AMPHIBIANS AND REPTILES
OF THE US–MEXICO BORDER
STATES

ANFIBIOS Y REPTILES DE
LOS ESTADOS DE LA FRONTERA
MÉXICO–ESTADOS UNIDOS

NUMBER FIFTY-TWO
W. L. Moody Jr. Natural History Series

Amphibians and Reptiles
of the US–Mexico Border States

Anfibios y reptiles
de los estados de la frontera
México–Estados Unidos

EDITED BY
JULIO A. LEMOS-ESPINAL

EDITADO POR
JULIO A. LEMOS ESPINAL

Texas A&M University Press
College Station, Texas

LIBRARY OF CONGRESS
CATALOGING-IN-PUBLICATION DATA

Amphibians and reptiles of the US–Mexico border
states = Anfibios y reptiles de los estados de la frontera
Mexico–Estados Unidos / edited by Julio A. Lemos-
Espinal. — First edition.
    pages cm — (W.L. Moody Jr. natural history series;
number fifty-two)
    In English and Spanish.
    Includes bibliographical references and index.
    ISBN 978-1-62349-306-6 (printed case: alk. paper)
— ISBN 978-1-62349-313-4 (ebook) 1. Amphibians—
Mexican-American Border Region. 2. Reptiles—
Mexican-American Border Region. I. Lemos-Espinal,
Julio A., editor. II. Title. III. Title: Anfibios y reptiles
de los estados de la frontera Mexico–Estados Unidos.
IV. Series: W.L. Moody Jr. natural history series; no. 52.
    QL651.A553 2015
    597.90972'1—dc23
                                              2015012200

WITH LOVE TO THE BEST OF MY LIFE,

*my wife Susy Sanoja-Sarabia and my
sons Julio and Sergio Lemos-Sanoja*

CON AMOR PARA LO MEJOR DE MI VIDA,

*mi esposa Susy Sanoja-Sarabia y mis
hijos Julio y Sergio Lemos-Sanoja*

Contents

Contenido

Acknowledgments

Agradecimientos

THIS STUDY WAS MADE possible through the generous support provided by Texas A&M University Press, Rusty Lizard Press, the Southwest Section of The Wildlife Society, the Tucson Herpetological Society, the Herpetology Department of the San Diego Natural History Museum, and the Department of Ecology and Evolutionary Biology at the University of Colorado. We are especially grateful to Shannon Davies from Texas A&M University Press, who supported and encouraged this book from the beginning.

We acknowledge the following people for their involvement in various stages and aspects of this book: George Bradley, Katie Duelm, Andrew T. Holycross, Louis Porras, and Susy Sanoja-Sarabia.

The majority of the photographs used to illustrate the amphibians and reptiles of the Border States region were taken by chapter authors (Randy Babb, Thomas C. Brennan, William Farr, Bradford D. Hollingsworth, Julio Lemos-Espinal, Charles W. Painter, and Jim Rorabaugh); however, a number of these photos were provided by Iván Ahumada Carrillo, Brad Alexander, Robert Bryson, Tim Burkhardt, Matthew Cage, Young Cage, Juan Alfonso Delgadillo Espinosa, Alan DeQueiroz, Erik Enderson, T. Fisher, Francisco García Camino, Daniel Garza-Tobón, Paul Hamilton, Peter Heimes, Terry Hibbitts, Toby J. Hibbitts,

ESTE ESTUDIO FUE POSIBLE a través del generoso apoyo proporcionado por Texas A&M University Press, Rusty Lizard Press, la Southwest Section of The Wildlife Society, la Tucson Herpetological Society, el Herpetology Department, San Diego Natural History Museum, y el Department of Ecology and Evolutionary Biology al University of Colorado. Estamos especialmente agradecidos con Shannon Davies de Texas A&M University Press, quien apoyó y alentó este libro desde su idea original.

Agradecemos a las siguientes personas por su participación en varias etapas y aspectos de este libro George Bradley, Katie Duelm, Andrew T. Holycross, Louis Parras, y Susy Sanoja-Sarabia.

La mayoría de las fotos que utilizamos en este libro para ilustrar los anfibios y reptiles de la región de los estados de la frontera pertenece a los autores de capítulos (Randy Babb, Thomas C. Brennan, William Farr, Bradford D. Hollingsworth, Julio Lemos-Espinal, Charles W. Painter y Jim Rorabaugh), sin embargo un número considerable de estas fotos pertenece a Iván Ahumada Carrillo, Brad Alexander, Tim Burkhardt, Robert Bryson, Matthew Cage, Young Cage, Juan Alfonso Delgadillo Espinosa, Alan DeQueiroz, Erik Enderson, T. Fisher, Francisco García Camino, Daniel Garza-Tobón, Paul Hamilton, Peter Heimes, Terry Hibbitts, Toby J. Hibbitts,

Troy D. Hibbitts, Marisa Ishimatsu, Randy Jennings, Kristopher Lappin, Carlos A. Luna Aranguré, Chris Mattison, Jimmy McGuire, Dana McLaughlin, Gary Nafis, Naturaleza y Cultura–Sierra Madre A. C., Barney Oldfield, Michael Patrikeev, Michael Price, Susy Sanoja-Sarabia, Jeff Servoss, Jack Sites, Robert G. Webb, William Wells, and John E. Werler. We are grateful to all of them for allowing us to use the photographs that appear in this book.

We are grateful to the administration of FES-Iztacala-UNAM, especially to the director, Patricia Dávila-Aranda, who has supported the studies of Julio Lemos-Espinal in northern Mexico.

Troy D. Hibbitts, Marisa Ishimatsu, Randy Jennings, Kristopher Lappin, Carlos A. Luna Aranguré, Chris Mattison, Jimmy McGuire, Dana McLaughlin, Gary Nafis, Naturaleza y Cultura – Sierra Madre A. C., Barney Oldfield, Michael Patrikeev, Michael Price, Susy Sanoja-Sarabia, Jeff Servoss, Jack Sites, Robert G. Webb, William Wells y John E. Werler, estamos muy agradecidos con todos ellos por permitirnos utilizar las fotografías que aparecen en este libro.

Agradecemos a la administración de la FES-Iztacala-UNAM, especialmente a su directora Patricia Dávila-Aranda, quien ha apoyado los estudios de Julio Lemos-Espinal en el norte de México.

Amphibians and Reptiles

of the US–Mexico Border States

Introduction:
How to Use This Book

JULIO A. LEMOS-ESPINAL

THE BIOLOGICAL DIVERSITY of the states on both sides of the United States–Mexico border attracted the attention of biologists even before the creation of the present border. Two of the most-studied classes of terrestrial vertebrates in this region are the amphibians and reptiles. Observations on their natural history and distribution have been accumulating since the United States and Mexican Boundary Survey was created in 1848 to determine the border between the two countries as defined in the Treaty of Guadalupe Hidalgo that ended the Mexican-American War. A number of monographic studies on amphibians and reptiles have been published for most of the 10 states that form the border between the United States and Mexico: Dixon (1987: Texas), Degenhardt et al. (1996: New Mexico), Grismer (2002: Baja California), Brennan and Holycross (2006: Arizona), Lemos-Espinal and Smith (2007a, 2007b: Chihuahua and Coahuila), and Stebbins and McGinnis (2012: California).

This book summarizes what is known about the distribution of amphibians and reptiles in each of the 10 states along the United States–Mexico border; provides a current list of amphibians and reptiles that have been recorded in each of these states; and analyzes which of those species are shared between the two countries,

which occur only north of the border in the United States, and which occur only south of the border in Mexico.

Chapter 2, "Nomenclatural and Distributional Notes," is a justification and discussion of the scientific and standard common names that are used throughout this book. In recent years there have been a number of changes in these names, and it is possible that the reader will not recognize the use of some names mentioned. The names used in this book have been standardized according to the names used in this chapter.

Chapters 3 through 12 are accounts of the herpetofauna for each of the 10 border states. Mexican border states are covered first from west to east (Baja California to Tamaulipas), and then the US border states, again from west to east (California to Texas). Each of these chapters follows the same format:

1. A short introduction that usually begins with the description of the general features of the state, including the amphibian and reptile species richness and the importance of these two vertebrate classes.
2. A summary of previous herpetological studies that usually begins with a description of work done by the early United States Boundary Commission expeditions to the border region

between 1848 and 1852 and then reviews subsequent work, and finishes with a discussion of present-day research and inventories. Also included is a table listing the type localities and authors for amphibians and reptiles described for the state.

3. Physiographic characteristics and their influence on the distribution of the herpetofauna. This section provides an overview of the biotic communities of the state as well as descriptions of geography, hydrology, climate, and the deep geological history that shaped the environments in which current-day amphibians and reptiles live. It also provides a context for the rich biodiversity of each state and fosters an understanding of the current patterns of herpetological distributions and affiliations. The discussion includes the number of amphibian and reptile species, number of endemic species in the state, species distributed only south (Mexico) or north (United States) of the border, as well as species shared by both countries, and species that may occur in the state. Conservation issues for amphibians and reptiles in the state are also discussed.

All the state chapters provide a current list of the amphibian and reptile species occurring within the state. All these lists are summarized in the appendix. Chapter 13, "Herpetofaunal Diversity of the United States–Mexico Border States," analyzes the data in the appendix by grouping the species by classes, orders, terrestrial species, and native species. It discusses the size of the state surface territory and environmental diversity with the species richness in amphibians and reptiles for individual states and the entire region.

Since the book is written in English and Spanish, it is divided in two parts. The first part is the English version, which is the language in which it was originally written; the second part is the Spanish version. These two parts are separated by photos that illustrate most of the species that inhabit the border region. Each photo is labeled with the scientific name of the species, English and Spanish standard name, and name of the photographer.

Nomenclatural and Distributional Notes

JULIO A. LEMOS-ESPINAL,
BRADFORD D. HOLLINGSWORTH,
AND WILLIAM FARR

Comments on Scientific Names

It has been argued that we are living in an era of taxonomic chaos (Pauly et al. 2009), with uncertainty of the stability of standard or even scientific names and the validity of various rationales for adoption of one name over another. To avoid contributing to such chaos, we have here been as consistent as possible with current literature. Nevertheless, in a few cases we have regarded it necessary in the interest of accuracy and understanding to adopt certain name changes. In this section we give short explanations of the reasons for use of scientific or standard English and Spanish names that appear in this book. Neither is firmly fixed, but usage will be an important factor in attaining consistency.

Batrachoseps altasierrae and B. bramei

Jockush et al. (2012) described two new species of slender salamanders (Batrachoseps altasierrae and B. bramei) from the region of the southern Sierra Nevada, California.

Siren lacertina

Flores-Villela and Brandon (1992) concluded two distinct species of Siren occur in the Rio Grande Valley of Texas and Tamaulipas, a smaller form that they assigned to S. intermedia nettingi (31–36 costal grooves, egg diameter no more than 2.5 mm) and a larger form assigned to S. lacertina (36–40 costal grooves, egg diameter reaches 3 mm), and placed S. i. texana in the synonymy of S. i. nettingi. They stated that although they could not find any differences separating the larger Rio Grande population from S. lacertina populations in the southeastern United States, their assignment of the larger Rio Grande specimens to S. lacertina was tentative, noting a hiatus in the species distribution between populations in Alabama and the Rio Grande Valley. Considering this distributional gap of well over 1000 km spanning numerous major drainage systems, we interpret the population of larger members of the genus Siren in the Rio Grande Valley as an undescribed taxon (Siren sp. indet.) rather than an isolated disjunct population of S. lacertina. A more definitive clarification of the status of the Rio Grande Siren genus will likely require collecting larger sample sizes for morphological and molecular analysis.

Incilius mccoyi and I. occidentalis

Recently Santos-Barrera and Flores-Villela (2011) described a new species (Incilius mccoyi), which formerly was regarded as part of *I. occidentalis*, from the Sierra Madre Occidental of Chihuahua.

Gastrophryne mazatlanensis and *G. olivacea*

In an analysis of the disjunct eastern and western clades of *Gastrophryne olivacea*, Streicher et al. (2012) elevated the western clade (Arizona, Sonora, and southwestern Chihuahua) to *G. mazatlanensis*. Elsewhere this anuran remains as *G. olivacea*.

Trachycephalus typhonius

Lavilla et al. (2010) reviewed the literature and newly available translations of the journal of Daniel Rolander (Hansen 2008), the collector of *Rana typhonia*, a species originally named by Linnaeus in 1758. They provided evidence demonstrating that this species is neither a bufonid nor an Asiatic ranid but the Neotropical hylid *Trachycephalus venulosus*. They regarded *R. typhonia* as an older synonym of *R. venulosa* (= *T. venulosus*) and redescribed its holotype under the new combination *T. typhonius*, which is the name used in this book.

Gopherus agassizii and *G. morafkai*

Murphy et al. (2011) investigated the problems associated with the identity of the Desert Tortoise (*Gopherus agassizii*), including date of publication of the original description, type locality, fate of the type series, description of other species from the Cape Region of Baja California Sur, allocation of the species to different genera, taxonomic arrangement of the species as a subspecies of the Bolson Tortoise, and existence of more than one species of Desert Tortoise. After analyzing all these issues and using genomic DNA, they found that a suite of characters serves to diagnose tortoises west and north of the Colorado River, the Mojavian population, from those east and south of the river in Arizona, Sonora, extreme southwestern Chihuahua, and Sinaloa. They concluded that species recognition is warranted for these two populations: *G. agassizii* for

the Mojavian population and *G. morafkai* for the Sonoran population. Thus, *G. agassizii* is a species found exclusively in the United States, and *G. morafkai* is a shared species between Mexico and the United States.

Gerrhonotus taylori

This is a close relative of *Gerrhonotus liocephalus*, which Good (1988) regarded a synonym without rank. However, the features noted by Tihen (1954) distinguish it from the latter species. It is possible that all of the isolated reports of *Gerrhonotus* from western Mexico (Good 1988) relate to *G. taylori*. The genus appears to be rare in that area, with insufficient material available for taxonomic allocation with confidence. Good (1994) elevated *G. infernalis* from synonymy with *G. liocephalus* and placed *G. taylori* in its synonymy. We regard the latter as a valid species and use it in this book.

Phyllodactylus nolascoensis

Two subspecies have been recognized for *Phyllodactylus homolepidurus*: *P. h. homolepidurus* from western Mexico and *P. h. nolascoensis* from Isla San Pedro Nolasco of the Sea of Cortés of Sonora. These two subspecies are diagnosable on the basis of differences in the number of paravertebral tubercles (31–41 in *P. h. homolepidurus* and 41–48 in *P. h. nolascoensis*) and longitudinal rows of dorsal tubercles (12–14 in *P. h. homolepidurus* and 14–16 in *P. h. nolascoensis*) (Dixon 1964). Grismer (1999) placed both subspecies in the synonymy of *P. homolepidurus* but commented on their potential to be different species. Because *P. h. nolascoensis* is diagnosable, restricted to an island, and has little chance of hybridizing with the mainland form, we regard it as a full species, *P. nolascoensis*, in this book.

Heloderma exasperatum

Building on the work of Douglas et al. (2010), Reiserer et al. (2013) elevated four subspecies of *Heloderma horridum* to species, including *H. horridum exasperatum* of southern Sonora and southwestern Chihuahua.

Sceloporus variabilis

Mendoza Quijano et al. (1998) reviewed the species boundaries and phylogenetic relationships of the *Sceloporus variabilis* species group, which included *S. v. variabilis* (southern Tamaulipas) and *S. v. marmoratus* (Nuevo León, Coahuila, Tamaulipas, Texas), using multilocus isozyme characters and regarded *S. marmoratus* as a full species. However, because *S. v. marmoratus*, a relatively wide-ranging taxon occurring from sea level to 1740 m in elevation, was represented by samples from only one remote locality in their review and morphological data supporting intergradation between the taxa, a number of authors have not followed this arrangement (e.g., Collins and Taggart 2009; Crother 2008; Dixon 2000; Köhler and Heimes 2002; Lavín-Murcio et al. 2005; Lemos-Espinal and Smith 2007; Uetz 2012). We do not recognize *S. v. marmoratus* as a distinct species from *S. v. variabilis* in this book.

Polychrotidae and Dactyloidae

Townsend et al. (2011) clarified iguanian evolutionary history by inferring phylogenies, using concatenated maximum-likelihood (ML) and Bayesian analyses of DNA sequence data from 29 nuclear protein-coding genes for 47 iguanian and 29 outgroup taxa. Within Pleurodonta, they found strong ML support for some interfamilial relationships and for monophyly of almost all families (except Polychrotidae). They corrected non-monophyly of Polychrotidae by recognizing the pleurodont genus *Anolis* (sensu lato) as a separate family (Dactyloidae), which includes the anoline genera *Anolis, Chamaeleolis, Chamaelinorops, Ctenonotus, Dactyloa, Phenacosaurus,* and *Semiurus,* and limited the name Polychrotidae to include only the genus *Polychrus.* Only the family Dactyloidae, with one genus (*Anolis*) and four species (one of them introduced), occurs in the Border States region.

Plestiodon brevirostris

We follow the recent revision of the *Plestiodon brevirostris* species group (Feria-Ortiz et al. 2011), elevating *P. b. bilineatus* of the Sierra Madre Occidental and *P. b. dicei* of the Sierra Madre Oriental to full species status. They noted their data was unable to be resolved if *P. b. pineus* is conspecific with *P. b. dicei* or if *P. b. dicei* is a paraphyletic. We tentatively refer to populations in Coahuila, Nuevo León, and parts of Tamaulipas (formerly *P. b. pineus* and *P. b. pineus* × *dicei* intergrades) as *P. dicei* until this relationship is clarified.

Scincella kikaapoa

García-Vázquez et al. (2010) described a new species of skink (*Scincella kikaapoa*) from the region of Cuatro Ciénegas, Coahuila.

Aspidoscelis neomexicana

It has been speculated that the New Mexico Whiptail (*Aspidoscelis neomexicana*) occurs in Chihuahua, Mexico (e.g., Liner and Casas-Andreu 2008), without the support of a voucher specimen. Several authors of chapters in this book have been working in northern Chihuahua along the southern side of the Rio Grande. None of us has found specimens of this species on the Mexican side. In addition, James Walker (pers. comm. 2011) has focused his attention on finding this species on the Mexican side but so far unsuccessfully. We regard the occurrence of *A. neomexicana* very likely in Chihuahua, but at this moment, without any voucher specimen, we regard it as a species occurring exclusively in the United States (New Mexico, Texas, and Arizona).

Aspidoscelis stictogramma

We follow Walker and Cordes (2011) in the elevation of *Aspidoscelis burti burti* and *A. b. stictogramma* to species.

Charina and Lichanura

Kluge (1993) argues for taxonomic equivalence of the closely related monotypic genera *Charina*, *Lichanura*, and *Calabaria*, with the latter two being placed in the synonymy of *Charina*. However, Crother (2008) and Grismer (2002) reasoned that their separation was warranted based on their long-established use and that both *Charina* and *Lichanura* are no longer regarded as monotypic. We accept the use of *Lichanura*, separated from *Charina*, in this book.

Lichanura trivirgata and L. orcutti

Based on mitochondrial DNA and preliminary morphological data, Wood et al. (2008) recognized two species within *Lichanura*: *L. trivirgata* and *L. orcutti*. *Lichanura orcutti* is found exclusively within the United States, while *L. trivirgata* occurs in both countries. Given further sampling along the border regions of the United States and Mexico, *L. orcutti* may also occur in both countries.

Adelphicos newmanorum

LaDuc (1996) determined that populations of *Adelphicos* in Tamaulipas and adjacent states in northeastern Mexico were diagnosable and allopatric from *A. quadrivirgatum*, which ranges from central Veracruz into Central America. Wilson and Johnson (2010) and Wilson et al. (2013) recognized *A. newmanorum* (Taylor 1950) as a full species without comment, although this was based largely on LaDuc. Future collecting could produce specimens within the hiatus between northeastern Hidalgo and central Veracruz, but we tentatively follow Taylor, LaDuc, Wilson and Johnson, and Wilson et al. in recognizing *A. newmanorum* as a distinct species pending new information resolving the issue. See Farr et al. (2013) for further discussion.

Arizona pacata

We follow Grismer (2002) in the elevation of the subspecies *Arizona elegans pacata* to full species.

Klauber (1946) distinguished *A. e. pacata* from all other subspecies of *Arizona* by having a lower number of body blotches, subcircular in shape. In addition, their distributions are separated by a distance of 10 km, with no sign of intergradation in specimens found closest to the gap (Grismer 2002). In this book we regard *A. pacata* to be a full species.

Contia longicaudae

Feldman and Hoyer (2010) described a new species of Sharp-tailed Snake (*Contia longicaudae*) from the outer Coast Ranges, Klamath Mountains, and portions of the Cascade Ranges in northern California and southern Oregon.

Hypsiglena

Mulcahy (2008) revised the arrangement of subspecies and species of the Common Nightsnake (*Hypsiglena torquata*) complex based upon mitochondrial DNA markers and recognized eight species of nightsnakes, one of which remains undescribed (Hooded Nightsnake). Four of these species occur in the Border States region: *H. chlorophaea* (Baja California, Sonora, Chihuahua, New Mexico, Arizona, and California), *H. jani* (Chihuahua, Coahuila, Nuevo León, Tamaulipas, Texas, New Mexico, and Arizona), *H. ochrorhyncha* (Baja California and California), and *H. slevini* (Baja California). We follow this arrangement in this book.

Lampropeltis getula

Pyron and Burbrink (2009) presented a systematic revision of the *Lampropeltis getula* group, based on a phylogeographic mtDNA analysis. They elevated the subspecies *L. g. californiae*, *L. g. holbrooki*, *L. g. niger*, and *L. g. splendida* to the rank of full species. Their samples were mostly from localities in the United States; however, and except for samples from the Baja California peninsula, only a single specimen was sampled from other areas of Mexico (Durango). Due to the lack of adequate sampling from Mexico, and in order to keep consistency among the 10 states, in this book we tentatively regard the taxon occurring in all 10 states as *L. getula*.

Lampropeltis knoblochi and L. pyromelana

Burbrink et al. (2011), applying Bayesian species delimitation methods and using three loci from samples of the Arizona Mountain Kingsnake (*Lampropeltis pyromelana*) taken throughout their range, found two species that are separated by low-elevation habitats between the Colorado Plateau / Mogollon Rim and the Sierra Madre Occidental: *L. pyromelana*, limited to the Colorado Plateau, in northern Arizona, western New Mexico, Utah, and Nevada; and *L. knoblochi* of the Sierra Madre Occidental of Sonora, Chihuahua, Sinaloa, and Durango, and the Madrean Sky Islands in southeastern Arizona and southwestern New Mexico. Therefore, *L. pyromelana* is a species exclusively of the United States, whereas *L. knoblochi* occurs in both countries.

Leptodeira maculata

Duellman (1958) described *Leptodeira annulata cussiliris*, and that name has been applied to the Tamaulipas population in 50 years of subsequent literature. Recent molecular-based studies of relationships within the genus *Leptodeira* have indicated that *L. annulata* is polyphyletic and elevated the former subspecies *L. a. cussiliris* to a full species (Daza et al. 2009; Mulcahy 2007). Daza et al. also found that *L. maculata* was nested within *L. a. cussiliris* and concluded that *L. maculata* "should therefore be synonymized" with *L. cussiliris* (in error). Wilson et al. (2013) and Uetz (2012) pointed out that *L. maculata* (Hallowell, 1861; as *Megalops maculatus*) has priority over *L. cussiliris*; therefore, *L. maculata* is the correct name for the species.

Masticophis

Utinger et al. (2005) proposed the synonymy of *Masticophis* with *Coluber* because of results indicating a paraphyletic *M. flagellum* when compared with *C. constrictor*. This arrangement, however, has not been accepted by a number of authors, including Burbrink et al. (2008), who found that *C. constrictor* is monophyletic and did not find any *M. flagellum* nested within the *C. constrictor* lineages, and thus questioned whether *Masticophis* should be synonymized with *Coluber*. Moreover, Collins and Taggart (2008) proposed retaining *Masticophis* as a separate genus based on the observation that Utinger et al. did not establish their findings on adequate sampling. Wilson and Johnson (2010) also concluded that Utinger et al. were deficient in sampling other populations of *C. constrictor* and, above all, because they did not sample different species of *Masticophis*, as only two were analyzed (*M. flagellum* and *M. taeniatus*). Until further evidence is available, we follow the conclusions of Collins and Taggart and Wilson and Johnson and regard *Masticophis* as a genus separate from *Coluber*.

Masticophis fuliginosus and M. flagellum

We follow Grismer (2002) in elevating *Masticophis flagellum fuliginosus* to full species. Both *M. fuliginosus* and *M. flagellum* have been found in sympatry at Bahía San Luis Gonzaga and Paseo de San Matías, Baja California, with no signs of intergradation (Grismer 1994; Grismer and Mellink 2005). *Masticophis fuliginosus* and *M. flagellum*, as separate species, occur in both the United States and Mexico.

Pliocercus bicolor

Some follow Savage and Crother (1989) in recognizing *Pliocercus elapoides* as a single, albeit highly variable, species. We follow Smith and Chiszar (2001a, 2001b) here in recognizing *P. bicolor*. This is in fact the most recent review and the arrangement followed by Liner and Casas-Andreu (2008). Smith and Chiszar's arrangement of the *P. elapoides* species group, recognizing multiple species and subspecies, has been criticized by some, who note that it was based solely on color-pattern variations correlated to geographic distribution (e.g., McCranie 2011). *Pliocercus bicolor* is unique among Smith and Chiszar's arrangement in having a distinguishing feature of scutellation (the posterior infralabial fused with the second labiogenial) and a relatively isolated, if not entirely allopatric, distribution. We remain impartial regarding the status of the southern extralimital members of the complex.

Thamnophis angustirostris

The taxonomic status of *Thamnophis angustirostris* is unclear. This species is known only from the lectotype specimen (USNM 959—Kennicott had not actually designated a holotype and apparently based his new taxon on more than one specimen [Rossman et al. (1996)]) collected in Parras, Coahuila. We regard it as a valid species since the Sierra de Parras and nearby localities have suitable environmental conditions for this gartersnake; eventually this species could be rediscovered with more intensive fieldwork in areas surrounding Parras.

Thamnophis rufipunctatus and *T. unilabialis*

Using coalescent and multilocus phylogenetic approaches, Wood et al. (2011) tested whether genetic diversification within the *Thamnophis rufipunctatus* species complex was consistent with two prevailing models of range fluctuation for species affected by Pleistocene climate changes. They identified three divergent clades within this species complex (*rufipunctatus, unilabialis*, and *nigronuchalis*) and suggested recognition of the two former subspecies (*T. r. rufipunctatus* and *T. r. unilabialis*) as full species: *T. rufipunctatus* (Mogollon Narrow-headed Gartersnake) of central and southeastern Arizona and southwestern New Mexico; and *T. unilabialis* (Madrean Narrow-headed Gartersnake) of the Sierra Madre Occidental of Sonora, Chihuahua, and Durango, and south-central Chihuahua and northern Durango.

Trimorphodon biscutatus

Devitt et al. (2008) examined patterns of geographic variation within *Trimorphodon biscutatus*, using multivariate statistical analyses of 33 morphological characters scored for 429 specimens. Principal components and discriminant analysis revealed six morphologically distinct groups that were concordant with lineages recovered in a phylogeographic analysis of mitochondrial DNA and with taxa traditionally recognized as species or subspecies. They concluded that *T. biscutatus* (sensu lato) comprises six evolutionary species (including *T. vilkinsonii*) and recommended elevating *T. biscutatus* (sensu stricto), *T. lambda, T. lyrophanes, T. paucimaculatus*, and *T. quadruplex* to the species level. Three of these species occur in the Border States region: Sonoran Lyresnake (*T. lambda* in Sonora, New Mexico, Arizona, and California), Baja California Lyresnake (*T. lyrophanes* in Baja California and California), and Texas Lyresnake (*T. vilkinsonii* in Chihuahua, Texas, and New Mexico).

Hydrophis platurus

We use *Hydrophis platurus* instead of *Pelamis platura* based on the work of Sanders et al. (2013), who combined *Pelamis* with several other seasnake genera into *Hydrophis*.

Rena and *Epictia*

Adalsteinsson et al. (2009) reviewed the phylogeny of the Leptotyphlopidae, using DNA sequences from nine mitochondrial and nuclear genes, and analyzed 91 individuals representing 34 recognized species of leptotyphlopids. Based on their results, they proposed a new classification of the family that recognizes two subfamilies, Epictinae (New World and Africa) and Leptotyphlopinae (Africa, Arabian Peninsula, and Southwest Asia), including 12 genera. Most of the genera corresponded to previously recognized species groups. Seven of those generic names were resurrected, whereas five others were newly named. A number of species formerly placed in *Leptotyphlops* were removed to *Rena*, a name originally proposed by Baird and Girard (1853). These included six species occurring in Mexico: *R. bressoni, R. dissecta,*

R. dulcis, R. humilis, R. maxima, and *R. myopica;* three of these also occur in the United States: *R. dissecta, R. dulcis,* and *R. humilis.* The other Mexican species (*L. goudotii*) was placed in the genus *Epictia,* a name originally proposed by Gray (1845). Adalsteinsson et al. also indicated that an unusually large number of species exist that are unrecognized and proposed that *R. h. boettgeri* should be recognized as a full species. In addition, Bocourt (1881) originally described *R. h. dugesii* as a species (*Catodon dugesii*), but Klauber (1940) regarded it as a subspecies of *Leptotyphlops* (= *Rena*) *humilis.* Later it was elevated again to species by Lemos-Espinal, Smith, and Chiszar (2004) on the basis of pigmentation and proximal caudal scale row count; Lemos-Espinal, Smith, Chiszar, and Woolrich-Piña (2004) split *R. h. segrega* off from *R. humilis* because of the number of scale rows around the base of the tail—*R. humilis* has 12 rows, and *R. segrega,* 10. We conclude that five species of *Rena* (*R. dissecta, R. dulcis, R. humilis, R. myopica,* and *R. segrega*) and one of *Epictia* (*E. goudotii*) occur in the Border States region.

Indotyphlops

We follow Hedges *et al.* (2014) in the use of the genus name *Indotyphlops* instead of *Ramphotyphlops.*

Comments on Standard English and Spanish Names

We adopted the English standard names proposed by Crother (2008) for those species that occur exclusively in the United States and for those that occur in the United States as well as in Mexico; the Spanish and English standard names proposed by Liner and Casas-Andreu (2008) for those species that occur only in Mexico; and the Spanish standard names for those that occur in Mexico as well as in the United States. For those species that occur exclusively in the United States, we translated the names proposed by Crother to Spanish.

The work done in these two publications represents excellent and extensive efforts to unify the use of standard names for amphibian and reptile species of both countries. However, if these names are wrongly applied, they can create confusion and contradict the meaning originally intended. Liner and Casas-Andreu (2008) proposed several names that would create such confusion; most of them are typographical errors, but we here give an explanation of why we differ in those few cases.

Batrachoseps major

The assigned name is Salamandra Pequeña, but that name contradicts the specific name, derived from the Latin word meaning "larger or greater," referring to the largest species of *Batrachoseps* (Beltz 2006). The name Salamandra Mayor is more appropriate for this species and is the one used in this book.

Ensatina eschscholtzii

The assigned name is Ensatina de Monterrey, but since Monterrey is a well-known city in Mexico (capital of Nuevo León), we suggest that the Spanish standard name should specify that it refers to "Monterrey Bay" and not to Monterrey City in Nuevo León. The name Ensatina de Bahía Monterrey seems more appropriate for this species and is the one used in this book.

Lithobates magnaocularis

The assigned name, Rana Leopardo del Noreste de México, is incorrect. The correct name should be Rana Leopardo del Noroeste de México since this species occurs in northwestern Mexico and is the one used in this book.

Alligator mississippiensis

The assigned name is Aligator Americano. The word "Aligator" does not exist in Spanish. The name Lagarto Americano seems more appropriate for this species and is the one used in this book.

Trachemys nebulosa

The assigned names Jicotea del Río and Baja California Slider are incorrect. The name Jicotea del Río is vague and does not define any special characteristic of this turtle; the name Baja California Slider is incorrect because it creates the wrong idea that this species occurs in Baja California, whereas it actually occurs only in Baja California Sur, Sonora, and Sinaloa. The name assigned in the reptile-database.org is more appropriate for the species, Black-bellied Slider, and the translation of it for the Spanish standard name, Jicotea de Vientre Negro. These are the names used in this book.

Gopherus

The assigned name for this genus of turtles is Galápago del Desierto, a name that nobody seems to use for them. This group of turtles is known as Tortugas de Desierto, which is the direct translation of the English standard name Desert Tortoise. We use the names Tortuga de Desierto (*Gopherus agassizii*), Tortuga de Berlandier (*G. berlandieri*), Tortuga del Bolsón de Mapimí (*G. flavomarginatus*), and Tortuga del Desierto de Sonora (*G. morafkai*) for the four species that occur in the Border States region.

Elgaria

The assigned name for this genus of lizards is Lagartos del Oeste. The name Lagarto refers to alligators, not lizards. We consider that the direct translation of the English standard name, Lagartijas Lagarto, should be used for them in order to avoid confusion. We use the names Lagartija Lagarto de Isla Cedros (*Elgaria cedrosensis*), Lagartija Lagarto del Norte de Estados Unidos (*E. coerulea*), Lagartija Lagarto de Montaña (*E. kingii*), Lagartija Lagarto del Suroeste de Estados Unidos (*E. multicarinata*—additional explanation in the following section), Lagartija Lagarto de Los Coronados (*E. nana*),

and Lagartija Lagarto de la Sierra Panamint (*E. panamintina*) for the six species that occur in the Border States region.

Elgaria multicarinata

The assigned name, Lagarto Meridional, is incorrect. The correct name should be Lagartija Lagarto del Suroeste de Estados Unidos since it is common in that region and enters northwestern Mexico through Baja California. The name proposed by Liner and Casas-Andreu (2008) erroneously indicates a southern distribution in Mexico.

Crotaphytus dickersonae

The assigned name, Sonoran Collared Lizard, has already been assigned to *Crotaphytus nebrius*. We therefore propose the name Dickerson's Collared Lizard.

Phrynosoma modestum

The assigned name, Tapayaxín, is a Nahuatl word meaning "those who cry blood," and it is used in some regions to refer to *Phrynosoma orbiculare* in relation to the blood-squirting from the eyes of this and other species, a habit that has not been reported for *P. modestum*. We suggest the name Camaleón Modesto, which is the meaning in Spanish of the specific name *modestum*, a Latin word meaning "modest, discreet, unassuming," in reference to the plain appearance of the species.

Sceloporus albiventris

This species was not considered by Liner and Casas-Andreu in their 2008 list. We suggest the name Bejore de Vientre Blanco and Western White-bellied Spiny Lizard, which are the ones used in this book.

Sceloporus cowlesi

This species was not considered by Liner and Casas-Andreu in their 2008 list. We suggest the name Lagartija de Cerca del Noroeste, which is the one used in this book.

Sceloporus graciosus

This species was not considered by Liner and Casas-Andreu in their 2008 list. We suggest the name Lagartija Común de las Salvias, which is the one used in this book.

Sceloporus vandenburgianus

The assigned name, Meridional de las Salvias, is incorrect because it refers to the distribution of this lizard in the United States, not in Mexico. The correct name should be Septentrional de las Salvias.

Ficimia olivacea

The assigned name, Nariz de Gancho Huaxtena, is incorrect. The name Huasteca is misspelled in the assigned name; it should be Huasteca. The correct name is Nariz de Gancho Huasteca.

Lampropeltis pyromelana

The assigned names, Culebra Real de Sonora and Sonoran Mountain Kingsnake, were changed by Burbrink et al. (2011) to Culebra Real de Arizona and Arizona Mountain Kingsnake.

Oxybelis aeneus

The assigned names are Bejuquilla Parda and Brown Vinesnake. Keiser (1982) recommended the English name Neotropical Vinesnake for the species, noting that this wide-ranging taxon is highly variable in color and pattern and in many regions of its distribution is not brown. We use the name proposed by Keiser and the translation of it for the Spanish standard name, Bejuquilla Neotropical.

Tantilla hobartsmithi

The assigned name is Culebra Cabeza Negra del Suroeste. The southwestern distribution implied in this name refers to the United States; moreover, the species has a wide distribution in the southern United States and northern Mexico and is not restricted to the southwestern United States. The correct name should be Culebra Cabeza Negra del Norte.

Thamnophis rufipunctatus

Since we are following Wood et al. (2011) in the classification of the Narrow-headed Garter-snake complex, we have adopted their proposed standard names, so the Spanish and English standard names for this species are Jarretera Cabeza-angosta del Mogollon and Mogollon Narrow-headed Gartersnake, respectively.

Thamnophis unilabialis

Since we are following Wood et al. (2011) in the classification of the Narrow-headed Gartersnake complex, we have adopted their proposed standard names. The Spanish and English standard names for this species are Jarretera Cabeza-angosta de Chihuahua and Madrean Narrow-headed Gartersnake, respectively.

Tropidodipsas fasciata

The assigned name, Caracolera Añillada, is incorrect. The correct name should be Caracolera Anillada.

Tropidodipsas repleta

This species was not considered by Liner and Casas-Andreu in their 2008 list. We suggest the names Caracolera Repleta and Banded Black-snake, which are used in this book.

Crotalus molossus

The assigned name, Cascabel Serrana, is applied to one of the most well-known rattlesnake species in Mexico. Most of the local people in northern Mexico refer to this species as Cola Prieta, which means "dark or blackish tail" and readily describes the species and differentiates it from other rattlesnake species. We consider the name Cola Prieta more appropriate for this snake, which has the added advantage of already being in widespread use by the local people.

Sistrurus catenatus

The assigned names are Cascabel de Massasauga and Massasauga Rattlesnake. For descriptive purposes and consistency with the Spanish standard name, instead of Massasauga as used by Crother (2008) and Liner and Casas-Andreu (2008), we use the combination Massasauga Rattlesnake.

Herpetofauna of Baja California

3

BRADFORD D. HOLLINGSWORTH,
CLARK R. MAHRDT, L. LEE GRISMER,
AND ROBERT E. LOVICH

Introduction

The study of the amphibians and reptiles within the state of Baja California has received considerable attention but usually within the context of the Baja California peninsula (Van Denburgh 1895a, 1895b; Schmidt 1922; Savage 1960; Murphy 1983b, 1983c; Grismer 1994b, 2002; Lovich et al. 2009; Murphy and Méndez de la Cruz 2010). Here we focus on the herpetofauna within the state of Baja California, which extends from the 28th parallel to the international border with the United States, covering the northern half of the Baja California peninsula. To the south, the state shares a 142 km border with Baja California Sur, while to the north it shares a 226 km border with California in the United States. In the northeastern corner of the state, a short border is shared with Arizona (36 km) and Sonora (112 km) following the historical drainage of the Rio Colorado. The remaining boundaries are with the Pacific Ocean and Sea of Cortés and include islands in both bodies of water. Altogether, the state is approximately 70,000 km² in area.

The state contains a diverse array of habitats, influenced by the extension of floral provinces from the north or the south and climatic effects from the Pacific Ocean and Sea of Cortés. The Peninsular Ranges are the dominant topographical feature of the state, running in a north-south orientation, with a gradual slope along the western side and an eastern face rising more abruptly, often presenting a precipitous escarpment. A series of mountains are contained within the Peninsular Ranges, with lower-elevation passes connecting regions from the west and east. Of these, the Sierra Juárez and San Pedro Mártir, located in the northern half of the state, represent a major zoogeographic barrier and a high-elevation ecosystem. Two national parks cap the highest elevations of these mountain ranges, Parque Nacional Constitución de 1857 in the Sierra Juárez and Parque Nacional Sierra San Pedro Mártir, which includes Picacho del Diablo, the highest peak (3095 m) in Baja California (Schad 1988). These mountains create a significant rain shadow effect between the relatively moist California Floristic Province along the Pacific Ocean and the Lower Colorado Desert to the east of the mountains (Shreve and Wiggins 1964; Hastings and Humphrey 1969).

The Peninsular Ranges in the southern half of the state include the Sierras de La Asamblea, Las Animas, San Borja, and La Libertad. Due to their lower elevations, these mountains have less effect in isolating the surrounding ecosystems. While the Sierra La Asamblea reaches nearly 1700 m in elevation and maintains an isolated mountaintop habitat, the southern portion of the state is dominated by the peninsular deserts of Baja California (Arriaga et al. 1997; Morrone 2005; Hafner and Riddle 1997), composed of the Vizcaíno and Central Gulf Coast Desert regions (Shreve and Wiggins 1964; Wiggins 1980). As a result, transitions between the California Floristic Province and Lower Colorado Desert in the north and the peninsular deserts in the south are less impacted by mountain ranges and intermix gradually across wider geographic areas (Grismer 2002).

A total of 119 amphibians and reptiles are found in the state of Baja California: 20 amphibians (4 salamanders, 16 frogs) and 99 reptiles (7 turtles, 51 lizards, 1 amphisbaenid, 40 snakes). Six species are introduced (*Xenopus laevis, Lithobates berlandieri, L. catesbeianus, Apalone spinifera, Hemidactylus turcicus,* and *Sauromalus varius*), and 113 are native. A total of 21 species are endemic to the state. Two species are endemic to the continental land masses (*Crotaphytus grismeri* and *Urosaurus lahtelai*), 2 are endemic to continental areas plus islands (*Elgaria cedrosensis* and *Anniella geronomensis*), and 17 are insular endemics (*Elgaria nana, Crotaphytus insularis, Phyllodactylus partidus, Sauromalus hispidus, Callisaurus splendidus, Petrosaurus slevini, Uta antiqua, U. encantadae, U. lowei, U. tumidarostra, Aspidoscelis cana, Lampropeltis herrerae, Pituophis insulanus, Crotalus angelensis, C. caliginis, C. lorenzoensis,* and *C. muertensis*).

The herpetofauna of Baja California is broadly shared with the United States. Of the 119 species found in the state, 88 (74%) also occur in the United States. All 6 of the state's marine species (*Caretta caretta, Chelonia mydas, Eretmochelys imbricata, Lepidochelys olivacea, Dermochelys coriacea,* and *Hydrophis platurus*) occur in coastal waters off California, and 5 of the 6 introduced species (the exception is *Sauromalus varius*) are also introduced into nonnative areas of the United States. The remaining 77 species have distributions that cross the northern border of the state with California (all 77) and Arizona (38 species).

There are 31 species of amphibians and reptiles from the state of Baja California that are endemic to Mexico, including the 21 species

endemic to the state plus 10 that have distributions that occur entirely within Mexico. *Elgaria cedrosensis, Petrosaurus repens, Phrynosoma cerroense, Sceloporus zosteromus, Aspidoscelis labialis, Bipes biporus, Arizona pacata, Hypsiglena slevini, Pituophis vertebralis,* and *Crotalus enyo* have distributions that extend southward into the state of Baja California Sur and are confined to the Baja California peninsula and its associated islands (Grismer 2002), while *Sauromalus varius* is an insular endemic from Sonora and introduced to Isla Roca Lobos, Baja California (Hollingsworth et al. 1997; Hollingsworth 1998).

Previous Herpetological Studies

The majority of works have focused on the faunal composition and diversity of the peninsula of Baja California, making it difficult to trace the history of herpetological research from within the boundaries of the state. Earlier explorations mainly describe amphibian and reptile observations from the peninsula, usually without establishing specific locations to discern whether the observation or collected material fell within the state of Baja California or Baja California Sur. Some exceptions are collections from the established port in Cabo San Lucas. For instance, P. E. Botta made natural history collections from Cabo San Lucas while onboard the French vessel *Le Héros* between 1827 and 1829 (Adler 1978). This included the type specimen of *Phyrnosoma coronatum* described by Blainville (1835) in *Reptiles de la Californie.* As the *Héros* made its way northward along the Pacific coast, it is unlikely Botta had the opportunity to collect within the state of Baja California before the ship made port in San Diego.

Earlier accounts of amphibian and reptile observations are reviewed in Grismer (2002), which includes details of the first written reports from the Baja California peninsula from the Jesuit missionary Clavijero in 1789. Besides a few earlier descriptions, herpetological observations were rare in the latter part of the eighteenth and first half of the nineteenth centuries (Grismer 2002). Following the Mexican-American War (1846–1848) and the Treaty of Guadalupe Hidalgo, the boundary between the two countries was revised. During this same period, the United States was exploring its newly acquired territory, and many expeditions were under way to document routes into the western frontier, including explorations into Baja California by the naturalists Spencer F. Baird, Lyman Belding, Paul E. Botta, Townsend S. Brandegee, Walter E. Bryant, Gustav Eisen, and William M. Gabb (Grismer 2002).

An expedition to Baja California in 1867, with the consent of Mexican authorities, was to detail the geological composition of the peninsula and would cross from coast to coast multiple times on the journey northward from Cabo San Lucas to Tijuana (Dall 1909). As part of the expedition team, William M. Gabb (1882), while documenting the geology of the peninsula, collected the type specimen of *Phyllorhynchus decurtatus* (Cope 1868, 311) in the "upper part of Lower California." In the account of the expedition, Gabb considered the peninsula to have three distinct districts: the Cape region, a middle section that covered the southern half of the peninsula to as far north as Santa Gertrudis located on the present-day state boundary, and a northern part above Santa Gertrudis. Smith and Taylor (1950, 322) list the restricted type locality as "San Fernando Misión," (between San Ignacio and) Baja California, Mexico. Here, the location of the type specimen is not exactly clear, but it can be interpreted as meaning the northern district based on Gabb's breakdown of the peninsula. If so, *P. decurtatus* represents the first species described from a specimen collected from the state of Baja California (see table 3.1).

The first review of the herpetofauna of Baja California was completed by John Van Denburgh in 1895, when he commented that "the peninsula of Lower California lies so far from usual routes of travel that few collections of its animals have found their way into museums." At this time, Van Denburgh (1895a, 1895b) reported on 61 species from the peninsula, plus an additional 7 species the following year (Van Denburgh 1896).

Twelve of these are confined to areas outside the state of Baja California. Mocquard (1899), with help from the collections of Léon Diguet, added another 11 species to the growing knowledge of the peninsular herpetofauna (Grismer 2002).

The Bureau of Biological Survey of the United States Department of Agriculture completed an expedition of the peninsula during 1905 and 1906 under the direction of Edward Nelson and assisted by Edward Goldman. The expedition began at the US border with the assistance of Frank Stephens of the San Diego Society of Natural History. The results, *Lower California and Its Natural Resources*, included accounts of amphibian and reptile discoveries, including the southern distribution of *Crotalus cerastes* (Nelson 1922). Nelson listed 76 currently recognized amphibians and reptiles from the peninsula, 62 of which are found in the state of Baja California.

In 1911, the Albatross Expedition led by Charles H. Townsend explored the Baja California peninsula and associated islands (Townsend 1916). Sponsored in part by the American Museum of Natural History and the United States National Museum, the expedition visited the islands of San Benito, Cedros, and Ángel de la Guarda and ported in Bahía San Francisquito in the state of Baja California. Specimens from the expedition led to the description of *Callisaurus splendidus* by Dickerson (1919), in addition to numerous other species from outside the state. The same material helped Schmidt (1922) produce another accounting of the herpetofauna of the peninsula and associated islands and treat the peninsula as a faunal unit (Grismer 2002).

Two of the most comprehensive works on the reptiles and amphibians of the Baja California region appeared in Van Denburgh (1922), *The Reptiles of Western North America*, and later in Joseph R. Slevin (1928), *The Amphibians of Western North America*. Much of the material used to expand on previous works came from the 1921 Silvergate Expedition to the Sea of Cortés under the leadership of Slevin (Hanna 1922; Slevin 1922). In addition, Van Denburgh and Slevin produced a number of checklists and accounts describing new species from the state (Van

Denburgh 1905; Van Denburgh and Slevin 1914, 1921a, 1921b, 1923).

Shortly after the Silvergate Expedition, the Museum of Vertebrate Zoology launched expeditions into Baja California from 1925 to 1931, which recorded additional, more detailed information on the habitat and ecology of the species found. The results of these expeditions were presented in Linsdale (1932).

Explorations continued into Baja California by both overland routes and sea. More detailed, comprehensive work was completed by Laurence M. Klauber from the San Diego Natural History Museum and Zoological Society of San Diego, who assembled specimens from a wide range of species that led to numerous taxonomic revisions (Klauber 1924, 1928, 1931a, 1931b, 1931c, 1931d, 1931e, 1932, 1934, 1935, 1936, 1940a, 1940b, 1940c, 1941, 1943a, 1943b, 1944, 1945, 1946a, 1946b, 1949a, 1949b, 1951, 1956, 1963). Frank S. Cliff explored the islands in the Sea of Cortés onboard the research ship *Orca*, which was made available to biologists by Joseph W. Sefton Jr. of the San Diego Natural History Museum and the Sefton Foundation of San Diego. As a result, a more detailed accounting of reptiles was made, with an emphasis on insular species of snakes (Cliff 1954a, 1954b). Richard G. Zweifel (1952a, 1958) provided additional notes on Pacific islands.

Charles H. Lowe Jr. and Kenneth S. Norris (1954) published studies on the biogeography and patterns of geographic variation. Jay M. Savage (1960, 1967) presented a dispersal-based account for the origin of the herpetofauna of the peninsula and discussed the evolution of the insular herpetofauna.

The Belvedere Expedition to the Sea of Cortés in the spring of 1962 produced a further advancement in our understanding of reptiles from islands in the Sea of Cortés and included visits to the Midriff Islands and Bahía de los Ángeles (Lindsay 1962). Herpetologists Charles E. Shaw, Donald Hunsaker, Dennis L. Bostic, and Michael Soulé were present. One notable publication included a study of biogeography of reptiles on islands, which analyzed endemism and species-area relationships (Soulé and Sloan 1966). In

addition, Shaw studied a number of reptiles, including *Anniella* and *Sauromalus* (Shaw 1940, 1945, 1949, 1953), while Bostic focused on reptiles and amphibians from the northern Vizcaíno Desert region (Bostic 1965, 1966a, 1966b, 1966c, 1966d, 1968, 1971, 1975).

John R. Ottley contributed to the taxonomy of *Lichanura*, natural history notes, and an updated island checklist compiled with Robert W. Murphy (Ottley 1978; Ottley et al. 1980; Ottley and Hunt 1981; Ottley and Jacobsen 1983; Murphy and Ottley 1984; Grismer and Ottley 1988). Ted J Case (1982, 1983) made contributions to island ecology and selection pressures with studies of *Sauromalus hispidus*. Hartwell H. Welsh (1988) studied the ecogeography of the amphibians and reptiles of the Sierra San Pedro Mártir with comments on the evolution of the herpetofauna and provided additions to the herpetofauna of the Lower Colorado Desert (Welsh and Bury 1984).

Robert W. Murphy has made several contributions to the origin and evolution of the herpetofauna of the Baja California peninsula (Murphy 1976, 1983b, 1983c; Murphy and Aguirre-Léon 2002b), checklists (Murphy 1983a; Murphy and Ottley 1984; Murphy and Aguirre-Léon, 2002a), taxonomy (Murphy 1974, 1975; Murphy and Smith 1985, 1991), phylogeny and historical biogeography (Murphy et al. 1995; Upton and Murphy 1997; Aguirre-Léon et al. 1999; Murphy et al. 2002; Lindell et al. 2005, 2008; Blair et al. 2009), and conservation (Murphy and Méndez de la Cruz 2010).

To date, the single most comprehensive work on the herpetofauna of the region is L. Lee Grismer's (2002) *Amphibians and Reptiles of Baja California Including Its Pacific Islands and the Islands in the Sea of Cortés*. Grismer has also contributed to the evolution of the herpetofauna of the peninsula (Grismer 1994a, 1994b; Grismer and Mellink 2005), checklists (Grismer 1999a, 2001), the regional herpetofauna (Grismer 1993; Grismer and McGuire 1993; Grismer and Hollingsworth 1996; Grismer et al. 1994; Grismer and Mellink 1994), taxonomy (Grismer 1990a, 1994c, 1999b), taxonomic revisions (Grismer 1988, 1990b, 1996, 1999c; McGuire and Grismer 1993; Grismer and McGuire 1996; Grismer and Hollingsworth 2001; Grismer et al. 2002), physiology (Hazard et al. 1998), and conservation (Lovich et al. 2009).

Studies by Robert E. Lovich and Clark R. Mahrdt document the mesic-adapted herpetofauna of the California Floristic Province, which includes notes on *Actinemys marmorata* (Lovich et al. 2005; Lovich et al. 2007), *Anaxyrus californicus* (Mahrdt et al. 2002; Mahrdt et al. 2003; Mahrdt and Lovich 2004), *Ensatina eschscholtzii* (Mahrdt 1975), and *E. klauberi* (Mahrdt et al. 1998). A summary of the terrestrial herpetofauna and conservation of Bahía de los Ángeles Biosphere Reserve was provided by Lovich and Mahrdt (2008) and for the state by Lovich et al. (2009).

Additional molecular and morphological studies of species groups with populations in the state of Baja California include *Trimorphodon* (Devitt et al. 2008), *Xantusia wigginsi* (Leavitt et al. 2007), *Phrynosoma* (Leaché and McGuire 2006; Leaché et al. 2009), *Sceloporus magister* (Leaché and Mulcahy 2007), *Hypsiglena* (Mulcahy 2008), *Lampropeltis zonata* (Rodríguez-Robles et al. 1999; Myers et al. 2013), *Pituophis* (Rodríguez-Robles 2000), *Ensatina klauberi* (Devitt et al. 2013), and *Anniella* (Papenfuss and Parham 2013).

In Baja California, 26 species of reptiles (19 lizards and 7 snakes) have their type localities within the state (table 3.1).

Physiographic Characteristics and Their Influence on the Distribution of the Herpetofauna

The topography within the state of Baja California is strongly associated with the formation of the Baja California peninsula and its origins from western mainland Mexico. A complex tectonic history and interactions between the Pacific and North American Plates resulted in the formation of the peninsula and the opening of the Sea of Cortés within the last seven to eight million years (Lonsdale 1989; Stock and Hodges

1989; Winker and Kidwell 1986; Carreño and Helenes 2002). In conjunction with the rifting of the peninsular landmass from western Mexico was the uplift of the Peninsular Range (Lonsdale 1989). These mountains extend the length of the Baja California peninsula and achieve their highest points in the northern half of Baja California. Depositional filling and changes in sea levels produced the flat, lower-elevation basins and plains in the Lower Colorado and Vizcaíno Deserts (Grismer 2002).

Amphibians and reptiles inhabit islands within the coastal waters of Baja California in both the Pacific Ocean and Sea of Cortés (Savage 1967; Murphy 1983a, 1983c; Murphy and Ottley 1984; Grismer 1993, 1999a, 1999b, 2001; Murphy and Aguirre-León 2002a, 2002b; Samaniego-Herrera et al. 2007). Isla Guadalupe, the westernmost point in Mexico, lies 241 km off the coast and is not inhabited by amphibians or reptiles, with the exception of sea turtles. Amphibians and reptiles inhabit the Pacific Ocean coastal islands of Islas Coronado, Todos Santos, San Martín, San Jerónimo, San Benito, and Cedros (Grismer 1993, 2001; Samaniego-Herrera et al. 2007). Islands in the Sea of Cortés belonging to the state of Baja California are inhabited by only reptiles. These islands include El Muerto, Coloradito, Encantada, Blancos, San Luis, Willard, Mejía, Granito, Ángel de la Guarda, Pond, Cardonosa Este, Partida Norte, Rasa, Salsipuedes, Roca Lobos, San Lorenzo Norte, San Lorenzo Sur, and numerous islands in Bahía de los Ángeles (Grismer 2002).

The climatic conditions of Baja California vary greatly because of its length, complex topography, elevation, and location between two very dissimilar bodies of water, the Pacific Ocean and the warm waters of the Sea of Cortés (Humphrey 1974). Naturalist-explorer Edward W. Nelson conveyed this condition in 1922 (p. 95): "The climate of Lower California as a whole is extremely hot and dry. There are, however, considerable local climatic variations according to the situation in relation to latitude, altitude, mountain ranges, and to the east and west coasts." With the exception of the northwestern and montane re-

TABLE 3.1. Type localities for amphibians and reptiles described from the state of Baja California, Mexico

Author(s)	Original name: Type locality
Lizards (19)	
Fitch (1934)	*Gerrhonotus* (= *Elgaria*) *cedrosensis*: **Cañon on southeast side of Cedros Island, Lower California, Mexico**
Fitch (1934)	*Gerrhonotus scincicauda nanus* (= *Elgaria nana*): **South Island, Los Coronados Islands, Lower California, Mexico**
Shaw (1940)	*Anniella geronimensis*: **San Geronimo Island, Lower California, Mexico**
McGuire (1994)	*Crotaphytus grismeri*: **Cañon David, a low pass that separates the contiguous Sierra de Los Cucapás and Sierra El Mayor, approximately 2 km W of Mex. Hwy 5 on the dirt road to the sulfur mine (turnoff at KM 49 S. Mexicali), Baja California, Mexico**
Van Denburgh & Slevin (1921b)	*Crotaphytus insularis*: **E coast Ángel de la Guarda Island seven miles north of Pond Island, Gulf of California, Mexico**
Smith & Tanner (1972)	*Crotaphytus insularis vestigium* (= *C. vestigium*): **Guadalupe Canyon, Juárez Mountains, Baja California**
Dixon (1966)	*Phyllodactylus partidus*: **Isla Partida, Baja California**
Stejneger (1891)	*Sauromalus hispidus*: **Ángel de la Guarda Island, Gulf of California**
Dickerson (1919)	*Callisaurus draconoides splendidus* (= *C. splendidus*): **Angel de la Guardia Island, Gulf of California, Baja California, Mexico**
Van Denburgh (1922)	*Uta* (= *Petrosaurus*) *slevini*: **Mejia Island, Gulf of California, Mexico**
Stejneger (1893)	*Phrynosoma cerroense*: **Cedros Island, Baja California, Mexico**[1]
Rau & Loomis (1977)	*Urosaurus lahtelai*: **4 km N parador Cataviña, elevation 564 m near Mexico Highway 1, State of Baja California, Mexico**
Ballinger & Tinkle (1968)	*Uta antiquus* (=*U. antiqua*): **Isla Salsipuedes, 28°44′ N, 112°59′ W, Golfo de California, Mexico**
Grismer (1994c)	*Uta encantadae*: **Isla Encantada, Gulf of California, Baja California, México**
Grismer (1994c)	*Uta lowei*: **Isla El Muerto, Gulf of California, Baja California, México**
Grismer (1994c)	*Uta tumidarostra*: **Isla Coloradito, Gulf of California, Baja California, México**
Van Denburgh & Slevin (1921b)	*Cnemidophorus canus* (= *Aspidoscelis cana*): **Sal Si Puedes Island, Gulf of California, Mexico**
Stejneger (1890)	*Cnemidophorus* (= *Aspidoscelis*) *labialis*: **Cerros Island, Lower California**[2]
Savage (1952)	*Xantusia vigilis wigginsi* (= *X. wigginsi*): **Arroyo 9 miles east of Miller's Landing, Distrito del Norte, Baja California, Mexico**
Snakes (7)	
Van Denburgh & Slevin (1923)	*Lampropeltis zonata herrerae* (= *L. herrerae*): **South Todos Santos Island, Lower California, Mexico**
Cope (1868)	*Phimothyra decurtata* (= *Phyllorhynchus decurtatus*): **San Fernando Misión (between San Ignacio and)**[1]
Klauber (1946b)	*Pituophis catenifer insulanus* (= *P. insulanus*): **Cedros (Cerros) Island off the west coast of Baja California, Mexico**
Klauber (1963)	*Crotalus mitchellii angelensis* (= *C. angelensis*): **About 4 miles southeast of Refugio Bay, at 1500 feet elevation, Isla Angel de la Guardia, Gulf of California, Mexico (near 29°29½′ N, 113°33′ W)**
Klauber (1949b)	*Crotalus viridis caliginis* (= *C. caliginis*): **South Coronado Island, off the northwest coast of Baja California, Mexico**
Radcliffe & Maslin (1975)	*Crotalus ruber lorenzoensis* (= *C. lorenzoensis*): **San Lorenzo Sur Island in the Gulf of California, Baja California Norte, Mexico**
Klauber (1949b)	*Crotalus mitchellii muertensis* (= *C. muertensis*): **El Muerto Island, Gulf of California, Mexico**

[1] Restricted type locality presented (Smith and Taylor 1950).

[2] Cochran (1961) reports type locality as "Cedros Island (given as 'Cerros Island' in the original description, but probably from San Quintín Bay), Baja California, Mexico."

gions, Baja California is generally characterized by relatively high annual mean temperatures with low amounts of precipitation (Grismer 2002). For almost the entire state the hottest months are July and August, and the coldest month is January (Markham 1972).

Two major climatic regimes, the Pacific Coastal Regime and Gulf Regime (Meigs 1966), originate in northern and southern latitudes, respectively, influencing weather patterns in both Baja California and neighboring southern California (Turner and Brown 1982). In the winter, northern-latitude cyclonic storms originating in the Gulf of Alaska move southward and bring rainfall to much of the northwest coast and montane regions. In southern latitudes during late summer and fall, hurricanes from the south develop in warm waters off the west coast of southern and central Mexico (Markham 1972). These storm systems occasionally deliver rainfall to the southern portions of the state. Large anticyclonic systems, generating wind speeds of over 200 kph, usually move north along the Pacific Ocean and hit the southwestern coastline of the state. Some Pacific hurricanes move north by several hundred kilometers from Cabo San Lucas, passing over the warm waters of the Sea of Cortés, increasing their intensity and wind speed. Smaller systems that develop locally in the Sea of Cortés also produce high winds, heavy rainfall, and rough seas. These Gulf systems, or chubascos, are usually short lived but are capable of causing severe damage once they make landfall (Grismer 2002).

In Baja California, most rivers are confined to the northwestern portion of the state. Waters originate in the high mountains and drain to the Pacific Ocean or Sea of Cortés. The Tijuana River drains the northwestern Sierra Juárez and southwestern San Diego County and empties into the Tijuana River Estuary in California. Rivers in Baja California are characterized by alluvial soils and intermittent flows depending on rainfall. Most no longer flow continuously to the ocean due to longer periods of drought and human diversions. Rivers draining to the Pacific Ocean include the Tijuana, Las Palmas, Guada-

lupe, Ensenada, Maneadero, Santo Tomas, San Vicente, Salado, San Rafael, San Telmo, Santo Domingo, Santa María, and El Rosario. South of the Río El Rosario, the desert conditions preclude continuous flow, although strong Pacific storms from the north or occasional hurricanes from the south cause episodic flooding within these usually dry river basins.

On the eastern side of the Peninsular Ranges, drainages are usually referred to by their canyon name, as their watercourses are ephemeral. In most years, the rainy season produces enough water to allow runoff to reach the desert floor. In the northern portion of the state, the Sierra Juárez drains into the basin Laguna Salada, which rarely fills high enough to drain to the Sea of Cortés. The Río Colorado, located along the state's boundary with Arizona and Sonora, historically formed a rich and productive delta but now is channeled and diverted for human use (Grismer 2002). South of the Sierra San Pedro Mártir, river drainages are dry, except in areas with spring-fed oases. In Baja California, spring-fed oases include San Fernando Velicatá, Cataviña, Santa María, Bahía de los Ángeles, San Borja, and Santa Gertrudis (Grismer and McGuire 1993).

The distributions of many of the native species in the state coincide with Baja California's phytogeographic regions (Grismer 1994b, 2002). As good indicators of natural biotic provinces, they likely are influenced by the same environmental characteristics that limit the distributions of amphibians and reptiles (Grismer 2002). Within the state of Baja California, five regions are present: (1) California, (2) Coniferous Forest, (3) Lower Colorado Desert, (4) Vizcaíno Desert, and (5) Central Gulf Coast Desert. A sixth region accounts for the marine species that spend the majority of their lives in the ocean (e.g., sea turtles and seasnakes). A comparison of the distributions of amphibians and reptiles and the boundaries of the phytogeographic regions reveals that 54 of the 113 (47.8%) native species are restricted to a single area. The remaining 59 species are generalists, with distributions occurring in two or more regions. These regions appear to be good indicators of the state's natural assemblages, despite broad transition zones between them.

The low percentage of amphibians native in the state (17/113, or 15%) compared to that of native reptiles (96/113, or 85%) is a result of the xeric conditions found throughout much of Baja California. Native amphibians are mainly confined to the California region and Coniferous Forests, but the Río Colorado and spring-fed oases also provide suitable habitats for these animals (Grismer and McGuire 1993; Grismer 1994b). Only two amphibian species (*Scaphiopus couchi* and *Anaxyrus punctatus*) have drought-tolerant life histories that allow them to range semicontinuously across the desert regions. The diversity of reptiles is more evenly distributed throughout the regions. The physical description of each region relies heavily on Grismer (1994b, 2002), which was adapted from Shreve and Wiggins (1964), Wiggins (1980), Turner and Brown (1982), and Cody et al. (1983).

California Region

Occupying the northwestern quarter of the state, the region extends 275 km from the US border to the vicinity of El Rosario along the Pacific coast, where it slowly intergrades into the Vizcaíno Desert region located to the south. The more southern regions in the state once supported communities similar to those now found throughout the California region, although successive periods of Pleistocene drying have replaced these with the Peninsular Deserts (Van Devender 1990). The result is intermixing over a large geographic area and the presence of remnants of the California region flora in the Vizcaíno Desert. To the east, the California region extends up the Sierra Juárez and Sierra San Pedro Mártir to an elevation as high as 2000 m at the start of the Jeffrey Pine Belt (Delgadillo 2004). Along the western slopes of the mountains, between 1500 and 2000 m, a meso-Mediterranean ecosystem of mixed chaparral-coniferous forest includes opens stands of Parry pine (*Pinus quadrifolia*) and montane chaparral

(Delgadillo 2004). The boundary of the California region usually extends to an elevation of about 1500 m, but sometimes higher. At lower elevations north of the Sierra Juárez, between the Sierras Juárez and San Pedro Mártir at Valle de la Trinidad and Paseo San Matías, and south of the San Pedro Mártir, the California region comes into contact and intermixes with the Lower Colorado Desert.

Within the Pacific Ocean, the California region is represented on four island groups: Islas Coronado (173.0 ha), Todos Santos (118.0 ha), San Martín (256.0 ha), and San Jerónimo (48.3 ha). Islas Coronado is located in coastal waters 13.0 km off Tijuana and is an archipelago of four main islands, two of which, Islas Coronado Norte and Coronado Sur, are inhabited by amphibians and reptiles (Grismer 1993, 2001; Samaniego-Herrera et al. 2007). Islas Todos Santos, located 17.3 km off Bahía de Ensenada, comprises two islands (Grismer 1993, 2001; Samaniego-Herrera et al. 2007). Isla San Martín is located 5.1 km off Bahía San Quintín, and Isla San Jerónimo is 9.7 km off the coast, just south of El Rosario (Samaniego-Herrera et al. 2007).

The California region is relatively cool most of the year due to prevailing northwesterly winds produced by the cold California Current, which blow inland off the waters of the Pacific Ocean (Hastings and Turner 1965; Meigs 1966). In summer months, temperatures average 20°C–25°C; cooler temperatures are found along the coastal margin, with warmer temperatures inland near the foothills of the Sierras Juárez and San Pedro Mártir. Winter temperatures average 10.0°C–12.5°C. The region receives most of its precipitation from Pacific winter storms originating in the Gulf of Alaska. Increased aridity occurs farther south in the vicinity of El Socorro. Here, the coast rarely receives any measurable rainfall from May to September (Humphrey 1974). In summer, low clouds and fog often cover the coastline, extending several kilometers inland and reaching 1000 m in elevation (Markham 1972).

This region is a southern extension of the coastal sage scrub and chaparral communities of southern California (Rebman and Roberts 2012). The dominant plant species in coastal areas are California sagebrush (*Artemisia californica*), white and black sage (*Salvia apiana* and *S. mellifera*), California buckwheat (*Eriogonum fasciculatum*), and coastal agave (*Agave shawii*). Chaparral occurs primarily within canyons in areas nearer to the coast and at higher elevations inland (Delgadillo 2004). Dominant chaparral species include chamise (*Adenostoma fasciculatum*), hoary leaf-lilac (*Ceanothus crassifolia*), toyon (*Heteromeles arbutifolia*), laurel sumac (*Malosma laurina*), and lemonadeberry (*Rhus integrifolia*).

CALIFORNIA REGION ASSEMBLAGE

Of the 54 amphibian and reptile species native to the California region, 8 are limited to the area: *Aneides lugubris, Ensatina eschscholtzii, Elgaria nana, Anniella geronimensis, A. stebbinsi, Diadophis punctatus, Lampropeltis herrerae,* and *Crotalus caliginis*.

The assemblage has 7.1% (8/113) of the state's total herpetofauna, representing three amphibians and five reptiles. Within Baja California, *Aneides lugubris, Ensatina eschscholtzii, Anniella stebbinsi,* and *Diadophis punctatus* have their southern distribution limits in the California region and enter Baja California from the United States. The two salamanders are confined to Mediterranean habitats in the southern part of their range. The southernmost distribution of *E. eschscholtzii* occurs at 22 km south of Ensenada along the coast (Peralta-García and Valdez-Villavicencio 2004), while the distribution of *Aneides lugubris* ends in the vicinity of Valle Santo Tomás (Lynch and Wake 1974).

The distribution of *Anniella stebbinsi* ranges widely throughout the California region (Bury 1983; Papenfuss and Parham 2013). In the north, it is found as far east as La Rumorosa, north of the Sierra Juárez (Hunt 1983). Farther south, it likely extends up sandy arroyos to the base of the Sierra San Pedro Mártir (Welsh 1988). Its southernmost distribution is found at Arroyo Pabellon, approximately 17 km south of San Quintín (Hunt 1983). The southernmost distri-

bution of *Diadophis punctatus* occurs on Isla San Martín located off the coast of San Quintín. On the peninsula, *D. punctatus* has been reported from Rancho San José and Arroyo San Telmo, 64 km to the north (Welsh 1988). No specimens have been reported from the intervening area, although these secretive snakes are likely more common than the paucity of specimens reflects (Grismer 2002).

Four assemblage members are endemic to Baja California. *Anniella geronimensis* is found on Islas San Jerónimo, San Martín, and coastal areas from approximately 6 km north of Colonia Guerrero to just south of Punta Baja at the northern edge of Bahía Rosario (Shaw 1949; Hunt 1983; Sánchez-Pacheco and Mellink 2001). Portions of its distribution likely occur within the ecotone between the California and Vizcaíno Desert regions, although its preferred habitat appears to be confined to within 4 km of the coast and does not range extensively into the desert transition (Shaw 1953; Grismer 2002).

The remaining three members, *Elgaria nana, Lampropeltis herrerae,* and *Crotalus caliginis,* are insular endemics. *Lampropeltis herrerae,* related to the California Mountain Kingsnake (*L. multifasciata*), is found on Isla Todos Santos Sur (Zweifel 1975; Samaniego-Herrera et al. 2007). Myers et al. (2013), in a passing remark, treated *L. herrerae* as a synonym of *L. multifasciata* in their revision of *L. zonata* but failed to discuss their reasoning. Grismer (1993) noted that its presence on the island represents a relictual distribution and is likely a result of diminishing habitat between the coast and the montane distribution of its nearest relative, *L. multifasciata,* in the Sierras Juárez and San Pedro Mártir.

Both *E. nana* and *C. caliginis* are endemic to Islas Coronados. *Elgaria nana* is found on both Islas Coronado Norte and Coronado Sur, while *C. caliginis* is found only on Isla Coronado Sur (Grismer 1993). Both appear to have smaller adult body sizes than those of their mainland relatives. *Elgaria nana* was found to be 42 mm smaller in snout-vent length than *E. multicarinata* (Fitch 1934). This led Grismer (2002) to elevate the originally described subspecies to species. Similarly, *C. caliginis* was originally described as a subspecies of the *C. viridis* (= *C. oreganus*) and elevated by Grismer (2002) based on its smaller adult body size.

A total of 10 amphibians and 44 reptiles live in the region, representing 47.8% (54/113) of the state's herpetofauna. In addition to the assemblage members, the region supports generalists, which occur in two or more of the phytogeographic regions of the state: *Batrachoseps major, Ensatina klauberi, Anaxyrus boreas, A. californicus, Pseudacris cadaverina, P. hypochondriaca, Rana draytonii, Spea hammondii, Actinemys marmorata, Elgaria multicarinata, Crotaphytus vestigium, Gambelia copeii, Coleonyx variegatus, Callisaurus draconoides, Phrynosoma blainvillii, Sceloporus occidentalis, S. orcutti, S. zosteromus, Urosaurus nigricaudus, Uta stansburiana, Plestiodon gilberti, P. skiltonianus, Aspidoscelis hyperythra, A. labialis, A. tigris, Xantusia henshawi, X. wigginsi, Lichanura trivirgata, Arizona elegans, Chilomeniscus stramineus, Hypsiglena ochrorhyncha, Lampropeltis getula, Masticophis fuliginosus, M. lateralis, Pituophis catenifer, P. vertebralis, Rhinocheilus lecontei, Salvadora hexalepis, Sonora semiannulata, Tantilla planiceps, Thamnophis hammondii, Trimorphodon lyrophanes, Rena humilis, Crotalus mitchellii, C. oreganus,* and *C. ruber.*

Batrachoseps major is found throughout the northern half of the California region but is more isolated in populations farther south. Grismer (1982) reports the southernmost locality to be Arroyo Grande, 24 km southeast of El Rosario. This species also occurs outside the California region in the Sierra San Pedro Mártir (Welsh 1988). Populations in the Sierra San Pedro Mártir are highly divergent from other populations, based on molecular data (Mártinez-Solano et al. 2012).

A number of species or subspecies have been described from the California region islands but have since been synonymized with their mainland relatives. *Aspidoscelis tigris vivida,* found on both Islas Coronado Norte and Coronado Sur, was originally described by Walker (1981) but recognized as a pattern class of *A. tigris* by Grismer (2002). *Pituophis catenifer coronalis,* found on Isla Coronado Sur, was originally described by Klauber (1946b) but later synonymized with its mainland counterparts by Grismer (2002). This snake is known from four museum specimens, but virtually nothing is known of its natural history (Oberbauer 1992; Grismer 2002). *Diadophis anthonyi,* found on Isla Todos Santos Sur, was originally described by Van Denburgh and Slevin (1923) but synonymized by Grismer (2001) due to lack of discrete diagnostic characters. *Elgaria multicarinata ignava* was described by Van Denburgh (1905) but recognized as a pattern class of *E. multicarinata* (Grismer 2002).

Much of the habitat in the California region has been replaced by urban development in metropolitan areas of Tijuana, Ensenada, and San Quintín. Smaller coastal communities are growing and in many areas have now merged with one another to form a sprawling coastal zone of urbanization. If not urbanized, most of the remaining flat areas have been converted to agriculture, including Valle Ojos Negros, Valle La Trinidad, and the San Quintín plain. These areas are water stressed as a result of groundwater extraction from the surrounding foothill canyons to support the agricultural industry. If the areas have not been drained, much of the original willow riparian forest of this region has been replaced by nonnative, highly invasive vegetation (*Arundo donax, Tamarix* spp., etc.). Illegal reptile collecting of high-value species (e.g., *Lichanura trivirgata, Lampropeltis herrerae*) has been documented (Mellink 1995). Mellink discovered collectors on Isla Todos Santos Sur attempting to take large numbers of reptiles off this small island. Cattle and other livestock are grazed throughout this region as well. The California region contains the southern third of one of five Mediterranean ecosystems on earth and is considered a biodiversity hot spot (Myers et al. 2001). Conservation areas along the coastal area are small and few. Ongoing planning is in progress to help protect Bahía San Quintín from development (The Nature Conservancy 2007). Conservation planning initiatives are in place to protect the fragile ecosystem of Colonet Mesa from future development (Clark et al. 2008).

Coniferous Forest Region

In the state of Baja California, prominent coniferous forests occur in both the Sierras Juárez and San Pedro Mártir. This forested ecosystem is the southernmost extension of the Sierran Montane Conifer Forest (Pase 1982) and supports cool mesic communities (Grismer 2002). The pine belt begins at 1500 m with mixed montane chaparral, pinyon, and Parry pine forests (Delgadillo 2004). The Sierra Juárez has more extensive forest coverage (342,113 ha) than the Sierra San Pedro Mártir (149,366 ha; SARH 1977). Extensive meadows exist at the mid-elevations of the mountains, which fill with water after heavy snowmelts (Welsh 1988).

Both mountains are central components of the Peninsular Range, oriented in a north-south direction, with gradually rising western slopes and steep eastern escarpments that drop precipitously to the desert floor. The eastern slope transitions are abrupt. At Picacho del Diablo, a 2700 m elevation gain is obtained within a 16 km distance from the Lower Colorado Desert to the top of the peak. The region extends south from the US-Mexican border approximately 300 km to near Cerro Matomi (Grismer 2002).

This region has the most reliable precipitation in the state, falling in winter and spring, and receives more rain than any other area in Baja California (Hastings and Turner 1965). Approximately 70% of the precipitation occurs in the winter from cold northern storms. Freezing temperatures are common; snow begins to fall around late November. The remainder of the precipitation falls in the summer during July to September (Roberts 1984) when warm, moist air from the Sea of Cortés moves in a north-westerly direction and is uplifted and cooled as it approaches the high elevations of the sierras (Humphrey 1974). Thunderstorms in July and August are not infrequent. Rainfall amounts can be highly variable, with a mean annual precipitation of 17.8 cm and the highest annual precipitation of 33.4 cm in the state (Hastings and Turner 1965). Drier conditions prevail at lower elevations on the east and west slopes. Mean monthly air temperatures for any given year may vary from −0.2°C to 17°C.

The floristic composition is relatively diverse, defined mainly by large shrubs and trees. Dominant species include pinyon pine (*Pinus quadrifolia*), Jeffrey pine (*P. jeffreyi*), lodgepole pine (*P. contorta*), sugar pine (*P. lambertiana*), California black oak (*Quercus kelloggii*), white fir (*Abies concolor*), quaking aspen (*Populus tremuloides*), and incense cedar (*Calocedrus decurrens*). These forests tend to have an open understory that includes greenleaf manzanita (*Arctostaphylos patula*).

CONIFEROUS FOREST REGION ASSEMBLAGE

Of the 41 species of amphibians and reptiles native to the region, 4 are limited to this area: *Rana boylii*, *Sceloporus vandenburgianus*, *Lampropeltis multifasciata*, and *Thamnophis elegans*.

This assemblage has 3.5% (4/113) of the state's total herpetofauna, representing one frog, a lizard, and two snakes. All four are montane-adapted species that prefer cool, mesic habitats. The distributions of each of the four species represent the southernmost extent of their ranges and are disjunct from their nearest populations in the United States (Grismer 2002). Both *S. vandenburgianus* and *L. multifasciata* are found in the Sierras Juárez and San Pedro Mártir, and the nearest populations to the north are found in the Laguna and Cuyamaca Mountains in San Diego County. Following Censky (1986) and references therein, Grismer (2002) regarded *S. vandenburgianus* to be absent from the Sierra Juárez. Based on museum records, this species occurs in both the Sierras Juárez and San Pedro Mártir. In the recent study of *L. zonata* by Myers et al. (2013), the populations from the Sierras Juárez and San Pedro Mártir were placed in *L. multifasciata*. Both populations had been previously referred to as the subspecies *L. z. agalma* (Zweifel 1952b, 1975), but Rodríguez-Robles et al. (1999) found the Sierra Juárez populations to be more closely related to populations in southern California, while the Sierra San Pedro Mártir population is more closely related to *L. herrerae* from Isla Todos Santos Sur. Myers et al. (2013) did not discuss *L. z. agalma* or *L. herrerae* in detail and treated both as synonyms of *L. multifasciata*.

Thamnophis elegans is found only in the Sierra San Pedro Mártir and is separated by 400 km from the nearest population found in the San Bernardino Mountains in California (Grismer 2002). Within the mountains, it occurs at elevations above 1820 m on the central slope (Welsh 1988).

The presence of *R. boylii* in the Sierra San Pedro Mártir needs additional research. The evidence for its presence in Baja California is based on two lost specimens. Loomis (1965) reported on specimens from Long Beach State College (1080 and 1081), collected by Elbert L. Sleeper in 1961 from the lower end of La Grulla meadow at 2000 m in the southern portion of the main scarp. The specimens were lost while being shipped (presumably between Long Beach State College and the University of California, Berkeley) but not before they were identified by R. C. Stebbins and R. G. Zweifel (Grismer 2002). Additional evidence of this species has not been found, leading Welsh (1988) and Lovich et al. (2009) to conclude that the species may be extirpated. If still present, the population of *R. boylii* represents a 480 km disjunction from the nearest, previously known population in the San Gabriel Mountains in California (Loomis 1965), which is also believed to be extirpated.

A typically montane species, *Ensatina klauberi* is found in isolated populations in both the Sierras Juárez and San Pedro Mártir. In the Sierra Juárez, it was found near Ranchos Las Cuevitas and Baja Largo del Sur, both approximately 19 km south of Laguna Hanson (Heim et al. 2005) and in the Sierra San Pedro Mártir at La Tasajera (Mahrdt et al. 1998). Recently, this species has been discovered in an extremely small area within an atypical habitat along the coast at San Quintín (Devitt et al. 2013). Until this discovery, *E. klauberi* would have been considered one of the best examples of a member of the Coniferous Forest region assemblage.

Despite the small number of species confined to the Coniferous Forest region, this area makes up an important mesic ecosystem within the state. A total of 9 amphibians and 32 reptiles live in the region, representing 36.3% (41/113) of the state's herpetofauna. In addition to the assemblage members, the region supports *Batrachoseps major*, *Ensatina klauberi*, *Anaxyrus boreas*, *A. californicus*, *Pseudacris cadaverina*, *P. hypochondriaca*, *Rana draytonii*, *Spea hammondii*, *Actinemys marmorata*, *Elgaria multicarinata*, *Gambelia copeii*, *Phrynosoma blainvillii*, *P. cerroense*, *Sceloporus occidentalis*, *S. orcutti*, *S. zosteromus*, *Uta stansburiana*, *Plestiodon gilberti*, *P. skiltonianus*, *Aspidoscelis tigris*, *Xantusia henshawi*, *X. wigginsi*, *Arizona elegans*, *Hypsiglena ochrorhyncha*, *Lampropeltis getula*, *Masticophis fuliginosus*, *M. lateralis*, *Pituophis catenifer*, *Rhinocheilus lecontei*, *Salvadora hexalepis*, *Tantilla planiceps*, *Thamnophis hammondii*, *Trimorphodon lyrophanes*, *Rena humilis*, *Crotalus mitchellii*, *C. oreganus*, and *C. ruber*.

Conservation threats to the area include climate change, cattle ranching, and logging. The latter is heavily managed and regulated to preserve the forest's integrity (SARH 1977). Two national parks are found in this region: Parque Nacional Constitución de 1857 (4950 ha) and Parque Nacional Sierra San Pedro Mártir (72,909 ha; Lovich et al. 2009).

Lower Colorado Desert Region

The desert regions of the northeastern quarter of the state contain the sedimentary deposits of the Río Colorado. Late Miocene to Late Pliocene sediments of the Bouse Formation represent a time when the region was submerged beneath a series of lakes filled by river water, which linked these bodies of water to the Sea of Cortés (Spencer and Pearthree 2001). As a result, the region is characterized by its broad, expansive basins lower than 400 m in elevation. The Lower Colorado Desert is the largest subregion of the Sonoran Desert and extends across southeastern California, southwestern Arizona, northwestern Sonora, and northeastern Baja California (Shreve 1951).

In the state of Baja California, the Lower Colorado Desert region extends 450 km from the US border to the vicinity of Bahía de los Ángeles. To the west, it is bounded by the steep slopes of the Sierras Juárez and San Pedro Mártir, and to the east, by the Arizona and Sonora border in the north and the Sea of Cortés in the southeast. Below Puertecitos, the region narrows to less than 13 km from the coast and in the south intergrades widely with the Gulf Coast Desert region (Peinado et al. 1994).

Despite its arid condition, the Río Colorado once provided a rich aquatic ecosystem, which included an expansive delta and estuary. Today, water no longer flows from the river to the Sea of Cortés. Extensive canal systems bring most of the water from the lower portions of the river to agricultural and urban areas in California, Arizona, and Baja California. Other sources of water within this region come from the eastern slopes of the Sierras Juárez and San Pedro Mártir.

Besides the eastern slopes of the sierras, the region contains a number of mountain ranges. Isolated between Laguna Salada and the Río Colorado delta, the Sierras de los Cucapás and El Mayor are contiguous but isolated within the region. Other mountain ranges connect to the Peninsular Range and include the Sierras Las Tinajas, Las Pintas, and San Felipe.

Within the Sea of Cortés, just north of Bahía San Luis Gonzaga, six islands occur within the region: El Muerto, Coloradito, Encantada, Blancos, San Luis, and Willard. The first five are a part of the Islas Las Encantadas archipelago, while Isla Willard is a nearshore mountain connected to the peninsula by a sand spit during extreme low tides.

The Lower Colorado Desert region, lying in the rain shadow of the Sierras Juárez and San Pedro Mártir, is the hottest and most barren desert in Baja California. As a result of low elevation and little topographic relief, it receives less than 5 cm of rainfall annually and is considered the driest desert region in North America (Meigs 1953; Hastings and Turner 1965). Most precipitation is unreliable and received during the summer months (Humphrey 1974), and on rare occasions the region is affected by hurricanes (Grismer 2002). However, these anticyclonic storms have come ashore as far north as San Felipe (Markham 1972). Mean temperatures for July and August are above 32.5°C, and the hottest monthly mean on record is 38.2°C; winter mean temperature is 12.5°C (Markham1972).

The vegetation of the region is dominated by creosote bush (*Larrea tridentata*). The northern portion lies directly within the rain shadow of the mountains and is characterized by low-lying, flat, sandy, hot, and arid conditions (Meigs 1953). Farther to the south, the region becomes more rugged, with large volcanic flows, mountain ranges, and dry washes. Dominant plants include white bursage (*Ambrosia dumosa*), ocotillo (*Fouquieria splendens*), brittlebush (*Encelia farinosa*), and desert agave (*Agave deserti*). In the arroyos and rugged foothill areas, mesquite (*Prosopis glandulosa*), smoke tree (*Psorothamnus spinosa*), little-leaf palo verde (*Cercidium microphyllum*), ironwood (*Olneya tesota*), and chollas (*Cylindropuntia* sp.) are found (Grismer 2002).

LOWER COLORADO DESERT REGION ASSEMBLAGE

Of the 56 species of amphibians and reptiles native to this region, 22 are limited to this region: *Anaxyrus cognatus, A. woodhousii, Incilius alvarius, Lithobates yavapaiensis, Crotaphytus grismeri, Gambelia wislizenii, Phrynosoma mcallii, P. platyrhinos, Sceloporus magister, Uma notata, Urosaurus graciosus, U. ornatus, Uta encantadae, U. lowei, U. tumidarostra, Chionactis occipitalis, Hypsiglena chlorophaea, Masticophis flagellum, Thamnophis marcianus, Crotalus atrox, C. cerastes,* and *C. muertensis.*

This assemblage has 19.5% (22/113) of the state's total herpetofauna, representing 4 amphibians and 18 reptiles. Of the six regions within the state, the Lower Colorado Desert region contains the most species confined to its boundaries. Of the 22 species, 5 are associated with the Río Colorado and its delta, 12 are arid-adapted and have their southernmost distributions in the region, and 5 are regional endemics.

Assemblage species that extend southward from the United States into the Lower Colorado Desert include *Gambelia wislizenii*, *Phrynosoma mcallii*, *P. platyrhinos*, *Sceloporus magister*, *Uma notata*, *Urosaurus graciosus*, *U. ornatus*, *Chionactis occipitalis*, *Hypsiglena chlorophaea*, *Masticophis flagellum*, *Crotalus atrox*, and *C. cerastes*. *Gambelia wislizenii* and *S. magister* are widely distributed across the region. Along the western edge of their range they are known to extend up Paseo San Matías and overlap with the distributions of *G. copeii* and *S. zosteromus*, respectively, with no signs of intergradation (McGuire 1996; Grismer and McGuire 1996). This same pattern also occurs with *M. flagellum* and *M. fuliginosus* (Grismer 1994b) and likely occurs in many other species as well (Grismer and Mellink 2005).

The distributions of *P. platyrhinos*, *Urosaurus graciosus*, *Chionactis occipitalis*, *H. chlorophaea*, and *Crotalus cerastes* occur broadly across the region. *Chionactis occipitalis* extends 34 km south of San Felipe (Grismer 1989a). Both *U. graciosus* (Grismer 1989b) and *P. platyrhinos* (Grismer 2002) extend southward to Bahía San Luis Gonzaga. *Crotalus cerastes* occurs as far south as Arroyo Calamajué (Nelson 1922) but is suspected to occur south of Bahía de los Ángeles (Grismer 2002). *Hypsiglena chlorophaea* has been newly circumscribed by Mulcahy (2008), using genetic data, and is believed to occur throughout the Lower Colorado Desert.

More limited in their southward ranges into the region are *P. mcallii*, *Uma notata*, *Urosaurus ornatus*, and *Crotalus atrox*. *Phrynosoma mcallii* extends only 80 km below the border (Funk 1981), while *U. notata* is found as far south as the northern end of the Sierra Las Pintas (Pough 1977). *Crotalus atrox* occurs in the extreme northeastern corner of the state and extends westward to the western side of the Sierra de los Cucapás (Grismer 2002). *Urosaurus ornatus* is strictly associated with the riparian vegetation along the Río Colorado (Grismer 2002).

Anaxyrus cognatus, *A. woodhousii*, *Incilius alvarius*, *Lithobates yavapaiensis*, and *Thamnophis marcianus* are associated with the Río Colorado, with *A. cognatus*, *A. woodhousii*, and *T. marcianus*

expanding their ranges in association with the agricultural industry and its network of canals and irrigation ditches. *Incilius alvarius* is believed to narrowly occur in Baja California, but little is known of the extent of its distribution. Brattstrom (1951) reported the only known specimen, which was found 48 km south of Mexicali and 7.6 km north of El Mayor. Brattstrom also reported on the occurrence of *T. marcianus* in the same location at 7.7 km north of El Mayor. *Lithobates yavapaiensis* has not been seen in the lower Río Colorado for more than three decades and is believed to be extirpated as the result of the introduction of *L. berlandieri* and *L. catesbeianus* (Vitt and Ohmart 1978; Platz and Frost 1984; Clarkson and de Vos 1986; Platz et al. 1990; Mellink and Ferreira-Bartrina 2000).

Of the five endemic species to the region, four are insular endemics on islands in the Encantada archipelago. *Crotalus muertensis* was described by Klauber (1949b) and is found only on Isla El Muerto, the northernmost island in the archipelago. *Uta lowei* is also endemic to Isla El Muerto and was described by Grismer (1994c), who also described *U. tumidarostra* from Isla Coloradito and *U. encantadae* from Isla Encantada and Islote Blancos. All the *Uta* from the Encantada archipelago have hypertrophied nasal glands and inflated rostrums as a result of intertidal habits and salty diets (Grismer 1994c).

McGuire (1994) described *Crotaphytus grismeri*, which is endemic to the Sierras de los Cucapás and El Mayor. The distribution of *C. grismeri* is hardly known, with the only records coming from the type series (McGuire 1994). These mountains appear to have been isolated from the Peninsular Ranges to the south and west for their entire history (Barnard 1970).

A total of 9 amphibians and 47 reptiles are native to the region, representing 49.6% (56/113) of the state's herpetofauna. In addition to the assemblage members, the region supports *Anaxyrus boreas*, *A. punctatus*, *Pseudacris cadaverina*, *P. hypochondriaca*, *Scaphiopus couchii*, *Crotaphytus vestigium*, *Coleonyx switaki*, *C. variegatus*, *Phyllodactylus nocticolus*, *Dipsosaurus dorsalis*, *Sauromalus ater*, *Callisaurus draconoides*,

Petrosaurus mearnsi, *Sceloporus orcutti*, *Urosaurus nigricaudus*, *Uta stansburiana*, *Plestiodon gilberti*, *Aspidoscelis tigris*, *Xantusia henshawi*, *X. wigginsi*, *Lichanura trivirgata*, *Arizona elegans*, *Bogertophis rosaliae*, *Hypsiglena chlorophaea*, *Lampropeltis getula*, *Phyllorhynchus decurtatus*, *Pituophis catenifer*, *Rhinocheilus lecontei*, *Salvadora hexalepis*, *Sonora semiannulata*, *Tantilla planiceps*, *Trimorphodon lyrophanes*, *Rena humilis*, *Crotalus mitchellii*, and *C. ruber*.

Severe threats to the northeastern region include extensive agricultural expansion and urbanization from the city of Mexicali. Little water flows in the natural river course of the Río Colorado. Throughout the region, racing organizations (e.g., SCORE International) sponsor off-road races, which promote extensive off-road recreational use of the desert in unregulated areas. The only protected regions are contained in the Alto Golfo de California Biosphere Reserve (Lovich et al. 2009; Murphy and Méndez de la Cruz 2010).

Vizcaíno Desert Region

This region lies in the southwestern corner of the state of Baja California and is separated from the Sea of Cortés to the east by the Central Gulf Coast Desert region, which extends from near El Rosario to Laguna San Ignacio in Baja California Sur. The western boundary is the Pacific Ocean; its eastern extent varies greatly but generally is limited by the Peninsular Range. The southern portion of the Vizcaino Desert region is much flatter and extends inland to the low, rocky foothills of the mountains. In Baja California, the flatter regions extend for only a short distance north of the state boundary with Baja California Sur to 20 km north of Villa Jesús María, while the northern portions consist of smaller mountain ranges, mesas, and dry washes (Bostic 1971).

In the Pacific Ocean, this region includes two island groups, Islas Cedros and San Benito, both located in the extreme southwestern area of the state. Isla Cedros is the largest Pacific island in the state, covering 34,827 ha, reaching 1204 m in elevation, and lying 100 km from the nearest point on the peninsula (Samaniego-Herrera et al.

2007). Islas San Benito is composed of an archipelago of three closely positioned islands (Islas San Benito Oeste, Medio, and Este) that cover approximately 600 ha, reach 212 m in elevation, and lie 140 km off the coast (Samaniego-Herrera et al. 2007).

The region is characterized as a "fog-type" temperate desert climate with little winter and summer precipitation (Meigs 1966) and is considered a part of the Peninsular Deserts (Hafner and Riddle 1997). The mild climatic condition is greatly influenced by prevailing westerly winds that move the cold California Current of the Pacific Ocean into coastal waters. The marine influence generates layers of cool, moist air that move beneath the dry, descending air, resulting in fog, low clouds, and the lack of precipitation (Bostic 1971). Heavy fog may extend inland for 10 km (Grismer 2002). Maximum precipitation occurs in winter (December–February) and averages 5.5 cm. Precipitation in the spring (March–May) and summer (June–August) progressively decreases, with summer receiving an average of 1 cm of rainfall (Hastings and Turner 1965). Mean air temperatures in the summer (July–September) and winter (January–February) are between 23°C and 28°C and 15°C and 18°C (Markham 1972), respectively, producing a mild, temperate climate.

A number of spring-fed oases, including San Fernando Velicatá, Cataviña, San Borja, and Santa Gertrudis, are found throughout the region, which support more mesic communities (Grismer and McGuire 1993).

In much of the region the vegetation is open, stunted, widely spaced, and depauperate because of continuous onshore winds from the Pacific Ocean (Grismer 2002). In areas protected from the winds, plant diversity increases sharply. The dominant plants include blue agave (*Agave cerulata*), datilillo (*Yucca valida*), bursages (*Ambrosia dumosa* and *A. camphorata*), cirio (*Fouquieria columnaris*), palo adán (*F. diguetii*), cardón (*Pachycereus pringlei*), elephant tree (*Pachycormus discolor*), pitahaya agria (*Stenocereus gummosus*), little-leaf palo verde (*Cercidium microphyllum*), and mesquite (*Prosopis glandulosa*).

VIZCAÍNO DESERT REGION ASSEMBLAGE

Of the 52 species of amphibians and reptiles native to this region, 5 are limited to this area: *Elgaria cedrosensis, Urosaurus lahtelai, Bipes biporus, Arizona pacata,* and *Pituophis insulanus.*

This assemblage has 4.4% (5/113) of the state's total herpetofauna, representing two lizards, one amphisbaenid, and two snakes. Three species represent endemics to the state of Baja California. *Pituophis insulanus* is endemic to Isla Cedros (Klauber 1946b) and is found throughout the island from the coast to mountaintops (Grismer 2002). *Elgaria cedroensis* is also a state endemic found on both Isla Cedros and the adjacent coastal areas of the peninsula from 30 km south of El Rosario to 20 km north of Villa Jesús María (Bostic 1971; Lais 1976; Grismer 1988; Grismer and Hollingsworth 2001). And *U. lahtelai,* described by Rau and Loomis (1977), is found in the vicinity of Cataviña, Las Arrastras de Arriola, and Las Palmas (Grismer 2002). The evolution of *U. lahtelai* is interesting because it appears to be associated with the exposure of the granitic batholith underlying this region and is surrounded by its sister species, *U. nigricaudus* (Rau and Loomis 1977; Grismer 1994b).

A number of species that occur in the Vizcaíno Desert region represent northern distribution extensions from the southern peninsula. Only *B. biporus* and *A. pacata* are confined to the region. *Bipes biporus* is the only amphisbaenid on the peninsula and extends across the state border to approximately 17 km north of Villa Jesús María (Papenfuss 1982). *Arizona pacata* prefers cool coastal deserts and extends northward to approximately the turnoff to Bahía de los Ángeles along Mex. Hwy 1 (Grismer 2002). Other species that follow a similar distributional pattern but range partly into other phytogeographic regions include *Petrosaurus repens, Aspidoscelis labialis, Chilomeniscus stramineus, Hypsiglena slevini,* and *Crotalus enyo.*

A total of 5 amphibians and 47 reptiles are native to the region, representing 46% (52/113) of the state's herpetofauna. In addition to the assemblage members, the region supports

Anaxyrus boreas, A. punctatus, Pseudacris cadaverina, P. hypochondriaca, Scaphiopus couchii, Elgaria multicarinata, Crotaphytus vestigium, Gambelia copeii, Coleonyx switaki, C. variegatus, Phyllodactylus nocticolus, Dipsosaurus dorsalis, Sauromalus ater, Callisaurus draconoides, Petrosaurus mearnsi, P. repens, Phrynosoma cerroense, Sceloporus occidentalis, S. orcutti, S. zosteromus, Urosaurus nigricaudus, Uta stansburiana, Aspidoscelis hyperythra, A. labialis, A. tigris, Xantusia wigginsi, Lichanura trivirgata, Bogertophis rosaliae, Chilomeniscus stramineus, Hypsiglena ochrorhyncha, H. slevini, Lampropeltis getula, Masticophis fuliginosus, M. lateralis, Phyllorhynchus decurtatus, Pituophis vertebralis, Rhinocheilus lecontei, Salvadora hexalepis, Sonora semiannulata, Tantilla planiceps, Thamnophis hammondii, Trimorphodon lyrophanes, Rena humilis, Crotalus enyo, C. mitchellii, C. oreganus, and *C. ruber.*

A few species would not be found in the arid conditions of the Vizcaíno Desert region if not for the presence of spring-fed oases. The presence of *Pseudacris cadaverina, P. hypochondriaca,* and *Thamnophis hammondii* in oases is believed to represent evidence of a more mesic ecological past (Grismer and McGuire 1993; McGuire and Grismer 1993). Likewise, mountaintop refugia have been found for *Elgaria multicarinata* in the Sierra La Asamblea and for *Sceloporus occidentalis* at the higher elevations of Isla Cedros (Grismer and Mellink 1994).

The region is mostly unpopulated and lacks large urbanized areas. Cattle ranching, small plots of agriculture, and disturbance in and around the water sources of the oases appear to be the greatest conservation threats. Salt mining just across the border in Baja California Sur has created the need for an industrial port on Isla Cedros. Valle de los Cirios is a protected area that extends from the southern state boundary to approximately El Rosario and includes 2,521,776 ha, or 35% of the state.

Central Gulf Coast Desert Region

A long, narrow stretch of desert bordering the Sea of Cortés, the Central Gulf Coast Desert region begins just south of Bahía de los Ángeles and extends southward into Baja California Sur to the Isthmus of La Paz (Shreve and Wiggins 1964; Wiggins 1980; Turner and Brown 1982; Cody et al. 1983; Cody et al. 2002). Some authors believe the Central Gulf Coast Desert begins farther north in the vicinity of Bahía Calamajué (Peinado et al. 1994), with the difference in opinions highlighting the broad intermixing that occurs between this region and the Lower Colorado Desert. To the east, the region is bounded by the Sea of Cortés, and to the west, by the Peninsular Range and Vizcaíno Desert region.

Numerous islands in the state of Baja California occur within the midriff region of the Sea of Cortés (Carreño and Helenes 2002). The Midriff Islands include Islas Mejía, Granito, Ángel de la Guarda, Pond, Cardonosa Este, Partida Norte, Rasa, Salsipuedes, San Lorenzo Norte, and San Lorenzo Sur. Numerous coastal islands occur within Bahía de los Ángeles, including Islas Bota, Cabeza de Caballo, Cerraja, Flecha, La Ventana, Mitlán, Pata, Piojo, and Smith (Grismer 2002).

This very hot, arid region receives nearly all of its precipitation in summer and fall. Severe droughts occur during March–June with a mean precipitation of 0.2 cm (Hastings and Turner 1965; Humphrey 1974). The majority of the Central Gulf Coast Desert rainfall originates from southern convectional storms and appears as runoff from the bordering Peninsular Range. Occasionally, the region receives rainfall as the result of chubascos that originate from the warm waters of the Sea of Cortés. Despite periods of severe drought, annual mean rainfall can reach 16.8 cm (Hastings and Turner 1965), provided there has been an active hurricane season. Temperatures in the Central Gulf Coast Desert are extremely hot. Mean temperatures for the warmest months in July–August are above 30°C; mean temperatures for the coldest month in January are 15°C (Markham 1972).

The vegetation characterized by Grismer (2002) includes leatherplant (*Jatropha cuneata*), lomboy (*J. cinerea*), palo adán (*Fouquieria diguetii*), copal (*Bursera hindsiana*), elephant tree (*B. microphylla*), cardón (*Pachycereus pringlei*), and little-leaf palo verde (*Cercidium microphyllum*).

CENTRAL GULF COAST DESERT REGION ASSEMBLAGE

Of the 45 species of amphibians and reptiles native to this region, 9 are limited to this area: *Crotaphytus insularis, Phyllodactylus partidus, Sauromalus hispidus, Callisaurus splendidus, Petrosaurus slevini, Uta antiqua, Aspidoscelis cana, Crotalus angelensis,* and *C. lorenzoensis.*

This assemblage has 8% (9/113) of the state's total herpetofauna, representing seven lizards and two snakes. All are insular endemics to the Midriff Islands. *Sauromalus hispidus* is the most widely distributed on these islands and is found on Islas Mejía, Granito, Ángel de la Guarda, Pond, Salsipuedes, San Lorenzo Norte, and San Lorenzo (Hollingsworth 1998). It is believed to have been introduced onto Islas Cabeza de Caballo, Flecha, La Ventana, Piojo, and Smith in Bahía de los Ángeles by the Seri Indians (Grismer et al. 1995; Hollingsworth 1998). *Sauromalus hispidus* has also been reported on Isla Rasa (Velarde et al. 2008).

Three species, *Crotaphytus insularis, Callisaurus splendidus,* and *Crotalus angelensis,* are endemic to Isla Ángel de la Guarda, while *Petrosaurus slevini* is endemic to Isla Ángel de la Guarda and its northern satellite Isla Mejía (McGuire 1996; Grismer 1999b; Murphy and Méndez de la Cruz 2010). Another three species are found on the San Lorenzo block, which includes Islas Salsipuedes, San Lorenzo Norte, and San Lorenzo Sur. *Uta antiqua* and *Aspidoscelis cana* occur on all three islands, while *Crotalus lorenzoensis* is limited to Isla San Lorenzo Sur (Grismer 2002; Murphy and Méndez de la Cruz 2010). *Phyllodactylus partidus* is endemic to two small islands, Islas Cardonosa Este and Partida Norte (Dixon 1964; Grismer 1999b).

A total of 2 amphibians and 43 reptiles are native to the region, representing 39.8% (45/113) of the state's herpetofauna. In addition to the assemblage members, the region supports *Ana-*

xyrus punctatus, Scaphiopus couchii, Crotaphytus vestigium, Gambelia copeii, Coleonyx switaki, C. variegatus, Phyllodactylus nocticolus, Dipsosaurus dorsalis, Sauromalus ater, Callisaurus draconoides, Petrosaurus repens, Phrynosoma cerroense, Sceloporus orcutti, S. zosteromus, Urosaurus nigricaudus, Uta stansburiana, Aspidoscelis hyperythra, A. tigris, Xantusia wigginsi, Lichanura trivirgata, Bogertophis rosaliae, Chilomeniscus stramineus, Hypsiglena ochrorhyncha, H. slevini, Lampropeltis getula, Masticophis fuliginosus, Phyllorhynchus decurtatus, Pituophis vertebralis, Salvadora hexalepis, Sonora semiannulata, Tantilla planiceps, Trimorphodon lyrophanes, Rena humilis, Crotalus enyo, C. mitchellii, and *C. ruber.*

Conservation threats to assemblage members are associated with their confinement on islands and the potential for introduced predators to greatly diminish population sizes (Murphy and Méndez de la Cruz 2010). All the islands within this assemblage are protected within the 387,956 ha Reserva de la Biósfera Bahía de Los Ángeles, Canales de Ballenas y de Salsipuedes (Lovich and Mahrdt 2008; Lovich et al. 2009). The coastal areas of the Central Gulf Coast Desert region within the state of Baja California are protected within Valle de los Cirios (Lovich et al. 2009).

Marine Region

The waters off the coasts of Baja California represent two dissimilar bodies. The Pacific Ocean is dominated by the north-to-south flow of the California Current, which brings cool water down the west coast of the state. In the Sea of Cortés, the current flows northward along the mainland coast and southward down the peninsula in the summer months and the reverse direction in the winter. Temperatures in the northern portion of the sea average 8.2°C in December and 32.6°C in August (Álvarez-Borrego 2002).

MARINE REGION ASSEMBLAGE

There are six species limited to this assemblage: *Caretta caretta, Chelonia mydas, Eretmochelys*

imbricata, Lepidochelys olivacea, Dermochelys coriacea, and *Hydrophis platurus.*

Five of these are sea turtles, and one is a seasnake. *Caretta caretta* is not as common as the other sea turtles (Grismer 2002). Despite its rarity in the Sea of Cortés, it has been observed as far north as Bahía Ometepec near the mouth of the Río Colorado (Shaw 1947). In the Pacific Ocean, it is known from as far north as the Islas Coronado (Caldwell 1962) and Isla Guadalupe (Smith and Smith 1979). *Chelonia mydas* is widespread in the Sea of Cortés but less common off the Pacific coast (Grismer 2002). It is also known to enter the Río Hardy and Río Colorado (Smith and Smith 1979) and was reported as far north as El Mayor, 80 km north of the mouth of the Río Colorado (Cliffton et al. 1982).

Eretmochelys imbricata is not commonly seen much farther north than Isla Cedros in the Pacific Ocean (Smith and Smith 1979) and has been reported as far north as Bahía de los Ángeles in the Sea of Cortés (Caldwell 1962). *Lepidochelys olivacea* is rare in the Sea of Cortés, although it has been found at San Felipe (Caldwell 1962). In the Pacific Ocean, it ranges along the entire west coast of Baja California (Zug et al. 1998). *Dermochelys coriacea* is considered rare in the Sea of Cortés but has been found as far north as Puerto Peñasco, Sonora, Mexico (Smith and Smith 1979). In the Pacific Ocean, its pelagic habits make sightings rare. *Hydrophis platurus* ranges throughout all areas of the Sea of Cortés, but is rarely observed in waters off the state of Baja California (Pickwell et al. 1983; Campbell and Lamar 2004).

Generalists

Of the 113 species native to the state of Baja California, 59 (52.2%) are considered generalists, with distributions extending into two or more phytogeographic regions.

GENERALIST ASSEMBLAGE

The generalists include the following species: *Batrachoseps major, Ensatina klauberi, Anaxyrus* boreas, *A. californicus, A. punctatus, Pseudacris cadaverina, P. hypochondriaca, Rana draytonii, Scaphiopus couchii, Spea hammondii, Actinemys marmorata, Elgaria multicarinata, Crotaphytus vestigium, Gambelia copeii, Coleonyx switaki, C. variegatus, Phyllodactylus nocticolus, Dipsosaurus dorsalis, Sauromalus ater, Callisaurus draconoides, Petrosaurus mearnsi, P. repens, Phrynosoma blainvillii, P. cerroense, Sceloporus occidentalis, S. orcutti, S. zosteromus, Urosaurus nigricaudus, Uta stansburiana, Plestiodon gilberti, P. skiltonianus, Aspidoscelis hyperythra, A. labialis, A. tigris, Xantusia henshawi, X. wigginsi, Lichanura trivirgata, Arizona elegans, Bogertophis rosaliae, Chilomeniscus stramineus, Hypsiglena ochrorhyncha, H. slevini, Lampropeltis getula, Masticophis fuliginosus, M. lateralis, Phyllorhynchus decurtatus, Pituophis catenifer, P. vertebralis, Rhinocheilus lecontei, Salvadora hexalepis, Sonora semiannulata, Tantilla planiceps, Thamnophis hammondii, Trimorphodon lyrophanes, Rena humilis, Crotalus enyo, C. mitchellii, C. oreganus,* and *C. ruber.*

Records of introduced species are best documented for *Lithobates berlandieri* along the Río Colorado, which is believed to be replacing *L. yavapaiensis* (Clarkson and Rorabaugh 1989; Platz et al. 1990; Rorabaugh et al. 2002). Also introduced into the Río Colorado is *Apalone spinifera* (Linsdale and Gressitt 1937; Miller 1946), although references to specific records are absent. *Lithobates catesbeianus* is known to occur in numerous watersheds within the California region (Clarkson and de Vos 1986; Grismer 2002) and the Río Colorado (Lovich et al. 2009), but not in oases in the southern portion of the state (Grismer and McGuire 1993). *Xenopus laevis* has been reported to occur between Ensenada and Tijuana (Ruíz-Campos and Valdez-Villavicencio 2012). *Hemidactylus turcicus* is reported from Ensenada (Martínez-Isaac and Valdez-Villavicencio 2000). Murphy and Méndez de la Cruz (2010) list *Taricha torosa* as introduced into Baja California, but no specific information is provided and its presence as an introduced species is doubtful. No records of this species are known from Baja California (Stebbins 1962; see also Slevin 1928). The southernmost distribution of this species occurs at Boulder Creek in San Diego County, California.

Herpetofauna of Sonora

JULIO A. LEMOS-ESPINAL
AND JAMES C. RORABAUGH

Introduction

The study of the amphibians and reptiles of Sonora has received a great deal of attention since the middle of the nineteenth century, undoubtedly because of the nearby presence of Arizona and California, with large population centers that include many specialists and amateurs interested in the rich herpetofauna of those areas. The continuity of the desert habitats of Arizona and Sonora and the 14 primary Sonoran islands in the Sea of Cortés represents a challenge for studies of biogeography, speciation, and island ecology and inevitably led to early exploration of parts of this state from more northern centers. The rugged terrain of the Sierra Madre Occidental of Sonora and its continuity with the tropical/subtropical environments of southern Sonora have also attracted the attention of numerous herpetologists to such regions as Álamos, Yécora, and the sky islands associated with the Sierra Madre Occidental, which have been extensively studied. The terrain and climate of northern, central, and coastal Sonora are also conducive to development of roads, allowing ready access to those parts of the state and promoting the development of herpetological studies.

Initially, samples of the amphibians and reptiles of Sonora were obtained from near the border with the states of Arizona and New Mexico by the Mexican–United States Boundary Commission. Although the commission did not focus exclusively on collections of herpetofaunal species, the number, identity, and collection localities of the specimens obtained were important and covered a considerable area of the state of Sonora.

The United States National Museum (USNM) played an important role in the study of the amphibians and reptiles of the state through the explorations of Nelson and Goldman at the end of the nineteenth century. The American Museum of Natural History (AMNH) promoted the study of the amphibians and reptiles of the islands of the Sea of Cortés and the Álamos-Güirocoba region of extreme southeastern Sonora. The California Academy of Sciences (CAS) supported the study of the herpetofauna of the extreme western part of the state through the work of Van Denburgh and Slevin in the early twentieth century.

At present several institutions such as the Universidad Nacional Autónoma de México, the University of Arizona, the US Fish and Wildlife Service, Sky Island Alliance, the Drylands Institute, and the Arizona-Sonora Desert Museum have been working on the study of the amphibians and reptiles of several parts of Sonora.

In general, most of the studies of the amphibians and reptiles of Sonora have focused on the islands in the Sea of Cortés and the coast, the border with Arizona and New Mexico, the Álamos-Güirocoba region, and the Yécora region. A large number of collections have been taken from along the main highways that cross the state, such as Mex. Hwys 2 (San Luis Río Colorado–Sonoyta–Caborca), 8 (Sonoyta–Puerto Peñasco), 15 (Estación Don–Nogales), and 16 (Maycoba–Hermosillo) and the road between Navojoa and Álamos. Some areas still have not been studied because of poor access and harsh environmental conditions. An example is the large desert area northeast of Puerto Libertad, south and east of Caborca, and west of that portion of Hwy 15 that runs between Hermosillo and Santa Ana.

We know that at present Sonora has a total of 195 herpetofaunal species: 38 amphibians (3 salamanders, 35 anurans) and 157 reptiles (16 turtles, 69 lizards, 72 snakes). Seven of these species are introduced: *Lithobates berlandieri*, *L. catesbeianus*, *Apalone spinifera*, *Hemidactylus frenatus*, *H. turcicus*, *Sauromalus hispidus*, and *Indotyphlops braminus*. Six are oceanic: *Caretta caretta*, *Chelonia mydas*, *Eretmochelys imbricata*, *Lepidochelys olivacea*, *Dermochelys coriacea*, and *Hydrophis platurus*. Another 13 are insular: *Phyllodactylus nocticolus*, *P. nolascoensis*, *Ctenosaura conspicuosa*, *C. nolascensis*, *Sauromalus hispidus*, *S. varius*, *Uta nolascensis*, *U. palmeri*, *Aspidoscelis baccata*, *A. estebanensis*, *A. martyris*, *Masticophis slevini*, and *Crotalus estebanensis*.

The topographically complex transition between the Neotropical and Nearctic zones, the highly diverse ecosystems of the Sierra Madre Occidental, the equally highly heterogeneous and diverse Sonoran Desert, the extensive coastline of the state, and its 14 primary islands with unique biota have resulted in a richness of amphibian and reptile species among the greatest in any state of northern Mexico. We are confident that the total number of Sonoran herpetofaunal species is actually even larger. The main additions would likely be found in the subtropical region of the extreme southeastern part of the state and the northeastern grasslands; they remain a lure for future herpetological studies.

Previous Herpetological Studies

The first herpetological collections made in the state of Sonora were obtained by the Mexican–United States Boundary Commission from 1849 to 1851. Collections were made in all of the states along the Mexico-US border, including Sonora. All the specimens collected by the commission are deposited in the USNM. Parts of these collections were reported by Baird (1859), Cope (1867, 1900), and Kellogg (1932).

Later, Lovewell and Heiligbrodt made a collection that was sent to Washburn College, Topeka, Kansas. Francis Whittemore Cragin reported this collection in 1884 and included information on *Sceloporus clarkii*, *Urosaurus ornatus*, *Thamnophis cyrtopsis*, and *Micruroides euryxanthus*.

A number of specimens collected by the Mexican–United States Boundary Commission became the types of several taxa originally described and reported for the state of Sonora. However, after the James Gadsden Purchase on December 30, 1853 (Sale of La Mesilla), a region of present-day southern Arizona and southwestern New Mexico became part of the United States. Hobart M. Smith and Edward H. Taylor (1950) restricted the involved type localities to Arizona or New Mexico. Those taxa include *Hyla affinis* (= *H. arenicolor*, Santa Rita Mountains, Arizona); *Kinosternon sonoriense* (Tucson, Arizona); *Heloderma suspectum* (Sierra de Morena, Arizona); *Sceloporus clarkii* (Santa Rita Mountains, Arizona); *S. poinsettii* (Río San Pedro of the Río Grande del Norte, New Mexico); *Eutaenia megalops* (= *Thamnophis eques megalops*, Tucson, Arizona); *Ophibolus splendida* (= *Lampropeltis getula splendida*, Santa Rita Mountains, Arizona); *Salvadora grahamiae* (Santa Rita Mountains, Arizona); and *Sonora semiannulata* (Santa Rita Mountains, Arizona).

Carl Lumholtz and his expedition explored the states of Sonora and Chihuahua from 1890 to 1892. The results of the birds and mammals collected in this expedition were reported by J. A. Allen in 1893. In this expedition F. Robinette collected the first specimen of what later became, along with two other specimens collected by a

Mr. Eustace in 1897, the type series of *Phryno-soma ditmarsi*, described by Stejneger in 1906. For a period of 73 years after the original description of this lizard species, there were no records of it until its rediscovery in 1970 by Vincent D. Roth, as documented by Lowe et al. (1971) and Roth (1997).

In 1892 Gustav Eisen and Walter Bryant made a collection of reptiles that was sent to the CAS. Van Denburgh reported this collection in 1898.

From October 27, 1898, to January 31, 1899, Edward Goldman and Edward Nelson of the USNM visited the extreme southeastern part of the state in the towns of Camoa, Batamotal, and Álamos, collecting and describing the wildlife of this region; Kellogg (1932) and Goldman (1951) reported the results of this expedition.

From February 23 to April 28, 1911, the Albatross Expedition to Lower California explored different points of the coast and islands of the Peninsula de Baja California and Sonora, under the auspices of the US Bureau of Fisheries, the AMNH, the New York Zoological Society, the New York Botanical Garden, and the USNM. Charles Haskins Townsend was in charge of this expedition, which collected 448 specimens of reptiles and 7 amphibians, all deposited in the USNM. In addition, a number of living reptiles were sent to the New York Zoological Park. The Sonoran localities visited by the expedition included Guaymas and the islands of San Esteban and Tiburón. Mary C. Dickerson (1919) and Karl P. Schmidt (1922) reported the herpetological results of the expedition. These two publications included the original descriptions of the following Sonoran taxa: *Aspidoscelis estebanensis, A. tigris disparilis, A. t. punctilinealis, Callisaurus draconoides inusitatus, Ctenosaura conspicuosa, Sauromalus ater townsendi*, and *S. varius* (Dickerson 1919); and *Crotaphytus dickersonae* (Schmidt 1922).

In 1922 John Van Denburgh included a list of the reptiles of Sonora in his *Reptiles of Western North America*. Joseph R. Slevin included a list of the amphibians of Sonora in his 1928 report, *Amphibians of Western North America*.

In June–July 1932 Morrow J. Allen, Jean Piatt,

and John Scofield of the Museum of Zoology of the University of Michigan collected 326 specimens—59 amphibians and 267 reptiles—near the cities of Puerto, Noria, Hermosillo, and Guaymas. Allen (1933) reported part of this collection, which included 4 amphibians, 12 lizards, 6 snakes, and 1 turtle species. This report includes the original description of *Dipsosaurus dorsalis sonoriensis*.

From June 19 to July 16, 1934, Edward Taylor visited Sonora, collecting amphibians and reptiles in the localities previously visited by Allen, Piatt, and Schofield and in other places such as Empalme. Taylor collected specimens of 5 species of amphibians, 18 lizards, 16 snakes, and 2 turtles. Part of this collection was reported by Hobart M. Smith (1935a, 1935b, 1935c, 1972) and included the original descriptions of *Phyllodactylus homolepidurus* from Hermosillo, *Uta taylori* (= *U. stansburiana taylori*) from Guaymas, and *Ctenosaura hemilopha macrolopha* (= *C. macrolopha*) from La Posa, San Carlos Bay. Taylor (1936b) compiled a complete report of the collection and included the original description of *Cnemidophorus* (= *Aspidoscelis*) *burti*.

In 1941 John W. Hilton collected amphibians and reptiles in the Güirocoba region. Charles M. Bogert, William Riemer, and Charles H. Lowe continued that study in 1942 in the mining town of Álamos (29 km northwest of Güirocoba). The collections made in these two years represent 15 species of amphibians, 9 turtles, 38 lizards, and 48 snakes, which were reported in Carr (1942, description of *Pseudemys scripta hiltoni* [= *Trachemys nebulosa hiltoni*]) and Bogert and Oliver (1945); this last report constitutes one of the best-known comprehensive publications on the herpetofauna of Sonora.

Richard G. Zweifel and Kenneth S. Norris of the Museum of Vertebrate Zoology, Berkeley, California, reported on collections they made during 1950 in the Güirocoba region of southeastern Sonora and points north of the international boundary (Zweifel and Norris 1955). They also reported on collections made by Laurence M. Klauber, A. A. Allanson, and Sterling Bunnell. They reported 10 species or subspecies of anurans, 18 lizards, 31 snakes, and 3 turtles and described *Lampropeltis getula nigritus* (= *L. g. nigrita*) and *Micruroides euryxanthus australis*.

Philip W. Smith and M. Max Hensley (1958) of the Illinois Natural History Survey and the Department of Zoology, Michigan State University, respectively, reported on a small collection of 3 amphibian species and 26 reptile species from the Pinacate region in northwestern Sonora, an area also investigated by Alberto González-Romero and Sergio Álvarez-Cárdenas during 1980–1983. In the latter study, the authors documented 42 species of amphibians and reptiles (González-Romero and Álvarez-Cárdenas 1989).

During August 1–22, 1964, the National Science Foundation (NSF) funded the Biology of the Insular Lizards of the Gulf Expedition (BILGE), carried out in the Sea of Cortés by a team of five professors, including Charles Carpenter (University of Oklahoma), Robert Clarke (Emporia State University, Kansas), James Dixon (New Mexico State University), Paul Maslin (University of Colorado), and Donald Tinkle (Texas Tech University), and five of their students: William C. Mc-Grew III, Louis J. Bussjaeger, Stanley Taft, Jack McCoy, and Orlando Cuellar. They visited numerous islands of the Sea of Cortés, including those assigned to Sonora (San Esteban, San Pedro Mártir, San Pedro Nolasco, and Tiburón). They collected many lizard specimens, which were deposited in several North American museums, especially at the University of Colorado. Reports of these collections and observations made on the trip are found in Dixon (1966), Ballinger and Tinkle (1972), and Smith (1972). The last publication included the original description of *Ctenosaura hemilopha nolascensis* (= *C. nolascensis*).

In the summers of 1964 and 1966 two private collections of amphibians and reptiles were made in the vicinity of Álamos. The specimens were deposited in the Herpetological Collection of Arizona State University. Part of this collection was reported by Max A. Nickerson and H. L. Heringhi (1966), including information on *Mastigodryas cliftoni*, *Sonora aemula*, and *Sympholis lippiens rectilimbus*. Additional information about these collections is found in Heringhi (1969).

In March 1965 and 1966 and in November 1966, Donald Tinkle and four of his students (Royce Ballinger, Charles McKinney, Gary Ferguson, and Orlando Cuellar) visited the town of Guaymas, San Carlos Bay, and the islands of San Pedro Nolasco, San Esteban, and Tiburón. Ballinger and Tinkle (1972) and McKinney (1969) reported the results of these trips.

Since 1974 Robert W. Murphy of the University of Toronto and the Royal Ontario Museum has been studying the herpetofauna of the islands of the Sea of Cortés (including those of Sonora) and several parts of continental Sonora and its coastline. The published work pertaining to Sonora includes Murphy (1976, 1983a, 1983b), Murphy and Aguirre-León (2002a, 2002b), and Davy et al. (2011).

In 1998 Gary P. Nabhan visited the towns of Punta Chueca and Desemboque de los Seris, collecting ethnobiological and herpetological information. Nabhan (2003) published the results of this research.

In 2000, Cecil R. Schwalbe and Charles H. Lowe published on the herpetofauna of the Álamos region, based on previous studies and the authors' own fieldwork, spanning about 50 years. The Álamos region includes the Sierra de Álamos north to the Río Mayo, south and east to the Río Cuchujaqui, and west and south to Masiaca. They reported 15 species of amphibians (all of them anurans) and 63 reptiles (6 turtles, 21 lizards, and 36 snakes).

Grismer's (2002) important publication listed herpetofaunal species that inhabit the islands of the Sea of Cortés, including those of Sonora, and provided detailed descriptions, identification keys, and illustrations of the taxa that occur on these islands.

More recently, Philip C. Rosen (2007) reported on the amphibians and reptiles of northwestern Sonora, in which he listed 2 amphibian species and 36 species of reptiles for the Pinacate-Gran Desierto region. Rosen and his coauthor, Cristina Melendez, sampled aquatic herpetofaunal species at 28 localities in Sonora and reported finding 14 species (Rosen and Melendez 2010). Seminoff and Nichols (2007) described the plight of sea turtles in the Alto Golfo region of the Sea of Cortés and included descriptions and species accounts for all five species found in Sonora's marine waters. James C. Rorabaugh (2010) discussed conservation of amphibians and reptiles of northwestern Sonora and southwestern Arizona and included a list of species with distributional notes. Rorabaugh et al. (2011) reported on the herpetofauna of the Northern Jaguar Reserve and vicinity northeast of Sahuaripa. Forty species of reptiles and 11 amphibian species were reported, including several range extensions. Rorabaugh et al. (2013) reported on 11 amphibian and 35 reptile species from two ranches in northeastern Sonora. Enderson et al. (2014) documented 20 amphibian and 73 reptile species in the Yécora region. Important distributional information has been collected by the Tucson-based Sky Island Alliance's Madrean Archipelago Biodiversity Assessment (MABA) during field trips to numerous mountain ranges in northeastern Sonora during 2009–2014, which was reported on in part by Van Devender et al. (2013). MABA has developed an online database of observations and collections it has made as well as records from several museums (http://www.madrean.org/maba/symbfauna/), which has become a valuable resource for herpetological and other animal and plant groups.

Distributional records, natural history notes, and ecological information have been reported in Avila et al. (2008: *Masticophis mentovarius*); Bonine et al. (2006: *Imantodes geminstratus*); Bury et al. (2002: *Gopherus agassizii* [= *morafkai*]); Dibble et al. (2007: *Callisaurus draconoides*; 2008: *Dipsosaurus dorsalis*); Enderson et al. (2006: *Gyalopion canum*; 2007: *Kinosternon integrum*); Enderson and Bezy (2007a: *Geophis dugesii*; 2007b: *Lampropeltis triangulum sinaloae*; 2007c: *Leptodeira splendida ephippiata*; 2007d: *Pseudoficimia frontalis*; 2007e: *Syrrhophus* [= *Eleutherodactylus*] *interorbitalis*); Enderson (2010: *Crotalus viridis*); Goldberg (1997: *Micruroides euryxanthus*); Lara-Góngora (1986: *Sceloporus grammicus disparilis* [= *S. lemosespinali*]); Lemos-Espinal, Chiszar, et al. (2004: lizards); Lemos-Espinal, Smith,

Hartman, and Chiszar (2004: amphibians and turtles); Lemos-Espinal, Smith, Chiszar, and Woolrich-Piña (2004: snakes); Lowe and Howard (1975: *Phrynosoma ditmarsi*); Macedonia et al. (2009: *Crotaphytus dickersonae*); Maldonado-Leal et al. (2009: *Hyla wrightorum*); O'Brien et al. (2006: *Drymarchon corais* and *Kinosternon integrum*); Palacio-Baez and Enderson (2012: *Incilius marmoreus*); Perrill (1983: *Phrynosoma ditmarsi*); Quijada-Mascareñas and Enderson (2007: *Ramphotyphlops* [= *Indotyphlops*] *braminus*); Quijada-Mascareñas et al. (2007: *Plestiodon obsoletus*); Recchio et al. (2007: *Geophis dugesii*); Rorabaugh (2013: *Trimorphodon tau*); Rorabaugh and King (2013: *Apalone spinifera*); Rorabaugh et al. (2008: *A. spinifera*; 2013: *Micrurus distans*); Rorabaugh and Servoss (2006: *Lithobates berlandieri*); Rosen and Quijada-Mascareñas (2009: *Aspidoscelis xanthonota*); Sherbrooke et al. (1998: *Phrynosoma ditmarsi*); Smith, Lemos-Espinal, and Heimes (2005: amphibians and lizards); Smith, Lemos-Espinal, and Chiszar (2005: amphibians and reptiles); Van Devender, Holm, and Lowe (1989: *Pseudoeurycea bellii sierraoccidentalis*); Van Devender, Lowe, and Holm (1989: *Pseudoeurycea bellii sierraoccidentalis*); Van Devender et al. (1994: *Oxybelis aeneus*); Van Devender and Enderson (2007: *Micrurus distans*); Villa et al. (2007: *Crotalus willardi*); Wiewandt et al. (1972: *Hypopachus variolosus*); and Winter et al. (2007: *Leptodactylus melanonotus*).

Recent original descriptions of new species are reported in Lara-Góngora (2004: *Sceloporus lemosespinali*); Smith, Lemos-Espinal, Hartman, and Chiszar (2005: *Tropidodipsas repleta*); Bezy et al. (2008: *Xantusia jaycolei*); Murphy et al. (2011: *Gopherus morafkai*); and Wood et al. (2011, *Thamnophis unilabialis*).

Other important studies are those by James C. Rorabaugh (2008), who reported the herpetofauna of continental Sonora, mentioning the distribution, protection status, and common names for all the species of amphibians and reptiles, and including some photos. The list of the herpetofauna reported in this publication includes 37 amphibian and 139 reptile species.

In 2009, Rorabaugh and Enderson provided a review of the lizard species of the state, including those of the islands of the Sea of Cortés. They reported a total of 64 lizard species in 21 genera and 11 families.

Enderson et al. (2009) provided a list of the amphibians and reptiles of the state of Sonora and compared it with those of the adjoining states. They reported a total of 187 species: 35 amphibians (3 salamanders, 32 anurans) and 152 reptiles (1 crocodile, 15 turtles, 64 lizards, and 72 snakes). Enderson et al. (2010) published another report of the herpetofauna of Sonora in which they listed the amphibians and reptiles currently known for continental Sonora and analyzed the conservation priorities for the amphibians and reptiles of the state.

Since 2003, Hobart M. Smith and Julio Lemos-Espinal have been studying the amphibians and reptiles of Sonora. Through this work they published in 2009 keys to the amphibians and reptiles and a list for the state (also Chihuahua and Coahuila). The list included a total of 192 species: 37 amphibians (3 salamanders, 34 anurans) and 155 reptiles (16 turtles, 65 lizards, 74 snakes). Twelve of the 192 species reported are insular, 6 are oceanic, and 174 are distributed in continental Sonora.

More recently, Lemos-Espinal et al. (forthcoming) published on the herpetofauna of the states of Sonora, Chihuahua, and Coahuila, presenting a list of all amphibian and reptile species known to occur in these three states; keys to those species; photographic illustrations; an account and spot distributional maps for each species; and a gazetteer including the coordinates and municipality for 2987 localities of collection for the species reported. The list for Sonora included 193 species: 37 amphibians (3 salamanders, 34 anurans) and 156 reptiles (16 turtles, 68 lizards, 72 snakes).

In addition to the studies mentioned previously we have now determined that there are 67 scientific names of amphibians (2 salamanders) and reptiles (7 turtles, 37 lizards, 21 snakes) with type localities in the state of Sonora (table 4.1).

Physiographic Characteristics and Their Influence on the Distribution of the Herpetofauna

The distribution of amphibians and reptiles is limited in part by their physiological and behavioral needs, which in turn are constrained by the environments in which they live. Unlike many birds, insects, mammals, and fishes, most amphibians and reptiles do not migrate or make only small seasonal movements within their home ranges (in Sonora, important exceptions include the marine reptiles, represented by five sea turtles and one seasnake, which migrate as far as thousands of kilometers between the Sea of Cortés and their breeding grounds). Nonmigratory species must be able to survive and reproduce within the range of environmental conditions present in their habitats, which may vary considerably from season to season and year to year. The climate in Sonora varies from very hot and arid in the northwestern deserts to cool and moist in the high eastern sierras, which is reflected in a diversity of biotic communities (Brown 1982a; Russell and Monson 1998). Seasonal or longer-term variations in temperature and precipitation impose temporal changes in biotic communities that can be visually dramatic and biologically significant. For example, thornscrub and tropical deciduous forest communities transform from a dry, unappealing brown thicket of thorny branches to tropical lushness in a matter of two to three weeks after the onset of the summer rains in June or July.

All 38 species of amphibians in Sonora need at least moist conditions for reproduction, and all but 5 species (in the genera *Pseudoeurycea*, *Eleutherodactylus*, and *Craugastor*) have an aquatic larval stage that requires ephemeral or permanent surface water for development. Although water is a major limiting factor for amphibians in Sonora, they are a remarkably adaptive group—only the driest valleys of the Gran Desierto de Altar in the northwestern portion of the state are devoid of amphibians. In contrast, reptiles are much less limited by water

TABLE 4.1 Type localities for amphibians and reptiles described from the state of Sonora, Mexico

Author(s)	Original name: Type locality
Salamanders (2)	
Shannon (1951)	*Ambystoma rosaceum sonoraensis* (= *A. rosaceum*): **32 mi south of the Arizona border, Sonora**
Lowe et al. (1968)	*Pseudoeurycea bellii sierraoccidentalis*: **ca. 11 mi (rd) E Sta. Ana, on old road to Yécora**
Turtles (7)	
Garman (1884)	*Sphargis* (= *Dermochelys*) *coriacea schlegelii*: **Tropical Pacific and Indian Ocean (restricted to Guaymas¹)**
Carr (1942)	*Pseudemys scripta hiltoni* (= *Trachemys nebulosa hiltoni*): **Güirocoba**
Legler & Webb (1970)	*Pseudemys scripta yaquia* (= *Trachemys yaquia*): **Río Mayo, Conicarit**
Bogert (1943)	*Terrapene klauberi* (= *T. nelsoni klauberi*): **Güirocoba, 18 mi SE of Álamos**
Berry & Legler (1980)	*Kinosternon alamosae*: **Rancho Carrizal, 7.2 km N and 11.5 km W of Alamos**
Hartweg (1938)	*Kinosternon flavescens stejnegeri* (= *K. arizonense*): **Llano**
Iverson (1981)	*Kinosternon sonoriense longifemorale*: **Sonoyta (31°51′ N, 112°50′ W)**
Lizards (37)	
Klauber (1945)	*Coleonyx variegatus sonoriense* (*sonoriensis*): **5 mi SE of Hermosillo**
Smith (1935b)	*Phyllodactylus homolepidurus*: **5 mi SW of Hermosillo**
Dixon (1964)	*Phyllodactylus homolepidurus nolascoensis* (= *P. nolascoensis*): **Isla San Pedro Nolasco**
Schmidt (1922)	*Crotaphytus dickersonae*: **Isla Tiburón**
Axtell & Montanucci (1977)	*Crotaphytus nebrius*: **14 km by road N of Rancho Cieneguita. 28°30′30″ N, 111°2′30″W**
Smith (1972)	*Ctenosaura hemilopha macrolopha* (= *C. macrolopha*): **La Posa, San Carlos Bay**
Dickerson (1919)	*Ctenosaura conspicuosa*: **Isla San Esteban**
Smith (1972)	*Ctenosaura nolascensis*: **Isla San Pedro Nolasco**
Allen (1933)	*Dispsosaurus dorsalis sonoriense* (= *D. d. sonoriensis*): **Hermosillo**
Dickerson (1919)	*Sauromalus townsendi* (= *S. ater townsendi*): **Isla Tiburón**
Dickerson (1919)	*Sauromalus varius*: **Isla San Esteban**
Bogert & Dorson (1942)	*Callisaurus draconoides brevipes*: **Güirocoba, 18 mi SE of Álamos**
Dickerson (1919)	*Callisaurus inusitatus* (= *C. draconoides inusitatus*): **Isla Tiburón**
Barbour (1921)	*Holbrookia thermophila* (= *H. elegans thermophila*): **San José de Guaymas**
Stejneger (1906)	*Phrynosoma ditmarsi*: **state of Sonora, not far from the boundary of Arizona**
Stejneger (1893)	*Phrynosoma goodei*: **coast deserts of Sonora (restricted to Puerto Libertad¹)**
Girard (1858)	*Phrynosoma regale* (= *P. solare*): **Sierra de la Nariz, near Zuñi**
Lara-Góngora (2004)	*Sceloporus lemosespinali*: **Yécora**
Smith (1938)	*Sceloporus undulatus virgatus* (= *S. virgatus*): **above Sta. María Mine, Tigre Mountains**
Heifetz (1941)	*Uma notata cowlesi* (= *U. rufopunctata*): **shores of Tepoca Bay**
Van Denburgh & Slevin (1921)	*Uta nolascensis*: **Isla San Pedro Nolasco**
Baird (1859a [1858])	*Uta ornata* var. *linearis* (= *Urosaurus ornatus linearis*): **Los Nogales**
Stejneger (1890)	*Uta palmeri*: **Isla San Pedro Mártir**
Baird (1859a [1858])	*Uta schottii* (= *Urosaurus ornatus schotti*): **Sta. Magdalena**

TABLE 4.1 *Continued*

Author(s)	Original name: Type locality
Smith (1935a)	*Uta taylori* (= *U. stansburiana taylori*): **10 mi NW of Guaymas**
Taylor (1933)	*Eumeces* (= *Plestiodon*) *parviauriculatus*: **Near Álamos**
Van Denburgh & Slevin (1921)	*Cnemidophorus baccatus* (= *Aspidoscelis baccata*): **Isla San Pedro Nolasco**
Taylor (1936b)	*Cnemidophorus* (= *Aspidoscelis*) *burti*: **La Posa, 10 mi NW of Guaymas**
Dickerson (1919)	*Cnemidophorus disparilis* (= *Aspidoscelis tigris disparilis*): **Isla Tiburón**
Dickerson (1919)	*Cnemidophorus* (= *Aspidoscelis*) *estebanensis*: **Isla San Esteban**
Stejneger (1891)	*Cnemidophorus* (= *Aspidoscelis*) *martyris*: **Isla San Pedro Mártir**
Wright (1967)	*Cnemidophorus* (= *Aspidoscelis*) *opatae*: **5.5 mi (by road) S of Oputo**
Dickerson (1919)	*Cnemidophorus punctilinealis* (= *Aspidoscelis tigris punctilinealis*): **Isla Tiburón**
Zweifel (1959)	*Cnemidophorus sacki* (= *Aspidoscelis costata*) *barrancarum*: **Rancho Güirocoba**
Zweifel (1959)	*Cnemidophorus sacki* (= *Aspidoscelis costata*) *griseocephalus*: **11.4 mi E of Navojoa**
Cope (1900)	*Cnemidophorus tessellatus* (= *Aspidoscelis tigris*) *aethiops*: **Hermosillo**
Bezy et al. (2008)	*Xantusia jaycolei*: **near Desemboque del Río San Ignacio**

Snakes (21)

Cope (1861)	*Chilomeniscus cinctus* (= *C. stramineus*): **near Guaymas**
Baird & Girard (1853)	*Diadophis regalis* (= *D. punctatus regalis*): **Sonora**
Taylor (1936a)	*Ficimia desertorum* (= *Gyalopion quadrangulare desertorum*): **about 12 km NW of Guaymas**
Tanner (1981)	*Hypsiglena torquata tiburonensis* (= *H. chlorophaea*): **Isla Tiburón**
Zweifel & Norris (1955)	*Lampropeltis getulus nigritus* (= *L. getula nigrita*): **30.6 road mi S of Hermosillo**
Smith & Tanner (1944)	*Leptodeira ephippiata* (= *L. splendida ephippiata*): **Agua Marin, 8.3 mi W-NW of Álamos**
Jan (1863)	*Masticophis bilineatus*: **western Mexico (restricted to Guaymas[1])**
Lowe & Woodin (1954)	*Masticophis flagellum cingulum*: **Moctezuma**
Lowe & Norris (1955)	*Masticophis slevini*: **Isla San Esteban**
Bogert & Oliver (1945)	*Phyllorhynchus browni fortitus* (= *P. browni*): **Álamos**
Smith & Langebartel (1951)	*Phyllorhynchus decurtatus norrisi* (= *P. decurtatus*): **45.1 mi S of Sta. Ana**
Bogert & Oliver (1945)	*Pseudoficimia hiltoni* (= *P. frontalis*): **Güirocoba**
Klauber (1937)	*Sonora* (= *Chionactis*) *palarostris*: **6 mi S of Hermosillo**
Taylor (1937)	*Tantilla hobartsmithi*: **near La Posa, 10 mi NW of Guaymas**
Tanner (1959)	*Thamnophis melanogaster chihuahuense*: **Río Bavispe below Three Rivers, Chihuahua-Sonora line[3]**
Cope (1886)	*Trimorphodon lambda*: **Guaymas**
Smith, Lemos-Espinal, Hartman, & Chiszar (2005)	*Tropidodipsas repleta*: **km 236.2, Hwy 16 Chihuahua-Hermosillo**
Kennicott (1860)	*Elaps* (= *Micrurus*) *euryxanthus*: **Sonora (restricted to Guaymas[1,2])**
Zweifel & Norris (1955)	*Micruroides euryxanthus australis*: **Güirocoba**
Kennicott (1861)	*Caudisona atrox sonoraensis* (= *Crotalus atrox*): **Sonora (restricted to Guaymas[1])**
Klauber (1949)	*Crotalus estebanensis*: **Isla San Esteban**

[1] Restrictions appeared in Smith and Taylor (1950).

[2] Restriction not accepted by Roze (1974).

[3] Originally reported for the state of Chihuahua by Tanner (1959); restricted here to the state of Sonora.

availability, and many species thrive even in the heart of Sonora's driest deserts. Sonora is rich in reptilian species because of their leathery eggs or live birth, scaly exterior that resists water loss, and an array of physiological and behavioral adaptations or exaptations to arid conditions (Gould and Vrba 1982; Pianka 1986).

As ectotherms, amphibians and reptiles depend on environmental heating and cooling for physiological processes. Although behaviorally adept at taking advantage of favorable microsites and conditions, when their habitats become exceedingly warm or cold, amphibians and reptiles seek underground or other retreats where temperatures do not exceed their tolerances, and many remain dormant through the hot pre-summer before the rains begin and during the cold of winter, particularly in the north and at high elevations. Cold is most limiting in the highest, northeastern sierras where winter temperatures occasionally drop below −10°C (Felger et al. 2001). Most amphibians and reptiles are not able to survive more than brief periods of subzero temperatures (Zug et al. 2001); hence, in cold periods, they avoid exposure and become dormant in underground burrows, at the bottom of ponds, or in other places that do not experience freezing temperatures. Cold winters and short growing seasons also influence reproductive strategies. For instance, the three high-elevation *Phrynosoma* species occurring in Sonora (*P. ditmarsi*, *P. hernandesi*, and *P. orbiculare*) are all live bearing (viviparous; Sherbrooke 2003). Oviparity (eggs are laid and develop underground or amid protective debris) does not provide for rapid enough embryonic development to allow hatching within a season. Rather, the females of these three species carry the embryos internally and keep them warm through basking and other behaviors. In contrast, all of the lower-elevation *Phrynosoma* in Sonora are oviparous. This pattern of viviparity, being more common at higher latitudes and higher elevation, generally holds true for lizards and snakes.

In the northwestern deserts, air temperatures exceeding 40°C and surface temperatures greater than 60°C are not uncommon in sum-

mer. Most diurnal desert reptiles have voluntary maximum body temperatures of 38°C–46°C and retreat to underground burrows or rock shelters when their body temperatures approach or meet those thresholds (Cowles and Bogert 1944; Brattstrom 1963). Thus, desert species tend to be active in the relative coolness of morning or late afternoon during summer, although many species—especially snakes—are active at night when temperatures have moderated.

Environment is the medium in which species take root, adapt, and grow. Thus, an understanding of the climate, topography, and biotic communities of Sonora lays a foundation for grasping the underpinning mechanisms that drive distributions and biodiversity of Sonora's rich herpetofauna.

Second in size (185,430 km²) to Chihuahua among the Mexican states, Sonora encompasses 9.5% of Mexico's land area and lies between latitudes 26.2° and 32.5° N and longitudes 115.0° and 108.8° W. Its eastern boundary with Chihuahua is situated mostly west of the Continental Divide in the Sierra Madre Occidental, while most of its western boundary is in the Sea of Cortés and includes 14 primary and numerous smaller islands. At approximately 1200 km², Tiburón is the largest of Sonora's islands and is also the largest island in Mexico. The mountains in the east form a more or less continuous chain of highlands and rugged peaks from the Sierra San Luis and the headwaters of the Río Bavispe near the US border south into Sinaloa. However, a basin and range topography west of the main mass of the Sierra Madre Occidental create a series of biogeographically distinct "sky islands," the most striking of which are in the northeast and characterized by islands of oaks and conifers separated from other such sky islands by seas of semidesert grasslands, plains grassland, or Chihuahuan desert scrub (Brown and Lowe 1980). The highest peak in Sonora (2625 m) is in one of those sky islands—the Sierra Los Ajos southeast of Cananea. Peaks nearly as high occur in the Sierra San Luis and the northern Sierra Madre Occidental near Mesa Tres Ríos (Felger et al. 2001).

Basin and range topography, with mountain ranges running generally northwest-southeast, continues westward to the coast. However, the sierras decline in elevation from east to west—most along or near the coast do not exceed 700 m. These lower sierras also form sky islands, but the differences in biotic communities between the peaks and valleys are less conspicuous than in eastern Sonora. Nonetheless, western montane flora and fauna tend to be richer than in the valleys. In northwestern Sonora, montane islands of the Arizona upland subdivision of Sonoran desert scrub may be surrounded by valleys of the Colorado subdivision of that same desert scrub. Farther south, thornscrub may occur in the mountains with Sonoran desert scrub in the valleys (Brown and Lowe 1980).

Almost all of Sonora drains to the Sea of Cortés. Major river drainages from north to south include the Ríos Colorado, Sonoyta, Concepción, Sonora, Yaqui, Matape, Mayo, and Fuerte. Only a small portion of northeastern Sonora drains east of the Continental Divide and into the Río Casas Grandes basin of Chihuahua. The Ríos San Pedro and Santa Cruz between Nogales and Naco take a circuitous route to the Sea of Cortés by flowing north to the Gila River in Arizona and then west into the Río Colorado. Most rivers in Sonora do not currently reach the Sea of Cortés except during floods because their flows have been diverted or dammed, or they simply run dry in the desert before reaching the sea (Minckley and Marsh 2009).

The present topography of Sonora began evolving about 35 million years ago when the volcanic Sierra Madre Occidental started to rise up from what had been a broad plain (Ferrusquia-Villafranca and Gonzáles-Guzmán 2005). Mountain building in eastern Sonora and western Chihuahua was followed by movements of the Pacific Plate 8 to 25 million years ago, which in effect pulled the western portions of Sonora along the San Andreas and associated faults to the north and west, creating the current basin and range topography and ultimately tearing Baja California from the mainland, which resulted in the initial formation of the

Sea of Cortés and its many islands (Scarborough 2000). Roughly 5 million years ago, the Sea of Cortés widened and elongated, attaining its current size and shape (Larson 1972). The sand sea that is the Gran Desierto de Altar was formed by sediment deposition from the Río Colorado, which began flowing into the Sea of Cortés about 5.5 million years ago after it was diverted south and east by mountain-building activity in the Sierra Nevada and coastal ranges of present-day California (Merriam and Bandy 1965; Howard 1996).

Current Sonoran climates are a product of local topography, latitude, and westerly winds that bring cool, moisture-laden air inland from the Pacific Ocean. Mountain ranges cause atmospheric uplift, cooling of air, and precipitation but also form rain shadows and subsequent dry zones opposite the source of moisture. Most rainfall in Sonora falls during the summer months—July to September—and results primarily from high pressure and rising air masses over a hot landscape that draws moist air in from the Pacific. Swirling moisture-laden air masses create violent thunderstorms in the summer that are often quite localized and visually spectacular. Debate is ongoing in regard to how much of the moisture driving these summer thunderstorms originates in the Gulf of Mexico; however, most climatologists believe the majority of summer moisture is derived from the Pacific and that the Sierra Madre Occidental has a rain shadow effect with regard to moisture from the Gulf of Mexico (Ingram 2000; Brito-Castillo et al. 2010). Rainfall in the winter months is less common but occurs from storm systems that originate in the eastern Pacific and travel eastward across the western United States. If low pressure develops in northwestern Mexico, these storms can swing far enough south to produce significant precipitation in Sonora. Winter storms tend to be gentler, of longer duration, and extend over larger areas than their summer counterparts (Brito-Castillo et al. 2010). The percentage of winter precipitation in annual rainfall totals increases to the north and west in Sonora. The driest region of Sonora is the northwest; mean annual precipitation is only 37 mm at San Luis Río Colorado. In contrast, Yécora in the Sierra Madre Occidental receives an annual average of 913 mm of precipitation (Felger et al. 2001).

Temperatures tend to decrease with elevation and latitude but also become more seasonally variable with latitude (Brito-Castillo et al. 2010). Locales in the Gran Desierto de Altar are known to have the most extreme summer temperatures; for example, a temperature of 56.7°C was recorded in the Sierra Blanca region in 1971 (May 1973). However, winters in the northwest are relatively cool. As a result, San Luis Río Colorado has a lower mean annual temperature (22.8°C) than the southeastern foothills at Álamos (23.5°C; Burquez et al. 1999). Freezing temperatures, which limit the distributions of many plants and animals, are rare in the southern lowlands and along the coast of Sonora but become increasingly common to the north and at higher elevations. Snow commonly covers the highest northeastern peaks for at least a few weeks in winter.

Present and geologically recent climate and topography are major forces driving the nature of biotic communities. In Sonora, a diversity of landscapes over a broad latitudinal and elevational range has engendered remarkable diversity in the herpetofauna and other animal and plant groups. A strong tropical influence from the south mingles with temperate biomes from the Rocky Mountains and Great Plains in the north and the Chihuahuan Desert primarily to the east. The largest community in terms of land area in Sonora is the Sonoran Desert, which is quite variable and has been described with four or five major subdivisions in the state (Shreve 1951; Turner and Brown 1982; Felger et al. 2001; Martínez-Yrizar et al. 2010). It occurs in the lowlands from San Luis Río Colorado south to Guaymas and on all of Sonora's islands in the Sea of Cortés. Generally, this community is characterized by a relatively open association of drought-resistant shrubs (e.g., *Larrea tridentata*, *Ambrosia* sp., *Lycium* sp.), trees (e.g., *Parkinsonia* sp., *Acacia* sp., *Prosopis* sp.), and succulents (e.g., *Carnegiea gigantea*, *Cylindropuntia* sp., *Opuntia*

sp., *Fouquieria* sp.). To the south and east, the plant community increases in stature, density, and richness until it grades into Sinaloan thornscrub (Brown 1982c), which is often divided into coastal and foothill variants (Burquez et al. 1992; Martínez-Yrizar et al. 2010). Thornscrub is composed of a tangle of primarily drought-deciduous and spiny shrubs, trees, and succulents, many of which store water in fleshy stems or tuberous roots. Representative plants include *Acacia* sp., *Bursera* sp., *Fouquieria* sp., *Ipomea arborescens*, *Jatropha* sp., *Lysiloma* sp., *Stenocereus thurberi*, *Havardia* sp., and *Cylindropuntia* sp. In the southeastern foothills predominantly south of Mex. Hwy 16, foothill thornscrub grades into tropical deciduous forest, which is a drought-deciduous, multilayered woodland community with strong tropical derivations and many frost-sensitive species (e.g.; *Ambrosia cordifolia*, *Bursera* sp., *Ceiba acuminata*, *Haematoxylum brasiletto*, *Lysiloma divaricata*, *Pachycereus pectin-aboriginum*, *Senna atomaria*, and *Stenocereus* sp.; Martin et al. 1998; Van Devender et al. 2000; Felger et al. 2001). Above and east of the tropical deciduous forest, as well as to the north in the lower mountains, oak woodland is found, which contains a diversity of mostly evergreen oaks (e.g., *Quercus arizonica*, *Q. chihuahuensis*, *Q. emoryi*, *Q. oblongifolia*), often in open woodlands mixed with species of the tropical deciduous forest at the lower elevations and latitudes, and in denser woodlands with scattered pines and junipers at the higher elevations and to the north. The lower limits of pine-oak woodland, which is typically above oak woodland, rarely extend below 1000 m but more characteristically are at 1300 m or higher (Martin et al. 1998). In this woodland community, various species of *Quercus* mingle with a variety of pines (e.g., *Pinus arizonica*, *P. ayacahuite*, *P. englemannii*, *P. oocarpa*, *P. strobiformis*). Above about 2135 m in the Sierras Los Ajos, El Tigre, San José, and San Luis, as well as on the Mesa del Campanero and in small patches in other mountain ranges in northeastern Sonora, mixed-conifer forests of pines and firs (*Abies* sp., *Pinus* sp., *Pseudotsuga menziesii*), sometimes with patches of aspen

(*Populus tremuloides*), can be found (Felger et al. 2001).

In the northeastern valleys, grasslands and Chihuahuan Desert scrublands predominate. Plains grasslands, which occur in the higher valleys, mostly above 1500 m east of the Río Santa Cruz along the US border to the Sierra San Luis, is a shortgrass prairie that frequently has been degraded by many decades of livestock grazing. Typical grass genera of the plains grassland include *Aristida*, *Bothriochloa*, *Bouteloua*, *Eragrostis*, *Leptochloa*, *Panicum*, and *Sporobolus*. Semidesert grasslands occur downslope in lower valleys to about 1100 m but have largely been invaded by shrubs and cacti characteristic of Sonoran desert scrub, a process likely driven by overgrazing, exclusion of fire, and changing climate but for which identification of specific drivers at specific sites is often elusive (Humphrey 1958; Peters and Havstad 2006). Fine examples of semidesert grasslands can be found on the bajadas of the Sierra Azul at Rancho El Aribabi and at other scattered locations in the northeast. In valleys east of the Sierra Los Ajos and north of the Río Bavispe, Chihuahuan Desert scrublands are found. This high desert is visually dominated by creosote bush (*Larrea tridentata*) and other shrubs or small trees (e.g., *Acacia neovernicosa*, *Flourensia cernua*, *Prosopis glandulosa*), as well as a variety of succulents (e.g., *Yucca elata*, *Dasylirion wheeleri*, *Fouquieria splendens*, *Cylindropuntia* sp.). Grasses are often scant to lacking where overgrazing has occurred.

Other biotic communities found within the major divisions just described include mangrove swamps and thickets on the coast and islands from Puerto Lobos south, and various riparian and wetland communities along rivers and streams. The latter tend to be biologically very diverse and provide crucial greenbelts and water through arid landscapes that are often physiologically challenging for many species, at least seasonally (Felger et al. 2001). The marine waters of the Sea of Cortés are yet another very different biome, which currently provides habitat for six marine reptiles.

Adaptations to the diverse environments of Sonora have given rise to varied and more or less distinct assemblages of amphibians and reptiles that are associated with the major biotic communities. Rorabaugh and Enderson (2009) described five lizard assemblages, including Sonoran Desert, Tropical/Subtropical, Montane, Grassland, and Chihuahuan Desert. Two species were considered generalists. Enderson et al. (2009) categorized Sonora's amphibians and reptiles into nine assemblages: Chihuahuan, Eastern Temperate, Generalist, Great Plains, Madrean, Marine, North American Desert Generalist, Sonoran, and Tropical. We use a delineation consistent with other chapters in this volume but fully recognize that any such classification is somewhat arbitrary. Herein we recognize seven assemblages.

SONORAN DESERT ASSEMBLAGE

Forty-six species have been assigned to this assemblage, which includes all 11 endemic island species: *Anaxyrus retiformis*, *Lithobates berlandieri*, *Kinosternon arizonense*, *Gopherus morafkai*, *Crotaphytus dickersonae*, *C. nebrius*, *Gambelia wislizenii*, *Coleonyx variegatus*, *Phyllodactylus homolepidurus*, *P. nocticolus*, *P. nolascoensis*, *Ctenosaura conspicuosa*, *C. nolascensis*, *Dipsosaurus dorsalis*, *Sauromalus ater*, *S. hispidus*, *S. varius*, *Callisaurus draconoides*, *Holbrookia elegans*, *Phrynosoma goodei*, *P. mcallii*, *P. solare*, *Sceloporus magister*, *Uma rufopunctata*, *Urosaurus graciosus*, *Uta nolascensis*, *U. palmeri*, *U. stansburiana*, *Aspidoscelis baccata*, *A. estebanensis*, *A. martyris*, *A. xanthonota*, *Xantusia jaycolei*, *X. vigilis*, *Lichanura trivirgata*, *Arizona elegans*, *Chilomeniscus stramineus*, *Chionactis occipitalis*, *C. palarostris*, *Masticophis slevini*, *Phyllorhynchus browni*, *P. decurtatus*, *Salvadora hexalepis*, *Crotalus cerastes*, *C. estebanensis*, and *C. mitchellii*.

Much of the Sonoran Desert is uninhabited and too dry to support much industry or agriculture, and as a result, vast expanses of this biome are relatively undisturbed, particularly in the arid northwest. The primary exceptions include agricultural and rural or urban development along the Río Colorado Valley, the valley of the Ríos

Magdalena, Asunción, and Concepción from about Magdalena to El Desemboque, and from Hermosillo west and southwest to Bahía Kino and Boca Tastiota. Cattle grazing is widespread and locally or seasonally heavy in most of the Sonoran Desert, although the drier areas, such as the Gran Desierto de Altar, support few cattle except after wet winters when the desert produces luxurious displays of annual herbaceous plants. Buffelgrass (*Pennisetum ciliare*), introduced from Africa or the Middle East, has been planted extensively for livestock forage in the central and southern portions of Sonoran desert scrub and has spread to many places in the Sonoran Desert. Where it has been planted, most of the native flora is often bulldozed, severely altering the environment. However, even in areas to which it has spread, the native plant community may be much altered due to competition as well as increased fire frequency facilitated by buffelgrass (Burquez et al. 2002; Franklin and Molina-Freaner 2010). Harvest of *Prosopis* spp. for fuelwood and *Olneya tesota* for wood carvings has affected vegetation structure and composition in some areas.

Climate change is anticipated to have significant effects on Sonoran Desert flora (Weiss and Overpeck 2005), as it is likely to have elsewhere in Sonora (IPCC 2007). Over a million hectares of Sonoran Desert are provided protection in the Reserva de la Biósfera Pinacate y Gran Desierto de Altar and the Reserva de la Biósfera Alto Golfo de California y Delta del Río Colorado. Most of Sonora's islands are included in the Reserva de la Biósfera Islas del Golfo de California.

SUBTROPICAL ASSEMBLAGE

Thirty-five species are included in this assemblage: *Incilius marmoreus, Rhinella marina, Craugastor occidentalis, Pachymedusa dacnicolor, Smilisca baudinii, Tlalocohyla smithi, Leptodactylus melanonotus, Hypopachus variolosus, Lithobates forreri, L. pustulosus, Trachemys nebulosa, Rhinoclemmys pulcherrima, Phyllodactylus tuberculosus, Heloderma exasperatum, Sceloporus albiventris, S. nelsoni, Urosaurus bicarinatus, Anolis nebulosus, Plestiodon parviauriculatus, Drymarchon melanurus, Drymobius margaritiferus, Imantodes gemmistratus, Lampropeltis triangulum, Leptodeira punctata, L. splendida, Leptophis diplotropis, Mastigodryas cliftoni, Pseudoficimia frontalis, Sonora aemula, Sympholis lippiens, Thamnophis validus, Trimorphodon tau, Micrurus distans, Agkistrodon bilineatus,* and *Crotalus basiliscus.*

These are species that have the heart of their distributions in subtropical or tropical areas. Most records for these species in Sonora are from the southeastern corner of the state at moderate to low elevation in tropical deciduous forest and/or coastal thornscrub. Much of the thornscrub in southern Sonora has been cleared for agriculture, which dominates the lowlands from Ciudad Obregón south to Yavaros. In the less accessible foothills, tropical deciduous forest is more nearly intact, although plantations of buffelgrass are not uncommon, a history of mining has scarred the landscape locally, and some harvest of fuelwood is in evidence, as well as harvest of other forest products such as the tree *Croton fantzianus*, which is used to make support stakes for vegetable crops. Cattle and other livestock are grazed throughout most of this region. The Área de Protección de Flora y Fauna Sierra de Álamos-Río Cuchujaqui, which includes the Reserva Monte Mojino, protects some of the best examples of tropical deciduous forest in Sonora.

MONTANE/FOOTHILLS ASSEMBLAGE

Sixty-five species have been placed in this assemblage: *Ambystoma rosaceum, Pseudoeurycea bellii, Anaxyrus kelloggi, A. mexicanus, Incilius mazatlanensis, I. mccoyi, Craugastor augusti, C. tarahumaraensis, Eleutherodactylus interorbitalis, Hyla arenicolor, H. wrightorum, Smilisca fodiens, Gastrophryne mazatlanensis, Lithobates chiricahuensis, L. magnaocularis, L. tarahumarae, L. yavapaiensis, Terrapene nelsoni, Trachemys yaquia, Kinosternon alamosae, K. integrum, Elgaria kingii, Crotaphytus collaris, Coleonyx fasciatus, Ctenosaura macrolopha, Holbrookia approximans, Phrynosoma ditmarsi, P. hernandesi, P. orbiculare, Sceloporus clarkii, S. jarrovii, S. lemosespinali, S. poinsettii, S. virgatus, Plestiodon callicephalus, P. obsoletus, Aspidoscelis burti, A. costata, A. opa-*

tae, *A. sonorae, A. stictogramma, Boa constrictor, Diadophis punctatus, Geophis dugesii, Gyalopion quadrangulare, Lampropeltis knoblochi, Masticophis mentovarius, Oxybelis aeneus, Pituophis deppei, Salvadora bairdii, S. deserticola, S. grahamiae, Senticolis triaspis, Storeria storerioides, Tantilla hobartsmithi, T. wilcoxi, T. yaquia, Thamnophis cyrtopsis, T. eques, T. melanogaster, T. unilabialis, Tropidodipsas repleta, Crotalus lepidus, C. pricei,* and *C. willardi.*

These are species characteristic of the foothills and mountains of eastern Sonora and may be found from lower-elevation foothills thornscrub or tropical deciduous forest well into conifer forests or be quite specialized in habitat use or localized in distribution. Much of this region is difficult to access, and most of the few paved or improved roads into the foothills and sierras have been in existence only in the last 20 years. Nonetheless, eastern Sonora has been subjected to many anthropogenic impacts, including widespread and often heavy livestock grazing, buffelgrass invasion, and plantations in the foothills and lower elevations; copper mining in the Cananea and Nacozari regions; timber and fuelwood harvest; and illicit drug production (Stoleson et al. 2005). Introduced aquatic species, including *Lithobates catesbeianus*, crayfish (*Orconectes virilis*), black bullhead (*Ameiurus melas*), green sunfish (*Lepomis cyanellus*), and others are becoming increasingly common, particularly in the foothills and river valleys in the northeast, and threaten native fishes, frogs, gartersnakes, and turtles in the region. The Bosque Nacional y Refugio de Vida Silvestre Los Ajos-Bavispe provides protection for eight key mountain ranges in the northeast. Conservation on this federally designated reserve is complemented by private reserves at the Northern Jaguar Reserve and Ranchos El Aribabi and Los Fresnos, as well as conservation ranching on other properties in northeastern Sonora.

GRASSLAND ASSEMBLAGE

We have placed seven species in this group, although all but *Crotalus viridis* occur in Chihuahuan Desert or other community types: *Ambystoma mavortium, Spea multiplicata, Terrapene ornata, Sceloporus cowlesi, S. slevini, Aspidoscelis uniparens,* and *Crotalus viridis.* Included are species characteristic of plains and semidesert grasslands (Brown 1982b). These biotic communities have been a focus of intense cattle grazing, often for well more than a century. Much of the semidesert grasslands have been invaded by desert shrubs, cacti, and trees. Heavy grazing removes fine fuels, thereby reducing fire frequency and favoring shrub and cacti invasion, which are less tolerant of fire than grasses. As mentioned previously, the causes of shrub and cacti invasion are not always clear. Cold winters in the higher-elevation plains grasslands are limiting for many of the desert shrubs and cacti that invade the lower, semidesert grasslands; hence, community type conversion is less likely to occur in plains grasslands. Rancho Los Fresnos, a private reserve owned by Naturalia, exhibits the best example of intact, well-managed plains grassland in Sonora.

CHIHUAHUAN DESERT ASSEMBLAGE

Seven species are included in this assemblage, although most occur in grassland communities as well, and a few have portions of their distributions in the mountains or foothills: *Anaxyrus debilis, Cophosaurus texanus, Phrynosoma cornutum, P. modestum, Aspidoscelis exsanguis, Gyalopion canum,* and *Heterodon kennerlyi.* Cattle grazing is the primary human activity in this community type. This shrub-dominated community is nearly devoid of grasses where grazing has been intense.

MARINE ASSEMBLAGE

The six species in this assemblage are exclusively wide-ranging, sea-dwelling reptiles that come to shore only to lay eggs. Most do not breed in Sonora: *Caretta caretta, Chelonia mydas, Eretmochelys imbricata, Lepidochelys olivacea, Dermochelys coriacea,* and *Hydrophis platurus.* Prior to its extirpation in the latter half of the twentieth century, *Crocodylus acutus* was a member of this assemblage and inhabited estuaries and mangrove swamps on the coast of Sonora

(Rorabaugh 2008). *Crocodylus acutus* was presumably eliminated by hunting and destruction or alteration of coastal estuaries and mangrove thickets. All five sea turtles have been protected by Mexican law since 1990, but their numbers have declined dramatically because of previously legal and now illegal harvest, as well as bycatch in the nets of commercial fishing boats and shrimp trawlers in the Sea of Cortés (Seminoff and Nichols 2007). The marine life of the Sea of Cortés has been much altered by intensive fishing well in excess of sustainability, including trawling for shrimp, which essentially scrapes the sea bottom and results in much capture and death of non-target species (Stoleson et al. 2005; Hastings and Findley 2007). Management and protection in the Reserva de la Biósfera Alto Golfo de California y Delta del Río Colorado and the Reserva de la Biósfera Islas del Golfo de California have mitigated some of the impacts to marine environments.

GENERALIST ASSEMBLAGE

Twenty-six species are distributed over a variety of biotic communities and cannot be logically assigned to any one type: *Anaxyrus cognatus, A. punctatus, A. woodhousii, Incilius alvarius, Lithobates catesbeianus, Scaphiopus couchii, Kinosternon sonoriense, Apalone spinifera, Heloderma suspectum, Urosaurus ornatus, Aspidoscelis tigris, Hypsiglena chlorophaea, Lampropeltis getula, Masticophis bilineatus, M. flagellum, Pituophis catenifer, Rhinocheilus lecontei, Sonora semiannulata, Thamnophis marcianus, Trimorphodon lambda, Micruroides euryxanthus, Rena humilis, Crotalus atrox, C. molossus, C. scutulatus,* and *C. tigris.*

Three additional species, *Hemidactylus frenatus, H. turcicus,* and *Indotyphlops braminus,* all introduced to Sonora, have been found only around human habitations or in city parks.

We know that at present Sonora has a total of 195 herpetofaunal species: 38 amphibians (3 salamanders, 35 anurans), and 157 reptiles (16 turtles, 69 lizards, 72 snakes). The low percentage of amphibian species in the state (19.5%, or 38/195) is a result of the strong dependence of this group on wet and moist places to live, which are limited to the eastern and northeastern mountains, the basins of rivers that run through the state, and the lush subtropics of the southeastern region and the foothills of the western slope of the Sierra Madre Occidental. Nevertheless, several amphibian species of Sonora are adapted to the arid-semiarid conditions of the Sonoran Desert. Of the three species of salamanders of Sonora, only the Barred Tiger Salamander (*Ambystoma mavortium*) has been recorded in the Sonoran Desert (municipal well of Puerto Peñasco) and was almost certainly introduced there. The other two salamander species are limited to the highlands of the eastern and northeastern mountains. *Ambystoma rosaceum* is commonly found with abundant populations in streams above 1000 m in elevation in these mountain ranges, and *Pseudoeurycea bellii* is known only from a couple of localities in the environs of Yécora.

Anurans are much better represented in the state. The tuberculated skin of the members of the family Bufonidae allows them to occupy arid-semiarid habitats such that abundant populations of *Anaxyrus cognatus, A. kelloggi, A. punctatus, A. retiformis,* and *Incilius alvarius* occur in the Sonoran Desert, and *A. debilis* occurs at the southern edge of the Chihuahuan Desert of northeastern Sonora. Members of other anuran families such as Microhylidae (*Gastrophryne mazatlanensis*), Hylidae (*Smilisca fodiens*), and Scaphiopodidae (*Scaphiopus couchii* and *Spea multiplicata*) are also commonly found in the arid-semiarid conditions of the Sonoran Desert. Other species, such as those of the families Craugastoridae and Eleutherodactylidae, are adapted to reproduce outside bodies of water. They lay their eggs in moist places, the embryo develops inside the egg, and hatchlings have the same morphology as the adults. However, all the Sonoran species of these families are commonly found in the eastern mountains and their foothills or in the canyons and lowlands of the southeastern part of the state. Most other species of Bufonidae, Hylidae, Microhylidae, and Ranidae are limited to this same region.

According to Rorabaugh (2008) and Lemos-

Espinal and Smith (2009), it is possible that three additional anuran species occur in Sonora but have not yet been recorded. The Pacific Stream Frog (*Craugastor vocalis*) may occur in extreme southeastern Sonora; Hardy and Mc-Diarmid (1969) reported this species in extreme northeastern Sinaloa, a region with environmental conditions similar to those of southeastern Sonora. Brennan and Holycross (2006) reported the presence of the Plains Leopard Frog (*Lithobates blairi*) in southeastern Arizona; it is likely that this species occurs in the open grasslands of extreme northeastern Sonora, which offer a suitable habitat. Another anuran species likely to occur in Sonora is the Plains Spadefoot (*Spea bombifrons*). There are numerous grasslands and Chihuahuan Desert shrublands in northeastern Sonora that could host populations of this species; however, it has not yet been recorded.

The number of turtle species in Sonora (16) represents 50% of the total number of turtle species that occur in Mexico (Liner and Casas-Andreu 2008). Five of the six Mexican sea turtles are found along the coast of Sonora. Three native species of terrestrial turtles occur in Sonora (*Gopherus morafkai, Terrapene nelsoni, T. ornata*), and the Spiny Softshell (*Apalone spinifera*) has been introduced in the northern part of the state (Rorabaugh et al. 2008; Rorabaugh and King 2013). Most of the other turtle species are commonly found in streams and ponds of the eastern mountains and extreme southeastern Sonora and in the rivers that run through the Sonoran Desert.

The Yellow Mud Turtle (*Kinosternon flavescens*) may occur in pools and cattle tanks of extreme northeastern Sonora. Brennan and Holycross (2006) reported it from extreme southeastern Arizona, near adjacent Sonora.

The lizards represent 35.4% (69/195) of the total number of herpetofaunal species in Sonora. Except for the marine habitats, they occur in practically all the available habitats in the state. Although their highest richness is found in the Sonoran Desert, they are also successful in the mountains of eastern and northeastern parts of the state and in the subtropical area of south-

eastern Sonora. We regard two lizard species as highly likely to occur in Sonora: the Twin-spotted Spiny Lizard (*Sceloporus bimaculosus*) and the Two-lined Short-nosed Skink (*Plestiodon bilineatus*). The Twin-spotted Spiny Lizard (*Sceloporus bimaculosus*) was mapped by Brennan (2008) in southeastern Arizona, close to the border with Sonora. The Two-lined Short-nosed Skink (*Plestiodon bilineatus*) was recorded in extreme western Chihuahua by Lemos-Espinal and Smith (2007) in a number of localities west of the Continental Divide, some of them close to the Sonora state line; the mountains of eastern Sonora provide suitable habitats for this lizard.

Snakes are the herpetofaunal group with the largest percentage of representation in Sonora: 36.9% (72/195) of the total number of amphibian and reptile species. It is highly likely that at least five more species occur within the state, three of them in northeastern Sonora (*Tantilla nigriceps, Rena dissecta,* and possibly *Sistrurus catenatus*); one in the eastern mountains of the state (*Conopsis nasus*); and another in extreme southeastern Sonora (an undescribed species of *Hypsiglena*).

The Plains Black-headed Snake (*T. nigriceps*) was mapped by Brennan (2008) in extreme southeastern Arizona; it may occur in grassland valleys of northeastern Sonora. The New Mexico Threadsnake (*R. dissecta*) also occurs in southeastern Arizona (Brennan and Holycross 2006) and was reported by Lemos-Espinal and Smith (2007) from extreme northwestern Chihuahua, in the Pradera de Janos (Rancho San Francisco), very close to the Sonora state line; it may occur in the grasslands west of the Sierra San Luis of extreme northeastern Sonora. And the Massasagua Rattlesnake (*S. catenatus*) was mapped by Brennan (2008) in scattered localities of southeastern Arizona; it may also occur in the grasslands of northeastern Sonora.

The Large-nosed Earthsnake (*C. nasus*) was reported by Lemos-Espinal and Smith (2007) from several localities in extreme western Chihuahua west of the Continental Divide and close to the state line with Sonora. Coniferous forest of the mountains of eastern Sonora provides suitable habitats for this colubrid snake, and we are confident of the presence of this species in Sonora. However, there are no records yet of this species within the state.

We have found a number of *Hypsiglena* in extreme southwestern Chihuahua (Chínipas-Milpillas region) and southeastern Sonora (Álamos-Güirocoba region) with morphological characteristics of *H. torquata* (e.g., white nape) along with typical *H. chlorophaea* individuals. *Hypsiglena* with a white nape from this region have long been considered to be intermediate between typical *torquata* and *chlorophaea* (Bogert and Oliver 1945; Dixon and Dean 1986). Mulcahy et al. (2014) recognized this white-naped form as a distinct species, but it has not yet been formally described.

Lemos-Espinal and Smith (2009) reported the possible occurrence of an amphisbaenid (*Bipes* sp.) in San Carlos Bay based on an unconfirmed report by Royce E. Ballinger. The sandy beaches of San Carlos Bay would provide the environmental requirements for this amphisbaenid to occur in Sonora.

Of the herpetozoan species of Sonora 120 (62%) are shared with the United States, and 75 species (38%) are not shared between the two countries. Of the 120 shared species, 75 are species that occur in the North American Mojave, Sonoran, and/or Chihuahuan Deserts. Most of them are limited to these arid-semiarid regions with their southern limits at the Mexican Plateau. In general, these are species that also occur in the North American states of California, Arizona, and/or New Mexico. Another 25 of these 120 shared species commonly occur in temperate woodlands of mountain ranges in southeastern Arizona, southwestern New Mexico, and the Sierra Madre Occidental and its associated sky islands of northeastern Sonora. Most of these species range from southern Arizona and New Mexico to northern Durango, Sinaloa, Zacatecas, or southern Chihuahua (e.g., *Hyla wrightorum, Lithobates chiricahuensis, L. tarahumarae, Sceloporus virgatus, Lampropeltis knoblochi,* and *Crotalus willardi*). Three of these species have southernmost limits in Sonora (e.g., *Lithobates yavapaiensis, Kinosternon arizonense,* and *Aspidoscelis*

sonorae). Some others have a widespread distribution in semiarid and temperate regions (e.g., *Hyla arenicolor, Thamnophis cyrtopsis, T. eques, Crotalus lepidus,* and *C. molossus*). The populations of one other species are distributed in the northern part of the Sierras Madres Occidental and Oriental (*Crotalus pricei*), with the western population ranging from southern Arizona to Nayarit. Another ranges from southern Canada to the Transvolcanic Belt of Mexico (*Diadophis punctatus*). Six of the 120 shared species are introduced in Sonora (*Lithobates berlandieri, L. catesbeianus, Apalone spinifera, Hemidactylus frenatus, H. turcicus,* and *Indotyphlops braminus*). Six are marine species with a widespread distribution through the world's oceans (*Caretta caretta, Chelonia mydas, Eretmochelys imbricata, Lepidochelys olivacea, Dermochelys coriacea,* and *Hydrophis platurus*). Seven species occur in the tropics and subtropics of the Americas on both coasts and enter the United States just at the southern tip of Texas (*Rhinella marina, Smilisca baudinii, Hypopachus variolosus, Drymarchon melanurus,* and *Drymobius margaritiferus*), or just at the southeastern tip of Arizona (*Oxybelis aeneus, Senticolis triaspis*). One of the 120 shared species ranges from Canada and Montana to South America, occurring in a wide variety of environmental conditions (*Lampropeltis triangulum*).

Sixty-seven of the 75 species that are not shared between the two countries are endemic to Mexico, 16 of them to Sonora: 11 to the islands of the Sea of Cortés (*Phyllodactylus nolascoensis, Ctenosaura conspicuosa, C. nolascensis, Sauromalus varius, Uta nolascensis, U. palmeri, Aspidoscelis baccata, A. estebanensis, A. martyris, Masticophis slevini,* and *Crotalus estebanensis*), and 5 to the continental part of the state (*Trachemys yaquia, Crotaphytus dickersonae, Phrynosoma ditmarsi, Aspidoscelis opatae,* and *Xantusia jaycolei*). One of the 67 endemic species was likely introduced to Alcatraz island of Kino Bay by the Seri Indians (*Sauromalus hispidus*). This species is originally endemic to the islands of the Sea of Cortés of Baja California. One more of these 67 endemic species is widespread in the Chihuahuan Desert (*Holbrookia approximans*); it is found only in the extreme northeastern part of Sonora and may occur in adjacent parts of Arizona and New Mexico. Another 13 of the 67 endemic species are characteristic of the temperate woodlands of the Sierra Madre Occidental. The range of some of them extends northwest from the Transvolcanic Belt of Mexico and reaches its northernmost limit in the state of Sonora (e.g., *Pseudoeurycea bellii, Phrynosoma orbiculare, Salvadora bairdii, Storeria storerioides,* and *Thamnophis melanogaster*). Others are limited to the northern part of the Sierra Madre Occidental in states such as Sonora, Chihuahua, Sinaloa, Durango, and Zacatecas (e.g., *Ambystoma rosaceum, Anaxyrus mexicanus, Incilius mccoyi, Craugastor tarahumaraensis, Sceloporus lemosespinali, Plestiodon parviauriculatus,* and *Thamnophis unilabialis*). Another 36 of the 67 endemic species are characteristic of the tropics and subtropics of Mexico. These species have ranges extending from southern Mexico through the canyons and foothills of the western slope of the Sierra Madre Occidental and reach their northernmost distribution in the lowlands or foothills of Sonora (e.g., *Pachymedusa dacnicolor, Tlalocohyla smithii, Kinosternon integrum,* and *Leptodeira punctata*). A few more are limited to the subtropics from north of the Río Santiago in Nayarit and also have their northernmost limit in the lowlands or foothills of Sonora (e.g., *Anaxyrus kelloggi, Incilius mazatlanensis, Lithobates magnaocularis, Coleonyx fasciatus, Heloderma exasperatum, Sceloporus nelsoni,* and *Sonora aemula*). The 8 species that are not endemic to Mexico but do not occur in the United States (*Leptodactylus melanonotus, Lithobates forreri, Rhinoclemmys pulcherrima, Phyllodactylus tuberculosus, Boa constrictor, Imantodes gemmistratus, Masticophis mentovarius,* and *Agkistrodon bilineatus*) are widespread species in the tropics and subtropics of Mexico, Central America, and even South America.

Herpetofauna of Chihuahua

JULIO A. LEMOS-ESPINAL
AND HOBART M. SMITH

Introduction

Knowledge of the extent of diversity of the herpetofauna of the state of Chihuahua has grown
very slowly. Up until about 1900 lack of knowledge was largely a result of difficulty of access
from the centers of academic activity in the
central parts of Mexico and universities and museums in the United States. Early explorations in
the state were limited to the Chihuahuan Desert
in the eastern and central parts, which were not
inspiring for intensive investigation, as they
remain to a certain extent even now. As the only
practical means of access to the western mountains, the desert constituted a bottleneck from
the south for exploration of the Sierra Madre
Occidental that covers most of the western third
of the state and has an excitingly diverse fauna.

For example, the "father" of Mexican herpetology, Alfredo Dugès (1896), who spent more
than 60 years of his studies in Mexico, in his
summary of the herpetofauna of Mexico reported but one species in Chihuahua (*Xerobates
polyphemus*), unrecognized then as a new species
(*Gopherus flavomarginatus*; Legler 1959b). The
first significant herpetological explorations in the
state came from the north, through surveys close
to the US border. Cope (1887) reported 37 species
in the state, and even the Smith and Taylor
checklists (1945, 1948, 1950a) reported only 102,
with relatively few from the Sierra Madre Occi-

dental. Tanner's summaries (1985, 1987, 1989) of
his and other studies in the western mountains
revealed an unexpectedly diverse herpetofauna
there of 120 species. Recently, several comprehensive publications on the diversity of the
amphibians and reptiles in this region have appeared. These include lists of all amphibian and
reptile species known to occur in Chihuahua,
keys to those species, photographs, an account
and distributional dot maps for each species, and
a gazetteer including the coordinates and municipality for the collection localities for the species
reported (Lemos-Espinal, Smith, Chiszar, and
Woolrich-Piña 2004; Lemos-Espinal et al., forthcoming; Lemos-Espinal and Smith 2007, 2009).

Further studies since about 1985, particularly
by our own group, continued the increasing
trend of known herpetofauna of the state, so we
can now list 174 species. The trend will obviously
continue, as indicated by the list herewith of possible occurrence of species in the state, judging
from their distribution in adjacent areas. The
real extent of the herpetofauna of the state probably exceeds 200—a lure for future workers.

Previous Herpetological Studies

Evidence of pre-Columbian interest in reptiles
and amphibians in Chihuahua exists in the
numerous petroglyphs in various parts of the

Chihuahuan Desert, for example, in the Sierra de Samalayuca, Sierra de Los Muertos, Cerros Colorados, and Arroyo Los Monos in the Sierra Madre Occidental.

Within historical times, the first herpetological investigation in Chihuahua appears to have been that of Thomas H. Webb, secretary and surgeon for the John Russell Bartlett Expedition exploring the vicinity of the US-Mexican border. Bartlett's group left El Paso, Texas, October 7, 1852, en route to the Ringgold Barracks of southern Texas, passing through the states of Chihuahua, Durango, Coahuila, and Nuevo León. The expedition arrived in Ciudad Chihuahua on October 22, 1852, passing through Guadalupe, Carrizal, Encinillas, and Saucillos on the way. There they remained for 10 days, during which time Webb collected various animals, including the first amphibians and reptiles from Mexico contributed to the National Museum of Natural History (USNM) (Baird 1859; Kellogg 1932).

Two years later, John Potts sent a collection of amphibians and reptiles from Chihuahua to USNM, where it was received April 7, 1854. In 1873 Edward Wilkinson spent two years in the mining town of Batopilas and returned in 1885 to spend most of his time near Ciudad Chihuahua. On both trips he collected amphibians and reptiles that were distributed to various North American museums, mostly the USNM. The collections were reported by Cope (1879, 1885).

In 1897 and 1899 Edward Goldman and Edward Nelson of USNM traversed much of the Sierra Tarahumara, collecting all manner of tetrapod vertebrates and describing the ecology of the area. Goldman (1951) published a detailed travelogue of all their travels in Mexico. Between 1900 and 1934 numerous naturalists visited Chihuahua, among them S. B. Benson, C. S. Brimley, H. H. Brimley, W. W. Brown, and W. B. Richardson, whose collections were deposited in various North American museums.

In June 1934, David Dunkle and Hobart M. Smith collected along the main north-south highways of Chihuahua, sampling the Samalayuca sand dunes and areas around Ciudad Chihuahua, Río San Pedro, the upper Río Conchos, Presa de Boquilla, and Escalón near the

state of Durango. That collection became part of the personal collection of Edward H. Taylor, later sold to the University of Illinois at Champaign-Urbana and Field Museum of Natural History. In October 1938 Hobart M. Smith and his wife, Rozella, collected 526 specimens of 29 species under a scholarship from the USNM at scattered localities in the state, including the municipalities of Ahumada, Ascensión, Buenaventura, Camargo, Casas Grandes, Chihuahua, Jiménez, and Juárez.

Between 1938 and 1940 Irving Knobloch made important collections in the vicinity of Maguarichic and Mojarachic. Portions of these collections were reported by Taylor and Knobloch (1940), including 3 species of amphibians and 19 reptiles. Taylor (1940a, 1940b, 1941, 1944) described several new species from the Knobloch collections, including *Eleutherodactylus tarahumaraensis*, *Lampropeltis knoblochi*, *Ambystoma rosaceum*, *A. fluvinatum*, and *Crotalus semicornutus*. All specimens were in his private collection, sold to the Field Museum of Natural History and the University of Illinois.

Important collections from the mountains of western Chihuahua were also made by James Anderson from 1952 to 1966 in the vicinities of Yahuarichic, Chuhuichupa, and Sierra del Nido. He died before his collections could be studied in full, but he published on parts of it (1960, 1961, 1962a, 1962b, 1962c, 1972). These works included distributional and natural history information on *Storeria storerioides*, *Plestiodon brevilineatus*, and *Sceloporus slevini*, as well as the original description of *Crotalus willardi amabilis* from the Sierra del Nido.

By far one of the largest collections from Chihuahua was amassed from 1956 to 1972 at Brigham Young University by Wilmer W. Tanner and his students, most important, Gerald W. Robison. Tanner published numerous publications at intervals after 1956, some with Robison, based on those collections, which were summarized in Tanner and Robison (1959) and Tanner (1957, 1985, 1987, 1989). The collection totaled 120 species, including 21 amphibians (2 salamanders, 19 anurans) and 99 reptiles (5 turtles, 42 lizards, 52 snakes). New taxa described in these works include *Plestiodon multilineatus, Sceloporus clarkii uriquensis, S. nelsoni coeruleus, S. poinsettii robisoni, Rena humilis chihuahuensis, R. dulcis supraocularis, Thamnophis melanogaster chihuahuensis*, and *T. rufipunctatus unilabialis* (= *T. unilabialis*).

Kenneth L. Williams and Pete S. Chrapliwy collected in various parts of northern Mexico in the summer of 1959, including Chihuahua, as reported by Williams, Chrapliwy, and Smith (1959, 1960); Williams, Smith, and Chrapliwy (1960); and Chrapliwy et al. (1961). Included was the original description of *Uma paraphygas*. The specimens are in the University of Illinois Museum of Natural History (UIMNH).

In August 1962 Kenneth L. Williams, Edward O. Moll, Francois Vuilleumier, and John E. Williams took canoes down the Río Conchos some 300 km, from Julimes to Ejido El Álamo (= Cañón de Barrera), collecting 282 specimens of 27 species (6 anurans, 3 turtles, 13 lizards, and 5 snakes). This material likewise was deposited in UIMNH and was reported by Smith et al. (1963).

Thomas R. Van Devender and Charles H. Lowe collected in the vicinities of Babícora, Temosachic, and Yepómera during the summers of 1971–1973. They obtained 204 specimens of amphibians (11 species) and 453 reptiles (29 species), now in the University of Arizona museum. The results were reported by Van Devender (1973a, 1973b), Van Devender and Van Devender (1985), and Van Devender and Lowe (1977).

A publication by Domínguez et al. (1977) reported a collection of amphibians and reptiles made at some time prior to 1974 under the auspices of the Dirección General de la Fauna Silvestre at 45 scattered localities, mostly west of Mex. Hwy 45 between Ciudad Juárez and Ciudad Chihuahua. Nine species of amphibians and 27 reptiles were included. All material is in the museum of the Instituto Politécnico Nacional.

Between 1975 and 1977 Robert Reynolds studied the relationship between rainfall and the activity of snakes along Mex. Hwy 16 (Ciudad Chihuahua-Ojinaga) between Aldama and Rancho El Pastor. The 418 snakes observed belonged to 20 colubrid and viperid species (Reynolds 1982).

In May 1982 Bruce Baker, Leroy Banicki, Rodolfo Corrales, and Robert Webb collected terrestrial vertebrates in the vicinity of Cerro Mohinora, which as reported by Webb and Baker (1984) included 11 amphibian and 5 reptile species, all more or less expected in that area except for an unidentified species of *Lithobates* in the *pipiens* complex that may represent *L. berlandieri* (Smith and Chiszar 2003). If so, it was the first record of this species of largely eastern distribution in the western mountains of the state. Those specimens are in the museum of the University of Texas at El Paso.

Bradley Shaffer and M. McKnight (1996) reported *Ambystoma velasci* (= *A. silvense*) from near El Vergel and Temosachic on the basis of specimens previously collected by Shaffer. The specimens are at the University of California at Davis.

Frost and Bagnara (1976), Legler (1959a), Tanner (1988, 1990), Webb et al. (2001), and Williams (1994) are some of the other works that have contributed to knowledge of the herpetofauna of Chihuahua.

Between 1993 and 2006 Hobart Smith and Julio Lemos-Espinal collected in many scattered localities throughout Chihuahua. That material has been reported in various works, including Chiszar et al. (1995), Lara-Góngora (2004), Larson et al. (1998), Lemos-Espinal and Smith (2007, 2009), Lemos-Espinal et al. (forthcoming), Taylor et al. (2003), and in several papers by Lemos-Espinal and colleagues and Smith and colleagues published between 1994 and 2009. Included in these publications was the first record for these taxa in the state: *Ambystoma mavortium, Eleutherodactylus interorbitalis, E. marnockii, Pachymedusa dacnicolor, Tlalocohyla smithii, Hypopachus variolosus, Lithobates forreri, L. magnaocularis, L. pustulosus, Terrapene nelsoni klauberi, Rhinoclemmys pulcherrima, Gopherus morafkai, Barisia ciliaris, Hemidactylus turcicus turcicus, Heloderma horridum exasperatum* (= *H. exasperatum*), *Sceloporus merriami annulatus, Boa constrictor imperator, Gyalopion quadrangulare, Imantodes gemmistratus latistratus, Masticophis flagellum cingulum, M. f. testaceus,*

Mastigodryas cliftoni, Thamnophis validus validus, Tropidodipsas repleta, Agkistrodon bilineatus bilineatus, and *Crotalus basiliscus.* Included were original descriptions of *Lithobates lemosespinali, Sceloporus edbelli* (= *S. consobrinus*), *S. lemosespinali, S. merriami sanojae, S. m. williamsi,* and *S. undulatus speari* (= *S. consobrinus*).

Recently, Georgina Santos-Barrera and Oscar Flores-Villela (2011) described a new species of toad (*Incilius mccoyi*) from the Sierra Madre Occidental of Chihuahua and Sonora. This species represents the toad populations from the Sierra Madre Occidental of Chihuahua formerly regarded as *I. occidentalis.* Other recent additions to the herpetofauna of Chihuahua are *Rhadinaea laureata* from the municipality of Guadalupe y Calvo (Villa et al. 2012) and *Nerodia erythrogaster bogerti* from near Ciudad Ojinaga (Uriarte-Garzón and García-Vázquez 2014).

In addition to the studies mentioned previously we have now determined that there are 39 scientific names of amphibians (3 salamanders, 5 anurans) and reptiles (15 lizards, 16 snakes) with type localities assigned to the state of Chihuahua (table 5.1).

Physiographic Characteristics and Their Influence on the Distribution of the Herpetofauna

Amphibians and reptiles possess characteristics that limit their distribution and activity. With the exception of some members of the Plethodontidae, Craugastoridae, and Eleutherodactylidae, all amphibians require bodies of water for reproduction. All amphibians and reptiles depend on the environment for regulation of temperature. Extremely low or high temperatures limit their activity to a greater degree than in endotherm vertebrates. Accordingly, it is necessary to know ambient physical characteristics in order to understand the nature of the herpetofauna of any given locality.

Chihuahua is the largest state in the Mexican Republic; its 245,612 km² (lying between the coordinate extremes of latitudes 25°38′ and 31°47′ N and longitudes 103°18′ and 109°7′ W) represent 12.6% of the total territory of the nation. Its extensive surface is physiographically complex and affects the distribution of the herpetofauna in equally complex ways.

Almost all of the western part of the state is occupied by the Sierra Madre Occidental, along the crest of which the Continental Divide passes, separating the Pacific drainages from the Atlantic. The highest altitude in the state is on Cerro Mohinora in the extreme southwestern part, at 3300 m (latitude 25°57′ N, longitude 107°3′ W). Other major elevations are Sierra de Gasachic (3060 m; latitude 28°15′ N, longitude 107°36′ W) and Cerro Güirichique (2740 m; latitude 26°52′ N, longitude 107°30′ W).

The altitude of the Sierra Madre diminishes to the north, terminating at about latitude 31° N, where the Sierra de San Luis extends northward to the border with New Mexico and reaches an altitude of about 2100 m.

The extreme southwestern Pacific slopes of the Sierra Madre are characterized by deep canyons that drop down to approximately 250 m in the Barranca del Septentrión/Cañón de Chínipas (INEGI 2004). The Sierra Madre varies in width from about 130 to 160 km in the south (west of Hidalgo de Parral) to about 65 to 80 km in the north (west of Casas Grandes) (Tanner 1985; Lemos-Espinal and Smith 2007, 2009).

Over half of the state of Chihuahua, east of the Sierra Madre, is represented by high plains at elevations of approximately 1200–1700 m. However, from these plains arise a large number of small to medium-sized isolated sierras, some of which reach altitudes of over 2000 m. Some are high enough to support coniferous forests, constituting continental "islands" surrounded by a "sea" of semiarid plains, where differentiation among populations is enhanced by isolation.

In the extreme northeast, deep canyons, analogous to those on the Pacific side of the Sierra Madre, cut into the edge of the high plains and support their own distinct herpetofaunal assemblage. Among them is the great Cañón de Santa Elena, in the Zona de Protección de Flora y Fauna Silvestre Cañón de Santa Elena, an ex-

TABLE 5.1 Type localities for amphibians and reptiles described from the state of Chihuahua, Mexico

Author(s)	Original name: Type locality
Salamanders (3)	
Taylor (1941)	*Ambystoma fluvinatum* (= *A. rosaceum*): **Mojárachic**
Taylor (1941)	*Ambystoma rosaceum:* **Mojárachic**
Owen (1844)	*Axolotes maculata* (= *Ambystoma rosaceum*): **Sierra Madre 26°6′N; 106°50′W**
Anurans (5)	
Girard (1854)	*Bufo insidior* (= *Anaxyrus debilis insidior*): **Chihuahua**
Santos-Barrera & Flores-Villela (2011)	*Incilius mccoyi:* **42 km S of Creel, on the Creel-Guachochi road**
Taylor (1940a)	*Eleutherodactylus tarahumaraensis:* **Mojárachic**
Smith & Chiszar (2003)	*Rana* (= *Lithobates*) *lemosespinali:* **between Creel and San Rafael**
Boulenger (1917)	*Rana* (= *Lithobates*) *tarahumarae:* **Yoquivo**[1]
Lizards (15)	
Stejneger (1890)	*Barisia levicollis:* **Chihuahua without defined locality**[1]
Tihen (1954)	*Gerrhonotus liocephalus taylori* (= *G. taylori*): **Clarines Mine, about 5 miles west of Santa Barbara, Chihuahua, Mexico, at an elevation of about 6800 feet**
Smith, Lemos-Espinal, Chiszar, & Ingrasci (2003)	*Elgaria usafa* (= *E. kingii*): **Ruinas of Rancho El Mesteño Chiquito**
Smith (1935b)	*Holbrookia bunkeri* (= *H. maculata bunkeri*): **15 mi S of Juárez**
Tanner & Robison (1959)	*Sceloporus clarkii uriquensis* (= *S. c. clarkii*): **Urique**
Smith, Chiszar, & Lemos-Espinal (1995)	*Sceloporus undulatus speari* (= *S. consobrinus*): **1.6 km N Hwy 2, 107°11′W, 31°33′N, 1250 m, on a side road intersecting Hwy. 2, 3.6 km E Microondas Duna, 18.6 km E San Martín, northern central Chihuahua**
Lemos-Espinal *in* Smith, Lemos-Espinal, & Chiszar (2003)	*Sceloporus merriami sanojae:* **Rancho Peñoles (27°7′49.6″N, 103°48′45.0″W), 1194 m**
Lemos-Espinal, Chiszar, & Smith (2000)	*Sceloporus merriami williamsi:* **El Fortín, 51 airline km straight W of the Río Grande**
Tanner & Robison (1959)	*Sceloporus nelsoni caeruleus* (= *S. n. barrancorum*): **Urique**
Tanner (1987)	*Sceloporus poinsettii robisoni* (= *S. p. macrolepis*): **Cuiteco**
Williams et al. (1959)	*Uma paraphygas:* **0.7 mi E of Carrillo, Chihuahua**
Mittleman (1942)	*Urosaurus unicus* (= *Urosaurus bicarinatus tuberculatus*): **Batopilas**
Smith (1935a)	*Uta caerulea* (= *Urosaurus ornatus caeruleus*): **30 mi N of Colonia García**
Tanner (1957)	*Eumeces* (= *Plestiodon*) *multilineatus:* **3 mi N of Colonia Chuhuichupa**
Wright & Lowe (1993)	*Cnemidophorus* (= *Aspidoscelis*) *inornata chihuahuae:* **8.1 miles east of Ciudad Chihuahua, 1440 meters elev.**
Snakes (16)	
Klauber (1946)	*Arizona elegans expolita:* **Casas Grandes**
Tanner (1961)	*Conopsis nasus labialis* (= *C. nasus*): **2 mi SE Creel**
Taylor (1940b)	*Lampropeltis knoblochi:* **Mojaráchic**
Smith (1941)	*Masticophis flagellum lineatulus:* **11 mi S of Buenaventura**
Baird & Girard (1852)	*Churchillia bellona* (= *Pituophis catenifer*): **Presidio del Norte, Chihuahua**[2]

TABLE 5.1 *Continued*

Author(s)	Original name: Type locality
Cope (1879)	*Procinura aemula*: **Batopilas**
Smith (1942a)	*Tantilla yaquia*: **Guasaremos, Río Mayo**
Smith (1942b)	*Thamnophis ordinoides errans* (= *T. errans*): **Colonia García**
Tanner (1985)	*Thamnophis rufipunctatus unilabialis* (= *T. rufipunctatus*): **0.5 mi SW of Bocoyna**
Cope (1886)	*Trimorphodon vilkinsoni*: **Chihuahua City**
Kennicott (1860)	*Micrurus diastema distans* (= *M. distans distans*): **Batosegachic**
Tanner (1985)	*Leptotyphlops humilis chihuahuensis* (= *Rena segrega*): **10.7 km NW of Ciudad Chihuahua**
Tanner (1985)	*Leptotyphlops humilis supraocularis* (= *Rena dissecta*): **Colonia Juárez**
Anderson (1962c)	*Crotalus willardi amabilis* (= *C. amabilis*): **Arroyo Mesteño**
Taylor (1944)	*Crotalus semicornutus* (= *C. lepidus klauberi*): **Mojárachic**
Klauber (1949)	*Crotalus willardi silus*: **Río Gavilán, 7 mi SW of Pacheco**

[1] Restrictions appeared in Smith & Taylor (1950b).

[2] Although the original type locality reported by Baird & Girard (1852) and Smith & Taylor (1950b, 1966) is Presidio del Norte, Chihuahua, this is in error; the correct type locality is Presidio del Norte, Texas.

tension of Big Bend National Park of the United States. It has an elevation of less than 800 m.

This complex topography has fundamental effects on the distribution of vegetation types and upon human activities. In the Sierra Madre, as well as in the highest parts of the ranges throughout the rest of the state, dense coniferous forests occur, dominated by Chihuahua pine (*Pinus leiophylla*), Apache pine (*P. engelmannii*), Mexican white pine (*P. ayacahuite*), Durango pine (*P. durangensis*), and white oak (*Quercus chihuahuensis*), among others. This type of vegetation represents 29.42% of the surface area of the state.

The deep barrancas of the west and southwest support tropical deciduous forests that have been relatively undisturbed due to the low density of human occupation. Notable plants are mauto (*Lysiloma divaricata*), tree morning glory (*Ipomoea arborescens*), guacimo (*Guazuma ulmifolia*), and fragrant bursera (*Bursera fagaroides*). This type of vegetation covers 2.38% of the surface area of the state.

East of the Sierra Madre the terrain is dominated by xerophytic shrubs such as creosote bush (*Larrea tridentata*), American tarwort (*Flou-rensia cernua*), mesquite (*Prosopis glandulosa*), Texas sotol (*Dasylirion texanum*), and ocotillo (*Fouquieria splendens*), among other species.

The principal grasslands are in the central and northeastern parts of the state. Some of the dominant species are blue grama (*Bouteloua gracilis*), hairy grama (*B. hirsute*), three-awn (*Aristida* sp.), sacaton (*Sporobolus airoides*), and speargrass (*Heteropogon contortus*). These grasslands make up 23.89% of the surface area of the state, and all that is suitable for grazing is grazed.

Cultivated parts of the state account for 7.38% of its area. The primary crops are maize (*Zea mays*), bean (*Phaseolus vulgaris*), alfalfa (*Medicago sativa*), sorghum (*Sorghum vulgare*), and apple (*Malus sylvestris*).

The climate of over half of the state (all east of the Sierra Madre) is reflected by its name, which was derived from the Nahuatl word *xicuacua*, meaning "a dry, sandy place." Three grades of aridity can be recognized: (1) very dry (precipitation less than 100 mm/yr, and an average annual temperature 16°C–20°C), characteristic of most of the northern and eastern parts of the state (parts of the municipalities of Ahumada, Ascensión, Buenaventura, Coyame, Guadalupe,

Juárez, Manuel Benavides, and Ojinaga), and to a lesser degree in the extreme southwest (part of the municipalities of Camargo and Jiménez); (2) dry (precipitation 300–600 mm/yr, mean annual temperature 15°C–25°C); and (3) semidry (precipitation 400–600 mm/yr, mean annual temperature 12°C–25°C).

The rest of the state has more humid climates, with no less precipitation than about 500 mm/yr. These areas may be regarded as (1) cool (precipitation no more than approximately 600 mm/yr, mean annual temperature 10°C–12°C) in the highest parts of the Sierra Madre; (2) temperate (precipitation no more than 600 mm/yr, mean annual temperature about 14°C) at moderate elevations in the Sierra Madre; (3) warm (precipitation 800–1400 mm/yr, mean annual temperature 20°C–24°C) in the barrancas in the southwestern part of the state; and (4) hot (precipitation 700–1000 mm/yr, mean annual temperature about 24°C) in the lowlands of the extreme southwest.

The rainy season extends from May to September, with a peak in July and August.

The major features of the topography of Chihuahua were established by the beginning of the Pleistocene epoch, during which several glaciation and interglacial events occurred that required extensive adjustments at repeated intervals in the ranges of the state's species of amphibians and reptiles. In recent times, the state has undergone considerable desertification.

These processes have eliminated some species from the state, restricted the range of others, expanded the range of some, and led to taxonomic differentiation of a few isolated populations formerly continuous with others, such as the subspecies *Crotalus willardi*. Some species, for example, *Chrysemys picta* and *Plestiodon multivirgatus*, now verge on extinction, with severely limited ranges. *Gopherus flavomarginatus* has a very small relictual range and is not likely to survive without protracted, intensive protection of both young and adults. Evidence indicates that *Sternotherus odoratus* once occurred in the state.

Most of the land in the state drains eastward into the Gulf of Mexico. A relatively small but very important region herpetologically on the state's western boundary drains into the Pacific Ocean in the Sea of Cortés. In the north-central portion of the state some rivers have only interior drainages that end in landlocked lakes and playas.

The Ríos Florido and San Pedro drain the central and southern parts of the state east of the Continental Divide and form the Río Conchos, which joins the Río Bravo, flowing eastward to the Gulf of Mexico. These rivers provide irrigation for southern, central, northern, and eastern parts of the state.

Drainage into the Pacific Ocean is provided by the Ríos El Fuerte, Mayo, Sinaloa, and Yaqui west of the Continental Divide. The Ríos Batopilas, Chínipas, Oteros, San Ignacio, and Urique unite as the Río El Fuerte; the Ríos Moris and Agua Caliente form the Río Mayo; the Río Pentatlán contributes to the Río Sinaloa; and the Río Papigochi flows into the Río Yaqui. Irrigation is provided for the Barrancas del Cobre by the El Fuerte; the Moris region, by the Mayo; the extreme southwest, by the Sinaloa; and the Miñaca, Yepómera, and Madera regions, by the Yaqui. The ranges of a number of tropical species, such as *Tlalocohyla smithi*, *Hypopachus variolosus*, *Pachymedusa dacnicolor*, *Heloderma exasperatum*, *Boa constrictor*, and *Oxybelis aeneus*, extend northward into the lowlands drained by these rivers.

Interior drainages are present mostly in the northwestern parts of the state. The Río Casas Grandes and Río Santa María empty into the Laguna de Guzmán, reduced in recent years to a few ponds. The Río El Carmen feeds the Laguna de Los Patos. Laguna El Barreal in the north and the central Laguna El Cuervo are depressions that form shallow swamps through the rainy season but dry completely in the dry season. They are very difficult to access while they hold water.

Geological formations over most of the state (91.39%) are of Cenozoic origin; only 8.23% are of Mesozoic origin, and 0.38%, Paleozoic (INEGI 2004). The Sierra Madre consists mostly of extrusive igneous rocks of Tertiary origin. Heavy rains in summers produce voluminous torrents that have rapidly cut into the western face of the escarpment, producing the famous, characteristically steep-sided, deep canyons of that exposure. At the same time erosive processes have exposed large deposits of minerals, leading to a long history of mining, especially for silver, gold, and copper. The most famous of these canyons, the Barrancas del Cobre, derived its name from its copper deposits.

The distribution and behavior of the amphibians and reptiles of Chihuahua have been greatly influenced by the varied ecologies arising from its diverse physical properties. The major barrier to a more or less uniform distribution is the Sierra Madre Occidental, which prevents most eastern species from reaching western slopes, and vice versa, and supports an extensive intrinsic biota. The lowland species that do occur on both sides of the Sierra Madre (e.g., *Anaxyrus cognatus, A. punctatus, Scaphiopus couchii, Urosaurus ornatus, Uta stansburiana, Salvadora deserticola, Crotalus atrox*) did not cross the divide but extended southward on both sides from the north.

The presence of western species in Chihuahua's herpetofauna results from connectivity through Pacific slope drainages and canyons that primarily drain to the Ríos El Fuerte and Mayo. Most species are restricted to low altitudes.

The activity of most reptiles and amphibians in Chihuahua is limited to April–October, but peak times vary according to altitude, timing of rainfall, and the adaptations of a given species. The activity season begins earlier and lasts longer in the tropical southwest and is most limited in the higher mountains. The temperature cycle does not match the precipitation cycle, so different species respond differently according to their adaptations.

The reptiles that begin their activity in April (e.g., *Aspidoscelis costata, Sceloporus albiventris, S. clarkii, S. lemosespinali, S. nelsoni, Phrynosoma orbiculare, Storeria storerioides, Pituophis catenifer*) are minimally active. Juveniles, which can warm up quickly, are the first to become active. Montane amphibians living in or near permanent water such as *Ambystoma rosaceum* and

various species of *Lithobates* tend to breed early and therefore may reach their maximum activity in April; activity of other montane amphibians as well as reptiles in all habitats tends to peak during the rainy season in June, July, August, and early September, then gradually taper off so that from mid-November through March there is virtually no activity. Only on warm days during winter will some heliophilic species emerge briefly.

Adaptations of the Chihuahua herpetofauna to these environmental variables have given rise to four distinct assemblages. (1) Subtropical: species confined to deep canyons with tropical deciduous forests that occur also in extreme southeastern Sonora and northeastern Sinaloa; (2) Montane: species occurring along the Sierra Madre Occidental and Sierra de San Luis, but also isolated sierras, such as the Sierra del Nido, Sierra del Pajarito, and Sierra Azul that rise from the adjacent high plains east of the Sierra Madre Occidental; (3) Chihuahuan Desert: species adapted to the arid/semiarid environment of the Chihuahuan Desert east of the Sierra Madre and its associated sierras; (4) Generalist: species that occur in more than one of the major habitats of the state and are not characteristic of any one of them.

SUBTROPICAL ASSEMBLAGE

Forty-one species are included in this assemblage: *Incilius mazatlanensis, Rhinella marina, Craugastor augusti, Pachymedusa dacnicolor, Smilisca baudinii, Tlalocohyla smithi, Hypopachus variolosus, Lithobates forreri, L. magnaocularis, Rhinoclemmys pulcherrima, Terrapene nelsoni, Gopherus morafkai, Phyllodactylus tuberculosus, Heloderma exasperatum, Ctenosaura macrolopha, Sceloporus albiventris, S. nelsoni, Urosaurus bicarinatus, Anolis nebulosus, Aspidoscelis costata, Boa constrictor, Drymarchon melanurus, Drymobius margaritiferus, Gyalopion quadrangulare, Imantodes gemmistratus, Lampropeltis triangulum, Leptodeira splendida, Leptophis diplotropis, Masticophis mentovarius, Mastigodryas cliftoni, Oxybelis aeneus, Senticolis triaspis, Sonora aemula, Sym-*

pholis lippiens, *Thamnophis validus*, *Trimorphodon tau*, *Micruroides euryxanthus*, *Micrurus distans*, *Rena humilis*, *Agkistrodon bilineatus*, and *Crotalus basiliscus*.

The latter two species are exceptional among these species in being typically tropical throughout their ranges, but they have also populated temperate areas, as they have on the Sierra de Milpillas, which constitutes a barrier for some species between the subtropical areas of southeastern Sonora and southwestern Chihuahua.

The density of human populations is quite low in these tropical zones, and as a result anthropogenic alteration of the environment is also quite low and aids in conservation. In contrast, in the immediate vicinity of the villages of Batopilas and Chínipas any snake encountered by a local resident is killed. An exception in Batopilas is the "limacoa" (*Boa constrictor*), which is considered beneficial through its predation on rodents. Some, among them *Lampropeltis triangulum*, *Oxybelis aeneus*, *Sonora aemula*, and *Sympholis lippiens*, are generally considered venomous and deadly by local inhabitants. Near Chínipas *Heloderma exasperatum* has been especially persecuted. Fortunately, Martín Velducea Avendaño and a few other residents have recently been promoting protection for these lizards, as well other reptiles, and apparently have diminished their persecution.

MONTANE ASSEMBLAGE

About 57 species occur in this high-altitude forested area, of which 40 are limited to this area and constitute the Montane assemblage: *Ambystoma rosaceum*, *A. silvense*, *Pseudoeurycea bellii*, *Anaxyrus mexicanus*, *Incilius mccoyi*, *Craugastor tarahumaraensis*, *Eleutherodactylus interorbitalis*, *Hyla wrightorum*, *Lithobates chiricahuensis*, *L. lemosespinali*, *L. tarahumarae*, *L. yavapaiensis*, *Barisia ciliaris*, *B. levicollis*, *Elgaria kingii*, *Gerrhonotus taylori*, *Phrynosoma hernandesi*, *P. orbiculare*, *Sceloporus jarrovii*, *S. lemosespinali*, *S. slevini*, *S. virgatus*, *Plestiodon bilineatus*, *P. callicephalus*, *P. multilineatus*, *P. parviauriculatus*, *Aspidoscelis sonorae*, *Conopsis nasus*, *Diadophis punctatus*, *Geophis dugesii*, *Lampropeltis knoblochi*, *Pituophis deppei*, *Rhadinaea laureata*, *Salvadora bairdii*, *Storeria storerioides*, *Thamnophis errans*, *T. melanogaster*, *Tropidodipsas repleta*, *Crotalus pricei*, and *C. willardi*.

The principal human activity in the Sierra Madre is timber harvest, mostly of pines, and oaks to a lesser extent. The three municipalities most important in this respect are Madera, Guadalupe y Calvo, and Guachochi. Although in some places a serious degradation of forest resources has occurred, most are reasonably intact. Large areas of rugged topography exist that are essentially inaccessible and therefore protected from exploitation of the dense forests of pine and oak and their wildlife.

The ancillary Sierra de San Luis and Sierra del Nido are also largely intact. The former is the northernmost limit of distribution of the jaguar (*Panthera onca*) in Mexico, according to Lafont-Terrazas (pers. comm. 2012), and the only place in the state where *Crotalus willardi obscurus* occurs. The Sierra del Nido is likewise exceptionally undisturbed and supports viable populations of four kinds of rattlesnakes (*Crotalus lepidus*, *C. molossus*, *C. pricei*, and *C. willardi*), and certain lizards (*Barisia levicollis*, *Sceloporus poinsettii*, *S. jarrovii*, among others) are abundant.

One of the best indicators of ecological stability in much of the forested area is the abundance of *Ambystoma rosaceum*, which is widely distributed and often very abundant in streams. Where streams have been degraded by logging, diversions, or other activities, this species is absent or scarce. The forest treefrogs (*Hyla arenicolor*, *H. wrightorum*) are also abundant throughout most of the Sierra Madre.

CHIHUAHUAN DESERT ASSEMBLAGE

Although the Chihuahuan Desert is characterized by a definite climatic continuity, it nevertheless has considerable diversity of habitats. Extensive grasslands occur in the central and northwestern parts of the state; sand dunes, both live and stabilized, occur to the north (Bolsón Cabeza de Vaca, Samalayuca system) and to the south (Bolsón de Mapimí); extensive barreales (shallow basins that hold water temporarily during the rainy season but are dry at other times) in extreme northern and central parts; ecotones of mixed xerophytic and wooded areas around isolated sierras in the middle of the desert; and riparian vegetation along the Ríos Conchos, Bravo, and Santa María.

Some 77 species inhabit these areas, and 57 are exclusive to it, within the boundaries of the state: *Ambystoma mavortium, Incilius alvarius, Anaxyrus cognatus, A. debilis, Eleutherodactylus marnockii, Gastrophryne olivacea, Lithobates catesbeianus, Scaphiopus couchii, Spea bombifrons, Chrysemys picta, Terrapene ornata, Trachemys gaigeae, Kinosternon durangoense, K. flavescens, K. hirtipes, Gopherus flavomarginatus, Apalone spinifera, Crotaphytus collaris, Gambelia wislizenii, Coleonyx brevis, Hemidactylus turcicus, Cophosaurus texanus, Holbrookia maculata, Phrynosoma cornutum, P. modestum, Sceloporus bimaculosus, S. consobrinus, S. cowlesi, S. merriami, Uma paraphygas, Uta stansburiana, Plestiodon multivirgatus, P. obsoletus, Aspidoscelis gularis, A. inornata, A. marmorata, A. tessellata, A. uniparens, Arizona elegans, Bogertophis subocularis, Gyalopion canum, Heterodon kennerlyi, Lampropeltis getula, Masticophis taeniatus, Nerodia erythrogaster, Pantherophis emoryi, Rhinocheilus lecontei, Sonora semiannulata, Tantilla hobartsmithi, T. nigriceps, T. wilcoxi, Thamnophis marcianus, Rena dissecta, R. segrega, Agkistrodon contortrix, Crotalus atrox,* and *C. viridis.*

Unfortunately, considerable anthropogenic environmental degradation has occurred. Stock ranches, especially of cattle, have drastically affected the face of the Chihuahuan Desert, where in the mid-nineteenth century the extensive grasslands were relatively free of bushes. Rivers traversing the region were bordered by dense gallery forests and frequently overflowed into large barreales that retained a relatively high level of humidity in the region. At present, cattle ranching has contributed to the domination of much of the former prairie grassland by shrubby vegetation. The barreales now retain water scarcely beyond the rainy season itself. Extensive

desertification has greatly reduced the range of the graminivorous *Gopherus flavomarginatus* and possibly exterminated *Sternotherus odoratus* in the vicinity of El Sauz. Formerly abundant populations of prairie dogs (*Cynomys ludovicialis*), wolves (*Canis lupus baileyi*), grizzly bears (*Ursus horribilis*), pronghorns (*Antilocapra americana*), aplomado falcons (*Falco femoralis*), and others that occurred in the Chihuahuan Desert have now been eliminated or much reduced in numbers and distribution. The wolves and bears no longer occur anywhere in the state; the prairie dogs are limited to the extreme northwest (Pradera de Janos); and the falcon, to the Desierto Coyamense.

In addition, intensive agricultural practices have destroyed thousands of hectares of native vegetation. Overuse of subterranean aquifers and reservoirs for irrigation and the needs of burgeoning cities like Chihuahua, Juárez, and Cuauhtémoc has seriously affected the flow of rivers such as the Bravo, Conchos, Florido, and San Pedro and hastened the desertification of the Chihuahuan Desert (Dinerstein et al. 2000).

The herpetofauna of the area has inevitably also suffered from increasing human activity. Perhaps the greatest impact has been upon amphibians because of their dependence on water for breeding and on high humidity at other times. The present scarcity, wide separation, and increasingly transient nature of bodies of water have greatly decreased amphibian survival. The Tiger Salamander (*Ambystoma mavortium*) was well known to farmers and ranchers in the municipalities of Ahumada and Coyame because of their abundance some two decades ago in the relatively numerous ponds and other permanent bodies of water there. Our survey of all accessible bodies of water in those regions revealed their presence only in the ponds on the La Bamba and Agua Zarca ranches, municipality of Coyame.

Aquatic reptiles have also been severely affected by desiccation in recent years. The Spiny Softshell (*Apalone spinifera*) has practically disappeared from the northwestern part of the state. It was once common in Laguna de Guzmán and Ojos de Santa María but now is rare.

The direct persecution, particularly of the reptiles of the state, by humans varies somewhat within its three major habitats, correlated in general with density of population. We have already mentioned it in connection with the Subtropical assemblage, but it is far worse in the more densely populated Chihuahuan Desert. With few exceptions snakes are regarded as venomous and deadly so are killed regardless of species. The actually venomous copperhead (*Agkistrodon contortrix*) and rattlesnakes (*Crotalus atrox, C. lepidus, C. molossus, C. scutulatus,* and *C. viridis*) cannot be distinguished by most people from harmless kinds, and therefore all are killed. The absence of a rattle is even more suspicious than the presence of one, so that such snakes (e.g., *Heterodon kennerlyi, Lampropeltis getula*) are regarded as especially dangerous.

In most areas the flesh of rattlesnakes is considered to have a beneficial effect on a wide variety of illnesses, from infections of the skin to diabetes and cancer. Such beliefs have led at times to a deliberate search for rattlesnakes not only to eat but also to dry out and grind into powder, which is sprinkled like salt on food.

GENERALIST ASSEMBLAGE

There are presently 30 species whose ranges in Chihuahua substantially involve more than one of the three major habitats of the state: *Anaxyrus punctatus, A. speciosus, A. woodhousii, Hyla arenicolor, Gastrophryne mazatlanensis, Lithobates berlandieri, Spea multiplicata, Kinosternon integrum, K. sonoriense, Holbrookia approximans, Sceloporus clarkii, S. poinsettii, Urosaurus ornatus, Plestiodon tetragrammus, Aspidoscelis exsanguis, Hypsiglena chlorophaea, H. jani, Masticophis bilineatus, M. flagellum, Pituophis catenifer, Salvadora deserticola, S. grahamiae, Thamnophis cyrtopsis, T. eques, T. unilabialis, T. sirtalis, Trimorphodon vilkinsonii, Crotalus lepidus, C. molossus,* and *C. scutulatus.*

A total of 174 herpetozoan species occur in Chihuahua: 38 amphibians (4 salamanders, 34 anurans) and 136 reptiles (13 turtles, 51 lizards, 72 snakes). The low percentage of amphibians in

the state (21.8%, or 38/174) is due to the dominance of semiarid environments in Chihuahua. Most of the amphibians are concentrated in the Sierra Madre Occidental and its associated sky islands, such as Sierra del Nido and Cumbres de Majalca. Three of the 18 ambystomatid salamanders of Mexico are found in Chihuahua, but two of them (*Ambystoma rosaceum, A. silvense*) are limited to these mountain ranges; only *A. mavortium* is adapted to live in the seasonal, still waters of the Chihuahuan Desert. Populations of these three salamanders are locally abundant; in fact, *A. rosaceum* is the Chihuahuan species best represented in herpetological collections. It is abundant in almost any small stream along the Sierra Tarahumara. The only plethodontid salamander in the state, *P. bellii*, is known from one locality (6 km west-northwest of Ocampo, in the baseball field El Águila); this poor representation of plethodontids is due to the relatively low humidity in the Sierra Madre Occidental and the absence in this region of bromeliad species, which are known to provide a sanctuary for plethodontids in the dry season.

The anurans of Chihuahua are also concentrated in the Sierra Madre Occidental and associated mountain ranges, although the families Bufonidae, Microhylidae, and Scaphiopodidae are well represented in the Chihuahuan Desert. The most conspicuous Chihuahuan Desert anurans in Chihuahua are those of the genera *Anaxyrus, Incilius, Gastrophryne, Spea,* and *Scaphiopus*. However, species of the families Hylidae (*Hyla, Pachymedusa, Smilisca, Tlalocohyla*) and Ranidae (*Lithobates*) are abundant and widespread in the woodlands of the Sierra Madre Occidental. Eight of the nine species of frogs of *Lithobates* occur in the mountains in the western part of the state. The only missing species is *L. catesbeianus*, which in Chihuahua has been introduced in the northern part of the state.

Lemos-Espinal and Smith (2007) stated that four additional anuran species are likely to occur in Chihuahua, three of them in the deep canyons and lowlands of the extreme southwestern part of the state. The Pacific Stream Frog (*Craugastor vocalis*) was recorded by Hardy and McDiarmid

(1969) in extreme northeastern Sinaloa, 16 km north-northeast of Choix near the state line with Chihuahua at an elevation of 520 m. The Sabinal Frog (*Leptodactylus melanonotus*) was recorded by Bogert and Oliver (1945) from Güirocoba and Álamos, Sonora, only about 25 and 35 km, respectively, from the Chihuahua border. Smith et al. (2005) reported it also from the Río Mayo at the floodgates of Presa Mocuzari, Sonora. Hardy and McDiarmid (1969) mapped localities for this species (as *L. occidentalis*, a junior synonym) from throughout the lowlands of Sinaloa, including a locality in the extreme northeastern corner. And the Lowland Burrowing Treefrog (*Smilisca fodiens*) has been recorded close to Chihuahua by Hardy and McDiarmid (1969) for Sinaloa, Bogert and Oliver (1945) for Sonora, and Trueb (1969) and Duellman (2001) for both states.

Another anuran species likely to occur in extreme northeastern Chihuahua is the Gulf Coast Toad (*Incilius nebulifer*). This species is represented by isolated populations at the southern extremity of the Big Bend region of Texas, adjacent to Coahuila (Conant and Collins 1998).

On the other hand, the low percentage of turtles in the state in relation to the total number of herpetozoan species (7.5%, or 13/174) is a reflection of the total number of turtles in the country (32; Liner and Casas-Andreu 2008). Chihuahua is host to 41% (13/32) of the turtle species found in Mexico, and it is likely that at least four other turtle species occur in the state. Three species have been recorded close the state line with Sonora and Sinaloa, in the extreme southwestern part of the state. *Kinosternon alamosae* has been recorded in the vicinity of Álamos, Sonora, about 35 km from the Chihuahua border. *Trachemys hiltoni* has been recorded from Güirocoba, about 25 km from Chihuahua, and from extreme northern Sinaloa. Legler and Webb (1970) stated that the species is limited to the Río El Fuerte drainage and that *T. yaquia* is limited to the drainages of the Río Mayo, Río Sonora, and Río Yaqui; however, Seidel (2002) mapped the range of both these species into Chihuahua, but only conjecturally.

In addition, the Common Snapping Turtle (*Chelydra serpentina*) occurs in the Rio Grande at least in New Mexico (Degenhardt et al. 1996) and may well occur farther south in extreme northeastern Chihuahua, where little turtle trapping has been done.

The lizards represent 29.3% (51/174) of the total number of herpetozoan species in Chihuahua. They are well represented through the state, and as indicated by Lemos-Espinal and Smith (2007), the diversity of this group in Chihuahua is even greater. They found that there are at least nine lizard species not yet recorded but likely to occur in the state: four of them in the deep canyons and lowlands of extreme southwestern Chihuahua; three in the extreme northeastern part of the state; and two in the extreme northwestern part.

Bogert and Oliver (1945) recorded the Zebra-tailed Lizard (*Callisaurus draconoides*) from Güirocoba and Álamos, Sonora (approximately 25 and 35 km, respectively, from the Chihuahua border), and Hardy and McDiarmid (1969) spotted it at several localities in extreme northeastern Sinaloa. The Black Banded Gecko (*Coleonyx fasciatus*) has been recorded from five localities along the foothills of the Sierra Madre Occidental of eastern Sonora: three of these localities are in the Álamos region, and one is fairly close to Chihuahua. Its habitat suggests that it might occur in some of the deep canyons of southwestern Chihuahua. The Regal Horned Lizard (*Phrynosoma solare*) ranges from southern Arizona through almost all of Sonora and into northern Sinaloa. Hardy and McDiarmid (1969) and Bogert and Oliver (1945) recorded it from very near Chihuahua in both Sinaloa and Sonora. It is a species of arid and semiarid habitats on plains, hills, and low mountain slopes. And the Desert Spiny Lizard (*Sceloporus magister*) shows a range similar to that of the preceding species. East of the Sea of Cortés, it is the western representative of the eastern *S. bimaculosus*.

In northeastern Chihuahua the presence of three additional lizard species is likely. Wright (1971) indicated that the New Mexico Whiptail (*Aspidoscelis neomexicana*) is known from only central New Mexico and extreme southwestern Texas; almost all records are from near the Rio Grande on both sides. He projected its range into Chihuahua along the Río Bravo; no records exist from there, but occurrence is highly likely. Conant and Collins (1998) depicted the southern part of the Big Bend region of Texas as part of the range of the Reticulate Banded Gecko (*Coleonyx reticulatus*). It may be expected in adjacent parts of Chihuahua. According to SEMARNAP (1997) this species was said to have been observed in a Benavides-Cañón de Santa Elena transect; however, no voucher material is available. Also, Conant and Collins (1998) projected the range of the Texas Alligator Lizard (*Gerrhonotus infernalis*) to include the southern part of the Big Bend region of Texas, southward through eastern Chihuahua, most of Coahuila, and other states to the south. According to SEMARNAP (1997) this species was said to have been observed (under the name of *G. liocephalus* [*sic*]) in the Benavides-Cañón de Santa Elena area transect; however, no voucher material is available.

In northwestern Chihuahua the presence of the Western Banded Gecko (*Coleonyx variegatus*) is expected. As indicated in Stebbins (2003) and Degenhardt et al. (1996), this species occurs in extreme southwestern New Mexico and probably also in adjacent northwestern Chihuahua. The Gila Monster (*Heloderma suspectum*) is also expected to occur in this part of the state. The known occurrence of this species in Sonora, Arizona, and New Mexico close to the Chihuahua border indicates that occurrence in Chihuahua is likely.

The snakes are the herpetozoan group with the largest percentage in Chihuahua (41.4%, or 72/174). It is highly likely that nine more snake species occur within the state: two in southwestern Chihuahua (*Phyllorhynchus browni, Pseudoficimia frontalis*); four in northeastern Chihuahua (*Coluber constrictor, Lampropeltis alterna, Pantherophis bairdi, Tantilla cucullata*); two in the northwestern part of the state (*Crotalus tigris, Sistrurus catenatus*); and one in extreme southeastern Chihuahua (*Tantilla atriceps*).

Bogert and Oliver (1945) recorded the Saddled Leaf-nosed Snake (*Phyllorhynchus browni*) from Álamos, about 35 km from the Chihuahua border; and Hardy (1972) reviewed the distribution of the False Ficimia (*Pseudoficimia frontalis*), citing specimens from near Álamos and Güirocoba, Sonora, about 35 and 25 km from the Chihuahua border, respectively.

The North American Racer (*Coluber constrictor*) is rare in Mexico, with only three records. Two are from Coahuila, including one from the extreme northwestern corner, in the Sierra del Carmen (Wilson 1966). Occurrence in Chihuahua seems likely. The Gray-banded Kingsnake (*Lampropeltis alterna*) is well known in the Big Bend of Texas and elsewhere in that state, as well as in Coahuila and other adjacent states in Mexico, but it has never been found in Chihuahua, although it almost certainly occurs there. Baird's Ratsnake (*Pantherophis bairdi*) occurs in western Texas, including all of the Big Bend, as well as northern Coahuila (Conant and Collins 1998); it is highly likely to occur in adjacent Chihuahua. The Trans-Pecos Black-headed Snake (*Tantilla cucullata*) is known only in Texas, in the Big Bend and immediate vicinity (Dixon et al. 2000); occurrence in adjacent Coahuila and Chihuahua is to be expected.

In northwestern Chihuahua the occurrence of the Tiger Rattlesnake (*Crotalus tigris*) is expected. Stebbins (2003) indicates occurrence of this species in the extreme southeastern corner of Arizona and in eastern Sonora near the Chihuahua border. An inhabitant of arid and semiarid foothill deserts, it may enter the latter state in some of its semiarid valleys. The Massasagua Rattlesnake (*Sistrurus catenatus*) is known from southern New Mexico (Degenhardt et al. 1996) and southeastern Arizona (Brennan and Holycross 2006); it likely occurs in adjacent Chihuahua.

In extreme southeastern Chihuahua the occurrence of the Mexican Black-headed Snake (*Tantilla atriceps*) is expected. The known range of this species comes close to the southeastern corner of the state (Conant and Collins 1998; Cole and Hardy 1981).

With the addition of these 26 species to the herpetofauna of Chihuahua the richness of amphibians and reptiles in the state would be 200, the highest for any state in northern Mexico, a result of the equally high richness of environments and spatial heterogeneity of the state.

Two of the 174 herpetozoan species currently known to occur in Chihuahua were introduced (*Lithobates catesbeianus, Hemidactylus turcicus*). Of Chihuahua's species 65% (113/174) are shared with the United States, and 35% (61/174) are not shared. Of the 113 shared species 72 occur in the North American deserts, such as the Sonoran, Chihuahuan, and Mojave. Some of them range even from southern Canada to the southern end of the Chihuahuan Desert in the Mexican states of San Luis Potosí and Querétaro (e.g., *Anaxyrus cognatus, A. debilis, Kinosternon hirtipes, Phrynosoma modestum, Heterodon kennerlyi, Masticophis taeniatus*). Others range to even more southern states, such as Oaxaca, Puebla, and southern Veracruz (e.g., *Ambystoma mavortium, Apalone spinifera, Masticophis taeniatus, Crotalus molossus, C. scutulatus*). Another 19 of these 113 shared species occur in temperate woodlands of mountainous habitats; they are distributed in the mountain ranges of central and southeastern Arizona and southwestern New Mexico, and their ranges extend southward through the Sierra Madre Occidental to Sinaloa, Durango, or Jalisco (e.g., *Hyla wrightorum, Lithobates chiricahuensis, L. tarahumarae, Elgaria kingii, Sceloporus slevini, S. virgatus, Lampropeltis knoblochi, Crotalus willardi*). Another 8 are species with subtropical affinities that enter the United States just at the southern tip of Texas (e.g., *Rhinella marina, Smilisca baudinii, Hypopachus variolosus, Drymarchon melanurus, Drymobius margaritiferus*), just at the southeastern tip of Arizona (*Oxybelis aeneus, Senticolis triaspis*), or in southeastern Arizona and southwestern New Mexico (*Tantilla yaquia*). The remaining 14 species (including the 2 introduced species) occur in a variety of habitats that show a widespread distribution, most of them from the southern United States (but they can range from southern Canada) to Central America or even South America (e.g., *Lampropeltis triangulum, Thamnophis cyrtopsis*) or in several types of habitats in northern Mexico and the southwestern United States (e.g., *Holbrookia elegans, Sceloporus poinsettii, Urosaurus ornatus, Opheodrys vernalis, Crotalus lepidus*).

Of the 61 species that are not shared, 3 are endemic to Chihuahua (*Lithobates lemosespinali, Barisia levicollis,* and *Plestiodon multilineatus*). Of these 61 species 23 are limited to woodlands of the Sierra Madre Occidental and associated mountain ranges (including the 3 endemic to Chihuahua). These are species of temperate affinities, and all of them are endemic to Mexico. Another 34 species are from the deep canyons and lowlands of the extreme southwestern part of the state; these are species of subtropical affinities that extend their ranges west and northwest through the woodlands and foothills of the Sierra Madre Occidental or enter the state eastward through the Río El Fuerte drainages. Of these 34 species 27 are endemic to Mexico. The 7 species that are not endemic (*Lithobates forreri, Rhinoclemmys pulcherrima, Phyllodactylus tuberculosus, Boa constrictor, Imantodes gemmistratus, Masticophis mentovarius, Agkistrodon bilineatus*) are widespread species of the tropics and subtropics of Mexico, Central America, and even South America. The remaining 4 species are limited to the Chihuahuan Desert of Mexico. Three (*Kinosternon durangoense, Gopherus flavomarginatus, Uma paraphygas*) are confined to the Bolsón de Mapimí in extreme southeastern Chihuahua and adjacent areas of Durango and Coahuila. The other (*Holbrookia approximans*) is widespread in the Chihuahuan Desert of northern Mexico, and it is highly likely that it occurs in adjacent areas of Arizona and New Mexico.

Herpetofauna of Coahuila

JULIO A. LEMOS-ESPINAL
AND HOBART M. SMITH

Introduction

In 1999, we initiated a study of the amphibians and reptiles of Coahuila in the western part of the state, which had received little attention previously. Subsequently we visited the state sporadically in areas from the extreme northwest to the southwest, including the Sierra Mojada, Químicas del Rey, Ejido El Alicante, Dunas Magnéticas, and Dunas de Bilbao. More recently we sampled the herpetofauna of the state more widely, visiting southeastern, southern, and central parts as well as other areas in the west.

Through this work and a study of the literature, it has become evident that knowledge of the herpetofauna of Coahuila is very limited, as noted in such early contributions as Gloyd and Smith (1942), Schmidt and Owens (1944), and Schmidt and Bogert (1947). This is true not only in the western part of the state but in almost all parts.

The rugged topography of the state accounts in part for the paucity of specimens available. For example, the high elevations of the Sierras de Burro and El Carmen are for the most part accessible only on foot. Isolated ranges in the north and west are equally inaccessible, and many areas of the desert are hazardous to travel either by foot or by car. Regions of relatively easy access have received the most attention, partic-

ularly along Mex. Hwy 40 between Torreón and Saltillo and Hwy 57 between Saltillo and Piedras Negras, as well as various side roads. Even those areas have not really been sampled well. However, large unsampled areas exist elsewhere, including Sierra El Carmen, Sierra El Burro, Llano El Guaje, the playa of the Río Bravo, and the high elevations along the southern border of the state.

The most intensive survey in the state has been carried out in the Cuatro Ciénegas Basin. Indeed, about 25% of all specimens from Coahuila, in 20 museums in the United States, one in Canada, and one in Mexico, are from that basin. The Sierra de Arteaga has also been a focus of attention, to a considerable extent through expansion of studies in adjacent Nuevo León. Work in other areas has been scattered, and most of the material in museums has not been reported.

We find that at present Coahuila is known to have 133 species of amphibians and reptiles. Represented are 4 salamander species, 20 anurans, 11 turtles, 49 lizards, and 49 snakes. All are native to the state except the introduced *Hemidactylus turcicus* and possibly *Rhinella marina*, which may have been introduced (as has been suggested) but which may be relictual in occurrence, as are some other species in the Cuatro Ciénegas Basin. The fully known herpetofauna will be considerably greater because large, critical areas of the state have not been sampled at all or very inadequately.

Previous Herpetological Studies

The first collector to reach Coahuila was Jean Louis Berlandier, who was the naturalist with the French/Mexican Boundary Commission in the field from 1827 to 1829 and who continued to work as a naturalist with other commissions until at least 1834. As reported in Berlandier and Chovell (1850), he collected briefly in the Saltillo area, where he reported *Lithobates* (= *Rana*) *catesbeianus*. Unfortunately, none of his voucher specimens from Coahuila are known.

The earliest herpetological collections of specimens now in museums from Coahuila were made by Thomas H. Webb, who accompanied an expedition in 1852 that traversed Chihuahua, Durango, Nuevo León, and Coahuila; in the latter state he collected at Parras and Saltillo (Baird 1859b). Shortly thereafter in 1853, Lt. Darius N. Couch, under the auspices of the Smithsonian Institution, collected along a transect through Tamaulipas, Nuevo León, Coahuila, and Durango. No synopsis of the collections was ever published, but numerous independent publications have been based on his material, including descriptions of several new species, some bearing his name. The localities he visited in Coahuila include Río Nazas, Parras, Castañuelas (Estación Seguí), Patos (General Cepeda]), Agua Nueva, Buena Vista, and Saltillo.

The seminal biological explorations in Mexico by Edward Nelson and Edward Goldman for the US National Museum (USNM) included forays into Coahuila in May 1895, May–August 1896, and April–May 1902. Among the species collected were *Spea multiplicata, Gopherus berlandieri, Cophosaurus texanus, Holbrookia approximans, Phrynosoma cornutum, Sceloporus cautus, S. grammicus, S. goldmani,* and *S. undulatus* (= *S. consobrinus*). A detailed itinerary, map, and list of species collected are in Goldman (1951).

In August 1932, E. H. Taylor and H. M. Smith traversed Mex. Hwy 40 from Saltillo to Torreón, collecting at various localities along the way. No summary of the collection was ever published, but all specimens are presently in the Field Museum of Natural History and the University of Illinois Museum of Natural History.

In 1937, Robert S. Sturgis collected 15 herpetozoa, all now in the USNM, in the Sierra El Carmen. In the same sierra, F. W. Millar, E. M. Dealey, and Raymond Foy collected 22 specimens in October and November 1940 and deposited them in the Chicago Academy of Sciences. Gloyd and Smith (1942) reported the two collections, which represented 10 species (1 anuran, 4 lizards, and 5 snakes). In late October and early November 1938, Hobart and Rozella Smith traveled the length of the highway from Torreón to Saltillo, collecting near Torreón, San Pedro de las Colonias, Parras, Jaral, San Hipólito, Saltillo, Arteaga, and Zapalinamé. Species collected included *Anaxyrus cognatus, Aspidoscelis gularis, Sceloporus oberon, S. ornatus, S. parvus, S. poinsettii, Uta stansburiana,* and *Masticophis taeniatus.* Between 1938 and 1939, Ernest G. Marsh collected plants and vertebrates in various localities in northern and central Coahuila for the Field Museum of Natural History; among them were some 800 specimens of amphibians and reptiles. That collection was reported by Schmidt and Owens (1944) and included 64 species (11 anurans, 6 turtles, 22 lizards, and 25 snakes).

In August 1946, Karl P. Schmidt and Charles M. Bogert collected the type series of *Uma exsul*, as well as other reptiles, 19.2 km north of San Pedro de las Colonias, as reported by Schmidt and Bogert (1947). Between 1950 and 1954, Albert A. Alcorn, J. R. Alcorn, and Robert W. Dickerman collected in various states of Mexico, including Coahuila. Chrapliwy (1956) reported the first record of *Bufo speciosus* for the state from that collection. Robert Webb in 1955 found a specimen of *Plestiodon brevirostris* (= *P. dicei*) at 20.8 km east of San Antonio de las Alazanas, municipality of Arteaga, that Axtell (1960) described as the holotype of *P. b. pineus*. Fugler and Webb (1956) reported a collection of amphibians and reptiles from central and southern Coahuila, obtained by Robert W. Dickerman and other members of an expedition from the University of Kansas.

In the summer of 1956, Pete S. Chrapliwy and Kenneth L. Williams collected in various parts of northern Mexico, including Coahuila. The results were reported by Williams, Chrapliwy, and Smith (1959, 1960); Williams, Smith, and Chrapliwy (1961); and Chrapliwy et al. (1961). All specimens are in the University of Illinois Museum of Natural History.

Richard G. Zweifel collected near Cuatro Ciénegas in June 1957, and from that collection subsequently published a first record for the state of *P. tetragrammus* (Zweifel 1958). He added that species to a list he compiled of other species of a more eastern range that have been discovered as isolated populations in the Cuatro Ciénegas Basin. Unfortunately, *Drymobius margaritiferus* was included erroneously in that list, leading to inclusion of that area as part of the range of that species in Conant (1975) and Conant and Collins (1998). No confirmation of that snake's existence in Coahuila exists to our knowledge, and occurrence there seems highly unlikely in view of the known range and habitat preference of the species.

In July 1961, Ralph W. Axtell and Michael D. Sabath (1963) collected the first known specimen of *Crotalus pricei miquihuanus* from Coahuila and a gartersnake described by Rossman (1969) as the holotype of *Thamnophis exsul*. In June 1966, Clarence J. McCoy and A. V. Bianculli found the first known specimen of *Sistrurus catenatus* in Mexico in the Cuatro Ciénegas Basin. In August, two years later, William L. Minckley and Ralph K. Minckley found another in the same basin. They were reported in McCoy and Minckley (1969). In June 1967, Ralph W. Axtell collected specimens of *Sceloporus* 800 m north of Cuesta Muralla that became the types of *S. cyanostictus* (originally described as *S. jarrovii cyanostictus*; Axtell and Axtell 1971). In July 1968, Max A. Nickerson and John N. Rinne collected a lizard at the extreme northern end of Sierra de San Marcos in the Cuatro Ciénegas Basin. It became the holotype of *Gerrhonotus lugoi*. Between 1969 and

1974, Ernest A. Liner, Jerry Gautreaux, Richard M. Johnson, and Allan H. Cheney collected herpetozoa near Múzquiz and Boquillas del Carmen, representing 49 taxa (9 anurans, 3 turtles, 17 lizards, and 20 snakes). All were reported in Liner et al. (1977).

In March 1970, Joseph T. Collins and Richard Montanucci obtained the first examples of *Sonora semiannulata* and *Tantilla gracilis* known from Coahuila in the vicinity of Don Martín and south of Nueva Rosita. They were reported by Savitzky and Collins (1971a, 1971b). Montanucci (1971) reported the presence of *Crotaphytus reticulatus* in the vicinity of Múzquiz and 1.6 km north of Nava. In 1977, Frederick H. Wills reported, in an unpublished dissertation, a study on the distribution, geographic variation, and natural history of *Sceloporus parvus*, including several specimens from Coahuila. During the summers of 1977, 1979, 1980, and 1982, Eric C. Axtell, Ralph W. Axtell, and Robert G. Webb collected in Sierra Texas, southwestern Coahuila. Axtell and Webb (1995) described *Crotaphytus antiquus*, which they obtained there.

In May 1983, Robert Powell, Nicholas A. Laposha, Donald D. Smith, and John S. Parmerlee explored the headwaters of the Río Sabinas, whose herpetology had been previously unknown because of the absence of roads. Powell et al. (1984) reported the collection made then, representing 5 anuran species, 3 turtles, and 4 snakes. In 1984, Clarence J. McCoy reported a total of 66 species in the herpetofauna of the Cuatro Ciénegas Basin, based primarily on his explorations there for more than two decades. Included were 8 anuran species, 4 turtles, 23 lizards, and 31 snakes. Between July and September 1987, Arturo Contreras-Arquieta visited the Cuatro Ciénegas Basin, collecting 392 specimens representing 38 species and observing 7 others. The totals included 4 anuran species, 4 turtles, 21 lizards, and 16 snakes (Contreras-Arquieta 1989). In May 1988, Ernest A. Liner, Richard R. Montanucci, Arturo González-Alonso, and Fernando Mendoza-Quijano collected herpetozoa between Múzquiz and Boquillas del Carmen and along the road to La Linda (on the east side of the

Sierra del Carmen); in the Sierra La Encandada; and in the Chihuahuan Desert surrounding these mountains. Their collection represented 18 taxa (2 anurans, 13 lizards, and 3 snakes), reported by González-Alonso et al. (1988), Liner et al. (1993), and Mendoza-Quijano et al. (1993). Among these was the first record of *Sceloporus cyanogenys* for the state. McGuire (1996) reported the presence of *Crotaphytus antiquus* in the Sierras de San Lorenzo, Texas, and Solis in extreme southwestern Coahuila.

Recently Héctor Gadsden-Esparza and his collaborators have studied the herpetofauna of the municipality of Viesca, southwestern Coahuila; results have been published in Gadsden-Esparza et al. (1993); Gadsden-Esparza et al. (1997); Gadsden-Esparza, López-Corrujedo, et al. (2001); Gadsden-Esparza, Palacios-Orona, and Cruz-Soto (2001); and Gadsden-Esparza et al. (2006). Gamaliel Castañeda-Gaytán and his collaborators have also been working in the same area, principally in the Dunas de Bilbao (Castañeda-Gaytán et al. 2004).

Our own collections from various parts of Coahuila were reported in Lemos-Espinal, Smith, and Ballinger (2004); Lemos-Espinal, Smith, Chiszar, and Woolrich-Piña (2004); Lemos-Espinal and Smith (2007, 2009); Lemos-Espinal et al. (forthcoming); Smith, Lemos-Espinal, and Chiszar (2003); and Smith et al. (2005a, 2005b). Javier Banda-Leal and Daniel Garza-Tobón of the Museo del Desierto in Saltillo have been very active in central and southeastern parts of the state. Through their endeavors they have discovered the first specimens known in Coahuila of *Ambystoma mavortium*, *Pseudoeurycea scandens*, and *Eleutherodactylus longipes*.

Despite the existence of these and other publications on the herpetofauna of Coahuila, the state remains one of the least known herpetologically of all the states of Mexico. It offers an unparalleled opportunity for additions to current knowledge. Its unexplored mountain chains very likely are inhabited by taxa completely unknown, which could have evolved in isolation for thousands of years, surrounded by the completely different habitat of extensive deserts.

Several species known in adjacent areas but not in Coahuila probably occur there. Many known distributional limits remain to be extended into new municipalities. Studies of geographic variation are a pressing need for most of the widespread species.

We have now determined that in addition to the studies mentioned previously, there are 31 scientific names of amphibians (2 anurans) and reptiles (5 turtles, 17 lizards, 7 snakes) with type localities assigned to the state of Coahuila (table 6.1).

Physiographic Characteristics and Their Influence on the Distribution of the Herpetofauna

Coahuila is the third-largest state of Mexico, encompassing 151,571 km², between latitudes 24°32′ S and 29°53′ N and between longitudes 99°51′ E and 103°58′ W. It represents 7.74% of the total area of the country. This large area, combined with its rugged topography of numerous mountain masses scattered throughout the lowlands of the state, creates disjunct habitats that result in disjunct or limited distributions of amphibian and reptile species occurring there. One result of such habitat fragmentation is the relatively large number of species (8) endemic to the state.

Extensive sierras in the northern part of the state appear to form a single mountain mass, although it is actually composed of three ranges: Sierra El Carmen, the western third; Sierra El Burro, the eastern third; and Sierra de Santa Rosa, the southern third. The greatest altitude (2120 m) is reached in the Sierra de Santa Rosa (latitude 28°18′ N, longitude 102°4′ W). These sierras constitute about 40%–50% of the northern part of the state; the rest of the northern part consists of plains whose average elevation is 1000 m.

Access to this northern region is provided by Coah. Hwy 93, between Melchor Múzquiz and La Cuesta de Malena; an unimproved road continues from there to Boquillas del Carmen. Hwy 93 skirts the southern bases of the Sierra

TABLE 6.1 Type localities for amphibians and reptiles described from the state of Coahuila, Mexico

Author(s)	Original name: Type locality
Anurans (2)	
Zweifel (1956)	*Eleutherodactylus* (= *Craugastor*) *augusti fuscofemora*: **Sacatón, 5 miles south of Cuatro Ciénegas**
Cope (1863)	*Scaphiopus rectifrenis* (= *S. couchii*): **Río Nazas**
Turtles (5)	
Schmidt & Owens (1944)	*Terrapene coahuila*: **Cuatro Ciénegas**
Ward (1984)	*Pseudemys concinna gorzugi* (= *P. gorzugi*): **3.5 mi W Jiménez, Río San Diego**
Legler (1960)	*Pseudemys scripta taylori* (= *Trachemys taylori*): **16 km S Cuatro Ciénegas**
Iverson (1981)	*Kinosternon hirtipes megacephalum*: **3.2 km SE Viesca (25°21′N, 102°48′W)**
Webb & Legler (1960)	*Trionyx ater* (= *Apalone spinifera ater*): **16 km S Cuatro Ciénegas**
Lizards (17)	
Smith (1942)	*Gerrhonotus levicollis ciliaris* (= *B. ciliaris*): **Sierra de Guadalupe**
McCoy (1970)	*Gerrhonotus lugoi*: **N tip of Sierra de San Marcos, ca. 11 km SW of Cuatro Ciénegas, ca. 800 m elev.**
Axtell & Webb (1995)	*Crotaphytus antiquus*: **2.1 km N–1.7 km E Vizcaya (25°46′4″N, 10311′48″W. elev. 1100 ± m) in the Sierra Texas**
Schmidt (1921)	*Holbrookia maculata dickersonae* (= *H. approximans*): **Castañuelas**
Smith (1938)	*Sceloporus cautus*: **30 mi N of El Salado, San Luis Potosí**
Williams et al. (1960)	*Sceloporus merriami australis*: **15.6 mi (25.1 km) E Cuatro Ciénegas**
Smith & Brown (1941)	*Sceloporus jarrovii oberon* (= *S. oberon*): **Arteaga**
Smith (1936)	*Sceloporus ornatus caeruleus*: **5 mi S of San Pedro de las Colonias**
Baird (1859a [1858])	*Sceloporus ornatus*: **Patos (= General Cepeda)**
Axtell & Axtell (1971)	*Sceloporus jarrovii cyanostictus* (= *S. cyanostictus*): **800 m N Cuesta La Muralla**
Schmidt & Bogert (1947)	*Uma exsul*: **sand dunes 12 mi N of San Pedro de las Colonias**
Axtell (1960)	*Eumeces* (= *Plestiodon*) *dicei pineus*: **20.8 km E San Antonio de las Alazanas**
García-Vázquez et al. (2010)	*Scincella kikaapoa*: **4 km SE Cuatro Ciénegas, Poza el Mojarral, 26°55′11.9″N; 100°6′53.2″W, 739 m elev.**
Cope (1892)	*Cnemidophorus gularis semifasciatus* (*Aspidoscelis gularis*): **Agua Nueva**
Duellaman & Zweifel (1962)	*Cnemidophorus septemvittatus pallidus* (= *Aspidoscelis gularis pallida*): **3 miles west of Cuatro Ciénegas**
Wright & Lowe (1993)	*Cnemidophorus* (= *Aspidoscelis*) *inornata cienegae*: **13.9 km SW of Cuatro Cienegas de Carranza, point of San Marcos Mtn.**
Cope (1892)	*Cnemidophorus variolosus* (= *Aspidoscelis marmorata variolosa*): **Parras**
Snakes (7)	
Cope (1861)	*Arizona jani* (= *Pituophis deppei jani*): **Buenavista**
Garman (1883)	*Rhinocheilus lecontei tessellatus*: **Monclova**
Kennicott (1860)	*Eutaenia* (= *Thamnophis*) *angustirostris* (*nomen dubium*): **Parras**
Kennicott (1860)	*Eutaenia cyrtopsis* (= *Thamnophis cyrtopsis*): **Rinconada**
Rossman (1969)	*Thamnophis exsul*: **17.6 km E and 5.6 km S San Antonio de las Alazanas**
Rossman (1963)	*Thamnophis proximus diabolicus*: **Río Nadadores, 5 km W Nadadores**
Garman (1887)	*Crotalus tigris palmeri* (= *C. lepidus lepidus*): **Monclova**

El Burro and Sierra El Carmen and the northern margin of Sierra de Santa Rosa. An unimproved road extends from the confluence of the three sierras, passing northward along the east side of Sierra El Carmen and the west side of Sierra El Burro, reaching La Linda. No roads extend into the higher parts and canyons of these sierras, and therefore access is very difficult; as a result these areas are very poorly known. Current knowledge of the herpetofauna of northern Coahuila is limited to the fringes of these mountains, a large part of which is totally unexplored. Important distributional information remains to be revealed there and very likely additional previously unknown taxa.

In the extreme western part of the state, isolated, relatively small sierras, oriented north to south, arise abruptly from the arid-semiarid plains. The principal ones are Sierra Las Cruces, Sierra Mojada, Sierra El Pino, and Sierra de Tlahualilo. The highest of these is Sierra Mojada (latitude 27°16′ N, longitude 103°42′ W), with a maximum altitude of 2450 m. All are difficult to access, especially Sierra Mojada.

Around these mountains the plains, at an average altitude of 1250 m, are dominated by areas of sand dunes: one group lies between Estación Sabaneta and an area east of Jaco (Chihuahua), a part of the Bolsón de Mapimí; one is on the plains of Aguanaval east of the Sierra de Tlahualilo (Dunas Magnéticas), part of the Zona de Silencio; and another is on the plains of the municipalities of Matamoros and Viesca, located in the extreme southwestern part of the area known as the Laguna de Mayrán. The areas to the east of Sierra Las Cruces, to the west of Sierra El Pino, and to the north of the trail from Químicas del Rey to Ocampo have not been surveyed despite its large size.

The extreme south-central and southeastern parts of the state are characterized by a series of east-west crustal folds that form several sierras, notably the Sierra de Arteaga, Sierra La Concordia, and Sierra de Parras, contiguous to the east with the Sierra Madre Oriental. Cerro La Nopalero (latitude 25°8′ N, longitude 103°14′ W),

at 3120 m, is the highest elevation in the area. Toward the southwest these ranges are continuous with those that form the southern limit of the Laguna de Mayrán.

The eastern part of the state is mostly flat, broken by several isolated, low ranges extending north-south, notably the Sierra Pájaros Azules (latitude 27°00′ N, longitude 100°53′ W), reaching an altitude of 1930 m, and Sierra La Gloria.

In the central part of the state is a small, low (about 750 m) valley of 120 km² surrounded by mountain ranges with elevations of up to 2500 m. For tens of thousands of years this valley had strictly internal drainage, fed by waters from several arroyos, which created a wide variety of aquatic habitats, including streams, wells, lakes, and marshes. Its isolation and antiquity led to a high degree of endemicity there (McCoy 1984). At present this valley is known as the Cuatro Ciénegas Basin.

Most of the geological structure of Coahuila is sedimentary rock, mostly marine but also continental, originating between the Paleozoic and Quaternary. The more typical of these are the limestones of the Mesozoic. These deposits have been folded by intense pressures, producing depressions and upthrusts. The orientation of the folds is east-west in the southern part of the state, northwest-southeast elsewhere in the state.

In diverse zones of the state, igneous rock outcrops occur, varying from Triassic to Quaternary. In some cases the more recent intrusions form the highest parts of the mountains. Small intrusions have been exposed due to erosion of the enclosing sedimentary rock and in some cases have been mineralized. In many places conglomerates of Tertiary origin form the extensive foothills of the mountains. Small exposures of Paleozoic metamorphic rock are scattered over the surface of the state.

Alluvial deposits derived from erosion of rock formations are the most recent geological alterations in the state. They cover most of the plains and in some areas have become several hundred meters deep.

Four major hydrological regions occur in

Coahuila: Bravo-Conchos, covering much of the state, at 95,236.33 km² in area; Mapimí in the west, 29,456.26 km²; Nazas-Aguanaval in the south-southwest, 21,908.22 km²; and El Salado in the southeast, 4977.56 km². The superficial and underground waters are treated separately in the following discussion.

The Bravo-Conchos region consists mainly of plains with altitudes of 1000–1800 m. It is an arid territory, increasing in aridity northward. Most of the rivers flow into the Río Bravo, although the Tortuguillas and Chancaplio Lakes drain internally.

No rivers or permanent natural reservoirs occur in the Mapimí region. It is arid and has no elevations of significance. All drainage is internal. Low areas are flooded when occasional torrential rain occurs, but the water completely evaporates or is absorbed by the ground in weeks or months. The average annual rainfall is less than 10 mm.

The part of the Nazas-Aguanaval region in Coahuila (it also extends into Durango and Zacatecas) is known as the Región Lagunera and is fed by the dead-ended great Ríos Nazas and Aguanaval. The Región Lagunera is the most important agricultural area in the state and supplies several cities, including Gómez Palacio and Lerdo in Durango, as well as Torreón, Matamoros, and San Pedro de las Colonias in Coahuila.

The El Salado region is the smallest in the state, although it is one of its most important internal drainage areas.

Much of Coahuila lies within the Chihuahuan Desert. The highlands in the extreme southeastern corner, including the Sierra de Arteaga, are an exception and constitute the extreme northern end of the Sierra Madre Oriental. The vegetative cover of the state is made up of six types of vegetation (Chihuahuan Desert Scrub; Tamaulipan Thornscrub; Submontane Scrub; Montane Forest; Sacatal Grassland; and Riparian, Subaquatic, and Aquatic Vegetation) and 12 plant communities that basically correspond to three floral provinces: the Mexican Plateau, the Coastal Plain of the Northeast, and the Sierra Madre Oriental (Rzedowski 1978).

Chihuahuan Desert Scrub

Chihuahuan Desert Scrub occupies great expanses of western and southern Coahuila. This vegetation type represents approximately 63% of the state and includes a series of plant communities (Rosetophilous Desert Scrubland, Creosote Bush Scrub, Bajada Dominated by Yuccas, Halophylous and Gypsophylous Scrublands) formed by scattered, perennial xerophilous shrubs and ephemeral herbaceous elements that occupy the driest habitats in the state, at 800–2600 m. Soil variation and topographic characteristics determine the diversity of the plant communities within this type of vegetation.

Rosetophilous Desert Scrubland occupies altitudes between 1000 and 2500 m, in rolling hills and mountain slopes where the greatest solar radiation is present or in the most exposed areas of canyons. It develops on shallow soils, usually stony and with good drainage, and covers approximately 6% of the state. The dominant species are short shrubs with thorny, evergreen leaves grouped in rosettes. The most representative species are *Agave lechuguilla*, *A. scabra*, *A. striata*, *Hechtia texensis*, and *Dasylirion palmeri*, associated with *Viguieria stenoloba*, *Parthenium argentatum*, *Euphorbia antisyphilitica*, and *Mimosa zygophylla*. Frequently this community displays a high diversity of cactus, of which the most common are *Opuntia microdasys*, *Echinocactus platyacanthus*, *Ferocactus pilosus*, *F. uncinatus*, *Echinocereus conglomeratus*, *E. pectinatus*, and *Thelocactus bicolor*. The most common herbaceous plants are *Bahia absinthifolia*, *Tiquilia canescens*, *T. greggii*, *Castilleja lanata*, and *Notholaena sinuata* and grass species such as *Bouteloua ramosa* and *Erioneuron avenaceum*.

Creosote Bush Scrub is one of the most characteristic plant communities, occupying approximately 48% of the state. It grows at elevations between 600 and 1500 m, in valleys and rolling hills with deep soils and few stones, and is dominated by shrubs 0.3–1.5 m tall with small leaves, as well as some individuals taller than 1.5 m. This plant community varies from low to high density, depending on the depth and humidity of the soil on which it develops. The most representative species are *Larrea tridentata*, *Flourensia cernua*, and *Parthenium incanum*, associated with *Lycium berlandieri*, *Opuntia rastrera*, *O. imbricata*, *Koeberlinia spinosa*, and *Fouquieria splendens*. Occasionally there are some arborescent individuals of *Yucca filifera* and *Prosopis glandulosa*. The most common herbaceous plants are *Bouteloua trifida*, *B. gracilis*, *Sporobolus airoides*, *Dasyochloa pulchella*, *Scleropogon brevifolius*, *Mühlenbergia porteri*, *Psilostrophe gnaphalodes*, and *Tiquilia canescens*.

The Bajada Dominated by Yuccas is characterized by dominance of arborescent plants of the genus *Yucca*, with heights between 8 and 10 m that give the aspect of a small forest or palmar. It occupies around 3% of the state and is distributed over 1500 m of altitude on deep, alluvial soils, sandy valleys, and mountain slopes, frequently mixed with elements of the Creosote Bush or Rosetophilous Desert Scrubland. The most common species are *Yucca filifera* and *Y. carnerosana*, associated with *Larrea tridentata*, *Flourensia cernua*, *Opuntia imbricata*, *O. rastrera*, *Agave lechuguilla*, *Dasylirion cedrosanum*, *Viguiera stenoloba*, and *Parthenium incanum*. The most common herbaceous plants are *Bouteloua gracilis*, *Psilostrophe gnaphalodes*, and *Zaluzania triloba*.

Halophylous and Gypsophylous Scrublands are present in central and southern Coahuila at elevations between 1000 and 2000 m, occupying about 6% of the state. The presence of these scrublands is determined by local soil conditions, but most are located in isolated river basins with internal drainage on which the accumulation of salty sediments is favorable, forming a series of salty valleys or temporary lakes. This scrubland usually is represented by species such as *Atriplex canescens*, *A. acanthocarpa*, *Allenrolfea occidentalis*, and *Suaeda palmeri*, with frequent infiltrations of tolerant species such as *Larrea tridentata*, *Prosopis glandulosa*, and *Opuntia imbricata*; the herbaceous layer is represented by *Pleuraphis mutica*, *Sporobolus airoides*, *S. spiciformis*, *Distichlis spicata*, and several species of *Machranthera*.

The gypsum deposits (hydrated calcium sulfate) were formed by precipitation and a subsequent evaporation of ancestral seas exposed in valleys and hills or forming dunes since the Tertiary period (Villarreal and Valdés 1992–1993). This community consists of scattered bushes or shrubs. Usually they present a rich flora of endemic species, the most common of which are *Frankenia johnstonii*, *Condalia ericoides*, *Koeberlinia spinosa*, *Lycium pallidum*, and *Viguiera dentata*, associated with *Prosopis glandulosa*, *Larrea tridentata*, and *Opuntia rastrera*. The herbaceous layer is dominated by the following grasses: *Mühlenbergia villiflora*, *M. arenicola*, and *Scleropogon brevifolius*. In the wettest months of the year the diversity of this layer increases, and the following plant species are present: *Hoffmanseggia watsonii*, *Zinnia acerosa*, *Lesquerella fendeleri*, *Lepidium montanum*, *Dicranocarpus parviflorus*, and *Sartwellia mexicana*.

Eighty-one species of amphibians and reptiles inhabit the Chihuahuan Desert Scrub, and 57 are exclusive to this vegetation type: *Anaxyrus cognatus*, *A. debilis*, *A. punctatus*, *A. speciosus*, *A. woodhousii*, *Incilius nebulifer*, *Eleutherodactylus marnockii*, *Gastrophryne olivacea*, *Scaphiopus couchii*, *Spea multiplicata*, *Gopherus flavomarginatus*, *Gerrhonotus lugoi*, *Crotaphytus antiquus*, *C. collaris*, *Gambelia wislizenii*, *Coleonyx brevis*, *C. reticulatus*, *Cophosaurus texanus*, *Holbrookia approximans*, *H. lacerata*, *Phrynosoma cornutum*, *P. modestum*, *Sceloporus bimaculosus*, *S. cautus*, *S. consobrinus*, *S. cyanostictus*, *S. maculosus*, *S. merriami*, *S. ornatus*, *Uma exsul*, *U. paraphygas*, *Uta stansburiana*, *Plestiodon obsoletus*, *Scincella kikaapoa*, *Aspidoscelis gularis*, *A. inornata*, *A. marmorata*, *Arizona elegans*, *Bogertophis subocularis*, *Gyalopion canum*, *Heterodon kennerlyi*, *Hypsiglena jani*, *Lampropeltis alterna*, *L. getula*, *Masticophis taeniatus*, *Rhinocheilus lecontei*, *Salvadora grahamiae*, *Sonora semiannulata*, *Tantilla atriceps*, *T. hobartsmithi*, *T. nigriceps*, *Thamnophis marcianus*, *Rena dissecta*, *R. segrega*, *Agkistrodon contortrix*, *Crotalus viridis*, and *Sistrurus catenatus*.

Most of these 57 species are widely distributed in the arid-semiarid regions of the United States and Mexico; they are adapted to sites with limited water availability and with well-defined dry and rainy seasons. The rainy season extends from late May to mid-September, with a peak in July and August. This constrains the activity of most of the amphibians and reptiles to the period of April to October, with a hibernation period from October to March. Most of these species emerge from their hibernacula, which could be mammal burrows, deep cracks in the ground, or rocks, in mid-April. However, when environmental conditions allow, it is possible to observe active individuals in the winter months. Also, the heliothermic nature of these species results in little to no activity on cloudy and cold days in the spring and summer.

All of the amphibians that inhabit the Chihuahuan Desert Scrub of Coahuila depend on bodies of water to reproduce, which is triggered by the first rains of late May–early June. Large concentrations of anurans can be observed in the temporary ponds formed at this time, on which large choruses of males, sometimes deafening, try to attract females. These concentrations can extend as long as water is available, resulting in opportunistic reproduction that sometimes is very risky due to high evaporation rates in the Chihuahuan Desert of Coahuila. In mid-August 2008, we observed large concentrations of *Anaxyrus debilis* in the temporary ponds on the sides of Hwy Parras-General Cepeda; these concentrations were formed by amplectant pairs. In the bottom of the ponds were a large number of eggs attached to the vegetation. Three weeks later, in September, we came back to check the tadpoles, but the ponds were dry and on their muddy bottoms were hundreds of dead tadpoles.

Most of the species of reptiles that inhabit this vegetation type in Coahuila are oviparous and can produce more than one clutch during their activity period, usually between late May and early September. However, some species, such as the snakes of the genera *Agkistrodon*, *Crotalus*, and *Sistrurus*, are viviparous; in general these species mate between April and May and give birth in June.

Nine of these 57 species (*Gopherus flavomarginatus*, *Gerrhonotus lugoi*, *Crotaphytus antiquus*, *Sceloporus cyanostictus*, *S. maculosus*, *S. ornatus*, *Uma exsul*, *U. paraphygas*, and *Scincella kikaapoa*) have small distributions: some are limited to the sand dunes of Chihuahua, Coahuila, and Durango (*Gopherus flavomarginatus*, *Uma exsul*, *U. paraphygas*); others, to the Cuatro Ciénegas Basin of central Coahuila (*Gerrhonotus lugoi* and *Scincella kikaapoa*); still others, to some isolated localities in extreme western Coahuila (*Crotaphytus antiquus*), western Coahuila and northeastern Durango (*Sceloporus maculosus*), the southern half of Coahuila (*S. cyanostictus*), and central and southern Coahuila (*S. ornatus*). These species are evidence of the great diversity of conditions present in the Chihuahuan Desert of Coahuila.

Four of the remaining 24 species that inhabit the Chihuahuan Desert Scrub (*Rhinella marina*, *Craugastor augusti*, *Smilisca baudinii*, and *Opheodrys aestivus*) have discontinuous distributions, suggesting that their presence in Coahuila could be relictual. Another, *Hemidactylus turcicus*, is an introduced species whose native range is the Mediterranean. This species has been very successful in colonizing the places to which it is introduced through a variety of means, but principally from ships and subsequently dispersing. The remaining 19 species (*Ambystoma mavortium*, *Lithobates berlandieri*, *S. minor*, *S. parvus*, *S. poinsettii*, *Urosaurus ornatus*, *Plestiodon tetragrammus*, *Coluber constrictor*, *Masticophis flagellum*, *Pantherophis bairdi*, *P. emoryi*, *Pituophis catenifer*, *Thamnophis cyrtopsis*, *T. proximus*, *Micrurus tener*, *Crotalus atrox*, *C. lepidus*, *C. molossus*, and *C. scutulatus*) are able to inhabit arid-semiarid, temperate, and/or subtropical environs.

Tamaulipan Thornscrub

Tamaulipan Thornscrub is the representative vegetation type of the floristic province of the Coastal Plain of the Northeast, which includes northern and northeastern Coahuila, in a series of plains and rolling hills located east of the Sierras El Carmen, La Babia, Santa Rosa, La Purisima, and La Gavia. It covers about 19% of the state, at elevations between 240 and 850 m, on deep, sandy, and gravel soils in the valleys and rocky and shallow soils in the rolling hills, usually with good drainage. The vegetation is composed of shrubs less than 2 m tall, a mixture of creosote bush, thorny, and thornless vegetation. Small communities of low trees are also frequent, concentrated in the most humid sites.

The most common associations are those of *Acacia rigidula–Leucophyllum frutescens* and *Prosopis glandulosa–Opuntia lindheimeri*, with elements of *Lippia graveolens*, *Agave lechuguilla*, and *Flourensia cernua* in the southern and western regions and *Colubrina texensis* in the northern region. Other bushes or small trees include *Karwinskia humboldtiana*, *Guaiacum angustifolium*, *Cercidium texanum*, *Ziziphus obtusifolia*, *Castela erecta*, *Opuntia leptocaulis*, *Citharexylum brachyanthum*, *Acacia berlandieri*, *A. farnesiana*, *A. constricta*, *A. greggii*, and *Diospyros texana*. Grasses are the principal component in the herbaceous layer, in open areas or among bushes; the most frequent are *Bouteloua trifida*, *B. curtipendula*, *Aristida purpurea*, *Tridens muticus*, *T. texanus*, *Panicum hallii*, *Pleuraphis mutica*, *Hilaria belangeri*, and *Pennisetum ciliare*. Other common herbaceous plants are *Gnaphalopsis micropoides* and *Ruellia nudiflora*.

Thirty-one species of amphibians and reptiles have been recorded in the Tamaulipan Thornscrub of Coahuila: *Anaxyrus punctatus*, *A. speciosus*, *Incilius nebulifer*, *Lithobates berlandieri*, *Scaphiopus couchii*, *Spea multiplicata*, *Gopherus berlandieri*, *Crotaphytus reticulatus*, *Sceloporus cyanogenys*, *S. olivaceus*, *S. variabilis*, *Scincella lateralis*, *Drymarchon melanurus*, *Lampropeltis triangulum*, *Leptodeira septentrionalis*, *Masticophis flagellum*, *M. schotti*, *Pantherophis bairdi*, *P. emoryi*, *Pituophis catenifer*, *Tantilla atriceps*, *T. gracilis*, *T. nigriceps*, *Thamnophis proximus*, *Micrurus tener*, *Rena dulcis*, *R. segrega*, *Crotalus atrox*, *C. lepidus*, *C. molossus*, and *C. scutulatus*. Only three of these species (*Gopherus berlandieri*, *Crotaphytus reticulates*, and *Sceloporus olivaceus*) might be considered limited to this vegetation type. The range of *G. berlandieri* is restricted to southern Texas and eastern Coahuila southward to east of the Sierra Madre Oriental through

most of Nuevo León and Tamaulipas to extreme northern Veracruz. *Crotaphytus reticulatus* is limited to southern Texas and eastern Coahuila to northwestern Tamaulipas, including northern Nuevo León. The distribution of these two reptile species lies within the Tamaulipan Thornscrub. *Sceloporus olivaceus* is distributed from north-central Texas southward to the Coastal Plains of the Gulf of Mexico to southern Tamaulipas, westward to near the Big Bend area of Texas and eastern Coahuila. The remaining 28 taxa are species with a wide distribution and the option to occupy other vegetation types.

Submontane Scrub

Submontane Scrub is found in the foothills or gentle slopes on the northern or northwestern faces of the mountain ranges of the state, as well as in canyons with good humidity, shallow soils, and rocks with little organic material. It covers about 5% of the state and is characterized by the dominance of shrubs and trees 1–3 m high, most of them thornless, sclerophyllous, and deciduous. The composition of this vegetation type changes with the climatic conditions of the area; for example, in the mountains of the Mexican Plateau, where temperatures are low, the dominant species are *Purshia plicata*, *Amelanchier denticulata*, *Lindleya mespiloides*, *Cercocarpus montanus*, *C. fothergilloides*, *Rhus virens*, *Berberis trifoliolata*, *Sophora secundiflora*, *Fraxinus cuspidata*, *Quercus pringlei*, *Q. striatula*, and *Q. intricata*, with an herbaceous layer composed of *Mühlenbergia rigida*, *M. emersleyi*, *M. dubia*, *Loeselia scariosa*, and *Leptochloa dubia*. However, in the mountain chains near the Coastal Plains of northeastern Mexico, where the temperatures are higher, the common species are *Helietta parvifolia*, *Havardia pallens*, *Amiris madrensis*, *Fraxinus greggii*, *Gochnatia hypoleuca*, *Acacia rigidula*, *A. berlandieri*, *Zanthoxylum fagara*, *Lippia graveolens*, and *Leucophyllum frutescens*, along with herbaceous plants including *Ruellia nudiflora*, *Croton fruticulosus*, *Abutilon wrightii*, *Bouteloua curtipendula*, *Erioneuron avenaceum*, *Tridens texanus*, *T. muticus*, and *Digitaria cognata*.

The most representative herpetozoan species of the Submontane Scrub of Coahuila include *Eleutherodactylus guttilatus, Lithobates berlandieri, Scaphiopus couchii, Spea multiplicata, Gerrhonotus infernalis, Sceloporus oberon, S. variabilis, Lampropeltis mexicana, Pituophis catenifer, P. deppei, Tantilla wilcoxi, Micrurus tener,* and *Crotalus lepidus.* However, none of these species is limited to the Submontane Scrub of Coahuila; all the amphibians and reptiles that inhabit this vegetation type are also present in other vegetation types.

Montane Forest

Montane Forest is the vegetation type that develops above the Submontane Scrub in the mountainous portions of the provinces of the Sierra Madre Oriental and the Mexican Plateau. It grows in the canyons and highest parts of the mountains, where the highest humidity occurs and the climate is temperate-semiarid to temperate-subhumid. It is formed by the following tree communities.

Oak forest is a tree community of humid and temperate sites, frequently found in the canyons and intermountain valleys above the Submontane Scrub, intermixing with this and the pine forest at elevations between 1200 and 2500 m. It covers about 1% of the state; the dominant species are trees 3–8 m high, with rounded tops in scattered or dense populations, usually very homogeneous. The common species are *Quercus gravesii, Q. grisea, Q. laceyi, Q. hypoxantha, Q. saltillensis,* and *Q. greggii,* associated with *Prunus serotina, Arbutus xalapensis, Pinus arizonica, P. cembroides,* and *Juniperus flaccida.* The most common shrubs are *Salvia regla, Garrya ovata,* and *Ageratina saltillensis,* along with herbaceous plants such as *Piptochaetium fimbriatum, Achillea millefolium, Bromus carinatus, Cologania pallida,* and *Pleopeltis guttata.*

Pine forest includes communities dominated by species of the pine genus, found on slopes with high humidity on the highest parts of the main mountain ranges of the state, at elevations between 1200 and 3000 m where the climate is temperate-subhumid. It covers about 2% of

the state. The pine forest mixed with species of arid and semiarid affinities is the most common tree community, defined by short trees with rounded tops, knotted branches, and trunks less than 30 cm in diameter. The dominant species are *Pinus cembroides* and *P. pinceana,* with scattered individuals of *Yucca filifera, Agave gentri, Opuntia stenopetala, Lindleya mespiloides, Juniperus saltillensis,* and *J. erythrocarpa,* associated with herbaceous plants such as *Chrysactinia mexicana, Piptochaetium fimbriatum, Bouteloua hirsuta,* and *B. dactyloides.* Other communities of this type are dominated by *Pinus arizonica* var. *stormiae,* which typically has low densities and is frequently mixed with species of the oak forest. These are found through most of the mountain ranges of central and northern Coahuila; in the southeastern part of the state (Sierra de Arteaga) the pine forest is composed of *Pinus teocote, P. greggii, P. hartwegii, P. strobiformis, P. ayacahuite* var. *brachyptera,* and *P. pseudostrobus,* associated with species of *Quercus, Arbutus, Ceanothus, Prunus,* and *Fraxinus,* among others.

Fir forest is a scattered, well-known community that covers approximately 2% of the state. It is found only in small areas in the Sierras de Arteaga, Zapalinamé, El Jabalí, La Madera, and Maderas del Carmen, at elevations between 2500 and 3400 m on high slopes, canyons, and mountaintops with northern exposure, where precipitation is higher than 600 mm per year and average temperature varies between 12°C and 14°C, on shallow soils with abundant organic matter. Trees of this community are 10–20 m tall; the most common are *Pseudotsuga menziesii* and species of the genus *Abies* (*A. vejarii, A. durangensis,* and *A. mexicana*), along with *Pinus strobiformis, P. hartwegii, Cupressus arizonica, Quercus greggii,* and *Arbutus xalapensis* (a low-density population of *Picea mexicana* is present in the fir forest of the Sierra de Arteaga). The most common species in the bush stratum are *Arctostaphylos pungens, Ceanothus coeruleus, Paxistima myrsinites, Symphoricarpos microphylus, Garrya ovata,* and *Sambucus nigra.* The most common herbaceous plants are *Pleopeltis guttata, Achillea millefolium, Alchemilla vulcanica, Geranium crenatifolium,*

Koeleria macrantha, Brachypodium mexicanum, Trisetum spicatum, Festuca macrantha, F. pinetorum, and *F. valdesii.*

Alpine and subalpine vegetation is found in the highest parts of the mountain ranges of southeastern Coahuila (La Marta, Las Vigas, Potreros de Abrego, and El Coauilón) above 3400 m of elevation, where the average temperature is lower than 10°C and frosts are common. It represents the highest limit of the Montane Forest, covers less than 1% of the state, and is characterized by short bushes (0.5–1.0 m tall) and herbaceous plants with some isolated trees of *Pinus culminicola* and *P. hartwegii.* The most common bush species are *Ceanothus buxifolius, C. greggii, Symphoricarpos microphyllus, Agave montana, Arctostaphylos pungens,* and *Quercus greggii.* The herbaceous layer is represented by *Lupinus cacuminis, Senecio lorathifolius, S. coahuilenses, Bromus carinatus, Penstemon leonensis, Arenaria lanuginosa,* and *Brachypodium mexicanum.*

Although the Montane Forest of Coahuila covers just 6% of the state, it hosts unique amphibian and reptile species for the state, as well as species with a wide distribution in Coahuila. Of the 29 herpetozoan species found in this vegetation type in Coahuila, 16 are restricted to this vegetation type: *Chiropterotriton priscus, Pseudoeurycea galeanae, P. scandens, Eleutherodactylus longipes, Ecnomiohyla miotympanum, Barisia ciliaris, Phrynosoma orbiculare, Sceloporus couchi, S. oberon, Plestiodon dicei, Scincella silvicola, Diadophis punctatus, Storeria hidalgoensis, Tantilla wilcoxi, Thamnophis exsul,* and *Crotalus pricei.* Most of these species have ranges extending southward through the Sierra Madre Oriental. *Sceloporus couchii* and *Thamnophis exsul* are limited to the Sierra Madre Oriental of Coahuila and Nuevo León. Five species (*Barisia ciliaris, Phrynosoma orbiculare, Diadophis punctatus, Tantilla wilcoxi, Crotalus pricei*) have discontinuous distributions with populations in the Sierra Madre Occidental as well as in the Sierra Madre Oriental, with a large gap represented by the Chihuahuan Desert. Populations of these species in both sierras might be different species, although at present they are regarded as a single

species. Nine more species are limited to the Sierra Madre Oriental (*Chiropterotriton priscus, Pseudoeurycea galeanae, P. scandens, Eleutherodactylus longipes, Ecnomiohyla miotympanum, Sceloporus oberon, Plestiodon dicei, Scincella silvicola, Storeria hidalgoensis*). The three salamander species (*Chiropterotriton priscus, Pseudoeurycea galeanae, P. scandens*) belong to Plethodontidae, the members of which do not have lungs and generally do not depend on bodies of water to reproduce. In all except one Mexican species the eggs are laid in moist places, and all embryonic development occurs inside the egg; hatchlings are identical to adults except in size.

The remaining 13 species of amphibians and reptiles from the Montane Forest of Coahuila (*Hyla arenicolor, Lithobates berlandieri, Gerrhonotus infernalis, Sceloporus grammicus, S. minor, S. spinosus, Lampropeltis mexicana, Pituophis catenifer, P. deppei, Thamnophis cyrtopsis, Crotalus lepidus, C. molossus, C. scutulatus*) have a broad distribution that can include arid-semiarid, temperate, and/or subtropical environments.

Sacatal Grassland

The Sacatal Grassland includes communities dominated by grasses; it is mainly found in valleys of moderate depth, as well as on gentle slopes and flatlands at elevations between 800 and 2500 m, frequently mixed with pine forest and Chihuahuan Desert Scrub. It covers approximately 8% of the state. The most characteristic community of this vegetation type is found in a series of valleys and flatlands in southeastern Coahuila, where it is a moderately open grassland associated with xerophilous plants. The dominant grass species are *Bouteloua gracilis, B. curtipendula, B. dactyloides,* and *Aristida divaricata,* with scattered bushes of *Opuntia imbricata, O. rastrera, Larrea tridentata, Flourensia cernua, Yucca carnerosana,* and *Prosopis glandulosa.* In some areas with gypsum-laden soils in central and southern Coahuila, gypsum-tolerant grasses are present; the common species are *Bouteloua chasei, Mühlenbergia villiflora, M. gypsophila, M. arenicola, Achnatherum editorum,* and

Scleropogon brevifolius. In soils with an accumulation of salt the following species are common: *Sporobolus airoides, S. wrightii, S. criptandrus, S. spiciformis, Pleuraphis mutica, Hilaria belangeri, Distichlis spicata,* and *D. litorales,* mixed with *Atriplex, Prosopis,* and *Suaeda.*

Sacatal Grassland communities are also found in the intermountain valleys of southern Coahuila; the most common species are *Mühlenbergia emersleyi, M. dubia, M. setifolia, Nasella tenuissima,* and *Stipa eminens,* mixed with bushes and shrubs of Submontane Scrub or Rosetophilous Scrub. In flatlands and open sites at high altitudes of the Montane Forest one finds grasslands dominated by *Stipa robusta* or *S. ichu* associated with species of *Pinus;* other common herbaceous plants are *Hymenoxys insignis, Gridelia grandiflora,* and *Senecio madrensis.*

In Coahuila, there are 46 herpetozoan species representative of the Sacatal Grassland, 3 of which are limited to this type of vegetation (*Sceloporus goldmani, S. samcolemani, Coluber constrictor*). The ranges of *S. goldmani* and *S. samcolemani* are limited to the grasslands of southeastern Coahuila and adjacent parts of Nuevo León and San Luis Potosí (this last state only for *S. goldmani*). Apparently, most of the populations of *S. goldmani* have been extirpated or are almost extirpated. *Coluber constrictor* has a wide distribution from southern Canada southward through most of the United States and the Atlantic slope of Mexico to Guatemala. In Coahuila, it is known only from the grasslands of the Chihuahuan Desert.

The remaining 43 species have a strong affinity for grasslands, although they can occupy other vegetation types: *Anaxyrus cognatus, A. debilis, A. woodhousii, Rhinella marina, Hyla arenicolor, Gastrophryne olivacea, Lithobates berlandieri, Scaphiopus couchii, Spea multiplicata, Gopherus berlandieri, G. flavomarginatus, Barisia ciliaris, Gerrhonotus infernalis, Phrynosoma cornutum, P. modestum, Sceloporus cautus, S. consobrinus, Plestiodon obsoletus, P. tetragrammus, Scincella kikaapoa, Aspidoscelis gularis, A. inornata, A. marmorata, Arizona elegans, Gyalopion canum, Heterodon kennerlyi, Hypsiglena jani, Lampropeltis*

getula, Masticophis flagellum, Pituophis catenifer, P. deppei, Rhinocheilus lecontei, Salvadora grahamiae, Sonora semiannulata, Tantilla gracilis, T. hobartsmithi, T. nigriceps, T. wilcoxi, Thamnophis marcianus, Crotalus atrox, C. scutulatus, C. viridis, and *Sistrurus catenatus.*

Riparian, Subaquatic, and Aquatic Vegetation

The communities of Riparian, Subaquatic, and Aquatic Vegetation are limited to rivers, streams, and bodies of water, which are concentrated in northern, northeastern, and central Coahuila at elevations between 300 and 1400 m. Parts of the riparian vegetation of the northern part of the state (Ríos Bravo, San Diego, San Rodrigo, Escondido, and Arroyo de las Vacas) are isolated tree communities of *Quercus fusiformis* and *Carya illinoensis,* frequently associated with individuals of *Prosopis glandulosa, Platanus glabrata, Diospyros texana, Ulmus crassifolia, Salix nigra, Celtis reticulata,* and *Morus celtidifolia.* Among the bushes are *Opuntia lindheimeri* and *Castela erecta;* and the dominant herbaceous plants are *Ruellia nudiflora, Sanvitalia ocymoides, Viguiera dentata, Allowissadula holosericea,* and *Malvastrum coromandelianum.* In the Río Sabinas one finds narrow galleries of *Taxodium mucronatum, Fraxinus berlanderiana, Morus celtidifolia, Salix nigra, Prosopis glandulosa,* and *Acacia farnesiana;* the last two are the dominant species in the Río Salado-Nadadores. The banks of these rivers frequently host dense populations of *Baccharis salicifolia* and *Arundo donax.* In the seasonal streams next to the mountain ranges are isolated trees of *Chilopsis linearis, Juglans microcarpa,* and *Dodonaea viscosa.* The subaquatic plant species grow in the permanent moist soils of rivers and bodies of water. The most common species are *Cyperus odoratus, Eleocharis cellulosa, E. geniculata, Fuirena simplex, Paspalum pubiflorum, Spartina spartinae, Setaria geniculata, Echinochloa colonum,* and *Poligonum punctatum.* The aquatic vegetation is represented by plants that grow in flooded places or underwater. The locality with the greatest amount of aquatic plants is Cuatro Ciénegas, where the following plant species have been reported: *Nymphaea ampla, Utricularia vulgaris, Najas marina, Ruppia maritima, Potamogeton nodosus, Nastartium officinale, Najar guadalupensis, Chara* spp., and *Zanichellia palustris* (Pinkava 1979). The following plant species are frequently found in the main rivers: *Nuphar luteum, Ceratophyllum demersum,* and *Heteranthera dubia*, with some populations of *Eichhornia crassipes.*

Thirty-seven species of amphibians and reptiles are found in the Riparian, Subaquatic, and Aquatic Vegetation types of Coahuila, 16 of which are amphibians that depend on bodies of water for their reproduction: *Ambystoma mavortium, Anaxyrus cognatus, A. debilis, A. punctatus, A. speciosus, A. woodhousii, Incilius nebulifer, Rhinella marina, Acris crepitans, Ecnomiohyla miotympanum, Hyla arenicolor, Smilisca baudinii, Lithobates berlandieri, L. catesbeianus, Scaphiopus couchii,* and *Spea multiplicata.* Only two, *Acris crepitans* and *Lithobates catesbeianus*, require permanent bodies of water to survive. In Mexico, the only known population of *A. crepitans* occurs in the higher part of the Río Sabinas, Coahuila, although this species is found from the Great Lakes region southwestward through the Great Plains to the Rio Grande and eastward from Big Bend. *Lithobates catesbeianus* also has a wide distribution in the United States, although the western populations are introduced. Although it was recorded in Coahuila more than 165 years ago (Smith et al. 2003a), recently it was recorded in Rancho La Burra, municipality of Guerrero, Coahuila, KM 92, Nuevo Laredo–Piedras Negras Highway (Garza-Tobón and Lemos-Espinal, 2013a, 2013b). This population is within the natural range of the species, thus should be regarded as a native species/population of Coahuila. The other 14 amphibian species that occur in this vegetation type inhabit sites that hold enough moisture (under logs, rocks, debris, etc.), so they can be found in places with a well-defined dry season and/or far away from bodies of water.

Another nine species that inhabit this vegetation type are turtles with a strong dependence on bodies of water for survival, including *Pseudemys gorzugi, Terrapene coahuila, Trachemys gaigeae,* *T. scripta, T. taylori, Kinosternon durangoense, K. flavescens, K. hirtipes,* and *Apalone spinifera.* The only strictly aquatic turtle is *A. spinifera.* The remaining eight species might spend long periods of time out of water, although they are always close to water.

Three more species are lizards with a strong affinity for riparian vegetation (*Sceloporus couchii, S. cyanogenys, Aspidoscelis tesselata*). *Sceloporus couchii* and *S. cyanogenys* are scansorial lizards mainly found in trees along riverbanks. *Aspidoscelis tesselata* is a species that lives among the bushes, grasses, and open spaces on or near riverbanks.

The remaining nine species are snakes (*Nerodia erythrogaster, N. rhombifer, Opheodrys aestivus, Thamnophis angustirostris* [description of the morphology of the type series of this species suggests strong aquatic habits], *T. cyrtopsis, T. exsul, T. marcianus, T. proximus, Agkistrodon contortrix*) that are rarely found far from bodies of water. All of them include aquatic prey in their diets.

The environment of Coahuila forms a pattern of harsh extremes in both climate and topography; of ecological islands in seas of homogeneity; of extensive deformation in fairly recent geological history, effecting successive isolations and merges of terrain; and of marginal availability of water in extensive areas of internal drainage (over a third of the state). The result, herpetologically, is a depauperate fauna of diverse origins, dominated by species adapted to harsh extremes of weather (like those of *Sceloporus*), but also with oases of astonishingly high endemism in patches of favorable environmental conditions (e.g., the Cuatro Ciénegas Basin, the habitat of four of the eight endemic species of the state) and islandlike distributions at higher elevations in widely separated mountain masses (e.g., *Gerrhonotus infernalis, Sceloporus grammicus, Pituophis deppei*).

Of course, the harsh environmental conditions have also restricted exploration, as discussed previously. Thus, knowledge of the herpetofauna of the state—its composition, distribution, taxonomy, natural history—is woefully deficient. Nevertheless, the environmental limitations here reviewed are constant or are more likely to worsen than improve.

As stated previously, a total of 133 species is known from Coahuila. Of these, the two amphibian orders Caudata and Anura are the groups least represented, 3% (4/133) and 15% (20/133), respectively, because of the dependence of amphibians on bodies of water and/or moist places. Most of Coahuila consists of arid-semiarid environments with a limited availability of water and moist areas, which restricts the presence of these groups in the state. Three of the four salamander species that inhabit Coahuila are limited to a small Montane Forest of the Sierra Madre Oriental of extreme southeastern Coahuila. The other salamander (*Ambystoma mavortium*) is the only one in the state adapted to live in arid-semiarid regions of North America. These salamanders live in ponds and areas where nonflowing water is available in the Chihuahuan Desert; in dry periods they can bury themselves to considerable depths to survive, following the receding moisture level.

It is likely that the Texas Salamander (*Eurycea neotenes*) inhabits extreme northern Coahuila; Dixon (2000) indicated the occurrence of this species at several localities in Texas adjacent to the extreme northern border of the state.

Similarly, the low proportion of anuran species is due to the limited availability of water and moisture in the state. The dominant genus in Coahuila (*Anaxyrus*) represents organisms with a tuberculated skin that allows them to survive under conditions of limited water availability. Species of other genera, such as *Incilius, Rhinella, Craugastor, Eleutherodactylus, Gastrophryne, Lithobates, Scaphiopus,* and *Spea*, require greater water or moisture availability.

It is possible that the Plains Spadefoot (*Spea bombifrons*) inhabits extreme northern Coahuila; Dixon (2000) indicated the occurrence of this species in a few counties in Texas just north of the Rio Grande.

However, in reptiles the scaled skin, amniotic eggs, and presence of lungs make them more broadly independent of water-essential activities.

Therefore, turtles, lizards, and snakes are well represented in the state.

Although the proportion of turtles is just 8% (11/133) of the total number of herpetozoan species in the state, it is more than twice the proportion of turtles in the country (3.3%, or 32/973; Liner and Casas-Andreu 2008). This gives us an idea of the large representation that this group of reptiles has in the state. All of the turtle species of Coahuila inhabit the arid-semiarid region. No turtle species has been recorded in the temperate region of southeastern Coahuila or in the high mountains scattered there. Furthermore, the number of turtle species in Coahuila probably is larger than reported here; species such as *Terrapene ornata* very likely inhabit the Chihuahuan Desert of Coahuila, although as yet there are no records of the species in the state.

Lizards are well represented in Coahuila and make up 37% (49/133) of the total number of herpetozoan species present in the state. By far the most speciose genus of reptiles (or amphibians) in Coahuila is the lizard genus *Sceloporus*. The number of *Sceloporus* species in the state (19) is one of the largest for any Mexican state and is likely a product of the large diversity of environments available for these lizards. However, the current list of lizards known for Coahuila may eventually prove to be woefully incomplete. Lemos-Espinal and Smith (2007) indicated that species such as the Common Lesser Earless Lizard (*Holbrookia maculata*), the Hernández Short-horned Lizard (*Phrynosoma hernandesi*), and the Chihuahuan Spotted Whiptail (*Aspidoscelis exsanguis*) possibly inhabit extreme northwestern Coahuila. The Pygmy Alligator Lizard (*Gerrhonotus parvus*) may occur in the pine forests of the Sierra de Arteaga, and the Eastern Spiny Lizard (*Sceloporus spinosus*) in the semiarid region of the extreme southeastern part of the state. The Green Anole (*Anolis carolinensis*), Laredo Striped Whiptail (*Aspidoscelis laredoensis*), and Six-lined Racerunner (*A. sexlineata*) probably occur in extreme northeastern Coahuila adjacent to Texas. The Torquate Lizard (*Sceloporus torquatus*), Bolson Night Lizard (*Xantusia bolsonae*), and Durangoan Night

Lizard (*X. extorris*) probably inhabit the extreme southwestern portion of the state.

Snakes also make up a large percentage of the Coahuilan herpetofauna (37%, or 49/133). The high diversity of snakes is the result of the variety of their adaptations to life in equally diverse habitats and ecosystems. According to Lemos-Espinal and Smith (2007), several species of snakes not recorded for Coahuila may inhabit the state. They suggested Taylor's Cantil (*Agkistrodon taylori*), the Tamaulipan Hook-nosed Snake (*Ficimia streckeri*), and the Red Black-headed Snake (*Tantilla rubra*) might occur in the southeastern portion of the state; the Tampico Threadsnake (*Rena myopica*) and Nuevo León Graceful Brown Snake (*Rhadinaea montana*) in the extreme eastern portion; Dekay's Brown-snake (*Storeria dekayi texana*) and the Trans-Pecos Black-headed Snake (*Tantilla cucullata*) in the extreme northeastern part; and the Big Bend Patch-nosed Snake (*Salvadora deserticola*) and Texas Lyresnake (*Trimorphodon vilkinsonii*) in extreme northwestern Coahuila. Rossman et al. (1996) indicated the presence of the Mexican Gartersnake (*Thamnophis eques*), the Mexican Black-bellied Gartersnake (*T. melanogaster*), and the Madrean Narrow-headed Gartersnake (*T. unilabialis*) in extreme southwestern Coahuila; however, we are not aware of specimens of any of these three species having been taken in the state. Thus, in the absence of records from Coahuila, we regard their occurrence there as conjectural.

Of the 133 species of herpetozoans that inhabit Coahuila, 98 are shared with the United States; most of these shared species (95%, or 93/98) occur in the Chihuahuan Desert and extend their ranges southward from the Great Plains of the United States to the southern tip of the Chihuahuan Desert in the Mexican states of San Luis Potosí and Querétaro. Only 4 of these shared species are characteristic of the American tropics and subtropics (*Rhinella marina*, *Smilisca baudinii*, *Drymarchon melanurus*, and *Leptodeira septentrionalis*). *Rhinella marina* has been recorded in the lowlands of central Coahuila, in the semiarid Cuatro Ciénegas Basin,

whereas the other 3 occur in the lowlands of northeastern Coahuila and the western foothills of the Sierra Madre Oriental. All 4 of the species with tropical affinities enter the United States only in the southern part of Texas. One of these 98 species is introduced, the Mediterranean Gecko (*Hemidactylus turcicus*), whose native range is the Mediterranean. In Coahuila it has a wide distribution, occupying several towns and ejidos (farming cooperatives) of the Chihuahuan Desert; it is widely distributed in the southern United States. Another of these shared species, *Coleonyx reticulatus*, is found almost exclusively in Coahuila, and except for one record on the Coahuila/Durango state line and records near Big Bend, Texas, just along the lower elevations near the Rio Grande, all the records of this species are from Coahuila.

All the 35 species that are not shared between Mexico and the United States are endemic to Mexico, 7 of them to Coahuila: *Terrapene coahuila*, *Trachemys taylori*, *Gerrhonotus lugoi*, *Crotaphytus antiquus*, *Uma exsul*, *Scincella kikaapoa*, and *Thamnophis angustirostris*. Three of these are limited to the Cuatro Ciénegas Basin (*T. coahuila*, *G. lugoi*, and *S. kikaapoa*); the other 4, to other localities in the Chihuahuan Desert of Coahuila. The taxonomic status of *Thamnophis angustirostris* is unclear. This species is known only from the lectotype specimen (USNM 959—Kennicott had not actually designated a holotype and apparently based his new taxon on more than one specimen [Rossman et al. (1996)]) collected in Parras, Coahuila. We regard it as a valid species since the Sierra de Parras and nearby localities have suitable environmental conditions for this gartersnake; eventually this species could be rediscovered with more intensive fieldwork in the areas around Parras.

Of the remaining 28 species endemic to Mexico, 15 (54%, or 15/28) are limited in eastern Mexico to the mountains and foothills of the Sierra Madre Oriental (*Chiropterotriton priscus*, *Pseudoeurycea galeanae*, *P. scandens*, *Eleutherodactylus longipes*, *Ecnomiohyla miotympanum*, *Barisia ciliaris*, *Phrynosoma orbiculare*, *Sceloporus minor*, *S. oberon*, *S. parvus*, *Plestiodon dicei*,

Scincella silvicola, Pituophis deppei, Storeria hidalgoensis, and *Thamnophis exsul*). They enter Coahuila only in the southeastern corner of the state. Three species are limited to scattered regions of northern Mexico: *Sceloporus couchi* to the northern Sierras of Coahuila and central western Nuevo León; *S. goldmani* to a small area in southeastern Coahuila, adjacent Nuevo León, and northeastern San Luis Potosí; and *S. maculosus* to the drainage of the Río Nazas in Durango and Coahuila. Three species are limited to the Mexican Plateau (*Sceloporus cautus, S. samcolemani,* and *Lampropeltis mexicana*). Another three species are limited to the small area of the Bolsón de Mapimí of southeastern Chihuahua, western Coahuila, and northeastern Durango (*Kinosternon durangoense, Gopherus flavomarginatus,* and *Uma paraphygas*). Two species (*Sceloporus cyanostictus* and *S. ornatus*) are limited to Coahuila and extreme western Nuevo León. *Sceloporus spinosus* is widely distributed from central Mexico northward through the Mexican Plateau, entering Coahuila only in the oak forests of the extreme southeastern part of the state in the Zapaliname Mountains. The last of these 35 endemic species (*Holbrookia approximans*) is limited to the Chihuahuan Desert of Mexico; however, it is highly likely that it occurs in the adjacent United States.

Herpetofauna of Nuevo León

JULIO A. LEMOS-ESPINAL
AND ALEXANDER CRUZ

Introduction

Nuevo León is one of the wealthiest Mexican states, it is highly industrialized, and the average inhabitant possesses a higher standard of living than the average inhabitant of the rest of Mexico. Its capital, Monterrey, is the third-largest city in Mexico. Unlike the rest of the Mexican states along the northern border, Nuevo León has a road network that facilitates access to almost every corner of the state. The flora and fauna of Nuevo León are very rich in species. Broadly speaking, it consists mainly of a group of species characteristic of the great deserts of North America, as well as species from the temperate forests of the Sierra Madre Oriental, and subtropical species that extend their distribution, in some cases even from Central or South America, through the lowlands of the Atlantic slope.

Even though Nuevo León has these characteristics, studies on the diversity and distribution of the species of amphibians and reptiles of the state are few, and those that have been developed are focused almost entirely on the forests of the Sierra Madre Oriental and satellite mountains to the north and northeast of Monterrey; the vast plains to the east of the Sierra Madre Oriental and the portion of the Mexican Plateau in the southwestern corner of the state remain mostly unstudied. Currently there is no program to systematically address the study of the state herpetofauna in different regions. Consequently, one of the main threats to the conservation of the richness of amphibians and reptiles of the state is the ignorance of it. This coupled with problems associated with the high demand for water, energy, and food to meet the needs of Monterrey, one of the largest and fastest-growing cities in Mexico, does not provide an encouraging outlook for the herpetofauna of Nuevo León.

To date, the presence of 132 species of amphibians (23 species) and reptiles (109 species) has been documented in the state. However, further herpetofaunal censuses, particularly in areas near the border with Tamaulipas, will likely considerably increase these numbers. In addition, studies on morphological variation and genetic composition of currently known populations, especially from those that are isolated, may demonstrate the existence of previously unknown species.

Previous Herpetological Studies

Herpetological work in northeastern Mexico began with the arrival of the French naturalist Jean Louis Berlandier. He came to Mexico in December 1826 at the age of 21 with the aim of assessing the natural resources of the border between

Texas and Mexico. By February 1828 Berlandier had begun his work at the border. Although Berlandier made most of his collections in Texas and Tamaulipas, it is likely that he also made collections of amphibians and/or reptiles in Nuevo León. However, records of such collections are difficult to obtain, and we could not find any that he developed in this state.

The first record of a herpetological collection in the state of Nuevo León is that made in 1852 by Thomas Webb, secretary and surgeon of Commissioner John Russel Bartlett's party of the Mexican–United States Boundary Commission. Upon completion of the Gila River survey, Bartlett's party left Tucson, Arizona, on July 17, 1852, for El Paso, Texas. Their route traversed northern Sonora and arrived in El Paso on August 18. The party left El Paso on October 7, proceeding to Ringgold Barracks, Texas, through the Mexican states of Chihuahua, Durango, Coahuila, and Nuevo León. They reached Santa Catarina on December 11, and Monterrey, the capital of Nuevo León, on December 12. They left Monterrey the next day by a route that passed through Marin, Carrizitos, and Cerralvo, and on December 19, they arrived at Mier y Noriega. The following day the party passed through Camargo and then across the Rio Grande to Ringgold Barracks (Kellogg 1932). All the specimens collected by the commission are deposited in the United States National Museum (USNM). Some of the specimens collected in Nuevo León were reported by Kellogg (1932).

After the collections made by Webb in 1852, Lt. Dario Nash Couch made a personal expedition to northeastern Mexico in 1853 to collect plants and animals and spent several months in the states of Coahuila, Nuevo León, and Tamaulipas. In Nuevo León he visited China, San Diego, Cadereyta, Monterrey, Boquilla, Villa de Santiago, Rinconada, and several localities in the Sierra Madre Oriental (Conant 1968). Specimens collected by Couch are deposited in the USNM. Several specimens that Couch collected in Nuevo León became part of the type series of new taxa, for example, *Bufo speciosus* (= *Anaxyrus speciosus*) collected in Pesquería Grande (= Villa García)

and described by Girard (1854); *Sceloporus couchii* collected in Santa Catarina and described by Baird (1859); *Cnemidophorus inornatus* (= *Aspidoscelis inornata*) collected in Pesquería Grande and described by Baird (1859); *Cnemidophorus octolineatus* (= *Aspidoscelis inornata*) collected in Pesquería Grande and described by Baird (1859); and *Eutaenia cyrtopsis* (= *Thamnophis cyrtopsis*) collected in Rinconada and described by Kennicott (1860). Other specimens collected by Couch were reported by Kellogg (1932).

Another of Couch's important contributions was the purchase of Berlandier's collection from his widow in 1853. He shipped part of Berlandier´s collection to Spencer F. Baird of the USNM and part of the plant collection to Switzerland (Beltz 2006).

In March, April, and June 1902, Edward Goldman and Edward Nelson of the USNM visited several localities in western Nuevo León: Cerro La Silla, Doctor Arroyo, Montemorelos, and Monterrey, where they collected vertebrates. Kellogg (1932) reported some of the anuran specimens collected.

In 1904 Hans Gadow and Seth Meek made a trip to Mexico to study the variation in the genus *Cnemidophorus*; during that trip Meek collected in the towns of San Juan (located south of Monterrey), Montemorelos, Garza Valdés, and La Cruz, all in the state of Nuevo León. Among the specimens collected was the type series of *C. gularis meeki* (= *Aspidoscelis gularis*). This material was reported by Gadow (1906).

The beginning of the twentieth century was characterized by a lack of herpetological work in Nuevo León. Edward H. Taylor and Hobart M. Smith did not visit Mexico for the first time until 1932, initiating one of the most prolific herpetological eras in Mexico. The same day that Hobart Smith graduated at the age of 20 with a bachelor's degree from Kansas State University, Edward Taylor picked him up to spend the entire summer collecting everywhere they could drive in Mexico, where they gathered more than 5000 specimens. On their trip down they stopped at Sabinas Hidalgo for several days and collected amphibians and reptiles there; among those specimens was the type of *Sceloporus parvus*, described by Hobart Smith in 1934. Six years later Smith was able to continue fieldwork on the herpetofauna of Mexico through a Walter Rathbone Bacon Traveling Scholarship. Between 1938 and 1940 he and his wife, Rozella, collected amphibians and reptiles throughout Mexico. On October 13 and 14, 1939, they stopped 24 km southeast of Galeana and collected amphibians and reptiles. On those two days they collected 19 specimens (all of them paratypes) of *Bolitoglossa galeanae* (= *Pseudoeurycea galeanae*) reported by Taylor (1941) and Taylor and Smith (1945); *Syrrophus smithii* (= *Eleutherodactylus guttilatus*) (the type and paratype), described and reported by Taylor (1940b); *Tantilla wilcoxi rubricata* (= *T. wilcoxi*), type described and reported by Smith (1942); and *Salvadora lineata* (= *S. grahamiae lineata*) reported by Taylor and Smith (1945). Taylor returned to Nuevo León in June 1936. He visited Sabinas Hidalgo (June 17) and Huasteca Canyon (June 20), 15 km west of Monterrey. Among the specimens collected at these two localities are the type and one of the paratypes of *Syrrhophus latodactylus* (= *Eleutherodactylus longipes*), which he described (Taylor 1940a).

Other important contributions of Hobart Smith to the knowledge of the herpetofauna of Nuevo León include the descriptions of *Leiolopisma caudaequinae* (= *Scincella silvicola caudaequinae*) (Smith 1951) and *Sceloporus scalaris samcolemani* (= *S. samcolemani*) (Smith and Hall 1974). J. P. Craig donated the type specimen of *Scincella silvicola caudaequinae* to the University of Illinois Museum of Natural History. He collected the specimen from Cascada Cola de Caballo, 40 km south of Monterrey on April 19, 1946. The type of *Sceloporus samcolemani*, collected by P. H. Litchfield between Providencia and La Paz, on July 16, 1960, is in the University of Michigan Museum of Zoology.

In 1934, Henry A. Pilsbry, Francis W. Pennell, and Cyril H. Harvey of the Academy of Natural Sciences of Philadelphia collected 141 specimens of amphibians and reptiles, representing 36 species from the states of Chihuahua, Coahuila, Durango, Mexico, Nuevo León, San Luis Potosí, and Zacatecas. In Nuevo León they visited Hacienda Pablillo, Cascada Cola de Caballo, Río Maurisco (25 km south of Monterrey), the trail between Pablillo and Alamar, the trail up El Infiernillo from the Hacienda Pablillo, and Cieneguillas (south of Galeana). Most of these locations are in the southwestern part of the state. They collected specimens of 12 species (1 salamander, 2 anurans, 7 lizards, and 2 snakes) in Nuevo León. Among the specimens collected was the type of *Sceloporus binocularis* (= *S. torquatus binocularis*). Dunn (1936) published the original description of this taxon along with the report of the specimens collected in Nuevo León.

On July 10, 1938, Radclyffe Roberts, a friend and field companion of Edward Taylor on several trips to Mexico, collected the type of *Bolitoglossa galeanae* (= *Pseudoeurycea galeanae*) in Pablillo, 24 km west of Galeana at 2133 m elevation. The original description of this taxon and the report of the specimen were published by Taylor (1941).

In the summers of 1938, 1939, and 1940, parties of students from the University of Illinois in Chicago under the leadership of Harry Hoogstraal collected a number of reptiles and amphibians in northeastern Mexico. Some of the specimens were incorporated into the collections of the Field Museum of Natural History (Smith 1944). Among the 53 snake specimens were a number from Nuevo León, including the type of *Rhadinaea montana*, collected by Harry Hoogstraal at Ojo de Agua, Galeana, and described by Hobart Smith in 1944. Some other specimens were sent to the Chicago Natural History Museum (now known as the Field Museum); among those is the type of *Chiropterotriton priscus*, collected in Cerro Potosí, near Galeana, on August 16, 1938, by E. J. Koestner, as well as the paratypes collected by Koestner and Hoogstraal at the same place. Originally, these specimens were mentioned by Taylor (1944, 217): "I have seen an undescribed species of this genus (*Chiropterotriton*) that was taken on Cerro Potosí near Galeana, Nuevo León." The report and description of the species were published by Rabb (1956).

In addition, Hobart Smith and Edward Taylor

published the checklist and key to the snakes of Mexico; the checklist and key to the amphibians of Mexico; the checklist and key to the reptiles of Mexico, exclusive of the snakes; and the type localities of Mexican amphibians and reptiles (Smith and Taylor 1945, 1948, 1950a, 1950b). In these publications they provided the checklists and keys to the amphibians and reptiles of Nuevo León and the list of type localities for this state. At the age of 100 Hobart Smith was still publishing on the herpetofauna of Mexico and supporting his fellow herpetologists with the same enthusiasm that had always characterized him.

One of the greatest contributors to the knowledge of the herpetofauna of Nuevo León was Ernest A. Liner, who started working in northeastern Mexico November 19–22, 1951. His first collecting trip took place in Nuevo León; his main objective for that trip was to collect *Pseudoeurycea galeanae*, which at that time was known only from the one type series near Galeana. That year he visited localities in the region of Monterrey, Arroyo Vaquerías (16 km north of Monterrey), Cascada Cola de Caballo, Santa Rosa Canyon, Linares, Galeana, Rancho La Montaña, and Doctor Arroyo. Although on that trip he did not collect *P. galeana*, he was able to gather several species of amphibians and reptiles and became familiar with the herpetofauna of the state. In the following years he visited Mexico as many times as he could (December 21, 1952–January 3, 1953; July 6–13, 1954; July 13–22, 1957; July 26–August 8, 1958; July 1960; and July 1961). On these trips he visited locations in eastern-northeastern Mexico, and during his 1960 trip he made a loop from the Atlantic to the Pacific coast, entering Mexico through Piedras Negras, Coahuila, and leaving it at Nogales, Sonora, traveling through the states of Coahuila, San Luis Potosí, Guanajuato, Aguascalientes, Jalisco, Nayarit, Sinaloa, and Sonora. Although he collected over a wide area in Mexico, his main collecting work was done in Nuevo León. The reports of his trips to Mexico, including all the places visited in Nuevo León and the specimens

he collected during this period are published in Liner (1964, 1966a, 1966b, 1991a, 1991b, 1992a, 1992b, 1993, 1994).

After 1961 Liner continued doing fieldwork in Nuevo León along with several other herpetologists, among them Allan H. Chaney, James R. Dixon, Richard M. Johnson, Douglas A. Rossman, and James F. Scudday. They published a number of distributional and natural history notes: *Rhadinaea montana* (July 18, 1972), from Ojo de Agua at Pablillo (Chaney and Liner 1986); *Crotalus pricei miquihuanus* (July 20, 1985), 1.8 km north of Los Mimbres Road (Chaney and Liner 1986); *Rhadinaea montana* (July 17, 1985), from 4.3 km north of Las Adjuntas (Chaney and Liner 1990); *Chiropterotriton priscus* (records of July and August 1980), from the surroundings of San Antonio Peña Nevada and La Encantada, southern Nuevo León (Chaney et al. 1982a); *Sceloporus poinsettii poinsettii* (records of July 1977 and August 1978), from Picacho Mountains at Rancho El Milagro (Chaney et al. 1982b); *Pseudoeurycea galeanae* (records of July and August 1980), from the area around San Antonio Peña Nevada and La Encantada, southern Nuevo León (Johnson et al. 1982a); *Sceloporus couchi* (records of July 1977 and August 1978), from Picacho Mountains at Rancho El Milagro (Johnson et al. 1982b); *Barisia ciliaris* (May 24, 1973), from Ojo de Agua near Pablillo (Liner et al. 1973); *Tantilla nigriceps fumiceps* (July 13, 1976), from Rancho El 86, 9.5 km south-southwest of Cerralvo (Liner et al. 1978); *Bogertophis subocularis* (July 28, 1981), from 35.4 km southwest of Bustamante (Liner, Chaney, and Johnson 1982); *Tantilla rubra rubra* (August 16, 1978), from Picacho Mountains at Rancho El Milagro (Liner, Johnson, and Chaney 1982); *Thamnophis cyrtopsis pulchrillatus* (records of August 1978 and June 1981), 5 km north of Las Crucitas and San Antonio Puerto Peña Nevada (Liner et al. 1990); *Crotalus lepidus lepidus* (July 18, 1972), Ojo de Agua at San Pablillo (Liner and Chaney 1986); *Rhadinaea montana* (July 18, 1972), Ojo de Agua at San Pablillo (Liner and Chaney 1987); *Tantilla rubra rubra* (August 16, 1978), 9.8 km southwest and 16.7 km west of

Cerralvo, on Rancho Milagro, Picacho Mountains (Liner and Chaney 1990b); *Sceloporus torquatus mikeprestoni* (records of July and August 1980), from the surroundings of San Antonio Peña Nevada (Liner and Chaney 1990a); *Aspidoscelis inornata inornata* (July 20, 1975), from 5.6 km west of Aramberri (Liner and Chaney 1995a); *A. i. paulula* (July 20, 1974), from 7.2 km north of Mier and Noriega at Rancho San Roberto (Liner and Chaney 1995b); and *Storeria occipitomaculata hidalgoensis* (= *S. hidalgoensis*) (July 5, 1965, and July 25, 1973), from Ojo de Agua above Pablillo and 8.4 km west of Ejido 14 de Marzo on Cerro Potosí (Liner and Johnson 1973).

Other publications authored or coauthored by Ernest Liner on the amphibians and reptiles of Nuevo León include the report of adults of the lizard *Sceloporus torquatus binocularis* collected in July 1954 and 1957 from the area around Galeana (Liner and Olson 1973); distributional records for the state for *Ambystoma tigrinum velasci* (= *A. mavortium*), *Chiropterotriton priscus*, *Pseudoeurycea galeanae*, *Anaxyrus speciosus*, *Spea multiplicata*, *Kinosternon flavescens flavescens*, *Plestiodon brevirostris pineus* (= *P. dicei*), *Sceloporus oberon*, *S. spinosus spinosus*, *S. torquatus binocularis*, *Lampropeltis alterna*, and *Crotalus scutulatus scutulatus* (Liner et al. 1976); the redescription of *Thamnophis exsul* (Rossman et al. 1989); the original description of *Sceloporus chaneyi*, collected by Ernest Liner and Richard Johnson on July 17, 1980, at Rancho La Encantada (Liner and Dixon 1992); the accounts for the *Catalogue of American Amphibians and Reptiles* of *Sceloporus chaneyi* (Liner and Dixon 1994) and *Rhadinaea montana* (Liner 1996); and a report of the herpetological type material from Nuevo León (Liner 1996a). The information reported by Liner in this last summarized paper is in table 7.1. Liner's large personal collection of Mexican herpetozoa was donated to the American Museum of Natural History (AMNH) shortly before his death.

Other contributors to the knowledge of the herpetofauna of Nuevo León are Robert and William Reese from St. Edward's University, Austin, Texas, who on August 23–24, 1969, vis-ited Las Lajitas (9.6 km west and 11.3 km north of Doctor Arroyo), 9.6 km north of Ascensión, and 8 km east of San Roberto junction and collected *Ambystoma velasci* (= *A. mavortium*) (first record for Nuevo León), *Pseudoeurycea galeanae*, and *Crotalus scutulatus*. This collection is reported in Reese (1971). About 12 years later, on March 21, 1983, Alec Knight from Sul Ross State University, Alpine, Texas, collected the type specimen of *Gerrhonotus parvus* from 1 km south of Galeana. Alec Knight and James Scudday published the original description and report of the specimen in 1985.

The Universidad Autónoma de Nuevo León (UANL) has also contributed to the knowledge of the herpetofauna of Nuevo León, mainly through people such as Javier Banda-Leal, Arturo Contreras-Arquieta, Armando Contreras-Balderas, Armando Contreras-Lozano, David Lazcano-Villarreal, Manuel Nevares, and Carlos Treviño-Saldaña. Individually or in groups, and in collaboration with people such as Robert W. Bryson, Michael S. Price, and Gerard T. Salmon, they have published several articles on the amphibians and reptiles of the state. In 1988, Carlos Treviño-Saldaña published the original description of *Sceloporus jarrovii cyaneus* (= *S. minor cyaneus*), collected at Cañón de la Presa Boca, Santiago. More recently there have been a number of publications on the distribution and natural history of the herpetofauna of Nuevo León, for example, *Elgaria parva* (= *Gerrhonotus parvus*): range extension (Banda-Leal et al. 2002) and *G. parvus*: maximum size (Banda-Leal et al. 2005) and natural history account (Bryson and Lazcano-Villarreal 2005); list of amphibians and reptiles of Nuevo León, reporting 113 species (Contreras-Arquieta and Lazcano-Villarreal 1995); distribution of the herpetofauna of Sierra Picachos with regard to vegetation types and altitude range (Contreras-Lozano et al. 2007); herpetofauna list of Cerro El Potosí, Galeana (Contreras-Lozano et al. 2010); herpetofauna list of San Antonio Peña Nevada, Zaragoza (Lazcano-Villarreal et al. 2004); herpetofauna list of Sierra Cerro de la Silla and

TABLE 7.1 Type localities for amphibians and reptiles described from the state of Nuevo León, Mexico

Author(s)	Original name: Type locality
Salamanders (2)	
Rabb (1956)	*Chiropterotriton priscus*: **at an elevation of 8000 feet on Cerro Potosí, near Ojo de Agua, about 11 mi west-northwest of Galeana**
Taylor (1941)	*Pseudoeurycea galeanae*: **15 mi west of Galeana**
Anurans (2)	
Taylor (1940a)	*Syrrhophus smithi* (= *Eleutherodactylus guttilatus*): **15 mi west of Galeana**
Taylor (1940b)	*Syrrhophus latodactylus* (= *Eleutherodactylus longipes*): **Huasteca Cañón, about 15 km west of Monterrey, about 680 m elevation**
Lizards (12)	
Knight & Scudday (1985)	*Elgaria parva* (= *Gerrhonotus parvus*): **3 km SE Galeana**
Liner & Dixon (1994)	*Sceloporus chaneyi*: **11.1 mi (17.0 km) SW Zaragoza, at Rancho La Encantada, 9300 feet (2835 meters)**
Baird (1859 [1858])	*Sceloporus couchi*: **Santa Catarina**
Cope (1885)	*Sceloporus torquatus cyanogenys* (= *S. cyanogenys*): **Monterrey**
Treviño-Saldaña (1988)	*Sceloporus jarrovi cyaneus* (= *S. minor cyaneus*): **Cañón de la Presa Boca, Santiago**
Smith (1934)	*Sceloporus parvus*: **Hills about 5 mi west of Sabinas Hidalgo**
Smith & Hall (1974)	*Sceloporus samcolemani*: **Between Providencia and La Paz**
Dunn (1936)	*Sceloporus binocularis* (= *S. torquatus binocularis*): **Trail between Pablillo and Alamar**
Smith (1951)	*Leiolopisma caudaequinae* (= *Scincella silvicola caudaequinae*): **Horsetail Falls, 25 mi S Monterrey**
Gadow (1906)	*Cnemidophorus gularis meeki* (= *Aspidoscelis gularis*): **Montemorelos (lectotype)**
Baird (1859 [1858])	*Cnemidophorus inornatus* (= *Aspidoscelis inornata*): **Pesquería Grande (= Villa García)**
Baird (1859 [1858])	*Cnemidophorus octolineatus* (= *Aspidoscelis inornata*): **Pesquería Grande (= Villa García)**
Snakes (10)	
Günther (1893)	*Coronella leonis* (= *Lampropeltis mexicana leonis*): **Nuevo León**
Kennicott (1860)	*Nerodia couchii* (= *N. erythrogaster transversa*): **San Diego y Santa Catarina, restricted to Santa Catarina**[1]
Duméril et al. (1854)	*Pituophis mexicanus* (= *P. catenifer sayi*): **Mexico, restricted to Sabinas Hidalgo**[1]
Smith (1944)	*Rhadinaea montana*: **Ojo de Agua, Galeana**
Günther (1895)	*Homalocranium* (= *Tantilla*) *atriceps*: **Nuevo León**
Smith (1942)	*Tantilla wilcoxi rubricata*: **15 mi southeast of Galeana, corrected to west of Galeana**[2]
Kennicott (1860)	*Eutaenia* (= *Thamnophis*) *cyrtopsis*: **Rinconada, Coahuila, corrected to Rinconada, Nuevo León**[3]
Zertuche & Treviño (1978)	*Crotalus lepidus castaneus* (= *C. lepidus*): **Paraje las Huertas, municipality of Monterrey**
Gloyd (1940)	*Crotalus triseriatus miquihuana* (= *C. pricei miquihuanus*): **Cerro El Potosí, near Galeana**
Harris & Simmons (1978)	*Crotalus durissus neoleonensis* (= *C. totonacus*): **Las Adjuntas, Santiago**

Note: Modified from Liner 1996a.

[1] Restriction appeared in Smith & Taylor (1950).

[2] Correction appeared in Cochran (1961).

[3] Correction appeared in Conant (1968).

distribution with regard to vegetation type and altitude range (Lazcano-Villarreal et al. 2007); herpetofauna list of a *Juniperus* forest in San Juan y Puentes, Aramberri (Lazcano-Villarreal, Contreras-Lozano, et al. 2009); report of the snakes found either alive or dead on Nuevo León roads (Lazcano-Villarreal, Salinas-Camarena, and Contreras-Lozano 2009); natural history notes and captive reproduction in *T. exsul* (Lazcano-Villarreal et al. 2011); *Sceloporus cyanostictus* and *S. merriami australis* first records for the state (Price et al. 2010; Price and Lazcano-Villarreal 2010); and *Lampropeltis mexicana* and *L. alterna* range extension (Salmon et al. 2001; Salmon et al. 2004).

Other distributional and natural history notes on the herpetofauna of Nuevo León concern *Diadophis punctatus regalis*: distributional record (Cole 1965); *Lampropeltis mexicana thayeri*: natural history account (Mattison 1998); and *L. m. thayeri*: coloration (Osborne 1983).

Since 2000 Robert W. Bryson Jr. has been working on the ecology, systematics, and natural history of the Mexican herpetofauna. He has made a number of important contributions to these topics, such as work on the reptiles of Nuevo León, and he has encouraged people from Nuevo León to do research on the ecology and systematics of reptiles. His other contributions to the knowledge of the herpetofauna of Nuevo León include the phylogeny of *G. parvus* (Conroy et al. 2005) and the phylogeny of *L. mexicana*, on which he reported samples of *L. alterna* from Cerro de la Silla, and the *L. mexicana* complex from near Iturbide and north of Doctor Arroyo (Bryson 2005; Bryson et al. 2007).

The herpetological research in the state of Nuevo León has documented a total of 132 species: 23 amphibians (3 salamanders, 20 anurans) and 109 reptiles (6 turtles, 42 lizards, 61 snakes). Two species have been introduced to the state: the Mediterranean House Gecko (*Hemidactylus turcicus*) and the Brahminy Blindsnake (*Indotyphlops braminus*). Only two species are endemic to Nuevo León, the Pygmy Alligator Lizard (*G. parvus*) and the Nuevo León Graceful Brown Snake (*Rhadinaea montana*).

Physiographic Characteristics and Their Influence on the Distribution of the Herpetofauna

The state of Nuevo León is located in the northeastern part of Mexico, between longitude s98°26′ and 101°14′ W, and latitudes 23°11′ and 27°49′ N. It is bordered to the north by the US state of Texas and the Mexican states of Coahuila and Tamaulipas; to the west it borders the states of Coahuila, San Luis Potosí, and Zacatecas; to the south it is adjacent to San Luis Potosí and Tamaulipas, with which it shares its entire eastern border. The border between Texas and Nuevo León stretches 15 km. The state covers 64,220 km², and elevations range from 50 to more than 3710 m above sea level. It is divided into 51 municipal entities. Its capital, Monterrey, forms a large metropolitan area that extends over nine municipalities: Monterrey, San Pedro Garza García, Santa Catarina, García, Guadalupe, San Nicolás de los Garza, Apodaca, General Escobedo, and Juárez. It is located in the west-central part of the state, hosts approximately 88% of the Nuevo León human population, and is the third most populous city in Mexico with more than four million inhabitants.

The topography of the state is complex and varied, but three large regions can be easily differentiated. The first covers the center, east, north, and northwest parts of the state, with an average altitude ranging from 50 to 250 m above sea level. This is a flat region with the exception of a series of small, low, scattered hills. The eastern and northern portions of this region are in the physiographic province of the Great Plains of North America, which comprises the municipalities from Anáhuac south to China. This is a very large region, covering 23,138 km² (36% of the state's surface). The topography in this vast region is fairly homogeneous and represented by a great succession of rolling hills and plains that on rare occasions are interrupted by a low mountain range or valley. The central portion of this flat region is part of the Coastal Plain of the Gulf, which comprises the municipalities from Pesquería south to Linares. This is a flat region consisting of gentle, rolling hills and plains of considerable length, except the Sierra de las Mitras in northwestern Monterrey. The soils of this region are deep and dark, unlike those in the Great Plains of North America, which are predominantly light and shallow.

The second region easily identifiable by its topography is the Sierra Madre Oriental, which in Nuevo León is a strip located mainly in the western portion of the state, occupying the towns from Lampazos de Naranjo south to General Zaragoza. In Nuevo León, the Sierra Madre Oriental crosses the state from southeast to northwest and separates the low-rise plains of the central, eastern, and northwestern parts of the state from a high and flat region in the southwestern corner (Mexican Plateau). The Sierra Madre Oriental presents very rugged terrain in the form of mountain ranges that reach an average of 2000 m above sea level; in general the elevation of the mountains decreases from south to north, although the highest peaks are located in the west-central part of the state. The highest elevations are Cerro El Morro in Sierra La Marta (3710 m), municipality of Galeana, in the west-central part of the state near the Coahuila state line; Cerro El Potosí (3700 m), municipality of Galeana, 80 km south of Monterrey; Cerro Peña Nevada (3660 m), municipality of General Zaragoza, southeastern end of the state near the Tamaulipas state line; Sierra Potrero de Abrego (3460 m), municipality of Santiago, in the west-central part of the state near the Coahuila state line; and Cerro de la Ascension (3200 m), municipality of Aramberri, in the southern part of the state. Although most of the Sierra Madre Oriental of Nuevo León is a continuous mountain chain, its northernmost portion, at the city of Monterrey, is interrupted by extensive valleys that separate small mountain ranges such as the Sierras Las Gomas and Lampazos (1540 m) in the northwestern part of the state and Sierra Picachos (1520 m) in the north-central part of the state. These ranges to the north of Monterrey form an archipelago of mountain islands in the middle of arid and semiarid valleys. This archipelago of mountains does not exceed 1550 m above sea level.

The third region is the Mexican Plateau of Nuevo León, located in the southwestern corner of the state next to Coahuila, Zacatecas, and San Luis Potosí. It covers the western end of the municipalities of Galeana and Aramberri, a small portion of the southwestern part of the municipality of General Zaragoza, and the entire municipalities of Doctor Arroyo and Mier y Noriega. It has few mountain elevations and is located at an altitude that varies between 1500 and 2000 m. There are some hills and small, irregular mountain ranges, which rise between 200 and 500 m above the level of the plateau, but in general it is a flat and arid-semiarid area that presents an uneven distribution of moisture. Precipitation is slightly higher at its northern end at the edge of the Sierra Madre Oriental than at the very low southern part, where the mountains form a shadow that results in low rainfall (Alanís-Flores et al. 1996; Velazco-Macías et al. 2008; Velazco-Macías 2009; INEGI 2010).

These two distinctive topographical regions of western-southwestern Nuevo León, the Sierra Madre Oriental and the Mexican Plateau, are part of the physiographic province of the Sierra Madre Oriental.

Nuevo León's current topography began its formation approximately 145 million years ago during the last period of the Mesozoic era, the Lower Cretaceous (145–98 million years ago). During this period, in shallow waters, the relatively horizontal layers were altered by enormous geological forces that caused great breaks and foldings as well as lift movements that raised the seabed over the surface level of the water, favoring the compaction of calcareous sediments, phenomena that gave rise to very thick limestone, rich in fossils. These formations are common in the mountains surrounding Monterrey. Later, during the Upper Cretaceous (98–66 million years ago), the sea became deeper and clays were deposited, which resulted in shale rocks of soft, brittle, fragile, and erodible nature, very common in the foothills of the Sierra Madre Oriental. Toward the end of the Cretaceous (66 million years ago), movements were generated that corrugated layers of rocks that today make up the Sierra Madre Oriental. This orogenic process extended toward the beginning of the Cenozoic era, in the Paleogene period (65–34 million years ago), characterized by great volcanic activity resulting in intrusive igneous rocks that are located in only a few areas of the state. Nuevo León emerged as we know it today at the end of the Neogene period of the Cenozoic, about 5.3 million years ago. Much of the low-altitude areas are covered by the conglomerates of the Neogene and alluvium rocks of the Quaternary. Thus, much of the surface rock of the state is of sedimentary origin (Alanís-Flores et al. 1996).

The greater part of Nuevo León (68.6%) presents extreme, hot, dry climates of semiarid type. Throughout the year the climate is very hot, especially in the extensive lowland plains. Precipitation is low and generally does not exceed 500 mm annually in most of the state. Rains occur in the summer or are scarce throughout the year. The moisture that the state receives comes from the Gulf of Mexico, and the average relative humidity is 62%. However, in the northwestern corner of the state, in the municipalities of Mina and García neighboring extreme eastern Coahuila, located in the driest region of the state, the climate is very arid. These municipalities have a flat topography. Due to the proximity of these municipalities to the Chihuahuan Desert, extremely high temperatures of up to 47°C can be observed; the rainfall in most parts is less than 300 mm per year and may be less than 200 mm. The Mexican Plateau in the southwestern part of the state is also hot and very dry; the average annual rainfall can be less than 200 mm. This type of climate occurs in 5% of the state.

Another important variation occurs in the temperate-subhumid climates of the southern and western parts of the state in the Sierra Madre Oriental. The climate of this region contrasts sharply with that of the extensive plains of the rest of the state. In the Sierra Madre Oriental, the average temperature is between 18°C and 20°C, and the average annual precipitation varies from 600 to 900 mm. At altitudes greater than 3000 m, there are alpine and subalpine climates, which are limited to Peña Nevada and

Cerro El Potosí. Frosts, hailstorms, and snow are limited to locations in the Sierra Madre Oriental. Temperate climates occur in 7.31% of the state.

The greatest rainfall rates occur in the Sierra Madre Oriental. This mountainous region plays an important role in the regulation of meteorological processes and hydrological dynamics, since it is located at the head of the basins, giving rise to several rivers (Alanís-Flores et al. 1996).

Finally, the Coastal Plain of the Gulf in the central part of the state presents tropical, semidry, and subhumid climates, with summer rains. This type of climate occurs in 19.1% of the state. This region, where Nuevo León's main rivers are located, receives an amount of rainfall intermediate between that of the Sierra Madre Oriental and the Mexican Plateau (Alanís-Flores et al. 1996).

The state of Nuevo León is crossed by several rivers, including a number of temporary streams that carry water only during the rainy season. The flow of the rivers is generally erratic and unpredictable, caused by changing climatic factors and the abrupt topography. Most of the rivers have small catchment basins, and not all of them have water throughout the year.

All surface and seasonal flows arise from runoff from the upper parts of the Sierra Madre Oriental in the western part of the state and follow routes, many of them winding, through the canyons and valleys of the mountains heading toward the lower, flat parts of the state. Most of their tributaries empty into the Río Bravo, San Fernando, and Soto La Marina, all in the state of Tamaulipas. Many of these seasonal streams dry quickly or are lost to the porous soils, so their water flow lasts only a few hours.

The main rivers that run through Nuevo León are the Ríos Bravo, Salado, Sabinas, San Juan, and Pesquería. The Río Bravo runs only through 15 km of the Nuevo León border, in the municipality of Anáhuac. The Río Salado extends its basin in the northern part of the state, rising in the Sierra de Santa Rosa in extreme northwestern Coahuila, entering Nuevo León through the northwestern corner of the municipality of

Lampazos. It then heads east toward the towns of Anáhuac and Vallecillos, where it enters the state of Tamaulipas and finally drains into the Río Bravo at the Falcón Dam of northwestern Tamaulipas. This river is fed by a number of tributaries through its passage across northern Nuevo León, most of them small seasonal streams such as the Camarón and the Galameses, both in the municipality of Anáhuac. Shortly before flowing into the Falcón Dam, it receives the waters of its most important tributary, the Río Sabinas, which arises in eastern Coahuila and enters Nuevo León through the western edge of the municipality of Mina. It crosses the municipalities of Bustamante, Villaldama, Sabinas Hidalgo, Vallecillo, and Parás and finally enters the state of Tamaulipas in its northwestern corner, where it joins the Río Salado.

The longest river, and perhaps the most important in the state, is the Río San Juan, which originates at the eastern edge of Coahuila and flows through Nuevo León in the municipalities of Santa Catarina, Monterrey, China, General Bravo, Doctor Coss, and Los Aldama, penetrating into northern Tamaulipas, where it feeds the Marte Dam and finally empties into the Río Bravo. In eastern Nuevo León, it feeds the Cuchillo Dam in the municipality of China, which provides water to the growing city of Monterrey. Some of the main tributaries of the Río San Juan are the Ríos Pilón and Pesquería.

In the southern part of the state most of the rivers are intermittent; all of them are directed toward the east, entering Tamaulipas. The most important are the Río Blanco in the municipality of Aramberri, which flows into Tamaulipas to eventually download its waters at the Vicente Guerrero Dam; and Río El Potosí, which passes through the municipalities of Galeana and Linares and reaches the state line with Tamaulipas, where it becomes intermittent and is now called Río Conchos, to finally drain into the Río San Fernando (Tamaulipas) which drains into the Gulf of Mexico.

The physiography, topography, humidity, temperature, and soil, among other features, condition the associated types of vegetation, which in turn create landscapes and environments that determine which organisms are present. The dependence of amphibians on bodies of water or humid environs limits their distribution in time and space to the rainy season, riparian habitats, and other places that retain humidity. Nevertheless, important populations of this group of organisms are successful in desert habitats. Equally, reptiles are limited by their dependence on external temperatures to regulate their body temperature. Unlike amphibians, their shelled eggs give them independence from bodies of water for reproduction, which along with other adaptations makes them so successful in northern Mexico.

Nuevo León hosts six types of vegetation (Chihuahuan Desert Scrub; Tamaulipan Thornscrub; Submontane Scrub; Montane Forest; Grasslands; and Riparian, Subaquatic, and Aquatic Vegetation) and 11 plant communities, which correspond to three floristic provinces: the Mexican Plateau, the Coastal Plain of the Northeast, and the Sierra Madre Oriental (Rzedowski 1978).

Chihuahuan Desert Scrub

Chihuahuan Desert Scrub is characteristic of the xeric habitats of the state that receive 300 to 400 mm of average annual precipitation at altitudes higher than 1400 m; it comprises two plant communities (Rosetophilous Desert Scrubland and Microphilous Desert Scrubland). Its uneven distribution is mainly due to the pattern of available soil moisture. Soils of sandy to silt-sandy and clayey texture that have the ability to retain moisture sustain dense and closed communities, whereas stony soils that retain little water produce poor and very open plant communities.

Rosetophilous Desert Scrubland occurs on mountainsides and slopes of various elevations, where there are rocky outcrops or shallow soils, usually stony and with good drainage. It is mainly located in the southwestern corner of the state, in the region of the Mexican Plateau in the municipalities of Aramberri, Doctor Arroyo, Galeana, Mier y Noriega, and Zaragoza; an equally important presence occurs in the northwestern corner of the state in the municipalities of Abasolo, Bustamante, Carmen, Escobedo, García, Lampazos, Marin, Mina, Salinas Victoria, and Villaldama; and lesser isolated patches in the central portion of the state in the municipalities of Cerralvo, Doctor González, and Higueras. The most conspicuous plants have succulent leaves arranged in rosettes, some with terminal spines or mucrones.

The most common plant species in this community are *Agave bracteosa, A. lechuguilla, A. striata, Berberis trifoliolata, Dasylirion texanum, Hechtia glomerata, Euphorbia antisyphilitica, Echinocereus enneacanthus, Echinocactus platyacanthus, Ferocactus pilosus, Opuntia leptocaulis, O. microdasys,* and *Opuntia* spp.

Microphilous Desert Scrubland is characterized by the dominance of shrub species having leaves or small leaflets. There are abundant cacti with spherical or flat stems, as well as plants that are spherical or flat themselves. Izote yucca (*Yucca filifera*) and Samandoca palm (*Y. carnerosana*) are abundant in flat terrain or alluvial fans of hills or mountains in the Mexican Plateau of the southwestern portion of the state, in the municipalities of Doctor Arroyo, Galeana, Mina y Mier, and Noriega; in the northwestern corner of the state in the municipalities of Bustamante, García, and Santa Catarina; and in the northern part of the state in the municipality of Anáhuac.

The most conspicuous plant species in this community are *Acacia berlandieri, A. rigidula, Castela texana, Celtis pallida, Chilopsis linearis, Echinocactus platyacanthus, Ephedra aspera, Flourensia cernua, Fouquieria splendens, Larrea tridentata, Mortonia greggii, Opuntia imbricata, O. leptocaulis, O. rastrera, O. microdasys, Partheniurn argentatum, P. incanum, Prosopis glandulosa, Sporobolus tiroides, Viguiera stenoloba, Yucca filifera,* and *Y. carnerosana.*

Fifty-nine species of amphibians and reptiles inhabit the Chihuahuan Desert Scrub of Nuevo León. Many are typical of the Chihuahuan Desert but have managed to extend their distributions eastward to the Coastal Plain of the Gulf where the Tamaulipan Thornscrub develops, passing through the plains and inter-

mountain valleys in the northwestern corner of Nuevo León. Others are limited to the arid-semiarid portions of northwestern Nuevo León in the municipalities of Lampazos and Mina or to the southwestern corner of the state in the municipalities of Doctor Arroyo, Galeana, Mier y Noriega, and Zaragoza. And a few have a very wide distribution and can be found throughout the entire state. Only 4 of these 59 species are limited to Chihuahuan Desert Scrub in Nuevo León: *Anaxyrus cognatus, Sceloporus cyanostictus, S. merriami,* and *Bogertophis subocularis.* The other 55 species recorded in this and other types of vegetation in the state are 8 amphibians: *Pseudoeurycea galeanae, Anaxyrus debilis, A. punctatus, A. speciosus, Eleutherodactylus longipes, Gastrophryne olivacea, Scaphiopus couchii,* and *Spea multiplicata;* 21 lizards: *Crotaphytus collaris, Coleonyx brevis, Cophosaurus texanus, Holbrookia approximans, H. lacerata, Phrynosoma cornutum, P. modestum, Sceloporus cautus, S. consobrinus, S. couchii, S. cyanogenys, S. minor, S. ornatus, S. parvus, S. poinsettii, S. spinosus, Uta stansburiana, Plestiodon tetragrammus, Aspidoscelis marmorata, A. gularis,* and *A. inornata;* and 26 snakes: *Arizona elegans, Gyalopion canum, Heterodon kennerlyi, Hypsiglena jani, Lampropeltis alterna, L. getula, L. triangulum, Masticophis flagellum, M. taeniatus, Pantherophis bairdi, P. emoryi, Pituophis catenifer, Rhinocheilus lecontei, Salvadora grahamiae, Sonora semiannulata, Tantilla atriceps, T. hobartsmithi, T. nigriceps, T. wilcoxi, Thamnophis cyrtopsis, T. marcianus, Trimorphodon tau, Micrurus tener, Crotalus atrox, C. scutulatus,* and *C. molossus.*

Tamaulipan Thornscrub

The Tamaulipan Thornscrub is characteristic of the Coastal Plain of the Northeast. The thorny scrub vegetation and mesquite lands of this region show physiognomic variants; the species can be tall and thorny or medium-sized and thornless. The altitude varies from 240 to 500 m, with an average annual rainfall of 700–800 mm. In favorable soil and moisture conditions, the stems have well-defined shafts and can reach more than 6 m in height. The most conspicuous for their abundance and coverage are *Prosopis laevigata, Pithecellobium ebano, Acacia rigidula, Castela texana, Celtis pallida, Parkinsonia macrum, Randia laetevirens, Cordia boissieri, Leucophyllum frutescens, Porlieria angustifolia, Opuntia leptocaulis, O. engelmannii, Zanthoxylum fagara,* and *Bumelia celastrina. Yucca filifera* reaches up to 10 m in height; in the herbaceous stratum layer one can find *Lupinus texensis* (Alanís-Flores et al. 1996; Velazco-Macías 2009).

This vegetation type is located mainly in the north, northeast, east, southeast, and central parts of the state and in the flat areas that separate the mountain massif of the Sierra Madre Oriental from the mountain ranges of La Ropa, Bustamante, and Picachos of the northwestern and central parts of the state. These latter ranges include the metropolitan area of Monterrey and all of the small towns surrounding it.

Of the 63 species of amphibians and reptiles inhabiting the Tamaulipan Thornscrub of Nuevo León, only 3 are limited to this type of vegetation: the Burrowing Toad (*Rhinophrynus dorsalis*), Berlandier Texas Tortoise (*Gopherus berlandieri*), and Reticulated Collared Lizard (*Crotaphytus reticulatus*). The Mediterranean House Gecko (*Hemidactylus turcicus*) has been recorded only in human structures surrounded by this type of vegetation in Nuevo León; however, it is an introduced species and has the ability to colonize virtually anywhere in hot, dry climates, mainly in areas with human construction. Another 13 species inhabit both Tamaulipan Thornscrub and Chihuahuan Desert Scrub (*Anaxyrus speciosus, Crotaphytus collaris, Coleonyx brevis, Cophosaurus texanus, Phrynosoma modestum, Sceloporus consobrinus, S. cyanogenys, Aspidoscelis marmorata, Heterodon kennerlyi, Lampropeltis alterna, Pantherophis bairdi, P. emoryi,* and *Tantilla atriceps*). Three more have been recorded in the Montane Forest as well as in the Tamaulipan Thornscrub (*Rhinella marina, Leptodeira septentrionalis,* and *Rena dulcis*). *Ficimia streckeri* occurs in the Submontane Scrub and the Tamaulipan Thornscrub, and *Tantilla nigriceps,* in the Tamaulipan Thornscrub and Grasslands. The other 41

species occupy various vegetation types within the Tamaulipan Thornscrub. They include 6 anurans (*Anaxyrus debilis, A. punctatus, Incilius nebulifer, Gastrophryne olivacea, Scaphiopus couchii,* and *Spea multiplicata*); 14 lizards (*Holbrookia approximans, H. lacerata, Phrynosoma cornutum, Sceloporus couchii, S. grammicus, S. olivaceus, S. poinsettii, S. serrifer, S. spinosus, S. variabilis, Uta stansburiana, Plestiodon tetragrammus, Aspidoscelis gularis,* and *A. inornata*); and 21 snakes (*Arizona elegans, Gyalopion canum, Hypsiglena jani, Masticophis flagellum, M. schotti, M. taeniatus, Pantherophis emoryi, Pituophis catenifer, Rhinocheilus lecontei, Salvadora grahamiae, Senticolis triaspis, Sonora semiannulata, Tantilla hobartsmithi, Thamnophis eques, T. marcianus, T. proximus, Micrurus tener, Rena myopica, Agkistrodon taylori, Crotalus atrox,* and *C. scutulatus*).

Submontane Scrub

The Submontane Scrub is a vegetation type of shrubs and subarboreal plants rich in life-forms; the size and distribution of the dominant and co-dominant species depend on the availability of water in the soil and subsoil. Depending on the orientation of the slopes occupied and soil conditions, this vegetation occurs in a wide altitudinal range from 500 m to over 1200 m in sites with an average annual rainfall of 700 to 800 mm. It is located on the lower slopes of the mountains and forms a wide belt that separates the Chihuahuan Desert Scrub and the Tamaulipan Thornscrub from the Montane Forests of the Sierra Madre Oriental, including those of the archipelago of mountains of the northern and northwestern parts of the state (Sierras Las Gomas, Lampazos, and Picachos).

Although the region has morphological and ecological variants, in general the most representative plant species are *Helietta parvifolia, Cordia boissieri, Sophora secundiflora, Gochnatia hypoleuca, Neopringlea integrifolia, Decatropis bicolor, Fraxinus greggii, Pithecellobium pallens, Leucophyllum frutescens, Acacia rigidula, A. berlandieri, A. farnesiana, Caesalpinia mexicana, Prosopis glandulosa, Dyospiros virginiana, D. texana,* and

Cercidium macrum. In some areas with protected habitats with abundant moisture and deep soils, there are small clusters of oak (*Quercus virginiana*) (Alanís-Flores et al. 1996; Velazco-Macías 2009).

There are 60 species of amphibians and reptiles found in this vegetation type: 9 amphibians, including 1 salamander (*Pseudoeurycea galeanae*) and 8 anurans (*Anaxyrus punctatus, Incilius nebulifer, Craugastor augusti, Eleutherodactylus cystignathoides, E. guttilatus, E. longipes, Hypopachus variolosus,* and *Spea multiplicata*); 19 lizards (*Barisia ciliaris, Gerrhonotus infernalis, G. parvus, Sceloporus cautus, S. couchii, S. grammicus, S. minor, S. oberon, S. olivaceus, S. ornatus, S. parvus, S. poinsettii, S. serrifer, S. torquatus, S. variabilis, Plestiodon dicei, P. tetragrammus, Aspidoscelis gularis,* and *Lepidophyma sylvaticum*); and 32 snakes (*Adelphicos newmanorum, Amastridium sapperi, Coluber constrictor, Diadophis punctatus, Drymarchon melanurus, Ficimia streckeri, Gyalopion canum, Lampropeltis getula, L. mexicana, L. triangulum, Masticophis flagellum, M. schotti, M. taeniatus, Oxybelis aeneus, Pituophis catenifer, P. deppei, Rhadinaea montana, Senticolis triaspis, Storeria dekayi, S. hidalgoensis, Tantilla rubra, T. wilcoxi, Thamnophis cyrtopsis, T. eques, Trimorphodon tau, Micrurus tener, Rena myopica, Agkistrodon taylori, Crotalus lepidus, C. molossus, C. scutulatus,* and *C. totonacus*). None of these species is limited to this vegetation type; they are also present in the Montane Forest, Chihuahuan Desert Scrub, and Tamaulipan Thornscrub.

Montane Forest

The Montane Forest comprises the arboreal vegetation that occurs above the Submontane Scrub in the mountainous areas of the Sierra Madre Oriental and Mexican Plateau provinces; it extends to the canyons and higher elevations of the mountains, where there is greater humidity and the climate is temperate-semidry to temperate-subhumid and includes the following communities: oak forest, pine-oak forest, pine forest, and coniferous forest. In addition, there are small patches of cloud forest that have a relictual distribution and that, unfortunately, are disappearing for various reasons.

Oak forest is part of the temperate forest community in the Sierra Madre Oriental. It occurs at altitudes above the Submontane Scrub, from areas with thin, rocky soils to areas with deep, well-drained soils. Physiognomically it is composed of trees and shrubs that are 12–15 m tall, dominated by species of *Quercus*; the holm oaks are linked ecologically and floristically to mixed forests and pine forests. Characteristic oak species are *Q. canbyi, Q. intricada, Q. laeta, Q. laceyi, Q. polymorpha, Q. pungens, Q. rysophylla,* and *Q. vaseyana*.

Pine-oak forest is composed of broadleaf and needle forests in temperate to subhumid areas; it is distributed over the oak forest in the mountains of the Sierra Madre Oriental. Despite the wide range of climatic conditions in the region, the presence of this type of forest is more related to temperature than moisture. This forest occurs between 550 and 900 m. The dominant tree species are oaks; the most common species are *Q. rysophylla, Q. laeta, Q. polymorpha,* and *Q. canbyi*. Coexisting with the oaks are the pine species *Pinus teocote* and *P. pseudostrobus*. In general, the trees in this forest do not have large diameters (10–30 cm) or heights (10–14 m).

Pine forest may occur in temperate and humid locations, with temperature decreasing with altitude. This is a community with trees up to 22 m in height. It is located above the pine-oak forest of the Sierra Madre Oriental; it is commonly associated with a few species of oaks, *Quercus* spp. and madrone (*Arbutus xalapensis*). The characteristic pine species are *P. teocote, P. pseudostrobus, P. arizonica, P. ayacahuite,* and *P. hartwegii*. There are variants in these forests such as Mexican pinyon (*P. cembroides*), located in the south-southwest of the state; in the municipalities of San Pedro Garza García, Santiago, Montemorelos, Rayones, and Galeana; and also in very restricted areas in the northern part of Nuevo León in the Sierra de Lampazos. They thrive in areas of low rainfall between 2000 and 2600 m in elevation on shallow, rocky soils. This type of forest presents open spaces of short trees

(4–8 m tall), with rounded crowns and trunks with diameters of 30–40 cm at chest height. The main tree species in the forest of Mexican pinyon of Galeana are pines (*P. cembroides, P. greggii*), junipers (*Juniperus flaccida, J. deppeana*), and madrone (*A. xalapensis*). The same community is also located in neighboring areas of the Chihuahuan Desert Scrub of the Mexican Plateau and is often associated with species such as the samandoca palm (*Yucca carnerosana*), izote yucca (*Y. filifera*), agaves (*Agave* spp.), and various shrubs.

From 3000 m to the summit of Cerro El Potosí in Galeana, a special type of vegetation is found in the form of scrub, branched from the base of the stem. The area, bordering alpine meadows and forming a continuous strip in the eastern and southern slopes, occurs in two isolated patches on the southwestern and western portions of the summit. It is a dense and short community where the dwarf pine (*P. culminicola*), less than 2 m tall, is the dominant species.

Coniferous forests (*Pseudotsuga-Abies*) are a variant of the temperate forests, characterized by tree species of typically pyramidal shape that offer great scenic attraction. This type of vegetation consists of tall trees (15–25 m) located in protected canyons with cold and humid climates, generally between 2000 and 2500 m in elevation, in the southern and southwestern parts of the state (municipalities of Galeana and Iturbide, Aramberri, and Zaragoza). The most common species in the tree canopy are *Pinus greggii* and *P. hartwegii*, along with *Pseudotsuga menziesii, P. flahaulti, Abies vejarii,* and *Cupressus arizonica*.

There are small areas of cedar forest (*Cupressus* spp.) found in restricted areas, forming more or less pure patches. These areas are also found in the southern and southwestern parts of the state (municipalities of Iturbide, Galeana, and Zaragoza). The trees are not burly and reach between 10 and 15 m in height. The characteristic species is *C. arizonica*.

Another community is the *Juniperus* forest associated with pine-oak forests. Its life-form is a tree or shrub, depending on environmental conditions. These tree species thrive in rocky soils of exposed limestone and low humidity, and their growth is slow. The most common species include *J. monosperma* and *J. flaccida*. This type of forest is found from the northwestern (municipalities of Santa Catarina, San Pedro Garza García, and Monterrey) to the southwestern parts of the state (municipalities of Galeana and Iturbide) in the Sierra Madre Oriental.

Cloud forest is located in areas with shallow or clayey soils with abundant organic matter between 800 and 1400 m in elevation, with high relative humidity year-round. It is located in the municipality of Zaragoza in the region bordering the state of Tamaulipas. Floristically it is characterized by sweetgum (*Liquidambar styraciflua*), oaks (*Q. sartorii* and *Q. germana*), red pine (*Pinus patula*), magnolia (*Magnolia schiedeana*), and eastern redbud (*Cercis canadiensis*). This is a mixed forest with abundant vines and epiphytes, as well as ferns and fungi in the understory (Alanís-Flores et al. 1996; Velazco-Macías 2009).

Of the 68 species of amphibians and reptiles that inhabit the Montane Forest of Nuevo León, only 15 appear to be limited to this type of vegetation and are typical of the Sierra Madre Oriental. One of the plethodontid salamanders, *Chiropterotriton priscus*, is limited to this type of vegetation. The other salamander of this family (*Pseudoeurycea galeanae*) and two eleutherodactylids (*Eleutherodactylus cystignathoides* and *E. longipes*) have a strong affinity for this type of vegetation but can also be found in semiarid conditions, as reported by Taylor (1940) in the description of *Syrrhophus smithi* (= *E. longipes*) from 24 km west of Galeana.

These four amphibian species are able to reproduce outside bodies of water by direct transformation. They deposit their eggs in moist places, either on soil or on the leaves of trees, and the development of the larvae takes place inside the egg; when they hatch, the individuals are fully metamorphosed. Another species of amphibian, the Small-eared Treefrog (*Ecnomiohyla miotympanum*), is limited to the Montane Forest in Nuevo León. The other 10 species limited to this vegetation type are 3 lizards (*Phrynosoma orbiculare, Sceloporus chaneyi,* and *Scincella silvicola*) and 7 snakes (*Drymobius margaritiferus, Leptophis mexicanus, Opheodrys aestivus, Thamnophis exsul, T. pulchrilatus, Tropidodipsas sartorii,* and *Crotalus pricei*). These snakes, especially those of the genus *Thamnophis*, are associated with bodies of water within the Montane Forest.

Another 18 species that inhabit the Montane Forest of Nuevo León can also be found in Submontane Scrub. Included are three amphibians: the Barking Frog (*Craugastor augusti*), the Spotted Chirping Frog (*Eleutherodactylus guttilatus*), and the Sheep Frog (*Hypopachus variolosus*); the first two species also reproduce by direct transformation. Three lizards (*Sceloporus torquatus, Plestiodon dicei,* and *Lepidophyma sylvaticum*) and 12 snakes (*Adelphicos newmanorum, Amastridium sapperi, Diadophis punctatus, Drymarchon melanurus, Lampropeltis mexicana, Pituophis deppei, Rhadinaea montana, Storeria dekayi, S. hidalgoensis, Tantilla rubra, Crotalus lepidus,* and *C. totonacus*) also inhabit this region.

Three more species, all of them lizards of the Anguidae family, occur in the Montane Forest, Submontane Scrub, and Grasslands: *Barisia ciliaris, Gerrhonotus infernalis,* and *G. parvus. Gerrhonotus parvus* is one of the two endemic species to Nuevo León; it inhabits the forests of the Sierra Madre Oriental in the municipality of Galeana in the extreme southwestern part of the state.

The remaining 32 species that inhabit the Montane Forest can be found in other vegetation types: 5 anurans (*Anaxyrus punctatus, Incilius nebulifer, Rhinella marina, Smilisca baudinii,* and *Gastrophryne olivacea*); 9 lizards (*Sceloporus grammicus, S. minor, S. oberon, S. olivaceus, S. parvus, S. poinsettii, S. spinosus, S. variabilis,* and *Plestiodon tetragrammus*); and 18 snakes (*Gyalopion canum, Lampropeltis getula, L. triangulum, Leptodeira septentrionalis, Oxybelis aeneus, Pituophis catenifer, Salvadora grahamiae, Senticolis triaspis, Tantilla wilcoxi, Thamnophis cyrtopsis, T. eques, T. proximus, Trimorphodon tau, Micrurus*

tener, Rena myopica, Agkistrodon taylori, Crotalus molossus, and *C. scutulatus*).

Grasslands

The Grasslands community is characterized by the dominance of grasses with thin, elongated leaves, although they can be mixed with some other species of the families Asteraceae, Fabaceae, and Chenopodiaceae. Natural climax grasslands occupy small areas in open spaces within desert scrub regions and in specific edaphic situations in places that have poor drainage, are subject to floods, or have excessive amounts of salts or the presence of gypsum (Alanís-Flores et al. 1996; Velazco-Macías 2009).

There are 36 species of amphibians and reptiles in the Grasslands of Nuevo León. Those with the greatest affinity for this vegetation type include 4 amphibians (*Gastrophryne olivacea, Scaphiopus couchii, Spea bombifrons,* and *S. multiplicata*); 11 lizards (*Barisia ciliaris, Gerrhonotus infernalis, G. parvus, Holbrookia approximans, H. lacerata, Sceloporus goldmani, S. samcolemani, Plestiodon obsoletus, Aspidoscelis inornata, A. gularis,* and *A. marmorata*); and 1 snake (*Sistrurus catenatus*). The other 20 species include 2 amphibians (*Anaxyrus debilis* and *Leptodactylus fragilis*); 3 lizards (*Phrynosoma cornutum, Uta stansburiana,* and *Plestiodon tetragrammus*); and 15 snakes (*Arizona elegans, Hypsiglena jani, Masticophis flagellum, Pituophis catenifer, Rhinocheilus lecontei, Salvadora grahamiae, Sonora semiannulata, Tantilla hobartsmithi, T. nigriceps, T. wilcoxi, Thamnophis marcianus, T. proximus, Trimorphodon tau, Crotalus atrox,* and *C. scutulatus*).

Riparian, Subaquatic, and Aquatic Vegetation

This community is closely linked to the orographic formation of the Sierra Madre Oriental as a great basin catchment with its tributaries of rivers and streams. It includes the tree and shrub vegetation located on the banks of streams in the municipalities of Linares, Montemorelos, Allende, General Terán, Cadereyta Jiménez, Santiago, Monterrey, Guadalupe, Sabinas, Cerralvo, Lampazos, and Bustamante. These forests consist of *Adiantum capillus-veneris, Arundo donax, Lobelia cardinales, Platanus occidentales, Populus tremuloides, Salix nigra, Taxodium mucronatum,* and *Ulmus crassifolia,* mixed with abundant species of aquatic and semiaquatic herbaceous plants, vines, and epiphytic species, such as *Tillandsia usneoides.* These forests grow primarily along the shores of the rivers and on the flat and wide river surface drainages. These sites, wet by permanent or sporadic runoff for long periods of time, allow species to acquire dimensions in height and diameter much larger than those located in dry areas (Alanís-Flores et al. 1996; Velazco-Macías 2009).

There are 34 species of amphibians and reptiles strongly associated with the Riparian, Subaquatic, and Aquatic Vegetation of Nuevo León. One of the 17 amphibians, the Barred Tiger Salamander (*Ambystoma mavortium*), has been recorded only in the ponds of the southwestern corner of the state. This is the only strictly aquatic species in Nuevo León, although metamorphosed adults can move between bodies of water that are considerable distances apart; however, only neotenic adults that retain larval characteristics such as the presence of gills and that are limited to bodies of water have been observed in the state. The other 16 are anurans: *Anaxyrus cognatus, A. debilis, A. punctatus, A. speciosus, Incilius nebulifer, Rhinella marina, Ecnomiohyla miotympanum, Smilisca baudinii, Leptodactylus fragilis, Gastrophryne olivacea, Hypopachus variolosus, Lithobates berlandieri, Rhinophrynus dorsalis, Scaphiopus couchii, Spea bombifrons,* and *S. multiplicata.* Although all of these species have a high affinity for bodies of water, all can occur in places where the humidity is high, and often they can be seen crossing roads or moving between bodies of water.

Another 5 species that inhabit this vegetation type are turtles: *Pseudemys gorzugi, Trachemys scripta, Kinosternon flavescens, K. integrum,* and *Apalone spinifera.* Only the Spiny Softshell (*A. spinifera*) is strictly aquatic and limited to this vegetation type. Four others are lizards, 3 of them of arboreal and scansorial habits, living in the trees beside rivers or dams (*Sceloporus couchii, S. cyanogenys,* and *S. serrifer*), whereas *Scincella silvicola* lives under logs and rocks near bodies of water. The 8 remaining species are colubrid snakes that rarely move away from bodies of water: *Nerodia erythrogaster, N. rhombifer, Thamnophis cyrtopsis, T. eques, T. exsul, T. marcianus, T. proximus,* and *T. pulchrilatus.*

As mentioned in the introduction, it is known that 132 species of herpetozoans inhabit Nuevo León: 23 amphibians (3 salamanders and 20 frogs) and 109 reptiles (6 turtles, 42 lizards, and 61 snakes). These are included in 28 families: 11 amphibians (2 salamanders and 9 frogs) and 17 reptiles (4 turtles, 8 lizards, and 5 snakes). The 132 species also represent 73 genera: 17 amphibians (3 salamanders and 14 frogs) and 56 reptiles (5 turtles, 14 lizards, and 37 snakes).

The proportions of the different groups of herpetozoa in Nuevo León are a reflection of the environments available in the state. Amphibians are strongly limited to wet sites and aquatic habitats, which are scarce in most parts of the state; they are found mainly in the temperate forests of the Sierra Madre Oriental and mountains to the north and northeast of the city of Monterrey. The greatest richness of amphibians also occurs in these regions, and in most cases their activity throughout the state is limited to the wet season. Even species of amphibians that need no bodies of water for reproduction, such as the two species of plethodontid salamanders (*Chiropterotriton priscus* and *Pseudoeurycea galeanae*), and the frogs of the families Craugastoridae (*Craugastor augusti*) and Eleutherodactylidae (*Eleutherodactylus cystignathoides, E. guttilatus,* and *E. longipes*), depend on moist places for their survival. Most of the other amphibian species that live in the rest of the state are limited in their activity to the rainy season. They start their activity with the first rains of spring–summer, a period in which they emerge from their winter refuges to feed and reproduce. At the end of the rainy season they can remain active during a short period of time to later return to their winter havens. In the arid-semiarid regions of the state they bury themselves until reaching a depth where humid conditions are suitable for survival and come to the surface only occasionally when some light or torrential rain moistens the soil surface. Some species of amphibians can be found in bodies of standing water virtually year-round. An example of this is the Barred Tiger Salamander (*Ambystoma mavortium*), which can be active year-round in reservoirs in the southwestern part of state. These strong environmental constraints result in the low proportions of amphibians (salamanders, 2% [3/132], and anurans, 15% [20/132]) in relation to the total number of species of herpetozoa from the state.

However, the reptiles with their amniote eggs, leathery skin, and different modes of reproduction (oviparous and viviparous) are much more diverse in the state and occupy virtually all available environments. The six species of turtles represent only 5% (6/132) of the total state herpetofauna and 19% (6/32) of the species of turtles reported by Liner and Casas-Andreu (2008) for Mexico. Two of these are river turtles (*Pseudemys gorzugi* and *Trachemys scripta*), two more live in cattle ponds and other types of standing water (*Kinosternon flavescens* and *K. hirtipes*), another is a Desert Tortoise (*Gopherus berlandieri*), and another occupies permanent water bodies (*Apalone spinifera*).

The 42 lizard species that inhabit Nuevo León represent 32% of the total number of herpetozoan species of the state. In addition to being a group with high species richness in the state, the abundance of many of their populations is also high. The scansorial, saxicolous, and heliothermic habits of the majority of the lizard species make them very conspicuous, which contributes to the observation of species and populations of this group of reptiles. One of the two species endemic to the state, the Pygmy Alligator Lizard (*Gerrhonotus parvus*), as well as one of the two introduced species to Nuevo León, the Mediterranean House Gecko (*Hemidactylus turcicus*), belongs to this group.

The group of herpetozoan species best represented in Nuevo León is the snakes, with 61

species representing 46% of the total number of species for the state. The habits of snakes are generally secretive and/or nocturnal, so they are not as conspicuous as lizards. Consequently, there is a common perception that populations of snakes are not as abundant as those of lizards. However, sampling aimed at proper habitats, microhabitats, and hours of the day or night may reveal equally abundant populations of many snakes.

Of the 132 species that occur in Nuevo León, 92 (70%) are shared with the United States; 19 of these shared species are found in Mexico, mainly in the lowlands and Atlantic slopes of the Sierra Madre Oriental; 14 of them enter the United States just in Texas: *Anaxyrus speciosus*, *Eleutherodactylus cystignathoides*, *E. guttilatus*, *Pseudemys gorzugi*, *Gopherus berlandieri*, *Gerrhonotus infernalis*, *Crotaphytus reticulatus*, *Holbrookia lacerata*, *Sceloporus cyanogenys*, *S. olivaceus*, *Ficimia streckeri*, *Pantherophis bairdi*, *Storeria dekayi*, and *Rena dulcis*. Five more have an eastern distribution in Mexico and the United States: *Incilius nebulifer*, *Nerodia erythrogaster*, *N. rhombifer*, *Opheodrys aestivus*, and *Micrurus tener*. Another of the shared species, *Sceloporus grammicus*, also enters the United States just in Texas but is distributed all along the Sierra Madre Oriental and Transvolcanic Belt of central Mexico. Another 51 of the shared species are typical of the North American deserts, and most of these are widely distributed in the southern United States and northern Mexico: *Ambystoma mavortium*, *Anaxyrus cognatus*, *A. debilis*, *A. punctatus*, *Gastrophryne olivacea*, *Scaphiopus couchii*, *Spea bombifrons*, *S. multiplicata*, *Trachemys scripta*, *Kinosternon flavescens*, *Apalone spinifera*, *Crotaphytus collaris*, *Coleonyx brevis*, *Cophosaurus texanus*, *Phrynosoma cornutum*, *P. modestum*, *Sceloporus consobrinus*, *S. merriami*, *S. poinsettii*, *Uta stansburiana*, *Plestiodon obsoletus*, *P. tetragrammus*, *Aspidoscelis gularis*, *A. inornata*, *A. marmorata*, *Arizona elegans*, *Bogertophis subocularis*, *Gyalopion canum*, *Heterodon kennerlyi*, *Hypsiglena jani*, *Lampropeltis alterna*, *L. getula*, *Masticophis flagellum*, *M. schotti*, *M. taeniatus*, *Pantherophis emoryi*, *Pituophis catenifer*, *Rhinocheilus lecontei*, *Salvadora grahamiae*, *Sonora semiannulata*,

Tantilla atriceps, T. hobartsmithi, T. nigriceps, T. wilcoxi, Thamnophis cyrtopsis, T. eques, Crotalus atrox, C. lepidus, C. molossus, C. scutulatus, and *Sistrurus catenatus.*

Three of the shared species are widely distributed in the temperate forests of North America. Two of them (*Craugastor augusti* and *Crotalus pricei*) extend from the southern United States southward through the Sierra Madre Occidental and Sierra Madre Oriental. The other (*Diadophis punctatus*) ranges from southern Canada to central Mexico.

Of the remaining 18 shared species, 16 have a broad distribution from the central or southern United States (*Rhinella marina, Smilisca baudinii, Leptodactylus fragilis, Hypopachus variolosus, Lithobates berlandieri, Rhinophrynus dorsalis, Sceloporus variabilis, Drymarchon melanurus, Drymobius margaritiferus, Leptodeira septentrionalis, Oxybelis aeneus, Senticolis triaspis, Thamnophis marcianus,* and *T. proximus*) or southern Canada (*Coluber constrictor* and *Lampropeltis triangulum*) to Central or even South America. The last two of the shared species are introduced, the Mediterranean Gecko (*Hemidactylus turcicus*) native to southern Europe and the Brahminy Blindsnake (*Indotyphlops braminus*) native to India.

Of the 132 species that occur in Nuevo León, 40 (30%) are not shared between Mexico and the United States; 35 of these species are endemic to Mexico, and 2 of them, *Gerrhonotus parvus* and *Rhadinaea montana,* are endemic to the Montane Forest of Nuevo León. Eighteen more are limited to the highlands of the Sierra Madre Oriental (*Chiropterotriton priscus, Pseudoeurycea galeanae, Eleutherodactylus longipes, Ecnomiohyla miotympanum, Sceloporus chaneyi, S. minor, S. oberon, S. parvus, Plestiodon dicei, Scincella silvicola, Lepidophyma sylvaticum, Adelphicos newmanorum, Pituophis deppei, Storeria hidalgoensis, Thamnophis exsul, Rena myopica, Agkistrodon taylori,* and *Crotalus totonacus*). Of these species, 4 have a narrow distribution in southeastern-eastern Coahuila and adjacent Nuevo León (*Chiropterotriton priscus, Pseudoeurycea galeanae, Sceloporus oberon,* and *Thamnophis exsul*); 1 is limited to Nuevo León and adjacent Tamauli-

pas (*S. chaneyi*); 1 is limited to Coahuila, Nuevo León, and Tamaulipas (*Plestiodon dicei*); 11 range from Nuevo León and Tamaulipas southward to southern Veracruz and northern Oaxaca, mainly on the Atlantic slopes of the Sierra Madre Oriental (*Eleutherodactylus longipes, Ecnomiohyla miotympanum, S. minor, S. parvus, Scincella silvicola, L. sylvaticum, Adelphicos newmanorum, Storeria hidalgoensis, Rena myopica, A. taylori,* and *Crotalus totonacus*); another, *Pituophis deppei,* occurs in the Sierra Madre Oriental, the Mexican Plateau, the Transvolcanic Belt, and the Sierra Madre Occidental.

Two of the remaining 15 species endemic to Mexico are limited to Coahuila and extreme western Nuevo León (*Sceloporus cyanosticus* and *S. ornatus*). Two are limited to scattered regions of northern Mexico: *S. couchi,* in the northern Sierras of Coahuila and central-western Nuevo León; and *S. goldmani,* in a small area in southeastern Coahuila, adjacent Nuevo León, and northeastern San Luis Potosí. Three species are limited to the Mexican Plateau (*S. cautus, S. samcolemani,* and *Lampropeltis mexicana*). Four species (*Kinosternon integrum, S. spinosus, S. torquatus,* and *Trimorphodon tau*) are widely distributed from central Mexico through the Mexican Plateau and, in some cases (*T. tau*), on both coasts. One species has disjunct populations in the highlands of Mexico (*Thamnophis pulchrilatus*), and another is limited to the Chihuahuan Desert of Mexico (*Holbrookia approximans*). The other 2 species (*Barisia ciliaris* and *Phrynosoma orbiculare*) occur on the Sierra Madre Occidental and the Sierra Madre Oriental; *P. orbiculare* ranges even into the mountains of the Transvolcanic Belt of the central part of the country.

Five of the species that occur in Nuevo León and that are not shared between Mexico and the United States (*Sceloporus serrifer, Amastridium sapperi, Leptophis mexicanus, Tantilla rubra,* and *Tropidodipsas sartorii*) are species of tropical-subtropical affinities whose ranges extend from Central America northward through the mountains and lowlands of eastern Mexico. All of them have their northernmost distribution in Nuevo León or Tamaulipas.

Herpetofauna of Tamaulipas

WILLIAM FARR

Introduction

A complete and comprehensive review of the herpetology of Tamaulipas has never been published. In the 254 years since Linnaeus (1758), only three lists of herpetofauna from Tamaulipas have appeared: Velasco (1892); Smith and Taylor (1966); and Lavin-Murcio, Hinojosa-Falcón, et al. (2005). Excluding a number of brief notes, there are only a small handful of articles in which Tamaulipas and its herpetofauna are the primary subjects: Gaige (1937); Martin (1955a, 1958); Baker and Webb (1966); Brown and Brown (1967); Greene (1970); Smith et al. (1976, 2003); Lavin-Murcio, Hinojosa-Falcón, et al. (2005); Lavin-Murcio, Sampablo Angel, et al. (2005); Farr et al. (2007); Farr et al. (2009); Farr et al. (2013); and Lazcano et al. (2009). The majority of what is known about the amphibians and reptiles of Tamaulipas comes from publications addressing the taxonomy and ecology of species of which portions of their distributions incidentally range within the geographic-political boundaries of the state, but Tamaulipas is not the subject of these studies. Currently 179 species of amphibians and reptiles can be confirmed to occur in Tamaulipas. This number will undoubtedly grow with future fieldwork, for species known to occur near the borders of neighboring states are anticipated in Tamaulipas (Smith and Dixon 1987), and occasional descriptions of new species from the state continue to appear (Lavin-Murcio and Dixon 2004; Bryson and Graham 2010). Situated on the threshold of the Tropic of Cancer, Tamaulipas has a rich and complex range of habitats, from barrier islands and mangrove swamps on the Gulf Coast, to pine-oak forest at over 3200 m in the Sierra Madre Oriental, from Chihuahuan Desert scrub of the Mexican Plateau, to cloud forest in the Reserva de la Biósfera El Cielo, and much remains to be discovered about the herpetofauna of the state.

Previous Herpetological Studies

The following account chronicles Linnaean and post-Linnaean fieldwork and literature pertaining to the herpetology of Tamaulipas. Pre-Columbian and the early Spanish colonists' knowledge of the herpetofauna are not reviewed. A bias is evident here, in that the majority of the activities of collectors are based largely on data from American museums. No attempt was made to acquire data from European collections, and efforts to acquire data from Mexican collections were mostly unsuccessful.

Carl Linnaeus (1758, 1766) named 18 (10.0%) of the currently valid species now known to occur in Tamaulipas. These were mostly wide-ranging species, and only *Phrynosoma orbiculare* was based on a type from Mexico. Following Linnaeus, the period from 1761 to 1840 saw the descriptions of an additional 29 new Tamaulipas species by 17 European and American authors, but like Linnaeus, none of the type material originated from Tamaulipas (see the following list for species, authors, and dates). The earliest investigations into the herpetofauna of Tamaulipas were conducted by Jean Louis Berlandier, a French naturalist who traveled to Mexico in 1826 to participate in a Mexican Boundary Survey from 1828 to 1829 (Geiser 1948; Berlandier 1980). At the conclusion of the survey Berlandier made his home in Matamoros, Tamaulipas, where he continued to collect, research, and write prolifically. Among his numerous manuscripts is an unfinished, handwritten, 219-page review of the herpetofauna of the region (Smith et al. 2003; Smith and Chiszar 2003; Chiszar et al. 2003). Smith et al. concluded that Berlandier would have been the author of 17 (9.49%) of the state's currently valid taxa had his manuscripts been published at the time of his death. Tragically, however, he drowned in the Río San Fernando in 1851, having published only a few botanical papers. After the Mexican-American War (1846–1848), the US government organized the United States and Mexican Boundary Survey (1848–1855), during which Arthur C. V. Schott collected *Sceloporus* northwest of Camargo in 1852 (Baird 1859b). Concurrently, but independently of the Boundary Survey, Lt. Darius Nash Couch collected natural history specimens in Mexico. He arrived in January 1853 and spent one month collecting in the vicinity of Brownsville, Texas, and Matamoros, Tamaulipas. While in Matamoros, Couch located the widow of Berlandier and bought the collections and manuscripts of the late naturalist. Couch then traveled westward, collecting in northern Tamaulipas in February and March and then on to Coahuila (Conant 1968). Among the Couch and Berlandier specimens from Tamaulipas are syntypes of *Anaxyrus* (= *Bufo*) *debilis* (Girard 1854), *Gopherus berlandieri* (Agassiz 1857), and *Leptodeira septentrionalis* (Kennicott 1859); the lectotype and

paralectotypes of *Plestiodon tetragrammus* (Baird [1858] 1859a); the holotype of *Lampropeltis triangulum annulata* (Kennicott 1860); and paratypes of *Siren intermedia texana* (Goin 1957).

Half of the currently valid species now known to occur in Tamaulipas (91 species, 50.83% of the state's total) were described between 1850 and 1885. Unlike previous examples, a number of these new taxa were based on specimens from Tamaulipas. Spencer Fullerton Baird, his French colleague Charles F. Girard, and Robert Kennicott at the National Museum of Natural History (USNM) collectively described 36 species (20.1%) in 10 years (1852–1861). About this same time at the Academy of Natural Sciences, Philadelphia (ANSP), Edward Hallowell described 4 new species now known from the state. Hallowell's successor at the ANSP, Edward Drinker Cope, described 24 species (13.4%) between 1861 and 1885, including *Notophthalmus meridionalis* (Cope 1880), based in part on specimens from Tamaulipas. Samuel Garman at the Museum of Comparative Zoology (MCZ) described 5 species, including *Rena myopica* (Garman 1883) from Tamaulipas. At the Museum National d'Histoire Naturelle (MNHN) in Paris, Auguste Henri André Duméril; his father, André Marie Constant Duméril; and Gabriel Bibron collectively described 9 species during 1841–1854, based on material from southern Mexico and Guatemala. Ten other zoologists in Europe, the United States, and Mexico described an additional 16 new species now known to inhabit Tamaulipas in the period between 1849 and 1885. Velasco (1892) published the first list of Tamaulipan herpetofauna, which included 82 species. Several species that appeared in his list were in error (e.g., *Basiliscus vittatus*, *Phyllodactylus tuberculosus*, *Loxocemus bicolor*); in spite of its shortcomings it was a significantly more complete inventory of the state's diversity than other reviews of Mexican herpetofauna of the era (e.g., Duméril et al. 1870–1909; Garman 1883; Günther 1885–1902; Cope 1887; Dugès 1896).

After 1885 and the advances made in the mid-nineteenth century, there was relatively little activity until 1932, almost a half century later. Only five of the currently valid species now known to occur in the state were described in this period. The last quarter of the nineteenth century produced a few small field collections made in the vicinity of Tampico by Edward Palmer in 1879, J. M. Priour in 1888, and J. O. Snyder in 1899. William Allison Lloyd of the US Bureau of Biological Survey (USNM) collected mostly along the US-Mexican border in 1891. Edward Nelson and Edward Goldman (USNM) surveyed for vertebrates in every state in Mexico (1892–1906) and collected in Tamaulipas in 1898 and 1901–1902 (Kellogg 1932; Goldman 1951). They were the first to collect in the Sierra Madre Oriental in Tamaulipas and collected paratypes of *Masticophis schotti ruthveni* (Ortenburger 1923) and *Crotalus pricei miquihuanus* (Gloyd 1940). Opening the first three decades of the twentieth century, ichthyologist Seth E. Meek (FMNH) made small collections in 1903. W. W. Brown (MCZ) collected in the Sierra Madre Oriental on three annual trips to Tamaulipas (1922–1924), including the holotypes of *Lampropeltis mexicana thayeri* (Loveridge 1924), *Craugastor batrachylus* (Taylor 1940), and *Phrynosoma orbiculare orientale* (Horowitz 1955). In 1930 the University of Michigan organized an expedition to the Sierra San Carlos (Gaige 1937), where Lee Raymond Dice collected nearly 250 specimens, including a new skink, *Plestiodon dicei* (Ruthven and Gaige 1933). A Dr. Kallert of Hamburg, Germany, collected in the vicinity of Tampico in 1930, including types of two subspecies, *Notophthalmus meridionalis kallerti* (Wolterstorff 1930) and *Micrurus tener maculates* (Rose 1967).

The fourth decade of the twentieth century saw the beginning of a new era in Mexican herpetology when Edward H. Taylor began over 20 years of investigations (1932–1953). He published extensively on Mexico, including descriptions of eight new species now known to occur in Tamaulipas. Taylor conducted fieldwork in Tamaulipas from 1932 to 1941, including an extended trip to Mexico with his student Hobart Smith in 1932, and he collected the holotype of *Agkistrodon taylori* (Burger and Robinson 1951) and paratypes of *Ameiva undulata podarga* (Smith

and Laufe 1946) in the state. Hobart M. Smith received his PhD under Taylor in 1936 and collected frequently in Tamaulipas from 1932 to 1939, at times with David Dunkle. He described six new species and two subspecies now known from the state. The *Herpetology of Mexico* (Smith and Taylor 1945, 1948, 1950a, 1966) established a modern foundation for Mexican herpetology. Hobart Smith continued to publish prodigiously on Mexico well beyond the end of the twentieth century, and additional contributions to Tamaulipas are cited later. Other than those by Smith and Taylor, only two new species were described in the period 1930–1949, both from specimens collected south of Tamaulipas. Miscellaneous notes on field collections were reported by Burt (1932), Dunkle (1935), and Dunkle and Smith (1937); and two subspecies, *Nerodia rhombifera blanchardi* (Clay 1938) and *Hypsiglena torquata dunklei* (Taylor 1938), were described from Tamaulipas. Archie F. Carr Jr. made field trips each December from 1939 to 1941, and other incidental field collections from 1930 to 1945 were made by Walter Mosauer in the summer of 1935, Ernest G. Marsh in 1938, and Myron and Evelyn Gordon in 1939.

With the impetus and a solid foundation laid down by Smith and Taylor, a new generation of herpetologists began to explore Mexico in unprecedented numbers after the Second World War. Charles F. Walker at the University of Michigan Museum of Zoology (UMMZ) and his students were particularly active in Tamaulipas from 1945 to 1960, making the approximately 3600 specimens in the UMMZ the largest Tamaulipas collection. Perhaps the single most significant contribution to herpetology in Tamaulipas was the work of Paul S. Martin, who spent well over 365 days in the field (1948–1953) and recorded 39 species (21.78% of the state's total) in the state for the first time (Martin 1955a). His primary study area was the Sierra de Guatemala (= El Cielo), harboring the northernmost cloud forest in the Western Hemisphere, for which he provided a biogeography of the herpetofauna (Martin 1958). Additional publications addressed broader issues of biogeography and Pleistocene history, but elements of the Tamaulipas herpetofauna were discussed in each (Martin et al. 1954; Martin 1955b; Martin and Harrell 1957). Later, Martin (2005) recounted his experiences collecting in Tamaulipas. Charles Walker collected in the Sierra de Guatemala in 1950 and 1953 and described three new species from the state: the endemic *Thamnophis mendax* (Walker 1955a), *Pseudoeurycea scandens* (Walker 1955b), and *Lepidophyma micropholis* (Walker 1955c). George B. Rabb (1958) collected there in 1951 and 1952 and described *Chiropterotriton cracens*. James A. Peters (1948) reported on *Laemanctus serratus* from the state and collected in Tamaulipas in 1949. Other collectors from the UMMZ active from 1945 to 1950 included Rezneat M. Darnell, William Heed, James Mosimann, Priscilla Starrett, and Hellmuth Wagner.

Several herpetologists from Texas were active in Tamaulipas in the decades after the war. Bryce C. Brown collected in Mexico, including Tamaulipas, most years from 1941 to 1978 (Brown and Smith 1942; Brown and Brown 1967; Auth, Smith, Brown, and Lintz 2000). William B. Davis made incidental collections of herpetofauna from Tamaulipas (Davis 1953a, 1953b). James R. Dixon, often with the aid of colleagues and students, collected in Tamaulipas at irregular intervals from 1949 to 2006, making the Texas Cooperative Wildlife Collection (TCWC) the third-largest Tamaulipas collection (ca. 1750 specimens), and he published many systematic reviews of taxa occurring in the region (e.g., Dixon 1959, 1962, 1969; Dixon and Thomas 1974; Dixon and Dean 1986; Dixon et al. 2000; Dixon and Vaughan 2003). The TCWC conducted field surveys of the Sierra San Carlos from 1975 to 1977 (Hendricks and Landry 1976), and Jack W. Sites Jr. surveyed the Sierra de Tamaulipas in 1979 (Sites and Dixon 1981, 1982). John E. Werler made incidental collections while traveling through Tamaulipas (Werler and Darling 1950; Werler 1951; Werler and Smith 1952; Smith et al. 1952), and Robert G. Webb reported on collections made in northern Mexico (Webb 1960) and Tamaulipas (Baker and Webb 1966). John S. Mecham collected salamanders on field

trips conducted from 1965 to 1969 (Mecham 1968a, 1968b; Mecham and Mitchell 1983). Allan Chaney at Texas A&I University in Kingsville collected in Tamaulipas between 1971 and 1981, often in the company of Ernest Liner.

Many other herpetologists were active in Mexico after World War II (1945–1975) and made contributions to our knowledge of Tamaulipas. The University of Illinois Mexican Expedition was conducted by Phillip Smith, Leslie Burger, Robert Reese, Frederick Shannon, and Lester Firschein in 1949 (Shannon and Smith 1950; Reese and Firschein 1950; Smith and Burger 1950). Burger and Robertson (1951) described *Agkistrodon taylori* from specimens previously collected in Tamaulipas. Ralph Axtell collected in Tamaulipas as early as 1949 (Axtell and Wasserman 1953; Axtell 1958b) and described two new subspecies now known to occur in the state (Axtell 1960; Axtell and Webb 1995). Roger Conant collected in the state between 1949 and 1965 (Conant 1965, 1969). Ernest Liner worked in Tamaulipas in most years from 1952 to 1963, made sporadic collections as late as 1988 (Liner 1964, 1966a, 1966b), and described two taxa now known to occur in the state (Liner 1960; Liner and Dixon 1992). Douglas Rossman conducted fieldwork in Tamaulipas in 1960 with M. J. Fouquette Jr. (Rossman and Fouquette 1963) and in 1965 with Edmond Keiser, Earl Olson, and others and reported on the genus *Thamnophis* (Rossman 1969, 1992a, 1992b). Craig Nelson and Max Nickerson collected in Mexico in the summers of 1963–1964 and reported new distributional records from Tamaulipas (Nelson and Nickerson 1966). John Lynch, with M. J. Landy, collected mostly anurans in 1964, publishing notes (Lynch 1963, 1965) and the original descriptions of two taxa (Lynch 1967, 1970). Fred Thompson collected the type series of *Xenosaurus platyceps* in Tamaulipas in 1965, later described in a review of the genus (King and Thompson 1968), and made additional collections in 1990 and 1992. Morris Jackson reported a northern range extension for *Tropidodipsas fasciata* (Jackson 1971). Sherman and Madge Minton collected in Tamaulipas eight of the years between 1960 and

1991. Selander et al. (1962) reported on a survey of vertebrates on the barrier islands of Tamaulipas, including notes on three species of herpetofauna. Greene (1970) reviewed the modes of reproduction of the Squamata occurring in the Sierra de Guatemala, based largely on data from Martin (1958).

The primary nesting beaches of *Lepidochelys kempii* were discovered in Tamaulipas in 1947 near the town of Rancho Nuevo. Humberto Chávez, Martín Contreras, and Eduardo Hernández found the species to be in serious decline by 1966 (Chávez et al. 1967, 1968), and recovery efforts were initiated, which ultimately involved the Mexican and US governments and numerous institutions and individuals. Casas-Andreu (1974) reported on the nesting habitat of *L. kempii*, and Plotkin (2007) reviewed the biology and conservation. Additional publications related to *L. kempii* and other sea turtles in Tamaulipas are too extensive to address here. Smith and Smith (1979) and Wilson and Zug (1991) reviewed much of the literature.

In the final quarter of the twentieth century several herpetologists published reports pertaining to the distribution, ecology, and taxonomy of the herpetofauna after conducting fieldwork in Tamaulipas. Smith and Álvarez (1976) described *Sceloporus torquatus mikeprestoni*. John Iverson, Peter Meylan, Ron Magill, and Diderot F. Gicca collected from 1974 to 1983, and Iverson and Berry (1979) reported on the distributional ecology of *Kinosternon*. Jonathan Campbell collected in Tamaulipas at irregular intervals between 1970 and 2009 in the company of various students and associates and published extensively on the broader subject of Middle American herpetofauna. Patrick M. Burchfield reported on the ecology of *Agkistrodon taylori* and the conservation of *L. kempii* (Burchfield 1982, 2004). Robert L. Bezy (1984) collected mostly *Lepidophyma* from 1967 to 1984, accompanied by Charles J. Cole and Carl S. Lieb on various occasions. Smaller collections were made in the late 1970s by Diderot Gicca in 1978 and James F. Lynch and Stephen D. Busack in 1979. Stephen Reilly collected with Ronald Brandon and Gail John-

ston near San José de las Rusias and reviewed the genus *Notophthalmus* (Reilly 1990). In 1986 J. F. Copp, D. E. Breedlove, R. R. Riviere, and B. R. Anderson collected in the Sierra Madre Oriental and reported *Sceloporus chaneyi* from the state (Watkins-Colwell et al. 1996). Earl Olson reviewed several *Sceloporus* taxa (Olson 1987, 1990) and described *S. torquatus madrensis* (Olson 1986). John J. Wiens collected *S. minor* and reviewed systematics of the genus *Sceloporus*, of which 13 species occur in Tamaulipas (Wiens and Reeder 1997; Wiens and Penkrot 2002). *Aspidoscelis laredoensis* was reported from Tamaulipas and was the subject of numerous publications (Walker 1986, 1987a, 1987b; Walker et al. 1986; Walker et al. 1990; Paulissen and Walker 1998). Flores-Villela and Brandon (1992) reviewed the status of the genus *Siren* in Tamaulipas and South Texas. New distributional records were reported for *Trimorphodon tau* (Dundee and Liner 1997) and *Gastrophryne elegans* (Sampablo-Brito and Dixon 1998). Smith et al. (1998) described *Leptotyphlops myopicus iversoni* (= *Rena myopica iversoni*), and Auth, Smith, Brown, Lintz, and Chiszar (2000) provided further observations on the taxa. Seidel et al. (1999) reviewed the distribution and taxonomy of *Trachemys* in Tamaulipas and reported on the distribution of *Kinosternon flavescens* (Seidel 1976). John E. Joy Jr. made some incidental collections from 1975 to 1979.

In recent years the natural history of *Xenosaurus platyceps* was the subject of a series of articles (Lemos-Espinal et al. 1997, 2003; Ballinger et al. 2000; Lemos-Espinal and Rojas-González 2000; Rojas-Gonzalez et al. 2008). Jiménez-Ramos et al. (1999) collected *Aspidoscelis sexlineata* in the state, and Pérez-Ramos et al. (2010) reviewed its status in Mexico. Terán-Juárez (2008) reported *Ophisaurus incomptus* from the Sierra Madre Oriental; however, it was not a new state record as indicated. Bailey et al. (2008) reported on the conservation status of *Pseudemys gorzugi*. Flores-Villela and Pérez-Mendoza (2006) reviewed literature on Tamaulipas, and Flores-Benabib and Flores-Villela (2008) reported *Epictia goudotii* from the state. Pablo A. Lavin-Murcio received his PhD under

James Dixon (Lavin-Murcio 1998) and taught at the Instituto Tecnológico de Ciudad Victoria (ITCV) from 1999 to 2004 while building a large collection from the state (Lavin-Murcio and Dixon 2004; Lavin-Murcio, Hinojosa-Falcón, et al. 2005; Lavin-Murcio, Sampablo Angel, et al. 2005). David Lazcano Villarreal at the Universidad Autónoma de Nuevo León (UANL) made incidental collections (1977–2010), conducted statewide surveys (1996–1997), and acquired the ITCV collection in 2009, making the UANL the second-largest collection of Tamaulipas material, with over 2370 specimens from the state (Lazcano et al. 2009; Farr et al. 2013). William L. Farr conducted field surveys from 2003 to 2009 and reported new distributional records, including 14 new state records and other notes (Farr et al. 2007; Farr et al. 2009; Farr et al. 2010; Farr et al. 2011; Farr 2011; Farr and Lazcano 2011; Farr et al. 2013), and Bryson and Graham (2010) described *Gerrhonotus farri* from these collections. Eli García-Padilla reported notes on *Anelytropsis papillosus* (García-Padilla and Farr 2010) and *Pantherophis bairdi* (García-Padilla et al. 2011).

Through the history of Tamaulipas herpetology, 48 species and subspecies of amphibians (4 salamanders, 6 anurans) and reptiles (1 crocodile, 3 turtles, 16 lizards, 18 snakes) have had type localities designated or restricted within the state, including 37 currently valid taxa (table 8.1).

Physiographic Characteristics and Their Influence on the Distribution of the Herpetofauna

Continental drift, tectonic shifting, erosion, climatic changes, and glaciations are some of the forces that act to transform landscapes and environments through geological time. Species must adapt to these changes or face extinction. These changes are among the key factors that drive evolution. Plants and animals evolve together and are inextricably linked to one another and their landscapes to form ecosystems. Understanding these changes, adaptations, and ecosystem as a

whole provides deeper insights into any single species or region of interest.

Located in northeastern Mexico between latitudes 22.20° N and 27.67° N and longitudes 97.14° W and 100.11° W, Tamaulipas is the sixth-largest state in Mexico in area (ca. 80,175 km², 4.1%). It shares a 370 km international border with the state of Texas to the north, 420 km of coastline with the Gulf of Mexico to the east; its neighbors are Nuevo León to the west and San Luis Potosí and Veracruz to the south. In 2010 the population was 3,268,554 (88.0% urban, 12.0% rural) with nearly half the residents in the larger cities on the northern frontier: Reynosa, Matamoros, and Nuevo Laredo. Greater metropolitan Tampico, including Altamira and Ciudad Madero, had a population greater than 825,000. Principal land uses include extensive agriculture throughout the Coastal Plain (sorghum, citrus fruit, sugarcane, agave), ranching statewide, logging in the Sierra Madre Oriental, and urbanization and industrialization (shipping, petroleum, manufacturing) in the Tampico area and the northern frontier.

The following information on herpetofauna is based on literature, approximately 16,000 museum records (about 11,000 examined by author), and more than 2700 personal field records collected during 2001–2009. The geological history and characters of Tamaulipas follow Ferrusquía-Villafranca (1993) and others cited in the following discussion. The topography and hydrology were described using topographical maps by Instituto Nacional de Estadística Geografía Informática (INEGI), Google Earth, literature, and data collected with a handheld GPS unit. *Juniperus* and *Pinus* occurrences in Tamaulipas were derived from Adams (2011) and Perry (1991), respectively.

Geologically, Middle America is one of the more complex areas in the world, and many aspects of the geological history of the Sierra Madre Oriental in particular are poorly understood. The Precambrian and Paleozoic basement of the region was continental (probably cratonic) and later overrun by seas. Marine depo-

TABLE 8.1 Type localities for amphibians and reptiles described from the state of Tamaulipas, Mexico

Author(s)	Original name: Type locality
Salamanders (4)	
Rabb (1958)	*Chiropterotriton chondrostega cracens* (= *C. cracens*): **Agua Linda, about 7 milesWNW of Gómez Farías, Tamaulipas**
Walker (1955b)	*Pseudoeurycea scandens*: **from the wall of a cave at "Rancho del Ciello," on the forested slopes of the Sierra Madre Oriental in southern Tamaulipas, about five miles NW of Gómez Farías: elevation about 3500 feet**
Cope (1880)	*Diemictylus miniatus meridionalis* (*Notophthalmus meridionalis*): **Matamoros, Mexico . . . tributaries of the Medina River and southward (restricted to Matamoros, Tamaulipas[1,2])**
Wolterstorff (1930)	*Diemictylus kallerti* (*Notophthalmus meridionalis kallerti*): **Angeblich Tampico-Mexiko. Wahrscheinlich aus dem gebirgigen Hinterland**
Anurans (6)	
Girard (1854)	*Bufo* (= *Anaxyrus*) *debilis*: **lower part of the valley of the Río Bravo (Río Grande), and in the province of Tamaulipas**
Taylor (1940)	*Eleutherodactylus* (= *Craugastor*) *batrachylus*: **Miquihuana, Tamaulipas, eighty miles southwest of Victoria, Tamaulipas**
Lynch (1967)	*Eleutherodactylus* (= *Craugastor*) *decoratus purpurus*: **near Rancho del Cielo, 5 km NW Gomez Farías, Tamaulipas, Mexico**
Lynch (1970)	*Syrrhophus* (= *Eleutherodactylus*) *dennisi*: **cave near El Pachón, 8 km. N Antiguo Morelos, Tamaulipas, Mexico, 250 m**
Baird (1854)	*Scaphiopus couchii*: **Coahuila and Tamaulipas (restricted to Matamoros, Tamaulipas[3])**
Cope (1863)	*Scaphiopus rectifrenis* (*S. couchii*): **Tamaulipas and Coahuila (restricted to Rio, Nazas, Coahuila[3])**
Crocodiles (1)	
Bocourt (1869)	*Crocodylus mexicanus* (= *C. moreletti*): **Tampico, Tamaulipas**
Turtles (3)	
Stejneger (1925)	*Kinosternon herrerai*: **Xochimilco, Valley of Mexico [in error] (restricted to La Laja, Veracruz [in error][4]); (restricted to Tampico, Tamaulipas[5])**
Gray (1849)	*Ciatudo mexicana* (= *Terrapene carolina mexicana*): **Mexico (restricted to vicinity of Tampico, Tamaulipas[6])**
Günther (1885)	*Emys cataspila* (= *Trachemys venusta cataspila*): **Mexico (restricted to Alvarado, Veracruz [in error][3]); (restricted to Tampico, Tamaulipas[7])**
Lizards (16)	
Bryson & Graham (2010)	*Gerrhonotus farri*: **north of Magdaleno Cedillo, 27 km SW of Tula, Municipio Tula, Tamaulipas, Mexico**
Shaw (1802)	*Lacerta* (= *Ctenosaura*) *acanthura*: **Not given (restricted to Tampico, Tamaulipas, Mexico[8,9])**
Harlan (1825)	*Cyclura teres* (= *Ctenosaura acanthura*): **Tampico, Tamaulipas**
Baird (1859a [1858])	*Holbrookia approximans*: **Lower Río Grande [in error]; Tamaulipas[10] [in error], (restricted to near Álamo de Parras or Parras de la Fuente, 25°25′ N, 102°11′ W, Coahuila, Mexico,[11] *fide*[12])**
Horowitz (1955)	*Phrynosoma orbiculare orientale*: **Miquihuana, Tamaulipas, Mexico**
Sites & Dixon (1981)	*Sceloporus grammicus tamaulipensis*: **4.3 km by road S of Hacienda Acuña (45.3 km by road N of Gonzales) in the Sierra de Tamaulipas mountains in southern Tamaulipas, Mexico**
Martin (1952)	*Sceloporus serrifer cariniceps* (= *S. s. plioporus*): **5 mi NE of Gómez Farías at Rancho Pano Ayuctle along the Río Sabinas**
Olson (1986)	*Sceloporus torquatus madrensis*: **1740 m above Rancho del Cielo, 7 km. NW Gómez Farías, Tamaulipas, Mexico**

TABLE 8.1 *Continued*

Author(s)	Original name: Type locality
Smith & Álvarez (1976)	*Sceloporus torquatus mikeprestoni*: **Marcela, Tamaulipas, Mexico**
Ruthven & Gaige (1933)	*Eumeces dicei* (= *Plestiodon dicei*): **24°38′N–99°01′W, Marmolejo . . . Sierra San Carlos, Tamaulipas, Mexico**
Cope (1900)	*Eumeces tetragrammus funebrosus* (= *Plestiodon tetragrammus tetragrammus*): **Matamoros, Mexico**
Baird (1859a [1858])	*Plestiodon tetragrammus*: **Lower Río Grande (restricted to Matamoros, Tamaulipas, Mexico[13])**
Smith & Laufe (1946)	*Ameiva undulata podarga*: **Seven miles west of Ciudad Victoria, Tamaulipas**
Walker (1955c)	*Lepidophyma micropholis*: **cave at El Pachón, about 5 miles. NNE of Antigua Morelos, Tamaulipas, elevation 600–700 feet**
Walker (1955c)	*Lepidophyma flavimaculatum tenebrarum* (= *L. sylvaticum*): **± 4.5 miles NW (by road) of Gómez Farías, in the Sierra Madre Oriental at "Rancho del Cielo," ± 3600 feet**
King & Thompson (1968)	*Xenosaurus platyceps*: **Tamaulipas, 15.4 mi SSW Ciudad Victoria, on the road to Jaumave, 4500 feet elevation**
Snakes (18)	
Cope (1895)	*Zamenis conirostris* (*Coluber constrictor oaxaca*): **Matamoros, Mex.**
Kennicott (1859)	*Taeniophis* (= *Coniophanes*) *imperialis*: **Brownsville, Texas [USNM catalogue and tag, Matamoros, Tamaulipas, *fide*[14,15]]**
Jan (1863)	*Glaphyrophis lateralis* (= *Coniophanes imperialis* in part): **Not given (restricted to Tampico[16])**
Kennicott (1859)	*Dipsas* (= *Leptodeira*) *septentrionalis*: **Matamoros, Tamaulipas and Brownsville, Tex. (restricted to Brownsville, Texas[3])**
Taylor (1939 [1938])	*Hypsiglena torquata dunklei* (= *H. jani dunklei*): **Hacienda La Clementina, near Forlón, Tamaulipas**
Stejneger (1893)	*Hypsiglena ochrorhynchus texana* (= *H. jani texana*): **between Laredo & Camargo, Texas (restricted to Mier, Tamaulipas[3])**
Kennicott (1860)	*Lampropeltis annulata* (= *L. triangulum annulata*): **Matamoros, Tamaulipas, Mexico**
Loveridge (1924)	*Lampropeltis thayeri* (= *L. mexicana thayeri*): **Miquihuana, Tamaulipas, Mexico**
Clay (1938)	*Natrix* (= *Nerodia*) *rhombifera blanchardi*: **Tamaulipas, Mexico, within a radius of 85 miles of Tampico in the triangle formed by the Río Tamesí and the Río Pánuco**
Smith (1943)	*Pliocercus elapoides celatus* (= *P. bicolor bicolor*): **Ciudad Victoria, Tamaulipas [in error[17]]**
Walker (1955a)	*Thamnophis mendax*: **near La Joya de Salas, Tamaulipas, ± 6000 ft.**
Cope (1895)	*Zamenis conirostris* (= *Coluber constrictor oaxaca*): **Matamoros, Mex.**
Rose (1967)	*Micrurus fulvius maculatus* (= *M. tener maculatus*): **Tampico, Tamaulipas, Mexico**
Brown & Smith (1942)	*Micrurus fitzingeri microgalbineus* (= *M. tener microgalbineus*): **seven km south of Antiguo Morelos, Tamaulipas, Mexico**
Lavin-Murcio & Dixon (2004)	*Micrurus tamaulipensis*: **Sierra de Tamaulipas, Rancho La Sauceda, ca. 50 km N González, Tamaulipas, Mexico**
Garman (1883)	*Stenostoma myopicum* (= *Rena myopica*): **Savineto, near Tampico, Tamaulipas, Mexico**
Smith et al. (1998)	*Leptotyphlops dulcis iversoni* (= *Rena myopica iversoni*): **oak woodland in the vicinity of Hwy 101, 23.7 km SW Río San Marcos (in turn 8.1 km SW Cd. Victoria), Tamaulipas, Mexico, 36.5 km NE of Jaumave**
Burger & Robertson (1951)	*Agkistrodon bilineatus taylori* (= *A. taylori*): **KM 833, 21 km N of Villagran, Tamaulipas**
Baird & Girard (1853)	*Crotalophorus Edwardsii* (= *Sistrurus catenatus edwardsii*): **Tamaulipas . . . S. bank of Río Grande . . . Sonora**

Restriction appeared in [1] Stejneger & Barbour (1933); [2] Smith & Taylor (1948); [3] Smith & Taylor (1950b); [4] Smith & Taylor (1950a); [5] Smith & Brandon (1968); [6] Müller (1936); [7] Smith & Smith (1979); [8] J. W. Bailey (1928); [10] Baird (1859a [1858]); [11] Axtell (1958a); [12] Degenhardt et al. (1996); [13] Taylor (1935); [14] J. R. Bailey (1939); [15] Cochran (1961); [16] J. R. Bailey (1937); [17] Martin (1958).
Restriction not accepted by [9] De Queiroz (1995).

sitions continued through the Late Cretaceous with evidence suggesting tectonic activity and instability in the Middle and Late Jurassic. The seas subsided to the east beginning in the Late Cretaceous, apparently associated with the uplift of Mexico's mainland (plateau). Folding and faulting occurred in the Early Tertiary, forming the ranges of the Sierra Madre Oriental, which comprise sedimentary marine rocks largely of Cretaceous and Jurassic origin with sedimentary clastic Cenozoic rock and recent alluvial deposits in the valleys and basins. Isolated exposures of older Precambrian, Paleozoic, and Mesozoic-Triassic rock occur locally, notably west of Ciudad Victoria. The Coastal Plain of Tamaulipas (the emerging continental shelf of the Gulf of Mexico) was tilted and lifted by tectonic activity in the Cretaceous-Paleocene but remained mostly submerged, probably until the older Quaternary (Heim 1940). The substrates of the Coastal Plain consist of alluvial soils over Cenozoic sedimentary marine depositions of limestone, sandstone, siltstone, and claystone, with various Paleogene exposures all resting on a Cretaceous base, which is occasionally exposed as well. The isolated ranges and mesas occurring on the Coastal Plain are igneous rock bodies, plutons, and alkaline basalt flows emplaced in upland areas in the Middle Tertiary where the sea had already subsided.

The hydrology of Tamaulipas is entirely associated with the Gulf of Mexico, excluding portions of two arid interior municipalities, Bustamante and Tula, in the extreme southwest. In the Tamaulipan Province the Río Bravo and its tributaries, the Río Salado and Río San Juan, drain vast areas of the continental interior, while the entire watersheds of the Río San Fernando (= Río Conchos) and the Río Soto La Marina (and its tributaries, Río Pilón, Río Purificación, Río Corona, and Río Palmas) lie within Tamaulipas and adjacent areas of Nuevo León. In the Veracruzan Province coastal areas are drained by relatively small river systems, including Río Carrizo, Río San Vicente, Río Carrizal (and tributaries Río San Pedro and Río Las Lajas), Río Tigre, and Río Barberena, the latter two empt-

ing into the Laguna San Andrés. The interior of the Veracruzan Province and the majority of the Sierra Madre Oriental in Tamaulipas are in the watershed of the Río Guayalejo and its tributaries, Río El Alamar, Río Chiue, Río Frío, Río Sabinas, Río Santa María de Guadalupe, and Río Las Animas. Exceptions occur in the Sierra Madre Oriental, where the eastern slopes north of the Tropic of Cancer drain into the Río Soto La Marina, and in the extreme south, where the Río Gallos Grandes and Río Los Galos drain into the Río Tampaón-Pánuco in San Luis Potosí. The Río Guayalejo is a tributary of the Río Tamesi, which forms the border of the state with Veracruz, where extensive marshes and lagoons (i.e., Champayan, Josecito, La Salada) sustain large areas of *Cyperus* and *Typha*. Other natural freshwater lakes are Laguna Los Soldados, a volcanic crater lake in the Sierra Maratinez, and some closed basins in the Sierra Madre Oriental: Laguna La Escondida, Laguna Isadora, Laguna La Loca, and the Lagunas Las Hondas. Several larger human-made lakes on the Coastal Plain, including Falcón Lake, Presa Marte Gómez, Presa Vicente Guerrero, Laguna República Española, Presa Ramiro Caballero, and Presa Emilio Portes, were constructed in the twentieth century. Karstic environments occur in both the Sierra Madre Oriental and the Sierra de Tamaulipas, particularly on the eastern slop of each, where mountain streams often drain through subterranean waterways, emerging on the Coastal Plain as rivers and lagoons: for example, the Río Sabinas, Río Frío, Laguna La Loca of the Sierra Madre Oriental, and Cenote El Zacatón (source of the Río Tigre) and other sinkholes of the Sierra de Tamaulipas. The Laguna Madre on the northern coast is described in more detail later.

Encompassed within its boundaries Tamaulipas has portions of three Nearctic subregions (the Tamaulipan, the Sierra Madre Oriental, and the Chihuahua Desert) and one Neotropical subregion (the Veracruzan), each with distinctive climates, topographies, and vegetation zones (Álvarez 1963; Flores-Villela 1993; Morrone et al. 2002).

The Tamaulipan Province

The Tamaulipan Province occupies approximately 141,500 km² in South Texas, northeastern Coahuila, northern Nuevo León, and the northern two-thirds of Tamaulipas (Blair 1950). Its boundaries are sharply defined by the Gulf of Mexico to the east, the Sierra Madre Oriental to the southwest, and the Balcones Escarpment in the northwest. Its boundaries are less distinct in the vicinity of the San Antonio River of Texas to the northeast, where pedocal soils change to pedalfer soils (Blair 1950), the Chihuahuan Desert to the west, and the Tropic of Cancer to the south. For the discussion here I delineate the southern limit at the Tropic of Cancer, excluding the northern foothills of the Sierra de Tamaulipas and coastal areas east of the Bordas Escarpment/Sierra de Las Rusias up to the Río Soto La Marina north of the Tropic line. The Bordas Escarpment transects this province where it runs from an area near the northern termination of the Sierra de Las Rusias, just east of Soto La Marina in the southeast, northwestward from there to the vicinity of Camargo, and on into Texas (Johnston 1963). The topography is flat with deeper, sandier soils northeast of this escarpment and rolling hills to the southwest. Elevations range from sea level to 400 m at the base of the Sierra Madre Oriental, abruptly ascending above the plain. An exception is the Sierra San Carlos, an isolated mountain range about 1448 km² in area, where elevations are generally 400–1000 m; however, a few ranges attain higher elevations, principally the Sierra Chiquita (1794 m max.).

The climate of this province (excluding the Sierra San Carlos) is semiarid with precipitation ranging from 400 to 750 mm; however, tropical depressions and hurricanes occasionally impact the Coastal Plain June through November and can increase rainfall significantly. Summer temperatures are hot (32°C–35°C), occasionally exceeding 40°C; winters are mild (2°C–8°C); and freezes are uncommon. Vast areas of riparian and thornscrub vegetation have been cleared for crop production, particularly in the northeast. Three assemblages are recognized here, Tamaulipan Thornscrub, the Laguna Madre, and the Sierra San Carlos. Authors have recognized vegetation zones within the thornscrub of South Texas that undoubtedly occur in Tamaulipas, but Tunnell (2002) identified significant information gaps, and these areas have not been delineated in Tamaulipas.

TAMAULIPAN THORNSCRUB ASSEMBLAGE

Tamaulipan Thornscrub is characterized by deciduous and thorny plants, including trees and shrubs (*Acacia berlandieri, A. greggii, A. rigidula, A. tortuosa, A. wrightii, Celtis pallid, Condalia hookeri, Cordia boissieri, Leucophyllum frutescens, Parkinsonia aculeata, Prosopis glandulosa, Vachellia farnesiana, Yucca treculeana*); grasses (*Aristida roemeriana, Bouteloua hirsute, Cenchrus incertus*); and perennial forbs (*Thymophylla pentachaeta, Heliotropium confertifolium*) (Blair 1950; Johnston 1963; Little 1976). Thornscrub is often taller (2–3 m) and denser in the east with exposure to coastal humidity, and shorter and more open in the west (Blair 1950; Martin et al. 1954). Riparian zones often include *Taxodium mucronatum, Ebenopsis ebano, Sabal mexicana,* and *Ulmus crassifolia.* Elements of tropical deciduous forest extend northward from the south in ever-narrowing and increasingly discontinuous patches at the base of the Sierra Madre Oriental. This region includes records for 84 species of herpetofauna: *Notophthalmus meridionalis, Siren intermedia, Siren* sp. indet., *Eleutherodactylus cystignathoides, Anaxyrus debilis, A. punctatus, A. speciosus, Incilius nebulifer, Rhinella marina, Pseudacris clarkii, Smilisca baudinii, Leptodactylus fragilis, L. melanonotus, Gastrophryne olivacea, Hypopachus variolosus, Lithobates berlandieri, L. catesbeianus, Rhinophrynus dorsalis, Scaphiopus couchii, Spea bombifrons, Pseudemys gorzugi, Terrapene carolina, Trachemys scripta, T. venusta, Kinosternon flavescens, Gopherus berlandieri, Apalone spiniferus, Crocodylus moreletii, Crotaphytus reticulates, Coleonyx brevis, Hemidactylus frenatus,*

H. turcicus, Cophosaurus texanus, Phrynosoma cornutum, P. modestum, Sceloporus consobrinus, S. cyanogenys, S. grammicus, S. olivaceus, S. variabilis, Urosaurus ornatus, Anolis sericeus, Plestiodon obsoletus, P. tetragrammus, Ameiva undulata, Aspidoscelis gularis, A. laredoensis, Boa constrictor, Arizona elegans, Coluber constrictor, Coniophanes imperialis, Drymarchon melanurus, Drymobius margaritiferus, Ficimia streckeri, Heterodon kennerlyi, Hypsiglena jani, Lampropeltis getula, L. triangulum, L. septentrionalis, Leptophis mexicanus, Masticophis flagellum, M. schotti, Nerodia erythrogaster, N. rhombifera, Opheodrys aestivus, Pantherophis emoryi, Pituophis catenifer, Rhinocheilus lecontei, Salvadora grahamiae, Sonora semiannulata, Storeria dekayi, Tantilla nigriceps, Thamnophis marcianus, T. proximus, Tropidodipsas fasciata, T. sartorii, Trimorphodon tau, Micrurus tener, Rena dulcis, R. myopica, Agkistrodon taylori, Crotalus atrox, C. totonacus, and *Sistrurus catenatus.*

LAGUNA MADRE ASSEMBLAGE

The extensive wetlands of the Laguna Madre are partitioned from the Gulf of Mexico by about 250 km of barrier islands and peninsulas. Tunnell and Judd (2002) reviewed this ecosystem, from which most of the information on vegetation here was derived. Barrier island sand dunes are vegetated with *Ipomoea imperati, Schizachyrium scoparium, Sesuvium portulacastrum,* and *Uniola paniculata.* The hypersaline lagoon (one of only five in the world), harbors about 480 km² of seagrass meadows, including *Cymodocea filiforme, Halodule beauderrei, Halophila engelmannii, Ruppia maritime,* and *Thalassia testudinum.* Clay dunes occur on the mainland with extensive marshlands dominated by *Spartina spartinae.* Salinity decreases inland, and limited areas of freshwater marshes and lagoons occur in peripheral areas, Laguna La Nacha being the largest. Elevations range from sea level to 3 m. Included are at least 16 species of amphibians and reptiles, 6 recorded on the barrier islands and peninsulas: *Holbrookia propinqua, Sceloporus variabilis,*

Aspidoscelis gularis, A. sexlineata, Gopherus berlandieri, and *Masticophis flagellum;* and 5 additional species recorded on the mainland adjacent to the Laguna Madre and among the clay dunes, mudflats, and marshlands: *Phrynosoma cornutum, S. olivaceus, Thamnophis proximus, Rena dulcis,* and *Crotalus atrox.* Five species of sea turtles (*Caretta caretta, Chelonia mydas, Eretmochelys imbricata, Lepidochelys kempii,* and *Dermochelys coriacea*) have been reported to nest on the beaches of southern Tamaulipas and Texas (with significantly varying degrees of frequency) and to enter the Laguna Madre of Texas (Tunnell and Judd 2002). Presumably, sea turtles utilize these habitats in this region; however, no specific information is available.

SIERRA SAN CARLOS ASSEMBLAGE

The thornscrub of the plain ascends to approximately 600 m, growing somewhat thicker and taller with elevation, and transitions into the semiarid oak forest (*Quercus* sp.) at 600–900 m, which predominates these mountains; however, a few ranges support pine-oak forest with *Pinus teocote* at 900–1794 m (Dice 1937). The higher elevations have higher humidity and milder temperatures than the plain. At least 42 species of herpetofauna can be confirmed there. Among these, 8 (19.0%) species do not appear on the surrounding coastal plain, although all occur in the Sierra Madre Oriental 40–50 km to the west: *Eleutherodactylus guttilatus, Gerrhonotus infernalis, Sceloporus minor, S. parvus, Plestiodon dicei, Pantherophis bairdi, Tantilla atriceps,* and *T. rubra.* Additionally 34 (81.0%) Coastal Plain species are recorded here: *Eleutherodactylus cystignathoides, Anaxyrus punctatus, Incilius nebulifer, Rhinella marina, Smilisca baudinii, Lithobates berlandieri, Trachemys venusta* (<500 m), *Gopherus berlandieri, Cophosaurus texanus, Phrynosoma cornutum, Sceloporus cyanogenys, S. grammicus, S. olivaceus, S. variabilis, Plestiodon obsoletus, P. tetragrammus, Aspidoscelis gularis, Arizona elegans, Drymarchon melanurus, Drymobius margaritiferus, Ficimia streckeri, Leptodeira septentrionalis, Ma-*

sticophis flagellum, *M. schotti*, *Opheodrys aestivus*, *Pantherophis emoryi*, *Salvadora grahamiae*, *Storeria dekayi*, *Tantilla nigriceps*, *Thamnophis marcianus*, *T. proximus*, *Micrurus tener*, *Rena myopica*, and *Crotalus atrox*.

Collectively, 98 species of reptiles and amphibians occur in the Tamaulipan Province within the state of Tamaulipas: 3 salamanders (3.0%, 2 families, 2 genera); 18 anurans (18.3%, 8 families, 13 genera); 12 turtles (12.2%, 6 families, 11 genera); 1 crocodile (1.0%); 24 lizards (24.5%, 7 families, 12 genera); and 40 snakes (40.8%, 5 families, 30 genera). Among these 98 species, 8 (8.0%) are endemic to Mexico, 81 (82.6%) occur in Mexico and the United States, and 9 (9.0%) occur in Mexico and Central America. Twenty-two (22.4%) species occur in the United States, Mexico, and Central America. Two invasive species, *Hemidactylus frenatus* and *H. turcicus*, occur here. Several species have limited distributions in this province. *Phrynosoma modestum* is known from one isolated record (Farr et al. 2007). In the "panhandle" *Nerodia erythrogaster*, *Urosaurus ornatus*, and *Lampropeltis getula* are known from only one, two, and four records each, respectively, in the state; and in the extreme northeast, *Pseudacris clarkii* (known from six specimens, two localities) and *Siren* sp. indet. (known from five specimens, one locality) reach their southern distributional limits here. *Siren intermedia*, *Spea bombifrons*, and *Sistrurus catenatus* also approach southern distributional limits here but have been recorded from isolated localities (presumably relict populations) farther south. Other species reaching southern distributional limits here that are more widespread in this province include *Apalone spinifera*, *Heterodon kennerlyi*, *Sonora semiannulata*, and *Tantilla nigriceps*. Two species occurring in the north, *Trachemys scripta* and *Rena dulcis*, are represented by sister species *T. venusta* and *R. myopica*, respectively, in the south. Neotropical species reaching their northern distributional limits here include *Leptodactylus melanonotus*, *Crocodylus moreletii*, *Ameiva undulata*, *Anolis sericeus*, *Boa constrictor*, *Leptophis mexicanus*, *Trimorphodon tau*, *Tropi-*

dodipsas fasciata, *T. sartorii*, *Agkistrodon taylori*, and *Crotalus totonacus*. *Holbrookia propinqua* and *Aspidoscelis sexlineata* are restricted to the coast in this region. *Siren* sp. indet., *Pseudemys gorzugi*, and *Aspidoscelis laredoensis* are restricted to the Río Bravo and adjacent areas. *Terrapene carolina* and *Opheodrys aestivus* occur in relatively wide-ranging but isolated relict populations in this region. Several species are restricted to western areas of this region, some apparently analogues to the Bordas Escarpment, including *Anaxyrus punctatus*, *Hypopachus variolosus*, *Crotaphytus reticulatus*, *Coleonyx brevis*, *Cophosaurus texanus*, *Phrynosoma modestum*, *Sceloporus consobrinus*, and *Urosaurus ornatus*. There are no museum records of *Holbrookia lacerata* from Tamaulipas, although their occurrence is likely (Farr et al. 2013).

The Veracruzan Province

The Veracruzan Province occupies most of the Coastal Plain of the Gulf of Mexico in Mexico. Although authors agree this province enters southern Tamaulipas, there is seldom agreement about the exact northern (and southern) limits (e.g., Goldman and Moore 1945; Álvarez 1963; Flores-Villela 1993; Morrone et al. 2002). The northern border recognized here was defined previously in the section on the Tamaulipan Province. The topography is primarily flat but interrupted by several hills and mountain ranges. The Sierra de Tamaulipas (ca. 2500 km²) is the largest range, dominating the north-central part of this province, ranging from 300 to 1000 m (1400 m max.) in elevation. To the east, the Sierra de Las Rusias (400 km²), with a maximum elevation of 300 m, and the Sierra Maratinez (150 km²), with elevations of 150–400 m (600 m max.), are isolated ranges supporting *Quercus* forest, but these do not harbor herpetofaunal assemblages significantly distinct from those of the Coastal Plain. South of the Sierra de Tamaulipas is a broad (ca. 10,800 km²), semiarid plain (<100 m) spreading between the coast and the Sierra Cucharas. To the west, about 50 km at 150–200 m elevation separates the Sierra de

Tamaulipas from the Sierra Madre Oriental, with semiarid tropical thornscrub in the east of this gap, becoming increasingly humid with tropical deciduous forest and low mesas (300–400 m elevation) to the west. In the southwest of this province the Veracruzan flora and fauna extend westward beyond the front ranges of the Sierra Madre Oriental, over the Sierra Cucharas (= Sierra del Abra; 350–450 m, 720 m max.) and the Sierra Tamalave (500–800 m, 1100 m max.), two low, relatively narrow (4–8 km) anticlinal folds, into two synclinal valleys (Chamal and Ocampo). These two ranges extending south from the Sierra de Guatemala onto the plain pose a minimal barrier for coastal fauna and harbor a minimal spectrum of the Sierra Madre Oriental herpetofauna. To the west of the Ocampo Valley, the Sierra Madre Oriental range rises to 1000–1300 m (1800 m max.).

The climate is tropical-subhumid with an annual rainfall of 700–1600 mm (70.0% occurring June–September), with the plain receiving less rainfall and areas adjacent to the Sierra Madre Oriental receiving the most. Ciudad Mante has a mean annual temperature of 24.8°C, a mean January low of 18.4°C, and a mean June high of 29.1°C. Tampico has a mean temperature of 24.2°C, a mean January low of 13.4°C, and a mean August high of 31.8°C. Four assemblages are identified here: Coastal Plain, Laguna San Andrés and Gulf Coast, Sierra de Tamaulipas, and Tropical Deciduous Forest.

COASTAL PLAIN ASSEMBLAGE

This region includes areas from sea level to 300 m between the coast and the Sierra Madre Oriental (Sierra de Guatemala, Sierra Cucharas), excluding tropical deciduous forest associated with the Sierra de Tamaulipas and the Sierra Madre Oriental occurring below this elevation. The vegetation is variable, and significant anthropogenic influences began before it was well documented. Tropical thorn forest and scrub dominate the semiarid plain south of the Sierra de Tamaulipas, including *Vachellia* (*Acacia*) *farnesiana, A. rigidula, Condalia hookeri, Cordia*

boissieri, Parkinsonia aculeate, Prosopis juliflora, and *Yucca treculeana* (Little 1976, 1977). Riparian areas often support lush gallery forest with flanking tropical deciduous forest. *Sabal mexicana* occurs throughout this region, but it is most abundant between the coast and the Sierra de Tamaulipas, sometimes forming small forests. Grasslands appear in some areas and may have been more extensive in the past (Martin 1958). Berlandier (1980, 503) described "immense," and Goldman (1951, 260) described "tolerably heavy," forest of *Quercus oleoides* and *Brosimum alicastrum* with small prairies near Altamira and more open grassy plains with acacia, mesquite, and cacti to the north. Remnants of oak forest can still be found between Altamira and the Río Soto La Marina. Records of 71 species of amphibians and reptiles can be confirmed: *Notophthalmus meridionalis, Eleutherodactylus cystignathoides, Anaxyrus debilis, A. speciosus, Incilius nebulifer, Rhinella marina, Trachycephalus typhonius, Scinax staufferi, Smilisca baudinii, Leptodactylus fragilis, L. melanonotus, Gastrophryne elegans, G. olivacea, Hypopachus variolosus, Lithobates berlandieri, L. catesbeianus, Rhinophrynus dorsalis, Scaphiopus couchii, Terrapene carolina, Trachemys scripta, T. venusta, Kinosternon flavescens, K. herrerai, K. scorpioides, Gopherus berlandieri, Crocodylus moreletii, Ophisaurus incomptus, Laemanctus serratus, Hemidactylus frenatus, H. turcicus, Ctenosaura acanthura, Phrynosoma cornutum, Sceloporus olivaceus, S. serrifer, S. variabilis, Anolis sericeus, Plestiodon tetragrammus, Ameiva undulata, Aspidoscelis gularis, Boa constrictor, Coluber constrictor, Coniophanes imperialis, Drymarchon melanurus, Drymobius margaritiferus, Ficimia streckeri, Imantodes cenchoa, Lampropeltis triangulum, Leptodeira maculata , L. septentrionalis, Leptophis mexicanus, Masticophis flagellum, M. mentovarius, M. schotti, Nerodia rhombifera, Opheodrys aestivus, Oxybelis aeneus, Pantherophis emoryi, Pituophis catenifer, Pseudelaphe flavirufa, Spilotes pullatus, Storeria dekayi, Thamnophis marcianus, T. proximus, Tropidodipsas fasciata, T. sartorii, Micrurus tener, Rena myopica, Indotyphlops braminus, Agkistrodon taylori, Crotalus atrox,* and *C. totonacus.*

LAGUNA SAN ANDRÉS AND GULF
COAST ASSEMBLAGE

This region covers approximately 170 km of
coastline between Río Soto La Marina and Río
Pánuco, including beaches and barrier peninsulas (Rancho Nuevo, Barra del Trodo, Barra
Chavarría), lagoons (San Andrés, 136 km²; Las
Marismas, 55 km²), and wetlands. The interior marshlands are not extensive. Elevations
range from sea level to 3 m. Some lagoons are
lined with mangrove trees (*Rhizophora mangle*,
Laguncularia racemosa, *Avicennia germinans*,
Conocarpus erectus), totaling 24 km² in the state
(CONABIO 2008). At least 16 species of herpetofauna have been recorded among the beaches,
sand dunes, and mangroves: *Incilius nebulifer*,
Caretta caretta, *Chelonia mydas*, *Eretmochelys
imbricata*, *Lepidochelys kempii*, *Dermochelys coriacea*, *Gopherus berlandieri*, *Ctenosaura acanthura*,
Iguana iguana, *Holbrookia propinqua*, *Sceloporus
variabilis*, *Aspidoscelis gularis*, *Boa constrictor*, *Masticophis flagellum*, *Drymarchon melanurus*, and
Thamnophis proximus.

SIERRA DE TAMAULIPAS ASSEMBLAGE

The vegetation zones for this assemblage include tropical deciduous forest (ca. 300–700 m)
with *Bursera simaruba* and *Lysiloma divaricata*;
montane scrub (600–900 m); and pine-oak forest (>800 m) with *Quercus* sp. and *Pinus teocote*
interspersed with grassland (Martin et al. 1954;
Cram et al. 2006). Elevations range from 300 to
1400 m. Among the 46 species of herpetofauna
confirmed from these mountains, 11 have not
been reported from the surrounding thornscrub
and forest of the Coastal Plain: *Pseudoeurycea
cephalica*, *Craugastor augusti*, *Eleutherodactylus
guttilatus*, *Ecnomiohyla miotympanum*, *Gerrhonotus infernalis*, *Sceloporus grammicus*, *Plestiodon
dicei*, *Lepidophyma sylvaticum*, *Rhadinaea gaigeae*,
Senticolis triaspis, and *Micrurus tamaulipensis*.
Additional species include *Notophthalmus meridionalis*, *Eleutherodactylus cystignathoides*, *Incilius
nebulifer*, *Rhinella marina*, *Scinax staufferi*,
Smilisca baudinii, *Hypopachus variolosus*, *Lithobates berlandieri*, *Scaphiopus couchii*, *Terrapene*

carolina, *Trachemys venusta*, *Kinosternon herrerai*,
Sceloporus olivaceus, *S. serrifer*, *S. variabilis*, *Anolis
sericeus*, *Plestiodon tetragrammus*, *Ameiva undulata*, *Aspidoscelis gularis*, *Coniophanes imperialis*,
Drymarchon melanurus, *Drymobius margaritiferus*,
Ficimia streckeri, *Imantodes cenchoa*, *Leptodeira
maculata*, *L. septentrionalis*, *Leptophis mexicanus*, *Masticophis schotti*, *Oxybelis aeneus*, *Spilotes
pullatus*, *Thamnophis proximus*, *Micrurus tener*,
Rena myopica, *Agkistrodon taylori*, and *Crotalus
totonacus*.

TROPICAL DECIDUOUS FOREST ASSEMBLAGE

This region includes the tropical deciduous
forest occurring mostly at 200–500 m (occasionally exceeding these parameters) at the eastern
base of the Sierra Madre Oriental, the Sierra
Cucharas, Sierra Tamalave, and the Chamal
and Ocampo Valleys, with the two intermittent
valleys becoming increasingly arid southward
in the vicinity of Antiguo Morelos and Nuevo
Morelos. Martin (1958) described this area as
tropical deciduous forest, but he speculated that
past anthropogenic activities might have affected
the current composition. Cram et al. (2006) described most of this region as perturbed forests
and scrublands with a high diversity. Vegetation
includes *Bromelia penguin*, *Esenbeckia runyonii*,
Guazuma ulmifolia, *Nectandra salicifolia*, *Petrea
arborea*, *(Pseudo) bombax ellipticum*, and *Thouinia
villosa* (Martin 1958; Cram et al. 2006). Moderately extensive *Sabal mexicana* forest occurs in
some areas, particularly in the Chamal Valley.
There are 82 species of herpetozoa recorded
here: *Bolitoglossa platydactyla*, *Chiropterotriton
multidentatus*, *Notophthalmus meridionalis*,
Craugastor augusti, *Eleutherodactylus cystignathoides*, *E. dennisi*, *E. longipes*, *Anaxyrus debilis*,
A. punctatus, *Incilius nebulifer*, *Rhinella marina*,
Scinax staufferi, *Smilisca baudinii*, *Trachycephalus
typhonius*, *Leptodactylus fragilis*, *L. melanonotus*,
Gastrophryne elegans, *G. olivacea*, *Hypopachus
variolosus*, *Lithobates berlandieri*, *L. catesbeianus*,
Rhinophrynus dorsalis, *Scaphiopus couchii*, *Terrapene carolina*, *Trachemys venusta*, *Kinosternon
scorpioides*, *Crocodylus moreletii*, *Gerrhonotus infernalis*, *Laemanctus serratus*, *Anelytropsis papillosus*,

Hemidactylus frenatus, H. turcicus, Ctenosaura acanthura, Cophosaurus texanus, S. serrifer, S. variabilis, Anolis sericeus, Plestiodon lynxe, P. tetragrammus, Scincella silvicola, Ameiva undulata, Aspidoscelis gularis, Lepidophyma micropholis, L. sylvaticum, Boa constrictor, Coluber constrictor, Coniophanes imperialis, C. piceivittis, Drymarchon melanurus, Drymobius margaritiferus, Ficimia olivacea, F. streckeri, Hypsiglena jani, Imantodes cenchoa, Lampropeltis triangulum, Leptodeira maculata, L. septentrionalis, Leptophis mexicanus, Masticophis flagellum, M. schotti, Mastigodryas melanolomus, Nerodia rhombifera, Oxybelis aeneus, Pantherophis emoryi, Pseudelaphe flavirufa, Rhadinaea gaigeae, Scaphiodontophis annulatus, Senticolis triaspis, Spilotes pullatus, Storeria dekayi, Tantilla rubra, Thamnophis marcianus, T. proximus, Tropidodipsas fasciata, T. sartorii, Trimorphodon tau, Micrurus tener, Epictia goudotii, Rena myopica, Agkistrodon taylori, Bothrops asper, and *Crotalus atrox.*

Collectively, 108 species are currently known from the Veracruzan Province in Tamaulipas: 4 salamanders (3.70%, 2 families, 4 genera); 23 anurans (21.29%, 9 families, 15 genera); 12 turtles (11.10%, 5 families, 9 genera); 1 crocodile (0.92%); 24 lizards (22.22%, 10 families, 17 genera); and 44 snakes (40.74%, 6 families, 35 genera). Among these species, 24 (22.22 %) are endemic to Mexico and 59 (54.62%) occur in the United States. Twenty-five species (23.14%) occur in Mexico and Central America, and 25 species (23.14%) occur in the United States, Mexico, and Central America. Many Neotropical species reach northern distributional limits here, including *Bolitoglossa platydactyla, Plestiodon lynxe, Lepidophyma micropholis, Coniophanes piceivittis, Ficimia olivacea, Mastigodryas melanolomus, Scaphiodontophis annulatus, Epictia goudotii,* and *Bothrops asper* in tropical deciduous forest; *Trachycephalus typhonius, Kinosternon herrerai, K. scorpioides, Laemanctus serratus, Ctenosaura acanthura, Iguana iguana, Leptodeira maculata, Masticophis mentovarius,* and *Pseudoelaphe flavirufa* near the coast; *Imantodes cenchoa* and *Spilotes pullatus* in the northern foothills of the Sierra de Tamaulipas; and *Scinax staufferi, Ga-*

strophryne elegans, and *Sceloporus serrifer,* which are more widespread. Species reaching their southern distributional limits include *Anaxyrus debilis, A. speciosus, Cophosaurus texanus,* and *Phrynosoma cornutum. Micrurus tamaulipensis* is endemic to the Sierra de Tamaulipas. *Eleutherodactylus dennisi* and *Lepidophyma micropholis* are restricted to the Sierra Cucharas in Tamaulipas. *Eretmochelys imbricata* nests in Tamaulipas are rare, and *Dermochelys coriacea* nests are extremely rare. *Plestiodon lynxe* and *Scaphiodontophis annulatus* are each known from a single specimen in Tamaulipas, and *Iguana iguana* is known from only two localities near the extreme southern coast. Species primarily associated with the Sierra Madre Oriental that have been recorded at 420 m and lower in the Sierra Madre Oriental or the Sierra Cucharas include *Chiropterotriton multidentatus, Craugastor augusti, Eleutherodactylus longipes, Gerrhonotus infernalis, Scincella silvicola, Rhadinaea gaigeae, Senticolis triaspis,* and *Tantilla rubra.* Five invasive species, *Lithobates catesbeianus, Trachemys scripta, Hemidactylus frenatus, H. turcicus,* and *Indotyphlops braminus,* occur in this region.

The Sierra Madre Oriental

The Sierra Madre Oriental covers approximately 145,500 km² of northeastern Mexico, running from the Transvolcanic Belt (northern Hidalgo, Querétaro, and Puebla) north to the vicinity of Monterrey, Nuevo León, where an east-west component runs into southeastern Coahuila. Although they are geologically a separate development, some consider the ranges extending northwest into the Trans-Pecos region of Texas to be part of the Sierra Madre Oriental. Within Tamaulipas, the Sierra Madre Oriental occupies a comparatively limited but significant area in the southwestern part of the state. The maximum elevation in Tamaulipas is greater than 3400 m on the eastern slope of Sierra Peña Nevada on the Nuevo León border; from there, the Sierra Madre Oriental extends eastward over 80 km to the front of the range, near Ciudad Victoria. Although irregular, there is a generalized trend of decreasing elevations and narrowing (east to west) toward the south near the San Luis Potosí border, where the Ocampo Valley is separated from the Chihuahuan Desert by a few ranges only 18 km wide and less than 1200 m in elevation at some points.

Within the Sierra Madre Oriental is the semiarid Jaumave Valley (ca. 1130 km²), with an average annual temperature of 21°C–23°C (45°C max., 0°C min.) and an annual precipitation of 500–560 mm. The Río Guayalejo drains the valley through a narrow gorge eastward to the Coastal Plain, and to the west a semiarid canyon extends to the semiarid Palmillas Valley and onward to the desert plateau, virtually transecting the Sierra Madres into northern and southern components, roughly but not precisely, at the Tropic of Cancer. The Sierra Madre Oriental produces a rain shadow, and the climate varies significantly depending on the exposure and elevation, although specific information has been difficult to obtain. Martin (1958) estimated the annual cloud forest precipitation to be 2000–2500 mm (3200 mm max.), the highest in the state, with a mean annual temperature of 19.4°C in 1954. The higher elevations northwest of the Jaumave Valley are cooler and semiarid and occasionally receive snow, although not annually. More than 150 temperatures recorded with specimen data at 2400–3200 m, July–August 2003–2009 (from 9:44 to 19:00), ranged from 14.44°C to 31.67°C, but nighttime lows (estimated at 10°C) were not recorded. Surface water is minimal, but afternoon rains occurred on most field days during July–October 2003–2009. The town of Jaumave, population 5633, is the largest in this region, and few others exceed 1000 in population. Significant areas of mature forest are in this region, although many are secondary growth. Three assemblages are identified here: Northern Sierras and Interior Slopes, Jaumave Valley, and Southern Sierras and Eastern Slope.

NORTHERN SIERRAS AND INTERIOR SLOPES ASSEMBLAGE

This region includes the Sierra Madre Oriental north and west of the Jaumave Valley (excluding the easternmost exposures of the front range)

and the semiarid interior slopes south of the valley. Elevations range from 1250 to greater than 3400 m. The vegetation of the Sierra Madre Oriental, like the climate, varies significantly depending on the exposure and elevation. Lower elevations and interior slopes descending into the Jaumave Valley and Chihuahuan Desert are semiarid with *Juniperus angosturana, J. flaccida, Pinus arizonica* var. *stormiae, P. cembroides, P. estevezii, P. nelsonii, Yucca carnerosana,* a variety of cacti, and occasional chaparral vegetation zones (Martin 1958). Intermediate elevations support dry oak and dry oak-pine forest. Temperate Madrean pine-oak woodlands characterize higher altitudes (>2250 m) in the northwest, where some of the more conspicuous vegetation includes *Arbutus xalapensis, Maguey americana, M. asperrima, Juniperus monticola, Pinus arizonica, P. cembroides, P. hartwegii, P. johannis, P. nelsonii, Quercus* sp., and *Dasylirion* sp. Recorded here are 37 amphibian and reptile species: *Chiropterotriton* cf. *priscus, Pseudoeurycea galeanae, Craugastor augusti, Eleutherodactylus guttilatus, E. verrucipes, Anaxyrus punctatus, Incilius nebulifer, Ecnomiohyla miotympanum, Hyla eximia, Spea multiplicata, Barisia ciliaris, Gerrhonotus infernalis, Phrynosoma orbiculare, Sceloporus chaneyi, S. grammicus, S. minor, S. olivaceus, S. parvus, S. spinosus, S. torquatus, Plestiodon dicei, Scincella silvicola, Aspidoscelis gularis, Ficimia hardyi, Lampropeltis mexicana, Masticophis schotti, Pantherophis bairdi, Salvadora grahamiae, Senticolis triaspis, Storeria hidalgoensis, Tantilla rubra, Thamnophis exsul, T. pulchrilatus, Rena myopica, Crotalus lepidus, C. molossus,* and *C. pricei.*

JAUMAVE VALLEY ASSEMBLAGE

Elevations in the Jaumave Valley range from 550 to 1250 m. The vegetation of the valley, like the herpetofauna, has species representative of both the Coastal Plain and the Chihuahuan Desert, including *Jatropha dioica, Prosopsis glandulosa, Cylindropuntia leptocaulis, Koeberlinia spinosa, Agave funkiana, A. lechuguilla, Yucca treculeana,* and *Y. filifera* (Martin 1958). Herpetofauna known from the valley includes 35 species: *Anaxyrus punctatus, Incilius nebulifer, Rhinella* marina, *Ecnomiohyla miotympanum*, *Lithobates berlandieri*, *Spea multiplicata*, *Kinosternon integrum*, *K. scorpioides*, *Gerrhonotus infernalis*, *Crotaphytus collaris*, *Hemidactylus frenatus*, *H. turcicus*, *Cophosaurus texanus*, *Phrynosoma cornutum*, *Sceloporus cautus*, *S. olivaceus*, *S. parvus*, *S. variabilis*, *Plestiodon tetragrammus*, *Aspidoscelis gularis*, *Boa constrictor*, *Coluber constrictor*, *Drymarchon melanurus*, *Hypsiglena jani*, *Masticophis flagellum*, *M. schotti*, *Pantherophis emoryi*, *Rhinocheilus lecontei*, *Senticolis triaspis*, *Tantilla rubra*, *Thamnophis proximus*, *Trimorphodon tau*, *Rena myopica*, *Micrurus tener*, and *Crotalus atrox*.

SOUTHERN SIERRAS AND EASTERN SLOPE ASSEMBLAGE

This assemblage includes the Sierra Madre south of the Jaumave Valley (excluding the semiarid interior slopes of the Jaumave Valley and Chihuahuan Desert) and the easternmost front range in the north. Elevations range from 500 m on the eastern slope, 1250 m on the interior slope, up to 2785 m. The vegetation is predominantly dry oak forest on the eastern slopes and summits and dry oak-pine forest at the summits and interior slopes. Trees include *Quercus canbyi*, *Q. polymorpha*, *Q. rysophylla*, *Juniperus flaccida*, *Pinus montezumae*, *P. teocote*, and *Arbutus xalapensis*, often with agaves (*Maguey* sp.) and palmettos (*Sabal* sp.) in the understory (Martin 1958). At lower elevations on the interior slopes these forests transition into the northern and interior communities described here. On the coastal versant, the dry oak forest often transitions into tropical deciduous forest of the Coastal Plain at 300–900 m; however, in some areas two additional vegetation zones also appear, cloud forest and tropical evergreen forest, although they are fragmented and unevenly dispersed. Cloud forests occur at elevations of 900–1700 m, predominantly at 1000–1500 m in the Sierra de Guatemala, although smaller fragments occur to the north and south. Cloud forest vegetation includes *Chamaedorea radicalis*, *Clethra macrophylla*, *Liquidambar styraciflua*, *Magnolia schiedeana*, *Meliosma alba*, *Pinus patula*, *Quercus germane*, *Q. sartorii*, *Turpina occidentalis*, and

several species of *Clethra* and *Symplocos* (Martin 1958; Jones and Gorchov 2000; Alcántara et al. 2002). Tropical evergreen forest occurs unevenly at lower elevations (ca. 500–1000 m) on the eastern slope, most predominantly between the cloud forest and the tropical deciduous forest in the Sierra de Guatemala, although additional segments occur to the south. Tropical evergreen forests include *Dendropanax arboreus*, *Brosimum alicastrum*, *Iresine tomentella*, *Quercus germana*, *Spondias* sp., *Ungnadia speciosa*, and *Zamia* sp. (Martin 1958; Cram et al. 2006). At least 60 herpetofaunal species occur here: *Chiropterotriton cracens*, *C. multidentatus*, *Pseudoeurycea belli*, *P. cephalica*, *P. scandens*, *Craugastor augusti*, *C. decoratus*, *Eleutherodactylus cystignathoides*, *E. guttilatus*, *E. longipes*, *Incilius nebulifer*, *Rhinella marina*, *Ecnomiohyla miotympanum*, *Hyla eximia*, *Smilisca baudinii*, *Hypopachus variolosus*, *Lithobates berlandieri*, *Kinosternon integrum*, *K. scorpioides*, *Abronia taeniata*, *Gerrhonotus infernalis*, *Ophisaurus incomptus*, *Laemanctus serratus*, *Anelytropsis papillosus*, *Sceloporus cyanogenys*, *S. grammicus*, *S. minor*, *S. parvus*, *S. scalaris*, *S. torquatus*, *S. variabilis*, *Anolis sericeus*, *Plestiodon dicei*, *Scincella silvicola*, *Aspidoscelis gularis*, *Lepidophyma sylvaticum*, *Xenosaurus platyceps*, *Adelphicos newmanorum*, *Amastridium sapperi*, *Coniophanes fissidens*, *Drymarchon melanurus*, *Drymobius margaritiferus*, *Geophis latifrontalis*, *Leptodeira septentrionalis*, *Leptophis mexicanus*, *Masticophis schotti*, *Pantherophis bairdi*, *Pliocercus bicolor*, *Rhadinaea gaigeae*, *Salvadora grahamiae*, *Storeria hidalgoensis*, *Tantilla rubra*, *Thamnophis cyrtopsis*, *T. mendax*, *Trimorphodon tau*, *Tropidodipsas sartorii*, *Micrurus tener*, *Bothrops asper*, *Crotalus lepidus*, and *C. totonacus*.

Collectively, 93 species are known from the Sierra Madre Oriental of Tamaulipas: 7 salamanders (7.5%, 1 family, 2 genera); 15 anurans (16.1%, 7 families, 11 genera); 2 turtles (2.1%, 1 family, 1 genus); 30 lizards (32.2%, 10 families, 16 genera); and 39 snakes (41.9%, 5 families, 28 genera). Among these species 3 (3.2%) are endemic to Tamaulipas; 41 (44%) are endemic to Mexico; 41 (44.0%) occur in the United States; and 24 (25.8%) range into Central America.

Two invasive species, *Hemidactylus frenatus* and *H. turcicus*, occur here. The cloud forest of the Sierra de Guatemala and the adjacent lowlands to the east are the most thoroughly surveyed areas in Tamaulipas. Approximately 35.0% (5447) of all available museum specimens from Tamaulipas were collected in the Gómez Farías area. Within this region *Sceloporus olivaceus* occurs in the Jaumave Valley and its canyons (<1750 m). *Sceloporus spinosus* is restricted to the Palmillas Valley and interior Chihuahuan Desert slopes, and *S. cautus* is known from a single specimen from the Palmillas Valley. *Sceloporus cyanogenys* occurs on the easternmost slopes of the Sierra Madre up to 1050 m (Martin 1958); however, a few museum records from interior and western localities are problematic and might involve some confusion with *S. minor*, although this is unresolved. *Coluber constrictor* is known from a single specimen. Martin believed a single record of *Mastigodryas melanolomus* at 1050 m was unreliable. Although known only from north of the Jaumave Valley, *Pantherophis bairdi* occurs in both the easternmost and interior ranges.

The Chihuahuan Desert

The Chihuahuan Desert rests on an interior plateau delimited by the Sierra Madre Occidental to the west and the Sierra Madre Oriental to the east. It occupies 450,000 km² (Morafka 1977) from San Luis Potosí in the south and extending into West Texas and southern New Mexico in the north. Definitions of this desert and the geographic area that it occupies vary. I follow Henrickson and Straw (1976) and Gómez-Hinostrosa and Hernández (2000). Many elements of this ecosystem enter extreme southwest Tamaulipas. The topography of this region of Tamaulipas comprises desert flats in the south near the San Luis Potosí border (1025 m); to the north, the base elevation ascends through canyons and valleys up to the town of Miquihuana (1850 m). Rising above the base elevation are a number of arid ranges, including the Sierra Las Ventanas (2270 m), Sierra La Norita (2490 m), Sierra El Fierro (2900 m), and Sierra El Pinal (2940 m). The high Sierra Madre north of Miquihuana and the Sierra Mocha to the east, with chaparral and juniper trees on the interior slopes and pine-oak forest at the summits, delineate the limits of this region in Tamaulipas. Some elements of the Chihuahuan Desert ecosystem filter through lower elevations of canyons into the Palmillas and Jaumave Valleys. Geologically, these ranges are part of the Sierra Madre, with the basins and ranges becoming more widely spaced westward. Upper Cretaceous limestone is predominant, with Jurassic and older rocks occurring locally and recent alluvial deposits in the flats, canyons, and mountain valleys (Gómez-Hinostrosa and Hernández 2000).

The rain shadow of the Sierra Madre produces a profoundly different climate and vegetation in these mountains from those only a few kilometers to the east, which is hot and arid, with summer temperatures as high as 43°C and winter lows rarely reaching 0°C. In the summer, night and day temperature fluctuations of greater than 15°C are typical (e.g., July 17, 2006, 21.11°C–36.67°C). Precipitation, occurring mostly in the summer, ranges from less than 400 mm on the flats to 500–600 m in the mountains. The arroyos of this region seldom hold water for more than a few hours immediately following rainfall. Two basic assemblages occur in this region, Chihuahuan Desert Flats and Chihuahuan Desert Mountains. However, a collecting bias is reflected in this region of the state, and for practical purposes I have also included Chihuahuan Desert Foothills, Canyons, and Valleys and the Laguna Isadora area, each representing an ecotone. Among the approximately 16,000 museum records available from Tamaulipas, only 154 (0.96%) are from this province, and the majority of these were collected from Hwy 101, which transects only the desert flats and one canyon. The majority of my personal field records were collected in the flats, foothills, canyons, and valleys. The mountains of this region have not been adequately sampled.

CHIHUAHUAN DESERT FLATS ASSEMBLAGE

This vegetation is dominated by microphyllous scrubs, including *Acacia farnesiana*, *A. rigidula*,

Larrea tridentata, and *Prosopis juliflora*. Other conspicuous species include *Yucca filifera* and *Myrtillocactus geometrizans* (Gómez-Hinostrosa and Hernández 2000). Vestiges of grasslands are minimal, but agricultural areas in the region might have supported this vegetation type in the past. Sand and dry mud are exposed in some overgrazed areas, sometimes in patches of several acres. Elevations generally range from 1025 to 1300 m, although isolated segments occur in some larger valleys as high as 1800 m, where chaparral vegetation from adjacent mountains may form an ecotone in the far north and east. The 25 species of herpetofauna recorded are *Anaxyrus cognatus, A. debilis, A. punctatus, Incilius nebulifer, Smilisca baudinii, Lithobates berlandieri, Spea multiplicata, Kinosternon integrum, Gerrhonotus farri, Anelytropsis papillosus, Cophosaurus texanus, Phrynosoma modestum, Sceloporus cautus, S. minor, Plestiodon obsoletus, Aspidoscelis gularis, A. inornata, Masticophis flagellum, M. schotti, Pantherophis emoryi, Pituophis deppei, Rhinocheilus lecontei, Tantilla atriceps, Trimorphodon tau,* and *Crotalus atrox.*

CHIHUAHUAN DESERT MOUNTAIN ASSEMBLAGE

Saxicolous scrubs typify the vegetation on these slopes and include *Agave lechuguilla, A. striata, Euphorbia antisyphilitica, Hechitia glomerata, Jatropha dioica, Fouquieria splendens, Ferocactus pilosus,* and *Mannillaria candida* (Gómez-Hinostrosa and Hernández 2000). The microphyllous scrubs of the flats and *Yucca filifera* also appear on these slopes but in significantly less abundance, and vegetation here is typically more open. Elevations range from 1400 to 2940 m. Goat herders were encountered on occasion; otherwise, this habitat is largely uninhabited and undisturbed. Only 11 amphibian and reptile species can currently be confirmed: *Craugastor augusti, Eleutherodactylus verrucipes, Spea multiplicata, Cophosaurus texanus, Phrynosoma orbiculare, Sceloporus grammicus, S. minor, S. parvus, Masticophis schotti, Tantilla rubra,* and

Crotalus molossus. Future sampling will no doubt augment this assemblage.

CHIHUAHUAN DESERT FOOTHILLS, CANYONS, AND VALLEYS ASSEMBLAGE

The vegetation and the herpetofauna of this assemblage reflect an ecotone comprising the preceding two assemblages; however, the majority of field records were recorded in these areas. Elevations range from 1100 to 1850 m. There are 33 species recorded: *Craugastor augusti, C. batrachylus, Anaxyrus punctatus, Incilius nebulifer, Hyla eximia, Lithobates berlandieri, Spea multiplicata, Kinosternon integrum, Gerrhonotus infernalis, Crotaphytus collaris, Anelytropsis papillosus, Cophosaurus texanus, Phrynosoma modestum, P. orbiculare, Sceloporus cautus, S. grammicus, S. minor, S. parvus, S. spinosus, Aspidoscelis gularis, Hypsiglena jani, Lampropeltis mexicana, Leptodeira septentrionalis, Masticophis flagellum, M. schotti, Pituophis deppei, Rhinocheilus lecontei, Senticolis triaspis, Tantilla wilcoxi, Micrurus tener, Crotalus atrox, C. molossus,* and *C. scutulatus.*

LAGUNA ISADORA ASSEMBLAGE

This unique area (an ecotone) of approximately 50 km² lies in a closed basin east of Ciudad Tula, surrounded on two and a half sides by the Sierra Madre, but also open and at the same elevation (1200–1300 m) as the desert flats to the west and northwest. Laguna Isadora typically holds water year-round, and the desert scrub vegetation of the flats intermingles with *Juniperus angosturana* and other chaparral elements from the sierras. Seven species have been recorded there, some of which are more characteristic of the adjacent sierras: *Lithobates berlandieri, Kinosternon integrum, K. scorpioides, Aspidoscelis gularis, Salvadora grahamiae, Thamnophis proximus,* and *Tropidodipsas sartorii.*

Collectively, 49 species are currently known from this region of Tamaulipas: 11 anurans (22.4%, 6 families, 8 genera), 2 turtles (4.0%, 1 family, 1 genus), 16 lizards (32.6%, 7 families,

9 genera), and 20 snakes (40.8%, 3 families, 15 genera). Among these species, 15 (30.6%) are endemic to Mexico, 32 (65.3%) occur in the United States, and 7 (14.2%) range into Central America. One invasive species, *Hemidactylus turcicus*, has been collected in the region. Of these 49 species, Morafka (1977) listed 37 (75.51%) as desert species, 5 (10.20%) as peripheral species, and did not include *Craugastor batrachylus* in his analysis. The remaining 6 (12.24%) species (*Smilisca baudinii, Anelytropsis papillosus, Gerrhonotus farri, Leptodeira septentrionalis, Trimorphodon tau, Tropidodipsas sartorii*) were only recently recorded in this region (Farr et al. 2007; Farr et al. 2009, Farr et al. 2013, unpub. data 2006, 2009; Bryson and Graham 2010). Five species (*Hyla eximia, Smilisca baudinii, Kinosternon scorpioides, Leptodeira septentrionalis, Tropidodipsas sartorii*) are known from a single record or a single peripheral locality within this region and are not considered representative of the Chihuahuan Desert herpetofauna. One frog (*Craugastor batrachylus*) and two lizards (*Gerrhonotus farri, Plestiodon obsoletus*) are also known from a single specimen each, and their status in this ecosystem is unclear. Artificial water tanks are numerous in this area and might be responsible for creating an unnatural resource and abundance of anurans and *Kinosternon integrum*, which are quite common there.

Currently, 179 species are known from Tamaulipas: 11 salamanders (6.14%, 3 families, 5 genera); 31 anurans (17.31%, 9 families, 18 genera); 15 turtles (8.37%, 6 families, 11 genera); 1 crocodile (0.55%); 47 lizards (26.25%, 13 families, 23 genera); and 74 snakes (41.34%, 6 families, 45 genera). Among these species 6 (3.35%) are endemic to Tamaulipas; 54 (30.16%) are endemic to Mexico; 96 (53.63%) occur in the United States and Mexico; 29 (16.2%) occur in Mexico and Central America only; and 25 (13.96%) occur in the United States, Mexico, and Central America.

Many Nearctic and Neotropical species reach their distributional limits on the Coastal Plain of South Texas, Tamaulipas, and Veracruz. The Tropic of Cancer in southern Tamaulipas is a somewhat arbitrary line to delineate these two biogeographical realms. Based on the herpetofaunal distribution, there is no clear, single demarcation point between these two regions but instead a gradient that runs throughout Tamaulipas and beyond.

Although not evenly distributed or equally abundant, 13 (7.26%) species have been recorded in all four of the ecological provinces of Tamaulipas and are only absent from higher elevations: *Anaxyrus punctatus* (≤1840 m), *Incilius nebulifer* (≤1740 m), *Lithobates berlandieri* (≤2000 m), *Gerrhonotus infernalis* (150–2092 m), *Aspidoscelis gularis* (≤2000.5 m), *Hemidactylus turcicus* (≤1150 m), *Hypsiglena jani* (≤1833 m), *Masticophis flagellum* (≤1216 m), *M. schotti* (ca. ≤2000 m), *Pantherophis emoryi* (ca. ≤1225 m), *Trimorphodon tau* (≤1550 m), *Micrurus tener* (≤1120 m), and *Crotalus atrox* (≤1280 m). Five (2.79%) species have been recorded from one of the Coastal Plain provinces, the Sierra Madre, and the Chihuahuan Desert: *Craugastor augusti* (200–2000 m), *Anelytropsis papillosus* (330–1360 m), *Rhinocheilus lecontei* (≤1506 m), *Salvadora grahamiae* (≤1600 m), and *Senticolis triaspis* (386–1546 m). Additionally, 15 (8.37%) species from the Coastal Plain occur at lower elevations in the Sierra Madre and occupy interior and western slopes adjacent to the Chihuahuan Desert: *Eleutherodactylus cystignathoides* (≤1447 m), *Ecnomiohyla miotympanum* (≤120–1600 m), *Smilisca baudinii* (≤1334 m), *Hypopachus variolosus* (≤1007 m), *Kinosternon scorpioides* (≤1279 m), *Sceloporus olivaceus* (≤1750 m), *Anolis sericeus* (100–1211 m), *Coluber constrictor* (≤1060 m), *Drymarchon melanurus* (≤1386 m), *Drymobius margaritiferus* (≤1309 m), *Leptodeira septentrionalis* (≤1698 m), *Thamnophis proximus* (≤1238 m), *Tropidodipsas sartorii* (≤1680 m), *Rena myopica* (≤1701 m), and *Crotalus totonacus* (≤1680 m). These distributional patterns bring into question the extent to which the Sierra Madre Oriental represents a barrier for these species. Some species that are absent from the Sierra Madre Oriental but occur both to the east and west have apparently dispersed from the north (e.g., *Anaxyrus debilis, Phrynosoma modestum,* and *Kinosternon flavescens*). For many other species (18.43%), the mesic and arid environments of the Veracruz Province and the Chihuahuan Desert would seem to present greater barriers for these species than the elevations of the Sierra Madre Oriental.

The International Union for Conservation of Nature (IUCN) Red List of 2014 categorized the herpetofauna of Tamaulipas as follows: CR = Critically Endangered, 2 (1.1%) species (*Eretmochelys imbricata, Lepidochelys kempii,*); EN = Endangered, 10 (5.58%) species (*Chiropterotriton cracens, C. multidentatus, Notophthalmus meridionalis, Eleutherodactylus dennisi, Caretta caretta, Chelonia mydas, Sceloporus chaneyi, Xenosaurus platyceps, Ficimia hardyi, Thamnophis mendax*); V = Vulnerable, 11 (6.14%) species (*Pseudoeurycea belli, P. scandens, Craugastor decoratus, Eleutherodactylus longipes, E. verrucipes, Dermochelys coriacea, Terrapene carolina, Abronia taeniata, Crotaphytus reticulatus, Lepidophyma micropholis, Storeria hidalgoensis*); NT = Near Threatened, 7 (3.1%) species (*Bolitoglossa platydactyla, Chiropterotriton priscus, Pseudoeurycea cephalica, Pseudoeurycea galeanae, Ecnomiohyla miotympanum, Kinosternon herrerai, Pseudemys gorzugi*); DD = Data Deficient, 5 (2.79%) species (*Craugastor batrachylus, Ophisaurus incomptus, Geophis latifrontalis, Rhadinaea gaigeae, Micrurus tamaulipensis*); LC = Least Concern, 110 (61.45%) species; and NE = Not Evaluated, 34 (18.99%) species. The Mexican federal government ("NORMA Oficial Mexicana NOM-059-SEMARNAT-2010") categorized the herpetofauna of Tamaulipas in 2010 as follows: P = In Danger of Extinction, 7 (3.91%) species (*Notophthalmus meridionalis, Caretta caretta, Chelonia mydas, Eretmochelys imbricata, Lepidochelys kempii, Dermochelys coriacea, Ophisaurus incomptus*); A = Threatened, 31 (17.32%) species (*Pseudoeurycea belli, P. cephalica, P. galeanae, Siren intermedia, Pseudemys gorzugi, Gopherus berlandieri, Crotaphytus collaris, C. reticulatus, Anelytropsis papillosus, Cophosaurus texanus, Phrynosoma orbiculare, Scincella silvicola,*

Lepidophyma micropholis, Boa constrictor, Coluber constrictor, Lampropeltis getula, L. mexicana, L. triangulum, Leptophis mexicanus, Masticophis flagellum, M. mentovarius, Nerodia erythrogaster, Pituophis deppei, Pliocercus bicolor, Tantilla atriceps, Thamnophis cyrtopsis, T. exsul, T. marcianus, T. mendax, T. proximus, Agkistrodon taylori); Pr = Subject to Special Protection, 41 (22.9%) species (*Bolitoglossa platydactyla, Chiropterotriton multidentatus, C. priscus, Pseudoeurycea scandens, Anaxyrus debilis, Craugastor batrachylus, C. decorates, Eleutherodactylus dennisi, E. verrucipes, Gastrophryne elegans, G. olivacea, Lithobates berlandieri, Rhinophrynus dorsalis, Crocodylus moreletii, Terrapene carolina, Trachemys scripta, Kinosternon herrerai, K. integrum, K. scorpioides, Apalone spinifera, Abronia taeniata, Laemanctus serratus, Coleonyx brevis, Ctenosaura acanthura, Iguana iguana, Sceloporus grammicus, Plestiodon lynxe, Lepidophyma sylvaticum, Xenosaurus platyceps, Geophis latifrontalis, Imantodes cenchoa, Leptodeira maculata, Tantilla rubra, Tropidodipsas sartorii, Micrurus tener, Crotalus atrox, C. lepidus, C. molossus, C. pricei, C. scutulatus,* and *Sistrurus catenatus*); and 100 (55.86%) species are not listed. The major protected areas in Tamaulipas include Alta Cumbre (30,327 ha) and Reserva de la Biósfera El Cielo (144,530 ha) in the Sierra Madre Oriental; Laguna Madre y Delta del Río Bravo (572,808 ha) and Parras de la Fuente (Reserva de la Paloma de Ala Blanca) (21,948 ha) in the Tamaulipan Province; Bernal de Horcasitas (18,204 ha) and Playa de Rancho Nuevo (ca. 20 km of beaches) in the Veracruzan Province. Most of these lands are privately owned, and few resources are available and devoted to their management and enforcement of wildlife regulations. All of the regions in Tamaulipas would benefit greatly from additional sanctuaries, but on the Coastal Plain, where the clearing of vegetation for agriculture and livestock continues unabated, the need for additional refuges is perhaps the greatest.

Herpetofauna of California

BRADFORD D. HOLLINGSWORTH
AND CLARK R. MAHRDT

Introduction

California's amphibian and reptile fauna has been greatly influenced by both complex geological processes and extreme climatic gradients. As an example, the highest and lowest points within the lower contiguous United States lie only 136 km apart. Mount Whitney, a product of the Sierra Nevada uplift, reaches 4421 m, while Badwater Basin within Death Valley sinks to −86 m below sea level. Between these two sites, the boreal climate of the snow-covered peaks goes through a dramatic transition and culminates with 57°C maximum temperatures on the desert floor. The position of California along the continental margin is the primary cause of both its complex geological history and climatic gradients. As the Pacific Plate slides and subducts beneath the North American continent, uplift and subsidence shape and reconfigure California's extreme topography. Climatic gradients develop as the Pacific storm cycle moves southward from the Gulf of Alaska to the west coast of California. Storms are strongest in the Pacific Northwest, but as the systems pass over the mountains, rain shadows produce deserts on their eastern sides.

A general pattern emerges where the highest amphibian diversity is located in the Pacific Northwest and montane habitats, while reptile diversity peaks in the southwestern deserts. Overall, the native amphibian and reptile fauna of California is composed of 173 species, with 71 amphibians (45 salamanders, 26 frogs) and 102 reptiles (8 turtles, 48 lizards, 46 snakes). Of these, 45 (26.0%) are endemic to the state (33 amphibians and 12 reptiles). Endemism is unevenly distributed across taxonomic groups: 30 salamanders, 3 frogs, 10 lizards, and 2 snakes. And, within salamanders, 25 of the 30 endemics belong to the Plethodontidae, which contains the most species in the state at the taxonomic level of the family (35 species).

The distributions of the remaining species extend beyond the state boundaries. These external influences include the Pacific Northwest, the Great Basin Desert, the Sonoran Desert, and the Baja California peninsula. The distributions of 92 of California's amphibian and reptile species extend into Mexico, representing more than half of the state's herpetofauna. The majority (86 species) span the southern border with the state of Baja California, while a few (6 species) extend eastward through Arizona, then southward to mainland Mexico. The high degree of overlap between the two countries demonstrates the great influence each region has on the other in shaping their respective faunas.

Previous Herpetological Studies

The discovery of California's native amphibians and reptiles began in the early 1800s, with expeditions both by land and by sea. The first species described from material collected within the state was *Elgaria coerulea*. In 1815, the German naturalist Adelbert von Chamisso circumnavigated the world as the appointed botanist on the Russian ship *Rurik*. Also onboard was the entomologist Johann Friedrich von Eschscholtz. Their voyage took them to both South America and up the coast to California, where they collected in and around San Francisco Bay. Upon their return to Europe in 1818, most of their material was deposited at the Zoological Museum of Berlin. In 1828, Wiegmann's description of the type specimen of *E. coerulea* incorrectly listed the locality as "Brasilien," which was later corrected to "San Francisco, California," by Stejneger (1902) after examining the specimen and voyage logs.

From 1823 to 1826, Eschscholtz made a second voyage onboard the Russian warship *Predpriaeteë*, which retraced the route of the *Rurik*, but this time he served as the expedition's chief naturalist. He collected along the coast and inland to the Great Central Valley with discoveries leading to the descriptions of *Aneides lugubris*, *Batrachoseps attenuatus*, *Taricha torosa*, and *Dicamptodon ensatus* (Adler 2007).

Two expeditions followed those of Chamisso and Eschscholtz. In the years 1827–1829, P. E. Botta was onboard the French commercial vessel *Le Héros* (Adler 1978). Botta made natural history collections along the entire coastline of the state with extended ports in both San Diego and San Francisco. The results of these collections were described by Henri Marie Ducrotay de Blainville from the Muséum National d'Histoire Naturelle in *Reptiles de la Californie* in 1835 and included the descriptions of *Callisaurus draconoides*, *Charina bottae*, *Elgaria multicarinata*, *Lampropeltis getula*, *L. zonata*, *Pituophis catenifer*, and *Tantilla planiceps* (Blainville 1835). The first American survey by sea was the US Exploring Expedition (1838–1842). This scientific survey of six ships made extensive collections throughout the Pacific region, including California, Oregon, and Washington. While a number of amphibian and reptile descriptions resulted from the material collected, only four were included in the expedition's accounts: *Sceloporus occidentalis*, *Contia tenuis*, *Thamnophis ordinoides*, and *Actinemys marmorata* (Baird and Girard 1852b; Girard 1858).

The US government began expeditions and surveys overland in 1845 and 1846 as a result of the Mexican-American War and the need to better understand the boundaries between the two countries. One notable surveyor was William Hemsley Emory, whose primary task was to make maps but also to amass natural history collections and maintain contact with notable scientists in the East. In 1846, his team fought in the Battle of San Pasqual near San Diego, which demonstrates the ruggedness of these early surveyors and the multiple roles their members took on (Adler 1978). Most of the material made its way to the Smithsonian Institution and, in combination with specimens collected by the US Exploring Expedition, served as the basis for numerous species descriptions published by Spencer F. Baird and Charles F. Girard (1852a, 1853). For California, this was a time of discovery, and species descriptions mounted faster than at any other time in its history (figure 9.1).

As hostilities ceased between the two countries and California entered the Union in 1850, overland expeditions to California increased dramatically and the discovery of new species continued. From New Mexico, expeditions in 1851 to the Colorado River and eventually San Diego included naturalist Samuel Washington Woodhouse, who was part of a team to find new routes to the West (Adler 1978). His material was later studied by Edward Hallowell and included the descriptions of *Pseudacris cadaverina*, *Phrynosoma mcallii*, and *Masticophis taeniatus*.

In 1853, the US Congress began an extensive search to find the best overland routes for a transcontinental railroad. Most of the railroad survey teams included a naturalist, as recom-

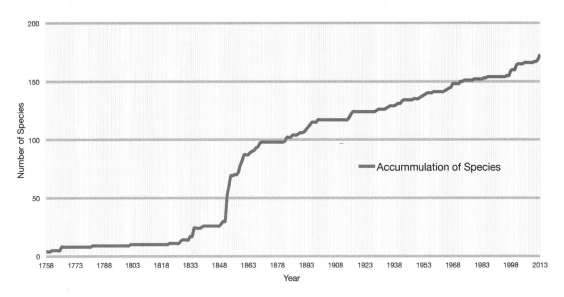

FIGURE 9.1 Accumulation of species descriptions for amphibians and reptiles found within California.

mended by Baird. Naturalists would return their collections back East for study and publication (Adler 1978). As a result, the high rate of species discoveries continued, and by the end of the decade nearly half of California's species had been described. Baird's assistants and protégés included Robert Kennicott and Edward D. Cope. Kennicott had described five of California's species from 1859 to 1861 but died a young man at the age of 30 while exploring Alaska. Cope was the most prolific author in American zoology and during the years 1860–1896 described 24 species that occur in California (Adler 1989).

By the end of the nineteenth century, routes to the West had been well established and herpetology began focusing on the exploration of specific geographic locations, detailing the distribution of species, and making accounts of the variation within species. Leonhard Stejneger, appointed by Baird at the Smithsonian Institution, eventually became curator of amphibians and reptiles. During his tenure, Stejneger began a new era of careful scholarship, detailed descriptions, and proper designations of specimens used in the analysis (Adler 1989). He described six species: *Ascaphus truei*, *Gambelia sila*, *Petrosaurus mearnsi*, *Sceloporus orcutti*, *Xantusia henshawi*, and *Lichanura orcutti*. In 1891, Stejneger led a survey to Death Valley, which included Frank Stephens from the San Diego Society of Natural History, who collected the type specimen of *Crotalus stephensi* (Stejneger 1893a; Klauber 1930).

The California Academy of Science, located in San Francisco, became the first institution in the state dedicated to natural history. Founded in 1852, the academy opened its first museum in 1874. In 1894, John Van Denburgh is credited with organizing its herpetology department and was appointed curator the following year (Slevin and Leviton 1956). By 1906, the collection had grown to 8100 specimens, primarily from western North America. However, in the same year the San Francisco earthquake destroyed all but 13 of these specimens and Van Denburgh was forced to rebuild the collection (Adler 1989). At the time, Joseph R. Slevin was in the Galápagos Islands and returned in November 1906 with 4506 specimens to start a new collection (Slevin and Leviton 1956).

Prior to the destruction of the collection, Van Denburgh published *The Reptiles of the Pacific Coast and Great Basin* (1897). After Van Denburgh had rebuilt the collections, he published his greatest work, *The Reptiles of Western North America* (1922), which has been compared to Holbrook's *North American Herpetology* (1842).

Six years later, Slevin released a companion volume, *The Amphibians of Western North America* (1928). Together, these volumes served as the modern foundation of taxonomic diversity of California and surrounding regions.

During the time of rebuilding the collection of the California Academy of Sciences, across the San Francisco Bay, the University of California, Berkeley, opened the Museum of Vertebrate Zoology in 1908 (Rodríguez-Robles et al. 2003). Its first director, Joseph Grinnell, quickly set out to build the collections of the museum, focused initially on regions primarily within California. Within 10 years, in 1917, Grinnell and his student Charles L. Camp published the first systematic review of California's amphibians and reptiles. Camp continued on in herpetology for a short time before focusing on paleontology. From 1915 to 1917, Camp described six species of California amphibians: *Anaxyrus californicus*, *A. canorus*, *Batrachoseps major*, *Hydromantes platycephalus*, *Rana muscosa*, and *R. sierrae*. Following Camp, Tracy I. Storer published *A Synopsis of the Amphibia of California* (1925).

During the early and mid-twentieth century, museums, colleges, and universities began to grow and ultimately include a herpetologist on staff. In 1926, the San Diego Society of Natural History appointed Laurence M. Klauber as their curator. Klauber's collections focused on the southwestern United States, and his numerous revisions of genera clarified much of the taxonomic confusion that embedded itself in the discipline. His greatest work was a two-volume set on rattlesnakes, which is still in print today (Klauber 1956, 1972).

Since the time of the boundary and railroad surveys of the 1850s, the accumulation of new species descriptions of amphibians and reptiles has continued to progress at a steady pace. The Museum of Vertebrate Zoology's early focus on the diversity of the state has made it one of the most productive centers. Robert C. Stebbins and David B. Wake, along with their students and colleagues, have described 14 species from California. In addition, Stebbins has produced a series of popular field guides starting in the early

1950s through today (Stebbins and McGinnis 2012). Today, productive herpetologists can be found at nearly every state-funded university and five California natural history museums.

Through the history of California herpetology, 102 species of amphibians (38 salamanders, 12 frogs) and reptiles (1 turtle, 30 lizards, 21 snakes) have their type localities within the state (table 9.1).

Physiographic Characteristics and Their Influence on the Distribution of the Herpetofauna

The distributions of amphibians and reptiles native to California are tied to both their evolutionary history and the region's geological progression (Rissler et al. 2006). Current distributions have strong tendencies to be correlated to the physiographic regions of the state, and patterns of faunal associations are evident. California is the third-largest state in the United States. Its area of 423,970 km^2 represents 4.3% of the country. The state extends 1240 km in length and 400 km in width, lying between latitudes 32°32′ N and 42°0′ N and longitudes 114°8′ W and 124°26′ W. It is positioned along the continental margin with a 1466 km coastline with the Pacific Ocean.

The combination of a high diversity of floral regions and a large number of endemic species makes California difficult to study as a single unit (Stebbins and Major 1965). Dividing the state by biotic provinces, life zones, plant communities, or by faunistic and floristic areas is one approach to better understand its complexity, but doing so has proved challenging due to the subdivisions' irregular, often complex boundaries. The rough, mountainous topography and diversity of climates make accurate delimitation problematic. As such, assigning specific groups to assemblages is inexact because even the core species may persist outside the defined boundaries. Stephens (1905) made one of the first attempts to explain the many different ecoregions within the state with the proposal to divide

TABLE 9.1 Type localities for amphibians and reptiles described from the state of California, USA

Author(s)	Original name: Type locality
Salamanders (38)	
Gray (1853)	*Ambystoma californiense*: **vicinity of San Francisco**[1]
Eschscholtz (1833)	*Triton ensatus* (= *Dicamptodon ensatus*): **probably near Fort Ross, Sonoma County**[2]
Strauch (1870)	*Plethodon flavipunctatus* (= *Aneides flavipunctatus*): **Californien (Neu-Albion)**
Cope (1883)	*Plethodon iĕcanus* (= *Aneides iecanus*): **Baird [McCloud River, Shasta County, California, USA]**[3]
Hallowell (1849)	*Salamandra lugubris* (= *Aneides lugubris*): **Monterey [Monterey County], Upper California**
Myers & Maslin (1948)	*Aneides niger*: **near the forks of Waddell Creek, Santa Cruz County, California**
Wake & Jackman (1999)	*Aneides vagrans*: **a point about 10 km S Maple Creek, Humboldt Co., California [. . .]**[4]
Eschscholtz (1833)	*Salamandrina attenuata* (= *Batrachoseps attenuatus*): **Umbegung der Bai St. Francisco auf Californien**
Jockusch et al. (2012)	*Batrachoseps altasierrae*: **1.5 mi [2.4 km] SE Alta Sierra [Greenhorn Mountains], Kern Co., California, USA**
Jockusch et al. (2012)	*Batrachoseps bramei*: **Packsaddle Canyon, adjacent to Kern River, 1137 m elevation, Tulare Co., California, USA**
Marlow et al. (1979)	*Batrachoseps campi*: **Long John Canyon, W slope of the Inyo Mountains, [. . .]**[4] **Inyo County, California, USA**[4]
Jockusch et al. (1998)	*Batrachoseps diabolicus*: **Hell Hollow, [. . .] Mariposa County, California**[4]
Wake (1996)	*Batrachoseps gabrieli*: **[. . .] Soldier Creek, [. . .] Los Angeles County, California [. . .]**[4]
Jockusch et al. (2001)	*Batrachoseps gavilanensis*: **0.5 miles (0.8 km) south of cement plant [. . .], San Benito Co., CA**[4]
Jockusch et al. (1998)	*Batrachoseps gregarius*: **Westfall Picnic Ground east of Highway 41, [. . .], Madera-Mariposa county line, California**[4]
Jockusch et al. (2001)	*Batrachoseps incognitus*: **[. . .] 14.7 km NE Highway 1 on San Simeon Creek Road, San Luis Obispo County, CA**[4]
Jockusch et al. (1998)	*Batrachoseps kawia*: **west side of the South Fork, Kaweah river, Tulare County, California**
Jockusch et al. (2001)	*Batrachoseps luciae*: **Don Dahvee Park, Monterey, Monterey Co., CA**
Camp (1915)	*Batrachoseps major*: **Sierra Madre, 1000 feet altitude, Los Angeles County, California**
Jockusch et al. (2001)	*Batrachoseps minor*: **from along the Santa Rita–Old Creek Road, [. . .], San Luis Obispo County, California [. . .]**[4]
Cope (1869)	*Batrachoseps nigriventris*: **Fort Tejon [Kern County], California**
Cope (1865)	*Hemidactylium pacificum* (= *Batrachoseps pacificus*): **one of the northern Channel Islands**[5]
Jockusch et al. (1998)	*Batrachoseps regius*: **south bank of the North Fork, Kings River, [. . .], Fresno county, California**[4]
Brame & Murray (1968)	*Batrachoseps relictus*: **[. . .] Kern River Canyon [. . .], Kern County, California**[4]
Wake et al. (2002)	*Batrachoseps robustus*: **[. . .] Kern Plateau, Tulare County, California [. . .]**[4]
Brame & Murray (1968)	*Batrachoseps simatus*: **[. . .] the south side of the Kern River Canyon [. . .], Kern County, California [. . .]**[4]
Brame & Murray (1968)	*Batrachoseps stebbinsi*: **[. . .] Piute Mountains, southern Sierra Nevada, Kern County, California [. . .]**[4]
Gray (1850)	*Ensatina eschscholtzii*: **Monterey**
Dunn (1929)	*Ensatina klauberi*: **Descanso, San Diego County, Calif[ornia]**
Gorman (1954)	*Hydromantes brunus*: **[. . .] 0.7 miles NNE Briceburg [. . .], Mariposa County, California**[4]
Camp (1916)	*Spelerpes platycephalus* (= *Hydromantes platycephalus*): **head of Lyell Cañon, [. . .], Yosemite National Park, California**[4]
Gorman & Camp (1953)	*Hydromantes shastae*: **[. . .] in the narrows of Low Pass Creek [. . .], Shasta County, California**[4]
Mead et al. (2005)	*Plethodon asupak*: **Muck-a-Muck Creek (41.774 N, 123.031 W) above Scott Bar, Siskiyou County California [. . .]**[4]
Van Denburgh (1916)	*Plethodon elongates*: **Requa, Del Norte County, California**
Stebbins & Lowe (1951)	*Rhyacotriton variegates*: **1.3 miles west of Burnt Ranch Post Office, Trinity County, California**
Twitty (1935)	*Triturus rivularis* (= *Taricha rivularis*): **Gibson Creek, about one mile west of Ukiah [Mendocino County], California**
Twitty (1942)	*Triturus sierrae* (= *Taricha sierrae*): **Cherokee Creek, in the hills above Chico, Butte County, California**
Rathke (1833)	*Triton torosa* (= *Taricha torosa*): **in der Umgebung der Bai St. Francisco auf Californien**

TABLE 9.1 *Continued*

Author(s)	Original name: Type locality
Frogs (12)	
Camp (1915)	*Bufo cognatus californicus* (= *Anaxyrus californicus*): **Santa Paula, 800 feet altitude, Ventura County, California**
Camp (1916)	*Bufo canorus* (= *Anaxyrus canorus*): **Porcupine Flat, 8100 feet, Yosemite National Park, Mariposa Co[unty], California**
Myers (1942)	*Bufo exsul* (= *Anaxyrus exsul*): **Deep Springs, Deep Springs Valley, Inyo County, California**
Girard (1859)	*Bufo alvarius* (= *Incilius alvarius*): **[old] Fort Yuma, Imperial County, California [. . .]**[4,6]
Cope (1866)	*Hyla cadaverina* (= *Pseudacris cadaverina*): **Tejon Pass**
Hallowell (1854)	*Hyla scapularis* var. *hypochondriaca* (= *Pseudacris hypochondriaca*): **Tejon Pass**
Jameson et al. (1966)	*Hyla regilla sierra* (= *Pseudacris sierra*): **1 1/4 miles SSE. of Tioga Pass Ranger Station [. . .]**[4]
Baird (1854)	*Rana boylii*: **vicinity of Coloma, along the South Fork of the American River, El Dorado County**[7]
Baird & Girard (1852)	*Rana draytonii*: **vicinity of San Francisco**[8]
Camp (1917)	*Rana muscosa*: **Arroyo Seco Cañon, at about 1300 feet altitude, near Pasadena, [Los Angeles County,] California**
Camp (1917)	*Rana sierrae*: **Matlack Lake, [. . .], two miles southeast of Kearsarge Pass, Sierra Nevada, Inyo County, California**[4]
Baird (1859a [1858])	*Scaphiopus hammondii* (= *Spea hammondii*): **Fort Reading, California**
Turtles (1)	
Cooper (1861)	*Xerobates agassizii* (= *Gopherus agassizii*): **mountains of California, near Fort Mojave**
Lizards (30)	
Wiegmann (1828)	*Gerrhonotus coeruleus* (= *Elgaria coerulea*): **San Francisco, California**[9]
Blainville (1835)	*Cordylus* (*Gerrhonotus*) *multi-carinatus* (= *Elgaria multicarinata*): **Monterey [Monterey Co., California]**[10]
Stebbins (1958)	*Elgaria panamintina*: **Surprise Canyon, [. . .], on the west side of the Panamint Mountains, Inyo County, California**[4]
Papenfuss & Parham (2013)	*Aniella alexanderae*: **Shale Rd., 1.3 km S (by road) junction with Hwy. 33, Kern County, California, U.S.A.**
Papenfuss & Parham (2013)	*Aniella campi*: **Big Spring, 5.8 km NW Junction Hwy. 14 (by Hwy. 178), Kern County, California, U.S.A.**
Papenfuss & Parham (2013)	*Aniella grinnelli*: **Jack Zaninovich Memorial Nature Trail, Sand Ridge Preserve, Kern County, California, U.S.A.**
Gray (1852)	*Aniella pulchra*: **Pinnacles National Monument, San Benito County, California**[11]
Papenfuss & Parham (2013)	*Aniella stebbinsi*: **El Segundo Dunes, Los Angeles International Airport, Los Angeles County, California, U.S.A.**
Stejneger (1890)	*Gambelia sila*: **Fresno, Cal.**
Baird (1859a [1858])	*Stenodactylus variegates* (= *Coleonyx variegatus*): **Winterhaven [Fort Yuma], Imperial County, California**[12]
Dixon (1964)	*Phyllodactylus nocticolus*: **Agua Caliente Hot Springs, San Diego County, California**
Baird & Girard (1852)	*Crotaphytus dorsalis* (= *Dipsosaurus dorsalis*): **Winterhaven [Fort Yuma], Imperial County**[12]
Stejneger (1894)	*Uta mearnsi* (= *Petrosaurus mearnsi*): **Summit of Coast Range, United States and Mexican boundary line, California**
Gray (1839)	*Phrynosoma blainvillii*: **California**
Hallowell (1852)	*Anota M'Callii* (= *Phrynosoma mcallii*): **close to the present town of Caléxico**[13]
Hallowell (1854)	*Sceloporus magister*: **Fort Yuma, Yuma County, Arizona**[12]
Baird & Girard (1852)	*Sceloporus occidentalis*: **Benicia**[14]
Stejneger (1893b)	*Sceloporus orcutti*: **the flat just east of Campo, San Diego Co., Calif.**[15]
Phelan & Brattstrom (1955)	*Sceloporus uniformis*: **Valyermo, Los Angeles, California**

TABLE 9.1 *Continued*

Author(s)	Original name: Type locality
Cope (1896)	*Sceloporus vandenburgianus*: **Campbell's Ranch, Laguna**[16]
Cope (1895)	*Uma inornata*: **Riverside County, California**[8]
Baird (1859a [1858])	*Uma notata*: **vicinity of Yuma, Arizona**[12]
Cope (1894)	*Uma scoparia*: **Mojave Desert, California**[8]
Hallowell (1854)	*Urosaurus graciosus*: **Winterhaven [Fort Yuma], Calif.**[12]
Van Denburgh (1896)	*Eumeces gilberti* (= *Plestiodon gilberti*): **Yosemite Valley, Mariposa County, California**
Grismer & Galvan (1986)	*Xantusia gracilis*: **Truckhaven Rocks in the Anza-Borrego Desert State Park, [. . .], San Diego County, California**[4]
Stejneger (1893b)	*Xantusia henshawi*: **Witch Creek, San Diego Co., California**
Cope (1883)	*Xantusia riversiana*: **California**
Bezy (1967)	*Xantusia sierrae*: **Granite Station, 9.1 mi (by rd) S Woody, 1700 ft, Kern Co., California**
Baird (1859a [1858])	*Xantusia vigilis*: **Fort Tejon, California**
Snakes (21)	
Blainville (1835)	*Tortrix bottae* (= *Charina bottae*): **Coast Range, opposite Monterey**[8]
Klauber (1943)	*Charina umbratica*: **Fern Valley, near Idyllwild, Riverside County, California**
Stejneger (1889)	*Lichanura orcutti*: **Colorado Desert, San Diego County, California**
Hallowell (1854)	*Rhinostoma occipitale* (= *Chionactis occipitalis*): **Mojave Desert**
Feldman & Hoyer (2010)	*Contia longicaudae*: **California, Mendocino County, 8.6 km E of junction with Highway 1 via State Route 128 [. . .]**[4]
Bocourt (1886)	*Coronella multifasciata* (= *Lampropeltis multifasciata*): **San Luis-Obispo, California**
Blainville (1835)	*Coluber Californiae* (= *Lampropeltis californiae*): **vicinity of Fresno, California**[8]
Lockington (1876) ex Blainville (1835)	*Coluber* (*Zacholus*) *zonatus* (= *Lampropeltis zonata*): **northern California**[8]
Hallowell (1853)	*Leptophis lateralis* (= *Masticophis* [= *Coluber*] *lateralis*): **San Diego**[8]
Blainville (1835)	*Pituophis catenifer*: **vicinity of San Francisco**[8]
Baird & Girard (1853)	*Rhinocheilus lecontei*: **San Diego**
Kennicott (1860a)	*Thamnophis atratus*: **California**
Kennicott (1859)	*Thamnophis couchii*: **Pitt River, California**
Baird & Girard (1853)	*Eutainia elegans* (= *Thamnophis elegans*): **El Dorado County, California**
Fitch (1940)	*Thamnophis gigas*: **Gadwall, Merced County, California**
Kennicott (1860)	*Eutaenia hammondii* (= *Thamnophis hammondii*): **San Diego, Cal.**
Baird & Girard (1853)	*Rena humilis*: **Upper Sonoran Life Zone of the Vallecito area**[17]
Hallowell (1854)	*Crotalus cerastes*: **borders of the Mohave river, and in the desert of the Mohave**
Cope (1892)	*Crotalus ruber*: **vicinity of San Diego, California**[8]
Kennicott (1861)	*Caudisona scutulata* (= *Crotalus scutulatus*): **Mojave Desert, California**[8]
Klauber (1930)	*Crotalus stephensi*: **two miles west of Jackass Springs, Panamint Mts., altitude 6200 ft., Inyo County, California**

[1] Restricted type locality presented (Schmidt 1953).
[2] Restricted type locality presented (Nussbaum 1976).
[3] Restricted type locality presented (Dunn 1926).
[4] Type locality abbreviated with [. . .]; refer to original description for the complete type locality.
[5] Restricted type locality presented (Van Denburgh 1905).
[6] Restricted type locality presented (Fouquette 1968).
[7] Restricted type locality presented (Jennings 1988).
[8] Restricted type locality presented (Schmidt 1953).
[9] Restricted type locality presented (Stejneger 1902).
[10] Restricted type locality presented (Fitch 1934).
[11] Restricted type locality presented (Murphy & Smith 1991).
[12] Restricted type locality presented (Smith & Taylor 1950).
[13] Restricted type locality presented (Klauber 1932).
[14] Restricted type locality presented (Grinnell & Camp 1917).
[15] Restricted type locality presented (Hall & Smith, 1979).
[16] Restricted type locality presented (Cochran 1961).
[17] Restricted type locality presented (Brattstrom 1953).

California into six major life zones with 17 faunal subdivisions. Many authors have further delimited and subdivided the state, either floristically (Jepson 1925; Wieslander 1935; Jensen 1947; Benson 1957; Munz and Keck 1959; Hickman 1993; Goudey and Smith 1994; Barbour et al. 2007) or faunistically (Van Dyke 1919, 1929; Mayer and Laudenslayer 1988; Schoenherr 1992).

While there is general agreement about the various biogeographic divisions, the floristic regions proposed by Hickman (1993) prove the most useful in understanding the distribution of California's herpetofauna. Of the state's 173 native amphibian and reptile species, 103 (59.5%) are confined to any one of Hickman's eight geographic regions. Because amphibians and reptiles are constrained more by water availability, temperature, and soil type than plant associations, it is not surprising that only a portion are confined to floristic regions. The remaining 70 (40.5%) species have distributions more broadly ranging, some of which span the entire state.

For those species confined to a single floristic area, forming assemblages helps reveal the emergent patterns and constraining elements that have shaped the biogeography of California's amphibians and reptiles. The eight regions are (1) Northwestern California, (2) Cascade Ranges, (3) Great Basin Province, (4) Central Western California, (5) Great Central Valley, (6) Sierra Nevada, (7) Southwestern California, and (8) Desert Province. Assemblages have as few as 2 species (Cascade Ranges) to as many as 35 species in the most specious (Desert Province). In addition to the floristic subdivisions of Hickman (1993), a marine division is added to account for the 5 species of sea turtle and 1 seasnake that occur in the Pacific Ocean along the California coastline. The remaining species are habitat generalists that occur in two or more subdivisions.

Northwestern California

Characterized by its cool, moist climate, mountainous terrain, and lush evergreen forests, this region extends from the Oregon border in the north to just above San Francisco Bay in the south. The Cascade Ranges and Great Central Valley bound this region to the east. Northwestern California contains approximately one-third of the coastline of the state and two prominent mountain ranges. The Klamath Ranges, located in the extreme northwest portion of the state, are composed predominantly of metamorphic and plutonic rocks. In the Cretaceous, the Klamath Ranges dismembered from the northern Sierra Nevada, shifting westward to fall in line with the North Coast Ranges (Schoenherr 1992). The Klamath Ranges consist of the Siskyou, Scott Bar, Salmon, Marble, and Scott Mountains and the Trinity Alps. Thompson Peak in the Trinity Alps is the highest at 2744 m. To the south lies the North Coast Ranges, which parallel the coastline. These mountains were formed by the uplifting of sedimentary rocks of the late Mesozoic (Schoenherr 1992). The major peaks rise above 1500 m, with many above 2000 m. In combination, these two mountain ranges share both climate and vegetation.

Northwestern California has the wettest climate in California. Average annual rainfall can exceed 200 cm. Most precipitation occurs in the winter months, and the cool, moist climate is the most predictable in the state. This favors the maintenance of old-growth forests, including the largest stands of the world's tallest tree, the coast redwood (*Sequoia sempervirens*), which reaches 115.0 m in height and 7.9 m in diameter. These forests rely on both heavy seasonal rains and coastal fog. The hot, dry summer months are moderated by coastal fogs, producing fog drip, which can account for 30% of the precipitation beneath the forest canopy. In the interior, the mountains produce a rain shadow, and annual rainfall drops to 50 cm.

In addition to redwood forests, Northwestern California is home to Douglas fir (*Pseudotsuga menziesii*), yellow pine (*Pinus ponderosa*), and mixed-evergreen and hardwood species. Forests in the Klamath Ranges include western hemlock (*Tsuga heterophylla*), grand fir (*Abies grandis*), or chinquapin (*Chrysolepis sempervirens*). Farther to the south, the Northern Coast Ranges transition to noble fir (*A. procera*) and red fir (*A. magnifica*) and lack hemlock. The eastern and southern interiors have more drought-tolerant species characterized by chaparral and pine-oak woodlands.

Soil types within the region are generally derived from either sedimentary or metamorphic rocks. The region is known for its extensive pockets of serpentine outcrops, consisting of soils with high magnesium and iron content. Serpentine soils have profound effects on the flora. Several hundred endemic plant species are restricted to this soil type (Kruckeberg 2006).

NORTHWESTERN CALIFORNIA ASSEMBLAGE

Of the 48 species that occur in this area, 15 are limited to the region: *Ambystoma gracile, Dicamptodon tenebrosus, Aneides ferreus, A. flavipunctatus, A. vagrans, Plethodon asupak, P. dunni, P. elongatus, P. stormi, Rhyacotriton variegatus, Taricha rivularis, Pseudacris regilla, Ascaphus truei, Rana aurora,* and *Thamnophis ordinoides.*

The region has the third-highest number of species confined to its boundaries, or 8.7% (15/173) of the state's total herpetofauna. Of these, 14 are amphibians (11 salamanders, 3 frogs) and 1 is a reptile. All are noted for their preference for cool, moist habitats, often shaded from sunlight beneath dense forest canopies.

The members of this assemblage show two distributional patterns. Nine are the southernmost distributions of species found farther to the north into Oregon, Washington, and in some cases, British Columbia, Canada: *Ambystoma gracile, Dicamptodon tenebrosus, Aneides ferreus, Plethodon dunni, Rhyacotriton variegatus, Pseudacris regilla, Ascaphus truei, Rana aurora,* and *Thamnophis ordinoides.* The remaining six are regional endemics: *Aneides flavipunctatus, A. vagrans, Plethodon asupak, P. elongatus, P. stormi,* and *Taricha rivularis.* Of these, *A. flavipunctatus, P. stormi,* and *P. elongatus* have distributions that cross into Oregon but are not far ranging outside the Northwestern California region.

Until recently, *A. ferreus* was considered to be farther ranging into California. At present, it is confined to the northwestern corner of the region, where it represents the southernmost distribution of the species. Wake and Jackman (1999) described *A. vagrans* from most of the populations pertaining to *A. ferreus* from California, with the contact zone between the two species located at the Smith River, just 34 km south of the Oregon border. The disjunct distribution of *A. vagrans* (or *A. ferreus*) on Vancouver Island, British Columbia, was long thought to be an introduction from northern California. Using both genetics and historical records from the logging industry, Jackman (1999) provided evidence that the Vancouver Island population originated from tan oak bark shipments from California. Once thought to reside only in fallen or downed logs, *A. vagrans* is now known to use the canopies of old-growth redwood forests as an ecological preference (Spickler et al. 2006).

Other members of the assemblage include recently circumscribed taxa. For instance, *Plethodon auspak* is confined to a small area along the Scott River and was recently described from populations previously recognized as either *P. stormi* or *P. elongatus* (Mead et al. 2005). Similarly, *Pseudacris regilla* was partitioned into multiple species across California, although the specific species boundaries or contact zones between the various groups have yet to be resolved (Recuero et al. 2006a, 2006b). Based on the available samples included in the study, it is presumed that *P. regilla* is confined to the northwestern region of state. The inclusion of *Rana aurora* is also the result of partitioning a more widespread species, having been found to be distinct from *R. draytonii* (Shaffer, Fellers, et al. 2004; Pauly et al. 2008).

While amphibians dominate the assemblage, the region supports a total of 48 species, of which 22 are reptiles. The reptile composition within the Northwestern California region is characterized by those that efficiently thermoregulate under limited sunlight conditions in well-shaded habitats, such as *Sceloporus graciosus* and *Contia longicaudae*, or function well in cool temperatures, such as *Elgaria coerulea, Charina bottae, Contia tenuis,* and *Lampropeltis zonata.* Others include species that prefer moist habitats or wetlands, such as *Diadophis punctatus, Thamnophis atratus, T. elegans, T. sirtalis,* and *Actinemys marmorata.*

A number of subspecies have been described within the Northwestern California region, although most are considered invalid and some may form the basis for the recognition of a more diverse fauna. For instance, two subspecies of *Ensatina eschscholtzii* are recognized from this area, *E. e. oregonensis* and *E. e. picta* (Moritz et al. 1992). Recent studies of *E. eschscholtzii* reveal a complex biogeographic scenario, with as many as four distinct populations within *E. e. oregonensis* and support for *E. e. picta* as a distinct species (Kuchta, Parks, and Wake 2009; Kuchta, Parks, Mueller, and Wake 2009). In other cases, the recognition of subspecies has not been supported: *Elgaria multicarinata scincicauda* is not distinct from *E. m. multicarinata* (Feldman and Spicer 2006); *Diadophis punctatus amabilis* is not distinct from *D. p. occidentalis* (Feldman and Spicer 2006; Fontanella et al. 2008); and *Lampropeltis zonata zonata* is not distinct from other *L. zonata* (Rodríguez-Robles et al. 1999; Myers et al. 2013).

Populations of *Ambystoma californiense* from southern Sonoma County are isolated and not formally described as a taxonomic unit, despite unpublished information that they are a distinct evolutionary lineage (Shaffer, Pauly, et al. 2004; Shaffer and Trenham 2005).

The Northwestern California region contains some of the most extensive stretches of undisturbed coastline in the state. The "Lost Coast" extends 120 km and has few roads or settlements. The majority is protected in the King Range National Conservation Area. Two additional parks exist in the old-growth forests. Humboldt Redwoods State Park protects 6880 ha, and Redwoods National Park includes 7948 ha. The major industries that affect land use include forestry, cattle ranching, and grape vineyards. Of these, logging has received the most attention

for practices that clear-cut the forests, increase erosion, and silt the creeks and rivers. Of the 48 species that occur in the region, more than one-third (35.4%) are currently designated as a conservation concern.

Cascade Ranges

A region dominated by prominent volcanic peaks, the Cascade Ranges extend from British Columbia to northern California. In California, the region is bounded to the north by Oregon, to the west by the metamorphic Klamath Ranges, to the southwest by the agricultural lands of the Great Central Valley, to the southeast by the Sierra Nevada, and to the east by the juniper savanna of the Modoc Plateau in the Great Basin Province (Hickman 1993). The Cascades began to uplift around seven million years ago as a result of a subduction-volcanic arc as the Juan de Fuca Plate beneath the Pacific Ocean pushed beneath the North American Plate (Schoenherr 1992). The most prominent features of the region are Mount Shasta and Mount Lassen, both having erupted in historical times (1786 and 1917, respectively). Mount Shasta reaches 4322 m in elevation and is the second-highest peak in the Cascade Ranges. Mount Lassen reaches 3189 m and is the southernmost peak of the ranges.

The region is typically divided into either the foothills or the highlands. While the high peaks are snow covered year-round, the bases of these mountains and their surrounding areas receive 150 cm of rain per year (Schoenherr 1992). Rain shadow effects occur between the boundaries of the Northwestern California region and along the eastern slopes of the Cascades.

Lower elevations within the Cascade Ranges consist of chaparral and pine-oak woodlands. Higher elevations are dominated by yellow pine (*Pinus ponderosa*) with damp forests on the west side of the slopes. On the east side of the Cascades, the habitat is typically drier and the trees are more widely spaced, with Great Basin plants growing in the understory. Other dominant forest species include sugar pine (*P. lam-*

bertiana), lodgepole pine (*P. contorta*), white fir (*Abies concolor*), and incense cedar (*Calocedrus decurrens*).

CASCADE RANGES ASSEMBLAGE

Of the 31 species that occur in this area, 2 are limited to the region: *Aneides iecanus* and *Hydromantes shastae*. The region has the fewest number of species confined to its boundaries, or 1.2% (2/173) of the state's total herpetofauna. Both are salamanders with preferences for cool, moist habitats. While the floristic region does not strongly correlate with amphibian and reptile endemism, it does represent a unique geological region of the state. Both species are endemic to the region. *Hydromantes shastae* has a narrow range around Shasta Reservoir and can be found in caves, limestone outcrops, and more rarely, within the leaf litter of the forest floor. *Aneides iecanus* was recently recognized as a cryptic species previously recognized as *A. flavipunctatus* (Rissler and Apodaca 2007) and assigned its available name by Frost (2011). *Aneides iecanus* is distributed around Shasta Reservoir as well but ranges more broadly than *H. shastae*.

Of the 31 species with distributions within this region, 67.7% (21/31) are reptiles and 32.3% (10/31) are amphibians. Of these, *Rana cascadae* is noted for having a distribution from British Columbia to northern California and is found predominantly within the Cascade Ranges. In California, it historically occurred throughout the Cascade region and in the Trinity Alps of the Northwestern California region. Today, it is believed to be mostly extirpated from the region, likely the result of the introduction of trout species for sport fishing, which are thought to eat their eggs, larvae, and froglets (Fellers and Drost 1993).

The Cascade Ranges region has a number of parks, including the 43,080 ha Lassen Volcanic National Park and a conglomerate of wildlife refuges, national forests, and recreational areas. About one-third (10/31) of the region's species are currently designated as a conservation concern.

Great Basin Province

In California, this region is represented as two distinct geographic areas: the Modoc Plateau and the eastern Sierra Nevada. The basin and range topography of the Great Basin Province extends into California from its central location in Nevada and makes up the eastern boundary of the northern half of the state. The province is bounded to the west by either the Cascade Mountains or the Sierra Nevada.

The Modoc Plateau, located in the northeastern corner of the state, is a 1500 m volcanic tableland formed approximately 25 million years ago, which borders the Great Basin to the east and shares the same floral characteristics. Great Basin sagebrush (*Artemisia tridentata*) is the dominant plant of the province, with the Modoc Plateau also containing western juniper (*Juniperus occidentalis*) savannas and extensive yellow (*Pinus ponderosa*) and Jeffrey pine (*P. jeffreyi*) forests (Hickman 1993). A number of large interior lakes provide abundant water to the region, including Goose and Eagle Lakes, which drain to the Great Central Valley by way of the Pit River (Schoenherr 1992). The Warner Mountains lie to the east of Modoc Plateau and represent a prominent range whose highest point is Eagle Peak at 3015 m.

The Great Basin Desert lies within the rain shadow of the Sierra Nevada and is composed of a series of north-south-trending mountains and valleys. This high desert has valley floors at approximately 1200–2000 m in elevation and rivers that drain to interior closed basins, often in dry or saline lakes. Great Basin sagebrush is the dominant plant, but saltbrush (*Atriplex confertifolia*) is more abundant at the center of the basins, and pinyon pine (*P. monophylla*), western juniper (*J. occidentalis*), and blackbrush (*Coleogyne ramosissima*) are found along the mountain slopes (Schoenherr 1992). In addition, this region contains extensive riparian corridors dominated by cottonwoods (*Populus fremontii* and *P. angustifolia*) and quaking aspen (*P. tremuloides*). The Inyo and White Mountains represent the most prominent range, whose highest points are Waucoba Mountain at 3390 m and White Mountain Peak at 4344 m, respectively. These mountains are noted for their stands of sub-alpine western bristlecone (*Pinus longaeva*) and limber pine (*P. flexilis*) communities (Hickman 1993). Western bristlecone pines are long-lived trees, with one individual dated as the oldest living tree in North America at 4700 years (Schoenherr 1992).

Both areas within the province are relatively high in elevation and lie within the rain shadows of the Cascade Mountains and Sierra Nevada. As such, the climate of the Great Basin Province is generally drier and cooler, with water flowing into the region from higher-elevation mountains. The province is a region of winter precipitation, mostly snow, totaling approximately 38–50 cm of rainfall annually.

GREAT BASIN PROVINCE ASSEMBLAGE

Of the 41 species that occur in this area, 6 are limited to the region: *Batrachoseps campi, Anaxyrus exsul, Lithobates pipiens, Rana pretiosa, Spea intermontana,* and *Phrynosoma douglasii.*

The region has 3.4% (6/173) of the state's total herpetofauna confined to its boundaries. Of these, 5 are amphibians and 1 is a reptile. All are noted for their preference for or tolerance of cool climates and, in the case of the amphibians, their reliance on moist habitats. Of the amphibians, 1 is a salamander and 4 are frogs.

The members of this assemblage show two distributional patterns. Four have the majority of their distributions outside California and extend in from Oregon or Nevada. These include *R. pretiosa* and *P. douglasii,* with distributions found on the Modoc Plateau or east of the Warner Mountains, and *L. pipiens* and *S. intermontana,* with distributions located on both the Modoc Plateau and east of the Sierra Nevada in the Great Basin Desert. It remains a possibility that *R. luteiventris* occurs in the extreme northeastern corner of California, east of the Warner Mountains (Camp 1917), with reports from Cope (1889)

and Grinnell and Camp (1917) placing it in Warner Valley and Pine Creek near Alturas, Modoc County, respectively. However, no recent studies have confirmed its presence in California, and the museum specimens from the Museum of Vertebrate Zoology cited in earlier works (MVZ 2098–99) have been assigned to *R. pretiosa.* The second distributional pattern is that of narrow endemics found in springs and streams of the White and Inyo Mountains, east of the Sierra Nevada. *Batrachoseps campi* is found along the eastern and western slopes of the Inyo Mountains in deep canyons with localized habitat (Hansen and Wake 2005). *Anaxyrus exsul* is found within approximately 15 ha of habitat, southeast of the White Mountains in Deep Springs Lake and its surrounding springs (Fellers 2005). These restricted distributions are tied to narrow aquatic habitats surrounded by wide-ranging desert.

Of the remaining 35 species found within the Great Basin Province, 7 are confined to this region and the Desert Province: *Elgaria panamintina, Crotaphytus bicinctores, Gambelia wislizenii, Callisaurus draconoides, Phrynosoma platyrhinos, Hypsiglena chlorophaea,* and *Sonora semiannulata.* All are broadly distributed across desert habitats, with the exception of *E. panamintina,* which has a similar distribution as the narrow endemics of the Great Basin Province. It is found in both the White and Inyo Mountains and would be included as part of this assemblage except for populations located farther to the south in the Panamint, Argus, and Cocos Ranges located within the Desert Province (Mahrdt and Beaman 2009).

The Modoc Plateau has been heavily transformed by agriculture (Schoenherr 1992). Reservoirs and irrigation systems have disrupted much of the watershed, with the remaining wetlands protected within a series of wildlife refuges and national forests. In the Great Basin Desert east of the Sierra Nevada, the Owens Valley has been dramatically altered. With the completion of the 359 km Los Angeles Aqueduct in 1913, nearly all of the Owens River was diverted to the Los Angeles metropolitan area. Of the 41 species that occur between the two regions of the Great Basin Province, 13 (31.7%) are currently designated as a conservation concern. The two ranid species restricted to the region, *L. pipiens* and *R. pretiosa*, appear to be nearly extirpated.

Central Western California

This region is bounded by the Pacific Ocean to the west and extends from just north of San Francisco Bay in the north to Point Conception in the south. The Great Central Valley bounds this region to the east. The San Francisco Bay region includes Mount Tamalpias, the Farallon Islands, the Santa Cruz Mountains, and the highest point on Mount Diablo at 1173 m. The Diablo Range extends southward, paralleling the coast, making up the principal mountains of the South Coast Ranges. The highest peak is San Benito Mountain at 1597 m. Smaller and more coastal mountains include the Gabilan Range, the Santa Lucia Range, and the Sierra Madre. The subduction of the Pacific Plate beneath the North American Plate is responsible for the uplift of this region; however, the presence of the San Andreas Fault system makes the geological history complicated (Schoenherr 1992).

The climate of the region is dominated by a winter storm cycle and hot, dry summer months moderated by coastal fogs. The region receives 38–50 cm of rain in the interior and more southern areas and 114–127 cm within the coastal mountains. The Santa Lucia Range supports the southernmost stands of the coast redwood (*Sequoia sempervirens*) and mixed-hardwood trees, which also can be found in pockets along the coast in the Santa Cruz Mountains and north of San Francisco Bay. The majority of the region contains blue oak (*Quercus douglasii*) and foothill pine (*Pinus sabiniana*) woodland and chaparral (Hickman 1993).

CENTRAL WESTERN CALIFORNIA ASSEMBLAGE

Of the 55 species that occur in this area, 5 are limited to the region: *Aneides niger, Batrachoseps gavilanensis, B. incognitus, B. luciae,* and *B. minor.* The region has 2.9% (5/173) of the state's total herpetofauna confined within its boundaries. All are salamanders that prefer cool, moist climates. Because this region is confined to the state, all 5 species represent endemics to the Coast Ranges.

Aneides niger has an isolated distribution within the Santa Cruz Mountains. It was recently elevated to species level using results that confirmed its separation from *A. flavipunctatus* by comparing molecular and morphological data and using niche modeling (Rissler and Apodaca 2007). The remaining 4 species are all Slender Salamanders. Of these, *B. incognitus, B. luciae,* and *B. minor* are found in different regions of the Santa Lucia Mountains and are distinguished by molecular evidence and small differences in morphology from other sympatric forms. *Batrachoseps gavilanensis* is more broadly distributed in the central Coast Ranges and is also distinguished on the basis of molecular evidence and small differences in morphology (Jockusch et al. 2001).

Of the remaining 50 species found in the region, a number have limited distributions, which include the Central Western region and regions to either the north or south. Two species are found between this region and Northwestern California. The newly described *Contia longicaudae* is found in the Santa Cruz Mountains and along the North Coast Ranges (Feldman and Hoyer 2010), while *Dicamptodon ensatus* is distributed from the Santa Cruz Mountains, around the San Francisco Bay, and as far north as Point Arena in Mendocino County (Bury 2005). Four species are found between Central Western and the Southwestern California regions: *B. nigriventris, Anaxyrus californicus, Pseudacris cadaverina,* and *Tantilla planiceps.*

A number of subspecies have been described within the Central Western California region, with some forming the basis to recognize a more diverse fauna. Three subspecies of *Ensatina eschscholtzii* are recognized from this area, *E. e. oregonensis, E. e. xanthoptica,* and *E. e. eschscholtzii* (Moritz et al. 1992). Recent studies of *E. eschscholtzii* reveal a complex biogeographic scenario, with as many as five regional lineages within *E. e. oregonensis* and two within *E. e. xanthoptica* (Kuchta, Parks, and Wake 2009; Kuchta, Parks, Mueller, and Wake 2009). Other subspecies represent extreme isolates. *Ambystoma macrodactylum croceum* occurs in the Santa Cruz Mountains, 250 km from the nearest population of *A. m. sigillatum* from the Sierra Nevada. In a recent study, Myers et al. (2013) found *Lampropeltis zonata* to be composed of two lineages, with their contact zone in the middle of the Central Western California region. As a result, *L. z. multifasciata* was elevated to a full species. The boundary between *L. zonata* and *L. multifasciata* is believed to be in the proximity of the Santa Cruz–Monterey County line.

In other cases, the recognition of subspecies has not been supported: *Diadophis punctatus vandenburgii* is not distinct from either *D. p. amabilis* or *D. p. occidentalis* (Feldman and Spicer 2006; Fontanella et al. 2008), and *Plestiodon gilberti cancellosus* is not distinct from other *P. gilberti* in the southwestern portion of the state (Richmond and Reeder 2002; Richmond and Jockusch 2007).

Several subspecies are in need of further investigation, including *Anniella pulchra nigra* (see Parham and Papenfuss 2009), *Sceloporus occidentalis bocourtii*, *Masticophis lateralis euryxanthus*, *Thamnophis atratus atratus*, *T. a. zaxanthus*, and *T. sirtalis tetrataenia* (= *T. s. infernalis*). While populations of *Ambystoma californiense* from northwestern Santa Barbara County are isolated and not formally described as a taxonomic unit, unpublished data suggest that they are a distinct evolutionary lineage (Shaffer, Pauly, et al. 2004; Shaffer and Trenham 2005). Similarly, populations of *Actinemys marmorata* from Santa Barbara County have been shown to be distinct (Spinks and Shaffer 2005). Support also exists for the recognition of three lineages of *Plestiodon skiltonianus* within this region (Richmond and Reeder 2002; Richmond and Jockusch 2007).

The Central Western California region is heavily urbanized in and around the San Francisco Bay area, which represents the second-largest metropolitan area within the state. Throughout the region, the Coast Ranges are used for ranching and agriculture. The region has a number of national forests, and wildlife refuges provide some protection to the region. The larger parks include Mount Diablo State Park with approximately 8000 ha and Pinnacles National Monument with more than 6500 ha. In addition, most coast redwood stands are preserved in smaller public parks and private reserves. Of the 55 species that occur within the Central Western California region, one-third (34.5%) are currently designated as a conservation concern.

Great Central Valley

Covering approximately 5.8 million ha, the Central Valley extends 650 km in length and 110 km in width. The region is flat with little topographical relief, and elevations range from below sea level to 150 m. The Central Valley is enclosed by the Sierra Nevada to the east and south, the Cascade Ranges to the north, and Coast Ranges to the west. Two major river systems and their watersheds divide the valley into northern and southern halves. The Sacramento River drains the northern half from the Cascade Ranges, and the San Joaquin River drains the southern half, both meeting centrally to form a delta system that empties into the San Francisco Bay.

The climate of the region is dominated by a winter storm cycle and hot, dry summers. The northern half of the valley receives as much as 50 cm of rainfall, while the southern half receives no more than 25 cm annually (Schoenherr 1992). Summer temperatures often exceed 32°C, and dense ground fog is known to form across large sections of the basin in the winter.

The deep sedimentary deposits of the Central Valley originated below sea level more than 145 million years ago, before the uplifting of the Coast Ranges (Schoenherr 1992). Over the last 5 million years, sediments have accumulated from alluvial deposits from the surrounding mountains (Kruckeberg 2006). Few topographical features are present, and water accumulates in marshes and within vernal pools.

Most of the native vegetation has long since been converted to agriculture, and few places represent a pristine condition. It is estimated that 99% of the valley has been transformed. Bunchgrasses had dominated the prairie habitats and included needlegrasses (*Stipa* sp.), triple-awned grasses (*Aristida* sp.), bluegrasses (*Poa* sp.), and ryegrasses (*Elymus* sp.), of which purple needlegrass (*S. pulchra*) was the most dominant (Schoenherr 1992). In addition, plants associated with marshes, vernal pools, riparian woodlands, and islands of oak savanna existed (Hickman 1993).

GREAT CENTRAL VALLEY ASSEMBLAGE

Of the 33 species that occur in this area, 4 are limited to the region: *Anniella alexanderae*, *A. grinnelli*, *Gambelia sila*, and *Thamnophis gigas*.

The region has one of the fewest number of species confined to its boundaries at 2.3% (4/173) of the state's total herpetofauna. Both *A. alexanderae* and *A. grinnelli* were recently described by Papenfuss and Parham (2013), and their ranges occur over a small region of the southern Central Valley. *Thamnophis gigas* historically existed in both northern and southern halves of the Central Valley, but today the species is confined to areas around the Sacramento River (Rossman et al. 1996). *Gambelia sila* historically existed within the drier, southern half of the Central Valley and adjacent areas along its southwestern edge (McGuire 1996). Today, *G. sila* is confined to scattered populations throughout its historical distribution in the Central Valley and the adjacent Carrizo and Elkhorn Plains and Cuyama Valley (Germano 2009).

Of the remaining 29 generalists with distributions within the Central Valley, *Sceloporus occidentalis biseriatus* and *Masticophis flagellum ruddocki* are in need of further investigation. In addition, the recently described species of *Anniella* from the southern half of the valley have disjunct distributions that are in need of further

analysis (Parham and Papenfuss 2009; Papenfuss and Parham 2013).

Few pristine areas within the Great Central Valley remain today. Agricultural use and metropolitan areas around Sacramento, Modesto, Fresno, and Bakersfield are the primary factors in the conversion of this prairie landscape. Small reserves exist throughout the valley, including numerous national wildlife refuges, mostly situated around semidisturbed wetlands. Of the 33 species that occur within the Great Central Valley, over one-third (33.3%) are currently designated as a conservation concern, including both species confined to the assemblage.

Sierra Nevada

The most prominent topographical feature in California, the Sierra Nevada rises to over 4250 m and is composed primarily of granitic rock, with overlying volcanics (Schoenherr 1992). It is bounded by the Cascade Ranges to the northwest, the Great Basin Province to the northeast and east, the Great Central Valley to the west, and the Desert and Southwestern California Provinces to the south. The range extends approximately 640 km in length and 80 km in width. The highest peak is Mount Whitney at 4421 m. In addition, this region includes the Tehachapi Mountains located at the southern terminus of the Sierra Nevada.

The climate varies dramatically depending on elevation and latitude but is characterized by a winter storm cycle bringing either rain to the foothills or snow to higher elevations, with the summer season being hot and dry. Annual rainfall can vary between 50 and 200 cm, and temperatures, from below freezing to above 32°C. In summer months, afternoon rain showers are not uncommon.

The Sierra Nevada granitic batholith formed in the Cretaceous as the result of the subduction of the Pacific Plate below the edge of the North American continent. Uplift of the batholith began 65 million years ago, with subsequent erosion to expose the granitic rock. At 20 million years ago, extensive volcanism covered the region. The present uplift began from 3 to 5 million years ago, resulting in a westward tilt. Uplifting at a faster rate in the southern end of the range has resulted in higher peaks. Rivers and glaciers cut deep canyons on both sides of the range, although the majority of watersheds flow into the Central Valley.

Floristically, the Sierra Nevada is divided between the Foothills and the High Sierra, and within each, a number of subregions have been described (Hickman 1993). Throughout most of the Foothills, blue oak (*Quercus douglasii*) and foothill pine (*Pinus sabiniana*) woodlands dominate, with intermixed serpentine soils. Higher in elevation, the forest transitions to mixed conifer-hardwood, including yellow pine (*P. ponderosa*), sugar pine (*P. lambertiana*), lodgepole pine (*P. contorta*), white fir (*Abies concolor*), incense cedar (*Calocedrus decurrens*), and California black oak (*Q. kelloggii*). The High Sierra is also home to the world's largest trees, the giant sequoia (*Sequoiadendron giganteum*), which exist in scattered groves on the western side of the mountains.

SIERRA NEVADA ASSEMBLAGE

Of the 53 species that occur in this area, 16 are limited to the region: *Batrachoseps altasierrae*, *B. bramei*, *B. diabolicus*, *B. gregarious*, *B. kawia*, *B. regius*, *B. relictus*, *B. robustus*, *B. simatus*, *B. stebbinsi*, *Hydromantes brunus*, *H. platycephalus*, *Anaxyrus canorus*, *Rana sierrae*, *Anniella campi*, and *Xantusia sierrae*.

The region has the second-highest number of species confined to its boundaries, or 9.2% (16/173) of the state's total herpetofauna. Of these, 14 are amphibians and 2 are reptiles. The assemblage is made up of 12 salamanders, 2 frogs, and 2 lizards. Most prefer cool, moist habitats, and many require secretive retreats in rocky outcrops.

The members of this assemblage are all endemic to this region. Two of the Slender Salamanders, *Batrachoseps diabolicus* and *B. gregarious*, are broadly distributed over the central and southern Sierra Nevada, respectively, while the remaining have narrow distributions in the extreme southern end of the range and Tehachapi Mountains. *Batrachoseps regius* is known from the Kings River, *B. kawia* from the Kaweah River, *B. relictus* and *B. simatus* from the lower Kern River Canyon, *B. robustus* from the Kern Plateau and eastern-facing slopes of the southeastern Sierra Nevada, and *B. stebbinsi* from the Tehachapi Mountains. Recently, Jockusch et al. (2012) described populations of *B. relictus* from the upper Kern River Canyon as *B. bramei* and those from Greenhorn Mountain to the Tule River drainage and upper elevations of the Little Kern River drainage as *B. altasierrae*. *Hydromantes brunus* and *H. platycephalus* prefer either limestone outcrops and slate slabs or talus slopes of granite, respectively (Wake and Papenfuss 2005a, 2005b). *Hydromantes brunus* is found in the foothills along the Merced River, while *H. platycephalus* is distributed more broadly across the Sierra Nevada but at higher elevations (1220 to 3600 m).

Of the two frogs restricted to this region, *Rana sierrae* has been recently removed from its synonymy with *R. muscosa* after genetic evidence demonstrated the populations in the vicinity of Kings Canyon were not closely related (Vredenburg et al. 2007). As a result, *R. sierrae* is distributed across much of the north and central Sierra Nevada and *R. muscosa* is distributed across the extreme southern end, with disjunct populations in the Transverse Ranges of Southwestern California. *Anaxyrus canorus* is found in the central Sierra Nevada at sites from 1950 to 3444 m in elevation (Davidson and Fellers 2005).

Papenfuss and Parham (2013) recently described *Anniella campi*, which is known only from three localities in a small region in the southern Sierra Nevada. One, Big Spring, contains only 2 ha of suitable habitat (Papenfuss and Parham 2013). *Xantusia sierrae* has a highly restricted distribution in the southern foothills of the Sierra Nevada. It is found along the western slopes of the Greenhorn Mountains at approximately 500 m in elevation and uses rock crevices in granitic boulders as its primary retreat (Bezy 2009).

Of the remaining 37 species, a number of subspecies have been described, with some forming the basis to recognize a more diverse fauna. Three subspecies of *Ensatina eschscholtzii* are recognized from the Sierra Nevada at present: *E. e. xanthoptica*, *E. e. platensis*, and *E. e. croceater* (Moritz et al. 1992). Recent studies of *E. eschscholtzii* reveal a complex biogeographic scenario, with as many as two regional lineages within *E. e. platensis* and weak evidence for *E. e. croceater* to be recognized as a distinct lineage (Kuchta, Parks, and Wake 2009; Kuchta, Parks, Mueller, and Wake 2009; Pereira and Wake 2009). Support exists for the separation of *Taricha sierrae* from *T. torosa*, as they are shown to hybridize along the Kaweah River (Kuchta 2007). Support also exists for the distinction of *Diadophis punctatus pulchellus* from the Sierra Nevada. Although its distributional pattern is similar to that of *T. torosa*, evidence suggests that members of coastal populations of *D. punctatus* migrated into the southern Sierra Nevada (Feldman and Spicer 2006; Fontanella et al. 2008). Further investigation of contact zones is warranted.

There are a number of undescribed populations that have been shown to be evolutionarily distinct. For instance, evidence for separate lineages for the Sierra Nevada populations have been provided for *Charina bottae* (Rodríguez-Robles et al. 2001; Feldman and Spicer 2006), *Contia tenuis* and *Elgaria multicarinata* (Feldman and Spicer 2006), and *Plestiodon gilberti* (Richmond and Reeder 2002; Richmond and Jockusch 2007). The discovery of more species of *Batrachoseps* from the evolutionarily complex southern portions of the range is likely (Jockusch et al. 2012).

In other cases, the recognition of *Lampropeltis zonata multicincta* does not appear to be warranted and is not distinct from other *L. zonata* (Rodríguez-Robles et al. 1999; Myers et al. 2013). Other subspecies are in need of further investigation, including *Sceloporus occidentalis taylori*.

The enormous size and rugged terrain of the Sierra Nevada provide some degree of environmental protection to the region. Still, the foothills are heavily ranched, the forests are logged, and the rivers have been dammed (Schoenherr 1992). Three large parks exist within the mountain, including the spectacular, glacial-carved Yosemite National Park and the contiguous Sequoia and Kings Canyon National Parks. Together, they amount to 576,920 ha of preserved land. Of the 53 species that occur within the Sierra Nevada, 21 (39.6%) are currently designated as a conservation concern.

Southwestern California

Characterized by its dry, moderate, coastal climate, the region extends from Point Conception to the US-Mexico border. It is bounded by Central Western California, the Great Central Valley, and the southern Sierra Nevada to the north; the Desert Province to the east; and the Pacific Ocean to the west. The region contains approximately one-third of the coastline of the state, two prominent mountain ranges, coastal basins and mesas, and the Channel Islands. The coastal influences are punctuated by sky island habitats of its mountains. The Transverse Ranges are positioned east to west and include the San Bernardino, San Gabriel, Santa Monica, and Santa Ynez Mountains. Of these, Mount San Gorgonio within the San Bernardino Mountains is the tallest peak, with an elevation of 3505 m. South of these ranges lie the north-to-south-positioned Peninsular Ranges, which include the San Jacinto, Palomar, Cuyamaca, and Laguna Mountains. Of these, Mount San Jacinto is the highest at 3302 m in elevation. The region also includes the Channel Islands, which are divided into two groups (Schoenherr et al. 1999). The Northern Channel Islands are positioned in an east-to-west line off Ventura and Santa Barbara Counties. They contain the Anacapa, Santa Cruz, Santa Rosa, and San Miguel Islands. The Southern Channel Islands are more dispersed and consist of Santa Catalina, Santa Barbara, San Nicolas, and San Clemente Islands.

A complex set of geological faults is associated with the northward movement of the Pacific Plate relative to the North American Plate along the San Andreas Fault (Schoenherr 1992). The Peninsular Ranges represent the northern end of the Baja California peninsula, which terminates at the San Jacinto Mountains. Across the San Gorgonio Pass to the north begin the Transverse Ranges with the San Bernardino Mountains. During Pliocene times, elevated sea levels inundated coastal regions and the Los Angeles Basin, possibly separating the two mountain ranges. The Channel Islands are a part of the continental borderland, formed by uplift as the result of the interactions of the region's complex fault system (Schoenherr et al. 1999).

Southwestern California experiences a classic Mediterranean climate with long, warm summers and moderately cool, wet winters, with snow in the higher elevations. This type of climate promotes vegetation with small, drought-tolerant leaves and waxy coatings (Schoenherr 1992). The area receives the tail end of winter storms originating in the northern Pacific Ocean, resulting in less precipitation than in more northern parts of the state. The coastal regions receive approximately 25 cm of annual rainfall. During the late spring and early summer months, warm temperatures are moderated by marine onshore breezes and fog. In the mid- to late summer, the region's mountains receive sporadic monsoon rains originating from tropical systems migrating up from the south.

Chaparral dominates the landscape of Southwestern California, with coastal sage scrub occurring at lower elevations and mixed conifer-hardwood forests at the higher elevations. The drought-tolerant chaparral is also noted for its association with periodic fires (Hanes 1971), which it requires for regrowth and germination (Quinn and Keeley 2006). Chaparral species include chamise (*Adenostoma fasciculatum*), red shank (*A. sparsifolium*), and laurel sumac (*Malosma laurina*). Large hardwood trees, such as the coast live oak (*Quercus agrifolia*) are interspersed across the landscape. Coastal sage scrub species include California sagebrush (*Artemisia californica*), California brittle-bush (*Encelia californica*),

black sage (*Salvia mellifera*), and white sage (*S. apiana*). Montane communities contain yellow pine (*Pinus ponderosa*), sugar pine (*P. lambertiana*), Coulter pine (*P. coulteri*), and California black oak (*Q. kelloggii*).

SOUTHWESTERN CALIFORNIA ASSEMBLAGE

Of the 56 species that occur in this area, 14 are limited to the region: *Batrachoseps gabrieli, B. major, B. pacificus, Ensatina klauberi, Gambelia copeii, Sceloporus orcutti, S. vandenburgianus, Aspidoscelis hyperythra, Xantusia henshawi, X. riversiana, Charina umbratica, Lichanura trivirgata, Masticophis fuliginosus,* and *Crotalus ruber.* The region has the fourth-highest number of species confined to its boundaries, or 8.1% (14/173) of the state's total herpetofauna. Of these, 4 are amphibians and 10 are reptiles.

The members of this assemblage represent either regional endemics or northern extensions of species distributed broadly across the Baja California peninsula. Eight are regional endemics, with varying forms of isolating mechanisms. Of these, *B. gabrieli* has the most limited distribution and is found in 13 discrete locations along the southern slopes of the San Gabriel and San Bernardino Mountains (Hansen, Goodman, and Wake 2005). *Batrachoseps major* is more broadly distributed across the region, with notable populations (*B. m. aridus*) on the desert slopes of the Peninsular Ranges. While *B. major* has a broader range, it is represented by at least two distinct lineages and is currently the subject of more research (Wake and Jockusch 2000).

Four of the regional endemics are confined to sky islands, with varying degree of elevational preferences. *Charina umbratica* is distributed from Mount Pinos to the San Jacinto Mountains and is confined to California (Rodríguez-Robles et al. 2001). *Ensatina klauberi, S. vandenburgianus,* and *X. henshawi* are distributed more widely across the region's mountains and include populations in northern Baja California (Grismer 2002; Lovich 2001).

The remaining two regional endemics are confined to the Channel Islands. *Xantusia river-*

siana is distributed across the Southern Channel Islands and is found on San Clemente, San Nicolas, and Santa Barbara, as well as Sutil Island, an islet off Santa Barbara Island (Fellers and Drost 2009). *Batrachoseps pacificus* is restricted to the Northern Channel Islands, including East Anacapa, Middle Anacapa, West Anacapa, Santa Cruz, Santa Rosa, and San Miguel (Hansen, Wake, and Fellers 2005).

Six species range across the Baja California peninsula to the south, and their northern distributions terminate in the Southwestern California region: *G. copeii, S. orcutti, A. hyperythra, L. trivirgata, M. fuliginosus,* and *Crotalus ruber* (Grismer 2002).

Of the remaining 42 species found in this region, 4 share their distributions with Central Western California (*Anaxyrus californicus, Pseudacris cadaverina, Tantilla planiceps,* and *Thamnophis hammondii*), and 4 share their distributions with the Desert Province (*Coleonyx variegatus, Lichanura orcutti, Trimorphodon lyrophanes,* and *Crotalus mitchellii*). *Rana muscosa* is also found in the southern Sierra Nevada. The remaining are more widely distributed generalists and occur in three or more of the regions.

A number of subspecies have been described within the Southwestern California region, with some forming the basis to recognize a more diverse fauna. For example, while *Diadophis punctatus modestus* and *D. p. similis* cannot be distinguished from each other, they were found to be distinct from other California subspecies of *Diadophis* (Feldman and Spicer 2006; Fontanella et al. 2008). Support also exists for the recognition of *Plestiodon skiltonianus interparietalis* and the possibility of additional lineages within *P. skiltonianus* (Richmond and Reeder 2002; Richmond and Jockusch 2007).

Other regional subspecies in need of further investigation include *Thamnophis sirtalis infernalis, Coleonyx variegatus abbotti, Aspidoscelis tigris stejnegeri, Xantusia riversiana reticulata,* and *X. r. riversiana.* For example, the contact zone between *Crotalus oreganus oreganus* and *C. o. helleri* occurs just north of the Santa Ynez Mountains at the boundary between this region and Central Western California (Schneider 1986). However, genetic studies have not consistently supported a clear distinction between the two at this location (Pook et al. 2000; Ashton and de Queiroz 2001; Douglas et al. 2002). Support for two Channel Island endemics, *Pituophis catenifer pumilis* and *Sceloporus occidentalis becki,* is in need of study, although Rodríguez-Robles and De Jesús-Escobar (2000) provided evidence that *P. c. pumulis* is nested within *P. catenifer.*

The Southwestern California region is the most heavily urbanized area of the state and struggles with population growth, road construction, and urban sprawl. Small, regional parks and reserves are present along the coast, but more than 90% of the natural habitat has been transformed into metropolitan areas. Preserves include the region's military bases, where development is limited due to the nature of their uses. Mountain habitat is protected in national forests and small state parks. Of the 56 species that occur within the region, 21 (37.5%) are currently designated as a conservation concern.

Desert Province

A surprisingly diverse ecosystem, the Desert Province is noted for its hot climate and low annual rainfall but also contains a number of aquatic systems and high-elevation mountain ranges. It is bounded by the Great Basin Province and Sierra Nevada to the north, Southwestern California to the west, Baja California to the south, and Arizona and Nevada to the east. Two deserts lie within the province. The more northern, high-elevation Mojave Desert has valley floors that range from 600 to 1200 m. The southern Lower Colorado Desert, a part of the larger Sonoran Desert, has valley floors from −60 to 100 m in elevation. Throughout the eastern Mojave Desert, mountain ranges provide extreme elevational gradients. The Panamint Range is the most dramatic, rising to 3369 m in elevation from valley floors found below sea level. Along the region's eastern boundary, the Colorado River has an average discharge through the desert of 402 m^3/s of water.

The Mojave Desert is a part of the basin and range system of spreading valley floors and uplifting mountains. The Lower Colorado Desert contains the Salton Trough, which has been successively flooded by either the northward extension of the Gulf of California in the Pliocene or by the Colorado River in Pleistocene and Holocene times. Drying in the last 10,000 years of both the river systems in the northern Mojave Desert and Lake Cahuilla in the Lower Colorado has made conditions ideal for the formation of windblown dune systems across much of the region (Schoenherr 1992).

Summer mean temperatures reach 43.5°C, while winter mean temperatures can drop as low 1.1°C. Annual rainfall ranges from 0 to 25 cm across most of the region, with high mountain peaks, some of which are often snowcapped, receiving up to 50 cm. Portions in the rain shadow of either the Sierra Nevada or the Peninsular Ranges receive proportionately less precipitation. The Lower Colorado Desert often receives sporadic monsoon rains in mid- to late summer.

Creosote bush (*Larrea tridentata*) is characteristic across both deserts, usually in association with burro bush (*Ambrosia dumosa*). The Mojave Desert also includes Great Basin sagebrush (*Artemisia tridentate*), Joshua tree (*Yucca brevifolia*), Mojave yucca (*Y. schidigera*), and a host of cholla cactus (*Cylindropuntia* sp.). The Lower Colorado Desert contains ocotillo (*Fouquieria splendens*), smoketree (*Psorothamnus spinosus*), ironwood (*Olneya tesota*), and palo verde (*Parkinsonia florida*), in addition to a rich community of cacti.

DESERT PROVINCE ASSEMBLAGE

Of the 63 species that occur in this area, 35 are limited to the subregion: *Anaxyrus cognatus, A. microscaphus, A. punctatus, A. woodhousii, Incilius alvarius, Lithobates yavapaiensis, Scaphiopus couchii, Crotaphytus vestigium, Coleonyx switaki, Phyllodactylus nocticolus, Heloderma suspectum, Dipsosaurus dorsalis, Sauromalus ater, Petrosaurus mearnsi, Phrynosoma mcallii, Sceloporus magister, Uma inornata, U. notata, U. scoparia, Urosaurus graciosus, U. nigricaudus, U. ornatus, Xantusia*

gracilis, X. wigginsi, Bogertophis rosaliae, Chionactis occipitalis, Phyllorhynchus decurtatus, Thamnophis marcianus, Trimorphodon lambda, Crotalus atrox, C. cerastes, C. scutulatus, C. stephensi, Kinosternon sonoriense, and *Gopherus agassizii*.

The region has the highest number of species confined to its boundaries, or 20.2% (35/173) of the state's total herpetofauna. Of these, 7 are amphibians and 28 are reptiles. Most prefer warm, dry habitats or have nocturnal or secretive habits to avoid the desert heat.

The members of this assemblage represent either regional endemics, northern extensions of species distributed broadly across the Baja California peninsula, species broadly distributed across the region's deserts, or those tied to the aquatic habitats centered on the Colorado River.

Regional endemics include species endemic to the Mojave or Lower Colorado Deserts. They include *Uma inornata, U. notata, U. scoparia, X. gracilis, Crotalus stephensi*, and *G. agassizii*. Of these, *U. inornata* and *X. gracilis* are narrowly endemic within the state. *Uma inornata* is confined to windblown sands of the Coachella Valley (Barrows and Fisher 2009), while *X. gracilis* is found in a relatively small sandstone formations within Anza-Borrego Desert State Park (Grismer and Galvan 1986). The remaining four species have distributions that extend beyond the state boundaries but do not broadly range across the region. *Uma scoparia* is confined to the Mojave Desert and includes an isolated population in the Bouse Dunes, Arizona (Espinoza 2009). *Crotalus stephensi* ranges across the Mojave Desert and portions of Nevada (Meik 2008), similar to the newly circumscribed *G. agassizii* (Murphy et al. 2011). *Uma notata* is found across the Lower Colorado Desert, which extends into northeastern Baja California (Grismer 2002).

Eight species confined to the California deserts also occur broadly across the Baja California peninsula: *Crotaphytus vestigium, Coleonyx switaki, Phyllodactylus nocticolus, Petrosaurus mearnsi, Phrynosoma mcallii, Urosaurus nigricaudus, Xantusia wigginsi*, and *Bogertophis rosaliae* (Grismer 2002). Another 12 species extend broadly across North American deserts: *Helo-*

derma suspectum, Dipsosaurus dorsalis, Sauromalus ater, Sceloporus magister, Urosaurus graciosus, U. ornatus, Chionactis occipitalis, Phyllorhynchus decurtatus, Trimorphodon lambda, Crotalus atrox, C. cerastes, and *C. scutulatus* (Stebbins 2003).

The remaining 9 species are confined to mesic habitats, most in association with the Colorado River. In addition, many have distributions within the Imperial Valley, likely as a result of the growing agricultural industry and the opening of the All-American Canal in 1940, which diverted waters from the Colorado River (Schoenherr 1992): *Anaxyrus cognatus, A. microscaphus, A. punctatus, A. woodhousii, Incilius alvarius, Lithobates yavapaiensis, Scaphiopus couchii, Thamnophis marcianus,* and *Kinosternon sonoriense* (Stebbins 2003).

The Desert Province contains a number of parks and reserves, including Death Valley and Joshua Tree National Parks, the Mojave National Preserve, and Anza-Borrego Desert State Park. Despite these conservation areas, the deserts are heavily utilized, with large metropolitan areas, military bases, agricultural regions, geothermal fields, dams along the Colorado River, and off-road vehicle use (Schoenherr 1992). More recently, large, open tracts of desert land are being developed for wind and solar energy generation. Of the 63 species that occur within the region, 16 (25.4%) are currently designated as a conservation concern.

Pacific Ocean

The continental shelf along the California coastline extends from 30 to 275 km offshore and averages 400 m in depth before plunging to depths of more than 3500 m. The cold California Current follows the coastline, flowing from north to south, but is disrupted at Point Conception, where the coastline extends to the east to form the Southern California Bight (Schoenherr et al. 1999). The south-to-north Southern California Countercurrent is the result of coastal eddies forming as the California Current flows past the bight. In summer, water temperatures in northern California average 14°C in the sum-

mer, while in southern California temperatures average 20°C.

PACIFIC OCEAN ASSEMBLAGE

There are 6 species limited to the subregion: *Hydrophis platurus, Caretta caretta, Chelonia mydas, Eretmochelys imbricata, Lepidochelys olivacea,* and *Dermochelys coriacea.* Of the 5 species of sea turtle that range into California waters, *C. mydas* is the most common and retreats into the San Diego Bay during late fall and winter months (Stinson 1984). The occurrence of *H. platurus* is rarer, but records have this species ranging as far north as San Clemente, Orange County (Campbell and Lamar 2004).

Generalists

Of the 173 species found in California, 70 (40.5%) are considered generalists, with distributions extending into two or more subregions.

GENERALIST ASSEMBLAGE

The Generalist assemblage includes these species: *Ambystoma californiense, A. macrodactylum, Dicamptodon ensatus, Aneides lugubris, Batrachoseps attenuatus, B. nigriventris, Ensatina eschscholtzii, Taricha granulosa, T. sierrae, T. torosa, Anaxyrus boreas, A. californicus, Pseudacris cadaverina, P. hypochondriaca, P. sierra, Rana boylii, R. cascadae, R. draytonii, R. muscosa, Spea hammondii, Elgaria coerulea, E. multicarinata, E. panamintina, Anniella pulchra, A. stebbinsi, Crotaphytus bicinctores, Gambelia wislizenii, Coleonyx variegatus, Callisaurus draconoides, Phrynosoma blainvillii, P. platyrhinos, Sceloporus graciosus, S. occidentalis, S. uniformis, Uta stansburiana, Plestiodon gilberti, P. skiltonianus, Aspidoscelis tigris, Xantusia vigilis, Charina bottae, Lichanura orcutti, Arizona elegans, Coluber constrictor, Contia longicaudae, C. tenuis, Diadophis punctatus, Hypsiglena chlorophaea, H. ochrorhyncha, Lampropeltis getula, L. multifasciata, L. zonata, Masticophis flagellum, M. lateralis, M. taeniatus, Pituophis catenifer, Rhinocheilus lecontei, Salvadora hexalepis, Sonora* semiannulata, *Tantilla hobartsmithi, T. planiceps, Thamnophis atratus, T. couchii, T. elegans, T. hammondii, T. sirtalis, Trimorphodon lyrophanes, Rena humilis, Crotalus mitchellii, C. oreganus,* and *Actinemys marmorata.*

The most broadly distributed of the generalists are *Anaxyrus boreas, Sceloporus occidentalis, Diadophis punctatus, Lampropeltis getula, Pituophis catenifer,* and *Crotalus oreganus.* These six species occur in all eight terrestrial regions.

California contains 18 islands that have amphibian and reptile records. Islands extend from the Southern Channel Islands north to the San Francisco Bay and include San Clemente, Santa Catalina, San Nicolas, Santa Barbara, Anacapa, Santa Cruz, Santa Rosa, San Miguel, Morro Rock, Año Nuevo, South and Southeast Farallon, Alcatraz, Angel, Brooks, East Marin, West Marin, Red Rock, and Yerba Buena (Schoenherr et al. 1999). Species inhabiting islands include *Aneides lugubris, Batrachoseps attenuatus, B. major, B. nigriventris, B. pacificus, Ensatina eschscholtzii, Pseudacris hypochondriaca, P. sierra, Rana draytonii, Elgaria coerulea, E. multicarinata, Sceloporus occidentalis, Uta stansburiana, Plestiodon skiltonianus, Xantusia riversiana, Charina bottae, Coluber constrictor, Diadophis punctatus, Hypsiglena ochrorhyncha, Lampropeltis getula, L. multifasciata, Pituophis catenifer, Thamnophis elegans, T. hammondii, Crotalus oreganus,* and *Actinemys marmorata.* In addition, three species have been introduced: *Lithobates catesbeianus, L. pipiens,*

and *Xantusia vigilis.* There are documented cases of species native to islands being moved from island to island or mainland to island. Some populations of *Uta stansburiana* and *Elgaria multicarinata* on the Channel Islands have been shown to be recently established (Mahoney et al. 2003).

As many as 21 species have been introduced into California. Some introductions pose severe ecological damage, such as *Lithobates catesbeianus* and *Nerodia* sp., while others are considered harmless. The persistence of each introduction is in need of more thorough documentation. Introduced species include *Ambystoma mavortium, Eleutherodactylus coqui, Xenopus laevis, Lithobates berlandieri, L. catesbeianus, L. sphenocephalus, Chamaeleo jacksonii, Gehyra mutilata, Hemidactylus turcicus, Tarentola annularis, T. mauritanica, Podarcis sicula, Nerodia fasciata, N. rhombifer, N. sipedon, Indotyphlops braminus, Apalone spinifera, Chelydra serpentina, Chrysemys picta, Graptemys pseudogeographica,* and *Trachemys scripta.*

A total of 89 subspecies have been recognized for 30 California species. However, many are in the process of being studied to verify their taxonomic validity. Amphibians belonging to 3 species, *Ambystoma macrodactylus, Anaxyrus boreas,* and *Ensatina eschscholtzii,* contain 10 subspecies between them. Reptiles have 79 subspecies recognized within 27 species.

Herpetofauna of Arizona

THOMAS C. BRENNAN
AND RANDALL D. BABB

Introduction

Arizona is one of the most herpetologically diverse states in the United States. Heterogeneous terrain combined with a wide altitudinal range, subtropical to temperate latitudes, and a bimodal rainfall regime conspire to endow Arizona with a multifarious and vibrant regional flora and fauna including six major biomes and a variety of riparian communities. Many species of plants and animals reach their northern distributional limits within the borders of Arizona. The Sky Island Province of southeastern Arizona is an epicenter of sorts for reptile diversity, boasting several species that are encountered nowhere else in the United States. The state's deserts also contribute significantly to its assortment of herpetozoan species. Arizona is the only state in which all four North American deserts are represented. These deserts, though low in endemic species, are great in their variety of reptile and amphibian fauna. The Sonoran Desert is the warmest, wettest, and most biologically diverse of these semiarid communities and contributes greatly to Arizona's collection of amphibian and reptile species.

Despite its abundance of flora and fauna, Arizona did not receive much attention from naturalists until the mid-1800s, largely because of its late induction into statehood and native peoples who fought hard to retain lands lived and hunted on. Early interest in the biota of the region was largely confined to exploratory and commercial forays by trappers, surveyors, miners, and boundary and military expeditions (Coues 1875; Wheeler 1875; Yarrow 1875; Davis 1982; Brown et al. 2009). Trappers and miners showed little interest in reptiles and amphibians, most likely because of lack of commercial value associated with these taxa. The exploratory and boundary survey period fostered a growing awareness and interest in this less exploitable wildlife (Coues 1875; Wheeler 1875; Yarrow 1875), and the type specimens of many species of reptiles and amphibians were collected during this time. General interest in the herpetofauna continued through the 1900s, which saw many new and unusual species discovered within the confines of the state. This interest has continued into the present. The relatively new science of genetic analysis has unearthed several species of reptiles and amphibians in recent years, and more are expected to come.

Previous Herpetological Studies

Arizona, when compared with the eastern United States, has a relatively short history of herpetological interest and research. Though

Spanish missionaries and explorers ventured far into what is now Arizona as early as the 1500s, they recorded essentially nothing regarding the reptiles and amphibians they encountered. Whether it was due to the unfavorable collecting seasons for ectotherms, the difficulty in finding and capturing reptiles and amphibians, or lack of interest, relatively little was documented regarding this state's herpetofauna. In like manner other early naturalists, such as the son of renowned naturalist John James Audubon who visited Arizona in the first half of the 1800s, mentioned reptiles and amphibians only in passing, as curiosities. Arizona was to remain largely a herpetological terra incognita until the last half of the 1800s. Explorations for the southern boundary surveys, western railway system, and the ensuing territorial years brought to light a diverse and fascinating reptile and amphibian fauna.

In these early years interest in documenting the state's wildlife was largely couched in the context of commercial exploitation and general exploration of the West (Davis 1982; Brown et al. 2009). Reptiles and amphibians were, more often than not, collected opportunistically and largely neglected in favor of the birds and mammals. In the mid-1800s this view of amphibians and reptiles began to change largely because of the broader interests and urgings of Spencer Fullerton Baird, who had taken the position of curator at the National Museum in 1850.

As a member of Col. J. D. Graham's 1851 expeditions of the Scientific Corps of the United States and Mexican Boundary Survey, John H. Clark collected amphibian and reptile specimens from present-day Arizona, including the type specimens of *Sceloporus clarkii*, *Lampropeltis splendida*, and *Salvadora grahamiae*. Clark's collection of the type specimen of *Sonora semiannulata* presumably occurred on this expedition as well (Stickel 1943; Degenhardt et al. 1996).

The Lorenzo Sitgreaves 1851 Expedition to explore and map the Zuni and Colorado Rivers traveled across the Colorado Plateau of northern Arizona, to the Colorado River west of Kingman, and then south through Yuma. Amphibian and reptile specimens, including the type specimens of *Anaxyrus woodhousii* and *Callisaurus draconoides ventralis,* were collected during this expedition. Expedition members included S. W. Woodhouse, for whom *A. woodhousii* is named.

Herpetozoan specimens collected in Arizona during W. H. Emory's 1855 US and Mexican boundary surveys included the type specimen of *Heloderma suspectum* from Morena Mountain on the international border southwest of San Miguel, Pima County (Bogert and Martín Del Campo 1956).

Army surgeons performed numerous roles during their deployment to the frontier, in addition to defense and medical duties; they were the designated naturalists and documented and collected wildlife. Famed ornithologist and collector C. Elliot Coues (pronounced "cows") was one of the first naturalists to earnestly collect reptiles and amphibians in Arizona and in turn authored the first in-depth treatment of the group. In 1863 Arizona was made a territory, and the 20-year-old C. E. Coues's first assignment, upon graduating from Columbia University, was to serve at Fort Whipple, situated near present-day Prescott. He was recommended for the position by Spencer F. Baird from the Smithsonian, who had earlier become acquainted with Coues's prowess as a naturalist. All along his journey to Fort Whipple Coues collected specimens. Lt. Charles A. Curtis, who was also attached to the expedition, wrote of Coues: "For creeping, crawling and wriggling things he had brought along a five-gallon keg of alcohol," and he often ventured far from the column in search of specimens, only returning at the end of the day with "his pockets and pouches filled with the trophies of his search." So keen was Coues's interest and dedication to obtaining specimens that he received a reprimand for firing his weapon to secure a particularly desirable bird during a touchy encounter with the native people of the area. The acquisitions made during his deployment to Arizona were jeopardized by the enlisted men who drank the alcohol in which the reptiles and amphibians were stored (Curtis 1902, 6–7). In a journal entry dated July 8, 1864, Coues describes

Ambystoma mavortium larvae encountered at Jacob's Well as "suspicious looking creatures" and as resembling catfish "barring the legs and long, fringe-like gills" (Coues 1871, 200). He arrived at his post later that month.

Though his stay at Fort Whipple was only 16 months, Coues collected an astonishing number of specimens and published many of the foundational faunal works in Arizona. Specimens collected during Coues's military service were deposited in the National Museum.

Coues is responsible for publishing one of the first scientific papers on the region's herpetofauna in 1875. "Synopsis of the Reptiles and Batrachians of Arizona: With Critical and Field Notes, and an Extensive Synonymy," which appeared in Wheeler's (1875) *Report upon Geographical Explorations and Surveys West of the One Hundredth Meridian, Volume V*, is the first thorough treatment of Arizona's reptiles and amphibians. In this report Coues discussed 83 species and numerous subspecies. His efforts brought to light the first specimens of *Crotalus cerberus*, then thought to be a subspecies of *C. lucifer* (= *C. oreganus*). Coues (1875) proposed the name *cerberus* for the subspecies he encountered in Arizona. During the execution of the work that was ultimately published in Coues's *Report*, many specimens were deposited in the National Museum, including notable Arizona species such as the first specimen of *Lampropeltis pyromelana*, along with commonly encountered species such as *Crotalus atrox, C. molossus, C. mitchellii, Crotaphytus collaris, Gambelia wislizenii*, and *Kinosternon sonoriense*.

Henry C. Yarrow served as the surgeon and chief zoologist on Lt. George M. Wheeler's expeditions to the West for the years of 1872 and 1874. He was head of a team of formidable field men, which included renowned ornithologist Henry W. Henshaw and paleontologist Edward D. Cope. Yarrow (1875) compiled a report on western herpetofauna similar to that of Coues, but of a broader, less focused context. In the introduction of his report, *Collections of Batrachians and Reptiles Made in Portions of Nevada, California, Colorado, Utah, New Mexico, and Arizona during the Years of 1871, 1872, 1873, and 1874*, Yarrow states that only in the year 1873 did Arizona along with New Mexico and Colorado serve as a focus for the expeditions' herpetological investigations. However, in 1874, two additional "natural history parties" were formed to work in areas of particular herpetological interest. One was under the leadership of Yarrow and worked in Colorado and New Mexico, while the other operated under the guidance of botanist and surgeon Joseph T. Rothrock and collected in Arizona and New Mexico as far south as the Mexican border. Cope's original descriptions of *Sceloporus jarrovii* and *Thamnophis rufipunctatus* can be found in Yarrow's 1875 work. Yarrow's report was also published in Wheeler's (1875) *Report* and immediately preceded Coues's in the text (Yarrow 1875).

Prominent among biologists working in the West during the late 1800s was the renowned naturalist and collector Edgar A. Mearns, who was stationed at Fort Verde in central Arizona from March 1884 through May 1888. He was a fervent collector and made many long explorations across the new territory. In his lifetime Mearns deposited numerous reptile specimens in the United States National Museum, though it has yet to be determined how many of these are from Arizona. In 1892 Mearns was assigned to the Mexican Boundary Surveys and reentered Arizona with collector Frank X. Holzner, who assisted in collecting specimens. This effort saw the acquisition of numerous additional specimens from Arizona (Mearns 1907; Richmond 1918).

The 1890s saw a proliferation of naturalists visiting and working in Arizona (Ewan 1981; Davis 1982; Brown et al. 2009). Army surgeons still played key roles as collectors of biological material in the territorial West, but professional collectors, biologists, and laypersons were now playing a greater role in securing specimens for institutions (Ewan 1981). The interests of most of these men were not in reptiles or amphibians but rather in natural history disciplines that incidentally brought them into contact with herpetozoan species. Several of the naturalists working in this time period collected type specimens of various reptiles and amphibians, and the names of these men would later become patronyms for creatures familiar to biologists working in the Southwest today. Army Surgeon General Timothy E. Wilcox, a botanist stationed at southeast Arizona's Fort Huachuca, collected reptiles extensively during his deployment, including the type specimen of the Chihuahuan Black-headed Snake (*Tantilla wilcoxi*) named after him (Stejneger 1902 [1903]). Wilcox also collected the westernmost specimen of *Sistrurus catenatus edwardsii* from the parade grounds on the fort (Holycross 2003). No additional *S. c. edwardsii* specimens have been documented on or near Fort Huachuca since. William W. Price of California, a tireless field man, collected many specimens from Arizona for John Van Denburgh at the California Academy of Sciences. Van Denburgh (1895) named a small new species of rattlesnake that Price collected *Crotalus pricei* in honor of the collector's efforts. Frank C. Willard was a schoolteacher and amateur ornithologist who lived in Tombstone and collected in southeastern Arizona. He sent a collection of about 40 reptiles and amphibians to the Field Museum in Chicago. Among these specimens was a single Ridge-nosed Rattlesnake (*Crotalus willardi*), which was described by ichthyologist Seth Meek in 1905 and named after Willard. Newspaperman Herbert Brown was an ornithologist/collector who worked in southern Arizona and was one of the founders of the herpetological and ornithological collections at the University of Arizona. So great was his dedication to natural history that his name pops up in nearly every acknowledgment of the period. The Saddled Leaf-nosed Snake (*Phyllorhynchus browni*) was named in Brown's honor in 1890 by L. Stejneger of the National Museum.

Numerous other biologists of note contributed specimens to collections during the late 1800s and early 1900s, mostly men who are more commonly associated with their work with other taxa. These important collections significantly advanced our understanding of Arizona's herpetofauna. Prominent among these men are Vernon O. Baily, J. A. Loring, E. W. Nelson, E. A. Goldman, Henry Skinner, and C. Birdseye.

Various local treatments of Arizona's herpetofauna began to appear in the 1890s and early 1900s. These reports often compiled the data and specimens procured from collecting efforts earlier that century, such as Leonard Stejneger's (1902 [1903]) "Reptiles of the Huachuca Mountains, Arizona." In this paper he addresses approximately 30 specimens secured by notable collectors T. Wilcox, W. Price, F. Holzner, A. Fisher, and Lieutenant Benson. He also provides the original description for *Tantilla wilcoxi* and dismisses *Crotalus scutulatus* as a regional variant of *C. atrox*. Other compendiums of this time were the products of personal efforts. Alexander G. Ruthven's 1907 "A Collection of Reptiles and Amphibians from Southern New Mexico and Arizona" provides a thorough treatment of the roughly 1000 specimens collected during his expedition to the area in 1906. It was notable because he paid particular attention to the habitat and context in which the animal was encountered, often documenting it in some detail, and photographs accompanied some of the accounts. In 1913 John Van Denburgh and Joseph R. Slevin published "A List of Reptiles and Amphibians of Arizona, with Notes on Species in the Collection of the Academy," which was the result of a personal field effort made by the authors. They list 86 species of reptiles and amphibians then thought to occur in the state and discuss specimens held at the California Academy. Several plates of *Crotalus* species follow the text. Van Denburgh published many papers on the regional herpetofauna during his time at the California Academy that had applications to Arizona.

Expeditions such as Henry A. Pilsbry's 1910 journey were continuing to ply Arizona for new natural wonders. Pilsbry, a malacologist, searched for mollusks in and around the Chiricahua, Mule, Dragoon, Santa Rita, Rincon, and Baboquivari Mountains. Though the expedition's focus was decidedly non-herpetological in nature, they collected several reptiles and amphibians incidentally, which were deposited in the Academy of Natural Sciences Museum in Philadelphia. This period also included US Army captain William L. Carpenter, an entomologist and naturalist, who collected a series of specimens in the Camp Verde and Prescott area for the National Museum in 1889. And banker and ornithologist John E. Law collected extensively in Arizona in 1919, particularly in the Chiricahua Mountains, and deposited numerous specimens in the University of California's Museum of Vertebrate Zoology.

In 1920 biologist Joseph R. Slevin, who followed J. Van Denburgh as curator of herpetology at the California Academy of Sciences, collected extensively in the Santa Rita and Huachuca Mountains. One small lizard he captured there would later bear his name. As a result of this trip, Slevin deposited well over 100 specimens in the collection he would later oversee. The 1920s and 1930s saw Charles T. Vorhies and Walter P. Taylor of the Biological Survey championing reptile and amphibian study and collection in Arizona. Aside from Herbert Brown, Arizona had no other resident biologists with herpetological leanings. Both collected numerous specimens, which were deposited in the University of Arizona's growing herpetological collection. Vorhies, an entomologist by training, showed a particular interest in reptiles and amphibians. He collected the first specimens of *Oxybelis aeneus* and found the first *Craugastor augusti* in Arizona, lobbied for the protection of the Gila Monster, and published miscellaneous short papers and articles on reptiles.

Interest in reptiles and amphibians increased during the ensuing years, and an awareness of Arizona's unique position in the national herpetological scene was slowly coming to light. By the 1930s and 1940s this new awareness had gained enough momentum to fuel research through many of the coming decades. This embrace of herpetology was driven in part by an increasing number of popular treatments of North American herpetofauna. The six-volume handbook series authored by Albert and Anna Wright, Archie Carr, Hobart Smith, and Sherman Bishop, though national in scope, addressed what was

known about Arizona's herpetofauna, providing detailed observations and behavior information. Many herpetologists of note were working, if only occasionally, in Arizona during this period, which saw two new species added to the state's herpetofauna (*Hyla wrightorum* and *Xantusia arizonae*) in addition to numerous subspecies. Renowned herpetologist Howard K. Gloyd, director of the Chicago Academy of Sciences, made several expeditions to Arizona to collect for that institution during this time.

During the 1940s Hebert Stanke became a public figure with the foundation of the Poisonous Animal Research Laboratory at Arizona State University. Though Stanke is best known for his work with scorpions, he lectured, consulted, and wrote on venomous reptiles and was known for having a post office–style drop box where students and the public could deposit specimens at his laboratory day or night. The focus of much of his herpetological writings was concerned with envenomation by rattlesnakes.

The 1950s witnessed an upsurge of herpetological activity in Arizona. Much of this was due to Charles H. Lowe's arrival at the University of Arizona in 1950. Charles H. Lowe was Arizona's first true resident herpetologist. Lowe's diverse interests are reflected by the varied topics of his numerous publications. During his 45-year tenure at the University of Arizona he, along with students such as Robert Bezy, J. Cole, and John Wright, grew the roughly 1000 specimen herpetological collection more than 50-fold, making it one of the most comprehensive assemblages of southwestern US herpetological material in the world. Lowe's presence brought numerous students to Arizona who studied under him and greatly expanded the knowledge about Arizona reptiles and amphibians. Many of these became authorities and mentors in their own right at diverse institutions. Among Lowe's many important contributions to Arizona herpetology are his numerous papers on the particularly vexing genus *Aspidoscelis*, often written in association with J. Wright and J. Cole. Charles Lowe authored and coauthored more than 160 popular and scientific publications, including a collaborative work with

C. R. Schwalbe and Terry Johnson, *Venomous Reptiles of Arizona*, the first semitechnical work of its kind dealing specifically with the state.

Another Arizona herpetologist of note, William H. Woodin, came to the state in 1952 to work at the renowned Arizona-Sonora Desert Museum. Two years later he became the director, finding himself in a position to influence the displays and programs of the institution. With the assistance of Lowe, Woodin helped increase understanding and to a good degree popularized reptiles and amphibians in the public's eye. Woodin also contributed to the regional body of work in publications such as his 1953 work, "Notes on Some Reptiles from the Huachuca Area of Southern Arizona," and his record of the first *Gyalopion quadrangulare* from Arizona (Woodin 1962), and through extensive collecting of museum material which would later be deposited at the University of Arizona.

In 1958 Arizona became home to another notable herpetologist. Howard K. Gloyd moved to Arizona to serve as a research associate and lecturer at the University of Arizona until his retirement in 1974, when he became emeritus professor. During his tenure at the University of Arizona he collaborated with and served as a mentor to the young Lowe.

Authors like Carl Kauffeld and Robert C. Stebbins, a friend and colleague of C. H. Lowe, also assisted in the popularization of Arizona reptiles with their books and research. Stebbins's 1954 publication of *Reptiles and Amphibians of Western North America* provided the first comprehensive guide to the West and therefore Arizona's herpetofauna, and Kauffeld's 1957 *Snakes and Snake Hunting* with chapters titled "Huachuca Heaven" and "Ajo Road," made Arizona a popular destination for amateur and professional reptile collectors alike.

The first book to be published solely on Arizona snakes was the 1965 work of the University of California's Jack A. Fowlie, *The Snakes of Arizona*, which covered all the recognized species and subspecies of Arizona. Interestingly, it would remain one of the very few books dedicated to Arizona's herpetofauna ever published.

However, throughout the 1960s, 1970s, and 1980s many herpetologists were working in Arizona on either particular species or genera and publishing outstanding papers on their research.

Wade C. Sherbrooke started his excellent work on the genus *Phrynosoma* during this period while attending the University of Arizona. Sherbrooke continued his research on the genus while serving as director of the American Museum of Natural History's Southwestern Research Station in Portal from 1982 to 2003 and still maintains his work there today.

The year 1983 saw the Arizona Game and Fish Department form the Nongame Branch, a division made up of dedicated biologists who oversaw the management of and research on state and federally protected wildlife and species of special concern. Cecil R. Schwalbe was appointed Arizona's first state herpetologist in 1984 and served until 1990. *Gopherus agassizii*, *G. morafkai*, *Lithobates* spp., and *Ambystoma mavortium stebbinsi* were some of the species monitored and researched. Work by the department with these and other species continues today. After leaving the Arizona Game and Fish Department, Schwalbe returned to the University of Arizona, working for the US Geological Survey. There he and his students continued work on many reptile and amphibians, including the aforementioned species.

Arizona State University built a formidable collection of preserved amphibian and reptile specimens during this period. Currently over 32,000 specimens, mostly from Arizona and the Southwest, reside there. This impressive collection reflects the fine work of the many noted herpetologists who attended and taught at this university. Laurie J. Vitt, James P. Collins, Arthur Hulse, James E. Platz, Brian K. Sullivan and later Andrew T. Holycross all made major contributions to the collection.

James C. Rorabaugh, now retired from the US Fish and Wildlife Service, contributed significantly throughout his career to the conservation of Arizona's reptiles and amphibians. He championed federal protection of *Lithobates chiricahuensis*, *Ambystoma mavortium stebbinsi*,

Phrynosoma mcallii, and *Crotalus willardi obscurus*. In addition, Rorabaugh and Rob Clarkson
were the first to note the precipitous decline of
Arizona's ranid frogs.

Though the ensuing years have seen numerous biologists who have worked on and contributed to unraveling the secrets of Arizona's reptiles and amphibians, few have assembled large
or diverse bodies of work specifically addressing
our state's herpetofauna. Among those who have
are James P. Collins and his students, who have
added much to understanding the state's only
native salamander species, *Ambystoma mavortium*. Brian K. Sullivan, once a student at Arizona State University and now a professor on the
West Campus, and his students have written on
reptiles and particularly amphibians, solving taxonomic and ecological riddles. James E. Platz, in
cooperation with others, was largely responsible
for deciphering identification issues that so long
plagued biologists working with leopard frogs in
the Southwest and described several new species
in doing so. Philip C. Rosen, a research scientist
at the University of Arizona, has been conducting and publishing research on a wide variety
of the state's herpetofauna for more than two
decades. Dale DeNardo and students at Arizona
State University have done much with *Crotalus
atrox* and *Heloderma suspectum*, and DeNardo
was one of the first investigators to discover and
document a *Heloderma* nest in the wild.

Sadly, any treatment of researchers and their
work in a state with a herpetofauna as diverse
and rich as that of Arizona will inevitably fall
far short of the mark and leave many deserving
people unmentioned. We recognize this and can
only apologize for our inadvertent omissions.
We wish to acknowledge Cecil R. Schwalbe for
his help and Andrew T. Holycross for generously
sharing his table of early snake collections in
Arizona.

Many of the type specimens of reptiles and
amphibians listed in table 10.1 were collected
prior to the 1853 Gadsden Purchase from Mexico
(Sale of La Mesilla). As a result, many of these
types have "Sonora, Mexico" listed as their
original locality. These localities have often been
revised over time by various authors to better reflect where the specimens came from.

Physiographic Characteristics and Their Influence on the Distribution of the Herpetofauna

Various characteristics of amphibians and reptiles limit their activity and distribution. Because
amphibians and reptiles depend on the environment for body-temperature regulation, they
generally possess a more narrow range of environmental tolerance than do endotherm vertebrates. Plant and animal community patterns are
formed in response to temperature and precipitation extremes as well as local conditions such
as elevation, longitude, slope exposure, geomass,
temperature inversions, cold air drainages, and
soil porosity (Lowe and Brown 1994). Therefore,
an understanding of the physical characteristics
of any given Arizona locality is necessary to understand the nature of its herpetofauna.

The state of Arizona encompasses a land area
of 295,523 km^2, making it the sixth-largest state
in the United States (Kearney and Peebles 1951;
DeLorme 2008). It lies between the coordinate extremes of roughly 31°20′ S, 36°60′ N,
−109°02′ E, and −114°49′ W (DeLorme 2008).
Elevation in the state ranges from a high of
3850 m at the top of Humphreys Peak in the San
Francisco Mountains near Flagstaff to 21.3 m
above sea level at the Colorado River near Yuma
(Chronic 1983; DeLorme 2008). The state can be
divided into three main physiographic provinces:
the Basin and Range Province of southern and
far western Arizona, the mountainous Transition Zone of central Arizona, and the Colorado
Plateau of northern Arizona.

The Basin and Range Province encompasses
most of southern Arizona and is characterized
by isolated, rocky mountains that rise abruptly
from alluvium-filled valleys and plains. In
the southeastern portion of this province the
Chiricahua, Huachuca, Pinaleño, Santa Catalina, and Santa Rita Ranges rise to over 2743 m
from valleys that range in elevation from about

TABLE 10.1 Type localities for amphibians and reptiles described from the state of Arizona, USA

Author(s)	Original name: Type locality
Anurans (7)	
Cope (1867)	*Bufo (= Anaxyrus) microscaphus*: **"Territory of Arizona, Upper Colorado River" = Fort Mohave, Mohave Co., Arizona Mohave County, Arizona**
Sanders & Smith (1951)	*Bufo debilis retiformis (= Anaxyrus retiformis)*: **14.4 miles south of Ajo, Pima County, Arizona**
Girard (1854)	*Bufo (= Anaxyrus) woodhousii*: **[Territory of] New Mexico, restricted to San Francisco Mountain, Coconino CO., Arizona**[1]
Cope (1866)	*Hyla arenicolor*: **Northern Sonora, Mexico, restricted to "Santa Rita Mts.," Pima and Santa Cruz counties, Arizona,**[2] **changed to Peña Blanca Springs, 10 miles northwest of Nogales, Santa Cruz County, Arizona**[3]
Taylor (1939 [1938])	*Hyla wrightorum*: **Eleven miles south of Springerville, Apache County, Arizona**
Platz & Mecham (1979)	*Rana (= Lithobates) chiricahuensis*: **Herb Martyr Lake, 6 km west of Portal, Coronado National Forest, Cochise Co., Arizona**
Platz & Frost (1984)	*Rana (= Lithobates) yavapaiensis*: **Tule Creek, Yavapai Co., Arizona**
Turtles (2)	
Le Conte (1854)	*Kinosternon sonoriense*: **Tucson, Pima Co., Arizona**
Murphy et al. (2011)	*Gopherus morafkai*: **Tucson (approximate location 32° 7′ N, 110° 56′ W, elevation 948 m), Pima County, Arizona**
Lizards (19)	
Gray (1838)	*Elgaria kingii*: **Mexico, restricted to Mojárachic Chihuahua,**[2] **changed to Huachuca Mountains Arizona**[4]
Klauber (1945)	*Coleonyx variegatus bogerti*: **Xavier, Pima Co., Arizona**
Cope (1869)	*Heloderma suspectum*: **Monument 146, Sierra de Moreno, Arizona/Sonora border.** See Bogert and Martín Del Campo (1956) for a complete discussion of the type locality.
Hallowell (1852)	*Callisaurus draconoides ventralis*: **New Mexico, restricted to Tucson, Pima Co., Arizona.**[2] See Degenhardt et al. (1996, 139–141) for further discussion on the type locality.
Schmidt (1921)	*Holbrookia maculata pulchra*: **Carr Canyon, 5,200 ft, Huachuca Mountains, Cochise Co., Arizona**
Gray (1845)	*Phrynosoma solare*: **California, changed to Tucson, [Pima Co.], Arizona**[4]
Baird & Girard (1852)	*Sceloporus clarkii*: **Province of Sonora, restricted to Santa Rita Mountains [Pima or Santa Cruz Co.], Arizona**[2]
Cope *in* Yarrow (1875)	*Sceloporus jarrovii*: **"southern Arizona," restricted to the "Huachuca Mts. Cochise Co., Arizona,**[2] **changed to "between Fort Grant and the Fort Bowie National Historic Site, Arizona, in an area encompassing the southeastern Pinaleño (Graham), Dos Cabezas, and northwestern Chiricahua mountains," Cochise Co., Arizona.**[5]
Hallowell (1854)	*Sceloporus magister*: **Near Fort Yuma, at Junction of Colorado and Gila, also near Tucson in Sonora, restricted to Fort Yuma, Yuma, Yuma Co., Arizona**[2]
Smith (1937)	*Sceloporus scalaris slevini (= S. slevini)*: **Miller Peak, Huachuca Mountains, Cochise Co., Arizona**
Cope (1895)	*Uma rufopunctata*: **Arizona, restricted to monument 200, Yuma Desert, Yuma County, Arizona**[6]
Van Denburgh (1896)	*Cnemidophorus (= Aspidoscelis) arizonae*: **Fairbank, Cochise County, Arizona**

TABLE 10.1 *Continued*

Author(s)	Original name: Type locality
Wright & Lowe (1993)	*Cnemidophorus (= Aspidoscelis) pai*: **Hermit Basin in Grand Canyon, 4800 ft, Coconino County, Arizona**
Lowe & Wright (1964)	*Cnemidophorus (= Aspidoscelis) sonorae*: **2 miles southwest of Oracle, 4,500 ft. elev., near the north base of the Santa Catalina Mountains, Pinal Co., Arizona**
Wright & Lowe (1965)	*Aspidoscelis uniparens*: **Fairbank, Cochise Co., Arizona**
Springer (1928)	*Cnemidophorus gularis velox (= Aspidoscelis velox)*: **Oraibi, Arizona, and Pueblo Bonito, New Mexico, restricted to Oraibi, Navajo Co., Arizona**[7]
Duellman & Lowe (1953)	*Cnemidophorus sacki xanthonotus (= Aspidoscelis xanthonota)*: **North fork of Álamo Canyon, Ajo Mountains, approximately 19 miles north of the international boundary in Organ Pipe Cactus National Monument, Pima Co., Arizona**
Klauber (1931)	*Xantusia arizonae*: **One mile south of Yarnell, 4940 ft, Yavapai County, Arizona**
Papenfuss et al. (2001)	*Xantusia bezyi*: **5.6 km S (by Highway 87) of Sunflower, elevation 948 m., Maricopa County, Arizona (33°49.48′ N, 111° 28.55′ W)**

Snakes (20)

Author(s)	Original name: Type locality
Klauber (1946)	*Arizona elegans philipi*: **10 miles east of Winslow, Navajo Co., Arizona**
Cope (1860)	*Gyalopion canum*: **Near Fort Buchanan Arizona**
Cope (1860)	*Hypsiglena chlorophaea*: **Fort Buchanan, Arizona**
Baird and Girard (1853)	*Lampropeltis getulus (= getula) splendida*: **Sonora, Mexico, restricted to Santa Rita Mountains, Arizona**[2]
Cope (1867)	*Ophibolus pyromelanus (= Lampropeltis pyromelana)*: **Fort Whipple [Yavapai County, Arizona]**[8]
Cope (1892)	*Masticophis (= Coluber) flagellum piceus*: **Camp Grant, Arizona, (= Fort Grant, Graham Co., Arizona)**[9]
Stejneger (1890)	*Phyllorhynchus browni*: **Tucson, Pima Co., Arizona**
Baird & Girard (1853)	*Salvadora grahamiae*: **Sonora, Mexico, restricted to Huachuca Mountains, Cochise Co., Arizona**[4]
Cope (1867)	*Salvadora hexalepis*: **Fort Whipple, Yavapai Co., Arizona**
Baird & Girard (1853)	*Sonora semiannulata*: **Sonora Mexico, restricted to the vicinity of the Santa Rita Mountains, Arizona**[10]
Stejneger (1902 [1903])	*Tantilla wilcoxi*: **"Fort Huachuca" Cochise Co., Arizona**
Kennicott (1860)	*Thamnophis macrostemma megalops (= T. eques megalops)*: **Tucson and St. Magdalena, restricted to Tucson, Pima Co., Arizona**[2]
Cope *in* Yarrow (1875)	*Chilopoma (= Thamnophis) rufipunctatus*: **"Southern Arizona," changed to the vicinity of Fort Apache, Apache Co., Arizona**[5]
Coues (1875)	*Crotalus cerberus*: **San Francisco Mountains, Coconino County, Arizona**
Gloyd (1936)	*Crotalus lepidus klauberi*: **Carr Canyon, Huachuca Mountains, Cochise Co., Arizona**
Cope (1867)	*Crotalus mitchellii pyrrhus*: **Canyon Prieto, Yavapai Co., Arizona**[2]
Van Denburgh (1895)	*Crotalus pricei*: **Huachuca Mountains, Cochise Co., Arizona**
Kennicott (1861)	*Caudisona scutulata (= Crotalus scutulatus)*: **Wickenburg, Maricopa Co., Arizona**[2]
Kennicott *in* Baird (1859)	*Crotalus tigris*: **Sasabe, Pima Co., Arizona**[2]
Meek (1905)	*Crotalus willardi*: **Tombstone, Arizona, revised to above Hamburg, middle branch of Ramsey Canyon, Huachuca Mountains (altitude about 7,000 ft.), Cochise County, Arizona**[11]

Restriction appeared in [1] Smith & Taylor (1948); [2] Smith & Taylor (1950); [3] Gorman (1960); [4] Schmidt (1953); [5] Webb & Axtell (1986); [6] Cochran (1961); [7] Lowe (1955); [8] Coues (1875); [9] Degenhardt et al. (1996); [10] Stickel (1943); [11] Swarth (1921).

914–1220 m. The north-south-oriented Babo-quivari Mountains at the center of Arizona's southern border form a rough dividing line between the southeastern portion of the Basin and Range Province and the lower-elevation western portion. The western basins vary in elevation from about 30 to 915 m. The rocky, steep, isolated ranges scattered throughout these flatlands include the Ajo, Cabeza Prieta, Comobabi, Gila, Gila Bend, Harcuvar, Harquahala, Kofa, Mohave, Quijotoa, Sand Tank, Sauceda, Sierra Estrella, Silver Bell, Tucson, and White Tank Mountains. Of these only the Harcuvar (1598 m), Harquahala (1732 m), and Mohave (1554 m) complexes exceed 1524 m in elevation.

The Transition Zone extends across central Arizona from just east of Kingman heading in an east-by-southeast direction through the White Mountains and into New Mexico, separating the Basin and Range from the Colorado Plateau. The Mogollon Rim, a nearly vertical escarpment of about 304 m in height and predominantly composed of Precambrian rock one to two billion years old, extends from the Verde River at the center of the state east by southeast into the White Mountains and forms the core of the Transition Zone (Kearney and Peebles 1951; Nations and Stump 1981; Chronic 1983). Several of this province's mountains reach well over 2286 m in elevation: Bradshaws, Black Hills, Mazatzals, Pinals, and Sierra Ancha. At an elevation of 3475.6 m, Baldy Peak in the White Mountains is the highest point in the Transition Zone.

The Colorado Plateau Province covers an area of approximately 116,550 km^2 across northern Arizona and is characterized by vast plains and isolated ranges. The plateaus here range from about 1270 to 2775 m in elevation and include Black Mesa and the Coconino, Hualapai, Kaibab, and Kanab Plateaus. Here, the state's most well-known physical feature, the Grand Canyon, and the state's highest point, the San Francisco Peaks stratovolcano, exist in stark contrast to the broad plateaus (Nations and Stump 1981). With a maximum depth of approximately 1.6 km, the Grand Canyon and associated upstream canyons serve as an effective barrier, isolating the plateaus to

the north (the land known as the Arizona Strip). A few of the ranges on the Arizona Strip are high enough in elevation to support coniferous forests, including the Kaibab Plateau (2791 m in elevation), Mount Logan, Mount Trumbull, and the Virgin Mountains. Distinctive ranges to the east in the Colorado Plateau Province include the Carrizo and Chuska Mountains near the border with New Mexico, both of which rise to over 2865 m in elevation and support coniferous forest.

With the exception of a few small creeks flowing south along the southern border, the entire area of the state is drained toward the west into the Colorado River, which flows south into Sonora, Mexico, en route to the Sea of Cortés. The major Arizona tributary of the Colorado is the Gila River, which, along with its major tributaries, drains most of the Transition Zone. The Little Colorado and Bill Williams Rivers drain most of the remainder of the state. From its headwaters in the White Mountains, the Little Colorado River flows west by northwest across the Colorado Plateau, where it is fed by north-flowing tributaries from the Transition Zone. This river is typically dry west of Winslow, but its bed carries floodwaters to the Colorado River in the eastern Grand Canyon. The western portion of the Transition Zone is drained by the Bill Williams River, which flows west into the Colorado River near the center of Arizona's western border. The Salt River, a tributary of the Gila, flows west across much of central Arizona and is fed by south-flowing tributaries from the Transition Zone, including central Arizona's Verde River; Tonto, Cherry, Canyon, and Cibeque Creeks; and the Black and White Rivers. The Gila River enters southeastern Arizona from New Mexico and flows west along the southern edge of the Transition Zone. Tributaries flowing south into the Gila from the eastern portion of the Transition Zone include the Blue River, San Francisco River, Eagle Creek, and San Carlos River. From the south the Gila is fed by the San Pedro River, which flows north from the state's southern border. Due to damming, diverting, and groundwater pumping, the once-perennial Gila currently has intermittent flow in most stretches in central and western Arizona (Brown et al. 1981). Here, the Gila's now typically dry bed passes near the southwest corner of Phoenix, where it is fed by the Salt River, which is also intermittent in the Phoenix area. Below the confluence with the Salt, the Gila's mostly dry bed continues west to Yuma, where it historically drained into the south-flowing Colorado River. The intermittent Santa Cruz River originates in the San Rafael Valley east of the Patagonia Mountains in southeastern Arizona, flows south into Mexico, makes a 180-degree turn, and then flows north along the western side of the Patagonia Mountains, through the western side of Tucson, and into the Santa Cruz Flats desert to the north, where it runs dry before reaching the Gila (Brown et al. 1981).

The diversity of topographical features, broad altitudinal range, and subtropical to temperate latitudes of Arizona contribute to the state's wide range of climatic conditions. Precipitation and temperature in the state vary greatly by season and location. Annual precipitation in the Transition Zone is usually 15–20 inches greater than that in the western deserts of the Basin and Range, and the state's high-elevation forests have an average annual temperature that is −6.6 to −3.8°C lower than that of the southwestern deserts (Sellers 1964).

The northeastern portion of Arizona's Colorado Plateau, the southwestern portion of the Basin and Range, and the San Simon Valley in southeastern Arizona receive less than 304.8 mm of precipitation in an average year (Sellers 1964; Sellers and Hill 1974; Garfin and Emanuel 2006). The basins of southwestern and far western Arizona are the most arid portions of the state and typically receive less than 127 mm of rain per year. In contrast, the highest portions of the Transition Zone, Colorado Plateau, and the highest southeastern ranges receive more than 635 mm of precipitation (rain and melted snow) in a typical year (Sellers 1964; Sellers and Hill 1974; Western Regional Climate Center 2001). Statewide, July through August is the wettest portion of the year. This summer moisture,

referred to as monsoon thunderstorms, often originates from the Gulf of Mexico, flowing into Arizona from the southeast. Arizona monsoon storms may also be associated with moist tropical air flowing into the state from the Sea of Cortés and the Pacific Ocean (Sellers 1964; Sellers and Hill 1974; Western Regional Climate Center 2001; Garfin and Emanuel 2006). Summer monsoon storms are typically in the form of brief, heavy, afternoon showers often preceded by blowing dust and accompanied by high winds, thunder, lightning, and flash flooding in washes and arroyos (Western Regional Climate Center 2001). Arizona normally experiences a second increase in precipitation from mid-November through mid-March, associated with large storms entering the continent from the Pacific Ocean. These cooler-month events are typically more subdued but longer in duration than the summer monsoon storms (Sellers 1964; Sellers and Hill 1974; Western Regional Climate Center 2001; Garfin and Emanuel 2006). Snow is a rarity in the western portion of the Basin and Range, but nearly all Arizona ranges rising to more than 2134 m in elevation receive more than 1.2 m of snow per year, and snow accounts for most of the winter precipitation in the highest portions of the state (Sellers 1964; Sellers and Hill 1974). May and June, a period in which southwestern Arizona frequently receives no measurable rain, are the driest months statewide (Sellers 1964).

Temperatures in Arizona vary as much as topography, altitude, and precipitation. Statewide, there is generally a −1.1°C to 4.4°C increase in temperatures between winter and summer, and daily maximum temperatures are occasionally more than 10°C higher than daily minimums (Sellers 1964; Sellers and Hill 1974; Western Regional Climate Center 2001). Subfreezing winter temperatures are common in the Transition Zone and on the Colorado Plateau, whereas in the basins and bajadas of southwestern Arizona winter daily low temperatures are usually above freezing and winter daily highs are near 21.0°C (Sellers 1964; Western Regional Climate Center 2001). In the Transition Zone summer daily temperatures typically range from lows near 4.4°C to high temperatures below 32.2°C (Sellers 1964; Western Regional Climate Center 2001). Summer daily high temperatures in southwestern Arizona are normally above 37.7°C, and lows are normally between 21.0°C and 26.6°C (Sellers 1964; Sellers and Hill 1974). Summer daily high temperatures in the valleys of southeastern Arizona are normally in the vicinity of 32.2°C, but above 2438 m on the adjacent mountains daily highs are usually below 29.4°C and morning low temperatures are often about 10.0°C (Sellers 1964). On the Colorado Plateau summer daily high temperatures regularly register above 32.2°C, except on the higher plateaus and mesas, and early-morning low temperatures are usually about 10.0°C to 15.5°C (Sellers 1964; Sellers and Hill 1974). The warmest period of the year usually occurs during late June and early July, before the onset of the summer monsoon rains (Sellers 1964).

The richly varied topography and climatic conditions of the state result in an equally diverse collection of communities. Arizona is home to 14 biotic communities ranging from hot and xeric deserts to frigid alpine tundra. These communities are intersected by numerous rivers and streams, which are home to many specialized organisms and contribute greatly to the biodiversity of the state.

Lower Colorado River Sonoran Desert Scrub

Lower Colorado River Sonoran Desert Scrub occurs on the flatlands of the southwestern quarter of the state at elevations ranging from 21 to 400 m. Creosote bush (*Larrea tridentata*), white bursage (*Ambrosia dumosa*), and desert saltbush (*Atriplex polycarpa*) characterize this community. Ironwood (*Olneya tesota*), blue palo verde (*Parkinsonia florida*), and smoketree (*Psorothamnus spinosus*) line the dry washes that transect these basins (Brown and Lowe 1980; Turner and Brown 1994). At the lowest elevations saguaros (*Carnegiea gigantea*) are largely confined to minor drainages and wash edges in this community.

Large portions of this community have been

lost or damaged as a result of agriculture, development, illegal cross-country operation of off-highway vehicles, and the overgrazing of cattle.

At least 42 amphibian and reptile species can be found in this community: *Anaxyrus cognatus, A. punctatus, A. retiformis, A. woodhousii, Incilius alvarius, Smilisca fodiens, Gastrophryne mazatlanensis, Scaphiopus couchii, Kinosternon arizonense, Gambelia wislizenii, Coleonyx variegatus, Heloderma suspectum, Dipsosaurus dorsalis, Callisaurus draconoides, Phrynosoma goodei, P. mcallii, P. platyrhinos, Sceloporus magister, Uma rufopunctata, U. scoparia, Urosaurus graciosus, U. ornatus, Uta stansburiana, Aspidoscelis tigris, Arizona elegans, Chilomeniscus stramineus, Chionactis occipitalis, Hypsiglena chlorophaea, Lampropeltis getula, Masticophis flagellum, Phyllorhynchus browni, P. decurtatus, Pituophis catenifer, Rhinocheilus lecontei, Salvadora hexalepis, Sonora semiannulata, Thamnophis marcianus* (near agricultural development), *Micruroides euryxanthus, Rena humilis, Crotalus atrox, C. cerastes,* and *C. scutulatus.*

Arizona Upland Sonoran Desert Scrub

Arizona Upland Sonoran Desert Scrub dominates the bajadas and slopes of the southwestern portion of the Basin and Range Province at elevations ranging from about 290 to 1050 m. It is a succulent-rich community with the saguaro (*Carnegiea gigantea*) being the most conspicuous of these. Dominant vegetation in this community also includes various cholla and prickly pear species (*Cylindropuntia* and *Opuntia,* respectively), fishhook pincushion (*Mammillaria grahamii*), fishhook barrel cactus (*Ferocactus wislizeni*), triangle-leaf bursage (*Ambrosia deltoidea*), brittlebush (*Encelia farinosa*), jojoba (*Simmondsia chinensis*), ocotillo (*Fouquieria splendens*), foothill palo verde (*Parkinsonia microphylla*), ironwood (*Olneya tesota*), mesquite (*Prosopis* sp.), and cat-claw acacia (*Acacia greggii*) (Brown and Lowe 1980; Turner and Brown 1994).

Invasive grasses and weeds such as buffelgrass (*Pennisetum ciliare*), Mediterranean grass (*Schismus barbatus*), red brome (*Bromus rubens*), and mustards (*Brassica* sp.) crowd out native Sonoran Desert plants, compete for water, and promote fire. Fires have no doubt always occurred in the Sonoran Desert, but the low density and relatively open spacing of native bunchgrasses kept fires low in intensity and of short duration. Nonnative grasses and forbs grow close together, often forming a continuous blanket of vegetation. Many Sonoran Desert species are not fire adapted and fare poorly during the high-intensity, extensive burns promoted by exotics. As a result, much of this community is being converted into a grass and scrubland largely devoid of succulents. Off-highway vehicle use, overgrazing, development, and habitat fragmentation are also major threats to this community.

At least 66 amphibian and reptile species occur in this community: *Ambystoma mavortium, Anaxyrus cognatus, A. microscaphus, A. punctatus, A. retiformis, A. woodhousii, Incilius alvarius, Hyla arenicolor, Smilisca fodiens, Gastrophryne mazatlanensis, Scaphiopus couchii, Spea multiplicata, Kinosternon arizonense, Gopherus morafkai, Crotaphytus bicinctores, C. collaris, C. nebrius, Gambelia wislizenii, Coleonyx variegatus, Heloderma suspectum, Dipsosaurus dorsalis, Sauromalus ater, Callisaurus draconoides, Cophosaurus texanus, Holbrookia elegans, Phrynosoma goodei, P. platyrhinos, P. solare, Sceloporus clarkii, S. magister, Urosaurus graciosus, U. ornatus, Uta stansburiana, Aspidoscelis stictogramma, A. tigris, A. xanthonota, Xantusia arizonae, X. bezyi, X. vigilis, Lichanura trivirgata, Arizona elegans, Chilomeniscus stramineus, Chionactis occipitalis, C. palarostris, Diadophis punctatus, Hypsiglena chlorophaea, Lampropeltis getula, Masticophis bilineatus, M. flagellum, Phyllorhynchus browni, P. decurtatus, Pituophis catenifer, Rhinocheilus lecontei, Salvadora hexalepis, Sonora semiannulata, Tantilla hobartsmithi, Thamnophis cyrtopsis, Trimorphodon lambda, Micruroides euryxanthus, Rena humilis, Crotalus atrox, C. cerastes, C. mitchellii, C. molossus, C. scutulatus,* and *C. tigris.*

Mohave Desert Scrub

Along the northern half of the state's western border at elevations ranging from about 275 to 1450 m, the predominant community is Mo-

have Desert Scrub. Conspicuous plants in this area include creosote bush (*Larrea tridentata*), desert holly (*Atriplex hymenelytra*), white burrobrush (*Hymenoclea salsola*), Mojave yucca (*Yucca schidigera*), blackbrush (*Coleogyne ramosissima*), shadscale (*Atriplex confertifolia*), and the nearly endemic Joshua tree (*Y. brevifolia*) (Brown and Lowe 1980; Turner 1994b).

This community is heavily scarred by the illegal cross-country operation of off-highway vehicles and has been damaged by overgrazing, development, and invasive vegetation.

At least 38 amphibian and reptile species occur in this community: *Anaxyrus cognatus, A. microscaphus, A. punctatus, A. woodhousii, Gopherus agassizii, G. morafkai, Crotaphytus bicinctores, Gambelia wislizenii, Coleonyx variegatus, Heloderma suspectum, Dipsosaurus dorsalis, Sauromalus ater, Callisaurus draconoides, Phrynosoma platyrhinos, Sceloporus magister, S. uniformis, Urosaurus graciosus, U. ornatus, Uta stansburiana, Aspidoscelis tigris, Xantusia vigilis, Lichanura trivirgata, Arizona elegans, Chionactis occipitalis, Hypsiglena chlorophaea, Lampropeltis getula, Masticophis flagellum, Phyllorhynchus decurtatus, Pituophis catenifer, Rhinocheilus lecontei, Salvadora hexalepis, Sonora semiannulata, Trimorphodon lambda, Rena humilis, Crotalus atrox, C. cerastes, C. mitchellii,* and *C. scutulatus.*

Chihuahuan Desert Scrub

Chihuahuan Desert Scrub is the predominant community in the lower valleys and bajadas of far southeastern Arizona at elevations ranging from about 900 to 1200 m. Common plants in this community include creosote bush (*Larrea tridentata*), whitethorn acacia (*Acacia constricta*), tarbush (*Flourensia cernua*), ocotillo (*Fouquieria splendens*), viscid acacia (*Acacia neovernicosa*), allthorn (*Koeberlinia spinosa*), mariola (*Parthenium incanum*), and honey mesquite (*Prosopis glandulosa*) (Brown and Lowe 1980; Brown 1994b).

At least 53 amphibian and reptile species can be found in this community: *Ambystoma mavortium, Anaxyrus cognatus, A. debilis, A. punctatus, A. woodhousii, Incilius alvarius, Hyla arenicolor, Scaphiopus couchii, Spea bombifrons, S. multi-*

plicata, Terrapene ornata, Kinosternon flavescens, Crotaphytus collaris, Gambelia wislizenii, Coleonyx variegatus, Heloderma suspectum, Callisaurus draconoides, Cophosaurus texanus, Holbrookia elegans, Phrynosoma cornutum, P. modestum, P. solare, Sceloporus bimaculosus, S. clarkii, S. cowlesi, S. magister, Urosaurus ornatus, Uta stansburiana, Aspidoscelis tigris, A. uniparens, Arizona elegans, Diadophis punctatus, Gyalopion canum, Heterodon kennerlyi, Hypsiglena jani, Lampropeltis splendida, Masticophis flagellum, Pituophis catenifer, Rhinocheilus lecontei, Salvadora deserticola, Sonora semiannulata, Tantilla hobartsmithi, T. nigriceps, Thamnophis cyrtopsis, T. marcianus, Trimorphodon lambda, Rena dissecta, R. humilis, Micruroides euryxanthus, Crotalus atrox, C. molossus, C. scutulatus, and C. tigris.

Great Basin Desert Scrub

Great Basin Desert Scrub occurs on the lower plateaus of the Colorado Plateau Province at elevations ranging from about 1190 to 1980 m. Conspicuous plants in this community include big sagebrush (Artemisia tridentata), Bigelow sagebrush (A. bigelovii), shadscale (Atriplex confertifolia), winterfat (Krascheninnikovia lanata), and blackbrush (Coleogyne ramosissima) (Brown and Lowe 1980; Turner 1994a).

At least 34 amphibian and reptile species occur in this community: Ambystoma mavortium, Anaxyrus cognatus, A. punctatus, A. woodhousii, Hyla arenicolor, Scaphiopus couchii, Spea bombifrons, S. intermontana, S. multiplicata, Crotaphytus bicinctores, C. collaris, Gambelia wislizenii, Coleonyx variegatus, Sauromalus ater, Holbrookia maculata, Phrynosoma platyrhinos, Sceloporus graciosus, S. magister, S. tristichus, S. uniformis, Urosaurus ornatus, Uta stansburiana, Aspidoscelis tigris, A. velox, Arizona elegans, Hypsiglena chlorophaea, H. jani, Lampropeltis getula, Pituophis catenifer, Rhinocheilus lecontei, Salvadora hexalepis, Sonora semiannulata, Crotalus oreganus, and C. viridis.

Semidesert Grassland

Semidesert Grassland exists at elevations ranging from about 1100 to 1700 m in many of the valleys and bajadas of the southeastern basins and in several locations along the southern edge of the Transition Zone. These communities are dominated by tobosa (Pleuraphis mutica), hairy grama (Bouteloua hirsuta), palmilla (Yucca rigida), sotol (Dasylirion wheeleri), soaptree yucca (Y. elata), various Agave species, cholla (Cylindropuntia sp.), and prickly pear (Opuntia sp.) (Brown and Lowe 1980; Brown 1994f).

At least 76 amphibian and reptile species can be found in this community: Ambystoma mavortium, Anaxyrus cognatus, A. debilis, A. microscaphus, A. punctatus, A. woodhousii, Incilius alvarius, Hyla arenicolor, Gastrophryne mazatlanensis, Scaphiopus couchii, Spea bombifrons, S. multiplicata, Terrapene ornata, Kinosternon flavescens, Gopherus morafkai, Elgaria kingii, Crotaphytus collaris, Gambelia wislizenii, Heloderma suspectum, Cophosaurus texanus, Holbrookia elegans, H. maculata, Phrynosoma cornutum, P. hernandesi, P. modestum, P. solare, Sceloporus bimaculosus, S. clarkii, S. cowlesi, S. magister, S. tristichus, Urosaurus ornatus, Uta stansburiana, Plestiodon obsoletus, Aspidoscelis arizonae, A. stictogramma, A. exsanguis, A. flagellicauda, A. sonorae, A. tigris, A. uniparens, Xantusia arizonae, X. bezyi, X. vigilis, Arizona elegans, Diadophis punctatus, Gyalopion canum, G. quadrangulare, Heterodon kennerlyi, Hypsiglena chlorophaea, H. jani, Lampropeltis getula, Masticophis bilineatus, M. flagellum, M. taeniatus, Oxybelis aeneus, Pituophis catenifer, Rhinocheilus lecontei, Salvadora deserticola, S. hexalepis, Senticolis triaspis, Sonora semiannulata, Tantilla hobartsmithi, Tantilla nigriceps, Thamnophis cyrtopsis, T. marcianus, Trimorphodon lambda, Micruroides euryxanthus, Rena dissecta, R. humilis, Crotalus atrox, C. cerberus, C. mitchellii, C. molossus, C. scutulatus, and Sistrurus catenatus.

Plains and Great Basin Grasslands

Plains and Great Basin Grasslands stretch across much of Arizona's Colorado Plateau and the San Rafael and Empire Valleys of southeastern Arizona. These communities occur at elevations ranging from about 1370 to 2300 m. The dominant grasses include blue grama (Bouteloua gracilis), galleta (*Pleuraphis jamesii*), Indian ricegrass (*Achnatherum hymenoides*), and alkali sacaton (*Sporobolus airoides*). Fire suppression and the overgrazing of cattle have disrupted the natural fire cycles in many of Arizona's grasslands, allowing the invasion of shrubs such as fourwing saltbush (*Atriplex canescens*), snakeweed (*Gutierrezia sarothrae*), and Mormon tea (*Ephedra viridis*) (Brown and Lowe 1980; Brown 1994e).

At least 51 amphibian and reptile species occur in Arizona's grassland communities: *Ambystoma mavortium, Anaxyrus cognatus, A. punctatus, A. woodhousii, Acris crepitans, Scaphiopus couchii, Spea bombifrons, S. intermontana, S. multiplicata, Terrapene ornata, Elgaria kingii, Crotaphytus bicinctores, C. collaris, Holbrookia elegans, H. maculata, Phrynosoma hernandesi, Sceloporus clarkii, S. cowlesi, S. graciosus, S. magister, S. slevini, S. tristichus, S. uniformis, Urosaurus ornatus, Uta stansburiana, Plestiodon multivirgatus, P. obsoletus, Aspidoscelis neomexicana, A. pai, A. sonorae, A. uniparens, A. velox, Arizona elegans, Coluber constrictor, Diadophis punctatus, Gyalopion canum, Heterodon kennerlyi, Hypsiglena chlorophaea, H. jani, Lampropeltis getula, L. triangulum, Masticophis taeniatus, Pituophis catenifer, Rhinocheilus lecontei, Salvadora deserticola, Sonora semiannulata, Thamnophis elegans, T. marcianus, Crotalus scutulatus, C. viridis,* and *Sistrurus catenatus.*

Interior Chaparral

Interior Chaparral occurs on many of the southern slopes of the Transition Zone at elevations between about 1000 and 2150 m. Common shrubs in this community include manzanita (*Arctostaphylos pungens*), shrub live oak (*Quercus turbinella*), birchleaf mountain mahogany (*Cercocarpus montanus*), skunkbush sumac (*Rhus trilobata*), silktassels (*Garrya sp.*), desert ceanothus (*Ceanothus greggii*), hollyleaf buckthorn (*Rhamnus crocea*), cliffrose (*Purshia stansburiana*), Arizona rosewood (*Vauquelinia californica*), and desert olive (*Forestiera shrevei*) (Brown and Lowe 1980; Pase and Brown 1994b).

At least 49 amphibian and reptile species can be found in this community: *Ambystoma*

mavortium, Anaxyrus microscaphus, A. punctatus, A. woodhousii, Hyla arenicolor, Spea multiplicata, Elgaria kingii, Crotaphytus bicinctores, C. collaris, Coleonyx variegatus, Heloderma suspectum, Sauromalus ater, Cophosaurus texanus, Holbrookia maculata, Phrynosoma hernandesi, Sceloporus clarkii, S. magister, S. tristichus, Urosaurus ornatus, Uta stansburiana, Plestiodon gilberti, P. obsoletus, Aspidoscelis flagellicauda, A. tigris, A. uniparens, A. velox, Xantusia arizonae, X. bezyi, X. vigilis, Lichanura trivirgata, Diadophis punctatus, Hypsiglena chlorophaea, H. jani, Lampropeltis pyromelana, Masticophis bilineatus, M. flagellum, M. taeniatus, Pituophis catenifer, Salvadora grahamiae, S. hexalepis, Sonora semiannulata, Tantilla hobartsmithi, Thamnophis cyrtopsis, Trimorphodon lambda, Micruroides euryxanthus, Crotalus atrox, C. cerberus, C. mitchellii, and *C. molossus.*

Madrean Evergreen Woodland

On the lower slopes and foothills of southeastern Arizona at elevations ranging from about 1280 to 2200 m the dominant community is Madrean Evergreen Woodland. Conspicuous plants in this landscape include Emory oak (*Quercus emoryi*), gray oak (*Q. grisea*), Arizona white oak (*Q. arizonica*), Mexican blue oak (*Q. oblongifolia*), alligator bark juniper (*Juniperus deppeana*), Mexican pinyon (*Pinus cembroides*), madrone (*Arbutus arizonica*), rainbow cactus (*Echinocereus rigidissimus*), Schott's yucca (*Yucca schottii*), banana yucca (*Y. baccata*), sotol (*Dasylirion wheeleri*), Palmer's century plant (*Agave palmeri*), and Parry's agave (*A. parryi*) (Brown and Lowe 1980; Brown 1994d).

At least 57 amphibian and reptile species occur in this community: *Ambystoma mavortium, Anaxyrus punctatus, A. woodhousii, Incilius alvarius, Craugastor augusti, Hyla arenicolor, H. wrightorum, Gastrophryne mazatlanensis, Spea multiplicata, Terrapene ornata, Elgaria kingii, Crotaphytus collaris, Heloderma suspectum, Cophosaurus texanus, Holbrookia elegans, H. maculata, Phrynosoma hernandesi, Sceloporus clarkii, S. jarrovii, S. slevini, S. tristichus, S. virgatus, Urosaurus ornatus, Plestiodon callicephalus, P. ob-*

soletus, Aspidoscelis stictogramma, A. exsanguis, A. flagellicauda, A. sonorae, A. uniparens, Diadophis punctatus, Hypsiglena chlorophaea, H. jani, Lampropeltis knoblochi, L. pyromelana, Masticophis bilineatus, Oxybelis aeneus, Pituophis catenifer, Salvadora deserticola, S. grahamiae, S. hexalepis, Senticolis triaspis, Sonora semiannulata, Tantilla hobartsmithi, T. wilcoxi, T. yaquia, Thamnophis cyrtopsis, Trimorphodon lambda, Micruroides euryxanthus, Rena dissecta, R. humilis, Crotalus atrox, C. cerberus, C. lepidus, C. molossus, C. pricei, and C. willardi.

Great Basin Conifer Woodland

Great Basin Conifer Woodland occurs in the middle elevations (1200–2300 m) of the Transition Zone and Colorado Plateau. This community is dominated by Rocky Mountain juniper (Juniperus scopulorum), Utah juniper (J. osteosperma), single-leaf pinyon (Pinus monophylla), and Rocky Mountain pinyon (P. edulis) (Brown and Lowe 1980; Brown 1994c).

At least 48 amphibian and reptile species occur in this community: Ambystoma mavortium, Anaxyrus microscaphus, A. punctatus, A. woodhousii, Hyla arenicolor, Spea intermontana, S. multiplicata, Elgaria kingii, Crotaphytus bicinctores, C. collaris, Heloderma suspectum, Cophosaurus texanus, Holbrookia maculata, Phrynosoma hernandesi, Sceloporus clarkii, S. cowlesi, S. magister, S. tristichus, S. uniformis, Urosaurus ornatus, Plestiodon gilberti, P. multivirgatus, P. obsoletus, P. skiltonianus, Aspidoscelis flagellicauda, A. pai, A. uniparens, A. velox, Xantusia bezyi, X. vigilis, Lichanura trivirgata, Diadophis punctatus, Hypsiglena chlorophaea, H. jani, Lampropeltis pyromelana, Masticophis taeniatus, Pituophis catenifer, Salvadora grahamiae, S. hexalepis, Sonora semiannulata, Tantilla hobartsmithi, Thamnophis cyrtopsis, T. elegans, Trimorphodon lambda, Crotalus cerberus, C. molossus, C. oreganus, and C. viridis.

Montane Conifer Forest

Montane Conifer Forest occurs on the high slopes and plateaus of Arizona at elevations ranging from about 2000 to 3050 m. The most conspicuous species in this community is the ponderosa pine (Pinus ponderosa). Other community members include Douglas fir (Pseudotsuga menziesii), Gambel oak (Quercus gambelii), and New Mexico locust (Robinia neomexicana) (Brown and Lowe 1980; Pase and Brown 1994c).

Years of active fire suppression and the clearing of fine fuels by overgrazing have resulted in an increase in the density of understory shrubs and young trees in this community. The regular fires that once maintained the forest, burning at ground level and with low intensity, are now fueled by dense undergrowth and have become catastrophic in nature.

At least 42 amphibian and reptile species can be found in this community: Ambystoma mavortium, Ensatina eschscholtzii, Anaxyrus microscaphus, A. punctatus, Hyla arenicolor, H. wrightorum, Pseudacris triseriata, Spea intermontana, S. multiplicata, Elgaria kingii, Phrynosoma hernandesi, Sceloporus cowlesi, S. jarrovii, S. slevini, S. tristichus, S. virgatus, Urosaurus ornatus, Plestiodon gilberti, P. multivirgatus, P. obsoletus, P. skiltonianus, Aspidoscelis exsanguis, A. flagellicauda, A. pai, A. velox, Diadophis punctatus, Hypsiglena chlorophaea, H. jani, Lampropeltis knoblochi, L. pyromelana, Masticophis taeniatus, Pituophis catenifer, Salvadora grahamiae, Senticolis triaspis, Tantilla wilcoxi, Thamnophis cyrtopsis, T. elegans, Crotalus cerberus, C. lepidus, C. molossus, C. pricei, and C. willardi.

Subalpine Grassland

High-elevation (2500–3500 m) flatlands, valleys, and rolling hills on the Kaibab Plateau, San Francisco Mountains, Chuska Mountains, and White Mountains are home to Subalpine Grassland. In this grass-forb community wild daisies, dandelions, larkspurs, asters, and clovers grow among bunchgrasses such as Arizona fescue (Festuca arizonica) (Brown and Lowe 1980; Brown 1994a).

At least 11 amphibian and reptile species occur in this community: Ambystoma mavortium, Hyla wrightorum, Pseudacris triseriata, Phrynosoma her-

nandesi, Sceloporus cowlesi, S. tristichus, Urosaurus ornatus, Plestiodon multivirgatus, Pituophis catenifer, Thamnophis elegans, and *Crotalus molossus.*

Subalpine Conifer Forest

The slopes of Arizona's highest mountains at elevations ranging from about 2450 to 3800 m are home to Subalpine Conifer Forest. Engelmann spruce (*Picea engelmannii*), corkbark fir (*Abies lasiocarpa*), bristlecone pine (*Pinus aristata*), aspen (*Populus tremuloides*), and Douglas fir (*Pseudotsuga menziesii*) are the most conspicuous plants in this community (Brown and Lowe 1980; Pase and Brown 1994d).

At least 17 amphibian and reptile species can be found in this community: *Pseudacris triseriata, Elgaria kingii, Phrynosoma hernandesi, Sceloporus cowlesi, S. jarrovii, S. slevini, S. tristichus, S. virgatus, Plestiodon skiltonianus, Lampropeltis knoblochi, L. pyromelana, Pituophis catenifer, Thamnophis elegans, Crotalus cerberus, C. lepidus, C. molossus,* and *C. pricei.*

Alpine Tundra

Alpine Tundra occurs atop Arizona's highest peaks, the San Francisco Peaks near Flagstaff, at elevations ranging from 3500 to 3862 m. There are no trees in this community, only lichens, mosses, herbaceous plants, and low-growing shrubs. Some of this community's members include Parry's lousewort (*Pedicularis parryi*), moonwort (*Botrychium* spp.), orange sneezeweed (*Helenium hoopesii*), American alpine speedwell (*Veronica wormskjoldii*), and San Francisco Peaks groundsel (*Packera franciscana*) (Brown and Lowe 1980; Pase and Brown 1994a).

Of herpetozoan species, only *Thamnophis elegans* has been observed in this community in Arizona. *Pituophis catenifer* has been observed just below the Alpine Tundra/Petran Subalpine Conifer Forest ecotone on San Francisco Mountain.

Riparian Communities

Numerous creeks and rivers stretch across the state, weaving together the various biotic communities. These riparian communities and other wetland communities provide critical habitat for numerous specialized organisms in Arizona. Near their origins, riparian communities are dominated by a variety of maples (*Acer* spp.), alders (*Alnus* spp.), and willows (*Salix* spp.). In the middle elevations the dominant vegetation includes cottonwoods (*Populus spp.*), willows (*Salix* spp.), sycamores (*Platanus* spp.), alders (*Alnus* spp.), walnut (*Juglans* spp.), and ash (*Fraxinus spp.*). At the lowest elevations, Arizona riparian communities are characterized by mesquites (*Prosopis* spp.), seep-willow (*Baccharis salicifolia*), desert willow (*Chilopsis linearis*), and cottonwoods (*Populus spp.*) (Brown and Lowe 1980; Minckley and Brown 1994).

Many of Arizona's riparian communities have been degraded and dried by overgrazing, damming, diverting, and groundwater pumping. In addition, invasive species such as crayfish, *Lithobates catesbeianus, L. berlandieri,* and various sportfish compete with native species. The introduced disease chytridiomycosis threatens Arizona's native aquatic amphibians.

At least 16 of Arizona's herpetozoan species are directly dependent on the state's riparian communities. Amphibian and reptile species that occur in Arizona's high-elevation riparian and wetland communities include *Lithobates catesbeianus, L. chiricahuensis, L. pipiens,* and *Thamnophis rufipunctatus.* The state's middle-elevation riparian and wetland communities are home to *L. catesbeianus, L. chiricahuensis, L. tarahumarae, L. yavapaiensis, Chrysemys picta, Kinosternon sonoriense, T. eques,* and *T. rufipunctatus.* Members of Arizona's low-elevation riparian and wetland communities include *Pseudacris hypochondriaca, L. berlandieri, L. blairi, L. catesbeianus, L. onca, L. tarahumarae, L. yavapaiensis, Chelydra serpentina, Chrysemys picta, Kinosternon sonoriense, Trachemys scripta, Apalone spinifera, Thamnophis eques,* and *T. rufipunctatus.*

Introduced species that may be dependent on areas of human development for their continued existence in Arizona include *Xenopus laevis,* which has been introduced to golf-course ponds in Tucson; *Chelydra serpentina,* which has been introduced to urban lakes and canals in the Phoenix metropolitan area; *Ctenosaura* spp. from two maternal lineages (*C. macrolopha* and *C. conspicuosa*) (Edwards et al. 2005), which have been introduced to the grounds of the Arizona-Sonora Desert Museum west of Tucson but have not spread beyond the immediate vicinity of the museum; *Hemidactylus turcicus*, which has been introduced to numerous areas of human habitation across western and southern Arizona; *Chalcides* sp., which has been introduced to the city of Mesa; *Cyrtopodion scabrum* currently known from the city of Mesa and Gila Bend, which may be much more widely distributed: and *Indotyphlops braminus*, which has been introduced to the Phoenix metropolitan area.

Presently a total of 153 (141 native, 13 introduced) herpetozoan species occur in Arizona, representing 30 families and 67 genera. Thirty of these species are amphibians (2 salamanders [1 introduced], 28 anurans [3 introduced]). The remaining 123 species are reptiles (10 turtles [3 introduced], 57 lizards [5 introduced], and 56 snakes [1 introduced]).

Amphibians make up a small percentage (20%, or 30/153) of Arizona's herpetozoan species, reflecting the group's dependence on moisture and the state's limited amount of mesic or hydric habitat. The Barred Tiger Salamander and most Arizona anurans are tied to moist habitat, which is generally limited to the mountainous Transition Zone, southeastern portion of the Basin and Range, and riparian communities. Nevertheless, several amphibians are adapted to the semiarid Sonoran Desert of the western Basin and Range Province, including *Anaxyrus punctatus, A. cognatus, A. retiformis, Incilius alvarius, Smilisca fodiens, Gastrophryne mazatlanensis,* and *Scaphiopus couchii*. The state's semiarid northern deserts are inhabited by *A. cognatus, A. woodhousii, Scaphiopus couchii, Spea multiplicata,* and *S. intermontana. Craugastor augusti,* which occurs in a handful of mesic ranges in southeastern Arizona, has adapted the ability to lay its eggs in moist places on land and is the only Arizona amphibian that is not dependent on bodies of water for reproduction.

Turtles also represent a small percentage (7%, or 10/153) of Arizona's herpetozoan species. Three of Arizona's native turtles (*Terrapene ornata, Gopherus agassizii,* and *G. morafkai*) are terrestrial, two (*Kinosternon arizonense* and *K. flavescens*) are semiaquatic, and two (*K. sonoriense* and *Chrysemys picta*) are aquatic. Three nonnative aquatic turtles (*Chelydra serpentina, Trachemys scripta,* and *Apalone spinifera*) have been introduced to Arizona. *Chelydra serpentina* has been introduced to urban lakes and canals in the Phoenix metropolitan area, whereas *T. scripta* and *A. spinifera* have been introduced to urban bodies in addition to natural creeks and rivers, reservoirs, and rural ponds in central, western, and southern Arizona.

The largest herpetozoan group in Arizona is the lizards, which make up 37% (57/153) of the state's species. With the exception of the Alpine Tundra community atop the San Francisco Mountains, every terrestrial habitat in the state has resident lizards. Some species, such as *Urosaurus ornatus* and *Uta stansburiana,* are found nearly statewide. Others such as the sand-flats-dwelling *Phrynosoma mcallii,* the dune-dwelling *Uma rufopunctata* and *U. scoparia,* and the rock-crevice-dwelling *Xantusia arizonae* and *X. bezyi* have relatively small ranges within specific habitats.

Snakes constitute 37% (56/153) of Arizona's herpetozoan species and occur in all of the state's terrestrial habitats, although *Thamnophis elegans* is the only species known to occur in Arizona's Alpine Tundra. *Pituophis catenifer* is found in all habitats except Alpine Tundra, while other species such as the high-elevation sky island–dwelling *Crotalus pricei* and the aquatic *T. rufipunctatus* have relatively small ranges within specific habitats. One species (*Coluber constrictor*) is documented from Arizona by a single specimen (BYU 100) collected in 1927 from "Eagar, Arizona." No additional *C. constrictor* specimens have been collected in Arizona. Through mtDNA analysis, D. G. Mulcahy (2006, 2008) discovered an undescribed species of *Hypsiglena* that is restricted to the Cochise Filter Barrier area of southeastern Arizona, adjacent New Mexico, and presumably adjacent Mexico. Upon description, this species would constitute Arizona's fifty-seventh species of snake.

Only four of the state's broad assortment of herpetotaxa are endemic to Arizona, and all are lizards (*Aspidoscelis arizonae, A. pai, Xantusia arizonae,* and *X. bezyi*).

Arizona shares 129 (84%) herpetozoan species with Mexico. Many of these shared species reach their northern distributional limits within or near the boundaries of the Sonoran Desert in Arizona: *Anaxyrus retiformis, Incilius alvarius, Smilisca fodiens, Kinosternon arizonense, Crotaphytus nebrius, Phrynosoma goodei, P. solare, Uma rufopunctata, Aspidoscelis stictogramma, A. xanthonota, Chilomeniscus stramineus, Chionactis palarostris, Masticophis bilineatus, Phyllorhynchus browni, Micruroides euryxanthus,* and *Crotalus tigris.* Most of these semiarid-region shared species reach their southern distributional limits in Sonora or Sinaloa. Many additional shared species reach their northern distributional limits in the sky island ranges and basins of southeastern Arizona: *Lithobates tarahumarae, Holbrookia elegans, Sceloporus jarrovii, S. slevini, S. virgatus, Plestiodon callicephalus, Gyalopion quadrangulare, Lampropeltis knoblochi, Oxybelis aeneus, Senticolis triaspis, Tantilla wilcoxi, T. yaquia, Crotalus pricei,* and *C. willardi.* The southern distributional limits of many of these southeastern Arizona species are on, or adjacent to, the Mexican Plateau. The range of *S. triaspis* extends south beyond Mexico to Costa Rica, and the exceptionally sizable range of *O. aeneus* extends from about 24 km north of Arizona's southern border to southeastern Brazil and Peru (Stebbins 2003).

Shared species that reach their northern distributional limits in or beyond the Transition Zone of central Arizona but within the boundaries of the state include *Hyla wrightorum, Lithobates chiricahuensis, L. yavapaiensis, Kinosternon sonoriense, Gopherus morafkai, Elgaria kingii, Cophosaurus texanus, Sceloporus clarkii, Aspidoscelis uniparens, Thamnophis eques,* and *Crotalus molossus.*

Shared species that reach their northern distributional limits outside the boundaries of Ar-

izona include *Ambystoma mavortium, Anaxyrus cognatus, A. debilis, A. punctatus, A. woodhousii, Craugastor augusti, Acris crepitans, Hyla arenicolor, Pseudacris hypochondriaca, Gastrophryne mazatlanensis, Lithobates catesbeianus, Scaphiopus couchii, Spea multiplicata, S. bombifrons, Chelydra serpentina, Chrysemys picta, Terrapene ornata, Trachemys scripta, Kinosternon flavescens, Apalone spinifera, Crotaphytus collaris, Gambelia wislizenii, Coleonyx variegatus, Hemidactylus turcicus, Heloderma suspectum, Dipsosaurus dorsalis, Sauromalus ater, Callisaurus draconoides, Holbrookia maculata, Phrynosoma cornutum, P. hernandesi, P. mcallii, P. modestum, P. platyrhinos, Sceloporus bimaculosus, S. cowlesi, S. magister, Urosaurus graciosus, U. ornatus, Uta stansburiana, Plestiodon gilberti, P. multivirgatus, P. obsoletus, P. skiltonianus, Aspidoscelis exsanguis, A. sonorae, A. tigris, Xantusia vigilis, Lichanura trivirgata, Arizona elegans, Chionactis occipitalis, Coluber constrictor, Diadophis punctatus, Gyalopion canum, Heterodon kennerlyi, Hypsiglena chlorophaea, H. jani, Lampropeltis getula, L. triangulum, Masticophis flagellum, M. taeniatus, Phyllorhynchus decurtatus, Pituophis catenifer, Rhinocheilus lecontei, Salvadora deserticola, S. grahamiae, S. hexalepis, Sonora semiannulata, Tantilla hobartsmithi, T. nigriceps, Thamnophis cyrtopsis, T. elegans, T. marcianus, Trimorphodon lambda, Rena dissecta, R. humilis, Crotalus atrox, C. cerastes, C. lepidus, C. mitchellii, C. oreganus, C. scutulatus, C. viridis,* and *Sistrurus catenatus.* Some of these shared species, including *Diadophis punctatus, Lampropeltis triangulum, Pituophis catenifer,* and *Crotalus viridis,* extend north into Canada.

Nine of the 129 species shared between Arizona and Mexico are introduced in Arizona (*Ensatina eschscholtzii, Xenopus laevis, Lithobates berlandieri, L. catesbeianus, Chelydra serpentina, Trachemys scripta, Apalone spinifera, Hemidactylus turcicus,* and *Indotyphlops braminus*).

In Arizona there are 17 native species that are not endemic but do not occur in Mexico. Of these, 4 are shared with only one neighboring state (*Aspidoscelis flagellicauda, Thamnophis rufipunctatus,* and *Crotalus cerberus* occur in Arizona and New Mexico; *Uma scoparia* occurs in Arizona and California). *Lithobates onca* may occur in three states (Arizona, Nevada, and possibly Utah). *Lampropeltis pyromelana* occurs in four states (Arizona, Nevada, New Mexico, and Utah). *Gopherus agassizii* occurs in four states (Arizona, California, Nevada, and Utah). Other native Arizona species that do not occur in Mexico are more broadly distributed in the western United States and, in some cases, western Canada (*Anaxyrus microscaphus, Spea intermontana, Crotaphytus bicinctores, Sceloporus graciosus, S. tristichus, S. uniformis,* and *Aspidoscelis velox*). A disjunct population of *Lithobates blairi* occurs in the Sulphur Springs Valley of far southeastern Arizona but apparently does not enter Mexico. The remainder of this anuran's broad distribution involves much of the south-central portion of the United States. The remaining two native species found in Arizona but not Mexico (*Pseudacris triseriata* and *Lithobates pipiens*) are widely distributed over large portions of the United States and Canada.

Arizona is one of the fastest-growing states in the United States. Residential, business, industrial, infrastructural, and agricultural development and the subsequent increased demand on the state's limited resources represent a major threat to natural habitats. Urban and suburban development directly consume natural areas; associated infrastructure such as roadways and canals fragment habitat; and increased pollution, waste, and human activity adversely affect ecosystems. The state is essentially bisected by the Central Arizona Project, a 336-mile system of canals, aqueducts, and pipelines that deliver Colorado River water from Lake Havasu near Parker to Pima, Pinal, and Maricopa Counties. An ever-growing, spider-web-like network of roadways blankets the state, and in many areas the habitat between these roadways is scarred by labyrinthine off-highway vehicle trails. These travel arteries act as "moving fences," forming substantial barriers to wildlife movements. New concerns on the horizon are "green energy" projects, which may require the employment of huge tracts of natural land for solar and wind farms.

Arizona has long had a vibrant cattle ranching industry, and there is virtually no place in the state in which grazing has not occurred. When done irresponsibly, cattle grazing can have dire impacts on native communities. Overgrazing has led to habitat conversion, promotion of noxious and exotic plant species, and loss of native forb species in many semiarid, lower-elevation portions of the state.

The construction of extensive physical barriers such as walls and fences along Arizona's southern border has disrupted or blocked the movement of many wildlife species, including reptiles and amphibians. Additionally increased human presence along these borderlands has led to a proliferation of roads, trash, and vehicle traffic. The impact of these activities on herpetofauna and other wildlife has yet to be explored.

Herpetofauna of New Mexico

CHARLES W. PAINTER
AND JAMES N. STUART

Introduction

New Mexico is a herpetologically diverse state with 136 species currently known, placing it within the top 10 in the United States in number of species (some variability in ranking can result depending on which taxonomy is used). Although New Mexico is drier than most states and has a paucity of salamander species, the anuran and turtle species are high for such a xeric state, and there is an abundance of lizards and snakes that represent 72% of the state's herpetofauna. Based on population size, New Mexico is the smallest of the United States–Mexico border states, and this is reflected in the small number of resident herpetologists active in the state. However, our current knowledge has benefited greatly from development of herpetology programs at the University of New Mexico, New Mexico State University, Western New Mexico University, and Eastern New Mexico University. In addition, the New Mexico Department of Game and Fish has developed an active non-game program during the past 25 years that has contributed significantly to the understanding of the natural history and distribution of the state's fauna, including amphibians and reptiles. The New Mexico Herpetological Society has also contributed significantly to this understanding. The most comprehensive treatise on the amphibians and reptiles of New Mexico was provided by Degenhardt et al. (1996), who reported 123 species from the state. The likelihood of new species being discovered within the borders is relatively low. However, as more species are investigated, using sophisticated genetic techniques, there is the possibility of cryptic, undescribed species being discovered (e.g., *Aspidoscelis*).

Previous Herpetological Studies

Much of the following material was adapted from the chapter "A Brief History of Herpetology in New Mexico" in Degenhardt et al. (1996), authored primarily by Andrew H. Price. The interested reader should consult that source for additional information.

The earliest students of herpetology in New Mexico were Native American peoples who created artwork such as petroglyphs, fetishes, and other ceremonial items that incorporated herpetological imagery or animal parts. These were significant totems in their everyday and spiritual lives (e.g., Whipple 1854 [1856]; Kennerly 1856; Henderson and Harrington 1914), and several species were seasonally used as food. Various species were iconographed by the Anasazi and

Mimbres cultures (Bettison et al. 1999; Davis 1995). Many of the Puebloan groups (Zia, Zuni, and Acoma) as well as the Apache, Navajo, and the Ute incorporated snakes, especially rattlesnakes, or turtles into their rituals (e.g., Cushing 1883; Klauber 1972; Fewkes 1986; Hough 2010).

Members of the Spanish colonial expedition led by Father Agustín Rodríguez and Francisco Sánchez Chamuscado in 1581 made the first written reference by Europeans to the herpetofauna of New Mexico, the rattlesnakes (Klauber 1972). After that, New Mexico remained relatively unexplored herpetologically until well into the nineteenth century. Following the War of 1812, the US Army created a division of topographical engineers, in part to explore the western frontier. The first exploring expedition of this division to enter what is now New Mexico was led by Maj. Stephen H. Long to the Rocky Mountains (James 1823). A small detachment of this expedition spent about two weeks of August 1820 exploring the Mora and Canadian River Valleys in northeastern New Mexico. During this portion of the journey they made reference to the Prairie Rattlesnake (*Crotalus viridis*) and to two forms of "orbicular lizards" (*Phrynosoma cornutum* and *P. hernandesi*), both said to be common (James 1823, 276).

Tension between Mexico and the United States provided the next opportunity for an addition to the printed record of New Mexico herpetology. In August and September 1845, Lt. James E. Abert of the Corps of Topographical Engineers spent about two weeks exploring the Canadian River Valley from Raton Pass to the Texas border. He commented on the abundant Prairie Rattlesnakes (*Crotalus viridis*) associated with prairie dog colonies along the way (Abert 1846). During 1846, both Lieutenant Abert and 1st Lt. William H. Emory joined the expedition of Col. Stephen W. Kearny against Mexico, which entered New Mexico along the Santa Fe Trail on August 6. During this expedition, only meager herpetological observations were recorded (Abert 1848; Emory 1848). First Lieutenant Emory noted the abundance of *Agama cornuta* (= *P. hernandesi*)

near the junction of the Vermejo and Canadian Rivers on August 9 and 11, and an encounter with a reddish *C. atrox* near Las Palomas on the Rio Grande on October 15. Lieutenant Abert remained in northern and central New Mexico, visiting various sites, including El Rito, Acoma, and Socorro, leaving New Mexico in mid-January 1847. He noted great numbers of horned lizards (*P. hernandesi*) near Lamy on September 29. He obtained a specimen of what was likely a Collared Lizard (*Crotaphytus collaris*) between Bernalillo and the mouth of the Jemez River on October 13. The disposition of this specimen is unknown. In 1846, Adolphus Wislizenus, an immigrant Swiss naturalist, embarked on one of the first privately funded natural history explorations of the West. He commented on the general abundance of lizards during his journey down the Santa Fe Trail and the Rio Grande Valley between June 23 and August 8 (Wislizenus 1848).

The decade prior to the Civil War was a period of increased herpetological activity in New Mexico. Following the end of the Mexican-American War in 1848, several military expeditions were dispatched to survey the international boundary with Mexico to locate the best route for a transcontinental railroad. Spencer Fullerton Baird, assistant secretary for the Smithsonian Institution, arranged to have naturalists accompany numerous of these expeditions (Dall 1914; Adler 1989). The material collected helped reveal the richness and diversity of the herpetofauna of the American Southwest, including New Mexico. The first expedition, led by Capt. Lorenzo Sitgreaves, explored the Zuni River and the Little Colorado River to its junction with the Colorado River and beyond, traveling about 1450 km from Santa Fe to Fort Yuma between August 15, 1851, and November 24, 1851 (Sitgreaves 1853). The expedition traveled up the Rio Puerco and the Rio San José from the Rio Grande to Laguna Pueblo and westward via Acoma Pueblo, Inscription Rock, and Zuni Pueblo to the Zuni River, leaving New Mexico along the course of the Zuni River on September 25. The United States and Mexican Boundary Survey, led by

Maj. William H. Emory, consisted of a series of separate expeditions conducted between 1851 and 1855 to determine the international boundary following the Treaty of Guadalupe Hidalgo in 1848 and the Gadsden Purchase in 1853 (Emory 1857, 1859). In 1853, Congress authorized surveys to find the best route for a transcontinental railroad from the Mississippi River to the Pacific coast. The results of these surveys were published as a 12-volume treatise with multiple authors between 1854 and 1859 (e.g., Pope 1854 [1855]). The expedition led by Lt. A. W. Whipple entered New Mexico along the Canadian River on September 21, 1853; traveled up Pajarito Creek and across Arroyo Cuerbito (= Cuervito) and the Gallinas River to the Pecos River at Anton Chico on September 27; and reached Albuquerque via Galisteo and Peña Blanca on October 5. On November 8, the expedition traveled south to Isleta Pueblo, crossed the Rio Grande, and followed the approximate route of the Sitgreaves Expedition, leaving New Mexico on November 28. Whipple (1854 [1856]) commented on the success of the zoological collections by members of his party. Another expedition, led by Lt. John G. Parke, crossed the Peloncillo Mountains into New Mexico through Stein's Pass on March 6, 1854, and traveled the old Boundary Commission Road past Fort Webster and Cooke's Spring, reaching the Rio Grande in the vicinity of Leasburg on March 12. Parke (1855) noted that many of his specimens were lost, as the containers they were in leaked. A third expedition, led by Bvt. Capt. John Pope, departed eastward from Doña Ana on February 12, 1854, and, passing through Soledad Canyon in the Organ Mountains on February 14, traveled via the Hueco, Alamo, and Cornudas Mountains through Guadalupe Pass to the Delaware River, then to its junction with the Pecos River on March 8. Pope (1854 [1855], 59) noted on March 7 that the expedition "killed a rattle-snake (the first we have yet seen) on a hill near camp. It was put in spirits and carried along."

Relatively little herpetological exploration took place in New Mexico during the period following the establishment of territorial boundaries in 1863 and prior to World War II. A series of government-sponsored expeditions to explore the geographic and geological features west of the 100th meridian, led by Lt. George M. Wheeler, US Army Corps of Engineers, were carried out between 1871 and 1874. The results were published in a six-volume treatise, one of which contained the zoology of the expeditions. Two chapters (Coues 1875; Yarrow 1875) treated New Mexico herpetology. Yarrow (1875) reported 1 salamander, 8 anurans, 21 lizards, 17 snakes, and 1 turtle from the New Mexico Territory. Other published references during this period (Cope 1883, 1896; Garman 1887; Townsend 1893; Stone and Rehn 1903; Bailey 1905; MacBride 1905; Ellis 1917) also presented information on specimens collected from New Mexico. Personnel associated with the US Biological Survey (1896–1939) also collected in New Mexico. Bailey (1913) listed 76 species, including 3 salamanders, 7 anurans, 31 lizards, 32 snakes, and 3 turtles, shortly after New Mexico became a state. He provided the first extensive statewide discussion of the ecology and zoogeography of individual species. Van Denburgh (1924) provided a checklist containing 85 species (2 salamanders, 12 anurans, 30 lizards, 36 snakes, and 5 turtles).

The founding of New Mexico State University in 1880 and the USDA's Jornada Experimental Range in 1912, along with the University of New Mexico in 1889, resulted in increased herpetological exploration in the Rio Grande Valley (e.g., Cockerel 1896; Herrick et al. 1899; Little and Keller 1937). Other noteworthy accounts during this period included that of Ruthven (1907), who collected between White Sands National Monument and Cloudcroft in the Sacramento Mountains. He provided information on 2 anurans, 13 lizards, 2 snakes, and 1 turtle. Mosauer (1932) spent three weeks in the Guadalupe Mountains and provided observations on 1 salamander, 1 anuran, 7 lizards, and 3 snakes.

After World War II, major military and supporting civilian installations were built that changed the face of New Mexico. These installations required new road construction, and the routine use of motorized vehicles made it easy to get to previously unexplored places and habitats and provided a direct mechanism for collecting specimens (e.g., Klauber 1939; Campbell 1953, 1956). Research programs and herpetological research collections were established at the state's colleges and universities. The arrival of William J. Koster at the University of New Mexico (UNM) in 1938 was especially important in helping build these collections. Although Koster was an ichthyologist, he was well acquainted with A. H. Wright and had a keen interest in herpetology. His collections of amphibians and reptiles resulted in over 5000 specimens. James S. Findley arrived at UNM in 1955, and although he was primarily a mammalogist, he and his students collected large numbers of amphibians and reptiles throughout New Mexico. William G. Degenhardt arrived at UNM from Texas A&M University in 1960 and developed the Division of Herpetology and graduate and undergraduate programs in herpetology. During this same time period other programs in herpetology developed at New Mexico State University (James R. Dixon, Walter G. Whitford, Joseph L. LaPointe), Western New Mexico University (Bruce J. Hayward), and Eastern New Mexico University (A. L. "Tony" Gennaro). This activity resulted in a significant increase in the knowledge of the distribution and natural history of the New Mexico herpetofauna, as well as in the number of publications similar to that seen in other western states and in Mexico (e.g., Webb 1970; Dixon 1987; Carpenter and Krupa 1989; Smith and Smith 1973, 1976, 1979 [1980], 1993). Noteworthy contributors to regional surveys during this period include Lewis (1950; Organ Mountains and Tularosa Basin, 34 species), Harris (1963; San Juan Basin, 21 species), Gehlbach (1965; Zuni and Zuni Mountains, 30 species), Jones (1970; Chaco Canyon National Monument, 18 species), Mecham (1979; Guadalupe Mountains, 50 species), and Best et al. (1983; Pedro Armendariz lava field, Sierra and Socorro Counties, 26 species). Important studies of select taxonomic groups during this period included those by Zweifel (1968), Creusere and Whitford (1976), and Woodward (1982, 1983, 1987) for anurans; Degenhardt and Christiansen (1974) for turtles; Tanner (1975),

Whitford and Creusere (1977), and Baltosser and Best (1990) for lizards; and Price and LaPointe (1990) for snakes.

The New Mexico Herpetological Society (NMHS) was formally established in October 1963, under the guidance of William G. Degenhardt and Ted L. Brown, as an organization dedicated to acquiring and sharing information about herpetology, herpetoculture, and the amphibians and reptiles of New Mexico. The society's initial activities included field trips to selected areas of New Mexico to investigate species peripheral to the state and those of special interest to New Mexico Department of Game and Fish (NMDGF). Since the establishment of the NMHS, members have contributed regional surveys that have added significantly to our understanding of the distribution of the herpetofauna in New Mexico. Included among many others are reports by T. Brown (2002, Mills Canyon, Harding and Mora Counties; 2003, Chloride Canyon, Black Range Sierra County); Brown and Stuart (1990, Canadian River Canyon, Harding, Mora, and San Miguel Counties), Stuart and Brown (1988, Dry Cimarron River Valley, Union County), and Stuart and White (1987, springs in Eddy County).

Numerous regional herpetofaunal surveys are available as unpublished reports, including those by Painter (1985, Gila and San Francisco River Valleys, southwestern New Mexico), Martin-Bashore (1997, Otero Mesa, McGregor Range, Fort Bliss), Krupa (1998, Rattlesnake Springs, Carlsbad Caverns National Park), Burkett and Black (1999, San Andres National Wildlife Refuge), Fox et al. (1999, Los Alamos County), Burkett (2000, White Sands Missile Range), Collins (2000, Kiowa National Grasslands of northeastern New Mexico), and Cummer et al. (2003, Valles Caldera National Preserve, Sandoval County).

Fritts et al. (1984) provided a complete review of the leopard frogs of New Mexico with distribution maps and notes on the morphology of adults and tadpoles and natural history. Altenbach and Painter (1998) published a complete bibliography of the state-endangered Jemez

Mountains Salamander (*Plethodon neomexicanus*). Beck (2005) published a thorough book on the natural history of the Gila Monster; much of his research for the book was conducted in southwestern New Mexico. Lannoo (2005 [and numerous authors therein]) reported on the conservation status and distribution of all amphibians in New Mexico.

Following the passage of the New Mexico Wildlife Conservation Act in 1974 and the creation of a nongame and endangered species program within the NMDGF, Charles W. Painter was hired as New Mexico's first state herpetologist in 1986 and retired from that position in 2013. He has conducted research and status surveys on numerous species of New Mexico's herpetofauna, particularly *Plethodon neomexicanus*, *Lithobates chiricahuensis*, *Aspidoscelis dixoni*, *Sceloporus arenicolus*, *Pseudemys gorzugi*, *Trachemys gaigeae*, and *Crotalus willardi obscurus*. He is a recognized authority on the barbaric practice of rattlesnake roundups.

By the early 1990s, sufficient information had been compiled to warrant preparation of a comprehensive publication on New Mexico's herpetofauna. The result, *Amphibians and Reptiles of New Mexico* (Degenhardt et al. 1996), provided the first major treatment of all the state's species and their geographic distribution and natural history and also identified gaps in our knowledge and topics for future research. The book subsequently stimulated additional reporting of new distribution and natural history information for the state, especially in the journal *Herpetological Review*. Genetic studies of herpetofaunal populations in New Mexico and elsewhere have resulted in many changes in taxonomy since 1996 (Crother 2008), as reflected in the current list of species known from the state. Many of these references concerning the New Mexico herpetofauna published since 1996 are cited in Stuart (2005).

In addition to the studies mentioned previously, we have now determined that there are 23 type localities of species and subspecies of amphibians (4 salamanders) and reptiles (15 lizards, 4 snakes) in the state of New Mexico (see table 11.1).

Physiographic Characteristics and Their Influence on the Distribution of the Herpetofauna

The following discussion was adapted in part from information in the chapter "A Physiographic Sketch of New Mexico" in Degenhardt et al. (1996), authored by Andrew H. Price. Additional information is available in New Mexico Department of Game and Fish (2006) and Western Regional Climate Center (2011).

New Mexico is the fifth-largest state in the United States, with a total area of 314,456 km². It is a square of approximately 563 km on each side and lies mostly between latitudes 32° and 37° N and longitudes 103° and 109° W. The state's topography consists mainly of high plateaus or mesas, with numerous mountain ranges, canyons, valleys, and normally dry arroyos. Average elevation is about 1433 m above sea level. The lowest point is just above Red Bluff Reservoir at 859 m where the Pecos River flows into Texas. The highest point is Wheeler Peak at 4011 m in the Sangre de Cristo Mountains.

The geology of New Mexico is complex and has been shaped by faults, uplifts, and volcanic activity. The Rio Grande Rift is a major fault that divides the state in half from north to south and contains the Rio Grande Gorge in northern New Mexico and the broad Rio Grande Valley in the central and southern parts of the state. The rift divides the southern Rocky Mountains of northern New Mexico into two major mountain complexes, the Sangre de Cristo Mountains to the east and the San Juan Mountains to the west. The Colorado Plateau and San Juan Basin lie west of the southern Rocky Mountains and are the major topographic features of northwestern New Mexico. The Jemez Mountains, a volcanic caldera, are disjunct from the southern Rocky Mountains but share many affinities with the more northern mountains of the state. Farther

TABLE 11.1 Type localities for amphibians and reptiles described from the state of New Mexico, USA

Author(s)	Original name: Type locality
Salamanders (4)	
Baird (1850)	*Ambystoma mavortia* (= *A. mavortium mavortium*): **"New Mexico"**
Hallowell (1852b [1853])	*Ambystoma nebulosum* (= *A. mavortium nebulosum*): **"New Mexico"**; this actually referred to the territory of New Mexico prior to the division of Arizona and New Mexico. Hallowell later restricted the type locality to **"San Francisco Mountain"** (= San Francisco Peaks, Coconino County, Arizona).
Taylor, 1941	*Plethodon hardii* (= *Aneides hardii*): **Sacramento Mountains at Cloudcroft (9000 ft.), New Mexico**
Stebbins & Riemer (1950)	*Plethodon neomexicanus*: **12 miles west and 4 miles south of Los Alamos, 8750± feet, Sandoval County, New Mexico**
Lizards (15)	
Baird & Girard (1852)	*Phrynosoma modestum*: the original description states the type material was **"brought from the valley of the Rio Grande west of San Antonio, by Gen. Churchill and from between San Antonio and El Paso del Norte, by Col. J.D. Graham"**; Smith and Taylor (1950) restricted the type locality to **"Las Cruces, Dona Ana County, New Mexico."** Axtell (1988) disagreed with this. Therefore, the actual type locality is in question (see Degenhardt et al. 1996 for additional detail).
Degenhardt & Jones (1972)	*Sceloporus graciosus arenicolous* (= *S. arenicolus*): **Mescalero Sands 3 1/2 miles N and 44 miles E Roswell, Chaves Co., New Mexico [T10S R31E]**
Phelan & Brattstrom (1955)	*Sceloporus magister bimaculosus* (= *S. bimaculosus*): **6.6 mi E. of San Antonio, Socorro County, New Mexico**
Lowe & Norris (1956)	*Sceloporus undulatus cowlesi* (= *S. cowlesi*): **White Sands, 3 miles northwest of the Monument headquarters, Otero County, New Mexico**
Smith et al. (1992)	*Sceloporus undulatus tedbrowni* (= *S. consobrinus*): **Chaves Co., New Mexico, 6 mi W Caprock, Lea Co., large dune, Waldrop Peak, 0.5 mi S Hwy 380**
Baird & Girard (1852)	*Sceloporus poinsetti*: original type locality was given as **"Rio San Pedro of the Rio Grande del Norte, and the province of Sonora"**; however, this was restricted to **"either the southern part of the Big Burrow [sic] Mountains, or the vicinity of Santa Rita, Grant Co., New Mexico"** by Webb (1988).
Cope *in* Yarrow (1875)	*Sceloporus undulatus tristichus* (= *S. tristichus*): **Taos [Taos County], New Mexico**
Stejneger (1890)	*Urosaurus ornatus levis*: **Tierra Amarilla [Rio Arriba County], New Mexico**
Schmidt (1921)	*Uta stansburiana stejnegeri*: **mouth of Dry Canyon, Alamogordo [Otero County], New Mexico**
Lowe (1956)	*Cnemidophorus exsanguis* (= *Aspidoscelis exsanguis*): **Socorro, Socorro County, New Mexico**
Lowe & Wright (1964)	*Cnemidohorus flagellicaudus* (= *Aspidoscelis flagellicauda*): **at San Francisco Hot Springs (Frisco Hot Springs), 4800 ft elev., Catron County, New Mexico**
Wright & Lowe (1993)	*Cnemidophorus inornatus gypsi* (= *Aspidoscelis gypsi*): **White Sands National Monument, at 3 miles (by road) NW Monument Headquarters, 4020 ft, Otero County, New Mexico**

TABLE 11.1 *Continued*

Author(s)	Original name: Type locality
Wright & Lowe (1993)	*Cnemidophorus inornatus juniperus* (= *Aspidoscelis inornata junipera*): **San Pedro Creek, 3 miles north and 2 miles east of San Antonio, 4550 ft, Bernalillo County, New Mexico**; Degenhardt et al. (1996) noted that "San Antonio" is actually San Antonito.
Wright & Lowe (1993)	*Cnemidophorus inornatus llanuras* (= *Aspidoscelis inornata llanuras*): **Carthage, 4990 ft, Socorro County, New Mexico**
Lowe & Zweifel (1952)	*Cnemidophorus neomexicanus* (= *Aspidoscelis neomexicana*): **McDonald ranch headquarters, 4800 feet elevation, 8.7 miles west and 22.8 miles south of New Bingham Post Office, Socorro County, New Mexico**
Snakes (4)	
Hallowell (1852a)	*Masticophis* (= *Coluber*) *taeniatus*: **New Mexico, west of the Rio Grande**; Smith and Taylor (1950) restricted to **Shiprock, San Juan County, New Mexico**
Hallowell (1852a)	*Pituophis melanoleucus affinis* (= *P. catenifer affinis*): the type locality **New Mexico** was restricted to **Zuni River, New Mexico**, by Hallowell *in* Sitgraves (1853); Smith and Taylor (1950) further restricted the type locality to **Zuni, McKinley County, New Mexico**
Baird & Girard (1853)	*Crotalus molossus*: **Fort Webster, St Rita del Cobre, Grant Co., New Mexico**
Harris & Simmons (1976)	*Crotalus willardi obscurus*: **Animas Mountains in Indian Creek Canyon near Animas Peak, Hidalgo County, New Mexico**

south is the Mogollon Plateau, which extends westward into Arizona. The Sacramento Highlands are the only major highlands east of the Rio Grande Valley in southern New Mexico and share some biological affinities with the more northerly mountains of the state. In addition, a number of smaller and isolated mountain ranges, separated by broad basins, comprise the Basin and Range topography of central and southern New Mexico on both sides of the Rio Grande Valley.

Much of the eastern one-third of New Mexico is characterized by level or gently rolling topography and includes the Llano Estacado, an elevated plain that extends into western Texas and has little natural aquatic habitat.

The state has approximately 606 km² of rivers, streams, lakes, and reservoirs that compose less than 1% of the state's total surface area. Although they are a small part of the state's landscape, rivers provide crucial habitat for many species that otherwise would not occur in the state's primarily xeric environment. The Con-

tinental Divide lies in the western half of New Mexico, where it is defined by high mountains in the northern part of the state and low ridge lines in the south. The state's major river systems found west of the Continental Divide include (1) the San Juan, which arises in southwestern Colorado and flows westward to join the Colorado River in Utah; (2) the Gila, which arises on the Mogollon Plateau and flows westward into Arizona to join the Colorado River; (3) the San Francisco, a major tributary of the Gila that arises in eastern Arizona; and (4) the Zuni, a tributary to the Little Colorado River that arises in western New Mexico. The major river systems found east of the Continental Divide are (1) the Rio Grande, which arises in southern Colorado and flows southward through New Mexico and along the United States–Mexico border to the Gulf of Mexico; (2) the Pecos, a major tributary of the Rio Grande that arises in northern New Mexico and flows southward through the eastern half of the state to join the Rio Grande in western Texas; (3) the Canadian, which arises in northern New Mexico and southern Colorado and flows eastward into Texas to join the Arkansas River and eventually the Mississippi; and (4) the Dry Cimarron, a small stream in northeastern New Mexico that is part of the Arkansas River drainage.

Several endorheic (closed) basins also occur in New Mexico, some of which support perennial streams or contain saline or gypsum soils that provide unique habitat for many species. The largest closed basins in the west are the Plains of San Agustin in Catron County and the Rio Mimbres Basin in Grant and Luna Counties. The Tularosa Basin in southern New Mexico is an intermountain area east of the Rio Grande Valley, best known for its extensive gypsum dune habitat.

Mean annual temperatures range from 17.8°C in the extreme southeast to 4.4°C or lower in high mountains and valleys of the north; elevation is a greater factor in determining the temperature of any specific locality than is latitude. This is shown by only a 1.7°C difference in mean temperature between stations at similar elevations, one in the extreme northeast and the other in the extreme southwest; however, at two stations only 24.1 km apart, but differing in elevation by 1433 m, the mean annual temperatures are 16.1°C and 7.2°C—a difference of 8.9°C or a little more than 1.7°C decrease in temperature for each 304.8 m increase in elevation.

During the summer months, daytime temperatures often exceed 37.8°C at elevations below 1524 m; but the average monthly maximum temperatures during July, the warmest month, range from slightly above 32.2°C at lower elevations to about 21.0°C at high elevations. Warmest days normally occur in June before the thunderstorm season sets in; during July and August, afternoon convective storms tend to decrease solar insolation, lowering temperatures before they reach their potential daily high. The highest temperature on record in New Mexico is 46.7°C at Orogrande on July 14, 1934, and at Artesia on June 29, 1918. A preponderance of clear skies and low relative humidity permits rapid cooling by radiation from the earth after sundown;

consequently, nights are usually comfortable in summer. The average range between daily high and low temperatures is 13.9°C and 19.4°C.

In January, the coldest month, average daytime temperatures range from about 10°C in the southern and central valleys to about –1°C in the higher elevations of the north. Minimum temperatures below freezing are common in all sections of the state during the winter, but subzero temperatures are rare except in the mountains. The lowest temperature recorded at regular observing stations in the state was –45.6°C at Gavilan on February 1, 1951. An unofficial low temperature of –49.4°C at Ciniza on January 13, 1963, was widely reported by the press.

The freeze-free season ranges from more than 200 days in the southern valleys to fewer than 80 days in the northern mountains, where some high mountain valleys freeze during the summer months.

The principal sources of moisture for the scant rains and snows that fall on New Mexico are the Pacific Ocean, 1288 km to the west, and the Gulf of Mexico, 804 km to the southeast. New Mexico has a mild, arid, or semiarid continental climate characterized by light precipitation totals, abundant sunshine, low relative humidity, and a relatively large annual and diurnal temperature range. The highest mountains have climate characteristics common to those of the Rocky Mountains.

Average annual precipitation ranges from less than 25.4 cm over much of the southern desert and the Rio Grande and San Juan Valleys to more than 50.8 cm at higher elevations in the state. A wide variation in annual totals is characteristic of arid and semiarid climates as illustrated by annual extremes of 7.5 cm and 86.2 cm at Carlsbad during a period of more than 71 years.

Summer rains fall almost entirely during brief but frequently intense thunderstorms. The general southeasterly circulation from the Gulf of Mexico brings moisture for these storms into the state, and strong surface heating combined with orographic lifting as the air moves over higher terrain causes air currents and condensation. July and August are the rainiest months over most of the state, with 30%–40% of the year's total moisture falling at that time. The San Juan Valley area is least affected by this summer circulation, receiving about 25% of its annual rainfall during July and August. During the warmest six months of the year, May through October, total precipitation averages from 60% of the annual total in the Northwestern Plateau to 80% of the annual total in the eastern plains.

Winter precipitation is caused mainly by frontal activity associated with the general movement of Pacific Ocean storms across the country from west to east. As these storms move inland, much of the moisture is precipitated over the coastal and inland mountain ranges of California, Nevada, Arizona, and Utah. Much of the remaining moisture falls on the western slope of the Continental Divide and over northern and high central mountain ranges. Winter is the driest season in New Mexico except for the portion west of the Continental Divide. This dryness is most noticeable in the Central Valley and on eastern slopes of the mountains.

Much of the winter precipitation falls as snow in the mountain areas, but it may occur as either rain or snow in the valleys. Average annual snowfall ranges from about 7.6 cm at the Southern Desert and Southeastern Plains stations to well over 2.54 m at Northern Mountain stations. Snowfall may exceed 7.62 m in the highest mountains of the north.

Average relative humidity in New Mexico is lower in the valleys but higher in the mountains because of the lower mountain temperatures. Relative humidity ranges from an average of near 65% about sunrise to near 30% in midafternoon; however, afternoon humidity in warmer months is often less than 20% and occasionally may be as low as 4%. The low relative humidity during periods of extreme temperatures eases the effect of summer and winter temperatures (Western Regional Climate Center 2011).

Biotic Communities

Based on the classification system developed by Brown and Lowe (1980) and D. Brown (1994b),

New Mexico supports at least 14 biotic communities ranging from high-elevation montane habitats that support few or no amphibians and reptiles to middle- and low-elevation grassland and desert communities with greater herpetofaunal diversity (Degenhardt et al. 1996). These communities are useful in understanding broad-scale patterns of herpetofaunal diversity but do not capture fine-scale landscape features such as soil type, topography, elevation, and availability of underground refugia that also greatly influence the distribution of herpetofauna. Also interspersed within these main biotic communities are various water bodies and riparian vegetation zones that, although uncommon in New Mexico, provide valuable habitat for many species.

Plains and Great Basin Grasslands

Plains and Great Basin Grasslands are common and widespread in New Mexico on plains and in valleys between 1200 and 2300 m in elevation. They extend from the northeast and east-central parts of the state westward through the Albuquerque area and into the San Juan Basin in the northwest, and southward to the Plains of San Agustin, near the Gila River, and the Animas Valley (Brown and Lowe 1980). They are absent in much of southern New Mexico. These grasslands consist mostly of short- or mixed-grass species (e.g., *Bouteloua* spp., *Achnatherum hymenoides*, *Pleuraphis jamesii*, *Sporobolus airoides*) occurring almost statewide and tall-grass stands (especially *Andropogon gerardi*, *Schizachyrium scoparium*) mostly limited to the eastern part of the state (D. Brown 1994h). Associated shrubs include *Atriplex canescens*, *Artemisia* spp., *Krascheninnikovia lanata*, *Yucca* spp., *Opuntia* spp., *Cylindropuntia* spp., and *Ericameria* spp. The oak *Quercus havardii* is found in this community in southeastern New Mexico. Overgrazing and fire suppression have impacted this community, leading to extensive encroachment by shrubs in many areas.

Associated herpetofauna (62 species) include *Ambystoma mavortium*, *Anaxyrus cognatus*, *A. debilis*, *A. microscaphus*, *A. punctatus*, *Pseudacris clarkii*, *P. maculata*, *Gastrophryne olivacea*, *Litho-*

bates blairi, Scaphiopus couchii, Spea bombifrons, S. multiplicata, Terrapene ornata, Kinosternon flavescens, Crotaphytus collaris, Gambelia wislizenii, Coleonyx brevis, Cophosaurus texanus, Holbrookia maculata, Phrynosoma cornutum, P. hernandesi, P. modestum, Sceloporus arenicolus, S. consobrinus, S. cowlesi, S. slevini, S. tristichus, Urosaurus ornatus, Uta stansburiana, Plestiodon multivirgatus, P. obsoletus, Aspidoscelis exsanguis, A. inornata, A. neomexicana, A. sexlineata, A. tesselata, A. tigris, A. uniparens, Arizona elegans, Coluber constrictor, Diadophis punctatus, Gyalopion canum, Heterodon nasicus, H. platirhinos, Hypsiglena jani, Lampropeltis getula, L. triangulum, Masticophis flagellum, M. taeniatus, Pantherophis emoryi, Pituophis catenifer, Rhinocheilus lecontei, Sonora semiannulata, Tantilla nigriceps, Thamnophis elegans, T. marcianus, T. radix, Tropidoclonion lineatum, Rena dissecta, Crotalus atrox, C. viridis, and *Sistrurus catenatus.*

Semidesert Grassland

Semidesert Grassland is typically found between 1100 and 1400 m and is closely associated with Chihuahuan Desert Scrub, with which it makes contact and intergrades. It occurs south of Albuquerque and Fort Sumner, bordering the Rio Grande and Pecos River Valleys, respectively, and across much of southern New Mexico, including near Roswell, Hobbs, Carrizozo, Las Cruces, Lordsburg, and most of Hidalgo County (Brown and Lowe 1980). Primary grass species include *Pleuraphis mutica, Bouteloua* spp., and *Aristida* spp. Cacti such as *Cylindropuntia* spp., *Opuntia* spp., and *Ferocactus wislizenii* are also important components. Although it was primarily a grass-dominated community prior to European settlement, livestock overgrazing has resulted in a disclimax grassland in many areas that is now dominated by shrubs such as *Prosopis glandulosa, Larrea tridentata, Flourensia cernua,* and *Gutierrezia* spp. (D. Brown 1994i).

Associated herpetofauna (72 species) include *Ambystoma mavortium, Anaxyrus cognatus, A. debilis, A. microscaphus, A. punctatus, A. speciosus, Incilius alvarius, Craugastor augusti, Gastrophryne olivacea, Lithobates blairi, Scaphiopus couchii,*

Spea bombifrons, S. multiplicata, Terrapene ornata, Kinosternon flavescens, Elgaria kingii, Crotaphytus collaris, Gambelia wislizenii, Coleonyx variegatus, Cophosaurus texanus, Holbrookia elegans, H. maculata, Phrynosoma cornutum, P. hernandesi, P. modestum, Sceloporus arenicolus, S. bimaculosus, S. clarkii, S. consobrinus, S. cowlesi, S. poinsettii, S. tristichus, Urosaurus ornatus, Uta stansburiana, Plestiodon obsoletus, Aspidoscelis exsanguis, A. flagellicauda, A. gularis, A. inornata, A. marmorata, A. neomexicana, A. sexlineata, A. sonorae, A. tesselata, A. uniparens, Arizona elegans, Diadophis punctatus, Gyalopion canum, Heterodon kennerlyi, H. nasicus, Hypsiglena jani, Lampropeltis getula, L. triangulum, Masticophis flagellum, M. taeniatus, Pantherophis emoryi, Pituophis catenifer, Rhinocheilus lecontei, Salvadora deserticola, S. grahamiae, Sonora semiannulata, Tantilla hobartsmithi, T. nigriceps, Thamnophis marcianus, Trimorphodon vilkinsonii, Micruroides euryxanthus, Rena dissecta, R. segrega, Crotalus atrox, C. scutulatus, C. viridis, and *Sistrurus catenatus.*

Chihuahuan Desert Scrub

Chihuahuan Desert Scrub is confined to the southern half of New Mexico between approximately 866 and 1500 m in elevation, where it is widespread in the Tularosa Basin and the Jornada del Muerto and at or near the White Sands National Monument and the towns of Socorro, Carlsbad, Artesia, Las Cruces, Deming, and Lordsburg (Brown and Lowe 1980). It is absent in the Bootheel. Much of this community is underlain by limestone, including caliche, and is typically present on gravel substrates and low, rocky hills and bajadas. The diverse topography within the region covered by Chihuahuan Desert Scrub contributes to the diversity of both floral and herpetofaunal species present in this biotic community. This desert scrub most frequently comes into contact with Semidesert Grassland, with which it shares many floral and faunal elements. Major shrub species include *Larrea tridentata, Flourensia cernua, Acacia* spp., and *Koeberlinia spinosa. Prosopis glandulosa* is abundant where sandy soils are present and where other plants generally associated with Semidesert Grassland may be present. Cacti are common, including species of *Opuntia, Cylindropuntia, Ferocactus,* and *Echinocactus* (D. Brown 1994c).

Associated herpetofauna (71 species) include *Ambystoma mavortium, Anaxyrus cognatus, A. debilis, A. punctatus, A. speciosus, Incilius alvarius, Craugastor augusti, Gastrophryne olivacea, Scaphiopus couchii, Spea bombifrons, S. multiplicata, Terrapene ornata, Kinosternon flavescens, Elgaria kingii, Crotaphytus collaris, Gambelia wislizenii, Coleonyx brevis, C. variegatus, Heloderma suspectum, Callisaurus draconoides, Cophosaurus texanus, H. maculata, Phrynosoma cornutum, P. modestum, Sceloporus bimaculosus, S. cowlesi, S. magister, S. poinsettii, Urosaurus ornatus, Uta stansburiana, Plestiodon obsoletus, Aspidoscelis dixoni, A. exsanguis, A. gularis, A. gypsi, A. inornata, A. marmorata, A. neomexicana, A. sonorae, A. tesselata, A. uniparens, Arizona elegans, Bogertophis subocularis, Diadophis punctatus, Gyalopion canum, Heterodon kennerlyi, H. nasicus, Hypsiglena jani, Lampropeltis alterna, L. getula, L. triangulum, Masticophis flagellum, M. taeniatus, Pituophis catenifer, Rhinocheilus lecontei, Salvadora deserticola, S. grahamiae, Sonora semiannulata, Tantilla hobartsmithi, T. nigriceps, T. yaqui, Thamnophis marcianus, Trimorphodon vilkinsonii, Micruroides euryxanthus, Rena dissecta, R. segrega, Crotalus atrox, C. lepidus, C. molossus, C. scutulatus,* and *C. viridis.*

Great Basin Desert Scrub

Great Basin Desert Scrub occurs intermittently in northwestern New Mexico, eastward to the Taos Plateau, and south to the area around Abiquiu Reservoir and near Gallup, generally at elevations between 1200 and 2200 m (Brown and Lowe 1980; Turner 1994). It is characterized by level plains or gently rolling terrain, typically on highly erodible soils dominated by several shrub species, especially *Artemisia* spp. and *Atriplex canescens,* along with *Chrysothamnus* spp., *Sarcobatus* spp., and *Tetradymia* spp.

Associated herpetofauna (19 species) include *Ambystoma mavortium, Spea bombifrons, S. mul-*

tiplicata, Crotaphytus collaris, Gambelia wislizenii, Sceloporus graciosus, S. magister, S. tristichus, Uta stansburiana, Aspidoscelis tigris, A. velox, Heterodon nasicus, Hypsiglena chlorophaea, Lampropeltis getula, L. triangulum, Masticophis taeniatus, Pituophis catenifer, Thamnophis elegans, and *Crotalus viridis.*

Madrean Evergreen Woodland

Madrean Evergreen Woodland is widely scattered in the southern part of the state, mainly between elevations of 1500 and 2200 m, with the most continuous occurrence found west of Silver City and into Arizona. It is present in the vicinity of the Animas, Peloncillo, Big Hatchet, Burro, Tres Hermanas, Organ, and Guadalupe Mountains (Brown and Lowe 1980). This encinal woodland is dominated by species of evergreen oaks (*Quercus* spp.) with pines (*Pinus* spp.) also present at higher elevations; it typically contacts grasslands or Interior Chaparral at its lower elevations (D. Brown 1994f).

Associated herpetofauna (45 species) include *Ambystoma mavortium, Anaxyrus punctatus, Hyla arenicolor, Terrapene ornata, Elgaria kingii, Crotaphytus collaris, Coleonyx brevis, Heloderma suspectum, Holbrookia elegans, Phrynosoma hernandesi, P. solare, Sceloporus clarkii, S. jarrovii, S. poinsettii, S. slevini, S. virgatus, Urosaurus ornatus, Plestiodon callicephalus, P. obsoletus, Aspidoscelis exsanguis, A. flagellicauda, A. sonorae, A. stictogramma, A. tesselata, Hypsiglena jani, Lampropeltis getula, L. knoblochi, Masticophis bilineatus, M. flagellum, M. taeniatus, Pituophis catenifer, Salvadora grahamiae, Senticolis triaspis, Tantilla hobartsmithi, T. nigriceps, T. yaquia, Thamnophis cyrtopsis, Trimorphodon lambda, T. vilkinsonii, Micruroides euryxanthus, Crotalus atrox, C. lepidus, C. molossus, C. viridis,* and *C. willardi.*

Great Basin Conifer Woodland

Great Basin Conifer Woodland, commonly known as pinyon-juniper or juniper woodland, is widely distributed in New Mexico, mainly between elevations of 1500 and 2300 m on bajadas, mesas, or foothills surrounding many of the state's mountain ranges, including the Sangre de Cristo, Jemez, Sandia, Manzano, Sacramento, and Guadalupe Mountains and in the Gila River drainage basin (Brown and Lowe 1980). It is absent from the southwestern corner of the state and the Llano Estacado of eastern New Mexico. This cold-adapted woodland is dominated by dwarf conifers, including one or more species of juniper (*Juniperus*) and pinyon (*Pinus edulis*). Other floral elements include shrubs such as *Fallugia*, *Rhus*, *Yucca*, *Atriplex*, and *Purshia* and grasses and forbs such as *Bouteloua*, *Bomus*, *Sphaeralcea*, and *Eriogonum* (D. Brown 1994d).

Although no herpetofauna is found exclusively in this habitat, 41 species may be present, including *Ambystoma mavortium*, *Anaxyrus microscaphus*, *A. punctatus*, *Spea multiplicata*, *Elgaria kingii*, *Crotaphytus collaris*, *Holbrookia maculata*, *Phrynosoma hernandesi*, *P. modestum*, *Sceloporus cowlesi*, *S. graciosus*, *S. tristichus*, *Urosaurus ornatus*, *Uta stansburiana*, *Plestiodon obsoletus*, *Aspidoscelis exsanguis*, *A. flagellicauda*, *A. inornata*, *A. neomexicana*, *A. tesselata*, *A. velox*, *Diadophis punctatus*, *Gyalopion canum*, *Heterodon nasicus*, *Hypsiglena chlorophaea*, *H. jani*, *Lampropeltis getula*, *L. triangulum*, *Masticophis flagellum*, *M. taeniatus*, *Pantherophis emoryi*, *Pituophis catenifer*, *Rhinocheilus lecontei*, *Salvadora grahamiae*, *Tantilla hobartsmithi*, *Trimorphodon lambda*, *Tropidoclonion lineatum*, *Rena dissecta*, *Crotalus atrox*, *C. molossus*, and *C. viridis*.

Interior Chaparral

Interior Chaparral is a shrub-dominated community that is common and widespread in Arizona but occurs only sparsely in New Mexico. It is found on slopes in or near the Burro, Florida, San Andres, Organ, and Guadalupe Mountains in the southern part of the state (Brown and Lowe 1980). Often found upslope from Semi-desert Grassland, Interior Chaparral is characterized by several shrub species, including *Garrya*, *Rhus trilobata*, *Purshia stansburiana*, *Forestiera pubescens*, *Berberis*, and occasionally shrub oaks (*Quercus*), often on granite or limestone sub-

strates at elevations of about 1700–2450 m (Pase and Brown 1994a).

Although lacking a well-defined herpetofaunal community, the 20 associated species include *Crotaphytus collaris*, *Sceloporus cowlesi*, *S. poinsettii*, *Urosaurus ornatus*, *Uta stansburiana*, *Plestiodon obsoletus*, *Aspidoscelis exsanguis*, *A. tesselata*, *Bogertophis subocularis*, *Hypsiglena jani*, *Masticophis taeniatus*, *Pituophis catenifer*, *Salvadora grahamiae*, *Tantilla hobartsmithi*, *Trimorphodon vilkinsonii*, *Rena segrega*, *Crotalus atrox*, *C. lepidus*, *C. molossus*, and *C. viridis*. *Lampropeltis alterna* also may occur in this community.

Great Basin Montane Scrubland

Great Basin Montane Scrubland has a limited distribution in New Mexico but becomes more common just north of the Colorado–New Mexico state line. Found in foothills and mountain areas at elevations of approximately 2300–2750 m, this scrubland occurs most extensively in the Raton Pass area east of the Sangre de Cristo Mountains and on the western slope of the Sandia-Manzano Mountain chain (Brown and Lowe 1980). This community is dominated by Gambel oak (*Quercus gambelii*) but also includes a variety of other shrubs such as *Cercocarpus*, *Robinia neomexicana*, *Amelanchier*, and *Symphoricarpus* (D. Brown 1994e).

Herpetofauna found in Montane Conifer Forest or Great Basin Desert Scrub also may occur in this community and include the following 13 species: *Crotaphytus collaris*, *Phrynosoma hernandesi*, *Urosaurus ornatus*, *Plestiodon multivirgatus*, *P. obsoletus*, *Aspidoscelis exsanguis*, *Masticophis flagellum*, *M. taeniatus*, *Opheodrys vernalis*, *Pituophis catenifer*, *Salvadora grahamiae*, *Crotalus molossus*, and *C. viridis*.

Rocky Mountain (Petran) and Madrean Montane Conifer Forest

Rocky Mountain (Petran) and Madrean Montane Conifer Forest is widespread and found in most uplands at elevations of about 2300–2900 m throughout New Mexico, including the Sangre de

Cristo, San Juan, Jemez, Gallinas, Sacramento, San Mateo, Magdalena, Mogollon, and Animas Mountains, and in the Black Range and near Mount Taylor (Brown and Lowe 1980). Included in this community are forests dominated by ponderosa or Chihuahuan pines (*Pinus ponderosa, P. leiophylla* in southwestern New Mexico), mixed conifer/Douglas fir–white fir (*Pseudotsuga menziesii–Abies concolor*), and aspen (*Populus tremuloides*). Other trees found include *Pinus strobiformis, Quercus gambelii,* and *Robinia neomexicana,* along with a variety of understory shrubs, forbs, and grasses (Pase and Brown 1994b).

Associated herpetofauna (35 species) include *Ambystoma mavortium, Aneides hardii, Plethodon neomexicanus, Anaxyrus microscaphus, Hyla arenicolor, Hyla wrightorum, Pseudacris maculata, Phrynosoma hernandesi, Sceloporus clarkii, S. cowlesi, S. graciosus, S. jarrovii, S. poinsettii, S. tristichus, S. virgatus, Plestiodon callicephalus, P. multivirgatus, P. obsoletus, Aspidoscelis exsanguis, A. velox, Coluber constrictor, Diadophis punctatus, Lampropeltis knoblochi, L. pyromelana, L. triangulum, Pantherophis emoryi, Pituophis catenifer, Salvadora grahamiae, Thamnophis elegans, Trimorphodon lambda, Crotalus cerberus, C. lepidus, C. molossus, C. viridis,* and *C. willardi.*

Montane Meadow Grasslands

Montane Meadow Grasslands are embedded within the Rocky Mountain (Petran) and Madrean Montane Conifer Forest and are characterized by a variety of grass and forb species, including wetland plants such as *Poa, Festuca, Carex, Cyperus, Eleocharis,* and *Juncus* (D. Brown 1994g). When associated with streams or saturated soils, this community provides habitat for six species of high-elevation herpetofauna such as *Ambystoma mavortium, Hyla wrightorum, Pseudacris maculata, Lithobates pipiens, Opheodrys vernalis,* and *Thamnophis elegans.*

Rocky Mountain (Petran) Subalpine Conifer Forest

Rocky Mountain (Petran) Subalpine Conifer Forest, often called spruce-fir forest, is restricted to high elevations at approximately 2500–3800 m

up to timberline in the major mountain ranges of the state, most notably in the Sangre de Cristo, San Juan, Jemez, and Sacramento Mountains (Pase and Brown 1994c). It is also present but less widespread in the Sandia, Magdalena, San Mateo, and Mogollon Mountains; the Black Range; and at Sierra Blanca. The 10 herpetofaunal species associated with this community include *Ambystoma mavortium, Aneides hardii, Plethodon neomexicanus, Anaxyrus boreas, Pseudacris maculata, Phrynosoma hernandesi, Lampropeltis pyromelana, Pituophis catenifer, Thamnophis elegans,* and *Crotalus molossus.*

Subalpine Scrub

The highest elevations in New Mexico include three biotic communities that have few species of herpetofauna. Subalpine Scrub is limited to the Sangre de Cristo Mountains in northern New Mexico at an elevation of approximately 3500 m, at and just below timberline (D. Brown 1994j). It is ecotonal with Rocky Mountain Subalpine Conifer Forests at somewhat lower elevations and may provide marginal habitat for species associated with the conifer forest community.

Alpine and Subalpine Grasslands

Alpine and Subalpine Grasslands are found in very limited areas of New Mexico, such as the higher elevations in the Sangre de Cristo, San Juan, Mogollon, Jemez, and Chuska Mountains and near Mount Taylor (Brown and Lowe 1980; D. Brown 1994a). Associated herpetofauna (6 species) may include *Ambystoma mavortium, Anaxyrus boreas* (San Juan Mountains), *Phrynosoma hernandesi, Pituophis catenifer, Pseudacris maculata,* and *Lithobates pipiens.*

Alpine Tundra

Alpine Tundra is confined to the highest elevations at approximately 3600–4000 m above timberline, including Wheeler Peak in the Sangre de Cristo Mountains of northern New Mexico and Sierra Blanca in the White Mountains of southern New Mexico (Pase 1994). Sedges (*Carex*) are a major component, and woody plants are limited to shrub willow (*Salix*) at some sites. One species, *Aneides hardii*, has been recorded from this community at an elevation of 3570 m in the White Mountains of New Mexico (Moir and Smith 1970).

Rivers and Wetlands

Although most herpetofauna (especially squamates) in New Mexico are entirely terrestrial and can be assigned to one or more of the previously discussed biotic communities, some species depend on free-standing water for at least part of their life history (e.g., reproduction, feeding). The 16 species that have distributions closely associated with one or more of the perennial river systems in New Mexico include *Acris crepitans*, *Lithobates berlandieri*, *L. catesbeianus*, *L. yavapaiensis*, *Chelydra serpentina*, *Apalone mutica*, *A. spinifera*, *Chrysemys picta*, *Pseudemys gorzugi*, *Trachemys gaigeae*, *T. scripta*, *Nerodia erythrogaster*, *Thamnophis eques*, *T. proximus*, *T. rufipunctatus*, and *T. sirtalis*.

Some semiaquatic herpetofaunal species are highly dependent on permanent or near-permanent wetlands such as spring pools, beaver ponds, cienegas, intermittent canyon streams, and some playa lakes, often with well-developed wetland flora. These sites are often isolated by drier habitats, and species that use these wetlands must be capable of dispersing between them. In grasslands and river valleys, agricultural irrigation has contributed to the creation of this habitat type or corridors for dispersal between these wetlands. Twelve species that use this habitat include *Anaxyrus boreas*, *A. woodhousii*, *Hyla arenicolor*, *H. wrightorum*, *Pseudacris clarkii*, *P. maculata*, *Lithobates blairi*, *L. chiricahuensis*, *L. pipiens*, *Kinosternon sonoriense*, *Thamnophis cyrtopsis*, and *T. elegans*.

Other amphibians and reptiles that are aquatic during at least part of their lives are primarily associated with ephemeral waters such as livestock ponds, arroyos, tinajas, and closed basins, mainly on the plains or in valleys. These sites typically lack wetland flora and have surface water only during summer monsoons. Herpeto-

fauna in this group are adapted to xeric environments, are often widespread in distribution, and may occur far from perennial waters. Representative species (11) are *Ambystoma mavortium*, *Scaphiopus couchii*, *Spea bombifrons*, *S. multiplicata*, *Anaxyrus cognatus*, *A. debilis*, *A. punctata*, *Incilius alvarius*, *Gastrophryne olivacea*, *Kinosternon flavescens*, and *Thamnophis marcianus*.

Three amphibians and one turtle in New Mexico do not require standing water for any part of their life history: *Aneides hardii*, *Plethodon neomexicanus*, *Craugastor augusti*, and *Terrapene ornata*.

A total of 136 herpetofaunal species are known to occur in New Mexico: 27 amphibians (3 salamanders, 24 anurans) and 109 reptiles (10 turtles, 47 lizards, 52 snakes). These species represent 24 families and 59 genera. The total number of species represents an increase of 13 from the 123 species recognized by Degenhardt et al. (1996). Two of these new additions since 1996 resulted from the discovery of species previously unknown in New Mexico (*Pseudacris clarkii*, *Heterodon platirhinos*); all others are due to recent taxonomic changes and the recognition of several subspecies as full species. In addition, through mtDNA analysis, Mulcahy (2006, 2008) discovered a species of *Hypsiglena* (as yet undescribed) that is restricted to the Cochise Filter Barrier area of southwestern New Mexico, adjacent Arizona, and presumably adjacent Mexico.

A majority of the taxa in New Mexico also occur in Mexico: 24 families (100%), 57 genera (57/59 = 97%), and 113 species (113/136 = 83%) occur on both sides of the border. The high percentage of taxa shared between New Mexico and northern Mexico presumably reflects the continuity of biotic communities and river drainage systems between these two regions.

Only three species of amphibians and reptiles are endemic to New Mexico: two salamanders (*Aneides hardii* and *Plethodon neomexicanus*) and one lizard (*Aspidoscelis gypsi*). Two additional species, *Sceloporus arenicolus* and *Aspidoscelis dixoni*, have very limited ranges in New Mexico. Populations of *S. arenicolus* occur in West Texas, adjacent to the populations in extreme southeastern New Mexico (Fitzgerald and Painter 2009). *Aspidoscelis dixoni* occurs in two widely disjunct populations in the Chinati Mountains of Presidio County in West Texas and at Antelope Pass in the Peloncillo Mountains of southwestern Hidalgo County, New Mexico. These populations are separated by approximately 500 km, and the population in New Mexico is distinctive and represents a distinct clone (Painter 2009).

American Bullfrogs (*Lithobates catesbeianus*) were introduced into New Mexico and now occur, sometimes in dense populations, throughout the state in perennial waters at elevations lower than about 2103 m (Degenhardt et al. 1996). They are extremely voracious and are thought to have contributed to population declines in select aquatic species (e.g., *Lithobates* spp., *Thamnophis rufipunctatus*). Until recently the American Bullfrog was considered a game species in New Mexico, and a hunting license was required for take. However, they were reclassified, and there are now no limits on numbers or methods of take.

The introduced Mediterranean Gecko (*Hemidactylus turcicus*) was first reported from New Mexico by Painter et al. (1992). The species appears to be expanding its range in urban areas of New Mexico and is currently known from Alamogordo, Albuquerque, Las Cruces, and Truth or Consequences (Byers et al. 2007). *Hemidactylus turcicus* is a commensal with humans, and populations are found on buildings and natural habitats that have been disturbed or altered through human activities.

Sonoran Desert Tortoise (*Gopherus morafkai*), recently separated from the Desert Tortoise (*G. agassizii*) (Murphy et al. 2011), is occasionally reported from Hidalgo County. Although it is not expected this is a viable population, there have been numerous encounters with specimens in remote areas where escaped or liberated individuals would not be expected to occur. It is uncertain, but doubtful, that reports from near Rodeo in the Peloncillo Mountains (B. Tomberlin, pers. comm.) or from near Cloverdale, southwest Hidalgo County (R. Turner and P. Melhlop, pers. comm.), actually represent a breeding

population of this species. Other sightings of *G. morafkai* occur statewide (e.g., Santa Fe), although these are considered escaped or liberated individuals.

Aquatic turtles are frequently introduced by humans in the river systems of New Mexico (e.g., Stuart 2000) although no nonnative turtle species is known to be established in the state. However, three native turtles are known or likely to have become established in New Mexico river systems where they naturally did not occur. *Trachemys scripta* is thought to be native only to the Pecos and Canadian River drainages; however, it is currently well established throughout the lower and middle Rio Grande drainage as far north as Bernalillo in Sandoval County (Stuart 2000). *Chelydra serpentina* is native to the Pecos, Canadian, and Dry Cimarron River drainages, although it has frequently been reported from the central Rio Grande drainage (Degenhardt et al. 1996; Stuart 2000). Stuart and Painter (1988) reported a specimen collected during summer 1978 that was the first *C. serpentina* known from the Rio Grande drainage. *Apalone spinifera* is native to the Rio Grande and Pecos, Canadian, and Dry Cimarron River drainages, although the species is currently well established in the lower Gila River in New Mexico, and individuals are infrequently seen in the San Francisco River. Miller (1946) suggested the Gila River population was likely the result of human introduction.

Four additional species occur in proximity to the southwestern corner of New Mexico, in Chihuahua, Sonora, or Arizona, and eventually might be found in the state. The Tarahumara Salamander (*Ambystoma rosaceum*) has been found in the Sierra San Luis of Sonora and Chihuahua, Mexico, as close as 27 km to the southern border of New Mexico (P. Warren and C. Painter, pers. obs.). Although this species is primarily a stream-dwelling species, it may be found in the higher-elevation stock tanks in the Animas or San Luis Mountains in Hidalgo County, New Mexico. However, as climate change continues, finding this species in New Mexico becomes less likely. Lemos-Espinal's Spiny Lizard (*Sceloporus lemosespinali*) has been collected in Sonora, Mexico, approximately 4.8–8.0 km south of the New Mexico border in the Sierra San Luis, based on specimens at the University of Texas at Arlington and the University of New Mexico. *Sceloporus lemosespinali* is expected to occur in the encinal woodland of the foothills of the Sierra San Luis in Hidalgo County, although repeated efforts by C. Painter and H. M. Smith failed to document the species in that area. The Chihuahuan Black-headed Snake (*Tantilla wilcoxi*) has been collected in Sonora, Mexico, approximately 4.8–8.0 km south of the New Mexico border in the Sierra San Luis, based on a specimen at the University of Texas at Arlington. *Tantilla wilcoxi* is expected to occur in the encinal woodland of the foothills of the Sierra San Luis in Hidalgo County, although repeated efforts by C. Painter failed to document the species in that area. Another species that likely occurs in southwestern New Mexico is the Tiger Rattlesnake (*Crotalus tigris*), which is known from only 0.7 km west of the New Mexico state line in Guadalupe Canyon, Cochise County, Arizona (University of Arizona Collection of Herpetology). An additional specimen was taken along Geronimo Trail on August 11, 1993 (Painter and Milensky 1993). It is reasonable to expect this species will be discovered in the rocky desert shrub or canyon habitat in extreme southwestern Hidalgo County, New Mexico.

The herpetofauna of New Mexico has experienced a variety of threats and impacts in the last few centuries following the arrival of settlers of both Mexican and American origin. Industries such as livestock grazing, timber harvesting, mining, fur trapping, and farming have modified the landscape in myriad ways. Overgrazing, suppression of wildfires, and the introduction of nonnative plants have changed the structure and composition of the state's grasslands, forests, and riparian woodlands. The demand for water for agricultural and urban uses has been met by damming and diversion of rivers, modification of springs, groundwater pumping, and removal of beavers and beaver dams, often resulting in the drying of cienegas and floodplains and severe erosion of stream channels. More recently, the introduction of nonnative fishes, crayfish, and American Bullfrogs has caused further degradation of aquatic environments and, along with the arrival of diseases such as chytridiomycosis and ranavirus, poses a direct threat to the state's amphibians (especially *Lithobates* spp.). Catastrophic wildfires in the montane forests of New Mexico threaten many upland species with restricted distribution (e.g., *Aneides hardii*, *Plethodon neomexicanus*, *Crotalus willardi*) and may become more frequent due to drought, climate change, and past fire suppression. Oil and gas development is a growing threat to wildlife in many rural areas in the state, and herbicide use to reduce shrub cover in rangelands is altering habitat for some terrestrial species (e.g., *Sceloporus arenicolus*). Perhaps most significantly, the rapid growth of New Mexico's human population since the late twentieth century has resulted in increased urbanization, road development, and demand for limited water resources, placing further pressure on the state's herpetofauna and other wildlife.

One species of herpetofauna in New Mexico is believed to be extirpated, *Anaxyrus boreas*, although a reintroduction program for this species is currently under way in Rio Arriba County. Species that have declined significantly in the state in recent years include *Plethodon neomexicanus*, *Lithobates chiricahuensis*, *L. pipiens*, *L. yavapaiensis*, and *Thamnophis eques*. Other species that are experiencing significant threats to their limited habitat in New Mexico include *Aneides hardii*, *Sceloporus arenicolus*, *Trachemys gaigeae*, and *Crotalus willardi*. *Trachemys gaigeae* is known to hybridize with nonnative *T. scripta* in the Middle Rio Grande Valley, which ultimately may cause elimination of the pure genotype of *T. gaigeae* in New Mexico. A similar threat may exist for some native populations of *Ambystoma mavortium* due to the introduction of nonnative conspecifics imported to the state as fish bait.

The commercial trade of amphibians and reptiles for pets, food, and skins and other novelties has experienced rapid growth worldwide, and New Mexico is no exception. During 2001–2008, 71,626 individual amphibians and reptiles representing 70 species were legally taken with commercial collecting licenses in New Mexico (NMDGF, unpubl. data). NMDGF regulates the legal trade of amphibians and reptiles through the sale of commercial collecting licenses, although the magnitude of the illegal trade and its impact on native populations of amphibians and reptiles are unknown (Fitzgerald et al. 2004).

Organized rattlesnake roundups in New Mexico are hopefully a thing of the past. Until around 2005 there was a single roundup in Alamogordo, Otero County, which traded approximately 227–680 kg of *Crotalus atrox* per year, depending upon the market value and demand. *Crotalus molossus* and *C. viridis* were also collected for the roundup, but the numbers traded were very low (Fitzgerald and Painter 2000; C. Painter, unpubl. data). This organized roundup no longer occurs in Alamogordo, although there are numerous hunters who still provide rattlesnakes to the commercial market.

Herpetofauna of Texas

JAMES R. DIXON

Introduction

Texas has a very diverse herpetofauna distributed over about 7% of the total land and water mass of the United States. Texas contains a land and water area of 692,247.5 km² — 678,357 km² of land and 13,890 km² of water. Texas is 1289 km (generally north to south) in length, and its greatest east-west distance is 1244 km.

The elevation in Texas is highly variable. Guadalupe Peak is 2675 m above sea level, and a twin peak, El Capitan, is at 2472 m, both in the Guadalupe Mountains National Park in Culberson County; and some 1290 km south is the Gulf of Mexico. In general, most of Trans-Pecos Texas has an elevation greater than 917 m, and the Panhandle varies from 612 to 1224 m. The landmass generally slopes from the northwest to the southeast across Texas.

Texas varies widely in the amount of precipitation across the state. At the city of Orange in the southeastern corner of the state, average rainfall measures about 1450 mm annually. At El Paso, the most western corner of the state, rainfall annually averages 20.3 mm. The recorded highest temperature was 53.6°C in Seymour in northwest Texas. The coldest recorded temperature was −11.4°C at Tulia in northwest-central Texas.

Interest in Texas herpetology probably began in the 1500s when the first Spanish settlers landed on Padre Island and encountered the Diamond-backed Rattlesnake. Since that time, many stories of the encounters with that snake have been told and retold so many times that the truth of the encounters has been lost in the imagination of the "teller." These stories continue today, as well as encounters with the snake.

Franz H. Troschel published the first scientific record of a new reptile from the relatively new state of Texas in 1852, the Greater Earless Lizard (*Cophosaurus texanus*). However, in the same year, Spencer Baird and Charles Girard published Texas records of new species descriptions for the Red-spotted Toad (*Bufo* [= *Anaxyrus*] *punctatus*), Crayfish Frog (*Rana* [= *Lithobates*] *areolata*), Keeled Lesser Earless Lizard (*Holbrookia propinqua*), Crevice Spiny Lizard (*Sceloporus poinsettii*), Ornate Tree Lizard (*Urosaurus ornatus*), Great Plains Skink (*Plestiodon obsoletus*), Texas Spotted Whiptail (*Aspidoscelis gularis*), and Marbled Whiptail (*A. marmorata*). In 1853, Baird and Girard followed with the descriptions of the Texas snakes Schott's Whipsnake (*Masticophis schotti*), Ornate Whipsnake (*M. ornatus* [= *M. taeniatus*]), Plains Hog-nosed Snake (*Heterodon nasicus*), Saltmarsh Watersnake (*Regina* [= *Nerodia*] *clarki*), Graham's Crayfish Snake (*R. grahamii*), Flat-headed Snake (*Tantilla gracilis*), Texas Threadsnake (*Rena dulcis*), Texas Coralsnake (*Elaps* [= *Micrurus*]) *tener*), and Western Diamond-backed Rattlesnake (*Crotalus atrox*).

Between 1853 and 1880, 1 salamander, 6 frogs, 3 turtles, 6 lizards, and 8 snakes were described from Texas. By the time the first herpetological checklist was published by John Strecker in 1915, the number of amphibians and reptiles in the state was 163. By 1950, Bryce Brown had recorded 181 species in his annotated checklist. Gerald Raun and Fredrick Gehlbach published a checklist of amphibians and reptiles in Texas in 1972 and recorded 199 species, followed by Robert Thomas's checklist for the Texas Department of Wildlife and Fisheries at 204 species in 1974 and 202 species in 1976. James Dixon's 1987 *Amphibians and Reptiles of Texas* recorded 204 species, and his revised 2000 edition provided a list of 219 species.

The state's list of 219 native species is currently composed of 28 salamanders, 41 frogs and toads, 1 alligator, 30 turtles, 45 lizards, and 74 snakes. Of these, 128 species are shared with Mexico: 4 salamanders, 23 frogs and toads, 15 turtles, 36 lizards, and 50 snakes. There are 14 salamanders, 1 toad, 3 turtles, and 2 snakes endemic to Texas.

The exotic species that maintain breeding populations in Texas are the Greenhouse Frog (*Euhyas planirostris*), Florida Red-bellied Turtle (*Pseudemys nelsoni*), Rough-scaled Gecko (*Cyrtopodion scabrum*), Common House Gecko (*Hemidactylus frenatus*), Endo-Pacific Gecko (*H. garnoti*), Mediterranean Gecko (*H. turcicus*), Mexican Spiny-tailed Iguana (*Ctenosaura pectinata*), Brown Anole (*Anolis sagrei*), Florida Watersnake (*Nerodia fasciata pictiventris*), and Brahminy Blindsnake (*Indotyphlops braminus*).

The accidental and/or released nonbreeding pets that have been found swimming, walking, or crawling around the byways of Texas are African Clawed Frog (*Xenopus laevis*), Boa Constrictor (*Boa constrictor*), Burmese Python (*Python molurus*), Reticulated Python (*P. reticulatus*), Ball Python (*P. regius*), Brown Tree Snake (*Boiga irregularis*), Mangrove Snake (*B. dendrophila*), Wandering Gartersnake (*Thamnophis elegans vagrans*), Florida Tortoise (*Gopherus polyphemus*), California Tortoise (*G. agassizii*), and the Burmese Brown Tortoise (*Manouria emys*).

The large landmass of Texas and its six biotic provinces with their relevant geology, climate, and plant communities account for the highest number of species of any state in the contiguous United States. However, the land is rapidly becoming modified by human growth and the necessity to feed and clothe the masses. Clearing of forests for pasture, cultivation, and urban growth tends to create loss of habitat for native species, for example, the Houston Toad, an endangered species. With human growth comes the demand for more and more water. These demands lower the water table, springs cease their flow, and aquatic salamanders disappear. Most humans have the "Ostrich Syndrome": what they cannot see cannot hurt them. Nearly all spring salamanders are threatened with low spring flows, and some are now on the endangered species list. It seems to me that Rachael Carson's (1962) *Silent Spring* is not far away.

Previous Herpetological Studies

About 1100 scholarly publications concerning Texas amphibians and reptiles were written between 1852 and 1970. By the end of 1987, more than 2000 additional publications had appeared. Over 3140 articles on various aspects of the Texas herpetofauna had been published by early 1998, and more than 1000 articles from 1999 to 2010. The literature includes only a few popular articles, newspaper accounts, and reports to federal and state agencies. The Texas herpetological literature has increased 38% following the publication of Gerald Raun and Fredrick Gehlbach's 1972 treatise. The first 115 years of Texas herpetological literature averaged only 10 articles per year, while the period from 1987 to 2000 averaged 48 articles per year, and the last 10-year period (2000–2009) averaged 77 articles per year.

By 1987, approximately 717 primary authors had utilized 292 publication resources. Seven of these, *Copeia* (239), *Herpetologica* (208), the

Texas Journal of Science (149), *Herpetological Review* (133), *Southwestern Naturalist* (116), *Catalogue of American Amphibians and Reptiles* (87), and *Journal of Herpetology* (63), contained about 50% of all published articles on Texas amphibians and reptiles. About 541 (27%) of all published articles were coauthored, with the bulk of coauthorship occurring after 1945. The coauthorship rate has doubled since 1945, and of the 573 coauthors, 259 (45%) never appeared as senior authors. There were about 719 primary authors, of which 439 appeared only once, and 614 (85%) appeared as authors fewer than five times. Of the 719 primary authors, 105 have published 53% (1070) of the total number of articles. Twenty-two authors have published 20 or more articles on Texas herpetology. They made up only 2% of all authors and 22% of all publications. Nine of the 22 authors are deceased, and 12 continue to publish on Texas herpetology. The names of those deceased, their Texas publications record, and their active period are John Strecker (58), 1902–1935; W. Frank Blair (39), 1949–1976; Donald W. Tinkle (26), 1951–1979; Edward D. Cope (25), 1859–1900; William W. Milstead (25), 1951–1978; Charles E. Burt (20), 1928–1938; Edward H. Taylor (20), 1931–1950; Bryce C. Brown (20), 1937–1967; and Roger Conant (20), 1942–1998. Strecker, a resident of Texas, achieved a publication record of 58 articles in 22 years, while Cope, a nonresident, published 25 articles over 41 years. Resident herpetologists have a distinct advantage concerning knowledge of their local herpetofauna.

The primary authors active through 2010 and their publication records are R. W. Axtell (51), 1950–2005; James R. Dixon (41), 1952–2009; Carl J. Franklin (23), 1996–2009; Fredrick R. Gehlbach (24), 1956–1991; Toby J. Hibbitts (31), 1997–2008; J. Alan. Holman (22), 1962–2003; Flavius C. Killebrew (26), 1975–1995; Chris T. McAllister (108), 1982–2008; Francis L. Rose (28), 1959–2010; Hobart M. Smith (85), 1933–2007; Thomas Vance (28), 1978–2001; and J. Martin Walker (27), 1980–2008.

Of active herpetologists, Hobart M. Smith's 85 published articles over a 74-year span are a monument to his excellence as a herpetologist and his desire to create herpetological activity among his colleagues. Only the young herpetologists of today will have the opportunity to be as productive as H. M. Smith. One of these has achieved this goal. The most prolific writer of Texas herpetofauna during the 1982–2010 era was Chris McAllister. He and his colleagues published nearly 108 articles in a 27-year period, surpassing Smith's record.

Brief History of Texas Herpetology

Some of the early writers on Texas herpetology neither resided in nor visited Texas. Most of their material came from early collectors of natural history lore such as Spanish explorers, botanists, and military survey engineers.

Spanish explorers during the period 1650–1700 recorded encounters with rattlesnakes in Texas. The French botanist Jean Louis Berlandier was probably the earliest science writer on amphibians and reptiles of Texas. Although Berlandier was French, he resided in Mexico and was one of the first scientists to write extensively on his Texas travels. Berlandier made several expeditions on horseback and on foot across the southern half of Texas between 1828 and 1834. His major expeditions were from Laredo to Bexar (February 20–March 1, 1828), from Bexar–San Felipe–Trinity and back to Bexar (April 13–June 18, 1828), Aransas-Goliad several times (May 1829), Bexar to Laredo (July 14–28, 1829), Bexar to Goliad (February 2–5, 1830), Matamoros to Goliad (April 1834), and Bexar–Eagle Pass–Laredo–Matamoros (June 10–July 28, 1834).

On his first trip, Berlandier encountered the Texas Tortoise and bullfrogs along the route. As he crossed the Nueces River, he encountered alligators and softshell turtles and remarked how common *Trionyx* [= *Apalone*] and *Rana* [= *Lithobates*] were around Bexar. Throughout Berlandier's trip to San Felipe, Trinity, and return trip to Bexar, he encountered box turtles, salamanders, Western Diamond-backed and Timber Rattlesnakes, *Rana* [= *Lithobates*], and alligators. Berlandier's trips between Copano Bay and

Goliad were frequently dangerous because of his contact with Timber Rattlesnakes. On one of his trips during 1829, he encountered the Texas Horned Lizard for the first time. Occasionally, Berlandier mentioned encounters with treefrogs but failed to give the scientific names. Because of his general observations and collecting, his name is associated in the literature with the Texas Tortoise, Rio Grande Leopard Frog, and many plant species (see Berlandier 1980).

John Bartlett and William Emery's 1850–1854 boundary surveys and their subsequent collections of amphibians and reptiles resulted in a partial list of Texas specimens through the publications of Spencer F. Baird and Charles F. Girard (1852–1854), and C. F. Girard (1852–1859). Baird and Girard recorded a total of 86 species in the area between Indianola and El Paso.

Several noted herpetologists began describing Texas species following the field collections of Berlandier, Emery, Bartlett, and others. Among these were E. D. Cope, G. A. Boulenger, Louis Agassiz, Albert Günther, Robert Kennicott, and Spencer F. Baird. By 1900, John Strecker had begun collecting in Texas, and by 1915, he had published the first definitive list (163 species) of reptiles and amphibians for the state. By 1933, Strecker had accumulated the largest collection of Texas reptiles and amphibians, and he housed the collection at Baylor University. By the time of his death, Strecker had written 58 papers on the herpetofauna of Texas. Brown (1950) produced the next major checklist of Texas amphibians and reptiles, recognizing 182 species, followed by Gerald Raun and Frederick Gehlbach (1972), who found 199 species. The most recent published state list by Robert Thomas (1976) records 203 species. The major herpetologists working with Texas species between Strecker's era and World War II were Frank Blanchard, Bryce Brown, Charles Burt, Roger Conant, Howard Gloyd, Laurence Klauber, Stanley Mulaik, Karl P. Schmidt, Hobart Smith, Edward Taylor, and Alan H. Wright. Only B. C. Brown and H. M. Smith actually resided in Texas.

There are 78 scientific names of amphibians (14 salamanders, 14 anurans) and reptiles (6

turtles, 20 lizards, 24 snakes) with type localities in the state of Texas (see table 12.1).

Physiographic Characteristics and Their Influence on the Distribution of the Herpetofauna

Texas contains 229 native and exotic species of amphibians and reptiles. To some extent, these taxa are restricted to particular vegetation communities, soil types, or water sources. Although the various environments of the state appear to have their own particular herpetofauna, a number of generalists tend to mask the uniqueness of some of these environments. Texas is situated at the junction of four major physiographic divisions of North America: the Rocky Mountains, Great Plains, Eastern Forests, and the Southwestern Deserts. Each of the latter divisions is further divided into biotic provinces.

Blair (1950) provides a map illustrating six biotic provinces in the state of Texas: the Kansan, Balconian, Texan, Austroriparian, Tamaulipan, and Chihuahuan.

Kansan Biotic Province

In Texas, the Kansas Biotic Province includes the eastern part of the Texas Panhandle and mixed-grass prairies. This province is continuous with the red plains south of the Red River. Moisture is low with a decrease in moisture from east to west, and the region is often classified as semiarid. The grasses are principally species of beardgrass and buffalograss. Sand dune areas are characterized by shin oak and sand sage. There are mesquite plains with various species of trees, such as mesquite, oak, elm, hackberry, and maple. The principal herpetofauna includes the turtles *Terrapene ornata* and *Kinosternon flavescens*; the lizards *Crotaphytus collaris*, *Cophosaurus texanus*, *Holbrookia maculata*, *Phrynosoma cornutum*, *P. modestum*, *Sceloporus consobrinus*, *Plestiodon obsoletus*, *Aspidoscelis gularis*, and *A. sexlineata*; and 38 species of snakes, for example, *Arizona elegans*, *Lampropeltis getula*, *Masticophis*

TABLE 12.1 Type localities for amphibians and reptiles described from the state of Texas, USA

Author(s)	Original name: Type locality
Salamanders (14)	
Matthes (1855)	*Salamandra texana* (= *Ambystoma texanum*): **Cummings Creek, Fayette Co.**
Chippindale et al. (2000)	*Eurycea chisholmensis*: **Main Salado Springs, Bell Co.**
Smith & Potter (1946)	*Eurycea latitans*: **Cascade Cavern 4.6 mi SE Boerne, Kendall Co.**
Bishop (1941)	*Eurycea nana*: **Headwaters, San Marcos River, San Marcos, Hays Co.**
Bishop & Wright (1937)	*Eurycea neotenes*: **creek, 5 mi N. Helotes, Bexar Co.**
Chippindale et al. (2000)	*Eurycea naufragia*: **head springs, Buford Hollow, trib. San Gabriel River, Below Lake Georgetown, Williamson Co.**
Stejneger (1896)	*Typhlomolge* (= *Eurycea*) *rathbuni*: **San Marcos, Hays Co.**
Stejneger (1896)	*Typhlomolge* (= *Eurycea*): **beneath Blanco River, 5 km NE San Marcos, Hays Co.**
Chippindale et al. (1993)	*Eurycea sosorum*: **Barton Springs, Zilker Park, Austin, Travis Co.**
Chippindale et al. (2000)	*Eurycea tonkawae*: **Stillhouse Hollow Springs, Travis Co.**
Mitchell & Reddell (1965)	*Eurycea tridentifera*: **Honey Creek Cave, Comal Co.**
Baker (1957)	*Eurycea troglodytes*: **pool near valdena farms sinkhole, Valdena Farms, Medina Co.**
Hillis et al. (2001)	*Eurycea waterlooensis*: **Barton Springs, Zilker Park, Austin, Travis Co.**
Grobman (1944)	*Plethodon albagula*: **20 miles N San Antonio, Bexar Co.**
Anurans (14)	
Girard (1854)	*Bufo* (= *Anaxyrus*) *debilis*: **Lower Río Grande**
Sanders (1953)	*Bufo* (= *Anaxyrus*) *houstonensis*: **Fairbanks, Harris Co.**
Baird & Girard (1852)	*Bufo* (= *Anaxyrus*) *punctatus*: **Río San Pedro of the Río Grande (= Devils River), Val Verde Co.**
Girard (1854)	*Bufo* (= *Anaxyrus*) *speciosus*: **Ringgold Barracks, Río Grande City, Starr Co.**
Bragg & Sanders (1951)	*Bufo woodhousii velatus* (= *Anaxyrus velatus*): **Elkhart, Anderson Co.**
Stejneger (1915)	*Syrrhophus campi* (= *Eleutherodactylus cystignathoides*): **Brownsville, Cameron Co.**
Cope (1878)	*Syrrhophus* (= *Eleutherodactylus*) *marnocki*: **near San Antonio, Bexar Co.**
Strecker (1910)	*Hyla versicolor chrysoscelis* (= *H. chryososcelis*): **Dallas, Dallas Co.**
Baird (1854)	*Helocaetes* (*Pseudacris*) *clarki*: **Galveston and Indianola**
Wright & Wright (1933)	*Pseudacris streckeri*: **unknown but likely Somerset, Bexar Co.**
Baird & Girard (1852)	*Rana* (= *Lithobates*) *areolata*: **Indianola, Calhoun Co.**
Baird (1859b)	*Rana* (= *Lithobates*) *berlandieri*: **Brownsville, Cameron Co.**
Strecker (1910)	*Scaphiopus hurteri*: **3.5 mi E Waco, McClellan Co.**
Cope (1863)	*Scaphiopus* (= *Spea*) *bombifrons*: **Llano Estacado**
Turtles (6)	
Stejneger (1925)	*Graptemys versa*: **Austin, Travis Co.**
Baur (1893)	*Pseudemys texana*: **San Antonio, Bexar Co.**
Hartweg (1939)	*Pseudemys scripta gaigeae* (= *Trachemys gaigeae*): **Boquillas, Río Grande, Brewster Co.**
Agassiz (1857)	*Platythyra* (= *Kinosternon*) *flavescens*: **Río Blanco**
Agassiz (1857)	*Xerobates* (= *Gopherus*) *berlandieri*: **Lower Río Grande**
Agassiz (1857)	*Aspidonectes* (= *Apalone*) *emoryi*: **Brownsville, Cameron Co.**
Lizards (20)	
Baird (1859a [1858])	*Gerrhonotus infernalis*: **Devils River, Val Verde Co.**
Baird (1859a [1858])	*Crotaphytus reticulatus*: **Laredo and ringgold barracks, Starr Co.**
Stejneger (1893)	*Coleonyx brevis*: **Helotes, Bexar Co.**
Davis & Dixon (1858)	*Coleonyx reticulatus*: **Black Gap Wildlife Management Area, Brewster Co.**

TABLE 12.1 *Continued*

Author(s)	Original name: Type locality
Troschel (1852)	*Cophosaurus texanus*: **Guadalupe River at New Braunfels, Comal Co.**
Cope (1880)	*Holbrookia lacerata*: **Helotes, Bexar Co.**
Baird & Girard (1852)	*Holbrookia propinqua*: **between San Antonio and Indianola**
Girard (1852)	*Phrynosoma modestum*: **between San Antonio and El Paso**
Stejneger (1916)	*Sceloporus disparilis* (= *S. grammicus disparilis*): **Lomita Ranch, 6 mi N. Hidalgo, Hidalgo Co.**
Wiegmann (1834)	*Sceloporus marmoratus* (= *variabilis*): **San Antono, Bexar Co.**
Stejneger (1904)	*Sceloporus merriami*: **East Painted Cave, near mouth Pecos River**
Smith (1934)	*Sceloporus olivaceus*: **near lower end of Arroyo Los Lomos, 3 mi SE Rio Grande City, Starr Co.**
Baird & Girard (1852)	*Sceloporus poinsettii*: **Río San Pedro of the Río Grande (= Devils River), Val Verde Co.**
Baird & Girard (1852)	*Urosaurus ornatus*: **Río San Pedro of the Río Grande (= Devils River), Val Verde Co.**
Baird & Girard (1852)	*Plestiodon obsoletum*: **Río San Pedro of the Río Grande (= Devils River), Val Verde Co.**
Bocourt (1879)	*Eumeces obtusirostris* (= *Plestiodon septentrionalis obtusirostris*): **unknown locality in Texas**
Cope (1880)	*Eumeces brevilineatus* (= *Plestiodon tetragrammus*): **Helotes Creek, Bexar Co.**
Baird & Girard (1852)	*Cnemidophorus* (=*Aspidoscelis*) *gularis*: **Indianola, Calhoun Co.**
McKinney et al. (1973)	*Cnemidophorus* (=*Aspidoscelis*) *laredoensis*: **Chacon Creek, at hwy 83, Laredo, Webb Co.**
Baird & Girard 1852	*Cnemidophorus* (=*Aspidoscelis*) *marmoratus*: **between San Antonio and El Paso**

Snakes (24)

Author(s)	Original name: Type locality
Kennicott *in* Baird (1859b)	*Arizona elegans*: **Lower Río Grande Valley**
Brown (1901b)	*Coluber* (=*Bogertophis*) *subocularis*: **Davis Mountains, Jeff Davis Co.**
Kennicott *in* Baird (1859b)	*Taeniophis* (= *Coniophanes*) *imperialis*: **Brownsville, Cameron Co.**
Smith (1941)	*Drymarchon erebennus* (= *D. melanurus erebennus*): **Eagle Pass, Maverick Co.**
Taylor (1941)	*Ficimia streckeri*: **3 mi E Río Grande City, Starr Co.**
Kennicott (1860)	*Heterodon kennerlyi*: **Lower Río Grande**
Baird & Girard (1853)	*Heterodon nasicus*: **Río Grande**
Stejneger (1893)	*Hypsiglena ochrorhyncha texana* (= *H. jani*): **between Laredo and Camargo**
Brown (1901a)	*Ophibolus alternus* (*Lampropeltis alterna*): **Davis Mountans, Jeff Davis Co.**
Kennicott *in* Baird (1859b)	*Dipsas* (= *Leptodeira*) *septentrionalis*: **Brownsville, Cameron Co.**
Baird & Girard (1853)	*Masticophis schotti*: **Eagle Pass, Maverick Co.**
Baird & Girard (1853)	*Masticophis ornatus* (= *M. taeniatus*): **between San Antonio and El Paso**
Baird & Girard (1853)	*Regina* (= *Nerodia*) *clarki*: **Indianola, Calhoun Co.**
Trapido (1941)	*Natrix* (= *Nerodia*) *harteri*: **Brazos River N of Palo Pinto, Palo Pinto Co.**
Yarrow (1880)	*Coluber* (= *Pantherophis*) *bairdi*: **near Ft. Davis, Jeff Davis Co.**
Baird & Girard (1852)	*Churchillia bellona* (= *Pituophis catenifer*): **Presidio del Norte, Texas**[1]
Baird & Girard (1853)	*Regina grahamii*: **Río Salado, 4 mi from San Antonio, Bexar Co.**
Baird & Girard (1853)	*Tantilla gracilis*: **Indianola, Calhoun Co.**
Kennicott (1860)	*Tantilla nigriceps*: **Indianola to Nueces**
Baird & Girard (1853)	*Elaps* (= *Micrurus*) *tener*: **Río San Pedro of the Río Grande (= Devils River), Val Verde Co.**
Baird & Girard (1853)	*Rena dulcis*: **between San Pedro and Commanche Springs**
Klauber (1939)	*Leptotyphlops humilis segeregus* (= *Rena segrega*): **Chalk Draw, Brewster Co.**
Baird & Girard (1853)	*Crotalus atrox*: **Indianola, Calhoun Co.**
Kennicott (1861)	*Caudisoma lepida* (= *Crotalus lepidus*): **Presidio del Norte and Eagle Pass, Maverick Co.**

[1] Although the original type locality reported by Baird & Girard (1852) and Smith & Taylor (1950, 1966) is Presidio del Norte, Chihuahua, this is in error; the correct type locality is Presidio del Norte, Texas.

flagellum, *Nerodia erythrogaster*, *Pantherophis emoryi*, *Tantilla nigriceps*, *Thamnophis radix*, *Rena dissecta*, *Crotalus atrox*, and *C. viridis*.

The only known salamander is the widely known *Ambystoma mavortium*. Representative amphibians are *Anaxyrus cognatus*, *A. debilis*, *A. punctatus*, *A. speciosus*, *A. woodhousii*, *Acris crepitans*, *Lithobates blairi*, *Scaphiopus couchi*, *Spea bombifrons*, and *S. multiplicata*.

Balconian Biotic Province

The Balconian Biotic Province is unique because of its geology, with surface flows of Comanchean Cretaceous limestone, igneous upthrusts of granite, and sediments as old as the Precambrian scattered across the region. At least eight major rivers are outflows from or form cuts through this region. The southern edge of this province forms the boundary with the northern boundary of the Tamaulipan Province. Its eastern boundary is the Texan boundary, the western boundary is the Chihuahuan, and the northern boundary is the Kansan. The rainfall of this province decreases from east to west. Its vegetation consists largely of short-tree forests of live oak, Texas oak, mesquite, and juniper, with an understory of Texas persimmon, beargrass, and gramagrasses. The riparian forests along the rivers often consist of elm, hackberry, pecan, and large live oaks. Along the southern edge of the river forests, large cypress trees (*Taxodium distichum*) occur.

Characteristic salamanders are all species of the genus *Eurycea*, except *E. quadridigitata*. One *Plethodon*, the Western Slimy Salamander (*P. albagula*), is also present. There are approximately 17 species of frogs and toads found in the Balconian Province, but only 2, *Craugastor augusti* and *Eleutherodactylus marnocki*, appear to be consistently associated with the province. Both species occur more widely to the south, primarily in Mexico. Representative species are *Anaxyrus debilis*, *A. punctatus*, *A. speciosus*, *A. woodhousii*, *Acris crepitans*, *Hyla chysoscelis*, *Pseudacris streckeri*, *Lithobates berlandieri*, *L. catesbeianus*, and *Scaphiopus couchi*.

Some 11 species of turtles occur in the Balconian Province. Only 3 aquatic and 1 terrestrial species are common: *Graptemys versa*, *Pseudemys texana*, *Terrapene ornata*, and *Apalone spinifera*.

Of the 16 lizards found in the Balconian Province, the majority of the species occur widely to the west, but a few species filter into the region from the Austroriparian and Tamaulipan Provinces. Characteristic species are *Gerrhonotus liocephalus*, *Cophosaurus texanus*, and *Sceloporus poinsettii*. Other commonly observed species are *Crotaphytus collaris*, *Coleonyx brevis*, *S. consobrinus*, *S. olivaceus*, *Urosaurus ornatus*, *Plestiodon obsoletus*, *P. tetragrammus*, and *Aspidoscelis gularis*.

There are 43 species of snakes that reach the Balconian Province. None are restricted to the region, but 27 are reasonably common: *Coluber constrictor*, *Diadophis punctatus*, *Gyalopion canum*, *Heterodon platirhinos*, *Hypsiglena jani*, *Lampropeltis getula*, *Masticophis flagellum*, *M. taeniatus*, *Nerodia erythrogaster*, *N. rhombifer*, *Pantherophis emoryi*, *P. obsoletus*, *Pituophis catenifer*, *Rhinocheilus lecontei*, *Salvadora grahamiae*, *Sonora semiannulata*, *Storeria dekayi*, *Tantilla nigriceps*, *Thamnophis cyrtopsis*, *T. marcianus*, *T. proximus*, *Rena dulcis*, *Micrurus tener*, *Agkistrodon contortrix*, *A. piscivorus*, *Crotalus atrox*, and *C. molossus*.

Texan Biotic Province

Dice (1943) described this province as the broad ecotone between the forests to the east and grasslands to the west. In Texas, this province is wedged between the Kansan, Balconian, and Tamaulipan Provinces to the west and the Austroriparian Province to the east. It is a mixture of forest and prairies, with an abundance of rainfall to the east and an almost deficient amount of moisture to the west. Sandy soils support forests of post oak, blackjack oak, and hickory, whereas clay soils support tallgrass prairie. The latter is now cultivated for various crops, while the forest lands now support some planted pine timber for harvest and/or cleared for pasture and cattle. The major rivers that drain the Texan Province to the south are the Brazos, Colorado, San Marcos, and Guadalupe, and the Red and Trinity Rivers drain the northern part of the province before entering the Austroriparian Province.

The herpetofauna represents a mixture of eastern and western species. The characteristic eastern species of amphibians are three species of salamanders (*Ambystoma texana, A. tigrinum,* and *Siren intermedia*), and eight species of frogs and toads (*Hyla cinerea, H. squirella, H. versicolor, Pseudacris crucifer, P. fouquettei, Gastrophryne carolinensis, Lithobates sphenocephala,* and *Scaphiopus hurteri*). Western species entering the province are *Anaxyrus nebulifer, Pseudacris clarki, P. streckeri, Gastrophryne olivacea,* and *Scaphiopus couchi*.

The reptile fauna also shows a mixture of eastern and western forms, with such eastern turtles as *Chelydra serpentina, Graptemys pseudogeographica, Terrapene carolina, Kinosternon subrubrum,* and *Sternotherus odoratus* and western forms such as *Pseudemys texana, T. ornata,* and *Apalone spinifera*. The only lizard that seems to belong as a characteristic species is *Plestiodon septentrionalis*. Western lizards that occur in the province are such species as *Crotaphytus collaris, Phrynosoma cornutum, Sceloporus olivaceus,* and *Plestiodon obsoletus*, while eastern species are *Anolis carolinensis, P. fasciatus, P. laticeps,* and *Scincella lateralis*.

Of the 41 species of snakes occurring in the province, only 14 are eastern and the remainder (27) are western. The only characteristic species are *Lampropeltis calligaster, Nerodia harteri, Regina grahamii, Tantilla gracilis, Tropidoclonion lineatum,* and *Virginia valeriae*.

Austroriparian Biotic Province

According to Blair, Dice (1943) configured this province from the Atlantic seaboard to East Texas. Blair (1950) states that the western boundary of the province is approximated by a line from western Harris County north to western Red River County. Blair further indicates that the western boundary is the western edge of the main body of pine and hardwood forests, but there is no physiographic break to limit the western edge of the forests.

The vegetation of this province is composed of the same species of hardwoods and pines that characterize the vegetation eastward to the Atlantic Ocean. Longleaf pine is the dominant pine in the Texas part of the province. The hardwoods of the southeastern segment of the forest (Big Thicket) are composed of sweetgum, magnolia, white oak, water oak, tupelo, cypress, wax myrtle, dogwood, beech, and an understory of palmetto and Spanish moss. The northern forest is less swampy and contains loblolly pine, yellow pine, red oak, post oak, and blackjack oak. There are about 13 species of salamanders, such as *Amphiuma tridactylus, Necturus beyeri, Siren intermedia*; several species of lowland *Ambystoma*; and *Notophthalmus viridescens*. The common frogs and toads are *Anaxyrus fowleri, A. nebulifer, Acris blanchardi, Hyla cinerea, H. squirella, H. versicolor, Pseudacris crucifer, P. fouquettei, Gastrophryne carolinensis, Lithobates catesbeianus,* and *L. clamitans*.

The characteristic turtle fauna consists of *Graptemys ouachitensis, Macrochelys temminckii, Pseudemys concinna, Terrapene carolinensis, Kinosternon subrubrum,* and *Sternotherus carinatus*. Typical lizards are *Ophisaurus attenuatus, Anolis carolinensis, Plestiodon fasciatus, P. laticeps,* and *Scincella lateralis*.

Characteristic snakes include *Cemophora coccinea, Farancia abacura, Lampropeltis triangulum, Nerodia cyclopion, N. erythrogaster, N. fasciata, N. rhombifer, Opheodrys aestivus, Regina rigida, Micrurus tener, Agkistrodon contortrix, A. piscivorus, Crotalus horridus,* and *Sistrurus miliarius*.

Tamaulipan Biotic Province

This province is limited by Blair (1950) by its consistent vegetation, climate, and rainfall, as well as its attendant vertebrate fauna. The eastern boundary begins near San Antonio and along a rather distinctive vegetation type from San Antonio south-southeast to the Gulf Coast near Rockport. The vegetation is predominantly thorn brush, with its northern boundary being the Balcones Fault zone from Del Rio to San Antonio and its southern boundary near the Gulf of Mexico. The vegetation is found mostly on caliche soils and consists of mesquite, various species of *Acacia* and *Mimosa*, cenizo, white brush, granjeno, tasajillo, prickly pear, *Condalia*, and *Castela*.

The sandy soil zone from near Corpus Christi to Raymondsville contains mottes of live oak and grasses that are not considered part of the Tamaulipan Province.

One salamander (*Notophthalmus meridionalis*) is considered endemic to the Tamaulipan Biotic Province. A possible endemic is the giant *Siren*, and *Ambystoma mavortium* is common. The population of giant *Siren* inhabiting the Rio Grande Valley in Brownsville, Texas, and Matamoros, Tamaulipas, was tentatively assigned to *S. lacertina* by Flores-Villela and Brandon (1992); however, this population actually seems to represent an undescribed species, since it, at this moment, is cited as *Siren* sp. indet. Some of the more characteristic frogs and toads are *Anaxyrus nebulifer, Rhinella marina, Eleutherodactylus cystignathoides, Smilisca baudinii, Leptodactylus fragilis, Hypopachus variolosus, Rhinophrynus dorsalis,* and *Scaphiopus couchi.*

Reptiles, especially some of the turtles and lizards, are unique to the Tamaulipan Province. The Texas Gopher Tortoise is endemic to this province, and both Spiny Softshells and Rio Grande Cooter are characteristic. Typical lizards are *Crotaphytus reticulatus, Coleonyx brevis, Holbrookia lacerata, H. propinqua, Phrynosoma cornutum, Sceloporus cyanogenys, S. grammicus, S. olivaceus, S. variabilis,* and *Aspidoscelis laredoensis.* Of 44 snakes found in the Tamaulipan Province, 7 species (*Coluber schotti, Coniophanes imperialis, Drymobius margaritiferus, Drymarchon melanurus, Ficimia streckeri, Leptodeira septentrionalis,* and *Tantilla atriceps*) are somewhat restricted to the province. In addition, some 17 species are typical of the province: *Arizona elegans, Coluber constrictor, Elaphe emoryi, Hypsiglena jani, Lampropeltis getula, L. triangulum, Masticophis flagellum, Pituophis catenifer, Rhinocheilus lecontei, Salvadora grahamiae, Sonora semiannulata, Storeria dekayi, Thamnophis marcianus, T. proximus, Micrurus tener, Crotalus atrox,* and *Sistrurus catenatus.*

Chihuahuan Biotic Province

Dice (1943) restricts the Chihuahuan Biotic Province to Trans-Pecos Texas, excluding the Guadalupe Mountains. It reaches its northern limit in southern New Mexico. According to Blair (1950), the eastern limit is the eastern Toyah Basin rim, south to Crockett County, where it follows the Pecos River south to its confluence with the Rio Grande. The western boundary is in Mexico. The province has more physiographic features than any other province. The northeastern part of the province is apparently an ancient bolson, now drained by the Pecos River. South and southeast of the Toyah Basin is the Stockton Plateau, which is the Trans-Pecos extension of the Comanchean Cretaceous limestone of the Edwards Plateau. This plateau comprises a distinct physiographic feature in southern Pecos, eastern Brewster, and most of Terrell Counties. According to Blair, the old Rio Grande embayment enters Trans-Pecos Texas in southern Terrell County. The entire course of the Rio Grande in Texas, down to Boquillas Canyon, is an ancient lake basin (see Sellards and Baker 1934). The highest peak in the Chihuahuan Province is Mount Livermore (2563 m) in the Davis Mountains. Emory Peak in the Chisos Mountains and Chinati Peak in the Chinati Mountains exceed 2154 m.

The climate of the Chihuahuan Province is classified as arid, with a −40 to −60 moisture deficiency index. The vegetation is typical Chihuahuan Desert plants and highly diversified. The grasses are buffalo, black gramma, tobosa, and galleta. The common shrubs are creosote bush, catclaw, blackbrush, and various species of prickly pear. In some shallow canyons huisache, mesquite, oaks, pinyon, and cedar predominate in the landscape.

The only salamander usually occurs in shallow ranch ponds (*Ambystoma mavortium*), and characteristic anurans are *Anayxrus cognatus, A. debilis, A. punctatus, A. speciosus, Scaphiopus couchi,* and *Spea multiplicata*; and in canyons with springs one can find *Hyla arenicolor.*

The turtle fauna consists of seven turtles; two are semiterrestrial, *Terrapene ornata* and *Kinoster-*

non flavescens. The Pecos River and Rio Grande contain four aquatic turtles: *Pseudemys gorzugi, Trachemys gaigeae, T. scripta* (captive releases), and *Apalone spinifera.* One of the semipermanent tributaries to the Rio Grande (Alamito Creek) contains the only population of *K. hirtipes* in the province.

About 33 species of lizards occur in the Texas portion of the Chihuahuan Biotic Province. Of these, 2 *Coleonyx* are characteristic, as are 3 *Phrynosoma,* 4 *Sceloporus,* 2 *Plestiodon,* 9 *Aspidoscelis, Crotaphytus collaris, Gambelia wislizenii, Cophosaurus texanus, Urosaurus ornatus,* and *Uta stansburiana.* Snake species number about 42. Common poisonous species are *Crotalus atrox, C. lepidus, C. molossus, C. scutulatus,* and *C. viridis.* Other typical species are bullsnakes (*Pituophis*), nightsnakes (*Hypsiglena*), kingsnakes (*Lampropeltis*), patch-nosed snakes (*Salvadora*), coachwhips and whipsnakes (*Coluber* and *Masticophis*), long-nosed snakes (*Rhinocheilus*), ratsnakes (*Pantherophis*), and glossy snakes (*Arizona*). Rare snakes include *Tantilla cucullata, Trimorphodon vilkinsoni,* and *Rena segrega.*

Of the 219 native herpetofaunal species in Texas, 91 (41%) are not shared with Mexico. Of those not shared with Mexico, 24 (86%) are salamanders, 18 (44%) are frogs and toads, 15 (50%) are turtles, 1 (100%) is an alligator, 9 (20%) are lizards, and 24 (32%) are snakes.

The status of the 13 unshared species of the spring salamanders (*Eurycea*) spread out along the escarpment of the southern edge of the Edwards Plateau is unresolved. The genes of some species show intermediacy, while others do not. Some species of blind taxa are very distinct, while others are not. I do not expect to see a solution to this problem soon. The non-spring salamander of the genus *Eurycea* in Texas, *E. quadridigitata,* is found only in the Austroriparian Biotic Province. Its range extends eastward to Florida and up the east coast to the Virginia border. The Western Slimy Salamander (*Plethodon albagula*) is a Texas endemic, found primarily along the canyons and stream bottoms near the escarpment of the Edwards Plateau.

The Spotted Dusky Salamander (*Desmognathus conanti*) is known only from a few localities in the Austroriparian Biotic Province in Texas but extends eastward to the Florida Panhandle. Most of the remaining unshared salamanders (*Ambystoma maculatum, A. opacum, A. talpoidium, A. texanum, A. tigrinum, Amphiuma tridactylum, Necturus beyeri,* and *Notophthalmus viridescens*) are eastern or southeastern salamanders of the United States and generally restricted to the Austroriparian Province in Texas.

The four salamanders shared with Mexico are Black-spotted Newt (*Notophthalmus meridionalis*), Barred Tiger Salamander (*Ambystoma mavortium*), Lesser Siren (*Siren intermedia*), and *Siren* sp. indet. The Black-spotted Newt is composed of two subspecies. The southern race may have evolved as a relic of a Pleistocene dispersal event and is known only from northern Veracruz and southern Tamaulipas. The northern subspecies occurs from about the southern third of the Texas coast to the southern limit of the Tamaulipan Biotic Province boundary in Mexico, very near the Tropic of Cancer, south of Llera, Tamaulipas. The second species, Barred Tiger Salamander, may be a remnant of a once widespread species, Tiger Salamander, whose distribution has been broken into several isolated segments, each evolving into independent lineages over time. Its distribution includes most of northwestern Texas, western New Mexico north to most of Nebraska, south through most of Kansas, and Oklahoma. Southwestward, the species extends into the Mexican states of Coahuila and Chihuahua. The third species shared with Mexico is *S. intermedia.* This species has a southeastern US distribution, ranging up the east coast as far as Virginia and up the Mississippi River Valley to Illinois and Indiana. In Texas the salamander is found in the Austroriparian and the southern coastal areas of the Texan and Tamaulipan Biotic Provinces. According to Petranka (1998), there are two taxa in Texas, *S. i. nettingi,* and *S. i. texana.* There is considerable controversy on the status of *S. i. texana.* The latter's distribution extends a short distance into Mexico below Brownsville and up the Texas coast as far as Kingsville. Currently, its genetics are similar to those of *intermedia*, yet sympatric with it in several South Texas localities, suggesting it may be a species. The taxonomic status of this taxon is yet to be resolved. In the particular case of this book and to be consistent with the Tamaulipas chapter, the *Siren* species not yet described is regarded as the fourth shared species with Mexico under the name of *Siren* sp. indet.

The anurans are almost equally split between those unshared with Mexico (44%) and those shared (56%). Of those unshared, one is endemic (*Anaxyrus houstonensis*) and also endangered. Of the 17 remaining unshared species, 8 are found in the Austroriparian Province: Crawfish Frog (*Lithobates areolatus*), Green Frog (*L. clamitans*), Pickerel Frog (*L. palustris*), Southern Leopard Frog (*L. sphenocephalus*), Eastern Narrow-mouthed Frog (*Gastrophryne carolinensis*), East Texas Toad (*A. velatus*), American Toad (*A. americanus*), and the Spring Peeper (*Pseudacris crucifer*). Seven of the remaining 9 species are generally Austroriparian in distribution but range farther to the west into the Texan Province: *Hyla chrysoscelis*, *H. cinerea*, *H. squirella*, *H. versicolor*, *Pseudacris fouquettei*, *P. streckeri*, and *Scaphiopus hurteri*. *Lithobates grylio* is restricted to the southeastern coastal marshes in Texas, and *L. blairi* is found in northwest Texas.

Of the 23 native species that are shared with Mexico, 8 are primarily found in the Tamaulipan Province: *Incilius nebulifer*, *Rhinella marina*, *Eleutherodactylus cystignathoides*, *Smilisca baudinii*, *Leptodactylus fragilis*, *Hypopachus variolosus*, *Lithobates berlandieri*, and *Rhinophrynus dorsalis*.

Of the remaining 15 taxa, one is a generalist but also Austroriparian, *Lithobates catesbeianus*. It has been introduced into Arizona, New Mexico, Mexico, and the Chihuahuan Desert in Texas, where it is considered an invasive species. Two species are primarily North American in distribution. *Anaxyrus woodhousii* is from eastern North America but extends into the Great Plains and the Chihuahuan Desert. *Anaxyrus cognatus* is principally from central North America but is also found in the Chihuahuan Desert grasslands at higher elevations. The remaining 12 species

all have affinities to the southwestern deserts: *A. debilis*, *A. punctatus*, *A. speciosus*, *Craugastor augusti*, *Eleutherodactylus guttilatus*, *E. marnocki*, *Hyla arenicolor*, *Pseudacris clarki*, *Gastrophryne olivacea*, *Scaphiopus couchi*, *Spea bombifrons*, and *S. multiplicata*. Many of the latter species also occur in the Balconian, Chihuahuan, and Tamaulipan Biotic Provinces.

The only crocodilian occurring in Texas is the American Alligator (*Alligator mississippiensis*), which has a southeastern US distribution, where it occurs from coastal North Carolina south to the tip of Florida, west to the tip of Texas. There are teeth from an alligator from Matamoros, Mexico, in the United States National Museum and a specimen reported from Brownsville, Texas (USNM 3184) by Baird (1859b, 5). Currently, alligators are common in areas of Cameron County and have been observed in the Rio Grande near Brownsville and along the Rio Grande as far up the river as 85 km east of Port Isabel. There are no records of a specimen from Mexico, but the alligator is in the river, so it is only a matter of time before one is taken from the Mexico side of the river. Alligators have also been observed at the mouth of Santa Elena Canyon, Big Bend National Park. The latter individual is a release from a small exhibit in Terlingua, Texas. It is interesting to note that *Crocodylus moreletii* occurs in the Río San Fernando, Tamaulipas, only 150 km south of the Rio Grande.

The 30 species of unshared/shared turtles are difficult to assign to a biotic province or even to a plant formation in Texas. Five species are marine turtles, 21 are freshwater turtles, 1 is a marine marsh turtle, and 3 are terrestrial turtles.

Of the unshared turtles, 3 are endemic to Texas: Cagle's Map Turtle (*Graptemys caglei*), Texas Map Turtle (*G. versa*), and Texas Cooter (*Pseudemys texana*). Cagle's Map Turtle is restricted to the Guadalupe River Basin, the Texas Map Turtle is restricted to the Colorado River Basin, and the Texas Cooter is restricted to the Colorado and Brazos River Basins. Three turtles are found only in the Mississippi River Basin: *G. ouachitensis*, *G. pseudogeographica*, and *Apalone mutica*. Four are found in the southeastern part of United States: *Macrochelys temminckii*, *Chrysemys dorsalis*, *Deirochelys reticularia*, and *Kinosternon subrubrum*. *Chelydra serpentina*, *Pseudemys concinna*, and *Sternotherus odoratus* are found throughout the eastern United States and at least throughout the Balconian Province. *Sternotherus carinatus* is found throughout the Austroriparian Province and into the states of Louisiana, Arkansas, and Oklahoma. *Malaclemys terrapin* occurs in Texas tidal marsh from Corpus Christi to the Sabine River.

The shared 15 species of turtles include the 5 species of marine turtles. Those that are shared with Mexico also include 3 turtles that range throughout the eastern half of the United States: *Pseudemys concinna*, *Trachemys scripta*, and *Apalone spinifera*. Two, *Trachemys gaigeae* and *Pseudemys gorzugi*, are restricted to the Rio Grande Basin. One species, *Kinosternon hirtipes*, is restricted to a tributary (Alameda Creek) of the Rio Grande in West Texas. Another species, *K. flavescens*, is found throughout Central and West Texas. *Chrysemys picta* is found in the Rio Grande, Pecos River, and extreme northeastern Texas. Three turtles are terrestrial: *Gopherus berlandieri* is restricted to the Tamaulipan Province; *Terrapene ornata* is a Great Plains species but is distributed to the west and occurs in Mexico; *T. carolina* is principally an eastern US species but is distributed southwestward in Texas and enters Mexico in Tamaulipas.

Of the 45 species of native lizards in Texas, only 9 are not shared with Mexico. Of these unshared species, 2 have very small distributions in Texas and New Mexico: Dunes Sagebrush Lizard (*Sceloporus arenicolus*) is known only from the Mescalero Sands of southeastern New Mexico and adjacent Texas; and Gray-checkered Whiptail (*Aspidoscelis dixoni*) occurs only along the Rio Grande terraces in Presidio County, Texas, and Antelope Pass in southwestern New Mexico. The Dunes Sagebrush Lizard is a habitat specialist, occurring only in or near dune blowouts where the apex of the dunes are somewhat stabilized by the shinnery oak–sand sage plant community. The Gray-checkered Whiptail is a parthenogenetic species produced from a

hybridization event between *A. marmorata* and *A. scalaris*. Three skinks, Coal Skink (*Plestiodon anthracinus*), Broad-headed Skink (*P. laticeps*), and Five-lined Skink (*P. fasciatus*), are found in the eastern half of the United States and are generally restricted to the Austroriparian Province in Texas. One other species, Slender Glass Lizard (*Ophisaurus attenuatus*), is found mainly in prairies, both coastal and inland, throughout the eastern half of the United States. One skink (*P. septentrionalis*) is restricted to the Great Plains Biome and is more restricted in its distribution to the Texan Province in Texas. The last lizard, the Green Anole (*Anolis carolinensis*), occurs throughout the southeastern United States and in the Texan, Austroriparian, Balconian, and parts of the Tamaulipan and Chihuahuan Biotic Provinces. This species appears to be easily transported, common in the pet trade, and is likely to be found in Mexico as an escaped pet.

Of the 36 species of native lizards shared with Mexico, more than half (21) are found in the Chihuahuan Biotic Province: *Gerrhonotus infernalis*, *Crotaphytus collaris*, *Gambelia wislizenii*, *Coleonyx brevis*, *C. reticulatus*, *Cophosaurus texanus*, *Phrynosoma cornutum*, *P. hernandesi*, *P. modestum*, *Sceloporus bimaculosus*, *S. consobrinus*, *S. cowlesi*, *S. merriami*, *S. poinsettii*, *Urosaurus ornatus*, *Uta stansburiana*, *Aspidoscelis exsanguis*, *A. inornata*, *A. marmorata*, *A. tesselata*, and *A. uniparens*.

Seven lizards are found in the Tamaulipan Biotic Province: *Crotaphytus reticulatus*, *Holbrookia propinqua*, *Sceloporus cyanogenys*, *S. grammicus*, *S. variabilis*, *Plestiodon tetragrammus*, and *Aspidoscelis laredoensis*. Of the remaining 8, *Holbrookia maculata* is distributed in the Chihuahuan and Kansan Provinces, and *H. lacerata* is found only in the Tamaulipan and Balconian Provinces. *Sceloporus olivaceus*, *Aspidoscelis gularis*, and *A. sexlineata* are generalists and found in all biotic provinces except the Chihuahuan. *Scincella lateralis* occurs in all Texas biotic provinces except the Kansan and Chihuahuan. *Plestiodon obsoletus* occurs in all biotic provinces except the Texan and Austroriparian, and *P. multivirgatus* occurs only in the Kansan Province.

Seventy-four snakes are native to Texas: 24 are

not shared with Mexico, and 50 are shared. The unshared species are generally found in the eastern part of the United States and East Texas, but some may occur in all biotic provinces except the Chihuahuan and/or Tamaulipan. Species more or less restricted to the Austroriparian Province are *Carphophis vermis, Farancia abacura, Nerodia cyclopion, N. fasciata, N. sipedon, Pantherophis guttatus, Pituophis ruthveni, Regina rigida,* and *Storeria occipitomaculata.* Those that occur in every biotic province except the Chihuahuan are *Heterodon platirhinos, Pantherophis obsoletus, Regina grahamii,* and *Virginia striatula.* Two species are endemic to Texas: *Tantilla cucullata* and *Nerodia harteri.* The former occurs only in the Chihuahuan Province, and *N. harteri* is restricted to riffle systems in the Brazos and the Colorado River Basins. Three species, *Lampropeltis calligaster, Crotalus horridus,* and *Sistrurus miliarius,* are found in the Austroriparian and the Texan Biotic Provinces. One species, the *Nerodia clarkii,* is found only along the Gulf Coast, from the Sabine River south to near Corpus Christi. One species, *Cemophora coccinea,* with a specialized reptilian egg diet, contains two isolated populations: one is restricted to the Austroriparian Province, and the other, to the Tamaulipan Province. *Thamnophis radix* is known only from the Kansan Province, while *Agkistrodon piscivorus* occurs over most of Texas except the Chihuahuan Province. The final two species, *Tropidoclonion lineatum* and *Virginia valeriae,* have very spotty distributions in Texas. Both species occupy parts of the Austroriparian and Texan Provinces.

Of the 50 species shared with Mexico, 15 are generalists in their distribution patterns and occur over most of Texas: *Coluber constrictor, Diadophis punctatus, Lampropeltis getula, L. triangulum, Masticophis flagellum, Nerodia erythrogaster, N. rhombifer, Opheodrys aestivus, Pituophis catenifer, Storeria dekayi, Thamnophis proximus, T. sirtalis, Micrurus tener, Agkistrodon contortrix,* and *Crotalus atrox.*

There are 21 shared snakes that occur in the Chihuahuan Biotic Province: *Bogertophis subocularis, Gyalopion canum, Hypsiglena jani, Lampropeltis alterna, Masticophis taeniatus,* *Rhinocheilus lecontei, Salvadora deserticola, S. grahamiae, Sonora semiannulata, Tantilla atriceps, T. hobartsmithi, T. nigriceps, Thamnophis cyrtopsis, T. marcianus, Trimorphodon vilkinsoni, Rena dissecta, R. dulcis, R. segrega, Crotalus lepidus, C. molossus,* and *C. scutulatus.* Some of these snakes also occur jointly in parts of other Texas biotic provinces, such as the Balconian and Tamaulipan. Some, however, are found only in the Tamaulipan Province: *Coniophanes imperialis, Drymarchon melanurus, Drymobius margaritiferus, Ficimia streckeri, Leptodeira septentrionalis,* and *Masticophis schotti.* The 4 remaining species are members of the Great Plains Biome: *Crotalus viridis* occurs only in northwestern Texas, primarily in the Kansan Province; *Tantilla gracilis* occurs in every biotic province in Texas except the Chihuahuan; *Heterodon kennerlyi* occurs in all provinces except the Balconian; and *Sistrurus catenatus* has a very spotty distribution in Texas and occurs in every province in Texas except the Austroriparian and the Balconian but is considered uncommon in all the other provinces.

Composition (by Political Units)

An analysis of species densities by counties (political units of unequal size) reveals more about resident herpetologists and intensive herpetofaunal inventories than about the distribution of the amphibians and reptiles in Texas. The highest numbers of recorded species occur in such counties as Bexar, Brazos, Dallas, Harris, Hays, McLennan, Tarrant, and Travis. Each of these counties contains a major state university and one or more resident herpetologists. Other species-rich counties may be the result of incidental collection of species while in pursuit of more desirable "target" species or can be attributed to their diverse environments. For example, Bexar County contains 97 amphibian and reptile species, the highest species density in the state and 48% of the total herpetofauna of Texas; Bexar County also has one of the most diverse environments in the state, and several active herpetologists reside there.

All major museums in the United States and smaller museums in Texas were asked to supply Texas records of their holdings of amphibians and reptiles. About 90% responded with more than 13,000 county records and more than 110,000 individual localities. Questionable identifications and isolated distributional records were verified whenever possible, either by visiting the museum or by borrowing the material for examination.

The more than 13,000 species records for 254 Texas counties reveal that 49 species occur in 60% of all Texas counties, and 50 species occur in only 1.5% of all counties. The distribution of species suggests that 25% of the taxa are abundant, 25% are rare, and 50% are of average distribution. Salamanders make up 4.2% of the total number of records; frogs and toads, 22.9%; turtles, 11.6%; lizards, 20.3%; and snakes, 41.0%. About 34% (4488) of the 13,284 records have been accumulated during the past 47 years. I have not totaled the last 12 years of records, but the majority are found in the journal *Herpetological Review* (see Dixon 2000).

Taxonomic Issues

I would be remiss not to discuss the excessive amount of taxonomic change that has taken place since 1987. The new "tools," such as karyology, immunology, electrophoresis, mtDNA, and nuclear DNA, change one's perspective of traditional taxonomy. These new tools, in conjunction with computer programs that can view the "whole evidence" in a matter of seconds, create a taxonomic delusion for "old-school" taxonomists. DNA biologists can develop thousands of scenarios for the evolutionist, cladist, and pheneticist to analyze with each segment of the gene. Who is correct? What happens to any technique if misapplied? For an interesting discussion of these problems the reader should study James Lazell's article "Taxonomic Tyranny and the Exoteric" (Lazell 1992, 14) or Pauly et al. (2009), "Taxonomic Freedom" statements. Their comments mirror my thoughts perfectly.

Conservation Issues

For the first time in the history of the Texas Parks and Wildlife Department (TPWD), the commissioners have taken a stand for the regulation and control of nongame wildlife in the state. In 1998, the TPWD issued new regulations concerning the collection, possession limits, take, and bartering of nongame wildlife, which includes amphibians and reptiles. Texas is far behind other states in taking action to ensure protection of nongame wildlife.

One issue is commercial collecting, which if left unchecked, could easily wipe out entire populations of turtles, lizards, and salamanders in certain areas of the state. Because of increased demand for certain species of turtles, tortoises, and other reptiles and amphibians in Europe and Asia for pets and food, many of our native species have undergone a sharp decrease in numbers.

I am reasonably happy to say that the TPWD is making an effort to stem the tide. We also can help accomplish this goal. In 1983, the TPWD created the Special Nongame and Endangered Species Conservation Fund. This fund may be used for nongame wildlife and endangered species research and conservation, habitat acquisition and development, and dissemination of information pertaining to these species. Money for the fund is obtained through private donations and sale of nongame wildlife art prints, decals, and stamps. For more information on the fund or endangered species, contact the TPWD.

In addition, the TPWD was directed by its commissioners to implement a ban on the collection of native Texas turtles in public waters. The department allowed the continued commercial collection of Red-ear Sliders, Common Snapping Turtles, and two species of softshell turtles in private waters. The commissioners also initiated a five-year study of Texas turtles to determine the extent of migration of these turtles between public and private waters.

Another issue is the lack of public concern for the protection of habitat for certain species. The Houston Toad is an excellent example. Exploitation of Houston Toad habitat for golf courses, housing, and agricultural developments slowly has undermined the ability of the toad to sustain itself. This issue applies to other Texas species as well: the Horned Lizard, Reticulate Collared Lizard, Rio Grande Chirping Frog, Black-spotted Newt, Indigo Snake, Cagle's Map Turtle, and the Texas Tortoise.

An increase in the number of motor vehicles has resulted in multiple deaths of slow-moving reptiles and amphibians. Another result of increased traffic is more and wider highways, all at the expense of habitat once used by various species of wildlife. Slowly and predictably, there is less and less green space, fewer animals, and few that mourn their passing. This applies to sea turtles as well. An increased demand for seafood has brought additional boats and nets and an increase in drowned sea turtles. Philosophically, one must ask where it will all end. Are we doing too little too late? Is Rachel Carson's (1962) *Silent Spring* coming into view?

Herpetofaunal Diversity of the United States–Mexico Border States

GEOFFREY R. SMITH
AND JULIO A. LEMOS-ESPINAL

THE 3169 KM BORDER that has separated Mexico and the United States for the last 160 years had its origins in the Mexican-American War, a war that began over a border dispute. In 1846, Mexico declared that the Nueces River was its northern border with Texas. The United States held that the border was actually the more southern Rio Grande. After Mexican troops attacked American troops who had ventured south of the Nueces in May 1846, the United States declared war on Mexico. After a year and a half of fighting, the United States ultimately prevailed and demanded an enormous swath of northern Mexico. On February 2, 1848, the Treaty of Guadalupe Hidalgo ended the war between the United States and Mexico and created a boundary line separating the two countries. The treaty compelled Mexico to relinquish 26,418,000 km² of its northern frontier, over half its territory, to the United States for $15 million. Today this territory comprises the states of California, Arizona, New Mexico, and parts of Texas, Nevada, Colorado, and Utah. The international border between the United States and Mexico runs from Imperial Beach, California, and Tijuana, Baja California, in the west to Matamoros, Tamaulipas, and Brownsville, Texas, in the east, following the middle of the Rio Grande (Río Bravo) "along the deepest channel" from its mouth on the Gulf of Mexico a distance of 2019 km to a point just upstream of El Paso and Ciudad Juárez. It then follows an overland line westward that is marked by monuments for a distance of 858 km to the Colorado River, during which it reaches its highest elevation at the intersection with the Continental Divide in the Sierra de San Luis of Chihuahua, Mexico. It then follows the middle of that river northward a distance of 38 km, and then it again follows a westward overland line marked by monuments a distance of 226 km in a straight line along the division of California and Baja California to the Pacific Ocean (Hughes 2007; Levanetz 2008).

The region along the border is characterized by deserts, rugged mountains, abundant sunshine, and two major rivers, the Colorado and the Rio Grande, which provide life-giving waters to the largely arid but fertile lands along the rivers in both countries. The US states along the border, from west to east, are California, Arizona, New Mexico, and Texas. The Mexican states are Baja California, Sonora, Chihuahua, Coahuila, Nuevo León, and Tamaulipas. In the United States, Texas has the longest border with Mexico of any state, and California has the shortest. In Mexico, Chihuahua has the longest border with the United States, and Nuevo León has the shortest. Texas borders four Mexican states (Tamaulipas, Nuevo León, Coahuila, and Chihuahua), the most of any US state. New

Mexico and Arizona each border two Mexican states (Chihuahua and Sonora; Sonora and Baja California, respectively). California borders only Baja California.

The surface area of the four US states totals 1,726,665 km^2 (= 17.6% of the total surface area of the United States); the six Mexican states total 799,288 km^2 (= 40.5% of the total surface area of Mexico). These 10 border states cover a total surface area of 2,525,953 km^2, lying between latitudes 22°12' and 42°00' N, and between longitudes 93°31' and 124°26' W, and range from below sea level (sinkholes in Sonora) to 4421 m elevation (Mount Whitney, California) (table 13.1).

This vast territory, combined with the wide latitudinal, longitudinal, and altitudinal ranges, encompasses a great deal of diversity, where all six major terrestrial biomes of North America are represented (Tundra, Coniferous Forest, Prairie, Deciduous Forest, Desert, and Tropical Rain Forest), and an incredible richness of wildlife is found. Amphibians and reptiles are two of the best-represented vertebrate classes in the border states. The species richness of herpetozoans in the 10 border states, including their oceanic islands, totals 648 species: 186 amphibians (85 salamanders, 101 anurans); 462 reptiles (2 croco-

diles, 50 turtles, 213 lizards, 1 amphisbaenid, and 196 snakes). Seven of these species are marine: 6 turtles (*Caretta caretta, Chelonia mydas, Eretmochelys imbricata, Lepidochelys kempi, L. olivacea,* and *Dermochelys coriacea*), and 1 snake (*Hydrophis platurus*). The species richness of each of the 10 border states is presented in tables 13.2 and 13.3.

Twelve of these 648 species were introduced to the region from Africa, Asia, or Europe: *Xenopus laevis* (introduced to Baja California, California, and Arizona); *Chamaeleo jacksonii* (California); *Cyrtopodion scabrum* (Texas); *Gehyra mutilata* (Baja California and California); *Hemidactylus frenatus* (Baja California, Sonora, Tamaulipas, and Texas); *H. garnoti* (Texas); *H. turcicus* (all 10 border states); *Tarentola annularis* and *T. mauritanica* (California); *Podarcis sicula* (California); *Chalcides* sp. (Arizona); and *Indotyphlops braminus* (Sonora, Nuevo León, Tamaulipas, Texas, Arizona, and California). Another 3 species were introduced to the region from the Caribbean: *Eleutherodactylus coqui* (introduced to California); *Euhyas planirostris* (Texas); and *Anolis sagrei* (Texas).

Nineteen more species that are native to the border states region, the southeastern United States, or southern Mexico have been intro-

TABLE 13.1 Comparison of the land surface area of the 10 border states

	LSA	%TSA	LN max	LN min	LW max	LW min
California	423,970	16.8	42°	32°32'	−124°26'	−114°8'
Arizona	295,253	11.7	37°	31°20'	−114°49'	−109°2'
New Mexico	315,194	12.5	37°	31°20'	−109°3'	−103°
Texas	692,248	27.4	36°30'	25°50'	−106°39'	−93°31'
Baja California	71,576	2.8	32°50'	27°42'	−117°7'	−112°45'
Sonora	185,430	7.3	32°3'	26°12'	−115°	−108°48'
Chihuahua	245,612	9.7	31°47'	25°38'	−109°7'	−103°18'
Coahuila	151,571	6.0	29°53'	24°32'	−103°58'	−99°51'
Nuevo León	64,924	2.6	27°49'	23°11'	−101°14'	−98°26'
Tamaulipas	80,175	3.2	27°40'	22°12'	−100°8'	−97°8'
TOTAL	2,525,953	100.0				

Note: Land surface area (LSA in km^2); percentage of total surface area (%TSA) in relation to the total amount of land in the 10 border states; maximum and minimum north latitude (LN max and LN min, respectively); and maximum and minimum west longitude (LW max and LW min, respectively).

TABLE 13.2 Species richness of the 10 border states

	S	E	I	CAU	ANU	CRO	TES	LAC	AMPH	OPH
California	194	45	21	46[1] (30)	31[5] (3)	—	13[5]	54[6] (10)	—	50[4] (2)
Arizona	153	4	12	2[1]	28[3]	—	10[3]	57[4] (4)	—	56[1]
New Mexico	136	3	2	3 (2)	24[1]	—	10	47[1] (1)	—	52
Texas	229	20	10	28 (14)	42[1] (1)	1	31[1] (3)	51[6]	—	76[2] (2)
Baja California	119	21	6	4	16[3]	—	7[1]	51[2] (15)	1	40 (6)
Sonora	195	16	7	3	35[2]	—	16[1] (1)	69[3] (13)	—	72[1] (2)
Chihuahua	174	3	2	4	34[1] (1)	—	13	51[1] (2)	—	72
Coahuila	133	7	1	4	20	—	11 (2)	49[1] (4)	—	49 (1)
Nuevo León	132	2	2	3	20	—	6	42[1] (1)	—	61 (1)[1]
Tamaulipas	179	6	3	11 (2)	31	1	15	47[2] (2)	—	74[1] (2)

Note: S = species richness; E = endemic; I = introduced; CAU = caudata; ANU = anurans; CRO = crocodilia; TES = testudines; LAC = lacertilia; AMPH = amphisbaenids; OPH = ophidia. Superscripts indicate number of introduced species within that group. Numbers in parentheses indicate number of endemic species within that group.

duced to some of the border states: *Ambystoma mavortium* (introduced to California); *Ensatina eschscholtzii* (Arizona); *Lithobates berlandieri* (Baja California, Sonora, Arizona, and California); *L. catesbeianus* (Baja California, Sonora, Chihuahua, New Mexico, Arizona, and California); *L. sphenocephalus* (California); *Chelydra serpentina* (Arizona and California); *Chrysemys picta* (California); *Graptemys pseudogeographica* (California); *Pseudemys nelsoni* (Texas); *Trachemys scripta* (Arizona and California); *Apalone spinifera* (Baja California, Sonora, Arizona, and California); *Ctenosaura* sp. (Arizona); *C. pectinata* (Texas); *Sauromalus hispidus* (Sonora); *S. varius* (Baja California); *Aspidoscelis neomexicana* (Arizona); *Nerodia fasciata* (California); *N. rhombifer* (California); and *N. sipedon* (California).

The species richness of the region, disregarding introduced species, totals 630 species; of these, 233 (37%) are shared between the two countries; 190 (30.2%) occur only north of the border in the United States, including 72 species endemic to one of the four US border states; and 207 (32.8%) occur only south of the border in Mexico, including 55 species endemic to one of the six Mexican border states (table 13.4).

We constructed a checklist for the United States–Mexico border states, using each of the state lists in this book (see appendix). In order to examine the patterns of shared species among

the border states, we used Jaccard hierarchical clustering analyses (see Enderson et al. 2009). These analyses allowed us to determine which states show the greatest similarity. We ran these analyses for several taxonomic groups: all species, amphibians, anurans, salamanders, reptiles, lizards, snakes, and turtles. We repeated these analyses for the appropriate groups with a data set that excluded marine species (six sea turtles and a seasnake) and a data set that excluded both marine species and nonnative species. We also ran a similar analysis on the biotic provinces found in each state (taken from Brown et al. 2007) to consider how similar states were in the habitats or environments available.

For the analysis of all species, there were several distinct clusters of states (figure 13.1). Texas clusters with Tamaulipas, Nuevo León, and Coahuila. Arizona and New Mexico cluster with Chihuahua and Sonora. Baja California and California cluster with each other. Removing marine species had no qualitative effect on the clustering of these states and had very little quantitative effect. The same held true for removing marine species and nonnative species.

For amphibians, the pattern of clustering differed from that found for all species combined (figure 13.2A). The biggest difference was in the placement of Texas, Baja California, and California. Baja California groups with a larger

TABLE 13.3. Species Richness by Class, Order, and in the 10 Border States

		CA	AZ	NM	TX	BC	SO	CH	CO	NL	TA
CLASS AMPHIBIA											
Order Caudata	85	46	2	3	28	4	3	4	4	3	11
Ambystomatidae	11	4	1	1	6		2	3	1	1	
Amphiumidae	1				1						
Dicamptodontidae	2	2									
Plethodontidae	61	35	1	2	16	4	1	1	3	2	8
Proteidae	1				1						
Rhyacotritonidae	1	1									
Salamandridae	6	4			2						1
Sirenidae	2				2						2
Order Anura	101	31	28	24	42	16	35	34	20	20	31
Bufonidae	22	9	7	8	10	6	12	10	7	6	6
Craugastoridae	5		1	1	1		3	2	1	1	3
Eleutherodactylidae	9	1			4		1	2	3	3	5
Hylidae	25	4	6	5	10	2	6	5	4	2	6
Leiopelmatidae	1	1									
Leptodactylidae	2				1		1			1	2
Microhylidae	5		1	1	3		2	3	1	2	3
Pipidae	1	1	1			1					
Ranidae	24	12	8	6	8	5	8	9	2	1	2
Rhinophrynidae	1				1					1	1
Scaphiopodidae	6	3	4	3	4	2	2	3	2	3	3
CLASS REPTILIA											
Order Crocodilia	2				1						1
Crocodylidae	2				1						1
Order Testudines	50	13	10	10	31	7	16	13	11	6	15
Chelonidae	5	4			4	4	4				4
Chelydridae	2	1	1	1	2						
Dermochelyidae	1	1			1	1	1				1
Emydidae	23	4	3	5	16	1	4	4	5	2	4
Geoemydidae	1						1	1			
Kinosternidae	12	1	3	2	5		4	5	3	2	4
Testudinidae	4	1	2		1		1	2	2	1	1
Trionychidae	2	1	1	2	2	1	1	1	1	1	1
Order Squamata											
Suborder Lacertilia	213	54	57	47	51	51	69	51	49	42	47
Anguidae	16	3	1	1	2	3	1	4	3	3	5
Anniellidae	6	5				2					
Chamaeleonidae	1	1									

TABLE 13.3 *Continued*

	CA	AZ	NM	TX	BC	SO	CH	CO	NL	TA
Corytophanidae	1									1
Crotaphytidae	12	5	4	2	3	5	4	2	4	2	2
Dactyloidae	4				2		1	1			1
Dibamidae	1									1
Eublepharidae	5	2	1	2	2	2	2	1	2	1	1
Gekkonidae	12	5	1	1	4	3	6	2	1	1	2
Helodermatidae	2	1	1	1			2	1			
Iguanidae	11	2	3		1	4	7	1			2
Lacertidae	1	1									
Phrynosomatidae	81	19	26	22	19	24	29	24	29	27	19
Scincidae	19	2	6	3	8	2	3	7	6	4	5
Teiidae	29	2	11	15	10	4	12	8	4	3	5
Xantusidae	11	6	3			2	2			1	2
Xenosauridae	1										1
Suborder Amphisbaenia	1					1					
Bipedidae	1					1					
Suborder Serpentes	196	50	56	52	76	40	72	72	49	61	74
Boidae	5	4	1			1	2	1			1
Colubridae	149	36	38	41	61	27	52	56	37	49	58
Elapidae	5	1	1	1	1	1	3	2	1	1	2
Leptotyphlopidae	6	1	2	2	3	1	1	3	3	2	3
Typhlopidae	1	1	1		1		1			1	1
Viperidae	30	7	13	8	10	10	13	10	8	8	9
Total	648	194	153	136	229	119	195	174	133	132	179

Note: CA = California; AR[4] = Arizona; NM = New Mexico; TX = Texas; BC = Baja California; SO = Sonora; CH = Chihuahua; CO = Coahuila; NL = Nuevo León; TA = Tamaulipas.

TABLE 13.4 Number of native species of the 10 border states by taxonomic group

	Shared	Not shared		Total
		US	Mexico	
Salamanders	8	67 (46)	10 (2)	85
Anurans	38	35 (4)	25 (1)	98
Crocodilians	—	1	1	2
Turtles	20	16 (3)	13 (3)	49
Lizards	79	33 (15)	88 (37)	200
Amphisbaenids	—	—	1	1
Snakes	88	38 (4)	69 (12)	195
Total	233	190	207	630

Note: Numbers in parentheses represent the number of endemic species in one of the four US or six Mexican border states.

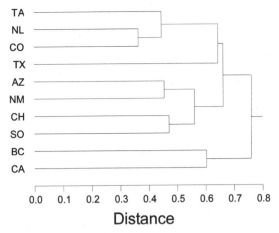

FIGURE 13.1 The cluster tree generated for all species of amphibians and reptiles in the border states. See text for methodology. Abbreviations: AZ = Arizona, BC = Baja California, CA = California, CH = Chihuahua, CO = Coahuila, NL = Nuevo León, NM = New Mexico, SO = Sonora, TA = Tamaulipas, and TX = Texas.

cluster of Tamaulipas, Nuevo León, and Coahuila; New Mexico and Arizona; and Sonora and Chihuahua. California then clusters with this larger group, and then Texas joins. Excluding nonnative species (no amphibians were marine species) resulted in a major change in the results of the cluster analysis (figure 13.2B). In this analysis, Baja California and California form a pair that is grouped with a cluster consisting of Tamaulipas, Nuevo León, and Coahuila; New Mexico and Arizona; and Sonora and Chihuahua paired with Texas.

The results of the analysis for all reptile species is nearly identical to the analysis for all species combined (figure 13.3A). This pattern of similarity remains the same when marine species are excluded from the analysis, as well as when nonnative species and marine species are both excluded from the analysis.

In the analysis of the distributions of anurans, the resulting cluster diagram places Arizona and New Mexico together as a pair of states that is grouped with the pair of Chihuahua and Sonora (figure 13.2C). This grouping is then clustered with a group consisting of Tamaulipas, Nuevo León, and Coahuila. Texas groups with

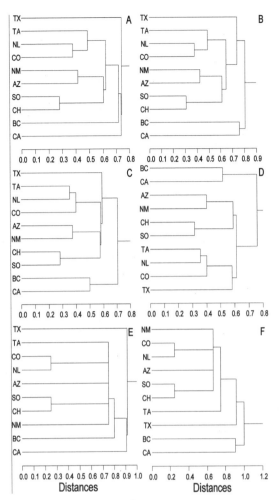

FIGURE 13.2 Cluster trees generated for (A) all amphibians, (B) all amphibians with nonnative species excluded, (C) anurans, (D) anurans with nonnative species excluded, (E) salamanders, and (F) salamanders with nonnative species excluded. See text for methodology. See figure 13.1 for abbreviations.

this cluster, and Baja California and California form a pair distinct from the remaining states. When nonnative species are excluded, the only change is that Texas moves to being clustered more closely with Tamaulipas, Nuevo León, and Coahuila (figure 13.2D).

The analysis of salamanders produces a unique cluster diagram (figure 13.2E). Coahuila and Nuevo León form a distinct pair, as do Sonora and Chihuahua. Beyond that there is a large cluster of equally similar states that includes Tamaulipas, the Coahuila and Nuevo León pair,

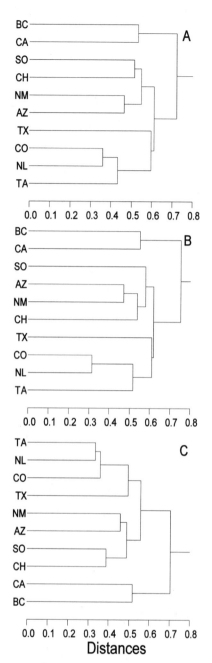

FIGURE 13.3 Cluster trees generated for (A) all reptiles, (B) lizards, and (C) snakes. See text for methodology. See figure 13.1 for abbreviations.

Arizona, the Sonora and Chihuahua pair, and New Mexico. This large cluster then pairs with Baja California and then California, with Texas joining last. There are several changes when nonnative salamanders are excluded from the analysis (figure 13.2F). In general there is greater resolution, with a group consisting of New Mex-

ico, the Coahuila and Nuevo León pair, Arizona, and the pair of Sonora and Chihuahua forming a group with Tamaulipas. Texas now groups with this cluster, and Baja California and California form their own pairing.

The analysis of lizard species is generally similar to the analysis for all reptile species with the exception of Chihuahua clustering more closely with the pair of Arizona and New Mexico, and Sonora forming a group with the cluster of Chihuahua, Arizona, and New Mexico (Figure 13.3B). Exclusion of nonnative lizards did not change the results of the analysis.

Snakes follow the same clustering pattern as for all reptiles and for lizards (figure 13.3C). This pattern holds for analyses excluding marine species and excluding marine species and nonnative species.

The analysis for turtles results in a group of states including Chihuahua, the pair of New Mexico and Arizona, the pair of Baja California and California, and Sonora (figure 13.4A). Another cluster of states that includes Tamaulipas, Nuevo León, and Coahuila pairs with the other large grouping of states. Texas stands alone. The exclusion of marine species restructures the results of the analysis (figure 13.4B). There is a cluster of states that includes a group of California and the pair of Arizona and New Mexico with the pair of Chihuahua and Sonora; another cluster includes Tamaulipas, Nuevo León, and Coahuila and Texas. Baja California now sits outside the other clusters. There are still further changes to the results when both marine species and nonnative species are excluded (figure 13.4C). Now, Baja California and California pair together and are separate from the rest of the states. The other states group into a subcluster of Tamaulipas, Nuevo León, and Coahuila; and a subcluster of Texas with New Mexico, Chihuahua, and Sonora; with Arizona grouping with these two subclusters.

The analysis of the biotic provinces results in several distinct clusters (figure 13.5). Texas clusters with New Mexico and Arizona, and Sonora and Chihuahua cluster with each other. These two clusters are included in a larger group.

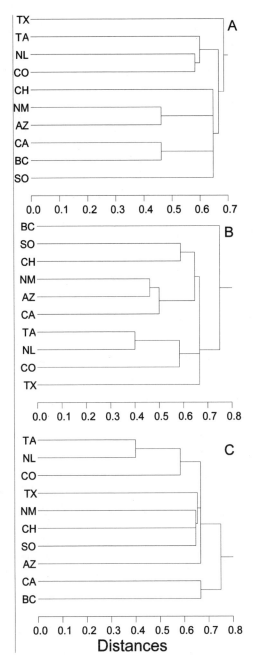

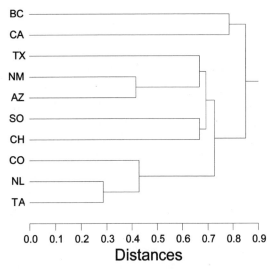

FIGURE 13.5 Cluster tree generated for biotic provinces found in the border states as defined and mapped by Brown et al. (2007). See text for methodology. See figure 13.1 for abbreviations.

FIGURE 13.4 Cluster trees generated for (A) all turtles, (B) turtles with marine species excluded, and (C) turtles with marine species and nonnative species excluded. See text for methodology. See figure 13.1 for abbreviations.

Coahuila, Nuevo León, and Tamaulipas form another cluster that is grouped with the Texas, New Mexico, Arizona, Sonora, and Chihuahua group. Baja California and California form a pair separated from the other states.

Our results generated several patterns. First, it is clear that, in general, certain sets of states tend to be more similar in species composition than others. In particular, Arizona and New Mexico group together frequently, as do Sonora and Chihuahua. This pattern of grouping for these states was also found in the analysis of the states surrounding Sonora by Enderson et al. (2009). This is as expected since Arizona and New Mexico share several biotic provinces in their northern parts, and Sonora and Chihuahua also share the Sierra Madre Occidental habitats, including the lowlands of the Barrancas del Cobre. These two pairs of states also tend to cluster together, as would be expected given that they all share the Sonoran Desert, Chihuahuan Desert, and sky island habitats. Tamaulipas, Coahuila, and Nuevo León are also generally found to be similar, and this group tends to cluster with Texas for many of the taxonomic groups we examined. Again, this clustering is expected given their geographic proximity and similarities in habitats. In addition, Tamaulipas, Coahuila, and Nuevo León share the Sierra Madre Oriental habitats, as well as part of the Coastal Plains, which are also shared with Texas. The final common pairing is between Baja California and California. Given the relatively isolated nature of Baja California and the fact it shares a common border and much of its habitat with California, this is not unexpected. One hypothesis for why such patterns are commonly seen across taxonomic groups is that they simply follow environmental conditions. As emphasized in many of the chapters of this book, environmental conditions and habitats found in each state appear to strongly drive the distribution and presence of the amphibian and reptile species found in these states. Indeed, many of the common pairings and clusters we observed in our taxonomic analyses also appear in the analysis of the similarities in biotic provinces found in these states (see figure 13.5). Thus, states that share biotic provinces will tend to share similar species, as would be expected. There are of course some exceptions; but in general it appears the distribution of habitat conditions across the border states drives many of the observed similarities in biodiversity across these states.

While many of the most similar groupings occur within either the United States (e.g., Arizona and New Mexico) or Mexico (Sonora and Chihuahua; Tamaulipas, Coahuila, and Nuevo León), there are important linkages across the border in terms of shared diversity, as is also highlighted in each of the chapters addressing each state's herpetofauna. The cross-border sharing of biodiversity indicates a strong need for habitats, and thus organisms, to be conserved as part of cross-border solutions, particularly since rapid urbanization along the United States–Mexico border can impact land cover and hydrology (Biggs et al. 2010) and the region is expected to become drier due to climate change (Seager and Vecchi 2010).

In addition, such cross-border sharing of biodiversity raises concerns about the construction of a barrier fence designed to limit the movement of people across the border. Such barriers may influence the herpetofauna if not properly designed and may lead to the isolation of populations that cross the border. Indeed, Lasky et al. (2011) identified 14 amphibian species and 21 reptile species that are potentially at risk from current border dispersal barriers, suggesting expansion might increase the number of species at risk. The consequences of the border fence is not only a concern for the herpetofauna. There is evidence that such a barrier could affect jaguars (*Panthera onca*) along the United States–Mexico border (McCain and Childs 2008). Studies along other border fences demonstrate negative effects on gene flow and access to good habitat across the border for some species (e.g., Daleszczyk and Bunevich 2009; Kaczensky et al. 2011). In addition, the extensive use of lighting along the border fence might affect the activity of nocturnal animals (e.g., Grigione and Mrykalo 2004), and given that many species of amphibians and reptiles are active at night, there could be consequences from artificial lighting for these species.

Second, for most of the taxonomic groups we examined, the exclusion of nonnative species had relatively little impact on the existing patterns of similarities in biodiversity of the herpetofauna. For the reptiles, except turtles, the removal of nonnative species had no effect on the patterns of biodiversity sharing we observed. There were relatively minor shifts in the observed patterns when we excluded nonnative amphibians from the analyses of all amphibians, anurans, and salamanders. In general, removing nonnative amphibians changed the placement of Texas; in all three cases, the removal of nonnative amphibians resulted in Texas moving from outside large clusters to inside these clusters and the pattern becoming more similar to the pattern found for the analysis of all species combined (see figure 13.1). The general lack of major effects of nonnative species in our analyses may be due to the relatively low numbers of nonnative reptiles and amphibians that have been introduced into Mexico, especially compared to the relatively high diversity of native herpetofauna. However, this does not mean that the further introduction of nonnative species will have no effect on the diversity and similarity of diversity in these border states (see Lavín-Murcio and Lazcano 2010). For example, all it takes is one very aggressive invasive species to reduce the native diversity of a particular state (e.g., American

Bullfrogs: Moyle 1973; Lawler et al. 1999; Casas-Andreu et al. 2001; Luja and Rodríguez-Estrella 2010; see also chapters by Hollingsworth and Mahrdt and by Painter and Stuart in this book). In addition, even without the negative effects of an aggressive invader species, increased anthropogenic movement of species around the world can lead to homogenization of diversity, as has been shown for amphibians and reptiles (Smith 2006) and fish (Rahel 2000; Villeger et al. 2011) in other areas of the world.

In our analyses, turtles as a group behaved differently. The overall turtle analysis generated a cluster tree relatively unique compared to all the other analyses. We believe this arises from the inclusion of sea turtles in this particular analysis. Once marine species are removed, the turtle cluster tree begins to become more similar to trees for the other reptiles, although there are still some significant differences, in particular the placement of California in a pairing with Arizona and New Mexico, a pattern that occurs for no other taxonomic group we examined. Further, removing nonnative turtles and marine species increases the similarity of the results to those of the other taxonomic groups, although there are still differences. However, the removal of nonnative species recaptures the general pairing of Baja California and California. It thus appears that of all the taxonomic groups we examined, nonnative species have a relatively important role in driving the observed patterns of similarity in turtles. This may arise from the relatively small number of turtles in these border states combined with the widespread introduction of one or two turtles (e.g., *Apalone spinifera*, *Trachemys scripta*).

In conclusion, the similarities of the herpetofauna among the border states likely reflect the distributions of shared environments among geographically proximate states. There are many herpetofaunal links between states across the United States–Mexico border because of these shared environments; however, these links may be threatened by the border fence erected by the US government (Lasky et al. 2011) as well as other threats encroaching on the region (reviewed in Lavín-Murcio and Lazcano 2010). Further empirical studies on the effects of changes along the United States–Mexico border on amphibians and reptiles are needed to fully understand the consequences and to develop potential remedial efforts.

Color Plates

1. *Ambystoma californiense*, California Tiger Salamander, Salamandra Tigre de California (Photo by Bradford Hollingsworth)

2. *Ambystoma gracile*, Northwestern Salamander, Salamandra del Noroeste (Photo by Bradford Hollingsworth)

3. *Ambystoma maculatum*, Spotted Salamander, Ajolote Manchado (Photo by Toby J. Hibbitts)

4. *Ambystoma mavortium*, Barred Tiger Salamander, Ajolote Tigre Rayado (Photo by Gary Nafis)

5. *Ambystoma mavortium*, Barred Tiger Salamander, Ajolote Tigre Rayado (Photo by Julio Lemos-Espinal)

6. *Ambystoma opacum*, Marbled Salamander, Ajolote Marmoleado (Photo by Toby J. Hibbitts)

7. *Ambystoma rosaceum*, Tarahumara Salamander, Ajolote Tarahumara (Photo by Iván Ahumada Carrillo)

8. *Ambystoma rosaceum*, Tarahumara Salamander, Ajolote Tarahumara (Photo by Iván Ahumada Carrillo)

9. *Ambystoma silvense*, Pine Woods Salamander, Salamandra del Bosque de Pino (Photo by Robert G. Webb)

10. *Ambystoma talpoideum*, Mole Salamander, Ajolote Topo (Photo by Toby J. Hibbitts)

11. *Ambystoma texanum*, Small-mouthed Salamander, Ajolote Texano (Photo by Toby J. Hibbitts)

12. *Dicamptodon ensatus*, California Giant Salamander, Salamandra Gigante de California (Photo by Bradford Hollingsworth)

13. *Dicamptodon tenebrosus*, Coastal Giant Salamander, Salamandra Gigante de la Costa (Photo by Bradford Hollingsworth)

14. *Aneides flavipunctatus*, Black Salamander, Salamandra Negra (Photo by Bradford Hollingsworth)

15. *Aneides lugubris*, Arboreal Salamander, Salamandra Arborícola (Photo by Bradford Hollingsworth)

16. *Aneides vagrans*, Wandering Salamander, Salamandra Vagabunda (Photo by Bradford Hollingsworth)

17. *Batrachoseps attenuatus,* California Slender Salamander, Salamandra Delgada de California (Photo by Bradford Hollingsworth)

18. *Batrachoseps luciae,* Santa Lucia Mountains Slender Salamander, Salamandra Delgada de las Montañas Santa Lucia (Photo by Bradford Hollingsworth)

19. *Batrachoseps major,* Garden Slender Salamander, Salamandra Mayor (Photo by Bradford Hollingsworth)

20. *Bolitoglossa platydactyla,* Broad-footed Salamander, Achoque de Tierra (Photo by Peter Heimes)

21. *Chiropterotriton priscus,* Primeval Flat-footed Salamander, Salamandra Primitiva (Photo by Tim Burkhardt)

22. *Desmognathus conanti,* Spotted Dusky Salamander, Salamandra de Conant (Photo by Toby J. Hibbitts)

23. *Ensatina eschscholtzii,* Monterey Ensatina, Ensatina de Bahía Monterey (Photo by Bradford Hollingsworth)

24. *Ensatina klauberi,* Large-blotched Ensatina, Ensatina Manchasgrandes (Photo by Bradford Hollingsworth)

25. *Eurycea chisholmensis*, Salado Salamander, Salamandra Salada (Photo by Toby J. Hibbitts)

26. *Eurycea latitans*, Cascade Caverns Salamander, Salamandra Cascada de Cavernas (Photo by Toby J. Hibbitts)

27. *Eurycea nana*, San Marcos Salamander, Salamandra de San Marcos (Photo by Toby J. Hibbitts)

28. *Eurycea neotenes*, Texas Salamander, Salamandra de Texas (Photo by Toby J. Hibbitts)

29. *Eurycea pterophila*, Fern Bank Salamander, Salamandra de Helechos (Photo by Toby J. Hibbitts)

30. *Eurycea quadridigitata*, Dwarf Salamander, Salamandra Enana (Photo by William Farr)

31. *Eurycea rathbuni*, Texas Blind Salamander, Salamandra Ciega Texana (Photo by Toby J. Hibbitts)

32. *Eurycea sosorum*, Barton Springs Salamander, Salamandra de Arroyos de Barton (Photo by Toby J. Hibbitts)

33. *Eurycea tonkawae*, Jollyville Plateau Salamander, Salamandra de Tonkawe (Photo by Toby J. Hibbitts)

34. *Eurycea troglodytes*, Valdina Farms Salamander, Salamandra Agujero-morador (Photo by William Farr)

35. *Eurycea waterlooensis*, Austin Blind Salamander, Salamandra Ciega de Austin (Photo by Toby J. Hibbitts)

36. *Plethodon albagula*, Western Slimy Salamander, Salamandra Babosa de Occidente (Photo by Toby J. Hibbitts)

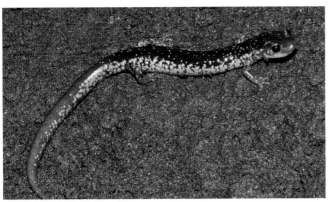

37. *Plethodon albagula*, Western Slimy Salamander, Salamandra Babosa de Occidente (Photo by William Farr)

38. *Pseudoeurycea bellii*, Bell's Salamander, Ajolote de Tierra (Photo by Michael Patrikeev)

39. *Pseudoeurycea cephalica*, Chunky False Brook Salamander, Babosa (Photo by William Farr)

40. *Pseudoeurycea galeanae*, Galeana False Brook Salamander, Tlaconete Neoleonense (Photo by William Farr)

41. *Pseudoeurycea scandens*, Tamaulipan False Brook Salamander, Tlaconete Tamaulipeco (Photo by William Farr)

42. *Necturus beyeri*, Gulf Coast Waterdog, Salamandra de la Costa del Golfo (Photo by Toby J. Hibbitts)

43. *Notophthalmus meridionalis*, Black-spotted Newt, Tritón de Manchas Negras (Photo by Gary Nafis)

44. *Notophthalmus meridionalis*, Black-spotted Newt, Tritón de Manchas Negras (Photo by William Farr)

45. *Notophthalmus viridescens*, Eastern Newt, Tritón de Oriental (Photo by William Farr)

46. *Notophthalmus viridescens*, Eastern Newt, Tritón de Oriental (Photo by Toby J. Hibbitts)

47. *Taricha rivularis*, Red-bellied Newt, Tritón de Vientre-rojo (Photo by Bradford Hollingsworth)

48. *Taricha torosa*, California Newt, Tritón de California (Photo by Bradford Hollingsworth)

49. *Anaxyrus americanus,* American Toad, Sapo Americano (Photo by Toby J. Hibbitts)

50. *Anaxyrus boreas,* Western Toad, Sapo Occidental (Photo by Bradford Hollingsworth)

51. *Anaxyrus californicus,* Arroyo Toad, Sapo de Arroyo (Photo by Bradford Hollingsworth)

52. *Anaxyrus cognatus,* Great Plains Toad, Sapo de Espuelas (Photo by William Farr)

53. *Anaxyrus debilis,* Green Toad, Sapo Verde (Photo by Bradford Hollingsworth)

54. *Anaxyrus houstonensis,* Houston Toad, Sapo de Houston (Photo by Toby J. Hibbitts)

55. *Anaxyrus kelloggi,* Little Mexican Toad, Sapo Mexicano (Photo by Bradford Hollingsworth)

56. *Anaxyrus mexicanus,* Mexican Madre Toad, Sapo Pie de Pala (Photo by Bradford Hollingsworth)

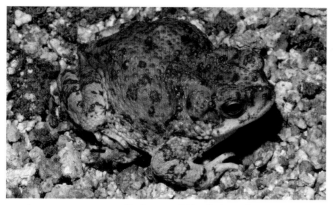

57. *Anaxyrus punctatus*, Red-spotted Toad, Sapo de Puntos Rojos (Photo by Bradford Hollingsworth)

58. *Anaxyrus retiformis*, Sonoran Green Toad (female and male), Sapo Sonorense (hembra y macho) (Photo by William Wells)

59. *Anaxyrus speciosus*, Texas Toad, Sapo Texano (Photo by Brad Alexander)

60. *Anaxyrus velatus*, East Texas Toad, Sapo del Este de Texas (Photo by Toby J. Hibbitts)

61. *Anaxyrus woodhousii*, Woodhouse's Toad, Sapo de Woodhouse (Photo by William Wells)

62. *Anaxyrus woodhousii*, Woodhouse's Toad, Sapo de Woodhouse (Photo by William Wells)

63. *Incilius alvarius*, Sonoran Desert Toad, Sapo del Desierto de Sonora (Photo by Bradford Hollingsworth)

64. *Incilius marmoreus*, Marbled Toad, Sapo Marmoleado (Photo by Matthew Cage)

65. *Incilius mazatlanensis*, Sinaloa Toad, Sapo de Mazatlán (Photo by Bradford Hollingsworth)

66. *Incilius mccoyi*, McCoy's Toad, Sapo de McCoy (Photo by Erik Enderson)

67. *Incilius nebulifer*, Gulf Coast Toad, Sapo Nebuloso (Photo by William Farr)

68. *Rhinella marina*, Cane Toad, Sapo Gigante (Photo by Iván Ahumada Carrillo)

69. *Craugastor augusti*, Barking Frog (adult), Sapo Ladrador (adulto) (Photo by Erik Enderson)

70. *Craugastor augusti*, Barking Frog (juvenile), Sapo Ladrador (juvenil) (Photo by Iván Ahumada Carrillo)

71. *Craugastor batrachylus*, Tamaulipan Arboreal Robber Frog, Rana Arborícola Tamaulipeca (Photo by William Farr)

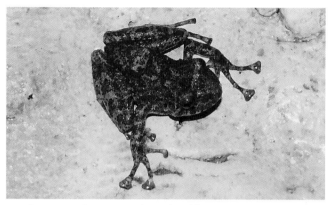

72. *Craugastor decoratus*, Adorned Robber Frog, Rana Labradora (Photo by William Farr)

73. *Craugastor tarahumaraensis*, Tarahumara Barking Frog, Rana Ladradora Amarilla (Photo by Peter Heimes)

74. *Craugastor tarahumaraensis*, Tarahumara Barking Frog, Rana Ladradora Amarilla (Photo by Peter Heimes)

75. *Eleutherodactylus cystignathoides*, Rio Grande Chirping Frog, Ranita Chirriadora del Río Bravo (Photo by William Farr)

76. *Eleutherodactylus dennisi*, Dennis's Chirping Frog, Rana Chirradora (Photo by William Farr)

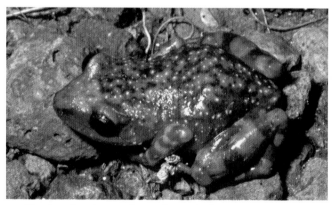

77. *Eleutherodactylus guttilatus*, Spotted Chirping Frog, Ranita Chirriadora de Manchas (Photo by Troy D. Hibbitts)

78. *Eleutherodactylus interorbitalis*, Spectacled Chirping Frog, Ranita de Lentes (Photo by Tim Burkhardt)

79. *Eleutherodactylus longipes*, Long-footed Chirping Frog, Ranita Chirriadora de la Huasteca (Photo by Tim Burkhardt)

80. *Eleutherodactylus marnockii*, Cliff Chirping Frog, Ranita Chirriadora de Escarpes (Photo by Tim Burkhardt)

81. *Eleutherodactylus marnockii*, Cliff Chirping Frog, Ranita Chirriadora de Escarpes (Photo by Troy D. Hibbitts)

82. *Eleutherodactylus verrucipes*, Big-eared Chirping Frog, Ranita Orejona (Photo by William Farr)

83. *Eleutherodactylus verrucipes*, Big-eared Chirping Frog, Ranita Orejona (Photo by Terry Hibbitts)

84. *Acris crepitans*, Northern Cricket Frog, Rana Grillo Norteña (Photo by Terry Hibbitts)

85. *Ecnomiohyla miotympanum*, Small-eared Treefrog, Calate Arborícola (Photo by William Farr)

86. *Hyla arenicolor*, Canyon Treefrog, Ranita de las Rocas (Photo by William Wells)

87. *Hyla arenicolor*, Canyon Treefrog, Ranita de las Rocas (Photo by Bradford Hollingsworth)

88. *Hyla chrysoscelis*, Cope's Gray Treefrog, Ranita Gris de Cope (Photo by Toby J. Hibbitts)

89. *Hyla cinerea*, Green Treefrog, Ranita Verde (Photo by Toby J. Hibbitts)

90. *Hyla cinerea*, Green Treefrog, Ranita Verde (Photo by William Farr)

91. *Hyla eximia*, Mountain Treefrog, Ranita de Montaña (Photo by Terry Hibbitts)

92. *Hyla squirella*, Squirrel Treefrog, Ranita Ardilla (Photo by Toby J. Hibbitts)

93. *Hyla versicolor*, Gray Treefrog, Ranita Gris (Photo by Toby J. Hibbitts)

94. *Hyla wrightorum*, Arizona Treefrog, Ranita de Wright (Photo by Erik Enderson)

95. *Pachymedusa dacnicolor*, Mexican Leaf Frog, Ranita Verduzca (Photo by Bradford Hollingsworth)

96. *Pseudacris cadaverina*, California Treefrog, Rana de Coro de California (Photo by Bradford Hollingsworth)

97. *Pseudacris crucifer*, Spring Peeper, Rana Pimienta (Photo by Toby J. Hibbitts)

98. *Pseudacris fouquettei*, Upland Chorus Frog, Rana de Coro de Tierra Alta (Photo by Toby J. Hibbitts)

99. *Pseudacris fouquettei*, Upland Chorus Frog, Rana de Coro de Tierra Alta (Photo by William Farr)

100. *Pseudacris hypochondriaca*, Baja California Treefrog, Rana de Coro de Baja California (Photo by Bradford Hollingsworth)

101. *Pseudacris streckeri*, Strecker's Chorus Frog, Rana de Coro de Strecker (Photo by Toby J. Hibbitts)

102. *Scinax staufferi*, Stauffer's Treefrog, Rana Arborícola Trompuda (Photo by William Farr)

103. *Smilisca baudinii*, Mexican Treefrog, Rana Arborícola Mexicana (Photo by Erik Enderson)

104. *Smilisca fodiens*, Lowland Burrowing Treefrog, Rana Chata (Photo by Bradford Hollingsworth)

105. *Tlalocohyla smithi*, Dwarf Mexican Treefrog, Ranita Enana Mexicana (Photo by Iván Ahumada Carrillo)

106. *Tlalocohyla smithi*, Dwarf Mexican Treefrog, Ranita Enana Mexicana (Photo by Iván Ahumada Carrillo)

107. *Trachycephalus typhonius*, Veined Treefrog, Rana Venulosa (Photo by William Farr)

108. *Leptodactylus fragilis*, White-lipped Frog, Ranita de Hojarasca (Photo by William Farr)

109. *Leptodactylus melanonotus*, Sabinal Frog, Rana del Sabinal (Photo by Jim Rorabaugh)

110. *Gastrophryne carolinensis*, Eastern Narrow-mouthed Frog, Ranita de Oriente (Photo by William Farr)

111. *Gastrophryne elegans*, Elegant Narrow-mouthed Toad, Sapito Elegante (Photo by William Farr)

112. *Gastrophryne olivacea*, Western Narrow-mouthed Toad, Ranita Olivo (Photo by Julio Lemos-Espinal)

113. *Hypopachus variolosus*, Sheep Frog, Rana Manglera (Photo by Iván Ahumada Carrillo)

114. *Hypopachus variolosus*, Sheep Frog, Rana Manglera (Photo by William Farr)

115. *Lithobates areolatus*, Crawfish Frog, Rana Cangrejo de Río (Photo by Toby J. Hibbitts)

116. *Lithobates berlandieri*, Rio Grande Leopard Frog, Rana Leopardo de El Río Bravo (Photo by Erik Enderson)

117. *Lithobates berlandieri*, Rio Grande Leopard Frog, Rana Leopardo de El Río Bravo (Photo by William Farr)

118. *Lithobates catesbeianus*, American Bullfrog, Rana Toro (Photo by Bradford Hollingsworth)

119. *Lithobates chiricahuensis*, Chiricahua Leopard Frog, Rana Leopardo Chiricahua (Photo by Erik Enderson)

120. *Lithobates clamitans*, Green Frog, Rana Verde (Photo by Toby J. Hibbitts)

121. *Lithobates forreri*, Forrer's Leopard Frog, Rana Leopardo de Forrer (Photo by Bradford Hollingsworth)

122. *Lithobates magnaocularis*, Northwest Mexico Leopard Frog, Rana Leopardo del Noroeste de México (Photo by Erik Enderson)

123. *Lithobates palustris*, Pickerel Frog, Rana de Ciénega (Photo by Toby J. Hibbitts)

124. *Lithobates pustulosus*, White-striped Frog, Rana Rayas Blancas (Photo by Iván Ahumada Carrillo)

125. *Lithobates sphenocephalus*, Southern Leopard Frog, Rana Leopardo del Sur de Estado Unidos (Photo by Toby J. Hibbitts)

126. *Lithobates tarahumarae*, Tarahumara Frog, Rana Tarahumara (Photo by Erik Enderson)

127. *Lithobates yavapaiensis*, Lowland Leopard Frog, Rana Leopardo de Yavapai (Photo by Jim Rorabaugh)

128. *Rana boylii*, Foothill Yellow-legged Frog, Rana Patas-amarillas (Photo by Bradford Hollingsworth)

129. *Rana draytonii*, California Red-legged Frog, Rana Patas-rojas de California (Photo by Bradford Hollingsworth)

130. *Rana muscosa*, Southern Mountain Yellow-legged Frog, Rana Patas-amarillas de las Montañas del Sur (Photo by Dana McLaughlin)

131. *Rhinophrynus dorsalis*, Burrowing Toad, Sapo de Madriguera (Photo by William Farr)

132. *Scaphiopus couchi*, Couch's Spadefoot, Cavador (Photo by Brad Alexander)

133. *Scaphiopus hurteri*, Hurter's Spadefoot, Sapo de Espuelas de Hurter (Photo by Toby J. Hibbitts)

134. *Spea bombifrons*, Plains Spadefoot, Sapo de Espuelas de los Llanos (Photo by William Farr)

135. *Spea hammondii*, Western Spadefoot, Sapo de Espuelas Occidental (Photo by Bradford Hollingsworth)

136. *Spea multiplicata*, Mexican Spadefoot, Sapo de Espuelas Mexicano (Photo by Bradford Hollingsworth)

137. *Alligator mississippiensis*, American Alligator, Lagarto Americano (Photo by William Farr)

138. *Crocodylus moreletii*, Morelet's Cocodile, Crocodrilo de Pantano (Photo by William Farr)

139. *Caretta caretta*, Loggerhead Sea Turtle, Caguama (Photo by Peter Heimes)

140. *Chelonia mydas*, Green Sea Turtle, Tortuga Blanca de Mar (Photo by Peter Heimes)

141. *Eretmochelys imbricata,* Hawksbill Sea Turtle, Carey (Photo by Peter Heimes)

142. *Lepidochelys kempii*, Kemp's Ridley Sea Turtle, Tortuga Golfina (Photo by William Farr)

143. *Lepidochelys olivacea*, Olive Ridley Sea Turtle, Tortuga Golfina (Photo by Peter Heimes)

144. *Macrochelys temminckii*, Alligator Snapping Turtle, Tortuga Lagarto de Temminck (Photo by Toby J. Hibbitts)

145. *Actinemys marmorata*, Western Pond Turtle, Tortuga Occidental de Estanque (Photo by Bradford Hollingsworth)

146. *Chrysemys picta*, Painted Turtle, Tortuga Pinta (Photo by Thomas C. Brennan)

147. *Deirochelys reticularia*, Chicken Turtle, Tortuga Pollo (Photo by Toby J. Hibbitts)

148. *Graptemys versa*, Texas Map Turtle, Tortuga Mapa Texana (Photo by William Farr)

149. *Malaclemys terrapin*, Diamond-backed Terrapin, Tortuga Espalda Diamante (Photo by Toby J. Hibbitts)

150. *Malaclemys terrapin*, Diamond-backed Terrapin, Tortuga Espalda Diamante (Photo by William Farr)

151. *Pseudemys concinna*, River Cooter, Tortuga Oriental de Río (Photo by William Farr)

152. *Pseudemys gorzugi*, Rio Grande Cooter, Tortuga de Oreja Amarilla (Photo by Troy D. Hibbitts)

153. *Pseudemys texana*, Texas Cooter, Jicotea Texana (Photo by Toby J. Hibbitts)

154. *Terrapene coahuila*, Coahuilan Box Turtle, Caja de Cuatro-ciénegas (Photo by Michael Price)

155. *Terrapene nelsoni*, Spotted Box Turtle, Caja de Manchas (Photo by Young Cage)

156. *Terrapene ornata*, Ornate Box Turtle, Caja Ornamentada (Photo by Thomas C. Brennan)

157. *Trachemys gaigeae*, Mexican Plateau Slider, Jicotea de la Meseta Mexicana (Photo by Charles W. Painter)

158. *Trachemys scripta*, Pond Slider, Jicotea de Estanque (Photo by Bradford Hollingsworth)

159. *Trachemys taylori*, Cuatrociénegas Slider, Jicotea de Cuatrociénegas (Photo by Peter Heimes)

160. *Trachemys venusta*, Mesoamerican Slider, Jicotea de Agua (Photo by William Farr)

161. *Trachemys yaquia*, Yaqui Slider, Jicotea del Yaqui (Photo by Jim Rorabaugh)

162. *Rhinoclemmys pulcherrima*, Painted Wood Turtle, Casco Rojo (Photo by Iván Ahumada Carrillo)

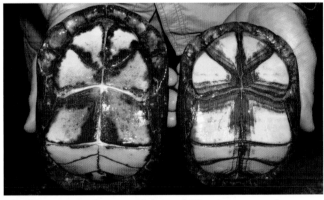

163. *Kinosternon alamosae* (right) and *Kinosternon sonoriense* (left), Álamos Mud Turtle, Casquito de Álamos, Sonora Mud Turtle, Casquito de Sonora (Photo by Jim Rorabaugh)

164. *Kinosternon alamosae*, Álamos Mud Turtle, Casquito de Álamos (Photo by Jim Rorabaugh)

165. *Kinosternon arizonense*, Arizona Mud Turtle, Casquito de Arizona (Photo by Thomas C. Brennan)

166. *Kinosternon durangoense*, Durango Mud Turtle, Casquito de Durango (Photo by Julio Lemos-Espinal)

167. *Kinosternon flavescens*, Yellow Mud Turtle, Casquito Amarillo (Photo by Thomas C. Brennan)

168. *Kinosternon herrerai*, Herrera's Mud Turtle, Casquito de Herrera (Photo by William Farr)

169. *Kinosternon hirtipes*, Rough-footed Mud Turtle, Casquito de Pata Rugosa (Photo by Terry Hibbitts)

170. *Kinosternon integrum*, Mexican Mud Turtle, Casquito de Fango Mexicana (Photo by William Farr)

171. *Kinosternon scorpioides*, Scorpion Mud Turtle, Tortuga Escorpión (Photo by William Farr)

172. *Kinosternon sonoriense*, Sonora Mud Turtle, Casquito de Sonora (Photo by Thomas C. Brennan)

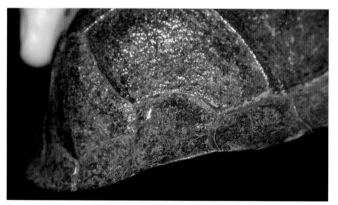

173. *Kinosternon sonoriense*, Sonora Mud Turtle, Casquito de Sonora (Photo by Jim Rorabaugh)

174. *Sternotherus odoratus,* Eastern Musk Turtle, Tortuga Almizclera (Photo by Toby J. Hibbitts)

175. *Gopherus agassizii,* Desert Tortoise, Tortuga de Desierto (Photo by Bradford Hollingsworth)

176. *Gopherus berlandieri,* Texas Tortoise, Tortuga de Berlandier (Photo by William Farr)

177. *Gopherus flavomarginatus,* Bolson Tortoise, Tortuga del Bolsón de Mapimí (Photo by Julio Lemos-Espinal)

178. *Gopherus morafkai,* Sonoran Desert Tortoise, Tortuga del Desierto de Sonora (Photo by Bradford Hollingsworth)

179. *Gopherus morafkai,* Sonoran Desert Tortoise, Tortuga del Desierto de Sonora (Photo by Bradford Hollingsworth)

180. *Apalone mutica,* Smooth Softshell, Tortuga de Concha Blanda Lisa (Photo by Toby J. Hibbitts)

181. *Apalone spinifera,* Spiny Softshell (adult), Tortuga de Concha Blanda (adulto) (Photo by William Farr)

182. *Apalone spinifera*, Spiny Softshell (juvenile), Tortuga de Concha Blanda (juvenil) (Photo by Susy Sanoja-Sarabia)

183. *Abronia taeniata*, Bromeliad Arboreal Alligator Lizard (adult), Escorpión Arborícola de Bandas (adulto) (Photo by Julio Lemos-Espinal)

184. *Abronia taeniata*, Bromeliad Arboreal Alligator Lizard (juvenile), Escorpión Arborícola de Bandas (juvenil) (Photo by Julio Lemos-Espinal)

185. *Barisia ciliaris*, Northern Alligator Lizard (female), Escorpión de Montaña (hembra) (Photo by Michael Price)

186. *Barisia ciliaris*, Northern Alligator Lizard (female and male), Escorpión de Montaña (hembra y macho) (Photo by Daniel Garza-Tobón)

187. *Barisia ciliaris*, Northern Alligator Lizard (juvenile), Escorpión de Montaña (juvenil) (Photo by Michael Price)

188. *Barisia levicollis*, Chihuahuan Alligator Lizard (female), Escorpión de Chihuahua (hembra) (Photo by Marisa Ishimatsu)

189. *Barisia levicollis*, Chihuahuan Alligator Lizard (male), Escorpión de Chihuahua (macho) (Photo by Peter Heimes)

190. *Barisia levicollis,* Chihuahuan Alligator Lizard (juvenile), Escorpión de Chihuahua (juvenil) (Photo by Peter Heimes)

191. *Elgaria cedrosensis,* Isla Cedros Alligator Lizard, Lagartija Lagarto de Isla Cedros (Photo by Bradford Hollingsworth)

192. *Elgaria kingii,* Madrean Alligator Lizard, Lagartija Lagarto de Montaña (Photo by Bradford Hollingsworth)

193. *Elgaria multicarinata,* Southern Alligator Lizard, Lagartija Lagarto del Suroeste de los Estados Unidos (Photo by Bradford Hollingsworth)

194. *Elgaria nana,* Islas Los Coronados Alligator Lizard, Lagartija Lagarto de Los Coronados (Photo by Bradford Hollingsworth)

195. *Gerrhonotus farri,* Farr's Alligator Lizard, Lagartija Escorpión de Farr (Photo by William Farr)

196. *Gerrhonotus infernalis,* Texas Alligator Lizard, Cantil de Tierra (Photo by William Farr)

197. *Gerrhonotus lugoi,* Lugo's Alligator Lizard, Lagartija Escorpión de Lugo (Photo by Peter Heimes)

198. *Ophisaurus attenuata*, Slender Glass Lizard, Lagartija de Cristal Delegada (Photo by Toby J. Hibbitts)

199. *Ophisaurus incomptus*, Plain-necked Glass Lizard, Lagartija de Cristal de Cuello Plano (Photo by Carlos A. Luna Aranguré)

200. *Ophisaurus incomptus*, Plain-necked Glass Lizard, Lagartija de Cristal de Cuello Plano (Photo by Carlos A. Luna Aranguré)

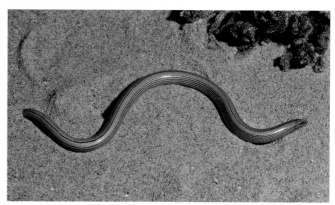

201. *Anniella geronimensis*, Baja California Legless Lizard, Lagartija Sin Patas (Photo by Bradford Hollingsworth)

202. *Anniella pulchra*, California Legless Lizard, Lagartija Apoda (Photo by Bradford Hollingsworth)

203. *Laemanctus serratus*, Serrated Casque-headed Lizard, Lemancto Coronado (Photo by Alfonso Delgadillo Espinosa)

204. *Crotaphytus antiquus*, Venerable Collared Lizard (female), Cachorón de Coahuila (hembra) (Photo by Jimmy Mcguire)

205. *Crotaphytus antiquus*, Venerable Collared Lizard (male), Cachorón de Coahuila (macho) (Photo by Jimmy Mcguire)

206. *Crotaphytus collaris*, Eastern Collared Lizard (female), Cachorón de Collar (hembra) (Photo by Julio Lemos-Espinal)

207. *Crotaphytus collaris*, Eastern Collared Lizard (male), Cachorón de Collar (macho) (Photo by William Wells)

208. *Crotaphytus dickersonae*, Dickerson's Collared Lizard (female), Cachorón Azul de Collar (hembra) (Photo by Barney Oldfield)

209. *Crotaphytus dickersonae*, Dickerson's Collared Lizard (male), Cachorón Azul de Collar (macho) (Photo by Paul Hamilton)

210. *Crotaphytus dickersonae*, Dickerson's Collared Lizard (female and male), Cachorón Azul de Collar (hembra y macho) (Photo by William Wells)

211. *Crotaphytus grismeri*, Sierra de las Cucupas Collared Lizard, Cachorón de Sierra de las Cucupas (Photo by Bradford Hollingsworth)

212. *Crotaphytus insularis*, Isla Ángel de la Guarda Collared Lizard, Cachorón de Collar Negro (Photo by Jimmy Mcguire)

213. *Crotaphytus insularis*, Isla Ángel de la Guarda Collared Lizard, Cachorón de Collar Negro (Photo by Jimmy Mcguire)

214. *Crotaphytus nebrius*, Sonoran Collared Lizard (female), Cachorón de Sonora (hembra) (Photo by Barney Oldfield)

215. *Crotaphytus nebrius*, Sonoran Collared Lizard (male), Cachorón de Sonora (macho) (Photo by Jimmy Mcguire)

216. *Crotaphytus nebrius*, Sonoran Collared Lizard (juvenile), Cachorón de Sonora (juvenil) (Photo by Julio Lemos-Espinal)

217. *Crotaphytus reticulatus*, Reticulate Collared Lizard (female), Reticulada de Collar (hembra) (Photo by Kristopher Lappin)

218. *Crotaphytus reticulatus*, Reticulate Collared Lizard (male), Reticulada de Collar (macho) (Photo by Jimmy Mcguire)

219. *Crotaphytus vestigium*, Baja California Collared Lizard, Cachorón de Baja California (Photo by Bradford Hollingsworth)

220. *Gambelia copeii*, Baja California Leopard Lizard, Cachorón Leopardo de Baja California (Photo by Bradford Hollingsworth)

221. *Gambelia wislizenii*, Long-nosed Leopard Lizard, Lagartija Mata Caballo (Photo by Bradford Hollingsworth)

222. *Anolis carolinensis*, Green Anole, Abaniquillo Verde (Photo by William Farr)

223. *Anolis carolinensis*, Green Anole, Abaniquillo Verde (Photo by Toby J. Hibbitts)

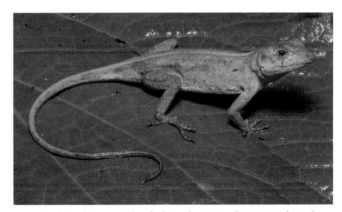

224. *Anolis nebulosus*, Clouded Anole, Roño de Paño (Photo by Iván Ahumada Carrillo)

225. *Anolis nebulosus*, Clouded Anole, Roño de Paño (Photo by Iván Ahumada Carrillo)

226. *Anolis sagrei*, Brown Anole, Abaniquillo Costero Maya (Photo by William Farr)

227. *Anolis sericeus*, Silky Anole, Abaniquillo Punto Azul (Photo by William Farr)

228. *Anelytropsis papillosus*, Mexican Blind Lizard, Lagartija Ciega Mexicana (Photo by William Farr)

229. *Coleonyx brevis*, Texas Banded Gecko (adult), Salamanquesa del Desierto (adulto) (Photo by Brad Alexander)

230. *Coleonyx brevis*, Texas Banded Gecko (juvenile), Salamanquesa del Desierto (juvenil) (Photo by William Farr)

231. *Coleonyx fasciatus*, Black Banded Gecko, Geco de Bandas Negras (Photo by Matthew Cage)

232. *Coleonyx fasciatus*, Black Banded Gecko, Geco de Bandas Negras (Photo courtesy of Naturaleza y Cultura—Sierra Madre A. C.)

233. *Coleonyx reticulatus*, Reticulate Banded Gecko, Geco Reticulado (Photo by Troy D. Hibbitts)

234. *Coleonyx variegatus*, Western Banded Gecko, Geco de Bandas Occidental (Photo by Thomas C. Brennan)

235. *Hemidactylus frenatus*, Common House Gecko, Besucona (Photo by William Farr)

236. *Hemidactylus turcicus*, Mediterranean House Gecko, Geco del Mediterráneo (Photo by Brad Alexander)

237. *Phyllodactylus homolepidurus*, Sonoran Leaf-toed Gecko, Salamanquesa de Sonora (Photo by Matthew Cage)

238. *Phyllodactylus nolascoensis*, San Pedro Nolasco Gecko, Salamanquesa de San Pedro Nolasco (Photo by Tim Burkhardt)

239. *Phyllodactylus partidus*, Isla Partida Norte Leaf-toed Gecko, Salamanquesa de Isla Partida Norte (Photo by Bradford Hollingsworth)

240. *Phyllodactylus tuberculosus*, Yellow-bellied Gecko, Geco Panza Amarilla (Photo by Jim Rorabaugh)

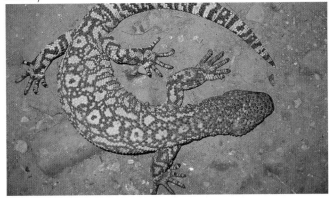

241. *Heloderma exasperatum*, Rio Fuerte's Beaded Lizard, Escorpión del Rió Fuerte's (Photo by Julio Lemos-Espinal)

242. *Heloderma suspectum*, Gila Monster, Monstruo de Gila (Photo by Julio Lemos-Espinal)

243. *Ctenosaura acanthura*, Mexican Spiny-tailed Iguana (male), Garrobo de Noreste (macho) (Photo by William Farr)

244. *Ctenosaura conspicuosa*, Isla San Esteban Spiny-tailed Iguana (male), Garrobo de Isla San Esteban (macho) (Photo by Bradford Hollingsworth)

245. *Ctenosaura conspicuosa*, Isla San Esteban Spiny-tailed Iguana (male), Garrobo de Isla San Esteban (macho) (Photo by Bradford Hollingsworth)

246. *Ctenosaura macrolopha*, Sonora Spiny-tailed Iguana (female), Garrobo de Sonora (hembra) (Photo by Jim Rorabaugh)

247. *Ctenosaura macrolopha*, Sonora Spiny-tailed Iguana (male), Garrobo de Sonora (macho) (Photo by Francisco García Camino)

248. *Ctenosaura macrolopha*, Sonora Spiny-tailed Iguana (juvenile), Garrobo de Sonora (juvenil) (Photo by Jim Rorabaugh)

249. *Ctenosaura nolascensis*, Isla San Pedro Nolasco Spiny-tailed Iguana (male), Garrobo de Isla San Pedro Nolasco (macho) (Photo by Bradford Hollingsworth)

250. *Ctenosaura nolascensis*, Isla San Pedro Nolasco Spiny-tailed Iguana (male), Garrobo de Isla San Pedro Nolasco (macho) (Photo by Bradford Hollingsworth)

251. *Dipsosaurus dorsalis*, Northern Desert Iguana, Cachorón Güero (Photo by Bradford Hollingsworth)

252. *Sauromalus ater*, Common Chuckwalla (male), Cachorón de Roca (macho) (Photo by William Wells)

253. *Sauromalus hispidus*, Spiny Chuckwalla, Iguana Espinosa de Pared (Photo by Bradford Hollingsworth)

254. *Sauromalus varius*, Piebald Chuckwalla, Iguana de Pared de Piebald (Photo by Chris Mattison)

255. *Sauromalus varius*, Piebald Chuckwalla, Iguana de Pared de Piebald (Photo by Chris Mattison)

256. *Callisaurus draconoides*, Zebra-tailed Lizard, Cachora Arenera (Photo by Thomas C. Brennan)

257. *Callisaurus splendidus*, Ángel de la Guarda Zebra-tailed Lizard, Cachora de Ángel de la Guarda (Photo by Bradford Hollingsworth)

258. *Cophosaurus texanus*, Greater Earless Lizard (male), Lagartijón Sordo (macho) (Photo by Julio Lemos-Espinal)

259. *Cophosaurus texanus*, Greater Earless Lizard (male), Lagartijón Sordo (macho) (Photo by Julio Lemos-Espinal)

260. *Holbrookia approximans*, Speckled Earless Lizard (female), Lagartija Sorda Manchada (hembra) (Photo by Julio Lemos-Espinal)

261. *Holbrookia elegans*, Elegant Earless Lizard, Lagartija Sorda Elegante (Photo by Bradford Hollingsworth)

262. *Holbrookia lacerata*, Spot-tailed Earless Lizard, Lagartija Cola Punteada (Photo by Michael Price)

263. *Holbrookia maculata*, Common Lesser Earless Lizard, Lagartija Sorda Pequeña (Photo by Tim Burkhardt)

264. *Holbrookia propinqua*, Keeled Earless Lizard, Lagartija Sorda Carinada (Photo by William Farr)

265. *Petrosaurus mearnsi*, Banded Rock Lizard, Lagarto de Roca Rayado (Photo by Bradford Hollingsworth)

266. *Petrosaurus repens*, Central Baja California Banded Rock Lizard, Lagarto de Roca de Hocico Corto (Photo by Bradford Hollingsworth)

268. *Phrynosoma cerroense*, Cedros Island Horned Lizard, Camaleón de Isla Cedros (Photo by Bradford Hollingsworth)

267. *Petrosaurus slevini*, Slevin's Banded Rock Lizard, Lagarto de Roca Bandeado (Photo by Bradford Hollingsworth)

269. *Phrynosoma cornutum*, Texas Horned Lizard, Camaleón Común (Photo by William Farr)

270. *Phrynosoma ditmarsi*, Rock Horned Lizard, Camaleón de Roca (Photo by Erik Enderson)

271. *Phrynosoma goodei*, Goode's Desert Horned Lizard, Camaleón de Sonora (Photo by Thomas C. Brennan)

272. *Phrynosoma hernandesi*, Greater Short-horned Lizard, Camaleón Cuernitos de Hernández (Photo by Erik Enderson)

273. *Phrynosoma mcallii*, Flat-tailed Horned Lizard, Camaleón de Cola Plana (Photo by Thomas C. Brennan)

274. *Phrynosoma modestum*, Round-tailed Horned Lizard, Camaleón Modesto (Photo by Michael Price)

275. *Phrynosoma orbiculare*, Mountain Horned Lizard, Camaleón de Montaña (Photo by T. Fisher)

276. *Phrynosoma platyrhinos*, Desert Horned Lizard, Camaleón del Desierto (Photo by Bradford Hollingsworth)

277. *Phrynosoma solare*, Regal Horned Lizard, Camaleón Real (Photo by Bradford Hollingsworth)

278. *Sceloporus albiventris*, Western White-bellied Spiny Lizard, Bejore de Vientre Blanco (Photo by Iván Ahumada Carrillo)

279. *Sceloporus bimaculosus*, Twin-spotted Spiny Lizard, Cachora (Photo by Julio Lemos-Espinal)

280. *Sceloporus cautus*, Shy Spiny Lizard, Espinosa Llanera (Photo by T. Fisher)

281. *Sceloporus chaneyi*, Chaney's Bunchgrass Lizard, Espinosa de Chaney (Photo by William Farr)

282. *Sceloporus clarkii*, Clark's Spiny Lizard (male), Bejori de Clark (macho) (Photo by William Wells)

283. *Sceloporus consobrinus*, Prairie Lizard, Lagartija de las Cercas (Photo by William Farr)

284. *Sceloporus couchii*, Couch's Spiny Lizard, Espinosa de las Rocas (Photo by Tim Burkhardt)

285. *Sceloporus cowlesi*, Southwestern Fence Lizard, Lagartija de Cerca del Noroeste (Photo by Bradford Hollingsworth)

286. *Sceloporus cyanogenys*, Bluechinned Rough-scaled Lizard, Lagartija de Barba Azul (Photo by Michael Price)

287. *Sceloporus cyanostictus*, Blue Spiny Lizard (male), Lagarto con Espinas Azules (macho) (Photo by Julio Lemos-Espinal)

288. *Sceloporus goldmani*, Goldman's Bunchgrass Lizard, Lagartija de Pastizal de Goldman (Photo by Jack Sites)

289. *Sceloporus grammicus*, Graphic Spiny Lizard (male), Lagartija de Árbol (macho) (Photo by Terry Hibbitts)

290. *Sceloporus jarrovii*, Yarrow's Spiny Lizard, Lagartija de Yarrow (Photo by Peter Heimes)

291. *Sceloporus jarrovii*, Yarrow's Spiny Lizard (female and male), Lagartija de Yarrow (hembra y macho) (Photo by William Wells)

292. *Sceloporus lemosespinali*, Lemos-Espinal's Spiny Lizard, Lagartija de Lemos-Espinal (Photo by Tim Burkhardt)

293. *Sceloporus lemosespinali*, Lemos-Espinal's Spiny Lizard (female), Lagartija de Lemos-Espinal (hembra) (Photo by Tim Burkhardt)

294. *Sceloporus lemosespinali*, Lemos-Espinal's Spiny Lizard, Lagartija de Lemos-Espinal (Photo by Marisa Ishimatsu)

295. *Sceloporus lemosespinali*, Lemos-Espinal's Spiny Lizard (male), Lagartija de Lemos-Espinal (macho) (Photo by Tim Burkhardt)

296. *Sceloporus maculosus*, Northern Snub-nosed Lizard (male), Lagartija Chata del Norte (macho) (Photo by Tim Burkhardt)

297. *Sceloporus maculosus,* Northern Snub-nosed Lizard (male), Lagartija Chata del Norte (macho) (Photo by Tim Burkhardt)

298. *Sceloporus magister,* Desert Spiny Lizard (female), Lagartija del Desierto (hembra) (Photo by William Wells)

299. *Sceloporus magister,* Desert Spiny Lizard (male), Lagartija del Desierto (macho) (Photo by Jim Rorabaugh)

300. *Sceloporus merriami,* Canyon Lizard (male), Lagartija de Cañón (macho) (Photo by Brad Alexander)

301. *Sceloporus merriami,* Canyon Lizard (male), Lagartija de Cañón (macho) (Photo by Brad Alexander)

302. *Sceloporus minor,* Minor Lizard (male), Lagartija Menor (macho) (Photo by Brad Alexander)

303. *Sceloporus nelsoni,* Nelson's Spiny Lizard, Espinosa de Nelson (Photo by Iván Ahumada Carrillo)

304. *Sceloporus nelsoni,* Nelson's Spiny Lizard, Espinosa de Nelson (Photo by Iván Ahumada Carrillo)

305. *Sceloporus oberon*, Royal Lesser Minor Lizard (male), Lagartija Menor Negra (macho) (Photo by Julio Lemos-Espinal)

306. *Sceloporus oberon*, Royal Lesser Minor Lizard (male), Lagartija Menor Negra (macho) (Photo by Julio Lemos-Espinal)

307. *Sceloporus occidentalis*, Western Fence Lizard, Bejori de Cerca Occidental (Photo by Bradford Hollingsworth)

308. *Sceloporus olivaceus*, Texas Spiny Lizard, Espinosa de los Árboles (Photo by Troy D. Hibbitts)

309. *Sceloporus olivaceus*, Texas Spiny Lizard, Espinosa de los Árboles (Photo by Troy D. Hibbitts)

310. *Sceloporus orcutti*, Granite Spiny Lizard, Espinosa del Granito (Photo by Bradford Hollingsworth)

311. *Sceloporus ornatus*, Ornate Spiny Lizard, Espinosa Ornamentada (Photo by Jack Sites)

312. *Sceloporus parvus*, Blue-bellied Lizard, Lagartija de Panza Azul (Photo by Terry Hibbitts)

313. *Sceloporus parvus*, Blue-bellied Lizard, Lagartija de Panza Azul (Photo by Terry Hibbitts)

314. *Sceloporus poinsettii*, Crevice Spiny Lizard, Lagartija de las Grietas (Photo by Brad Alexander)

315. *Sceloporus samcolemani*, Coleman's Bunchgrass Lizard, Lagartija de Coleman (Photo by Terry Hibbitts)

316. *Sceloporus scalaris,* Light-bellied Bunchgrass Lizard, Lagartija Escamosa (Photo by Terry Hibbitts)

317. *Sceloporus serrifer,* Rough-scaled Lizard, Escamosa del Ocote (Photo by William Farr)

318. *Sceloporus slevini,* Slevin's Bunchgrass Lizard, Espinosa de Pastizal de Slevin (Photo by Thomas C. Brennan)

319. *Sceloporus slevini,* Slevin's Bunchgrass Lizard, Espinosa de Pastizal de Slevin (Photo by Thomas C. Brennan)

320. *Sceloporus spinosus*, Eastern Spiny Lizard, Chintete Espinoso (Photo by William Farr)

321. *Sceloporus torquatus*, Torquate Lizard (male), Espinosa de Collar (macho) (Photo by Julio Lemos-Espinal)

322. *Sceloporus torquatus*, Torquate Lizard (male), Espinosa de Collar (macho) (Photo by Julio Lemos-Espinal)

323. *Sceloporus uniformis*, Yellow-backed Spiny Lizard, Lagartija Espinosa de Espalda-amarilla (Photo by Bradford Hollingsworth)

324. *Sceloporus vandenburgianus*, Southern Sagebrush Lizard, Septentrional de las Salvias (Photo by Bradford Hollingsworth)

325. *Sceloporus variabilis*, Rose-bellied Lizard (female), Panza Azul-Rosada (hembra) (Photo by Troy D. Hibbitts)

326. *Sceloporus variabilis*. Rose-bellied Lizard (female), Panza Azul-Rosada (hembra) (Photo by Julio Lemos-Espinal)

327. *Sceloporus variabilis,* Rose-bellied Lizard (male), Panza Azul-Rosada (macho) (Photo by William Farr)

328. *Sceloporus variabilis,* Rose-bellied Lizard (male), Panza Azul-Rosada (macho) (Photo by Julio Lemos-Espinal)

329. *Sceloporus virgatus,* Striped Plateau Lizard, Lagartija Rayada de la Meseta (Photo by Thomas C. Brennan)

330. *Sceloporus zosteromus,* Baja California Spiny Lizard, Bejori (Photo by Bradford Hollingsworth)

331. *Uma exsul,* Fringe-toed Sand Lizard (male), Lagartija Arenera (macho) (Photo by Julio Lemos-Espinal)

332. *Uma inornata,* Coachella Fringe-toed Lizard, Arenera del Valle Coachella (Photo by Bradford Hollingsworth)

333. *Uma notata,* Colorado Desert Fringe-toed Lizard, Arenera del Desierto del Colorado (Photo by Bradford Hollingsworth)

334. *Uma paraphygas,* Chihuahua Fringe-toed Lizard, Arenera de Chihuahua (Photo by Julio Lemos-Espinal)

335. *Uma rufopunctata,* Yuman Desert Fringe-toed Lizard, Arenera de Manchas Laterales (Photo by Thomas C. Brennan)

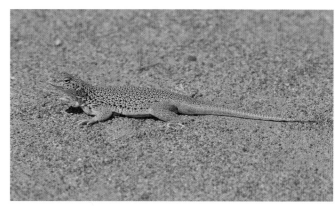

336. *Uma scoparia,* Mohave Fringe-toed Lizard, Arenera del Desierto de Mohave (Photo by Bradford Hollingsworth)

337. *Urosaurus bicarinatus,* Tropical Tree Lizard, Roñito Arborícola (Photo by Susy Sanoja-Sarabia)

338. *Urosaurus graciosus,* Long-tailed Brush Lizard, Roñito de Matorral (Photo by Thomas C. Brennan)

339. *Urosaurus graciosus,* Long-tailed Brush Lizard, Roñito de Matorral (Photo by Thomas C. Brennan)

340. *Urosaurus lahtelai,* Baja California Brush Lizard, Roñito de Matorral Bajacaliforniano (Photo by Bradford Hollingsworth)

341. *Urosaurus nigricaudus,* Black-tailed Brush Lizard, Roñito de Matorral Cola-negra (Photo by Bradford Hollingsworth)

342. *Urosaurus ornatus,* Ornate Tree Lizard, Roñito Ornado (Photo by Bradford Hollingsworth)

343. *Uta nolascensis*, Isla San Pedro Nolasco Lizard, Mancha-lateral de San Pedro Nolasco (Photo by Bradford Hollingsworth)

344. *Uta palmeri*, Isla San Pedro Mártir Side-blotched Lizard, Mancha-lateral de San Pedro Mártir (Photo by Bradford Hollingsworth)

345. *Uta stansburiana*, Common Side-blotched Lizard, Mancha-lateral Común (Photo by Thomas C. Brennan)

346. *Uta stansburiana*, Common Side-blotched Lizard, Mancha-lateral Común (Photo by Thomas C. Brennan)

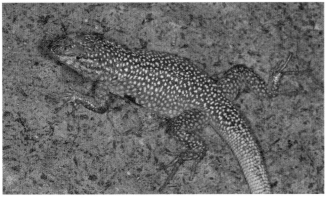

347. *Uta tumidarostra*, Swollen-nosed Side-Blotched Lizard, Mancha-lateral Narigona (Photo by Bradford Hollingsworth)

348. *Plestiodon anthracinus*, Coal Skink, Lincer del Carbón (Photo by Toby J. Hibbitts)

349. *Plestiodon dicei*, Dice's Short-nosed Skink, Alicante de Cola Azul (Photo by Tim Burkhardt)

350. *Plestiodon callicephalus*, Mountain Skink, Lincer de Barranco (Photo by Iván Ahumada Carrillo)

351. *Plestiodon fasciatus*, Five-lined Skink, Lincer de Cinco Líneas (Photo by Toby J. Hibbitts)

352. *Plestiodon fasciatus*, Five-lined Skink, Lincer de Cinco Líneas (Photo by Toby J. Hibbitts)

353. *Plestiodon laticeps*, Broad-headed Skink, Lincer de Cabeza-ancha (Photo by William Farr)

354. *Plestiodon lynxe*, Oak Forest Skink, Lincer de los Encinos (Photo by Terry Hibbitts)

355. *Plestiodon multivirgatus*, Many-lined Skink, Lincer Variable (Photo by Troy D. Hibbitts)

356. *Plestiodon multivirgatus*, Many-lined Skink, Lincer Variable (Photo by Troy D. Hibbitts)

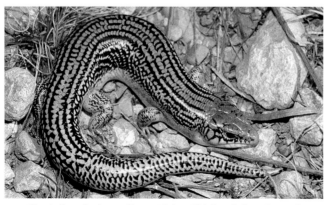

357. *Plestiodon obsoletus*, Great Plains Skink (adult), Lincer de Llanura (adulto) (Photo by Troy D. Hibbitts)

358. *Plestiodon obsoletus*, Great Plains Skink (subadult), Lincer de Llanura (subadulto) (Photo by Jim Rorabaugh)

359. *Plestiodon obsoletus*, Great Plains Skink (juvenile), Lincer de Llanura (juvenil) (Photo by Jeff Servoss)

360. *Plestiodon parviauriculatus*, Northern Pygmy Skink, Lincer Pigmeo Norteño (Photo by Peter Heimes)

361. *Plestiodon septentrionalis*, Prairie Skink, Lincer Norteño (Photo by Toby J. Hibbitts)

362. *Plestiodon skiltonianus*, Western Skink, Lincer Occidental (Photo by Bradford Hollingsworth)

363. *Plestiodon tetragrammus*, Four-lined Skink (adult), Lincer de Cuatro Líneas (adulto) (Photo by Brad Alexander)

364. *Plestiodon tetragrammus*, Four-lined Skink (juvenile), Lincer de Cuatro Líneas (juvenil) (Photo by William Farr)

365. *Scincella lateralis*, Little Brown Skink, Escíncela de Tierra (Photo by Troy D. Hibbitts)

366. *Scincella silvicola*, Taylor´s Ground Skink, Correlón (Photo by William Farr)

367. *Ameiva undulata*, Rainbow Ameiva, Lagartija Metálica (Photo by William Farr)

368. *Aspidoscelis baccata*, San Pedro Nolasco Whiptail, Huico de San Pedro Nolasco (Photo by Bradford Hollingsworth)

369. *Aspidoscelis burti*, Canyon Spotted Whiptail, Huico Manchado del Cañón (Photo by Bradford Hollingsworth)

370. *Aspidoscelis costata*, Western Mexico Whiptail, Huico Llanero (Photo by Bradford Hollingsworth)

371. *Aspidoscelis dixoni*, Gray-checkered Whiptail, Huico Gris (Photo by Toby J. Hibbitts)

372. *Aspidoscelis exsanguis*, Chihuahuan Spotted Whiptail, Corredora de Chihuahua (Photo by Peter Heimes)

373. *Aspidoscelis gularis*, Texas Spotted Whiptail, Corredora Pinta Texana (Photo by William Farr)

374. *Aspidoscelis hyperythra*, Orange-throated Whiptail, Huico Garganta-naranja (Photo by Bradford Hollingsworth)

375. *Aspidoscelis inornata*, Little Striped Whiptail, Huico Liso (Photo by Michael Price)

376. *Aspidoscelis labialis*, Baja California Whiptail, Huico de Baja California (Photo by Bradford Hollingsworth)

377. *Aspidoscelis marmorata*, Western Marbled Whiptail, Huico Marmóreo (Photo by Troy D. Hibbitts)

378. *Aspidoscelis opatae*, Opata Whiptail, Huico Opata (Photo by Bradford Hollingsworth)

379. *Aspidoscelis sonorae*, Sonoran Spotted Whiptail, Huico Manchado de Sonora (Photo by Thomas C. Brennan)

380. *Aspidoscelis tesselata*, Common Checkered Whiptail, Huico Teselado (Photo by Julio Lemos-Espinal)

381. *Aspidoscelis tigris*, Tiger Whiptail, Huico Tigre (Photo by Thomas C. Brennan)

382. *Aspidoscelis tigris*, Tiger Whiptail, Huico Tigre (Photo by Thomas C. Brennan)

383. *Aspidoscelis uniparens*, Desert Grassland Whiptail, Huico de la Pradera del Desierto (Photo by Thomas C. Brennan)

384. *Lepidophyma sylvaticum*, Madrean Tropical Night Lizard, Lagartija Nocturna de Montaña (Photo by Terry Hibbitts)

385. *Xantusia gracilis*, Sandstone Night Lizard, Nocturna de Arenisca (Photo by Bradford Hollingsworth)

386. *Xantusia henshawi*, Granite Night Lizard, Nocturna de Granito (Photo by Bradford Hollingsworth)

387. *Xantusia riversiana*, Island Night Lizard, Nocturna Insular (Photo by Bradford Hollingsworth)

388. *Xantusia vigilis*, Desert Night Lizard, Nocturna del Desierto (Photo by Bradford Hollingsworth)

389. *Xantusia wigginsi*, Wiggins' Night Lizard, Nocturna de Wiggins (Photo by Bradford Hollingsworth)

390. *Xenosaurus platyceps*, Flathead Knob-scaled Lizard, Xenosauro de Cabeza-plana (Photo by Susy Sanoja-Sarabia)

391. *Bipes biporus*, Five-toed worm Lizard, Dos Manos de Cinco Dedos (Photo by Bradford Hollingsworth)

392. *Boa constrictor*, Boa constrictor, Mazacoatl (Photo by Julio Lemos-Espinal)

393. *Lichanura orcutti*, Desert Rosy Boa, Boa del Desierto (Photo by Bradford Hollingsworth)

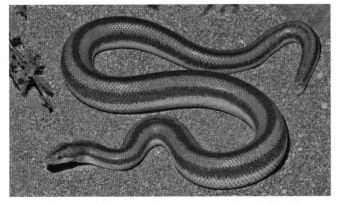

394. *Lichanura trivirgata*, Mexican Rosy Boa, Boa del Desierto Mexicana (Photo by Bradford Hollingsworth)

395. *Arizona elegans*, Glossy Snake, Brillante Arenícola (Photo by William Farr)

396. *Arizona pacata*, Baja California Glossy Snake, Brillante Peninsular (Photo by Bradford Hollingsworth)

397. *Bogertophis rosaliae*, Baja California Ratsnake, Ratonera de Baja California (Photo by Bradford Hollingsworth)

398. *Bogertophis subocularis*, Trans-Pecos Ratsnake, Ratonera de Trans-Pecos (Photo by Brad Alexander)

399. *Cemophora coccinea*, Scarletsnake, Serpiente Escarlata (Photo by Toby J. Hibbitts)

400. *Chilomeniscus stramineus*, Variable Sandsnake, Arenera de Modelo Variable (Photo by Bradford Hollingsworth)

401. *Chionactis occipitalis*, Western Shovel-nosed Snake, Rostro de Pala Occidental (Photo by Thomas C. Brennan)

402. *Chionactis palarostris*, Sonoran Shovel-nosed Snake, Rostro de Pala de Sonora (Photo by Thomas C. Brennan)

403. *Coluber constrictor*, North American Racer, Corredora (Photo by Michael Price)

404. *Coniophanes imperialis*, Regal Black-striped Snake, Culebra de Raya Negra (Photo by William Farr)

405. *Conopsis nasus*, Large-nosed Earthsnake, Culebra de Nariz Grande (Photo by Terry Hibbitts)

406. *Diadophis punctatus*, Ring-necked Snake, Culebra de Collar (Photo by William Wells)

407. *Drymarchon melanurus*, Central American Indigo Snake, Palancacóatls (Photo by Iván Ahumada Carrillo)

408. *Drymobius margaritiferus*, Speckled Racer, Petatillo (Photo by William Farr)

409. *Farancia abacura*, Red-bellied Mudsnake, Serpiente de Fango de Vientre-rojo (Photo by William Farr)

410. *Ficimia olivacea*, Brown Hook-nosed Snake, Nariz de Gancho Huasteca (Photo by William Farr)

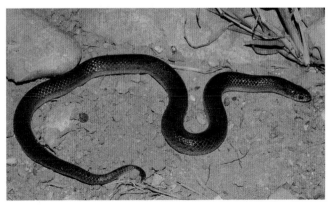

411. *Ficimia streckeri*, Tamaulipan Hook-nosed Snake, Narez de Gancho Tamaulipeca (Photo by William Farr)

412. *Geophis latifrontalis*, Potosí Earth Snake, Minadora de San Luis Potosí (Photo by William Farr)

413. *Gyalopion canum*, Chihuahuan Hook-nosed Snake, Naricilla Chihuahuense (Photo by John E. Werler)

414. *Gyalopion quadrangulare.* Thornscrub Hook-nosed Snake, Naricilla del Desierto (Photo by Thomas C. Brennan)

415. *Heterodon kennerlyi*, Mexican Hog-nosed Snake, Trompa de Cerdo Mexicana (Photo by Peter Heimes)

416. *Heterodon kennerlyi*, Mexican Hog-nosed Snake, Trompa de Cerdo Mexicana (Photo by Peter Heimes)

417. *Heterodon platirhinos*, Eastern Hog-nosed Snake, Trompa de Cerdo Oriental (Photo by Toby J. Hibbitts)

418. *Heterodon platirhinos*, Eastern Hog-nosed Snake, Trompa de Cerdo Oriental (Photo by Toby J. Hibbitts)

419. *Hypsiglena chlorophaea*, Desert Nightsnake, Nocturna Verde-oscuro (Photo by Bradford Hollingsworth)

420. *Hypsiglena jani*, Chihuahuan Nightsnake, Nocturna de Chihuahua (Photo by Thomas C. Brennan)

421. *Hypsiglena ochrorhyncha*, Coast Nightsnake, Nocturna Moteada (Photo by Bradford Hollingsworth)

422. *Hypsiglena slevini*, Slevin's Nightsnake, Nocturna de Baja California (Photo by Bradford Hollingsworth)

423. *Imantodes cenchoa*, Blunthead Tree Snake, Cordelilla Manchada (Photo by William Farr)

424. *Imantodes gemmistratus*, Central American Tree Snake, Cordelilla Escamuda (Photo by Iván Ahumada Carrillo)

425. *Imantodes gemmistratus*, Central American Tree Snake, Cordelilla Escamuda (Photo by Iván Ahumada Carrillo)

426. *Lampropeltis alterna*, Gray-banded Kingsnake, Culebra Real Gris (Photo by Brad Alexander)

427. *Lampropeltis alterna*, Gray-banded Kingsnake, Culebra Real Gris (Photo by Brad Alexander)

428. *Lampropeltis alterna*, Gray-banded Kingsnake, Culebra Real Gris (Photo by John E. Werler)

429. *Lampropeltis calligaster*, Prairie Kingsnake, Culebra Real de Pradera (Photo by Toby J. Hibbitts)

430. *Lampropeltis calligaster*, Prairie Kingsnake, Culebra Real de Pradera (Photo by Toby J. Hibbitts)

431. *Lampropeltis getula*, Common Kingsnake, Barila (Photo by Bradford Hollingsworth)

432. *Lampropeltis getula*, Common Kingsnake, Barila (Photo by Bradford Hollingsworth)

433. *Lampropeltis getula*, Common Kingsnake, Barila (Photo by Brad Alexander)

434. *Lampropeltis getula*, Common Kingsnake, Barila (Photo by John E. Werler)

435. *Lampropeltis knoblochi*, Chihuahuan Mountain Kingsnake, Culebra Real de Chihuahua (Photo by William Wells)

436. *Lampropeltis knoblochi,* Chihuahuan Mountain Kingsnake, Culebra Real de Chihuahua (Photo by Peter Heimes)

437. *Lampropeltis mexicana,* San Luis Potosí Kingsnake, Culebra Real Roja (Photo by Brad Alexander)

438. *Lampropeltis pyromelana,* Arizona Mountain Kingsnake, Culebra Real de Arizona (Photo by Thomas C. Brennan)

439. *Lampropeltis triangulum,* Milksnake, Falsa Coralillo (Photo by John E. Werler)

440. *Lampropeltis triangulum,* Milksnake, Falsa Coralillo (Photo by Brad Alexander)

441. *Lampropeltis triangulum,* Milksnake, Falsa Coralillo (Photo by Matthew Cage)

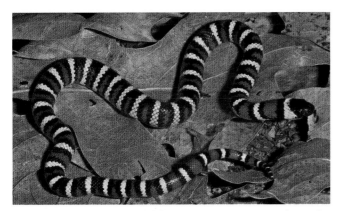

442. *Lampropeltis zonata,* California Mountain Kingsnake, Culebra Real de California (Photo by Bradford Hollingsworth)

443. *Leptodeira maculata,* Southwestern Cat-eyed Snake, Escombrera del Suroeste Mexicano (Photo by William Farr)

444. *Leptodeira punctata*, Western Cat-eyed Snake, Escombrera del Occidente (Photo by Bradford Hollingsworth)

445. *Leptodeira septentrionalis*, Northern Cat-eyed Snake, Escombrera Manchada (Photo by William Farr)

446. *Leptodeira splendida*, Splendid Cat-eyed Snake, Escombrera Sapera (Photo by Iván Ahumada Carrillo)

447. *Leptophis diplotropis*, Pacific Coast Parrot Snake, Ratonera de la Costa del Pacífico (Photo by Peter Heimes)

448. *Leptophis mexicanus*, Mexican Parrot Snake, Ranera Mexicana (Photo by Julio Lemos-Espinal)

449. *Masticophis bilineatus*, Sonoran Whipsnake, Látigo de Sonora (Photo by Bradford Hollingsworth)

450. *Masticophis bilineatus*, Sonoran Whipsnake, Látigo de Sonora (Photo by Brad Alexander)

451. *Masticophis flagellum*, Coachwhip, Chirrionera (Photo by Troy D. Hibbitts)

452. *Masticophis flagellum*, Coachwhip, Chirrionera (Photo by Troy D. Hibbitts)

453. *Masticophis fuliginosus*, Baja California Coachwhip, Chirrionera de Baja California (Photo by Bradford Hollingsworth)

454. *Masticophis lateralis*, Striped Racer, Corredora Rayada (Photo by Bradford Hollingsworth)

455. *Masticophis mentovarius*, Neotropical Whipsnake, Sabanera (Photo by Iván Ahumada Carrillo)

456. *Masticophis schotti*, Schott's Whipsnake, Látigo de Schott (Photo by William Farr)

457. *Masticophis taeniatus*, Striped Whipsnake, Látigo Rayada (Photo by John E. Werler)

458. *Mastigodryas cliftoni*, Clifton's Lizard, Eater, Lagartijera de Clifton (Photo by Iván Ahumada Carrillo)

459. *Nerodia clarki*, Saltmarsh Watersnake, Culebra de Agua de Clark (Photo by Toby J. Hibbitts)

460. *Nerodia cycopion*, Mississippi Green Watersnake, Culebra de Agua del Mississippi (Photo by Toby J. Hibbitts)

461. *Nerodia erythrogaster*, Plain-bellied Watersnake, Culebra de Agua de Vientre Plano (Photo by John E. Werler)

462. *Nerodia fasciata*, Southern Watersnake, Culebra de Agua del Sur de Estados Unidos (Photo by William Farr)

463. *Nerodia harteri*, Harter's Watersnake, Culebra de Agua de Harter (Photo by Toby J. Hibbitts)

464. *Nerodia rhombifer*, Diamond-backed Watersnake, Culebra de Agua Diamantada (Photo by John E. Werler)

465. *Opheodrys aestivus*, Rough Greensnake, Estival Rugosa (Photo by William Farr)

466. *Oxybelis aeneus*, Neotropical Vinesnake, Bejuquilla Neotropical (Photo by Peter Heimes)

467. *Oxybelis aeneus*, Neotropical Vinesnake, Bejuquilla Neotropical (Photo by Thomas C. Brennan)

468. *Pantherophis bairdi,* Baird's Ratsnake, Ratonera de Bosque (Photo by John E. Werler)

469. *Pantherophis emoryi,* Great Plains Ratsnake, Ratonera de Emory (Photo by William Farr)

470. *Pantherophis guttatus,* Red Cornsnake, Ratonera Roja (Photo by Toby J. Hibbitts)

472. *Pantherophis obsoletus,* Texas Ratsnake, Ratonera Texana (Photo by Toby J. Hibbitts)

471. *Phyllorhynchus browni,* Saddled Leaf-nosed Snake, Culebra Ensillada (Photo by Bradford Hollingsworth)

473. *Phyllorhynchus browni,* Saddled Leaf-nosed Snake, Culebra Ensillada (Photo by Young Cage)

474. *Phyllorhynchus decurtatus,* Spotted Leaf-nosed Snake, Culebra Nariz Moteada (Photo by Bradford Hollingsworth)

475. *Phyllorhynchus decurtatus*, Spotted Leaf-nosed Snake, Culebra Nariz Moteada (Photo by Thomas C. Brennan)

476. *Pituophis catenifer*, Gopher Snake, Cincuate Casero (Photo by Thomas C. Brennan)

477. *Pituophis deppei*, Mexican Bullsnake, Cincuate Mexicano (Photo by William Farr)

478. *Pituophis insulanus*, Isla Cedros Gophersnake, Cincuate de Isla Cedros (Photo by Bradford Hollingsworth)

479. *Pituophis ruthveni*, Louisiana Pinesnake, Cincuate de Louisiana (Photo by Toby J. Hibbitts)

480. *Pituophis vertebralis*, Baja California Gophersnake, Cincuate de San Lucas (Photo by Bradford Hollingsworth)

481. *Pliocercus bicolor*, Northern False Coral Snake, Falsa Coral del Norte (Photo by William Farr)

482. *Pseudelaphe flavirufa*, Tropical Ratsnake, Ratonera del Trópico (Photo by William Farr)

483. *Pseudoficimia frontalis,* False Ficimia, Ilamacoa (Photo by Matthew Cage)

484. *Pseudoficimia frontalis,* False Ficimia, Ilamacoa (Photo by Matthew Cage)

485. *Regina grahami,* Graham's Crayfish Snake, Regina de Graham (Photo by Toby J. Hibbitts)

486. *Regina rigida,* Glossy Crayfish Snake, Regina Brillante (Photo by Toby J. Hibbitts)

487. *Rhadinaea gaigeae,* Gaige's Pine Forest Snake, Hojaras-quera de Gaige (Photo by William Farr)

488. *Rhinocheilus lecontei,* Long-nosed Snake, Culebra Nariz-larga (Photo by Peter Heimes)

489. *Rhinocheilus lecontei,* Long-nosed Snake, Culebra Nariz-larga (Photo by William Farr)

490. *Rhinocheilus lecontei,* Long-nosed Snake, Culebra Nariz-larga (Photo by Bradford Hollingsworth)

491. *Rhinocheilus lecontei*, Long-nosed Snake, Culebra Nariz-larga (Photo by Jim Rorabaugh)

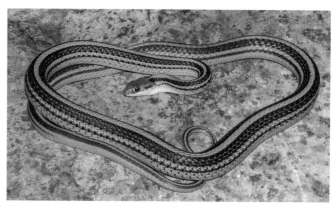

492. *Salvadora bairdii*, Baird's Patch-nosed Snake, Culebra Chata de Baird (Photo by Iván Ahumada Carrillo)

493. *Salvadora bairdii*, Baird's Patch-nosed Snake, Culebra Chata de Baird (Photo by Iván Ahumada Carrillo)

494. *Salvadora deserticola*, Big-Bend Patch-nosed Snake, Cabestrillo del Big-Bend (Photo by Troy D. Hibbitts)

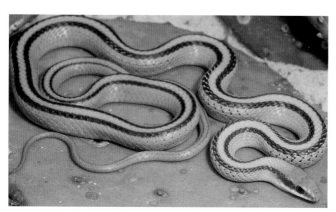

495. *Salvadora grahamiae*, Eastern Patch-nosed Snake, Culebra Chata de Montaña (Photo by John E. Werler)

496. *Salvadora grahamiae*, Eastern Patch-nosed Snake, Culebra Chata de Montaña (Photo by John E. Werler)

497. *Salvadora hexalepis*, Western Patch-nosed Snake, Cabestrillo (Photo by Bradford Hollingsworth)

498. *Scaphiodontophis annulatus*, Guatemala Neck-banded Snake, Culebra Añadida de Guatemala (Photo by Peter Heimes)

499. *Senticolis triaspis*, Green Ratsnake, Culebra Ratonera Verde (Photo by William Farr)

500. *Sonora aemula*, File-tailed Grand Snake, Culebra de Tierra Cola Plana(Photo by Young Cage)

501. *Sonora aemula*, Fire-tailed Grand Snake, Culebra de Tierra Cola Plana (Photo by Julio Lemos-Espinal)

502. *Sonora semiannulata*, Western Groundsnake, Culebrilla de Tierra (Photo by Brad Alexander)

503. *Sonora semiannulata*, Western Groundsnake, Culebrilla de Tierra (Photo by Troy D. Hibbitts)

504. *Sonora semiannulata*, Western Groundsnake, Culebrilla de Tierra (Photo by Young Cage)

505. *Sonora semiannulata*, Western Groundsnake, Culebrilla de Tierra (Photo by John E. Werler)

506. *Sonora semiannulata*, Western Groundsnake, Culebrilla de Tierra (Photo by John E. Werler)

507. *Sonora semiannulata*, Western Groundsnake, Culebrilla de Tierra (Photo by William Wells)

508. *Sonora semiannulata*, Western Groundsnake, Culebrilla de Tierra (Photo by Michael Price)

509. *Sonora semiannulata*, Western Groundsnake, Culebrilla de Tierra (Photo by Brad Alexander)

510. *Sonora semiannulata*, Western Groundsnake, Culebrilla de Tierra (Photo by Troy D. Hibbitts)

511. *Sonora semiannulata*, Western Groundsnake, Culebrilla de Tierra (Photo by William Wells)

512. *Sonora semiannulata*, Western Groundsnake, Culebrilla de Tierra (Photo by Bradford Hollingsworth)

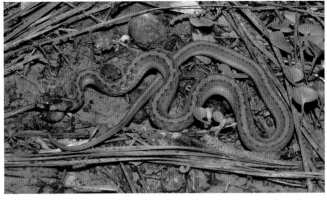

513. *Storeria dekayi*, Dekay's Brownsnake, Culebra Parda (Photo by William Farr)

514. *Storeria hidalgoensis*, Mexican Yellow-bellied Brownsnake, Culebra de Cuello Blanco (Photo by Tim Burkhardt)

515. *Storeria occipitomaculata*, Red-bellied Snake, Serpiente de Vientre-Rojo (Photo by Toby J. Hibbitts)

516. *Storeria storerioides*, Mexican Brownsnake, Culebra Parda Mexicana (Photo by Julio Lemos-Espinal)

517. *Sympholis lippiens*, Mexican Short-tailed Snake, Culebra Cola Corta Mexicana (Photo by Iván Ahumada Carrillo)

518. *Tantilla atriceps*, Mexican Black-headed Snake, Culebrilla de Cabeza Negra (Photo by John E. Werler)

519. *Tantilla gracilis*, Flat-headed Snake, Culebra Cabeza-plana (Photo by Troy D. Hibbitts)

520. *Tantilla hobartsmithi*, Smith's Black-headed Snake, Culebra Cabeza Negra del Norte (Photo by Randy Babb)

521. *Tantilla nigriceps*, Plains Black-headed Snake, Culebra Cabeza Negra de los Llanos (Photo by John E. Werler)

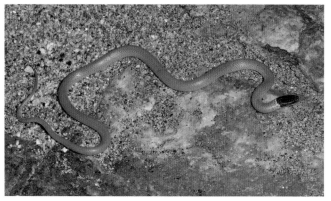

522. *Tantilla planiceps*, Western Black-headed Snake, Culebra Cabeza Negra Occidental (Photo by Bradford Hollingsworth)

523. *Tantilla planiceps*, Western Black-headed Snake, Culebra Cabeza Negra Occidental (Photo by Brad Alexander)

524. *Tantilla rubra*, Red Black-headed Snake, Rojilla (Photo by Terry Hibbitts)

525. *Tantilla wilcoxi*, Chihuahuan Black-headed Snake, Centipedívora de Chihuahua (Photo by Thomas C. Brennan)

526. *Tantilla yaquia*, Yaqui Black-headed Snake, Culebra Cabeza Negra Yaqui (Photo by Brad Alexander)

527. *Thamnophis cyrtopsis*, Black-necked Gartersnake, Jarretera Cuello-negro (Photo by Peter Heimes)

528. *Thamnophis elegans*, Terrestrial Gartersnake, Jarretera Terrestre Occidental (Photo by Alan DeQueiroz)

529. *Thamnophis eques*, Mexican Gartersnake, Jarretera Mexicana (Photo by Bradford Hollingsworth)

530. *Thamnophis exsul*, Exile Mexican Gartersnake, Jarretera Mexicana Exiliada (Photo by Tim Burkhardt)

531. *Thamnophis exsul*, Exile Mexican Gartersnake, Jarretera Mexicana Exiliada (Photo by Tim Burkhardt)

532. *Thamnophis hammondii*, Two-striped Gartersnake, Jarretera de Hammond (Photo by Bradford Hollingsworth)

533. *Thamnophis marcianus*, Checkered Gartersnake, Sochuate (Photo by Thomas C. Brennan)

534. *Thamnophis melanogaster*, Mexican Black-bellied Gartersnake, Jarretera Vientre-negro Mexicana (Photo by Erik Enderson)

535. *Thamnophis proximus*, Western Ribbonsnake, Jarretera Occidental (Photo by William Farr)

536. *Thamnophis pulchrilatus*, Mexican Highland Gartersnake, Jarretera Mexicana del Altiplano (Photo by William Farr)

537. *Thamnophis sirtalis*, Common Gartersnake, Jarretera Común (Photo by Toby J. Hibbitts)

538. *Thamnophis unilabialis*, Madrean Narrow-headed Gartersnake, Jarretera Cabeza-angosta de Chihuahua (Photo by Randy Jennings)

539. *Thamnophis validus*, Mexican West Coast Gartersnake, Jarretera Mexicana del Pacífico (Photo by Francisco García Camino)

540. *Trimorphodon lambda*, Sonoran Lyresnake, Ilimacoa de Sonora (Photo by Bradford Hollingsworth)

541. *Trimorphodon lyrophanes*, California Lyresnake, Ilamacoa de California (Photo by Bradford Hollingsworth)

542. *Trimorphodon tau*, Mexican Lyresnake, Falsa Nauyaca Mexicana (Photo by Brad Alexander)

543. *Tropidoclonion lineatum*, Lined Snake, Serpiente Lineada (Photo by Toby J. Hibbitts)

544. *Tropidodipsas fasciata*, Banded Snail Sucker, Caracolera Anillada (Photo by William Farr)

545. *Tropidodipsas repleta*, Banded Blacksnake, Caracolera Repleta (Photo by Young Cage)

546. *Tropidodipsas sartorii*, Terrestrial Snail Sucker, Caracolera Terrestre (Photo by William Farr)

547. *Virginia striatula*, Rough Earthsnake, Serpiente Rugosa de Tierra (Photo by William Farr)

548. *Micruroides euryxanthus*, Sonoran Coralsnake, Coralillo Occidental (Photo by Randy Babb)

549. *Micrurus distans*, West Mexican Coralsnake, Coralillo Bandas Claras (Photo by Tim Burkhardt)

550. *Micrurus tamaulipensis*, Sierra de Tamaulipas Coralsnake, Coralillo de la Sierra de Tamaulipas (Photo by William Farr)

551. *Micrurus tener*, Texas Coralsnake, Coralillo Texano (Photo by Brad Alexander)

552. *Hydrophis platurus,* Yellow-bellied Seasnake, Culebra del Mar (Photo by Peter Heimes)

553. *Rena dissecta,* New Mexico Threadsnake, Culebrilla Ciega de Nuevo México (Photo by Thomas C. Brennan)

554. *Rena dissecta,* New Mexico Threadsnake, Culebrilla Ciega de Nuevo México (Photo by Thomas C. Brennan)

555. *Rena dissecta,* New Mexico Threadsnake, Culebrilla Ciega de Nuevo México (Photo by Thomas C. Brennan)

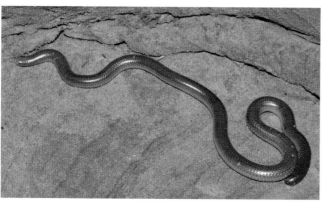

556. *Rena dulcis,* Texas Threadsnake, Culebrilla Ciega Texana (Photo by William Farr)

557. *Rena dulcis,* Texas Threadsnake, Culebrilla Ciega Texana (Photo by Troy D. Hibbitts)

558. *Rena humilis,* Western Threadsnake, Culebra Lombriz (Photo by Peter Heimes)

559. *Rena humilis,* Western Threadsnake, Culebra Lombriz (Photo by Bradford Hollingsworth)

560. *Rena humilis*, Western Threadsnake, Culebra Lombriz (Photo by Brad Alexander)

561. *Rena myopica*, Tampico Threadsnake, Culebrilla Ciega de Tampico (Photo by William Farr)

562. *Rena segrega*, Chihuahuan Threadsnake, Culebra Lombriz de Trans-Pecos (Photo by John E. Werler)

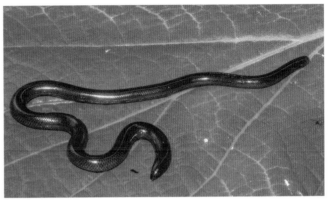

563. *Indotyphlops braminus*, Brahminy Blindsnake, Culebrilla Ciega (Photo by Iván Ahumada Carrillo)

564. *Indotyphlops braminus*, Brahminy Blindsnake, Culebrilla Ciega (Photo by Iván Ahumada Carrillo)

565. *Agkistrodon bilineatus*, Cantil, Cantil (Photo by Iván Ahumada Carrillo)

566. *Agkistrodon contortrix*, Copperhead, Mocasín (Photo by Brad Alexander)

567. *Agkistrodon piscivorus*, Cottonmouth, Boca de Algodón (Photo by Toby J. Hibbitts)

568. *Agkistrodon taylori*, Taylor's Cantil, Metapil (Photo by William Farr)

569. *Agkistrodon taylori*, Taylor's Cantil, Metapil (Photo by Alfonso Delgadillo Espinosa)

570. *Bothrops asper*, Terciopelo, Cuatro Narices (Photo by Peter Heimes)

571. *Crotalus angelensis*, Isla Ángel de la Guarda Rattlesnake, Cascabel de Isla Ángel de la Guarda (Photo by Bradford Hollingsworth)

572. *Crotalus atrox*, Western Diamond-backed Rattlesnake, Cascabel de Diamantes (Photo by Bradford Hollingsworth)

573. *Crotalus basiliscus*, Mexican West Coast Rattlesnake, Saye (Photo by Iván Ahumada Carrillo)

574. *Crotalus caliginis*, Islas Coronado Rattlesnake, Cascabel de Isla Coronado (Photo by Bradford Hollingsworth)

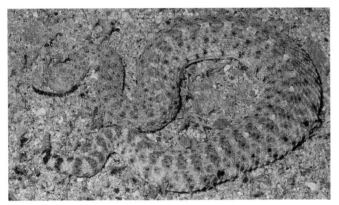

575. *Crotalus cerastes*, Sidewinder, Víbora Cornuda (Photo by Bradford Hollingsworth)

576. *Crotalus cerastes*, Sidewinder, Víbora Cornuda (Photo by Barney Oldfield)

577. *Crotalus enyo*, Baja California Rattlesnake, Cascabel de Baja California (Photo by Bradford Hollingsworth)

578. *Crotalus estebanensis*, Isla San Esteban Black-tailed Rattlesnake, Cascabel de San Esteban (Photo by Bradford Hollingsworth)

579. *Crotalus horridus*, Timber Rattlesnake, Cascabel de Bosque (Photo by William Farr)

580. *Crotalus lepidus*, Rock Rattlesnake, Cascabel Verde (Photo by William Farr)

581. *Crotalus lepidus*, Rock Rattlesnake, Cascabel Verde (Photo by Iván Ahumada Carrillo)

582. *Crotalus lorenzoensis*, San Lorenzo Island Rattlesnake, Cascabel de San Lorenzo (Photo by Bradford Hollingsworth)

583. *Crotalus mitchellii*, Speckled Rattlesnake, Víbora Blanca (Photo by Bradford Hollingsworth)

584. *Crotalus molossus*, Black-tailed Rattlesnake, Cola Prieta (Photo by Michael Price)

585. *Crotalus molossus*, Black-tailed Rattlesnake, Cola Prieta (Photo by Iván Ahumada Carrillo)

586. *Crotalus oreganus*, Western Rattlesnake, Serpiente de Cascabel Occidental (Photo by Bradford Hollingsworth)

587. *Crotalus pricei*, Twin-spotted Rattlesnake, Cascabel de Manchas-gemelas (Photo by Brad Alexander)

588. *Crotalus pricei*, Twin-spotted Rattlesnake, Cascabel de Manchas-gemelas (Photo by Iván Ahumada Carrillo)

589. *Crotalus scutulatus*, Mojave Rattlesnake, Chiauhcoatl (Photo by Julio Lemos-Espinal)

590. *Crotalus tigris*, Tiger Rattlesnake, Cascabel Tigre (Photo by Jim Rorabaugh)

591. *Crotalus totonacus*, Totonacan Rattlesnake, Tepocolcoatl (Photo by William Farr)

592. *Crotalus viridis*, Prairie Rattlesnake, Serpiente de Cascabel de la Pradera (Photo by Julio Lemos-Espinal)

593. *Crotalus willardi*, Ridge-nosed Rattlesnake, Cascabel de Nariz-afilada (Photo by Robert Bryson)

594. *Crotalus willardi*, Ridge-nosed Rattlesnake, Cascabel de Nariz-afilada (Photo by Marisa Ishimatsu)

595. *Crotalus willardi*, Ridge-nosed Rattlesnake, Cascabel de Nariz-afilada (Photo by Young Cage)

596. *Crotalus willardi*, Ridge-nosed Rattlesnake, Cascabel de Nariz-afilada (Photo by Matthew Cage)

597. *Crotalus willardi,* Ridge-nosed Rattlesnake, Cascabel de Nariz-afilada (Photo by Young Cage)

598. *Sistrurus catenatus,* Massasauga Rattlesnake, Cascabel de Massasauga (Photo by Thomas C. Brennan)

599. *Sistrurus catenatus,* Massasauga Rattlesnake, Cascabel de Massasauga (Photo by John E. Werler)

600. *Sistrurus miliarius,* Pygmy Rattlesnake, Cascabel Pigmea (Photo by Toby J. Hibbitts)

Anfibios y reptiles

de los estados de la frontera

México–Estados Unidos

Introducción:
Cómo Usar Este Libro

JULIO A. LEMOS-ESPINAL

La diversidad biológica de los estados a ambos lados de la frontera México-Estados Unidos ha llamado la atención de biólogos aun antes de la creación de frontera como hoy se conoce. Dos de las clases de vertebrados terrestres más estudiadas en esta región son los anfibios y reptiles. Observaciones sobre su historia natural y distribución han sido acumuladas desde que la United States and Mexican Boundary Survey fue creada en 1848 para determinar la frontera entre los dos países como lo definió el Tratado de Guadalupe Hidalgo que terminó la Guerra México-Estados Unidos. Varios estudios monográficos sobre anfibios y reptiles han sido publicados para la mayoría de los 10 estados que forman la frontera entre México y los Estados Unidos: Dixon (1987: Texas), Degenhardt et al. (1996: Nuevo México), Grismer (2002: Baja California), Brennan y Holycross (2006: Arizona), Lemos-Espinal y Smith (2007a, 2007b: Chihuahua y Coahuila), y Stebbins y McGinnis (2012: California).

Este libro resume lo que se conoce sobre la distribución de anfibios y reptiles en cada uno de los 10 estados a lo largo de la frontera México-Estados Unidos; proporciona una lista actualizada de anfibios y reptiles que han sido registrados en cada uno de estos estados; y analiza cuáles de estas especies se comparten entre los dos países, cuáles habitan únicamente al norte

de la frontera en los Estados Unidos, y cuáles habitan únicamente al sur de la frontera en México.

El Capítulo 2, "Notas sobre Nomenclatura y Distribución," es una justificación y discusión de los nombres científicos y estándar que son utilizados en este libro. En años recientes ha habido un número de cambios en estos nombres, y es posible que el lector no reconozca el uso de algunos nombres mencionados. Los nombres utilizados en este libro han sido estandarizados de acuerdo a los nombres utilizados en este capítulo.

Los capítulos del 3 al 12 son las descripciones de la herpetofauna para cada uno de los 10 estados de la frontera. Los estados de la frontera mexicana se encuentran primero de oeste a este (Baja California a Tamaulipas), seguidos por los estados de la frontera estadounidense, también de oeste a este (California a Texas). Cada uno de estos capítulos sigue el mismo formato:

1. Una pequeña introducción que generalmente inicia con la descripción de las características generales del estado, incluyendo la riqueza específica de anfibios y reptiles y la importancia de estas dos clases de vertebrados.
2. Una sinopsis de los estudios herpetológicos previos que generalmente comienza con la descripción del trabajo realizado por las

expediciones de la United States Boundary
Commission a la región de la frontera entre
los años 1848 y 1852 para posteriormente
revisar el trabajo subsecuente, y termina con
una discusión de las investigaciones e inventarios actuales. También incluye un cuadro
que enlista las localidades tipo y autores de
anfibios y reptiles descritos para el estado.

3. Características fisiográficas y su influencia
   en la distribución de la herpetofauna. Esta
   sección proporciona una revisión de las
   comunidades bióticas del estado así como
   descripciones de geografía, hidrología, clima
   e historia geológica que moldearon los ambientes en los cuales los anfibios y reptiles
   actuales viven. También proporciona un contexto para la rica biodiversidad de cada estado
   y promueve un entendimiento de los patrones
   actuales de distribución herpetológica y ayuda
   al entendimiento de los patrones actuales de
   distribución de la herpetofauna y sus afinidades. La discusión incluye el número de
   especies de anfibios y reptiles, número de
   especies endémicas al estado, especies que
   se distribuyen sólo al sur (México) o al norte
   (Estados Unidos) de la frontera, así como
   especies que se comparten por ambos países,

y especies que pueden habitar en el estado.
Problemas de conservación para anfibios y
reptiles en el estado también son discutidos.

Todos los capítulos de los estados proporcionan
una lista de los anfibios y reptiles que habitan
en el estado. Todas esas listas están resumidas
en el anexo. El Capítulo 13, "Diversidad de la
Herpetofauna de los Estados de la Frontera
México-Estados Unidos," analiza los datos del
anexo agrupando a las especies por clases,
órdenes, especies terrestres y especies nativas.
Éste discute el tamaño de la superficie territorial
de los estados y la diversidad ambiental con la
riqueza de especies de anfibios y reptiles para
cada estado y la región entera.

Debido a que este libro está escrito en inglés y
español, está dividido en dos partes. La primera
parte es la versión en inglés, que es el idioma
en que este libro fue escrito originalmente; la
segunda parte es la versión en español. Estas dos
partes están separadas por fotos que ilustran la
mayoría de las especies que habitan la región de
la frontera. Cada foto se presenta con el nombre
científico de la especie, los nombres estándar en
inglés y español, y el nombre del fotógrafo.

JULIO A. LEMOS-ESPINAL,
BRADFORD D. HOLLINGSWORTH,
Y WILLIAM FARR

Comentarios sobre Nombre Científicos

Se ha argumentado que estamos viviendo en una época de caos taxonómico (Pauly et al. 2009), con la incertidumbre en la estabilidad de nombres estándar o aun científicos y la validez de diversas justificaciones para la adopción de un nombre u otro. Para evitar contribuir a este caos, en este libro hemos sido tan consistentes como ha sido posible con la literatura actual. Sin embargo, en algunos casos hemos considerado necesario en el interés de la precisión y el entendimiento adoptar ciertos cambios de nombre. En esta sección damos una explicación corta de las razones para el uso de nombres científicos o estándar en inglés y español que aparecen en este libro. Ninguno está totalmente aceptado, pero su uso será un factor importante para lograr la constancia en su aceptación.

Batrachoseps altasierrae y B. bramei

Jockush et al. (2012) describieron dos especies nuevas de salamandras delgadas (*Batrachoseps altasierrae* y *B. bramei*) de la región del sur de Sierra Nevada, California.

Siren lacertina

Flores-Villela y Brandon (1992) concluyeron que dos especies distintas de *Siren* habitan en el Valle del Río Bravo de Texas y Tamaulipas, una forma pequeña que asignaron a *S. intermedia nettingi* (31–36 surcos costales, diámetro de huevo no mayor a 2.5 mm), y una forma más grande que asignaron a *S. lacertina* (36–40 surcos costales, diámetro de huevo alcanzando 3 mm), y colocaron a *S. intermedia texana* en la sinonimia de *S. intermedia nettingi*. Flores-Villela y Brandon (1992) indicaron que aunque no pudieron encontrar ninguna diferencia para separar a la población de individuos más grandes del Río Bravo de la población de *S. lacertina* del sureste de los Estados Unidos, su asignación de la población de individuos más grandes del Río Bravo a *S. lacertina* fue tentativa, señalando un vacío en la distribución de la especie entre poblaciones en Alabama y el Valle del Río Bravo. Considerando este vacío distribucional de más de 1000 kilómetros que se expande sobre varios sistemas ribereños importantes, nosotros interpretamos que la población de individuos más grandes del género *Siren* en el Valle del Río Bravo es un taxón aún no descrito (*Siren* sp. indet.) en lugar de una población aislada de *S. lacertina*. Una mejor clarificación del estatus del género *Siren* del Río Bravo probablemente requerirá la recolecta de muestras grandes para ser analizadas morfológica y molecularmente.

Incilius mccoyi e I. occidentalis

Recientemente Santos-Barrera y Flores-Villela (2011) describieron una nueva especie (*Incilius mccoyi*) que anteriormente era considerada como parte de *I. occidentalis*, de la Sierra Madre Occidental de Chihuahua.

Gastrophryne mazatlanensis y G. olivacea

En un análisis de los clados disjuntos del este y del oeste de *Gastrophryne olivacea*, Streicher et al. (2012) elevaron el clado del oeste (Arizona, Sonora y suroeste de Chihuahua) a *G. mazatlanensis*. En el resto de su distribución este anuro permanece como *G. olivacea*.

Trachycephalus typhonius

Lavilla et al. (2010) revisaron la literatura y una traducción reciente del diario de Daniel Rolander (Hansen 2008), el colector de *Rana typhonia*, una especie originalmente nombrada por Linnaeus en 1758. Proporcionaron evidencia que demuestra que esta especie no es ni un bufónido ni un ranido asiático, sino el hylido neotropical *Trachycephalus venulosus*. Consideraron a *Rana typhonia* como una sinonimia mayor de *Rana venulosa* (= *Trachycephalus venulosus*), y re describieron su holotipo bajo la nueva combinación *Trachycephalus typhonius*, el cual es el nombre utilizado en este libro.

Gopherus agassizii y G. morafkai

Murphy et al. (2011) investigaron los problemas asociados con la identidad de la Tortuga de Desierto (*G. agassizii*), incluyendo: fecha de publicación de la descripción original, localidad tipo, destino de la serie tipo, descripción de otras especies para la región del Cabo de Baja California Sur, asignación de la especie a diferentes géneros, arreglo taxonómico de la especie como una subespecie de la Tortuga del Bolsón de Mapimí, y la existencia de más de una especie de Tortuga de Desierto. Después de analizar todos estos problemas usando ADN genómico, encontraron que un conjunto de características sirve para diferenciar las tortugas al oeste y norte del Río Colorado, la población del Desierto de Mojave, de las tortugas del este y sur de este río en Arizona, Sonora, extremo suroeste de Chihuahua y Sinaloa. Concluyeron que el rango de especie se justifica para estas dos poblaciones: *G. agassizii* para la población del Desierto de Mojave; y *G. morafkai* para la población del Desierto de Sonora. Por lo que *G. agassizii* es una especie que se encuentra exclusivamente en los Estados Unidos, y *G. morafkai* es una especie compartida entre México y los Estados Unidos.

Gerrhonotus taylori

Ésta es una especie cercanamente relacionada con *Gerrhonotus liocephalus*, de la que fue considerada como una sinonimia sin rango por Good (1988). Sin embargo, las características señaladas por Tihen (1954) distinguen a estos dos taxa. Es posible que todos los reportes aislados de *Gerrhonotus* para el oeste de México (Good 1988) sean de *G. taylori*. El género parece ser raro en esa área, con insuficiente material disponible para asignarlos taxonómicamente con confianza. Good (1994) elevó *G. infernalis* de la sinonimia con *G. liocephalus*, y colocó a *G. taylori* en su sinonimia. Consideramos a *G. taylori* como una especie válida en este libro.

Phyllodactylus nolascoensis

Dos subespecies han sido reconocidas para *Phyllodactylus homolepidurus*: *P. h. homolepidurus* del occidente de México y *P. h. nolascoensis* de la Isla San Pedro Nolasco del Mar de Cortés de Sonora. Estas dos subespecies se pueden distinguir sobre la base de diferencias en el número de tubérculos paravertebrales (31–41 en *P. h. homolepidurus* y 41–48 en *P. h. nolascoensis*), e hileras longitudinales de tubérculos dorsales (12–14 en *P. h. homolepidurus* y 14–16 en *P. h. nolascoensis*) (Dixon 1964). Grismer (1999) colocó a ambas subespecies en la sinonimia de *P. homolepidurus*, pero comentó sobre su potencial de ser especies diferentes. Debido a que *P. h. nolascoensis* tiene características únicas, está restringido a una isla y tiene poca probabilidad de hibridar con la forma continental, consideramos a éste con el rango de especie, *P. nolascoensis*.

Heloderma exasperatum

Basándose en el estudio de Douglas et al. (2010), Reiserer et al. (2013) elevaron cuatro subespecies de *Heloderma horridum* al rango de especie, incluyendo *H. horridum exasperatum* del sur de Sonora y suroeste de Chihuahua.

Sceloporus variabilis

Mendoza Quijano et al. (1998) revisaron los límites de la especie y relaciones filogenéticas de la especie-grupo *Sceloporus variabilis*, el cual incluye entre otros a *S. v. variabilis* (sur de Tamaulipas) y *S. v. marmoratus* (Nuevo León, Coahuila, Tamaulipas, Texas) y consideraron a *S. marmoratus* con el rango de especie. Sin embargo, debido a que *S. v. marmoratus*, un taxón con una distribución relativamente amplia que habita desde el nivel del mar hasta 1740 m de altitud, estuvo representado únicamente por muestras de una localidad remota en su revisión, y debido a que los datos morfológicos apoyan la intergradación entre los taxa de este grupo, varios autores no han seguido este arreglo (ejem. Collins y Taggart 2009; Crother 2008; Dixon 2000; Köhler 2002; Lavín-Murcio et al. 2005; Lemos-Espinal y Smith 2007; Utez 2012). En este libro no reconocemos a *S. v. marmoratus* como una especie distinta de *S. v. variabilis*.

Polychrotidae y Dactyloidae

Townsend et al. (2011) aclararon la historia evolutiva de iguánidos infiriendo filogenias utilizando análisis concatenados de máxima probabilidad (MP) y bayesianos de secuencias de ADN para 29 genes nucleares que codifican proteínas para 47 taxa de iguánidos y 29 de grupos externos.

En Pleurodontos encontraron una fuerte MP apoyada por algunas relaciones interfamiliares y por monofilia para casi todas las familias (excepto Polychrotidae). Corrigieron la ausencia de monofilia en Polychrotidae reconociendo al género pleurodonto *Anolis* (sensu lato) como una familia separada (Dactyloidae) la cual incluye los géneros *Anolis*, *Chamaeleolis*, *Chamaelinorops*, *Ctenonotus*, *Dactyloa*, *Phenacosaurus* y *Semiurus*, y limitaron el nombre Polychrotidae que incluye únicamente al género *Polychrus*. Únicamente la familia Dactyloidae con un género (*Anolis*) y cuatro especies (una de ellas introducida) habita en la región de los Estados de la Frontera.

Plestiodon brevirostris

Feria-Ortiz et al. (2011) revisaron la especie-grupo *Plestiodon brevirostris* elevando *Plestiodon b. bilineatus* de la Sierra Madre Occidental y *Plestiodon b. dicei* de la Sierra Madre Oriental al rango de especies. Ellos señalaron que sus datos no pudieron resolver si *Plestiodon b. pineus* es conespecífico con *P. b. dicei* o si *P. b. dicei* es parafilético. Tentativamente nos referimos a las poblaciones de Coahuila, Nuevo León y parte de Tamaulipas (anteriormente consideradas como *P. b. pineus* y como intergrados de *P. b. pineus* × *dicei*) como *P. dicei* hasta que esta relación sea clarificada.

Scincella kikaapoa

García-Vázquez et al. (2010) describieron una especie nueva de scincido (*Scincella kikaapoa*) para la región de Cuatro Ciénegas, Coahuila.

Aspidoscelis neomexicana

Se ha especulado que el Huico de Nuevo México (*Aspidoscelis neomexicana*) habita en Chihuahua, México (ejem. Liner y Casas-Andreu 2008), sin el apoyo de un registro curatorial. Varios autores de capítulos de este libro han trabajado en el norte de Chihuahua a lo largo del lado sur del Río Bravo. Ninguno de nosotros ha encontrado individuos de esta especie en el lado mexicano. Además, James Walker (comunicación personal 2011) ha enfocado su atención a encontrar esta especie en el lado mexicano pero hasta ahora sin éxito. Consideramos altamente probable la presencia de *A. neomexicana* en Chihuahua, pero en este momento, sin un registro curatorial, consideramos a esta especie exclusiva de los Estados Unidos (Nuevo México y Texas e introducida en Arizona).

Aspidoscelis stictogramma

En este libro consideramos *Aspidoscelis burti burti* y *A. b. stictogramma* con el rango de especie de acuerdo a Walker y Cordes (2011).

Charina y Lichanura

Kluge (1993) argumentó equivalencia taxonómica de los géneros monotípicos cercanamente relacionados *Charina*, *Lichanura* y *Calabaria*, con los últimos dos estando colocados en la sinonimia de *Charina*. Sin embargo, Crother (2008) y Grismer (2002) argumentaron que su separación está garantizada en el uso que estos nombres han tenido durante mucho tiempo y en que *Charina* y *Lichanura* ya no son considerados monotípicos. En este libro aceptamos el uso de *Lichanura* como un género diferente de *Charina*.

Lichanura trivirgata y L. orcutti

Basándose sobre ADN mitocondrial y datos morfológicos preliminares, Wood et al. (2008) reconocieron dos especies en *Lichanura*: *L. trivirgata* y *L. orcutti*. *Lichanura orcutti* se encuentra exclusivamente en los Estados Unidos, mientras que *L. trivirgata* habita en ambos países. Muestreos adicionales a lo largo de la región fronteriza de los Estados Unidos y México, podrían probar la presencia de *L. orcutti* en ambos países.

Adelphicos newmanorum

LaDuc (1996) determinó que las poblaciones de *Adelphicos* en Tamaulipas y estados adyacentes en el noreste de México fueron diferentes y alopátricas de *A. quadrivigatum* que se distribuye desde el centro de Veracruz hasta Centroamérica. Wilson y Johnson (2010) y Wilson et al. (2013) reconocieron *A. newmanorum* (Taylor 1950) con el rango de especie sin hacer comentarios al respecto, aunque en gran parte se basaron sobre LaDuc. Colectas futuras podrían registrar especímenes entre el noreste de Hidalgo y centro de Veracruz, pero, tentativamente seguimos a Taylor, LaDuc, Wilson y Johnson, y Wilson et al. en reconocer *A. newmanorum* como una especie distinta esperando nueva información que resuelva este problema. Ver Farr et al. (2013) para una discusión adicional.

Arizona pacata

Seguimos a Grismer (2002) en la elevación de la subespecie *A. elegans pacata* al rango de especie. Klauber (1946) diferenció *A. e. pacata* de todas las otras especies de *Arizona* por tener un número menor de manchas corporales de forma subcircular. Además, sus distribuciones están separadas por una distancia de 10 km, sin señales de intergradación en especímenes encontrados cerca de este vacío (Grismer, 2002). En este libro consideramos a *Arizona pacata* con el rango de especie.

Contia longicaudae

Feldman y Hoyer (2010) describieron una especie nueva de Culebra del Bosque de Cola-afilada (*Contia longicaudae*) para las Sierras Costeras, Montañas Klamath, y porciones de las Montañas Cascada en el norte de California y sur de Oregón.

Hypsiglena

Mulcahy (2008) revisó el arreglo de subespecies y especies del complejo de la Serpiente Nocturna de Collar (*Hypsiglena torquata*) basándose sobre marcadores de ADN mitocondrial, y reconoció ocho especies de Serpientes Nocturnas, una de las cuales permanece sin describir (Nocturna de Capucha). Cuatro de estas especies habitan en la región de los Estados de la Frontera. *Hypsiglena chlorophaea* (Baja California, Sonora, Chihuahua, Nuevo México, Arizona y California), *H. jani* (Chihuahua, Coahuila, Nuevo León, Tamaulipas, Texas, Nuevo México y Arizona), *H. ochrorhyncha* (Baja California y California), y *H. slevini* (Baja California). Éste es el arreglo que seguimos en este libro.

Lampropeltis getula

Pyron y Burbrink (2009) presentaron una revisión de la sistemática del grupo *Lampropeltis getula*, basándose sobre una análisis filogeográfico de ADN mitocondrial. Elevaron a las subespecies *L. g. californiae*, *L. g. holbrooki*, *L. g. niger* y *L. g. splendida* al rango de especies. Sus muestras fueron principalmente de localidades de los Estados Unidos; sin embargo, y excepto por muestras de la Península de Baja California únicamente un espécimen de otras áreas de México (Durango) fue incluido en el análisis. Debido a la carencia de muestras adecuadas para México, y para mantener la consistencia entre los diez estados, tentativamente consideramos al taxón que habita en los 10 estados de la frontera México—Estados Unidos como *Lampropeltis getula*.

Lampropeltis knoblochi y L. pyromelana

Burbrink et al. (2011), aplicando métodos bayesianos de delimitación de especies, utilizando tres loci de muestras de la Culebra Real de Arizona (*Lampropeltis pyromelana*) tomadas sobre toda su distribución, encontraron dos especies que están separadas por hábitats de baja altitud entre la Meseta de Colorado / Borde Mogollón (= Mogollon Rim) y la Sierra Madre Occidental: *Lampropeltis pyromelana*, limitada a la Meseta de Colorado, en el norte de Arizona, oeste de Nuevo México, Utah y Nevada; y *Lampropeltis knoblochi* de la Sierra Madre Occidental de Sonora, Chihuahua, Sinaloa y Durango, y el archipiélago de islas continentales en el sureste de Arizona y suroeste de Nuevo México. Por lo que *L. pyromelana* es una especie que habita exclusivamente

en los Estados Unidos, y *L. knoblochi* habita en ambos países.

Leptodeira maculata

Duellman (1958) describió *Leptodeira annulata cussiliris* y ese nombre fue aplicado a las poblaciones de Tamaulipas en los 50 años siguientes a su descripción. Estudios moleculares recientes de las relaciones en el género *Leptodeira* indicaron que *L. annulata* es polifilética y elevaron *L. a. cussiliris* al rango de especie (Daza et al. 2009; Mulcahy 2007). Daza et al. también encontraron que *L. maculata* estaba anidada en *L. a. cussiliris* y concluyeron que *L. maculata* "debe ser considerada una sinonimia" de *L. cussiliris* (en error). Wilson et al. (2013) y Uetz (2013) señalaron que *L. maculata* (Hallowell 1861; con el nombre *Megalops maculatus*) tiene prioridad sobre *L. cussiliris*, por lo que *L. maculata* es el nombre correcto para esta especie.

Masticophis

Utinger et al. (2005) propusieron la sinonimia de *Masticophis* con *Coluber* debido a que sus resultados indicaron que *Masticophis flagellum* es parafilética cuando ésta es comparada con *Coluber constrictor*. Sin embargo, este arreglo no ha sido aceptado por varios autores, incluyendo Burbrink et al. (2008), quienes encontraron que *Coluber constrictor* es monofilética y no encontraron ninguna *M. flagellum* anidada en los linajes de *C. constrictor*, cuestionando si *Masticophis* debe considerarse una sinonimia de *Coluber*. Además, Collins y Taggart (2008) propusieron retener *Masticophis* como un género separado basándose sobre la observación de que Utinger et al. no fundamentaron sus resultados en un muestreo adecuado. Wilson y Johnson (2010) también concluyeron que Utinger et al. tuvieron un muestreo deficiente de poblaciones de *C. constrictor* y, sobre todo, no incluyeron muestras de diferentes especies de *Masticophis*, pues únicamente dos fueron analizadas (*M. flagellum* y *M. taeniatus*). Hasta que haya más evidencia disponible, seguimos las conclusiones de Collins y Taggart y Wilson y Johnson y consideramos a *Masticophis* como un género separado de *Coluber*.

Masticophis fuliginosus y *M. flagellum*

Seguimos a Grismer (2002) quien elevó *Masticophis flagellum fuliginosus* al rango de especie. Tanto *M. fuliginosus* como *M. flagellum* se han encontrado en simpatría en la Bahía San Luis Gonzaga y en Paseo de San Matías, Baja California, sin muestras de intergradación (Grismer, 1994; Grismer y Mellink, 2005). *Masticophis fuliginosus* y *M. flagellum*, como especies separadas, habitan tanto en Estados Unidos como en México.

Pliocercus bicolor

Algunos autores siguen a Savage y Crother (1989) en reconocer *Pliocercus elapoides* como una sola especie altamente variable. En este libro, seguimos a Smith y Chiszar (2001a, 2001b). De hecho ésta es la revisión más reciente y el arreglo seguido por Liner y Casas-Andreu (2008). El arreglo de Smith y Chiszar de la especie-grupo *Pliocercus elapoides* que reconoce múltiples especies y subespecies, ha sido criticado por algunos autores señalando que está basado únicamente sobre variaciones del patrón de coloración correlacionadas con la distribución geográfica (ejem. McCranie 2011). *Pliocercus bicolor* es única entre las especies del arreglo de Smith y Chiszar en tener características distintivas en las escamas (la infralabial posterior está fusionada con la segunda labiogenial) y una distribución relativamente aislada, sino enteramente alopátrica. Permanecemos imparciales respecto al estatus de los miembros que habitan el extremo sur de la distribución de este complejo.

Thamnophis angustirostris

El estatus taxonómico de *T. angustirostris* no es claro. Esta especie es conocida sólo por el lectotipo (USNM 959—Kennicott no designó un ho-

lotipo y aparentemente basó su nuevo taxón en más de un espécimen [Rossman et al. (1996)]) recolectado en Parras, Coahuila. Consideramos a esta como una especie válida ya que la Sierra de Parras y localidades de alrededor tiene condiciones ambientales adecuadas para esta especie de serpiente; eventualmente esta especie puede ser redescubierta con un trabajo de campo más intenso en los alrededores de Parras.

Thamnophis rufipunctatus y *T. unilabialis*

Utilizando enfoques filogenéticos y de genética poblacional, Wood et al. (2011) probaron si la diversidad genética en el complejo de especies de *Thamnophis rufipunctatus* era consistente con dos modelos prevalecientes de fluctuación distribucional para especies afectadas por los cambios climáticos del Pleistoceno. Identificaron tres clados divergentes dentro de este complejo (*rufipunctaus*, *unilabialis* y *nigronuchalis*) y propusieron reconocer a las dos primeras que anteriormente tenían el rango de subespecies (*T. r. rufipunctatus* y *T. r. unilabialis*) con el rango de especie: *Thamnophis rufipunctatus* (Jarretera Cabeza-angosta del Mogollón) del centro de Arizona y sureste y suroeste de Nuevo México; y *T. unilabialis* (Jarretera Cabeza-angosta de Chihuahua) de la Sierra Madre Occidental de Sonora, Chihuahua y Durango y el centro-sur de Chihuahua y norte de Durango.

Trimorphodon biscutatus

Devitt et al. (2008) examinaron patrones de variación geográfica en *T. biscutatus* usando análisis estadístico multivariado de 33 características morfológicas en 429 especímenes. Análisis de componentes principales y discriminantes revelaron seis grupos morfológicamente distintos que fueron concordantes con los linajes recuperados en un análisis filogeográfico de ADN mitocondrial y con taxa tradicionalmente reconocidos como especies o subespecies. Concluyeron que *T. biscutatus* (sensu lato) comprende seis especies evolutivas (incluyendo *T. vilkinsonii*) y recomendaron elevar *T. biscutatus* (sensu stricto), *T. lambda*, *T. lyrophanes*, *T. paucimaculatus* y *T. quadruplex* al rango de especies. Tres de estas especies habitan en la región de los Estados de la Frontera: Ilimacoa de Sonora (*T. lambda* en Sonora, Nuevo México, Arizona y California), Ilimacoa de Baja California (*T. lyrophanes* en Baja California y California), y Serpiente de Tetalura (*T. vilkinsonii* en Chihuahua, Texas y Nuevo México).

Hydrophis platurus

Utilizamos el nombre *Hydrophis platurus* en lugar de *Pelamis platura* basándonos sobre el trabajo de Sanders et al. (2012), quienes consideraron a *Pelamis* junto con otros géneros de serpientes de mar en el género *Hydrophis*.

Rena y Epictia

Adalsteinsson et al. (2009) revisaron la filogenia de Leptotyphlopidae usando secuencias de ADN de nueve genes mitocondriales y nucleares, analizando 91 individuos representando 34 especies reconocidas de leptotyphlopidos. Basándose sobre sus resultados, propusieron una nueva clasificación de la familia la cual reconoce dos subfamilias, Epictinae (América y África) y Leptotyphlopinae (África, Península Arábica y suroeste de Asia), incluyendo 12 géneros. La mayoría de los géneros corresponden a especies grupo reconocidas previamente. Siete de estos nombres de géneros fueron revividos mientras que cinco de ellos fueron nuevos. Varias de las especies anteriormente colocadas en *Leptotyphlops* fueron reubicadas en el género *Rena*, un nombre originalmente propuesto por Baird y Girard (1853). Estas incluyen seis especies que habitan en México, *R. bressoni, R. dissecta, R. dulcis, R. humilis, R. maxima* y *R. myopica*; tres de las cuales también habitan en los Estados Unidos (*R. dissecta, R. dulcis* y *R. humilis*). La otra especie mexicana (*Leptotyphlops goudotii*) fue colocada en el género *Epictia*, un nombre originalmente propuesto por Gray (1845).

Adalsteinsson et al. (2009) también indicaron que existe un número inusualmente grande de especies que no han sido reconocidas y propusieron que *R. humilis boettgeri* debe ser reconocida con el rango de especie. Además, Bocourt (1881) originalmente describió *Rena humilis dugesii* como una especie (*Catodon dugesii*), pero Klauber (1940) consideró a ésta como una subespecie de *Leptotyphlops* (= *Rena*) *humilis*. Posteriormente ésta fue de nuevo elevada al rango de especie por Lemos-Espinal, Smith y Chiszar (2004) sobre la base de pigmentación y número proximal de hileras caudales; Lemos-Espinal, Smith, Chiszar y Woolrich-Piña (2004) separaron *R. h. segrega* de *R. humilis* por el número de hileras de escamas alrededor de la base de la cola, *R. humilis* teniendo 12 hileras, *R. segrega* 10. Concluimos que cinco especies de *Rena* (*R. dissecta, R. dulcis, R. humilis, R. myopica* y *R. segrega*), y una de *Epictia* (*E. goudotii*) habitan en la región de los Estados de la Frontera.

Indotyphlops

Seguimos a Hedges et al. (2014) en el uso del nombre genérico *Indotyphlops* en lugar de *Ramphotyphlops*.

Comentarios sobre los nombres estándar en inglés y español

En este libro hemos adoptado los nombre estándar en ingles propuestos por Crother (2008) para las especies que habitan exclusivamente en los Estados Unidos así como para las que habitan tanto en Estados Unidos como en México, y los nombres estándar en español e inglés propuestos por Liner y Casas-Andreu (2008) para las especies que habitan únicamente en México, y los nombres en español para las especies que habitan tanto en México como en Estados Unidos. Para las especies que habitan exclusivamente en los Estados Unidos tradujimos a español los nombres propuestos por Crother (2008).

El trabajo de estas dos publicaciones representa excelentes y extensos esfuerzos para unificar el uso de nombres estándar para las especies de anfibios y reptiles de ambos países. Sin embargo, si estos nombres se aplican erróneamente pueden crear confusión y contradecir el significado original para el cual fueron propuestos. Liner y Casas-Andreu (2008) propusieron varios nombres que podrían crear tal confusión; la mayoría de ellos son errores tipográficos, aquí proporcionamos una explicación de porqué diferimos en el uso de algunos de ellos.

Batrachoseps major

El nombre asignado es Salamandra Pequeña, pero este nombre contradice al nombre específico, derivado de la palabra en latín que significa "más grande," refiriéndose a la especie más grande de *Batrachoseps* (Beltz 2006). El nombre Salamandra Mayor es más apropiado para esta especie y es el que utilizamos en este libro.

Ensatina eschscholtzii

El nombre asignado es Ensatina de Monterrey, pero como Monterrey es una ciudad bien conocida en México (capital de Nuevo León), sugerimos que el nombre en español debe especificar que éste se refiere a Bahía Monterrey y no a la ciudad de Monterrey en Nuevo León. El nombre Ensatina de Bahía Monterrey parece más apropiado para esta especie y es el que utilizamos en este libro.

Lithobates magnaocularis

El nombre asignado es Rana Leopardo del Noreste de México, éste es incorrecto. El nombre correcto debe ser Rana Leopardo del Noroeste de México ya que esta especie habita en el noroeste de México y es el que utilizamos en este libro.

Alligator mississippiensis

El nombre asignado es Aligator Americano. La palabra "Aligator" no existe en español. El nombre Lagarto Americano parece más apropiado

para esta especie y es el que utilizamos en este libro.

Trachemys nebulosa

Los nombres Jicotea del Río y Baja California Slider son incorrectos. El nombre Jicotea del Río es vago y no define ninguna característica especial de esta Tortuga, y el nombre Baja California Slider es incorrecto porque crea la idea errónea de que esta especie habita en Baja California cuando en realidad habita en Baja California Sur, Sonora y Sinaloa. El nombre asignado en http://www.reptile-database.org es más apropiado para esta especie Black-bellied Slider y su traducción para el nombre en español, Jicotea de Vientre Negro. Estos nombres son los que utilizamos en este libro.

Gopherus

El nombre asignado para este género de tortugas es Galápago del Desierto un nombre que nadie parece utilizar al referirse a ellas. Este grupo de tortugas es conocido como Tortugas de Desierto que es la traducción directa del nombre estándar en inglés Desert Tortoise. Aquí utilizamos los nombres: Tortuga de Desierto (*Gopherus agassizii*), Tortuga de Berlandier (*Gopherus berlandieri*), Tortuga del Bolsón de Mapimí (*Gopherus flavomarginatus*), y Tortuga del Desierto de Sonora (*Gopherus morafkai*), para las cuatro especies que habitan en la región de los Estados de la Frontera.

Elgaria

El nombre asignado para este género de lagartijas es Lagartos del Oeste. El nombre Lagarto se refiere a Caimanes no a Lagartijas. Consideramos que la traducción directa del nombre estándar en inglés, Lagartijas Lagarto debe ser utilizada para evitar confusiones. Aquí utilizamos los nombres: Lagartija Lagarto de Isla Cedros (*E. cedrosensis*), Lagartija Lagarto del Norte de Estados Unidos (*E. coerulea*), Lagartija Lagarto de Montaña (*E. kingii*), Lagartija Lagarto del Suroeste de Estados Unidos (*E. multicarinata*—se proporciona una explicación adicional en el siguiente párrafo), Lagartija Lagarto de Los Coronado (*E. nana*), y Lagartija Lagarto de la Sierra Panamint (*E. panamintina*), para las seis especies que habitan en la región de los Estados de la Frontera.

Elgaria multicarinata

El nombre asignado, Lagarto Meridional, es incorrecto. El nombre correcto debe ser Lagartija Lagarto del Suroeste de los Estados Unidos ya que ésta es común en esta región y entra al noroeste de México a través de Baja California. El nombre propuesto por Liner y Casas-Andreu (2008) erróneamente indica una distribución sureña en México.

Crotaphytus dickersonae

El nombre asignado, Sonoran Collared Lizard, ya fue asignado a *Crotaphytus nebrius*, aquí proponemos y usamos el nombre Dickerson's Collared Lizard.

Phrynosoma modestum

El nombre asignado, Tapayaxín, es una palabra Náhuatl que significa "aquellos que lloran sangre," y es utilizado en algunas regiones para referirse a *Phrynosoma orbiculare* en relación al hábito de expulsar sangre a través de los ojos que presenta esta y otra especies del género, hábito que no ha sido reportado en *P. modestum*. Aquí sugerimos y utilizamos el nombre Camaleón Modesto que es el significado en español del nombre específico *modestum*, una palabra en latín que significa "modesto o discreto," en referencia a la apariencia normal (sin escamas en forma de grandes espinas) de esta especie.

Sceloporus albiventris

Esta especie no fue considerada por Liner y Casas-Andreu en su listado del 2008. Sugerimos el nombre Bejore de Vientre Blanco y Western

White-bellied Spiny Lizard, los cuales utilizamos en este libro.

Sceloporus cowlesi

Esta especie no fue considerada por Liner y Casas-Andreu en su listado del 2008. Sugerimos el nombre Lagartija de Cerca del Noroeste, el cual utilizamos en este libro.

Sceloporus graciosus

Esta especie no fue considerada por Liner y Casas-Andreu en su listado del 2008. Sugerimos el nombre Lagartija Común de las Salvias, el cual utilizamos en este libro.

Sceloporus vandenburgianus

El nombre asignado, Meridional de las Salvias, es incorrecto debido a que se refiere a la distribución de esta lagartija en los Estados Unidos, no en México. El nombre correcto debe ser Septentrional de las Salvias, el cual utilizamos en este libro.

Ficimia olivacea

El nombre asignado, Nariz de Gancho Huaxteca, es incorrecto. El nombre Huasteca está mal escrito en el nombre asignado, éste debe ser Huasteca. El nombre correcto es Nariz de Gancho Huasteca.

Lampropeltis pyromelana

Los nombres asignados, Culebra Real de Sonora y Sonoran Mountain Kingsnake, fueron modificado por Burbrink et al. (2011) a Culebra Real de Arizona y Arizona Mountain Kingsnake, éstos son los que utilizamos en este libro.

Oxybelis aeneus

Los nombres asignados son Bejuquilla Parda y Brown Vinesnake. Keiser (1982) recomendó el nombre en inglés de Neotropical Vinesnake para esta especie, señalando que este taxón de distribución amplia tiene una variación muy alta en patrón de coloración y en muchas regiones de su distribución no es parda. Aquí utilizamos el nombre propuesto por Keiser (1982) y su traducción para el nombre estándar en español, Bejuquilla Neotropical.

Tantilla hobartsmithi

El nombre asignado es Culebra Cabeza Negra del Suroeste. La distribución en el suroeste implícita en este nombre se refiere a los Estados Unidos; además, esta especie tiene una distribución amplia en el sur de Estados Unidos y norte de México, y no se limita al suroeste de los Estados Unidos. El nombre correcto debe ser Culebra Cabeza Negra del Norte.

Thamnophis rufipunctatus

Debido a que aquí seguimos a Wood et al. (2011) en la clasificación del complejo de la Jarretera Cabeza-angosta, hemos adoptado los nombres estándar propuestos por ellos: Jarretera Cabeza-angosta del Mogollón y Mogollon Narrowheaded Gartersnake.

Thamnophis unilabialis

Debido a que aquí seguimos a Wood et al. (2011) en la clasificación del complejo de la Jarretera Cabeza-angosta, hemos adoptado los nombres estándar propuestos por ellos: Jarretera Cabeza-angosta de Chihuahua y Madrean Narrowheaded Gartersnake.

Tropidodipsas fasciata

El nombre asignado Caracolera Añillada es incorrecto. El nombre correcto debe ser Caracolera Anillada.

Tropidodipsas repleta

Esta especie no fue incluida por Liner y Casas-Andreu en su listado del 2008. Sugerimos los nombres de Caracolera Repleta y Banded Blacksnake.

Crotalus molossus

El nombre asignado, Cascabel Serrana, se aplica a una de las especies de víboras de cascabel mejor conocidas en México. La mayoría de la gente local en el norte de México se refiere a esta especie como "Cola Prieta," que significa "Dark or Blackish Tail" y que describe fácilmente a esta especie diferenciándola de otras especies de víboras de cascabel. Consideramos que el nombre Cola Prieta es más apropiado para esta serpiente y tiene la ventaja de ser ampliamente utilizado por la gente local.

Sistrurus catenatus

Cascabel de Massasauga, Massasauga Rattlesnake. Para propósitos descriptivos y para ser consistentes con el nombre estándar en español, aquí utilizamos la combinación Massasauga Rattlesnake, en lugar de Massasauga utilizado por Crother (2008) y Liner y Casas-Andreu (2008).

BRADFORD D. HOLLINGSWORTH,
CLARK R. MAHRDT, L. LEE GRISMER,
Y ROBERT E. LOVICH

Introducción

El estudio de los anfibios y reptiles del estado de Baja California ha recibido considerable atención, pero por lo general en el contexto de la Península de Baja California (Van Denburgh 1895a, 1895b; Schmidt 1922; Savage 1960; Murphy 1983b, 1983c; Grismer 1994b, 2002; Lovich et al. 2009; Murphy y Méndez de la Cruz 2010). En este trabajo nos enfocamos en la herpetofauna del estado de Baja California, que se extiende desde el paralelo 28 hasta la frontera internacional con los Estados Unidos, cubriendo la mitad norte de la Península de Baja California. Al sur, Baja California comparte 142 km de frontera estatal con Baja California Sur, mientras que al norte comparte 226 km de frontera con el estado norteamericano de California. En el extremo noreste del estado se comparten 36 km con Arizona, y 112 km con Sonora siguiendo el cauce histórico del Río Colorado. El resto del estado de Baja California limita con el Océano Pacífico y con el Mar de Cortés, e incluye islas en estos dos cuerpos de agua. En total, el estado de Baja California comprende un área de aproximadamente 70,000 km².

El estado contiene una amplia gama de hábitats, la cual tiene influencias de las Provincias Florísticas que se extienden desde el norte o desde el sur, y por los climas del Océano Pacífico y el Mar de Cortés. La Cordillera Peninsular es la característica topográfica dominante en el estado, ésta corre en dirección norte a sur, con una pendiente gradual a lo largo de su vertiente oeste, y un levantamiento más abrupto en su vertiente este, frecuentemente presentando una escarpa precipitada. En la Cordillera Peninsular hay una serie de montañas con valles de baja altitud que conectan las regiones del oeste con las del este. La Sierra Juárez y la Sierra San Pedro Mártir, localizadas en la mitad norte del estado, representan barreras zoogeográficas así como ecosistemas de elevaciones altas. Dos parques nacionales poseen las elevaciones más altas de estas sierras, el Parque Nacional Constitución de 1857, en la Sierra Juárez y el Parque Nacional Sierra San Pedro Mártir, que incluye el Picacho del Diablo, y que representa el punto más alto con 3095 m en Baja California (Schad 1988). Estas montañas crean un efecto importante de sombra de lluvia entre la relativamente húmeda Provincia Florística de California y el Desierto Inferior del Colorado al este de estas montañas (Shreve y Wiggins 1964; Histings y Humphrey 1969).

La Cordillera Peninsular en la mitad sur del estado incluye las Sierras de La Asamblea, Las Animas, San Borja y La Libertad. Debido a que estas montañas son más bajas, éstas tienen un menor efecto de aislar los ecosistemas circundantes. Mientras la Sierra de La Asamblea alcanza casi 1,700 m de altitud y mantiene un hábitat aislado de cima de montaña, la parte sur del estado está dominada por el Desierto de la Península de Baja California (Arriaga et al. 1997; Morrone 2005; Hafner y Riddle 1997), formado por las regiones desérticas del Vizcaíno y la Costa Central del Golfo (Shreve y Wiggins 1964; Wiggins 1980). Como resultado, la transición entre la provincia florística de California y el Desierto Inferior del Colorado en el norte y los desiertos de la península en el sur se ven menos afectadas por estas sierras y se entremezclan gradualmente a través de áreas geográficas más amplias (Grismer 2002).

Un total de 119 especies de anfibios y reptiles habitan en el estado de Baja California. De estos, 20 son anfibios (4 salamandras, 16 anuros) y 99 son reptiles (7 tortugas, 51 lagartijas, 1 amphisbenido y 40 serpientes). Seis especies son introducidas (*Xenopus laevis, Lithobates berlandieri, L. catesbeianus, Apalone spinifera, Hemidactylus turcicus* y *Sauromalus varius*) y 113 son nativas. Un total de 21 especies son endémicas al estado. Dos especies son endémicas de la parte continental (*Crotaphytus grismeri* y *Urosaurus lahtelai*), 2 del área continental y de las islas del estado (*Elgaria cedrosensis* y *Anniella geronomensis*) y 17 son endémicas de las islas del estado (*Elgaria nana, Crotaphytus insularis, Phyllodactylus partidus, Sauromalus hispidus, Callisaurus splendidus, Petrosaurus slevini, Uta antiqua, U. encantadae, U. lowei, U. tumidarostra, Aspidoscelis cana, Lampropeltis herrerae, Pituophis insulanus, Crotalus angelensis, C. caliginis, C. lorenzoensis* y *C. muertensis*).

La herpetofuana de Baja California se comparte ampliamente con los Estados Unidos. De las 119 especies halladas en el estado, 88 (74%) también habitan en los Estados Unidos. Las 6 especies marinas del estado (*Caretta caretta, Chelonia mydas, Eretmochelys imbricata, Lepidochelys olivacea, Dermochelys coriacea* e *Hydrophis platurus*) también habitan en las aguas costeras de California y 5 de las 6 especies introducidas (la excepción es *Sauromalus varius*) también han sido introducidas en áreas no nativas de los Estados Unidos. Las restantes 77 especies tienen distribuciones que cruzan la frontera norte con los estados de California (todas las 77 especies) y Arizona (38 de las 77 especies).

Las restantes 31 especies de anfibios y reptiles que habitan el estado de Baja California son endémicas de México. Éstas incluyen 21 especies endémicas al estado, y 10 más cuyas distribuciones se limitan a México. *Elgaria cedrosensis Petrosaurus repens, Phrynosoma cerroense, Sceloporus zosteromus, Aspidoscelis labialis, Bipes biporus, Arizona pacata, Hypsiglena slevini, Pituophis vertebralis* y *Crotalus enyo* tienen distribuciones que se extienden hacia el sur en el estado de Baja California Sur y se limitan a la Península de Baja California y sus islas asociadas (Grismer 2002), mientras que *Sauromalus varius* es una especie endémica a islas del Mar de Cortés pertenecientes a Sonora y que ha sido introducida a la Isla Roca Lobos, Baja California (Hollingsworth et al. 1997; Hollingsworth 1998).

Estudios Herpetológicos Previos

La mayoría de los trabajos se han enfocado en la composición faunística y diversidad de la Península de Baja California, lo que dificulta rastrear la historia de la investigación herpetológica dentro de los límites del estado. Las primeras exploraciones describen observaciones de anfibios y reptiles de la península, generalmente sin establecer lugares específicos para discernir si la observación o el material recolectado se refieren a Baja California o a Baja California Sur. Algunas excepciones son las colectas realizadas en el Puerto de Cabo San Lucas. Por ejemplo, Paul E. Botta hizo recolectas de historia natural en Cabo San Lucas cuando formaba parte de la tripulación del barco francés, *Le Héros*, entre 1827 y 1829 (Adler 1978). Éstas incluyeron el espécimen tipo de *Phyrnosoma coronatum* descrito por Blainville (1835) en *Reptiles de La Californie*. Cuando *Le Héros* se dirigió al norte a lo largo de la costa del Pacífico, es poco probable que Botta haya tenido la oportunidad de recolectar en el estado de Baja

California antes de que el barco llegará al Puerto de San Diego.

Los reportes iniciales sobre observaciones de anfibios y reptiles se encuentran en Grismer (2002), éstos incluyen el primer reporte detallado de la Península de Baja California por el misionero Jesuita Francisco Javier Clavijero escrito en 1789. Además de los reportes iniciales, las observaciones herpetológicas fueron raras en la última parte del siglo 18 y la primera mitad del siglo 19 (Grismer 2002). Después de la Guerra México—Estados Unidos (1846–1848) y del Tratado de Guadalupe Hidalgo, se reformó el límite entre los dos países. Durante este mismo periodo los Estados Unidos exploraron el territorio recientemente adquirido y se realizaron varias expediciones para encontrar rutas de acceso al occidente de Estados Unidos. Esto incluyó exploraciones a Baja California por los naturalistas Spencer F. Baird, Lyman Belding, Paul E. Botta, Townsend S. Brandegee, Walter E. Bryant, Gustav Eisen y William M. Gabb (Grismer 2002).

En 1867 se realizó una expedición a Baja California, con el consentimiento de las autoridades mexicanas, para detallar la composición geológica de la península, esta expedición cruzó varias veces de costa a costa en sus viajes hacia el norte desde Cabo San Lucas hasta Tijuana (Dall 1909). Como parte del equipo de esta expedición, William M. Gabb, mientras documentaba la geología de la península (Gabb 1882), recolectó el espécimen tipo de *Phyllorhynchus decurtatus* (Cope 1868: p. 311) en "la parte superior de Baja California." En el reporte de esta expedición Gabb consideró tres distritos diferentes para la península, incluyendo la Región del Cabo, una sección central que cubría la mitad sur de la península y hacia el norte hasta Santa Gertrudis localizada en el actual límite estatal, y una parte norte por encima de Santa Gertrudis. Smith y Taylor (1950: p. 322) restringieron la localidad tipo como "Misión de San Fernando (entre San Ignacio y)," Baja California, México. La ubicación del espécimen tipo de *P. decurtatus* no es clara, pero se puede interpretar como proveniente del distrito norte basándose en la división de Gabb.

De ser así, *P. decurtatus* representa la primera especie descrita a partir de un espécimen recolectado en el estado de Baja California (ver Cuadro 3.1).

La primera revisión de la herpetofauna de Baja California fue completada por John Van Denburgh en 1895, cuando comentó "La Península de Baja California está tan lejos de las rutas habituales de viajes que pocas colecciones de animales de este estado han sido depositadas en museos." En ese entonces, Van Denburgh (1895a, 1895b) reportó 61 especies para la península, más 7 especies adicionales al año siguiente (Van Denburgh 1896). Doce de éstas limitadas a áreas fuera del estado de Baja California. Mocquard (1899), con la ayuda de las colectas realizadas por León Diguet, añadió otras 11 especies al creciente conocimiento de la herpetofauna peninsular (Grismer 2002).

La Oficina del Censo Biológico del Departamento de Agricultura de los Estados Unidos realizó una expedición a la península de 1905 a 1906 bajo la dirección de Edward Nelson y asistido por Edward Goldman. La expedición comenzó en la frontera de Estados Unidos con la ayuda de Frank Stephens de la Sociedad de Historia Natural de San Diego. Los resultados fueron publicados como (*La Península de*) *Baja California y sus Recursos Naturales*, e incluyeron los reportes de los anfibios y reptiles descubiertos, así como el reporte de la distribución sureña de *Crotalus cerastes* (Nelson 1922). Nelson reportó 76 especies de anfibios y reptiles actualmente reconocidos para la península, de los cuales 62 se encuentran en el estado de Baja California.

En 1911, la expedición de *El Albatros* liderada por Charles H. Townsend exploró la Península de Baja California y sus islas asociadas (Townsend 1916). Patrocinada en parte por el American Museum of Natural History y el United States National Museum, esta expedición visitó las islas San Benito, Cedros y Ángel de la Guarda y la Bahía San Francisquito todas en el estado de Baja California. Especímenes recolectados en esta expedición llevaron a la descripción de *Callisaurus splendidus* por Dickerson (1919), además de muchas otras especies de fuera del estado. Este mismo material ayudó a Schmidt (1922) a escribir sobre la herpetofauna de la península y sus islas asociadas y a considerar a la península como una unidad faunística (Grismer 2002).

Dos de los trabajos más completos de los reptiles y anfibios de la región de Baja California fueron los realizados por Van Denburgh (1922), *Los Reptiles del Oeste de Norteamérica*, y más tarde por Joseph R. Slevin (1928), *Los Anfibios del Oeste de Norteamérica*. Gran parte del material utilizado para expandir los trabajos previos fueron obtenidos en 1921 por la expedición *Silvergate* al Mar de Cortés bajo el liderazgo de Slevin (Hanna 1922; Slevin 1922). Adicionalmente, Van Denburgh y Slevin produjeron varios listados y reportes en donde describieron especies nuevas para el estado (Van Denburgh 1905; Van Denburgh y Slevin 1914, 1921a, 1921b, 1923).

Poco después el Museum of Vertebrate Zoology envió expediciones a Baja California entre 1925 y 1931, que registraron información adicional y más detallada sobre el hábitat y la ecología de las especies halladas. Los resultados de estas expediciones se presentaron en Linsdale (1932).

Las exploraciones a Baja California continuaron por ambas rutas por tierra y por mar. Trabajos más detallados y completos fueron realizados por Laurence M. Klauber, del San Diego Natural History Museum y la Zoological Society of San Diego, que reunieron especímenes de un amplio número de especies y que llevaron a numerosas revisiones taxonómicas (Klauber 1924, 1928, 1931a, 1931b, 1931c, 1931d, 1931e, 1932, 1934, 1935, 1936, 1940a, 1940b, 1940c, 1941, 1943a, 1943b, 1944, 1945, 1946a, 1946b, 1949a, 1949b, 1951, 1956 y 1963). Frank S. Cliff exploró las islas en el Mar de Cortés a bordo del buque de investigación *Orca*, que fue puesto a disposición de la investigación biológica por Joseph W. Sefton, Jr. del San Diego Natural History Museum y la Sefton Foundation of San Diego. Como resultado, se obtuvo un reporte más detallado de los reptiles, con énfasis en especies de serpientes insulares (Cliff 1954a, 1954b). Notas adicionales sobre las islas del Pacífico fueron proporcionadas por Richard G. Zweifel (1952a, 1958).

Estudios sobre la biogeografía y patrones de variación geográfica fueron publicados por Charles H. Lowe, Jr. y Kenneth S. Norris (1954). Jay M. Savage (1960) presentó un reporte sobre el origen de la herpetofauna de la península basándose en datos de dispersión y discutió la evolución de la herpetofauna insular (Savage 1967).

La expedición *Belvedere* al Mar de Cortés en la primavera de 1962 produjo un avance mayor en el entendimiento de los reptiles de las Islas del Mar de Cortés, e incluyó visitas a las Islas Midriff y a la Bahía de los Ángeles (Lindsay 1962). Los herpetólogos Charles E. Shaw, Donald Hunsaker, Dennis L. Bostic y Michael Soulé formaron parte de esta expedición. Una publicación notable derivada de esta expedición incluye un estudio sobre la biogeografía de los reptiles de las islas, en la cual se analizan las relaciones de endemismos y el área ocupada por las especies (Soulé y Sloan 1966). Adicionalmente, Shaw estudió varias especies de reptiles incluyendo *Anniella* y *Sauromalus* (Shaw 1940, 1945, 1949 y 1953), mientras que Bostic se enfocó a reptiles y anfibios del norte de la región del Desierto del Vizcaíno (Bostic 1965, 1966a, 1966b, 1966c, 1966d, 1968, 1971 y 1975).

John R. Ottley contribuyó a la taxonomía de *Lichanura*, notas de historia natural, y un listado actualizado de la herpetofauna de las islas compilado en colaboración con Robert W. Murphy (Ottley 1978; Ottley et al. 1980; Ottley y Hunt 1981; Ottley y Jacobsen 1983; Murphy y Ottley 1984; y Grismer y Ottley 1988). Ted J. Case hizo contribuciones importantes a estudio de la ecología de islas y presiones de selección con estudios sobre *Sauromalus hispidus* (Case 1982, 1983). Hartwell H. Welsh estudió la ecogeografía de los anfibios y reptiles de la Sierra San Pedro Mártir con comentarios sobre la evolución de la herpetofauna (Welsh 1988) y reportó adiciones a la herpetofauna del Desierto Inferior del Colorado (Welsh y Bury 1984).

Robert W. Murphy ha hecho varias contribuciones al origen y evolución de la herpetofauna de la Península de Baja California (Murphy 1976, 1983b, 1983c; Murphy y Aguirre-Léon 2002b), listados (Murphy 1983a; Murphy y Ottley 1984; Murphy y Aguirre-Léon 2002a), taxonomía (Murphy 1974, 1975; Murphy y Smith 1985, 1991), filogenia y biogeografía histórica (Murphy et al. 1995; Upton y Murphy 1997; Aguirre-Léon et al. 1999; Murphy et al. 2002; Lindell et al. 2005, 2008; Blair et al. 2009), y conservación (Murphy y Méndez de la Cruz 2010).

A la fecha, el trabajo más completo sobre la herpetofauna de la región es el libro de L. Lee Grismer publicado en 2002, *Anfibios y Reptiles de Baja California incluyendo sus Islas del Pacífico y las Islas del Mar de Cortés*. Grismer también ha contribuido al conocimiento de la evolución de la herpetofauna de la península (Grismer 1994a, 1994b; Grismer y Mellink 2005), listados (Grismer 1999a, 2001), herpetofauna regional (Grismer 1993; Grismer y McGuire 1993; Grismer y Hollingsworth 1996; Grismer et al. 1994; Grismer y Mellink 1994), taxonomía (Grismer 1990a; Grismer 1994c; Grismer 1999b), revisiones taxonómicas (Grismer 1988; Grismer 1990b, 1996, 1999c; McGuire y Grismer 1993; Grismer y McGuire 1996; Grismer y Hollingsworth 2001; Grismer et al. 2002), fisiología (Hazard et al. 1998) y conservación (Lovich et al. 2009).

Estudios realizados por Robert E. Lovich y Clark R. Mahrdt han documentado la herpetofauna adaptada a ambientes templados de la Provincia Florística de California, los cuales incluyen notas sobre *Actinemys marmorata* (Lovich et al. 2005, 2007), *Anaxyrus californicus* (Mahrdt et al. 2002, 2003; Mahrdt y Lovich 2004), *Ensatina eschscholtzii* (Mahrdt 1975) y *E. klauberi* (Mahrdt et al. 1998). Un estudio sobre la herpetofauna terrestre y la conservación de la Reserva de la Biosfera Bahía de los Ángeles fue proporcionados por Lovich y Mahrdt (2008) y para el estado por Lovich et al. (2009).

Estudios moleculares y morfológico adicionales de grupos de especies con poblaciones en el estado de Baja California incluyen *Trimorphodon* (Devitt et al. 2008), *Xantusia wigginsi* (Leavitt et al. 2007), *Phrynosoma* (Leaché y McGuire 2006; Leaché et al. 2009), *Sceloporus magister* (Leaché y Mulcahy 2007), *Hypsiglena* (Mulcahy 2008), *Lampropeltis zonata* (Rodríguez-Robles

et al. 1999; Myers et al. 2013), y *Pituophis* (Rodríguez-Robles 2000), *Ensatina klauberi* (Devitt et al. 2013) y *Anniella* (Papenfuss y Parham 2013).

En Baja California, 26 especies de reptiles (19 lagartijas y 7 serpientes) tienen su localidad tipo en el estado (Cuadro 3.1).

Características Fisiográficas y su Influencia sobre la Herpetofauna

La topografía del estado de Baja California está fuertemente asociada con la formación de la Península de Baja California y sus orígenes a partir del occidente continental de México. Una compleja historia de tectónica de placas e interacciones entre las placas de América del Norte y del Pacífico dio lugar a la formación de la Península y la apertura del Mar de Cortés en los últimos siete a ocho millones de años (Lonsdale 1989; Stock y Hodges 1989; Winker y Kidwell 1986; Carreño y Helenes 2002). Junto con la separación de la península del occidente continental de México se produjo el levantamiento de la Cordillera Peninsular (Lonsdale 1989). Estas montañas se extienden a lo largo de la Península de Baja California y alcanzan sus puntos más altos en la mitad norte de Baja California. El depósito de sedimentos y cambios en los niveles del mar produjeron la cuenca de baja altitud y planicies de los desiertos del Vizcaíno y parte Inferior del Colorado (Grismer 2002).

Las islas costeras de Baja California en el Océano Pacífico y el Mar de Cortés están habitadas por anfibios y reptiles (Savage 1967; Murphy 1983a, 1983c; Murphy y Ottley 1984; Grismer 1993, 1999a, 1999b, 2001; Murphy y Aguirre-León 2002a, 2002b; Samaniego-Herrera et al. 2007). La Isla Guadalupe, que representa el punto más occidental de México, se encuentra 241 kilómetros de la costa y no está habitada por anfibios o reptiles, con excepción de las tortugas marinas. Las islas costeras del Océano Pacífico que están habitadas por reptiles y anfibios incluyen las Islas Coronado, Todos Santos, San Martín, San Jerónimo, San Benito y Cedros

CUADRO 3.1. Localidades tipo para anfibios y reptiles descritos para el estado de Baja California, México.

Autor(es)	Nombre original: Localidad tipo
Lagartijas (19)	
Fitch (1934)	*Gerrhonotus (= Elgaria) cedrosensis*: **Cañon on southeast side of Cedros Island, Lower California, Mexico**
Fitch (1934)	*Gerrhonotus scincicauda nanus (= Elgaria nana)*: **South Island, Los Coronados Islands, Lower California, Mexico**
Shaw (1940)	*Anniella geronimensis*: **San Geronimo Island, Lower California, Mexico**
McGuire (1994)	*Crotaphytus grismeri*: **Cañon David, a low pass that separates the contiguous Sierra de Los Cucapás and Sierra El Mayor, approximately 2 km W of Mex. Hwy 5 on the dirt road to the sulfur mine (turnoff at KM 49 S. Mexicali), Baja California, Mexico**
Van Denburgh & Slevin (1921b)	*Crotaphytus insularis*: **E coast Ángel de la Guarda Island seven miles north of Pond Island, Gulf of California, Mexico**
Smith & Tanner (1972)	*Crotaphytus insularis vestigium (= C. vestigium)*: **Guadalupe Canyon, Juárez Mountains, Baja California**
Dixon (1966)	*Phyllodactylus partidus*: **Isla Partida, Baja California**
Stejneger (1891)	*Sauromalus hispidus*: **Ángel de la Guarda Island, Gulf of California**
Dickerson (1919)	*Callisaurus draconoides splendidus (= C. splendidus)*: **Angel de la Guardia Island, Gulf of California, Baja California, Mexico**
Van Denburgh (1922)	*Uta (= Petrosaurus) slevini*: **Mejia Island, Gulf of California, Mexico**
Stejneger (1893)	*Phrynosoma cerroense*: **Cedros Island, Baja California, Mexico**[1]
Rau & Loomis (1977)	*Urosaurus lahtelai*: **4 km N parador Cataviña, elevation 564 m near Mexico Highway 1, State of Baja California, Mexico**
Ballinger & Tinkle (1968)	*Uta antiquus (=U. antiqua)*: **Isla Salsipuedes, 28°44′ N, 112°59′ W, Golfo de California, Mexico**
Grismer (1994c)	*Uta encantadae*: **Isla Encantada, Gulf of California, Baja California, México**
Grismer (1994c)	*Uta lowei*: **Isla El Muerto, Gulf of California, Baja California, México**
Grismer (1994c)	*Uta tumidarostra*: **Isla Coloradito, Gulf of California, Baja California, México**
Van Denburgh & Slevin (1921b)	*Cnemidophorus canus (= Aspidoscelis cana)*: **Sal Si Puedes Island, Gulf of California, Mexico**
Stejneger (1890)	*Cnemidophorus (= Aspidoscelis) labialis*: **Cerros Island, Lower California**[2]
Savage (1952)	*Xantusia vigilis wigginsi (= X. wigginsi)*: **Arroyo 9 miles east of Miller's Landing, Distrito del Norte, Baja California, Mexico**
Serpientes (7)	
Van Denburgh & Slevin (1923)	*Lampropeltis zonata herrerae (= L. herrerae)*: **South Todos Santos Island, Lower California, Mexico**
Cope (1868)	*Phimothyra decurtata (= Phyllorhynchus decurtatus)*: **San Fernando Misión (between San Ignacio and)**[1]
Klauber (1946b)	*Pituophis catenifer insulanus (= P. insulanus)*: **Cedros (Cerros) Island off the west coast of Baja California, Mexico**
Klauber (1963)	*Crotalus mitchellii angelensis (= C. angelensis)*: **About 4 miles southeast of Refugio Bay, at 1500 feet elevation, Isla Angel de la Guardia, Gulf of California, Mexico (near 29°29½′ N, 113°33′ W)**
Klauber (1949b)	*Crotalus viridis caliginis (= C. caliginis)*: **South Coronado Island, off the northwest coast of Baja California, Mexico**
Radcliffe & Maslin (1975)	*Crotalus ruber lorenzoensis (= C. lorenzoensis)*: **San Lorenzo Sur Island in the Gulf of California, Baja California Norte, Mexico**
Klauber (1949b)	*Crotalus mitchellii muertensis (= C. muertensis)*: **El Muerto Island, Gulf of California, Mexico**

[1] Restricción de la localidad tipo presentada por Smith y Taylor (1950).

[2] Cochran (1961) reportó la localidad tipo como "Cedros Island ("Cerros Island" en la descripción original, pero probablemente proveniente de la Bahía de San Quintín), Baja California, México.

(Grismer 1993, 2001; Samaniego-Herrera et al. 2007). Las islas del Mar de Cortés, pertenecientes al estado de Baja California están habitadas sólo por reptiles. Estas islas son El Muerto, Coloradito, Encantada, Blancos, San Luis, Willard, Mejía, Granito, Ángel de la Guarda, Pond, Cardonosa Este, Partida Norte, Rasa, Salsipuedes, Roca Lobos, San Lorenzo Norte, San Lorenzo Sur, y varias islas en Bahía de los Ángeles (Grismer 2002).

Las condiciones climáticas de Baja California varían mucho, debido a su longitud, topografía compleja, altitud y su ubicación entre dos cuerpos de agua muy disímiles, el Océano Pacífico y las cálidas aguas del Mar de Cortés (Humphrey 1974). Esta misma condición fue mencionada en 1922 (p. 95) por el naturalista-explorador Edward W. Nelson: "el clima de Baja California como un todo es extremadamente cálido y seco. Hay, sin embargo, importantes variaciones climáticas locales de acuerdo a la situación en relación con la latitud, altitud, sierras y a las costas este y oeste." Con la excepción del extremo noroeste y las regiones montañosas, Baja California se caracteriza generalmente por temperaturas medias anuales relativamente altas con bajas cantidades de precipitación (Grismer 2002). Para casi todo Baja California, los meses más cálidos son julio y agosto y el mes más frío es enero (Markham 1972).

Dos grandes regímenes climáticos, el de la Costa del Pacífico y el del Mar de Cortés (Meigs 1966), se originan en latitudes norte y sur, respectivamente, éstos influyen en los patrones climáticos tanto en Baja California como en el sur de California (Turner y Brown 1982). En el invierno, las tormentas ciclónicas de latitud norte se originan en el Golfo de Alaska desplazándose hacia el sur llevando lluvia a gran parte de la costa noroeste y regiones montañosas. En latitudes sur, a fines del verano y otoño, huracanes del sur se desarrollan en aguas cálidas en la costa occidental del sur y centro de México (Markham 1972). Estos sistemas de tormenta ocasionalmente llevan precipitaciones a la porción sur del estado. Grandes sistemas anticiclónicos, generando vientos de más de 200 km/h, suelen moverse hacia el norte a lo largo del Océano Pacífico y golpean la costa suroeste del estado. Algunos huracanes del Pacífico se mueven hacia el norte cientos de kilómetros, pasando por las cálidas aguas del Mar de Cortés, incrementando su intensidad y velocidad de viento. Sistemas más pequeños que se desarrollan localmente en el Mar de Cortés también producen fuertes vientos, tormentas y mar picado. Estos sistemas de golfo o chubascos son generalmente de corta duración, pero son capaces de causar graves daños una vez que tocan tierra (Grismer 2002).

En Baja California, la mayoría de los ríos se encuentran en la parte noroeste del estado. Las aguas se originan en las montañas altas y drenan al Océano Pacífico o al Mar de Cortés. El Río Tijuana drena el noroeste de Sierra Juárez y el condado de San Diego y desemboca en el estuario del Río Tijuana en California. Los ríos en Baja California se caracterizan por presentar suelos aluviales y flujos intermitentes dependiendo de las lluvias. La mayoría ya no fluye continuamente al océano debido a largos períodos de sequía y a canalizaciones y desviaciones hechas por humanos. Los ríos que drenan hacia el Océano Pacífico son Tijuana, Las Palmas, Guadalupe, Ensenada, Maneadero, Santo Tomás, San Vicente, Salado, San Rafael, San Telmo, Santo Domingo, Santa María y El Rosario. Al sur del Río El Rosario, las condiciones desérticas impiden el flujo continuo, aunque tormentas fuertes del Pacífico desde el norte o huracanes ocasionales provenientes del sur causan inundaciones episódicas dentro de estas cuencas normalmente secas.

En el lado oriental de la Cordillera Peninsular, los arroyos son generalmente nombrados por el nombre del cañón pues sus flujos son efímeros. En la mayoría de los años, la temporada de lluvias produce suficiente agua para permitir que los escurrimientos de agua alcancen el suelo del desierto. En la parte norte del estado, las corrientes de la Sierra Juárez desembocan en la cuenca de la Laguna Salada, que rara vez se llena lo suficientemente como para drenar en el Mar de Cortés. El Río Colorado, ubicado a lo largo de la frontera del estado con Arizona y Sonora, históricamente ha formado un delta rico y productivo, pero ahora este río se ha canalizado y desviado para uso humano (Grismer 2002). Al sur de la Sierra San Pedro Mártir, los cauces de ríos están secos, excepto en áreas alimentadas con aguas de oasis. En Baja California, los ríos alimentados por oasis incluyen San Fernando Velicatá, Cataviña, Santa María, Bahía de los Ángeles, San Borja y Santa Gertrudis (Grismer y McGuire 1993).

Las distribuciones de muchas de las especies nativas del estado coinciden con las regiones fitogeográficas de Baja California (Grismer 1994b, 2002). Como buenas indicadoras de provincias bióticas naturales, ellas probablemente están influenciadas por las mismas características del medio ambiente que limitan la distribución de anfibios y reptiles (Grismer 2002). Dentro del estado de Baja California, existen cinco regiones: (1) California, (2) Bosques de Coníferas, (3) Desierto Inferior del Colorado, (4) Desierto del Vizcaíno, y (5) Desierto de la Costa Central del Golfo. Adicionalmente, hemos añadido una sexta región para tener en cuenta las especies marinas que pasan la mayor parte de sus vidas en el mar (por ejemplo, tortugas marinas y serpientes marinas). Una comparación de las distribuciones de anfibios y reptiles y los límites de las regiones fitogeográficas revela que 54 de las 113 (47.8%) especies nativas están restringidas a una sola región. Las restantes 59 especies son generalistas, con distribuciones que ocupan dos o más regiones. Estas regiones parecen ser un buen indicador del estado natural de los ensambles, a pesar las amplias zonas de transición entre ellas.

El bajo porcentaje de anfibios nativos en el estado (17/113 o 15%), en comparación con los reptiles nativos (96/113 o 85%) es un resultado de las condiciones de sequía encontradas a lo largo de Baja California. Los anfibios nativos se limitan principalmente a la región de California y a los Bosques de Coníferas, mientras que el Río Colorado y los oasis también proporcionan un hábitat adecuado para estos animales (Grismer y McGuire 1993; Grismer 1994b). Sólo dos especies de anfibios (*Scaphiopus couchi* y *Anaxyrus punctatus*) tienen historias de vida tolerantes a la sequía que les permiten distribuirse casi de forma continua en todas las regiones desérticas. La diversidad de reptiles está más uniformemente distribuida a través de las regiones de Baja California. La descripción física de cada región fue tomada en gran medida de Grismer (1994b, 2002), que fue adaptada de Shreve y Wiggins (1964), Wiggins (1980), Turner y Brown (1982) y Cody et al. (1983).

Región de California

Ocupando el cuarto noroeste del estado, esta región se extiende 275 km desde la frontera de Estados Unidos a las cercanías de El Rosario a lo largo de la costa del Pacífico donde lentamente se incorpora en la región del Desierto del Vizcaíno situado al sur. Las regiones más meridionales del estado alguna vez hospedaron comunidades similares a las que ahora se encuentran en toda la Región de California, aunque los periodos sucesivos de sequía durante el Pleistoceno sustituyeron a éstas con los Desiertos Peninsulares (Van Devender 1990). El resultado está entremezclado sobre una gran área geográfica y la presencia de los restos de la flora de la Región de California en el Desierto del Vizcaíno. Al este, la Región de California se extiende hasta la Sierra Juárez y la Sierra San Pedro Mártir a una altitud de hasta 2000 m donde inicia el cinturón de Pino de Jeffrey (Delgadillo 2004). A lo largo de la ladera oeste de las montañas, entre 1500 y 2000 m, un ecosistema meso-mediterráneo de bosque mixto de coníferas y chaparral que incluye bosques abiertos de pino de Parry (*Pinus quadrifolia*) y chaparral de montaña (Delgadillo 2004). El límite de la Región de California normalmente se extiende a unos 1500 m, pero a veces a altitudes superiores. Áreas en elevaciones más bajas al norte de la Sierra Juárez, entre la Sierra Juárez y la Sierra San Pedro Mártir en Valle de la Trinidad y Paseo San Matías, y al sur de San Pedro Mártir, permiten a la Región de California entrar en contacto y entremezclarse con el Desierto Inferior del Colorado.

En el Océano Pacífico, la Región de California está representada por cuatro grupos de islas: Islas Coronado (173 hectáreas), Todos Santos (118 hectáreas), San Martín (256 hectáreas) y San Jerónimo (48.3 hectáreas). Las Islas Coronado se encuentran en aguas costeras a 13.0 km de Tijuana, éstas forman un archipiélago de cuatro islas principales, de las cuales, la Coronado Norte y la Coronado Sur hospedan anfibios y reptiles (Grismer 1993, 2001; Samaniego-Herrera et al. 2007). Las Islas Todos Santos se encuentran a 17.3 km de la Bahía de Ensenada y están compuestas por dos islas (Grismer 1993, 2001; Samaniego-Herrera et al. 2007). La Isla San Martín está a 5.1 km de la Bahía San Quintín, mientras que la Isla San Jerónimo está situada a 9.7 km de la costa, al sur de El Rosario (Samaniego Herrera et al. 2007).

La Región de California es relativamente fresca la mayor parte del año debido a que prevalecen los vientos del noroeste que se producen por la fría Corriente de California, que sopla fuera de las aguas del Océano Pacífico (Hastings y Turner 1965; Meigs 1966). En los meses de verano, las temperaturas promedian entre 20 y 25°C; temperaturas más frías se encuentran a lo largo de la margen costera, con temperaturas más cálidas cerca de las estribaciones de las sierras Juárez y San Pedro Mártir. Las temperaturas invernales promedian entre 10.0 y 12.5°C. La región recibe la mayor parte de su precipitación de las tormentas de invierno del Pacífico que se originan en el Golfo de Alaska. El incremento de aridez se produce más al sur en las cercanías de El Socorro. Aquí, la costa rara vez recibe cualquier precipitación medible de mayo a septiembre (Humphrey 1974). En verano, niebla y nubes bajas a menudo cubre la costa, extendiéndose varios kilómetros tierra adentro y alcanzando 1000 m de altitud (Markham 1972).

Esta región es una extensión al sur de las comunidades costeras de matorral de Salvia y chaparral del sur de California (Rebman y Roberts 2012). Las especies de plantas dominantes en las zonas costeras son artemisa de California (*Artemisia californica*), salvia blanca y salvia negra (*Salvia apiana* y *S. mellifera*), alforfón de California (*Eriogonum fasciculatum*) y agave costero (*Agave shawii*). El chaparral se encuentra principalmente en los cañones en áreas más cercanas a la costa y a elevaciones más altas tierra adentro (Delgadillo 2004). Las especies dominantes del chaparral incluyen chamiso (*Adenostoma fasciculatum*), hoja-lila (*Ceanothus crassifolia*), toyon (*Heteromeles arbutifolia*), laurel sumac (*Malosma laurina*) y mora limonada (*Rhus integrifolia*).

ENSAMBLE DE LA REGIÓN DE CALIFORNIA

De las 54 especies de anfibios y reptiles nativas de la región de California, 8 están limitada a esta región: *Aneides lugubris, Ensatina eschscholtzii, Elgaria nana, Anniella geronimensis, A. stebbinsi, Diadophis punctatus, Lampropeltis herrerae* y *Crotalus caliginis.*

Este ensamble tiene 7.1% (8/113) de la herpetofauna total del Estado, representando tres anfibios y cinco reptiles. En Baja California, *Aneides lugubris, Ensatina eschscholtzii, Anniella stebbinsi* y *Diadophis punctatus* tienen sus límites de distribución sur en la Región de California y entran a Baja California desde los Estados Unidos. Las dos salamandras están confinadas a hábitats mediterráneos en la parte sur de su distribución. La distribución más austral de *E. eschscholtzii* se encuentra a 22 km al sur de Ensenada, a lo largo de la costa (Peralta-García y Valdez-Villavicencio 2004), mientras que la distribución *A. lugubris* llega hasta las inmediaciones del Valle de Santo Tomás (Lynch y Wake 1974).

La distribución de *Anniella stebbinsi* varía ampliamente en la Región de California (Bury 1983; Papenfuss y Parham 2013). En el norte, su distribución más oriental se encuentra en La Rumorosa, al norte de la Sierra Juárez (Hunt 1983). Más al sur, probablemente se extiende hasta arroyos arenosos en la base de la Sierra San Pedro Mártir (Welsh 1988). Su distribución más austral se encuentra en el Arroyo Pabellón, aproximadamente a 17 km al sur de San Quintín (Hunt 1983). La distribución más austral de *Diadophis punctatus* se encuentra en la Isla San Martín, ubicada frente a la costa de San Quintín.

En la península, *D. punctatus* ha sido registrada desde el Rancho San José y el Arroyo San Telmo, 64 km al norte (Welsh 1988). Ningún espécimen ha sido reportado en la zona que queda entre estas dos localidades, aunque estas serpientes de hábitos secretivos probablemente son más comunes que lo reflejado por los pocos especímenes recolectados (Grismer 2002).

Cuatro miembros de este ensamble son endémicos de Baja California. *Anniella geronimensis* se encuentra en la Isla San Jerónimo, Isla San Martín y la costa 6 km al norte de la Colonia Guerrero ligeramente al sur de Punta Baja en la orilla norte de la Bahía Rosario (Shaw 1949; Hunt 1983; Sánchez-Pacheco y Mellink 2001). Probablemente habita en el ecotono entre las regiones de los desiertos de California y el Vizcaíno, aunque su hábitat preferido parece estar limitado a 4 km de la costa y no se distribuye extensamente en el desierto de transición (Shaw 1953; Grismer 2002).

Las tres especies restantes, *Elgaria nana*, *Lampropeltis herrerae* y *Crotalus caliginis* son endémicas a islas. *Lampropeltis herrerae*, relacionada a la Serpiente Real de California (*L. multifasciata*), se encuentra en la Isla Todos Santos Sur (Zweifel 1975; Samaniego-Herrera et al. 2007). Myers et al. (2013), consideraron a *L. herrerae* como una sinonimia de *L. multifasciata* en su revisión de *L. zonata*, pero no discutieron su razonamiento. Grismer (1993) señaló que su presencia en esta isla representa una distribución relictual y probablemente es el resultado de la disminución de hábitat entre la costa y las montañas que habita su pariente más cercano, *L. multifasciata*, en las sierras Juárez y San Pedro Mártir.

Elgaria nana y *Crotalus caliginis* son endémicas a las Islas Coronado. *Elgaria nana* se encuentra en las Islas Coronado Norte y Coronado Sur, mientras que *C. caliginis* sólo se encuentra en la Isla Coronado Sur (Grismer 1993). Ambas parecen tener tamaños corporales más pequeños en comparación con sus parientes del continente. *Elgaria nana* es 42 mm más pequeña en longitud hocico cloaca que *Elgaria multicarinata* (Fitch 1934). Sobre esta base

Grismer (2002) elevó al taxón originalmente descrito como subespecie al rango de especie. Asimismo, *Crotalus caliginis* originalmente fue descrita como una subespecie de la especie continental *C. viridis* (= *C. oreganus*) y fue elevada por Grismer (2002) sobre la base del pequeño tamaño corporal que alcanzan los adultos de la Isla Coronado Sur.

Un total de 10 especies de anfibios y 44 de reptiles viven en esta región representando 47% (54/113) de la herpetofauna del estado. Adicionalmente a las especies de este ensamble, la región también hospeda especies generalistas, las cuales habitan en dos o más regiones fitogeográficas del estado: *Batrachoseps major*, *Ensatina klauberi*, *Anaxyrus boreas*, *A. californicus*, *Pseudacris cadaverina*, *P. hypochondriaca*, *Rana draytonii*, *Spea hammondii*, *Actinemys marmorata*, *Elgaria multicarinata*, *Crotaphytus vestigium*, *Gambelia copeii*, *Coleonyx variegatus*, *Callisaurus draconoides*, *Phrynosoma blainvillii*, *Sceloporus occidentalis*, *S. orcutti*, *S. zosteromus*, *Urosaurus nigricaudus*, *Uta stansburiana*, *Plestiodon gilberti*, *P. skiltonianus*, *Aspidoscelis hyperythra*, *A. labialis*, *A. tigris*, *Xantusia henshawi*, *X. wigginsi*, *Lichanura trivirgata*, *Arizona elegans*, *Chilomeniscus stramineus*, *Hypsiglena ochrorhyncha*, *Lampropeltis getula*, *Masticophis fuliginosus*, *M. lateralis*, *Pituophis catenifer*, *P. vertebralis*, *Rhinocheilus lecontei*, *Salvadora hexalepis*, *Sonora semiannulata*, *Tantilla planiceps*, *Thamnophis hammondii*, *Trimorphodon lyrophanes*, *Rena humilis*, *Crotalus mitchellii*, *C. oreganus* y *C. ruber*.

Batrachoseps major se encuentra en toda la mitad norte de la Región de California, pero en poblaciones más aisladas hacia la porción sur. Grismer (1982) reportó la localidad más austral en Arroyo Grande, 24 km al sureste de El Rosario. Esta especie también habita fuera de la Región de California en la Sierra San Pedro Mártir (Welsh 1988). Las poblaciones de la Sierra San Pedro Mártir son altamente divergentes de otras poblaciones sobre la base de datos moleculares (Mártinez-Solano et al. 2012).

Varias especies o subespecies han sido descritas para las islas de la Región de California, pero han sido sinonimizadas con sus parientes del continente. *Aspidoscelis tigris vivida*, se encuentra en las Islas Coronado Norte y Coronado Sur, fue originalmente descrita por Walker (1981), pero reconocida como una clase de patrón de *A. tigris* por Grismer (2002). *Pituophis catenifer coronalis*, que habita en la Isla Coronado Sur, fue originalmente descrita por Klauber (1946b), pero más tarde sinonimizada con sus homólogos del continente por Grismer (2002). Esta serpiente es conocida por cuatro especímenes de museo y no se sabe prácticamente nada de su historia natural (Oberbauer 1992; Grismer 2002). *Diadophis anthonyi*, que habita en la Isla Todos Santos Sur, originalmente fue descrita por Van Denburgh y Slevin (1923), pero sinonimizada por Grismer (2001) debido a la falta de caracteres diagnósticos. *Elgaria multicarinata ignava* fue descrita por Van Denburgh (1905), pero posteriormente fue reconocida como una clase de patrón de *E. multicarianata* (Grismer 2002).

Gran parte del hábitat en la Región de California ha sido sustituido por el desarrollo urbano en las áreas metropolitanas de Tijuana, Ensenada y San Quintín. Pequeñas comunidades costeras están creciendo y en muchas áreas se han fusionado entre ellas para formar una extensa zona costera urbanizada. Si no se han urbanizado, la mayoría de las áreas planas restantes han sido convertidas a la agricultura, incluyendo el Valle de Ojos Negros, Valle de La Trinidad y la llanura de San Quintín. Estas áreas tienen una baja disponibilidad de agua, como resultado de la extracción de aguas subterráneas de los cañones de estribaciones cercanas para apoyar al sector agrícola. Si no se han drenado, gran parte de los bosque ribereños de sauce de esta región han sido sustituidos por vegetación no nativa, altamente invasiva (*Arundo donax, Tamarix* spp., etc.). La recolecta ilegal de especies de reptiles de alto valor económico (por ejemplo, *Lichanura trivirgata, Lampropeltis herrerae*) ha sido documentada (Mellink 1995). Mellink encontró colectores en la Isla Todos Santos Sur tratando de recolectar cantidades grandes de reptiles. El ganado vacuno junto con otros tipos de ganado pastorea en toda esta región. La Región de California contiene el tercio sur de uno de los

cinco ecosistemas mediterráneos en la tierra y es considerada una región de alta biodiversidad (Myers et al. 2001). Las áreas naturales protegidas a lo largo de la zona costera son pequeñas y escasas. En la actualidad se están impartiendo cursos de planificación para ayudar a proteger la Bahía San Quintín a través de un desarrollo ordenado (The Natural Conservancy 2007). Se han llevado a cabo iniciativas de planes de conservación para proteger el frágil ecosistema de Colonet Mesa de desarrollos futuros (Clark et al. 2008).

Región del Bosque de Coníferas

En el estado de Baja California, prominentes bosques de coníferas se desarrollan en las Sierras Juárez y San Pedro Mártir. Este ecosistema boscoso es la extensión más austral de los bosques de coníferas de montaña (Pase 1982) y hospedan comunidades de afinidades templadas (Grismer 2002). El cinturón de pino comienza a 1500 m con una mezcla de pino de Parry, pino piñonero y chaparral (Delgadillo 2004). La Sierra Juárez tiene la mayor extensión de bosques con una cobertura de 34,2113 hectáreas, en comparación con la Sierra San Pedro Mártir con tan sólo 149,366 hectáreas (SARH 1977). Extensas praderas se presentan en las elevaciones medias de las montañas, las cuales se inundan del agua producto del derretimiento de nieve de las partes altas de las montañas (Welsh 1988).

Estas dos sierras son un componente central de la Cordillera Peninsular, orientada en dirección norte-sur, con una elevación gradual en la vertiente occidental y pendientes empinadas en la vertiente oriental que caen estrepitosamente hasta el suelo del desierto. Las transiciones de la vertiente oriental son abruptas. En el Picacho del Diablo, se presenta un gradiente de 2700 m en tan sólo 16 km del Desierto Inferior del Colorado a la cima del pico. Esta región se extiende hacia el sur desde la frontera México—Estados Unidos a aproximadamente 300 km hasta aproximadamente cerca del Cerro Matomi (Grismer 2002).

Esta región recibe la mayor precipitación en

el Estado, con lluvias en invierno y primavera (Hastings y Turner 1965). Aproximadamente 70% de esta precipitación se produce en el invierno a través de tormentas frías del norte. Temperaturas bajo cero son comunes; la nieve comienza a caer a finales de noviembre. El otro 30% de la precipitación cae en el verano entre los meses de julio a septiembre (Roberts 1981) cuando corrientes de aire cálido y húmedo provenientes del Mar de Cortés se mueven en dirección noroeste, elevándose y enfriándose a medida que se aproximan a las elevaciones altas de las sierras (Humphrey 1974). Las tormentas en julio y agosto son comunes. La cantidad de precipitación puede ser altamente variable, con un promedio anual de 17.8 cm y un máximo anual de 33.4 cm para el estado (Hastings y Turner 1965). Condiciones más secas prevalecen a elevaciones más bajas en ambas vertientes (occidental y oriental). El promedio mensual de temperatura del aire para cualquier año dado puede variar desde −0.2°C hasta 17°C.

La composición florística es relativamente diversa, ésta está definida principalmente por arbustos grandes y árboles. Especies dominantes incluyen pino piñonero (*Pinus quadrifolia*), pino de Jeffrey (*Pinus jeffreyi*), pino costero (*Pinus contorta*), pino de azúcar (*Pinus lambertiana*), encino negro de California (*Quercus kelloggii*), abeto (*Abies concolor*), aspen (*Populus tremuloides*) y cedro de incienso (*Calocedrus decurrens*). Estos bosques tienen un sotobosque abierto que incluye manzanita de hoja-verde (*Arctostaphylos patula*).

ENSAMBLE DE LA REGIÓN DEL BOSQUE DE CONÍFERAS

De las 41 especies de anfibios y reptiles nativas de la región, 4 están limitadas a esta área: *Rana boylii*, *Sceloporus vandenburgianus*, *Lampropeltis multifasciata* y *Thamnophis elegans*.

Este ensamble tiene 3.5% (4/113) de la herpetofauna total del estado, que representa un anuro, una lagartija, y dos serpientes. Estas cuatro son especies adaptadas a ambientes montañosos, prefieren hábitats templados fríos. La distribu-

ción de cada una de ellas representa el punto más austral de su distribución y se encuentra separada de la distribución de las poblaciones más cercanas de estas especies en los Estados Unidos (Grismer 2002). *Sceloporous vandenburgianus* y *Lampropelits multifasciata* habitan en las Sierras Juárez y San Pedro Mártir y las poblaciones más cercanas al norte se encuentran en Laguna y en las Montañas Cuyamaca en el Condado de San Diego. De acuerdo con Censky (1986) y haciendo referencia a él, Grismer (2002) consideró a *S. vandenburgianus* ausente de la Sierra Juárez. Sobre la base de registros de museos, se sabe que esta especie habita en las Sierras Juárez y San Pedro Mártir. En el reciente estudio de *L. zonata* por Myers et al. (2013), las poblaciones de las Sierras Juárez y San Pedro Mártir fueron colocadas en *L. multifasciata*. Ambas poblaciones de *L. zonata* han sido consideradas por algunos autores como la subespecie *L. z. agalma* (Zweifel 1952b, 1975), pero Rodríguez-Robles et al. (1999) encontraron que las poblaciones de Sierra Juárez están más cercanamente relacionadas a las poblaciones del sur de California, mientras que la población de la Sierra San Pedro Mártir está más cercanamente relacionada a *L. herrerae* de la Isla Todos Santos Sur. Myers et al. (2013) no discutieron ni *L. z. agalma*, ni *L. herrerae*, en detalle y trataron a ambas como sinonimias de *L. multifasciata*.

Thamnophis elegans habita sólo en la Sierra San Pedro Mártir y está separada por 400 km de la población más cercana que se encuentra en las Montañas de San Bernardino, California (Grismer 2002). En estas montañas, se distribuye por encima de 1820 m sobre la ladera central (Welsh 1988).

La presencia de *Rana boylii* en la Sierra San Pedro Mártir necesita investigación adicional. La evidencia de su presencia en Baja California está basada sobre dos ejemplares perdidos. Loomis (1965) reportó estos especímenes del California State College en Long Beach (1080 y1081), recolectados por Elbert L. Sleeper en 1961 en el extremo inferior de la Pradera La Grulla a 2000 m en la parte sur de la escarpa principal. Los especímenes se perdieron cuando fueron enviados en préstamo (presumiblemente entre Long Beach State College y la Universidad de California, Berkeley), pero no antes de que fueran identificados por R. C. Stebbins y R. G. Zweifel (Grismer 2002). No se ha encontrado evidencia adicional de la presencia de esta especie en el estado, por lo que Welsh (1988) y Lovich et al. (2009) concluyeron que esta especie podría haber sido extirpada. Si aún persiste en Baja California, esta población de *R. boylii* representa una separación de 480 km de la población más cercana, previamente conocida en las Montañas de San Gabriel en California (Loomis 1965), que también se cree ha sido extirpada.

Una especie típicamente de montaña, *Ensatina klauberi* se encuentra en poblaciones aisladas en las Sierras Juárez y San Pedro Mártir. En la Sierra Juárez, habita cerca de los ranchos Las Cuevitas y Baja Largo del Sur, ambos aproximadamente a 19 km al sur de Laguna Hanson (Heim et al. 2005) y en la Sierra San Pedro Mártir en La Tasajera (Mahrdt et al. 1998). Recientemente, esta especie ha sido descubierta en un área muy pequeña dentro de un hábitat atípico para ella a lo largo de la costa de San Quintín (Devitt et al. 2013). Hasta este descubrimiento, *Ensatina klauberi* había sido considerada uno de los mejores ejemplos de especies del ensamble de la Región del Bosque de Coníferas.

A pesar del pequeño número de especies confinadas a la Región del Bosque de Coníferas, esta área constituye un importante ecosistema templado dentro del estado. Un total de 9 anfibios y 32 reptiles viven en la región, representando el 36.3% (41/113) de herpetofauna del estado. Además de los miembros de ensamble, la región también hospeda: *Batrachoseps major, Ensatina klauberi, Anaxyrus boreas, A. californicus, Pseudacris cadaverina, P. hypochondriaca, Rana draytonii, Spea hammondii, Actinemys marmorata, Elgaria multicarinata, Gambelia copeii, Phrynosoma blainvillii, P. cerroense, Sceloporus occidentalis, S. orcutti, S. zosteromus, Uta stansburiana, Plestiodon gilberti, P. skiltonianus, Aspidoscelis tigris, Xantusia henshawi, X. wigginsi, Arizona elegans, Hypsiglena ochrorhyncha, Lampropeltis getula, Masticophis fuliginosus, M. lateralis, Pituophis catenifer, Rhinocheilus lecontei, Salvadora hexalepis, Tantilla planiceps, Thamnophis hammondii, Trimorphodon lyrophanes, Rena humilis, Crotalus mitchellii, C. oreganus* y *C. ruber.*

Amenazas de conservación de esta región incluyen el cambio climático, la ganadería y la tala de bosques. Esta última se maneja y regula para preservar la integridad del bosque (SARH 1977). Dos parques nacionales se encuentran en esta región, el Parque Nacional Constitución de 1857 que comprende 4950 hectáreas y el Parque Nacional Sierra San Pedro Mártir que comprende 72,909 hectáreas (Lovich et al. 2009).

Región del Desierto Inferior del Colorado

Las regiones desérticas del cuarto noreste del estado contienen los depósitos sedimentarios del Río Colorado. Sedimentos del Mioceno Tardío al Plioceno Tardío de la formación Bouse representan un tiempo cuando la región estuvo sumergida bajo una serie de lagos llenos del agua de los ríos que unían a estos cuerpos de agua con el Mar de Cortés (Spencer y Pearthree 2001). Como resultado, la región se caracteriza por sus amplias cuencas expansivas por debajo de 400 m de altitud. El Desierto Inferior del Colorado es la subregión más grande del Desierto de Sonora y se extiende a través del sureste de California, suroeste de Arizona, noroeste de Sonora y noreste de Baja California (Shreve 1951).

En el estado de Baja California, la Región del Desierto Inferior del Colorado se extiende 450 km desde la frontera con los Estados Unidos hasta las inmediaciones de Bahía de los Ángeles. Al oeste, limita con las empinadas laderas de las Sierras Juárez y San Pedro Mártir, al norte y al este con la frontera de Arizona y Sonora, y al sureste por el Mar de Cortés. Por debajo de Puertecitos, la región reduce su anchura a menos de 13 km de la costa, y en el sur se entremezcla ampliamente con la Región del Desierto de la Costa del Golfo (Peinado et al. 1994).

A pesar de su condición de zonas árida, el Río Colorado alguna vez proporcionó un rico ecosistema acuático, que incluía un delta expansivo y un estuario. Actualmente, el agua ya no fluye de este río al Mar de Cortés. Un extenso sistema de canales proporciona la mayor cantidad de agua desde las partes inferiores del río a las áreas agrícolas y urbanas en California, Arizona y Baja California. Otras fuentes de agua dentro de esta región provienen de las laderas orientales de las Sierras Juárez y San Pedro Mártir.

Además de las laderas orientales de las sierras, la región posee varias cadenas montañosas. Aisladas entre Laguna Salada y el Delta del Río Colorado, se encuentran las Sierras de Los Cucapás y El Mayor. Otras cadenas montañosas se conectan a la Cordillera Peninsular e incluyen las Sierras Las Tinajas, Las Pintas y San Felipe.

En el Mar de Cortés, ligeramente al norte de Bahía San Luis Gonzaga, hay seis islas pertenecientes a esta región: El Muerto, Coloradito, Encantada, Blancos, San Luis y Willard. Las cinco primeras forman parte del archipiélago de Islas Las Encantadas, mientras que la Isla Willard es una montaña cerca de la costa conectada a la península por un banco de arena durante mareas extremadamente bajas.

La Región del Desierto Inferior del Colorado, se localiza en la sombra de lluvia de las Sierras Juárez y San Pedro Mártir, éste es el desierto más caluroso y más estéril en Baja California. Como resultado de la baja altitud y el poco relieve topográfico, recibe menos de 5 cm de lluvia al año y es considerada la región de desierto más seca de América del Norte (Meigs 1953; Hastings y Turner 1965). La mayoría de la precipitación es incipiente y cae durante los meses de verano (Humphrey 1974) y en raras ocasiones se ve afectada por huracanes (Grismer 2002). Sin embargo, estas tormentas anticiclónicas han tocado tierra tan al norte como San Felipe (Markham 1972). Las temperaturas promedio para julio y agosto están por encima de la 32.5°C y la promedio máxima mensual es un registro de 38.2°C; la temperatura promedio de invierno es 12.5°C (Markham 1972).

La vegetación de la región está dominada por arbustos de gobernadora (*Larrea tridentata*). La parte norte se encuentra directamente dentro de la sombra de lluvia de las montañas y se caracteriza por ser de baja altitud, plana, arenosa,

caliente y de condiciones áridas (Meigs 1953). Más al sur, la región es más accidentada con grandes flujos de lava seca, cadenas montañosas y cauces secos. Las plantas dominantes incluyen hierba de burro (*Ambrosia dumosa*), ocotillo (*Fouquieria splendens*), incienso (*Encelia farinosa*) y agave de desierto (*Agave deserti*). En los arroyos y zonas escarpadas de las estribaciones, mezquite (*Prosopis glandulosa*), árbol de humo (*Psorothamnus spinosa*), palo verde de hoja chica (*Cercidium tetragonum*), palo fierro (*Olneya tesota*) y chollas (*Cylindropuntia* sp.) (Grismer 2002).

ENSAMBLE DE LA REGIÓN DEL DESIERTO INFERIOR DEL COLORADO

De las 56 especies de anfibios y reptiles nativos a esta región, 22 están limitadas a ésta: *Anaxyrus cognatus*, *A. woodhousii*, *Incilius alvarius*, *Lithobates yavapaiensis*, *Crotaphytus grismeri*, *Gambelia wislizenii*, *Phrynosoma mcallii*, *P. platyrhinos*, *Sceloporus magister*, *Uma notata*, *Urosaurus graciosus*, *U. ornatus*, *Uta encantadae*, *U. lowei*, *U. tumidarostra*, *Chionactis occipitalis*, *Hypsiglena chlorophaea*, *Masticophis flagellum*, *Thamnophis marcianus*, *Crotalus atrox*, *C. cerastes* y *C. muertensis*.

Este ensamble tiene 19.5% (22/113) de la herpetofauna total del estado, representado por 4 especies de anfibios y 18 de reptiles. De las seis regiones del estado, la Región del Desierto Inferior del Colorado hospeda al mayor número de especies confinada a sus límites. De las 22 especies, 5 están asociadas con el Río Colorado y su delta, 12 son especies adaptadas a condiciones áridas y tienen sus distribuciones más australes en esta región, y 5 son endémicas a la región.

Las especies de este ensamble que se extienden hacia el sur desde los Estados Unidos hasta el Desierto Inferior del Colorado incluyen *Gambelia wislizenii*, *Phrynosoma mcallii*, *P. platyrhinos*, *Sceloporus magister*, *Uma notata*, *Urosaurus graciosus*, *U. ornatus*, *Chionactis occipital*, *Hypsiglena chlorophaea*, *Masticophis flagellum*, *Crotalus atrox* y *C. cerastes*. *Gambelia wislizenii* y *S. magister* están ampliamente distribuidas en toda la región. A lo largo del borde occidental de su distribución se sabe que se extienden hasta Paseo San Matías y se sobreponen con las distribuciones de *Gambelia copeii* y *Sceloporus zosteromus*, respectivamente, sin muestras de intergradación (McGuire 1996; Grismer y McGuire 1996). Este mismo patrón también se presenta con *Masticophis flagellum* y *M. fulginosus* (Grismer 1994b) y es probable que se presente en muchas otras especies (Grismer y Mellink 2005).

Phrynosoma platyrhinos, Urosaurus graciosus, Chionactis occipitalis, Hypsiglena chlorophaea y *Crotalus cerastes* se distribuyen ampliamente en toda la región. *Chionactis occipitalis* se extiende 34 km al sur de San Felipe (Grismer 1989a). *Urosaurus graciosus* (Grismer 1989b) y *Phrynosoma platyrhinos* (Grismer 2002) se extienden hacia el sur hasta la Bahía San Luis Gonzaga. *Crotalus cerastes* se distribuye hasta el Arroyo Calamajué (Nelson 1922), pero se cree que habita al sur de Bahía de los Ángeles (Grismer 2002). La distribución de *Hypsiglena chlorophaea* fue recientemente delimitada por Mulcahy (2008) utilizando datos genéticos y se considera que habita a través del Desierto Inferior del Colorado.

Especies con distribuciones más limitadas que no se extienden tan al sur son *Phrynosoma mcallii, Uma notata Urosaurus ornatus* y *Crotalus atrox*. *Phrynosoma mcallii* se extiende a sólo 80 km al sur de la frontera (Funk 1981), mientras que *U. notata* presenta su distribución más austral en el extremo norte de la Sierra Las Pintas (Pough 1977). *Crotalus atrox* habita en el extremo noreste del estado y se extiende hacia el oeste hasta el lado occidental de la Sierra de Los Cucapás (Grismer 2002). *Urosaurus ornatus* está estrictamente asociada con la vegetación ribereña a lo largo del Río Colorado (Grismer 2002).

Anaxyrus cognatus, A. woodhousii, Incilius alvarius, Lithobates yavapaiensis y *Thamnophis marcianus* están asociadas con el Río Colorado; *A. cognatus, A. woodhousii* y *T. marcianus* han ampliado sus distribuciones en asociación con la industria agrícola y diques de su red de canales y riego. Se cree que *I. alvarius* habita en Baja California pero se sabe poco de su distribución en el estado. El único espécimen conocido fue reportado por Brattstrom (1951)

48 km al sur de Mexicali y 7.6 km al norte de El Mayor. Brattstrom también reportó la presencia de *T. marcianus* en la misma localidad a 7.7 km al norte de El Mayor. Por más de tres décadas, *L. yavapaiensis* no ha sido registrada en la parte inferior del Río Colorado y se piensa que ha sido extirpada como resultado de la introducción de *L. berlandieri* y *L. catesbeianus* (Vitt y Ohmart 1978; Platz y Frost 1984; Clarkson y de Vos 1986; Platz et al. 1990; Mellink y Ferreira-Bartrina 2000).

De las cinco especies endémicas a la región, cuatro son endémicas a islas del Archipiélago Las Encantadas. *Crotalus muertensis* fue descrita por Klauber (1949b) y se encuentra sólo en la Isla El Muerto, la isla más septentrional del archipiélago. *Uta lowei* también es endémica de la Isla El Muerto y fue descrita por Grismer (1994c), quien también describió *Uta tumidarostra* de la Isla Coloradito y *U. encantadae* de la Isla Encantada y el Islote Blancos. Todas las especies de *Uta* del archipiélago Las Encantadas tienen glándulas nasales hipertrofiadas y la porción del rostro está inflada como resultado de sus hábitos intermareales y dietas saladas (Grismer 1994c).

McGuire (1994) describió *Crotaphytus grismeri*, que es una especie endémica a las Sierras de los Cucapás y El Mayor. La distribución de *C. grismeri* es poco conocida, los únicos registros que se tienen para esta especie provienen de la serie tipo (McGuire 1994). Al parecer, desde su origen, estas montañas han estado aisladas de la Cordillera Peninsular ubicada al sur y al oeste (Barnard 1970).

Un total de 9 anfibios y 47 reptiles son nativos a esta región representando 49.6% (56/113) de la herpetofauna del estado. Además de estas especies, la región hospeda: *Anaxyrus boreas, A. punctatus, Pseudacris cadaverina, P. hypochondriaca, Scaphiopus couchii, Crotaphytus vestigium, Coleonyx switaki, C. variegatus, Phyllodactylus nocticolus, Dipsosaurus dorsalis, Sauromalus ater, Callisaurus draconoides, Petrosaurus mearnsi, Sceloporus orcutti, Urosaurus nigricaudus, Uta stansburiana, Plestiodon gilberti, Aspidoscelis tigris, Xantusia henshawi, X. wigginsi, Lichanura trivirgata, Arizona elegans, Bogertophis rosaliae, Hypsiglena chlorophaea, Lampropeltis getula, Phyllorhynchus decurtatus, Pituophis catenifer, Rhinocheilus lecontei, Salvadora hexalepis, Sonora semiannulata, Tantilla planiceps, Trimorphodon lyrophanes, Rena humilis, Crotalus mitchellii* y *C. ruber.*

Amenazas graves a esta región del noreste incluyen una amplia expansión agrícola y la urbanización de la ciudad de Mexicali. Poca agua fluye en el curso natural del Río Colorado. En toda la región, varias organizaciones (por ejemplo, SCORE International) patrocinan carreras todo terreno, que promueven el extenso uso recreativo de vehículos doble tracción en áreas no reguladas del desierto. Las únicas regiones protegidas se encuentran en la Reserva de la Biosfera del Alto Golfo de California (Lovich et al. 2009; Murphy y Méndez de la Cruz 2010).

Región del Desierto del Vizcaíno

Esta región se encuentra en la esquina suroeste del estado de Baja California y está separada del Mar de Cortés ubicado al este por la Región del Desierto Central de la Costa del Golfo. La región se extiende desde cerca de El Rosario hasta la Laguna San Ignacio en Baja California Sur. Su límite occidental es el Océano Pacífico, mientras que la extensión de su límite oriental varía mucho, pero en general está limitada por la Cordillera Peninsular. La porción meridional de la región del Desierto del Vizcaíno es mucho más plana y se extiende hacia el interior hasta las faldas y estribaciones rocosas de las montañas. En Baja California, las regiones más planas se extienden sólo a pocos kilómetros al norte de la frontera estatal con Baja California Sur a 20 km al norte de Villa Jesús María, mientras que las porciones del norte de la Región del Desierto del Vizcaíno consisten de cadenas montañosas más pequeñas, mesas y cauces secos (Bostic 1971).

En el Océano Pacífico, esta región incluye dos grupos de islas, Isla Cedros e Islas San Benito, ambos ubicados en el extremo suroeste del estado. La Isla Cedros es la más grande del Pacífico en el estado, cubriendo 34,827 hectáreas, con una altitud de 1204 m y situada a 100 km del punto más cercano en la península (Samaniego-Herrera et al. 2007). Las Islas San Benito son un archipiélago de tres islas ubicadas cercanamente entre sí (Isla San Benito Oeste, Medio y Este). Estas tres islas cubren aproximadamente 600 hectáreas, alcanzan una altitud de 212 m y se encuentran a 140 km de la costa (Samaniego-Herrera et al. 2007).

La región está caracterizada como un clima desértico templado "tipo niebla" con poca precipitación de invierno y verano (Meigs 1966) y es considerada como una parte de los desiertos de la península (Hafner y Riddle 1997). Estas condiciones climáticas someras están muy influenciadas por los vientos del oeste que llevan la Corriente Fría de California del Océano Pacífico a las aguas de la costa. Esta influencia marina genera capas de aire frío y húmedo que se mueven bajo el aire seco descendiente, lo que resulta en niebla, nubes bajas y falta de precipitación (Bostic 1971). Niebla densa se puede extenderse hasta 10 km hacia el interior (Grismer 2002). La máxima precipitación ocurre en invierno y su promedio es de 5.5 cm (diciembre-febrero). La precipitación en primavera (marzo-mayo) y verano (junio-agosto) disminuye progresivamente, con el verano recibiendo un promedio de 1 cm de precipitación (Hastings y Turner 1965). La temperatura promedio del aire en el verano (julio-septiembre) e invierno (enero-febrero) va de 23°C a 28°C y de 15°C a 18°C respectivamente, produciendo un clima templado agradable (Markham 1972).

En la región se encuentran varios arroyos alimentados por oasis, los cuales hospedan comunidades más templadas (Grismer y McGuire 1993). Éstos incluyen San Fernando Velicatá, Cataviña, San Borja y Santa Gertrudis.

En gran parte de la región la vegetación es abierta, raquítica, ampliamente espaciada y depauperada debido a la presencia de vientos continuos en la costa provenientes del Océano Pacífico (Grismer 2002). En áreas protegidas de los vientos, la riqueza específica de planta aumenta bruscamente. Las plantas dominantes incluyen agave azul (*Agave cerulata*), datilillo

(*Yucca valida*), hierbas de burro (*Ambrosia dumosa* y *A. camphorata*), cirio (*Fouquieria columnaris*), palo Adán (*Fouquieria californicus*), cardón (*Pachycereus pringlei*), árbol de elefante (*Pachycormus discolor*), pitahaya agria (*Stenocereus gummosus*), palo verde de pocas hojas (*Cercidium microphyllum*) y mezquite (*Prosopis glandulosa*).

ENSAMBLE DE LA REGIÓN DEL DESIERTO DEL VIZCAÍNO

De las 52 especies de anfibios y reptiles nativos de esta región, 5 están limitadas a ésta: *Elgaria cedrosensis, Urosaurus lahtelai, Bipes biporus, Arizona pacata* y *Pituophis insulanus.*

Este ensamble tiene 4.4% (5/113) de la herpetofauna total del estado, representando dos serpientes, un amphisbenido y dos lagartijas. Tres de estas especies son endémicas al estado de Baja California. *Pituophis insulanus* es endémica a Isla Cedros (Klauber 1946b) y se encuentra en toda la isla desde su costa hasta la cima de las montañas (Grismer 2002). *Elgaria cedroensis* también es endémica a Isla Cedros y áreas costeras adyacentes de la península desde 30 km al sur de El Rosario hasta 20 km al norte de Villa Jesús María (Bostic 1971; Lais 1976; Grismer 1988; Grismer y Hollingsworth 2001). *Urosaurus lahtelai,* descrita por Rau y Loomis (1977), habita en las proximidades de Cataviña, Las Arrastras de Arriola y Las Palmas (Grismer 2002). La evolución de *Urosaurus lahtelai* es interesante porque parece estar asociado con la exposición del batolito granítico subyacentes a esta región y está rodeada de poblaciones de su especie hermana, *U. nigricaudus* (Rau y Loomis, 1977; Grismer 1994b).

Varias de las especies que habitan en la Región del Desierto del Vizcaíno tienen su distribución más septentrional en esta región y se extienden desde el sur de la península. Sólo *Bipes biporus* y *Arizona pacata* están confinadas a esta región. *Bipes biporus* es el único amphisbenido en la península y se extiende a través de la frontera del estado hasta a unos 17 km al norte de Villa Jesús María (Papenfuss 1982). *Arizona pacata* prefiere la costa fría del desierto y se

extiende hacia el norte hasta aproximadamente la desviación a Bahía de los Ángeles a lo largo de la carretera Méx. 1 (Grismer 2002). Otras especies que siguen un patrón similar de distribución pero que se distribuyen en otras regiones fitogeográficas incluyen *Petrosaurus repens, Aspidoscelis labialis, Chilomeniscus stramineus, Hypsiglena slevini* y *Crotalus enyo.*

Un total de 5 anfibios y 47 reptiles son nativos a esta región representando 46% (52/113) de la herpetofauna estatal. Además de las especies pertenecientes a este ensamble, la región también hospeda: *Anaxyrus boreas, A. punctatus, Pseudacris cadaverina, P. hypochondriaca, Scaphiopus couchii, Elgaria multicarinata, Crotaphytus vestigium, Gambelia copeii, Coleonyx switaki, C. variegatus, Phyllodactylus nocticolus, Dipsosaurus dorsalis, Sauromalus ater, Callisaurus draconoides, Petrosaurus mearnsi, P. repens, Phrynosoma cerroense, Sceloporus occidentalis, S. orcutti, S. zosteromus, Urosaurus nigricaudus, Uta stansburiana, Aspidoscelis hyperythra, A. labialis, A. tigris, Xantusia wigginsi, Lichanura trivirgata, Bogertophis rosaliae, Chilomeniscus stramineus, Hypsiglena ochrorhyncha, H. slevini, Lampropeltis getula, Masticophis fuliginosus, M. lateralis, Phyllorhynchus decurtatus, Pituophis vertebralis, Rhinocheilus lecontei, Salvadora hexalepis, Sonora semiannulata, Tantilla planiceps, Thamnophis hammondii, Trimorphodon lyrophanes, Rena humilis, Crotalus enyo, C. mitchellii, C. oreganus* y *C. ruber.*

Algunas especies no se podrían encontrar en las condiciones áridas de la Región del Desierto del Vizcaíno si no fuera por la presencia de arroyos alimentados por oasis. La presencia de *P. cadaverina, P. hypochondriaca* y *Thamnophis hammondii* en oasis se cree que constituyen evidencia de un pasado ecológico más templado (Grismer y McGuire 1993; McGuire y Grismer 1993). Asimismo, se han encontrado una población de *Elgaria multicarinata* en la cima de las montañas de la Sierra La Asamblea y una población de *Sceloporus occidentalis* en las elevaciones más altas de la Isla Cedros (Grismer y Mellink 1994).

La región está en su mayoría despoblada y

carece de grandes áreas urbanizadas. Ganadería, parcelas agrícolas pequeñas y áreas perturbadas en y alrededor de los oasis parecen ser las principales amenazas de conservación. La minería de sal al otro lado de la frontera estatal en Baja California Sur ha creado la necesidad de un puerto industrial en la Isla Cedros. El Valle de los Cirios es un área protegida que se extiende desde el sur de la frontera estatal hasta aproximadamente El Rosario e incluye 2,521,776 hectáreas o el 35% de la superficie estatal.

Región del Desierto Central de la Costa del Golfo

Esta región es un tramo largo y angosto de desierto bordeando el Mar de Cortés, que comienza ligeramente al sur de Bahía de los Ángeles y se extiende hacia el sur en Baja California Sur hasta el Istmo de La Paz (Shreve y Wiggins 1964; Wiggins 1980; Turner y Brown 1982; Cody et al. 1983, 2002). Algunos autores consideran que el Desierto Central de la Costa del Golfo comienza aún más al norte en las inmediaciones de Bahía Calamajué (Peinado et al. 1994), con la diferencia de opiniones destacando la amplia zona donde se entremezclan esta región y el Desierto Inferior del Colorado. Al este, la región está limitada por el Mar de Cortés, mientras que al oeste por la Cordillera Peninsular y la Región del Desierto del Vizcaíno.

Una cantidad considerable de islas del estado de Baja California se encuentran en la región Midriff del Mar de Cortés (Carreño y Helenes 2002). Las islas de esta región incluyen Mejía, Granito, Ángel de la Guarda, Pond, Cardosa Este, Partida Norte, Rasa, Salsipuedes, San Lorenzo Norte y San Lorenzo Sur. Varias islas costeras se localizan en la Bahía de los Ángeles, incluyendo Bota, Cabeza de Caballo, Cerraja, Flecha, La Ventana, Mitlán, Pata, Piojo y Smith (Grismer 2002).

Esta región es muy caliente, árida y recibe casi la totalidad de sus precipitaciones en verano y otoño. Sequías severas ocurren durante marzo-junio, con una precipitación media de 0.20 cm. (Hastings y Turner 1965; Humphrey 1974). La mayoría de las lluvias del Desierto Central de la Costa del Golfo se origina de las tormentas convencionales del sur y aparece como escorrentías procedentes de la Cordillera Peninsular que representa el límite oeste de esta región. En ocasiones, la región recibe precipitaciones como resultado de chubascos que se originan en las cálidas aguas del Mar de Cortés. A pesar de los períodos de sequía severa, el promedio anual de precipitación puede llegar a 16.8 cm (Hastings y Turner 1965), siempre y cuando haya habido una temporada de huracanes activa. Las temperaturas en el Desierto Central de la Costa del Golfo son claramente calientes. Las temperaturas promedio de los meses más cálidos en julio y agosto están por encima de 30°C; la temperatura promedio del mes más frío en enero es de 15°C (Markham 1972).

La vegetación de esta región caracterizada por Grismer (2002) incluye planta de piel (*Jatropha cuneata*), sangregado (*Jatropha cinerea*), palo Adán (*Fouquieria californicus*), copal (*Bursera hindsiana*), árbol de elefante (*B. microphylla*), cardón (*Pachycereus pringlei*) y palo verde de hoja pequeña (*Cercidium microphyllum*).

ENSAMBLE DEL DESIERTO CENTRAL DE LA COSTA DEL GOLFO

De las 45 especies de anfibios y reptiles nativas a esta región, 9 están limitadas a ésta: *Crotaphytus insularis, Phyllodactylus partidus, Sauromalus hispidus, Callisaurus splendidus, Petrosaurus slevini, Uta antiqua, Aspidoscelis cana, Crotalus angelensis* y *C. lorenzoensis*.

Este ensamble tiene 8% (9/113) de la herpetofauna total del estado, representando siete lagartijas y dos serpientes. Todas son endémicas a islas de la región Midriff. *Sauromalus hispidus* es la más ampliamente distribuida en estas islas y se encuentra en las islas Mejía, Granito, Ángel de la Guarda, Pond, Salsipuedes, San Lorenzo Norte y San Lorenzo (Hollingsworth 1998). Se cree que ha sido introducida a las islas Cabeza de Caballo, Flecha, La Ventana, Piojo y Smith en Bahía de los Ángeles por la etnia Seri (Grismer et al. 1995; Hollingsworth 1998). *Sauromalus hispidus* también ha sido reportada en la Isla Rasa (Velarde et al. 2008).

Tres especies, *Crotaphytus insularis*, *Callisaurus splendidus* y *Crotalus angelensis* son endémicas a la Isla Ángel de la Guarda, mientras que *Petrosaurus slevini* es endémica a la Isla Ángel de la Guarda y a su isla satelital al norte de ésta, Isla Mejía (McGuire 1996; Grismer 1999b; Murphy y Méndez de la Cruz 2010). Otras tres especies se encuentran en el bloque San Lorenzo, que incluye las islas Salsipuedes, San Lorenzo Norte, San Lorenzo Sur. *Uta antiqua* y *Aspidoscelis cana* habitan en estas tres islas, mientras que *Crotalus lorenzoensis* está limitada a la Isla San Lorenzo Sur (Grismer 2002; Murphy y Méndez de la Cruz 2010). *Phyllodactylus partidus* es endémica a dos islas pequeñas, Cardonosa Este y Partida Norte (Dixon 1964; Grismer 1999b).

Un total de 2 anfibios y 43 reptiles son nativos a esta región, representando 39.8% (45/113) de la herpetofauna estatal. Además de las especies de este ensamble, la región también hospeda: *Anaxyrus punctatus*, *Scaphiopus couchii*, *Crotaphytus vestigium*, *Gambelia copeii*, *Coleonyx switaki*, *C. variegatus*, *Phyllodactylus nocticolus*, *Dipsosaurus dorsalis*, *Sauromalus ater*, *Callisaurus draconoides*, *Petrosaurus repens*, *Phrynosoma cerroense*, *Sceloporus orcutti*, *S. zosteromus*, *Urosaurus nigricaudus*, *Uta stansburiana*, *Aspidoscelis hyperythra*, *A. tigris*, *Xantusia wigginsi*, *Lichanura trivirgata*, *Bogertophis rosaliae*, *Chilomeniscus stramineus*, *Hypsiglena ochrorhyncha*, *H. slevini*, *Lampropeltis getula*, *Masticophis fuliginosus*, *Phyllorhynchus decurtatus*, *Pituophis vertebralis*, *Salvadora hexalepis*, *Sonora semiannulata*, *Tantilla planiceps*, *Trimorphodon lyrophanes*, *Rena humilis*, *Crotalus enyo*, *C. mitchellii* y *C. ruber*.

Amenazas para la conservación de las especies de este ensamble están asociadas a su distribución limitada a las islas y el potencial para introducir especies depredadoras que disminuirían grandemente sus tamaños poblacionales (Murphy y Méndez de la Cruz 2010). Todas las islas de este ensamble están protegidas dentro de las 387,956 hectáreas de la Reserva de la Biósfera Bahía de Los Ángeles, Canales de Ballenas y de Salsipuedes (Lovich y Mahrdt 2008; Lovich et al. 2009). Las áreas costeras de la Región del Desierto Central de la Costa del Golfo en el estado de Baja California están protegidas dentro del Área Natural Protegida Valle de los Cirios (Lovich et al. 2009).

Región Marina

Las aguas frente a las costas de Baja California representan dos cuerpos de agua diferentes. El Océano Pacífico está dominado por el flujo de norte a sur de la Corriente de Baja California, que lleva agua fría a la costa oeste del estado. En el Mar de Cortés, la corriente fluye hacia el norte a lo largo de la costa continental (Sonora) y hacia el sur a lo largo de la península en los meses de verano; la dirección de esta corriente se invierte en los meses de invierno. Las temperaturas en la porción norte del Mar de Cortés son de 8.2°C en diciembre y de 32.6°C en agosto (Álvarez-Borrego 2002).

ENSAMBLE DE LA REGIÓN MARINA

Hay seis especies limitadas a este ensamble: *Caretta caretta*, *Chelonia mydas*, *Eretmochelys imbricata*, *Lepidochelys olivacea*, *Dermochelys coriacea* e *Hydrophis platurus*.

Cinco de éstas son tortugas marinas, y otra es una serpiente de mar. *Caretta caretta* no es tan común como las otras tortugas marinas (Grismer 2002). A pesar de su rareza en el Mar de Cortés, ha sido observada hasta la Bahía Ometepec cerca de la boca del Río Colorado (Shaw 1947). En el Océano Pacífico, se sabe que llega hasta las Islas Coronado (Caldwell 1962) y a la Isla Guadalupe (Smith y Smith 1979). *Chelonia mydas* está ampliamente distribuida en el Mar de Cortés, pero es menos común en la costa del Pacífico (Grismer 2002). También se sabe que entra a los Ríos Hardy y Colorado (Smith y Smith 1979) y fue reportada en El Mayor, 80 km al norte de la boca del Río Colorado (Cliffton et al. 1982).

Eretmochelys imbricata no ha sido observada más allá de uno pocos kilómetros al norte de Isla Cedros en el Océano Pacífico (Smith y Smith 1979) y su registro más septentrional en el Mar de Cortés es en Bahía de los Ángeles (Caldwell 1962). *Lepidochelys olivacea* es rara en el Mar de Cortés, aunque se ha reportado en San Felipe (Caldwell 1962). En el Océano Pacífico, se distribuye a lo largo de toda la costa oeste de Baja California (Zug et al. 1998). *Dermochelys coriacea* es considerada rara en el Mar de Cortés, su registro más septentrional en este mar es en Puerto Peñasco, Sonora, México (Smith y Smith 1979). En el Océano Pacífico, es de hábitats pelágicos haciendo que su observación sea rara. La serpiente de mar *Hydrophis platurus* se distribuye en todo el Mar de Cortés, pero raramente se le observa en aguas del estado de Baja California (Pickwell et al. 1983; Campbell y Lamar 2004).

Generalistas

De las 113 especies nativas al estado de Baja California, 59 (52.2%) son consideradas generalistas, con distribuciones que se extienden en dos o más regiones fitogeográficas.

ENSAMBLE DE GENERALISTAS

El ensamble de generalistas incluye las siguientes especies: *Batrachoseps major, Ensatina klauberi, Anaxyrus boreas, A. californicus, A. punctatus, Pseudacris cadaverina, P. hypochondriaca, Rana draytonii, Scaphiopus couchii, Spea hammondii, Actinemys marmorata, Elgaria multicarinata, Crotaphytus vestigium, Gambelia copeii, Coleonyx switaki, Coleonyx variegatus, Phyllodactylus nocticolus, Dipsosaurus dorsalis, Sauromalus ater, Callisaurus draconoides, Petrosaurus mearnsi, P. repens, Phrynosoma blainvillii, P. cerroense, Sceloporus occidentalis, S. orcutti, S. zosteromus, Urosaurus nigricaudus, Uta stansburiana, Plestiodon gilberti, P. skiltonianus, Aspidoscelis hyperythra, A. labialis, A. tigris, Xantusia henshawi, X. wigginsi, Lichanura trivirgata, Arizona elegans, Bogertophis rosaliae, Chilomeniscus stramineus, Hypsiglena ochrorhyncha, H. slevini, Lampropeltis getula, Masticophis fuliginosus, M. lateralis, Phyllorhynchus decurtatus, Pituophis catenifer, P. vertebralis, Rhinocheilus lecontei, Salvadora hexalepis, Sonora semiannulata, Tantilla planiceps, Thamnophis hammondii, Trimorphodon lyrophanes, Rena humilis, Crotalus enyo, C. mitchellii, C. oreganus y C. ruber.*

Registros bien documentados para especies introducidas han sido reportados para *Lithobates berlandieri* a lo largo del Río Colorado, la cual se piensa está desplazando a *L. yavapaiensis* (Clarkson y Rorabaugh 1989; Platz et al. 1990; Rorabaugh et al. 2002). Otra especie introducida al Río Colorado es *Apalone spinifera* (Lindsdale y Gressitt 1937; Miller 1946), aunque hay una carencia de referencias sobre registros específicos a este respecto. *Lithobates catesbeianus* habita en varios cauces de la Región de California (Clarkson y de Vos 1986; Grismer 2002), el Río Colorado (Lovich et al. 2009), pero no en oasis de la porción sur del estado (Grismer y McGuire 1993). *Xenopus laevis* ha sido registrada entre Ensenada y Tijuana (Ruíz-Campos y Valdez-Villavicencio 2012). *Hemidactylus turcicus* ha sido reportada para Ensenada (Martínez-Isaac y Valdez-Villavicencio 2000). Murphy y Méndez de la Cruz (2010) reportaron *Taricha torosa* como una especie introducida en Baja California, pero no proporcionaron información específica y su presencia como especie introducida es dudosa. No se conoce ningún registro de esta especie para Baja California (Stebbins 1962; ver también Slevin 1928). La distribución más austral de esta especie se encuentra en Boulder Creek en el Condado de San Diego, California.

JULIO A. LEMOS-ESPINAL
Y JAMES C. RORABAUGH

Introducción

El estudio de los anfibios y reptiles de Sonora
ha recibido gran atención desde mediados del
siglo diecinueve, indudablemente debido a la
proximidad de los estados norteamericanos
de Arizona y California, con grandes centros
poblacionales donde viven muchos especialistas
y amateurs interesados en la rica herpetofauna
de este estado mexicano. La continuidad de los
hábitats de desierto de Arizona y Sonora y las 14
islas principales del Mar de Cortés pertenecientes a Sonora representan un reto para realizar
estudios de biogeografía, especiación, ecología
de islas, etc., e inevitablemente han resultado en
la exploración temprana de partes de Sonora por
pobladores de Arizona y California. El escabroso
terreno de la Sierra Madre Occidental de Sonora
y su continuidad con los ambientes tropicales/
subtropicales del sur de México también ha
atraído la atención de muchos herpetólogos a
regiones como Álamos, Yécora y las islas continentales asociadas a la Sierra Madre Occidental,
las cuales han sido extensamente estudiadas.
El terreno y clima del norte, centro y costa de
Sonora también ha propiciado el desarrollo
de caminos permitiendo el fácil acceso a estas
partes del estado y promoviendo el desarrollo de
estudios herpetológicos.

Inicialmente, muestras de anfibios y reptiles
de Sonora fueron obtenidas cerca de la frontera
con los estados de Arizona y Nuevo México por
la Mexico-United States Boundary Commission,
aunque ésta no estaba enfocada exclusivamente
a la recolección de especies de anfibios y reptiles,
el número, identidad y localidades de colecta de
los especímenes obtenidos fue importante y cubrió un área considerable del estado de Sonora.

El United States National Museum (USNM)
jugó un papel importante en el estudio de los
anfibios y reptiles del estado a través de las exploraciones de Nelson y Goldman a finales del siglo
diecinueve. El American Museum of Natural
History (AMNH) promovió el estudio de los anfibios y reptiles de las islas del Mar de Cortés y la
región de Álamos-Güirocoba del extremo sureste
de Sonora. La California Academy of Sciences
(CAS) apoyó el estudio de los herpetozoarios del
extremo oeste del estado a través de los estudios
de Van Denburgh y Slevin a principios del siglo
veinte.

En la actualidad varias instituciones como la
Universidad Nacional Autónoma de México, la
University of Arizona, el US Fish and Wildlife
Service, el Drylands Institute, y el Arizona-
Sonora Desert Museum, entre otras han estado
trabajando sobre el estudio de los anfibios y
reptiles de varias partes de Sonora.

En general, la mayoría de los estudios de anfibios y reptiles de Sonora ha estado enfocada a

las islas del Mar de Cortés y la costa del estado, la frontera con Arizona y Nuevo México, la región de Álamos-Güirocoba, y la región de Yécora. Se ha recolectado una cantidad considerable de especímenes a lo largo de las principales carreteras que cruzan el estado, tales como la Méx. 2 (San Luis Río Colorado—Sonoyta—Caborca), Méx. 8 (Sonoyta—Puerto Peñasco), Méx. 15 (Estación Don—Nogales), Méx. 16 (Maycoba—Hermosillo) y la carretera entre Navojoa y Álamos. Aún hay áreas que no han sido estudiadas debido a la falta de caminos y a condiciones ambientales extremas. Un ejemplo es la gran área desértica al noreste de Puerto Libertad, sur y este de Caborca, y oeste de la porción de la carretera Méx. 15 que corre entre Hermosillo y Santa Ana.

En la actualidad sabemos que Sonora tiene un total de 195 especies de anfibios y reptiles: 38 anfibios (3 salamandras, 35 anuros), y 157 reptiles (16 tortugas, 69 lagartijas, 72 serpientes). Siete de estas especies son introducidas: *Lithobates berlandieri*, *L. catesbeianus*, *Apalone spinifera*, *Hemidactylus frenatus*, *H. turcicus*, *Sauromalus hispidus* e *Indotyphlops braminus*. Seis son marinas: *Caretta caretta*, *Chelonia mydas*, *Eretmochelys imbricata*, *Lepidochelys olivacea*, *Dermochelys coriacea* e *Hydrophis platurus*. Otras 13 son isleñas: *Phyllodactylus nocticolus*, *P. nolascoensis*, *Ctenosaura conspicuosa*, *C. nolascensis*, *Sauromalus hispidus*, *S. varius*, *Uta nolascensis*, *U. palmeri*, *Aspidoscelis baccata*, *A. estebanensis*, *A. martyris*, *Masticophis slevini*, y *Crotalus estebanensis*.

La compleja transición topográfica entre las zonas Neártica y Neotropical, la alta diversidad de ecosistemas de la Sierra Madre Occidental, la igualmente alta heterogeneidad y diversidad del Desierto de Sonora, la extensa costa del estado, y sus 14 islas principales con biotas únicas, han resultado en una de las más grandes riquezas de especies de anfibios y reptiles en cualquier estado del norte de México. Estamos seguros que el número total de especies de anfibios y reptiles de Sonora es aún más grande. Las principales adiciones posiblemente se encuentre en la región subtropical del extremo sureste del estado, éstas permanecen como un atrayente para estudios herpetológicos futuros.

Estudios Herpetológicos Previos

Las primeras colectas de anfibios y reptiles hechas en el estado de Sonora se obtuvieron por la Mexico-United States Boundary Commission entre 1849 y 1851. Estas colectas se hicieron en todos los estados de la frontera entre México y Estados Unidos, incluyendo Sonora. Todos los especímenes recolectados por esta Comisión están depositados en el USNM. Parte de esta colección fue reportada por Baird (1859), Cope (1867, 1900) y Kellogg (1932).

Posteriormente, Lovewell y Heiligbrodt hicieron una colección de herpetozoarios que fue enviada al Washburn College, Topeka, Kansas. Esta colección fue reportada por Francis Whittemore Cragin en 1884 e incluyó información sobre *Sceloporus clarkii*, *Urosaurus ornatus*, *Thamnophis cyrtopsis* y *Micruroides euryxanthus*.

Muchos de los especímenes recolectados por la Mexico-United States Boundary Commission son los tipos de varios taxa originalmente descritos y reportados para el estado de Sonora. Sin embargo, después de la Compra de James Gadsden en diciembre 30, 1853 (Venta de la Mesilla), una región que actualmente se encuentra en el sur de Arizona y suroeste de Nuevo México se volvió parte de los Estados Unidos. Hobart M. Smith y Edward H. Taylor (1950) restringieron las localidades tipo de estos taxa a Arizona y Nuevo México. Estos taxa incluyen *Hyla affinis* (= *H. arenicolor* Santa Rita Mountains, Arizona); *Kinosternon sonoriense* (Tucson, Arizona); *Heloderma suspectum* (Sierra de Morena, Arizona); *Sceloporus clarkii* (Santa Rita Mountains, Arizona); *Sceloporus poinsettii* (Río San Pedro del Río Grande del Norte, Nuevo México); *Eutaenia megalops* (= *Thamnophis eques megalops*, Tucson, Arizona); *Ophibolus splendida* (= *Lampropeltis getula splendida*, Santa Rita Mountains, Arizona); *Salvadora grahamiae* (Santa Rita Mountains, Arizona); *Sonora semiannulata* (Santa Rita Mountains, Arizona).

Carl Lumholtz y su equipo de expedición exploraron los estados de Sonora y Chihuahua entre 1890 y 1892. Las aves y mamíferos recolectados en esta expedición fueron reportados por

Allen en 1893. En esta expedición F. Robinette recolectó el primer espécimen de lo que más tarde se volvió, junto con otros dos especímenes recolectados por el Señor Eustace en 1897, la serie tipo de *Phrynosoma ditmarsi*, especie descrita por Stejneger en 1906. Durante un periodo de 73 años después de la descripción original de esta especie de lagartija, no hubo registros de ella hasta su redescubrimiento en 1970 por Vincent D. Roth, el cual fue documentado por Lowe et al. (1971) y Roth (1997).

En 1892 Gustav Eisen y Walter Bryant hicieron una colección de reptiles que fue enviada a la CAS. Esta colección fue reportada por Van Denburgh en 1898.

Entre octubre 27, 1898 y enero 31, 1899, Edward Goldman y Edward Nelson del USNM visitaron el extremo sureste del estado en los poblados de Camoa, Batamotal, y Álamos, recolectando y describiendo la vida silvestre de esta región; los resultados de esta expedición fueron reportados por Kellogg (1932) y Goldman (1951).

Entre febrero 23 y abril 28 de 1911, la *Albatross* Expedition a Baja California se llevó a cabo en diferentes puntos de la costa e islas de la Península de Baja California y Sonora, bajo los auspicios del United States Bureau of Fisheries, el American Museum of Natural History, la New York Zoological Society, el New York Botanical Garden, y el USNM. Charles Haskins Townsend estuvo encargado de esta expedición en la cual 448 especímenes de reptiles y 7 de anfibios fueron recolectados. Adicionalmente, una cantidad de reptiles vivos fue enviada al New York Zoological Park. Las localidades de Sonora visitadas por esta expedición incluyeron Guaymas y las islas San Esteban y Tiburón. Los resultados herpetológicos de esta expedición fueron reportados por Mary C. Dickerson en 1919 y Karl P. Schmidt en 1922. Estas dos publicaciones incluyen la descripción original de los siguientes taxa de Sonora: *Aspidoscelis estebanensis, Aspidoscelis tigris disparilis, Aspidoscelis tigris punctilinealis, Callisaurus draconoides inusitatus, Ctenosaura conspicuosa, Sauromalus ater townsendi* y *Sauromalus varius* (Dickerson 1919); y *Crotaphytus dickersonae* (Schmidt 1922).

En 1922 John Van Denburgh incluyó un listado de reptiles de Sonora en su libro de *Los Reptiles del Oeste de Norteamérica*. Una lista de anfibios de Sonora fue incluida por Joseph R. Slevin en su reporte de 1928 sobre *Los Anfibios del Oeste de Norteamérica*.

En el verano de 1932 (junio-julio) Morrow J. Allen, Jean Piatt y John Scofield del Museo de Zoología de la Universidad de Michigan recolectaron 326 especímenes: 59 anfibios y 267 reptiles, cerca de las ciudades de Puerto, Noria, Hermosillo y Guaymas. Allen (1933) reportó parte de esta colección la cual incluyó 4 especies de anfibios, 12 de lagartijas, 6 de serpientes y 1 de tortugas. Este reporte incluye la descripción original de *Dipsosaurus dorsalis sonoriensis*.

Entre junio 19 y julio 16 de 1934, Edward Taylor visitó Sonora, recolectando anfibios y reptiles en las localidades previamente visitadas por Allen, Piatt y Schofield y algunas otras como Empalme. Taylor recolectó especímenes de 5 especies de anfibios, 18 de lagartijas, 16 de serpientes y 2 de tortugas. Parte de esta colección fue reportada por Hobart M. Smith (1935a, 1935b, 1935c, 1972) estas publicaciones incluyen la descripción original de *Phyllodactylus homolepidurus* de Hermosillo, *Uta taylori* (= *U. stansburiana taylori*) de Guaymas, y *Ctenosaura hemilopha macrolopha* (= *C. macrolopha*) de La Posa, Bahía de San Carlos. Un reporte completo de esta colección fue publicado por Taylor (1936b) éste incluye la descripción original de *Cnemidophorus* (= *Aspidoscelis*) *burti*.

En 1941 John W. Hilton recolectó anfibios y reptiles en la región de Güirocoba. Este estudio fue continuado en 1942 por Charles M. Bogert, William Riemer y Charles H. Lowe en el pueblo minero de Álamos (29 km al noroeste de Güirocoba). Durante estos dos años recolectaron especímenes de 15 especies de anfibios, 9 de tortugas, 38 de lagartijas, y 48 de serpientes. Esta colección fue reportada por Carr (1942: descripción de *Pseudemys scripta hiltoni* [= *Trachemys nebulosa hiltoni*]), y Bogert y Oliver (1945), ésta última constituye una de las publicaciones más completas y mejor conocidas sobre la herpetofauna de Sonora.

En 1950 Richard G. Zweifel y Kenneth S. Norris del Museum of Vertebrate Zoology, Berkeley, California, recolectaron anfibios y reptiles en la región de Güirocoba y en varios puntos hacia el norte hasta la frontera con Estados Unidos (Zweifel y Norris 1955). En esta publicación también reportaron una colección hecha por Laurence M. Klauber, A. A. Allanson y Sterling Bunnell. Este reporte menciona 10 especies y subespecies de anuros, 18 de lagartijas, 31 de serpientes y 3 de tortugas, e incluye la descripción original de *Lampropeltis getula nigritus* (= *L. getula nigrita*) y *Micruroides euryxanthus australis*.

Philip W. Smith y M. Max Hensley del Illinois Natural History Survey y el Department of Zoology, Michigan State University, respectivamente, reportaron una pequeña colección de 3 especies de anfibios y 26 de reptiles de la región del Pinacate en el noroeste de Sonora (Smith y Hensley 1958), un área también investigada por Alberto González-Romero y Sergio Álvarez-Cárdenas durante 1980–1983. En este último estudio los autores reportaron 42 especies de anfibios y reptiles (González-Romero y Álvarez-Cárdenas 1989).

En agosto 1–22 de 1964, la National Science Foundation (NSF) auspició la expedición Biology of the Insular Lizards of the Gulf Expedition (BILGE), llevada a cabo en el Mar de Cortés por un equipo de cinco profesores Charles Carpenter (University of Oklahoma), Robert Clarke (Emporia State University, Kansas), James Dixon (New Mexico State University), Paul Maslin (University of Colorado), y Donald Tinkle (Texas Tech. University), y cinco de sus estudiantes: William C. McGrew III, Louis J. Bussjaeger, Stanley Taft, Jack McCoy y Orlando Cuellar. Visitaron varias islas del Mar de Cortés, incluyendo algunas pertenecientes a Sonora (San Esteban, San Pedro Mártir, San Pedro Nolasco, Tiburón). Recolectaron especímenes de varias especies de lagartijas los cuales fueron depositados en varios museos norteamericanos, especialmente en la University of Colorado. Reportes de esta colección y observaciones sobre la expedición se pueden encontrar en Dixon (1966), Ballinger y Tinkle (1972), y Smith (1972). Ésta última publicación

incluye la descripción original de *Ctenosaura hemilopha nolascensis* (= *C. nolascensis*).

En los veranos de 1964 y 1966 dos colecciones diferentes de anfibios y reptiles fueron hechas por colectores privados en los alrededores de Álamos. Los especímenes recolectados fueron depositados en la Herpetological Collection of Arizona State University. Parte de estas colecciones fue reportada por Max A. Nickerson y H. L. Heringhi (1966) incluyendo información sobre *Mastigodryas cliftoni*, *Sonora aemula* y *Sympholis lippiens rectilimbus*. Información adicional sobre estas colecciones se encuentra en Heringhi (1969).

En marzo 1965 y 1966, y noviembre 1966, Donald Tinkle y cuatro de sus estudiantes (Royce Ballinger, Charles McKinney, Gary Ferguson y Orlando Cuellar) visitaron el poblado de Guaymas y la Bahía de San Carlos, y las islas de San Pedro Nolasco, San Esteban y Tiburón. Los resultados de estos viajes fueron reportados por Ballinger y Tinkle (1972), y McKinney (1969).

Desde 1974 Robert W. Murphy de la University of Toronto y el Royal Ontario Museum ha estado estudiando la herpetofauna de las islas del Mar de Cortés (incluyendo las pertenecientes a Sonora) así como de la parte continental de Sonora. Información sobre la herpetofauna de Sonora derivada de este trabajo ha sido publicada en Murphy (1976, 1983a, 1983b), Murphy y Aguirre-León (2002a, 2002b), y Davy et al. (2011).

En 1998 Gary P. Nabhan visitó los poblados de Punta Chueca y Desemboque de los Seris, recolectando información etnobiológica y herpetológica. Los resultados de esta investigación fueron publicados por Nabhan (2003).

En el año 2000, Cecil R. Schwalbe y Charles H. Lowe publicaron sobre la herpetofauna de la región de Álamos, esta publicación se basó sobre una revisión de estudios anteriores y el propio trabajo de campo de los autores, el cual se extendía 50 años atrás. Recopilaron la lista de especies de anfibios y reptiles de la región de Álamos, la cual incluyó la Sierra de Álamos al norte del Río Mayo, al sur y al este del Río Cuchujaqui y al oeste y sur de Masiaca. En total reportaron 15 especies de anfibios (todas ellas anuros) y 63 de reptiles (6 tortugas, 21 lagartijas y 36 serpientes).

Una publicación importante es la de Grismer (2002) quien proporcionó un listado de los anfibios y reptiles que habitan las islas del Mar de Cortés, incluyendo las pertenecientes a Sonora, así como una descripción detallada, claves de identificación y fotos de los taxa que habitan en estas islas.

Más recientemente, Philip C. Rosen reportó sobre los anfibios y reptiles del noroeste de Sonora, proporcionando una lista de 2 especies de anfibios y 26 de reptiles para la región de El Pinacate—Gran Desierto de Altar (Rosen 2007). Posteriormente, Rosen y Cristina Melendez muestrearon la herpetofauna de 28 localidades acuáticas de Sonora reportando un total de 14 especies (Rosen y Melendez 2010). Seminoff y Nichols (2007) describieron la situación actual de las tortugas marinas de la región del Alto Golfo del Mar de Cortés, e incluyeron fichas diagnósticas para cada una de las cinco especies que se desarrollan en territorio Sonorense. James C. Rorabaugh discutió la conservación de los anfibios y reptiles del noroeste de Sonora y suroeste de Arizona, e incluyó una lista de especies con notas distribucionales (Rorabaugh 2010). Rorabaugh et al. (2011) reportaron sobre la herpetofauna de la Northern Jaguar Reserve y sus alrededores al noreste de Sahuaripa. Registraron 40 especies de reptiles y 11 de anfibios, incluyendo varias extensiones en la distribución de estas especies. Rorabaugh et al. (2013) reportaron 11 especies de anfibios y 35 de reptiles registradas en dos ranchos del noreste de Sonora. Enderson et al. (2014) reportaron 20 especies de anfibios y 73 de reptiles para la región de Yécora. Información distribucional importante ha sido recolectada por la Sky Island Alliance's Madrean Archipelago Biodiversity Assessment (MABA) de Tucson, Arizona, a través del trabajo de campo desarrollado entre 2009 y 2014 en varias cadenas montañosas del noreste de Sonora, parte de esta información ha sido reportada por Van Devender et al. (2013). La MABA ha desarrollado una base de datos de libre acceso en línea que contiene

registros de observaciones y colectas que el personal de esta organización ha hecho así como registros curatoriales de varios museos (http://www.madrean.org/maba/symbfauna/). Esta base de datos se ha convertido en un valioso recurso de información sobre herpetofauna y otros grupos de plantas y animales.

Registros distribucionales, notas de historia natural e información ecológica han sido reportados en Ávila et al. (2008: *Masticophis mentovarius*); Bonine et al. (2006: *Imantodes geminstratus*); Bury et al. (2002: *Gopherus agassizii* [= *morafkai*]); Dibble et al. (2007: *Callisaurus draconoides*; 2008: *Dipsosaurus dorsalis*); Enderson et al. (2006: *Gyalopion canum*; 2007 *Kinosternon integrum*); Enderson y Bezy (2007a: *Geophis dugesii*; 2007b: *Lampropeltis triangulum sinaloae*; 2007c: *Leptodeira splendida ephippiata*; 2007d: *Pseudoficimia frontalis*; 2007e: *Syrrhophus* [= *Eleutherodactylus*] *interorbitalis*); Enderson (2010: *Crotalus viridis*); Goldberg (1997: *Micruroides euryxanthus*); Lara-Góngora (1986 *Sceloporus grammicus disparilis* [= *S. lemosespinali*]); Lemos-Espinal, Chiszar et al. (2004: lagartijas); Lemos-Espinal, Smith, Hartman y Chiszar (2004: anfibios y tortugas); Lemos-Espinal, Smith, Chiszar y Woolrich-Piña (2004: serpientes); Lowe y Howard (1975: *Phrynosoma ditmarsi*); Macedonia et al. (2009: *Crotaphytus dickersonae*); Maldonado-Leal et al. (2009: *Hyla wrightorum*); O'Brien et al. (2006: *Drymarchon corais* y *Kinosternon integrum*); Palacio-Baez y Enderson (2012: *Incilius marmoreus*); Perrill (1983: *Phrynosoma ditmarsi*); Quijada-Mascareñas y Enderson (2007: *Ramphotyphlops* [= *Indotyphlops*] *braminus*); Quijada-Mascareñas et al. (2007: *Plestiodon obsoletus*); Recchio et al. (2007: *Geophis dugesii*); Rorabaugh (2013: *Trimorphodon tau*); Rorabaugh y King (2013: *Apalone spinifera*); Rorabaugh et al. (2008: *Apalone spinifera*; 2013: *Micrurus distans*); Rorabaugh y Servoss (2006: *Lithobates berlandieri*); Rosen y Quijada-Mascareñas (2009: *Aspidoscelis xanthonota*); Sherbrooke et al. (1998: *Phrynosoma ditmarsi*); Smith, Lemos-Espinal y Heimes (2005: anfibios y lagartijas); Smith, Lemos-Espinal y Chiszar (2005: anfibios y reptiles); Van Devender, Holm y Lowe (1989:

Pseudoeurycea bellii sierraoccidentalis); Van Devender, Lowe y Holm (1989: *Pseudoeurycea bellii sierraoccidentalis*); Van Devender et al. (1994: *Oxybelis aeneus*); Van Devender y Enderson (2007: *Micrurus distans*); Villa et al. (2007: *Crotalus willardi*); Wiewandt et al. (1972: *Hypopachus variolosus*); y Winter et al. (2007: *Leptodactylus melanonotus*).

Descripciones originales recientes de especies nuevas están reportadas en Lara-Góngora (2004: *Sceloporus lemosespinali*); Smith, Lemos-Espinal, Hartman y Chiszar (2005: *Tropidodipsas repleta*); Bezy et al. (2008: *Xantusia jaycolei*); y Murphy et al. (2011: *Gopherus morafkai*); y Wood et al. (2011: *Thamnophis unilabialis*).

Otros estudios importantes son los de James C. Rorabaugh (2008), quien reportó la herpetofauna de la parte continental de Sonora mencionando la distribución, estatus de protección y nombres comunes de todas las especies de anfibios y reptiles, con fotos de algunos de ellos. La lista de la herpetofauna reportada en esta publicación incluye 37 especies de anfibios y 139 de reptiles.

En 2009, Rorabaugh y Enderson proporcionaron una revisión de las especies de lagartijas del estado, incluyendo las de las islas de Mar de Cortés pertenecientes a Sonora. En total reportaron 64 especies de lagartijas en 21 géneros y 11 familias.

Enderson et al. (2009) proporcionaron un listado de anfibios y reptiles del estado de Sonora y compararon éste con listados de los estados contiguos. Reportaron un total de 187 especies: 35 anfibios (3 salamandras, 32 anuros) y 152 reptiles (1 cocodrilo, 15 tortugas, 64 lagartijas y 72 serpientes). Otro reporte de la herpetofauna de Sonora fue publicado por Enderson et al. (2010) en el cual proporcionaron una lista de los anfibios y reptiles conocidos para la parte continental de Sonora y analizaron las prioridades de conservación para la herpetofauna de este estado.

Desde 2003, Hobart M. Smith y Julio Lemos-Espinal han estado estudiando los anfibios y reptiles de Sonora. A través de este trabajo en 2009 publicaron unas claves para la identifica-

ción y una lista de los anfibios y reptiles de este estado (incluyendo también a los estados de Chihuahua y Coahuila). Este listado incluyó un total de 192 especies: 37 anfibios (3 salamandras, 34 anuros), y 155 reptiles (16 tortugas, 65 lagartijas, 74 serpientes). Doce de las 192 especies reportadas son isleñas, 6 son marinas, y 174 se distribuyen en la parte continental del estado.

Más recientemente, Lemos-Espinal et al. (en prensa) publicaron sobre la herpetofauna de los estados de Sonora, Chihuahua y Coahuila, presentando una lista de todos los anfibios y reptiles que se sabe habitan estos tres estados; claves para la identificación de estas especies; ilustraciones fotográficas de ellos; una ficha diagnostica y un mapa puntual sobre la distribución de cada una de las especies reportadas; y un gacetero que incluye las coordenadas geográficas y municipio al que pertenecen un total de 2987 localidades en donde han sido recolectadas las especies reportadas. La lista para Sonora incluye un total de 193 especies: 37 anfibios (3 salamandras, 34 anuros), y 156 reptiles (16 tortugas, 68 lagartijas, 72 serpientes).

Adicionalmente a los estudios mencionados previamente hemos determinado que hay 67 nombres científicos de anfibios (2 salamandras) y reptiles (7 tortugas, 37 lagartijas, 21 serpientes) con localidad tipo en el estado de Sonora (Cuadro 4.1).

Características Fisiográficas y su Influencia sobre la Herpetofauna

La distribución de los anfibios y reptiles está limitada en parte por sus necesidades fisiológicas y conductuales, las cuales a su vez están limitadas por los ambientes en que viven. A diferencia de muchas aves, insectos, mamíferos y peces, la mayoría de los anfibios y reptiles no migran o únicamente realizan movimientos estacionales pequeños dentro de sus ámbitos hogareños (en Sonora, excepciones importantes incluyen a los reptiles marinos, representados por cinco tortugas y una serpiente, los cuales migran miles de kilómetros entre el Mar de Cortés y sus sitios

Autor(es)	Nombre original: Localidad tipo
Salamandras (2)	
Shannon (1951)	*Ambystoma rosaceum sonoraensis* (= *A. rosaceum*): **32 mi south of the Arizona border, Sonora**
Lowe et al. (1968)	*Pseudoeurycea bellii sierraoccidentalis*: **ca. 11 mi (rd) E Sta. Ana, on old road to Yécora**
Tortugas (7)	
Garman (1884)	*Sphargis* (= *Dermochelys*) *coriacea schlegelii*: **Tropical Pacific and Indian Ocean (restringida a Guaymas[1])**
Carr (1942)	*Pseudemys scripta hiltoni* (= *Trachemys nebulosa hiltoni*): **Güirocoba**
Legler & Webb (1970)	*Pseudemys scripta yaquia* (= *Trachemys yaquia*): **Río Mayo, Conicarit**
Bogert (1943)	*Terrapene klauberi* (= *T. nelsoni klauberi*): **Güirocoba, 18 mi SE of Álamos**
Berry & Legler (1980)	*Kinosternon alamosae*: **Rancho Carrizal, 7.2 km N and 11.5 km W of Alamos**
Hartweg (1938)	*Kinosternon flavescens stejnegeri* (= *K. arizonense*): **Llano**
Iverson (1981)	*Kinosternon sonoriense longifemorale*: **Sonoyta (31°51′ N, 112°50′ W)**
Lagartijas (37)	
Klauber (1945)	*Coleonyx variegatus sonoriense* (*sonoriensis*): **5 mi SE of Hermosillo**
Smith (1935b)	*Phyllodactylus homolepidurus*: **5 mi SW of Hermosillo**
Dixon (1964)	*Phyllodactylus homolepidurus nolascoensis* (= *P. nolascoensis*): **Isla San Pedro Nolasco**
Schmidt (1922)	*Crotaphytus dickersonae*: **Isla Tiburón**
Axtell & Montanucci (1977)	*Crotaphytus nebrius*: **14 km by road N of Rancho Cieneguita. 28°30′30″ N, 111°2′30″W**
Smith (1972)	*Ctenosaura hemilopha macrolopha* (= *C. macrolopha*): **La Posa, San Carlos Bay**
Dickerson (1919)	*Ctenosaura conspicuosa*: **Isla San Esteban**
Smith (1972)	*Ctenosaura nolascensis*: **Isla San Pedro Nolasco**
Allen (1933)	*Dispsosaurus dorsalis sonoriense* (= *D. d. sonoriensis*): **Hermosillo**
Dickerson (1919)	*Sauromalus townsendi* (= *S. ater townsendi*): **Isla Tiburón**
Dickerson (1919)	*Sauromalus varius*: **Isla San Esteban**
Bogert & Dorson (1942)	*Callisaurus draconoides brevipes*: **Güirocoba, 18 mi SE of Álamos**
Dickerson (1919)	*Callisaurus inusitatus* (= *C. draconoides inusitatus*): **Isla Tiburón**
Barbour (1921)	*Holbrookia thermophila* (= *H. elegans thermophila*): **San José de Guaymas**
Stejneger (1906)	*Phrynosoma ditmarsi*: **state of Sonora, not far from the boundary of Arizona**
Stejneger (1893)	*Phrynosoma goodei*: **coast deserts of Sonora (restringida a Puerto Libertad[1])**
Girard (1858)	*Phrynosoma regale* (= *P. solare*): **Sierra de la Nariz, near Zuñi**
Lara-Góngora (2004)	*Sceloporus lemosespinali*: **Yécora**
Smith (1938)	*Sceloporus undulatus virgatus* (= *S. virgatus*): **above Sta. María Mine, Tigre Mountains**
Heifetz (1941)	*Uma notata cowlesi* (= *U. rufopunctata*): **shores of Tepoca Bay**
Van Denburgh & Slevin (1921)	*Uta nolascensis*: **Isla San Pedro Nolasco**
Baird (1859a [1858])	*Uta ornata* var. *linearis* (= *Urosaurus ornatus linearis*): **Los Nogales**
Stejneger (1890)	*Uta palmeri*: **Isla San Pedro Mártir**
Baird (1859a [1858])	*Uta schottii* (= *Urosaurus ornatus schotti*): **Sta. Magdalena**
Smith (1935a)	*Uta taylori* (= *U. stansburiana taylori*): **10 mi NW of Guaymas**

Autor(es)	Nombre original: Localidad tipo
Taylor (1933)	*Eumeces (= Plestiodon) parviauriculatus*: **near Álamos**
Van Denburgh & Slevin (1921)	*Cnemidophorus baccatus (= Aspidoscelis baccata)*: **Isla San Pedro Nolasco**
Taylor (1936b)	*Cnemidophorus (= Aspidoscelis) burti*: **La Posa, 10 mi NW of Guaymas**
Dickerson (1919)	*Cnemidophorus disparilis (= Aspidoscelis tigris disparilis)*: **Isla Tiburón**
Dickerson (1919)	*Cnemidophorus (= Aspidoscelis) estebanensis*: **Isla San Esteban**
Stejneger (1891)	*Cnemidophorus (= Aspidoscelis) martyris*: **Isla San Pedro Mártir**
Wright (1967)	*Cnemidophorus (= Aspidoscelis) opatae*: **5.5 mi (by road) S of Oputo**
Dickerson (1919)	*Cnemidophorus punctilinealis (= Aspidoscelis tigris punctilinealis)*: **Isla Tiburón**
Zweifel (1959)	*Cnemidophorus sacki (= Aspidoscelis costata) barrancarum*: **Rancho Güirocoba**
Zweifel (1959)	*Cnemidophorus sacki (= Aspidoscelis costata) griseocephalus*: **11.4 mi E of Navojoa**
Cope (1900)	*Cnemidophorus tessellatus (= Aspidoscelis tigris) aethiops*: **Hermosillo**
Bezy et al. (2008)	*Xantusia jaycolei*: **near Desemboque del Río San Ignacio**
Serpientes (21)	
Cope (1861)	*Chilomeniscus cinctus (= C. stramineus)*: **near Guaymas**
Baird & Girard (1853)	*Diadophis regalis (= D. punctatus regalis)*: **Sonora**
Taylor (1936a)	*Ficimia desertorum (= Gyalopion quadrangulare desertorum)*: **about 12 km NW of Guaymas**
Tanner (1981)	*Hypsiglena torquata tiburonensis (= H. chlorophaea)*: **Isla Tiburón**
Zweifel & Norris (1955)	*Lampropeltis getulus nigritus (= L. getula nigrita)*: **30.6 road mi S of Hermosillo**
Smith & Tanner (1944)	*Leptodeira ephippiata (= L. splendida ephippiata)*: **Agua Marin, 8.3 mi W-NW of Álamos**
Jan (1863)	*Masticophis bilineatus*: **western Mexico (restringida a Guaymas[1])**
Lowe & Woodin (1954)	*Masticophis flagellum cingulum*: **Moctezuma**
Lowe & Norris (1955)	*Masticophis slevini*: **Isla San Esteban**
Bogert & Oliver (1945)	*Phyllorhynchus browni fortitus (= P. browni)*: **Álamos**
Smith & Langebartel (1951)	*Phyllorhynchus decurtatus norrisi (= P. decurtatus)*: **45.1 mi S of Sta. Ana**
Bogert & Oliver (1945)	*Pseudoficimia hiltoni (= P. frontalis)*: **Güirocoba**
Klauber (1937)	*Sonora (= Chionactis) palarostris*: **6 mi S of Hermosillo**
Taylor (1937)	*Tantilla hobartsmithi*: **near La Posa, 10 mi NW of Guaymas**
Tanner (1959)	*Thamnophis melanogaster chihuahuense*: **Río Bavispe below Three Rivers, Chihuahua-Sonora line[3]**
Cope (1886)	*Trimorphodon lambda*: **Guaymas**
Smith, Lemos-Espinal, Hartman, & Chiszar (2005)	*Tropidodipsas repleta*: **km 236.2, Hwy 16 Chihuahua-Hermosillo**
Kennicott (1860)	*Elaps (= Micrurus) euryxanthus*: **Sonora (restringida a Guaymas[1,2])**
Zweifel & Norris (1955)	*Micruroides euryxanthus australis*: **Güirocoba**
Kennicott (1861)	*Caudisona atrox sonoraensis (= Crotalus atrox)*: **Sonora (restringida a Guaymas[1])**
Klauber (1949)	*Crotalus estebanensis*: **Isla San Esteban**

[1] Restricción reportada en Smith y Taylor (1950)

[2] Restricción no aceptada por Roze (1974)

[3] Originalmente reportada para el estado de Chihuahua por Tanner (1959), restringida aquí al estado de Sonora

de reproducción). Las especies que no migran deben ser capaces de sobrevivir y reproducirse dentro del intervalo de condiciones ambientales presentes en sus hábitats, los cuales pueden presentar variaciones estacionales y anuales. El clima en Sonora varía desde muy caliente y árido en los desiertos del noroeste hasta fresco y húmedo en las sierras altas del este, lo cual se refleja en una diversidad de comunidades bióticas (Brown 1982a; Russell y Monson 1998). Variaciones estacionales o a largo plazo en temperatura y precipitación imponen cambios temporales en comunidades bióticas que pueden ser visualmente dramáticos y biológicamente significativos. Por ejemplo las comunidades del matorral espinoso y el bosque tropical caducifolio se transforman desde un matorral de ramas espinosas, seco, de color café y poco atractivo a un matorral tropical exuberante en tan sólo dos o tres semanas después del inicio de las lluvias de verano en junio y julio.

Las 38 especies de anfibios de Sonora necesitan por lo menos condiciones húmedas para reproducirse, y todas excepto 5 especies de los géneros *Pseudoeurycea, Eleutherodactylus* y *Craugastor*, tienen un estadio larval acuático que requiere cuerpos de agua estacionales o permanentes para su desarrollo. Aunque el agua es un factor limitante para los anfibios de Sonora, éstos son un grupo remarcablemente adaptado, únicamente los valles más secos del Gran Desierto de Altar en la porción noroeste del estado carecen de anfibios. En contraste, los reptiles están mucho menos limitados por la disponibilidad de agua y muchas especies habitan aun en el corazón de los desiertos más secos de Sonora. La alta riqueza de reptiles de Sonora se debe al huevo coriáceo o condición vivípara, escamas externas que resisten la perdida de agua, y al matiz de adaptaciones o exaptaciones fisiológicas y conductuales a condiciones áridas que esta clase de vertebrados presenta (Gould y Vrba 1982; Pianka 1986).

Como ectotermos, anfibios y reptiles dependen del ambiente para los procesos fisiológicos de calentarse y enfriarse. Aunque conductualmente adeptos a tomar ventaja de condiciones

y micrositios favorables, cuando sus hábitats se vuelven extremadamente calientes o fríos, anfibios y reptiles se refugian bajo el suelo u otros refugios donde las condiciones no exceden sus tolerancias a la temperatura, y muchos permanecen inactivos durante la temporada caliente previa a las lluvias de verano y, particularmente en el norte y a elevaciones altas, durante el invierno. El frío es más limitante en las sierras altas del noreste donde las temperaturas de invierno ocasionalmente caen por debajo de –10°C (Felger et al. 2001). La mayoría de los anfibios y reptiles no pueden sobrevivir más que periodos breves a temperaturas bajo cero (Zug et al. 2001); por lo tanto, en periodos fríos, evitan salir y permanecen inactivos en madrigueras bajo el suelo, en el fondo de charcas, o en otros lugares donde pueden evitar las temperaturas de congelamiento. Inviernos fríos y periodos cortos de crecimiento también influyen en las estrategias de reproducción. Por ejemplo, las tres especies de *Phrynosoma* que habitan a elevaciones altas en Sonora (*P. ditmarsi, P. hernandesi* y *P. orbiculare*) son vivíparas (Sherbrooke 2003). La oviparidad, en la cual se depositan huevos y éstos se desarrollan bajo el suelo o alguna cubierta protectora, no proporciona un desarrollo embrionario suficientemente rápido que permita eclosionar en una estación. Por lo que, las hembras de estas tres especies llevan los embriones dentro de sus cuerpos y los mantienen calientes a través de asolearse y otras conductas donde captan calor. En contraste, todas las especies de *Phrynosoma* de altitudes bajas en Sonora son ovíparas. Este patrón donde la viviparidad es más común a elevaciones y latitudes altas se da para lagartijas y serpientes en general.

En los desiertos del noroeste, las temperaturas del aire exceden 40°C y las temperaturas del sustrato mayores a 60°C no son raras en verano. La mayoría de las especies de reptiles diurnas de desierto tienen una temperatura corporal máxima voluntaria de 38°C a 46°C, y se refugian en madrigueras bajo el suelo o rocas cuando sus temperaturas se aproximan o llegan a este umbral (Cowles y Bogert 1944; Brattstrom 1963). Por lo que, especies de desierto tienden a estar activas en el frío relativo de la mañana o las últimas horas de la tarde durante el verano, aunque muchas especies, especialmente serpientes, están activas en la noche cuando las temperaturas son moderadas.

El ambiente es el medio en el cual las especies habitan, se adaptan y crecen. Debido a esto, el entendimiento del clima, topografía y comunidades bióticas de Sonora establece los fundamentos que sustentan los mecanismos que condicionan la distribución y diversidad de la rica herpetofauna de Sonora.

Sonora es el segundo estado más grande en México, sus 185,430 km² quedan comprendidos entre los 26.2° y 32.5° de latitud norte, y 115.0° y 108.8° de longitud oeste, y representan el 9.5% de la superficie territorial del país. Su límite este con Chihuahua está situado mayormente al oeste de la División Continental en la Sierra Madre Occidental, mientras que la mayor parte de su límite oeste está representado por el Mar de Cortés e incluye 14 islas principales y numerosos islotes pequeños. Con aproximadamente 1200 km², Tiburón es la isla más grande de Sonora, de hecho, es la isla más grande de todo México. Las montañas del este forman una sierra más o menos continua de tierras altas y picos escabrosos desde la Sierra de San Luis y la cabecera del Río Bavispe cerca de la frontera con Estados Unidos hasta el sur del estado en la frontera con Sinaloa. Sin embargo, una cuenca y topografía montañosa al oeste de la masa principal de la Sierra Madre Occidental crea una serie de islas continentales biogeográficamente distintas, la mayoría en el noreste del estado caracterizadas por tener islas de encinos y coníferas separadas de otras islas similares por mares de pastizales semiáridos, planicies, o matorral desértico chihuahuense (Brown y Lowe, 1980). La montaña más alta en Sonora (2625 m) es la Sierra de los Ajos al sureste de Cananea, ésta representa una de esas islas continentales. Montañas casi tan altas como la anterior se encuentran en la Sierra de San Luis y en el extremo norte de la Sierra Madre Occidental cerca de la Mesa de Tres Ríos (Felger et al. 2001).

La topografía de valles y montañas, con sierras generalmente corriendo en dirección noroeste-sureste, continua hacia el oeste hasta la costa. Sin embargo, las sierras disminuyen su altitud de este a oeste, la mayoría a lo largo o cerca de la costa no exceden 700 m de altitud. Estas sierras bajas también forman islas continentales, pero las diferencias en comunidades bióticas entre los picos y valles son menos conspicuas que en el este de Sonora. La flora y fauna de las montañas del oeste tiende a ser más rica que la de los valles. En el noroeste de Sonora, las islas continentales de la subdivisión de las tierras altas de Arizona del matorral desértico sonorense pueden estar rodeadas por valles de la subdivisión Colorado de este mismo matorral desértico. Más hacia el sur, el matorral espinoso puede estar presente en los valles de las montañas con matorral desértico sonorense (Brown y Lowe 1980).

Casi todos los ríos de Sonora desembocan en el Mar de Cortés. De norte a sur, los principales ríos de Sonora son Colorado, Sonoyta, Concepción, Sonora, Yaqui, Matape, Mayo y Fuerte. Únicamente una porción pequeña del noreste de Sonora desemboca al este de la División Continental en la cuenca del Río Casas Grandes de Chihuahua. Los Ríos San Pedro y Santa Cruz entre Nogales y Naco toman una ruta sinuosa hasta el Mar de Cortés fluyendo hacia el norte hasta el Río Gila en Arizona y de ahí hacia el oeste hasta el Río Colorado. Actualmente la mayoría de los ríos en Sonora no llegan al Mar de Cortés excepto cuando se desbordan pues han sido desviados, convertidos en presas, o simplemente se secan en el desierto antes de alcanzar el mar (Minckley y Marsh 2009).

La presente topografía de Sonora empezó a evolucionar hacer aproximadamente 35 millones de años cuando la Sierra Madre Occidental comenzó a elevarse desde lo que era una amplia planicie (Ferrusquia-Villafranca y Gonzáles-Guzmán 2005). El surgimiento de montañas en el este de Sonora y oeste de Chihuahua fue seguido por movimientos de la placa tectónica del Pacífico hace 8–25 millones de años, los cuales empujaron las porciones del oeste de Sonora junto con la Falla de San Andrés y fallas asociadas a ésta hacia el norte y oeste, creando la topografía actual de cuencas y sierras y finalmente separando a Baja California de la parte continental, lo que resultó en la formación inicial del Mar de Cortés y sus islas (Scarborough 2000). Aproximadamente hace 5 millones de años el Mar de Cortés se ensanchó y alargó, adquiriendo su forma y tamaño actual (Larson 1972). La arena marina del Gran Desierto de Altar se formó por el depósito de sedimentos del Río Colorado, los cuales comenzaron a fluir al Mar de Cortés hace aproximadamente 5.5 millones de años después de que este río fue desviado hacia el sur y el este por la creación de montañas en la Sierra Nevada y sierras costeras de la actual California (Howard 1996; Merriam y Bandy 1965).

Los climas de Sonora son un producto de la topografía local, latitud, y vientos del oeste que traen aire frío y húmedo del Océano Pacífico a la parte continental. Las sierras causan un levantamiento atmosférico, enfrían el aire y provocan precipitación, pero también forman barreras para la lluvia y subsecuentemente zonas secas opuestas a la fuente de humedad. En Sonora, la mayoría de las lluvias caen en los meses de verano (julio a septiembre) y principalmente son el resultado de las altas presiones y levantamiento de masas de aire sobre un paisaje caliente que recibe aire húmedo del Pacífico. Remolinos de masas de aire húmedo crean tormentas violentas en el verano que frecuentemente son muy localizadas y visualmente espectaculares. Existe un debate sobre qué tanto de la humedad de verano que provoca estas tormentas se origina en el Golfo de México; sin embargo, muchos de los climatólogos consideran que la mayor parte de la humedad de verano se deriva del Pacífico, y que la Sierra Madre Occidental tiene un efecto de barrera sobre la humedad del Golfo de México (Ingram 2000; Brito-Castillo et al. 2010). Las lluvias en invierno son menos comunes, pero se dan a través de los sistemas de lluvias originados en el Pacífico oriental que se desplazan hacia

el este a través del oeste de los Estados Unidos. Si se desarrollan presiones bajas en el suroeste de Estados Unidos, estas tormentas pueden moverse lo suficientemente al sur para producir precipitaciones importantes en Sonora. Las lluvias de invierno tienden a ser menos fuertes, de mayor duración, y se extienden sobre áreas más grandes que las lluvias de verano (Brito-Castillo et al. 2010). El porcentaje de lluvias invernales en el total de precipitación anual incrementa en el norte y oeste de Sonora. La región más seca de Sonora es la noroeste; la precipitación media anual es únicamente de 37 mm en San Luis Río Colorado. En contraste, Yécora en la Sierra Madre Occidental, recibe un promedio de precipitación anual de 913 mm (Felger et al. 2001).

Las temperaturas tienden a disminuir con la elevación y la latitud, pero también se vuelven más variables estacionalmente con la latitud (Brito-Castillo et al. 2010). Se sabe que el Gran Desierto de Altar presenta las temperaturas más extremosas en el verano; por ejemplo una temperatura de 56.7°C fue registrada en la región de Sierra Blanca en 1971 (May 1973). Sin embargo, los inviernos en el noroeste son relativamente frescos. Como resultado, San Luis Río Colorado tiene una temperatura anual media (22.8°C) más baja que las estribaciones de la Sierra Madre Occidental del sureste en el poblado de Álamos (23.5°C; Burquez et al. 1999). Temperaturas de congelamiento, las cuales limitan la distribución de muchas plantas y animales, son raras en las tierras bajas del sur y a lo largo de la costa de Sonora, pero se vuelven más comunes en el norte y las montañas altas del este. La nieve comúnmente cubre las montañas más altas del noreste durante al menos unas cuantas semanas en invierno.

Cambios presentes y geológicos en clima y topografía son las principales fuerzas que influyen en la naturaleza de las comunidades bióticas. En Sonora, una diversidad de paisajes sobre un amplio intervalo latitudinal y altitudinal han engendrado una remarcable diversidad de herpetofauna y otros grupos de animales y plantas. Una fuerte influencia tropical del sur se mezcla con los biomas templados de las Montañas Rocallosas y las Grandes Planicies en el norte, y con el Desierto de Chihuahua principalmente en el este. La comunidad más grande en términos de territorio ocupado en Sonora es el Desierto Sonorense, el cual es muy variable y ha sido descrito con cuatro o cinco subdivisiones principales en el estado (Shreve 1951; Turner y Brown 1982; Felger et al. 2001; Martínez-Yrizar et al. 2010). Éste ocupa las tierras bajas desde San Luis Río Colorado hacia el sur hasta Guaymas, y todas las islas del Mar de Cortés. Generalmente esta comunidad está caracterizada por una asociación relativamente abierta de matorrales resistentes a sequía (ejem: *Larrea tridentata*, *Ambrosia* sp., *Lycium* sp.), árboles (ejem: *Parkinsonia* sp., *Acacia* sp., *Prosopis* sp.), y suculentas (ejem: *Carnegiea gigantean*, *Cylindropuntia* sp., *Opuntia* sp., *Fouquieria* sp.). Hacia el sur y el este, las comunidades incrementan en estatura, densidad y riqueza hasta que cambia gradualmente a matorral espinoso de Sinaloa (Brown 1982c), el cual frecuentemente está delineado en vertiente costera y estribaciones (Burquez et al. 1992, Martínez-Yrizar et al. 2010). El matorral espinoso está compuesto de una mezcla principalmente de matorrales espinosos y caducifolios resistentes a sequía, árboles y suculentas, muchas de las cuales pueden almacenar agua en sus tallos carnosos o raíces tuberosas. Plantas representativas incluyen *Acacia* sp., *Bursera* sp., *Fouquieria* sp., *Ipomea arborescens*, *Jatropha* sp., *Lysiloma* sp., *Stenocereus thurberi*, *Havardia* sp. y *Cylindropuntia* sp. En las estribaciones del sureste predominantemente al sur de la carretera Méx. 16, el matorral espinoso cambia a bosque tropical caducifolio, el cual es un bosque deciduo resistente a sequía, de múltiples capas con fuertes derivaciones tropicales y muchas especies sensitivas a heladas (ejem: *Ambrosia cordifolia*, *Bursera* sp., *Ceiba acuminata*, *Haematoxylum brasiletto*, *Lysiloma divaricata*, *Pachycereus pectin-aboriginum*, *Senna atomaria* y *Stenocereus* sp., Martin et al. 1998; Van Devender et al. 2000; Felger et al. 2001). Arriba y al este del bosque tropical caducifolio, así como al norte de las montañas más bajas, se encuentra el bosque de encino, el cual contiene una diversidad principalmente de encinos perennes (ejem: *Quercus arizonica*, *Q. chihuahensis*, *Q. emoryi*, *Q. oblongifolia*), frecuentemente en bosques abiertos mezclados con especies del bosque tropical caducifolio a altitudes y latitudes más bajas, y en los bosques más densos con pinos y táscates esparcidos a elevaciones más altas y hacia el norte. El límite inferior del bosque de pino–encino, que típicamente se desarrolla por encima del bosque de encino, raramente se extiende debajo de los 1000 m, más comúnmente se encuentra a 1300 o más metros de altitud (Martin et al. 1998). En esta comunidad boscosa, varias especies de *Quercus* se mezclan con una variedad de especies de pinos (ejem: *Pinus arizonica*, *P. ayacahuite*, *P. englemannii*, *P. oocarpa*, *P. strobiformis*). Aproximadamente por encima de 2135 m en las Sierras Los Ajos, El Tigre, San José y San Luis, así como en la Mesa del Campanero y en parches pequeños en otras montañas y otras sierras en el noreste de Sonora, se pueden encontrar bosques mixtos de coníferas de pinos y abetos (*Abies* sp., *Pinus* sp., *Pseudotsuga menziesii*), algunas veces con parches de aspen (*Populus tremuloides*) (Felger et al. 2001).

En los valles del noreste, predominan matorrales y pastizales del Desierto de Chihuahua. El pastizales de planicies, que se da en los valles más altos, principalmente por encima de 1500 m al este del Río Santa Cruz a lo largo de la frontera con Estados Unidos hasta la Sierra de San Luis, es una pradera de pastos cortos que frecuentemente ha sido degradada por muchas décadas de pastoreo de ganado. Los géneros comunes de pastos de este pastizal incluyen *Aristida*, *Bothriochloa*, *Bouteloua*, *Eragrostis*, *Leptochloa*, *Panicum* y *Sporobolus*. Pastizales de semi-desierto se dan pendiente abajo en los valles inferiores a aproximadamente 1100 m, éstos han sido grandemente invadidos por matorrales y cactus característicos del matorral desértico sonorense, proceso probablemente causado por sobrepastoreo, fuego, y cambio climático, frecuentemente la identificación de las causas específicas de este proceso en sitios específicos son alusivas (Humphrey 1958; Peters y Havstad 2006). Ejemplos de pastizales

de semi-desierto se pueden encontrar en las
bajadas de la Sierra Azul en Rancho El Aribabi
y en otras localidades esparcidas en el noreste.
En valles al este de la Sierra de Ajos y al norte
de Río Bavispe, se desarrollan matorrales del
Desierto de Chihuahua. Este desierto alto está
visualmente dominado por gobernadora (*Larrea
tridentata*) y otros matorrales o árboles pequeños
(ejem: *Acacia neovernicosa, Flourensia cernua,
Prosopis glandulosa*), así como por una variedad
de suculentas (ejem: *Yucca elata, Dasylirion
wheeleri, Fouquieria splendens, Cylindropuntia* sp.).
Frecuentemente donde ha habido sobrepastoreo
los pastos son escasos o están ausentes.

Otras comunidades bióticas que se encuentran dentro de las principales divisiones
descritas anteriormente incluyen manglares y
matorrales sobre la costa e islas desde el sur de
Puerto Lobos, y varias comunidades riparias y de
pantano a lo largo de ríos. Éstas últimas tienden
a ser muy diversas y proporcionan importantes cinturones de vegetación verde y de agua a
través de los paisajes áridos que frecuentemente
presentan limitantes importantes en la disponibilidad de alimento, agua, cubierta, etc., para
varias especies, por lo menos estacionalmente
(Felger et al. 2001). Las aguas marinas del Mar
de Cortés representan otro bioma muy diferente,
el cual actualmente proporciona hábitat para seis
reptiles marinos.

Adaptaciones a los diversos ambientes de
Sonora han formado una variedad de ensambles
más o menos distintivos de anfibios y reptiles
que están asociados con las principales comunidades bióticas. Rorabaugh y Enderson (2009)
describieron cinco ensambles de lagartijas,
incluyendo Desierto de Sonora, Tropical/
Subtropical, de Montaña, Pastizal y Desierto de
Chihuahua. Dos especies fueron consideradas
generalistas. Enderson et al. (2009) categoriza-
ron a los anfibios y reptiles en nueve ensambles:
Chihuahuense, Templado del Este, Generalista,
Grandes Planicies, Sierra Madre, Marino, Generalista de Desiertos de Norteamérica, Sonorense
y Tropical. Nosotros utilizamos una división
consistente con otros capítulos de este libro, pero
un reconocimiento total de cualquiera de estas

clasificaciones es arbitrario. En este capítulo
reconocemos siete ensambles.

ESPECIES DEL DESIERTO DE SONORA

Cuarenta y seis especies han sido asignadas a
este ensamble, el cual incluye las 11 especies
endémicas a islas: *Anaxyrus retiformis, Lithobates berlandieri, Kinosternon arizonense, Gopherus
morafkai, Crotaphytus dickersonae, C. nebrius,
Gambelia wislizenii, Coleonyx variegatus, Phyllodactylus homolepidurus, P. nocticolus, P. nolascoensis, Ctenosaura conspicuosa, C. nolascensis,
Dipsosaurus dorsalis, Sauromalus ater, S. hispidus,
S. varius, Callisaurus draconoides, Holbrookia
elegans, Phrynosoma goodei, P. mcallii, P. solare,
Sceloporus magister, Uma rufopunctata, Urosaurus
graciosus, Uta nolascensis, U. palmeri, U. stansburiana, Aspidoscelis baccata, A. estebanensis,
A. martyris, A. xanthonota, Xantusia jaycolei,
X. vigilis, Lichanura trivirgata, Arizona elegans,
Chilomeniscus stramineus, Chionactis occipitalis,
C. palarostris, Masticophis slevini, Phyllorhynchus
browni, P. decurtatus, Salvadora hexalepis, Crotalus cerastes, C. estebanensis, C. mitchellii*. Mucho
del Desierto de Sonora está deshabitado por humanos y es demasiado seco para el desarrollo de
industrias o agricultura, como resultado, vastas
regiones de este bioma están relativamente in-
alteradas, particularmente el noroeste árido. Las
principales excepciones incluyen los desarrollos
agrícolas, rurales y urbanos a lo largo del Valle
del Río Colorado, el valle de los Ríos Magdalena,
Asunción y Concepción desde las cercanías
de Magdalena hasta El Desemboque, y desde
Hermosillo hacia el oeste y suroeste hasta Bahía
Kino y Boca Tastiota. El sobrepastoreo de ganado
está muy extendido y local o estacionalmente
es muy fuerte en la mayor parte del Desierto
de Sonora, aunque áreas más secas, tales como
el Gran Desierto de Altar, tienen poco ganado
excepto después de inviernos húmedos cuando
el desierto produce lujosas exhibiciones de
herbáceas anuales. El pasto africano (*Pennisetum
ciliare*), introducido de África o el Medio Oriente,
ha sido plantado extensamente para forraje de
ganado en las partes centro y sur ocupadas por

el matorral desértico sonorense. Donde ha sido plantado, frecuentemente la vegetación nativa fue removida con tractores, alterando severamente el ambiente. Sin embargo, aun en áreas donde se ha propagado, la comunidad de plantas nativas puede estar muy alterada debido a la competencia así como por un incremento en la frecuencia de fuegos facilitada por el pasto africano (Burquez et al. 2002; Franklin y Molina-Freaner 2010). La cosecha de *Prosopis* sp. para leña o carbón y *Olneya tesota* para artesanías ha afectado la estructura y composición de la vegetación en algunas áreas. Se ha anticipado que el cambio climático tendrá un efecto importante en la flora del Desierto de Sonora (Weiss y Overpeck 2005), así como probablemente en el resto de Sonora (IPCC 2007). Más de un millón de hectáreas del Desierto de Sonora están protegidas en la Reserva de la Biósfera Pinacate y Gran Desierto de Altar y la Reserva de la Biósfera Alto Golfo de California y Delta del Río Colorado. La mayoría de las islas de Sonora están incluidas en la Reserva de la Biósfera Islas del Golfo de California.

ESPECIES DE AFINIDADES SUBTROPICALES

Treinta y cinco especies están incluidas en este ensamble: *Incilius marmoreus, Rhinella marina, Craugastor occidentalis, Pachymedusa dacnicolor, Smilisca baudinii, Tlalocohyla smithi, Leptodactylus melanonotus, Hypopachus variolosus, Lithobates forreri, L. pustulosus, Trachemys nebulosa, Rhinoclemmys pulcherrima, Phyllodactylus tuberculosus, Heloderma exasperatum, Sceloporus albiventris, S. nelsoni, Urosaurus bicarinatus, Anolis nebulosus, Plestiodon parviauriculatus, Drymarchon melanurus, Drymobius margaritiferus, Imantodes gemmistratus, Lampropeltis triangulum, Leptodeira punctata, L. splendida, Leptophis diplotropis, Mastigodryas cliftoni, Pseudoficimia frontalis, Sonora aemula, Sympholis lippiens, Thamnophis validus, Trimorphodon tau, Micrurus distans, Agkistrodon bilineatus, Crotalus basiliscus.* Éstas son especies que se distribuyen principalmente en áreas subtropicales o tropicales. La mayoría de los registros para estas especies en

Sonora provienen del extremo sureste del estado a elevaciones bajas o moderadas en el bosque tropical caducifolio y/o matorral espinoso de la costa. La mayoría del matorral espinoso en el sur de Sonora ha sido removido por agricultura, que domina las tierras bajas desde Ciudad Obregón hacia el sur hasta Yavaros. En las estribaciones menos accesibles, el bosque tropical caducifolio está más intacto, aunque plantaciones de pasto africano no son raras, una larga historia minera ha marcado el paisaje local, y algunas cosechas de madera para leña son evidentes, así como cosechas de otros productos forestales tales como el árbol *Croton fantzianus*, el cual es utilizado para hacer estacas guía y de soporte para cultivos vegetales. Ganado vacuno y otros tipos de ganado forrajean a través de la mayor parte de esta región. El Área de Protección de Flora y Fauna Sierra de Álamos-Río Cuchujaqui, la cual incluye la Reserva Monte Mojino, protege algunos de los mejores ejemplos del bosque tropical caducifolio de Sonora.

ENSAMBLE DE MONTAÑA Y ESTRIBACIONES

Sesenta y cinco especies han sido colocadas en este ensamble: *Ambystoma rosaceum, Pseudoeurycea bellii, Anaxyrus kelloggi, A. mexicanus, Incilius mazatlanensis, I. mccoyi, Craugastor augusti, C. tarahumaraensis, Eleutherodactylus interorbitalis, Hyla arenicolor, H. wrightorum, Smilisca fodiens, Gastrophryne mazatlanensis, Lithobates chiricahuensis, L. magnaocularis, L. tarahumarae, L. yavapaiensis, Terrapene nelsoni, Trachemys yaquia, Kinosternon alamosae, K. integrum, Elgaria kingii, Crotaphytus collaris, Coleonyx fasciatus, Ctenosaura macrolopha, Holbrookia approximans, Phrynosoma ditmarsi, P. hernandesi, P. orbiculare, Sceloporus clarkii, S. jarrovii, S. lemosespinali, S. poinsettii, S. virgatus, Plestiodon callicephalus, P. obsoletus, Aspidoscelis burti, A. costata, A. opatae, A. sonorae, A. stictogramma, Boa constrictor, Diadophis punctatus, Geophis dugesii, Gyalopion quadrangulare, Lampropeltis knoblochi, Masticophis mentovarius, Oxybelis aeneus, Pituophis deppei, Salvadora bairdii, S. deserticola, S. grahamiae,*

Senticolis triaspis, Storeria storerioides, Tantilla hobartsmithi, T. wilcoxi, T. yaquia, Thamnophis cyrtopsis, T. eques, T. melanogaster, T. unilabialis, Tropidodipsas repleta, Crotalus lepidus, C. pricei y *C. willardi.* Éstas son especies características de las estribaciones y montañas del este de Sonora, y se pueden encontrar desde altitudes bajas en el matorral espinoso de las estribaciones o el bosque tropical caducifolio hasta elevaciones altas en el bosque de coníferas, o pueden ser muy especializadas en el uso de hábitat o con una distribución muy localizada y específica. Mucha de esta región es de acceso difícil y la mayoría de los pocos caminos pavimentados o terracerías de las estribaciones y sierras fueron construidos hace apenas 20 años. Sin embargo, el este de Sonora ha estado sujeto a gran cantidad de impactos antrópicos, incluyendo el generalizado y frecuentemente excesivo pastoreo del ganado, la invasión y plantación de pasto africano en las estribaciones y altitudes más bajas, la minería de cobre en las regiones de Cananea y Nacozari, la tala de bosques para madera y leña, y la producción ilícita de drogas (Stoleson et al. 2005). Especies acuáticas introducidas, incluyendo a *Lithobates catesbeianus,* el cangrejo de río (*Orconectes virilis*), el cabeza de toro negro (*Ameiurus melas*), el pez sol verde (*Lepomis cyanellus*) y otras, se han vuelto más comunes, particularmente en las estribaciones y valles de ríos en el noreste, y amenazan a los peces, anuros, culebras y tortugas de la región. El Bosque Nacional y Refugio de Vida Silvestre Los Ajos-Bavispe proporciona protección para ocho sierras importantes del noreste del estado. La conservación en esta reserva federal está complementada por reservas privadas en la Northern Jaguar Reserve y Ranchos El Aribabi y Los Fresnos, así como la conservación a través de crianza en ranchos y otras propiedades en el noreste de Sonora.

ESPECIES DE PASTIZALES

Hemos colocado a siete especies en este grupo, aunque todas excepto *Crotalus viridis* habitan en el Desierto de Chihuahua o en otros tipos de comunidades: *Ambystoma mavortium, Spea*

multiplicata, Terrapene ornata, Sceloporus cowlesi, S. slevini, Aspidoscelis uniparens, Crotalus viridis. Algunas de estas especies son características de planicies y pastizales de semi-desierto (Brown 1982b). Estas comunidades bióticas han sido el foco de un intenso pastoreo de ganado, frecuentemente durante más de un siglo. Muchos de estos pastizales de semi-desierto han sido invadidos por matorrales de desierto, cactus y árboles. El sobrepastoreo remueve la fuente de combustible, reduciendo la frecuencia de fuego y favoreciendo la invasión de matorrales y cactus, los cuales son menos tolerantes al fuego que los pastos. Sin embargo, las causas de la invasión de matorrales y cactus no son siempre claras. Inviernos fríos en los pastizales de las planicies más altas están limitados por muchos de los matorrales y cactus de desierto que invaden los pastizales inferiores de semi-desierto, por lo tanto la conversión en el tipo de comunidad vegetal es menos probable de ocurrir en los pastizales de planicies altas. El Rancho Los Fresnos, una reserva privada de la ONG Naturalia, es el mejor ejemplo en Sonora de un pastizal de planicie intacto y bien manejado.

ESPECIES DEL DESIERTO DE CHIHUAHUA

Siete especies están incluidas en este ensamble, aunque la mayoría habita también en comunidades de pastizal, y algunas llegan a ocupar las montañas y estribaciones: *Anaxyrus debilis, Cophosaurus texanus, Phrynosoma cornutum, P. modestum, Aspidoscelis exsanguis, Gyalopion canum* y *Heterodon kennerlyi.* El pastoreo de ganado vacuno es la principal actividad humana en este tipo de comunidad. En esta comunidad dominada por matorrales, cuando el pastoreo ha sido muy intenso, casi no se observan pastos.

ENSAMBLE MARINO

Las seis especies de este ensamble son exclusivamente reptiles habitantes de mar con una distribución amplia que salen a la orilla del mar sólo para depositar sus huevos. La mayoría de ellos no se reproduce en Sonora: *Caretta caretta, Che-*

lonia mydas, Eretmochelys imbricata, Lepidochelys olivacea, Dermochelys coriacea, Hydrophis platurus. Antes de su extirpación a finales de la segunda mitad del siglo veinte, *Crocodylus acutus* era un miembro de este ensamble y habitaba estuarios y manglares sobre la costa de Sonora (Rorabaugh 2008). *Crocodylus acutus* fue presumiblemente eliminado por la caza excesiva y la destrucción o alteración de los estuarios y manglares costeros. Las 5 tortugas marinas han estado protegidas por la ley mexicana desde 1990, pero sus números han disminuido dramáticamente debido a su captura anteriormente legal y ahora ilegal, así como a la captura accidental en redes de botes pesqueros comerciales y redes de arrastre para camarón en el Mar de Cortés (Seminoff y Nichols 2007). La vida marina del Mar de Cortés ha sido muy alterada por la pesca intensiva que excede por mucho el nivel de sustentabilidad, incluyendo la pesca de arrastre de camarón, que esencialmente rasguña el fondo marino y resulta en mucha captura accidental de especies diferentes al camarón (Stoleson et al. 2005; Hastings y Findley 2007). El manejo y protección en la Reserva de la Biósfera Alto Golfo de California y Delta del Río Colorado y la Reserva de la Biósfera Islas del Golfo de California han mitigado algunos de los impactos a los ambientes marinos.

ENSAMBLE GENERALISTA

Veintiséis especies están distribuidas sobre una variedad de comunidades bióticas y no pueden ser asignadas a un ensamble específico: *Anaxyrus cognatus, A. punctatus, A. woodhousii, Incilius alvarius, Lithobates catesbeianus, Scaphiopus couchii, Kinosternon sonoriense, Apalone spinifera, Heloderma suspectum, Urosaurus ornatus, Aspidoscelis tigris, Hypsiglena chlorophaea, Lampropeltis getula, Masticophis bilineatus, M. flagellum, Pituophis catenifer, Rhinocheilus lecontei, Sonora semiannulata, Thamnophis marcianus, Trimorphodon lambda, Micruroides euryxanthus, Rena humilis, Crotalus atrox, C. molossus, C. scutulatus* y *C. tigris.*

Tres especies adicionales, *Hemidactylus frenatus, H. turcicus* e *Indotyphlops braminus,* son introducidas en Sonora y únicamente se les ha encontrado alrededor de habitaciones humanas o en parques de ciudades.

Sabemos que en la actualidad Sonora tiene un total de 195 especies de anfibios y reptiles: 38 anfibios (3 salamandras, 35 anuros), y 157 reptiles (16 tortugas, 69 lagartijas, 72 serpientes). El bajo porcentaje de especies de anfibios en el estado (19.5% = 38/195) es resultado de la fuerte dependencia de este grupo por lugares húmedos para vivir, el cual está limitado a las montañas del este y noreste, los cauces de ríos que corren a través del estado, y los exuberantes subtrópicos de la región sureste y estribaciones de la pendiente oeste de la Sierra Madre Occidental. Sin embargo, algunas especies de anfibios de Sonora están adaptadas a las condiciones áridas/semiáridas del Desierto de Sonora. De las tres especies de salamandras de Sonora, únicamente el Ajolote Tigre Rayado (*Ambystoma mavortium*) ha sido registrada en el Desierto de Sonora (pozo municipal de Puerto Peñasco) y casi seguramente fue introducido ahí. Las otras dos especies de salamandras están limitadas a las tierras altas de las montañas del este y noreste. Poblaciones abundantes de *Ambystoma rosaceum* se encuentran comúnmente en arroyos por encima de 1000 m de altitud en estas montañas, y *Pseudoeurycea bellii* se conoce únicamente para localidades en los alrededores de Yécora.

Los anuros están mucho mejor representados en el estado. La piel tuberculada de los miembros de la familia Bufonidae les permite ocupar hábitats áridos-semiáridos tal que poblaciones abundantes de *Anaxyrus cognatus, A. kelloggi, A. punctatus, A. retiformis* e *Incilius alvarius,* habitan en el Desierto de Sonora, y *Anaxyrus debilis,* habita en la punta del Desierto de Chihuahua del noreste de Sonora. Miembros de otras familias de anuros tales como Microhylidae (*Gastrophryne mazatlanensis*), Hylidae (*Smilisca fodiens*), y Scaphiopodidae (*Scaphiopus couchii* y *Spea multiplicata*), también se encuentran comúnmente en condiciones áridas-semiáridas del Desierto de Sonora. Otras especies, como las de las familias Craugastoridae y Eleutherodactylidae, están adaptadas para reproducirse fuera de cuerpos de agua. Depositan sus huevos en lugares húmedos, los embriones se desarrollan dentro del huevo, y las crías tienen la misma morfología que los adultos. Sin embargo, todas las especies de estas familias en Sonora se encuentran comúnmente en las montañas del este y sus estribaciones, o en los cañones y tierras bajas de la parte sureste del estado. La mayoría de las otras especies de Bufonidae, Hylidae, Microhylidae y Ranidae están limitadas a esta misma región.

De acuerdo con Rorabaugh (2008) y Lemos-Espinal y Smith (2009), es posible que tres especies adicionales de anuros habiten en Sonora pero aún no han sido registradas. La Rana de Arroyo del Pacífico (*Craugastor vocalis*) pueden habitar en el extremo sureste de Sonora; Hardy y McDiarmid (1969) reportaron a esta especie en el extremo noreste de Sinaloa, una región con condiciones ambientales similares a las del sureste de Sonora. Brennan y Holy-cross (2006) reportaron la presencia de la Rana Leopardo de Blair (*Lithobates blairi*) en el sureste de Arizona; probablemente esta especie habita en los pastizales abiertos del extremo noreste de Sonora, los cuales ofrecen un hábitat adecuado para esta especie de rana. Otra especie de anuro que probablemente habita en Sonora es el Sapo de Espuelas de los Llanos (*Spea bombifrons*). Hay una cantidad considerable de pastizales y matorrales del Desierto de Chihuahua en el noreste de Sonora que pueden hospedar poblaciones de esta especie; sin embargo, aún no ha sido registrada.

El número de especies de tortugas de Sonora (16) representa 50% del número total de especies de tortugas que habitan en México (Liner y Casas-Andreu 2008). Cinco de las seis Tortugas Marinas Mexicanas se encuentran en las costas de Sonora. Tres especies terrestres de tortugas nativas habitan en Sonora (*Gopherus morafkai*, *Terrapene nelsoni* y *T. ornata*), y la Tortuga de Concha Blanda (*Apalone spinifera*) ha sido introducida en la parte norte del estado (Rorabaugh et al. 2008; Rorabaugh y King 2013). La mayoría de las otras especies de tortugas se encuentra comúnmente en arroyos y pozas de las montañas del este y extremo sureste, y en los ríos que corren a través del Desierto de Sonora.

Probablemente la Casquito Amarilla (*Kinosternon flavescens*) habita en las pozas y aguajes del extremo noreste de Sonora. Brennan y Holycross (2006) reportaron a esta tortuga en el extremo sureste de Arizona, cerca de la parte adyacente de Sonora.

Las lagartijas representan 35.4% (69/195) del número total de especies de anfibios y reptiles de Sonora. Excepto por los hábitats marinos, éstas habitan en prácticamente todos los hábitats disponibles en el estado. Aunque su mayor riqueza de especies se encuentra en el Desierto de Sonora, también son abundantes en las montañas del este y noreste del estado y en el área subtropical del sureste de Sonora. Consideramos que es muy probable que otras dos especies de lagartijas habiten en Sonora: la Cachora (*Sceloporus bimaculosus*), y el Alicante de Dos Líneas (*Plestiodon bilineatus*). La distribución de *S. bimaculosus* fue reportada por Brennan (2008) en el sureste de Arizona, cerca de la frontera con Sonora. *Plestiodon bilineatus* fue registrado en el extremo oeste de Chihuahua por Lemos-Espinal y Smith (2007) en varias localidades al oeste de la División Continental, algunas de ellas cerca de la línea estatal con Sonora; las montañas del este de Sonora ofrecen los mismos hábitats observados por ellos en Chihuahua.

Las serpientes son el grupo herpetofaunístico con el mayor porcentaje en Sonora; ellas representan el 36.9% (72/195) del número total de especies de anfibios y reptiles. Es muy probable que por lo menos otras 5 especies habiten en el estado, tres de ellas en el noreste de Sonora (*Tantilla nigriceps*, *Rena dissecta*, y posiblemente *Sistrurus catenatus*); una en las montañas del este del estado (*Conopsis nasus*); y otra en el extremo sureste de Sonora (una especie de *Hypsiglena* que todavía no ha sido descrita).

La Culebrilla de Cabeza Negra (*T. nigriceps*) fue registrada por Brennan (2008) en el extremo sureste de Arizona; ésta podría habitar en los pastizales de las planicies del noreste de Sonora. La Culebrilla Ciega de Nuevo México (*R. dissecta*) fue reportada por Lemos-Espinal y Smith (2007) en el extremo noroeste de Chihuahua, en la Pradera de Janos (Rancho San Francisco), muy cerca de la línea estatal con Sonora; ésta podría habitar en los pastizales al oeste de la Sierra de San Luis del extremo noreste de Sonora. Y la Cascabel de Massasagua (*S. catenatus*) fue reportada por Brennan (2008) en localidades esparcidas en el sureste de Arizona; ésta podría habitar en los pastizales del noreste de Sonora.

La Culebra de Nariz Grande (*C. nasus*) fue reportada por Lemos-Espinal y Smith (2007) en varias localidades del extremo oeste de Chihuahua al oeste de la División Continental y cerca de la línea estatal con Sonora. Los bosques de coníferas de las montañas del este de Sonora ofrecen un hábitat adecuado para esta culebra y estamos seguros de su presencia en Sonora. Sin embargo, a la fecha no se ha registrado en este estado.

Hemos encontrado varias *Hypsiglena* en el extreme suroeste de Chihuahua (región de Chínipas-Milpillas) y sureste de Sonora (región de Álamos-Güirocoba) con características morfológicas de *H. torquata* (por ejemplo nuca blanca) junto con individuos típicos de *H. chlorophaea*. Las *Hypsiglena* con nuca blanca de esta región han sido consideradas desde hace mucho como intermedios entre individuos típicos de *torquata* y *chlorophaea* (Bogert y Oliver1945; Dixon y Dean 1986). Mulcahy et al. (2014) reconocieron a esta forma de nuca blanca como una especie distinta, pero aún no ha sido formalmente descrita.

Lemos-Espinal y Smith (2009) reportaron la posible presencia de un anfisbénido (*Bipes* sp.) en la Bahía de San Carlos basándose sobre un reporte sin confirmar por Royce E. Ballinger. Las playas arenosas de Bahía de San Carlos pueden proporcionar los requerimientos ambientales para que este anfisbénido habite en Sonora.

Ciento veinte de las especies de anfibios y reptiles de Sonora (= 62%) se comparten con los Estados Unidos, y 75 especies (= 38%) no se comparten entre estos dos países. Setenta y cinco de las 120 especies compartidas son especies que habitan en los desiertos norteamericanos de Mojave, Sonora y/o Chihuahua. La mayoría de ellas está limitada a estas regiones áridas/semiáridas con sus límites más sureños en el Altiplano Mexicano. En general, éstas son especies que también habitan en los estados norteamericanos de California, Arizona, y/o Nuevo México. Otras 25 de estas 120 especies compartidas son especies que comúnmente habitan en bosques templados de las montañas del sureste de Arizona, suroeste de Nuevo México, y de la Sierra Madre Occidental y sus islas continentales del noreste de Sonora. La mayoría de estas especies se distribuyen desde el sur de Arizona y Nuevo México hasta el norte de Durango, Sinaloa, Zacatecas o sur de Chihuahua (ejem. *Hyla wrightorum, Lithobates chiricahuensis, L. tarahumarae, Sceloporus virgatus, Lampropeltis knoblochi* y *Crotalus willardi*, entre otras). Tres especies tienen su límite más sureño en Sonora (ejem. *Lithobates yavapaiensis, Kinosternon arizonensis* y *Aspidoscelis sonorae*). Algunas otras tienen una distribución amplia en regiones semiáridas y templadas (ejem. *Hyla arenicolor, Thamnophis cyrtopsis, T. eques, Crotalus lepidus* y *C. molossus*). Las poblaciones de una más están distribuidas en la parte norte de las Sierras Madres Occidental y Oriental (*Crotalus pricei*), con las poblaciones del oeste distribuyéndose desde el sur de Arizona hasta Nayarit. Otra más se distribuye desde el sur de Canadá hasta el Eje Transvolcánico de México (*Diadophis punctatus*). Seis de las 120 especies compartidas son introducidas en Sonora (*Lithobates berlandieri, L. catesbeianus, Apalone spinifera, Hemidactylus frenatus, H. turcicus* e *Indotyphlops braminus*). Seis son especies marinas con una distribución amplia a través de los océanos del mundo (*Caretta caretta, Chelonia mydas, Eretmochelys imbricata, Lepidochelys olivacea, Dermochelys coriacea* e *Hydrophis platurus*). Siete son especies que habitan en los trópicos y subtrópicos de América y que se distribuyen sobre ambas costas entrando a los Estados Unidos únicamente en la punta sur de Texas (*Rhinella marina, Smilisca baudinii, Hypopachus variolosus, Drymarchon melanurus, Drymobius margaritiferus*), o sólo en la punta sureste de Arizona (*Oxybelis aeneus, Senticolis triaspis*). Y una de las 120 especies compartidas se distribuye desde Canadá y Montana hasta Sudamérica, ocupando una amplia variedad de condiciones ambientales (*Lampropeltis triangulum*).

Sesenta y siete de las 75 especies que no se comparten entre los dos países son endémicas a México, 16 de ellas a Sonora: 11 a las islas del Mar de Cortés (*Phyllodactylus nolascoensis, Ctenosaura conspicuosa, C. nolascensis, Sauromalus varius, Uta nolascensis, U. palmeri, Aspidoscelis baccata, A. estebanensis, A. martyris, Masticophis slevini* y *Crotalus estebanensis*), y 5 a la parte continental del estado (*Trachemys yaquia, Crotaphytus dickersonae, Phrynosoma ditmarsi, Aspidoscelis opatae* y *Xantusia jaycolei*). Una de las 67 especies endémicas probablemente fue introducida a la Isla Alcatraz de Bahía Kino por la etnia Seri (*Sauromalus hispidus*). Esta especie es originalmente endémica a las islas del Mar de Cortés de Baja California. Una más de estas 67 especies endémicas está ampliamente distribuida en el Desierto de Chihuahua (*Holbrookia approximans*); ésta se encuentra únicamente en el extremo noreste de Sonora, y es muy probable que habite en las partes adyacentes de Arizona y Nuevo México. Otras 13 de las 67 especies endémicas son especies características de los bosques templados de la Sierra Madre Occidental. La distribución de algunas de ellas se extiende hacia el noroeste desde el Eje Transvolcánico de México y alcanzan su distribución más norteña en el estado de Sonora (ejem. *Pseudoeurycea bellii, Phrynosoma orbiculare, Salvadora bairdii, Storeria storerioides* y *Thamnophis melanogaster*). Otras están limitadas a la parte norte de la Sierra Madre Occidental en estados tales como Sonora, Chihuahua, Sinaloa, Durango y Zacatecas (ejem. *Ambystoma rosaceum, Anaxyrus mexicanus, Incilius mccoyi, Craugastor tarahumaraensis, Sceloporus lemosespinali, Plestiodon parviauriculatus* y *Thamnophis unilabialis*). Otras 36 de las 67 especies endémicas son especies características de los trópicos y subtrópicos de México. Estas especies se distribuyen desde el sur de México a través de los cañones y estribaciones de la vertiente oeste de la Sierra Madre Occidental y alcanzan su distribución más norteña en las tierras bajas y estribaciones de Sonora (ejem. *Pachymedusa dacnicolor, Tlalocohyla smithii, Kinosternon integrum* y *Leptodeira punctata*, entre otras especies). Algunas más están limitadas a los subtrópicos desde el norte del Río Santiago en Nayarit y también alcanzan su distribución más norteña en las tierras bajas y estribaciones de Sonora (ejem. *Anaxyrus kelloggi, Incilius mazatlanensis, Lithobates magnaocularis, Coleonyx fasciatus, Heloderma exasperatum, Sceloporus nelsoni* y *Sonora aemula*, entre otras especies). Las 8 especies que no son endémicas a México, y que no habitan en los Estados Unidos (*Leptodactylus melanonotus, Lithobates forreri, Rhinoclemmys pulcherrima, Phyllodactylus tuberculosus, Boa constrictor, Imantodes gemmistratus, Masticophis mentovarius,* y *Agkistrodon bilineatus*), son especies ampliamente distribuidas en los trópicos y subtrópicos de México, Centroamérica y hasta Sudamérica.

Herpetofauna de Chihuahua

JULIO A. LEMOS ESPINAL Y HOBART M. SMITH

5

Introducción

El conocimiento de la diversidad de la herpetofauna del estado de Chihuahua ha incrementado
muy lentamente, hasta el año de 1900 esto se
debió principalmente a la dificultad para acceder
al estado desde las instituciones de educación/
investigación del centro de México y universidades y museos de los Estados Unidos. Las
primeras exploraciones en el estado estuvieron
limitadas al Desierto de Chihuahua en las partes
este y centro, éstas no fueron inspiradas para
ser investigaciones intensas, hasta cierto punto
éste ha sido el caso aún en la actualidad. Ya que
el único acceso a las montañas del oeste, era el
desierto el cual constituía un cuello de botella
desde el sur para poder explorar la Sierra Madre
Occidental que cubre la mayor parte del tercio
oeste del estado y posee una diversidad faunística extraordinaria.

Por ejemplo, Alfredo Dugès (1896), "padre"
de la herpetología mexicana se dedicó más de
60 años al estudio de la herpetofauna de México,
en su compendio de la herpetofauna de México,
reportó solamente una especie para Chihuahua
(*Xerobates polyphemus*), no reconocida en ese
entonces como una especie nueva (*Gopherus flavomarginatus* Legler, 1959b). De hecho la primera
exploración herpetológica en el estado vino del

norte, por un reconocimiento cerca de la frontera con los Estados Unidos. Cope (1887) reportó
37 especies para el estado y los listados de Smith
y Taylor (1945, 1948, 1950a) reportaron sólo 102,
y relativamente pocas de éstas para la Sierra
Madre Occidental. Los compendios de Tanner
(1985, 1987, 1989) de sus propios estudios en
las montañas del oeste, así como estudios de
otros investigadores en el estado, revelaron una
diversidad inesperada de herpetofauna de 120
especies. Recientemente se han publicado varios
compendios sobre el total de la diversidad de anfibios y reptiles de Chihuahua, éstos incluyen el
listado de especies, fichas diagnosticas para cada
especie, claves ilustradas para la identificación
de las especies, fotos, mapas distribucionales y
gacetero con las localidades de colecta (Lemos-
Espinal et al. 2004d; Lemos-Espinal et al. en
prensa; Lemos-Espinal y Smith 2007, 2009).

Estudios posteriores a aproximadamente
1985, particularmente los realizados por nosotros, continuaron la tendencia de incrementar
el conocimiento de la herpetofauna del estado,
de tal forma que ahora podemos enlistar a 174
especies. Obviamente esta tendencia continuará,
el número real de la riqueza herpetofaunística
del estado probablemente excede 200 especies,
lo que representa un atractivo para investigaciones futuras.

Estudios Herpetológicos Previos

Las evidencias del interés precolombino en anfibios y reptiles en Chihuahua están plasmadas en numerosos petroglifos en varias partes del Desierto Chihuahuense, por ejemplo en la Sierra de Samalayuca, Sierra de Los Muertos, Cerros Colorados, Arroyo Los Monos y en la Sierra Madre Occidental.

La primera investigación herpetológica en Chihuahua parece haber sido el trabajo de Thomas H. Webb, secretario y cirujano de la Mexico-United States Boundary Commission comandada por John Russell Bartlett de los Estados Unidos, que exploraba la frontera mexicana. El grupo de Bartlett dejó la ciudad de El Paso, Texas el 7 de octubre de 1852 dirigiéndose a Ringgold Barracas, Texas a través de los estados mexicanos de Chihuahua, Durango, Coahuila y Nuevo León. La expedición llegó a la ciudad de Chihuahua el 22 de octubre de 1852 pasando a través de los poblados de Guadalupe, Carrizal, Encinillas y Saucillos. Este grupo permaneció diez días en la ciudad de Chihuahua, tiempo durante el cual Webb recolectó diversos tipos de fauna, constituyendo la primera colección de anfibios y reptiles de México depositados en el United States National Museum (USNM) (Baird 1859; Kellogg 1932).

Dos años más tarde, John Potts envió una colección de anfibios y reptiles de Chihuahua al USNM donde fue recibida el 7 de abril de 1854. En 1873 el naturalista Edward Wilkinson visitó México, pasando dos años en el pueblo minero de Batopilas, donde recolectó varios anfibios y reptiles. En 1885 regresó al estado, en donde pasó la mayor parte del tiempo en Ciudad Chihuahua, recolectando anfibios y reptiles en los alrededores. Su colección fue distribuida en varios museos norteamericanos, principalmente el USNM, y fue reportada parcialmente por Cope (1879, 1885).

En 1897 y 1899 Edward Goldman y Edward Nelson del USNM viajaron a través de la mayor parte de la Sierra Tarahumara, recolectando y describiendo la historia natural de la vida silvestre de esta región; los resultados de esta expedición fueron reportados por Goldman (1951). Posteriormente a las exploraciones de Goldman y Nelson, un número considerable de botánicos, mastozoólogos y naturalistas visitaron el estado de Chihuahua entre 1900 y 1938; en estas expediciones ocasionalmente se recolectaban anfibios y reptiles, los cuales fueron depositados en colecciones norteamericanas por S. B. Benson, C. S. Brimley, H. H. Brimley, W. W. Brown, y W. B. Richardson entre otros.

En junio de 1934 Hobart M. Smith y David Dunkle recolectaron a lo largo de los principales caminos que conectaban el norte y sur del estado de Chihuahua, muestreando los Médanos de Samalayuca, y áreas alrededor de Ciudad Chihuahua, Río San Pedro, la parte superior del Río Conchos, Presa de la Boquilla y el poblado de Escalón cerca de la línea divisoria con Durango. Esa colección se volvió parte de la colección personal de Edward H. Taylor, y posteriormente fue vendida a la Universidad de Illinois en Champaign-Urbana y al Field Museum of Natural History. En octubre de 1938 Hobart M. Smith y su esposa Rozella B. Smith recolectaron 526 especímenes de 29 especies bajo el auspicio del USNM en varias localidades esparcidas en el estado, incluyendo los municipios de Ahumada, Ascensión, Buenaventura, Camargo, Casas Grandes, Chihuahua, Jiménez y Juárez.

Entre 1938 y 1940 Irving Knobloch estudió la herpetofauna de los alrededores de Maguarichic y Mojarachic. Parte de los resultados obtenidos fueron reportados por Taylor y Knobloch (1940), éstos incluyen 3 especies de anfibios y 19 de reptiles. Otros resultados se encuentran reportados en Taylor (1940a, 1940b, 1941, 1944) éstos contienen las descripciones originales de *Eleutherodactylus* (= *Craugastor*) *tarahumaraensis*, *Lampropeltis knoblochi*, *Ambystoma rosaceum*, *A. fluvinatum* y *Crotalus semicornutum*. Todos los especímenes fueron depositados en la colección privada de Edward H. Taylor, y posteriormente vendidos al Field Museum of Natural History y a la University of Illinois.

Colecciones importantes de las montañas del

oeste de Chihuahua fueron hechas por James Anderson entre 1952 y 1966 en los alrededores de Yahuarichic, Chuhuichupa y Sierra del Nido. Anderson murió antes de que sus colecciones pudieran ser reportadas completamente, pero información importante sobre ellas aparece en algunas de sus publicaciones: Anderson (1960, 1961, 1962a, 1962b, 1962c, 1972). En ellas reportó información sobre la distribución e historia natural de *Storeria storerioides*, *Plestiodon brevilineatus* y *Sceloporus slevini*, incluyendo la descripción de *Crotalus willardi amabilis* de Sierra del Nido.

Por mucho una de las colecciones más grande de anfibios y reptiles de Chihuahua fue acumulada entre 1956 y 1972 en Brigham Young University por Wilmer W. Tanner y sus estudiantes, el principal Gerald W. Robison. Una gran cantidad de publicaciones de Tanner y Robison apareció en intervalos después de 1956, basada en esta colección, la cual está resumida en Tanner y Robison (1959) y Tanner (1957, 1985, 1987, 1989). Esta colección totaliza 120 especies, incluyendo 21 anfibios (2 salamandras, 19 anuros) y 99 reptiles (5 tortugas, 42 lagartijas y 52 serpientes). Taxa nuevos descritos en estos trabajos incluyen a *Plestiodon multilineatus*, *Sceloporus clarkii uriquensis*, *S. nelsoni coeruleus*, *S. poinsettii robisoni*, *Rena humilis chihuahuaensis*, *R. dulcis supraocularis*, *Thamnophis melanogaster chihuahuaensis* y *T. rufipunctatus unilabialis* (= *T. unilabialis*).

En el verano de 1959 Kenneth L. Williams y Pete S. Chrapliwy recolectaron en varias partes del norte de México, incluyendo Chihuahua. Estos resultados fueron reportados por Williams, Chrapliwy y Smith (1959, 1960); Williams, Smith y Chrapliwy (1960); y Chrapliwy et al. (1961). Entre ellos se encuentra la descripción original de *Uma paraphygas*. Estos especímenes se encuentran depositados en el University of Illinois Museum of Natural History (UIMNH).

En agosto de 1962 Kenneth L. Williams, Edward O. Moll, Francois Vuilleumier y John E. Williams viajaron a través del Río Conchos en la parte comprendida entre Julimes y Ejido El

Álamo (= Cañón de Barrera). Recorrieron una distancia de 300 km navegando sobre el río y recolectaron 282 especímenes de 27 especies: 6 anuros, 3 tortugas, 13 lagartijas y 5 serpientes. Este material se encuentra depositado en el UIMNH, y fue reportado por Smith et al. (1963).

En los veranos de 1971–1973 Thomas R. Van Devender y Charles H. Lowe recolectaron anfibios y reptiles en la región de Babícora, Temósachic y Yepómera. Recolectaron 204 especímenes de 11 especies de anfibios y 453 de 29 especies de reptiles, los cuales depositaron en el museo de la University of Arizona. Estos resultados fueron reportados por Van Devender (1973a, 1973b), Van Devender y Van Devender (1985), y Van Devender y Lowe (1977).

Una publicación de Domínguez et al. (1977) reportó una colección de anfibios y reptiles realizada en 1974 bajo el auspicio de la Dirección General de Fauna Silvestre en 45 localidades esparcidas mayormente al oeste de la carretera Méx. 45 entre Ciudad Juárez y Ciudad Chihuahua. Nueve especies de anfibios y 27 de reptiles fueron reportadas. Todo este material está depositado en el museo del Instituto Politécnico Nacional.

Entre 1975 y 1977 Robert P. Reynolds estudió la relación entre la precipitación pluvial y la actividad de las serpientes a lo largo de la carretera Méx. 16 (Ciudad Chihuahua-Ojinaga) en el tramo comprendido de Aldama al Rancho El Pastor. Los resultados que obtuvo incluyen observaciones sobre 418 organismos pertenecientes a 20 especies de las familias Colubridae y Viperidae (Reynolds 1982).

En mayo de 1982 Bruce Baker, Leroy Banicki, Rodolfo Corrales y Robert Webb recolectaron vertebrados terrestres en los alrededores del Cerro Mohinora, reportando 11 especies de anfibios y 5 de reptiles (Webb y Baker 1984), todos ellos más o menos esperados para esta área excepto por una especie no identificada de *Lithobates* del complejo *pipiens* que puede representar a *Lithobates berlandieri* (Smith y Chiszar 2003). Si así fuera, éste fue el primer registro de esta especie de distribución amplia en el este de las montañas del oeste del estado. Estos especímenes se encuentran en el museo de la University of Texas at El Paso.

Bradley Shaffer y M. McNight (1996) registraron la presencia de *Ambystoma velasci* (= *A. silvense*) cerca de los poblados de El Vergel y Temósachic sobre la base de especímenes recolectados previamente por Shaffer. Estos especímenes se encuentran depositados en la University of California at Davis.

Frost y Bagnara (1976), Legler (1959a), Tanner (1988, 1990), Webb et al. (2001) y Williams (1994) son algunos de varios otros autores que han contribuido al conocimiento de la herpetofauna de Chihuahua.

Entre 1993 y 2006 Hobart M. Smith y Julio Lemos-Espinal recolectaron en diferentes localidades esparcidas sobre todo el estado de Chihuahua Los resultados de este trabajo están reportados en Chiszar et al. (1995), Lara-Góngora (2004), Larson et al. (1998), Lemos-Espinal y Smith (2007, 2009), Lemos-Espinal et al. (en prensa), Taylor et al. (2003), y varios artículos de Lemos-Espinal y colaboradores y Smith y colaboradores publicados entre 1994 y 2009. Estos resultados contienen el primer registro para el estado de los taxa: *Ambystoma mavortium*, *Eleutherodactylus interorbitalis*, *E. marnockii*, *Pachymedusa dacnicolor*, *Tlalocohyla smithii*, *Hypopachus variolosus*, *Lithobates forreri*, *L. magnaocularis*, *L. pustulosus*, *Terrapene nelsoni klauberi*, *Rhinoclemmys pulcherrima*, *Gopherus morafkai*, *Barisia ciliaris*, *Hemidactylus t. turcicus*, *Heloderma horridum exasperatum* (= *H. exasperatum*), *Sceloporus merriami annulatus*, *Boa constrictor imperator*, *Gyalopion quadrangulare*, *Imantodes gemmistratus latistratus*, *Masticophis flagellum cingulum*, *M. f. testaceus*, *Mastigodryas cliftoni*, *Thamnophis v. validus*, *Tropidodipsas repleta*, *Agkistrodon b. bilineatus* y *Crotalus basiliscus* entre otros. Asimismo la descripción original de *Lithobates lemosespinali*, *Sceloporus edbelli* (= *S. consobrinus*), *S. lemosespinali*, *S. merriami sanojae*, *S. m. williamsi* y *S. undulatus speari* (= *S. consobrinus*).

Recientemente, Georgina Santos-Barrera y Oscar Flores-Villela (2011) describieron una nueva especie de sapo (*Incilius mccoyi*) de la Sierra Madre Occidental de Chihuahua y Sonora. Esta especie representa las poblaciones de sapos de la Sierra Madre Occidental de Chihuahua anteriormente consideradas como *Incilius occidentalis*. Otras adiciones recientes a la herpetofauna de Chihuahua son *Rhadinaea laureata* del municipio de Guadalupe y Calvo (Villa et al. 2012), y *Nerodia erythrogaster bogerti* de cerca de Ciudad Ojinaga (Uriarte-Garzón y García-Vázquez 2014).

Adicionalmente a los estudios mencionados previamente, hemos determinado que hay 39 nombres científicos de anfibios (3 salamandras, 5 anuros) y reptiles (15 lagartijas, 16 serpientes) con localidades tipo asignadas al estado de Chihuahua (Cuadro 5.1).

Características Fisiográficas y su Influencia sobre la Herpetofauna

Los anfibios y reptiles poseen características que limitan su distribución y actividad. A excepción de los miembros de las familias Plethodontidae, Craugastoridae y Eleutherodactylidae, todos los demás anfibios necesitan reproducirse en cuerpos de agua. Todos los anfibios y reptiles dependen del medio externo para poder regular su temperatura. Temperaturas extremadamente frías o calientes impiden su actividad. Debido a esto es necesario conocer cuáles son las características del ambiente físico para así poder entender la distribución de la herpetofauna de un determinado lugar.

Chihuahua es el estado más grande de la República Mexicana; sus 245,612 km² de extensión (comprendidos dentro de las coordenadas extremas 25°38' al sur; 31°47' al norte; 103°18' al este; 109°7' al oeste) representan el 12.6% de la superficie del territorio nacional. Esta superficie tan extensa alberga una fisiografía compleja la cual condiciona la distribución de los anfibios y reptiles en el estado.

Casi toda la porción oeste del estado de Chi-

CUADRO 5.1 Localidades tipo para anfibios y reptiles descritos para el estado de Chihuahua, México

Autor(es)	Nombre original: Localidad tipo
Salamandras (3)	
Taylor (1941)	*Ambystoma fluvinatum* (= *A. rosaceum*): **Mojárachic**
Taylor (1941)	*Ambystoma rosaceum*: **Mojárachic**
Owen (1844)	*Axolotes maculata* (= *Ambystoma rosaceum*): **Sierra Madre 26°6′N; 106°50′W**
Anuros (5)	
Girard (1854)	*Bufo insidior* (= *Anaxyrus debilis insidior*): **Chihuahua**
Santos-Barrera & Flores-Villela (2011)	*Incilius mccoyi*: **42 km S of Creel, on the Creel-Guachochi road**
Taylor (1940a)	*Eleutherodactylus tarahumaraensis*: **Mojárachic**
Smith & Chiszar (2003)	*Rana* (= *Lithobates*) *lemosespinali*: **between Creel and San Rafael**
Boulenger (1917)	*Rana* (= *Lithobates*) *tarahumarae*: **Yoquivo**[1]
Lagartijas (15)	
Stejneger (1890)	*Barisia levicollis*: **Chihuahua without defined locality**[1]
Tihen (1954)	*Gerrhonotus liocephalus taylori* (= *G. taylori*): **Clarines Mine, about 5 miles west of Santa Barbara, Chihuahua, Mexico, at an elevation of about 6800 feet**
Smith, Lemos-Espinal, Chiszar, & Ingrasci (2003)	*Elgaria usafa* (= *E. kingii*): **Ruinas of Rancho El Mesteño Chiquito**
Smith (1935b)	*Holbrookia bunkeri* (= *H. maculata bunkeri*): **15 mi S of Juárez**
Tanner & Robison (1959)	*Sceloporus clarkii uriquensis* (= *S. c. clarkii*): **Urique**
Smith, Chiszar, & Lemos-Espinal (1995)	*Sceloporus undulatus speari* (= *S. consobrinus*): **1.6 km N Hwy 2, 107°11′W, 31°33′N, 1250 m, on a side road intersecting Hwy. 2, 3.6 km E Microondas Duna, 18.6 km E San Martín, northern central Chihuahua**
Lemos-Espinal *in* Smith, Lemos-Espinal, & Chiszar (2003)	*Sceloporus merriami sanojae*: **Rancho Peñoles (27°7′49.6″N, 103°48′45.0″W), 1194 m**
Lemos-Espinal, Chiszar, & Smith (2000)	*Sceloporus merriami williamsi*: **El Fortín, 51 airline km straight W of the Río Grande**
Tanner & Robison (1959)	*Sceloporus nelsoni caeruleus* (= *S. n. barrancorum*): **Urique**
Tanner (1987)	*Sceloporus poinsettii robisoni* (= *S. p. macrolepis*): **Cuiteco**
Williams et al. (1959)	*Uma paraphygas*: **0.7 mi E of Carrillo, Chihuahua**
Mittleman (1942)	*Urosaurus unicus* (= *Urosaurus bicarinatus tuberculatus*): **Batopilas**
Smith (1935a)	*Uta caerulea* (= *Urosaurus ornatus caeruleus*): **30 mi N of Colonia García**
Tanner (1957)	*Eumeces* (= *Plestiodon*) *multilineatus*: **3 mi N of Colonia Chuhuichupa**
Wright & Lowe (1993)	*Cnemidophorus* (= *Aspidoscelis*) *inornata chihuahuae*: **8.1 miles east of Ciudad Chihuahua, 1440 meters elev.**
Serpientes (16)	
Klauber (1946)	*Arizona elegans expolita*: **Casas Grandes**
Tanner (1961)	*Conopsis nasus labialis* (= *C. nasus*): **2 mi SE Creel**
Taylor (1940b)	*Lampropeltis knoblochi*: **Mojárachic**
Smith (1941)	*Masticophis flagellum lineatulus*: **11 mi S of Buenaventura**
Baird & Girard (1852)	*Churchillia bellona* (= *Pituophis catenifer*): **Presidio del Norte, Chihuahua**[2]

CUADRO 5.1 *Continued*

Autor(es)	Nombre original: Localidad tipo
Cope (1879)	*Procinura aemula*: **Batopilas**
Smith (1942a)	*Tantilla yaquia*: **Guasaremos, Río Mayo**
Smith (1942b)	*Thamnophis ordinoides errans* (= *T. errans*): **Colonia García**
Tanner (1985)	*Thamnophis rufipunctatus unilabialis* (= *T. rufipunctatus*): **0.5 mi SW of Bocoyna**
Cope (1886)	*Trimorphodon vilkinsoni*: **Chihuahua City**
Kennicott (1860)	*Micrurus diastema distans* (= *M. distans distans*): **Batosegachic**
Tanner (1985)	*Leptotyphlops humilis chihuahuensis* (= *Rena segrega*): **10.7 km NW of Ciudad Chihuahua**
Tanner (1985)	*Leptotyphlops humilis supraocularis* (= *Rena dissecta*): **Colonia Juárez**
Anderson (1962c)	*Crotalus willardi amabilis* (= *C. amabilis*): **Arroyo Mesteño**
Taylor (1944)	*Crotalus semicornutus* (= *C. lepidus klauberi*): **Mojárachic**
Klauber (1949)	*Crotalus willardi silus*: **Río Gavilán, 7 mi SW of Pacheco**

[1] Restricción reportada en Smith y Taylor (1950b).

[2] Aunque la localidad tipo reportada por Baird & Girard (1852) y Smith & Taylor (1950b, 1966) es Presidio del Norte, Chihuahua, esto es un error; la localidad tipo correcta es Presidio del Norte, Texas.

huahua está representada por la Sierra Madre Occidental donde se localizan las mayores altitudes del estado y a través de la cual corre longitudinalmente de norte a sur la División Continental, la cual separa los sistemas hidrológicos del Pacífico de los del Atlántico. El punto más alto en Chihuahua es el Cerro Mohinora ubicado en el extremo suroeste a 3300 m de altitud (25°57′N, 107°3′O). Otras elevaciones importantes son la Sierra de Gasachic (3060 m de altitud; 28°15′N, 107°36′O) y el Cerro Güirichique (2740 m de altitud; 26°52′N, 107°30′O).

La altitud de la Sierra Madre Occidental disminuye gradualmente de sur a norte. Hacia el norte esta sierra termina a aproximadamente los 31° de latitud norte en donde nace la Sierra de San Luis la cual se extiende hasta la frontera con Nuevo México. La Sierra de San Luis deja el territorio nacional en un punto de aproximadamente 2100 m de altitud.

El extremo suroeste de la Sierra Madre Occidental está caracterizado por barrancas profundas las cuales llegan a bajar hasta aproximadamente 250 m de altitud en la Barranca del Septentrión/Cañón de Chínipas (INEGI 2004). La Sierra Madre Occidental varía en anchura

de aproximadamente 130 a 160 kms en la parte sur (al oeste de Hidalgo del Parral), y de 65 a 80 kms en la parte norte (al oeste de Casas Grandes) (Tanner 1985; Lemos-Espinal y Smith 2007, 2009).

La mayor parte del estado de Chihuahua al este de la Sierra Madre Occidental está representada por un altiplano extenso que varía de aproximadamente 1200 a 1700 m de altitud. Sin embargo, sobre este altiplano surge una cantidad grande de sierras pequeñas y medianas que en algunas ocasiones llegan a rebasar los 2000 m de altitud. En las partes altas de estas sierras se desarrollan bosques de coníferas, los cuales forman islas continentales que se encuentran rodeadas de condiciones ambientales totalmente diferentes a las que se presentan en ellas. Esta situación ha propiciado el aislamiento y diferenciación de las poblaciones que ahí se desarrollan.

En el extremo noreste, cañones profundos, similares a los de la vertiente oeste de la Sierra Madre Occidental, atraviesan las planicies altas y hospedan sus propios ensambles herpetofaunísticos. Entre ellos está la Zona de Protección de Flora y Fauna Silvestre Cañón de Santa Elena, una extensión del Big Bend National Park de los

Estados Unidos. Éste se encuentra a una altitud de 800 m.

Esta topografía tan accidentada ha condicionado los tipos de vegetación presentes en el estado así como las actividades productivas del ser humano. En la Sierra Madre Occidental, así como en las partes altas de las serranías al norte, sur, centro y este del estado, se presentan bosques densos de coníferas dominados por: pino cheguis *(Pinus leiophylla)*, pino real *(P. engelmanni)*, pino huicoyo *(P. ayacahuite)*, pino colorado *(P. durangensis)* y encino blanco *(Quercus chihuahuensis)*, entre otras especies. Este tipo de vegetación representa el 29.42% de la superficie del estado.

Las barrancas del oeste y suroeste del estado poseen bosques tropicales caducifolios que afortunadamente se encuentran en condiciones favorables debido a la densidad baja de la población humana que ahí habita. Algunas especies presentes en este tipo de vegetación son: mauto *(Lysiloma divaricata)*, palo blanco *(Ipomoea arborescens)*, guácima *(Guazuma ulmifolia)* y copal *(Bursera fagaroides)*; este tipo de vegetación representa el 2.38% de la superficie del estado.

Al este de la Sierra Madre Occidental domina el matorral xerófito el cual se entremezcla con pastizales naturales e inducidos en las porciones centro y noroeste y superficies para la agricultura en casi la totalidad de su extensión. Las especies más conspicuas del matorral xerófito son: gobernadora *(Larrea tridentata)*, hojasén *(Flourensia cernua)*, mezquite *(Prosopis glandulosa)*, sotol *(Dasylirion texanum)* y ocotillo *(Fouquieria splendens)*, entre otras especies; este tipo de vegetación comprende el 32.41% del total de la superficie estatal. Los principales pastizales se encuentran en la parte centro y noroeste del estado.

Algunas de las especies dominantes de éstos son: navajita *(Bouteloua gracilis)*, navajita velluda *(Bouteloua hirsuta)*, zacate tres barbas *(Aristida sp.)*, zacatón *(Sporobolus airoides)* y zacate colorado *(Heteropogon contortus)*; este tipo de vegetación comprende el 23.89% de la superficie total del estado.

Las zonas agrícolas comprenden el 7.38% de la superficie total del estado, y los principales cultivos son: maíz *(Zea mays)*, fríjol *(Phaseolus vulgaris)*, alfalfa *(Medicago sativa)*, sorgo *(Sorghum vulgare)*, y manzana *(Malus sylvestris)*.

La palabra Chihuahua proviene del Náhuatl *xicuacua* que significa lugar seco y arenoso describiendo así las condiciones de más de la mitad del estado (al este de la Sierra Madre Occidental) que presenta climas de tipo seco en su diferentes variantes: muy seco (precipitación menor de 100 mm al año, con una temperatura media anual entre 16°C y 20°C), presente en su mayoría en las porciones norte y este (parte de los municipios de Ahumada, Ascensión, Buenaventura, Coyame, Guadalupe, Juárez, Manuel Benavides y Ojinaga entre otros), y en menor grado en el extremo sureste (parte de los municipios de Camargo y Jiménez); seco (precipitación de 300 a 600 mm al año, con una temperatura media anual entre 15°C y 25°C); y semiseco (precipitación de 400 a 600 mm al año, con una temperatura media anual de 12°C a 25°C).

El resto del estado presenta climas: semifrío subhúmedo (precipitación de 500 a 600 mm al año, con una temperatura media anual de 10°C a 12°C) en las partes más altas de la Sierra Madre Occidental y Sierra del Nido; templado subhúmedo (precipitación de 500 a 600 mm al año, con una temperatura media anual de 14°C), sobre la parte media de la Sierra Madre Occidental; semicálido subhúmedo (precipitación de 800 a 1400 mm al año, con una temperatura media anual de 20°C a 24°C), en la región de las barrancas al suroeste del estado; y una porción pequeña del estado con clima cálido subhúmedo (precipitación entre 700 y 1000 mm al año, con temperatura media anual de 24°C), al suroeste del estado.

El periodo de lluvias en el estado es de mayo a septiembre; la máxima cantidad de estas ocurre en los meses de julio y agosto.

Las principales características de la topografía de Chihuahua se establecieron desde el comienzo del Pleistoceno, durante el cual varias glaciaciones y periodos interglaciares ocurrieron y requirieron de ajustes extensos a intervalos repetidos en las distribuciones ocupadas por las especies de anfibios y reptiles que habitan el estado. En tiempos recientes, el estado ha experimentado una desertificación considerable.

Estos procesos removieron algunas especies del estado, restringieron la distribución de otras, expandieron la de algunas más, y resultaron en la diferenciación taxonómica de algunas poblaciones aisladas que anteriormente tenían una distribución continua con otras poblaciones, tales como las subespecies de *Crotalus willardi*. Algunos ejemplos son *Chrysemys picta* y *Plestiodon multivirgatus*, que en la actualidad están al borde de la extinción, con distribuciones severamente limitadas. *Gopherus flavomarginatus* tienen una distribución relictual muy pequeña y es improbable que sobreviva sin una protección intensiva y a largo plazo tanto de individuos jóvenes como de adultos. Hay evidencias que indican que *Sternotherus odoratus* alguna vez habitó en el estado.

La mayoría de los ríos del estado fluyen hacia el este y desembocan en el Golfo de México. Los ríos de una región relativamente pequeña pero herpetológicamente importante en el extremo oeste del estado drenan en el Mar de Cortés del Océano Pacífico. En la porción norte-centro del estado algunos ríos son cuencas cerradas que vierten sus aguas en lagos.

Los Ríos Florido y San Pedro corren a través de las partes centro y sur del estado al este de la División Continental y forman el Río Conchos, el cual se une al Río Bravo, fluyendo hacia el este hasta el Golfo de México. Estos ríos irrigan las partes sur, centro, norte y este del estado.

El flujo hacia el Océano Pacífico es proporcionado por los Ríos El Fuerte, Mayo, Sinaloa y Yaqui al oeste de la División Continental. Los Ríos Batopilas, Chínipas, Oteros, San Ignacio y Urique se unen al Río El Fuerte; los Ríos Moris y Agua Caliente forman el Río Mayo; el Río Pentatlán vierte sus aguas al Río Sinaloa; y el Río Papigochi se une al Río Yaqui. El Río El Fuerte irriga las Barrancas del Cobre; el Río Mayo la región de Moris; el Río Sinaloa el extremo suroeste; y el Río Yaqui las regiones de Miñaca, Yepómera y Madera. La distribución de un número de especies tropicales, tales

como *Tlalocohyla smithi*, *Hypopachus variolosus*, *Pachymedusa dacnicolor*, *Heloderma exasperatum*, *Boa constrictor* y *Oxybelis aeneus* se extiende hacia el norte hasta las tierras bajas de Chihuahua a través de estas cuencas.

Cuencas cerradas se presentan principalmente en la parte noroeste del estado. Los Ríos Casas Grandes y Santa María vacían sus aguas en la Laguna de Guzmán, reducida en años recientes a unas cuantas pozas. El Río El Carmen alimenta la Laguna de Los Patos. La Laguna El Barreal en el norte y la Laguna El Cuervo en el centro del estado son depresiones que forman pantanos someros durante la estación de lluvias pero están completamente secas en la estación seca. Cuando contienen agua, el acceso a ellas es muy difícil.

Las formaciones geológicas de la mayoría del estado (91.39%) son de origen Cenozoico; únicamente 8.23% son de origen Mesozoico, y 0.38% de origen Paleozoico (INEGI 2004). La Sierra Madre Occidental consiste de rocas ígneas extrusivas de origen Terciario. Durante el verano ocurre una cantidad grande de lluvias torrenciales las cuales llenan ríos y arroyos que corren rápidamente ejerciendo su acción erosiva sobre los suelos. Este tipo de erosión ha resultado en los cañones profundos del suroeste del estado y ha dejado al descubierto una cantidad grande de depósitos minerales entre los que destacan plata, oro y cobre. La abundancia de este último en la región de las barrancas es la que da origen al nombre de Barrancas del Cobre.

La distribución y actividad de los anfibios y reptiles en el estado de Chihuahua se ve fuertemente influenciada por las condiciones anteriormente mencionadas. La topografía del estado representa barreras importantes en la distribución de estos. La porción oeste del estado representada por la Sierra Madre Occidental es una barrera que impide el flujo libre de especies entre los estados de Chihuahua y Sonora. Asimismo, esta sierra evita que las especies que habitan el Desierto Chihuahuense ubicado al este de la misma, puedan ocupar el extremo oeste del estado. La cima de la Sierra Madre Occidental marca la División Continental la cual divide al continente en las regiones occidental (Pacífico) y oriental (Atlántico). Igualmente, la porción de las barrancas del extremo suroeste de Chihuahua se caracteriza por la presencia de especies con afinidades tropicales las cuales ingresaron al estado de Chihuahua principalmente a través de los afluentes del Río El Fuerte. Algunas especies como *Crotalus basiliscus* parecen haber sido capaces de pasar a través de sierras altas, como la de Milpillas en el municipio de Chínipas, ya que a esta especie la hemos registrado en la cima de Milpillas así como en la vertiente occidental de la misma la cual corresponde al estado de Sonora, y en la vertiente oriental que corresponde al estado de Chihuahua. Adicionalmente esta serpiente es abundante tanto en el valle alrededor del poblado de Chínipas así como en las partes altas de las sierras que rodean a este valle. Algunas especies se distribuyen ampliamente a ambos lados de la Sierra Madre Occidental (Chihuahua y Sonora), tales como *Anaxyrus cognatus, A. punctatus, Scaphiopus couchii, Urosaurus ornatus, Uta stansburiana, Salvadora deserticola, Crotalus atrox* entre otras especies.

La actividad de la mayoría de los anfibios y reptiles en Chihuahua está limitada al periodo abril-octubre, pero los picos de actividad varían de acuerdo a la altitud, época de lluvias, y adaptaciones de una especie dada. La estación de actividad inicia antes y dura más en el suroeste tropical del estado y está más limitada en las montañas altas. Los ciclos de temperatura no están relacionados con el ciclo de precipitación, por lo que especies diferentes responden diferentemente de acuerdo a sus adaptaciones.

Los reptiles que inician su actividad en abril (ejem: *Aspidoscelis costata, Sceloporus albiventris, S. clarkii, S. lemosespinali, S. nelsoni, Phrynosoma orbiculare, Storeria storerioides, Pituophis catenifer*) se observan mínimamente activos. Individuos juveniles que se pueden calentar rápidamente, son los primeros que se observan activos. Los anfibios que habitan en montañas y que viven en o cerca de cuerpos de agua permanente tales como *Ambystoma rosaceum* y varias especies de *Lithobates* ser reproducen en abril y alcanzan su máxima

actividad en este mes; la actividad de otros anfibios de montaña así como de los reptiles en todos los hábitats su máximo durante la estación de lluvias en junio, julio, agosto y principios de septiembre, después disminuye gradualmente tal que desde mediados de noviembre hasta marzo virtualmente no hay actividad de ellos. Únicamente durante días calientes de invierno algunas especies heliofílicas salen de sus refugios brevemente.

A grandes rasgos la herpetofauna del estado de Chihuahua puede dividirse en cuatro grupos grandes, que a continuación se mencionan. (1) Subtropical: especies con afinidades tropicales distribuidas en las barrancas profundas del suroeste del estado y asociadas a la selva baja caducifolia de la región. Estas especies se comparten con las especies presentes en el extremo noreste del estado de Sinaloa y el extremo sureste del estado de Sonora. (2) Montaña: especies de afinidades templadas las cuales se distribuyen en la porción oeste del estado a lo largo de la Sierra Madre Occidental y Sierra de San Luis. Muchas de ellas ocupan las partes altas de las diferentes sierras que se elevan sobre el altiplano ubicado al este de la Sierra Madre Occidental, como: Sierra del Nido, Sierra del Pajarito, Sierra Azul, etc. (3) Desierto de Chihuahua: especies de afinidades áridas/semiáridas, y características del Desierto Chihuahuense, ubicado al este de la Sierra Madre Occidental. (4) Generalistas: especies que pueden presentarse en más de un tipo de hábitat y que no son características de ninguno de ellos.

ENSAMBLE SUBTROPICAL

Cuarenta y un especies están incluidas en este ensamble: *Incilius mazatlanensis, Rhinella marina, Craugastor augusti, Pachymedusa dacnicolor, Smilisca baudinii, Tlalocohyla smithi, Hypopachus variolosus, Lithobates forreri, L. magnaocularis, Rhinoclemmys pulcherrima, Terrapene nelsoni, Gopherus morafkai, Phyllodactylus tuberculosus, Heloderma exasperatum, Ctenosaura macrolopha, Sceloporus albiventris, S. nelsoni, Urosaurus bicarinatus, Anolis nebulosus, Aspidoscelis costata,*

Boa constrictor, Drymarchon melanurus, Drymobius margaritiferus, Gyalopion quadrangulare, Imantodes gemmistratus, Lampropeltis triangulum, Leptodeira splendida, Leptophis diplotropis, Masticophis mentovarius, Mastigodryas cliftoni, Oxybelis aeneus, Senticolis triaspis, Sonora aemula, Sympholis lippiens, Thamnophis validus, Trimorphodon tau, Micruroides euryxanthus, Micrurus distans, Rena humilis, Agkistrodon bilineatus y *Crotalus basiliscus.*

Las últimas dos especies son excepcionales entre las demás ya que son típicamente tropicales a través de sus distribuciones, pero también han logrado ocupar áreas templadas, como la Sierra de Milpillas, la cual constituye una barrera para algunas especies entre las áreas subtropicales del sureste de Sonora y del suroeste de Chihuahua.

Esta región presenta una densidad baja de población humana por lo que la mayor parte de ella se encuentra poco o nada alterada, lo que ha ayudado a la conservación de la herpetofauna del lugar. Sin embargo, en las cercanías de los poblados de Batopilas y Chínipas el encuentro con casi cualquier especie de serpiente resulta en la muerte de la serpiente. En Batopilas una de las pocas especies que se respeta es la Limacoa (*Boa constrictor*) la cual se considera benéfica para los cultivos por su acción depredadora sobre roedores. Otras serpientes como *Lampropeltis triangulum, Oxybelis aeneus, Sonora aemula* y *Sympholis lippiens*, entre otras, son consideradas muy venenosas por lo que en general se les mata. En el municipio de Chínipas el Escorpión (*Heloderma exasperatum*) era fuertemente atacado, afortunadamente Martín Velducea Avendaño junto con varios pobladores de la localidad han estado promoviendo la protección de esta lagartija así como de las serpientes en general por lo que aparentemente se ha disminuido la perdida de estos reptiles por ataques del ser humano.

ENSAMBLE DE MONTAÑA

Aproximadamente 57 especies habitan en las áreas boscosas de elevaciones altas, de estas 40

se presentan sólo en estos bosques y constituyen el Ensamble de Montaña, éstas son: *Ambystoma rosaceum, A. silvense, Pseudoeurycea bellii, Anaxyrus mexicanus, Incilius mccoyi, Craugastor tarahumaraensis, Eleutherodactylus interorbitalis, Hyla wrightorum, Lithobates chiricahuensis, L. lemosespinali, L. tarahumarae, L. yavapaiensis, Barisia ciliaris, B. levicollis, Elgaria kingii, Gerrhonotus taylori, Phrynosoma hernandesi, P. orbiculare, Sceloporus jarrovii, S. lemosespinali, S. slevini, S. virgatus, Plestiodon bilineatus, P. callicephalus, P. multilineatus, P. parviauriculatus, Aspidoscelis sonorae, Conopsis nasus, Diadophis punctatus, Geophis dugesii, Lampropeltis knoblochi, Pituophis deppei, Rhadinaea laureata, Salvadora bairdii, Storeria storerioides, Thamnophis errans, T. melanogaster, Tropidodipsas repleta, Crotalus pricei* y *C. willardi.*

La principal actividad productiva de esta región es el aprovechamiento forestal maderable, éste se enfoca principalmente al aprovechamiento de pinos y en menor medida al de encinos. Los tres municipios principales en donde se da esta producción son, en orden de importancia: Madera, Guadalupe y Calvo y Guachochi. Aunque existen sitios en donde se puede observar una degradación seria del ambiente, la mayor parte de esta sierra presenta condiciones buenas. A lo largo de ella se observan bosques densos de pino y encino; la inaccesibilidad a un porcentaje grande de estos bosques ha promovido el desarrollo adecuado de las especies de fauna silvestre que ahí habitan.

Otros sistemas montañosos como la Sierra de San Luis y la Sierra del Nido están intactos por lo que presentan condiciones óptimas para el desarrollo de la vida silvestre. La Sierra de San Luis ubicada en el extremo noroeste del estado es quizá la localidad más norteña en donde se distribuye el jaguar (*Panthera onca*) en México (Lafont-Terrazas, comunicación personal, 2012). En esta misma sierra se desarrolla la única población de *Crotalus willardi obscurus* de la República Mexicana. Igualmente, la Sierra del Nido presenta condiciones excepcionales, las poblaciones de víboras de cascabel (*Crotalus lepidus, C. molossus, C. pricei* y *C. willardi*) y lagartijas

270

CAPÍTULO 5

(*Barisia levicollis*, *Sceloporus poinsettii*, *S. jarrovii*, etc.) son abundantes.

Uno de los mejores indicadores del buen estado en que se encuentran estos bosques es la cantidad grande de sitios en que se ha registrado a la salamandra tarahumara (*Ambystoma rosaceum*); en la mayoría de estos sitios las poblaciones de esta salamandra muestran una densidad alta. En arroyos que han sido alterados por tala del bosque, desviación de cauces, u otras actividades esta especie está ausente o es rara. Asimismo, las ranas arborícolas (*Hyla arenicolor* e *H. wrightorum*) presentan una distribución bastante amplia a lo largo de estas sierras.

ENSAMBLE DEL DESIERTO DE CHIHUAHUA

Aunque el Desierto de Chihuahua está caracterizado por una definitiva continuidad climática, éste tiene una considerable diversidad de hábitats. Hay pastizales extensos en las partes centro y noroeste del estado; sistemas de médanos de arena, tanto activos como estabilizados, se presentan en el norte (Bolsón Cabeza de Vaca–Médanos de Samalayuca) y en el sur (Bolsón de Mapimí); extensos barreales (grandes áreas inundables que mantienen aguas someras temporalmente durante la época de lluvias pero que están secas el resto del año) en el extremo norte y la parte central del estado; ecotonos en donde se mezcla la vegetación xerófita y boscosa en la base de las sierras que nacen en forma aislada en medio del desierto; vegetación riparia a lo largo de los Ríos Conchos, Bravo y Santa María.

Setenta y siete especies habitan estas áreas y 57 son exclusivas a ellas dentro de los límites del estado: *Ambystoma mavortium*, *Incilius alvarius*, *Anaxyrus cognatus*, *A. debilis*, *Eleutherodactylus marnockii*, *Gastrophryne olivacea*, *Lithobates catesbeianus*, *Scaphiopus couchii*, *Spea bombifrons*, *Chrysemys picta*, *Terrapene ornata*, *Trachemys gaigeae*, *Kinosternon durangoense*, *K. flavescens*, *K. hirtipes*, *Gopherus flavomarginatus*, *Apalone spinifera*, *Crotaphytus collaris*, *Gambelia wislizenii*, *Coleonyx brevis*, *Hemidactylus turcicus*, *Cophosaurus texanus*, *Holbrookia maculata*, *Phrynosoma cornutum*, *P. modestum*, *Sceloporus bimaculosus*, *S. consobrinus*, *S. cowlesi*, *S. merriami*, *Uma paraphygas*, *Uta stansburiana*, *Plestiodon multivirgatus*, *P. obsoletus*, *Aspidoscelis gularis*, *A. inornata*, *A. marmorata*, *A. tessellata*, *A. uniparens*, *Arizona elegans*, *Bogertophis subocularis*, *Gyalopion canum*, *Heterodon kennerlyi*, *Lampropeltis getula*, *Masticophis taeniatus*, *Nerodia erythrogaster*, *Pantherophis emoryi*, *Rhinocheilus lecontei*, *Sonora semiannulata*, *Tantilla hobartsmithi*, *T. nigriceps*, *T. wilcoxi*, *Thamnophis marcianus*, *Rena dissecta*, *R. segrega*, *Agkistrodon contortrix*, *Crotalus atrox* y *C. viridis*.

Desafortunadamente las actividades productivas del ser humano han causado una degradación importante de esta región. La actividad pecuaria enfocada principalmente a la ganadería extensiva de vacunos ha provocado un cambio drástico en el paisaje del Desierto Chihuahuense, el cual a mediados del siglo diecinueve estaba caracterizado por pastizales extensos relativamente libres de arbustos. Los ríos que corrían a través de esta región poseían bosques densos de galería; estos ríos frecuentemente desembocaban en zonas grandes de barreal en donde llegaban a formar humedales extensos. En la actualidad la ganadería extensiva ha provocado que los arbustos dominen áreas anteriormente ocupadas por pastizales así como los hábitats riparios de la región. Las zonas de barreal que hasta hace poco contenía agua durante la mayor parte del año, actualmente no llegan a captar la suficiente cantidad para permanecer con agua por periodos mayores a la época de lluvias. Los cambios en la humedad de esta región podrían explicar la presente ausencia de la tortuga almizclera *Sternotherus odoratus* que en 1903 fue registrada por Meek en la localidad de El Sauz. Igualmente, en esta región se presentaban poblaciones abundantes del perrito de las praderas (*Cynomys ludovicianus*), lobo mexicano (*Canis lupus baileyi*), oso grisli (*Ursus horribilis*), berrendo (*Antilocapra americana*) y halcón plomado (*Falco femoralis*), entre otros. En la actualidad estas poblaciones han sido extirpadas del estado de Chihuahua (lobo mexicano y oso grisli), o son escasas y se limitan a áreas pequeñas como el perrito de las praderas en el extremo noroeste del estado (Pradera de Janos), y el halcón plomado en la parte central del estado (Desierto Coyamense).

Asimismo, la agricultura de riego ha destruido miles de hectáreas de esta región. La sobreutilización del agua subterránea para usos agrícolas y urbanos de ciudades crecientes como Chihuahua, Juárez y Cuauhtémoc, entre otras, ha afectado seriamente el flujo normal de ríos como Bravo, Conchos, Florido, San Pedro, etc. Esta sobreutilización junto con la ganadería extensiva ha provocado una degradación considerable de cuerpos de agua, hábitats riparios, pastizales y matorrales de la parte desértica del estado de Chihuahua (Dinerstein et al. 2000).

La herpetofauna de esta región también ha sufrido los embates de la población humana. Quizá el mayor impacto se ha dado en los anfibios ya que la reducción en la disponibilidad de agua ha disminuido poblaciones de esta clase de tetrápodo. La Salamandra Tigre (*Ambystoma mavortium*) es un habitante bien conocido por los pobladores del estado de Chihuahua, quienes cuentan que hace aproximadamente veinte años esta salamandra era abundante en los aguajes de los municipios de Ahumada y Coyame, sin embargo, la escasez de agua ha resultado en la desaparición de esta salamandra de la mayoría de estos aguajes. Durante el presente estudio revisamos minuciosamente la mayor cantidad posible de aguajes y presones de estos dos municipios, únicamente registramos su presencia en los aguajes de los ranchos La Bamba y Agua Zarca, municipio de Coyame.

Reptiles acuáticos también han sido afectados por la desecación en años recientes. La Tortuga de Concha Blanca (*Apalone spinifera*) prácticamente ha desaparecido de la parte noroeste del estado. La Laguna de Guzmán y los Ojos de Santa María, en el municipio de Ascensión, albergaban poblaciones abundantes de esta tortuga, pero en la actualidad la desecación de La Laguna de Guzmán y alteraciones de los Ojos de

Santa María han provocado que ésta se presente en números muy bajos.

La persecución directa, en particular de los reptiles del estado, por los humanos varía en los tres grandes hábitats, en general está correlacionada con la densidad de la población humana. Anteriormente mencionamos esto para el ensamble subtropical, pero es peor en el Desierto de Chihuahua que está más densamente poblado. Con pocas excepciones las serpientes son consideradas como venenosas y mortales tal que, independientemente de la especie, son atacadas en cuanto se les ve. El Cantil (*Agkistrodon contortrix*) y las serpientes de cascabel (*Crotalus atrox, C. lepidus, C. molossus, C. scutulatus* y *C. viridis*) no pueden ser diferenciadas de las especies inofensivas por lo que todas son atacadas. La ausencia de un cascabel es aún más sospechosa que la presencia de uno, tal que serpientes como *Heterodon kennerlyi* y *Lampropeltis getula*) son consideradas especialmente peligrosas.

En la mayoría de las áreas la carne de serpiente de cascabel es considerada con efectos benéficos sobre una amplia variedad de enfermedades, desde infecciones de la piel hasta diabetes y cáncer. En algunas ocasiones, estas creencias resultan en la búsqueda deliberada de serpientes de cascabel no sólo para comer pero también para secarlas y molerlas para convertirlas en polvo el cual se le pone a la comida en forma de sal.

ENSAMBLE DE GENERALISTAS

Hay 30 especies cuyas distribuciones en Chihuahua substancialmente involucran más de uno de los tres hábitats principales del estado: *Anaxyrus punctatus, A. speciosus, A. woodhousii, Hyla arenicolor, Gastrophryne mazatlanensis, Lithobates berlandieri, Spea multiplicata, Kinosternon integrum, K. sonoriense, Holbrookia approximans, Sceloporus clarkii, S. poinsettii, Urosaurus ornatus, Plestiodon tetragrammus, Aspidoscelis exsanguis, Hypsiglena chlorophaea, H. jani, Masticophis bilineatus, M. flagellum, Pituophis catenifer, Salvadora deserticola, S. grahamiae, Thamnophis cyrtopsis,*

T. eques, T. unilabialis, T. sirtalis, Trimorphodon vilkinsonii, Crotalus lepidus, C. molossus y *C. scutulatus.*

Un total de 174 especies de anfibios y reptiles habitan en Chihuahua: 38 anfibios (4 salamandras, 34 anuros), y 136 reptiles (13 tortugas, 51 lagartijas, 72 serpientes). El bajo porcentaje de especies de anfibios en el estado (21.8% o 38/174) se debe a la dominancia de ambientes semiáridos en Chihuahua. La mayoría de los anfibios está concentrada en la Sierra Madre Occidental y sus islas continentales asociadas, tales como la Sierra del Nido y Cumbres de Majalca. Tres de las 18 especies de salamandras ambystomatidas de México se encuentran en Chihuahua, dos de ellas (*Ambystoma rosaceum* y *A. silvense*) están limitadas a sistemas montañosos; únicamente *A. mavortium* está adaptada para vivir en aguas estacionales y estancadas en el Desierto de Chihuahua. Poblaciones de estas tres salamandras son localmente abundantes, de hecho, *A. rosaceum* es la especie de Chihuahua mejor representada en colecciones herpetológicas. Ésta es abundante en casi cualquier arroyo pequeño de la Sierra Tarahumara. La única salamandra plethodontida en el estado, *P. bellii*, es conocida únicamente en una localidad (6 km al oeste-noroeste de Ocampo, en el campo de béisbol El Águila); esta pobre representación de plethodontidos es debida al relativamente bajo porcentaje de humedad en la Sierra Madre Occidental y la ausencia en esta región de especies de bromelias las cuales se sabe proporcionan un refugio para plethodontidos en la estación seca.

Los anuros de Chihuahua también están concentrados en la Sierra Madre Occidental y cadenas montañosas asociadas, aunque las familias Bufonidae, Microhylidae y Scaphiopodidae están bien representadas en el Desierto de Chihuahua. Los anuros más conspicuos del Desierto de Chihuahua son las especies de los géneros *Anaxyrus, Incilius, Gastrophryne, Spea* y *Scaphiopus*. Por otra parte, miembros de las familias Hylidae (*Hyla, Pachymedusa, Smilisca, Tlalocohyla*) y Ranidae (*Lithobates*) son abundantes y están ampliamente distribuidos en los

bosques de la Sierra Madre Occidental. Ocho de las nueve especies de ranas del género *Lithobates* habitan en las montañas de la parte oeste del estado. La única especie ausente en esta región es *L. catesbeianus*, la cual en Chihuahua ha sido introducida en la parte norte del estado.

Lemos-Espinal y Smith (2007) señalaron que cuatro especies adicionales de anuros probablemente habitan en Chihuahua; tres de ellas en las barrancas y tierras bajas del extremo suroeste del estado. La Rana de Arroyo del Pacífico (*Craugastor vocalis*) fue registrada por Hardy y McDiarmid (1969) en el extremo noreste de Sinaloa, a 16 km al norte-noreste de Choix, 520 m de altitud, cerca de la línea estatal con Chihuahua. La Rana del Sabinal (*Leptodactylus melanonotus*) fue registrada por Bogert y Oliver (1945) en Güirocoba y Álamos, Sonora, aproximadamente a 25 y 35 km respectivamente de la frontera con Chihuahua. Smith et al. (2005) también reportaron a esta especie para el Río Mayo en la compuerta de la Presa Mocuzari, Sonora. Hardy y McDiarmid (1969) reportaron localidades para esta especie (como *L. occidentalis*, una sinonimia menor) a través de las tierras bajas de Sinaloa, incluyendo una localidad en la esquina noreste de este estado. Y la Rana Chata (*Smilisca fodiens*) ha sido registrada cerca de Chihuahua por Hardy y McDiarmid (1969) para Sinaloa, Bogert y Oliver (1945) para Sonora, y Trueb (1969) y Duellman (2001) para ambos estados.

Otra especie de anuro que probablemente habita en el extremo noreste de Chihuahua es el Sapo Nebuloso (*Incilius nebulifer*). Esta especie de sapo está representada por poblaciones aisladas en el extremo sur de la región del Big Bend de Texas, adyacente a Coahuila (Conant y Collins 1998).

Por otra parte, el bajo porcentaje de tortugas en el estado en relación con el número total de especies de anfibios y reptiles (7.5% o 13/174) es un reflejo del número total de tortugas en el país (32; Liner y Casas-Andreu 2008). En Chihuahua habita el 41% (13/32) de las especies del país. Aún más, es probable que por lo menos otras cuatro especies habiten en el estado. Tres especies han sido registradas cerca de la línea estatal con Sonora y Sinaloa, en el extremo suroeste del estado. *Kinosternon alamosae* ha sido registrada en los alrededores de Álamos, Sonora, aproximadamente 35 km de la frontera con Chihuahua. *Trachemys hiltoni* ha sido registrada en Güirocoba, aproximadamente a 25 km de la línea estatal con Chihuahua, y en el extremo norte de Sinaloa. Seidel (2002) reportó la distribución de esta especie incluyendo Chihuahua, pero únicamente conjeturalmente. Legler y Webb (1970) señalaron que esta especie está limitada a la cuenca del Río El Fuerte. Estos últimos autores mencionaron que *Trachemys yaquia* está limitada a la cuenca del Río Mayo, Río Sonora y Río Yaqui, sin embargo, Seidel (2002) reportó la distribución de esta especie incluyendo Chihuahua, pero únicamente conjeturalmente.

Adicionalmente, la Tortuga Lagarto (*Chelydra serpentina*) habita en el Río Bravo por lo menos en Nuevo México (Degenhardt et al. 1996), y también puede habitar más hacia el sur en el extremo noreste de Chihuahua, donde se han hecho poco muestreos de tortugas.

Las lagartijas representan el 29.3% (51/174) del número total de especies de anfibios y reptiles de Chihuahua. Están bien representadas a través del estado y como lo señalan Lemos-Espinal y Smith (2007), la riqueza de especies de este grupo en Chihuahua es aún más grande. Ellos mencionan que por lo menos hay nueve especies de lagartijas que aún no han sido registradas en el estado y que es muy probable que habiten ahí; cuatro de ellas en las barrancas y tierras bajas del extremo suroeste de Chihuahua; tres en el extremo noreste del estado; y dos en el extremo noroeste.

La Cachora Arenera (*Callisaurus draconoides*) fue registrada por Bogert y Oliver (1945) en Güirocoba y Álamos, Sonora (aproximadamente a 25 y 35 km respectivamente de la frontera con Chihuahua), y Hardy y McDiarmid (1969) la reportaron para varias localidades en el extremo noreste de Sinaloa. El Geco de Bandas Negras (*Coleonyx fasciatus*) ha sido registrado en cinco localidades a lo largo de las estribaciones de la Sierra Madre Occidental del este de Sonora, tres de estas localidades están en la región de Álamos, una de ellas muy cerca de Chihuahua. Su hábitat sugiere que ésta puede habitar en las barrancas del suroeste de Chihuahua. El Camaleón Real (*Phrynosoma solare*) se distribuye desde el sur de Arizona a través de la mayor parte de Sonora, hasta el norte de Sinaloa. Hardy y McDiarmid (1969) y Bogert y Oliver (1945) registraron a esta especie muy cerca de Chihuahua en Sinaloa y Sonora. Ésta es una especie de hábitats áridos y semiáridos en planicies, colinas y montañas de pendiente baja. Y la Lagartija del Desierto (*Sceloporus magister*) muestra una distribución similar a la de la especie anterior. Al este del Mar de Cortés, ésta es la representante occidental de *S. bimaculosus*.

En el noreste de Chihuahua es probable la presencia adicional de tres especies de lagartijas. Wright (1971) indicó que el Huico de Nuevo México (*Aspidoscelis neomexicana*) es conocido únicamente para el centro de Nuevo México y extremo suroeste de Texas; casi todos los registros de esta especie provienen de cerca del Río Bravo. Proyectó la distribución de esta especie incluyendo Chihuahua a lo largo del Río Bravo; no existe ningún registro para esta región, sin embargo, su presencia ahí es muy probable. Conant y Collins (1998) incluyeron la parte sur de la región del Big Bend de Texas como parte de la distribución del Geco Reticulado (*Coleonyx reticulatus*). Éste puede habitar la parte adyacente de Chihuahua. De acuerdo con SEMARNAP (1997) esta especie fue observada en un transecto de Benavides al Cañón de Santa Elena; sin embargo, no se presentó ningún material de apoyo a esta observación. Conant y Collins (1998) también reportaron la distribución del Cantil de Tierra (*Gerrhonotus infernalis*) incluyendo la parte sur de la región del Big Bend de Texas, extendiéndose hacia el sur a través del este de Chihuahua, la mayor parte de Coahuila y otros estados hacia el sur. De acuerdo con SEMARNAP (1997) esta especie fue observada (bajo el nombre de *G. leiocephalus* [sic]) en un transecto de Benavides al Cañón de Santa Elena; sin embargo, no se presentó ningún material de apoyo a esta observación.

En el noroeste de Chihuahua puede habitar el Geco de Bandas Occidental (*Coleonyx variegatus*). Como lo indican Stebbins (2003) y Degenhardt et al. (1996), esta especie habita el extremo suroeste de Nuevo México, y probablemente también la parte adyacente del noroeste de Chihuahua. El Monstruo del Gila (*Heloderma suspectum*) también puede habitar está parte del estado. La presencia de esta especie en Sonora y Nuevo México cerca de la frontera con Chihuahua indica que ésta puede habitar en este último estado.

Las serpientes son el grupo de herpetozoarios con el mayor porcentaje en Chihuahua, con 41.4% o 72/174). Aún más, es muy probable que 9 especies adicionales habiten en el estado: 2 de ellas en el suroeste de Chihuahua (*Phyllorhynchus browni* y *Pseudoficimia frontalis*); 4 en el noreste (*Coluber constrictor, Lampropeltis alterna, Pantherophis bairdi* y *Tantilla cucullata*); 2 en la parte noroeste del estado (*Crotalus tigris* y *Sistrurus catenatus*); y 1 en el extremo sureste de Chihuahua (*Tantilla atriceps*).

La Culebra Ensillada (*Phyllorhynchus browni*) fue registrada por Bogert y Oliver (1945) en Álamos, aproximadamente a 35 km de la frontera con Chihuahua; y Hardy (1972) revisó la distribución de la Ilamacoa (*Pseudoficimia frontalis*), citando especímenes de cerca de Álamos y Güirocoba, Sonora, aproximadamente a 35 y 25 km de la frontera con Chihuahua, respectivamente.

La Corredora (*Coluber constrictor*) es una especie rara en México, con únicamente tres registros. Dos de ellos para Coahuila, incluyendo uno para el extremo noroeste, en la Sierra del Carmen (Wilson 1966), su presencia en Chihuahua es muy probable. La Culebra Real Gris (*Lampropeltis alterna*) es bien conocida en el Big Bend de Texas, y en el resto del estado, así como en Coahuila y otros estados adyacentes en México, pero nunca ha sido registrada en Chihuahua, aunque casi seguramente habita ahí. La Ratonera de Bosque (*Pantherophis bairdi*) habita en el oeste de Texas, incluyendo todo el Big Bend, así como el norte de Coahuila (Conant y Collins 1998); es muy probable que habite en la parte adyacente de Chihuahua. La Serpiente de Cabeza Negra de Trans-Pecos (*Tantilla cucul-*

lata) es conocida únicamente en el Big Bend y sus alrededores, Texas (Dixon et al. 2000); su presencia en las partes adyacentes de Coahuila y Chihuahua es probable.

En el noroeste de Chihuahua es probable que habite la Cascabel Tigre (*Crotalus tigris*). Stebbins (2003) señaló la presencia de esta especie en el extremo sureste de Arizona, y en el este de Sonora cerca de la frontera con Chihuahua. Ésta es una habitante de las estribaciones áridas-semiáridas de desiertos, puede entrar a Chihuahua en alguno de sus valles semiáridos. Otra víbora de cascabel, la Cascabel de Massasagua (*Sistrurus catenatus*), es conocida para el sur de Nuevo México (Degenhardt et al. 1996) y sureste de Arizona (Brennan y Holycross 2006); probablemente habita en la parte adyacente de Chihuahua.

En el extremo sureste de Chihuahua la presencia de la Culebrilla de Cabeza Negra (*Tantilla atriceps*) es probable. La distribución conocida de esta especie se acerca a la esquina sureste del estado (Conant y Collins 1998; Cole y Hardy 1981).

Con la adición de estas 26 especies a la herpetofauna de Chihuahua la riqueza de anfibios y reptiles en el estado podría ser de 200, la mayor para cualquier estado del norte de México, un resultado de la igualmente alta riqueza de ambientes y heterogeneidad espacial del estado.

Dos de las 174 especies de herpetozoarios conocidas para Chihuahua fueron introducidas (*Lithobates catesbeianus* y *Hemidactylus turcicus*). Sesenta y cinco por ciento (113/174) de estas especies se comparten con los Estados Unidos, y 35% (61/174) son especies que no se comparten. Setenta y dos de las 113 especies compartidas habitan en los desiertos de Norteamérica de Sonora, Chihuahua y Mojave. Algunas de ellas se distribuyen desde el sur de Canadá hasta el extremo sur del Desierto de Chihuahua en los estados mexicanos de San Luis Potosí y Querétaro (ejem. *Anaxyrus cognatus, A. debilis, Kinosternon hirtipes, Phrynosoma modestum, Heterodon kennerlyi, Masticophis taeniatus*). Algunas otras se distribuyen más al sur hasta los estados de Oaxaca, Puebla y Veracruz (ejem. *Ambystoma mavortium, Apalone spinifera, Masticophis taenia-*

tus, *Crotalus molossus, C. scutulatus*). Diecinueve más de estas 113 especies compartidas son especies que habitan en los bosques templados de hábitats montañosos; éstas se distribuyen en las cadenas montañosas del centro y sureste de Arizona y suroeste de Nuevo México extendiéndose hacia el sur a través de la Sierra Madre Occidental hasta Sinaloa, Durango o Jalisco (ejem. *Hyla wrightorum, Lithobates chiricahuensis, L. tarahumarae, Elgaria kingii, Sceloporus slevini, S. virgatus, Lampropeltis knoblochi, Crotalus willardi*). Otras 8 son especies de afinidades subtropicales que entran a los Estados Unidos sólo en la punta sur de Texas (*Rhinella marina, Smilisca baudinii, Hypopachus variolosus, Drymarchon melanurus, Drymobius margaritiferus*) o sólo en el extremo sureste de Arizona (*Oxybelis aeneus* y *Senticolis triaspis*) o sureste de Arizona y suroeste de Nuevo México (*Tantilla yaquia*). Las restantes 14 especies (incluyendo las 2 especies introducidas) son especies que habitan en una variedad de hábitats y muestran una distribución amplia, la mayoría de ellas en los estados del sur de Estados Unidos (pero algunas se distribuyen desde el sur de Canadá) hasta Centroamérica o Sudamérica (*Lampropeltis triangulum, Thamnophis cyrtopsis*), o que habitan en varios tipos de hábitats en el norte de México y suroeste de Estados Unidos (*Holbrookia elegans, Sceloporus poinsettii, Urosaurus ornatus, Opheodrys vernalis, Crotalus lepidus*).

Tres de las 61 especies que no se comparten son endémicas a Chihuahua (*Lithobates lemose-* spinali, Barisia levicollis y *Plestiodon multilineatus*). Veintitrés de estas 61 especies están limitadas a los bosques de la Sierra Madre Occidental y montañas asociadas (incluyendo las 3 endémicas a Chihuahua). Éstas son especies de afinidades templadas, y todas ellas son endémicas a México. Otras 34 especies habitan las barrancas y tierras bajas del extremo suroeste del estado; éstas son especies de afinidades subtropicales que extienden su distribución hacia el oeste y noroeste a través de los bosques y estribaciones de la Sierra Madre Occidental, o que entran al estado extendiéndose hacia el este a través de la cuenca del Río El Fuerte. Veintisiete de estas 34 especies son éndemicas a México. Las 7 especies que no son endémicas (*Lithobates forreri, Rhinoclemmys pulcherrima, Phyllodactylus tuberculosus, Boa constrictor, Immantodes gemmistratus, Masticophis mentovarius, Agkistrodon bilineatus*), son especies de distribución amplia en los trópicos y subtrópicos de México, Centroamérica y, algunas de ellas, hasta Sudamérica. Las 4 especies restantes están limitadas al Desierto de Chihuahua de México. Tres de ellas (*Kinosternon durangoense, Gopherus flavomarginatus, Uma paraphygas*) al Bolsón de Mapimí en el extremo sureste de Chihuahua y áreas adyacentes de Durango y Coahuila. La otra (*Holbrookia approximans*) está ampliamente distribuida en el Desierto de Chihuahua del norte de México y es muy probable que también habite en las áreas adyacentes de Arizona y Nuevo México.

JULIO A. LEMOS-ESPINAL Y HOBART M. SMITH

Introducción

En 1999 iniciamos un estudio prospectivo sobre la herpetofauna de Coahuila enfocándonos exclusivamente a la porción oeste de este estado, la cual había sido poco explorada anteriormente. Posteriormente visitamos en forma esporádica esta región, haciendo recorridos desde el extremo noroeste al suroeste del estado, revisando diversas localidades cercanas a Sierra Mojada, Químicas del Rey, Ejido El Alicante, Dunas Magnéticas y Dunas de Bilbao. Recientemente hemos revisado más ampliamente la herpetofauna del estado, visitando las porciones sureste, sur y centro así como otras áreas en el oeste.

A través de este trabajo y de la revisión de literatura, se hace evidente que el conocimiento de la herpetofauna de Coahuila está muy limitado, como fue señalado en las primeras publicaciones de Gloyd y Smith (1942), Schmidt y Owens (1944), y Schmidt y Bogert (1947). Esto es cierto no únicamente para la parte oeste del estado sino para casi todo el estado.

La accidentada topografía del estado es una de las causas de la ausencia de trabajos en Coahuila. Por ejemplo, las montañas de las sierras de El Burro y El Carmen son en su mayor parte accesibles sólo a pie. Sierras aisladas en el norte y oeste del estado son igualmente inaccesibles, y muchas áreas del desierto son inseguras cuando se cruzan a pie o camioneta. Regiones de acceso relativamente fácil han recibido la mayor atención, particularmente a lo largo de las autopistas Méx. 40 entre Torreón y Saltillo, y Méx. 57 entre Saltillo y Piedras Negras, así como varias carreteras que conectan a estas autopistas. Sin embargo, aun estas áreas no han sido bien muestreadas. Existen extensiones de proporciones considerablemente grandes sobre prácticamente todo el territorio estatal, las cuales incluyen Sierra El Carmen, Sierra El Burro, Llano El Guaje, playas del Río Bravo, y las montañas a lo largo de la frontera sur del estado.

Los muestreos más intensivos en el estado han sido llevados a cabo en el Bolsón de Cuatro Ciénegas. De hecho, aproximadamente el 25 % de todos los especímenes de Coahuila, en 20 museos de los Estados Unidos, uno de Canadá, y uno de México, son de este Bolsón. La Sierra de Arteaga también ha sido el foco de una atención considerable, mucha de ésta a través de la expansión de estudios en las partes adyacentes de Nuevo León. El trabajo en otras áreas ha sido escaso, y la mayoría del material depositado en colecciones y museos no ha sido reportado.

En la actualidad sabemos que Coahuila está habitado por 133 especies de anfibios y reptiles: 4 salamandras, 20 anuros, 11 tortugas, 49 lagartijas y 49 serpientes. Todas ellas nativas del estado excepto por *Hemidactylus turcicus*, que es una especie introducida, y posiblemente *Rhinella marina*, la cual pudo haber sido introducida (como ha sido sugerido), pero que posiblemente presente una distribución relictual, al igual que algunas otras especies en el Bolsón de Cuatro Ciénegas. La lista de anfibios y reptiles del estado es seguramente mucho mayor, ya que áreas grandes e importantes del estado no han sido muestreadas o lo han sido inadecuadamente.

Estudios Herpetológicos Previos

El primer colector de anfibios y reptiles que trabajó en Coahuila fue Jean Louis Berlandier, quien fue el naturalista de campo asociado al censo Francés/Mexicano de la frontera mexicana de 1827 a 1829, y quien continuó trabajando como naturalista en otras comisiones por lo menos hasta 1834. Como se reportó en Berlandier y Chovell (1850), él recolectó brevemente en el área de Saltillo, en donde registró *Lithobates catesbeianus*. Desafortunadamente ningún espécimen de sus recolectas en Coahuila se conoce.

El primer registro que se tiene de un estudio herpetológico desarrollado en el estado de Coahuila se refiere a la recolecta realizada por Thomas H. Webb quien en 1852 viajó a través de los estados de Chihuahua, Durango, Coahuila y Nuevo León. En el estado de Coahuila Webb recolectó anfibios y reptiles en los poblados de Parras y Saltillo (Baird, 1859b). Posteriormente, el Teniente Darius N. Couch en 1853, bajo el apoyo del Instituto Smithsoniano realizó un viaje que incluyó los estados de Durango, Coahuila, Nuevo León y Tamaulipas. En este viaje recolectó varios ejemplares tipo de especies nuevas que posteriormente serían nombradas en honor a él. Todas las localidades visitadas por Couch en 1853 se ubican en la porción sur de Coahuila (Río Nazas, Parras, Castañuelas [Estación Seguí], Patos [General Cepeda], Agua Nueva, Buena Vista y Saltillo).

Posteriormente, en mayo de 1895, mayo-agosto de 1896, y abril-mayo de 1902 Edward Goldman y Edward Nelson del US National Museum (USNM), visitaron localidades en el este de Coahuila. En estos viajes recolectaron *Spea multiplicata, Gopherus berlandieri, Cophosaurus texanus, Holbrookia approximans, Phrynosoma cornutum, Sceloporus cautus, S. grammicus, S. goldmani* y *S. undulatus* (= *S. consobrinus*), entre otras especies. Sus resultados se encuentran reportados en Goldman (1951).

En agosto de 1932, Edward H. Taylor y Hobart M. Smith viajaron a través de la carretera Méx. 40, desde Saltillo hasta Torreón, recolectando en varias localidades a lo largo de esta carretera. Ningún reporte de esta colecta fue alguna vez publicado, pero los especímenes recolectados se encuentran depositados en el Field Museum y el University of Illinois Museum of Natural History, Estados Unidos.

En 1937 Robert S. Sturgis recolectó 15 especímenes de anfibios y reptiles provenientes de la Sierra El Carmen, los cuales fueron depositados en el USNM. Posteriormente, F. W. Millar, E. M. Dealey y Raymond Foy recolectaron 22 especímenes en octubre y noviembre de 1940, y 11 en octubre de 1941. Los especímenes recolectados por ellos se encuentran depositados en el museo de la Chicago Academy of Sciences. Los resultados de estas recolectas (1 anuro, 4 lagartijas y 5 serpientes) están reportados en Gloyd y Smith (1942). A finales de octubre y hasta mediados de noviembre de 1938 Hobart y Rozella Smith recolectaron anfibios y reptiles en los alrededores de Torreón, San Pedro de las Colonias, Parras, Jaral, San Hipólito, Saltillo, Arteaga y Zapalinamé, y a lo largo de la carretera Saltillo-Torreón. En ese viaje recolectaron especímenes de *Anaxyrus cognatus, Aspidoscelis gularis, Sceloporus oberon, S. ornatus, S. parvus, S. poinsetti, Uta stansburiana* y *Masticophis taeniatus* entre otras especies. Entre 1938 y 1939 Ernest G. Marsh recolectó en varias localidades del norte y centro de Coahuila, aproximadamente 800 especímenes de anfibios y reptiles, entre otros grupos de vertebrados terrestres, peces y planta, todos ellos depositados en el Field Museum of Natural History. Los resultados sobre los anfibios y reptiles recolectados están reportados en Schmidt y Owens (1944), éstos incluyen 11 especies de anuros, 6 de tortugas, 22 de lagartijas, y 25 de serpientes.

En agosto de 1946 Karl P. Schmidt y Charles M. Bogert estuvieron recolectando a 19.2 km al N de San Pedro de Colonias. Entre los organismos recolectados se encuentra la serie tipo de *Uma exsul*, especie endémica a Coahuila que fue descrita por Schmidt y Bogert (1947). Entre 1950 y 1954 Albert A. Alcorn, J. R. Alcorn, y Robert W. Dickerman recolectaron especímenes de anfibios y reptiles en varios estados de la República Mexicana incluyendo Coahuila. Entre éstos se encuentra el primer registro de *Anaxyrus speciosus* para el estado, el cual fue reportado por Chrapliwy (1956). En 1955 Robert Webb recolectó un espécimen de *Plestiodon brevirostris* (= *P. dicei*) a 20.8 km al E de San Antonio de las Alazanas, municipio de Arteaga, el cual representa el holotipo de *P. b. pineus* descrito por Ralph Axtell en 1960. Fugler y Webb (1956) reportaron una colección de anfibios y reptiles del centro y sur de Coahuila obtenida por Robert W. Dickerman en 1954 y por un grupo de investigadores de la Universidad de Kansas.

En el verano de 1956 Pete S. Chrapliwy y Kenneth Williams recolectaron en varias partes del norte de México, incluyendo Coahuila. Sus resultados están reportados en Williams, Chrapliwy y Smith (1959, 1960); Williams, Smith y Chrapliwy (1961); y Chrapliwy et al. (1961). Estos especímenes se encuentran depositados en el University of Illinois Museum of Natural History.

En junio de 1957 Richard G. Zweifel estuvo recolectando en los alrededores del poblado de Cuatro Ciénegas, en ese periodo registró el primer ejemplar de *Plestiodon tetragammus* para el estado de Coahuila. Este hallazgo está reportado en Zweifel (1958). Desafortunadamente, *Drymobius margaritiferus* fue incluida erróneamente en ese listado, lo que resultó en la inclusión de esa área como parte de la distribución de esta especie en Conant (1975) y Conant y Collins (1998). Hasta donde sabemos, no existe ninguna confirmación de la presencia de esta especie en Coahuila, y su presencia ahí parece altamente improbable en vista de la distribución conocida y preferencias de hábitat de *D. margaritiferus*.

En julio de 1961 Michael D. Sabath recolectó un espécimen de *Crotalus pricei miquihuanus* a 18.1 km al este-sureste de San Antonio de las Alazanas. Éste corresponde al primer registro de esta subespecie para el estado de Coahuila, y fue reportado por Axtell y Sabath (1963). Ese mismo día Michael D. Sabath y Ralph Axtell recolectaron una culebra del género *Thamnophis* a 17.6 km al este y 5.6 km al sur de San Antonio de las Alazanas. Este espécimen representa el holotipo de *Thamnophis exsul*, el cual fue descrito por Rossman (1969). En junio de 1966 Clarence J. McCoy y A. V. Bianculli recolectaron en el Bolsón de Cuatro Ciénegas el primer espécimen de *Sistrurus catenatus edwardsi* para la República Mexicana. Posteriormente, en agosto de 1968 William. L. Minckley y Ralph K. Minckley recolectaron otro espécimen de esta subespecie en este mismo bolsón. Estos resultados se encuentran reportados en McCoy y Minckley (1969). En junio de 1967 Ralph Axtell recolectó lagartijas del género *Sceloporus* a 800 m al N de la cuesta La Muralla, estas lagartijas constituyen la serie tipo de *Sceloporus cyanostictus*, originalmente descrita como una subespecie de *Sceloporus jarrovii* por Axtell y Axtell (1971). En julio de 1968 Max A. Nickerson y John N. Rinne recolectaron una lagartija del género *Gerrhonotus* en el extremo norte de la Sierra de San Marcos, Bolsón de Cuatro Ciénegas. Ese ejemplar representa el holotipo de *Gerrhonotus lugoi* descrito por McCoy en 1970. Entre 1969 y 1974 Ernest A. Liner, Jerry Gautreaux, Richard M. Johnson y Allan H. Chaney recolectaron anfibios y reptiles en Múzquiz y Boquillas de Carmen. En total recolectaron especímenes de 49 taxa (9 anuros, 3 tortugas, 17 lagartijas, 20 serpientes). Estos resultados se encuentran publicados en Liner et al. (1977).

En marzo de 1970 Joseph T. Collins y Richard Montanucci, estuvieron recolectando serpientes en los alrededores de Don Martín y al sur de Nueva Rosita, ambas localidades en Coahuila. Entre los especímenes recolectados se encuentran los primeros registros de *Sonora s. semiannulata* y *Tantilla gracilis* para el estado de Coahuila. Estos resultados fueron reportados por Savitzky y Collins (1971a y 1971b). Montanucci (1971) reportó la presencia de *Crotaphytus reticulatus* en los alrededores de Múzquiz y a 1.6 km al norte de Nava, Coahuila. En 1977 Frederick H. Wills reportó un estudio sobre la distribución, variación geográfica e historia natural de *Sceloporus parvus*, incluyendo varios registros para Coahuila. Durante los veranos de 1977, 1979, 1980 y 1982 Eric C. Axtell, Ralph W. Axtell y Robert G. Webb, recolectaron lagartijas del género *Crotaphytus* en la Sierra Texas, suroeste de Coahuila. Estos ejemplares constituyen la serie tipo de *Crotaphytus antiquus*, descrita por Axtell y Webb (1995).

En mayo de 1983 Robert Powell, Nicholas A. Laposha, Donald D. Smith y John S. Parmerlee visitaron la parte superior de la cuenca del Río Sabinas, la cual hasta ese entonces había permanecido sin estudiar debido a la ausencia de caminos para acceder a ella. En ese viaje registraron 5 especies de anuros, 3 de tortugas y 4 de serpientes. Estos resultados se encuentran reportados en Powell et al. (1984). En 1984 Clarence J. McCoy reportó un total de 66 especies nativas del Bolsón de Cuatro Ciénegas (8 anfibios, 4 tortugas, 23 lagartijas y 31 serpientes) y dos especies introducidas (1 lagartija y 1 sapo). De julio a septiembre de 1987 Arturo Contreras-Arquieta visitó el Bolsón de Cuatro Ciénegas en donde recolectó 392 especímenes de 38 especies de anfibios y reptiles, y observó otras 7. El total de las especies registradas incluyó: 4 anuros, 4 tortugas, 21 saurios y 16 serpientes (Contreras-Arquieta 1989). En mayo de 1988 Ernest A. Liner, Richard R. Montanucci, Arturo González-Alonso y Fernando Mendoza-Quijano, recolectaron anfibios y reptiles entre Múzquiz y Boquillas del Carmen (lado oeste de Sierra de El Carmen), y a lo largo de la brecha a La Linda (lado este de Sierra de El Carmen); Sierra La Encantada; y, sobre la parte del Desierto Chihuahuense que rodea a estas montañas. En total recolectaron 18 taxa (2 anuros, 3 serpientes y 13 lagartijas). Entre éstos se encuentra: el primer registro de *Sceloporus cyanogenys* para el estado (González-Alonso et al. 1988; Liner et al. 1993; y Mendoza-Quijano et al. 1993). McGuire (1996), reportó la presencia de *Crotaphytus antiquus* en las Sierras de San Lorenzo, Texas y Solis, en el extremo suroeste de Coahuila.

Recientemente Héctor Gadsden-Esparza y colaboradores, han estado trabajando en el municipio de Viesca, suroeste de Coahuila; los resultados que han obtenidos están publicados en Gadsden-Esparza et al. (1993); Gadsden-Esparza et al. (1997); Gadsden-Esparza, López-Corrujedo et al. (2001); Gadsden-Esparza, Palacios-Orona y Cruz-Soto (2001); y Gadsden-Esparza et al. (2006). Igualmente, Gamaliel Castañeda-Gaytán y colaboradores han estado trabajando en esta porción de Coahuila, enfocándose principalmente a la región de las Dunas de Bilbao. Viesca. Los resultados que han obtenido se encuentran publicados en Castañeda-Gaytán et al. (2004).

Los resultados del trabajo que hemos desarrollado en el estado de Coahuila se encuentran reportados en Lemos-Espinal, Smith y Ballinger (2004); Lemos-Espinal, Smith, Chiszar y Woolrich-Piña (2004); Lemos-Espinal y Smith (2007, 2009); Lemos-Espinal et al. (en prensa); Smith, Lemos-Espinal y Chiszar (2003); y Smith et al. (2005a, 2005b, 2005c). Javier Banda-Leal y Daniel Garza-Tobón del Museo del Desierto en Saltillo han estado trabajando intensamente sobre las porciones centro y sureste de Coahuila. A través de estos esfuerzos han podido registrar la presencia de *Ambystoma mavortium*, *Pseudoeurycea scandens* y *Eleutherodactylus longipes* en el estado.

Aunque existen estas y otras publicaciones sobre la herpetofauna de Coahuila, éste es uno de los estados mexicanos menos conocidos herpetológicamente. Éste ofrece una oportunidad única para incrementar el conocimiento actual de los anfibios y reptiles del norte de México. Sus cadenas montañosas aún inexploradas muy probablemente están habitadas por taxa completamente desconocidos, los cuales pueden haber evolucionado aisladamente durante miles de años, rodeadas de hábitats desérticos completamente diferentes. Varias especies que se conocen para regiones adyacentes a Coahuila probablemente también habitan en este estado. La distribución conocida para muchas especies en Coahuila se pueden extender a límites aún desconocidos. Estudios sobre la variación geográfica son necesarios para la mayoría de las especies que presentan una distribución amplia.

Adicionalmente a los estudios mencionados previamente, hemos determinado que hay 31 nombres científicos asignados a anfibios (2 anuros) y reptiles (5 tortugas, 17 lagartijas, 7 serpientes) con localidades tipo asignadas al estado de Coahuila (Cuadro 6.1).

Características Fisiográficas y su Influencia sobre la Herpetofauna

Coahuila es la tercera entidad de la República Mexicana en extensión territorial, sus 151,571 km² (comprendidos dentro de las coordenadas extremas 24°32' al sur; 29°53' al norte; 99°51' al este; 103°58' al oeste) representan el 7.74% de la superficie total de nuestro país. Esta extensión de territorio se combina con una topografía accidentada representada por una gran cantidad de macizos montañosos esparcidos sobre todo el territorio coahuilense los cuales le dan una apariencia rugosa al estado, y producen distribuciones discretas para varios de los taxa de anfibios y reptiles que ahí habitan. Uno de los resultados de esta fragmentación del hábitat es el relativamente alto número de especies endémicas al estado (8).

La parte norte del estado presenta sierras extensas que parecen formar un sólo macizo montañoso el cual recibe tres nombres diferentes: Sierra El Carmen (en el tercio oeste); Sierra El Burro (en el tercio este); y, Sierra de Santa Rosa (en el tercio sur). El punto más alto en este conjunto de serranías es la Sierra de Santa Rosa (28°18'N, 102°4'O), con 2120 de altitud. Estas sierras representan aproximadamente el 40–50% de la porción norte de Coahuila, el otro 50–60% está representado en su mayoría por planicies extensas cuya altitud promedio es aproximadamente 1000 m.

Esta región se comunica a través de la carretera Coah. 93, que va de Melchor Múzquiz a La Cuesta de Malena, continuando como camino de terracería hasta el poblado de Boquillas del Carmen. La carretera Coah. 93 corre a lo largo de la base sur de la Sierra del Burro y Sierra El Carmen, y norte de la Sierra de Santa Rosa. Otro camino de terracería sale del punto de confluencia de esta tres sierras, corriendo entre el lado este de la Sierra El Carmen y el oeste de la Sierra del Burro, y llega hasta el poblado de La Linda. El acceso a las partes altas y cañones de estas sierras es difícil y ha resultado en un conocimiento muy pobre de esta área. El conocimiento actual sobre el norte de Coahuila se limita a la base de estas serranías. En estas sierras existe una cantidad considerable de áreas totalmente desconocidas que seguramente contienen taxa que representan registros distribucionales nuevos o hasta especies nuevas.

Sobre el extremo oeste del estado corren longitudinalmente de norte a sur sierras aisladas y pequeñas que surgen de repente sobre las planicies áridas/semiáridas que las rodean. Las principales son Sierra Las Cruces, Sierra Mojada, Sierra El Pino y Sierra de Tlahualilo. El punto más alto en estas sierras lo representa Sierra Mojada (27°16'N, 103°42'O) con una altitud de 2450 m. Esta área también es de difícil acceso, principalmente Sierra Mojada.

En esta misma región se presentan condiciones únicas donde predominan médanos de arena (médanos entre el extremo este de Jaco [Chihuahua] y Estación Sabaneta), los cuales son parte del Bolsón de Mapimí; médanos de Aguanaval al este de la Sierra de Tlahualilo (Dunas Magnéticas) que son parte de la Zona del Silencio; y médanos de los municipios de Matamoros y Viesca, ubicados en el extremo suroeste del área conocida como Desierto Laguna de Mayrán. La altitud promedio de las zonas bajas del oeste de Coahuila es aproximadamente 1250 m. La parte al este de la Sierra Las Cruces, al oeste de la Sierra El Pino y al norte de brecha entre Químicas del Rey y Ocampo, conocida como Llano El Guaje, no ha sido muestreada a pesar de ser un área plana de proporción considerable.

El sur del estado se caracteriza por plegamientos orográficos representados por varias sierras dispuestas transversalmente que corren de este a oeste. Éstas son la continuación del extremo norte de la Sierra Madre Oriental, siendo

CUADRO 6.1 Localidades tipo para anfibios y reptiles descritos para el estado de Coahuila, México

Autor(es)	Nombre original: Localidad tipo
Anuros (2)	
Zweifel (1956)	*Eleutherodactylus* (= *Craugastor*) *augusti fuscofemora*: **Sacatón, 5 miles south of Cuatro Ciénegas**
Cope (1863)	*Scaphiopus rectifrenis* (= *S. couchii*): **Río Nazas**
Tortugas (5)	
Schmidt & Owens (1944)	*Terrapene coahuila*: **Cuatro Ciénegas**
Ward (1984)	*Pseudemys concinna gorzugi* (= *P. gorzugi*): **3.5 mi W Jiménez, Río San Diego**
Legler (1960)	*Pseudemys scripta taylori* (= *Trachemys taylori*): **16 km S Cuatro Ciénegas**
Iverson (1981)	*Kinosternon hirtipes megacephalum*: **3.2 km SE Viesca (25°21′N, 102°48′W)**
Webb & Legler (1960)	*Trionyx ater* (= *Apalone spinifera ater*): **16 km S Cuatro Ciénegas**
Lagartijas (17)	
Smith (1942)	*Gerrhonotus levicollis ciliaris* (= *B. ciliaris*): **Sierra de Guadalupe**
McCoy (1970)	*Gerrhonotus lugoi*: **N tip of Sierra de San Marcos, ca. 11 km SW of Cuatro Ciénegas, ca. 800 m elev.**
Axtell & Webb (1995)	*Crotaphytus antiquus*: **2.1 km N–1.7 km E Vizcaya (25°46′4″N, 10311′48″W. elev. 1100 ± m) in the Sierra Texas**
Schmidt (1921)	*Holbrookia maculata dickersonae* (= *H. approximans*): **Castañuelas**
Smith (1938)	*Sceloporus cautus*: **30 mi N of El Salado, San Luis Potosí**
Williams et al. (1960)	*Sceloporus merriami australis*: **15.6 mi (25.1 km) E Cuatro Ciénegas**
Smith & Brown (1941)	*Sceloporus jarrovii oberon* (= *S. oberon*): **Arteaga**
Smith (1936)	*Sceloporus ornatus caeruleus*: **5 mi S of San Pedro de las Colonias**
Baird (1859a [1858])	*Sceloporus ornatus*: **Patos (= General Cepeda)**
Axtell & Axtell (1971)	*Sceloporus jarrovii cyanostictus* (= *S. cyanostictus*): **800 m N Cuesta La Muralla**
Schmidt & Bogert (1947)	*Uma exsul*: **sand dunes 12 mi N of San Pedro de las Colonias**
Axtell (1960)	*Eumeces* (= *Plestiodon*) *dicei pineus*: **20.8 km E San Antonio de las Alazanas**
García-Vázquez et al. (2010)	*Scincella kikaapoa*: **4 km SE Cuatro Ciénegas, Poza el Mojarral, 26°55′11.9″N; 100°6′53.2″W, 739 m elev.**
Cope (1892)	*Cnemidophorus gularis semifasciatus* (*Aspidoscelis gularis*): **Agua Nueva**
Duellaman & Zweifel (1962)	*Cnemidophorus septemvittatus pallidus* (= *Aspidoscelis gularis pallida*): **3 miles west of Cuatro Ciénegas**
Wright & Lowe (1993)	*Cnemidophorus* (= *Aspidoscelis*) *inornata cienegae*: **13.9 km SW of Cuatro Ciénegas de Carranza, point of San Marcos Mtn.**
Cope (1892)	*Cnemidophorus variolosus* (= *Aspidoscelis marmorata variolosa*): **Parras**
Serpientes (7)	
Cope (1861)	*Arizona jani* (= *Pituophis deppei jani*): **Buenavista**
Garman (1883)	*Rhinocheilus lecontei tessellatus*: **Monclova**
Kennicott (1860)	*Eutaenia* (= *Thamnophis*) *angustirostris* (*nomen dubium*): **Parras**
Kennicott (1860)	*Eutaenia cyrtopsis* (= *Thamnophis cyrtopsis*): **Rinconada**
Rossman (1969)	*Thamnophis exsul*: **17.6 km E and 5.6 km S San Antonio de las Alazanas**
Rossman (1963)	*Thamnophis proximus diabolicus*: **Río Nadadores, 5 km W Nadadores**
Garman (1887)	*Crotalus tigris palmeri* (= *C. lepidus lepidus*): **Monclova**

las principales Sierra de Arteaga, Sierra La Concordia y Sierra de Parras. El punto más alto de esta región lo representa el Cerro La Nopalera (25°8′N, 103°14′O), con 3120 m de altitud. Sobre el extremo sureste este conjunto de sierras se observa de forma continua, limitando en su parte norte al Desierto Laguna de Mayrán.

La parte este del estado es en su mayoría plana presentándose unas cuantas sierras aisladas, tales como: Sierra Pájaros Azules y Sierra La Gloria. El punto más alto en estas sierras lo representa la Sierra Pájaros Azules (27°0′N, 100°53′O), con 1930 m de altitud.

En la parte centro se presenta un valle pequeño de aproximadamente 120 km^2 rodeado de montañas de hasta 2500 m de altitud, cuya base tiene una altitud promedio de aproximadamente 750 m. Anteriormente este valle era una cuenca cerrada de varios arroyos que alimentaban a un sistema hidrológico de un amplio espectro de hábitats acuáticos, representados por ríos, pozas, lagos y pantanos. Estos hábitats acuáticos son de una antigüedad considerable. El aislamiento de este valle junto con la gran variedad de hábitats presentes en él ha resultado en el desarrollo de taxa endémicos a la región (McCoy 1984). En el presente estudio nos referimos a este valle como el Bolsón de Cuatro Ciénegas.

El territorio coahuilense está constituido en su mayor parte por rocas de origen sedimentario, tanto marino como continental cuyas edades van desde el Paleozoico hasta el Cuaternario. Las más típicas de ellas son las calizas del Mesozoico. Estas rocas han sido afectadas por intensos plegamientos, así como afallamientos e intrusiones relacionadas a ellos. La orientación de los plegamientos es en dirección este-oeste en el sur del estado, y noroeste-sureste en el resto de él. Así las sierras se orientan también preferentemente en tales direcciones.

En diversas zonas del estado se encuentran rocas ígneas cuyas edades varían desde el Triásico hasta el Cuaternario. Las extrusivas son las más jóvenes de ellas, y forman, en algunos casos, las partes más altas de las sierras, mientras que las intrusivas han quedado expuestas en cuerpos pequeños debido a la erosión de las rocas sedimentarias a las cuales intrusionaron y en algunos casos mineralizaron. En muchos lugares afloran conglomerados continentales terciarios, que constituyen lomeríos y las extensas bajadas de las sierras.

Los aluviones son los depósitos más recientes y están constituidos por detritos de las diversas rocas mencionadas. Ellos cubren la mayor parte de los llanos y alcanzan en algunos casos espesores de varios cientos de metros. Por último, las rocas metamórficas Paleozoicas afloran en pequeñas áreas dispersas por varias zonas en la entidad.

El estado de Coahuila queda comprendido en parte de las regiones hidrológicas: "Bravo-Conchos" que abarca gran parte del estado con 95,236.33 km^2; Mapimí en la porción oeste con 29,456.26 km^2; Nazas-Aguanaval en la parte sur-suroeste con 21,908.22 km^2 y finalmente la región El Salado con una área muy reducida en la parte sureste con 4977.56 km^2.

La región Bravo-Conchos, en su mayoría está constituida por tierras planas, con altitud media de 1000 m a 1800 m. Es una región árida cuya sequedad se agudiza en el norte. La mayoría de las corrientes del norte desaguan en el Río Bravo, y hay además algunas cuencas endorreicas, como las de las lagunas Tortuguillas y Chancaplio, dentro del área de la cuenca del Río Conchos.

La región del Bolsón de Mapimí no presenta corrientes o almacenamientos perennes, se caracteriza por su aridez y la ausencia casi total de elevaciones importantes. Todas las cuencas que la conforman son endorreicas. Los cauces son temporales y se forman a consecuencia de alguna precipitación intensa, borrándose semanas o meses después. En toda la región los índices de escurrimiento superficial son muy bajos, menores a los 10 mm anuales que, por supuesto, van a dar al fondo de los bolsones, en donde se llegan a formar lagunas intermitentes.

La parte coahuilense de la región Nazas-Aguanaval se encuentra al suroeste de la entidad y abarca también partes de los estados de Durango y Zacatecas, se le conoce con el nombre de Región Lagunera y corresponde a las cuencas cerradas de los grandes Ríos Nazas y Aguanaval. Estos ríos alimentan a la zona agrícola más importante de la Entidad, la Comarca Lagunera, y a varias de las ciudades que en ella se enmarcan, tanto en Durango (Gómez Palacio y Lerdo) como en Coahuila (Torreón, Matamoros y San Pedro de las Colonias).

De todas las regiones hidrológicas comprendidas dentro del estado de Coahuila, la de El Salado es la que menor área ocupa. Es, sin embargo, una de las vertientes interiores más importantes del país. Está integrada por un conjunto hidrográfico de cuencas cerradas de dimensiones muy diferentes.

Prácticamente todo el estado de Coahuila queda incluido dentro de la zona biogeográfica conocida como Desierto Chihuahuense, a excepción de una parte pequeña en el extremo sureste del estado, donde se encuentra la Sierra de Arteaga, la cual representa el extremo norte de la Sierra Madre Oriental. La cubierta vegetal del estado se compone por seis tipos de vegetación (Matorral Desértico Chihuahuense; Matorral Tamaulipeco; Matorral Submontano; Bosque de Montaña; Zacatal; y Vegetación Acuática, Subacuática y Riparia) y 12 comunidades vegetales, que básicamente corresponden a tres provincias florísticas: Altiplano Mexicano, Planicie Costera del Noreste y Sierra Madre Oriental (Rzedowski 1978).

Matorral Desértico Chihuahuense

Ocupa grandes extensiones del oeste y sur del Coahuila. Este tipo de vegetación representa aproximadamente el 63% de la superficie estatal y comprende una serie de comunidades vegetales (matorral rosetófilo, matorral micrófilo, izotal, matorral halófilo y gipsófilo) constituidas por arbustos xerófilos esparcidos, perennes y elementos herbáceos efímeros, que se presentan en los hábitats más xéricos del estado, en altitudes que van de los 800 a los 2600 m, las variaciones en las características edáficas y topográficas, son las causantes en determinar las diversas asociaciones vegetales dentro de este tipo de vegetación.

El Matorral Rosetófilo se presenta en altitudes que varían de los 1000 a 2500 m, se desarrolla en lomeríos y laderas de montañas donde incide una mayor radiación solar o bien en las áreas más expuestas de los cañones. Se presenta en sitios con suelos someros, usualmente pedregosos y con buenas condiciones de drenaje, cubre aproximadamente el 6% del estado. Las especies dominantes son arbustos bajos con hojas agrupadas en forma de roseta, espinosos y perennifolios, las especies más representativas son: *Agave lechuguilla, A. scabra, A. striata, Hechtia texensis* y *Dasylirion palmeri*, asociadas con *Viguieria stenoloba, Parthenium argentatum, Euphorbia antisyphillitica* y *Mimosa zygophylla*. A menudo esta comunidad presenta una riqueza alta de cactáceas, las más frecuentes son: *Opuntia microdasys, Echinocactus platyacanthus, Ferocactus pilosus, F. uncinatus, Echinocereus conglomeratus, E. pectinatus* y *Thelocactus bicolor*; las herbáceas más comunes son: *Bahia absinthifolia, Tiquilia canescens, T. greggii, Castilleja lanata, Notholaena sinuata* y especies de gramíneas como *Bouteloua ramosa* y *Erioneuron avenaceum*.

El Matorral Micrófilo es una de las comunidades más características del estado, ocupa cerca del 48% de la superficie, se presenta entre los 600 y 1500 m de altitud, en valles y lomeríos con suelos profundos y poca pedregosidad, está dominada por arbustos micrófilos de 0.30–1.5 m de altura, así como algunos individuos arbóreos mayores a 1.5 m de alto, las comunidades varían de poco densas a densas en relación a la profundidad del suelo y la humedad disponible, el componente principal lo constituyen: *Larrea tridentata, Flourensia cernua* y *Parthenium incanum*, asociados con *Lycium berlandieri, Opuntia rastrera, O. imbricata, Koeberlinia spinosa* y *Fouquieria splendens*, en algunos casos se presentan individuos arborescentes de *Yucca filifera* y *Prosopis glandulosa*; las herbáceas que ocurren con mayor frecuencia son: *Bouteloua trifida, B. gracilis, Sporobolus airoides, Dasyochloa pulchella, Scleropogon brevifolius, Mühlenbergia porteri, Psilostrophe gnaphalodes* y *Tiquilia canescens*.

El Izotal se caracteriza por la alta dominancia de plantas arborescentes del género *Yucca*, con alturas entre 8 y 10 m que dan el aspecto de un bosquecillo o palmar, ocupa cerca del 3% del área de estudio y se distribuye por encima de los 1500 m de altitud, sobre suelos aluviales, profundos y arenosos de valles y laderas de sierras, a menudo, mezclados con elementos del matorral micrófilo o rosetófilo. Las especies más frecuentes son: *Yucca filifera* y *Y. carnerosana*, asociadas con *Larrea tridentata*, *Flourensia cernua*, *Opuntia imbricata*, *O. rastrera*, *Agave lechuguilla*, *Dasylirion cedrosanum*, *Viguiera stenoloba* y *Parthenium incanum*; las herbáceas más comunes son: *Bouteloua gracilis*, *Psilostrophe gnaphalodes* y *Zaluzania triloba*.

Los Matorrales Halófilo y Gypsófilo se presentan en el centro y sur de la entidad, entre los 1000 y 2000 m de altitud, ocupan el 6% de la superficie estatal, su presencia está determinada por condiciones edáficas locales. El primero de ellos se localiza en cuencas aisladas con drenaje interno (cuencas endorreicas) donde se propicia la acumulación de sedimentos salinos formando una serie de valles salinos o lagunas temporales, este matorral se compone usualmente por especies como *Atriplex canescens*, *A. acanthocarpa*, *Allenrolfea occidentalis* y *Suaeda palmeri*, con frecuentes infiltraciones de especies tolerantes como *Larrea tridentata*, *Prosopis glandulosa* y *Opuntia imbricata*; el estrato herbáceo se compone por *Pleuraphis mutica*, *Sporobolus airoides*, *S. spiciformis*, *Distichlis spicata* y especies de *Machranthera*.

Los depósitos de yeso (sulfato de calcio hidratado) se forman por precipitación y una evaporación subsecuente de mares antiguos expuestos en valles y colinas o formando dunas desde el tiempo del terciario (Villarreal y Valdés 1992–1993). Las comunidades están formadas por plantas arbustivas o subarbustivas, cespitosas y esparcidas, usualmente presentan una flora rica en endemismos, las especies más frecuentes son: *Frankenia johnstonii*, *Condalia ericoides*, *Koeberlinia spinosa*, *Lycium pallidum* y *Viguiera dentata*, asociadas con *Prosopis glandulosa*, *Larrea tridentata* y *Opuntia rastrera*; en el estrato herbáceo dominan las gramíneas: *Muhlenbergia villiflora*, *M. arenicola* y *Scleropogon brevifolius*, en

las épocas de mayor humedad la diversidad del estrato aumenta y se presentan: *Hoffmanseggia watsonii*, *Zinnia acerosa*, *Lesquerella fendeleri*, *Lepidium montanum*, *Dicranocarpus parviflorus* y *Sartwellia mexicana* (www.gramineasdecoahuila .com).

Existen 81 especies de anfibios y reptiles que habitan el Matorral Desértico Chihuahuense y 57 son exclusivas de este tipo de vegetación: *Anaxyrus cognatus*, *A. debilis*, *A. punctatus*, *A. speciosus*, *A. woodhousii*, *Incilius nebulifer*, *Eleutherodactylus marnockii*, *Gastrophryne olivacea*, *Scaphiopus couchii*, *Spea multiplicata*, *Gopherus flavomarginatus*, *Gerrhonotus lugoi*, *Crotaphytus antiquus*, *C. collaris*, *Gambelia wislizenii*, *Coleonyx brevis*, *C. reticulatus*, *Cophosaurus texanus*, *Holbrookia approximans*, *H. lacerata*, *Phrynosoma cornutum*, *P. modestum*, *Sceloporus bimaculosus*, *S. cautus*, *S. consobrinus*, *S. cyanostictus*, *S. maculosus*, *S. merriami*, *S. ornatus*, *Uma exsul*, *U. paraphygas*, *Uta stansburiana*, *Plestiodon obsoletus*, *Scincella kikaapoa*, *Aspidoscelis gularis*, *A. inornata*, *A. marmorata*, *Arizona elegans*, *Bogertophis subocularis*, *Gyalopion canum*, *Heterodon kennerlyi*, *Hypsiglena jani*, *Lampropeltis alterna*, *L. getula*, *Masticophis taeniatus*, *Rhinocheilus lecontei*, *Salvadora grahamiae*, *Sonora semiannulata*, *Tantilla atriceps*, *T. hobartsmithi*, *T. nigriceps*, *Thamnophis marcianus*, *Rena dissecta*, *R. segrega*, *Agkistrodon contortrix*, *Crotalus viridis* y *Sistrurus catenatus*.

La mayoría de estas 57 especies presentan una distribución amplia en las regiones áridas/semiáridas de Estados Unidos y México, están adaptadas para habitar sitios con limitada disponibilidad de agua, con una marcada estacionalidad entre la época de lluvias y la seca. La época de lluvias se extiende desde finales de mayo hasta mediados de septiembre, con la máxima incidencia de éstas en los meses de julio y agosto. Esto limita la actividad de la mayoría de los anfibios y reptiles ahí presentes al periodo de abril a octubre, con un periodo de hibernación de octubre a marzo. La mayoría de estas especies salen de sus refugios de hibernación, los que pueden ser madrigueras de mamíferos, grietas profundas en suelo o rocas, etc., a mediados de abril. Sin embargo, cuando las condiciones ambientales lo permiten, se puede observar individuos activos aun en los meses de invierno. Asimismo, la naturaleza heliotérmica de estas especies resulta en una baja o nula actividad en días nublados y fríos de primavera-verano.

Todos los anfibios característicos del Matorral Desértico Chihuahuense dependen de cuerpos de agua para su reproducción, la cual inician con las primeras lluvias de finales de mayo—principios de junio. Durante este periodo se pueden observar grandes concentraciones de anuros en diferentes cuerpos de agua, donde se llega a oír el canto, algunas veces ensordecedor, de los machos tratando de atraer a las hembras para realizar el amplexus. Estas concentraciones se pueden extender mientras haya cuerpos de agua disponibles, presentándose una reproducción oportunista que en ocasiones es muy riesgosa debido a la alta tasa de evaporación que llega a haber en el Desierto Chihuahuense de Coahuila. A mediados de agosto del 2008, observamos concentraciones grandes de *Anaxyrus debilis* en las charcas que se forman a los lados de la carretera Parras—General Cepeda, estas concentraciones estaban formadas de pares amplectantes. En el fondo de las charcas pudimos observar gran cantidad de huevos adheridos a la vegetación de las charcas. Tres semanas después, ya en el mes de septiembre, regresamos para registrar renacuajos, sin embargo, las charcas estaban secas y en sus fondos lodosos se observaban cientos de renacuajos muertos.

La mayoría de las especies de reptiles que habitan el Matorral Desértico Chihuahuense es ovípara, por lo que pueden dejar más de una puesta durante su periodo de actividad. Por lo común esto lo hacen de finales de mayo a principios de septiembre. Sin embargo, algunas especies, como las serpientes de los géneros *Agkistrodon*, *Crotalus* y *Sistrurus*, son vivíparas, generalmente estas especies copulan entre abril y mayo y las crías nacen a partir del mes de junio.

Nueve de estas 57 especies (*Gopherus flavomarginatus*, *Gerrhonotus lugoi*, *Crotaphytus antiquus*, *Sceloporus cyanostictus*, *S. maculo-*sus, *S. ornatus*, *Uma exsul*, *Uma paraphygas* y *Scincella kikaapoa*) presentan una distribución considerablemente pequeña, éstas se limitan a sistemas de médanos de arena en Chihuahua, Coahuila y Durango (*Gopherus flavomarginatus*, *Uma exsul*, *Uma paraphygas*), al Bolsón de Cuatro Ciénegas del centro de Coahuila (*Gerrhonotus lugoi*, *Scincella kikaapoa*), o a algunas localidades aisladas en el extremo oeste de Coahuila (*Crotaphytus antiquus*), oeste de Coahuila y noreste de Durango (*Sceloporus maculosus*), mitad sur de Coahuila (*Sceloporus cyanostictus*), y centro y sur de Coahuila (*Sceloporus ornatus*). Estas especies de distribución limitada son una muestra de la gran variedad de condiciones presentes en el Desierto Chihuahuense de Coahuila.

Cuatro de las restantes veinticuatro especies que ocupan el Matorral Desértico Chihuahuense (*Rhinella marina*, *Craugastor augusti*, *Smilisca baudinii* y *Opheodrys aestivus*), tienen distribuciones discontinuas que sugieren que su presencia en Coahuila podría ser de tipo relictual. Una más (*Hemidactylus turcicus*), es una especie introducida cuya distribución original es el Mediterráneo. Ésta ha sido muy exitosa en colonizar los lugares a los que llega a través de diferentes medios de transporte, principalmente barcos, dispersándose desde los puertos de llegada a prácticamente cualquier lugar. Las diecinueve especies restantes (*Ambystoma mavortium*, *Lithobates berlandieri*, *S. minor*, *S. parvus*, *S. poinsettii*, *Urosaurus ornatus*, *Plestiodon tetragrammus*, *Coluber constrictor*, *Masticophis flagellum*, *Pantherophis bairdi*, *P. emoryi*, *Pituophis catenifer*, *Thamnophis cyrtopsis*, *T. proximus*, *Micrurus tener*, *Crotalus atrox*, *C. lepidus*, *C. molossus*, *C. scutulatus*), son especies que tienen la facultad de habitar en ambientes templados, desérticos y/o subtropicales.

Matorral Tamaulipeco

Éste es el tipo de vegetación característico de la provincia florística de la Planicie Costera del Noreste, la cual abarca el norte y noreste del estado, en una serie de planicies y lomeríos ubicados al este de la Sierra del Carmen, La Babia, Santa Rosa, La Purísima y La Gavia, ocupa una extensión aproximada del 19% de la superficie estatal. Las altitudes varían desde los 240 hasta los 850 m, los suelos son gravosos, arenosos y profundos en los valles a pedregosos y someros en lomeríos, usualmente con buen drenaje; la vegetación se integra por extensos matorrales de menos de 2 m de altura, compuestos por una mezcla de arbustos micrófilos, espinosos e inermes, también son frecuentes pequeñas comunidades de árboles bajos que se concentran en los sitios más húmedos.

Las asociaciones más representativas son las de *Acacia rigidula - Leucophyllum frutescens* y *Prosopis glandulosa - Opuntia lindheimeri*, con elementos de *Lippia graveolens, Agave lechuguilla* y *Flourensia cernua* en las regiones de la porción sur y oeste, y *Colubrina texensis* en las regiones del norte. Otros arbustos o árboles pequeños son: *Karwinskia humboldtiana, Guaiacum angustifolium, Cercidium texanum, Ziziphus obtusifolia, Castela erecta, Opuntia leptocaulis, Citharexylum brachyanthum, Acacia berlandieri, A. farnesiana, A. constricta, A. greggii* y *Diospyros texana*, en el estrato herbáceo las gramíneas son el componente principal, se presentan en sitios abiertos o protegidas entre los arbustos, las más frecuentes son: *Bouteloua trifida, B. curtipendula, Aristida purpurea, Tridens muticus, T. texanus, Panicum hallii, Pleuraphis mutica, Hilaria belangeri* y *Pennisetum ciliare*, otras herbáceas comunes son: *Gnaphalopsis micropoides* y *Ruellia nudiflora*.

Treinta y un especies de anfibios y reptiles han sido registradas en el Matorral Tamaulipeco de Coahuila: *Anaxyrus punctatus, Anaxyrus speciosus, Incilius nebulifer, Lithobates berlandieri, Scaphiopus couchii, Spea multiplicata, Gopherus berlandieri, Crotaphytus reticulatus, Sceloporus cyanogenys, S. olivaceus, S. variabilis, Scincella lateralis, Drymarchon melanurus, Lampropeltis triangulum, Leptodeira septentrionalis, Masticophis flagellum, M. schotti, Pantherophis bairdi, P. emoryi, Pituophis catenifer, Tantilla atriceps, T. gracilis, T. nigriceps, Thamnophis proximus, Micrurus tener, Rena dulcis, R. segrega, Crotalus atrox, C. lepidus, C. molossus* y *C. scutulatus*. Únicamente tres de estas especies (*Gopherus*

berlandieri, Crotaphytus reticulatus y *Sceloporus olivaceus*) se pueden considerar exclusivas de este tipo de vegetación. *Gopherus berlandieri* tiene una distribución que se limita desde el sur de Texas y este de Coahuila extendiéndose hacia el sur al este de la Sierra Madre Oriental a través de la mayor parte de Nuevo León y Tamaulipas hasta el extremo norte de Veracruz. *Crotaphytus reticulatus* se limita desde el sur de Texas y este de Coahuila hasta el noroeste de Tamaulipas, incluyendo el norte de Nuevo León. La distribución de estas dos especies de reptiles queda totalmente incluida en el Matorral Tamaulipeco. Por su parte, *Sceloporus olivaceus* se distribuye desde el norte central de Texas extendiéndose hacia el sur a través de las planicies costeras del Golfo de México hasta el sur de Tamaulipas, hacia el oeste cerca del área del Big Bend de Texas y este de Coahuila. Las 28 restantes son especies que tienen una distribución amplia y que tienen la facultad de habitar otros tipos de vegetación.

Matorral Submontano

Este tipo de vegetación se presenta en el pie de monte o en laderas medias de los macizos montañosos del estado con exposiciones topográficas norte o noroeste, así como en cañones con buenas condiciones de humedad, los suelos son someros y pedregosos con poca materia orgánica, ocupa el 5% de la entidad y se caracteriza por la dominancia de arbustos y árboles de 1 a 3 m de altura, la mayoría inermes, esclerófilos y caducifolios, la composición del matorral difiere en relación a las condiciones climáticas prevalecientes, de esta manera en las montañas del altiplano, donde predominan las bajas temperaturas las especies dominantes son: *Purshia plicata, Amelanchier denticulata, Lindleya mespiloides, Cercocarpus montanus, C. footergiroides, Rhus virens, Berberis trifoliolata, Sophora secundiflora, Fraxinus cuspidata, Quercus pringlei, Q. striatula* y *Q. intricata*, con un estrato herbáceo compuesto por *Mühlenbergia rigida, M. emersleyi, M. dubia, Loeselia scariosa,* y *Leptochloa dubia*; mientras que en las serranías próximas a la Planicie Costera del Noreste, donde el clima es más cálido, se presentan *Helietta parvifolia*, *Havardia pallens*, *Amiris madrensis*, *Fraxinus greggii*, *Gochnatia hypoleuca*, *Acacia rigidula*, *A. berlandieri*, *Zanthoxylum fagara*, *Lippia graveolens* y *Leucophyllum frutescens*, acompañadas por herbáceas como: *Ruellia nudiflora*, *Croton fruticulosus*, *Abutilon wrigthii*, *Bouteloua curtipendula*, *Erioneuron avenaceum*, *Tridens texanus*, *T. muticus* y *Digitaria cognata*.

Las especies de herpetozoarios más representativas de Matorral Submontano de Coahuila son trece: *Eleutherodactylus guttilatus*, *Lithobates berlandieri*, *Scaphiopus couchii*, *Spea multiplicata*, *Gerrhonotus infernalis*, *Sceloporus oberon*, *S. variabilis*, *Lampropeltis mexicana*, *Pituophis catenifer*, *P. deppei*, *Tantilla wilcoxi*, *Micrurus tener* y *Crotalus lepidus*. Sin embargo, ninguna de estas especies es exclusiva del Matorral Submontano de Coahuila, todos los anfibios y reptiles que habitan este tipo de vegetación se presentan en otros tipos de vegetación.

Bosque de Montaña

Éste comprende la vegetación arbórea que se desarrolla por encima del matorral submontano, en las porciones montañosas de las provincias Sierra Madre Oriental y Altiplano Mexicano, crece en los cañones y partes altas de montañas, donde hay mayor humedad y el clima es templado semiseco a templado subhúmedo, se integra por las siguientes comunidades.

Bosque de Encino es una comunidad propia de sitios húmedos y templados, con frecuencia se presenta en cañones y valles intermontanos, por encima del matorral submontano, entremezclándose con éste y el bosque de pino, entre los 1200 y 2500 m de altitud, ocupa aproximadamente el 1% de la superficie estatal; las especies dominantes son árboles pequeños de entre 3 y 8 m de altura, con copas redondeadas en poblaciones esparcidas o densas, usualmente muy homogéneas, los componentes más frecuentes son: *Quercus gravesii*, *Q. grisea*, *Q. laceyi*, *Q. hypoxantha*, *Q. saltillensis*, *Q. greggii*, asociados

con *Prunus serotina*, *Arbutus xalapensis*, *Pinus arizonica*, *Pinus cembroides* y *Juniperus flaccida*, las arbustivas más comunes son: *Salvia regla*, *Garrya ovata* y *Ageratina saltillensis*, el estrato herbáceo con frecuencia presenta las siguientes especies: *Piptochaetium fimbriatum*, *Achillea milleifolium*, *Bromus carinatus*, *Cologania pallida* y *Pleopeltis guttata*

Bosque de Pino es una comunidad dominada por especies del género *Pinus*, se distribuye sobre laderas con mayor humedad y partes altas de los principales macizos montañosos del estado, en altitudes que van de los 1200 a los 3000 m, donde el clima es templado sub-húmedo, cubre aproximadamente el 2% de la entidad. El bosque de pino piñonero mezclado con especies de naturaleza xerica es la comunidad más frecuente, se trata de un bosque con árboles bajos, copas redondeadas, ramas nudosas y troncos con un diámetro menor a 30 cm, las especies dominantes son: *Pinus cembroides* y *Pinus pinceana*, con individuos esparcidos de *Yucca filifera*, *Agave gentri*, *Opuntia stenopetala*, *Lindleya mespiloides*, *Juniperus saltillensis* y *J. erythrocarpa*, asociados con herbáceas como: *Chrysactinia mexicana*, *Piptochaetium fimbriatum*, *Bouteloua hirsuta* y *B. dactyloides*. Otras comunidades de este tipo son las formadas por *Pinus arizonica* var. *stormiae*, las cuales son poco densas y a menudo se mezclan con especies del bosque de encino, se distribuyen a través de la mayoría de los sistemas montañosos del centro y norte del estado; en el sur de la entidad (Sierra de Arteaga) son frecuentes bosques de pino compuestos por *Pinus teocote*, *P. greggii*, *P. hartweguii*, *P. strobiformis*, *P. ayacahuite* var. *brachyptera* y *P. pseudostrobus*, asociados con especies de *Quercus*, *Arbutus*, *Ceanothus*, *Prunus* y *Fraxinus*, entre otras.

Bosque de Oyamel. Esta comunidad cubre aproximadamente el 2% de la superficie estatal, su distribución es dispersa y muy localizada, sólo se presenta en pequeñas porciones de las sierras de Arteaga, Zapalinamé, el Jabalí, La Madera y Maderas del Carmen, entre los 2500 y 3400 m de altitud, en laderas altas, cañones y cimas con exposición norte, donde la precipitación es superior a los 600 mm anuales y la temperatura media oscila entre 12°C y 14°C, los suelos son someros y con abundante materia orgánica; la comunidad presenta una altura que varía de 10 a 20 m y está dominada por *Pseudotsuga menziesii* y especies de *Abies* (*A. vejarii, A. durangensis* y *A. mexicana*), con *Pinus strobiformis, P. hartweguii, Cupressus arizonica, Quercus greggii* y *Arbutus xalapensis*, (como parte de este bosque también se presentan pequeñas poblaciones de *Picea mexicana* en la Sierra de Arteaga), en el estrato arbustivo son frecuentes *Arctostaphylos pungens, Ceanothus coeruleus, Paxistima myrsinites, Symphoricarpos microphylus, Garrya ovata* y *Sambucus nigra*, entre las herbáceas más comunes se encuentran: *Pleopeltis guttata, Achillea milleifolium, Alchemilla vulcanica, Geranium crenatifolium, Koeleria macrantha, Brachypodium mexicanum, Trisetum spicatum, Festuca macrantha, F. pinetorum* y *F. valdesii.*

Vegetación Alpina-Subalpina. Se presenta en las partes más altas de las sierras del sur del estado (La Marta, Las Vigas, Potrero de Abrego y El Coahuilón), por encima de los 3400 m de altitud, donde la temperatura media es menor a los 10°C y las heladas son frecuentes, constituye el límite superior del bosque de montaña y ocupa menos del 1% del estado, se caracteriza por la dominancia de arbustos bajos (0.50–1 m de altura) y de herbáceas, con algunos árboles aislados de *Pinus culminicola* y *P. hartweguii*; las especies arbustivas más frecuentes son: *Ceanothus beauxifolius, C. greggii, Symphoricarpos microphyllus, Agave montana, Arctostaphylos pungens* y *Quercus greggii*; como parte del estrato herbáceo se presentan las siguientes especies: *Lupinus cacuminis, Senecio lorathifolius, S. coahuilenses, Bromus carinatus, Penstemon leonensis, Arenaria lanuginosa* y *Brachypodium mexicanum* (www.gramineasdecoahuila.com).

Aunque el Bosque de Montaña de Coahuila representa sólo el 6% de la superficie estatal, éste hospeda especies de anfibios y reptiles únicas para el estado, así como especies que presentan una distribución amplia en Coahuila. Veintinueve especies de herpetozoarios son representantes del Bosque de Montaña de

Coahuila. Dieciseis de estas 29 especies son exclusivas de este tipo de vegetación: *Chiropterotriton priscus, Pseudoeurycea galeanae, P. scandens, Eleutherodactylus longipes, Ecnomiohyla miotympanum, Barisia ciliaris, Phrynosoma orbiculare, Sceloporus couchi, S. oberon, Plestiodon dicei, Scincella silvicola, Diadophis punctatus, Storeria hidalgoensis, Tantilla wilcoxi, Thamnophis exsul* y *Crotalus pricei*. La mayoría de estas especies extienden su distribución hacia el sur a través de la Sierra Madre Oriental. Dos de ellas (*Sceloporus couchii* y *Thamnophis exsul*) se encuentra limitada a la Sierra Madre Oriental de Coahuila y Nuevo León. Otras cinco (*Barisia ciliaris, Phrynosoma orbiculare, Diadophis punctatus, Tantilla wilcoxi* y *Crotalus price*), presentan distribuciones discontinuas con poblaciones tanto en la Sierra Madre Occidental como en la Oriental con un gran vacío en su distribución representado por el Desierto Chihuahuense. Poblaciones de estas especies en ambas sierras pueden ser especies diferentes, sin embargo, a la fecha siguen siendo consideradas de la misma especie. Nueve más se distribuyen sólo en la Sierra Madre Oriental (*Chiropterotriton priscus, Pseudoeurycea galeanae, P. scandens, Eleutherodactylus longipes, Ecnomiohyla miotympanum, Sceloporus oberon, Plestiodon dicei, Scincella silvícola* y *Storeria hidalgoensis*). Las tres salamandras aquí presentes (*Chiropterotriton priscus, Pseudoeurycea galeanae* y *P. scandens*), pertenecen a la familia Plethodontidae, estas salamandras tienen pulmones y no dependen de cuerpos de agua para su reproducción. Los huevos los depositan en lugares húmedos, todo el desarrollo embrionario lo realizan dentro del huevo, por lo que no presentan metamorfosis, las crías salidas del huevo son idénticas a individuos adultos.

Las trece especies restantes del Bosque de Montaña de Coahuila (*Hyla arenicolor, Lithobates berlandieri, Gerrhonotus infernalis, Sceloporus grammicus, S. minor, S. spinosus, Lampropeltis mexicana, Pituophis catenifer, P. deppei, Thamnophis cyrtopsis, Crotalus lepidus, C. molossus* y *C. scutulatus*), tienen una distribución muy amplia, que puede incluir ambientes áridos/semiáridos, templados y/o subtropicales.

Zacatal

El Zacatal comprende las comunidades dominadas por gramíneas; se desarrollan principalmente en valles con suelos moderadamente
profundos, así como en laderas poco inclinadas
y mesetas, entre los 800 y 2500 m de altitud,
con frecuencia se mezclan con bosques de pino
piñonero y comunidades del Matorral Desértico
Chihuahuense, ocupan aproximadamente el 8%
del estado. La comunidad más característica se
presenta en una serie de valles y planicies del sureste del estado, se trata de un zacatal mediano
abierto en asociación con plantas xerófilas, las
gramíneas dominantes son: *Bouteloua gracilis,
B. curtipendula, B. dactyloides* y *Aristida divaricata*, con arbustos dispersos de *Opuntia imbricata, O. rastrera, Larrea tridentata, Flourensia cernua, Yucca carnerosana* y *Prosopis glandulosa*. En
algunas áreas con vegetación gypsófila del centro
y sur del estado se presenta el zacatal gypsófilo,
donde son frecuentes: *Bouteloua chasei, Mühl-
enbegia villiflora, M. gypsophila, M. arenicola, Achnatherum editorum* y *Scleropogon brevifolius*. En
suelos con acumulación de sales son comunes
colonias de *Sporobolus airoides, S. wrigthii, S. criptandrus, S. spiciformis, Pleuraphis mutica, Hilaria
belanjeri, Distichlis spicata* y *D. litorales*, mezclados con especies de *Atriplex, Prosopis* y *Suaeda*.
En valles intermontanos del sur del estado se
desarrollan comunidades de gramíneas ama-
colladas, las más frecuentes son: *Mühlenbergia
emersleyi, M. dubia, M. setifolia, Nasella tenuissima* y *Stipa eminens*, mezcladas con arbustos del
matorral submontano o del matorral rosetófilo.
En los sitios de alta montaña como mesetas y
claros de bosque se presentan zacatales de *Stipa
robusta* o *S. ichu* asociadas con especies de *Pinus*,
otras herbáceas frecuentes son: *Hymenoxys
insignis, Gridelia grandiflora* y *Senecio madrensis*
(www.gramineasdecoahuila.com).

En Coahuila, las especies de herpetozoarios
características del Zacatal son 47, tres de estas
especies están limitadas a este tipo de vegetación
(*Sceloporus goldmani, S. samcolemani* y *Coluber
constrictor*). La distribución de *Sceloporus goldmani* y *S. samcolemani*, se limita a zacatales del

sureste de Coahuila, partes adyacentes de Nuevo
León y noreste de San Luis Potosí (este último
estado únicamente para *S. goldmani*). Al parecer
la mayoría de las poblaciones de *S. goldmani* han
sido extirpadas o están a punto de extirparse.
Coluber constrictor presenta una distribución
amplia que va desde el sur de Canadá, hacia el
sur a través de gran parte de los Estados Unidos
y la vertiente del Atlántico de México, hasta Guatemala. En Coahuila se conoce únicamente para
áreas de zacatal del Desierto Chihuahuense.

Las 43 especies restantes tienen una fuerte
afinidad por pastizales, sin embargo, éstas
pueden habitar otros tipos de vegetación, éstas
son las siguientes: *Anaxyrus cognatus, A. debilis,
A. woodhousii, Rhinella marina, Hyla arenicolor,
Gastrophryne olivacea, Lithobates berlandieri,
Scaphiopus couchii, Spea multiplicata, Gopherus
berlandieri, G. flavomarginatus, Barisia ciliaris,
Gerrhonotus infernalis, Phrynosoma cornutum,
P. modestum, Sceloporus cautus, S. consobrinus,
Plestiodon obsoletus, P. tetragrammus, Scin-
cella kikaapoa, Aspidoscelis gularis, A. inornata,
A. marmorata, Arizona elegans, Gyalopion canum,
Heterodon kennerlyi, Hypsiglena jani, Lampropeltis
getula, Masticophis flagellum, Pituophis cateni-
fer, P. deppei, Rhinocheilus lecontei, Salvadora
grahamiae, Sonora semiannulata, Tantilla gracilis,
T. hobartsmithi, T. nigriceps, T. wilcoxi, Tham-
nophis marcianus, Crotalus atrox, C. scutulatus,
C. viridis* y *Sistrurus catenatus*.

Vegetación Riparia, Subacuática y Acuática

Estas comunidades se restringen a los ríos, arroyos y cuerpos de agua presentes en el estado, las
cuales se concentran en el norte, noreste y centro
de la entidad, en altitudes que van de 300 a 1400
m. Como parte de la vegetación riparia del norte
del estado (Ríos Bravo, San Diego, San Rodrigo,
Escondido y Arroyo de las Vacas) se presentan
comunidades arbóreas aisladas de *Quercus
fusiformis* y *Carya illinoensis*, a menudo asociados
con individuos de *Prosopis glandulosa, Platanus
glabrata, Diospyros texana, Ulmus crassifolia,
Salix nigra, Celtis reticulata* y *Morus celtidifolia*,
entre las arbustivas figuran: *Opuntia lindheimeri*

y *Castela erecta*, las herbáceas dominantes son: *Ruellia nudiflora*, *Sanvitalia ocymoides*, *Viguiera dentata*, *Allowissadula holosericea* y *Malvastrum coromandelianum*, mientras que en el Río Sabinas son frecuentes galerías estrechas de *Taxodium mucronatum*, *Fraxinus berlanderiana*, *Morus celtidifolia*, *Salix nigra*, *Prosopis glandulosa* y *Acacia farnesiana*, las últimas dos especies son dominantes en el Río Salado-Nadadores, con frecuencia las riveras de éstos presentan colonias densas de *Baccharis salicifolia* y *Arundo donax*. En los arroyos intermitentes próximos a serranías son comunes árboles aislados de *Chilopsis linearis*, *Juglans microcarpa* y *Dodonaea viscosa*. Las plantas subacuáticas crecen en suelos permanentemente húmedos, de ríos y cuerpos de agua, las especies más frecuentes son: *Cyperus odoratus*, *Eleocharis cellulosa*, *E. geniculata*, *Fuirena simplex*, *Paspalum pubiflorum*, *Spartina spartinae*, *Setaria geniculata*, *Echinochloa colonum* y *Poligonum punctatum*; la vegetación acuática se conforma por plantas que crecen en lugares inundados o dentro del agua, la localidad con mayor cantidad de plantas acuáticas es Cuatro Ciénegas, de donde se reportan las siguientes especies: *Nymphaea ampla*, *Utricularia vulgaris*, *Najas marina*, *Ruppia maritima*, *Potamogeton nodosus*, *Nastartium officinale*, *Najar guadalupensis*, *Chara* spp. y *Zanichellia palustris* (Pinkava 1979, 1984) en los principales ríos son frecuentes: *Nuphar luteum*, *Ceratophyllum demersum* y *Heteranthera dubia*, con algunas infestaciones de *Eichhornia crassipes* (www.gramineasdecoahuila.com).

Treinta y siete especies habitan la Vegetación Riparia, Subacuática y Acuática de Coahuila. De éstas 16 son anfibios que dependen del agua para su reproducción: *Ambystoma mavortium*, *Anaxyrus cognatus*, *A. debilis*, *A. punctatus*, *A. speciosus*, *A. woodhousii*, *Incilius nebulifer*, *Rhinella marina*, *Acris crepitans*, *Ecnomiohyla miotympanum*, *Hyla arenicolor*, *Smilisca baudinii*, *Lithobates berlandieri*, *L. catesbeianus*, *Scaphiopus couchii* y *Spea multiplicata*. Únicamente dos de estas especies (*Acris crepitans* y *Lithobates catesbeianus*) requieren de cuerpos de agua permanente para sobrevivir. En México, la única población

conocida de *A. crepitans* es la de la parte superior del Río Sabinas, Coahuila, sin embargo, ésta se distribuye desde la región de los Grandes Lagos extendiéndose hacia el suroeste a través de las Grandes Planicies hasta el Río Bravo y hacia el este desde el Big Bend. *Lithobates catesbeianus* presenta una distribución muy amplia en Estados Unidos, aunque las poblaciones del oeste son introducidas, y en México habita en la costa noreste. Hace 165 años fue registrada en Saltillo, Coahuila (Smith et al. 2003a), pero desde entonces no había habido ningún reporte sobre su presencia en Coahuila hasta que recientemente (Garza-Tobón y Lemos-Espinal 2013a, 2013b) fue registrada en el Rancho La Burra, km 92, carretera Nuevo Laredo—Piedras Negras. Esta población cae dentro de la distribución natural de la especie por lo que no es considerada como una población introducida en Coahuila. Las otras 14 especies de anfibios que ocupan este tipo de vegetación pueden habitar sitios que retienen suficiente cantidad de humedad (bajo troncos, rocas, hojarasca, etc.), por lo que se les puede encontrar en lugares con una marcada estación seca y/o alejadas de cuerpos de agua.

Otras nueve especies que ocupan este tipo de vegetación son tortugas con una fuerte dependencia de cuerpos de agua para su sobrevivencia, éstas son: *Pseudemys gorzugi*, *Terrapene coahuila*, *Trachemys gaigeae*, *T. scripta*, *T. taylori*, *Kinosternon durangoense*, *K. flavescens*, *K. hirtipes* y *Apalone spinifera*. De éstas, la única estrictamente acuática es *A. spinifera*. Las otras ocho especies pueden pasar periodos de tiempo considerablemente grandes fuera del agua.

Otras tres especies son lagartijas con una gran afinidad por vegetación riparia (*Sceloporus couchii*, *S. cyanogenys* y *Aspidoscelis tesselata*). *Sceloporus couchii* y *S. cyanogenys* son lagartijas trepadoras que se encuentran principalmente en árboles a los lados de cauces de ríos. *Aspidoscelis tesselata* es una lagartija que vive entre los arbustos, maleza y claros a los lados de ríos.

Las nueve especies restantes (*Nerodia erythrogaster*, *N. rhombifer*, *Opheodrys aestivus*, *Thamnophis angustirostris* [la descripción de la morfología de la serie tipo de esta especie sugiere hábitos acuáticos], *T. cyrtopsis, T. exsul, T. marcianus, T. proximus* y *Agkistrodon contortrix*), son serpientes que raramente se encuentran alejadas de cuerpos de agua. Todas ellas incluyen en sus dietas presas acuáticas.

La topografía y características mencionadas anteriormente forma un patrón de extremos rigurosos en clima y topografía; de islas ecológicas en mares de homogeneidad; de una deformación extensa en la reciente historia geológica, efectuando aislamientos sucesivos y surgimientos de terreno; y de disponibilidad marginal de agua en áreas extensas de cuencas internas (más de un tercio del estado). El resultado, herpetológicamente, es una fauna depauperada de orígenes diversos, dominada por especies adaptadas a extremos rigurosos de clima (como las lagartijas del género *Sceloporus*), pero también con oasis de endemismos asombrosamente altos (ejem. Bolsón de Cuatro Ciénegas), en parches de condiciones ambientales favorables, y distribuciones en forma de islas en las cimas de macizos montañosos separados ampliamente (ejem. *Gerrhonotus infernalis, Sceloporus grammicus* y *Pituophis deppei*).

Por supuesto, las condiciones ambientales rigurosas también han limitado la exploración, como se discutió previamente. El conocimiento de la herpetofauna del estado (su composición, distribución, taxonomía, historia natural) es tristemente deficiente. No obstante, las limitantes ambientales revisadas aquí son constantes o más probables de empeorar que de mejorar.

Como se mencionó anteriormente, un total de 133 especies se conocen para Coahuila. De este total los dos órdenes de anfibios: caudata y anura, son los grupos menos representados 3% (4/133) y 15% (20/133) respectivamente. Esto se debe a la dependencia que los anfibios tienen por cuerpos de agua y/o lugares húmedos. La mayor parte de Coahuila está ocupada por ambientes áridos/semiáridos con una limitada disponibilidad de agua y sitios húmedos, igualmente esto limita la presencia de estos grupos en el estado. Tres de las cuatro salamandras que habitan Coahuila se limitan a la pequeña

porción del Bosque de Montaña de la Sierra Madre Oriental que entra al extremo sureste de Coahuila. La otra salamandra (*Ambystoma mavortium*), es la única en el estado que está adaptada para habitar regiones áridas/semiáridas del norte de América. Esta salamandra ocupa aguajes y cuerpos de agua estancada y temporal del Desierto Chihuahuense, en periodos secos se puede enterrar a una profundidad en la que encuentra la suficiente cantidad de humedad para sobrevivir.

Es posible que la Salamandra Texana (*Eurycea neotenes*) habite en el extremo norte de Coahuila, Dixon (2000) indicó la presencia de esta especie en varias localidades en la parte de Texas adyacente al extremo norte de Coahuila.

Igualmente, la baja proporción de anuros se debe a la limitada disponibilidad de agua y humedad en el estado. El género de anuros dominante en el estado de Coahuila (*Anaxyrus*), representa organismos de piel tuberculada que les permite sobrevivir en condiciones de limitada disponibilidad de agua. Especies de otros géneros, como *Incilius, Rhinella, Craugastor, Eleutherodactylus, Gastrophryne, Lithobates, Scaphiopus* y *Spea*, requieren de una mayor cantidad de agua o humedad disponible.

Posiblemente el Sapo de Espuelas de los Llanos (*Spea bombifrons*) habita en el extremo norte de Coahuila, Dixon (2000) reportó la presencia de esta especie en localidades de Texas a pocos kilómetros de la frontera con Coahuila. Las condiciones ambientales del Desierto de Coahuila son adecuadas para el desarrollo de poblaciones de esta especie de anuro.

Por otra parte, la piel coriácea carente de glándulas, huevo amniota y presencia de pulmones en los reptiles les brinda independencia del agua para el desarrollo de las actividades esenciales. De tal forma que tortugas, lagartijas y serpientes están muy bien representadas en el estado.

Aunque la proporción de tortugas es de tan sólo el 8% (11/133) del total de la herpetofauna estatal, ésta es más del doble de la proporción de tortugas a nivel nacional (3.3% o 32/973, Liner y Casas-Andreu 2008). Lo que nos da una idea de la alta representación que este grupo de reptiles tiene en el estado. Todas las especies de tortugas de Coahuila se desarrollan en la región árida/semiárida del estado. Ninguna especie de tortuga se ha registrado en la región templada del sureste o de las cimas de las cadenas montañosas dispersas en el territorio Coahuilense. Más aún, el número de especies de tortugas en Coahuila seguramente es mayor al que actualmente se tiene registrado, ya que especies como *Terrapene ornata* muy posiblemente habitan el Desierto Chihuahuense de Coahuila, sin embargo, a la fecha no se ha registrado.

Las lagartijas están bien representadas en Coahuila, la proporción de éstas en relación al total de herpetozoarios presentes en el estado es de 37% (49/133). Por mucho el género de herpetozoarios más diversificado de Coahuila es *Sceloporus*. El número de especies de este género presentes en el estado (19) es uno de los más grandes para un estado mexicano, muestra de la diversidad de ambientes disponibles para estas lagartijas. Adicionalmente, el número de especies de lagartijas que habitan el estado seguramente es mayor, Lemos-Espinal y Smith (2007) reportaron que especies como la Lagartija Sorda Pequeña (*Holbrookia maculata*), el Camaleón Cuernitos de Hernández (*Phrynosoma hernandesi*), y la Corredora de Chihuahua (*Aspidoscelis exsanguis*), posiblemente habitan el extremo noroeste de Coahuila. El Lagarto Pigmeo (*Gerrhonotus parvus*) los bosques de pino de la Sierra de Arteaga, el Chintete Espinoso (*Sceloporus spinosus*), la parte semiárida del extremo sureste del estado. El Anolis Verde (*Anolis carolinensis*), la Llanera de Laredo (*Aspidoscelis laredoensis*), y el Huico de Seis Líneas (*Aspidoscelis sexlineatus*), el extremo noreste de Coahuila adyacente a Texas. Y la Espinosa de Collar (*Sceloporus torquatus*), la Nocturna del Bolsón (*Xantusia bolsonae*), y la Nocturna de Durango (*Xantusia extorris*), el extremo suroeste del estado.

Igualmente las serpientes están bien representadas en Coahuila con 37% (49/133). La alta diversidad de serpientes es el resultado de la igualmente gran diversidad de adaptaciones que este suborden presenta para habitar ambientes áridos/semiáridos y templados. De acuerdo con Lemos-Espinal y Smith (2007), varias especies de serpientes aún no registradas para Coahuila pueden habitar este estado. Ellos consideran que en el extremo sureste del estado pueden habitar la Metapil (*Agkistrodon taylori*), la Nariz de Gancho Tamaulipeca (*Ficimia streckeri*), y la Rojilla (*Tantilla rubra*). En el extremo este la Culebrilla Ciega de Tampico (*Rena myopica*), y la Hojarasquera de Nuevo León (*Rhadinaea montana*). En el extremo noreste la Culebra Parda (*Storeria dekayi texana*), y la Culebra Cabeza-negra de Trans-Pecos (*Tantilla cucullata*). En el extremo noroeste el Cabestrillo del Big Bend (*Salvadora deserticola*), y la Serpiente de Tetalura (*Trimorphodon vilkinsonii*). Por otra parte, Rossman et al. (1996) reportaron la distribución de la Jarretera Mexicana (*Thamnophis eques*), la Jarretera Vientre Negro Mexicana (*Thamnophis melanogaster*), y la Jarretera Cabeza-angosta de Chihuahua (*Thamnophis unilabialis*), en el extremo suroeste de Coahuila, sin embargo, no hay registros de especímenes de ninguna de estas especies recolectados en esta región, por lo que no las incluimos en la herpetofauna del estado.

Noventa y ocho de las 133 especies de anfibios y reptiles que habitan Coahuila se comparten con los Estados Unidos; la mayoría de estas especies compartidas (95% o 93/98) habitan en el Desierto de Chihuahua y extienden su distribución hacia el sur desde las Grandes Planicies de los Estados Unidos hasta el extremo sur del Desierto de Chihuahua en los estados mexicanos de San Luis Potosí o Querétaro. Únicamente 4 de estas especies compartidas son características de los trópicos y subtrópicos de América (*Rhinella marina, Smilisca baudinii, Drymarchon melanurus* y *Leptodeira septentrionalis*). La primera de ellas ha sido registrada en las tierras bajas del centro de Coahuila, en el semiárido Bolsón de Cuatro Ciénegas. Las otras tres especies habitan en las tierras bajas y noreste de Coahuila y en las estribaciones del oeste de la Sierra Madre Oriental. Estas 4 especies de afinidades tropicales entran a los Estados Unidos únicamente en la parte sur de Texas. Una de estas 98 especies es una especie introducida, el Geco del Mediterráneo (*Hemidactylus turcicus*) cuya distribución original es el Mediterráneo. En Coahuila tiene una distribución amplia ocupando varios pueblos y ejidos del Desierto de Chihuahua; éste está ampliamente distribuido en el sur de los Estados Unidos. Otra de estas especies compartidas, *Coleonyx reticulatus*, tiene una distribución casi exclusiva en Coahuila, excepto por un registro en la frontera entre Coahuila y Durango, y por una pequeña población aislada en la parte más baja del Big Bend de Texas cerca del Río Bravo, todos los demás registros provienen de Coahuila.

Todas las 35 especies que no se comparten entre México y los Estados Unidos son endémicas a México; 7 de ellas a Coahuila: *Terrapene coahuila, Trachemys taylori, Gerrhonotus lugoi, Crotaphytus antiquus, Uma exsul, Scincella kikaapoa* y *Thamnophis angustirostris*.

Tres de estas siete están limitadas al Bolsón de Cuatro Ciénegas (*T. coahuila, G. lugoi* y *S. kikaapoa*), las otras 4 a otras localidades del Desierto de Chihuahua de Coahuila. El estatus taxonómico de una de ellas, *Thamnophis angustirostris*, no es claro. Esta especie es conocida únicamente por el espécimen lectotipo (USNM 959—Kennicott no designó un holotipo y aparentemente tenía más de un espécimen sobre él que basó la descripción de este taxón [Rossman et al. (1996)]) recolectado en Parras, Coahuila. Consideramos a ésta como una especie valida ya que la Sierra de Parras y localidades de alrededor tienen las condiciones ambientales adecuadas para esta especie; eventualmente esta especie podría ser redescubierta con un trabajo de campo más intenso en los alrededores de Parras.

Quince (54% o 15/28) de las restantes 28 especies endémicas a México están limitadas en el este de México a las montañas y estribaciones de la Sierra Madre Oriental (*Chiropterotriton priscus, Pseudoeurycea galeanae, P. scandens, Eleutherodactylus longipes, Ecnomiohyla miotympanum, Barisia ciliaris, Phrynosoma orbiculare, Sceloporus minor, S. oberon, S. parvus, Plestiodon dicei, Scincella silvicola, Pituophis deppei, Storeria hidalgoensis* y *Thamnophis exsul*). Éstas entran a Coahuila únicamente en el extremo sureste del estado. Tres especies más están limitadas a regiones esparcidas del norte de México: *Sceloporus couchi* a las sierras del norte de Coahuila y centro oeste de Nuevo León; *Sceloporus goldmani* a un área pequeña en el sureste de Coahuila, parte adyacente de Nuevo León y noreste de San Luis Potosí; y *Sceloporus maculosus* a los afluentes del Río Nazas en Durango y Coahuila. Tres especies más están limitadas al Altiplano Mexicano (*Sceloporus cautus, S. samcolemani* y *Lampropeltis mexicana*). Otras tres especies están limitadas a un área pequeña del Bolsón de Mapimí de sureste Chihuahua, oeste Coahuila y noreste Durango (*Kinosternon durangoense, Gopherus flavomarginatus* y *Uma paraphygas*). Dos especies más (*Sceloporus cyanostictus* y *S. ornatus*) se distribuyen únicamente en Coahuila y el extremo oeste de Nuevo León. Una más, *Sceloporus spinosus*, se distribuye ampliamente desde el centro de México y se extiende hacia el norte sobre el Altiplano Mexicano entrando a Coahuila únicamente a los bosques de encino del extremo sureste del estado en la Sierra de Zapaliname. Y la última de estas 35 especies endémicas (*Holbrookia approximans*) está limitada al Desierto de Chihuahua de México; sin embargo, es altamente probable que habite en la parte adyacente de los Estados Unidos.

JULIO A. LEMOS-ESPINAL Y ALEXANDER CRUZ

Introducción

Nuevo León es uno de los estados mexicanos de mayor riqueza económica, está altamente industrializado y el promedio de sus habitantes poseen una calidad de vida superior a la del resto del país. Su capital, la ciudad de Monterrey, es la tercera ciudad más grande de México. A diferencia del resto de los estados mexicanos de la frontera norte, Nuevo León posee una red de carreteras que hace posible el ingreso a prácticamente cualquier rincón del estado. La flora y fauna de Nuevo León es muy rica en especies, a grandes rasgos ésta se compone principalmente de un arreglo de especies características de los grandes desiertos de Norteamérica, así como especies de los bosques templados de la Sierra Madre Oriental, y especies subtropicales que extienden su distribución, en algunos casos desde Centro o Sudamérica, a través de las tierras bajas de la vertiente del Atlántico.

A pesar de que Nuevo León tiene estas características, los estudios sobre la diversidad y distribución de las especies de anfibios y reptiles del estado son muy escasos y los pocos que ahí se han desarrollado están enfocados casi en su totalidad a los bosques de la Sierra Madre Oriental y montañas satelitales al norte y noreste de la ciudad de Monterrey, las extensas planicies al este de la Sierra Madre Oriental permanecen en su

mayor parte sin estudiar al igual que la porción del Altiplano Mexicano del extremo suroeste del estado. En la actualidad no existe un programa que aborde sistemáticamente el estudio de la herpetofauna estatal en sus diferentes regiones, tal que una de las principales amenazas para la conservación de la riqueza de anfibios y reptiles del estado es el desconocimiento que se tiene sobre ella. Esto aunado a problemas de una gran demanda de agua, energía y recursos alimenticios para cubrir las necesidades de la ciudad de Monterrey, muestra un panorama poco alentador para la herpetofauna de Nuevo León.

A la fecha se ha documentado la presencia de 132 especies de anfibios (23 especies) y reptiles (109 especies) en el estado, sin embargo, el desarrollo de inventarios herpetofaunísticos, particularmente en áreas cercanas a la frontera con Tamaulipas, pueden incrementar considerablemente este número. Asimismo, estudios sobre las variaciones morfológicas y análisis genético de las poblaciones actualmente conocidas, especialmente de aquellas que se encuentran aisladas, podrán probar la existencia de especies nuevas.

Estudios Herpetológicos Previos

El trabajo herpetológico en el noreste de México comenzó con la llegada del naturalista francés Jean Louis Berlandier. Llegó a México en diciembre de 1826 a la edad de 21 años con el objetivo de evaluar los recursos naturales de la frontera entre Texas y México. En febrero de 1828 Berlandier había comenzado su trabajo en la frontera. Aunque Berlandier hizo la mayoría de sus recolectas en Texas y Tamaulipas, es probable que también haya recolectado anfibios y/o reptiles en Nuevo León. Sin embargo, los registros de sus colecciones son difíciles de obtener y no hemos podido encontrar ninguno desarrollado en este estado.

El primer registro de una colección herpetológica en el estado de Nuevo León es el que hizo en 1852 Thomas Webb, Secretario y Cirujano del equipo del Comisionado John Russel Bartlett para la United States and Mexican Boundary Commission. Al finalizar la evaluación del Río Gila, el equipo de Bartlett salió de Tucson, Arizona, el 17 de julio de 1852, para dirigirse a El Paso, Texas. Su ruta incluyó el norte de Sonora y continuó hacia El Paso, Texas, llegando a este último punto el 18 de agosto de 1852. Dejaron El Paso, Texas, el 7 de octubre de 1852 procediendo hacia Ringgold Barracks, Texas a través de los estados mexicanos de Chihuahua, Durango, Coahuila y Nuevo León. Llegaron a Santa Catarina el 11 de diciembre, y a Monterrey, capital de Nuevo León, el 12 de diciembre. Dejaron Monterrey al siguiente día tomando una ruta que pasó por Marin, Carrizitos y Cerralvo, y el 19 de diciembre de 1852 llegaron a Mier y Noriega. Los siguientes días pasaron a través de Camargo y cruzaron el Río Bravo para Ringgold Barracks, Texas (Kellogg 1932). Todos los especímenes recolectados por la United States and Mexican Boundary Commission están depositados en el United States National Museum (USNM). Algunos de los especímenes recolectados en Nuevo León fueron reportados por Kellogg (1932).

Después de las recolectas hechas por Thomas Webb en 1852, el Teniente Dario Nash Couch hizo una expedición personal al noreste de México en 1853 para recolectar plantas y animales, pasó varios meses en los estados de Coahuila, Nuevo León y Tamaulipas. En Nuevo León visitó China, San Diego, Cadereyta, Monterrey,

Boquilla, Villa de Santiago, Rinconada y varias localidades en la Sierra Madre Oriental (Conant 1968). Los especímenes recolectados por Couch están depositados en el USNM. Varios especímenes recolectados por Couch en Nuevo León se volvieron parte de la serie tipo de taxa nuevos. Algunos de éstos son: *Bufo speciosus* (= *Anaxyrus speciosus*), recolectado en Pesquería Grande (= Villa García) y descrito por Girard (1854); *Sceloporus couchii*, recolectado en Santa Catarina y descrito por Baird (1859); *Cnemidophorus inornatus* (= *Aspidoscelis inornata*), recolectado en Pesquería Grande y descrito por Baird (1859); *Cnemidophorus octolineatus* (= *Aspidoscelis inornata*), recolectado en Pesquería Grande y descrito por Baird (1859); *Eutaenia cyrtopsis* (= *Thamnophis cyrtopsis*), recolectado en Rinconada y descrito por Kennicott (1860). Otros especímenes recolectados por Couch fueron reportados por Kellogg (1932).

Otra contribución importante de Couch fue la compra de la colección de Berlandier a la viuda de éste en 1853. Envió parte de esta colección a Spencer F. Baird del USNM y parte de la colección de plantas a Suiza (Beltz 2006).

En marzo, abril y junio de 1902, Edward Goldman y Edward Nelson del USNM visitaron varias localidades en el oeste de Nuevo León: Cerro La Silla, Doctor Arroyo, Montemorelos y Monterrey. Recolectaron vertebrados en estas localidades; algunos especímenes de anuros recolectados por ellos fueron reportados por Kellogg (1932).

En 1904 Hans Gadow y Seth Meek hicieron un viaje a México para estudiar la variación en el género *Cnemidophorus*, en ese viaje Meek recolectó en los poblados de San Juan localizado al sur de Monterrey, Montemorelos, Garza Valdés y La Cruz, todos en el estado de Nuevo León. Entre los especímenes recolectados estaba la serie tipo de *Cnemidophorus gularis meeki* (= *Aspidoscelis gularis*). Este material fue reportado por Gadow (1906).

El inicio del siglo 20 estuvo caracterizado por una carencia de trabajo herpetológico en Nuevo León. Hasta que en 1932 Edward H. Taylor y Hobart M. Smith visitaron México por primera vez iniciando una de las eras herpetológicas más

prolíficas en México. El mismo día que Hobart
Smith se graduó de la licenciatura a la edad de
20 años en la Kansas State University, Edward
Taylor lo recogió para pasar todo el verano recolectando en cualquier sitio que pudieran visitar
en México, obteniendo más de 5000 especímenes. En su viaje de ida pararon en Sabinas
Hidalgo por varios días y recolectaron anfibios y
reptiles en esa localidad; entre los especímenes
recolectados estaba el tipo de *Scelporus parvus*,
descrito por Hobart Smith en 1934. Seis años
más tarde Hobart Smith pudo continuar su
trabajo de campo para estudiar la herpetofauna
de México a través de la Walter Rathbone Bacon
Traveling Scholarship. Él y su esposa Rozella
estuvieron recolectando anfibios y reptiles entre
1938 y 1940 en todo México. En octubre 13 y 14
de 1939, Rozella y Hobart Smith pararon 24 km
al sureste de Galeana y recolectaron anfibios y
reptiles. En esos dos días recolectaron 19 especímenes (todos ellos paratipos) de *Bolitoglossa
galeanae* (= *Pseudoeurycea galeanae*), reportados
por Taylor (1941) y Taylor y Smith (1945); *Syrrophus smithii* (= *Eleutherodactylus guttilatus*) (el
tipo y paratipo), descritos y reportados por Taylor
(1940b); *Tantilla wilcoxi rubricata* (= *T. wilcoxi*) el
tipo descrito y reportado por Smith (1942); y *Salvadora lineata* (= *S. grahamiae lineata*) reportada
por Taylor y Smith (1945). Taylor regresó a Nuevo
León en junio de 1936. Visitó Sabinas Hidalgo
(junio 17), y el Cañón de la Huasteca (junio 20),
15 km al oeste de Monterrey. Entre los especímenes recolectados en estas dos localidades
están el tipo y uno de los paratipos de *Syrrhophus
latodactylus* (= *Eleutherodactylus longipes*), descrito
por él (Taylor 1940a).

Otra importante contribución de Hobart
Smith al conocimiento de la herpetofauna de
Nuevo León son las descripciones de *Leiolopisma
caudaequinae* (= *Scincella silvicola caudaequinae*)
(Smith 1951), y *Sceloporus scalaris samcolemani*
(= *S. samcolemani*) (Smith y Hall 1974). El especí-
men tipo de *S. silvicola caudaequinae* fue recibido
como un regalo al University of Illinois Museum
of Natural History por parte del Señor J. P. Craig,
quien lo recolectó en la Cascada Cola de Caballo, 40 km al sur de Monterrey el 19 de abril de

1946. El tipo de *Sceloporus samcolemani* está en el
University of Michigan Museum of Zoology, fue
recolectado por P. H. Litchfield entre Providencia y La Paz, el 16 de julio de 1960.

En 1934, Henry A. Pilsbry, Francis W. Pennell
y Cyril H. Harvey, de la Academy of Natural
Sciences of Philadelphia recolectaron 141 especímenes de anfibios y reptiles, representando
36 especies para los estados de Chihuahua,
Coahuila, Durango, México, Nuevo León, San
Luis Potosí y Zacatecas. En Nuevo León visitaron
Hacienda Pablillo, Cascada Cola de Caballo, Río
Maurisco 25 km al sur de Monterrey, el camino
entre Pablillo y Alamar, el camino arriba de El
Infiernillo de la Hacienda Pablillo, y Cieneguillas
al sur de Galeana. La mayoría de estas localidades se encuentran en la parte suroeste del
estado. Recolectaron especímenes de 12 especies
(1 salamandra, 2 anuros, 7 lagartijas y 2 serpientes) en Nuevo León. Entre los especímenes
recolectados estaba el tipo de *Sceloporus binocularis* (= *S. torquatus binocularis*). La descripción
original de este taxón junto con el reporte de los
especímenes recolectados en Nuevo León fue
publicado por Dunn (1936).

El 10 de julio de 1938 Radclyffe Roberts,
amigo y compañero de campo de Edward Taylor
en varios viajes a México, recolectó el tipo de
Bolitoglossa galeanae (= *Pseudoeurycea galeanae*)
en Pablillo, 24 km al oeste de Galeana a 2133 m
de altitud. La descripción original de este taxón
junto con el reporte del espécimen fueron publicados por Taylor (1941).

En los veranos de 1938, 1939 y 1940 grupos
de estudiantes de la University of Illinois in
Chicago bajo la dirección de Harry Hoogstraal
recolectaron varios anfibios y reptiles en el
noreste de México. Algunos de los especímenes
recolectados fueron incorporados al Field Museum of Natural History (Smith 1944). Entre los
53 especímenes de serpientes recolectadas había
varios especímenes de Nuevo León, incluyendo
el tipo de *Rhadinaea montana*, recolectada por
Hoogstraal en Ojo de Agua, Galeana, y descrita
por Hobart Smith en 1944. Otros especímenes
fueron enviados al Chicago Natural History
Museum; entre esos especímenes estaba el

tipo de *Chiropterotriton priscus*, recolectado en Cerro Potosí, cerca de Galeana el 16 de agosto de 1938, por E. J. Koestner, así como los paratipos recolectados por Koestner y Hoogstraal en el mismo lugar. Originalmente, estos especímenes fueron mencionados por Taylor (1944, 217) "He visto una especie aún no descrita de este género (*Chiropterotriton*) que fue recolectada en Cerro Potosí cerca de Galeana, Nuevo León." El reporte y descripción de la especie fueron publicados por Rabb (1956).

Adicionalmente, Hobart Smith y Edward Taylor publicaron: el listado y claves para las serpientes de México; el listado y claves para los anfibios de México; el listado y clave para los reptiles de México, excluyendo la serpientes; y las localidades tipo de anfibios y reptiles mexicanos (Smith y Taylor 1945, 1948, 1950a, 1950b). En estas publicaciones proporcionaron los listados y claves para los anfibios y reptiles de Nuevo León, y la lista de localidades tipo para este estado. A los 100 años de edad Hobart Smith todavía publicaba sobre la herpetofauna de México y apoyaba a sus compañeros herpetólogos con el mismo entusiasmo que siempre lo caracterizó.

Una de las mayores contribuciones al conocimiento de la herpetofauna de Nuevo León fue la realizada por Ernest A. Liner; él comenzó a trabajar en el noreste de México en noviembre 19–22 de 1951. Su primer viaje de colecta lo hizo a Nuevo León; su principal objetivo en este viaje era recolectar *Pseudoeurycea galeanae* la cual en ese tiempo era conocida sólo por la serie tipo proveniente de cerca de Galeana. Ese año visitó localidades en los alrededores de Monterrey, Arroyo Vaquerías (16 km al norte de Monterrey), Cascada Cola de Caballo, Cañón de Santa Rosa, Linares, Galeana, Rancho La Montaña y Doctor Arroyo. Aunque en ese viaje no recolectó *P. galeana*, pudo recolectar varias especies de anfibios y reptiles y se familiarizo con la herpetofauna del estado. En los siguientes años visitó México tantas veces como pudo (diciembre 21, 1952—enero 3, 1953; julio 6–13, 1954; julio 13–22, 1957; julio 26–agosto 8, 1958; julio 1960; y julio 1961). En estos viajes visitó localidades en el este—noreste de México y en su viaje de 1960 dio toda la vuelta

desde el Atlántico hasta el Pacífico entrando a México a través de Piedras Negras, Coahuila y saliendo en Nogales, Sonora, viajando a través de los estados de Coahuila, San Luis Potosí, Guanajuato, Aguascalientes, Jalisco, Nayarit, Sinaloa y Sonora. Aunque recolectó sobre un área muy amplia en México su principal trabajo fue realizado en Nuevo León. El reporte de sus viajes a México, incluyendo todas las localidades visitadas en Nuevo León así como los especímenes recolectados durante este periodo de tiempo están reportados en Liner (1964, 1966a, 1966b, 1991a, 1991b, 1992a, 1992b, 1993, 1994).

Después de 1961 Liner continuo haciendo trabajo de campo en Nuevo León junto con otros herpetólogos, entre ellos Allan H. Chaney, James R. Dixon, Richard M. Johnson, Douglas A. Rossman y James F. Scudday. Publicaron varias notas distribucionales y de historia natural: *Rhadinaea montana* (julio018, 1972) del Ojo de Agua en Pablillo (Chaney y Liner 1986); *Crotalus pricei miquihuanus* (julio 20, 1985) 1.8 km al norte del camino Los Mimbres (Chaney y Liner 1986); *Rhadinaea montana* (julio 17, 1985) de 4.3 km al norte de Las Adjuntas (Chaney y Liner 1990); *Chiropterotriton priscus* (registros de julio y agosto 1980) de los alrededores de San Antonio Peña Nevada y La Encantada, sur de Nuevo León (Chaney et al. 1982a); *Sceloporus p. poinsettii* (registros de julio 1977 y agosto 1978) de Sierra Picacho en Rancho El Milagro (Chaney et al. 1982b); *Pseudoeurycea galeanae* (registros de julio y agosto 1980) de los alrededores de San Antonio Peña Nevada y La Encantada, sur de Nuevo León (Johnson et al. 1982a); *Sceloporus couchi* (registros de julio 1977 y agosto 1978) de Sierra Picacho en Rancho El Milagro (Johnson et al. 1982b); *Barisia ciliaris* (mayo 24, 1973) del Ojo de Agua cerca de Pablillo (Liner et al. 1973); *Tantilla nigriceps fumiceps* (julio 13, 1976) del Rancho El 86, 9.5 km sur-suroeste de Cerralvo (Liner et al. 1978); *Bogertophis subocularis* (julio 28, 1981) de 35.4 km al suroeste de Bustamante (Liner, Chaney y Johnson 1982); *Tantilla r. rubra* (agosto 16, 1978) de Sierra Picacho en Rancho El Milagro (Liner, Johnson y Chaney 1982); *Thamnophis cyrtopsis pulchrillatus* (registros de agosto 1978 y junio 1981) 5 km al N de Las Crucitas y San Antonio Puerto Peña Nevada (Liner et al. 1990); *Crotalus1. lepidus* (julio 18, 1972) Ojo de Agua en San Pablillo (Liner y Chaney 1986); *Rhadinaea montana* (julio 18, 1972) Ojo de Agua en San Pablillo (Liner y Chaney 1987); *Tantilla r. rubra* (agosto 16, 1978) 9.8 km al suroeste de Cerralvo y 16.7 km al oeste, en Rancho Milagro, Sierra Picacho (Liner y Chaney 1990b); *Sceloporus torquatus mikeprestoni* (registros de julio y agosto de 1980) de los alrededores de San Antonio Peña Nevada (Liner y Chaney 1990a); *Aspidoscelis i. inornata* (julio 20, 1975) de 5.6 km al oeste de Aramberri (Liner y Chaney 1995a); *Aspidoscelis inornata paulula* (julio 20, 1974) de 7.2 km al norte de Mier y Noriega en Rancho San Roberto (Liner y Chaney 1995b); *Storeria occipitomaculata hidalgoensis* (= *S. hidalgoensis*) (julio 5, 1965 y julio 25, 1973) de Ojo de Agua arriba de Pablillo y 8.4 km al oeste de Ejido 14 de Marzo en Cerro Potosí (Liner y Johnson 1973).

Otras publicaciones de autoría o coautoría por Ernest Liner sobre los anfibios y reptiles de Nuevo León incluyen el reporte de individuos adultos de la lagartija *Sceloporus torquatus binocularis* recolectados en Julio 1954 y 1957 en los alrededores de Galeana (Liner y Olson 1973); registros distribucionales para el estado de *Ambystoma tigrinum velasci* (= *A. mavortium*), *Chiropterotriton priscus*, *Pseudoeurycea galeanae*, *Anaxyrus speciosus*, *Spea multiplicata*, *Kinosternon f. flavescens*, *Plestiodon brevirostris pineus* (= *P. dicei*), *Sceloporus oberon*, *S. s. spinosus*, *S. torquatus binocularis*, *Lampropeltis alterna* y *Crotalus s. scutulatus* (Liner et al. 1976); la redescripción de *Thamnophis exsul* (Rossman et al. 1989); la descripción original de *Sceloporus chaneyi*, especímenes recolectados por Ernest Liner y Richard Johnson en julio 17, 1980, en Rancho La Encantada (Liner y Dixon 1992); las fichas para el Catalogue of American Amphibians and Reptiles de *Sceloporus chaneyi* (Liner y Dixon 1994), y *Rhadinaea montana* (Liner 1996); y un reporte del material herpetológico tipo de Nuevo León (Liner 1996a). La información reportada por Liner en este último artículo se encuentra resumida en el Cuadro 7.1.

La extensa colección personal de Liner de anfibios y reptiles mexicanos fue donada poco antes de su muerte al American Museum of Natural History (AMNH).

Otras contribuciones importantes al conocimiento de la herpetofauna de Nuevo León son las de Robert y William Reese de St. Edward's University, Austin, Texas, quienes en agosto 23–24, 1969 visitaron las localidades de Las Lajitas (9.6 km al oeste y 11.3 km al norte de Doctor Arroyo), 9.6 km al norte de Ascensión, y 8 km al este de la intersección San Roberto, y recolectaron *Ambystoma velasci* (= *A. mavortium*) (primer registro para Nuevo León), *Pseudoeurycea galeanae* y *Crotalus scutulatus*. Esta colección fue reportada por Reese (1971). Aproximadamente 12 años más tarde, el 21 de marzo de 1983 Alec Knight de Sul Ross State University, Texas, recolectó el espécimen tipo de *Gerrhonotus parvus* a 1 km al sur de Galeana. La descripción original y reporte del espécimen fueron publicadas por Alec Knight y James Scudday en 1985.

La Universidad Autónoma de Nuevo León (UANL) también ha contribuido al conocimiento de la herpetofauna de Nuevo León, principalmente a través de Javier Banda-Leal, Arturo Contreras-Arquieta, Armando Contreras-Balderas, Armando Contreras-Lozano, David Lazcano-Villarreal, Manuel Nevares y Carlos Treviño-Saldaña. Individualmente o en grupos, y en colaboración con gente como Robert W. Bryson, Michael S. Price y Gerard T. Salmon, han publicado varios artículos sobre los anfibios y reptiles del estado. En 1988, Carlos Treviño-Saldaña publicó la descripción original de *Sceloporus jarrovii cyaneus* (= *S. minor cyaneus*), en el Cañón de la Presa Boca, Santiago. Más recientemente ha habido varias publicaciones sobre la distribución e historia natural de la herpetofauna de Nuevo León: *Elgaria parva* (= *Gerrhonotus parvus*): extensión de su distribución (Banda-Leal et al. 2002); *Gerrhonotus parvus*: talla máxima (Banda-Leal et al. 2005); *Gerrhonotus parvus*: ficha sobre historia natural (Bryson y Lazcano-Villarreal 2005); lista de anfibios y reptiles de Nuevo León, reportando 113 especies (Contreras-Arquieta y Lazcano-

Localidades tipo para anfibios y reptiles descritos para el estado de Nuevo León, México

Autor(es)	Nombre original: Localidad tipo
Salamandras (2)	
Rabb (1956)	*Chiropterotriton priscus*: **at an elevation of 8000 feet on Cerro Potosí, near Ojo de Agua, about 11 mi west-northwest of Galeana**
Taylor (1941)	*Pseudoeurycea galeanae*: **15 mi west of Galeana**
Anuros (2)	
Taylor (1940a)	*Syrrhophus smithi* (= *Eleutherodactylus guttilatus*): **15 mi west of Galeana**
Taylor (1940b)	*Syrrhophus latodactylus* (= *Eleutherodactylus longipes*): **Huasteca Cañón, about 15 km west of Monterrey, about 680 m elevation**
Lagartijas (12)	
Knight & Scudday (1985)	*Elgaria parva* (= *Gerrhonotus parvus*): **3 km SE Galeana**
Liner & Dixon (1994)	*Sceloporus chaneyi*: **11.1 mi (17.0 km) SW Zaragoza, at Rancho La Encantada, 9300 feet (2835 meters)**
Baird (1859 [1858])	*Sceloporus couchi*: **Santa Catarina**
Cope (1885)	*Sceloporus torquatus cyanogenys* (= *S. cyanogenys*): **Monterrey**
Treviño-Saldaña (1988)	*Sceloporus jarrovi cyaneus* (= *S. minor cyaneus*): **Cañón de la Presa Boca, Santiago**
Smith (1934)	*Sceloporus parvus*: **Hills about 5 mi west of Sabinas Hidalgo**
Smith & Hall (1974)	*Sceloporus samcolemani*: **Between Providencia and La Paz**
Dunn (1936)	*Sceloporus binocularis* (= *S. torquatus binocularis*): **Trail between Pablillo and Alamar**
Smith (1951)	*Leiolopisma caudaequinae* (= *Scincella silvicola caudaequinae*): **Horsetail Falls, 25 mi S Monterrey**
Gadow (1906)	*Cnemidophorus gularis meeki* (= *Aspidoscelis gularis*): **Montemorelos (lectotype)**
Baird (1859 [1858])	*Cnemidophorus inornatus* (= *Aspidoscelis inornata*): **Pesquería Grande (= Villa García)**
Baird (1859 [1858])	*Cnemidophorus octolineatus* (= *Aspidoscelis inornata*): **Pesquería Grande (= Villa García)**
Serpientes (10)	
Günther (1893)	*Coronella leonis* (= *Lampropeltis mexicana leonis*): **Nuevo León**
Kennicott (1860)	*Nerodia couchii* (= *N. erythrogaster transversa*): **San Diego y Santa Catarina, restricted to Santa Catarina**[1]
Duméril et al. (1854)	*Pituophis mexicanus* (= *P. catenifer sayi*): **Mexico, restringida a Sabinas Hidalgo**[1]
Smith (1944)	*Rhadinaea montana*: **Ojo de Agua, Galeana**
Günther (1895)	*Homalocranium* (= *Tantilla*) *atriceps*: **Nuevo León**
Smith (1942)	*Tantilla wilcoxi rubricata*: **15 mi southeast of Galeana, corregida al oeste de Galeana**[2]
Kennicott (1860)	*Eutaenia* (= *Thamnophis*) *cyrtopsis*: **Rinconada, Coahuila, corregida a Rinconada, Nuevo León**[3]
Zertuche & Treviño (1978)	*Crotalus lepidus castaneus* (= *C. lepidus*): **Paraje las Huertas, municipio de Monterrey**
Gloyd (1940)	*Crotalus triseriatus miquihuana* (= *C. pricei miquihuanus*): **Cerro El Potosí, near Galeana**
Harris & Simmons (1978)	*Crotalus durissus neoleonensis* (= *C. totonacus*): **Las Adjuntas, Santiago**

Note: Modificado de Liner 1996a.

[1] Restricción reportada en Smith & Taylor (1950).

[2] Corrección reportada en Cochran (1961).

[3] Corrección reportada en Conant (1968).

Villarreal 1995); distribución de la herpetofauna de Sierra Picachos con respecto a los tipos de vegetación y gradiente altitudinal (Contreras-Lozano et al. 2007); lista de la herpetofauna del Cerro El Potosí, Galeana (Contreras-Lozano et al. 2010); lista de la herpetofauna de San Antonio Peña Nevada, Zaragoza (Lazcano-Villarreal et al. 2004); lista de la herpetofauna del Cerro de la Silla y distribución con respecto a los tipos de vegetación y gradiente altitudinal (Lazcano-Villarreal et al. 2007); lista de la herpetofauna de un bosque de *Juniperus* en San Juan y Puentes, Aramberri (Lazcano-Villarreal, Contreras-Lozano, et al. 2009); reporte de las serpientes encontradas vivas o muertas en carreteras de Nuevo León (Lazcano-Villarreal, Salinas-Camarena y Contreras-Lozano 2009); notas de historia natural y reproducción en cautiverio en *T. exsul* (Lazcano-Villarreal et al. 2011); primeros registros para el estado de *Sceloporus cyanostictus* y *S. merriami australis* (Price et al. 2010; Price y Lazcano-Villarreal 2010); extensión de la distribución de *Lampropeltis mexicana* y *L. alterna* (Salmon et al. 2001; Salmon et al. 2004).

Otras notas que se han publicado sobre distribución e historia natural de la herpetofauna de Nuevo León son las de *Diadophis punctatus regalis*: registro distribucional (Cole 1965); *Lampropeltis mexicana thayeri*: ficha de historia natural (Mattison 1998); y *L. m. thayeri*: coloración (Osborne 1983).

Desde el año 2000 Robert W. Bryson ha estado trabajando sobre la ecología, sistemática e historia natural de la herpetofauna mexicana. Ha publicado varias contribuciones importantes en estos temas; parte de su trabajo ha sido realizado sobre los reptiles de Nuevo León, y ha incentivado a la gente de Nuevo León a hacer investigación sobre la ecología y sistemática de reptiles. Los registros distribucionales y notas de historia natural que ha publicado con la gente de la UANL fueron mencionados en anteriormente. Otras contribuciones importantes al conocimiento de la herpetofauna de Nuevo León son la filogenia de *Gerrhonotus parvus* (Conroy et al. 2005); y la filogenia de *Lampropeltis mexicana* en la cual reporta muestras de *L. alterna*

recolectadas en Cerro de la Silla y, el complejo *L. mexicana* con muestras de cerca de Iturbide y del norte de Doctor Arroyo (Bryson 2005; Bryson et al. 2007).

La investigación herpetológica hecha en el estado de Nuevo León ha documentado un total de 132 especies: 23 anfibios (3 salamandras, 20 anuros), y 109 reptiles (6 tortugas, 42 lagartijas, 61 serpientes). Dos de estas 132 especies han sido introducidas al estado: el Geco del Mediterráneo (*Hemidactylus turcicus*), y la Culebrilla Ciega de Brahminy (*Indotyphlops braminus*). Únicamente 2 de estas 132 especies son endémicas a Nuevo León, la Lagartija Pigmea (*G. parvus*) y la Hojarasquera de Nuevo León (*Rhadinaea montana*).

Características Fisiográficas y su Influencia sobre la Herpetofauna

El estado de Nuevo León se localiza en la parte noreste de México, entre los 98°26′ y 101°14′ de longitud oeste, y los 23°11′ y 27°49′ de latitud norte. Limita al norte con el estado norteamericano de Texas, y los estados mexicanos de Coahuila y Tamaulipas; al occidente limita con los estados de Coahuila, San Luís Potosí y Zacatecas, hacia el sur colinda con San Luis Potosí y Tamaulipas, con éste último también comparte la totalidad de su límite oriental. La frontera entre Nuevo León y Texas se extiende a lo largo de 15 km. La superficie estatal es de 64,220 km², las elevaciones van desde los 50 hasta los 3710 m sobre el nivel del mar. Está dividido en 51 entidades municipales. Su capital es la ciudad de Monterrey la cual forma una gran área metropolitana que se extiende sobre nueve municipios: Monterrey, San Pedro Garza García, Santa Catarina, García, Guadalupe, San Nicolás de los Garza, Apodaca, General Escobedo y Juárez. Ésta se ubica en el oeste central del estado y hospeda a aproximadamente el 88% de la población humana de Nuevo León, lo que coloca a ésta como la tercera ciudad más poblada de México con más de cuatro millones de habitantes.

La topografía del estado es compleja y muy variada, sin embargo se pueden diferenciar fácilmente tres grandes regiones en el estado. La primera de ellas abarca el centro, este, norte y noroeste del estado, con un promedio de altitud que varía de 50 a 250 m sobre el nivel del mar. Ésta es una región plana a excepción de una serie de lomeríos y cerros de poca altura que se encuentran esparcidos en ella. El este y norte de esta región forma parte de la Provincia Fisiográfica Gran Llanura de Norteamérica la cual comprende los municipios, de norte a sur, desde Anáhuac hasta China. Ésta es una región muy extensa dentro del estado, ocupando 23,138 km² de Nuevo León (= 36% de la superficie estatal). La topografía en esta extensa región es bastante homogénea, está representada por una gran sucesión de lomeríos y llanuras, que en raras ocasiones se ven interrumpidos por una sierra baja o un valle. La parte centro de esta región plana de Nuevo León forma parte de otra Provincia Fisiográfica llamada Planicie Costera del Golfo la cual, de norte a sur, comprende los municipios desde Pesquería hasta Linares. Ésta es una región plana constituida por lomeríos suaves con bajadas y llanuras de extensión considerable, a excepción por la Sierra de las Mitras al noroeste de Monterrey. Los suelos de esta región son profundos y de color oscuro, a diferencia de los que se presentan en la provincia de la Gran Llanura de Norteamérica en donde los suelos que predominan son de color claro.

La segunda región fácilmente identificable por su topografía es la Sierra Madre Oriental la cual en Nuevo León forma una franja ubicada principalmente en la porción oeste del estado, ocupando los municipios, de norte a sur, desde Lampazos de Naranjo hasta General Zaragoza.

En Nuevo León la Sierra Madre Oriental cruza el estado de sureste a noroeste y separa a las planicies de poca altura del centro, este y noroeste de una región alta y plana del extremo suroeste del estado (Altiplano Mexicano). La Sierra Madre Oriental presenta terrenos muy accidentados en forma de sierras que alcanzan un promedio de 2000 m de altura sobre el nivel del mar, en general la altitud de esta sierra disminuye en dirección sur a norte, aunque los picos más altos se ubican en el oeste central del estado. Las

máximas elevaciones son: Cerro El Morro en Sierra La Marta (3710 m de altitud) municipio de Galeana, en el oeste central del estado cerca de la división estatal con Coahuila; Cerro El Potosí (3700 m de altitud), a 80 km al sur de Monterrey, también en el municipio de Galeana; Cerro Peña Nevada (3660 m de altitud), municipio General Zaragoza, extremo sureste del estado cerca de la división estatal con Tamaulipas; Sierra Potrero de Abrego (3460 m de altitud), municipio Santiago, en el oeste central del estado cerca de la división estatal con Coahuila; y Cerro de la Ascensión (3200 m de altitud), municipio de Aramberri, en el sur del estado. Aunque la mayor parte de la Sierra Madre Oriental de Nuevo León forma un macizo montañoso continuo, el extremo norte de ésta, representado por Monterrey, se ve interrumpido por valles extensos que separan cadenas montañosas pequeñas como las sierras Las Gomas y Lampazos (1540 m de altitud) del noroeste del estado, y Sierra Picachos (1520 m de altitud) del centro-norte del estado. Estas sierras al norte de Monterrey se presentan en forma de un archipiélago siendo éstas verdaderas islas de montaña en medio de los valles áridos y semiáridos del noreste de México. Las montañas de este archipiélago no sobrepasan los 1550 m de altitud.

La tercera región es el Altiplano Mexicano de Nuevo León, éste se ubica en el extremo suroeste del estado, queda pegado a Coahuila, Zacatecas y San Luis Potosí. Abarca el extremo oeste de los municipios de Galeana y Aramberri, una porción pequeña del suroeste del municipio de General Zaragoza, y la totalidad de los municipios de Doctor Arroyo y Mier y Noriega. Tiene pocas elevaciones montañosas, y se localiza a una altura que varía entre 1500 y 2000 m. Sobre ella aparecen cerros y serranías irregulares, que se elevan entre los 200 y 500 m sobre el nivel del Altiplano pero en general es una zona plana y desértica que presenta una distribución irregular de la humedad. Ésta es ligeramente alta en su parte norte en el límite con la Sierra Madre Oriental, y es muy baja en la parte sur, donde la sierra forma una sombra de montaña, que induce a bajas precipitaciones (Alanís-Flores et al. 1996, Velazco-Macías et al. 2008; Velazco-Macías 2009, INEGI 2010).

Estas dos distintivas regiones topográficas del oeste-suroeste de Nuevo León, Sierra Madre Oriental y Altiplano Mexicano, forman parte de la Provincia Fisiográfica de la Sierra Madre Oriental.

La topografía que actualmente muestra Nuevo León inició su formación hace aproximadamente 145 millones de años durante el último periodo de la Era Mesozoica, el Cretácico Inferior (145–98 millones de años). Durante este periodo, en condiciones de aguas poco profundas, las capas, relativamente horizontales, fueron alteradas por enormes fuerzas geológicas que provocaron grandes rupturas y plegamientos así como movimientos ascensionales que elevaron el fondo marino por encima del nivel de la superficie del agua, se favoreció la compactación de sedimentos calcáreos, fenómeno que dio origen a calizas de gran espesor, ricas en fósiles. Estas formaciones son comunes en las montañas que rodean Monterrey. Posteriormente, durante el Cretácico Superior (hace 98–66 millones de años) el mar se hizo más profundo y se depositaron arcillas; el resultado consistió en rocas lutitas, de naturaleza suave, quebradiza, frágiles y erosionables, muy comunes en las estribaciones de la Sierra Madre Oriental. Hacia fines del Cretácico Superior (hace 66 millones de años) se generaron movimientos que corrugaron las capas de rocas que hoy configuran la Sierra Madre Oriental. Este proceso orogénico se extendió hacia el inicio de la Era Cenozoica, en el periodo Paleógeno (hace 65–34 millones de años), durante el cual se dio una gran actividad volcánica que originó rocas ígneas intrusivas, las cuales se localizan sólo en unas cuantas áreas del estado. Nuevo León emergió totalmente a finales del periodo Neógeneo de la Era Cenozoica, hace aproximadamente 5.3 millones de años. Gran parte de sus áreas bajas están cubiertas por rocas de conglomerado del Neógeneo y aluviones del Cuaternario. Estos eventos geológicos explican el hecho de que gran parte de las rocas superficiales del estado sean de origen sedimentario (Alanís-Flores et al. 1996).

La mayor parte de Nuevo León (68.6% de la superficie estatal) presenta climas extremosos, calientes y secos de tipo semiárido. A lo largo de casi todo el año el clima es muy caliente, especialmente en las extensas planicies de baja altitud. La precipitación es baja y por lo general no sobrepasa los 500 mm anuales en la mayor parte del estado. Las lluvias se presentan en el verano o son escasas a lo largo de todo el año. La humedad que recibe el estado llega a través de los vientos del Golfo de México y muestra un promedio del 62%. Sin embargo, existen variaciones importantes a este tipo de condición climática. En el extremo noroeste del estado, en los municipios de Mina y García que colindan con el extremo este de Coahuila, se presenta un clima muy árido. Estos municipios tienen una topografía plana. Debido a la proximidad de estos municipios con el Desierto Chihuahuense, temperaturas extremadamente altas hasta de 47°C se llegan a observar, la precipitación en la mayor parte de ellos es menor a 300 mm anuales y puede ser menor a 200 mm. Estos dos municipios representan la zona más seca del estado. Igualmente en la región del Altiplano Mexicano, suroeste del estado, el clima es cálido y muy seco, la precipitación anual promedio puede llegar a ser menor a 200 mm, siendo aquí donde se presenta la menor precipitación pluvial. Este tipo de clima se presenta en el 5% de la superficie estatal.

Otra variante importante son los climas templados subhúmedos del sur y oeste del estado, presentes en la Sierra Madre Oriental. El clima de esta región contrasta fuertemente con el de las extensas planicies del resto del estado. En la Sierra Madre Oriental la temperatura oscila alrededor de 18°C-20°C y el promedio de precipitación anual varía de 600 a 900 mm. En las partes más altas de la sierra, a altitudes superiores a 3000 m sobre el nivel del mar, se presentan climas de tipo alpino y subalpino, limitados a Peña Nevada y Cerro El Potosí. Igualmente los fenómenos de heladas, granizadas y nevadas se limitan a localidades de la Sierra Madre Oriental. El clima templado se presenta en 7.31% de la superficie estatal.

En la Sierra Madre Oriental se presentan las

mayores precipitaciones pluviales. Esta región montañosa desempeña un papel importante en la regulación de los procesos meteorológicos y la dinámica hidrológica, ya que actúa como cabecera de las cuencas, dando origen a varios ríos (Alanís-Flores et al. 1996).

Finalmente, la Planicie Costera del Golfo en la parte centro del estado, presenta climas de tipo tropical, semicálidos y subhúmedos, con lluvias en verano. Este tipo de climas se presenta en 19.1% de la superficie estatal. Aquí se presenta una precipitación pluvial intermedia y se localizan las principales corrientes permanentes del estado (Alanís-Flores et al. 1996).

El territorio de Nuevo León es atravesado por varios ríos, incluyendo un número de corrientes temporales que se presentan únicamente durante la época de lluvias. En términos generales se puede decir que el flujo de los ríos de Nuevo León es errático e impreciso, causado por los cambiantes factores climáticos y la abrupta topografía. La mayoría de los ríos tienen pequeñas cuencas de captación, y no todos llevan agua durante todo el año.

Todas las corrientes superficiales y estacionales nacen de los escurrimientos de las partes altas de la Sierra Madre Oriental del oeste del estado y siguen rutas, en muchas ocasiones sinuosas, a través de los cañones y valles de esta sierra dirigiéndose hacia las partes bajas y planas del estado, en su gran mayoría éstas descargan sus afluentes en los Ríos Bravo (= Río Grande), San Fernando y Soto La Marina, todos ellos en el estado de Tamaulipas. Muchas de estas corrientes estacionales se secan rápidamente o se pierden en los suelos porosos, por lo que el flujo de sus aguas dura tan sólo unas pocas horas.

Los principales ríos que corren a través de Nuevo León son el Bravo, Salado, Sabinas, San Juan, Pesquería. El Río Bravo únicamente corre a través de los 15 km de frontera de Nuevo León, en el municipio de Anáhuac. El Río Salado extiende su cuenca en la parte norte del estado, nace en la Sierra de Santa Rosa del extremo noroeste de Coahuila, internándose a Nuevo León a través del extremo noroeste del municipio de Lampazos y dirigiéndose hacia el este a través de los

municipios de Anáhuac y Vallecillos e internándose en el estado de Tamaulipas para finalmente desembocar en el Río Bravo en la Presa Falcón del noroeste de Tamaulipas. Este río es alimentado por numerosos tributarios a través de su paso por el norte de Nuevo León, la mayoría de ellos pequeños arroyos estacionales como El Camarón y El Galameses, ambos del municipio de Anáhuac. Poco antes de desembocar en la Presa Falcón recibe las aguas de su afluente más importante, el Río Sabinas. Este último río también corre a través del norte de Nuevo León, se origina en el este de Coahuila entrando a Nuevo León a través del extremo oeste del municipio de Mina y cruzando los municipios de Bustamante, Villaldama, Sabinas Hidalgo, Vallecillo y Parás, para finalmente internarse en el extremo noroeste de Tamaulipas y unirse en esta región al Río Salado.

El río de mayor longitud, y quizás el más importante del estado, es el Río San Juan, éste se origina en el extremo este de Coahuila y fluye a través de Nuevo León por los municipios de Santa Catarina, Monterrey, China, General Bravo, Dr. Coss y Los Aldama, internándose en el norte de Tamaulipas donde alimenta a la Presa Marte para finalmente desembocar en el Río Bravo. En el este de Nuevo León alimenta la Presa El Cuchillo en el municipio de China, la cual proporciona agua a la creciente ciudad de Monterrey. Algunos de los principales tributarios del Río San Juan son los ríos Pilón y Pesquería.

En el sur del estado la mayoría de las corrientes son de tipo intermitente, todas ellas se dirigen hacia el este para internarse a Tamaulipas. Las más importantes son las del Río Blanco en el municipio de Aramberri, éste se interna en Tamaulipas para finalmente desembocar en la Presa Vicente Guerrero; Río El Potosí que atraviesa los municipios de Galeana y Linares hasta alcanzar la división estatal con Tamaulipas en donde se vuelve una corriente intermitente y cambia su nombre a Río Conchos, para finalmente desembocar en el Río San Fernando (Tamaulipas) que desagua en el Golfo de México.

La interacción entre organismos y ambiente a través del tiempo produce un cambio gradual que resulta en la distribución actual de las formas vivientes y explica su diversidad biológica. Es necesario conocer la complejidad de características resumidas en los párrafos anteriores para poder entender la diversidad biológica del estado. La fisiografía, topografía, cantidad de humedad, temperatura y tipo de suelo entre otras características, condicionan la presencia de un determinado tipo de vegetación, los cuales a su vez crean paisajes y ambientes que condicionan la presencia de otros organismos. La dependencia de los anfibios por cuerpos de agua y/o lugares húmedos limita su distribución en el tiempo y el espacio a los meses de lluvias y hábitats riparios o lugares que retienen humedad. Aun así, poblaciones importantes de este grupo de organismos son exitosas en hábitats desérticos. Igualmente, los reptiles se ven limitados por su dependencia de la temperatura externa para regular su temperatura corporal. A diferencia de los anfibios, su huevo con cascara les proporciona independencia de cuerpos de agua para su reproducción, lo que junto con otras adaptaciones los hace una de las clases de vertebrados más exitosa en el norte de México.

Nuevo León hospeda seis tipos de vegetación (Matorral Desértico Chihuahuense; Matorral Tamaulipeco; Matorral Submontano; Bosque de Montaña; Zacatal; y Vegetación Acuática, Subacuática y Riparia) y 11 comunidades vegetales, que corresponden a tres provincias florísticas: Altiplano Mexicano, Planicie Costera del Noreste y Sierra Madre Oriental (Rzedowski 1978).

Matorral Desértico Chihuahuense

Es característico de los hábitats xéricos del estado que reciben de 300 a 400 mm de precipitación anual promedio, a altitudes superiores a 1400 m sobre el nivel del mar; comprende dos comunidades vegetales (Matorral Rosetófilo y Matorral Micrófilo). Su distribución irregular obedece principalmente al patrón de humedad disponible en el suelo. Este factor es otra condicionante que influye en la localización y desarrollo de sus especies. Los suelos de textura arenosa a limoarenosa y arcillosa, que tienen la capacidad de re-

tener humedad, sustentan comunidades densas y cerradas, mientras que los suelos pedregosos que retienen poca agua presentan comunidades vegetales pobres y muy abiertas.

Matorral Rosetófilo. Esta comunidad se presenta sobre flancos montañosos y taludes de varias elevaciones, en los cuales existen afloramientos rocosos o suelos esqueléticos de litosoles. Se localizan principalmente en el extremo suroeste del estado, en la parte correspondiente al Altiplano Mexicano en los municipios de Aramberri, Dr. Arroyo, Galeana, Mier y Noriega y Zaragoza. Con una presencia igualmente importante en el extremo noroeste del estado en los municipios de Abasolo, Bustamante, Carmen, Escobedo, García, Lampazos, Marín, Mina, Salinas Victoria, Villaldama. Y en manchones aislados en la porción centro del estado en los municipios de Cerralvo, Dr. González e Higueras. Los elementos más conspicuos presentan hojas suculentas agrupadas en rosetas, algunas con espinas terminales o mucrones.

Las especies de plantas más comunes son *Agave bracteosa*, *A. lechuguilla*, *A. striata*, *Berberis trifoliolata*, *Dasylirion texanum*, *Hechtia glomerata*, *Euphorbia antisyphilitica*, *Echinocereus enneacanthus*, *Echinocactus platyacanthus*, *Ferocactus pilosus*, *Opuntia leptocaulis*, *O. rnicrodasys* y *Opuntia* spp.

Matorral Micrófilo. Este tipo de vegetación se caracteriza par la dominancia de especies tipo arbustivo, con hojas o foliolos pequeños y a menudo olorosos. Se encuentran abundantes cactáceas de tallos esféricos o planos; asimismo plantas tipo Palma China o Palma Ixtlera se localizan abundantemente en terrenos planos o en abanicos aluviales de lomas o cerros en el Altiplano Mexicano de la porción suroeste del estado, en los municipios de Dr. Arroyo, Galeana, Mina y Mier y Noriega; en el extremo noroeste del estado en los municipios de Bustamante, García y Santa Catarina; y en el norte del estado en el municipio de Anáhuac.

Las especies características son *Acacia berlandieri*, *A. rigidula*, *Castela texana*, *Celtis pallida*, *Chilopsis linearis*, *Echinocactus platyacanthus*, *Ephedra aspera*, *Flourensia cernua*, *Fouquieria* splendens, *Larrea tridentata*, *Mortonia greggii*, *Opuntia imbricata*, *O. leptocaulis*, *O. rastrera*, *O. rnicrodasys*, *Partheniurn argentatum*, *P. incanum*, *Prosopis glandulosa*, *Sporobolus tiroides*, *Viguiera stenoloba*, *Yucca filifera* y *Y. carnerosana*.

Cincuenta y nueve especies de anfibios y reptiles habitan el Matorral Desértico Chihuahuense de Nuevo León. Muchas de estas especies son típicas del Desierto Chihuahuense y han logrado extender su distribución hacia el este hasta la Planicie Costera del Golfo donde el Matorral Tamaulipeco se desarrolla, pasando a través de las planicies del extremo noroeste de Nuevo León y los valles intermontanos que se forman entre el archipiélago de montañas de esta parte del estado. Otras más se limitan a las porciones áridas-semiáridas del noroeste de Nuevo León en los municipios de Lampazos y Mina o al extremo suroeste en los municipios de Dr. Arroyo, Galeana, Mier y Noriega y Zaragoza. Y unas cuantas tienen una distribución muy amplia pudiéndose encontrar en prácticamente todo el estado. Únicamente 4 de estas 59 especies están limitadas al Matorral Desértico Chihuahuense de Nuevo León, éstas son *Anaxyrus cognatus*, *Sceloporus cyanostictus*, *S. merriami* y *Bogertophis subocularis*. Las otras 55 especies que se han registrado en este y otros tipos de vegetación en el estado son 8 anfibios: *Pseudoeurycea galeanae*, *Anaxyrus debilis*, *A. punctatus*, *A. speciosus*, *Eleutherodactylus longipes*, *Gastrophryne olivacea*, *Scaphiopus couchii*, *Spea multiplicata*; 21 lagartijas: *Crotaphytus collaris*, *Coleonyx brevis*, *Cophosaurus texanus*, *Holbrookia approximans*, *H. lacerata*, *Phrynosoma cornutum*, *P. modestum*, *Sceloporus cautus*, *S. consobrinus*, *S. couchii*, *S. cyanogenys*, *S. minor*, *S. ornatus*, *S. parvus*, *S. poinsettii*, *S. spinosus*. *Uta stansburiana*, *Plestiodon tetragrammus*, *Aspidoscelis marmorata*, *A. gularis*, *A. inornata*; y 26 serpientes: *Arizona elegans*, *Gyalopion canum*, *Heterodon kennerlyi*, *Hypsiglena jani*, *Lampropeltis alterna*, *L. getula*, *L. triangulum*, *Masticophis flagellum*, *M. taeniatus*, *Pantherophis bairdi*, *P. emoryi*, *Pituophis catenifer*, *Rhinocheilus lecontei*, *Salvadora grahamiae*, *Sonora semiannulata*, *Tantilla atriceps*, *T. hobartsmithi*, *T. nigriceps*, *T. wilcoxi*,

Thamnophis cyrtopsis, *T. marcianus*, *Trimorphodon tau*, *Micrurus tener*, *Crotalus atrox*, *C. scutulatus* y *C. molossus*.

Matorral Tamaulipeco

Éste es el tipo de vegetación característico de la provincia florística de la Planicie Costera del Noreste, el matorral espinoso y los mezquitales de esta región presentan variantes fisonómicas, las especies pueden ser altas espinosas o medianas subinermes. Las altitudes varían desde 240 hasta 500 m sobre el nivel del mar, en localidades con una precipitación anual promedio de 700–800 mm. En condiciones de suelo y humedad favorables, los tallos poseen fustes bien definidos y se presentan formas arbóreas de más de 6 metros de altura, entre las que destacan por abundancia y cobertura *Prosopis laevigata*, *Pithecellobium ebano*, *Acacia rigidula*, *Castela texana*, *Celtis pallida*, *Cercidium macrum*, *Randia laetevirens*, *Cordia boissieri*, *Leucophyllum frutescens*, *Porlieria angustifolia*, *Opuntia leptocaulis*, *O. engelmannii*, *Zanthoxylum fagara* y *Bumelia celastrina*, destacando *Yucca filifera* con hasta de 10 metros de altura. En el estrato herbáceo destaca *Lupinus texensis* (Alanís-Flores et al. 1996; Velazco-Macías 2009).

Este tipo de vegetación se localiza principalmente en el norte, noreste, este, sureste y centro del estado, y en las zonas planas que interrumpen y separan el macizo montañoso de la Sierra Madre Oriental de las sierras de La Ropa, Bustamante y Picacho del noroeste y centro del estado. Éstas últimas incluyen el área metropolitana de Monterrey y todos los pueblos pequeños que la rodean.

Sesenta y tres especies de anfibios y reptiles habitan el Matorral Tamaulipeco de Nuevo León, únicamente 3 de ellas están limitadas a este tipo de vegetación: el Sapo de Madriguera (*Rhinophrynus dorsalis*), la Tortuga de Desiérto de Berlandier (*Gopherus berlandieri*), y la lagartija Reticulada de Collar (*Crotaphytus reticulatus*). Otra de estas 63 especies sólo ha sido registrada en construcciones humanas rodeadas por este tipo de vegetación en Nuevo León, sin embargo, es una especie introducida y tiene la capacidad de habitar prácticamente en cualquier lugar, principalmente dentro de construcciones humanas, esta especie es el Geco del Mediterráneo (*Hemidactylus turcicus*). Otras 13 especies habitan las partes áridas-semiáridas del estado ocupando el Matorral Tamaulipeco y el Matorral Desértico Chihuahuense (*Anaxyrus speciosus, Crotaphytus collaris, Coleonyx brevis, Cophosaurus texanus, Phrynosoma modestum, Sceloporus consobrinus, S. cyanogenys, Aspidoscelis marmorata, Heterodon kennerlyi, Lampropeltis alterna, Pantherophis bairdi, P. emoryi* y *Tantilla atriceps*). Tres más han sido registradas en el Bosque de Montaña y el Matorral Tamaulipeco (*Rhinella marina, Leptodeira septentrionalis* y *Rena dulcis*). Otra habita el Matorral Submontano y el Matorral Tamaulipeco (*Ficimia streckeri*), y otra más el Matorral Tamaulipeco y los pastizales (*Tantilla nigriceps*). Las otras 41 especies ocupan varios tipos de vegetación incluyendo el Matorral Tamaulipeco, éstas son 6 anuros (*Anaxyrus debilis, Anaxyrus punctatus, Incilius nebulifer, Gastrophryne olivacea, Scaphiopus couchii* y *Spea multiplicata*), 14 lagartijas (*Holbrookia approximans, H. lacerata, Phrynosoma cornutum, Sceloporus couchii, S. grammicus, S. olivaceus, S. poinsettii, S. serrifer, S. spinosus, S. variabilis, Uta stansburiana, Plestiodon tetragrammus, Aspidoscelis gularis* y *A. inornata*), y 21 serpientes (*Arizona elegans, Gyalopion canum, Hypsiglena jani, Masticophis flagellum, M. schotti, M. taeniatus, Pantherophis emoryi, Pituophis catenifer, Rhinocheilus lecontei, Salvadora grahamiae, Senticolis triaspis, Sonora semiannulata, Tantilla hobartsmithi, Thamnophis eques, T. marcianus, T. proximus, Micrurus tener, Rena myopica, Agkistrodon taylori, Crotalus atrox* y *C. scutulatus*).

Matorral Submontano

Es una formación arbustiva y subarbórea muy rica en formas de vida; el vigor, talla y distribución de las especies dominantes y codominantes dependen de la disponibilidad de agua en el suelo y subsuelo. Dependiendo de la orientación de las pendientes que ocupa así como de las condiciones del suelo se presenta en un amplio intervalo altitudinal desde los 500 hasta los 1200 m sobre el nivel del mar, en localidades con una precipitación anual promedio de 700 a 800 mm. Se ubican en los taludes inferiores de las montañas de hecho forman un amplio cinturón que separa el Matorral Desértico Chihuahuense del suroeste y noroeste y al Matorral Tamaulipeco del norte, este, sureste y centro del estado de los Bosques de Montaña de la Sierra Madre Oriental, incluyendo los de los macizos montañosos aislados del norte y noroeste del estado (sierras de Gomas, Lampazos y Picachos).

Aunque tiene variantes morfológicas y ecológicas, en términos generales en este matorral las especies vegetales más representativas son *Helietta parvifolia, Cordia boissieri, Sophora secundiflora, Gochnatia hypoleuca, Neopringlea integrifolia, Decatropis bicolor, Fraxinus greggii, Pithecellobium pallens, Leucophyllum frutescens, Acacia rigidula, Acacia berlandieri, Acacia farnesiana, Caesalpinia mexicana, Prosopis glandulosa, Dyospiros virginiana, Dyospiros texana* y *Cercidium macrum*. En algunas áreas con hábitats protegidos con abundante humedad y suelos profundos, se pueden encontrar agrupaciones pequeñas del encino *Quercus virginiana*. El matorral submontano ocupa localidades en la Sierra Madre Oriental y archipiélago de montañas del norte y noroeste del estado (Alanís-Flores et al. 1996; Velazco-Macías 2009).

Sesenta especies ocupan este tipo de vegetación, éstas son: nueve anfibios, una salamandra (*Pseudoeurycea galeanae*) y ocho anuros (*Anaxyrus punctatus, Incilius nebulifer, Craugastor augusti, Eleutherodactylus cystignathoides, E. guttilatus, E. longipes, Hypopachus variolosus* y *Spea multiplicata*); 19 lagartijas (*Barisia ciliaris, Gerrhonotus infernalis, G. parvus, Sceloporus cautus, S. couchii, S. grammicus, S. minor, S. oberon, S. olivaceus, S. ornatus, S. parvus, S. poinsettii, S. serrifer, S. torquatus, S. variabilis, Plestiodon dicei, P. tetragrammus, Aspidoscelis gularis* y *Lepidophyma sylvaticum*); y 32 serpientes (*Adelphicos quadrivirgatum, Amastridium sapperi, Coluber constrictor, Diadophis punctatus, Drymarchon melanurus, Ficimia streckeri, Gyalopion canum,*

Lampropeltis getula, L. mexicana, L. triangulum, Masticophis flagellum, M. schotti, M. taeniatus, Oxybelis aeneus, Pituophis catenifer, P. deppei, Rhadinaea montana, Senticolis triaspis, Storeria dekayi, S. hidalgoensis, Tantilla rubra, T. wilcoxi, Thamnophis cyrtopsis, T. eques, Trimorphodon tau, Micrurus tener, Rena myopica, Agkistrodon taylori, Crotalus lepidus, C. molossus, C. scutulatus y C. totonacus). Ninguna de estas especies está limitada a este tipo de vegetación. Las especies que aquí habitan se presentan en el Bosque de Montaña o en alguno de los otros matorrales del estado, Desértico Chihuahuense y/o Tamaulipeco.

Bosque de Montaña

Éste comprende la vegetación arbórea que se desarrolla por encima del matorral submontano, en las porciones montañosas de las provincias Sierra Madre Oriental y Altiplano Mexicano, crece en los cañones y partes altas de montañas, donde hay mayor humedad y el clima es templado semiseco a templado subhúmedo, se integra por las siguientes comunidades: bosque de encino, bosque de pino-encino, bosque de pino y bosque de coníferas. Adicionalmente, en el estado se presentan pequeños manchones de bosque mesófilo de montaña que presentan una distribución relictual y que desafortunadamente están a punto de desaparecer debido a diversos factores.

El bosque de encino forma parte de la comunidad forestal templada sobre el área de la Sierra Madre Oriental, se presenta a altitudes por encima del Matorral Submontano; desde áreas con suelos delgados y rocosos, hasta en sitios de suelos profundos y bien drenados; fisonómicamente se componen de árboles y arbustos que van de 12 a 15 metros de altura, dominando las especies del género *Quercus*; los encinares están muy ligados ecológica y florísticamente con los bosques mixtos y bosques de pinos. Las especies de encinos características son *Quercus canbyi, Q. intricada, Q. laeta, Q. laceyi, Q. polymorpha, Q. pungens, Q. rysophylla y Q. vaseyana*.

El bosque de pino—encino se integra por bosques de especies de hoja ancha y en agujas,

sobre áreas templadas a subhúmedas, se distribuye por encima del bosque de encino en las montañas de la Sierra Madre Oriental. A pesar de la amplia gama de condiciones climáticas en la región, la presencia de este tipo de bosque obedece más al factor térmico que a la humedad.

Este bosque se presenta entre una altitud de 550 a 900 m. Dominan principalmente los encinos entre los que destacan *Quercus rysophylla, Quercus laeta, Quercus polymorpha y Quercus canbyi*. Conviviendo con los encinos se encuentran los pinos *Pinus teocote* y *P. pseudostrobus*. Los árboles en general no presentan grandes tallas en diámetro o en altura, siendo éstas de 10 a 30 cm, y 10 a 14 m respectivamente.

El bosque de pino se puede presentar en lugares templados y húmedos, con el factor térmico decreciente a medida que se asciende en altitud. Se trata de una comunidad con árboles hasta 22 m de alto. Se localizan por encima de los bosques de pino-encino de la Sierra Madre Oriental; están asociados comúnmente con algunas especies de encinos, *Quercus* spp. y madroño, *Arbutus xalapensis*. Las especies de pinos características son *Pinus teocote, P. pseudostrobus, P. arizonica, P. ayacahuite y P. hartwegii*. Se pueden encontrar variantes de estos bosques como son los de pino piñonero (*P. cembroides*), que se localizan en el sur—suroeste del estado, en los municipios de San Pedro Garza García, Santiago, Montemorelos, Rayones y Galeana; asimismo en áreas muy restringidas en la parte norte de Nuevo León en la Sierra de Lampazos. Prosperan en áreas de baja precipitación entre los 2000 a 2600 msnm, sobre suelos someros y rocosos. Este tipo de bosque presenta espacios abiertos de árboles bajos de 4 a 8 m de altura, con copas redondeadas y troncos con diámetros de 30 a 40 cm a la altura del pecho. Las principales especies arbóreas del bosque de pino piñonero de Galeana son *Pinus cembroides, P. greggii*, táscates (*Juniperus flaccida y J. deppeana*) y madroño (*Arbutus xalapensis*). Se localiza también en áreas limítrofes del matorral desértico del altiplano y con frecuencia se asocian especies como la palma samandoca (*Yucca carnerosana*), palma china (*Yucca filifera*), magueyes (*Agave* spp.) y arbustos diversos.

Desde los 3000 metros de altitud hasta la cima del cerro El Potosí en Galeana, se localiza un tipo especial de vegetación en forma de matorral, ramificados desde la base del tallo, de menos de 2 m de altura, de pinos enanos (*Pinus culminicola*) que colindando con el prado alpino, forma una franja continua en el declive oriental y sur, presentándose además en dos manchones aislados al suroeste y oeste de la cima. Ésta se presenta como una comunidad densa y baja en la que el pino enano es dominante.

Bosque de Coníferas. Los bosques de *Pseudotsuga—Abies* son una variante de los bosques templados que se caracteriza por especies de gran atractivo escénico con árboles de formas típicas piramidales. Este tipo de vegetación está constituido por árboles altos de 15 a 25 m, que se localizan en cañones protegidos con climas fríos y húmedos, generalmente entre 2000 a 2500 m de altitud en el sur y suroeste del estado (municipios de Galeana, Iturbide, Aramberri y Zaragoza). Las especies que sobresalen dominando en el dosel arbóreo son *Pinus greggii* y *P. hartwegii*, junto con *Pseudotsuga menziesii*, *P. flahaulti*, *Abies vejarii* y *Cupressus arizonica*.

Hay otro tipo de bosque en áreas restringidas, formando manchones más o menos puros en algunas pequeñas áreas, de bosques de cedro (*Cupressus* spp.) También se encuentra en el sur y suroeste del estado (municipios de Iturbide, Galeana y Zaragoza). Los árboles no son corpulentos y alcanzan entre 10 y 15 m de altura, la especie característica es *C. arizonica*.

Otra comunidad es la de bosque de *Juniperus* spp., que se asocia a los bosques de pinos y encinos. Su forma de vida es arbórea o arbustiva según sean las condiciones ambientales. Se sabe que estas especies de árboles prosperan en lugares de suelos rocosos de calizas expuestas y baja humedad y su crecimiento es lento. Las especies más comunes son *Juniperus monosperma* y *J. flaccida*. Se encuentra desde la parte noroeste del estado (municipios de Santa Catarina, San Pedro Garza García y Monterrey), hasta la porción suroeste (municipios de Iturbide y Galeana) en la Sierra Madre Oriental.

El bosque mesófilo de montaña se localiza en áreas con suelos poco profundos o en áreas de suelos arcillosos con abundante materia orgánica. Se ubica entre 800 y 1400 m de altitud, con alta humedad relativa todo el año y se localiza en el municipio de Zaragoza en la región limítrofe con el estado de Tamaulipas. Florísticamente está caracterizado por liquidámbar (*Liquidambar styraciflua*), encinos (*Quercus sartorii* y *Q. germana*), pino rojo (*Pinus patula*), magnolia (*Magnolia schiedeana*) y duraznillo (*Cercis canadiensis*). Es un bosque mixto con abundantes enredaderas y epífitas, así como helechos y hongos en el sotobosque (Alanís-Flores et al. 1996; Velazco-Macías 2009).

Sesenta y ocho especies de anfibios y reptiles habitan los Bosques de Montaña de Nuevo León, sin embargo, únicamente 15 de estas 68 especies parecen estar limitadas a este tipo de vegetación. En Nuevo León éstas son especies típicas de la Sierra Madre Oriental. Una de las salamandras plethodontidas del estado está limitada a este tipo de vegetación: *Chiropterotriton priscus*. La otra salamandra de esta familia: *Pseudoeurycea galeanae*, así como dos eleutherodactylos: *Eleutherodactylus cystignathoides* y *E. longipes*, tienen una fuerte afinidad por este tipo de vegetación pero se pueden encontrar en condiciones semiáridas como las reportadas por Taylor (1940) en la descripción de *Syrrhophus smithi* (= *Eleutherodactylus longipes*) a 24 km al oeste de Galeana.

Estas cuatro especies de anfibios tienen la facultad de poder reproducirse fuera de cuerpos de agua. Sus huevos los depositan en lugares húmedos, ya sea en la tierra o en las hojas de árboles, y el desarrollo de la larva se lleva a cabo dentro del huevo, tal que al eclosionar los individuos están totalmente metamorfoseados. Otra especie de anfibio, el Calate Arborícola, *Ecnomiohyla miotympanum*, está limitado a los Bosque de Montaña de Nuevo León. Las otras 10 especies limitadas a este tipo de vegetación son 3 lagartijas (*Phrynosoma orbiculare*, *Sceloporus chaneyi* y *Scincella slivicola*) y 7 serpientes (*Drymobius margaritiferus*, *Leptophis mexicanus*, *Opheodrys aestivus*, *Thamnophis exsul*, *T. pulchrilatus*, *Tropidodipsas sartorii* y *Crotalus pricei*). Estas serpientes, especialmente las del género *Tham-*

nophis, están asociadas a cuerpos de agua dentro de los Bosques de Montaña.

Otras 18 especies que habitan en los Bosques de Montaña de Nuevo León se pueden encontrar también en el Matorral Submontano, éstas son 3 anfibios el Sapo Ladrador (*Craugastor augusti*) que al igual que los plethodontidos y eleuherodactylus no necesita de cuerpos de agua para reproducirse, la Ranita Chirriadora de Manchas (*Eleutherodactylus guttilatus*), y la Rana Manglera (*Hypopachus variolosus*); 3 lagartijas (*Sceloporus torquatus*, *Plestiodon dicei* y *Lepidophyma sylvaticum*); y 12 serpientes (*Adelphicos newmanorum*, *Amastridium sapperi*, *Diadophis punctatus*, *Drymarchon melanurus*, *Lampropeltis mexicana*, *Pituophis deppei*, *Rhadinaea montana*, *Storeria dekayi*, *S. hidalgoensis*, *Tantilla rubra*, *C. lepidus* y *Crotalus totonacus*).

Tres especies más, todas ellas lagartijas de la familia Anguidae, ocupan Bosques de Montaña, Matorral Submontano y Pastizales, estás son *Barisia ciliaris*, *Gerrhonotus infernalis* y *G. parvus*. Ésta última es la única especie endémica a Nuevo León, habita en los bosques de la Sierra Madre Oriental del municipio de Galeana, en el extremo suroeste del estado.

Las otras 32 especies que habitan los Bosques de Montaña del estado, se pueden encontrar en otros tipos de vegetación, éstas incluyen 5 anfibios anuros (*Anaxyrus punctatus*, *Incilius nebulifer*, *Rhinella marina*, *Smilisca baudinii* y *Gastrophryne olivacea*); 9 lagartijas (*Sceloporus grammicus*, *S. minor*, *S. oberon*, *S. olivaceus*, *S. parvus*, *S. poinsettii*, *S. spinosus*, *S. variabilis* y *Plestiodon tetragrammus*); y 18 serpientes (*Gyalopion canum*, *Lampropeltis getula*, *L. triangulum*, *Leptodeira septentrionalis*, *Oxybelis aeneus*, *Pituophis catenifer*, *Salvadora grahamiae*, *Senticolis triaspis*, *Tantilla wilcoxi*, *Thamnophis cyrtopsis*, *T. eques*, *T. proximus*, *Trimorphodon tau*, *Micrurus tener*, *Rena myopica*, *Agkistrodon taylori*, *Crotalus molossus* y *C. scutulatus*).

Pastizales

Esta comunidad se caracteriza porque sobresalen herbáceas graminoides con hojas delgadas y alargadas, aunque pueden combinarse en algunas otras con especies de las familias Asteraceae, Fabaceae y Chenopodiaceae. Los pastizales clímax naturales ocupan áreas reducidas en espacios abiertos dentro de los matorrales desérticos así como en situaciones edáficas específicas en lugares con mal drenaje, inundables o con excesivas sales o existencia de yeso (Alanís-Flores et al. 1996; Velazco-Macías, 2009).

Treinta y seis especies de anfibios y reptiles se pueden encontrar en los pastizales de Nuevo León. Los que mayor afinidad tienen por este tipo de vegetación son 4 anfibios (*Gastrophryne olivacea*, *Scaphiopus couchii*, *Spea bombifrons* y *S. multiplicata*), 11 lagartijas (*Barisia ciliaris*, *Gerrhonotus infernalis*, *G. parvus*, *Holbrookia approximans*, *H. lacerata*, *Sceloporus goldmani*, *S. samcolemani*, *Plestiodon obsoletus*, *Aspidoscelis inornata*, *A. gularis*, *A. marmorata*) y una serpiente de cascabel (*Sistrurus catenatus*). Las otras 20 especies que ocupan este tipo de vegetación son 2 anfibios (*Anaxyrus debilis* y *Leptodactylus fragilis*), 3 lagartijas (*Phrynosoma cornutum*, *Uta stansburiana* y *Plestiodon tetragrammus*), y 15 serpientes (*Arizona elegans*, *Hypsiglena jani*, *Masticophis flagellum*, *Pituophis catenifer*, *Rhinocheilus lecontei*, *Salvadora grahamiae*, *Sonora semiannulata*, *Tantilla hobartsmithi*, *T. nigriceps*, *T. wilcoxi*, *Thamnophis marcianus*, *T. proximus*, *Trimorphodon tau*, *Crotalus atrox* y *C. scutulatus*).

Vegetación Riparia, Subacuática y Acuática

Esta comunidad se encuentra estrechamente ligada a la formación orográfica de la Sierra Madre Oriental como una gran cuenca de captación con sus afluentes de ríos y arroyos, agrupa tanto a la vegetación arbórea como a la arbustiva que se encuentra en las riberas de las corrientes de agua en los municipios de Linares, Montemorelos, Allende, General Terán, Cadereyta Jiménez, Santiago, Monterrey, Guadalupe, Sabinas, Cerralvo, Lampazos y Bustamante. Estos bosques se componen por *Adiantum capillus-veneris*, *Arundo donax*, *Lobelia cardinales*, *Platanus occidentales*, *Populus tremuloides*, *Salix nigra*, *Taxodium mucronatum* y *Ulmus crassifolia* encontrándose abundantes especies herbáceas acuáticas o semiacuáticas, enredaderas y especies epifitas como *Tillandsia usneoides*.

Estos bosques crecen principalmente en las orillas de los ríos, así como en riberas planas y amplias con drenaje superficial. Dichos sitios, humedecidos por escurrimientos perennes o esporádicos, durante largo tiempo, permiten que las especies adquieran dimensiones en altura y diámetro mucho mayor que las que se ubican en zonas secas (Alanís-Flores et al. 1996; Velazco-Macías 2009).

Treinta y cuatro especies de anfibios y reptiles están fuertemente asociadas a la vegetación Riparia, Subacuática y Acuática de Nuevo León. Éstas son 17 anfibios: el Ajolote Tigre Rayado (*Ambystoma mavortium*) que únicamente ha sido registrada en aguajes del extremo suroeste del estado, ésta es la única especie que en Nuevo León es estrictamente acuática, aunque individuos adultos totalmente metamorfoseados pueden salir del agua y desplazarse entre aguajes a distancias considerables, en el estado sólo se han observado individuos neoténicos (adultos que conservan las características larvales como presencia de branquias y que están limitados a cuerpos de agua); y 16 anuros (*Anaxyrus cognatus, A. debilis, A. punctatus, A. speciosus, Incilius nebulifer, Rhinella marina, Ecnomiohyla miotympanum, Smilisca baudinii, Leptodactylus fragilis, Gastrophryne olivacea, Hypopachus variolosus, Lithobates berlandieri, Rhinophrynus dorsalis, Scaphiopus couchii Spea bombifrons y S. multiplicata*). Aunque todas estas especies tienen gran afinidad por cuerpos de agua, todas ellas pueden habitar lugares donde la humedad es alta, y frecuentemente se les llega a ver cruzando caminos o desplazándose entre cuerpos de agua.

Otras cinco especies son tortugas que habitan este tipo de vegetación, éstas son *Pseudemys gorzugi, Trachemys scripta, Kinosternon flavescens, K. integrum y Apalone spinifera*. De éstas únicamente la Tortuga de Concha Blanda (*Apalone spinifera*), es estrictamente acuática y está limitada a este tipo de vegetación. Otras 4 son lagartijas, 3 de ellas de hábitos trepadores y arborícolas, viven en los árboles de los lados de ríos o

presas (*Sceloporus couchii, S. cyanogenys y S. serrifer*). Y *Scincella silvicola*, es una lagartija que vive debajo de troncos y rocas cerca de cuerpos de agua. Las 8 especies restantes son serpientes de la familia colubridae que rara vez se alejan de cuerpos de agua, éstas son *Nerodia erythrogaster, N. rhombifer, Thamnophis cyrtopsis, T. eques, T. exsul, T. marcianus, T. proximus y T. pulchrilatus*.

En la actualidad se sabe que en Nuevo León habitan 132 especies de herpetozoarios: 23 anfibios (3 salamandras y 20 anuros), y 109 reptiles (6 tortugas, 42 lagartijas y 61 serpientes). Estas especies representan 28 familias: 11 de anfibios (2 de salamandras y 9 de anuros), y 17 de reptiles (4 de tortugas, 8 de lagartijas y 5 de serpientes). Las 132 especies también representan 73 géneros; 17 de anfibios (3 de salamandras y 14 de anuros), y 56 de reptiles (5 de tortugas, 14 de lagartijas y 37 de serpientes).

Las proporciones de los diferentes grupos de herpetozoarios en Nuevo León son un reflejo de los ambientes disponibles en el estado. Los anfibios están fuertemente limitados a sitios húmedos y hábitats acuáticos, los cuales son escasos en la mayor parte de Nuevo León, éstos se encuentran principalmente en los bosques templados de la Sierra Madre Oriental y montañas al norte y noreste de la ciudad de Monterrey. Igualmente la mayor riqueza de anfibios se presentan en estas regiones, y en la mayoría de los casos su actividad en todo el estado está limitada a la época de lluvias. Aun especies de anfibios que no necesitan de cuerpos de agua para su reproducción, como las dos especies de salamandras plethodontidas (*Chiropterotriton priscus y Pseudoeurycea galeanae*), y las ranas de las familias Craugastoridae (*Craugastor augusti*), y Eleutherodactylidae (*E. cystignathoides, E. guttilatus y E. longipes*), dependen de lugares húmedos para su sobrevivencia. La mayoría de las otras especies de anfibios que habitan en el resto del estado están limitados a la época de lluvias, su actividad la inician con las primeras lluvias de primavera - verano, periodo en el que salen de sus refugios invernales para alimentarse y reproducirse, al término de las lluvias pueden permanecer activos durante un periodo de tiempo corto para posteriormente regresar a sus refugios invernales. En las regiones áridas-semiáridas del estado se entierran hasta alcanzar una profundidad en donde las condiciones de húmeda son adecuadas, para salir a la superficie sólo ocasionalmente cuando alguna lluvia ligera o torrencial humedece la superficie del suelo. En cuerpos de agua permanente, se pueden encontrar especies de anfibios prácticamente durante todo el año. Un ejemplo de esto es el Ajolote Tigre Rayado (*Ambystoma mavortium*) que puede estar activo durante todo el año en aguajes permanentes del suroeste del estado. Estas fuertes limitantes ambientales resultan en las proporciones tan bajas de anfibios (salamandras 2% [3/132] y anuros 15% [20/132]) en relación al total de especies de herpetozoarios del estado.

Por otro lado, los reptiles con su huevo amniota, su piel coriácea y sus diferentes modo de reproducción (ovíparo y vivíparo) son mucho más diversos en el estado y ocupan prácticamente todos ambientes disponibles. Las 6 especies de tortugas representan sólo el 5% (6/132) de total de la herpetofauna estatal y el 19% (6/32) de las especies de tortugas reportadas por Liner y Casas-Andreu (2008) para México. Dos de éstas son tortugas de río (*Pseudemys gorzugi* y *Trachemys scripta*), otras dos de aguajes y charcas (*Kinosternon flavescens* y *K. hirtipes*), otra es una Tortuga de Desierto (*Gopherus berlandieri*) y otra más ocupa cuerpos de agua permanente (*Apalone spinifera*).

Las 42 especies de lagartijas que habitan Nuevo León representan el 32% del número total de herpetozoarios del estado. Además de ser un grupo con una riqueza de especies alta en el estado la abundancia de muchas de sus poblaciones es igualmente alta. Los hábitos escansoriales, saxícolas y heliotérmicos de la mayoría de las lagartijas las hacen individuos muy conspicuos fáciles de observar, lo cual contribuye a un mayor registro de las especies y poblaciones de este grupo de reptiles. A este grupo pertenece una de las dos especie endémica al estado, el Lagarto Pigmeno (*Gerrhonotus parvus*), y una de las

dos especie introducidas a Nuevo León, el Geco del Mediterráneo (*Hemidactylus turcicus*).

El grupo de herpetozoarios mejor representado en Nuevo León es el de las serpientes, con 61 especies que representan el 46% de total estatal. A diferencia de las lagartijas, los hábitos de las serpientes son por lo general secretivos y/o nocturnos, por lo que no son tan conspicuos como los lacertilios. Debido a esto se tiene la idea de que las poblaciones de estos organismos no son tan abundantes como las de las lagartijas, sin embargo, muestreos dirigidos a los hábitats, microhábitats y horas adecuadas del día o la noche pueden revelar poblaciones igualmente abundantes de serpientes.

Noventa y dos (70%) de las 132 especies que habitan en Nuevo León se comparten con los Estados Unidos; 19 de estas especies compartidas se encuentran principalmente en las tierras bajas y pendientes del Atlántico de la Sierra Madre Oriental, 14 de ellas entran a los Estados Unidos únicamente en Texas: *Anaxyrus speciosus, Eleutherodactylus cystignathoides, E. guttilatus, Pseudemys gorzugi, Gopherus berlandieri, G. infernalis, Crotaphytus reticulatus, Holbrookia lacerata, Sceloporus cyanogenys, S. olivaceus, Ficimia streckeri, Storeria dekayi, Pantherophis bairdi* y *Rena dulcis*. Cinco más tienen una distribución oriental en México y los Estados Unidos: *Incilius nebulifer, Nerodia erythrogaster, N. rhombifer, Opheodrys aestivus* y *Micrurus tener*. Una más de las especies compartidas, *Sceloporus grammicus*, también entra a los Estados Unidos solamente en Texas, pero se distribuye a todo lo largo de la Sierra Madre Oriental y el Eje Transvolcánico del centro de México. Otras 51 de las especies compartidas son típicas de los Desierto de Norteamérica, la mayoría de ellas tienen una distribución amplia en el sur de los Estados Unidos y norte de México, éstas son *Ambystoma mavortium, Anaxyrus cognatus, A. debilis, A. punctatus, Gastrophryne olivacea, Scaphiopus couchii, Spea bombifrons, S. multiplicata, Trachemys scripta, Kinosternon flavescens, Apalone spinifera, Crotaphytus collaris, Coleonyx brevis, Cophosaurus texanus, Phrynosoma cornutum, P. modestum, Sceloporus consobrinus, S. merriami, S. poinsettii, Uta stan-*

sburiana, Plestiodon obsoletus, P. tetragrammus, Aspidoscelis gularis, A. inornata, A. marmorata, Arizona elegans, Bogertophis subocularis, Gyalopion canum, Heterodon kennerlyi, Hypsiglena jani, Lampropeltis alterna, L. getula, Masticophis flagellum, M. schotti, M. taeniatus, Pantherophis emoryi, Pituophis catenifer, Rhinocheilus lecontei, Salvadora grahamiae, Sonora semiannulata, Tantilla atriceps, T. hobartsmithi, T. nigriceps, T. wilcoxi, Thamnophis cyrtopsis, T. eques, Crotalus atrox, C. lepidus, C. molossus, C. scutulatus y *Sistrurus catenatus*.

Tres de las especies compartidas tienen una distribución amplia en los bosques templados de Norteamérica, dos de ellas (*Craugastor augusti* y *Crotalus price*), desde el sur de los Estados Unidos extendiéndose hacia el sur a través de la Sierra Madre Occidental y Sierra Madre Oriental, una más (*Diadophis punctatus*), desde el sur de Canadá hasta el centro de México.

Dieciséis de las restantes 18 especies compartidas tienen una distribución amplia desde el sur de los Estados Unidos (*Rhinella marina, Smilisca baudinii, Leptodactylus fragilis, Hypopachus variolosus, Lithobates berlandieri, Rhinophrynus dorsalis, Sceloporus variabilis, Drymarchon melanurus, Drymobius margaritiferus, Leptodeira septentrionalis, Oxybelis aeneus, Senticolis triaspis* y *Thamnophis marcianus*) o desde el sur de Canadá (*Coluber constrictor* y *Lampropeltis triangulum*) hasta Centro o Sudamérica. Las dos últimas especies compartidas son especies introducidas, el Geco del Mediterráneo (*Hemidactylus turcicus*) originario del sur de Europa, y la Culebrilla Ciega (*Indotyphlops braminus*) originaria de la India.

Cuarenta (30%) de las 132 especies que habitan en Nuevo León no se comparten entre México y los Estados Unidos; 35 de estas especies son endémicas a México; 2 de ellas, *Gerrhonotus parvus* y *Rhadinaea montana*, al Bosque de Montaña de Nuevo León. Dieciocho más a las partes altas de la Sierra Madre Oriental (*Chiropterotriton priscus, Pseudoeurycea galeanae, Eleutherodactylus longipes, Ecnomiohyla miotympanum, Sceloporus chaneyi, S. minor, S. oberon, S. parvus, Plestiodon dicei, Scincella slivicola, Lepidophyma sylvaticum, Pituophis deppei, Rhadinaea montana, Storeria*

hidalgoensis, Rena myopica, Thamnophis exsul, Agkistrodon taylori y *Crotalus totonacus*). Cuatro de estas especies tienen una distribución limitada al este-sureste de Coahuila y partes adyacentes de Nuevo León (*Chiropterotriton priscus, Pseudoeurycea galeanae, Sceloporus oberon* y *Thamnophis exsul*); una está limitadas a Nuevo León y partes adyacentes de Tamaulipas (*S. chaneyi*); una está limitada a Coahuila, Nuevo León y Tamaulipas (*Plestiodon dicei*); otras 11 se distribuyen desde Nuevo León y Tamaulipas extendiéndose hasta el sur de Veracruz y norte de Oaxaca, principalmente sobre la vertiente del Atlántico de la Sierra Madre Oriental (*Eleutherodactylus longipes, Ecnomiohyla miotympanum, S. minor, S. parvus, Scincella silvicola, L. sylvaticum, Adelphicos newmanorum, Storeria hidalgoensis, Rena myopica, A. taylori* y *Crotalus totonacus*); y una más, *Pituophis deppei*, habita en la Sierra Madre Oriental, el Altiplano Mexicano, el Eje Transvolcánico, y la Sierra Madre Occidental.

Dos de las restantes 15 especies endémicas a México están limitadas a Coahuila y el extremo oeste de Nuevo León (*Sceloporus cyanosticus* y *S. ornatus*). Dos más están limitadas a regiones aisladas en el norte de México: *S. couchi* a las sierras del norte de Coahuila y centro-oeste de Nuevo León; y *S. goldmani* a un área pequeña en el sureste de Coahuila, parte adyacente de Nuevo León y noreste de San Luis Potosí. Otras tres especies endémicas a México están limitadas al Altiplano Mexicano (*S. cautus, S. samcolemani* y *Lampropeltis mexicana*). Cuatro más (*Kinosternon integrum, S. spinosus, S. torquatus* y *Trimorphodon tau*) están distribuidas ampliamente desde el centro de México a través del Altiplano Mexicano, y en algunos casos (*T. tau*) sobre ambas costas. Una más tiene poblaciones aisladas en las partes altas de México (*Thamnophis pulchrilatus*), otra más está limitada al Desierto Chihuahuense de México (*Holbrookia approximans*). Las otras 2 especies endémicas a México (*Barisia ciliaris* y *Phrynosoma orbiculare*), habitan en la Sierra Madre Occidental y Sierra Madre Oriental, una de ellas, *P. orbiculare*, se distribuye desde las montañas del Eje Transvolcánico de la parte central del país.

Cinco de las especies que habitan en Nuevo León y que no se comparten entre México y los Estados Unidos (*Sceloporus serrifer, Amastridium sapperi, Leptophis mexicanus, Tantilla rubra* y *Tropidodipsas sartorii*) son especies de afinida-des tropicales-subtropicales que extienden su distribución hacia el norte desde Centroamérica a través de las montañas y tierras bajas del este de México. Todas ellas presentan su distribución más septentrional en Nuevo León o Tamaulipas.

WILLIAM FARR

Introducción

Nunca se ha publicado una revisión completa y exhaustiva de la herpetología de Tamaulipas. En los 254 años que han pasado desde Linnaeus (1758), sólo tres listas de la herpetofauna de Tamaulipas han aparecido, Velasco (1892), Smith y Taylor (1966) y Lavin-Murcio, Hinojosa-Falcón, et al. (2005). Con excepción de una serie de notas breves, hay sólo un pequeño número de artículos en los que Tamaulipas y su herpetofauna son los principales temas: Gaige (1937); Martin (1955a, 1958); Baker y Webb (1966); Brown y Brown (1967); Greene (1970); Smith et al. (1976, 2003); Lavin-Murcio, Hinojosa-Falcón, et al. (2005); Lavin-Murcio, Sampablo Angel, et al. (2005); Farr et al. (2007); Farr et al. (2009); y Lazcano et al. (2009). La mayoría de lo que se conoce acerca de los anfibios y reptiles de Tamaulipas proviene de publicaciones que abordan la taxonomía y ecología de especies cuya distribución incidentalmente incluye los límites geográficos-políticos del estado pero en los que Tamauliipas no es el objeto de estos estudios. Actualmente 179 especies de anfibios y reptiles pueden confirmarse en Tamaulipas. Sin duda, este número irá creciendo con trabajos de campo en el futuro, para especies que se sabe habitan cerca de los límites estatales de los estados vecinos (Smith y Dixon 1987), y ocasionalmente descripciones de nuevas especies para el estado continuarán apareciendo (Lavin-Murcio y Dixon 2004; Bryson y Graham 2010). Situado en el umbral del Trópico de Cáncer, Tamaulipas tiene una rica y compleja variedad de hábitats, desde islas y manglares en la costa del golfo, hast bosques de pino-encino que se desarrollan a más de 3200 m de altitud en la Sierra Madre Oriental, desde el Matorral Desértico Chihuahuense del Altiplano Mexicano, hasta el Bosque Mesófilo de Montaña de la Reserva de la Biosfera El Cielo, y aún queda mucho por descubrir acerca de la herpetofauna del estado.

Estudios Herpetológicos Previos

La siguiente es una crónica del trabajo de campo realizado desde Linnaeus así como de la literatura relativa a la herpetología de Tamaulipas. El conocimiento precolombino así como el de los primeros colonizadores españoles sobre la herpetofauna de Tamaulipas no es mencionado aquí. En esta sección hay un sesgo evidente ya que la mayoría de las actividades de los recolectores está basada en mucho en datos de museos norteamericanos. No se intentó adquirir datos de colecciones europeas y los esfuerzos para adquirir datos de colecciones mexicanas en su mayoría fueron infructuosos. Carl Linnaeus (1758, 1766) nombró 18 (10.0%) de las especies actualmente válidas que ahora se sabe que habitan en Tamaulipas. Éstas son en su mayoría especies de distribución amplia y sólo la descripción de *Phrynosoma orbiculare* tiene la serie tipo en México. Después de Linnaeus, el período de 1761 a 1840 vio la descripción de 29 especies nuevas cuya distribución incluye Tamaulipas, estas descripciones fueron hechas por 17 autores europeos y norteamericanos, pero al igual que Linnaeus, ninguna de estas descripciones incluyó material de Tamaulipas (ver Cuadro 8.1). Las primeras investigaciones sobre la herpetofauna de Tamaulipas fueron realizadas por Jean Louis Berlandier, un naturalista francés que llegó a México en 1826 para participar en un censo de la frontera del norte de México entre 1828 y 1829 (Geiser 1948; Berlandier 1980). Una vez concluido este estudio Berlandier se quedó a vivir en Matamoros, Tamaulipas, donde continuó recolectando, investigando y escribiendo prolíficamente. Entre sus numerosos manuscritos se encuentra uno sin terminar de 219 páginas sobre la herpetofauna de la región (Chiszar et al. 2003; Smith et al. 2003; Smith y Chiszar 2003). Smith y colaboradores concluyeron que Berlandier podría haber sido el autor de 17 (9.49 %) de los taxa actualmente válidos para el estado si sus manuscritos hubieran sido publicados antes de su muerte. Sin embargo, trágicamente se ahogó en el Río San Fernando en 1851, habiendo publicado únicamente unos cuantos artículos sobre botánica. Después de la Guerra México-Estados Unidos (1846–48), el gobierno de los Estados Unidos organizó la United States-Mexican Boundary Survey (1848–55) durante la cual Arthur C. V. Schott recolectó *Sceloporus* al noroeste de Camargo en 1852 (Baird 1859b). Simultáneamente, pero independientemente de la Boundary Survey, el Teniente Darius Nash Couch recolectó especímenes en México. Couch llegó en enero de 1853 y pasó un mes recolectando en los alrededores de Brownsville, Texas y Matamoros, Tamaulipas. Estando en Matamoros, Couch localizó a la viuda de Berlandier y le compró la colección y manuscritos del naturalista. Posteriormente, Couch se dirigió al oeste recolectando

en el norte de Tamaulipas en febrero y marzo y luego en Coahuila (Conant 1968). Entre los especímenes recolectados por Couch y Berlandier en Tamaulipas se encuentran los sintipos de *Anaxyrus* (= *Bufo*) *debilis* (Girard 1854), *Gopherus berlandieri* (Agassiz 1857), y *Leptodeira septentrionalis* (Kennicott 1859), el lectotipo y paralectotipos de *Plestiondon tetragrammus* (Baird 1859a), el holotipo de *Lampropeltis triangulum annulata* (Kennicott 1860), y los paratipos de *Siren intermedia texana* (Goin 1957).

La mitad de las especies actualmente válidas que actualmente se sabe que habitan Tamaulipas (91 especies, 50.83%) fueron descritas entre 1850 y 1885. A diferencia de los ejemplos anteriores, un número de estos taxa nuevos se basaron en muestras de Tamaulipas. Spencer Fullerton Baird, su colega francés Charles F. Girard y Robert Kennicott del United States National Museum (USNM) describieron colectivamente 36 especies (20.1%) en diez años (1852–1861). Aproximadamente en estos mismos años en la Academy of Natural Sciences of Philadelphia (ANSP), Edward Hallowell describió 4 especies nuevas que ahora se sabe habitan en el estado. El sucesor de Hallowell en la ANSP, Edward Drinker Cope describió 24 especies (13.4% del total para el estado) entre 1861 y 1885 incluyendo *Notophthalmus meridionalis* (Cope 1880) basándose en parte sobre especímenes de Tamaulipas. Samuel Garman del Museum of Comparative Zoology (MCZ), describió cinco especies incluyendo *Rena myopica* (Garman 1884 [1883]) para Tamaulipas. En el Museum National d´ Histoire Naturelle (MNHN) en Paris, Francia, Auguste Henri André Duméril, su padre André Marie Constant Duméril, y Gabriel Bibron describieron colectivamente 9 especies entre 1841 y 1854, basándose sobre material del sur de México y Guatemala. Otros 10 zoólogos en Europa, Estados Unidos y México describieron 16 especies nuevas que ahora se sabe habitan en Tamaulipas entre 1849 y 1885. Velasco (1892) publicó el primer listado de la herpetofauna de Tamaulipas el cual incluyó 82 especies. Varias especies que aparecieron en su listado fuero erróneamente reportadas (ejem. *Basiliscus vittatus, Phyllodactylus tuberculosus, Loxocemus bicolor*). Sin embargo, a pesar de sus deficiencias fue el inventario significativamente más completo de la diversidad del estado en comparación con otras revisiones de la herpetofauna mexicana de esa época (ejem. Duméril et al. 1870–1909; Garman 1884; Günther 1885–1902; Cope 1887; Dugès 1896).

Después de 1885 y los progresos realizados a mediados del siglo diecinueve, hubo relativamente poca actividad hasta 1932, casi medio siglo después. Sólo cinco de las especies que actualmente se sabe habitan en el estado fueron descritas en este período. El último cuarto del siglo diecinueve produjo algunas colecciones pequeñas de campo realizadas en las cercanías de Tampico por Edward Palmer en 1879, J. M. Priour en 1888 y J. O. Snyder en 1899. William Lloyd Allison, del US Bureau of Biological Survey (USNM) recolectó principalmente a lo largo de la frontera mexicana con Estados Unidos en 1891. Edward Nelson y Edward Goldman (USNM) recolectaron vertebrados en cada estado de México (1892–1906) y específicamente en Tamaulipas en 1898 y 1901–1902 (Kellogg 1932; Goldman 1951). Fueron los primeros en recolectar en la Sierra Madre Oriental de Tamaulipas y obtuvieron los paratipos de *Masticophis schotti ruthveni* (Ortenburger 1923) y *Crotalus pricei miquihuanus* (Gloyd 1940). En las tres primeras décadas del siglo veinte el ictiólogo Seth E. Meek, del Field Museum of Natural History (FMNH), hizo una pequeña recolecta en 1903. W. W. Brown (MCZ) recolectó en la Sierra Madre Oriental en tres viajes anuales a Tamaulipas (1922–1924) incluyendo los holotipos de *Lampropeltis mexicana thayeri* (Loveridge 1924), *Craugastor batrachylus* (Taylor 1940), y *Phrynosoma orbiculare orientale* (Horowitz 1955). En 1930 la University of Michigan organizó una expedición a la Sierra de San Carlos (Gaige 1937), donde Lee Raymond Dice recolectó casi 250 especímenes incluyendo un scíncido nuevo, *Plestiodon dicei* (Ruthven y Gaige 1933). E. Kallert de Hamburgo, Alemania recolectó en las cercanías de Tampico en 1930 incluyendo los tipos de dos subespecies, *Notophthalmus meridionalis kallerti*

(Wolterstorff 1930) y *Micrurus tener maculatus* (Rose 1967).

La cuarta década del siglo veinte vio el comienzo de una nueva era en herpetología mexicana cuando Edward H. Taylor inició más de 20 años (1932–1953) de investigaciones. Publicó extensamente sobre México incluyendo descripciones de ocho especies nuevas que ahora se sabe habitan en Tamaulipas. Taylor realizó trabajo de campo en Tamaulipas de 1932 a 1941, incluyendo un extenso viaje a México con su estudiante Hobart M. Smith y recolectó el holotipo de *Agkistrodon taylori* (Burger y Robinson 1951) y paratipos de *Ameiva undulata podarga* (Smith y Laufe 1946) en el estado. Hobart M. Smith recibió su doctorado bajo la dirección de Taylor en 1936 y recolectó con frecuencia en Tamaulipas de 1932 a 1939, a veces con David Dunkle. Describió seis nuevas especies y dos subespecies que ahora se sabe habitan en el estado. Los libros de *Herpetology of Mexico* (Smith y Taylor 1945, 1948, 1950a, 1966) establecieron una base moderna sobre la herpetología mexicana. Smith continuó publicando prodigiosamente sobre México después del fin del siglo veinte, y sus contribuciones adicionales sobre Tamaulipas se citan en las siguientes líneas. Aparte de las especies descritas por Smith y Taylor, únicamente dos especies nuevas fueron descritas en el periodo de 1930–1949, ambas a partir de especímenes recolectados en el sur de Tamaulipas. Varias notas sobre colecciones de campo fueron reportadas por Burt (1932), Dunkle (1935) y Dunkle y Smith (1937), y dos subespecies, *Nerodia rhombifera blanchardi* (Clay 1938) e *Hypsiglena torquata dunklei* (Taylor 1938) fueron descritas para Tamaulipas. Archie F. Carr hizo trabajo de campo cada diciembre desde 1939 hasta 1941 y otras colectas incidentales de campo entre 1930 y 1945 fueron realizadas por Walter Mosauer en el verano de 1935, Ernest G. Marsh en 1938, Myron y Evelyn Gordon en 1939.

Con el ímpetu así como con una sólida base establecida por Hobart Smith y Edward Taylor, una nueva generación de herpetólogos comenzó a explorar México en números sin precedentes después de la Segunda Guerra Mundial. Charles F. Walker del University of Michigan Museum of Zoology (UMMZ) y sus estudiantes estuvieron particularmente activos en Tamaulipas desde 1945 hasta 1960 haciendo al UMMZ la mayor colección de anfibios y reptiles de Tamaulipas con cerca de 3600 especímenes. Tal vez la contribución más importante a la Herpetología de Tamaulipas sea la obra de Paul S. Martin. Martin pasó más de 365 días en el campo (1948–1953) y registró 39 especies para el estado por primera vez (Martin 1955a). Su área de estudio principal fue la Sierra de Guatemala (= El Cielo), que alberga el bosque mesófilo de montaña más septentrional en el hemisferio occidental, para el cual proporcionó el estudio de la biogeografía de la herpetofauna (Martin 1958). Otras publicaciones que abordan cuestiones más amplias sobre biogeografía e historia del Pleistoceno pero, en las que elementos de la herpetofauna de Tamaulipas se discuten son Martin et al. (1954), Martin (1955b), Martin y Harrell (1957). Más tarde, Martin (2005) relató su experiencia de campo en Tamaulipas. Charles Walker recolectó en la Sierra de Guatemala en 1950 y 1953, y describió tres especies nuevas para el estado la culebra endémica *Thamnophis mendax* (Walker 1955a), *Pseudoeurycea scandens* (Walker 1955b), y *Lepidophyma micropholis* (Walker 1955c). George B. Rabb recolectó en Tamaulipas en 1951 y 1952 y describió *Chiropterotriton cracens* (Rabb 1958). James A. Peters escribió sobre una población de Tamaulipas de *Laemanctus serratus* (Peters 1948) y recolectó en este estado en 1949. Otros investigadores del UMMZ activos de 1945 a 1950 incluyen a Rezneat M. Darnell, William Heed, James Mosimann, Priscilla Starrett y Hellmuth Wagner.

Varios herpetólogos de Texas, Estados Unidos estuvieron activos en Tamaulipas en las décadas posteriores a la Segunda Guerra Mundial. Bryce C. Brown recolectó en México, incluyendo Tamaulipas, la mayoría de los años de 1941 a 1978 (Brown y Smith, 1942; Brown y Brown, 1967; Auth, Smith, Brown y Lintz 2000). William B. Davis hizo recolectas incidentales de herpetofauna en Tamaulipas (Davis 1953a; Davis 1953b). James R. Dixon, frecuentemente con la ayuda de colegas y estudiantes, recolectó en Tamaulipas a intervalos irregulares desde 1949 hasta 2006, haciendo la Texas Cooperative Wildlife Collection (TCWC) la tercera mayor colección de Tamaulipas (cerca de 1750 especímenes) y publicó muchas revisiones sistemáticas de taxa que habitan en la región (ejem. Dixon 1959, 1962, 1969; Dixon y Thomas 1974; Dixon y Dean 1986; Dixon et al. 2000; Dixon y Vaughan 2003). La TCWC llevó a cabo estudios de campo sobre la Sierra San Carlos desde 1975 a 1977 (Hendricks y Landry 1976) y Jack W. Sites estudio la Sierra de Tamaulipas en 1979 (Sites y Dixon 1981, 1982). John E. Werler hizo recolectas incidentales viajando a través de Tamaulipas (Werler y Darling 1950; Werler 1951; Werler y Smith 1952; Smith et al. 1952) y Robert G. Webb reportó una recolecta realizada en el norte de México (Webb 1960) y Tamaulipas (Baker y Webb 1966). John S. Mecham recolectó salamandras en varios viajes a campo llevados a cabo de 1965 a 1969 (Mecham 1968a, 1968b; Mecham y Mitchell 1983). Allan Chaney de Texas A & I University en Kingsville recolectó en Tamaulipas entre 1971 y 1981, a menudo en compañía de Ernest Liner.

Muchos otros herpetólogos estuvieron activos en México después de la Segunda Guerra Mundial (1945–1975) e hicieron contribuciones importantes al conocimiento de Tamaulipas. La University of Illinois Mexican Expedition fue llevada a cabo por Phillip Smith, Leslie Burger, Robert Reese, Frederick Shannon y Lester Firschein en 1949 (Shannon y Smith 1950; Reese y Firschein 1950; Smith y Burger 1950). Burger y Robertson (1951) describieron *Agkistrodon taylori* de especímenes recolectadós previamente en Tamaulipas. Ralph Axtell recolectó en Tamaulipas a principios de 1949 (Axtell y Wasserman 1953; Axtell 1958b), y describió dos subespecies nuevas que ahora se sabe habitan en el estado (Axtell 1960; Axtell y Webb 1995). Roger Conant recolectó en el estado entre 1949 y 1965 (Conant 1965, 1969). Ernest Liner trabajó en Tamaulipas en la mayoría de los años entre 1952 y 1963 y realizó recolectas esporádicas hasta 1988 (Liner 1964, 1966a, 1966b) además describió dos taxa, que ahora se sabe habitan en Tamaulipas (Liner 1960; Liner y Dixon 1992). Douglas Rossman realizó trabajo de campo en Tamaulipas en 1960 con M. J. Fouquette (Rossman y Fouquette 1963) y en 1965 con Edmond Keiser, Earl Olson y otros, además escribió sobre el género *Thamnophis* (Rossman 1969, 1992a, 1992b). Craig Nelson y Max Nickerson recolectaron en México en los veranos de 1963 y 1964 y reportaron registros distribucionales nuevos para Tamaulipas (Nelson y Nickerson 1966). John Lynch junto con M. J. Landy, recolectaron principalmente anuros en 1964 y Lynch publicó sobre las especies recolectadas (Lynch 1963, 1965) incluyendo la descripción original de dos taxa (Lynch 1967, 1970). Fred Thompson recolectó la serie tipo de *Xenosaurus platyceps* en Tamaulipas en 1965, más tarde la describió en una revisión del género (King y Thompson 1968) e hizo recolectas adicionales en 1990 y 1992. Morris Jackson registró una extensión hacia el norte para *Tropidodipsas fasciata* (Jackson 1971). Sherman y Madge Minton recolectaron en Tamaulipas ocho años entre 1960 y 1991. Selander et al. (1962) reportaron un estudio de los vertebrados en las islas de Tamaulipas, incluyendo notas sobre tres especies de herpetofauna. Greene (1970) examinó los modos de reproducción de las lagartijas y serpientes que habitan la Sierra de Guatemala basándose mucho sobre datos obtenidos por Martin (1958).

Las principales playas de anidación de *Lepidochelys kempii* fueron descubriertas en Tamaulipas en 1947 cerca de la localidad de Rancho Nuevo. Humberto Chávez, Martín Contreras y Eduardo Hernández encontraron la especie en grave declive en 1966 (Chávez et al. 1967, 1968) y se iniciaron los esfuerzos de recuperación, que en última instancia incluyeron la participación de los gobiernos de México y Estados Unidos y numerosas instituciones privadas. Casas-Andreu (1974) reportó los habitos de anidación de *L. kempii* y Plotkin (2007) revisó la biología y conservación de esta especie. Publicaciones adicionales relacionadas con *L. kempii* y otras

tortugas marinas en Tamaulipas son demasiado extensas para abordar aquí. Smith y Smith (1979) y Wilson y Zug (1991) revisaron gran parte de la literatura para Tamaulipas.

En el último cuarto del siglo veinte varios herpetólogos publicaron sobre distibución, ecología y taxonomía de la herpetofauna tras realizar trabajo de campo en Tamaulipas. Smith y Álvarez (1976) describieron *Sceloporus torquatus mikeprestoni*. John Iverson, Peter Meylan, Ron Magill y Diderot F. Gicca recolectaron de 1974 a 1983 e Iverson y Berry (1979) publicaron sobre la distribución ecológica de *Kinosternon*. Jonathan Campbell recolectó en Tamaulipas a intervalos irregulars entre 1970 y 2009 en compañía de varios colegas y estudiantes y publicó extensamente sobre la herpetofauna de Mesoamérica. Patrick M. Burchfield publicó sobre la ecología de *Agkistrodon taylori* y la conservación de *L. kempii* (Burchfield 1982, 2004). Robert L. Bezy (1984) recolectó principalmente *Lepidophyma* de 1967 a 1984, en varias ocasiones acompañado por Charles J. Cole y Carl S. Lieb. Recolectas más pequeñas se hicieron en la década de 1970 por Diderot Gicca en 1978 y James F. Lynch y Stephen D. Busack en 1979. Stephen Reilly recolectó junto con Ronald Brandon y Gail Johnston cerca de San José de las Rusias y revisó el género *Notophthalmus* (Reilly 1990). En 1986 J. F. Copp, D. E. Breedlove, R. R. Riviere y B. R. Anderson recolectaron en la Sierra Madre Oriental de Tamaulipas y reportaron la presencia de *Sceloporus chaneyi* en el estado (Watkins-Colwell et al. 1996). Earl Olson examinó varios taxa de *Sceloporus* (Olson 1987, 1990) y describió *Sceloporus torquatus madrensis* (Olson 1986). John J. Wiens recolectó *Sceloporus minor* e hizo una revisión sitemática del género *Sceloporus*, del cual 13 especies habitan en Tamaulipas (Wiens y Reeder 1997; Wiens y Penkrot 2002). *Aspidoscelis laredoensis* fue reportada para el estado de Tamaulipas y fue objeto de numerosas publicaciones (Walker 1986, 1987a, 1987b; Walker et al. 1986; Walker et al. 1990; Paulissen y Walker 1998). Flores-Villela y Brandon (1992) revisaron el estatus del género *Siren* en Tamaulipas y el sur de Texas. Registros distribucionales nuevos fueron reportados para *Trimorphodon tau* (Dundee y Liner 1997) y *Gastrophryne elegans* (Sampablo-Brito y Dixon 1998). Smith et al. (1998) describieron *Leptotyphlops* (= *Rena*) *myopica iversoni* y Auth, Smith, Brown, Lintz y Chiszar (2000) proporcionaron otras observaciones sobre esta subespecie. Seidel et al. (1999) revisó la distribución y taxonomía de *Trachemys* en Tamaulipas y publicó sobre la distribución de *Kinosternon flavescens* (Seidel 1976). John E. Joy realizó alguna recolectas entre 1975 y 1979.

En años recientes la historia natural de *Xenosaurus platyceps* fue objeto de una serie de artículos (Lemos-Espinal et al. 1997, 2003; Ballinger et al. 2000; Lemos-Espinal y Rojas-González 2000; Rojas González et al. 2008). Ramos-Jiménez et al. (1999) recolectaron *Aspidoscelis sexlineata* en el estado y Pérez-Ramos et al. (2010) revisaron su estatus en México. Terán-Juárez (2008) reportó la presencia de *Ophisaurus incomptus* en la Sierra Madre Oriental de Tamaulipas. Bailey et al. (2008) publicaron sobre el estatus de conservación *Pseudemys gorzugi*. Flores-Villela y Pérez-Mendoza (2006) revisaron la literatura sobre la herpetofauna de Tamaulipas, y Flores-Benabib y Flores-Villela (2008) reportaron por primera vez *Epictia goudotii* para el estado. Pablo A. Lavin-Murcio recibió su doctorado bajo la dirección de James Dixon (Lavin-Murcio 1998) y enseñó en el Instituto Tecnológico de Ciudad Victoria (ITCV) de 1999 a 2004 acumulando una gan colección de anfibios y reptiles del estado (Lavin-Murcio y Dixon 2004; Lavin-Murcio, Hinojosa-Falcón, et al. 2005; Lavin-Murcio, Sampablo Angel, et al. 2005). David Lazcano Villarreal de la Universida Autónoma de Nuevo León (UANL) hizo recolectas incidentales (1977–2010), adicionalmente hizo estudios a nivel estatal sobre los anfibios y reptiles (1996–1997), y en el 2009 adquirió la colección del ITCV haciendo a la UANL la segunda mayor colección de Tamaulipas con más de 2730 especímenes de este estado (Lazcano et al. 2009; Farr et al. 2013). William L. Farr realizó estudios de campo de 2003 a 2009 y publicó registros distribucionales incluyendo 14 registros nuevos para el estado así como otras notas (Farr et al. 2007;

Farr et al. 2009; Farr et al. 2010; Farr et al. 2011; Farr 2011; Farr and Lazcano 2011; Farr et al. 2013), y Bryson y Graham (2010) describieron *Gerrhonotus farri* con ejemplares de esta colección. Eli García-Padilla reportó notas sobre *Anelytropsis papillosus* (García-Padilla y Farr 2010) y *Pantherophis bairdi* (García-Padilla et al. 2011).

A través de la historia de la herpetología de Tamaulipas, 48 especies y subespecies de anfibios (4 salamandras, 6 anuros) y reptiles (1 cocodrilo, 3 tortugas, 16 lagartijas, 18 serpientes) tienen localidades tipo designada o restringida al estado de Tamaulipas, incluyendo 37 taxa actualmente válidos (Cuadro 8.1).

Características Fisiográficas y su Influencia sobre la Herpetofauna

Deriva continental, desplazamiento tectónico, erosión, cambios climáticos y glaciaciones son algunas de las fuerzas que actúan para transformar los ambientes y paisajes a través del tiempo geológico. Las especies se deben adaptar a estos cambios o enfrentar la extinción. Estos cambios son algunos de los factores claves producen evolución. Plantas y animales evolucionan juntos y están estrechamente vinculados entre sí y sus paisajes formando los ecosistemas. Comprender estos cambios, adaptaciones y ecosistema en su conjunto, proporciona una visión más profunda de cualquier especie o región de interés.

Situado en el noreste de México entre latitud 22.20° N–27. 67° N y longitud 97.14° O–100.11° O, Tamaulipas es el sexto mayor estado en superficie territorial de México (aproximadamente 80,175 km², 4.1%). Comparte una frontera internacional de 370 km con el estado de Texas, Estados Unidos al norte, 420 km de la costa con el Golfo de México hacia el este, limita con Nuevo León al oeste y San Luis Potosí y Veracruz hacia el sur. En 2010 la población era de 3,268,554 (88.0% urbana, 12.0% rural) con casi la mitad de los residentes en las grandes ciudades de la frontera norte, Reynosa, Matamoros y Nuevo Laredo. La ciudad más grande es Tampico, que incluyendo Altamira y Ciudad Madero

Autor(es)	Nombre original: Localidad tipo
Salamandras (4)	
Rabb (1958)	*Chiropterotriton chondrostega cracens* (= *C. cracens*): **Agua Linda, about 7 milesWNW of Gómez Farías, Tamaulipas**
Walker (1955b)	*Pseudoeurycea scandens*: **from the wall of a cave at "Rancho del Ciello," on the forested slopes of the Sierra Madre Oriental in southern Tamaulipas, about five miles NW of Gómez Farías: elevation about 3500 feet**
Cope (1880)	*Diemictylus miniatus meridionalis* (*Notophthalmus meridionalis*): **Matamoros, Mexico . . . tributaries of the Medina River and southward (restringida a Matamoros, Tamaulipas[1,2])**
Wolterstorff (1930)	*Diemictylus kallerti* (*Notophthalmus meridionalis kallerti*): **Angeblich Tampico-Mexiko. Wahrscheinlich aus dem gebirgigen Hinterland**
Anuros (6)	
Girard (1854)	*Bufo* (= *Anaxyrus*) *debilis*: **lower part of the valley of the Río Bravo (Río Grande), and in the province of Tamaulipas**
Taylor (1940)	*Eleutherodactylus* (= *Craugastor*) *batrachylus*: **Miquihuana, Tamaulipas, eighty miles southwest of Victoria, Tamaulipas**
Lynch (1967)	*Eleutherodactylus* (= *Craugastor*) *decoratus purpurus*: **near Rancho del Cielo, 5 km NW Gómez Farías, Tamaulipas, Mexico**
Lynch (1970)	*Syrrhophus* (= *Eleutherodactylus*) *dennisi*: **cave near El Pachón, 8 km. N Antiguo Morelos, Tamaulipas, Mexico, 250 m**
Baird (1854)	*Scaphiopus couchii*: **Coahuila and Tamaulipas (restringida a Matamoros, Tamaulipas[3])**
Cope (1863)	*Scaphiopus rectifrenis* (*S. couchii*): **Tamaulipas and Coahuila (restringida a Rio, Nazas, Coahuila[3])**
Cocodrilos (1)	
Bocourt (1869)	*Crocodylus mexicanus* (= *C. moreletti*): **Tampico, Tamaulipas**
Tortugas (3)	
Stejneger (1925)	*Kinosternon herrerai*: **Xochimilco, Valley of Mexico [in error] (restringida a La Laja, Veracruz [en error][4]); (restringida a Tampico, Tamaulipas[5])**
Gray (1849)	*Ciatudo mexicana* (= *Terrapene carolina mexicana*): **Mexico (restringida a vicinity of Tampico, Tamaulipas[6])**
Günther (1885)	*Emys cataspila* (= *Trachemys venusta cataspila*): **Mexico (restringida a Alvarado, Veracruz [en error][3]); (restringida a Tampico, Tamaulipas[7])**
Lagartijas (16)	
Bryson & Graham (2010)	*Gerrhonotus farri*: **north of Magdaleno Cedillo, 27 km SW of Tula, Municipio Tula, Tamaulipas, Mexico**
Shaw (1802)	*Lacerta* (= *Ctenosaura*) *acanthura*: **Not given (restringida a Tampico, Tamaulipas, Mexico[8,9])**
Harlan (1825)	*Cyclura teres* (= *Ctenosaura acanthura*): **Tampico, Tamaulipas**
Baird (1859a [1858])	*Holbrookia approximans*: **Lower Río Grande [en error]; Tamaulipas[10] [en error], (restringida a near Álamo de Parras or Parras de la Fuente, 25°25′ N, 102°11′ W, Coahuila, Mexico,[11] *fide*[12])**
Horowitz (1955)	*Phrynosoma orbiculare orientale*: **Miquihuana, Tamaulipas, Mexico**
Sites & Dixon (1981)	*Sceloporus grammicus tamaulipensis*: **4.3 km by road S of Hacienda Acuña (45.3 km by road N of Gonzales) in the Sierra de Tamaulipas mountains in southern Tamaulipas, Mexico**
Martin (1952)	*Sceloporus serrifer cariniceps* (= *S. s. plioporus*): **5 mi NE of Gómez Farías at Rancho Pano Ayuctle along the Río Sabinas**
Olson (1986)	*Sceloporus torquatus madrensis*: **1740 m above Rancho del Cielo, 7 km. NW Gómez Farías, Tamaulipas, Mexico**

Autor(es)	Nombre original: Localidad tipo
Smith & Álvarez (1976)	*Sceloporus torquatus mikeprestoni*: **Marcela, Tamaulipas, Mexico**
Ruthven & Gaige (1933)	*Eumeces dicei* (= *Plestiodon dicei*): **24°38′N–99°01′W, Marmolejo . . . Sierra San Carlos, Tamaulipas, Mexico**
Cope (1900)	*Eumeces tetragrammus funebrosus* (= *Plestiodon tetragrammus tetragrammus*): **Matamoros, Mexico**
Baird (1859a [1858])	*Plestiodon tetragrammus*: **Lower Río Grande (restringida a Matamoros, Tamaulipas, Mexico[13])**
Smith & Laufe (1946)	*Ameiva undulata podarga*: **Seven miles west of Ciudad Victoria, Tamaulipas**
Walker (1955c)	*Lepidophyma micropholis*: **cave at El Pachón, about 5 miles. NNE of Antigua Morelos, Tamaulipas, elevation 600–700 feet**
Walker (1955c)	*Lepidophyma flavimaculatum tenebrarum* (= *L. sylvaticum*): **± 4.5 miles NW (by road) of Gómez Farías, in the Sierra Madre Oriental at "Rancho del Cielo," ± 3600 feet**
King & Thompson (1968)	*Xenosaurus platyceps*: **Tamaulipas, 15.4 mi SSW Ciudad Victoria, on the road to Jaumave, 4500 feet elevation**

Serpientes (18)

Kennicott (1859)	*Taeniophis* (= *Coniophanes*) *imperialis*: **Brownsville, Texas [USNM catálogo y etiqueta, Matamoros, Tamaulipas, fide[14,15]]**
Jan (1863)	*Glaphyrophis lateralis* (= *Coniophanes imperialis* en parte): **Not given (restringida a Tampico[16])**
Kennicott (1859)	*Dipsas* (= *Leptodeira*) *septentrionalis*: **Matamoros, Tamaulipas and Brownsville, Tex. (restringida a Brownsville, Texas[3])**
Taylor (1939 [1938])	*Hypsiglena torquata dunklei* (= *H. jani dunklei*): **Hacienda La Clementina, near Forlón, Tamaulipas**
Stejneger (1893)	*Hypsiglena ochrorhynchus texana* (= *H. jani texana*): **between Laredo & Camargo, Texas (restringida a Mier, Tamaulipas[3])**
Kennicott (1860)	*Lampropeltis annulata* (= *L. triangulum annulata*): **Matamoros, Tamaulipas, Mexico**
Loveridge (1924)	*Lampropeltis thayeri* (= *L. mexicana thayeri*): **Miquihuana, Tamaulipas, Mexico**
Clay (1938)	*Natrix* (= *Nerodia*) *rhombifera blanchardi*: **Tamaulipas, Mexico, within a radius of 85 miles of Tampico in the triangle formed by the Río Tamesí and the Río Pánuco**
Smith (1943)	*Pliocercus elapoides celatus* (= *P. bicolor bicolor*): **Ciudad Victoria, Tamaulipas [en error[17]]**
Walker (1955a)	*Thamnophis mendax*: **near La Joya de Salas, Tamaulipas, ± 6000 ft.**
Cope (1895)	*Zamenis conirostris* (= *Coluber constrictor oaxaca*): **Matamoros, Mex.**
Rose (1967)	*Micrurus fulvius maculatus* (= *M. tener maculatus*): **Tampico, Tamaulipas, Mexico**
Brown & Smith (1942)	*Micrurus fitzingeri microgalbineus* (= *M. tener microgalbineus*): **seven km south of Antiguo Morelos, Tamaulipas, Mexico**
Lavin-Murcio & Dixon (2004)	*Micrurus tamaulipensis*: **Sierra de Tamaulipas, Rancho La Sauceda, ca. 50 km N González, Tamaulipas, Mexico**
Garman (1883)	*Stenostoma myopicum* (= *Rena myopica*): **Savineto, near Tampico, Tamaulipas, Mexico**
Smith et al. (1998)	*Leptotyphlops dulcis iversoni* (= *Rena myopica iversoni*): **oak woodland in the vicinity of Hwy 101, 23.7 km SW Río San Marcos (in turn 8.1 km SW Cd. Victoria), Tamaulipas, Mexico, 36.5 km NE of Jaumave**
Burger & Robertson (1951)	*Agkistrodon bilineatus taylori* (= *A. taylori*): **KM 833, 21 km N of Villagran, Tamaulipas**
Baird & Girard (1853)	*Crotalophorus Edwardsii* (= *Sistrurus catenatus edwardsii*): **Tamaulipas . . . S. bank of Río Grande . . . Sonora**

Restricción reportada en[1] Stejneger & Barbour (1933); [2] Smith & Taylor (1948); [3] Smith & Taylor (1950b); [4] Smith & Taylor (1950a); [5] Smith & Brandon (1968); [6] Müller (1936); [7] Smith & Smith (1979); [8] J. W. Bailey (1928); [10] Baird (1859a [1858]); [11] Axtell (1958a); [12] Degenhardt et al. (1996); [13] Taylor (1935); [14] J. R. Bailey (1939); [15] Cochran (1961); [16] J. R. Bailey (1937); [17] Martin (1958).
Restricción no aceptada por [9] De Queiroz (1995).

tiene una población de más 825,000 habitantes. Los principales usos que se le da a la tierra incluyen agricultura extensiva a lo largo de la llanura costera (sorgo, cítricos, caña de azúcar, agave), la ganadería se practica a nivel estatal, la tala comercial se practica en la Sierra Madre Oriental, con la urbanización y la industrialización (transportación maritima, petróleo, manufactura) en la zona de Tampico y en la frontera norte.

La siguiente información sobre herpetofauna está basa en revisión de literatura, aproximadamente 16,000 registros de museo (cerca de 11,000 examinados por el autor), y más de 2700 registros personales de campo recolectados entre 2001 y 2009. La historia geológica y características de Tamaulipas se tomaron de Ferrusquía-Villafranca (1993) y otros investigadores que se citan a continuación. La topografía y la hidrología fueron descritas usando mapas topográficos del Instituto Nacional de Estadística Geografía Informática (INEGI), Google Earth, literatura y registros recopilados con una unidad GPS de mano. La distribución de *Juniperus* y *Pinus* en Tamaulipas se tomaron de Adams (2011) y Perry (1991), respectivamente.

Geológicamente, norteamerica es una de las áreas más complejas del mundo y muchos aspectos de la historia geológica de la Sierra Madre Oriental en particular están mal entendidos. Los basamentos del Precámbrico y Paleozoico de la región fueron continentales (probablemente cratónicos) y posterior saturados por mares. Deposiciones marinas continuaron durante el Cretácico Superior con evidencias que sugieren actividad tectónica e inestabilidad en el Jurásico Medio y Superior. Los mares cedieron en el este al comienzo del Cretácico Superior aparentemente asociados con la elevación del territorio continental de México (altiplano). En el Terciario Inferior se produjeron plegamientos y fallas formando las cordilleras de la Sierra Madre Oriental, que se componen de rocas sedimentarias marinas en gran medida de origen Cretácico y Jurásico con rocas sedimentarias clásticas del Cenozoico y depósitos aluviales recientes en los valles y cuencas. Exposiciones aisladas de rocas del Precámbrico, Paleozoico y Mesozoico-Triásico se encuentran a nivel local, en particular al oeste de Ciudad Victoria. La planicie costera de Tamaulipas (la emergente plataforma continental del Golfo de México) fue levantada e inclinada por la actividad tectónica en el Cretácico-Paleoceno pero principalmente permaneció sumergida, probablemente hasta el Pleistoceno (Heim 1940). Los sustratos de la planicie costera consisten de suelos aluviales sobre deposiciones marinas sedimentarias del Cenozoico de piedra caliza, piedra arenisca, pizarras y arcilla con diversas exposiciones del Paleógeno todas ellas reposando sobre una base del Cretácico que ocasionalmente se encuentra expuesta. Las sierras y mesetas aisladas que se observan en la planicie costera son cuerpos de roca ígnea, plutones y flujos de basalto alcalinos que en el Terciario Medio se encontraban en zonas altas y en donde el mar había desaparecido.

La hidrología de Tamaulipas está totalmente asociada con el Golfo de México, con excepción de dos municipios áridos interiores en el extremo suroeste. En la Provincia Tamaulipeca el Río Bravo (= Río Grande) y sus tributarios los Ríos Salado y San Juan, drenar zonas vastas del interior continental mientras que toda la cuenca del Río San Fernando (= Río Conchos) y el Río Soto La Marina (y sus tributarios los Ríos Pilón, Purificación, Corona y Palmas) se encuentran en el interior de Tamaulipas y áreas adyacentes de Nuevo León. En la Provincia Veracruzana las zonas costeras son drenadas por sistemas ribereños relativamente pequeños incluyendo los Ríos Carrizo, Carrizal (y sus tributarios los Ríos San Pedro y Las Lajas), San Vicente, Tigre y Barberena, estos dos últimos vacian sus aguas en la Laguna San Andrés. El interior de la Provincia Veracruzana y la mayor parte de la Sierra Madre Oriental de Tamaulipas son irrigadas por el Río Guayalejo y sus tributarios los Ríos El Alamar, Chiue, Frío, Sabinas, Santa María de Guadalupe y Las Animas, algunas excepciones se observan en la Sierra Madre Oriental, donde la vertiente oriental al norte del Trópico de Cáncer desemboca en el Río Soto La Mariana y en el extremo sur, donde los Ríos Gallos Grandes y Los Galos desaguan en el Río Tampaón-Pánuco en San

Luis Potosí. El Río Guayalejo es un tributario del Río Tamesi, que forma la frontera con el estado de Veracruz, donde extensos pantanos y lagunas (Lagunas Champayan, Josecito, La Salada) sustentan grandes áreas de *Cyperus* y *Typha*. Otros lagos naturales son Laguna Los Soldados, un lago cráter volcánico en la Sierra Maratinez, y algunas cuencas cerradas en la Sierra Madre Oriental: Laguna La Escondida, Laguna Isadora, Laguna La Loca y Lagunas Las Hondas. Varias presas sobre la Planicie Costera, incluyendo Falcón, Marte Gómez, Vicente Guerrero, República Española, Ramiro Caballero y Emilio Portes, fueron construidas en el siglo veinte. Ambientes kársticos se presentan tanto en la Sierra Madre Oriental como en la Sierra de Tamaulipas, particularmente en la vertiente este de cada una de ellas, donde los ríos de las montañas frecuentemente son subterráneos, emergiendo en la Planicie Costera como ríos o lagos, por ejemplo Río Sabinas, Río Frío, Laguna La Loca de la Sierra Madre Oriental y Cenote El Zacatón (que proporciona agua al Río Tigre) y otros sótanos de la Sierra de Tamaulipas. La Laguna Madre sobre la costa norte se describe con más detalle más adelante.

Comprendidos dentro de sus límites Tamaulipas tiene porciones de tres subregiones Neárticas, Tamaulipeca, Sierra Madre Oriental y Desierto Chihuahuense, y una subregión Neotropical, la Veracruzana, cada una con climas, topografías y zonas de vegetación distintivas (Álvarez 1963; Flores-Villela 1993; Morrone et al. 2002).

Provincia Tamaulipeca

Ocupa 141,500 km² en el sur de Texas, noreste de Coahuila, norte de Nuevo León y los dos tercios del norte de Tamaulipas (Blair 1950). Sus límites están muy definidos por el Golfo de México hacia el este, la Sierra Madre Oriental al suroeste y la Provincia de los Balcones en el noroeste. Sus límites están menos diferenciados en las proximidades del Río San Antonio de Texas hacia el noreste, donde los suelos pedocal cambiar a suelos pedalfers (Blair 1950), el Desierto

Chihuahuense al oeste y el Trópico de Cáncer al sur. Para la discusión realizada aquí, marqué el límite sur en el Trópico de Cáncer excluyendo las estribaciones del norte de la Sierra de Tamaulipas y las áreas costeras de la Escarpada de Reynosa/Sierra de las Rusias arriba del Río Soto La Marina, al norte de la línea del Trópico de Cáncer. La Escarpada de Reynosa divide a esta provincia, corriendo desde el área al norte del extremo norte de Sierra de Las Rusias, ligeramente al este de Soto La Marina en el sureste, extendiéndose hacia el noroeste desde ahí hasta las proximidades de Camargo, y hasta Texas, Estados Unidos (Johnston 1963). La topografía es plana con suelos arenosos profundos al noreste de esta escarpada y algunas colinas al suroeste. La altitud varía desde el nivel del mar hasta 400 m en la base de la Sierra Madre Oriental, ascendiendo abruptamente sobre la planicie. Una excepción es la Sierra de San Carlos, una sierra aislada de aproximadamente 1448 km² donde las altitudes son generalmente de 400 a 1000 m, sin embargo, algunas montañas alcanzan mayores alturas, principalmente la Sierra Chiquita (1794 m).

El clima de esta provincia (excluyendo la Sierra San Carlos) es semiárido con precipitaciones que van de 400 a 750 mm, sin embargo, huracanes y depresiones tropicales ocasionalmente afectan la planicie costera de junio a noviembre y pueden aumentar considerablemente la precipitación. Las temperaturas de verano son calientes (32°C-35°C) ocasionalmente excediendo 40°C, los inviernos son leves (2°C-8°C) y las heladas son raras. Áreas extensas de vegetación riparia y matorral espinoso han sido sustituidas para dar paso a cultivos agrícolas, particularmente en el noreste. Se pueden reconocer tres ensambles de anfibios y reptiles en esta provincia que incluyen Matorral Espinoso Tamaulipeco, Laguna Madre y Sierra de San Carlos. Varios autores reconocen diferentes tipos de vegetación dentro del Matorral Espinoso Tamaulipeco del sur de Texas que indudablemente se presentan en Tamaulipas pero Tunnell (2002) identifico importantes vacios de información y estas áreas no han sido delineadas en Tamaulipas.

ENSAMBLE DEL MATORRAL ESPINOSO TAMAULIPECO

Este matorral se caracteriza por plantas espinosas y deciduas, incluyendo árboles y arbustos (*Acacia berlandieri, A. greggii, A. rigidula, A. tortuosa, A. wrightii, Celtis pallid, Condalia hookeri, Cordia boissieri, Leucophyllum frutescens, Parkinsonia aculeata, Prosopis glandulosa, Vachellia farnesiana, Yucca treculeana*), pastos (*Aristida roemeriana, Bouteloua hirsute, Cenchrus incertus*) y plantas perennes (*Dyssodia berlandieri, Heliotropium confertifolium*) (Blair 1950; Johnston 1963; Little 1976). El Matorral Espinoso frecuentemente es más alto (2–3 m) y más denso en el este donde está expuesto a la humedad de la costa, y más corto y más abierto en el oeste (Blair 1950; Martin et al. 1954). Las zonas ribereñas a menudo incluyen *Taxodium mucronatum, Ebenopsis ebano, Sabal mexicana* y *Ulmus crassifolia*. Elementos del bosque tropical caducifolio se extienden hacia el norte desde el sur en parches que se hacen más angostos y discontinuos a medida que incrementa la latitud sobre la base de la Sierra Madre Oriental. Esta región incluye 84 especies de anfibios y reptiles: *Notophthalmus meridionalis, Siren intermedia, Siren* sp. indet., *Eleutherodactylus cystignathoides, Anaxyrus debilis, A. punctatus, A. speciosus, Incilius nebulifer, Rhinella marina, Pseudacris clarkii, Smilisca baudinii, Leptodactylus fragilis, L. melanonotus, Gastrophryne olivacea, Hypopachus variolosus, Lithobates berlandieri, L. catesbeianus, Rhinophrynus dorsalis, Scaphiopus couchii, Spea bombifrons, Pseudemys gorzugi, Terrapene carolina, Trachemys scripta, T. venusta, Kinosternon flavescens, Gopherus berlandieri, Apalone spiniferus, Crocodylus moreletii, Crotaphytus reticulates, Coleonyx brevis, Hemidactylus frenatus, H. turcicus, Cophosaurus texanus, Phrynosoma cornutum, P. modestum, Sceloporus consobrinus, S. cyanogenys, S. grammicus, S. olivaceus, S. variabilis, Urosaurus ornatus, Anolis sericeus, Plestiodon obsoletus, P. tetragrammus, Ameiva undulata, Aspidoscelis gularis, A. laredoensis, Boa constrictor, Arizona elegans, Coluber constrictor, Coniophanes imperialis, Drymarchon melanurus, Drymobius margaritiferus, Ficimia streckeri, Heterodon kennerlyi, Hypsiglena jani, Lampropeltis getula, L. triangulum, L. septentrionalis, Leptophis mexicanus, Masticophis flagellum, M. schotti, Nerodia erythrogaster, N. rhombifera, Opheodrys aestivus, Pantherophis emoryi, Pituophis catenifer, Rhinocheilus lecontei, Salvadora grahamiae, Sonora semiannulata, Storeria dekayi, Tantilla nigriceps, Thamnophis marcianus, T. proximus, Tropidodipsas fasciata, T. sartorii, Trimorphodon tau, Micrurus tener, Rena dulcis, R. myopica, Agkistrodon taylori, Crotalus atrox, C. totonacus* y *Sistrurus catenatus*.

ENSAMBLE DE LA LAGUNA MADRE

Los extensos humedales de la Laguna Madre se extienden desde el Golfo de México y están divididos por 250 km de islotes y penínsulas. Túnel y Judd (2002) revisaron este ecosistema y su publicación es la base de la mayoría de la información que se reporta aquí. Los islotes de médanos de arena están habitados por *Ipomoea imperati, Schizachyrium scoparium, Sesuvium portulacastrum* y *Uniola paniculata*. La laguna hipersalina (una de sólo cinco en el mundo), alberga aproximadamente 480 km² de praderas de pasto marino incluyendo *Cymodocea filiforme, Halodule beauderrei, Halophila engelmannii, Ruppia maritima* y *Thalassia testudinum*. Médanos de arcilla se presentan en la parte continental con extensas zonas pantanosas dominadas por *Spartina spartinae*. Al interior la salinidad disminuye y áreas pantanosas de agua dulce y lagunas se presentan en áreas periféricas, la más grande de ellas es la Laguna La Nacha. La altitud varía de nivel del mar a 3 m. Aquí habitan por lo menos 16 especies de anfibios y reptiles, 6 se han registrado en los islotes de médanos de arena y penínsulas: *Holbrookia propinqua, Sceloporus variabilis, Aspidoscelis gularis, A. sexlineata, Gopherus berlandieri* y *Masticophis flagellum*, y 5 especies adicionales se han registrado en la parte continental adyacente a la Laguna Madre, entre los médanos de arcilla, planicies lodosas y pantanos: *Phrynosoma cornutum, Sceloporus*

olivaceus, *Thamnophis proximus*, *Rena dulcis* y
Crotalus atrox. Cinco especies de tortugas marinas: *Caretta caretta*, *Chelonia mydas*, *Eretmochelys
imbricata*, *Lepidochelys kempii* y *Dermochelys coriacea* han sido registradas anidando en las playas
del sur de Tamaulipas y Texas (con una variación
y frecuencia considerable) y en la entrada a la
Laguna Madre de Texas (Tunnel y Judd 2002).
Presumiblemente, en esta región, las tortugas
marinas utilizan estos hábitats, sin embargo, no
existe información específica al respecto.

ENSAMBLE DE LA SIERRA DE SAN CARLOS

El matorral espinoso de la planicie asciende
hasta cerca de 600 m creciendo un poco más
grueso y alto con la altitud y en la zona de
transición con el bosque semiárido de encino
(*Quercus* sp.) a 600–900 m, que predomina en
estas montañas, sin embargo, algunas montañas
hospedan bosques de pino-encino con *Pinus
teocote* a 900–1794 m (Dice 1937). Las elevaciones más altas reciben mayor humedad y temperaturas más agradables comparadas con las de
la planicie. Por lo menos 42 especies de anfibios
y reptiles han sido registradas aquí. Entre éstas,
8 especies (19.0%) no habitan en los alrededores de la planicie costera, aunque todas ellas
habitan en la Sierra Madre Oriental 40–50 km
al oeste, incluyendo *Eleutherodactylus guttilatus*,
Gerrhonotus infernalis, *Sceloporus minor*, *S. parvus*, *Plestiodon dicei*, *Pantherophis bairdi*, *Tantilla
atriceps* y *T. rubra*. Adicionalmente 34 especies
(81.0%) de la Planicie Costera se han registrado
únicamente aquí incluyendo *Eleutherodactylus
cystignathoides*, *Anaxyrus punctatus*, *Incilius
nebulifer*, *Rhinella marina*, *Smilisca baudinii*,
Lithobates berlandieri, *Trachemys venusta* (<500
m), *Gopherus berlandieri*, *Cophosaurus texanus*,
Phrynosoma cornutum, *Sceloporus cyanogenys*,
S. grammicus, *S. olivaceus*, *S. variabilis*, *Plestiodon
obsoletus*, *P. tetragrammus*, *Aspidoscelis gularis*,
Arizona elegans, *Drymarchon melanurus*, *Drymobius margaritiferus*, *Ficimia streckeri*, *Leptodeira
septentrionalis*, *Masticophis flagellum*, *M. schotti*,
Opheodrys aestivus, *Pantherophis emoryi*, *Salva-

dora grahamiae, *Storeria dekayi*, *Tantilla nigriceps*,
Thamnophis marcianus, *T. proximus*, *Micrurus
tener*, *Rena myopica* y *Crotalus atrox*.

Colectivamente 98 especies habitan en la Provincia Tamaulipeca del Tamaulipas: 3 salamandras
(3.0%, 2 familias, 2 géneros); 18 anuros (18.3%, 8
familias, 13 géneros); 12 tortugas (12.2%, 6 familias, 11 géneros); 1 cocodrilo (1.0%); 24 lagartijas
(24.5%, 7 familias, 12 géneros); 40 serpientes
(40.8%, 5 familias, 30 géneros). Entre estas 98
especies, 8 (8.0%) son endémicas de México, 81
(82.6%) se distribuyen en México y los Estados
Unidos, y 9 (9.0%) se distribuyen en México
y Centroamérica. Veintidos (22.4%) especies
habitan en Estados Unidos, México y Centroamérica. Dos especies introducidas, *Hemidactylus
frenatus* y *H. turcicus* se encuentran aquí. Varias
especies tienen distribuciones limitadas en esta
provincia. *Phrynosoma modestum* es conocida por
registro aislado (Farr et al. 2007). En el extremo
noroeste *Urosaurus ornatus*, *Lampropeltis getula*
y *Nerodia erythrogaster* son conocidas por sólo
uno o dos registros o localidades para el estado,
al igual que *Pseudacris clarkii* y *Siren* sp. indet.
en el noreste, estas dos últimas alcanzan sus
limites australes aquí. *Siren intermedia*, *Spea
bombifrons* y *Sistrurus catenatus* también alcanzan
su distribución más austral aquí pero han sido
registradas en localidades aisladas (presumiblemente poblaciones relictuales) más al sur. Otras
especies que alcanzan sus limites de distribución
austral aquí están más ampliamente distribuidas
en esta provincia incluyendo *Apalone spinifera*,
Heterodon kennerlyi, *Sonora semiannulata* y
Tantilla nigriceps. Dos especies que habitan en
el norte, *Trachemys scripta* y *Rena dulcis*, están
representadas por especies hermanas *T. venusta*
y *R. myopica* respectivamente, en el sur. Especies neotropicales alcanzando su distribución
más septentrional aquí incluyen *Leptodactylus
melanonotus*, *Crocodylus moreletii*, *Anolis sericeus*,
Ameiva undulata, *Boa constrictor*, *Leptophis mexicanus*, *Trimorphodon tau*, *Tropidodipsas fasciata*,
T. sartorii, *Agkistrodon taylori* y *Crotalus totonacus*.
Holbrookia propinqua y *Aspidoscelis sexlineata*

están limitadas a la costa en esta región. *Siren* sp. indet., *Pseudemys gorzugi* y *Aspidoscelis laredoensis* están limitadas al Río Bravo y áreas adyacentes. *Terrapene carolina* y *Opheodrys aestivus* relativamente están ampliamente distribuidas pero en poblaciones relictuales en esta región. Varias especies están limitadas a las áreas del oeste de esta región, algunas aparentemente análogas a la Escarpada de Reynosa, incluyendo, *Anaxyrus punctatus, Hypopachus variolosus, Crotaphytus reticulatus, Coleonyx brevis, Cophosaurus texanus, Phrynosoma modestum, Sceloporus consobrinus* y *Urosaurus ornatus*. No existen registros confirmados de *Holbrookia lacerata* para Tamaulipas, aunque su ocurrencia en este estado es probable (Farr et al. 2013).

Provincia Veracruzana

Ocupa la mayor parte de la Planicie Costera del Golfo de México en México. Aunque varios autores coinciden en que esta provincia abarca el sur de Tamaulipas, el límite norte (y sur) exacto rara vez coinciden entre autores (ejem: Goldman y Moore 1945; Álvarez 1963; Flores-Villela 1993; Morrone et al. 2002). La frontera norte reconocida aquí fue definida anteriormente en la Provincia Tamaulipeca. La topografía es principalmente plana, pero se ve interrumpida por varias colinas y montañas. La Sierra de Tamaulipas (cerca de 2500 km²) es la más grande, dominando la parte norte-centro de esta provincia, su altitud varía de 300 a 1000 m (máxima 1400 m). Al este, la Sierra de Las Rusias (400 km²) con una elevación máxima de 300 m, y la Sierra de Maratinez (150 km²) con elevaciones de 150–400 m (máxima 600 m), son sierras aisladas que hospedan bosques de *Quercus* pero éstos no albergan ensambles significativamente distintos a los de la Planicie Costera. Al sur de la Sierra de Tamaulipas hay una amplia (cerca de 10,800 km²) planicie semiárida de menos de 100 m de altitud, expandiéndose entr la costa y la Sierra de Cucharas. Al oeste, aproximadamente 50 km de 150–200 m de altitud separan la Sierra de Tamaulipas de la Sierra Madre Oriental, con matorral espinoso semiárido y tropical al este de este tramo, volviéndose más húmedo en el bosque tropical caducifolio y mesas bajas (300–400 m de altitud) hacia el oeste. En el suroeste de la Provincia Veracruzana la flora y fauna se extienden hacia el oeste más allá de las montañas frontales de la Sierra Madre Oriental, sobre la Sierra de las Cucharas (= Sierra del Abra) (350–450 m, máxima 720 m) y la Sierra Tamalave (500–800 m, máxima 1100 m), dos pliegues anticlinales bajos relativamente angostos (4–8 km), en dos valles sinclinales (Chamal y Ocampo). Estas dos sierras que se extienden desde el sur desde la Sierra de Guatemala hasta la planicie representan una pequeña barrera para la fauna de la costa y albergan algunas especies de la herpetofauna de la Sierra Madre Oriental. Hacia el oeste del Valle de Ocampo, la Sierra Madre Oriental se eleva a 1000–1300 m (máxima 1800 m).

El clima es tropical subhúmedo con una precipitación anual de 700–1600 mm (70% entre junio y septiembre), con la planicie recibiendo la menor precipitación y la Sierra Madre Oriental recibiendo la mayor. Ciudad Mante tiene una media de temperatura anual de 24.8°C, la media inferior en enero es de 18.4°C, y la media superior en junio es de 29.1°C. Tampico tiene una temperatura media de 24.2°C, la media inferior en enero es de 13.4°C, y la media superior en agosto es de 31.8°C. Cuatro ensambles se identifican en esta región: Planicie Costera, Laguna San Andrés y Costa del Golfo, Sierra de Tamaulipas y Bosque Tropical Caducifolio.

ENSAMBLE DE LA PLANICIE COSTERA

Esta región incluye áreas desde el nivel del mar hasta 300 m entre la costa y la Sierra Madre Oriental (Sierra de Guatemala, Sierra Cucharas) excluyendo los bosques tropicales caducifolios asociados con la Sierra de Tamaulipas y la Sierra Madre Oriental que habitan por debajo de esta altitud. La vegetación es variable e influencias antropogénicas significativas comenzaron antes de que ésta fuera bien estudiada. El matorral tropical espinoso domina la planicie semiárida al sur de la Sierra de Tamaulipas incluyendo *Vachellia [Acacia] farnesiana, A. rigidula, Condalia* hookeri, *Cordia boissieri, Parkinsonia aculeate, Prosopis juliflora, Yucca treculeana,* (Little 1976, 1977). Zonas ribereñas hospedan a menudo bosques exuberantes de galería acompañados de bosque tropical caducifolio. *Sabal mexicana* habita en toda esta región, pero es más abundante entre la costa y la Sierra de Tamaulipas a veces formando bosques pequeños. Los pastizales se observan en algunas áreas y pueden haber sido más extensos en el pasado (Martin 1958). Berlandier (1980) describio "inmensos," y Goldman (1951) "tolerablemente densos," bosques de *Quercus oleoides* y *Brosimum alicastrum* con praderas pequeñas cerca de Altamira y pastizales más abiertos con acacias, mezquites y cactus en el norte. Aún se pueden encontrar remanentes de bosques de encino entre Altamira y el Río Soto La Marina. Se han confirmado registros para 71 especies de anfibios y reptiles en esta región: *Notophthalmus meridionalis, Eleutherodactylus cystignathoides, Anaxyrus debilis, A. speciosus, Incilius nebulifer, Rhinella marina, Trachycephalus typhonius, Scinax staufferi, Smilisca baudinii, Leptodactylus fragilis, L. melanonotus, Gastrophryne elegans, G. olivacea, Hypopachus variolosus, Lithobates berlandieri, L. catesbeianus, Rhinophrynus dorsalis, Scaphiopus couchii, Terrapene carolina, Trachemys scripta, T. venusta, Kinosternon flavescens, K. herrerai, K. scorpioides, Gopherus berlandieri, Crocodylus moreletii, Ophisaurus incomptus, Laemanctus serratus, Hemidactylus frenatus, H. turcicus, Ctenosaura acanthura, Phrynosoma cornutum, Sceloporus olivaceus, S. serrifer, S. variabilis, Anolis sericeus, Plestiodon tetragrammus, Ameiva undulata, Aspidoscelis gularis, Boa constrictor, Coluber constrictor, Coniophanes imperialis, Drymarchon melanurus, Drymobius margaritiferus, Ficimia streckeri, Imantodes cenchoa, Lampropeltis triangulum, Leptodeira maculata , L. septentrionalis, Leptophis mexicanus, Masticophis flagellum, M. mentovarius, M. schotti, Nerodia rhombifera, Opheodrys aestivus, Oxybelis aeneus, Pantherophis emoryi, Pituophis catenifer, Pseudelaphe flavirufa, Spilotes pullatus, Storeria dekayi, Thamnophis marcianus, T. proximus, Tropidodipsas fasciata, T. sartorii, Micrurus tener, Rena myopica, Indotyphlops braminus, Agkistrodon taylori, Crotalus atrox y C. totonacus.*

ENSAMBLE DE LA LAGUNA SAN ANDRÉS Y COSTA DEL GOLFO

Esta región cubre 170 km de la costa entre los Ríos Soto La Marina y Pánuco incluyendo playas y penínsulas de barrera (Rancho Nuevo, Barra del Trodo, Barra Chavarria), lagunas (San Andrés, 136 km²; Las Marismas, 55 km²) y humedales. Las zonas pantanosas interiores no son extensas. La altitud varía desde el nivel del mar hasta 3 m. Algunas lagunas están cubiertas de árboles de mangle (*Rhizophora mangle, Avicennia germinans, Laguncularia racemosa, Conocarpus erectus*) que suman un total de 24 km² en el estado (CONABIO 2008). Se han registrado al menos 16 especies de anfibios y reptiles entre las playas, dunas y manglares incluyendo *Incilius nebulifer, Caretta caretta, Chelonia mydas, Eretmochelys imbricata, Lepidochelys kempii, Dermochelys coriacea, Gopherus berlandieri, Ctenosaura acanthura, Iguana iguana, Holbrookia propinqua, Sceloporus variabilis, Aspidoscelis gularis, Boa constrictor, Masticophis flagellum, Drymarchon melanurus y Thamnophis proximus.*

ENSAMBLE DE LA SIERRA DE TAMAULIPAS

Los tipos de vegetación incluidos en esta región incluyen: bosque tropical caducifolio (300–700 m) con *Bursera simaruba* y *Lysiloma divaricata*; matorral de montaña (600–900 m de altitud); y bosques de pino-encino (>800 m de altitud) con *Quercus* sp. y *Pinus teocote* intercalados con pastizales (Martin et al. 1954; Cram et al. 2006). La altitud varía de 300 a 1400 m. Entre las 46 especies confirmadas para estas montañas, 11 no se han sido registradas en el matorral espinoso que las rodea ni en los bosques de la Planicie Costera. Las especies de anfibios y reptiles de esta sierra incluyen *Pseudoeurycea cephalica, Craugastor augusti, Eleutherodactylus guttilatus, Ecnomiohyla miotympanum, Gerrhonotus infernalis, Sceloporus grammicus, Plestiodon dicei, Lepidophyma sylvaticum, Rhadinaea gaigeae, Senticolis triaspis y Micrurus tamaulipensis.* Especies adicionales incluyen *Notophthalmus meridionalis, Eleutherodactylus cystignathoides, In-*

cilius nebulifer, Rhinella marina, Scinax staufferi, Smilisca baudinii, Hypopachus variolosus, Lithobates berlandieri, Scaphiopus couchii, Terrapene carolina, Trachemys venusta, Kinosternon herrerai, S. olivaceus, S. serrifer, S. variabilis, Anolis sericeus, P. tetragrammus, Ameiva undulata, Aspidoscelis gularis, Coniophanes imperialis, Drymarchon melanurus, Drymobius margaritiferus, Ficimia streckeri, Imantodes cenchoa, Leptodeira annulata, L. septentrionalis, Leptophis mexicanus, Masticophis schotti, Oxybelis aeneus, Spilotes pullatus, Thamnophis proximus, M. tener, Rena myopica, Agkistrodon taylori y *Crotalus totonacus.*

ENSAMBLE DEL BOSQUE TROPICAL CADUCIFOLIO

Esta región incluye los bosques tropicales caducifolios principalmente a 200–500 m de altitud (en ocasiones en altitudes superiores) en la base este de la Sierra Madre Oriental, Sierra Cucharas, Sierra Tamalave y los valles de Chamal y Ocampo, con los dos valles intermitentes volviéndose más áridos hacia el sur en las cercanías de Antiguo Morelos y Nuevo Morelos. Martin (1958) describió esta área como un bosque tropical caducifolio, pero especuló que actividades humanas pasadas podrían haber afectado la composición actual. Cram et al. (2006) describieron la mayoría de esta región como bosques perturbados y matorrales de alta diversidad. La vegetación incluye *Bromelia penguin, Esenbeckia runyonii, Guazuma ulmifolia, Nectandra salicifolia, Petrea arborea, [Pseudo] bombax ellipticum, Thouinia villosa* (Martin 1958; Cram et al. 2006). Bosques de *Sabal mexicana* moderadamente extensos se presentan en algunas áreas, particularmente en el Valle Chamal. Hay 82 especies de anfibios y reptiles registradas aquí, éstas incluyen *Bolitoglossa platydactyla, Chiropterotriton multidentatus, Notophthalmus meridionalis, Craugastor augusti, Eleutherodactylus cystignathoides, E. dennisi, E. longipes, Anaxyrus debilis, A. punctatus, Incilius nebulifer, Rhinella marina, Scinax staufferi, Smilisca baudinii, Trachycephalus typhonius, Leptodactylus fragilis, L. melanonotus, Gastrophryne elegans, G. olivacea, Hypopachus*

variolosus, Lithobates berlandieri, L. catesbeianus, Rhinophrynus dorsalis, Scaphiopus couchii, Terrapene carolina, Trachemys venusta, Kinosternon scorpioides, Crocodylus moreletii, Gerrhonotus infernalis, Laemanctus serratus, Anelytropsis papillosus, Hemidactylus frenatus, H. turcicus, Ctenosaura acanthura, Cophosaurus texanus, S. serrifer, S. variabilis, Anolis sericeus, Plestiodon lynxe, P. tetragrammus, Scincella silvicola, Ameiva undulata, Aspidoscelis gularis, Lepidophyma micropholis, L. sylvaticum, Boa constrictor, Coluber constrictor, Coniophanes imperialis, C. piceivittis, Drymarchon melanurus, Drymobius margaritiferus, Ficimia olivacea, F. streckeri, Hypsiglena jani, Imantodes cenchoa, Lampropeltis triangulum, Leptodeira maculata , L. septentrionalis, Leptophis mexicanus, Masticophis flagellum, M. schotti, Mastigodryas melanolomus, Nerodia rhombifera, Oxybelis aeneus, Pantherophis emoryi, Pseudelaphe flavirufa, Rhadinaea gaigeae, Scaphiodontophis annulatus, Senticolis triaspis, Spilotes pullatus, Storeria dekayi, Tantilla rubra, Thamnophis marcianus, T. proximus, Tropidodipsas fasciata, T. sartorii, Trimorphodon tau, Micrurus tener, Epictia goudotii, Rena myopica, Agkistrodon taylori, Bothrops asper y *Crotalus atrox.*

Colectivamente 108 especies de anfibios y reptiles se conocen actualmente para la Provincia Veracruzana en Tamaulipas, 4 salamandras (3.7%, 2 familias, 4 géneros); 23 anuros (21.29%, 9 familias, 15 géneros); 12 tortugas (11.1%, 5 familias, 9 géneros); 1 cocodrilo (0.92%); 24 lagartijas (22.22%, 10 familias, 17 géneros); 44 serpientes (40.74%, 6 familias, 35 géneros). Entre estas especies, 24 (22.22%) son endémicas a México y 59 (54.62%) se distribuyen en Estados Unidos. Veinticinco especies (23.14%) habitan en México y Centroamérica, y otras 25 (23.14%) habitan en Estados Unidos, México y Centroamérica. Muchas especies neotropicales alcanzan su límite septentrional aquí, incluyendo *Bolitoglossa platydactyla, Plestiodon lynxe, Lepidophyma micropholis, Coniophanes piceivittis, Ficimia olivacea, Mastigodryas melanolomus, Scaphiodontophis annulatus, Epictia goudotii* y *Bothrops asper* en el bosque tropical caducifolio; *Trachycephalus typhonius, Kinosternon herrerai, K. scorpioides, Laemanctus serratus, Ctenosaura acanthura, Iguana iguana, Leptodeira annulata, Masticophis mentovarius* y *Pseudoelaphe flavirufa* cerca de la costa; *Imantodes cenchoa* y *Spilotes pullatus* en las estribaciones del norte de la Sierra de Tamaulipas; y otras más distribuidas más ampliamente *Scinax staufferi, Gastrophryne elegans* y *Sceloporus serrifer*. Especies que alcanzan sus límites de distribución austral: *Anaxyrus debilis, A. speciosus, Cophosaurus texanus* y *Phrynosoma cornutum*. *Micrurus tamaulipensis* es endémica de la Sierra de Tamaulipas. *Eleutherodactylus dennisi* y *Lepidophyma micropholis* están limitadas a la Sierra Cucharas en Tamaulipas. Nidos de *Eretmochelys imbricata* en Tamaulipas son raros y nidos de *Dermochelys coriacea* son extremadamente raros. *Plestiodon lynxe* y *Scaphiodontophis annulatus* son conocidas en Tamaulipas por un solo espécimen e *Iguana iguana* es conocida únicamente para dos localidades cerca del extremo de la costa sur. Especies asociadas a la Sierra Madre Oriental que han sido registras a 420 m o menos de altitud en la Sierra Madre Oriental o Sierra Cucharas incluyen *Chiropterotriton multidentatus, Craugastor augusti, Eleutherodactylus longipes, Gerrhonotus infernalis, Scincella silvicola, Rhadinaea gaigeae, Senticolis triaspis* y *Tantilla rubra*. Cinco especies introducidas, *Lithobates catesbeianus, Trachemys scripta, Hemidactylus frenatus, H. turcicus* e *Indotyphlops braminus* habitan en esta región.

Sierra Madre Oriental

La Sierra Madre Oriental cubre cerca de 145,500 km² del noreste de México desde el Eje Transvolcánico cinturón (norte de Hidalgo, Querétaro y Puebla) extendiéndose hacia el norte hasta las cercanías de Monterrey, Nuevo León, donde un archipiélago de montañas asociadas a esta sierra corre de este a oeste hasta el sureste de Coahuila. Aunque geológicamente se desarrollaron independientemente, algunos consideran que las montañas que se extienden al noroeste y forman la región de Trans-Pecos también forman parte de la Sierra Madre Oriental. En Tamaulipas, la Sierra Madre Oriental ocupa un área relativamente limitada pero significativa en el suroeste del estado. Es la máxima elevación en Tamaulipas con más de 3400 m de altitud sobre la ladera oriental de la Sierra Peña Nevada en la frontera con Nuevo León, a partir de ahí la Sierra Madre se extiende hacia el este más de 80 km al frente de la cordillera, cerca de Ciudad Victoria. Aunque irregular, hay una tendencia generalizada de disminución de elevaciones y estrechamiento (de este a oeste) hacia el sur cerca de la frontera con San Luis Potosí, donde el Valle de Ocampo está separado del Desierto Chihuahuense por algunas montañas de tan sólo 18 km de anchura y menos de 1200 m de altitud en algunos puntos. Dentro de la Sierra Madre, se encuentra el Valle de Jaumave, una región semiárida de cerca de 1130 km² con una temperatura promedio anual de 21°C-23°C (máxima 45°C, mínima 0°C) y una precipitación anual de 500–560 mm. El Río Guayalejo drena el valle a través de un estrecho desfiladero que se dirige hacia el este hasta la planicie costera, y hacia el oeste un cañon semiárido se extiende hasta el Valle de Palmillas, también una región semiárida, hasta la planicie costera, y más adelante hasta la planicie desertica, virtualmente dividiendo la Sierra Madre Oriental en un componente norte y uno sur, aproximada pero no precisamente en el Trópico de Cáncer. La Sierra Madre Oriental produce una sombra de lluvia y el clima varía considerablemente en función de la exposición y la elevación, aunque información específica ha sido difícil de obtener. Martin (1958) estima que la precipitación anual en el bosque mesófilo de montaña es de 2000–2500 mm (máxima 3200 mm), la más alta en el estado, con una temperatura media anual de 19.4°C en 1954. Las elevaciones más altas al noroeste del Valle de Jaumave son más frescas, semiáridas y en ocasiones reciben nieve, aunque no anualmente. Más de 150 registros de temperatura de especímenes capturados entre 2400 y 3200 m de altitud, entre julio y agosto de los años 2003 a 2009, y entre 9:44 am y 7:00 pm, oscilaron entre 14.44°C y 31.67°C, pero la disminución durante la noche (estimada en 10°C) no fue registrada. El agua superficial es mínima, pero las lluvias por la tarde se produjeron en casi todos los días julio-octubre 2003–2009. El pueblo de Jaumave, con una población de 5633 habitantes, es el más grande en esta región y sólo algunos otros superan los 1000 habitantes. Importantes áreas de bosques maduros se desarrollan en esta región, aunque muchos son de crecimiento secundario.

ENSAMBLE DE LAS SIERRAS DEL NORTE Y LADERAS INTERIORES

Esta región incluye a la Sierra Madre Oriental al norte y al oeste del Valle de Jaumave (excluyendo las exposiciones más orientales de la cordillera frontal) y las laderas interiores al sur del Valle. La altitud varía de 1250 a más de 3400 m. La vegetación de la Sierra Madre Oriental al igual que su clima varía considerablemente en función de la exposición y la altitud. Elevaciones más bajas y laderas interiores que descienden hasta el Valle de Jaumave y el Desierto Chihuahuense son semiáridas con *Juniperus angosturana, J. flaccida, Pinus arizonica* var. *stormiae, P. cembroides, P. estevezii, P. nelsonii, Yucca carnerosana*, una variedad de cactus y ocasionalmente zonas de vegetación de chaparral (Martin 1958). Elevaciones intermedias hospedan bosques secos de encino y pino-encino. Bosques templados de enebros y pino-encino son característicos de elevaciones más altas (>2250 m) en el noroeste donde la vegetación más conspicua incluye *Arbutus xalapensis, Maguey americana, M. asperrima, Juniperus monticola, Pinus arizonica, P. cembroides, P. hartwegii, P. johannis, P. nelsonii, Quercus* sp. y *Dasylirion* sp. Se han registrado 37 especies de anfibios y reptiles en esta región, incluyendo *Chiropterotriton* cf. *priscus, Pseudoeurycea galeanae, Craugastor augusti, Eleutherodactylus guttilatus, E. verrucipes, Anaxyrus punctatus, Incilius nebulifer, Ecnomiohyla miotympanum, Hyla eximia, Spea multiplicata, Barisia ciliaris, Gerrhonotus infernalis, Phrynosoma orbiculare, Sceloporus chaneyi, S. grammicus, S. minor, S. olivaceus, S. parvus, S. spinosus, S. torquatus, Plestiodon dicei, Scincella silvicola, Aspidoscelis gularis, Ficimia*

hardyi, Lampropeltis mexicana, Masticophis schotti, Pantherophis bairdi, Salvadora grahamiae, Senticolis triaspis, Storeria hidalgoensis, Tantilla rubra, Thamnophis exsul, T. pulchrilatus, Rena myopica, Crotalus lepidus, C. molossus y *C. pricei.*

ENSAMBLE DEL VALLE DE JAUMAVE

La altitud varía de 550 a 1250 m. La vegetación del Valle, así como la herpetofauna, tiene especies representativas de la planicie costera y el Desierto Chihuahuense, entre ellas *Jatropha dioica, Prosopis glandulosa, Cylindropuntia leptocaulis, Koeberlinia spinosa, Agave funkiana, A. lecheguilla, Yucca treculeana* y *Y. filifera* (Martin 1958). La herpetofauna de esta región incluye 35 especies: *Anaxyrus punctatus, Incilius nebulifer, Rhinella marina, Ecnomiohyla miotympanum, Lithobates berlandieri, Spea multiplicata, Kinosternon integrum, K. scorpioides, Gerrhonotus infernalis, Crotaphytus collaris, Hemidactylus frenatus, H. turcicus, Cophosaurus texanus, Phrynosoma cornutum, Sceloporus cautus, S. olivaceus, S. parvus, S. variabilis, Plestiodon tetragrammus, Aspidoscelis gularis, Boa constrictor, Coluber constrictor, Drymarchon melanurus, Hypsiglena jani, Masticophis flagellum, M. schotti, Pantherophis emoryi, Rhinocheilus lecontei, Senticolis triaspis, Tantilla rubra, Thamnophis proximus, Trimorphodon tau, Rena myopica, Micrurus tener* y *Crotalus atrox.*

ENSAMBLE DE LAS SIERRAS DEL SUR Y LADERA ORIENTAL

Esta región incluye la Sierra Madre Oriental al sur del Valle de Jaumave (excluyendo la ladera interior semiárida del Valle de Jaumave y Desierto Chihuahuense) y la frente de la sierra que queda más hacia el oriente en el norte. La altitud varía desde 500 m en la vertiente oriental, 1250 m sobre la vertiente interior, hasta 2785 m. La vegetación dominante es el bosque seco de encino en las laderas orientales y cumbres, y bosque seco de pino-encino en las cumbres y laderas interiores. Los árboles incluyen *Quercus canbyi, Q. polymorpha, Q. rysophylla, Juniperus* flaccida, Pinus montezumae, P. teocote y Arbutus xalapensis frecuentemente con agaves (Maguey sp.) y palmetos (Sabal sp.) en el sotobosque (Martin 1958). En las elevaciones más bajas en la ladera interior estos bosques cambian a las comunidades del norte y del interior descritas anteriormente y en los párrafos siguientes. En la vertiente de la costa el bosque seco de encino frecuentemente cambia a bosque tropical caducifolio de la planicie costera a 300–900 m de altitud, sin embargo, en algunas áreas dos tipos adicionales de vegetación también se presentan, el bosque mesófilo de montaña y el bosque tropical perenifolio, aunque éstos están fragmentados y distribuidos desigualmente. El bosque mesófilo de montaña se encuentra a 900–1700 m de altitud, más comúnmente a 1000–1500 m en la Sierra de Guatemala, aunque fragmentos pequeños se observan en el norte y sur. La vegetación del bosque mesófilo de montaña incluye Chamaedorea radicalis, Clethra macrophtlla, Liquidambar styraciflua, Magnolia schiedeana, Meliosma alba, Pinus patula, Quercus germane, Q. sartorii, Turpina occidentalis y varias especies de Clethra y Symplocos (Martin 1958; Jones y Gorchov 2000; Alcántara et al. 2002). El bosque tropical perenifolio se presenta desigualmente a elevaciones más bajas (aproximadamente 500–1000 m) sobre la vertiente oriental, más comúnmente entre el bosque mesófilo de montaña y el bosque tropical caducifolio en la Sierra de Guatemala, aunque segmentos adicionales se presentan en el sur. El bosque tropical perenifolio incluye Dendropanax arboreus, Brosimum alicastrum, Iresine tomentella, Quercus germana, Spondias sp., Ungnadia speciosa, Zamia sp. (Martin 1958; Cram et al. 2006). Por lo menos 60 especies de anfibios y reptiles habitan aquí, incluyendo Chiropterotriton cracens, C. multidentatus, Pseudoeurycea belli, P. cephalica, P. scandens, Craugastor augusti, C. decoratus, Eleutherodactylus cystignathoides, E. guttilatus, E. longipes, Incilius nebulifer, Rhinella marina, Ecnomiohyla miotympanum, Hyla eximia, Smilisca baudinii, Hypopachus variolosus, Lithobates berlandieri, Kinosternon integrum, K. scorpioides, Abronia taeniata, Gerrhonotus infernalis, Ophisaurus incomptus, Laemanctus serratus, Anelytropsis papillosus, Sceloporus cyanogenys, S. grammicus, S. minor, S. parvus, S. scalaris, S. torquatus, S. variabilis, Anolis sericeus, Plestiodon dicei , Scincella silvicola, Aspidoscelis gularis, Lepidophyma sylvaticum, Xenosaurus platyceps, Adelphicos newmanorum , Amastridium sapperi , Coniophanes fissidens, Drymarchon melanurus, Drymobius margaritiferus, Geophis latifrontalis, Leptodeira septentrionalis, Leptophis mexicanus, Masticophis schotti, Pantherophis bairdi, Pliocercus bicolor, Rhadinaea gaigeae, Salvadora grahamiae, Storeria hidalgoensis, Tantilla rubra, Thamnophis cyrtopsis, T. mendax, Trimorphodon tau, Tropidodipsas sartorii, Micrurus tener, Bothrops asper, Crotalus lepidus y C. totonacus.

Colectivamente se conocen 93 especies de anfibios y repitles para la Sierra Madre Oriental de Tamaulipas: 7 salamandras (7.5%, 1 familia, 2 géneros); 15 anuros (16.1%, 7 familias, 11 géneros); 2 tortugas (2.1%, 1 familia, 1 género); 30 lagartijas (32.2%, 10 familias, 16 géneros); y 39 serpientes (41.9%, 5 familias, 28 géneros). De estas 93 especies: 3 (3.2%) son endémicas de Tamaulipas; 41 (44.0%) son endémicas de México; 41 (44.0%) habitan también en Estados Unidos; y 24 (25.8%) se encuentran también en Centroamérica. Dos especies introducidas, Hemidactylus frenatus y H. turcicus habitan aquí. El bosque mesófilo de montaña de la Sierra de Guatemala y tierras bajas adyacentes del este son las áreas más minuciosamente estudiadas en Tamaulipas. Aproximadamente 35.0% (5447) de todos los especímenes de museo disponible de Tamaulipas fueron recolectados en el área de Gómez Farías. Dentro de esta región Sceloporus olivaceus habita en el Valle de Jaumave y sus cañones (<1750 m de altitud). Sceloporus spinosus está restringida al Valle de Palmillas y laderas interiores del Desierto Chihuahuense y Sceloporus cautus se conoce por un solo espécimen para el Valle de Palmillas. Sceloporus cyanogenys habita en la ladera más oriental de la Sierra Madre Oriental hasta 1050 m de altitud (Martin 1958); sin embargo, algunos registros de museo en localidades del interior son problemáticos y podrían representar cierta confusión con *S. minor*, aunque esto no se ha resuelto. *Coluber constrictor* se conoce por un solo ejemplar. Martin (1958) consideró que el único registro de *Mastigodryas melanolomus* recolectado a 1050 m de altitud no era confiable. Aunque se conoce únicamente para el norte del Valle de Jaumave, *Pantherophis bairdi* habita tanto en las montañas del interior como en las localizada más al oriente en esta región.

Desierto Chihuahuense

Esta región descansa sobre una meseta interior delimitada por la Sierra Madre Occidental al oeste y la Sierra Madre Oriental al este. Ocupa 450,000 km² (Morafka 1977) desde San Luis Potosí, en el sur, extendiéndose hasta el oeste de Texas y el sur de Nuevo México en el norte. Las definiciones de este desierto y el área geográfica que ocupa varían. La definición aquí utilizada sigue los criterios de Henrickson y Straw (1976) y Gómez-Hinostrosa y Hernández (2000). Muchos elementos de este ecosistema entrar en el extremo suroeste de Tamaulipas. La topografía de esta región de Tamaulipas se compone de planicies desérticas en el sur cerca de la frontera con San Luis Potosí (1025 m), hacia el norte de esta región la elevación base asciende a través de cañones y valles hasta la ciudad de Miquihuana (1850 m). Elevandose por encima de la altitud de la base hay una serie de sierras áridas que incluyen las Sierras Las Ventanas (2270 m), La Norita (2490 m), El Fierro (2900 m) y El Pinal (2940 m). La alta Sierra Madre Oriental al norte de Miquihuana y la Sierra Mocha al este, con chaparrales y árboles de Táscate sobre la ladera interior, y bosques de pino-encino en la cima, delineando los límites de esta región en Tamaulipas. Algunos elementos del ecosistema del Desierto Chihuahuense se filtran a través de las elevaciones más bajas de los cañones en los valles de Palmillas y Jaumave. Geologicamente estas sierras son parte de la Sierra Madre Oriental descrita anteriormente, con los valles y sierras más ampliamente separados hacia el oeste. Rocas calcáreas del Cretácico Superior con

rocas del Jurásico y más antiguas presentándose localmente, y depósitos aluviales recientes en las planicies, cañones, y valles de montaña (Gómez-Hinostrosa y Hernández 2000).

La sombra de lluvia de la Sierra Madre Oriental produce una diferencia profunda en climas y vegetación en estas montañas del clima que se presenta pocos kilómetros al este, y que es caliente y árido, con temperaturas de verano hasta de 43°C y temperaturas bajas de invierno rara vez llegando a 0°C. Durante el verano la fluctuación de la temperatura del día con respecto a la noche es de más de 15°C (ejem. 17 de julio 2006, 21.11°C y -36.67°C) son típicas. La precipitación se da principalmente en el verano, ésta varía desde menos de 400 mm en las planicies hasta 500–600 mm en las montañas. Los arroyos de esta región rara vez contienen agua por más de unas pocas horas inmediatamente después de que cae una lluvia. Dos ensambles de anfibios y reptiles se presentan en esta región: Planicies Desérticas y Montañas. Sin embargo, existe un sesgo en las recolectas realizadas en esta región del estado y para propósitos practicos también he incluido ensambles de Estribaciones del Desierto Chihuahuense, Cañones y Valles, y el área de Laguna Isadora, cada uno representando un ecotono. Entre los cerca de 16,000 registros museográficos disponibles para Tamaulipas, únicamente 154 (0.96 %) pertenecen a esta provincia, y la mayoría de ellos fueron recolectados a lo largo de la carretera Méx. 101 la cual atraviesa únicamente por las planicies desérticas y un cañón. La mayoría de mis registros de campo fueron recolectados en planicies, estribaciones, cañones y valles. Las montañas de esta región no han sido adecuadamente muestreadas.

ENSAMBLE DE LAS PLANICIES DEL DESIERTO CHIHUAHUENSE

Aquí domina la vegetación del matorral microfilo incluyendo *Acacia farnesiana*, *Acacia rigidula*, *Larrea tridentata* y *Prosopis julifora*. Otras especies conspicuas incluyen *Yucca filifera* y *Myrtillocactus geometrizans* (Gómez-Hinostrosa y Hernández 2000). Hay vestigios pequeños de pastizales pero las áreas agrícolas de esta región podrían haber hospedado este tipo de vegetación en el pasado. Arena y barro seco están expuestos en algunas áreas donde ha habido sobrepastoreo, algunas veces en parches de varios acres. La altitud generalmente oscila entre 1025 y 1300 m, aunque segmentos aislados se presentan en algunos valles más altos de hasta 1800 m donde la vegetación de chaparral de las montañas adyacentes puede formar un ecotono en el extremo norte y el este. Las 25 especies de anfibios y reptiles de esta región incluyen: *Anaxyrus cognatus, A. debilis, A. punctatus, Incilius nebulifer, Smilisca baudinii, Lithobates berlandieri, Spea multiplicata, Kinosternon integrum, Gerrhonotus farri, Anelytropsis papillosus, Cophosaurus texanus, Phrynosoma modestum, Sceloporus cautus, S. minor, Plestiodon obsoletus, Aspidoscelis gularis, A. inornata, Masticophis flagellum, M. schotti, Pantherophis emoryi, Pituophis deppei, Rhinocheilus lecontei, Tantilla atriceps, Trimorphodon tau* y *Crotalus atrox.*

ENSAMBLE DE LAS MONTAÑAS DEL DESIERTO CHIHUAHUENSE

Matorrales saxicolas caracterizan la vegetación sobre estas pendientes e incluyen *Agave lecheguilla, A. striata, Euphorbia antisyphilitica, Hechitia glomerata, Jatropha dioica, Fouquieria splendens, Ferocactus pilosus* y *Mannillaria candida* (Gómez-Hinostrosa y Hernández 2000). Los matorrales microfilos de las planicies y *Yucca filifera* aparecen también en estas pendientes pero con una abundancia considerablemente menor y la vegetación aquí es normalmente más abierta. La altitud varía de 1400 a 2940 m. En algunas ocasiones pude observar rebaños de cabras, de lo contrario este hábitat está deshabitado y sin perturbar. Actualmente sólo 11 especies de anfibios y repitles se pueden confirmar para esta región incluyendo: *Craugastor augusti, Eleutherodactylus verrucipes, Spea multiplicata, Cophosaurus texanus, Phrynosoma orbiculare, Sceloporus grammicus, S. minor, S. parvus, Masticophis schotti, Tantilla rubra* y *Crotalus molossus.*

Sin duda alguna muestreos futuros aumentaran el número de especies de este ensamble.

ENSAMBLE DE LAS ESTRIBACIONES, CAÑONES Y VALLES DEL DESIERTO CHIHUAHUENSE

La vegetación y herpetofauna de este ensamble reflejan un ecotono compuesto de los dos ensambles descritos anteriormente, sin embargo, la mayoría de los registros de campo fueron obtenidos en estas áreas. La altitud varía de 1100 a 1850 m. Las 33 especies de anfibios y reptiles registradas aquí incluyen: *Craugastor augusti, C. batrachylus, Anaxyrus punctatus, Incilius nebulifer, Hyla eximia, Lithobates berlandieri, Spea multiplicata, Kinosternon integrum, Gerrhonotus infernalis, Crotaphytus collaris, Anelytropsis papillosus, Cophosaurus texanus, Phrynosoma modestum, P. orbiculare, Sceloporus cautus, S. grammicus, S. minor, S. parvus, S. spinosus, Aspidoscelis gularis, Hypsiglena jani, Lampropeltis mexicana, Leptodeira septentrionalis, Masticophis flagellum, M. schotti, Pituophis deppei, Rhinocheilus lecontei, Senticolis triaspis, Tantilla wilcoxi, Micrurus tener, Crotalus atrox, C. molossus* y *C. scutulatus.*

ENSAMBLE DE LAGUNA ISADORA

Esta área única (un ecotono) de cerca de 50 km² se encuentra en una cuenca cerrada al este de Ciudad Tula, rodeada en dos y un medio lados por la Sierra Madre Oriental, pero también abierta y a la misma altitud (1200–1300 m) que las planicies desérticas que se encuentran al oeste y noroeste de esta región. La Laguna Isadora normalmente tiene agua durante todo el año y la vegetación de matorral desértico de las planicies se entremezcla con *Juniperus angosturana* y otros elementos de chaparral de las sierras. Se han registrado siete especies de anfibios y reptiles, algunas de las cuales son más características de las sierras adyacentes, incluyendo *Lithobates berlandieri, Kinosternon integrum, K. scorpioides, Aspidoscelis gularis, Salvadora grahamiae, Thamnophis proximus* y *Tropidodipsas sartorii.*

Actualmente se conocen colectivamente 49 especies de anfibios y reptiles para esta región de Tamaulipas: 11 anuros (22.4%, 6 familias, 8 géneros), 2 tortugas (4.0%, 1 familia, 1 género), 15 lagartijas (32.6%, 7 familias, 9 géneros) y 20 serpientes (40.8%, 3 familias, 16 géneros). Entre estas 49 especies 15 (30.6%) son endémicas de México, 32 (65.3%) habitan también en Estados Unidos y 7 se distribuyen en México y Centroamérica (14.2%). Una especie introducida, *Hemidactylus turcicus*, ha sido recolectada en la región. De estas 49 especies, Morafka (1977) enlistó 37 (75.51%) como especie de desierto, 5 (10.20%) como especie periféricas, y no incluyó *Craugastor batrachylus* en su análisis. Las restantes 6 (12.24%) especies (*Smilisca baudinii, Anelytropsis papillosus, Gerrhonotus farri, Leptodeira septentrionalis, Trimorphodon tau, Tropidodipsas sartorii*) se registraron recientemente en esta región (Farr et al. 2007; Farr et al. 2009, Farr et al. 2013, datos sin publicar 2006, 2009; Bryson y Graham 2010). Cinco especies (*Hyla eximia, Smilisca baudinii, Kinosternon scorpioides, Leptodeira septentrionalis, Tropidodipsas sartorii*) se conocen por sólo un registro o para una localidad periférica a esta región y no son consideradas como representativas de la herpetofauna del Desierto Chihuahuense. Un anuro (*Craugastor batrachylus*) y dos lagartijas (*Gerrhonotus farri, Plestiodon obsoletus*) también son conocidas por un solo espécimen y su estatus en este ecosistema no es claro. Hay gran cantidad de aguajes construidos por el hombre en esta región y esto podría estar causando la creación de un recurso no natural y una abundancia de anuros y de *Kinosternon integrum* que no es normal aquí.

Actualmente se conocen 179 especies de anfibios y reptiles para Tamaulipas: 11 salamandras (6.14%, 3 familias, 5 géneros); 31 anuros (17.31%, 9 familias, 18 géneros); 15 tortugas (8.37%, 6 familias, 11 géneros); 1 cocodrilo (0.55%); 47 lagartijas (26.25%, 13 familias, 23 géneros); y 74 serpientes (41.34%, 6 familias, 45 géneros). Entre estas 179 especies: 6 (3.35%) son endémicas a Tamaulipas; 54 (30.16%) son endémicas a México; 96 (53.63%) habitan en Estados Unidos y México; 29 (16.2%) sólo se conocen para México

y Centroamérica; 25 (13.96%) habitan en Estados Unidos, México y Centroamérica.

Muchas especies neárticas y neotropicales alcanzan sus límites distribucionales en la Planicie Costera del sur de Texas, Tamaulipas y Veracruz. El Trópico de Cáncer en el sur de Tamaulipas es una línea algo arbitraria que separa a estas dos regiones biogeográficas. Con base a la distribución de la herpetofauna, no hay ningún punto de separación claro de estas dos regiones, en su lugar se observa un gradiente que corre sobre el territorio de Tamaulipas y más allá.

Aunque no igualmente distribuidas ni igualmente abundantes, 13 (7.26%) especies han sido registradas en las cuatro Provincias de Tamaulipas y sólo están ausentes en las elevaciones más altas (*Anaxyrus punctatus* [≤ 1840 m], *Incilius nebulifer* [≤ 1740 m], *Lithobates berlandieri* [≤ 2000 m], *Gerrhonotus infernalis* [150–2092 m], *Aspidoscelis gularis* [≤ 200.5 m], *Hemidactylus turcicus* [≤ 1150 m], *Hypsiglena jani* [≤ 1833 m], *Masticophis flagellum* [≤ 1216 m], *M. schotti* [≤ 2000 m], *Pantherophis emoryi* [≤ 1225 m], *Trimorphodon tau* [≤ 1550 m], *Micrurus tener* [≤ 1120 m] y *Crotalus atrox* [≤ 1280 m]. Cinco (2.79%) especies han sido registradas en la Planicie Costera del Golfo, La Sierra Madre Oriental y el Desierto Chihuahuense (*Craugastor augusti* [200–2000 m], *Anelytropsis papillosus* [330–1360 m], *Rhinocheilus lecontei* [≤ 1506 m], *Salvadora grahamiae* [≤ 1600 m] y *Senticolis triaspis* [386–1546 m]). Adicionalmente, 15 (8.37%) especies de la Planicie Costera habitan en las elevaciones más bajas de la Sierra Madre Oriental y ocupan las laderas interior y oeste adyacentes al Desierto Chihuahuense (*Eleutherodactylus cystignathoides* [≤ 1447 m], *Ecnomiohyla miotympanum* [120–1600 m], *Smilisca baudinii* [≤ 1334 m], *Hypopachus variolosus* [≤ 1007 m], *Kinosternon scorpioides* [≤ 1279 m], *Sceloporus olivaceus* [≤ 1750 m], *Anolis sericeus* [100–1211 m], *Coluber constrictor* [≤ 1060 m], *Drymarchon melanurus* [≤ 1386 m], *Drymobius margaritiferus* [≤ 1309 m], *Leptodeira septentrionalis* [≤ 1698 m], *Thamnophis proximus* [≤ 1238 m], *Tropidodipsas sartorii* [≤ 1680 m], *Rena myopica* [≤ 1701 m], *Crotalus totonacus* [≤ 1680 m]). Este patrón distribucional genera la pregunata ¿en qué medida la Sierra Madre Oriental representa una barrera para estas especies? Algunas especies que están ausentes de la Sierra Madre Oriental pero que habitan a ambos lados de ella aparentemente se han dispersado desde el norte (por ejemplo *Anaxyrus debilis, Kinosternon flavescens* y *Phrynosoma modestum*). Para muchas otras de las 33 especies (18.43%) los ambientes templados y áridos de las Provincias Veracruzana y del Desierto Chihuahuense podrían representan una barrera más importante que las elevaciones de la Sierra Madre Oriental.

La Lista Roja del 2011 de la Unión Internacional para la Conservación de la Naturaleza (UICN) categoriza la herpetofauna de Tamaulipas de la siguiente manera: CR = especies en peligro crítico, 2 (1.1%) especies (*Eretmochelys imbricata, Lepidochelys kempii*); EN = en peligro de extinción, 10 (5.58%) especies (*Chiropterotriton cracens, C. multidentatus, Notophthalmus meridionalis, Eleutherodactylus dennisi, Caretta caretta, Chelonia mydas, Sceloporus chaneyi, Xenosaurus platyceps, Ficimia hardyi, Thamnophis mendax*); V = vulnerables, 11 (6.14%) especies (*Pseudoeurycea belli, P. scandens, Craugastor decoratus, Eleutherodactylus longipes, E. verrucipes, Dermochelys coriacea, Terrapene carolina, Abronia taeniata, Crotaphytus reticulatus, Lepidophyma micropholis, Storeria hidalgoensis*); NT = cerca de estar amenazadas, 7 (3.91%) especies (*Bolitoglossa platydactyla, Chiropterotriton priscus, Pseudoeurycea cephalica, Pseudoeurycea galeanae, Ecnomiohyla miotympanum, Kinosternon herrerai, Pseudemys gorzugi*) ; DD = con información deficiente, 5 (2.79%) especies; LC = de preocupación menor, 110 (61.45%) especies (*Craugastor batrachylus, Ophisaurus incomptus, Geophis latifrontalis, Rhadinaea gaigeae, Micrurus tamaulipensis*); NE = no evaluadas, 34 (18.99%) especie. El gobierno federal de México (NORMA Oficial Mexicana NOM-059-SEMARNAT-2010), categorizó la

herpetofauna de Tamaulipas en el año 2010 de la siguiente manera: P = en peligro de extinción, 7 (3.91%) especies (*Notophthalmus meridionalis, Caretta caretta, Chelonia mydas, Eretmochelys imbricata, Lepidochelys kempii, Dermochelys coriacea, Ophisaurus incomptus*); A = Amenazadas, 31 (17.32%) especies (*Pseudoeurycea belli, P. cephalica, P. galeanae, Siren intermedia, Pseudemys gorzugi, Gopherus berlandieri, Crotaphytus collaris, C. reticulatus, Anelytropsis papillosus, Cophosaurus texanus, Phrynosoma orbiculare, Scincella silvicola, Lepidophyma micropholis, Boa constrictor, Coluber constrictor, Lampropeltis getula, L. mexicana, L. triangulum, Leptophis mexicanus, Masticophis flagellum, M. mentovarius, Nerodia erythrogaster, Pituophis deppei, Pliocercus bicolor, Tantilla atriceps, Thamnophis cyrtopsis, T. exsul, T. marcianus, T. mendax, T. proximus, Agkistrodon taylori*); Pr = sujetas a protección especial, 41 (22.9%) especies; y 94 (52.51%) especies sin estatus de protección. Las principales áreas naturales protegidas de Tamaulipas incluyen: Alta Cumbre (30,327 ha.) y Reserva de la Biosfera El Cielo (144,530 ha.) en la Sierra Madre Oriental; Laguna Madre y Delta del Río Bravo (572,808 ha.) y Parras de la Fuente (Reserva de la Paloma de Ala Blanca) (21,948 ha.) en la Provincia Tamaulipeca; Bernal de Horcasitas (18,204 ha.) y Playa de Rancho Nuevo (aproximadamente 20 km de playas) en la Provincia Veracruzana. La mayoría de estas áreas protegidas son de propiedad privada por lo que hay pocos recursos disponibles dedicados a su manejo y aplicación correcta de las leyes de protección de vida silvestre. Todas estas regiones de Tamaulipas se beneficiarían considerablemente de la creación de santuarios adicionales pero, en la Planicie Costera, donde continúa el aclareo de vegetación para agricultura y ganadería, la necesidad de refugios adicionales es quizás la mayor.

BRADFORD D. HOLLINGSWORTH
Y CLARK R. MAHRDT

Introducción

La fauna de anfibios y reptiles de California ha sido enormemente influenciada por procesos geológicos complejos y gradientes climáticos extremos. Como un ejemplo, los puntos más altos y más bajos dentro del área continental del estado se encuentran a tan sólo 136 km de distancia. El Monte Whitney, que forma parte de la Sierra Nevada, alcanza 4421 m de altitud, mientras que la cuenca Badwater en Valle de la Muerte se hunde a -86 m por debajo del nivel del mar. Entre estos dos sitios, el clima boreal de los picos nevados pasa por una transición dramática y culmina con 57°C de temperatura máxima en el suelo del desierto. La posición de California a lo largo del margen continental es la causa principal de su compleja historia geológica y gradientes climáticos. A medida que la Placa del Pacífico se desliza y hunde bajo el continente norteamericano, levantamientos y hundimientos forman y reconfiguran la topografía extrema de California. Gradientes climáticos se desarrollan a medida que el ciclo de tormentas del Pacífico se mueve hacia el sur desde el Golfo de Alaska hasta la costa oeste de California. Las tormentas son más fuertes en el noroeste del Pacífico, pero a media que éstas pasan y se disipan sobre las montañas, sombras de lluvia producen desiertos en sus lados orientales.

Un patrón general emerge donde se encuentra la mayor diversidad de anfibios en el noroeste del Pacífico y hábitats montañosos, mientras que la diversidad más alta de reptiles se encuentra en los desiertos del suroeste. En general, la fauna nativa de anfibios y reptiles de California está compuesta por 173 especies, con 71 anfibios (45 salamandras, 26 anuros) y 102 reptiles (8 tortugas, 48 lagartijas, 46 serpientes). De éstas, 45 (26.0%) son endémicas del estado (33 anfibios y 12 reptiles). Los endemismo se distribuye desigualmente entre los grupos taxonómicos: 30 salamandras, 3 anuros, 10 lagartijas y 2 serpientes. Y, dentro de las salamandras, 25 de las 30 endémicas pertenecen a la familia Plethodontidae, que contiene el mayor número de especies en el estado al nivel taxonómico de familia (35 especies).

Las distribuciones de las especies restantes se extienden más allá de los límites del Estado. Estas influencias externas incluyen el Noroeste del Pacífico, el Desierto de la Gran Cuenca, el Desierto de Sonora y la Península de Baja California. Las distribuciones de 92 especies de reptiles y anfibios de California se extienden a México, representando más de la mitad de la herpetofauna del estado. La mayoría (86 especies) tiene su límite sur de distribución en el estado de Baja California, mientras que pocas (6 especies) se extienden hacia el este a través de

Arizona, y de ahí hacia el sur hasta México continental. El alto grado de sobreposición entre los dos países demuestra la gran influencia que cada uno de ellos tiene sobre el otro en la formación de sus respectivas faunas.

Estudios Herpetológicos Previos

El descubrimiento de anfibios y reptiles nativos de California comenzó a principios del siglo diecinueve, con expediciones por tierra y por mar. La primera especie descrita del material recolectado en el estado fue *Elgaria coerulea*. En 1815, el naturalista alemán Adelbert von Chamisso circunnavegó el mundo como el botánico designado en el buque ruso *Rurik*. A bordo de este barco también estaba el entomólogo Johann Friedrich von Eschscholtz. Su viaje les llevó desde Sudamérica hasta la costa de California donde recolectaron en y alrededor de la Bahía de San Francisco. A su regreso a Europa en 1818, la mayor parte de su material fue depositado en el Zoological Museum of Berlin. En 1828, la descripción de Wiegmann del espécimen tipo de *E. coerulea* incorrectamente reportaba la localidad como "Brasil," que más tarde fue corregida a "San Francisco, California" por Stejneger (1902) después de examinar el ejemplar y los registros del viaje.

De 1823 a 1826 Eschscholtz hizo un segundo viaje a bordo del buque de guerra ruso *Predpriaeteë*, que recorrió la ruta del *Rurik*, pero en esa ocasión fungió como naturalista en jefe de la expedición (Adler 2007). Recolectó a lo largo de la costa y tierra adentro hasta el Gran Valle Central con descubrimientos que condujeron a las descripciones de *Aneides lugubris*, *Batrachoseps attenuatus*, *Taricha torosa* y *Dicamptodon ensatus* (Adler 2007).

Dos expediciones siguieron a las de Chamisso y Eschscholtz. En los años 1827–1829, P.E. Botta estuvo a bordo del buque comercial francés llamado *Le Héros* (Adler 1978). Botta recolectó especímenes a lo largo de toda la costa del estado incluyendo los puertos de San Diego y San Francisco. Los resultados de estas colecciones fueron descritos por Henri Marie Ducrotay de Blainville del Muséum National d'Histoire Naturelle en *Reptiles de La Californie* en 1835 e incluyen las descripciones de *Callisaurus draconoides*, *Charina bottae*, *Elgaria multicarinata*, *Lampropeltis getula*, *L. zonata*, *Pituophis catenifer* y *Tantilla planiceps* (Blainville 1835). La primera expedición estadounidense por mar fue la United States Exploring Expedition (1838–1842). Esta expedición científica de seis buques realizó recolectas de especímenes en toda la región del Pacífico, incluyendo California, Oregón y Washington. Aunque varias descripciones de anfibios y reptiles resultaron del material recolectado, únicamente cuatro fueron incluidas en el reporte de la expedición: *Sceloporus occidentalis*, *Contia tenuis*, *Thamnophis ordinoides* y *Actinemys marmorata* (Baird y Girard 1852b; Girard 1858).

Expediciones y estudios por tierra comenzaron en 1845 y 1846 por el gobierno de Estados Unidos como resultado de la Guerra México-Estados Unidos y la necesidad de entender mejor los límites entre los dos países. William Hemsley Emory, un notable topógrafo, cuya tarea principal era hacer mapas, pero también obtener notas y especímenes de historia natural y mantener contacto con científicos notables del oriente del país (Adler 1978). En 1846, su batallón luchó en la batalla de San Pascual, cerca de San Diego, lo que demuestra la robustez de estos primeros topógrafos y las múltiples funciones que desarrollaban (Adler 1978). La mayor parte del material recolectado por Emory fue depositado en el Smithsonian Institute y junto con los especímenes recolectados por la United States Exploring Expedition, sirvieron de base para numerosas descripciones de especies publicadas por Spencer F. Baird y Charles F. Girard (1852a, 1853). Para California, éste fue el momento de descubrimientos y descripciones de especies más rápido que cualquier otro momento de su historia (Fig. 9.1).

Cuando cesaron las hostilidades entre los dos países y California entró a la Unión en 1850, las expediciones por tierra en territorio de California aumentaron considerablemente y el descubrimiento de nuevas especies continuo. Desde

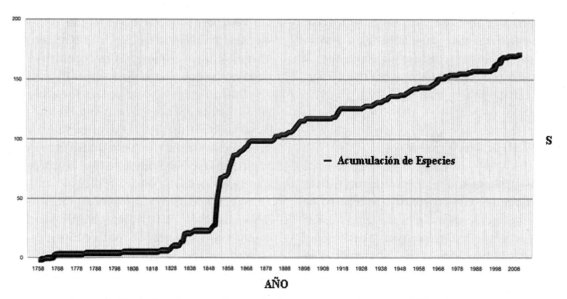

FIGURA 9.1 Acumulación de descripciones de especies de anfibios y reptiles para California.

Nuevo México, expediciones en 1851 al Río Colorado y eventualmente a San Diego, incluyeron al naturalista Samuel Washington Woodhouse, quien fue parte de un equipo que buscaba rutas nuevas hacia el occidente (Adler 1978). El material recolectado por él más tarde fue estudiado por Edward Hallowell e incluye las descripciones de *Pseudacris cadaverina*, *Phrynosoma mcallii* y *Masticophis taeniatus*.

En 1853, el Congreso de Estados Unidos comenzó una búsqueda exhaustiva para encontrar las mejores rutas terrestres para un ferrocarril transcontinental. La mayoría de los equipos de estudio del ferrocarril había incluido a un naturalista, según lo recomendado por Baird. Los naturalistas enviarían los especímenes recolectados al oriente del país para su estudio y publicación (Adler 1978). Como resultado, la alta tasa de descubrimientos de especies continuó y a finales de la década casi la mitad de las especies de California había sido descrita. Los asistentes y protegidos de Baird incluían a Robert Kennicott y Edward D. Cope. Kennicott había descrito cinco especies de California, entre 1859 y 1861, pero falleció a la edad de 30 años mientras exploraba Alaska. Cope fue el autor más prolífico en zoología estadounidense, entre 1860 y 1896 describió 24 especies que habitan en California (Adler 1989).

A finales del siglo diecinueve, las rutas hacia el oeste había quedado bien establecidas y la herpetología comenzó a enfocarse en la exploración de localidades geográficas específicas, detallando la distribución de especies y reportando la variación dentro de las especies. Leonhard Stejneger, nombrado por Baird del Smithsonian Institution, eventualmente fue curador de anfibios y reptiles. Durante su gestión, Stejneger comenzó una nueva era de estudios cuidadosos, descripciones detalladas y designaciones adecuadas de muestras utilizadas en el análisis (Adler 1989). Describió seis especies, incluyendo *Ascaphus truei*, *Gambelia sila*, *Petrosaurus mearnsi*, *Sceloporus orcutti*, *Xantusia henshawi* y *Lichanura orcutti*. En 1891, Stejneger lideró una expedición al Valle de la Muerte, que incluía a Frank Stephens, de la Sociedad de Historia Natural de San Diego y quien recolectó el espécimen tipo de *Crotalus stephensi* (Stejneger 1893; Klauber 1930).

La California Academy of Science, localizada en San Francisco, llegó a ser la primera institución en el estado dedicada a historia natural. Fundada en 1852, la academia abrió su primer museo en 1874. En 1894, John Van Denburgh organizó el primer departamento de herpetología y fue nombrado curador al año siguiente (Slevin y Leviton 1956). Para 1906, había crecido a 8100 especímenes, principalmente del oeste de Norteamérica. Sin embargo, el terremoto de San Francisco destruyó todos estos especímenes excepto 13 de ellos y Van Denburgh se vio forzado a reconstruir la colección (Adler 1989). En ese tiempo, Joseph R. Slevin estaba en las islas Galápagos, y regresó en noviembre de 1906 con 4506 especímenes para comenzar la reconstrucción.

Antes de la destrucción de la colección, Van Denburgh publicó el libro *Los Reptiles de la Costa del Pacífico y la Gran Cuenca* en 1897 (Adler 1989). Fue hasta que Van Denburgh había reconstruido las colecciones cuando él publicó sus trabajos más importantes *Los Reptiles del Oeste de Norteamérica* en 1922, que ha sido comparado con el libro de Holbrook *Herpetología Norteamericana* (1842). Seis años más tarde, Slevin publicó un libro complementario *Los Anfibios del Oeste de Norteamérica* en 1928. Juntos, estos dos libros sirvieron como la base moderna sobre la diversidad taxonómica de California y regiones de alrededor.

Durante la época de la reconstrucción de la colección de la California Academy of Sciences, al otro lado de la Bahía de San Francisco, la Universidad de California, Berkeley abrió el Museo de Zoología de Vertebrados en 1908 (Rodriguez-Robles et al. 2003). Su primer director, Joseph Grinnell, rápidamente hizo los arreglos para construir las colecciones del Museo, enfocándose inicialmente en regiones principalmente de California. En diez años, Grinnell y su alumno Charles L. Camp publicaron la primera revisión sistemática de anfibios y reptiles de California en 1917. Camp continuó trabajando en herpetología durante un corto período de tiempo antes de enfocarse en paleontología. Entre 1915 y 1917, Camp describió seis especies de anfibios de California: *Batrachoseps major, Anaxyrus californicus, A. canorus, Hydromantes platycephalus, Rana muscosa* y *R. sierrae*. Después de Camp, Tracy I. Storer publicó *Una Sinopsis de los Anfibios de California* en 1925.

Durante principios y mediados del siglo veinte, museos, escuelas y universidades comenzaron a crecer y en última instancia incluyeron a herpetólogos en el personal. En 1926, la San Diego Society of Natural History nombró a Laurence M. Klauber como su curador. El trabajo de Klauber se centró en el suroeste de Estados Unidos y sus numerosas revisiones de géneros aclararon gran parte de la confusión taxonómica que lo atrajo a trabajar en esta disciplina. Su principal obra fue una publicación de dos volúmenes sobre serpientes de cascabel, que aún se imprime en la actualidad (Klauber 1956, 1972).

Desde la época de las expediciones de la Boundary Commission y el Ferrocarril en la década de 1850, la acumulación de descripciones de especies nuevas de anfibios y reptiles ha seguido aumentando a un ritmo sostenido. El enfoque inicial del Museum of Vertebrate Zoology sobre la diversidad del estado hizo a éste uno de los más productivos. Robert C. Stebbins y David B. Wake, junto con sus estudiantes y colegas, han descrito 14 especies de California. Además, Stebbins ha elaborado una serie de guías de campo desde inicios de la década de 1950 hasta la actualidad (Stebbins y McGinnis 2012). Actualmente, se pueden encontrar herpetólogos productivos en casi todas las universidades estatales, así como en los cinco museos de historia natural de California.

A través de la historia de la herpetología de California, 102 especies de anfibios (38 salamandras, 12 anuros) y reptiles (1 tortuga, 30 lagartijas, 21 serpientes) tienen sus localidades tipo dentro del estado (Cuadro 9.1).

Características Fisiográficas y su Influencia sobre la Herpetofauna

Las distribuciones de anfibios y reptiles nativos de California están vinculadas a su historia evolutiva y evolución geológica de la región (Rissler et al. 2006). Las distribuciones actuales tienen fuertes tendencias a estar correlacionadas con las regiones fisiográficas del estado y patrones de asociaciones faunística son evidentes. California es el tercer estado más grande de los Estados Unidos. Sus 423,970 km² representan 4.3% del país. El estado tiene 1240 km de largo y 400 km de ancho, situado entre 32°32′ - 42°0′ de

CUADRO 9.1 Localidades tipo para anfibios y reptiles descritos para el estado de California, Estados Unidos

Autor(es)	Nombre original: Localidad tipo
Salamandras (38)	
Gray (1853)	*Ambystoma californiense*: vicinity of San Francisco[1]
Eschscholtz (1833)	*Triton ensatus* (= *Dicamptodon ensatus*): probably near Fort Ross, Sonoma County[2]
Strauch (1870)	*Plethodon flavipunctatus* (= *Aneides flavipunctatus*): Californien (Neu-Albion)
Cope (1883)	*Plethodon iëcanus* (= *Aneides iecanus*): Baird [McCloud River, Shasta County, California, USA][3]
Hallowell (1849)	*Salamandra lugubris* (= *Aneides lugubris*): Monterey [Monterey County], Upper California
Myers & Maslin (1948)	*Aneides niger*: near the forks of Waddell Creek, Santa Cruz County, California
Wake & Jackman (1999)	*Aneides vagrans*: a point about 10 km S Maple Creek, Humboldt Co., California [. . .][4]
Eschscholtz (1833)	*Salamandrina attenuata* (= *Batrachoseps attenuatus*): Umbegung der Bai St. Francisco auf Californien
Jockusch et al. (2012)	*Batrachoseps altasierrae*: 1.5 mi [2.4 km] SE Alta Sierra [Greenhorn Mountains], Kern Co., California, USA
Jockusch et al. (2012)	*Batrachoseps bramei*: Packsaddle Canyon, adjacent to Kern River, 1137 m elevation, Tulare Co., California, USA
Marlow et al. (1979)	*Batrachoseps campi*: Long John Canyon, W slope of the Inyo Mountains, [. . .][4] Inyo County, California, USA[4]
Jockusch et al. (1998)	*Batrachoseps diabolicus*: Hell Hollow, [. . .] Mariposa County, California[4]
Wake (1996)	*Batrachoseps gabrieli*: [. . .] Soldier Creek, [. . .] Los Angeles County, California [. . .][4]
Jockusch et al. (2001)	*Batrachoseps gavilanensis*: 0.5 miles (0.8 km) south of cement plant [. . .], San Benito Co., CA[4]
Jockusch et al. (1998)	*Batrachoseps gregarius*: Westfall Picnic Ground east of Highway 41, [. . .], Madera-Mariposa county line, California[4]
Jockusch et al. (2001)	*Batrachoseps incognitus*: [. . .] 14.7 km NE Highway 1 on San Simeon Creek Road, San Luis Obispo County, CA[4]
Jockusch et al. (1998)	*Batrachoseps kawia*: west side of the South Fork, Kaweah river, Tulare County, California
Jockusch et al. (2001)	*Batrachoseps luciae*: Don Dahvee Park, Monterey, Monterey Co., CA
Camp (1915)	*Batrachoseps major*: Sierra Madre, 1000 feet altitude, Los Angeles County, California
Jockusch et al. (2001)	*Batrachoseps minor*: from along the Santa Rita–Old Creek Road, [. . .], San Luis Obispo County, California [. . .][4]
Cope (1869)	*Batrachoseps nigriventris*: Fort Tejon [Kern County], California
Cope (1865)	*Hemidactylium pacificum* (= *Batrachoseps pacificus*): one of the northern Channel Islands[5]
Jockusch et al. (1998)	*Batrachoseps regius*: south bank of the North Fork, Kings River, [. . .], Fresno county, California[4]
Brame & Murray (1968)	*Batrachoseps relictus*: [. . .] Kern River Canyon [. . .], Kern County, California[4]
Wake et al. (2002)	*Batrachoseps robustus*: [. . .] Kern Plateau, Tulare County, California [. . .][4]
Brame & Murray (1968)	*Batrachoseps simatus*: [. . .] the south side of the Kern River Canyon [. . .], Kern County, California [. . .][4]
Brame & Murray (1968)	*Batrachoseps stebbinsi*: [. . .] Piute Mountains, southern Sierra Nevada, Kern County, California [. . .][4]
Gray (1850)	*Ensatina eschscholtzii*: Monterey
Dunn (1929)	*Ensatina klauberi*: Descanso, San Diego County, Calif[ornia]
Gorman (1954)	*Hydromantes brunus*: [. . .] 0.7 miles NNE Briceburg [. . .], Mariposa County, California[4]
Camp (1916)	*Spelerpes platycephalus* (= *Hydromantes platycephalus*): head of Lyell Cañon, [. . .], Yosemite National Park, California[4]
Gorman & Camp (1953)	*Hydromantes shastae*: [. . .] in the narrows of Low Pass Creek [. . .], Shasta County, California[4]

Autor(es)	Nombre original: Localidad tipo
Mead et al. (2005)	*Plethodon asupak*: Muck-a-Muck Creek (41.774 N, 123.031 W) above Scott Bar, Siskiyou County California [. . .][4]
Van Denburgh (1916)	*Plethodon elongates*: Requa, Del Norte County, California
Stebbins & Lowe (1951)	*Rhyacotriton variegates*: 1.3 miles west of Burnt Ranch Post Office, Trinity County, California
Twitty (1935)	*Triturus rivularis* (= *Taricha rivularis*): Gibson Creek, about one mile west of Ukiah [Mendocino County], California
Twitty (1942)	*Triturus sierrae* (= *Taricha sierrae*): Cherokee Creek, in the hills above Chico, Butte County, California
Rathke (1833)	*Triton torosa* (= *Taricha torosa*): in der Umgebung der Bai St. Francisco auf Californien
Anuros (12)	
Camp (1915)	*Bufo cognatus californicus* (= *Anaxyrus californicus*): Santa Paula, 800 feet altitude, Ventura County, California
Camp (1916)	*Bufo canorus* (= *Anaxyrus canorus*): Porcupine Flat, 8100 feet, Yosemite National Park, Mariposa Co[unty], California
Myers (1942)	*Bufo exsul* (= *Anaxyrus exsul*): Deep Springs, Deep Springs Valley, Inyo County, California
Girard (1859)	*Bufo alvarius* (= *Incilius alvarius*): [old] Fort Yuma, Imperial County, California [. . .][4,6]
Cope (1866)	*Hyla cadaverina* (= *Pseudacris cadaverina*): Tejon Pass
Hallowell (1854)	*Hyla scapularis* var. *hypochondriaca* (= *Pseudacris hypochondriaca*): Tejon Pass
Jameson et al. (1966)	*Hyla regilla sierra* (= *Pseudacris sierra*): 1 1/4 miles SSE. of Tioga Pass Ranger Station [. . .][4]
Baird (1854)	*Rana boylii*: vicinity of Coloma, along the South Fork of the American River, El Dorado County[7]
Baird & Girard (1852)	*Rana draytonii*: vicinity of San Francisco[8]
Camp (1917)	*Rana muscosa*: Arroyo Seco Cañon, at about 1300 feet altitude, near Pasadena, [Los Angeles County,] California
Camp (1917)	*Rana sierrae*: Matlack Lake, [. . .], two miles southeast of Kearsarge Pass, Sierra Nevada, Inyo County, California[4]
Baird (1859a [1858])	*Scaphiopus hammondii* (= *Spea hammondii*): Fort Reading, California
Tortugas (1)	
Cooper (1861)	*Xerobates agassizii* (= *Gopherus agassizii*): mountains of California, near Fort Mojave
Lagartijas (30)	
Wiegmann (1828)	*Gerrhonotus coeruleus* (= *Elgaria coerulea*): San Francisco, California[9]
Blainville (1835)	*Cordylus* (*Gerrhonotus*) *multi-carinatus* (= *Elgaria multicarinata*): Monterey [Monterey Co., California][10]
Stebbins (1958)	*Elgaria panamintina*: Surprise Canyon, [. . .], on the west side of the Panamint Mountains, Inyo County, California[4]
Papenfuss & Parham (2013)	*Anniella alexanderae*: Shale Rd., 1.3 km S (by road) junction with Hwy. 33, Kern County, California, U.S.A.
Papenfuss & Parham (2013)	*Anniella campi*: Big Spring, 5.8 km NW Junction Hwy. 14 (by Hwy. 178), Kern County, California, U.S.A.
Papenfuss & Parham (2013)	*Anniella grinnelli*: Jack Zaninovich Memorial Nature Trail, Sand Ridge Preserve, Kern County, California, U.S.A.
Gray (1852)	*Anniella pulchra*: Pinnacles National Monument, San Benito County, California[11]

Autor(es)	Nombre original: Localidad tipo
Papenfuss & Parham (2013)	*Anniella stebbinsi*: **El Segundo Dunes, Los Angeles International Airport, Los Angeles County, California, U.S.A.**
Stejneger (1890)	*Gambelia sila*: **Fresno, Cal.**
Baird (1859a [1858])	*Stenodactylus variegates* (= *Coleonyx variegatus*): **Winterhaven [Fort Yuma], Imperial County, California**[12]
Dixon (1964)	*Phyllodactylus nocticolus*: **Agua Caliente Hot Springs, San Diego County, California**
Baird & Girard (1852)	*Crotaphytus dorsalis* (= *Dipsosaurus dorsalis*): **Winterhaven [Fort Yuma], Imperial County**[12]
Stejneger (1894)	*Uta mearnsi* (= *Petrosaurus mearnsi*): **Summit of Coast Range, United States and Mexican boundary line, California**
Gray (1839)	*Phrynosoma blainvillii*: **California**
Hallowell (1852)	*Anota M'Callii* (= *Phrynosoma mcallii*): **close to the present town of Caléxico**[13]
Hallowell (1854)	*Sceloporus magister*: **Fort Yuma, Yuma County, Arizona**[12]
Baird & Girard (1852)	*Sceloporus occidentalis*: **Benicia**[14]
Stejneger (1893b)	*Sceloporus orcutti*: **the flat just east of Campo, San Diego Co., Calif.**[15]
Phelan & Brattstrom (1955)	*Sceloporus uniformis*: **Valyermo, Los Angeles, California**
Cope (1896)	*Sceloporus vandenburgianus*: **Campbell's Ranch, Laguna**[16]
Cope (1895)	*Uma inornata*: **Riverside County, California**[8]
Baird (1859a [1858])	*Uma notata*: **vicinity of Yuma, Arizona**[12]
Cope (1894)	*Uma scoparia*: **Mojave Desert, California**[8]
Hallowell (1854)	*Urosaurus graciosus*: **Winterhaven [Fort Yuma], Calif.**[12]
Van Denburgh (1896)	*Eumeces gilberti* (= *Plestiodon gilberti*): **Yosemite Valley, Mariposa County, California**
Grismer & Galvan (1986)	*Xantusia gracilis*: **Truckhaven Rocks in the Anza-Borrego Desert State Park, [. . .], San Diego County, California**[4]
Stejneger (1893b)	*Xantusia henshawi*: **Witch Creek, San Diego Co., California**
Cope (1883)	*Xantusia riversiana*: **California**
Bezy (1967)	*Xantusia sierrae*: **Granite Station, 9.1 mi (by rd) S Woody, 1700 ft, Kern Co., California**
Baird (1859a [1858])	*Xantusia vigilis*: **Fort Tejon, California**

Serpientes (21)

Blainville (1835)	*Tortrix bottae* (= *Charina bottae*): **Coast Range, opposite Monterey**[8]
Klauber (1943)	*Charina umbratica*: **Fern Valley, near Idyllwild, Riverside County, California**
Stejneger (1889)	*Lichanura orcutti*: **Colorado Desert, San Diego County, California**
Hallowell (1854)	*Rhinostoma occipitale* (= *Chionactis occipitalis*): **Mojave Desert**
Feldman & Hoyer (2010)	*Contia longicaudae*: **California, Mendocino County, 8.6 km E of junction with Highway 1 via State Route 128 [. . .]**[4]
Bocourt (1886)	*Coronella multifasciata* (= *Lampropeltis multifasciata*): **San Luis-Obispo, California**
Blainville (1835)	*Coluber Californiae* (= *Lampropeltis californiae*): **vicinity of Fresno, California**[8]
Lockington (1876) ex Blainville (1835)	*Coluber (Zacholus) zonatus* (= *Lampropeltis zonata*): **northern California**[8]
Hallowell (1853)	*Leptophis lateralis* (= *Masticophis* [= *Coluber*] *lateralis*): **San Diego**[8]
Blainville (1835)	*Pituophis catenifer*: **vicinity of San Francisco**[8]
Baird & Girard (1853)	*Rhinocheilus lecontei*: **San Diego**
Kennicott (1860a)	*Thamnophis atratus*: **California**

Autor(es)	Nombre original: Localidad tipo
Kennicott (1859)	*Thamnophis couchii*: **Pitt River, California**
Baird & Girard (1853)	*Eutainia elegans* (= *Thamnophis elegans*): **El Dorado County, California**
Fitch (1940)	*Thamnophis gigas*: **Gadwall, Merced County, California**
Kennicott (1860)	*Eutaenia hammondii* (= *Thamnophis hammondii*): **San Diego, Cal.**
Baird & Girard (1853)	*Rena humilis*: **Upper Sonoran Life Zone of the Vallecito area**[17]
Hallowell (1854)	*Crotalus cerastes*: **borders of the Mohave river, and in the desert of the Mohave**
Cope (1892)	*Crotalus ruber*: **vicinity of San Diego, California**[8]
Kennicott (1861)	*Caudisona scutulata* (= *Crotalus scutulatus*): **Mojave Desert, California**[8]
Klauber (1930)	*Crotalus stephensi*: **two miles west of Jackass Springs, Panamint Mts., altitude 6200 ft., Inyo County, California**

[1] Restricción de la localidad tipo reportada en (Schmidt 1953).
[2] Restricción de la localidad tipo reportada en (Nussbaum 1976).
[3] Restricción de la localidad tipo reportada en (Dunn 1926).
[4] Localidad tipo abreviada con [. . .]; ver la descripción original para localidad tipo completa.
[5] Restricción de la localidad tipo reportada en (Van Denburgh 1905).
[6] Restricción de la localidad tipo reportada en (Fouquette 1968).
[7] Restricción de la localidad tipo reportada en (Jennings 1988).
[8] Restricción de la localidad tipo reportada en (Schmidt 1953).
[9] Restricción de la localidad tipo reportada en (Stejneger 1902).
[10] Restricción de la localidad tipo reportada en (Fitch 1934).
[11] Restricción de la localidad tipo reportada en (Murphy and Smith 1991).
[12] Restricción de la localidad tipo reportada en (Smith & Taylor 1950).
[13] Restricción de la localidad tipo reportada en (Klauber 1932).
[14] Restricción de la localidad tipo reportada en (Grinnell & Camp 1917).
[15] Restricción de la localidad tipo reportada en (Hall & Smith, 1979).
[16] Restricción de la localidad tipo reportada en (Cochran 1961).
[17] Restricción de la localidad tipo reportada en (Brattstrom 1953).

latitud norte y 114°8′–124°26′ de longitud oeste, y colocado a lo largo de la margen continental con 1466 km de costa en el Océano Pacífico.

La combinación de una gran diversidad de regiones florísticas y un gran número de especies endémicas dificulta estudiar a California como una sola unidad (Stebbins y Major 1965). Dividir el estado por provincias bióticas, zonas de vida, comunidades vegetales, o por áreas faunística y florísticas es un enfoque para entender mejor su complejidad, pero hacer esto ha resultado difícil debido a que los límites de subdivisiones irregulares a menudo son complejos. La topografía montañosa e intrincada y diversidad de climas hace difícil una delimitación precisa. Como tal, la asignación de grupos específicos a ensambles es inexacta porque incluso las principales especies pueden persistir fuera de los límites definidos. Uno de los primeros intentos de explicar la variedad de ecorregiones diferentes dentro del estado fue hecho por Stephens (1905) con la propuesta de dividir California en seis de las principales zonas de vida con 17 subdivisio-

nes faunísticas. Posteriormente, muchos autores han delimitado y dividido el estado, florística (Jepson 1925; Wieslander 1935; Jensen 1947; Benson 1957; Munz y Keck 1959; Hickman 1993; Goudey y Smith 1994; Barbour et al. 2007) o faunísticamente (Van Dyke 1919, 1929; Mayer y Laudenslayer 1988; Schoenherr 1992).

Si bien existe un acuerdo general para las distintas divisiones biogeográficas, las regiones florísticas propuestas por Hickman (1993) han mostrado ser las más útiles en la comprensión de la distribución de la herpetofauna de California. De las 173 especies de anfibios y reptiles nativos del estado y especies de reptiles, 103 (59.5%) se limitan a alguna de las ocho regiones geográficas de Hickman. Debido a que los anfibios y reptiles están más limitados por la disponibilidad de agua, temperatura y tipo de suelo, más que por las asociaciones de plantas, no es sorprendente que sólo una parte esté limitada a las regiones florísticas. Las restantes 70 especies (40.5%) tienen distribuciones más amplias, algunas de las cuales abarcan todo el estado.

Para las especies que se limita a una sola área florística, conformar ensambles ayuda a revelar los patrones emergentes y elementos limitantes que han formado la biogeografía de anfibios y reptiles de California. Las ocho regiones son: (1) Noroeste de California; (2) Montañas Cascada; (3) Provincia de la Gran Cuenca; (4) Oeste Central de California; (5) Gran Valle Central; (6) Sierra Nevada; (7) Suroeste de California; y (8) Provincia del Desierto. Algunos ensambles tienen tan sólo dos especies (Sierra Cascada), comparados con 35 especies del ensamble con más especies (Provincia del Desierto). Además de las subdivisiones florísticas de Hickman (1993), una división marina es agregada para incluir a las cinco especies de tortugas marinas y una serpiente marina que habitan en el Océano Pacífico a lo largo de la costa de California. El resto de las especies es generalista en la ocupación del hábitat en dos o más subdivisiones.

Noroeste de California

Se caracteriza por su clima frío y húmedo, terreno montañoso y frondosos bosques perennes, esta región se extiende desde la frontera con Oregón en el norte hasta justo por encima de la Bahía de San Francisco en el sur. Las Montañas Cascada y el Gran Valle Central están pegados a esta región hacia el este. El noroeste de California contiene aproximadamente la tercera parte de la costa del estado y dos sierras prominentes. Las Montañas Klamath están situadas en el extremo noroeste del estado y están compuestas predominantemente de rocas metamórficas y plutónicas. En el Cretácico, las Montañas Klamath se separaron del norte de la Sierra Nevada, desplazándose hacia el oeste hasta quedar en línea con las Sierras de la Costa Norte (Schoenherr 1992). Las Montañas Klamath consisten de las Montañas Siskyou, Scott Bar, Salmon, Marble, Montañas Scott y de los Alpes Trinity. El Thompson Peak en los Alpes Trinity es el más alto con 2744 m. Hacia el sur se encuentra la Sierra de la Costa Norte que corre paralela a la costa. Estas montañas se formaron por el levantamiento de rocas sedimentarias del Mesozoico

Tardío (Schoenherr 1992). Sus principales picos se elevan por encima de 1500 m, con varios por encima de 2000 m. Combinadas, estas dos sierras comparten tanto clima como vegetación.

El noroeste de California tiene el clima más húmedo en el estado. La precipitación media anual puede superar los 200 cm. La mayor precipitación ocurre en los meses de invierno y el clima frío y húmedo es el más predecible en el estado. Esto favorece el mantenimiento de los bosques primarios, incluyendo los árboles más grandes del mundo, la secoya costera (*Sequoia sempervirens*), que alcanza hasta 115 m de altura y 7.9 m de diámetro. Estos bosques dependen de lluvias estacionales y niebla costera. Los meses de verano calurosos y secos son moderados por nieblas costeras, produciendo un rocío de niebla, que puede comprender hasta el 30% de la precipitación por debajo del dosel forestal. En el interior, las montañas producen una sombra de lluvia y la precipitación anual desciende a 50 cm.

Además de los bosques de secoyas, el noroeste de California hospeda bosques de abeto (*Pseudotsuga menziesii*) y pino amarillo (*Pinus ponderosa*) y una mezcla de especies perennes y de maderas duras. Las Montañas Klamath tienen bosques que incluyen *Tsuga heterophylla*, *Abies grandis* o *Chrysolepis sempervirens*. Más al sur, las Sierras de la Costa Norte cambian a bosques de *Abies procera* y *A. magnifica*, y carecen de *T. heterophylla*. Las partes internas del este y del sur tienen más especies tolerantes a sequía, caracterizadas por chaparral y bosques de pino-encino.

Los suelos dentro de la región generalmente derivan de rocas sedimentarias o metamórficas. La región es conocida por sus extensas áreas de afloramientos de serpentinas, que consisten en suelos con altos contenidos de hierro y magnesio. Los suelos de serpentina tienen efectos profundos sobre la flora. Varios cientos de especies de plantas endémicas están restringidas a este tipo de suelo (Kruckeberg 2006).

ENSAMBLE DEL NOROESTE DE CALIFORNIA

De las 48 especies que habitan en esta área, 15 están limitadas a la región: *Ambystoma gracile,*

Dicamptodon tenebrosus, Aneides ferreus, A. flavi-punctatus, A. vagrans, Plethodon asupak, P. dunni, P. elongatus, P. stormi, Rhyacotriton variegatus, Taricha rivularis, Pseudacris regilla, Ascaphus truei, Rana aurora y *Thamnophis ordinoides.*

La región tiene el tercer mayor número de especies confinadas a sus límites, u 8.7% (15/173) de la herpetofauna total del estado. De éstas, 14 son anfibios y uno es un reptil. Todos son conocidos por su preferencia por hábitats frescos y húmedos, a menudo protegidos de la luz solar bajo el denso dosel del bosque. De los anfibios, 11 son salamandras y tres son anuros.

Los miembros de este ensamble muestran dos patrones de distribución. Nueve especies encuentran su distribución más austral aquí y se extienden hacia el norte en Oregón, Washington y en algunos casos, Columbia Británica y Canadá. Éstos incluyen *Ambystoma gracile, Dicamptodon tenebrosus, Aneides ferreus, Plethodon dunni, Rhyacotriton variegatus, Pseudacris regilla, Ascaphus truei, Rana aurora* y *Thamnophis ordinoides.* Las seis restantes son endémicas a la región e incluyen *Aneides flavipunctatus, Aneides vagrans, Plethodon asupak, Plethodon elongatus, P. stormi* y *Taricha rivularis.* De éstas, *Aneides flavipunctatus, Plethodon stormi* y *P. elongatus* tienen distribuciones que llegan hasta Oregón, pero que no van más allá fuera de la región del Noroeste de California.

Hasta hace poco, *Aneides ferreus* era considerada con una distribución amplia en California. En la actualidad, está confinada a la esquina noroeste de la región, donde representa el límite más austral en la distribución de la especie. Wake y Jackman (1999) describieron *A. vagrans* de la mayoría de las poblaciones pertenecientes a *A. ferreus* de California, con la zona de contacto entre las dos especies, ubicada en el Río Smith, a 34 km al sur de la frontera de Oregón. La distribución aislada de *Aneides vagrans* (o *A. ferreus*) en la Isla de Vancouver, Columbia Británica, durante mucho tiempo fue considerada como una introducción del norte de California. Utilizando registros genéticos e históricos de la industria maderera, Jackman (1999) proporcionó evidencias que la población de la Isla de Vancouver se

originó de envíos de corteza de encino café desde California. Alguna vez considerada como residente sólo de troncos caídos o derribados, ahora se sabe que la preferencia ecológica de *A. vagrans* incluye los doseles de los bosques primarios de Secoya (Spickler et al. 2006).

Otros miembros de este ensamble incluyen taxa recientemente limitados a esta región. Por ejemplo, *Plethodon auspak* se limita a una pequeña área del Río Scott y recientemente fue descrito a partir de poblaciones previamente reconocidas como *P. stormi* o *P. elongatus* (Mead et al. 2005). Similarmente, *Pseudacris regilla* fue dividida en varias especies a través de California, aunque los límites de determinadas especies o las zonas de contactos entre los diversos grupos aún no han sido resueltos (Recuero et al. 2006a, 2006b). Basándose sobre las muestras disponibles incluidas en el estudio, se presume que *P. regilla* está limitada a la región noroeste del estado. La inclusión de *Rana aurora* también es el resultado de la separación de una especie más extendida, habiéndose demostrado que es distinta de *R. draytonii* (Shaffer, Fellers, et al. 2004; Pauly et al. 2008).

Mientras que los anfibios dominan este ensamble, la región hospeda un total de 48 especies, de las cuales 22 son reptiles. La composición de reptiles en la región del noroeste de California se caracteriza por especies que termorregulan eficientemente bajo condiciones limitadas de luz solar en hábitats sombreados, tales como *Sceloporus graciosus* y *Contia longicaudae* o que funcionan bien en temperaturas frescas, tales como *Elgaria coerulea, Charina bottae, Contia tenuis* y *Lampropeltis zonata.* Otros incluyen especies que prefieren hábitats húmedos o humedales, incluyendo *Diadophis punctatus, Thamnophis atratus, T. elegans, T. sirtalis* y *Actinemys marmorata.*

Varias subespecies han sido descritas dentro de la región Noroeste de California, aunque la mayoría han sido consideradas inválidas y algunas pueden formar la base para el reconocimiento de una fauna más diversa. Por ejemplo, se reconocen dos subespecies de *Ensatina eschscholtzii* para esta área, *E. e. oregonensis* y *E. e.*

picta (Moritz et al. 1992). Estudios recientes de *E. eschscholtzii* revelan un complejo escenario biogeográfico, con cuatro poblaciones distintas dentro de *E. e. oregonensis* y apoyan la idea que *E. e. picta* es una especie distinta (Kuchta, Parks y Wake 2009; Kuchta, Parks, Mueller y Wake 2009). En otros casos, el reconocimiento de subespecies no ha sido apoyado: *Elgaria multicarinata scincicauda* no es distinta de *E. m. multicarinata* (Feldman y Spicer 2006); *Diadophis punctatus amabilis* no es distinta de *D. p. occidentalis* (Feldman y Spicer 2006; Fontanella et al. 2008); y *Lampropeltis zonata zonata* no es distinta de otra *L. zonata* (Rodríguez-Robles et al. 1999; Myers et al. 2013).

Poblaciones de *Ambystoma californiense* del sur del Condado de Sonoma están aisladas y no han sido formalmente descritas como una unidad taxonómica, a pesar de información no publicada que las reportan como un linaje evolutivo distinto (Shaffer, Pauly, et al. 2004; Shaffer y Trenham 2005).

La región Noroeste de California contiene algunos de los tramos más extensos de costa sin perturbar en el estado. La "Costa Perdida" se extiende 120 km y tiene pocos caminos o asentamientos. La mayor parte de esta región está protegida en el King Range National Conservation Area. Existen dos parques adicionales en los bosques primarios. El Humboldt Redwoods State Park que protege 6880 hectáreas y el Redwoods National Park que incluye 7948 hectáreas. Las principales industrias que le dan uso a la tierra incluyen la silvicultura, la ganadería y el cultivo de uvas. De éstas, la silvicultura ha recibido la mayor atención por la tala de bosques, el incremento de erosión, y el incremento de cieno de los ríos y arroyos. De las 48 especies que habitan en la región, más de un tercio (35.4%) tienen el estatus de preocupación de conservación.

Montañas Cascada

Ésta es una región dominada por prominentes picos volcánicos, las Montañas Cascada se extienden desde Columbia Británica hasta el norte de California. En California, la región está rodeada por Oregón al norte, por las Montañas Klamath al oeste, por los cultivos agrícolas del Gran Valle Central al suroeste, por la Sierra Nevada al sureste, y por la sabana de enebros de la Meseta Modoc en la Provincia de la Gran Cuenca al este (Hickman 1993). Las Montañas Cascada comenzaron a elevarse hace alrededor de siete millones de años como resultado de un contacto activo entre dos placas como la placa Juan de Fuca debajo del Océano Pacífico, empujado por debajo de la placa Norteamericana (Schoenherr 1992). Los volcanes más característicos de la región son el Monte Shasta y el Monte Lassen, ambos hicieron erupción en tiempos históricos (1786 y 1917, respectivamente). El Monte Shasta alcanza 4322 m de altitud y es el segundo pico más alto en las Montañas Cascada. El Monte Lassen alcanza 3189 m y es el pico más austral de estas montañas.

La región se divide típicamente en estribaciones o tierras altas. Mientras que las montañas altas están cubiertas de nieve durante todo el año, las bases de estas montañas y las áreas de alrededor reciben 150 cm de lluvia al año (Schoenherr 1992). Los efectos de la sombra de lluvia se producen entre los límites de la región Noroeste de California y a lo largo de la vertiente oriental de las Montañas Cascada.

Las elevaciones inferiores de las Montañas Cascada contienen chaparrales y bosques de pino-encino. Las elevaciones más altas están dominadas por pino amarillo (*Pinus ponderosa*) con bosques húmedos en el lado oeste de las laderas. En el lado este de las Montañas Cascada, el hábitat es normalmente más seco y los árboles están más ampliamente espaciados con plantas de la Gran Cuenca que crecen en el sotobosque. Otras especies forestales dominantes incluyen el pino de azúcar (*Pinus lambertiana*), pino de costa (*Pinus contorta*), abeto blanco (*Abies concolor*) y cedro de incienso (*Calocedrus decurrens*).

ENSAMBLE DE LAS MONTAÑAS CASCADA

De las 31 especies que habitan en esta área, dos están limitadas a la región: *Aneides iecanus* e *Hydromantes shastae*. La región tiene el menor número de especies limitadas a ella, o 1.2% (2/173) de la herpetofauna total del estado. Ambas son salamandras con preferencias por hábitats frescos y húmedos. Mientras que la región florística no está fuertemente correlacionada con endemismo de anfibios y reptiles, ésta representa una región geológica única del estado. Ambas especies son endémicas de la región. *Hydromantes shastae* tiene una distribución limitada alrededor de la Presa Shasta y puede encontrarse en cuevas, afloramientos de piedra caliza, y más raramente, en la hojarasca del suelo del bosque. *Aneides iecanus* recientemente fue reconocida como una especie críptica previamente considerada como *A. flavipunctatus* (Rissler y Apodaca 2007) y fue nombrada por Frost (2011). *Aneides iecanus* también se distribuye alrededor de la Presa Shasta, pero se distribuye más ampliamente que *H. shastae*.

De las 31 especies que se distribuyen en esta región, 67.7% (21/31) son reptiles y 32.3% (10/31) son anfibios. De estas, *Rana cascadae* tiene una distribución desde Columbia Británica hasta el norte de California y se encuentra predominantemente en las Montañas Cascada. En California, se distribuía históricamente en toda la región de las Montañas Cascada, así como en los Alpes de Trinidad de la región noroeste de California. En la actualidad, se considera que ha sido extirpada en la mayor parte de la región, es probable que esto sea el resultado de la introducción de especies de trucha para pesca deportiva, se cree que estos peces consumen los huevos, larvas y renacuajos de la *R. cascadae* (Fellers y Drost 1993).

La región de las Montañas Cascada tiene varios parques incluyendo 43,080 hectáreas del Lassen Volcanic National Park y un conglomerado de refugios de vida silvestre, bosques nacionales, y áreas recreativas. Aproximadamente un tercio (10/31) de las especies de la región tiene actualmente un estatus de preocupación de conservación.

Provincia de la Gran Cuenca

En California, esta región está representada como dos áreas geográficas distintas: la Meseta Modoc y el este de Sierra Nevada. La topografía de cuenca y sierras de la Provincia de la Gran Cuenca se extiende hasta California desde su parte central en Nevada y forma el límite este de la mitad norte del estado. La provincia está limitada en el oeste por las Montañas Cascada o la Sierra Nevada.

La Meseta Modoc, situada en la esquina noreste del estado, es una meseta volcánica de 1500 m de altitud, formada hace aproximadamente 25 millones de años, que limita la Gran Cuenca al este y con quien comparte las mismas características florísticas. La artemisa de la Gran Cuenca (*Artemisia tridentata*) es la planta dominante de esta provincia, con la Meseta Modoc que también contiene sabanas de enebro occidental (*Juniperus occidentalis*) y extensos bosques de pino amarillo (*Pinus ponderosa*) y pino de Jeffrey (*Pinus jeffreyi*) (Hickman 1993). Varios lagos interiores y grandes proporcionan agua abundante a la región, incluyendo los lagos Eagle y Goose, los cuales irrigan el Gran Valle Centra a través del Río Pit (Schoenherr 1992). Las Montañas Warner se encuentran al este de la Meseta Modoc y representan una sierra prominente cuyo punto más alto es el Eagle Peak a 3015 m.

El Desierto de la Gran Cuenca se encuentra dentro de la sombra de lluvia de la Sierra Nevada y se compone de una serie de montañas y valles con orientación norte a sur. Este desierto alto tiene valles a aproximadamente 1200 a 2000 m de altitud y ríos que drenan a cuencas cerradas internas, frecuentemente en lagos secos o salinos. La artemisa de la Gran Cuenca es la planta dominante, pero el arbusto de sal (*Atriplex confertifolia*) es más abundante en el centro de las cuencas y el pino piñón (*Pinus monophylla*), el enebro occidental (*Juniperus occidentalis*) y el arbusto negro (*Coleogyne ramosissima*) se encuentra en las laderas de la montaña (Schoenherr 1992). Además, esta región contiene amplios corredores ribereños dominadas por álamos (*Populus fremontii* y *P. angustifolia*) y aspen (*P. tremuloides*). Las Montañas Inyo y White representan las sierras más prominentes cuyos puntos más altos son la Montaña Waucoba a 3390 m y el Pico de la Montaña White a 4344 m. Estas montañas son conocidas por sus bosques subalpinos de pino longevo (*Pinus longaeva*) y pino flexible (*Pinus flexilis*) (Hickman 1993). El pino longevo es una especie de árbol longeva, con el espécimen más longevo en Norteamérica de 4700 años (Schoenherr 1992).

Ambas áreas dentro de esta provincia son relativamente altas y se encuentran en la sombra de lluvia de las Montañas Cascada y de la Sierra Nevada. Como tal el clima de la Provincia de la Gran Cuenca es generalmente más seco y fresco, con agua fluyendo en la región desde las partes altas de las montañas. La provincia es una región de precipitación invernal, principalmente en forma de nieve, totalizando aproximadamente 38 a 50 cm de precipitación anual.

ENSAMBLE DE LA PROVINCIA DE LA GRAN CUENCA

De las 41 especies que habitan en esta región, 6 están limitadas a ella: *Batrachoseps campi, Anaxyrus exsul, Lithobates pipiens, Rana pretiosa, Spea intermontana* y *Phrynosoma douglasii*.

La región hospeda al 3.4% (6/173) de la herpetofauna total del estado. De éstas, 5 son anfibios y 1 es un reptil. Todas son conocidas por su preferencia o tolerancia a climas fríos, y en el caso de los anfibios, su dependencia por hábitats húmedos. De los anfibios, 1 es una salamandra y 4 son anuros.

Los miembros de este ensamble muestran dos patrones de distribución. Cuatro tienen la mayoría de sus distribuciones fuera de California en Oregón o Nevada. Éstas incluyen *Rana pretiosa* y *Phrynosoma douglassii* con distribuciones que se encuentran en la Meseta Modoc o al este de las Montañas Warner, y *Lithobates pipiens* y *Spea intermontana* que se distribuyen en la Meseta Modoc y al este de la Sierra Nevada, en el Desierto de la Gran Cuenca. Existe la posibilidad que *R. luteiventris* habite en la esquina del extremo noreste de California, al este de las Montañas Warner (Camp, 1917), con los reportes de Cope (1889) y Grinnell y Camp (1917) colocándola en el Valle Warner y Pine Creek cerca de Alturas, Condado de Modoc, respectivamente. Sin embargo, no hay estudios recientes

que confirmen su presencia en California y los especímenes de museo citados en trabajos anteriores (MVZ 2098–99) han sido asignados a *R. pretiosa*. El segundo patrón de distribución es el de distribuciones muy restringidas de especies endémicas encontradas en manantiales y arroyos de las Montañas Inyo y White, al este de Sierra Nevada. *Batrachoseps campi* se encuentra a lo largo de las laderas este y oeste de las Montañas Inyo en cañones profundos en hábitats muy localizados (Hansen y Wake 2005). *Anaxyrus exsul* se encuentra dentro de aproximadamente 15 hectáreas de hábitat, al sureste de las Montañas White en Deep Springs Lake y sus manantiales de alrededor (Fellers 2005). Estas distribuciones limitadas están vinculadas a pequeños hábitats acuáticos rodeados de extensos desiertos.

De las restantes 35 especies que habitan en la Provincia de la Gran Cuenca, 7 están limitadas a esta región y a la Provincia del Desierto. Éstas incluyen *Elgaria panamintina, Crotaphytus bicinctores, Gambelia wislizenii, Callisaurus draconoides, Phrynosoma platyrhinos, Hypsiglena chlorophaea* y *Sonora semiannulata*. Todas se distribuyen ampliamente a través de los hábitats de desierto, con la excepción de *Elgaria panamintina*, que tiene una distribución similar a la de especies endémicas de la Provincia de la Gran Cuenca. Se encuentra en las Montañas Inyo y White y puede ser incluida como parte de este ensamble excepto por las poblaciones localizadas más al sur en las sierras Panamint, Argus y Cocos ubicadas dentro de la Provincia del Desierto (Mahrdt y Beaman 2009).

La Meseta Modoc ha sido fuertemente transformada por la agricultura (Schoenherr 1992). Presas y sistemas de riego han afectado gran parte de la cuenca, con lo que queda de humedales protegidos dentro de una serie de refugios de vida silvestre y bosques nacionales. En el Desierto de la Gran Cuenca, al este de la Sierra Nevada, el Valle Owens ha sido alterado considerablemente. Con la finalización de los 359 km del acueducto de Los Angeles en 1913, casi todo el Río Owens fue desviado al área metropolitana de Los Angeles. De las 41 especies que habitan entre las dos regiones de la Provincia de la Gran Cuenca, 13 (31.7%) tienen un estatus de preocupación de conservación. Las dos especies de ranidos restringidas a la región, *Lithobates pipiens* y *Rana pretiosa*, parecen estar casi extirpadas.

Oeste Central de California

Esta región está limitada por el Océano Pacífico al oeste y se extiende desde justo al norte de la Bahía de San Francisco en el norte, hasta Punta Concepción en el sur. El Gran Valle Central delimita esta región hacia el este. La región de la Bahía de San Francisco incluye el Monte Tamalpias, las Islas Farallón, las Montañas Santa Cruz, y el punto más alto, el Monte Diablo con 1173 m de altitud. La Sierra Diablo se extiende hacia el sur, paralela a la costa, conformando las principales montañas de la Sierra Costera del Sur. El pico más alto es la Montaña San Benito con 1597 m. Montañas más pequeñas y más costeras incluyen la Sierra Gabilan, las Sierra Santa Lucía y las Montañas Sierra Madre. La subducción de la placa del Pacífico debajo de la placa de América del Norte es responsable del levantamiento de esta región, sin embargo, la presencia del sistema de la Falla de San Andrés hace la historia geológica complicada (Schoenherr 1992).

El clima de la región está dominado por un ciclo de tormenta de invierno y meses de verano calientes y secos moderados por nieblas costeras. La región recibe 38 a 50 cm de lluvia en las áreas interiores y más meridionales y 114 a 127 cm en las montañas costeras. La Sierra Santa Lucía hospeda el bosque más austral de secoya de costa (*Sequoia sempervirens*) y una mezcla de árboles de madera dura, los cuales pueden ser encontrados en bolsones a lo largo de la costa en las Montañas Santa Cruz y al norte de la Bahía de San Francisco. La mayor parte de esta región hospeda bosques de encino azul (*Quercus douglasii*) y pino de estribaciones (*Pinus sabiniana*) y chaparrales (Hickman 1993).

ENSAMBLE DEL OESTE CENTRAL DE CALIFORNIA

De las 55 especies que habitan en esta región, 5 están limitadas a ella: *Aneides niger*, *Batrachoseps gavilanensis*, *B. incognitus*, *B. luciae* y *B. minor*.

Esta región hospeda 2.9% (5/173) de la herpetofauna total del estado. Todas son salamandras que prefieren climas frescos y húmedos. Debido a que esta región se limita al estado, estas cinco especies son endémicas a las Sierras Costeras.

Aneides niger tiene una distribución aislada en las Montañas Santa Cruz. Recientemente fue elevada al rango de especie utilizando resultados que confirman su separación de *A. flavipunctatus* por comparación de datos moleculares y morfológicos y utilizando modelación de nicho (Rissler y Apodaca 2007). Las restantes cuatro especies son salamandras delgadas. De éstas, *Batrachoseps incognitus*, *B. luciae* y *B. minor* se encuentran en diferentes regiones de las Montañas Santa Lucía y se distinguen por evidencias moleculares y pequeñas diferencias en la morfología de las otras formas simpátricas (Jockusch et al. 2001). *Batrachoseps gavilanensis* está más ampliamente distribuida en las Sierras Costeras del centro y también sobre bases de evidencias moleculares y pequeñas diferencias en la morfología (Jockusch et al. 2001).

De las restantes 50 especies que se encuentran en la región, varias tienen distribuciones limitadas que incluyen la región del Oeste Central y regiones hacia el norte o hacia el sur. Dos especies han sido encontradas entre esta región y el Noroeste de California. La recientemente descrita *Contia longicaudae* se encuentra en las Montañas Santa Cruz y a lo largo de la Sierra de la Costa Norte (Feldman y Hoyer 2010), mientras que *Dicamptodon ensatus* se distribuye desde las Montañas Santa Cruz, alrededor de la Bahía de San Francisco, y hacia el norte hasta Punta Arena en el Condado Mendocino (Bury 2005). Cuatro especies se encuentran entre las regiones del Oeste Central y el Suroeste de California: *Batrachoseps nigriventris*, *Anaxyrus californicus*, *Pseudacris cadaverina* y *Tantilla planiceps*.

Varias subespecies han sido descritas dentro de la región del Oeste Central de California, con algunas formando la base para reconocer una fauna más diversa. Se reconocen tres subespecies de *Ensatina eschscholtzii* para esta área, *E. e. oregonensis*, *E. e. xanthoptica* y *E. e. eschscholtzii* (Moritz et al. 1992). Estudios recientes sobre *E. eschscholtzii* revelan un complejo escenario biogeográfico, con cinco linajes regionales dentro de *E. e. oregonensis* y dos dentro de *E. e. xanthroptica* (Kuchta, Parks y Wake 2009; Kuchta, Parks, Mueller y Wake 2009). Otras subespecies representan aislamientos extremos. *Ambystoma macrodactylum croceum* habita en las Montañas Santa Cruz, 250 km de la población más cercana de *A. m. sigillatum* de la Sierra Nevada. En un estudio reciente, Myers et al. (2013) encontraron que *Lampropeltis zonata* está compuesta de dos linajes con sus respectivas zonas de contacto a la mitad de la región Central del Oeste de California. Como un resultado de esto, *L. z. multifasciata* fue elevada al rango de especie. Se cree que la frontera en la distribución de *Lampropeltis zonata* y *L. multifasciata* está en las proximidades de la línea de separación entre los condados Santa Cruz y Monterey.

En otros casos, el reconocimiento de subespecies no ha sido apoyado: *Diadophis punctatus vandenburgii* no es distinta de *D. p. amabilis* o de *D. p. occidentalis* (Feldman y Spicer 2006; Fontanella et al. 2008) y *Plestiodon gilberti cancellosus* no es distinto de otro *P. gilberti* en la porción suroeste del estado (Richmond y Reeder 2002; Richmond y Jockusch 2007).

Se necesita más investigación en varias subespecies, incluyendo *Anniella pulcra nigra* (véase Parham y Papenfuss 2009), *Sceloporus occidentalis bocourtii*, *Masticophis lateralis euryxanthus*, *Thamnophis atratus atratus*, *T. atratus zaxanthus* y *T. sirtalis tetrataenia* (= *T. s. infernalis*). Mientras que las poblaciones de *Ambystoma californiense* del noroeste del Condado de Santa Bárbara están aisladas y no oficialmente descritas como una unidad taxonómica, datos no publicados sugieren que son un linaje evolutivo distinto (Shaffer, Pauly, et al. 2004; Shaffer y Trenham 2005). Del mismo modo, las poblaciones de *Actinemys marmorata* del Condado de Santa Bárbara han demostrado ser distintas (Spinks y Shaffer 2005). También existe suficiente evidencia para el reconocimiento de tres linajes de *Plestiodon skiltonianus* dentro de esta región (Richmond y Reeder 2002; Richmond y Jockusch 2007).

La región del Oeste Central de California está fuertemente urbanizada en y alrededor del área de la Bahía de San Francisco, que representa la segunda mayor área metropolitana dentro del estado. En toda la región, la Sierra Costera se utilizan para la agricultura y la ganadería. La región tiene varios bosques nacionales y refugios de vida silvestre que proporcionan cierta protección a los recursos naturales de la región. Los parques más grandes incluyen el Parque Estatal Mount Diablo con aproximadamente 8000 hectáreas y el Monumento Nacional Pinnacles con más de 6500 hectáreas. Además, la mayoría de los bosques de secoya se preservan en parques públicos y reservas privadas más pequeñas. De las 55 especies que habitan en la región del Oeste Central de California, un tercio (34.5%) tienen un estatus de preocupación de conservación.

Gran Valle Central

Con aproximadamente 5.8 millones de hectáreas, el Valle Central se extiende 650 km de largo y 110 km de ancho. La región es plana con poco relieve topográfico y sierras que se levantan desde abajo del nivel del mar hasta los 150 m. El Valle Central está rodeado por la Sierra Nevada en el este y el sur, la Montaña Cascada en el Norte y las Sierras Costeras en el oeste. Hay dos sistemas fluviales principales y sus cuencas dividen el valle en una mitad norte y una sur. El Río Sacramento drena la mitad septentrional de las Montañas Cascada y el Río San Joaquín drena la mitad meridional, ambos se unen en el centro para formar un sistema de delta que desemboca en la Bahía de San Francisco.

El clima de la región está dominado por un ciclo de tormentas de invierno y veranos calurosos y secos. La mitad norte del valle recibe hasta 50 cm de precipitación, mientras que la mitad sur recibe no más de 25 cm anualmente (Schoenherr 1992). Las temperaturas en verano a menudo superan 32°C y masas densas de niebla se forman a través de amplios sectores de la cuenca en el invierno.

Los depósitos sedimentarios profundos del Valle Central se originaron por debajo del nivel del mar hace más de 145 millones de años, antes del levantamiento de las Sierras de la Costa

(Schoenherr 1992). En los últimos 5 millones de años, se han acumulado sedimentos de depósitos aluviales de las montañas circundantes (Kruckeberg 2006). Existen pocas características topográficas y el agua se acumula en pantanos y pozas efímeras.

Desde hace mucho tiempo la mayor parte de la vegetación nativa se ha convertido a la agricultura y pocos lugares representan una condición original. Se estima que el 99% del valle ha sido transformado. El zacate amacollado había dominado los hábitats de la pradera e incluido pastos de agujas (*Stipa* sp.), hierba de tres-aristas (*Aristida* sp.), hierba azul (*Poa* sp.) y centeno (*Elymus* sp.), de los cuales, el pasto púrpura de agujas (*S. pulcra*) era el más común (Schoenherr 1992). Además, existían plantas asociadas con marismas, pozas efímeras, vegetación riparia y manchones de encino de sabana (Hickman 1993).

ENSAMBLE DEL GRAN VALLE CENTRAL

De las 33 especies que habitan en esta área, 4 están limitadas a esta región: *Anniella alexanderae, A. grinnelli, Gambelia sila* y *Thamnophis gigas*.

Esta región tiene uno de los menores números de especies confinadas dentro de sus límites con 2.3% (4/173) de la herpetofauna total del estado. Las dos especies de *Anniella*: *A. alexanderae* y *A. grinnelli* fueron recientemente descritas por Papenfuss y Parham (2013) y sus distribuciones ocupan una región pequeña del sur del Valle Central. *Thamnophis gigas* históricamente se distribuía en las mitades septentrional y meridional del Valle Central, pero en la actualidad, la especie se limita a áreas alrededor del Río Sacramento (Rossman et al. 1996). *Gambelia sila* históricamente se distribuía en la mitad más seca (sur) del Valle Central y las zonas adyacentes a lo largo de su borde suroeste (McGuire 1996). En la actualidad, *G. sila* se limita a poblaciones esparcidas a lo largo de su distribución histórica en el Valle Central y las mesetas adyacentes de Carrizo y Elkhorn y al Valle Cuyama (Germano 2009).

De las 29 especies restantes que se distribuyen en el Valle Central, *Sceloporus occidentalis*

biseriatus y *Masticophis flagellum ruddocki* necesitan ser investigadas más profundamente. Adicionalmente, las especies de *Annietta* recientemente descritas para la mitad sur del valle tienen distribuciones separadas que requieren de un análisis mas detenido (Parham y Papenfuss 2009; Papenfus y Parham 2013).

Quedan pocas áreas cuyas condiciones sean originales en el Gran Valle Central. El uso agrícola y urbanización alrededor de Sacramento, Modesto, Fresno y Bakersfield son los factores principales en la transformación de este paisaje de pradera. Existen pequeñas reservas en todo el valle, incluyendo varios Refugios Nacionales de Vida Silvestre, en su mayoría situados alrededor de los humedales semi-alterados. De las 33 especies que habitan en el Gran Valle Central, un tercio (33.3%) tienen el estatus de preocupación de conservación, incluyendo las dos especies endémicas a esta región.

Sierra Nevada

La característica topográfica más prominente en California es la Sierra Nevada, ésta se eleva a más de 4250 m y está compuesta principalmente de rocas de granito, con recubrimientos volcánicos (Schoenherr 1992). Está limitada por las Montañas Cascada al noroeste, la Provincia de la Gran Cuenca al noreste y este, el Gran Valle Central al oeste y por las provincias del Desierto y del Suroeste de California al sur. Esta sierra se extiende sobre 640 km de longitud y 80 km de ancho. El pico más alto es el Monte Whitney a 4421 m. Además, esta región incluye las Montañas Tehachapi ubicadas en el extremo sur de la Sierra Nevada.

El clima varía considerablemente en función de la latitud y altitud, pero se caracteriza por un ciclo de tormenta de invierno que trae lluvia a las colinas o nieve a las partes altas de las montañas, con veranos calientes y secos. La precipitación anual puede variar entre 50 a 200 cm y las temperaturas desde bajo cero hasta arriba de 32°C. En los meses de verano, las lluvias por la tarde no son raras.

El batolito granítico de la Sierra Nevada se

formó en el Cretácico como resultado de la subducción de la placa del Pacífico por debajo del
borde del continente norteamericano. El levantamiento del batolito comenzó hace 65 millones
de años, con erosión posterior que expuso la roca
granítica. Hace 20 millones de años, volcanes
extensos cubrían la región. El levantamiento actual comenzó entre 3 y 5 millones de años atrás,
resultando en una inclinación hacia el oeste. El
levantamiento a un ritmo más rápido en el extremo sur de la sierra ha resultados en picos más
altos. Ríos y glaciares forman cañones profundos
a ambos lados de la sierra, aunque la mayoría de
los ríos fluyen en el Valle Central.

Florísticamente, la Sierra Nevada está dividida
entre las Estribaciones y la Sierra Alta, y dentro
de cada una, varias subregiones han sido descritas (Hickman 1993). Sobre la mayor parte de las
estribaciones, bosques de encino azul (*Quercus
douglasii*) y de pino de estribaciones (*Pinus
sabiniana*) dominan, con suelos de serpentinas
entre mezclados. En las montañas más altas el
bosque cambia a bosque mixto de coníferas y de
maderas duras, incluyendo pino amarillo (*Pinus
ponderosa*), pino de azúcar (*Pinus lambertiana*),
pino de la costa (*Pinus contorta*), abeto (*Abies
concolor*), cedro (*Calocedrus decurrens*) y encino
negro de California (*Quercus kelloggii*). La Sierra
Alta también es hogar de los árboles más altos
del mundo, la secoya gigante (*Sequoiadendron
giganteum*), que se presenta en manchones esparcidos sobre el lado oeste de las montañas.

ENSAMBLE DE LA SIERRA NEVADA

De las 53 especies que habitan en esta área, 16 están limitadas a la región: *Batrachoseps altasierrae*,
B. bramei, *B. diabolicus*, *B. gregarious*, *B. kawia*,
B. regius, *B. relictus*, *B. robustus*, *B. simatus*,
B. stebbinsi, *Hydromantes brunus*, *H. platycephalus*, *Anaxyrus canorus*, *Rana sierrae*, *Anniella
campi* y *Xantusia sierrae*.

La región tiene el segundo mayor número
de especies confinadas a sus límites, o 9.2%
(16/173) de la herpetofauna total del estado. De
éstas, 14 son anfibios y 2 son reptiles. El ensamble se compone de 12 salamandras, 2 anuros, y 2

lagartijas. La mayoría prefiere hábitats frescos y
húmedos, y muchos requieren refugios escondi-
dos en afloramientos rocosos.

Todos los miembros de este ensamble son
endémicos a esta región. Dos de las Salamandras
Delgadas, *Batrachoseps diabolicus* y *B. gregarious*
están ampliamente distribuidas en las partes
centro y sur, respectivamente, de la Sierra Nevada, mientras que el resto de las especies tienen
distribuciones limitadas en el extremo sur de la
sierra y de las Montañas Tehachapi. *Batrachoseps
regius* es conocido para el Río Kings, *B. kawia*
para el Río Kaweah, *B. relictus* y *B. simatus* para
la parte inferior del Cañón del Río Kern, *B. robustus* para la Meseta Kern y las laderas orientales
del sureste de la Sierra Nevada, y *B. stebbinsi* para
las Montañas Tehachapi. Recientemente, Jockusch et al. (2012) describieron poblaciones de
B. relictus de la parte superior del Cañón del Río
Kern como *B. bramei* y a las poblaciones desde
las Montañas Greenhorn hasta el drenaje del
Río Tula y elevaciones superiores del drenaje del
Río Little Kern como *B. altasierrae*. *Hydromantes
brunus* y *H. platycephalus* prefieren afloramientos de piedra caliza y losas o taludes de laderas
de granito, respectivamente (Wake y Papenfuss
2005a, 2005b). *Hydromantes brunus* se encuentra en las estribaciones del Río Merced, mientras
que *H. platycephalus* está distribuido más
ampliamente a través de la Sierra Nevada, pero a
mayores altitudes (1220 a 3600 m).

De los dos anuros que se limitan a esta
región, *Rana sierrae* fue removida recientemente
de su sinonimia con *R. muscosa* después de que
pruebas genéticas demostraron que las poblaciones en los alrededores de Cañón Kings estaban
estrechamente relacionadas (Vredenburg et al.
2007). Como resultado, *R. sierrae* se distribuye
en gran parte del norte y centro de la Sierra Nevada y *R. muscosa* se distribuye en el extremo sur,
con poblaciones disjuntas en las Sierras Transversales del suroeste de California. *Anaxyrus
canorus* se encuentra en la parte central de Sierra
Nevada en sitios desde 1950 a 3444 m de altitud
(Davidson y Fellers 2005).

Anniella campi fue descrita recientemente
por Papenfuss y Parham (2013) y se conoce

únicamente para tres localidades en una región pequeña del sur de Sierra Nevada. Una de ellas, Big Spring, tiene sólo dos hectáreas de hábitat adecuado para esta especie (Papenfuss y Parham 2013). *Xantusia sierrae* tiene una distribución muy restringida en las estribaciones meridionales de la Sierra Nevada. Se encuentra a lo largo de la vertiente occidental de las Montañas Greenhorn a aproximadamente 500 m de altura y utiliza las grietas de rocas graníticas como su principal refugio (Bezy 2009).

De las restantes 37 especies, se han descrito varias subespecies, con algunas de ellas formando la base para reconocer una fauna más diversa. En la actualidad se reconocen tres subespecies de *Ensatina eschscholtzii* de la Sierra Nevada: *E. e. xanthoptica*, *E. e. platensis* y *E. e. croceater* (Moritz et al. 1992). Estudios recientes de *E. eschscholtzii* revelan un complejo escenario biogeográfico, con dos linajes regionales para *E. e. platensis* y evidencia débil para el reconocimiento de *E. e. croceater* como un linaje distinto (Kuchta, Parks y Wake 2009; Kuchta, Parks, Mueller y Wake 2009; Pereira y Wake 2009). Existen evidencias para la separación de *Taricha sierrae* de *T. torosa* aunque se ha mostrado que hibridan a lo largo del Río Kaweah (Kuchta 2007). También existe evidencia para la distinción de *Diadophis punctatus pulchellus* de la Sierra Nevada. Aunque similar al patrón de distribución de *T. torosa*, la evidencia sugiere que los miembros de las poblaciones costeras de *D. punctatus* migraron hacia el sur de la Sierra Nevada (Feldman y Spicer 2006; Fontanella et al. 2008). Una investigación adicional de zonas de contacto está garantizada.

Hay una serie de poblaciones no descritas que han mostrado ser evolutivamente distintas. Por ejemplo, se han proporcionado evidencias de linajes separados para las poblaciones de la Sierra Nevada de *Charina bottae* (Rodríguez-Robles et al. 2001; Feldman y Spicer 2006), *Contia tenuis* (Feldman y Spicer 2006), *Elgaria multicarinata* (Feldman y Spicer 2006), y *Plestiodon gilberti* (Richmond y Reeder 2002; Richmond y Jockusch 2007). Y es probable que se descubran más especies del complejo evolutivo de *Batra-*

choseps de la porción sur de esta sierra (Jockusch et al. 2012).

En otros casos el reconocimiento de *Lampropeltis zonata multicincta* no parece ser justificado y no es distinta de otra *L. zonata* (Rodríguez-Robles et al. 1999; Myers et al. 2013). Otras subespecies necesitan más investigación, incluyendo *Sceloporus occidentalis taylori*.

La gran extensión y terreno accidentado de la Sierra Nevada proporcionan cierto grado de protección ambiental a la región. Aun así, las estribaciones de la sierra están densamente pobladas de ranchos, sus bosques son talados, y sus ríos han sido convertidos en presas (Schoenherr 1992). Existen tres grandes parques en esta sierra, incluyendo el espectacular Parque Nacional Yosemite y los contiguos Parques Nacionales de Secoya y del Cañón Kings. Los tres juntos representan un total de 576,920 hectáreas de tierra preservada. De las 53 especies que habitan en la Sierra Nevada, 21(39.6%) tienen un estatus de preocupación de conservación.

Suroeste de California

Se caracteriza por su clima costero seco y moderado, la región se extiende desde Punta Concepción hasta la frontera de Estados Unidos y México. Está limitada por el Oeste Central de California, el Gran Valle Central y el sur de la Sierra Nevada al norte, por la Provincia de Desierto al este, y por el Océano Pacífico al oeste. La región contiene aproximadamente la tercera parte de la costa del estado, dos sierras costeras prominentes, cuencas y mesetas costeras, y las Islas del Canal del California. Las influencias costeras se ven interrumpidas por los hábitats de islas continentales de sus montañas. Las sierras transversales están ubicadas de este a oeste e incluyen las montañas San Bernardino, San Gabriel, Santa Mónica y Santa Ynez. De éstas, el Monte San Gorgonio que forma parte de las Montañas San Bernardino es el más alto, con una altitud de 3505 m. Al sur de estas sierras se encuentran ubicadas de norte a sur las Sierras Peninsulares, que incluyen las Montañas San Jacinto, Palomar, Cuyamaca y Laguna. De éstas, el Monte San Jacinto es el más alto con 3302 m de altitud. La región también incluye las Islas del Canal de California, las cuales están divididas en dos grupos (Schoenherr et al. 1999). Las Islas del Canal del Norte se ubican en una línea este—oeste en los Condados de Ventura y Santa Bárbara. Éstas están formadas por cuatro islas: Anacapa, Santa Cruz, Santa Rosa y San Miguel. Las Islas del Canal del Sur están más dispersas y están formadas por las islas: Santa Catalina, Santa Bárbara, San Nicolás y San Clemente.

Un complejo conjunto de fallas geológicas están asociados con el movimiento hacia el norte de la Placa del Pacífico relativo a la Placa de Norteamérica a lo largo de la Falla de San Andrés (Schoenherr 1992). Las Sierras Peninsulares representan el extremo norte de la Península de Baja California, que termina con las Montañas de San Jacinto. A través del Paso San Gorgonio al norte comienzan las Sierras Transversales con las Montañas San Bernardino. Durante el Plioceno, el nivel del mar se elevó inundando la región costera y la cuenca de Los Angeles, posiblemente separando las dos sierras. Las Islas del Canal de California son parte de la frontera continental, formada por levantamientos como resultado de la interacción del complejo sistema de fallas de la región (Schoenherr et al. 1999).

El Suroeste de California presenta un clásico clima mediterráneo con largos veranos cálidos e inviernos moderadamente frescos y húmedos, con nieve en las elevaciones más altas. Este tipo de clima promueve vegetación con hojas pequeñas y recubrimientos cerosos tolerantes a sequía (Schoenherr 1992). La zona recibe el final de las tormentas de invierno originadas en el norte del Océano Pacífico resultando en una menor precipitación comparada con las regiones más septentrionales del estado. Las regiones costeras reciben aproximadamente 25 cm de precipitación anual. Al final de la primavera y principios de verano, las temperaturas cálidas son moderadas por las brisas marinas y la niebla. De mediados a finales del verano, las montañas de la región reciben esporádicas lluvias de monzón procedentes de sistemas tropicales que migran desde el sur.

El chaparral domina el paisaje del Suroeste de California, con matorral costero que ocupa las elevaciones más bajas y bosques mixtos de coníferas y maderas duras en las elevaciones más altas. El chaparral tolerante a la sequía también es conocido por su asociación con los incendios periódicos (Hanes 1971), que requiere de este fenómeno para volver a brotar y germinar (Quinn y Keeley 2006). Las especies del chaparral incluyen chamiso (*Adenostoma fasiculatum*), espiga roja (*A. sparsifolium*) y lentisco (*Malosma laurina*). Árboles grandes de madera dura, tales como encino costero (*Quercus agrifolia*) están intercalados a través del paisaje. Las especies de matorral costero incluyen artemisa de California (*Artemisia californica*), encelia de California (*Encelia californica*), salvia negra (*Salvia mellifera*) y salvia blanca (*S. apiana*). Las comunidades de montaña contienen pino amarillo (*Pinus ponderosa*), pino de azúcar (*Pinus lambertiana*), pino de Coulter (*Pinus coulteri*) y encino negro de California (*Quercus kelloggii*).

ENSAMBLE DEL SUROESTE DE CALIFORNIA

De las 56 especies que habitan aquí, 14 están limitadas a esta región: *Batrachoseps gabrieli*, *B. major*, *B. pacificus*, *Ensatina klauberi*, *Gambelia copeii*, *Sceloporus orcutti*, *S. vandenburgianus*, *Aspidoscelis hyperythra*, *Xantusia henshawi*, *X. riversiana*, *Charina umbratica*, *Lichanura trivirgata*, *Masticophis fuliginosus* y *Crotalus ruber*. La región tiene el cuarto mayor número de especies confinadas a sus límites, u 8.1% (14/173) de la herpetofauna total del estado. De éstos, 4 son anfibios y 10 son reptiles.

Los miembros de este ensamble representan endemismos regionales o extensiones del norte de especies distribuidas ampliamente a través de toda la Península de Baja California. Ocho son endémicas de la región, con diferentes formas de mecanismos de aislamiento. De éstas, *Batrachoseps gabrieli* tiene la distribución más limitada y se encuentra en 13 localidades separadas a lo largo de la ladera sur de las Montañas San Gabriel y San Bernardino (Hansen, Goodman y Wake 2005). En comparación, *B. major* está más ampliamente distribuida en toda la región, con importantes poblaciones (*B. m. aridus*) en las laderas de desierto de las Sierras Peninsulares. Además de que *B. major* tiene una distribución más amplia, está representado por al menos dos linajes distintos y es actualmente objeto de más investigaciones (Wake y Jockusch 2000).

Cuatro de las especies endémicas de la región están limitadas a las islas continentales, con diferentes grados de preferencias altitudinales. *Charina umbratica* se distribuye desde el Monte Pinos hasta las Montañas San Jacinto y se limita a California (Rodríguez-Robles et al. 2001). *Ensatina klauberi*, *Sceloporus vandenburgianus* y *Xantusia henshawi* se distribuyen más ampliamente a través de las montañas de la región e incluyen las poblaciones en el norte de Baja California (Grismer 2002; Lovich 2001).

Las otras dos especies endémicas de la región se limitan a las Islas del Canal de California. *Xantusia riversiana* se distribuye a través de las Islas del Canal Sur y se encuentra en San Clemente, San Nicolás y Santa Bárbara, así como, en la Isla Sutil, un islote de la Isla Santa Bárbara (Fellers y Drost 2009). *Batrachoseps pacificus* está restringida a las Islas del Canal del Norte, incluyendo las tres Islas de Anacpa (Este, Centro y Oeste), y las Islas Santa Cruz, Santa Rosa y San Miguel (Hansen, Wake y Fellers 2005).

Seis especies se distribuyen a través de la Península de Baja California hasta la punta sur y tienen sus límites de distribución norte en la región del suroeste de California. Éstas incluyen *Gambelia copeii*, *Sceloporus orcutti*, *Aspidoscelis hyperythra*, *Lichanura trivirgata*, *Masticophis fuliginosus* y *Crotalus ruber* (Grismer 2002).

De las restantes 42 especies que se encuentran en esta región, 4 se comparten con el Oeste Central de California (*Anaxyrus californicus*, *Pseudacris cadaverina*, *Tantilla planiceps* y *Thamnophis hammondii*), y 4 se comparten con la Provincia del Desierto (*Coleonyx variegatus*, *Lichanura orcutti*, *Trimorphodon lyrophanes* y *Crotalus mitchellii*). *Rana muscosa* también se encuentra en el sur de la Sierra Nevada. El resto de las especies están más ampliamente distribuidas y habitan en tres o más de las regiones.

Varias subespecies han sido descritas para la región del Suroeste de California, con algunas de ellas formando la base para reconocer una fauna más diversa. Por ejemplo, mientras que *Diadophis punctatus modestus* y *D. p. similis* no puede distinguirse una de otra, se encontró que son distintas de otra subespecie de *Diadophis* de California (Feldman y Spicer 2006; Fontanella et al. 2008). También existen evidencias para el reconocimiento de *Plestiodon skiltonianus interparietalis* y la posibilidad de linajes adicionales dentro de *P. skiltonianus* (Richmond y Reeder 2002; Richmond y Jockusch 2007).

Otras subespecies regionales que necesitan más investigación incluyen *Coleonyx variegatus abbotti*, *Aspidoscelis tigris stejnegeri*, *Xantusia riversiana reticulata*, *X. r. riversiana* y *Thamnophis sirtalis infernalis*. Por ejemplo, la zona de contacto entre *Crotalus oreganus oreganus* y *C. o. helleri* se encuentra ligeramente al norte de las Montañas Santa Ynez, en el límite entre esta región y el Oeste Central de California (Schneider 1986). Sin embargo, los estudios genéticos no han mostrado evidencia consistente de una clara distinción entre estas dos en esta localidad (Pook et al. 2000; Ashton y de Queiroz 2001; Douglas et al. 2002). Se necesita más investigación sobre dos subespecies endémicas de las Islas del Canal de California, *Pituophis catenifer pumilis* y *Sceloporus occidentalis becki*, aunque Rodríguez-Robles y Jesús-Escobar (2000) proporcionaron evidencia que muestra que *P. c. pumulis* está anidada dentro de *P. catenifer*.

La región Suroeste de California es la zona más urbanizada del estado y enfrenta el crecimiento de la población, la construcción de carreteras y la expansión urbana. Hay parques y reservas regionales pequeñas a lo largo de la costa, pero más del 90% del hábitat natural ha sido transformado en áreas metropolitanas. Las áreas protegidas incluyen bases militares de la región donde el desarrollo es limitado debido a la naturaleza de sus usos. Los hábitats de montaña están protegidos en bosques nacionales y pequeños parques estatales. De las 56 especies que habitan en la región, 21 (37.5%) tienen un estatus de preocupación de conservación.

Provincia del Desierto

Éste es un ecosistema sorprendentemente diverso, se caracteriza por su clima cálido y escasas precipitaciones anuales, pero también contiene un número de sistemas acuáticos y altas sierras montañosas. Está limitado por la Provincia de la Gran Cuenca y la Sierra Nevada al norte, el Suroeste de California al oeste, Baja California al sur, y Arizona y Nevada al este. Dos desiertos se encuentran dentro de la provincia. El más septentrional es el Desierto de Mojave, un desierto de elevación alta que tiene valles que van desde 600 hasta 1200 m de altitud. El más meridional es el Desierto de la Parte Inferior del Colorado, que es parte del extenso Desierto de Sonora, éste tiene valles que van desde −60 a 100 m de altitud. Interrumpiendo el extremo este del Desierto de Mojave, se encuentran sierras que proporcionan gradientes altitudinales extremos. La Sierra de Panamint es la más espectacular, elevándose a 3369 m de altitud, desde los valles que se encuentran por debajo del nivel del mar. A lo largo de su frontera oriental, el Río Colorado tiene un caudal promedio a través del desierto de 402 m³ de agua por segundo.

El Desierto de Mojave es parte del sistema de cuencas y sierras de valles esparcidos y levantamientos montañosos. El Desierto de la Parte Inferior del Colorado contiene el Salton Trough que ha sido sucesivamente inundado por la extensión hacia el norte del Golfo de California en el Plioceno o por el Río Colorado en el Pleistoceno y Holoceno. Secándose en los últimos 10,000 años de los sistemas ribereños en el norte del Desierto de Mojave y del Lago Cahuilla en la Parte Inferior del Colorado, lo que ha creado condiciones ideales para la formación de los sistemas de dunas de viento en gran parte de la región (Schoenherr 1992).

La temperatura media de verano alcanza 43°C mientras que la temperatura media de invierno puede caer hasta 1.1°C. La precipitación anual varía desde 0 a 25 cm en la mayor parte de la región, con las cimas de las montañas, algunas de las cuales permanecen con nieve, recibiendo hasta 50 cm. Porciones en la sombra de lluvia de la Sierra Nevada o de las Sierras Peninsulares reciben proporcionalmente menos precipitación. El Desierto de la Parte Inferior del Colorado frecuentemente recibe lluvias esporádicas de monzón de mediados a fines del verano.

La gobernadora (*Larrea tridentata*) es característica de ambos desiertos, generalmente en asociación con el arbusto burro (*Ambrosia dumosa*). El desierto de Mojave también incluye artemisa de la Gran Cuenca (*Artemisia tridentate*), yuca (*Yucca brevifolia*), yuca de Mojave (*Y. schidigera*) y gran cantidad de cactus tipo cholla (*Cylindropuntia* sp.). El Desierto de la Parte Inferior del Colorado hospeda ocotillo (*Fouquieria splendens*), corona de cristo (*Psorothamnus spinosus*), palo fierro (*Olneya tesota*) y palo verde (*Parkinsonia florida*), además de una rica comunidad de cactus.

ENSAMBLE DE LA PROVINCIA DE DESIERTO

De las 63 especies que habitan aquí, 35 están limitadas a esta subregión: *Anaxyrus cognatus*, *A. microscaphus*, *A. punctatus*, *A. woodhousii*, *Incilius alvarius*, *Lithobates yavapaiensis*, *Scaphiopus couchii*, *Crotaphytus vestigium*, *Coleonyx switaki*, *Phyllodactylus nocticolus*, *Heloderma suspectum*, *Dipsosaurus dorsalis*, *Sauromalus ater*, *Petrosaurus mearnsi*, *Phrynosoma mcallii*, *Sceloporus magister*, *Uma inornata*, *U. notata*, *U. scoparia*, *Urosaurus graciosus*, *U. nigricaudus*, *U. ornatus*, *Xantusia gracilis*, *X. wigginsi*, *Bogertophis rosaliae*, *Chionactis occipitalis*, *Phyllorhynchus decurtatus*, *Thamnophis marcianus*, *Trimorphodon lambda*, *Crotalus atrox*, *C. cerastes*, *C. scutulatus*, *C. stephensi*, *Kinosternon sonoriense* y *Gopherus agassizii*.

La región tiene el mayor número de especies confinadas a sus límites, o el 20.2% (35/173) de la herpetofauna total del estado. De éstas, 7 son anfibios y 28 son reptiles. La mayoría prefiere hábitats cálidos y secos o tienen hábitos nocturnos o secretivos para evitar el calor del desierto.

Los miembros de este ensamble representan especies endémicas de la región, especies que se extienden hacia el norte a través de la Península de Baja California, especies que se distribuyen ampliamente a través de la región de los desiertos, o especies fuertemente ligadas a los hábitats acuáticos que se centran en el Río Colorado.

Las especies endémicas a la región incluyen especies endémicas al Desierto de Mojave o al Desierto de la Parte Inferior del Colorado. Éstas incluyen *Uma inornata, U. notata, U. scoparia, X. gracilis, Crotalus stephensi* y *G. agassizii*. De éstas, *U. inornata* y *Xantusia gracilis* son endémicas del estado. *Uma inornata* está limitada a las arenas arrastradas por viento del Valle Coachella (Barrows y Fisher 2009), mientras que *X. gracilis* se encuentra en formaciones relativamente pequeñas de rocas areniscas dentro del Parque Estatal Anza-Borrego (Grismer y Galvan 1986). Las restantes cuatro especies tienen distribuciones que se extienden más allá de los límites estatales, pero no se distribuyen ampliamente a través de la región. *Uma scoparia* está limitada al Desierto de Mojave e incluye poblaciones aisladas en las Dunas Bouse, Arizona (Espinoza 2009). *Crotalus stephensi* se distribuye a través del Desierto de Mojave y porciones de Nevada (Meik 2008), similar a la recientemente circunscrita *Gopherus agassizii* (Murphy et al. 2011). *Uma notata* se encuentra a través del Desierto de la Parte Inferior del Colorado, el cual se extiende hasta el noreste de Baja California (Grismer 2002).

Ocho especies confinadas a los desiertos de California también habitan ampliamente en toda la Península de Baja California e incluyen *Crotaphytus vestigium, Coleonyx switaki, Phyllodactylus nocticolus, Petrosaurus mearnsi, Phrynosoma mcallii, Urosaurus nigricaudus, Xantusia wigginsi* y *Bogertophis rosaliae* (Grismer 2002). Otras doce especies se extienden ampliamente a través de los desiertos de Norteamérica e incluyen *Heloderma suspectum, Dipsosaurus dorsalis, Sauromalus ater, Sceloporus magister, Urosaurus graciosus, U. ornatus, Chionactis occipitalis, Phyllorhynchus decurtatus, Trimorphodon lambda, Crotalus atrox, C. cerastes* y *C. scutulatus* (Stebbins 2003).

Las restantes 9 especies se limitan a hábitats templados, la mayoría en asociación con el Río Colorado. Además, muchos tienen distribuciones en el Valle Imperial, probablemente como resultado de la creciente industria agrícola y la apertura del All American Canal en 1940, que desvía las aguas del Río Colorado (Schoenherr 1992). Éstas incluyen *Anaxyrus cognatus, A. microscaphus, A. punctatus, A. woodhousii, Incilius alvarius, Lithobates yavapaiensis, Scaphiopus couchii, Thamnophis marcianus* y *Kinosternon sonoriense* (Stebbins 2003).

La Provincia de Desierto alberga varios parques y reservas, incluyendo los Parques Nacionales de Death Valley y de Joshua Tree, la Reserva Nacional Mojave, y el Parque Estatal Desierto de Anza-Borrego. A pesar de tener estas áreas de conservación, los desiertos están fuertemente utilizados con grandes áreas metropolitanas, bases militares, regiones agrícolas, campos geotérmicos, presas a lo largo del Río Colorado y uso de vehículos todo terreno (Schoenherr 1992). Más recientemente, grandes extensiones abiertas de tierras del desierto están siendo desarrolladas para generación de energía solar y eólica. De las 63 especies que habitan dentro de la región, 16 (25.4%) tienen un estatus de preocupación de conservación.

OCÉANO PACÍFICO

La plataforma continental a lo largo de la costa de California se extiende desde 30 hasta 275 km mar adentro y tiene una profundidad promedio de 400 m, antes de hundirse a profundidades de más de 3500 m. La fría Corriente de California sigue la costa, que fluye de norte a sur, pero se interrumpe en Punta Concepción, donde la línea costera se extiende hacia el este para formar la Ensenada del Sur de California (Schoenherr et al. 1999). De sur a norte la Corriente del Sur de California es el resultado de remolinos costeros que se forman a medida que la Corriente de California fluye a través de la ensenada. Las temperaturas del agua en el norte de California promedian 14°C en el verano, mientras que en el Sur de California promedian 20°C.

ENSAMBLE DEL OCÉANO PACÍFICO

Hay 6 especies limitadas a esta subregión: *Hydrophis platurus, Caretta caretta, Chelonia mydas, Eretmochelys imbricata, Lepidochelys olivacea* y *Dermochelys coriacea*. De las 5 especies de tortugas marinas que se distribuyen en las agua de California, *Chelonia mydas* es la más común y se queda en la Bahía de San Diego durante el final del otoño y meses de invierno (Stinson 1984). La presencia de *Hydrophis platurus* es rara, pero se tienen registros de esta especie en localidades tan al norte como San Clemente, Condado Orange (Campbell y Lamar 2004).

Generalistas

De las 173 especies que habitan en California, 70 (40.5%) son consideradas generalistas, con distribuciones que se extienden en dos o más subregiones.

ENSAMBLE GENERALISTA

Este ensamble incluye las siguientes especies: *Ambystoma californiense, A. macrodactylum, Dicamptodon ensatus, Aneides lugubris, Batrachoseps attenuatus, B. nigriventris, Ensatina eschscholtzii, Taricha granulosa, T. sierrae, T. torosa, Anaxyrus boreas, A. californicus, Pseudacris cadaverina, P. hypochondriaca, P. sierra, Rana boylii, R. cascadae, R. draytonii, R. muscosa, Spea hammondii, Elgaria coerulea, E. multicarinata, E. panamintina, Anniella pulchra, A. stebbinsi, Crotaphytus bicinctores, Gambelia wislizenii, Coleonyx variegatus, Callisaurus draconoides, Phrynosoma blainvillii, P. platyrhinos, Sceloporus graciosus, S. occidentalis, S. uniformis, Uta stansburiana, Plestiodon gilberti, P. skiltonianus, Aspidoscelis tigris, Xantusia vigilis, Charina bottae, Lichanura orcutti, Arizona elegans, Coluber constrictor, Contia longicaudae, C. tenuis, Diadophis punctatus, Hypsiglena chlorophaea, H. ochrorhyncha, Lampropeltis getula, L. multifasciata, L. zonata, Masticophis flagellum, Masticophis lateralis, Masticophis taeniatus, Pituophis catenifer, Rhinocheilus lecontei, Salvadora hexalepis, Sonora semiannulata, Tantilla hobartsmithi,* *T. planiceps, Thamnophis atratus, T. couchii, T. elegans, T. hammondii, T. sirtalis, Trimorphodon lyrophanes, Rena humilis, Crotalus mitchellii, C. oreganus* y *Actinemys marmorata.*

Las especies generalistas más ampliamente distribuidas son *Anaxyrus boreas, Sceloporus occidentalis, Diadophis punctatus, Lampropeltis getula, Pituophis catenifer* y *crotalus oreganus*. Estas cuatro especies habitan en las ocho regiones terrestres.

California posee 18 islas que tienen registros de anfibios y reptiles. Las islas se extienden desde las Islas del Canal Sur hacia el norte hasta la Bahía de San Francisco e incluyen las Islas San Clemente, Santa Catalina, San Nicolás, Santa Bárbara, Anacapa, Santa Cruz, Santa Rosa, San Miguel, Morro Rock, Año Nuevo, Farallón Sur y Sureste, Alcatraz, Ángel, Brooks, Marin Este, Marin Oeste, Red Rock y Yerba Buena (Schoenherr et al. 1999). Las especies que habitan estas islas incluyen *Aneides lugubris, Batrachoseps attenuatus, B. major, B. nigriventris, B. pacificus, Ensatina eschscholtzii, Pseudacris hypochondriaca, P. sierra, Rana draytonii, Elgaria coerulea, E. multicarinata, Sceloporus occidentalis, Uta stansburiana, Plestiodon skiltonianus, Xantusia riversiana, Charina bottae, Coluber constrictor, Diadophis punctatus, Hypsiglena ochrorhyncha, Lampropeltis getula, L. multifasciata, Pituophis catenifer, Thamnophis elegans, T. hammondii, Crotalus oreganus* y *Actinemys marmorata*. Además, tres especies han sido introducidas: *Lithobates catesbeianus, L. pipiens* y *Xantusia vigilis*. Hay casos documentados de especies nativas de las islas que han sido transportadas de isla a isla o de tierra continental a isla. Un ejemplo de esto es la presencia de *Uta stansburiana* y *Elgaria multicarinata* en varias Islas del Canal de California, algunas de las cuales se han establecido recientemente (Mahoney et al. 2003).

Veintiuna especies han sido introducidas a California. Algunas de éstas representan daños ecológicos severos, tales como *Lithobates catesbeianus* y *Nerodia* sp., mientras que otras son consideradas inofensivas. La persistencia de cada especie introducida necesita documentación más profunda. Especies introducidas incluyen

Ambystoma mavortium, Eleutherodactylus coqui, Xenopus laevis, Lithobates berlandieri, L. catesbeianus, L. sphenocephalus, Chamaeleo jacksonii, Gehyra mutilata, Hemidactylus turcicus, Tarentola annularis, T. mauritanica, Podarcis sicula, Nerodia fasciata, N. rhombifer, N. sipedon, Indotyphlops braminus, Apalone spinifera, Chelydra serpentina, Chrysemys picta, Graptemys pseudogeographica y *Trachemys scripta.*

Un total de 89 subespecies han sido reconocidas para 30 especies de California. Sin embargo, muchas están siendo estudiadas para comprobar su validez taxonómica. Los anfibios tienen tres especies, *Anaxyrus boreas, Ambystoma macrodactylus* y *Ensatina eschscholtzii* que contienen 10 subespecies entre ellos. Los reptiles tienen 79 subespecies reconocidas dentro de 27 especies.

THOMAS C. BRENNAN Y RANDALL D. BABB

Introducción

Arizona es uno de los estados más ricos en especies de anfibios y reptiles de los Estados Unidos. La heterogeneidad espacial del estado combinada con sus amplios intervalos altitudinal, latitudinal, y régimen de precipitación bimodal se unen para formar una diversa y llamativa flora y fauna regionales incluyendo seis biomas principales y una variedad de comunidades riparias. Muchas especies de plantas y animales alcanzan su límite norte de distribución en Arizona. La provincia de Islas Continentales del sureste de Arizona es un epicentro de arreglos para la diversidad de reptiles, albergando varias especies que no se encuentran en ninguna otra parte de los Estados Unidos. Los desiertos del estado también contribuyen significativamente a su diversidad herpetológica. Arizona es el único estado donde los cuatro Desiertos de Norteamérica están representados. Estos desiertos, aunque pobres en especies endémicas son ricos en su variada fauna de anfibios y reptiles. El Desierto de Sonora es el más caliente, húmedo, y con mayor diversidad biológica de estas comunidades semiáridas y contribuye grandemente a la diversidad de anfibios y reptiles de Arizona.

A pesar de su abundancia en especies de flora y fauna, Arizona no recibió mucha atención de naturalistas hasta mediados del siglo diecinueve. En gran parte esto se debió a su tardío estatus de estado y a la gente nativa que peleó fuertemente la tierra en que vivían y cazaban. El interés inicial en la biota de la región estuvo grandemente confinado a las incursiones exploratorias y comerciales de tramperos, inspectores, mineros y expediciones fronterizas y militares (Coues 1875; Wheeler 1875; Yarrow 1875; Davis 1982; Brown et al. 2009). Tramperos y mineros mostraron poco interés en reptiles y anfibios, probablemente debido a la carencia de valor comercial asociado con estos taxa. El periodo de exploración y evaluación de la frontera fomentó una creciente conciencia e interés en estas formas de vida silvestre menos aprovechables (Coues 1875; Wheeler 1875; Yarrow 1875) y los especímenes tipo de muchas especies de reptiles y anfibios fueron recolectados durante ese tiempo. El interés general en la herpetofauna continúo a través del siglo veinte, el cual vio muchas especies nuevas e inusuales descubiertas en el estado. Este interés ha continuado hasta la actualidad. La relativamente nueva ciencia de análisis genético ha desenterrado varias especies de reptiles y anfibios en años recientes y se espera que varias más sean descubiertas.

Estudios Herpetológicos Previos

Arizona, comparada con el este de los Estados Unidos, tiene una historia relativamente corta de interés e investigación herpetológica. Aunque los misioneros y exploradores españoles se aventuraron al territorio de lo que hoy conocemos como Arizona desde el siglo dieciséis, no registraron nada sobre sus encuentros con anfibios y reptiles. Ya sea que esto haya sido debido a las condiciones estacionales desfavorables para la recolecta de ectotérmos, a la dificultad para encontrar y capturar reptiles y anfibios o a la falta de interés; se había documentado relativamente poco sobre la herpetofauna de este estado. De igual manera otros naturalistas como el hijo del renombrado naturalista John James Audubon quien visitó Arizona en la primera mitad del siglo diecinueve, mencionaron anfibios y reptiles sólo anecdóticamente, como curiosidades. Arizona permaneció como terra incognita hasta la segunda mitad del siglo diecinueve. Las exploraciones de la evaluación de la frontera sur, el sistema ferroviario del oeste, y los consiguientes años territoriales sacaron a luz una diversa y fascinante fauna de reptiles y anfibios.

En estos primeros años el interés en documentar la vida silvestre del estado estuvo guiado en el contexto de explotación comercial y exploración del oeste (Davis 1982; Brown et al. 2009; Baird 1857). Reptiles y anfibios fueron comúnmente recolectados oportunistamente e ignorados a favor de fauna más conspicua como aves y mamíferos. A mediados del siglo diecinueve esta percepción hacia anfibios y reptiles comenzó a cambiar en gran medida debido a intereses más amplios y a las exhortaciones de Spencer Fullerton Baird quien en 1850 había ocupado la posición de curador del Museo Nacional de los Estados Unidos.

John H. Clark, como parte del equipo científico de la expedición de 1851 de la frontera México—Estados Unidos comandada por el Coronel J. D. Graham, recolectó especímenes de anfibios y reptiles en el territorio de Arizona incluyendo los ejemplares tipo de *Sceloporus clarkii*, *Lampropeltis getula splendida* y *Salvadora grahamiae*. Es probablemente que el ejemplar tipo de *Sonora semiannulata* recolectado por Clark, también haya sido capturado en esta expedición (Stickel 1943; Degenhardt et al. 1996).

En 1851 Lorenzo Sitgreaves realizó una expedición para explorar y mapear los ríos Colorado y Zuni viajando desde la Meseta de Colorado del norte de Arizona, al Río Colorado al oeste de Kingman, y hacia el sur a través de Yuma, Arizona. Recolectó anfibios y reptiles incluyendo los especímenes tipo de *Anaxyrus woodhousii* y *Callisaurus draconoides ventralis*. Entre los miembros de esta expedición se encontraba S. W. Woodhouse, *Anaxyrus woodhousii* fue nombrado en su honor.

En 1855 la expedición de la frontera México—Estados Unidos comandada por W. H. Emory recolectó especímenes de anfibios y reptiles incluyendo el ejemplar tipo de *Heloderma suspectum* en la Montaña Morena sobre la frontera internacional al suroeste de San Miguel, Condado Pima, Arizona (Bogert y Martín Del Campo 1956).

Los cirujanos de la armada de los Estados Unidos desempeñaron varias funciones durante su estancia en la frontera sur, además de sus obligaciones médicas y de defensa; ellos eran los naturalistas designados y documentaban y recolectaban vida silvestre. El famoso ornitólogo y colector C. Elliot Coues (que se pronuncia "cows"), fue uno de los primeros naturalistas que inicialmente recolectó anfibios y reptiles en Arizona y a su vez fue autor del primer libro detallado sobre estos grupos. En 1863 Arizona fue declarada un territorio, y C. E. Coues de 20 años de edad, después de graduarse de la University of Columbia, fue asignado para servir en Fort Whipple, situado cerca de lo que actualmente es Prescott. Fue recomendado para esta posición por Spencer F. Baird del Smithsonian, quien estaba familiarizado con las proezas de Coues como naturalista. Recolectó especímenes a todo lo largo del viaje a Fort Whipple. El teniente Charles A. Curtis, quien también estuvo en esta expedición, escribió sobre Coues "Él ha traído un barril de cinco galones de alcohol con criaturas rastreras y retorsientes" y frecuentemente se aventuró más allá de su columna (pelotón) en busca de especímenes, regresando únicamente al final del día con "sus bolsas y costales llenos con los trofeos de su búsqueda." Tan fuerte era el interés y dedicación de Coues para obtener especímenes, que recibió una reprimenda por disparar su arma para recolectar un ave durante un delicado encuentro con nativos del área. Las adquisiciones hechas durante su estancia en Arizona fueron puestas en peligro por algunas personas de su batallón quienes bebieron el alcohol en que los anfibios y reptiles estaban almacenados (Curtis 1902, 6–7). En una publicación fechada el 8 de julio de 1864, Coues describió larvas de *Ambystoma mavortium* encontradas en el Pozo Jacob, como "criaturas de apariencia sospechosa" y parecidas a un pez gato "poseyendo extremidades y branquias largas en forma de dedos" (Coues 1871:200). Coues llegó a su puesto más tarde ese mes.

Aunque su estancia en Fort Whipple duró sólo 16 meses, Coues recolectó una asombrosa cantidad de especímenes y publicó muchas de las bases sobre el trabajo faunístico en Arizona. Los especímenes recolectados durante el servicio militar de Coues fueron depositados en el US National Museum.

En 1875 Coues publicó uno de los primeros artículos científicos sobre la herpetofauna de la región. "Sinopsis de los reptiles y batracios de Arizona: con notas críticas y de campo, y una extensiva sinonimia," el cual apareció en el reporte de Wheeler (1875) titulado *Reporte de las exploraciones geográficas y evaluaciones al oeste del meridiano 100, volumen V.* Éste es el primer estudio detallado sobre los anfibios y reptiles de Arizona. En este reporte Coues discutió sobre 83 especies de anfibios y reptiles junto con numerosas subespecies. Sus esfuerzos sacaron a luz los primeros especímenes de *Crotalus cerberus*, en ese entonces considerada como una subespecie de *Crotalus lucifer = C. oreganus.* Coues propuso el nombre *cerberus* para la subespecie que encontró en Arizona (Coues 1875). Durante el desarrollo del trabajo que finalmente fue publicado en *Reporte de las exploraciones geográficas y evaluaciones al oeste del meridiano 100,* muchos especímenes fueron depositados en el US National Museum. Entre ellos estaban especies notables de Arizona como el primer espécimen de *Lampropeltis pryomelana,* junto con especies comúnmente observadas tales como *Crotalus atrox, C. molossus, C. mitchellii,*

Crotophytus collaris, Gambellia wislizenii y *Kinosternon sonoriense.*

Henry C. Yarrow sirvió como cirujano y zoólogo en jefe en la expedición al oeste del Teniente George M. Wheeler durante los años de 1872 y 1874. Fue el líder de un equipo de formidables hombres de campo, el cual incluía al renombrado ornitólogo Henry W. Henshaw y al paleontólogo Edward D. Cope. Yarrow compiló un reporte sobre la herpetofauna del oeste similar al de Coues, pero de un contexto más amplio y menos enfocado. En la introducción de este reporte titulada *Colección de Batracios y Reptiles hecha en porciones de Nevada, California, Colorado, Utah, Nuevo México y Arizona durante los años de 1871, 1872, 1873 y 1874,* Yarrow indicó que únicamente en el año de 1873 Arizona junto con Nuevo México y Colorado fueron el foco de sus investigaciones herpetológicas. Sin embargo, en 1874, se formaron dos grupos adicionales de historia natural para trabajar en áreas de particular interés herpetológico. Una fue liderada por Yarrow y trabajó en Colorado y Nuevo México mientras que la otra operó bajo la guía del botánico y cirujano Joseph T. Rothrock y recolectó en Arizona y Nuevo México hasta la frontera con México. En el trabajo de Yarrow de 1875 se encuentran las descripciones originales de Cope de *Sceloporus jarrovii* y *Thamnophis rufipunctatus.* El reporte de Yarrow también fue publicado en el reporte de Wheeler (1875) y precedía inmediatamente después del reporte de Coues (Yarrow 1875).

Uno de los biólogos sobresalientes trabajando en el oeste de Estados Unidos a finales del siglo diecinueve fue el renombrado naturalista y colector Edgar A. Mearns quien estuvo asignado a Fort Verde en el centro de Arizona desde marzo de 1884 hasta mayo de 1888. Mearns fue un ferviente colector e hizo muchas exploraciones largas a través de este territorio nuevo. Durante su vida Mearns depositó gran cantidad de especímenes de reptiles en el Smithsonian Museum, aunque aún está por determinarse cuántos de éstos son de Arizona. En 1892 Mearns fue asignado a la Mexican Boundary Surveys y reingresó a Arizona con el colector Frank X. Holzner quien lo asistió en la recolección de especímenes. A través de este esfuerzo se adquirieron numerosos especímenes adicionales para Arizona (Mearns 1907; Richmond 1918).

La década de 1890 vio una proliferación de naturalistas visitando y trabajando en Arizona (Ewan 1981; Davis 1982; Brown et al. 2009). Los cirujanos de la armada de los Estados Unidos jugaron un papel clave como colectores de material biológico en el territorio del oeste, pero colectores profesionales, biólogos, y religiosos jugaron un gran papel en la obtención de especímenes para instituciones (Ewan 1981). La mayoría de estas personas no estaba interesada en anfibios y reptiles sino en disciplinas de historia natural que los pusieron en contacto con estas clases de vertebrados. Un buen número de esto naturalistas recolectó los especímenes tipo de varios anfibios y reptiles y más tarde se volvieron los patronimios de criaturas familiares a biólogos trabajando en el suroeste de ahora. El cirujano de la armada General Timothy E. Wilcox, un botánico asignado a Fort Huachuca en el sureste de Arizona, recolectó reptiles extensamente durante su estancia ahí, incluyendo el espécimen tipo de la culebra Centipedívora de Chihuahua (*Tantilla wilcoxi*) nombrada en su honor (Stejneger 1902 [1903]). También recolectó el espécimen de *Sistrurus cantenatus edwardsi* registrado más al oeste en los terrenos de este fuerte (Holycross 2003). Ningún otro espécimen ha sido documentado en o cerca de Fort Huachuca desde entonces. William W. Price de California, un incansable hombre de campo, recolectó muchos especímenes para John Van Denburgh de la California Academy of Sciences. Van Denburgh (1895) nombró una nueva especie pequeña de cascabel recolectada por Price como *Crotalus pricei* en honor a los esfuerzos del naturalista en 1895. Frank C. Willard fue un maestro de escuela y ornitólogo amateur en Tombstone y recolectó en el sureste de Arizona. Envió una colección de aproximadamente 40 anfibios y reptiles al Field Museum en Chicago. Entre estos especímenes estaba una Cascabel de Nariz-afilada (*Crotalus willardi*), la cual fue descrita por el ictiólogo Seth Meeks en 1905 y nombrada en honor a Willard.

El periodista Herbert Brown fue un ornitólogo/colector quien trabajó en el sur de Arizona y fue uno de los fundadores de las colecciones herpetológica y ornitológica de la University of Arizona. Tan grande fue su dedicación a la historia natural, que su nombre aparece casi en cualquier sección de agradecimientos en publicaciones de ese periodo. La Culebra Ensillada (*Phyllorhynchus browni*) fue nombrada en honor a Brown en 1890 por Leonard Stejneger del US National Museum.

Muchos otros biólogos contribuyeron con especímenes para colecciones a finales del siglo diecinueve y principios del siglo veinte, hombres que son más comúnmente asociados con sus trabajos con otros taxa. Estas colecciones importantes incrementaron significativamente nuestro entendimiento de la herpetofauna de Arizona. Sobresalientes entre éstos fueron Vernon O. Baily, J. A. Loring, E. W. Nelson, E. A. Goldman, Henry Skinner y C. Birdseye.

Varias publicaciones locales sobre la herpetofauna del estado comenzaron a aparecer en la década de 1890 y principios de siglo veinte. Estos reportes frecuentemente compilaron los datos y especímenes adquiridos a través de esfuerzos de colecciones a principios de ese siglo, tales como *Reptiles de la Sierra Huachuca, Arizona* publicado por Leonard Stejneger en 1902 (1903). En esta publicación reporta aproximadamente 30 especímenes recolectados por los notables T. Wilcox, W. Price, F. Holzner, A. Fisher y Teniente Benson. También proporciona la descripción original de *Tantilla wilcoxi* y rechaza *Crotalus scutulatus* como una variante regional de *Crotalus atrox*. Otros compendios fueron el producto de esfuerzos personales. El artículo *Una Colección de Reptiles y Anfibios del Sur de Nuevo México y Arizona* publicado en 1907 por Alexander G. Ruthven es un estudio detallado de aproximadamente 1000 especímenes recolectados durante su expedición a esta región en 1906. Esta publicación fue notable en que dio particular atención al hábitat y contexto en el cual el animal fue encontrado, frecuentemente documentando al organismo en cierto detalle y en que se proporcionó una fotografía para algunas fichas diagnósticas. En 1913 John Van Denburgh y Joseph R. Slevin publicaron *Una Lista de Reptiles y Anfibios de Arizona, con notas sobre especies en la Colección de la Academía de Ciencias de California* que fue el resultado de esfuerzos de campo personales hechos por los autores. Ellos enlistaron 86 especies de reptiles y anfibios que se pensaba habitaban en el estado y discutieron los especímenes depositados en la California Academy. Varias imágenes de especies de *Crotalus* siguen al texto de esta publicación. Van Denburgh publicó muchos artículos sobre la herpetofauna regional durante el tiempo que estuvo en la California Academy of Sciences los cuales tienen aplicaciones para Arizona.

Expediciones tales como el viaje de Henry A. Pilsbry en 1910 enriquecieron a Arizona de nuevas maravillas naturales. Pilsbry, un malacólogo, buscaba moluscos en y alrededor de las montañas Chiricahua, Mule, Dragoon, Santa Rita, Rincón y Baboquivari. Aunque el enfoque de sus expediciones no era herpetológico, en ellas se recolectaron incidentalmente varios reptiles y anfibios, los cuales fueron depositados en la Academy of Natural Sciences Museum en Filadelfia. Este periodo también incluyó al Capitán de la armada de los Estados Unidos William L. Carpenter, un entomólogo y naturalista, quien recolectó una serie de especímenes en el área de Camp Verde y Prescott para el US National Museum en 1889. El banquero y ornitólogo John E. Law recolectó extensamente en Arizona en 1919, particularmente en las Montañas Chiricahua, y depositó numerosos especímenes en el Museum of Vertebrate Zoology de la University of California.

En 1920 el biólogo Joseph R. Slevin, sucesor de J. Van Denburgh como Curador de Herpetología de la California Academy of Sciences, recolectó extensamente en las Montañas Santa Rita y Huachuca. Una pequeña lagartija recolectada por él fue nombrada más tarde en su honor (*Sceloporus slevini*). Como uno de los resultados de este viaje, Slevin depositó más de 100 especímenes en la colección que posteriormente revisó. Las décadas de 1920 y 1930 vieron a Charles T. Vorhies y Walter P. Taylor de la

Biological Survey liderando el estudio y colecta de anfibios y reptiles en Arizona. Además de Herbert Brown, Arizona no contaba con otros biólogos residentes con inclinaciones herpetológicas. Vorhies, un entomólogo de formación, mostró un particular interés en anfibios y reptiles. Recolectó el primer espécimen de *Oxybelis aeneus* y encontró el primer *Craugastor augusti* en Arizona, presionó para la protección del Monstruo de Gila, y publicó un arreglo misceláneo de artículos sobre reptiles.

El interés en los anfibios y reptiles incrementó en los siguientes años y una conciencia de la posición única de Arizona en la escena nacional herpetológica fue lentamente surgiendo. Para las décadas de 1930 y 1940 esta nueva conciencia había ganado suficiente momentum para incentivar investigaciones en las siguientes décadas. Esta acogida de herpetología fue guiada en parte por un incremento en el número de publicaciones populares de la herpetofauna Norteamericana. La serie de manuales de 6 volúmenes escrita por Albert y Anna Wright, Archie Carr, Hobart Smith y Sherman Bishop, aunque de un enfoque nacional, incluyó lo que se conocía sobre la herpetofauna de Arizona, proporcionando observaciones detalladas e información conductual. Muchos herpetólogos notables estuvieron trabajando, aunque sólo ocasionalmente, en este estado. Este periodo vio dos nuevas especies añadirse a la herpetofauna estatal: *Hyla wrightorum* y *Xantusia arizonae* así como numerosas subespecies. El renombrado herpetólogo Howard K. Gloyd, director de la Chicago Academy of Sciences, hizo varias expediciones a Arizona para recolectar para esa institución durante este tiempo.

Durante la década de 1940 Hebert Stanke se volvió una figura pública con los fundamentos del Poisonous Animal Research Laboratory de la Arizona State University. Aunque Stanke es mejor conocido por su trabajo con escorpiones, impartía conferencias, daba consultas y escribía sobre reptiles venenosos y fue conocido por tener una oficina postal estilo abierto donde los estudiantes y el público podían depositar especímenes en su laboratorio durante el día o la noche. El enfoque de la mayoría de sus escritos herpetológicos era de envenenamiento por mordedura de cascabel.

La década de 1950 atestiguó un nuevo surgimiento de la actividad herpetológica en Arizona. Mucho de esto debido a la llegada de Charles H. Lowe a la University of Arizona en 1950. Charles H. Lowe fue el primer herpetólogo residente de Arizona. Los diversos intereses de Lowe están reflejados en la variedad de tópicos de sus numerosas publicaciones. Durante sus 45 años de trabajar con la University of Arizona él, junto con sus estudiantes tales como Robert Bezy, J. Cole y John Wright, incrementaron los aproximadamente 1000 especímenes de la colección herpetológica más de 50 veces, haciendo a ésta la colección más completa de material herpetológico del suroeste de los Estados Unidos en el mundo. La presencia de Lowe trajo numerosos estudiantes a Arizona quienes estudiaron bajo su dirección y expandieron grandemente el conocimiento de los reptiles y anfibios de Arizona. Muchos de éstos se volvieron autoridades y mentores por sí mismos en sus instituciones. Entre una de las muchas e importantes contribuciones de Lowe a la herpetología de Arizona están los numerosos artículos sobre el género *Apsidoscelis*, frecuentemente asociado con J. Wright y J. Cole. Charles Lowe fue autor y coautor de más de 160 publicaciones científicas y populares incluyendo un trabajo en colaboración con C. R. Schwalbe y Terry Johnson titulado *Reptiles Venenosos de Arizona*, el primer trabajo semi-técnico enfocado específicamente al estado.

Otro herpetólogo notable de Arizona, William H. Woodin, llegó al estado en 1952 para trabajar en el renombrado Arizona Sonoran Desert Museum. Dos años más tarde fue nombrado director, encontrándose en una posición para influir en los programas de la institución. Con la asistencia de Lowe, Woodin ayudó a incrementar el entendimiento y en buen grado popularidad de anfibios y reptiles a la vista del público. Woodin también contribuyó al trabajo regional con publicaciones tales como su trabajo de 1953 *Notas sobre algunos reptiles del área Huachuca del sur de Arizona*, y el registro del primer espécimen de

Gyalopion quadrangulare para Arizona (Woodin 1962), y a través de sus extensas colecciones de material museográfico que posteriormente fue depositado en la University of Arizona.

En 1958 Arizona fue el hogar de otro herpetólogo notable. Howard K. Gloyd se mudó a Arizona para ocupar la posición de investigador asociado y profesor de la University of Arizona hasta su retiro en 1974 año en que fue nombrado profesor emérito de esta universidad. Durante su permanencia en la University of Arizona colaboró y sirvió como mentor del joven Lowe.

Autores como Carl Kauffeld y Robert C. Stebbins, un amigo y colega de C. H. Lowe, también contribuyeron con la popularización de los reptiles de Arizona con sus libros e investigaciones. La publicación de Stebbins (1954) de *Reptiles y Anfibios del Oeste de Norteamérica* proporcionó la primera guía detallada del oeste y por lo tanto de la herpetofauna de Arizona. Mientras que la publicación de Kauffeld (1957) *Serpientes y Cazadores de Serpientes* con los capítulos "Cielo Huachuca" y "Camino Ajo" hicieron a Arizona un destino popular para colectores amateurs y profesionales por igual.

El primer libro publicado exclusivamente sobre las serpientes de Arizona apareció en 1965, éste fue escrito por Jack A. Fowlie de la University of California y se tituló *Las Serpientes de Arizona*. Éste cubrió todas las especies y subespecies de serpientes reconocidas para Arizona en un modesto libro de pasta dura. Interesantemente, éste puede permanecer como uno de los pocos libros dedicados a la herpetofauna de Arizona que se han publicado. Sin embargo, a través de las décadas de 1960, 1970 y 1980 muchos herpetólogos estuvieron trabajando en Arizona sobre especies o géneros particulares, y publicando excelentes artículos sobre sus investigaciones.

Wade C. Sherbroke comenzó su excelente trabajo sobre el género *Phyrnosoma* durante el tiempo que estuvo asistiendo a la University of Arizona. Sherbroke continuó su investigación sobre el género sirviendo como director de 1982 a 2003 del Southwestern Research Station en Portal perteneciente al American Museum of

Natural History y aún se mantiene trabajando ahí en la actualidad.

En 1983 el Arizona Game and Fish Department formó el Nongame Branch, una división creada por biólogos dedicados a supervisan a nivel estatal y federal el manejo e investigación de la vida silvestre y especies con estatus especial. Cecil R. Schwalbe fue nombrado el primer herpetólogo trabajando para el estado en 1984 y trabajó con esta institución hasta 1990. *Gopherus agassizii, G. morafkai, Lithobates* sp. y *Ambystoma marvortium stebbinsi* fueron algunas de las especies monitoreadas e investigadas por él. El trabajo del Departamento con estas y otras especies continúa en la actualidad. Después de dejar el Arizona Game and Fish Department Schwalbe regresó a la University of Arizona, trabajando para el U.S. Geological Survey. Ahí, él y sus estudiantes continuaron trabajando con muchas especies de reptiles y anfibios incluyendo las antes mencionadas.

Durante este periodo, la Arizona State University también creo una formidable colección de especímenes de anfibios y reptiles preservados. Actualmente más de 32,000 especímenes, la mayoría de Arizona y el suroeste de los Estados Unidos residen aquí. Esta impresionante colección es el reflejo del excelente trabajo de muchos herpetólogos notables que asistieron y enseñaron en esta universidad. Laurie J. Vitt, James P. Collins, Arthur Hulse, J. E. Platz, Brian K. Sullivan y recientemente Andrew T. Holycross han hecho contribuciones importantes a la colección.

James C. Rorabaugh, ahora retirado del U.S. Fish and Wildlife Service, contribuyó significativamente a través de su carrera a la conservación de los reptiles y anfibios de Arizona. Él es quien más ha hecho para la protección federal de *Lithobates chiricahuensis, Ambystoma mavortium stebbinsi, Phrynosoma mcallii* y *Crotalus willardi obscurus*. Adicionalmente, él y Rob Clarkson fueron los primeros en observar la precipidata disminución de ranidos de Arizona.

Aunque los siguientes años han visto numerosos biólogos que han trabajado y contribuido a revelar los secretos de los anfibios y reptiles de Arizona, pocos han acumulado grandes o diversas cantidades de trabajo enfocándose específicamente a la herpetofauna de nuestro estado. Entre los que lo han hecho están James P. Collins y sus estudiantes quienes han incrementado mucho del entendimiento de la única especie de salamandra nativa del estado, *Ambystoma mavortium*. Brian K. Sullivan, anteriormente estudiante de la Arizona State University, y actualmente profesor del Campus Oeste, y sus estudiantes han escrito sobre reptiles y particularmente sobre anfibios contestando cuestionamientos taxonómicos y ecológicos. J. E. Platz, en cooperación con otros, fue responsable de descifrar problemas de identificación del complejo de Ranas Leopardo en el suroeste de los Estados Unidos y de describir varias especies nuevas. Philip C. Rosen, un investigador de la University of Arizona, ha estado conduciendo y publicando investigaciones sobre una variedad increíble de herpetofauna estatal por más de dos décadas. Dale DeNardo y estudiantes de la Arizona State University han hecho mucho con *Crotalus atrox* y *Heloderma suspectum* y Dale fue uno de los primeros investigadores en descubrir y documentar los nidos de *Heloderma* en vida silvestre.

Tristemente, cualquier revisión sobre investigadores y sus trabajos sobre herpetofauna en un estado tan diverso y rico como Arizona inevitablemente queda incompleto y no menciona todos los trabajos que merecen ser mencionados. Reconocemos esto y únicamente podemos disculparnos por nuestras inadvertidas omisiones. Deseamos agradecemos a Cecil R. Schwalbe por su ayuda y a Andrew T. Holycross que generosamente compartió su cuadro sobre las colectas inciales de serpientes en Arizona.

Muchos de los especímenes tipo de anfibios y reptiles enlistados en el Cuadro 10.1 fueron recolectados antes de la Compra que Gadsden hizo a México en 1853 (Venta de la Mesilla). Como resultado muchos de estos tipos tienen como localidad original Sonora, México. Estas localidades frecuentemente han sido revisadas por varios autores a través del tiempo para reflejar mejor la localidad de donde provienen los especímenes.

Autor(es)	Nombre original: Localidad tipo
Anuros (7)	
Cope (1867)	*Bufo* (= *Anaxyrus*) *microscaphus*: "Territory of Arizona, Upper Colorado River" = Fort Mohave, Mohave Co., Arizona Mohave County, Arizona
Sanders & Smith (1951)	*Bufo debilis retiformis* (= *Anaxyrus retiformis*): 14.4 miles south of Ajo, Pima County, Arizona
Girard (1854)	*Bufo* (= *Anaxyrus*) *woodhousii*: [Territory of] New Mexico, restricted to San Francisco Mountain, Coconino CO., Arizona[1]
Cope (1866)	*Hyla arenicolor*: Northern Sonora, Mexico, restricted to "Santa Rita Mts.," Pima and Santa Cruz counties, Arizona,[2] changed to Peña Blanca Springs, 10 miles northwest of Nogales, Santa Cruz County, Arizona[3]
Taylor (1939 [1938])	*Hyla wrightorum*: Eleven miles south of Springerville, Apache County, Arizona
Platz & Mecham (1979)	*Rana* (= *Lithobates*) *chiricahuensis*: Herb Martyr Lake, 6 km west of Portal, Coronado National Forest, Cochise Co., Arizona
Platz & Frost (1984)	*Rana* (= *Lithobates*) *yavapaiensis*: Tule Creek, Yavapai Co., Arizona
Tortugas (2)	
Le Conte (1854)	*Kinosternon sonoriense*: Tucson, Pima Co., Arizona
Murphy et al. (2011)	*Gopherus morafkai*: Tucson (approximate location 32° 7′ N, 110° 56′ W, elevation 948 m), Pima County, Arizona
Lagartijas (19)	
Gray (1838)	*Elgaria kingii*: Mexico, restringida a Mojárachic Chihuahua,[2] modificada a Huachuca Mountains Arizona[4]
Klauber (1945)	*Coleonyx variegatus bogerti*: Xavier, Pima Co., Arizona
Cope (1869)	*Heloderma suspectum*: Monument 146, Sierra de Moreno, Arizona/Sonora border. Ver Bogert y Martín Del Campo (1956) para una discusión completa de la localidad tipo.
Hallowell (1852)	*Callisaurus draconoides ventralis*: New Mexico, restricted to Tucson, Pima Co., Arizona.[2] See Degenhardt et al. (1996, 139–141) for further discussion on the type locality.
Schmidt (1921)	*Holbrookia maculata pulchra*: Carr Canyon, 5,200 ft, Huachuca Mountains, Cochise Co., Arizona
Gray (1845)	*Phrynosoma solare*: California, modificada a Tucson, [Pima Co.], Arizona[4]
Baird & Girard (1852)	*Sceloporus clarkii*: Province of Sonora, restringida a Santa Rita Mountains [Pima or Santa Cruz Co.], Arizona[2]
Cope *in* Yarrow (1875)	*Sceloporus jarrovii*: "southern Arizona," restringida a "Huachuca Mts. Cochise Co., Arizona,"[2] modificada a "between Fort Grant and the Fort Bowie National Historic Site, Arizona, in an area encompassing the southeastern Pinaleño (Graham), Dos Cabezas, and northwestern Chiricahua mountains," Cochise Co., Arizona.[5]
Hallowell (1854)	*Sceloporus magister*: Near Fort Yuma, at Junction of Colorado and Gila, also near Tucson in Sonora, restringida a Fort Yuma, Yuma, Yuma Co., Arizona[2]
Smith (1937)	*Sceloporus scalaris slevini* (= *S. slevini*): Miller Peak, Huachuca Mountains, Cochise Co., Arizona
Cope (1895)	*Uma rufopunctata*: Arizona, restricted to monument 200, Yuma Desert, Yuma County, Arizona[6]
Van Denburgh (1896)	*Cnemidophorus* (= *Aspidoscelis*) *arizonae*: Fairbank, Cochise County, Arizona
Wright & Lowe (1993)	*Cnemidophorus* (= *Aspidoscelis*) *pai*: Hermit Basin in Grand Canyon, 4800 ft, Coconino County, Arizona

Autor(es)	Nombre original: Localidad tipo
Lowe & Wright (1964)	*Cnemidophorus* (= *Aspidoscelis*) *sonorae*: **2 miles southwest of Oracle, 4,500 ft. elev., near the north base of the Santa Catalina Mountains, Pinal Co., Arizona**
Wright & Lowe (1965)	*Aspidoscelis uniparens*: **Fairbank, Cochise Co., Arizona**
Springer (1928)	*Cnemidophorus gularis velox* (= *Aspidoscelis velox*): **Oraibi, Arizona, and Pueblo Bonito, New Mexico, restringida a Oraibi, Navajo Co., Arizona**[7]
Duellman & Lowe (1953)	*Cnemidophorus sacki xanthonotus* (= *Aspidoscelis xanthonota*): **North fork of Álamo Canyon, Ajo Mountains, approximately 19 miles north of the international boundary in Organ Pipe Cactus National Monument, Pima Co., Arizona**
Klauber (1931)	*Xantusia arizonae*: **One mile south of Yarnell, 4940 ft, Yavapai County, Arizona**
Papenfuss et al. (2001)	*Xantusia bezyi*: **5.6 km S (by Highway 87) of Sunflower, elevation 948 m., Maricopa County, Arizona (33°49.48′ N, 111° 28.55′ W)**

Serpientes (20)

Autor(es)	Nombre original: Localidad tipo
Klauber (1946)	*Arizona elegans philipi*: **10 miles east of Winslow, Navajo Co., Arizona**
Cope (1860)	*Gyalopion canum*: **Near Fort Buchanan Arizona**
Cope (1860)	*Hypsiglena chlorophaea*: **Fort Buchanan, Arizona**
Baird and Girard (1853)	*Lampropeltis getulus (= getula) splendida*: **Sonora, Mexico, restricted to Santa Rita Mountains, Arizona**[2]
Cope (1867)	*Ophibolus pyromelanus* (= *Lampropeltis pyromelana*): **Fort Whipple [Yavapai County, Arizona]**[8]
Cope (1892)	*Masticophis* (= *Coluber*) *flagellum piceus*: **Camp Grant, Arizona, (= Fort Grant, Graham Co., Arizona)**[9]
Stejneger (1890)	*Phyllorhynchus browni*: **Tucson, Pima Co., Arizona**
Baird & Girard (1853)	*Salvadora grahamiae*: **Sonora, Mexico, restringida a Huachuca Mountains, Cochise Co., Arizona**[4]
Cope (1867)	*Salvadora hexalepis*: **Fort Whipple, Yavapai Co., Arizona**
Baird & Girard (1853)	*Sonora semiannulata*: **Sonora Mexico, restringida a the vicinity of the Santa Rita Mountains, Arizona**[10]
Stejneger (1902 [1903])	*Tantilla wilcoxi*: **"Fort Huachuca" Cochise Co., Arizona**
Kennicott (1860)	*Thamnophis macrostemma megalops* (= *T. eques megalops*): **Tucson and St. Magdalena, restringida a Tucson, Pima Co., Arizona**[2]
Cope *in* Yarrow (1875)	*Chilopoma* (= *Thamnophis*) *rufipunctatus*: **"Southern Arizona," modificada a the vicinity of Fort Apache, Apache Co., Arizona**[5]
Coues (1875)	*Crotalus cerberus*: **San Francisco Mountains, Coconino County, Arizona**
Gloyd (1936)	*Crotalus lepidus klauberi*: **Carr Canyon, Huachuca Mountains, Cochise Co., Arizona**
Cope (1867)	*Crotalus mitchellii pyrrhus*: **Canyon Prieto, Yavapai Co., Arizona**[2]
Van Denburgh (1895)	*Crotalus pricei*: **Huachuca Mountains, Cochise Co., Arizona**
Kennicott (1861)	*Caudisona scutulata* (= *Crotalus scutulatus*): **Wickenburg, Maricopa Co., Arizona**[2]
Kennicott *in* Baird (1859)	*Crotalus tigris*: **Sasabe, Pima Co., Arizona**[2]
Meek (1905)	*Crotalus willardi*: **Tombstone, Arizona, modificada a above Hamburg, middle branch of Ramsey Canyon, Huachuca Mountains (altitude about 7,000 ft.), Cochise County, Arizona**[11]

Restricción reportada en [1] Smith & Taylor (1948); [2] Smith & Taylor (1950); [3] Gorman (1960); [4] Schmidt (1953); [5] Webb & Axtell (1986); [6] Cochran (1961); [7] Lowe (1955); [8] Coues (1875); [9] Degenhardt et al. (1996); [10] Stickel (1943); [11] Swarth (1921).

Características Fisiográficas y su Influencia sobre la Herpetofauna

Varias características de anfibios y reptiles limitan sus actividades y distribuciones. Debido a que anfibios y reptiles dependen del ambiente para regular su temperatura corporal ellos generalmente poseen un intervalo más estrecho de tolerancia ambiental que los vertebrados endotérmicos. Los patrones de comunidades de plantas y animales están formados en respuesta a temperaturas y precipitaciones extremas así como a condiciones locales tales como la altitud, longitud, pendiente de exposición, inversiones de temperatura, flujo de aire frío, porosidad del suelo, etc. (Lowe y Brown 1994). Por lo tanto, el entendimiento de las características físicas de cualquier localidad dada en Arizona es necesario para entender la naturaleza de su herpetofauna.

El estado de Arizona comprende una superficie de 295,253 km² que lo hace el sexto estado en extensión territorial de los Estados Unidos (Kearney y Peebles 1951; DeLorme 2008). Éste cae entre las coordenadas extremas de aproximadamente 31°20′ al sur, 36°60′ al norte, –109°02′ al este y –114°49′ al oeste (DeLorme 2008). La altitud en el estado varía desde 3850 m en la cima del Humphreys Peak en las Montañas San Francisco cerca de Flagstaff hasta 21.3 m sobre el nivel del mar en el Río Colorado cerca de Yuma (Chronic 1983; DeLorme 2008). El estado puede ser dividido en tres provincias fisiográficas principales: la Provincia de Sierra y Cuenca del sur y extremo oeste de Arizona, la Zona Montañosa de Transición del centro de Arizona, y la Meseta de Colorado del norte de Arizona.

La Provincia de Sierra y Cuenca comprende la mayor parte del sur de Arizona y está caracterizada por montañas rocosas aisladas que se surgen abruptamente desde planicies y valles aluviales. En la porción sureste de esta provincia las montañas Chiricahua, Pinaleno, Huachuca, Santa Catalina y Santa Rita se elevan hasta 2743 m desde los valles que varían en altitud desde 914 hasta 1220 m. Las Montañas Baboquivari orientadas de norte a sur, en el centro de la frontera sur de Arizona, forman una accidentada

línea divisoria entre la porción sureste de la Provincia de Sierra y Cuenca y la porción oeste de altitudes más bajas. Los valles del oeste varían en altitud desde aproximadamente 30 m hasta aproximadamente 915 m. Las rocosas y empinadas sierras esparcidas a través de estas tierras planas incluyen las Montañas Ajo, Cabeza Prieta, Comobabi, Gila, Gila Bend, Harcuvar, Harquahala, Kofa, Mohave, Quijotoa, Sand Tank, Sauceda, Sierra Estrella, Silver Bell, Tucson y White Tank. De éstas únicamente los complejos Mohave (1554 m), Harquahala (1732 m) y Harcuvar (1598 m) exceden 1524 m en altitud.

La Zona de Transición se extiende a través del centro de Arizona desde el este de Kingman dirigiéndose al este-sureste a través de las Montañas White, hasta Nuevo México, separando la Provincia de Sierra y Cuenca de la Meseta de Colorado. El Mogollon Rim, un acantilado casi vertical de aproximadamente 305 m de altura compuesto predominantemente de rocas precámbricas de 1–2 mil millones de años de antigüedad, se extiende desde el Verde River en el centro del estado hacia el sureste hasta las Montañas White y forma la parte central de la Zona de Transición (Kearney y Peebles 1951; Nations y Stump 1981; Chronic 1983). Varias de las montañas de esta provincia alcanzan una altitud de más de 2286 m sobre el nivel del mar. Éstas incluyen las Bradshaws, Black Hills, Mazatzals, Pinals y Sierra Ancha. A una altitud de 3475.6 m, Baldy Peak en las Montañas White, es el punto más alto en la Zona de Transición.

La provincia de la Meseta de Colorado cubre un área de aproximadamente 116,549.5 km² a través del norte de Arizona y se caracteriza por vastas planicies y sierras aisladas. Aunque las rocas del Precámbrico de esta provincia están expuestas a la vista en los acantilados y repisas, la mayoría de ellas están enterradas en una gruesa capa de sedimentos del Paleozoico y Mesozoico (Nations y Stump 1981; Chronic 1983). La altitud de estas mesetas varía de 1270 a 2775 m sobre el nivel del mar e incluyen la Hualapai, Coconino, Kaibab, Kanab y Black Mesa. Aquí, la característica física mejor conocida del estado, es el Gran Cañón, y el punto más alto del estado, el volcán Pico de San Francisco, contrasta fuertemente con las vastas mesetas (Nations y Stump 1981). Con una profundidad máxima de aproximadamente 1.6 km, el Gran Cañón y cañones asociados, sirve como una barrera efectiva, aislando las mesetas del norte (la parte conocida como la Arizona Strip). Unas cuantas de las sierras en la Arizona Strip son los suficientemente altas para hospedar bosques de coníferas. Éstas incluyen las Virgin Mountains, Mount Tumbull, Mount Logan y la Kaibab Plateau con 2791 m de altitud. Sierras importantes en el este de la Provincia Meseta de Colorado incluyen las Sierras Chuska y Carrizo cerca de la frontera con Nuevo México, ambas se elevan más de 2865 m y hospedan bosques de coníferas.

A excepción de algunos arroyos pequeños que fluyen hacia el sur a lo largo de la frontera sur, todos los ríos del estado fluyen hacia el oeste desembocando en el Río Colorado, el cual fluye hacia el sur hasta Sonora, México desembocando en el Mar de Cortés. El río más grande de Arizona tributario del Río Colorado es el Río Gila, el cual irriga la mayor parte de la Zona de Transición. Los ríos Little Colorado y Bill Williams irrigan la mayor parte del resto del estado. Desde su cabecera en las Montañas White, el Río Little Colorado fluye hacia el noroeste a través de la Meseta de Colorado donde es alimentado por los tributarios que fluyen hacia el norte en la Zona de Transición. Típicamente éste es un río seco al oeste de Winslow pero su cauce lleva aguas hasta el Río Colorado al este del Gran Cañón. La porción oeste de la Zona de Transición es irrigada por el Río Bill Williams, que fluye al oeste hasta el Río Colorado cerca del centro de la frontera oeste de Arizona. El Río Salt, un tributario del Gila, fluye hacia el oeste a través de la mayor parte del centro de Arizona y es alimentado por los tributarios que fluyen hacia el sur de la Zona de Transición incluyendo los ríos del centro de Arizona como Verde River, Tonto Creek, Cherry Creek, Canyon Creek, Cibeque Creek y los Ríos Black y White. El Río Gila entra al sureste de Arizona desde Nuevo México y fluye al oeste a lo largo de la orilla sur de la Zona de Transición. Los tributarios del Gila que fluyen hacia el sur desde la porción este de la Zona de

Transición incluyen al Blue River, San Francisco River, Eagle Creek y San Carlos River. En el sur el Gila es alimentado por el Río San Pedro, el cual fluye hacia el norte desde la frontera sur del estado. Debido a la construcción de reservorios, desviación de cauces, y bombeo de agua subterránea, el alguna vez permanente Río Gila actualmente tiene únicamente flujo intermitente en la mayoría de su porciones del centro y oeste de Arizona (Brown et al. 1981). Aquí, el cauce del Gila típicamente seco pasa cerca de la esquina suroeste de Phoenix donde es alimentado por el Río Salt, el cual también es intermitente en el área de Phoenix. Debajo de la unión con el Río Salt, el cauce del Gila mayormente seco continua al oeste hasta Yuma donde históricamente desemboca en el Río Colorado que fluye hacia el sur. El Río Santa Cruz se origina en el Valle de San Rafael al este de la Montañas Patagonia en el sureste de Arizona, fluye hacia el sur hasta México, haciendo un giro de 180°, y entonces fluye hacia el norte a lo largo del lado oeste de la Montañas Patagonia, a través del lado oeste de Tucson, y hasta la planicies desérticas de Santa Cruz en el norte donde se seca antes de alcanzar el Gila (Brown et al. 1981).

Las diversas características topográficas de Arizona, el gran intervalo altitudinal, y condiciones subtropicales y templadas contribuyen a la amplia variedad de condiciones climáticas del estado. La precipitación varía mucho entre estaciones y localidades. La precipitación anual en la Zona de Transición es generalmente 38–52 cm mayor que la de los desiertos del oeste de la Provincia de Sierra y Cuenca, y los bosques de las altas elevaciones del estado tienen un promedio de temperatura anual que es −6.7 a −3.9 °C inferior que la de los desiertos del suroeste (Seller 1964).

La porción noreste de la Meseta de Colorado de Arizona, la porción suroeste de la Sierra y Cuenca, y el Valle de San Simón en el sureste de Arizona reciben menos de 304.8 mm de precipitación anual promedio (Sellers 1964; Sellers y Hill 1974; Garfin y Emanuel 2006). Los valles del suroeste y extremo oeste de Arizona son las porciones más áridas del estado, y típicamente reciben menos de 127 mm de lluvia por año (Sellers 1964; Sellers y Hill 1974; Western Regional Climate Center 2001). En contraste, las porciones más altas de la Zona de Transición, Meseta del Colorado, y las montañas más altas del sureste reciben más de 635 mm de precipitación (lluvia y nieve derretida) al año (Sellers 1964; Sellers y Hill 1974; Western Regional Climate Center 2001). En todo el estado el periodo julio-agosto es el más húmedo del año. Esta humedad de verano, conocida como las tormentas del monzón, se puede originar desde el Golfo de México, fluyendo hasta Arizona desde el sureste. En Arizona, las tormentas del monzón también pueden estar asociadas con el surgimiento de aire tropical desde el Mar de Cortés y Océano Pacífico (Sellers 1964; Sellers y Hill 1974; Western Regional Climate Center 2001; Garfin y Emanuel 2006). Las lluvias del monzón del verano típicamente se presentan en forma de tormentas breves en la tarde frecuentemente acompañadas de vientos altos, lluvia, truenos, tormentas de arena, y arribazones en ríos y arroyos (Western Regional Climate Center 2001). La segunda estación de lluvias en Arizona se extiende desde mediados de noviembre hasta mediados de marzo y está asociada con tormentas ciclónicas de gran escala que entran al continente desde el Océano Pacífico. Estas tormentas de meses más fríos son relativamente de ligeras a moderadas en intensidad pero de duración más larga que las tormentas del monzón de verano (Sellers 1964; Sellers y Hill 1974; Western Regional Climate Center 2001; Garfin y Emanuel 2006). La nieve es rara en la porción oeste de la Provincia de Sierra y Cuenca pero casi todas las montañas de Arizona que se elevan a más de 2133.5 m reciben más de 1.2 m de nieve por año, la nieve representa la mayor cantidad de precipitación del invierno en las partes más altas del estado (Sellers 1964; Sellers y Hill 1974). El periodo de mayo-junio en el suroeste de Arizona frecuentemente no recibe cantidades medibles de lluvia, éstos son los meses más secos en todo el estado (Sellers 1964).

La temperatura en Arizona varía tanto como la topografía, altitud y precipitación. A nivel estatal, por lo general hay un incremento en temperatura de –1.1°C a 4.4°C entre las tempe-

raturas máximas diarias de invierno y verano y ocasionalmente más de 10.0°C en los mínimos diarios (Sellers 1964; Sellers y Hill 1974; Western Regional Climate Center 2001). Temperaturas invernales por debajo de cero grados centígrados son comunes en la Zona de Transición y en la Meseta de Colorado, mientras que en los valles y bajadas del suroeste de Arizona las temperaturas mínimas de un día de invierno comúnmente son por arriba de cero grados centígrados y las máximas cerca de 21.0°C (Seller 1964; Western Regional Climate Center 2001). En la Zona de Transición las temperaturas diarias de verano comúnmente varían de 4.4°C a 32.2°C (Seller 1964; Western Regional Climate Center 2001). El promedio de temperaturas diarias de verano en los valles del sureste de Arizona son regularmente arriba de 32.2°C, pero en las montañas adyacentes de más de 2438.5 m de altitud las máximas diarias son generalmente por debajo de 29.4°C y las temperaturas bajas de las mañanas son frecuentemente de alrededor de 10.0°C (Seller 1964). En la Meseta de Colorado las temperaturas máximas diarias de verano regularmente son por encima de los 32.2°C, excepto en las mesetas altas, y las temperaturas mínimas de las mañanas son de alrededor de 10.0°C o por debajo de 15.5°C (Sellers 1964; Sellers y Hill 1974). El periodo más caliente del año comúnmente ocurre de finales de junio a principios de julio, antes del inicio de las lluvias del monzón de verano (Sellers 1964).

La rica variedad de condiciones climáticas y topográficas del estado resulta en una igualmente colección de diversas comunidades bióticas. En Arizona se desarrollan 14 comunidades bióticas que van desde los desiertos calientes y xéricos hasta la fría tundra alpina. Estas comunidades son atravesadas por números ríos y arroyos, los cuales hospedan muchos organismos especializados y contribuyen grandemente a la biodiversidad del estado.

Matorral Desértico Sonorense del Bajo Río Colorado

Éste se presenta en las planicies del cuarto suroeste del estado a altitudes que varían desde 21 hasta 400 m. La gobernadora (*Larrea triden-*

tata), el bursage blanco (*Ambrosia dumosa*), y el arbusto de sal del desierto (*Atriplex polycarpa*) caracterizan a esta comunidad. El palo fierro (*Olneya tesota*), el palo verde azul (*Parkinsonia florida*), y el árbol de humo (*Psorothamnus spinosus*) bordean los cauces secos que atraviesan estas planicies (Brown y Lowe 1980; Turner y Brown 1994). En las altitudes más bajas los saguaros (*Carnegiea gigantea*) están confinados a los arroyos y orillas de cauces secos en esta comunidad.

Grandes porciones de esta comunidad se han perdido o han sido dañadas como resultado de la agricultura, desarrollo, uso ilegal de vehículos doble tracción, y sobrepastoreo de ganado.

Por lo menos 42 especies de anfibios y reptiles se pueden encontrar en esta comunidad incluyendo: *Anaxyrus cognatus, A. punctatus, A. retiformis, A. woodhousii, Incilius alvarius, Smilisca fodiens, Gastrophryne olivacea, Scaphiopus couchii, Kinosternon arizonense, Gambelia wislizenii, Coleonyx variegatus, Heloderma suspectum, Dipsosaurus dorsalis, Callisaurus draconoides, Phrynosoma goodei, P. mcallii, P. platyrhinos, Sceloporus magister, Uma rufopunctata, U. scoparia, Urosaurus graciosus, U. ornatus, Uta stansburiana, Aspidoscelis tigris, Arizona elegans, Chilomeniscus stramineus, Chionactis occipitalis, Hypsiglena chlorophaea, Lampropeltis getula, Masticophis flagellum, Phyllorhynchus browni, P. decurtatus, Pituophis catenifer, Rhinocheilus lecontei, Salvadora hexalepis, Sonora semiannulata, Thamnophis marcianus* (cerca de campos agrícolas), *Micruroides euryxanthus, Rena humilis, Crotalus atrox, C. cerastes* y *C. scutulatus.*

El Matorral Desértico Sonorense de la Tierras Altas de Arizona

Éste domina las bajadas y pendientes de la porción suroeste de la Cuenca y Provincia de Sierra a altitudes que varían desde aproximadamente 290 a 1050 m. Ésta es una comunidad rica en suculentas con el saguaro (*Carnegiea gigantea*), siendo la más conspicua de éstas. La vegetación dominante en esta comunidad también incluye, varias especies de chollas y tunas (*Cylindropuntia* y *Opuntia* respectivamente), así como cactus de

Graham (*Mammillaria grahamii*), barril (*Ferocactus wislizeni*), bursage de hojas triangulares (*Ambrosia deltoidea*), arbusto frágil (*Encelia farinosa*), jojoba (*Simmondsia chinensis*), ocotillo (*Fouquieria splendens*), palo verde (*Parkinsonia microphylla*), palo fierro (*Olneya tesota*), mezquite (*Prosopis* sp.) y uña de gato (*Acacia greggii*) (Brown y Lowe 1980; Turner y Brown 1994).

Pastos y hierbas invasivas tales como el bufel (*Pennisetum ciliare*), pasto del mediterráneo (*Schismus barbatus*), bromo rojo (*Bromus rubens*) y mostaza (*Brassica* sp.), crecen junto a las plantas del Desierto Sonorense, compiten por agua, y promueven incendios. Sin lugar a dudas los incendios siempre han ocurrido en el Desierto Sonorense, pero éstos han sido de baja intensidad y macollos nativos con un espaciamiento relativamente abierto mantenían los incendios a baja intensidad y corta duración. Pastos y hierbas no nativos crecen sin este espaciamiento formando una cubierta continua de vegetación. Con la colonización del suroeste por especies de pastos y hierbas exóticas, los incendios se han vuelto más frecuentes en las comunidades del desierto. Muchas especies del Desierto Sonorense no están adaptadas al fuego y no resisten los extensos incendios de alta intensidad promovidos por especies exóticas. Como resultado, mucho de esta comunidad está siendo convertida a un pastizal y matorral grandemente desprovisto de suculentas. Vehículos doble tracción de campo traviesa, sobrepastoreo, desarrollo, y fragmentación del hábitat también son amenazas importantes para esta comunidad.

Por lo menos 66 especies de anfibios y reptiles habitan esta comunidad incluyendo: *Ambystoma mavortium, Anaxyrus cognatus, A. microscaphus, A. punctatus, A. retiformis, A. woodhousii, Incilius alvarius, Hyla arenicolor, Smilisca fodiens, Gastrophryne mazatlanensis, Scaphiopus couchii, Spea multiplicata, Kinosternon arizonense, Gopherus morafkai, Crotaphytus bicinctores, C. collaris, C. nebrius, Gambelia wislizenii, Coleonyx variegatus, Heloderma suspectum, Dipsosaurus dorsalis, Sauromalus ater, Callisaurus draconoides, Cophosaurus texanus, Holbrookia elegans, Phrynosoma goodei, P. platyrhinos, P. solare,*

Sceloporus clarkii, S. magister, Urosaurus gracio-sus, U. ornątus, Uta stansburiana, Aspidoscelis stictogramma, A. tigris, A. xanthonota, Xantusia arizonae, X. bezyi, X. vigilis, Lichanura trivir-gata, Arizona elegans, Chilomeniscus stramineus, Chionactis occipitalis, C. palarostris, Diadophis punctatus, Hypsiglena chlorophaea, Lampropel-tis getula, Masticophis bilineatus, M. flagellum, Phyllorhynchus browni, P. decurtatus, Pituophis catenifer, Rhinocheilus lecontei, Salvadora hexale-pis, Sonora semiannulata, Tantilla hobartsmithi, Thamnophis cyrtopsis, Trimorphodon lambda, Micruroides euryxanthus, Rena humilis, Crotalus atrox, C. cerastes, C. mitchellii, C. molossus, C. scu-tulatus y *C. tigris.*

Matorral Desértico de Mohave

A lo largo de la mitad norte de la frontera oeste del estado a altitudes que varían desde apro-ximadamente 275 hasta 1450 m la comunidad predomínate es el Matorral Desértico de Mo-have. Plantas conspicuas en esta área incluyen gobernadora (*Larrea tridentata*), acebo de de-sierto (*Atriplex hymenelytra*), cepillo burro blanco (*Hymenoclea salsola*), yuca de Mohave (*Yucca schidigera*), cepillo negro (*Coleogyne ramosissima*), sábalo (*Atriplex confertifolia*), y la casi endémica Yuca (*Yucca brevifolia*) (Brown y Lowe 1980; Turner 1994b).

Esta comunidad profundas cicatrices deja-das por la operación ilegal de vehículos doble tracción y cross-country y ha sido dañada por sobrepastoreo, desarrollo y vegetación invasiva.

Por lo menos 38 especies de anfibios y reptiles habitan esta comunidad incluyendo: *Anaxyrus cognatus, A. microscaphus, A. punctatus, A. woodhousii, Gopherus agassizii, G. morafkai, Crotaphytus bicinctores, Gambelia wislizenii, Coleonyx variegatus, Heloderma suspectum, Dipsosaurus dorsalis, Sauromalus ater, Calli-saurus draconoides, Phrynosoma platyrhinos, Sceloporus magister, S. uniformis, Urosaurus graciosus, U. ornatus, Uta stansburiana, Aspido-scelis tigris, Xantusia vigilis, Lichanura trivirgata, Arizona elegans, Chionactis occipitalis, Hypsiglena chlorophaea, Lampropeltis getula, Masticophis*

flagellum, Phyllorhynchus decurtatus, Pituophis ca-tenifer, Rhinocheilus lecontei, Salvadora hexalepis, Sonora semiannulata, Trimorphodon lambda, Rena humilis, Crotalus atrox, C. cerastes, C. mitchellii y *C. scutulatus.*

Matorral Desértico Chihuahuense

Ésta es la comunidad dominante en los valles bajos y bajadas someras del extremo sureste de Arizona a elevaciones que varían desde aproxi-madamente 900 hasta 1200 m. Plantas comunes en esta comunidad incluyen *Larrea tridentata, Acacia constricta, Flourensia cernua, Fouquieria splendens, Acacia neovernicosa, Koeberlinia spi-nosa, Parthenium incanum* y *Prosopis glandulosa* (Brown y Lowe 1980; Brown 1994b).

Por lo menos 53 especies de anfibios y reptiles pueden ser encontradas en esta comunidad incluyendo: *Ambystoma mavortium, Anaxyrus cognatus, A. debilis, A. punctatus, A. woodhousii, Incilius alvarius, Hyla arenicolor, Scaphiopus couchii, Spea bombifrons, S. multiplicata, Terra-pene ornata, Kinosternon flavescens, Crotaphytus collaris, Gambelia wislizenii, Coleonyx variegatus, Heloderma suspectum, Callisaurus draconoides, Cophosaurus texanus, Holbrookia elegans, Phryno-soma cornutum, P. modestum, P. solare, Sceloporus bimaculosus, S. clarkii, S. cowlesi, S. magister, Urosaurus ornatus, Uta stansburiana, Aspidoscelis tigris, A. uniparens, Arizona elegans, Diadophis punctatus, Gyalopion canum, Heterodon kennerlyi, Hypsiglena jani, Lampropeltis getula, Masticophis flagellum, Pituophis catenifer, Rhinocheilus lecontei, Salvadora deserticola, Sonora semiannulata, Tantilla hobartsmithi, T. nigriceps, Thamnophis cyrtopsis, T. marcianus, Trimorphodon lambda, Micruroides euryxanthus, Rena dissecta, R. hu-milis, Crotalus atrox, C. molossus, C. scutulatus* y *C. tigris.*

Matorral Desértico de la Gran Cuenca

Esta comunidad se presenta en las planicies más bajas de la Provincia de la Meseta de Colorado a elevación que varían desde aproximadamente 1190 hasta 1980 m. Plantas conspicuas de esta comunidad incluyen *Artemisia tridentata, Artemisia bigelovii, Atriplex confertifolia, Krascheninnikovia lanata* y *Coleogyne ramosissima* (Brown y Lowe 1980; Turner 1994a).

Por lo menos 34 especies de anfibios y reptiles habitan esta comunidad incluyendo: *Ambystoma mavortium, Anaxyrus cognatus, A. punctatus, A. woodhousii, Hyla arenicolor, Scaphiopus couchii, Spea bombifrons, S. intermontana, S. multiplicata, Crotaphytus bicinctores, C. collaris, Gambelia wislizenii, Coleonyx variegatus, Sauromalus ater, Holbrookia maculata, Phrynosoma platyrhinos, Sceloporus graciosus, S. magister, S. tristichus, S. uniformis, Urosaurus ornatus, Uta stansburiana, Aspidoscelis tigris, A. velox, Arizona elegans, Hypsiglena chlorophaea, H. jani, Lampropeltis getula, Pituophis catenifer, Rhinocheilus lecontei, Salvadora hexalepis, Sonora semiannulata, Crotalus oreganus* y *C. viridis.*

Pastizal Semidesértico

El Pastizal Semidesértico se presenta a elevaciones que varían desde aproximadamente 1100 hasta 1700 m en muchos de los valles y bajadas del sureste y en varias localidades a lo largo de la orilla sur de la Zona de Transición. Estas comunidades están dominadas por tobosa (*Pleuraphis mutica*), macollo (*Bouteloua hirsuta*), palmilla (*Yucca rigida*), sotol (*Dasylirion wheeleri*), yuca de hoja delgada (*Yucca elata*), varias especies de *Agave*, cholla (*Cylindropuntia* sp.) y nopal tunero (*Opuntia* sp.) (Brown y Lowe 1980; Brown 1994f).

Por lo menos 76 especies de anfibios y reptiles pueden ser encontradas en esta comunidad incluyendo: *Ambystoma mavortium, Anaxyrus cognatus, A. debilis, A. microscaphus, A. punctatus, A. woodhousii, Incilius alvarius, Hyla arenicolor, Gastrophryne mazatlanensis, Scaphiopus couchii, Spea bombifrons, S. multiplicata, Terrapene ornata, Kinosternon flavescens, Gopherus morafkai, Elgaria kingii, Crotaphytus collaris, Gambelia wislizenii, Heloderma suspectum, Cophosaurus texanus, Holbrookia elegans, H. maculata, Phrynosoma cornutum, P. hernandesi, P. modestum, P. solare, Sceloporus bimaculosus, S. clarkii, S. cowlesi,*

S. magister, S. tristichus, Urosaurus ornatus, Uta stansburiana, Plestiodon obsoletus, Aspidoscelis arizonae, A. stictogramma, A. exsanguis, A. flagellicauda, A. sonorae, A. tigris, A. uniparens, Xantusia arizonae, X. bezyi, X. vigilis, Arizona elegans, Diadophis punctatus, Gyalopion canum, G. quadrangulare, Heterodon kennerlyi, Hypsiglena chlorophaea, H. jani, Lampropeltis getula, Masticophis bilineatus, M. flagellum, M. taeniatus, Oxybelis aeneus, Pituophis catenifer, Rhinocheilus lecontei, Salvadora deserticola, S. hexalepis, Senticolis triaspis, Sonora semiannulata, Tantilla hobartsmithi, Tantilla nigriceps, Thamnophis cyrtopsis, T. marcianus, Trimorphodon lambda, Micruroides euryxanthus, Rena dissecta, R. humilis, Crotalus atrox, C. cerberus, C. mitchellii, C. molossus, C. scutulatus y *Sistrurus catenatus.*

Pastizal de la Meseta y de la Gran Cuenca

Éste se presenta a través de la mayor parte de la Meseta de Colorado de Arizona y en los valles de San Rafael y Empire del sureste del estado. Estas comunidades se desarrollan a altitudes de 1370 a 2300 m. Los pastos dominantes incluyen *Bouteloua gracilis, Pleuraphis jamesii, Achnatherum hymenoides* y *Sporobolus airoides.* La supresión de incendios y el pastoreo de ganado han alterado el ciclo de natural de incendios en la mayoría de los pastizales de Arizona permitiendo la invasión de arbustos tales como *Atriplex canescens, Gutierrezia sarothrae* y *Ephedra viridis* (Brown y Lowe 1980; Brown 1994e).

Por lo menos 51 especies de anfibios y reptiles habitan en los pastizales de Arizona incluyendo: *Ambystoma mavortium, Anaxyrus cognatus, A. punctatus, A. woodhousii, Acris crepitans, Scaphiopus couchii, Spea bombifrons, S. intermontana, S. multiplicata, Terrapene ornata, Elgaria kingii, Crotaphytus bicinctores, C. collaris, Holbrookia elegans, H. maculata, Phrynosoma hernandesi, Sceloporus clarkii, S. cowlesi, S. graciosus, S. magister, S. slevini, S. tristichus, S. uniformis, Urosaurus ornatus, Uta stansburiana, Plestiodon multivirgatus, P. obsoletus, Aspidoscelis neomexicana, A. pai, A. sonorae, A. uniparens, A. velox, Arizona elegans, Coluber constrictor, Diadophis punctatus,*

Gyalopion canum, Heterodon kennerlyi, Hypsiglena chlorophaea, H. jani, Lampropeltis getula, L. triangulum, Masticophis taeniatus, Pituophis catenifer, Rhinocheilus lecontei, Salvadora deserticola, Sonora semiannulata, Thamnophis elegans, T. marcianus, Crotalus scutulatus, C. viridis y *Sistrurus catenatus.*

Chaparral Interior

El Chaparral Interior se presenta en muchas de las pendientes sur de la Zona de Transición a altitudes entre aproximadamente 1000 y 2150 m. Arbustos comunes en esta comunidad incluyen *Arctostaphylos pungens, Quercus turbinella, Cercocarpus montanus, Rhus trilobata, Garrya sp., Ceanothus greggii, Rhamnus crocea, Purshia stansburiana, Vauquelinia californica* y *Forestiera shrevei* (Brown y Lowe 1980; Pase y Brown 1994b).

Por lo menos 49 especies de anfibios y reptiles pueden ser encontradas en esta comunidad incluyendo: *Ambystoma mavortium, Anaxyrus microscaphus, A. punctatus, A. woodhousii, Hyla arenicolor, Spea multiplicata, Elgaria kingii, Crotaphytus bicinctores, C. collaris, Coleonyx variegatus, Heloderma suspectum, Sauromalus ater, Cophosaurus texanus, Holbrookia maculata, Phrynosoma hernandesi, Sceloporus clarkii, S. magister, S. tristichus, Urosaurus ornatus, Uta stansburiana, Plestiodon gilberti, P. obsoletus, Aspidoscelis flagellicauda, A. tigris, A. uniparens, A. velox, Xantusia arizonae, X. bezyi, X. vigilis, Lichanura trivirgata, Diadophis punctatus, Hypsiglena chlorophaea, H. jani, Lampropeltis pyromelana, Masticophis bilineatus, M. flagellum, M. taeniatus, Pituophis catenifer, Salvadora grahamiae, S. hexalepis, Sonora semiannulata, Tantilla hobartsmithi, Thamnophis cyrtopsis, Trimorphodon lambda, Micruroides euryxanthus, Crotalus atrox, C. cerberus, C. mitchellii* y *C. molossus.*

Encinar

Sobre las pendientes bajas y estribaciones del sureste de Arizona a altitudes que varían desde aproximadamente 1280 a 2200 m la comunidad dominante es el Encinar. Plantas conspicuas de esta comunidad incluyen encino de Emory (*Quercus emoryi*), encino gris (*Quercus grisea*), encino blanco de Arizona (*Quercus arizonica*), encino azul mexicano (*Quercus oblongifolia*), táscate (*Juniperus deppeana*), piñón mexicano (*Pinus cembroides*), madroño (*Arbutus arizonica*), cactus arcoíris (*Echinocereus rigidissimus*), yuca de Schott (*Yucca schottii*), yuca banana (*Yucca baccata*), sotol (*Dasylirion wheeleri*), agave de Palmer (*Agave palmeri*), y agave de Parry (*Agave parryi*) (Brown y Lowe 1980; Brown 1994d).

Por lo menos 57 especies de anfibios y reptiles habitan esta comunidad incluyendo: *Ambystoma mavortium, Anaxyrus punctatus, A. woodhousii, Incilius alvarius, Craugastor augusti, Hyla arenicolor, H. wrightorum, Gastrophryne mazatlanensis, Spea multiplicata, Terrapene ornata, Elgaria kingii, Crotaphytus collaris, Heloderma suspectum, Cophosaurus texanus, Holbrookia elegans, H. maculata, Phrynosoma hernandesi, Sceloporus clarkii, S. jarrovii, S. slevini, S. tristichus, S. virgatus, Urosaurus ornatus, Plestiodon callicephalus, P. obsoletus, Aspidoscelis stictogramma, A. exsanguis, A. flagellicauda, A. sonorae, A. uniparens, Diadophis punctatus, Hypsiglena chlorophaea, H. jani, Lampropeltis knoblochi, L. pyromelana, Masticophis bilineatus, Oxybelis aeneus, Pituophis catenifer, Salvadora deserticola, S. grahamiae, S. hexalepis, Senticolis triaspis, Sonora semiannulata, Tantilla hobartsmithi, T. wilcoxi, T. yaquia, Thamnophis cyrtopsis, Trimorphodon lambda, Micruroides euryxanthus, Rena dissecta, R. humilis, Crotalus atrox, C. cerberus, C. lepidus, C. molossus, C. pricei* y *C. willardi.*

Bosque de Coníferas de la Gran Cuenca

Este bosque se presenta a altitudes medias (1200 a 2300 m) de la Zona de Transición y Meseta de Colorado. Esta comunidad está dominada por táscate de las Montañas Rocallosas (*Juniperus scopulorum*), táscate de Utah (*Juniperus osteosperma*), pino de una hoja (*Pinus monophylla*), y pino de las Montañas Rocallosas (*Pinus edulis*) (Brown y Lowe 1980; Brown 1994c).

Por lo menos 48 especies de anfibios y reptiles habitan esta comunidad incluyendo: *Ambystoma mavortium, Anaxyrus microscaphus,*

A. punctatus, A. woodhousii, Hyla arenicolor,
Spea intermontana, S. multiplicata, Elgaria kingii,
Crotaphytus bicinctores, C. collaris, Heloderma
suspectum, Cophosaurus texanus, Holbrookia maculata, Phrynosoma hernandesi, Sceloporus clarkii,
S. cowlesi, S. magister, S. tristichus, S. uniformis,
Urosaurus ornatus, Plestiodon gilberti, P. multivirgatus, P. obsoletus, P. skiltonianus, Aspidoscelis
flagellicauda, A. pai, A. uniparens, A. velox, Xantusia bezyi, X. vigilis, Lichanura trivirgata, Diadophis
punctatus, Hypsiglena chlorophaea, H. jani,
Lampropeltis pyromelana, Masticophis taeniatus,
Pituophis catenifer, Salvadora grahamiae, S. hexalepis, Sonora semiannulata, Tantilla hobartsmithi,
Thamnophis cyrtopsis, T. elegans, Trimorphodon
lambda, Crotalus cerberus, C. molossus, C. oreganus
y C. viridis.

Bosque de Coníferas de Montaña

Este bosque se presenta en las pendientes altas
y mesetas de Arizona a altitudes desde aproximadamente 2000 hasta 3050 m. La especie más
conspicua en esta comunidad es el pino Ponderosa (Pinus ponderosa). Otros miembros de esta
comunidad incluyen el abeto de Douglas (Pseudotsuga menziesii), encino de Gambel (Quercus
gambelii) y robinia (Robinia neomexicana) (Brown
y Lowe 1980; Pase y Brown 1994c).

Años de una activa supresión de incendios y
el aclaramiento de la vegetación nativa a través
de pastoreo ha resultado en un incremento en
la densidad de malezas y árboles jóvenes en esta
comunidad. Los incendios regulares que alguna
vez mantuvieron el bosque, con quemas de baja
intensidad a nivel del suelo, son ahora alimentados por malezas y se han vuelto de naturaleza
catastrófica.

Por lo menos 42 especies de anfibios y
reptiles pueden encontrarse en esta comunidad
incluyendo: Ambystoma mavortium, Ensatina
eschscholtzii, Anaxyrus microscaphus, A. punctatus, Hyla arenicolor, H. wrightorum, Pseudacris
triseriata, Spea intermontana, S. multiplicata,
Elgaria kingii, Phrynosoma hernandesi, Sceloporus cowlesi, S. jarrovii, S. slevini, S. tristichus,
S. virgatus, Urosaurus ornatus, Plestiodon gilberti,

P. multivirgatus, P. obsoletus, P. skiltonianus, Aspi-
doscelis exsanguis, A. flagellicauda, A. pai, A. velox,
Diadophis punctatus, Hypsiglena chlorophaea,
H. jani, Lampropeltis knoblochi, L. pyromelana,
Masticophis taeniatus, Pituophis catenifer, Salva-
dora grahamiae, Senticolis triaspis, Tantilla wilcoxi,
Thamnophis cyrtopsis, T. elegans, Crotalus cerberus,
C. lepidus, C. molossus, C. pricei y C. willardi.

Pastizal Subalpino

Planicies, valles y colinas de elevaciones altas
(2500–3500 m) en la Meseta de Kaibab, y en las
Montañas San Francisco, Chuska y White hospedan el Pastizal Subalpino. En esta comunidad de
pastos y hierbas las margaritas silvestres, dientes
de león, tréboles, espuelas y asters crecen entre
macollos como la festuca de Arizona (Festuca
arizonica) (Brown y Lowe 1980; Brown 1994a).

Por lo menos 11 especies de anfibios y reptiles
habitan esta comunidad incluyendo: Ambystoma
mavortium, Hyla wrightorum, Pseudacris triseriata, Phrynosoma hernandesi, Sceloporus cowlesi,
S. tristichus, Urosaurus ornatus, Plestiodon multivirgatus, Pituophis catenifer, Thamnophis elegans y
Crotalus molossus.

Bosque Subalpino de Coníferas

Las pendientes de las montañas más altas de
Arizona que varían de 2450 a 3800 m hospedan al Bosque Subalpino de Coníferas. La picea
de Engelmann (Picea engelmannii), el abeto de
corteza de corcho (Abies lasiocarpa), el pino de
conos de cerda (Pinus aristata), el aspen (Populus
tremuloides) y el abeto de Douglas (Pseudotsuga
menziesii) son las especies de árboles más conspicuas en esta comunidad (Brown y Lowe 1980;
Pase y Brown 1994d).

Por lo menos 17 especies de anfibios y reptiles
habitan esta comunidad incluyendo: Pseudacris
triseriata, Elgaria kingii, Phrynosoma hernandesi, Sceloporus cowlesi, S. jarrovii, S. slevini,
S. tristichus, S. virgatus, Plestiodon skiltonianus,
Lampropeltis knoblochi, L. pyromelana, Pituophis
catenifer, Thamnophis elegans, Crotalus cerberus,
C. lepidus, C. molossus y C. pricei.

Tundra Alpina

La comunidad de Tundra Alpina se presenta en la cima de los picos más altos de Arizona, el Pico de San Francisco cerca de Flagstaff, con altitudes que varían de 3500 a 3862 m. No hay árboles en esta comunidad, únicamente líquenes, musgos, plantas herbáceas y arbustos de bajo crecimiento. Algunos miembros de esta comunidad incluyen *Pedicularis parryi, Botrychium* sp., *Helenium hoopesii, Veronica wormskjoldii,* y *Packera franciscana* (Brown y Lowe 1980; Pase y Brown 1994a).

La única especie de herpetozoario que se presenta aquí es *Thamnophis elegans. Pituophis catenifer* ha sido observada cerca del ecotono de Tundra Alpina/Bosque Petran de Coníferas Subalpino en las Montañas San Francisco.

Comunidades Riparias

Numerosos arroyos y ríos fluyen a través del estado corriendo sobre varias comunidades bióticas. Estas comunidades riparias y otras comunidades pantanosas proporcionan hábitats para numerosos organismos especializados. Cerca de sus orígenes, las comunidades riparias están dominadas por una variedad de maples (*Acer* sp.), ailes (*Alnus* sp.) y sauces (*Salix* sp.). A altitudes moderadas la vegetación dominante incluye álamos (*Populus sp.*), sauces (*Salix* sp.), sicomoros (*Platanus* sp.), ailes (*Alnus* sp.), nueces (*Juglans* sp.) y fresnos (*Fraxinus sp.*). En las elevaciones más bajas las comunidades riparias están caracterizadas por mezquites (*Prosopis* sp.), chila (*Baccharis salicifolia*), sauce de desierto (*Chilopsis linearis*), y álamos (*Populus sp.*) (Brown y Lowe 1980; Minckley y Brown 1994).

Muchas de las comunidades riparias de Arizona han sido secadas y degradadas por el sobrepastoreo, creación de reservorios, desviación de cauces y bombeo de aguas subterráneas. Adicionalmente, especies invasivas como el cangrejo de río, *Lithobates catesbeianus, L. berlandieri* y varios peces introducidos compiten con las especies nativas. La enfermedad introducida quitridiomicosis amenaza a los anfibios acuáticos nativos.

Por lo menos 16 especies de anfibios y reptiles dependen directamente de comunidades riparias. Especies de anfibios y reptiles que habitan en comunidades riparias o pantanos de elevaciones altas incluyen: *Lithobates catesbeianus, L. chiricahuensis, L. pipiens* y *Thamnophis rufipunctatus.* Comunidades riparias y pantanos de elevaciones medias hospedan a: *Lithobates catesbeianus, L. chiricahuensis, L. tarahumarae, L. yavapaiensis, Chrysemys picta, Kinosternon sonoriense, Thamnophis eques* y *Thamnophis rufipunctatus.* Especies de anfibios y reptiles que habitan en comunidades riparias y pantanos de elevaciones bajas incluyen: *Pseudacris hypochondriaca, Lithobates berlandieri, L. blairi, L. catesbeianus, L. onca, L. tarahumarae, L. yavapaiensis, Chelydra serpentina, Chrysemys picta, Kinosternon sonoriense, Trachemys scripta, Apalone spinifera, Thamnophis eques* y *T. rufipunctatus.*

Especies introducidas que pueden depender de áreas de desarrollos humanos para su continua existencia en Arizona incluyen: *Xenopus laevis,* que ha sido introducida a las pozas de los campos de golf en Tucson; *Chelydra serpentina,* que ha sido introducida a lagos y canales urbanos en el área metropolitana de Phoenix; *Ctenosaura* spp., de dos linajes maternos (*C. macrolopha* y *C. conspicuosa*) (Edwards et al. 2005) que ha sido introducida a los terrenos del Arizona Sonora Desert Museum en el oeste de Tucson pero no se ha dispersado a los alrededores del museo; *Hemidactylus turcicus,* que ha sido introducida a numerosas áreas de habitaciones humanas a través del oeste y sur de Arizona; *Chalcides* sp., que ha sido introducida a la Ciudad de Mesa; e *Indotyphlops braminus,* que ha sido introducida al área metropolitana de Phoenix.

Actualmente un total de 153 especies (141 nativas, 12 introducidas) de herpetozoarios habitan en Arizona, representando 30 familias y 67 géneros. Treinta de estas especies son anfibios (2 salamandras [1 introducida], 28 anuros [3 introducidos]). Las restantes 123 especies son reptiles (10 tortugas [3 introducidas], 57 lagartijas [4 introducidas], y 56 serpientes [1 introducida]).

Los anfibios representan un porcentaje

pequeño (20% o 30/153) de las especies de herpetozoarios de Arizona, reflejando la dependencia
de este grupo por lugares húmedos y por ambientes templados o acuáticos. El Ajolote Tigre
Rayado y la mayoría de los anuros de Arizona
están vinculados a habitats humedos, los cuales
en Arizona están generalmente limitados a las
montañas de la Zona de Transición, a la porción
sureste de la Provincia de Sierra y Cuenca, y a
las comunidades riparias. Sin embargo, varios
anfibios están adaptados a las condiciones se-
miaridas del Desierto de Sonora de la parte oeste
de la Provincia de Sierra y Cuenca. Éstas incluyen *Anaxyrus punctatus, A. cognatus, A. retiformis,
Incilius alvarius, Smilisca fodiens, Gastrophryne
mazatlanensis* y *Scaphiopus couchii*. Los desiertos
semiáridos del norte del estado están habitados
por *Anaxyrus cognatus, A. woodhousii, Scaphiopus
couchii, Spea multiplicata* y *S. intermontana. Craugastor augusti*, habita en varias sierras templadas
del sureste de Arizona, está adaptada para depositar sus huevos en el suelo de lugares humedos
y es el único anfibios de Arizona que no depende
de cuerpos de agua para su reproducción.

Las tortugas también representan un pequeño
porcentaje (7% or 10/153) de las especies de
herpetozoarios de Arizona. Tres de las especies
nativas de Arizona (*Terrapene ornata, Gopherus
agassizii* y *G. morafkai*) son terrestres, 2 (*Kinosternon arizonense* y *K. flavescens*) son semiacuáticas,
y 2 (*Kinosternon sonoriense* y *Chrysemys picta*) son
acuáticas. Tres especies de tortugas no nativas
(*Chelydra serpentina, Trachemys scripta* y *Apalone
spinifera*) han sido introducidas a Arizona. *Chelydra serpentina* ha sido introducida a los lagos y
canales urbanos del área metropolitan de Phoenix mientras que *Trachemys scripta* y *Apalone
spinifera* han sido introducidas a cuerpos de agua
urbanos, arroyos y ríos naturales, reservorios y
aguajes del centro, oeste y sur de Arizona.

El grupo de herpetozoarios mejor representado en Arizona es el de las lagartijas con 37%
(57/153) de las especies del estado. A excepción
de la comunidad de tundra alpina de la cima de
las Montañas San Francisco, todos los hábitats terrestres del estado hospedan lagartijas.
Algunas especies como *Urosaurus ornatus* y

Uta stansburiana se encuentran casi en todo el
estado. Otras como el Camaleón de Cola Plana
Phrynosoma mcallii, las areneras *Uma rufopunctata* y *U. scoparia*, y las lagartijas nocturnas
Xantusia arizonae y *X. bezyi* tienen distribuciones
relativamente pequeñas en hábitats específicos.

Las serpientes representan el 37% (56/153) de
las especies de herpotozoarios de Arizona y habitan en todos los hábitats terrestres del estado,
aunque *Thamnophis elegans* es la única especie
conocida que habita en la tundra alpina de Arizona. *Pituophis catenifer* se encuentra en todos
los habitats except por la tundra alpina, mientras
que otras especies tales como la Cascabel de
Manchas-gemelas *Crotalus pricei* y la serpiente
acuática *Thamnophis rufipunctatus* tienen distribuciones relativamente pequeñas en hábitats
específicos. Una especie (*Coluber constrictor*) ha
sido registrada para Arizona por sólo un espécimen (BYU 100) recolectado en 1927 en "Eagar,
Arizona." No hay especímenes adicionales de
C. constrictor recolectados en Arizona. A través
de análisis de ADN mitocondrial, D. G. Mulcahy
(2006, 2008) descubrió una especie no descrita
de *Hypsiglena* que se limita al área de las Cochise
Filter Barrier del sureste de Arizona, parte adyacente a Nuevo México, y presumiblemente parte
adyacente a México. Cuando sea descrita esta
especie podrá representar la especie 57 de las
serpientes de Arizona.

Únicamente cuatro del amplio arreglo de
herpetotaxa del estado son endémicas a Arizona
y todas ellas son lagartijas (*Aspidoscelis arizonae,
Aspidoscelis pai, Xantusia arizonae* y *Xantusia
bezyi*).

Ciento veintinueve (84%) de las especies de
herpetozoarios de Arizona se comparten con
México. Muchas de estas especies compartidas
alcanzan su límite norte de distribución en o
cerca de los límites del Desierto de Sonora de
Arizona. Estas especies incluyen *Anaxyrus retiformis, Incilius alvarius, Smilisca fodiens, Kinosternon
arizonense, Crotaphytus nebrius, Phrynosoma
goodei, P. solare, Uma rufopunctata, Aspidoscelis stictogramma, A. xanthonota, Chilomeniscus
stramineus, Chionactis palarostris, Masticophis
bilineatus, Phyllorhynchus browni, Micruroides*

euryxanthus y *Crotalus tigris*. La mayoría de estas especies de regiones semiáridas alcanzan su límite sur en Sonora o Sinaloa. La mayoría del resto de las especies compartidas alcanzan su límite norte de distribución en las sierras que forman islas continentales y cuencas del sureste de Arizona. Estas especies incluyen *Lithobates tarahumarae, Holbrookia elegans, Sceloporus jarrovii, S. slevini, S. virgatus, Plestiodon callicephalus, Gyalopion quadrangulare, Lampropeltis knoblochi, Oxybelis aeneus, Senticolis triaspis, Tantilla wilcoxi, T. yaquia, Crotalus pricei* y *C. willardi*. El límite sur de muchas de estas especies del sureste de Arizona se encuentra en o adyacente al Altiplano Mexicano. La distribución de *Senticolis triaspis* se extiende hasta Costa Rica, y la distribución excepcionalmente grande de *Oxybelis aeneus* se extiende desde aproximadamente 24 km al norte de la frontera sur de Arizona hasta el sureste de Brasil y Perú (Stebbins 2003).

Especies compartidas que alcanzan su límite norte de distribución en o más allá de la Zona de Transición del centro de Arizona pero dentro de las fronteras del estado incluyen *Hyla wrightorum, Lithobates chiricahuensis, L. yavapaiensis, Kinosternon sonoriense, Gopherus morafkai, Elgaria kingii, Cophosaurus texanus, Sceloporus clarkii, Aspidoscelis uniparens, Thamnophis eques* y *Crotalus molossus*.

Especies compartidas que alcanzan su límite norte de distribución fuera de las fronteras de Arizona incluyen *Ambystoma mavortium, Anaxyrus cognatus, A. debilis, A. punctatus, A. woodhousii, Craugastor augusti, Acris crepitans, Hyla arenicolor, Pseudacris hypochondriaca, Gastrophryne mazatlanensis, Lithobates catesbeianus, Scaphiopus couchii, Spea bombifrons, S. multiplicata, Chelydra serpentina, Chrysemys picta, Terrapene ornata, Trachemys scripta, Kinosternon flavescens, Apalone spinifera, Crotaphytus collaris, Gambelia wislizenii, Coleonyx variegatus, Hemidactylus turcicus, Heloderma suspectum, Dipsosaurus dorsalis, Sauromalus ater, Callisaurus draconoides, Holbrookia maculata, Phrynosoma cornutum, P. hernandesi, P. mcallii, P. modestum, P. platyrhinos, Sceloporus bimaculosus, S. cowlesi, S. magister, Urosaurus graciosus, U. ornatus, Uta* stansburiana, *Plestiodon gilberti, P. multivirgatus, P. obsoletus, P. skiltonianus, Aspidoscelis exsanguis, A. sonorae, A. tigris, Xantusia vigilis, Lichanura trivirgata, Arizona elegans, Chionactis occipitalis, Coluber constrictor, Diadophis punctatus, Gyalopion canum, Heterodon kennerlyi, Hypsiglena chlorophaea, H. jani, Lampropeltis getula, L. triangulum, Masticophis flagellum, M. taeniatus, Phyllorhynchus decurtatus, Pituophis catenifer, Rhinocheilus lecontei, Salvadora deserticola, S. grahamiae, S. hexalepis, Sonora semiannulata, Tantilla hobartsmithi, T. nigriceps, Thamnophis cyrtopsis, T. elegans, T. marcianus, Trimorphodon lambda, Rena dissecta, R. humilis, Crotalus atrox, C. cerastes, C. lepidus, C. mitchellii, C. oreganus, C. scutulatus, C. viridis* y *Sistrurus catenatus*. Algunas de estas especies compartidas, incluyendo *Diadophis punctatus, Lampropeltis triangulum, Pituophis catenifer* y *Crotalus viridis* se extienden hacia el norte hasta Canadá.

Nueve de las 129 especies compartidas entre Arizona y México son introducidas en Arizona (*Ensatina eschscholtzii, Xenopus laevis, Lithobates berlandieri, L. catesbeianus, Chelydra serpentina, Trachemys scripta, Apalone spinifera, Hemidactylus turcicus* e *Indotyphlops braminus*).

En Arizona hay 17 especies nativas que no son endémicas y que no habitan en México. De éstas, 4 se comparten con sólo un estado vecino (*Aspidoscelis flagellicauda, Thamnophis rufipunctatus* y *Crotalus cerberus* habitan en Arizona y Nuevo México; *Uma scoparia* habita en Arizona y California). *Lithobates onca* puede habitar en tres estados (Arizona, Nevada y posiblemente Utah). *Lampropeltis pyromelana* habita en cuatro estados (Arizona, Nevada, Nuevo México y Utah). *Gopherus agassizii* habita en cuatro estados (Arizona, California, Nevada y Utah). Otras especies nativas de Arizona que no habitan en México están más ampliamente distribuidas en el oeste de Estados Unidos y, en algunos casos, oeste de Canadá (*Anaxyrus microscaphus, Spea intermontana, Crotaphytus bicinctores, Sceloporus graciosus, S. tristichus, S. uniformis* y *Aspidoscelis velox*). Una población aislada de *Lithobates blairi* habita en el Sulphur Springs Valley del extremo sureste de Arizona pero aparentemente no entra a México. El resto de la amplia distribución de estos anuros incluye mucho de la porción sur-central de Estados Unidos. Las restantes dos especies nativas que se encuentran en Arizona pero no en México (*Pseudacris triseriata* y *Lithobates pipiens*) se distribuyen ampliamente sobre una porción grande de Estados Unidos y Canadá.

Arizona es uno de los estados que está creciendo más rápidamente en los Estados Unidos. Desarrollos residenciales, industriales, de infraestructura, de negocios, y agrícolas y el subsecuente incremento en la demanda de los limitados recursos del estado representan una amenaza mayor para los hábitats naturales. Desarrollos urbanos y suburbanos directamente consumen áreas naturales, la infraestructura asociada como caminos y canales fragmenta el hábitat, incrementan la contaminación, desechos, y actividades humanas que afectan adversamente el ecosistema. El estado está esencialmente partido en dos por el Proyecto Central de Arizona, un sistema de 541 km de canales, acueductos, y líneas de tubería que distribuyen las aguas del Río Colorado desde el Lake Havasu cerca de Parker hasta los condados de Pima, Pinal y Maricopa. Una siempre creciente red de caminos en forma de telaraña cubre el estado y en muchas áreas los hábitats entre estos caminos están cicatrizados por un laberinto de autopistas. Estas arterias de viaje actúan como cercas móviles formando barreras sustanciales para los movimientos de la vida silvestre. Nuevas amenazas en el horizonte son los proyectos de energía verde, los cuales pueden requerir el empleo de enormes superficies para la creación de granjas solares o eólicas.

Varias especies de plantas exóticas han sido introducidas a Arizona y muchas se han convertido en invasivas, amenazando las especies nativas del estado. Pastos y yerbas invasivas tales como bufel (*Pennisetum ciliare*), pasto del mediterranean (*Schismus barbatus*), bromo rojo (*Bromus rubens*) y mostaza (*Brassica* sp.), se establecen junto a plantas nativas, compiten por

agua, y promueven incendios. Especies de fauna invasiva como el cangrejo de río, la Rana Toro, peces y tortugas exóticas han tenido impactos severos en los lugares donde se establecen. Estas especies importadas compiten con las especies nativas por recursos, actúan como almacenadores y transmisores de enfermedades, y frecuentemente son depredadores importantes de la herpetofauna nativa.

Mucho del oeste de los Estados Unidos está siendo testigo de los impactos de casi 100 años de supresión de fuego en comunidades adaptadas a incendios. La supresión de este factor natural del ecosistema ha resultado en el crecimiento de una maleza densa en muchas comunidades boscosas. Los fuegos de baja intensidad que alguna vez mantuvieron estas comunidades ahora están alimentados por una densa maleza y se han convertido en fuegos de alta intensidad de naturaleza catastrófica. Los incendios catastróficos han resultado en una conversión masiva del hábitat en muchas partes de Arizona.

Arizona ha tenido una larga historia de industria ganadera vibrante y virtualmente no hay ningún lugar en el estado en el cual el pastoreo no haya ocurrido. Cuando éste se realiza irresponsablemente puede tener impactos terribles sobre las comunidades nativas. El sobrepastoreo ha provocado la conversión del hábitat, promovido especies exóticas y nocivas, y perdida de especies de herbáceas nativas en muchas partes áridas de elevaciones bajas del estado.

La construcción de extensas barreras fisicas tales como muros y cercas a lo largo de la frontera sur de Arizona ha interrumpido o bloqueado el movimiento de muchas especies de vida silvestre incluyendo anfibios y reptiles. Adicionalmente el incremento de la presencia humana a lo largo de estas fronteras ha provocado una proliferación de caminos, basura y tráfico de vehículos. Los impactos de estas actividades sobre la herpetofauna y otra vida silvestre aún no han sido evaluados.

CHARLES W. PAINTER Y JAMES N. STUART

Introducción

Nuevo México es un estado herpetológicamente diverso con 136 especies conocidas actualmente, colocándolo dentro de las 10 entidades de Estados Unidos con mayor número de especies (puede haber algunas variaciones en el orden dependiendo de la clasificación taxonómica que se utilice). Aunque Nuevo México es más seco que la mayoría del resto de los Estados Unidos y tiene pocas especies de salamandras, el número de especies de anuros y tortugas es muy alto para un estado tan xérico y las especies de lagartijas y serpientes representan el 72% de la herpetofauna del estado. Con base al tamaño de la población humana, Nuevo México es el estado de la frontera México—Estados Unidos más pequeño de los Estados Unidos, y esto se refleja en el reducido número de herpetólogos activos residentes en el estado. Sin embargo, nuestro conocimiento actual se ha beneficiado en gran medida por el desarrollo de programas de Herpetología en la University of New Mexico, New Mexico State University, Western New Mexico University y Eastern New Mexico University. Además, el New Mexico Department of Game and Fish ha desarrollado un programa activo de fauna no cinegética durante los últimos 25 años que ha contribuido significativamente al entendimiento de la historia natural y la distribución de la fauna del estado, incluyendo anfibios y reptiles. La New Mexico Herpetological Society también ha contribuido significativamente a este entendimiento. El libro más completo sobre los anfibios y reptiles de Nuevo México fue producido por Degenhardt et al. (1996) quienes reportaron 123 especies para el estado. La probabilidad de que nuevas especies sean descubiertas dentro de las fronteras de Nuevo México es relativamente baja. Sin embargo, a medida que más especies son investigadas utilizando técnicas genéticas existe la posibilidad de que especies cripticas aún no descritas sean descubiertas (ejem. *Aspidoscelis*).

Estudios Herpetológicos Previos

Gran parte del siguiente material fue adaptado del capítulo "Una breve historia de Herpetología en Nuevo México" de Degenhardt et al. (1996), escrito principalmente por Andrew H. Price. El lector interesado debe consultar esta fuente para obtener información adicional.

El interés por la herpetología de Nuevo México se inició en los pueblos nativos americanos quienes crearon petroglifos, fetiches, y otros objetos ceremoniales que incorporaban imágenes herpetológicas o partes de anfibios o reptiles. Éstos eran importantes totems en su vida cotidiana y espiritual (ejem. Whipple 1854 [1856]; Kennerly

1856; Henderson y Harrington 1914), y estacionalmente varias especies fueron utilizadas como alimento. Se crearon iconos de varias especies por las culturas Anasazi y Mimbres (Bettison et al. 1999; Davis 1995). Muchos de los grupos de la cultura Pueblo (Zia, Zuni y Acoma) así como las culturas Apache, Navajo y Ute incorporaron serpientes, especialmente de cascabel, o tortugas en sus rituales (ejem. Cushing 1883; Klauber 1972; Fewkes 1986; Hough 2010).

La primera referencia escrita sobre la herpetofauna de Nuevo México por europeos fue sobre las serpientes de cascabel escrita por los miembros de la expedición española dirigida por el Padre Agustín Rodríguez y Francisco Sánchez Chamuscado en 1581 (Klauber 1972). Después de esa contribución, Nuevo México permaneció relativamente inexplorado herpetológicamente hasta el siglo diecinueve. Tras la guerra de 1812, el ejército de Estados Unidos creó una división de ingenieros topográficos, en parte para explorar la frontera occidental. La primera expedición de exploración de esta división para entrar en lo que hoy es Nuevo México fue dirigida por el mayor Stephen H. Long a las Montañas Rocallosas (James 1823). Un pequeño destacamento de esta expedición pasó cerca de dos semanas de agosto de 1820 explorando los valles de los ríos Mora y Canadiense en el noreste de Nuevo México. Durante esta parte del viaje hizo referencia a la Serpiente de Cascabel de la Pradera (*Crotalus viridis*) y a dos formas de "lagartijas orbiculares" (*Phrynosoma cornutum* y *P. hernandesi*), ambas reportadas como comunes (James 1823: 276).

La tensión entre México y los Estados Unidos proporcionó la siguiente oportunidad para una adición al registro impreso de la herpetología de Nuevo México. En agosto y septiembre de 1845, el Teniente James E. Abert del Cuerpo de Ingenieros Topográficos pasó dos semanas explorando el Valle del Río Canadiense desde Paso Ratón hasta la frontera con Texas. Comentó sobre la abundante serpiente de cascabel (*Crotalus viridis*) asociada con las colonias del perrito de la pradera a lo largo del camino (Abert 1846). En 1846, el Teniente Abert y el Primer Teniente William H. Emory se unieron a la expedición del

Coronel Stephen W. Kearny contra México, que entró a Nuevo México a lo largo del camino de Santa Fe el 6 de agosto. Durante esta expedición, sólo se registraron algunas observaciones herpetológicas (Abert 1848; Emory 1848). El Primer Teniente Emory señaló la abundancia de "*Agama cornuta*" (= *P. hernandesi*) cerca de la confluencia de los Ríos Vermejo y Canadá los días 9 y 11 de agosto y un encuentro con una *Crotalus atrox* rojiza cerca de Las Palomas en el Río Grande en octubre 15. El Teniente Abert permaneció en el norte y centro de Nuevo México, visitando varios sitios incluyendo El Rito, Acoma y Socorro, dejando Nuevo México a mediados de enero de 1847. Señaló que había "un gran número de lagartijas cornudas" (*P. hernandesi*), cerca de Lamy en septiembre 29. Obtuvo un espécimen de una "lagartija singular" (*Crotaphytus collaris*?) entre Bernalillo y la boca del Río Jemez octubre 13. Se desconoce el lugar en donde este espécimen fue depositado. En 1846, Adolphus Wislizenus, un naturalista suizo emigrante, inició una de las primeras exploraciones apoyadas con fondos privados en el oeste. Comentó sobre la abundancia general de lagartijas durante su viaje por el camino de Santa Fe y el Valle de Río Grande entre junio 23 y agosto 8 (Wislizenus 1848).

La década previa a la Guerra Civil fue un período de mayor actividad herpetológica en Nuevo México. Tras el final de la Guerra México-Estados Unidos en 1848, se enviaron varias expediciones militares para el estudio de la frontera internacional con México para buscar la mejor ruta para un ferrocarril transcontinental. Spencer Fullerton Baird, secretario adjunto del Instituto Smithsoniano, hizo arreglos para que hubiera naturalistas acompañando varias de estas expediciones (Dall 1914; Adler 1989). El material recolectado ayudó a descubrir la riqueza y diversidad de la herpetofauna del suroeste estadounidense, incluyendo Nuevo México. La primera expedición, liderada por el Capitán Lorenzo Sitgraves, exploró los Ríos Zuni y Little Colorado hasta su confluencia con el Río Colorado y más allá, viajando a 1450 km desde Santa Fe hasta Fuerte Yuma entre agosto 15 y noviembre 24 de 1851 (Sitgraves 1853). La

expedición viajó hasta los Ríos Puerco y San
José, y desde el Río Grande hasta Laguna Pueblo
y hacia el oeste vía Acoma Pueblo, Inscription
Rock y Zuni Pueblo hasta el Río Zuni, saliendo
de Nuevo México a lo largo del curso del Río
Zuni en septiembre 25. La United States—
Mexico Boundary Survey, liderada por el Mayor
William H. Emory, consistió en una serie de
expediciones independientes realizada entre 1851
y 1855 para determinar la frontera internacional
tras el Tratado de Guadalupe Hidalgo de 1848
y la Compra de Gadsden en 1853 (Emory 1857,
1859). En 1853, el Congreso autorizó expediciones para encontrar la mejor ruta para un
ferrocarril transcontinental desde el río Misisipi
hasta la costa del Pacífico. Los resultados de estas
expediciones se publicaron como un tratado
de 12 volúmenes con varios autores entre 1854
y 1859 (ejem. Pope 1854 [1855]). La expedición
dirigida por el Teniente A. W. Whipple entró en
Nuevo México a lo largo del Río Canadiense en
septiembre 21 de 1853, viajó hasta Arroyo Pajarito
y a través del Arroyo Cuerbito (= Cuervito) y del
Río Gallinas hasta el Río Pecos en Anton Chico
en septiembre 27 llegando a Albuquerque vía
Galisteo y Peña Blanca en octubre 5. En noviembre 8, la expedición viajó hacia el sur hasta Isleta
Pueblo, atravesando el Río Grande y siguiendo
la ruta aproximada de la expedición de Sitgraves, dejando Nuevo México en noviembre 28.
Whipple (1854 [1856]) comentó sobre el éxito de
las colecciones zoológicas por miembros de su
equipo. Otra expedición, liderada por el Teniente
John G. Parke, cruzó las Montañas Peloncillo en
Nuevo México a través del Paso Stein en marzo
6 de 1854 y viajó a través del antiguo camino de
la Boundary Comission pasando Fort Webster
y Cooke´s Spring, llegando al Río Grande en
los alrededores de Leasburg en marzo 12. Parke
(1855) señaló que muchos de sus especímenes
se perdieron porque los contenedores donde
estaban almacenados se estaban derramando.
Una tercera expedición, liderada por el Capitán
Brevet John Pope, partió hacia el este de Doña
Ana en febrero 12 de 1854 y, pasando por Soledad
Canyon en las Montañas Organ, en febrero 14,
viajó a través de las Montañas Hueco, Álamo y

Cornuas a través del Paso Guadalupe hasta el
Río Delaware y, hasta su confluencia con el Río
Pecos en marzo 8. Pope (1854 [1855]: 59) señaló
en marzo 7 que la expedición "mató una serpiente de cascabel (la primera que hemos visto)
en una colina cerca al campamento. Fue puesta
en alcohol y transportada."

Relativamente poca exploración herpetológica
tuvo lugar en Nuevo México durante casi un
siglo tras el establecimiento de las fronteras territoriales entre 1863 y hasta antes de la Segunda
Guerra Mundial. Una serie de expediciones
patrocinadas por el gobierno para explorar las
características geográficas y geológicas al oeste
del meridiano 100, lideradas por el Teniente
George M. Wheeler, de la U.S. Army Corps
of Engineers, se llevó a cabo entre 1871 y 1874.
Los resultados fueron publicados en un tratado
de seis volúmenes, uno de los cuales contenía
la zoología de las expediciones. Dos capítulos
(Coues 1875; Yarrow 1875) fueron sobre la herpetología de Nuevo México. Yarrow (1875) reportó 1
salamandra, 8 anuros, 21 lagartijas, 17 serpientes
y 1 tortuga para el territorio de Nuevo México.
Otras referencias de publicaciones durante
este período (Cope 1883, 1896; Garman 1887;
Townsend 1893; Stone y Rehn 1903; Bailey 1905;
MacBride 1905; Ellis 1917) también presentaron
información sobre especímenes recolectados en
Nuevo México. Personal asociado con la U.S. Biological Survey (1896–1939) también recolectó en
Nuevo México. Bailey (1913) enlistó 76 especies,
incluyendo 3 salamandras, 7 anuros, 31 lagartijas,
32 serpientes y 3 tortugas poco después de que
Nuevo México se convirtió en un estado. Además escribió la primera discusión a nivel estatal
sobre la ecología y zoogeografía de especies. Van
Denburgh (1924) proporcionó un listado conte-
niendo 85 especies (2 salamandras, 12 anuros, 30
lagartijas, 36 serpientes y 5 tortugas).

La Fundación de la New Mexico State University en 1880 y Jornada Experimental Range del
U.S Department of Agriculture (USDA) en 1912,
junto con la University of New Mexico en 1889,
resultaron en un incremento de la exploración
herpetológica en el Valle de Río Grande (Cockerel 1896; Herrick et al. 1899; Little y Keller 1937).

Otros trabajos notables durante este período incluyen Ruthven (1907) que recolectó entre White Sands National Monument y Cloudcroft en las Montañas Sacramento. Proporcionó información sobre 2 anuros, 13 lagartijas, 2 serpientes y 1 tortugas. Mosauer (1932) pasó tres semanas en las Montañas Guadalupe y proporcionó observaciones sobre 1 salamandra, 1 anuros, 7 lagartijas y 3 serpientes.

Tras la Segunda Guerra Mundial, se construyeron grandes instalaciones militares y civiles que cambiaron la imagen de Nuevo México. Estas instalaciones requirieron la construcción de carreteras nuevas, y el uso rutinario de vehículos motorizados facilitó llegar a lugares y hábitats previamente inexplorados y proporcionaron un mecanismo directo para recolectar especímenes (ejem. Klauber 1939; Campbell 1953, 1956). En las universidades estatales se establecieron programas de investigación y colecciones de investigación herpetológica. La llegada de William J. Koster a la University of New Mexico (UNM) en 1938 fue especialmente importante en ayudar a crear estas colecciones. Aunque Koster era ictiólogo, conocía bien a A. H. Wright y tenía un gran interés en herpetología. Sus colecciones de anfibios y reptiles resultaron en más de 5000 ejemplares. James S. Findley, llegó a la UNM en 1955, y aunque fue principalmente mastozoólogo, él y sus estudiantes recolectaron gran cantidad de anfibios y reptiles en todo Nuevo México. William G. Degenhardt llegó a la UNM en 1960 proveniente de la Universidad de Texas A & M y creó la División de Herpetología y los programas de licenciatura y posgrado en herpetología. Durante este mismo período, otros programas en herpetología se desarrollaron en la New Mexico State University (James R. Dixon, Walter G. Whitford, Joseph L. LaPointe), Western New Mexico University (Bruce J. Hayward), y Eastern New Mexico University (A. L. "Tony" Gennaro). Esta actividad resultó en un aumento significativo en el conocimiento de la distribución y la historia natural de la herpetofauna de Nuevo México, así como en el número de publicaciones similares a las vistas en otros estados del oeste y en México (ejem. Webb 1970;

Dixon 1987; Carpenter y Krupa 1989; Smith y Smith 1973, 1976, 1979 [1980], 1993). Destacados colaboradores en los estudios regionales durante este período incluyen Lewis (1950; Organ Mountains y Tularosa Basin, 34 especies), Harris (1963; San Juan Basin, 21 especies), Gehlbach (1965; Zuni; Zuni Mountains, 30 especies), Jones (1970; Chaco Canyon National Monument, 18 especies), Mecham (1979; Guadalupe Mountains, 50 especies) y Best et al. (1983; Pedro Armendáriz lava field, Condados Sierra y Socorro, 26 especies). Importantes estudios de diferente grupos taxonómicos durante este período incluyeron Zweifel (1968), Creusere y Whitford (1976) y Woodward (1982, 1983, 1987) de anuros; Degenhardt y Christiansen (1974) de tortugas; Tanner (1975), Whitford y Creusere (1977) y Baltosser y Best (1990) de lagartijas; y Price y LaPointe (1990) de serpientes.

La New Mexico Herpetological Society (NMHS) se estableció formalmente en octubre de 1963, bajo la dirección de William G. Degenhardt y Ted L. Brown, como una organización dedicada a adquirir y compartir información acerca de herpetología, herpetocultura y los anfibios y reptiles de Nuevo México. Las actividades iniciales de la sociedad incluyeron visitas de campo a áreas seleccionadas de Nuevo México para investigar especies periféricas al estado y aquellas de especial interés para el New Mexico Department of Game and Fish (NMDGF). Desde el establecimiento de la NMHS, los miembros han contribuido a los estudios regionales que han incrementado significativamente nuestro entendimiento de la distribución de la herpetofauna en Nuevo México. Destacan los reportes de Brown (2002: Mills Canyon, condados Harding y Mora; 2003: Chloride Canyon, condado de Black Range Sierra), Brown y Stuart (1990: Canadian River Canyon, condados de Harding, Mora y San Miguel), Stuart y Brown (1988: Dry Cimarron River Valley, condado Unión), y Stuart y White (1987: manantiales en el condado Eddy).

Muchos estudios regionales sobre herpetofauna están disponibles como reportes sin publicar, incluyendo los de Painter (1985: Valles de los Ríos Gila y San Francisco, suroeste de

Nuevo México), Martin-Bashore (1997: Otero Mesa, McGregor Range, Fort Bliss), Krupa (1998: Rattlesnake Springs, Carlsbad Caverns National Park), Burkett y Black (1999, San Andres National Wildlife Refuge), Fox et al. (1999: condado Los Álamos), Burkett (2000: White Sands Missile Range), Collins (2000: Kiowa National Grasslands del noreste de Nuevo México), y Cummer et al. (2003: Valles Caldera National Preserve, condado Sandoval).

Fritts et al. (1984) proporcionaron una revisión completa de las Ranas Leopardo de Nuevo México con mapas de distribución y notas sobre la morfología de los adultos y renacuajos e historia natural. Altenbach y Painter (1998) publicaron la bibliografía completa de la Salamandra de las Montañas Jemez, *Plethodon neomexicanus*, la cual a nivel estatal se encuentra en peligro de extinción; Beck (2005) publicó un libro sobre la historia natural del Monstruo de Gila, gran parte de su investigación para el libro se llevó a cabo en el suroeste de Nuevo México. Lannoo (2005 [y muchos autores en ese libro]), reportaron sobre el estado de conservación y distribución de todos los anfibios en Nuevo México.

Tras la aprobación de la Ley de Conservación de Vida Silvestre de Nuevo México en 1974 y la creación del programa de especies no cinegéticas en peligro de extinción en el NMDGF, Charles W. Painter fue contratado como primer herpetólogo estatal de Nuevo México en 1986 y se retiró de ese puesto en 2013. Él ha desarrollado investigaciones y estudios sobre el estatus de conservación en numerosas especies de la herpetofauna de Nuevo México, particularmente *Plethodon neomexicanus*, *Lithobates chiricahuensis*, *Aspidoscelis dixoni*, *Sceloporus arenicolus*, *Pseudemys gorzugi*, *Trachemys gaigeae* y *Crotalus willardi obscurus*. Es considerado como una autoridad sobre la práctica barbárica de redadas de víboras de cascabel.

A principios de la década de 1990, se había reunido suficiente información para garantizar la publicación de un libro sobre la herpetofauna de Nuevo México. El resultado, *Anfibios y Reptiles de Nuevo México*, por Degenhardt et al. (1996), proporcionó el primer tratado importante de especies del estado y su distribución geográfica e historia natural y también identificó vacíos en el conocimiento y temas para investigaciones futuras. El libro estimuló posteriormente reportes adicionales de nueva información de distribución e historia natural para el estado, especialmente en la revista *Herpetological Review*. Los estudios genéticos de poblaciones de anfibios y reptiles en Nuevo México y en otros lugares han resultado en gran cantidad de cambios en taxonomía desde 1996 (Crother 2008), como se refleja en la lista actual de especies conocidas para el estado. Muchas de estas referencias sobre la herpetofauna de Nuevo México publicadas desde 1996 se citan en Stuart (2005).

Adicionalmente a los estudios mencionados previamente, hemos determinado que hay 21 localidades tipo de especies y subespecies de anfibios (4 salamandras) y reptiles (15 lagartijas y 4 serpientes) en el estado de Nuevo México (ver Cuadro 11.1).

Características Fisiográficas y su influencia en la distribución de la herpetofauna

Parte de la siguiente discusión fue adaptada de información del capítulo "Un esbozo fisiográfico de Nuevo México," de Degenhardt et al. (1996) escrito por Andrew H. Price. Información adicional está disponible en el NMDGF (2006) y Western Regional Climate Center (2011).

Nuevo México es el quinto estado más grande de los Estados Unidos, con un área total de 314,456 km². Es un cuadro de aproximadamente 563 km en cada lado, y se encuentra principalmente entre 32° y 37° de latitud norte y 103° y 109° de longitud oeste. La topografía del estado consiste principalmente de mesetas o mesas, con numerosas cadenas montañosas, cañones, valles y arroyos normalmente secos. La altitud promedio es aproximadamente 1433 m sobre el nivel del mar. El punto más bajo está justo por encima de la presa Red Bluff a 859 m donde el Río Pecos fluye en Texas. El punto más alto es Wheeler Peak a 4011 m en las Montañas Sangre de Cristo.

Autor(es)	Nombre original: Localidad tipo
Salamandras (4)	
Baird (1850)	*Ambystoma mavortia* (= *A. mavortium mavortium*): "**New Mexico**"
Hallowell (1852b [1853])	*Ambystoma nebulosum* (= *A. mavortium nebulosum*): "**New Mexico**"; ésta se refiere al territorio de Nuevo México antes de la división de Arizona y Nuevo México. Hallowell restringió la localidad tipo a "**San Francisco Mountain**" (= San Francisco Peaks, Coconino County, Arizona).
Taylor, 1941	*Plethodon hardii* (= *Aneides hardii*): **Sacramento Mountains at Cloudcroft (9000 ft.), New Mexico**
Stebbins & Riemer (1950)	*Plethodon neomexicanus*: **12 miles west and 4 miles south of Los Alamos, 8750± feet, Sandoval County, New Mexico**
Lagartijas (15)	
Baird & Girard (1852)	*Phrynosoma modestum*: La descripción original señala que el material tipo fue "**brought from the valley of the Rio Grande west of San Antonio, by Gen. Churchill and from between San Antonio and El Paso del Norte, by Col. J.D. Graham**"; Smith y Taylor (1950) restringieron la localidad tipo a "**Las Cruces, Dona Ana County, New Mexico.**" Axtell (1988) no estuvo de acuerdo con esto. Por lo tanto, la localidad tipo está en cuestión (ver Degenhardt et al. 1996 para detalles adicionales).
Degenhardt & Jones (1972)	*Sceloporus graciosus arenicolous* (= *S. arenicolus*): **Mescalero Sands 3 1/2 miles N and 44 miles E Roswell, Chaves Co., New Mexico [T10S R31E]**
Phelan & Brattstrom (1955)	*Sceloporus magister bimaculosus* (= *S. bimaculosus*): **6.6 mi E. of San Antonio, Socorro County, New Mexico**
Lowe & Norris (1956)	*Sceloporus undulatus cowlesi* (= *S. cowlesi*): **White Sands, 3 miles northwest of the Monument headquarters, Otero County, New Mexico**
Smith et al. (1992)	*Sceloporus undulatus tedbrowni* (= *S. consobrinus*): **Chaves Co., New Mexico, 6 mi W Caprock, Lea Co., large dune, Waldrop Peak, 0.5 mi S Hwy 380**
Baird & Girard (1852)	*Sceloporus poinsetti*: la localidad tipo original fue dada como "**Rio San Pedro of the Rio Grande del Norte, and the province of Sonora**"; sin embargo, ésta fue restringida a "**either the southern part of the Big Burrow [sic] Mountains, or the vicinity of Santa Rita, Grant Co., New Mexico**" por Webb (1988).
Cope *in* Yarrow (1875)	*Sceloporus undulatus tristichus* (= *S. tristichus*): **Taos [Taos County], New Mexico**
Stejneger (1890)	*Urosaurus ornatus levis*: **Tierra Amarilla [Rio Arriba County], New Mexico**
Schmidt (1921)	*Uta stansburiana stejnegeri*: **mouth of Dry Canyon, Alamogordo [Otero County], New Mexico**
Lowe (1956)	*Cnemidophorus exsanguis* (= *Aspidoscelis exsanguis*): **Socorro, Socorro County, New Mexico**
Lowe & Wright (1964)	*Cnemidohorus flagellicaudus* (= *Aspidoscelis flagellicauda*): **at San Francisco Hot Springs (Frisco Hot Springs), 4800 ft elev., Catron County, New Mexico**
Wright & Lowe (1993)	*Cnemidophorus inornatus gypsi* (= *Aspidoscelis gypsi*): **White Sands National Monument, at 3 miles (by road) NW Monument Headquarters, 4020 ft, Otero County, New Mexico**
Wright & Lowe (1993)	*Cnemidophorus inornatus juniperus* (= *Aspidoscelis inornata junipera*): **San Pedro Creek, 3 miles north and 2 miles east of San Antonio, 4550 ft, Bernalillo County, New Mexico**; Degenhardt et al. (1996) señalaron que "San Antonio" es San Antonito.

CUADRO II.I *Continued*

Autor(es)	Nombre original: Localidad tipo
Wright & Lowe (1993)	*Cnemidophorus inornatus llanuras* (= *Aspidoscelis inornata llanuras*): **Carthage, 4990 ft, Socorro County, New Mexico**
Lowe & Zweifel (1952)	*Cnemidophorus neomexicanus* (= *Aspidoscelis neomexicana*): **McDonald ranch headquarters, 4800 feet elevation, 8.7 miles west and 22.8 miles south of New Bingham Post Office, Socorro County, New Mexico**
Serpientes (4)	
Hallowell (1852a)	*Masticophis* (= *Coluber*) *taeniatus*: **New Mexico, west of the Rio Grande**; Smith y Taylor (1950) restringieron la localidad tipo a **Shiprock, San Juan County, New Mexico**
Hallowell (1852a)	*Pituophis melanoleucus affinis* (= *P. catenifer affinis*): la localidad tipo **New Mexico** fue restringida a **Zuni River, New Mexico,** por Hallowell *in* Sitgraves (1853); Smith y Taylor (1950) más adelante restringieron la localidad tipo a **Zuni, McKinley County, New Mexico**
Baird & Girard (1853)	*Crotalus molossus*: **Fort Webster, St Rita del Cobre, Grant Co., New Mexico**
Harris & Simmons (1976)	*Crotalus willardi obscurus*: **Animas Mountains in Indian Creek Canyon near Animas Peak, Hidalgo County, New Mexico**

La geología de Nuevo México es compleja y se ha formado por fallas, levantamientos y actividad volcánica. La grieta del Río Grande es importante ya que divide al estado en dos mitades de norte a sur y contiene la Garganta del Río Grande en el norte de Nuevo México y el amplio Valle del Río Grande en los partes centro y sur del estado. Esta grieta divide las Montañas Rocallosas del Sur del norte de Nuevo México en dos complejos de montañas más importantes, las Montañas Sangre de Cristo al este y las Montañas de San Juan al oeste. La Meseta de Colorado y la Cuenca de San Juan se encuentran al oeste de las Montañas Rocallosas del Sur y son las principales características topográficas del noroeste de Nuevo México. Las Montañas Jemez, una caldera volcánica, están alejadas de las Montañas Rocallosas del Sur pero comparten muchas afinidades con las montañas más septentrionales del estado. Más al sur se encuentra la Meseta del Mogollón, que se extiende hacia el oeste hasta Arizona. Las tierras altas de Sacramento son el único gran altiplano al este del Valle de Río Grande en el sur de Nuevo México y comparten algunas afinidades biológicas con las montañas más septentrionales del estado. Además, una serie de cadenas montañosas más pequeñas y aisladas, separadas por grandes cuencas, conforman la topografía de Cuenca y Sierra del centro y sur de Nuevo México a ambos lados del Valle del Río Grande.

La mayor parte del tercio oriental de Nuevo México se caracteriza por una topografía de colinas suaves e incluye el Llano Estacado, una planicie elevada que se extiende en el oeste de Texas y tiene un poco de hábitat acuático natural.

El estado tiene aproximadamente 606 km² de ríos, arroyos, lagos y embalses que comprenden menos del 1% de la superficie total del estado. Aunque son una pequeña parte del paisaje del estado, proporcionan hábitat crucial para muchas especies que de lo contrario no podrían vivir en un ambiente principalmente seco. La División Continental se encuentra en la mitad occidental de Nuevo México, donde se define por montañas altas en la parte norte del estado y crestas bajas en la parte sur. Los principales sistemas de ríos del estado se encuentran al oeste de la División Continental e incluyen: 1) San Juan, que surge en el suroeste de Colorado y fluye hacia el oeste para unirse el Río Colorado en Utah; 2) Gila, que nace en la Meseta de

Mogollón y fluye hacia el oeste hasta Arizona para unirse al Río Colorado; 3) San Francisco, un afluente del Gila que surge en la región oriental de Arizona; y 4) Zuni, un afluente del Río Little Colorado que surge en el oeste de Nuevo México. Los principales ríos al este de la División Continental son: 1) Grande, que surge en el sur de Colorado y fluye hacia el sur a través de Nuevo México y a lo largo de los Estados Unidos-frontera con México hasta el Golfo de México; 2) Pecos, un afluente del Río Grande que surge en el norte de Nuevo México y fluye hacia el sur a través de la mitad oriental del estado para unirse al Río Grande en el occidente de Texas; 3) Canadiense, que surge en el norte de Nuevo México y el sur de Colorado y fluye hacia el este hasta Texas para unirse al Río Arkansas y eventualmente al Mississippi; y 4) Dry Cimarron, un pequeño arroyo en el noreste de Nuevo México que forma parte de las aguas del Río Arkansas.

También existen varias cuencas endorreicas (cerradas) en Nuevo México, algunas de éstas con flujos perennes, o que contienen suelos salinos o de yeso que proporcionan un hábitat único para muchas especies. Las cuencas cerradas más grandes en el oeste son las Planicies de San Agustín en el condado de Catron y la cuenca del Río Mimbres en los condados de Grant y Luna. La Cuenca Tularosa en el sur de Nuevo México es un área entre montañas al este del Valle del Río Grande, conocido por su hábitat de extensas dunas de yeso.

El promedio de temperatura anual varía de 17.8°C en el extremo sureste a 4.4°C o menos en las montañas altas y valles del norte; la altitud es más importante que la latitud en la determinación de la temperatura de cualquier localidad específica. Esto se muestra por sólo una diferencia de 1.7°C en la temperatura promedio entre estaciones en elevaciones similares, una en el extremo noreste y otra en el extremo suroeste; sin embargo, en dos estaciones separadas por tan sólo 24.1 km de distancia, pero con una diferencia de altitud de 1433 m, las temperaturas medias anuales son 16.1°C y 7.2°C, una diferencia de 8.9°C o una disminución de un poco más

de 1.7°C de temperatura por cada 304.8 m de altitud.

Durante los meses de verano, las temperaturas diurnas a menudo superan 37.8°C en elevaciones por debajo de los 1524 m; pero el promedio de temperatura máxima mensual durante julio, el mes más caluroso, va desde ligeramente por encima de 32.2°C en elevaciones más bajas hasta por encima de los 21.1°C en elevaciones altas. Los días más cálidos normalmente ocurren en junio antes de la temporada de tormentas; durante julio y agosto ocurren tormentas convectivas por las tardes, que tienden a disminuir la insolación, disminuyendo las temperaturas antes de que alcancen su máximo potencial diario. La temperatura más alta que se ha registrado en Nuevo México es de 46.7°C en Orogrande en julio 14 de 1934 y en Artesia en junio 29 de 1918. Un predominio de cielos despejados y baja humedad relativa permite el rápido enfriamiento por radiación de la tierra después de la puesta del sol; en consecuencia, las noches son generalmente confortables en verano. El intervalo promedio entre temperaturas alta y baja es de 13.9°C a 19.4°C.

En enero, el mes más frío, el promedio de temperaturas diurnas varía desde alrededor de 10°C en el sur y los valles centrales hasta alrededor de –1°C en las elevaciones más altas del norte. Temperaturas mínimas inferiores al punto de congelamiento son comunes en todo el estado durante el invierno, pero temperaturas inferiores al punto de congelamiento son raras excepto en las montañas. La temperatura más baja registrada en una estación meteorológica del estado fue –45.6°C en Gavilán en febrero 1 de 1951. Una temperatura baja sin registro oficial de –49.4°C en Ciniza en enero 13 de 1963, fue ampliamente reportada por la prensa.

La temporada libre de congelación oscila entre más de 200 días en los valles del sur a menos de 80 días en las montañas del norte donde algunos valles de alta montaña se congelan durante los meses de verano.

Las principales fuentes de humedad por las escasas lluvias y nieves que caen sobre Nuevo

México son el Océano Pacífico, 1288 km al oeste, y el Golfo de México, 804 km al sureste. Nuevo México tiene un clima continental templado, árido o semiárido, caracterizado por ligeras precipitaciones totales, abundante sol, baja humedad y un intervalo de temperatura diurna y anual relativamente grande. Las montañas más altas tienen características climáticas similares a las de las Montañas Rocallosas.

La precipitación media anual oscila entre menos de 25.4 cm sobre gran parte del desierto del sur y los valles de San Juan y del Río Grande hasta más de 50.8 cm en las partes más altas del estado. Una amplia variación en los totales anuales es característica de los climas áridos y semiáridos como lo ilustran los extremos anuales de 7.5 cm y 86.2 cm en Carlsbad durante un período de más de 71 años.

Las lluvias de verano caen casi en su totalidad durante las breves, pero frecuentemente intensas tormentas. La corriente general del sureste proveniente del Golfo de México trae humedad con estas tormentas en el estado, y el fuerte calentamiento de la superficie combinado con los levantamientos orográficos hace que los movimientos del aire sobre terrenos más altos provoquen corrientes de aire y condensación. Julio y agosto son los meses más lluviosos en la mayor parte del estado, con 30–40 % de la humedad total del año cayendo en estos meses. La zona del Valle de San Juan es menos afectada por esta corriente de verano, recibiendo alrededor del 25 % de su precipitación anual durante julio y agosto. Durante los 6 meses más cálidos del año, de mayo a octubre, la precipitación total promedio de 60 % del total anual en la meseta del noroeste a 80 % del total anual en las planicies del este.

La precipitación de invierno es causada principalmente por la actividad frontal asociada con el movimiento general de las tormentas del Pacífico en todo el país de oeste a este. A medida que estas tormentas se mueven hacia el interior, gran parte de la humedad se precipita sobre las cordilleras de montañas costeras y del interior de California, Nevada, Arizona y Utah. Gran parte de la humedad restante cae en la ladera oeste de la División Continental y en el norte y las altas cordilleras centrales. El invierno es la temporada más seca en Nuevo México excepto en la parte oeste de la División Continental. Esta sequedad es más notable en el Valle Central y en la vertiente oriental de las montañas.

Gran parte de la precipitación de invierno cae como nieve en las zonas montañosas, pero se puede presentar como lluvia o nieve en los valles. El promedio anual de nevadas varía desde aproximadamente 7.6 cm en el desierto del sur y estaciones de las planicies del hasta más de 2.54 m en las estaciones de las montaña del norte. Las nevadas pueden superar 7.62 m en las montañas más altas del norte.

En Nuevo México el promedio de humedad relativa es más bajo en los valles y más alto en las montañas debido a temperaturas más bajas en las montañas. La humedad relativa oscila entre un promedio de casi 65 % cerca del amanecer hasta 30 % a media tarde; sin embargo, la humedad de la tarde en los meses más cálidos suele ser inferior al 20 % y ocasionalmente puede ser hasta de 4 %. La baja humedad relativa durante los períodos de temperaturas extremas facilita el efecto de las temperaturas de verano e invierno (Western Regional Climate Center 2011).

Comunidades Bióticas

Basado sobre el sistema de clasificación desarrollado por Brown y Lowe (1980) y Brown (1994a), Nuevo México hospeda al menos 14 comunidades bióticas que van desde hábitats de montaña de elevaciones altas que hospedan pocas o ninguna especie de anfibios y reptiles hasta pastizales de elevaciones medias y bajas y comunidades de desierto con una gran diversidad herpetofaunística (Degenhardt et al. 1996). Estas comunidades son útiles para el entendimiento a gran escala de los patrones de diversidad herpetofaunística, pero no detallan las características del paisaje a fina escala tales como el tipo de suelo, topografía, elevación y disponibilidad de los refugios subterráneos que influyen mucho en la distribución de la herpetofauna. También intercaladas dentro de estas comunidades bióticas hay varios cuerpos de agua y zonas de vegetación riparia que, aunque poco comunes en Nuevo México, proporcionan hábitats valiosos para muchas especies.

Planicies y Pastizales de la Gran Cuenca

Las Planicies y Pastizales de la Gran Cuenca son comunes y están ampliamente distribuidas en Nuevo México en las llanuras y valles entre 1200 y 2300 m de altitud. Se extienden desde las partes noreste y centro-este del estado hacia el oeste a través del área de Albuquerque y en la Cuenca de San Juan en el noroeste, y hacia el sur hasta las Planicies de San Agustín, cerca del Río Gila y Animas Valley (Brown y Lowe 1980). Están ausentes en gran parte del sur de Nuevo México. Estos pastizales consisten principalmente de especies de pastos cortos o mixtos (ejem. *Bouteloua* spp., *Achnatherum hymenoides, Pleuraphis jamesii* y *Sporobolus airoides*) habitando casi todo el estado, y pastos altos (especialmente *Andropogon gerardi, Schizachyrium scoparium*) principalmente limitado a la parte este del estado (D. Brown 1994h). Arbustos asociados incluyen *Atriplex canescens, Artemisia* spp., *Krascheninnikovia lanata, Yucca* spp., *Opuntia* spp., *Cylindropuntia* spp. y *Ericameria* spp. El encino *Quercus havardii* se encuentra en esta comunidad en el sureste de Nuevo México. El sobrepastoreo y la supresión de fuego han impactado esta comunidad, provocando una amplia invasión de arbustos en muchas áreas. La herpetofauna asociada incluye *Ambystoma mavortium, Anaxyrus cognatus, A. debilis, A. microscaphus, A. punctatus, Pseudacris clarkii, P. maculata, Gastrophryne olivacea, Lithobates blairi, Scaphiopus couchii, Spea bombifrons, S. multiplicata, Terrapene ornata, Kinosternon flavescens, Crotaphytus collaris, Gambelia wislizenii, Coleonyx brevis, Cophosaurus texanus, Holbrookia maculata, Phrynosoma cornutum, P. hernandesi, P. modestum, Sceloporus arenicolus, S. consobrinus, S. cowlesi, S. slevini, S. tristichus, Urosaurus ornatus, Uta stansburiana,*

Plestiodon multivirgatus, P. obsoletus, Aspidoscelis exsanguis, A. inornata, A. neomexicana, A. sexlineata, A. tesselata, A. tigris, A. uniparens, Arizona elegans, Coluber constrictor, Diadophis punctatus, Gyalopion canum, Heterodon nasicus, H. platirhinos, Hypsiglena jani, Lampropeltis getula, L. triangulum, Masticophis flagellum, M. taeniatus, Pantherophis emoryi, Pituophis catenifer, Rhinocheilus lecontei, Sonora semiannulata, Tantilla nigriceps, Thamnophis elegans, T. marcianus, T. radix, Tropidoclonion lineatum, Rena dissecta, Crotalus atrox, C. viridis y *Sistrurus catenatus.*

Pastizal Semidesértico

El Pastizal Semidesértico se encuentra normalmente entre 1100 y 1400 m y está estrechamente relacionado con el Matorral Desértico Chihuahuense con el que hace contacto y se entremezcla. Se encuentra al sur de Albuquerque y Fort Sumner, bordeando los valles de los Ríos Grande y Pecos, respectivamente, y en gran parte del sur de Nuevo México incluyendo Roswell, Hobbs, Carrizozo, Las Cruces, Lordsburg y la mayor parte del condado de Hidalgo (Brown y Lowe 1980). Las principales especies de pastos incluyen *Pleuraphis mutica, Bouteloua* spp. y *Aristida* spp. Cactus tales como *Cylindropuntia* spp., *Opuntia* spp. y *Ferocactus wislizenii* también son componentes importantes. Aunque originalmente era una comunidad dominada por pastos antes de la llegada de los europeos, el sobrepastoreo de ganado ha resultado en un pastizal clímax alterado en muchas áreas que ahora están dominadas por arbustos tales como *Prosopis glandulosa, Larrea tridentata, Flourensia cernua* y *Gutierrezia* spp. (D. Brown 1994i).

La herpetofauna asociada (72 especies) incluye *Ambystoma mavortium, Anaxyrus cognatus, A. debilis, A. microscaphus, A. punctatus, A. speciosus, Incilius alvarius, Craugastor augusti, Gastrophryne olivacea, Lithobates blairi, Scaphiopus couchii, Spea bombifrons, S. multiplicata, Terrapene ornata, Kinosternon flavescens, Elgaria kingii, Crotaphytus collaris, Gambelia wislizenii, Coleonyx variegatus, Cophosaurus texanus, Holbrookia elegans, H. maculata, Phrynosoma* cornutum, *P. hernandesi*, *P. modestum*, *Sceloporus arenicolus*, *S. bimaculosus*, *S. clarkii*, *S. consobrinus*, *S. cowlesi*, *S. poinsettii*, *S. tristichus*, *Urosaurus ornatus*, *Uta stansburiana*, *Plestiodon obsoletus*, *Aspidoscelis exsanguis*, *A. flagellicauda*, *A. gularis*, *A. inornata*, *A. marmorata*, *A. neomexicana*, *A. sexlineata*, *A. sonorae*, *A. tesselata*, *A. uniparens*, *Arizona elegans*, *Diadophis punctatus*, *Gyalopion canum*, *Heterodon kennerlyi*, *H. nasicus*, *Hypsiglena jani*, *Lampropeltis getula*, *L. triangulum*, *Masticophis flagellum*, *M. taeniatus*, *Pantherophis emoryi*, *Pituophis catenifer*, *Rhinocheilus lecontei*, *Salvadora deserticola*, *S. grahamiae*, *Sonora semiannulata*, *Tantilla hobartsmithi*, *T. nigriceps*, *Thamnophis marcianus*, *Trimorphodon vilkinsonii*, *Micruroides euryxanthus*, *Rena dissecta*, *R. segrega*, *Crotalus atrox*, *C. scutulatus*, *C. viridis* y *Sistrurus catenatus*.

Matorral Desértico Chihuahuense

El Matorral Desértico Chihuahuense está confinado a la mitad sur de Nuevo México entre aproximadamente 866 y 1500 m de altitud donde está ampliamente distribuido en la Cuenca Tularosa y la Jornada del Muerto y en o cerca del White Sands National Monument y las ciudades de Socorro, Carlsbad, Artesia, Las Cruces, Deming y Lordsburg (Brown y Lowe 1980). Éste está ausente en el Bootheel. Gran parte de esta comunidad se desarrolla sobre piedra caliza, incluyendo caliche, y está normalmente presente sobre sustratos de grava y sobre pequeñas colinas rocosas y bajadas. La diversa topografía en la región cubierta por Matorral Desértico contribuye a la diversidad de especies de flora y herpetofauna presentes en esta comunidad biótica. Este matorral desértico con más frecuencia entra en contacto con pastizales semidesérticos con los cuales comparte muchos elementos de flora y fauna. Las principales especies de arbustos incluyen *Larrea tridentata*, *Flourensia cernua*, *Acacia* spp. y *Koeberlinia spinosa*. *Prosopis glandulosa* es abundante en donde hay suelos arenosos y en donde otras plantas generalmente asociadas con pastizales semidesérticos pueden estar presentes. Los cactus más comunes, incluyen especies de *Opuntia*, *Cylindropuntia*, *Ferocactus* y *Echinocactus* (D. Brown 1994j).

La herpetofauna asociada (71 especies) incluye *Ambystoma mavortium*, *Anaxyrus cognatus*, *A. debilis*, *A. punctatus*, *A. speciosus*, *Incilius alvarius*, *Craugastor augusti*, *Gastrophryne olivacea*, *Scaphiopus couchii*, *Spea bombifrons*, *S. multiplicata*, *Terrapene ornata*, *Kinosternon flavescens*, *Elgaria kingii*, *Crotaphytus collaris*, *Gambelia wislizenii*, *Coleonyx brevis*, *C. variegatus*, *Heloderma suspectum*, *Callisaurus draconoides*, *Cophosaurus texanus*, *H. maculata*, *Phrynosoma cornutum*, *P. modestum*, *Sceloporus bimaculosus*, *S. cowlesi*, *S. magister*, *S. poinsettii*, *Urosaurus ornatus*, *Uta stansburiana*, *Plestiodon obsoletus*, *Aspidoscelis dixoni*, *A. exsanguis*, *A. gularis*, *A. gypsi*, *A. inornata*, *A. marmorata*, *A. neomexicana*, *A. sonorae*, *A. tesselata*, *A. uniparens*, *Arizona elegans*, *Bogertophis subocularis*, *Diadophis punctatus*, *Gyalopion canum*, *Heterodon kennerlyi*, *H. nasicus*, *Hypsiglena jani*, *Lampropeltis alterna*, *L. getula*, *L. triangulum*, *Masticophis flagellum*, *M. taeniatus*, *Pituophis catenifer*, *Rhinocheilus lecontei*, *Salvadora deserticola*, *S. grahamiae*, *Sonora semiannulata*, *Tantilla hobartsmithi*, *T. nigriceps*, *T. yaqui*, *Thamnophis marcianus*, *Trimorphodon vilkinsonii*, *Micruroides euryxanthus*, *Rena dissecta*, *R. segrega*, *Crotalus atrox*, *C. lepidus*, *C. molossus*, *C. scutulatus* y *C. viridis*.

Matorral Desértico de la Gran Cuenca

El Matorral Desértico de la Gran Cuenca habita intermitentemente en el noroeste de Nuevo México, extendiéndose hacia el este hasta la Meseta Taos, y al sur hasta el área alrededor de Abiquiu Reservoir y cerca de Gallup, generalmente a elevaciones entre 1200 y 2200 m (Brown y Lowe 1980; Turner 1994). Éste está caracterizado por planicies de nivel o colinas suaves, normalmente en suelos altamente erosionables, dominados por varias especies de arbustos, especialmente *Artemisia* spp. y *Atriplex canescens*, junto con *Chrysothamnus* spp., *Sarcobatus* spp. y *Tetradymia* spp.

La herpetofauna asociada (19 especies) incluye *Ambystoma mavortium*, *Spea bombifrons*, *S. mul-*

tiplicata, Crotaphytus collaris, Gambelia wislizenii, Sceloporus graciosus, S. magister, S. tristichus, Uta stansburiana, Aspidoscelis tigris, A. velox, Heterodon nasicus, Hypsiglena chlorophaea, Lampropeltis getula, L. triangulum, Masticophis taeniatus, Pituophis catenifer, Thamnophis elegans y Crotalus viridis.

Encinar

El Encinar está ampliamente esparcido en la parte sur del estado, principalmente entre 1500 y 2200 m de altitud, con una presencia más continua al oeste de Silver City y hasta Arizona. Está presente en los alrededores de las Ánimas, Peloncillo, Big Hatchet, Burro, Tres Hermanas, Organ y las Montañas Guadalupe (Brown y Lowe 1980). Este bosque de encinos está dominado por especies perennes de encino (*Quercus* spp.) con pinos (Pinus spp.) también presentes en las elevaciones más altas; normalmente hace contacto con los pastizales o el chaparral interior en el límite inferior de la distribución altitudinal de estas dos comunidades (D. Brown 1994c).

La herpetofauna asociada (45 especies) incluye *Ambystoma mavortium, Anaxyrus punctatus, Hyla arenicolor, Terrapene ornata, Elgaria kingii, Crotaphytus collaris, Coleonyx brevis, Heloderma suspectum, Holbrookia elegans, Phrynosoma hernandesi, P. solare, Sceloporus clarkii, S. jarrovii, S. poinsettii, S. slevini, S. virgatus, Urosaurus ornatus, Plestiodon callicephalus, P. obsoletus, Aspidoscelis exsanguis, A. flagellicauda, A. sonorae, A. stictogramma, A. tesselata, Hypsiglena jani, Lampropeltis getula, L. knoblochi, Masticophis bilineatus, M. flagellum, M. taeniatus, Pituophis catenifer, Salvadora grahamiae, Senticolis triaspis, Tantilla hobartsmithi, T. nigriceps, T. yaquia, Thamnophis cyrtopsis, Trimorphodon lambda, T. vilkinsonii, Micruroides euryxanthus, Crotalus atrox, C. lepidus, C. molossus, C. viridis y C. willardi.*

Bosques de Coníferas de la Gran Cuenca

Los Bosques de Coníferas de la Gran Cuenca, comúnmente conocidos como bosques de piñonero-juniperus o de juniperus, están ampliamente distribuidos en Nuevo México, principalmente entre 1500 y 2300 m de altitud sobre bajadas, mesas o estribaciones rodeando muchas de las sierras del estado incluyendo las de Sangre de Cristo, Jemez, Sandia, Manzano, Sacramento y las Montañas Guadalupe, y en la cuenca del Río Gila (Brown y Lowe1980). Está ausente desde la esquina suroeste del estado y el Llano Estacado del este de Nuevo México. Este bosque adaptado al frío está dominado por coníferas enanas incluyendo una o más especies de enebro (*Juniperus*) y piñón (*Pinus edulis*). Otros elementos florales incluyen arbustos como *Fallugia, Rhus, Yucca, Atriplex* y *Purshia* y pastos y herbáceas como *Bouteloua, Bomus, Sphaeralcea* y *Eriogonum* (D. Brown 1994b).

Aunque no hay una herpetofauna exclusiva de este hábitat, 41 especies puede estar presente, incluyendo *Ambystoma mavortium, Anaxyrus microscaphus, A. punctatus, Spea multiplicata, Elgaria kingii, Crotaphytus collaris, Holbrookia maculata, Phrynosoma hernandesi, P. modestum, Sceloporus cowlesi, S. graciosus, S. tristichus, Urosaurus ornatus, Uta stansburiana, Plestiodon obsoletus, Aspidoscelis exsanguis, A. flagellicauda, A. inornata, A. neomexicana, A. tesselata, A. velox, Diadophis punctatus, Gyalopion canum, Heterodon nasicus, Hypsiglena chlorophaea, H. jani, Lampropeltis getula, L. triangulum, Masticophis flagellum, M. taeniatus, Pantherophis emoryi, Pituophis catenifer, Rhinocheilus lecontei, Salvadora grahamiae, Tantilla hobartsmithi, Trimorphodon lambda, Tropidoclonion lineatum, Rena dissecta, Crotalus atrox, C. molossus* y *C. viridis.*

Chaparral Interior

El Chaparral Interior es una comunidad dominada por arbustos común y ampliamente distribuida en Arizona, pero escasa en Nuevo México. Se encuentra en laderas en o cerca del Burro, Florida, San Andrés, Organ y las Montañas Guadalupe en la parte sur del estado (Brown y Lowe 1980). A menudo se encuentra en la pendiente superior del Pastizal Semidesértico. El

Chaparral Interior está caracterizado por varias especies de arbustos incluyendo *Garrya, Rhus trilobata, Purshia stansburiana, Forestiera pubescens, Berberis* y ocasionalmente arbusto de encinos (*Quercus*) frecuentemente sobre sustratos de granito o piedra caliza en elevaciones aproximadas a 1700–2450 m (Pase y Brown 1994c).

Aunque carece de una comunidad herpetofaunística bien definida, las 20 especies asociadas incluyen *Crotaphytus collaris, Sceloporus cowlesi, S. poinsettii, Urosaurus ornatus, Uta stansburiana, Plestiodon obsoletus, Aspidoscelis exsanguis, A. tesselata, Bogertophis subocularis, Hypsiglena jani, Masticophis taeniatus, Pituophis catenifer, Salvadora grahamiae, Tantilla hobartsmithi, Trimorphodon vilkinsonii, Rena segrega, Crotalus atrox, C. lepidus, C. molossus* y *C. viridis. Lampropeltis alterna* también puede presentarse en esta comunidad.

Matorral de Montaña de la Gran Cuenca

El Matorral de Montaña de la Gran Cuenca tiene una distribución limitada en Nuevo México, pero se hace más común al norte de la división estatal entre Colorado y Nuevo México. Se encuentra en estribaciones y áreas montañosas a elevaciones de alrededor de 2300 a 2750 m, este matorral se encuentra más extensamente en el área de Paso Ratón al este de las Montañas Sangre de Cristo y sobre la pendiente oeste de la cadena montañosa Sandia-Manzano (Brown y Lowe 1980). Esta comunidad está dominada por el encino de Gamel (*Quercus gambelii*) pero también incluye una variedad de matorrales tales como *Cercocarpus, Robinia neomexicana, Amelanchier* y *Symphoricarpus* (D. Brown 1994e).

Especies de anfibios y reptiles del Bosque de Coníferas de Montaña o Matorral Desértico de la Gran Cuenca también se presentan en esta comunidad e incluyen las siguientes 13 especies *Crotaphytus collaris, Phrynosoma hernandesi, Urosaurus ornatus, Plestiodon multivirgatus, P. obsoletus, Aspidoscelis exsanguis, Masticophis flagellum, M. taeniatus, Opheodrys vernalis, Pituophis catenifer, Salvadora grahamiae, Crotalus molossus* y *C. viridis.*

Bosque de las Montañas Rocallosas (Petran) y el Bosque de Coníferas de Montaña

El Bosque de las Montañas Rocallosas (Petran) y el Bosque de Coníferas de Montaña está ampliamente distribuido y se encuentra en la mayoría de las tierras altas alrededor de 2300 y 2900 m de altitud a través de Nuevo México incluyendo las Montañas Sangre de Cristo, San Juan, Jemez, Gallinas, Sacramento, San Mateo, Magdalena, Mogollón y Ánimas, y el Black Range y cerca de Mount Taylor (Brown y Lowe 1980). Incluidos en esta comunidad están los bosques dominados por el pino Ponderosa o pino chihuahuense (*Pinus ponderosa, P. leiophylla* en el suroeste de Nuevo México), Coníferas Mixtas/Abeto de Douglas-Abeto Blanco (*Pseudotsuga menziesii–Abies concolor*) y aspen (*Populus tremuloides*). Otros árboles incluyen *Pinus strobiformis, Quercus gambelii* y *Robinia neomexicana*, junto con una variedad de matorrales, hierbas y pastos en el sotobosque (Pase y Brown 1994b).

La herpetofauna asociada (35 especies) incluye *Ambystoma mavortium, Aneides hardii, Plethodon neomexicanus, Anaxyrus microscaphus, Hyla arenicolor, Hyla wrightorum, Pseudacris maculata, Phrynosoma hernandesi, Sceloporus clarkii, S. cowlesi, S. graciosus, S. jarrovii, S. poinsettii, S. tristichus, S. virgatus, Plestiodon callicephalus, P. multivirgatus, P. obsoletus, Aspidoscelis exsanguis, A. velox, Coluber constrictor, Diadophis punctatus, Lampropeltis knoblochi, L. pyromelana, L. triangulum, Pantherophis emoryi, Pituophis catenifer, Salvadora grahamiae, Thamnophis elegans, Trimorphodon lambda, Crotalus cerberus, C. lepidus, C. molossus, C. viridis* y *C. willardi.*

Pastizales de Pradera de Montaña

Los Pastizales de Pradera de Montaña se encuentran en las Montañas Rocallosas (Petran) y Bosques de Coníferas de Montaña, y se caracterizan por una variedad de especies de pastos y hierbas, incluyendo plantas de humedales como *Poa, Festuca, Carex, Cyperus, Eleocharis* y *Juncus* (D. Brown 1994g). Cuando está asociada con arroyos o suelos saturados, esta comunidad proporciona hábitat para 6 especies de herpetofauna de elevaciones altas como *Ambystoma mavortium, Hyla wrightorum, Pseudacris maculata, Lithobates pipiens, Opheodrys vernalis* y *Thamnophis elegans.*

Bosque de Coníferas Subalpinos de las Montañas Rocallosas (Petran)

Los Bosque de Coníferas Subalpinos de las Montañas Rocallosas (Petran), frecuentemente llamado Bosque de picea-abeto, están restringidos a elevaciones altas de aproximadamente 2500 a 3800 m hasta el límite altitudinal de la vegetación arbórea en las principales cadenas montañosas del estado, sobre todo en las Sangre de Cristo, San Juan, Jemez y Sacramento (Pase y Brown 1994a). También está presente pero menos extendido en las montañas Sandia, Magdalena, San Mateo y Mogollón; el Black Range; y la Sierra Blanca. Las 10 especies de herpetofauna asociadas con esta comunidad incluyen *Ambystoma mavortium, Aneides hardii, Plethodon neomexicanus, Anaxyrus boreas, Pseudacris maculata, Phrynosoma hernandesi, Lampropeltis pyromelana, Pituophis catenifer, Thamnophis elegans* y *Crotalus molossus.*

Matorral Subalpino

Las elevaciones más altas en Nuevo México incluyen tres comunidades bióticas que tienen pocas especies de herpetofauna. El Matorral Subalpino se limita a las Montañas Sangre de Cristo en el norte de Nuevo México a elevaciones de alrededor de 3500 m de altitud, y justo por debajo del límite altitudinal de la vegetación arbórea (D. Brown 1994d). Forma un ecotono con los Bosques Subalpinos de Coníferas de las Montañas Rocallosas a elevaciones ligeramente inferiores y puede proporcionar hábitat marginal para especies asociadas con la comunidad del Bosque de Coníferas.

Pastizales Alpinos y Subalpinos

Los Pastizales Alpinos y Subalpinos se encuentran en zonas muy limitadas de Nuevo México,

tales como las elevaciones más altas de las Montañas Sangre de Cristo, San Juan, Mogollón, Jemez y Chuska y cerca de Monte Taylor (Brown y Lowe 1980; D. Brown 1994f). Especies asociadas (6 especies) pueden incluir *Ambystoma mavortium, Anaxyrus boreas* (Montañas San Juan), *Phrynosoma hernandesi, Pituophis catenifer, Pseudacris maculata* y *Lithobates pipiens.*

Tundra Alpina

La Tundra Alpina se limita a las elevaciones más altas entre aproximadamente 3600 y 4000 m. Se encuentra por encima del límite altitudinal de la vegetación arbórea incluyendo el Wheeler Peak en las Montañas Sangre de Cristo del norte de Nuevo México y Sierra Blanca en la Montañas White del sur de Nuevo México (Pase 1994). Las juncias (*Carex*) son un componente importante, las plantas leñosas están limitadas a sauces (*Salix*) en algunos sitios. Una especie, *Aneides hardii*, ha sido registrada en esta comunidad a 3570 m de altitud en las Montañas White de Nuevo México (Moir y Smith 1970).

Ríos y Humedales

Aunque en Nuevo México la mayoría de la herpetofauna (especialmente del orden squamata) es completamente terrestre y se pueden asignar a una o más de las comunidades bióticas mencionadas anteriormente, algunas especies dependen de cuerpos de agua para al menos parte de sus historias de vida (ejem. reproducción, alimentación). Las 16 especies que tienen distribuciones estrechamente asociadas con uno o más de los sistemas fluviales perennes en Nuevo México incluyen *Acris crepitans, Lithobates berlandieri, L. catesbeianus, L. yavapaiensis, Chelydra serpentina, Apalone mutica, A. spinifera, Chrysemys picta, Pseudemys gorzugi, Trachemys gaigeae, T. scripta, Nerodia erythrogaster, Thamnophis eques, T. proximus, T. rufipunctatus* y *T. sirtalis.*

Algunas especies semiacuáticas de herpetofauna son altamente dependientes de los humedales permanentes o casi permanente como ojos de agua, estanques castores, ciénagas, arroyos intermitentes de cañón y algunas playas de lagos, frecuentemente con flora de humedales bien desarrollada. Estos sitios frecuentemente están aislados de hábitats más secos y especies que utilizan estos humedales se pueden dispersar entre ellos. En las praderas y valles fluviales, el riego agrícola ha contribuido a la creación de este tipo de hábitat o corredores para la dispersión entre estos humedales. Doce especies que utilizan este hábitat incluyen *Anaxyrus boreas, A. woodhousii, Hyla arenicolor, H. wrightorum, Pseudacris clarkii, P. maculata, Lithobates blairi, L. chiricahuensis, L. pipiens, Kinosternon sonoriense, Thamnophis cyrtopsis* y *T. elegans.*

Otros anfibios y reptiles que son acuáticos durante al menos parte de sus vidas están principalmente asociados con aguas efímeras tales como aguajes para ganado, arroyos, tinajas y cuencas cerradas, principalmente en las planicies o valles. Estos sitios típicamente carecen de flora de humedales y tienen agua superficial sólo durante las lluvias del monzón de verano. La herpetofauna en este grupo está adaptada a ambientes xéricos, por lo común tienen distribuciones amplias, y pueden vivir lejos de aguas perennes. Las especies representativas (11) son *Ambystoma mavortium, Scaphiopus couchii, Spea bombifrons, S. multiplicata, Anaxyrus cognatus, A. debilis, A. punctata, Incilius alvarius, Gastrophryne olivacea, Kinosternon flavescens* y *Thamnophis marcianus.*

En Nuevo México tres anfibios y una tortuga no requieren de agua estancadas para ninguna parte de sus historias de vida: *Aneides hardii, Plethodon neomexicanus, Craugastor augusti* y *Terrapene ornata.*

Un total de 136 especies de herpetofauna habita en Nuevo México: 27 anfibios (3 salamandras, 24 anuros) y 109 reptiles (10 tortugas, 47 lagartos, 52 serpientes). Estas especies representan 24 familias y 59 géneros. El número total de especies representa un aumento de 13 de las 123 especies reconocidas por Degenhardt et al. (1996). Dos de estas especies que se añadieron después de 1996 resultaron del descubrimiento de especies anteriormente desconocidas en Nuevo México (*Pseudacris clarkii, Heterodon platirhinos*); el resto de las adiciones se han debido a cambios taxonómicos recientes y el actual reconocimiento del rango de especie para taxa anteriormente considerados como subespecies. Además, a través de análisis de ADN mitocondrial, Mulcahy (2006, 2008) descubrió una especie de *Hypsiglena* (todavía sin describir) que se limita al área de Cochise Filter Barrier del suroeste de Nuevo México, adyacente a Arizona, y presumiblemente adyacente a México.

La mayoría de los taxa que habitan en Nuevo México también se encuentran en México: 24 familias (100%), 57 géneros (57/59 = 97%) y 113 especies (113/136 = 83%) habitan en ambos lados de la frontera. El alto porcentaje de taxa compartidos entre Nuevo México y el norte de México, presumiblemente refleja la continuidad de las comunidades bióticas y sistemas fluviales entre estas dos regiones.

Sólo tres especies de anfibios y reptiles son endémicas a Nuevo México: dos salamandras, *Aneides hardii* y *Plethodon neomexicanus,* y una lagartija, *Aspidoscelis gypsi.* Dos especies adicionales, *Sceloporus arenicolus* y *Aspidoscelis dixoni* tienen distribuciones muy limitadas en Nuevo México. Las poblaciones de *S. arenicolus* se encuentran en el oeste de Texas, adyacente a las poblaciones del extremo sureste de Nuevo México (Fitzgerald y Painter 2009). *Aspidoscelis dixoni* se encuentra en dos poblaciones muy alejadas, una de ellas en las Montañas Chinati del condado de Presidio en el oeste de Texas y la otra en Antelope Pass en las Montañas Peloncillo del condado de Hidalgo, suroeste del estado. Estas poblaciones están separadas por aproximadamente 500 km, y la población de Nuevo México es distintiva y representa un clon distinto (Painter 2009).

La Rana Toro, *Lithobates catesbeianus,* es una especie introducida en Nuevo México y ahora se encuentra, algunas veces en poblaciones densas, en aguas perennes de todo el estado a alturas inferiores a 2103 m (Degenhardt et al. 1996). Son extremadamente voraces y se cree que han contribuido a la disminución de algunas especies acuáticas (ejem. *Lithobates* spp., *Thamnophis rufipunctatus*). Hasta hace poco la Rana Toro era considerada una especie cinegética en Nuevo México y se requería de una licencia de caza para capturarla. Sin embargo, fueron reclasificadas y ahora no hay límite en el número ni método de colecta.

El Geco del Mediterráneo, *Hemidactylus turcicus*, es otra especie introducida, el primer reporte para el estado fue publicado por Painter et al. (1992). La especie parece estar ampliando su distribución en zonas urbanas de Nuevo México y actualmente es conocida para Alamogordo, Albuquerque, Las Cruces y Truth or Consequences (Byers et al. 2007). *Hemidactylus turcicus* es un comensal con los seres humanos, y sus poblaciones se encuentran en construcciones y hábitats naturales que han sido alterados a través de las actividades humanas.

La Tortuga del Desierto de Sonora (*Gopherus morafkai*), recientemente separada de la Tortuga de Desierto (*Gopherus agassizii*) (Murphy et al. 2011), ocasionalmente ha sido reportada para el Condado de Hidalgo. Aunque no se espera que haya alguna población viable pues ha habido numerosos encuentros de especímenes en áreas remotas donde individuos han escapado o han sido liberados y donde no pueden habitar en forma natural. Es dudoso que reportes de cerca de Rodeo en las Montañas Peloncillo (B. Tomberlin, comunicación personal) o cerca de Cloverdale, suroeste del condado de Hidalgo (R. Turner; P. Melhlop, comunicación personal) representen poblaciones reproductivas de esta especie. Se han reportado otros avistamiento de *G. morafkai* para todo el estado (ejem. Santa Fe) aunque éstos son considerados individuos que han escapado o han sido liberados.

Frecuentemente los humanos han introducido tortugas acuáticas en los sistemas fluviales de Nuevo México (Stuart 2000), aunque no se conoce ninguna especie de tortuga no nativa que se haya establecido en el estado. Sin embargo, se sabe o es probable que tres tortugas nativas se hayan llegado a establecer en Nuevo México en donde no habitaban de forma natural. Se piensa que *Trachemys scripta* es nativa sólo en las cuencas de los Ríos Pecos y Canadiense, sin embargo, actualmente está bien establecida a través de la parte baja y media de la cuenca del Río Grande extendiéndose hacia el norte hasta Bernalillo en el condado Sandoval (Stuart 2000). *Chelydra serpentina* es nativa de los Ríos Pecos, Canadienses y Dry Cimarron, aunque con frecuencia se ha reportado desde la parte central de la cuenca del Río Grande (Degenhardt et al. 1996; Stuart 2000). Stuart y Painter (1988) reportaron un espécimen recolectado durante el verano de 1978, el primero de *C. serpentina* conocido para la cuenca del Río Grande. *Apalone spinifera* es nativa de los ríos Pecos, Grande, Canadiense y Dry Cimarron aunque la especie está actualmente bien establecida en la parte inferior del Río Gila en Nuevo México y son poco comunes en el Río San Francisco. Miller (1946) sugirió que la población del Río Gila probablemente fue el resultado de introducción por humanos.

Cuatro especies adicionales habitan en las proximidades de la esquina suroeste de Nuevo México, en Chihuahua, Sonora o Arizona, y eventualmente podrían ser registradas en el estado. El Ajolote Tarahumara, *Ambystoma rosaceum*, se ha registrado en la Sierra de San Luis de Sonora y Chihuahua, a tan sólo 27 km de la frontera sur de Nuevo México (P. Warren and C. Painter, observaciones personales). Aunque esta especie es principalmente un habitante de arroyos se podría encontrar en aguajes para ganado de elevaciones altas en las montañas Ánimas o San Luis en el condado de Hidalgo, Nuevo México. Sin embargo, a media que el cambio en el clima continúa, el encontrar a esta especie en Nuevo México se hace menos probable. La Lagartija de Lemos-Espinal, *Sceloporus lemosespinali*, ha sido recolectada en Sonora, aproximadamente 4.8–8 km al sur de la frontera sur de Nuevo México en la Sierra de San Luis de acuerdo a especímenes depositados en la University of Texas at Arlington y la University of New Mexico. Es probable que *Sceloporus lemosespinali* habite en el bosque de encinos de las estribaciones de la Sierra de San Luis en el condado de Hidalgo, aunque esfuerzos repetidos por C. Painter y H. M. Smith no han podido documentar la presencia de esta especie en el área. La serpiente Centipedívora de Chihuahua,

Tantilla wilcoxi, ha sido recolectada en Sonora, aproximadamente 4.8–8 km al sur de la frontera sur de Nuevo México en la Sierra de San Luis de acuerdo a especímenes depositados en la University of Texas at Arlington. Es probable que *Tantilla wilcoxi* habite el bosque de encinos de las estribaciones de la Sierra San Luis en el condado de Hidalgo, aunque esfuerzos repetidos por C. Painter han fallado en documentar la presencia de esta especie en el área. Otra especie que probablemente habita en el suroeste de Nuevo México es la Cascabel Tigre, *Crotalus tigris*, la cual se ha recolectado a tan sólo 700 m al oeste de Nuevo México en la línea estatal en el Cañón Guadalupe, condado de Cochise, Arizona (University of Arizona Collection of Herpetology). Un espécimen adicional fue recolectado a lo largo del Geronimo Trail en agosto 11 de 1993 (Painter y Milensky 1993). Es razonable esperar que esta especie sea descubierta en hábitats como el rocoso matorral desértico o cañones en el condado de Hidalgo, del extremo suroeste de Nuevo México.

La herpetofauna de Nuevo México ha experimentado una serie de amenazas e impactos en los últimos siglos después de la llegada de los colonos de origen mexicano y estadounidense. Industrias tales como pastoreo, tala, minería, curtidoras y agricultura han modificado el paisaje de innumerables maneras. El sobrepastoreo, y la supresión de incendios forestales y la introducción de plantas no nativas han cambiado la estructura y composición de las praderas, bosques y vegetación ribereña del estado. La demanda de agua para usos agrícolas y urbanos han canalizado y creado presas, y el desvío de ríos, modificación de manantiales, el bombeo de las aguas subterráneas y la remoción de castores y sus represas, a menudo han resultado en el secado de ciénegas y llanuras aluviales y en una severa erosión de canales de arroyos. Más recientemente, la introducción de peces no nativos, cangrejos de río y Rana Toro han causado una degradación adicional de los ambientes acuáticos y, junto con la llegada de enfermedades tales como quitridiomicosis y ranavirus, representan una amenaza directa para los anfibios del estado (especialmente *Lithobates* spp.). Incendios catastróficos en los bosques de las montañas de Nuevo México amenazan muchas de las especies de tierras altas que tienen distribuciones limitadas (ejem. *Aneides hardii*, *Plethodon neomexicanus*, *Crotalus willardi*) y pueden volverse más frecuentes debido a la sequía, cambio climático, y a las pasadas supresiones de fuegos. El desarrollo de de la industria del petróleo y el gas es una amenaza creciente a la fauna silvestre de muchas áreas rurales en el estado y el uso de herbicidas para reducir la cubierta arbustiva en los pastizales para el ganado está alterando el hábitat de muchas especies terrestre (ejem. *Sceloporus arenicolus*). Quizás aún más importante, el rápido crecimiento de la población humana de Nuevo México desde finales del siglo veinte ha resultado en un incremento de urbanización, desarrollo de carreteras, y mayor demanda de recursos hídricos limitados, poniendo más presión sobre la herpetofauna estatal y otras formas de vida silvestre.

Se cree que una de las especies de la herpetofauna de Nuevo México ha sido extirpada, *Anaxyrus boreas*, aunque en la actualidad se está llevando a cabo un programa de reintroducción para esta especie en el condado de Río Arriba. Especies que en años recientes han disminuido significativamente sus poblaciones en el estado incluyen: *Plethodon neomexicanus*, *Lithobates chiricahuensis*, *L. pipiens*, *L. yavapaiensis* y *Thamnophis eques*. Otras especies que han experimentado amenazas importantes a sus limitados hábitats en Nuevo México incluyen *Aneides hardii*, *Sceloporus arenicolus*, *Trachemys gaigeae* y *Crotalus willardi*. Se sabe que *Trachemys gaigeae* hibrida con *T. scripta*, especie no nativa, en el Valle Medio del Río Grande que en última instancia puede causar la eliminación del genotipo puro de *T. gaigeae* en Nuevo México. Una amenaza similar puede existir para algunas poblaciones nativas de *Ambystoma mavortium* debido a la introducción de poblaciones introducidas de esta misma especie importadas para el estado como carnada para peces.

El intercambio comercial de anfibios y reptiles para mascotas, alimento, pieles y otras novedades ha experimentado un rápido creci-miento a nivel mundial, y Nuevo México no es la excepción. Durante 2001–2008, 71,626 especí-menes de anfibios y reptiles representando 70 especies fueron legalmente recolectados con licencias de colecta comercial en Nuevo México (NMDGF datos sin publicar). El NMDGF regula el comercio legal de anfibios y reptiles a través de la venta de licencias comerciales, aunque la mag-nitud del comercio ilegal y sus impactos sobre las poblaciones nativas de anfibios y reptiles es desconocida (Fitzgerald et al. 2004).

Se tiene la esperanza que las persecuciones organizadas de serpientes de cascabel sean una cosa del pasado. Hasta alrededor del año 2005 había una sola persecución en Alamogordo, condado de Otero, que comercializaba apro-ximadamente entre 227 y 680 kg de *Crotalus atrox* por año, dependiendo del valor de mercado y de la demanda. *Crotalus molossus* y *C. viridis* también fueron recolectadas por las persecucio-nes pero los números de este comercio fueron muy bajos (Fitzgerald y Painter 2000; C. Painter datos sin publicar). Ya no existen estas persecu-ciones organizadas en Alamogordo, aunque hay numerosos cazadores que aún proporcionan serpientes de cascabel al mercado comercial.

JAMES R. DIXON

Introducción

Texas tiene una herpetofauna diversa distribuida sobre aproximadamente 7% de la superficie terrestre y acuática de los Estados Unidos. Texas tiene un área de tierras y aguas de 692,247.5 km²; 678,357 km² de superficie terrestre y 13,890 km² de superficie acuática. La longitud de Texas es de 1289 km (generalmente de norte a sur), y su mayor distancia de este a oeste es1244 km.

La altitud de Texas varía mucho. Las montañas más altas son el Pico de Guadalupe con 2675 m y el Pico Gemelo (El Capitán) con 2472 m, ambas en el Parque Nacional Montañas de Guadalupe, condado de Culberson; 1290 km al sur se encuentra el Golfo de México. En general la mayoría de las montañas de Trans-Pecos tienen una altitud mayor a 917 m, y las Panhandle varían de 612 a 1224 m. En general, la pendiente de la masa continental tiene una dirección de noroeste a sureste a través de Texas.

La precipitación de Texas varía ampliamente a través del estado. En la ciudad de Orange, en el extremo sureste del estado, el promedio anual de precipitación es aproximadamente 1450 mm. En El Paso, extremo oeste del estado, el promedio anual de precipitación es 20.3 mm. El registro de temperatura más alta es de 53.6°C en Seymour, noroeste de Texas. El registro de temperatura

más baja es de 11.4°C en Tulia, noroeste central de Texas.

El interés en la herpetología de Texas probablemente inició en el siglo catorce cuando los primeros colonizadores españoles llegaron a la Isla del Padre, y encontraron la Cascabel de Diamantes. Desde entonces, muchas historias sobre los encuentros con esta serpiente se han narrado y vuelto a narrar tal que algunos de ellas son tan sólo narraciones imaginarias. Estas historias continúan en la actualidad, así como los encuentros con esta serpiente.

El primer registro de una publicación científica sobre una especie nueva del relativamente estado nuevo de Texas fue el del Lagartijón Sordo (*Cophosaurus texanus*), por Franz H.Troschel en 1852. Sin embargo, ese mismo año, Spencer Baird y Charles Girard publicaron las descripciones de especies nuevas como el Sapo de Puntos Rojos (*Bufo* [*Anaxyrus*] *punctatus*), Rana Cangrejo de Río (*Lithobates areolata*), Lagartija Sorda Carinada (*Holbrookia propinqua*), Lagartija de las Grietas (*Sceloporus poinsettii*), Roñito Ornado (*Urosaurus ornatus*), Lincer de Llanura (*Plestiodon obsoletus*), Corredora Pinta Texana (*Aspidoscelis gularis*), y Huico Marmóreo (*Aspidoscelis marmorata*). En 1853, Baird y Girard continuaron con las descripciones de las serpiente texanas Látigo de Schott (*Masticophis schotti*), Látigo Rayada

(*Masticophis ornatus* [= *M. taeniatus*]), Trompa de Cerdo Occidental (*Heterodon nasicus*), Culebra de Agua de Clark (*Regina* [= *Nerodia*] *clarki*), Regina de Graham (*Regina grahamii*), Culebra Cabeza-plana (*Tantilla gracilis*), Culebrilla Ciega Texana (*Rena dulcis*), Coralillo Texano (*Elaps* [= *Micrurus*]) *tener*), y Cascabel de Diamantes (*Crotalus atrox*).

Entre 1853 y 1880, 1 salamandra, 6 anuros, 3 tortugas, 6 lagartijas y 8 serpientes fueron descritas para Texas. Por primera vez un listado herpetológico de Texas fue publicado por John Strecker en 1915, el número de anfibios y reptiles reportados fue 163. Para 1950, Bryce Brown había registrado 181 especies en su lista anotada. Gerald Raun y Fredrick Gehlbach publicaron un listado de anfibios y repiltes de Texas en 1972, y registraron 199 especies, posteriormente Robert Thomas del Texas Department of Wildlife and Fisheries reportó 204 especies en 1974, y 202 en 1976. En 1987, James Dixon, reportó 204 especies en su libro *Anfibios y Reptiles de Texas* y su edición revisada de año 2000 proporcionó una lista de 219 especies.

La lista estatal de 219 especies nativas está compuesta de 28 salamandras, 41 anuros, 1 cocodrilo, 30 tortugas, 45 lagartijas y 74 serpientes. De éstas, 128 se comparten con México. Por grupos, éstas son 4 salamandras, 23 anuros, 15 tortugas, 36 lagartijas y 50 serpientes. El número de especies endémicas a Texas consiste de 14 salamandras, 1 sapo, 3 tortugas y 2 serpientes.

Las especies exóticas que mantienen poblaciones reproductivas en Texas son Rana de Invernadero (*Euhyas planirostris*), Tortuga Vientre-rojo de Florida (*Pseudemys nelsoni*), Geco de Cola Rugosa (*Cyrtopodion scabrum*), Besucona (*Hemidactylus frenatus*), Geco del Indo-Pacífico (*H. garnoti*), Geco del Mediterráneo (*H. turcicus*), Garrobo de Roca (*Ctenosaura pectinata*), Abaniquillo Costero Maya (*Anolis sagrei*), Culebra de Agua de Florida (*Nerodia fasciata pictiventris*) y Culebrilla Ciega (*Indotyphlops braminus*).

Las mascotas liberadas accidentalmente y/o sin poblaciones reproductivas que han sido encontradas nadando, caminando o arrastrándose en áreas de Texas son: Rana Africana de Garras (*Xenopus laevis*), Boa Constrictor (*Boa constrictor*), Pitón Birmano (*Python molurus*), Pitón Reticulado (*Python reticulatus*), Pitón Regio (*Python regius*), Serpiente Arborícola Café (*Boiga irregularis*), Serpiente de Mangle (*Boiga dendrophila*), Jarretera Terrestre de Occidente (*Thamnophis elegans vagrans*), Galápago de Florida (*Gopherus polyphemus*), Tortuga del Desierto (*Gopherus agassizii*), y Tortuga Café Birmana (*Manouria emys*).

La extensa superficie de Texas, sus seis provincias bióticas con su relevante geología, clima, y comunidades vegetales, contribuyen a que Texas tenga la mayor riqueza de anfibios y reptiles de cualquier estado norteamericano contiguo a él. Sin embargo, esta superficie está siendo modificada rápidamente por el crecimiento de la población humana y la necesidad de alimentar y vestir a las masas. La tala de bosques para campos agrícolas, pastura, y crecimiento urbano tienden a crear perdida de hábitats de especies nativas, por ejemplo, el Sapo de Houston, un especie en peligro de extinción. Con el crecimiento de la población humana viene una demanda por más y más agua. Esta demanda baja los niveles de agua, cesa el flujo de arroyos, y las salamandras acuáticas desaparecen. La mayoría de los humanos tenemos el "Síndrome de Ostrich," lo que no podemos ver, no nos daña. Casi todas las salamandras de arroyo están amenazadas por el bajo flujo de arroyos y algunas de ellas están consideradas en peligro de extinción. Tal parece que *La Primavera Silenciosa* de Rachel Carlson (1962) se aproxima.

Estudios Herpetológicos Previos

Aproximadamente 1100 publicaciones sobre anfibios y reptiles de Texas fueron escritas entre 1852 y 1970. A finales de 1987, más de 2000 publicaciones adicionales habían aparecido. Más de 3140 artículos sobre varios aspectos de la herpetofauna de Texas habían sido publicados para 1998, y más de 1000 artículos aparecieron en 1999–2010. La literatura incluye únicamente algunos artículos de divulgación, notas de periódico, y reportes de agencias federales y estatales. La literatura herpetológica de Texas ha incrementado 38% desde la publicación de Gerald Raun y Fredrick Gehlbach de 1972. En los primeros 115 años de publicaciones herpetológicas sobre Texas se escribieron un promedio de 10 artículos por año, mientras que durante el periodo 1987–2000 el promedio fue de 48 artículos por año, y para los últimos diez años (2000–09) el promedio fue de 77 artículos por año.

Para 1987, aproximadamente 717 autores principales habían utilizado 292 recursos de publicación. Siete de estos: *Copeia* (239), *Herpetologica* (208), el *Texas Journal of Science* (149), *Herpetological Review* (133), *Southwestern Naturalist* (116), *Catalogue of American Amphibians and Reptiles* (87), y *Journal of Herpetology* (63), contenían aproximadamente 50% de todos los artículos publicados sobre los anfibios y reptiles de Texas. Aproximadamente 541 (27%) de todos los artículos publicados fueron en coautoría, con la mayoría de éstos publicados después de 1945. La tasa de coautoría fue duplicada desde 1945, y de los 573 autores, 259 (45%) nunca aparecieron como primeros autores. Hubo aproximadamente 719 autores principales, de los cuales 439 aparecieron únicamente una vez, y 614 (85%) aparecieron como autores menos de cinco veces. De los 719 autores principales, 105 han publicado 1070 (53%) del número total de artículos. Veintidós autores han publicado 20 o más artículos sobre la herpetología de Texas. Ellos representan tan sólo el 2% de todos los autores y 22% de todas las publicaciones. Nueve de los 22 autores han fallecido y doce continúan publicando sobre la herpetología de Texas. Los nombres de los que fallecieron, su record de publicaciones sobre Texas, y sus periodos de actividad son: John Strecker (58) 1902–35; W. Frank Blair (39) 1949–76; Donald W. Tinkle (26) 1951–79; Edward D. Cope (25) 1859–1900; William W. Milstead (25) 1951–78; Charles E. Burt (20) 1928–38; Edward H. Taylor (20) 1931–50; Bryce C. Brown (20) 1937–67; y Roger Conant (20) 1942–98. Strecker, un residente de Texas, tiene el record de publicar 58 artículos en 22 años, mientras que Cope, un no residente de Texas, publicó 25 artículos en

41 años. Los herpetólogos residentes tienen una gran ventaja en relación con el conocimiento de su herpetofauna local.

Los principales autores activos hasta 2010 y sus records de publicaciones son: R. W. Axtell (51) 1950–2005; James R. Dixon (41) 1952–2009; Carl J. Franklin (23) 1996–2009; Fredrick R. Gehlbach (24) 1956–91; Toby J. Hibbitts (31) 1997–2008; J. Alan. Holman (22) 1962–2003; Flavius C. Killebrew (26) 1975–95; Chris T. McAllister (108) 1982–2008; Francis L. Rose (28) 1959–2010; Hobart M. Smith (85) 1933–2007; Thomas Vance (28) 1978–2001; y J. Martin Walker (27) 1980–2008.

De los herpetólogos activos, los 85 artículos de Hobart M. Smith en un periodo de 74 años (1933–2007) son un monumento a su excelencia como herpetólogo y su deseo de crear actividad herpetológica entre sus colegas. Únicamente los herpetólogos jóvenes de la actualidad tendrán la oportunidad de ser tan productivos como Hobart M. Smith. Uno de éstos ha alcanzado esta meta. El escritor más prolífico sobre la herpetofauna de Texas durante el periodo 1982–2010 fue Chris McAllister. Él y sus colegas publicaron cerca de 108 artículos en un periodo de 27 años, sobrepasando el record de 85 publicaciones en un periodo de 74 años de Hobart M. Smith.

Breve Historia de la Herpetología de Texas

Algunos de los primeros escritores sobre la herpetología de Texas no residieron ni visitaron Texas. La mayoría de su material provino de colectores de historia natural tales como los primeros exploradores españoles y militares realizando exploraciones.

Durante el periodo 1650–1700 los exploradores españoles registraron encuentros con víboras de cascabel de Texas. El botánico francés, Jean Louis Berlandier, probablemente fue el primer escritor científico en reportar anfibios y reptiles de Texas. Aunque Berlandier era francés, residía en México y fue uno de los primeros científicos que escribió extensamente sobre sus viajes a Texas. Berlandier hizo varias expediciones a caballo y a pie a través de la mitad sur de Texas entre 1828 y 1834. Sus principales expediciones fueron de Laredo a Bexar (febrero 20–marzo 1, 1828), Bexar—San Felipe—Trinidad—Bexar (abril 13–junio 18, 1828), Aransas—Goliad varias veces (mayo 1829), Bexar—Laredo (julio 14–28, 1829), Bexar—Goliad (febrero 2–5, 1830), Matamoros—Goliad (abril 1834), Bexar—Eagle Pass—Laredo—Matamoros (junio 10–julio 28, 1834).

En su primer viaje, Berlandier encontró la Galápago de Berlandier y ranas toro en su ruta. Cuando cruzó el Río Nueces encontró lagartos y tortugas de concha blanda, y reportó que *Trionyx* (= *Apalone*) y *Rana* (= *Lithobates*) eran abundantes alrededor de Bexar. En el viaje de San Felipe—Trinidad—Bexar, encontró tortugas caja, salamandras, Cascabel de Diamante, Cascabel del Bosque, *Rana* (= *Lithobates*) y Lagartos. Sus viajes a Copano Bay y Goliad frecuentemente fueron peligrosos debido a su contacto con la Cascabel del Bosque. En uno de sus viajes de 1829, encontró al Camaleón Común por primera vez. Ocasionalmente, Berlandier mencionó encuentros con ranas arborícolas, pero no proporcionó sus nombres científicos. Debido a sus observaciones y recolectas generales, su nombre está asociado con la Galápago de Berlandier, la Rana Leopardo del Río Bravo, y muchas especies de plantas (ver Berlandier 1980).

Las exploraciones de la frontera México—Estados Unidos de John Bartlett y William Emery, y sus subsecuentes recolectas de anfibios y reptiles, resultaron en una lista parcial de especímenes de Texas publicada por Spencer F. Baird y Charles F. Girard (1852–1854), y C. F. Girard (1852–1859). Baird y Girard registraron un total de 86 especies en el área entre Indianola y El Paso.

Varios otros herpetólogos comenzaron a describir especies de Texas después de las colecciones de campo de Berlandier, Emery, Bartlett, y otros. Entre éstos están Edward D. Cope, G.A. Boulenger, Louis Agassiz, Albert Günther, Robert Kennicott, y Spencer F. Baird. Para 1900, John Strecker había comenzado a recolectar en Texas, y para 1915, había publicado su primer listado definitivo de 163 especies de anfibios y reptiles de Texas, adicionalmente

creó la colección de Baylor University. Cuando falleció, Strecker había escrito 58 artículos sobre la herpetofauna de Texas. Brown (1950) produjo el siguiente listado de anfibios y reptiles de Texas, reconociendo 182 especies, seguido por Gerald Raun y Frederick Gehlbach (1972) quienes reportaron 199 especies. Posteriormente Robert Thomas (1976) reportó 203 especies. Los principales herpetólogos trabajando con especies texanas entre la Era de Strecker y la Segunda Guerra Mundial fueron Frank Blanchard, Bryce Brown, Charles Burt, Roger Conant, Howard Gloyd, Laurence Klauber, Stanley Mulaik, Karl P. Schmidt, Hobart Smith, Edward Taylor, y Alan H. Wright. Únicamente B. C. Brown y H. M. Smith fueron residentes de Texas.

Hay 78 nombres científicos de anfibios (14 salamandras, 14 anuros) y reptiles (6 tortugas, 20 lagartijas, 24 serpientes) con localidades tipo en el estado de Texas (ver Cuadro 12.1).

Características Fisiográficas y su Influencia sobre la Herpetofauna

Texas posee 229 especies nativas y exóticas de anfibios y reptiles. En cierta medida estos taxa están limitados a comunidades vegetales particulares, tipos de suelos, o fuentes de agua. Aunque los varios ambientes del estado parecen tener sus propias herpetofaunas, un número de generalistas tienden a ocultar las características únicas de estos ambientes. Texas está situado en la unión de cuatro principales divisiones fisiográficas de Norteamérica: las Montañas Rocallosas, las Grandes Planicies, los Bosques del Este, y los Desiertos del Suroeste. Cada una de estas divisiones está dividida en provincias bióticas.

Blair (1950) proporcionó un mapa ilustrando las seis provincias bióticas en el estado de Texas. Éstas son Kansas, Balcones, Texana, Austroriparia, Tamaulipeca y Chihuahuense.

Provincia Biótica Kansas

En Texas, esta provincial incluye las partes este del Texas Panhandle y las praderas de pastos

Autor(es)	Nombre original: Localidad tipo
Salamandras (14)	
Matthes (1855)	*Salamandra texana* (= *Ambystoma texanum*): **Cummings Creek, Fayette Co.**
Chippindale et al. (2000)	*Eurycea chisholmensis*: **Main Salado Springs, Bell Co.**
Smith & Potter (1946)	*Eurycea latitans*: **Cascade Cavern 4.6 mi SE Boerne, Kendall Co.**
Bishop (1941)	*Eurycea nana*: **Headwaters, San Marcos River, San Marcos, Hays Co.**
Bishop & Wright (1937)	*Eurycea neotenes*: **creek, 5 mi N. Helotes, Bexar Co.**
Chippindale et al. (2000)	*Eurycea naufragia*: **head springs, Buford Hollow, trib. San Gabriel River, Below Lake Georgetown, Williamson Co.**
Stejneger (1896)	*Typhlomolge* (= *Eurycea*) *rathbuni*: **San Marcos, Hays Co.**
Stejneger (1896)	*Typhlomolge* (= *Eurycea*): **beneath Blanco River, 5 km NE San Marcos, Hays Co.**
Chippindale et al. (1993)	*Eurycea sosorum*: **Barton Springs, Zilker Park, Austin, Travis Co.**
Chippindale et al. (2000)	*Eurycea tonkawae*: **Stillhouse Hollow Springs, Travis Co.**
Mitchell & Reddell (1965)	*Eurycea tridentifera*: **Honey Creek Cave, Comal Co.**
Baker (1957)	*Eurycea troglodytes*: **pool near valdena farms sinkhole, Valdena Farms, Medina Co.**
Hillis et al. (2001)	*Eurycea waterlooensis*: **Barton Springs, Zilker Park, Austin, Travis Co.**
Grobman (1944)	*Plethodon albagula*: **20 miles N San Antonio, Bexar Co.**
Anuros (14)	
Girard (1854)	*Bufo* (= *Anaxyrus*) *debilis*: **Lower Río Grande**
Sanders (1953)	*Bufo* (= *Anaxyrus*) *houstonensis*: **Fairbanks, Harris Co.**
Baird & Girard (1852)	*Bufo* (= *Anaxyrus*) *punctatus*: **Río San Pedro of the Río Grande (= Devils River), Val Verde Co.**
Girard (1854)	*Bufo* (= *Anaxyrus*) *speciosus*: **Ringgold Barracks, Río Grande City, Starr Co.**
Bragg & Sanders (1951)	*Bufo woodhousii velatus* (= *Anaxyrus velatus*): **Elkhart, Anderson Co.**
Stejneger (1915)	*Syrrhophus campi* (= *Eleutherodactylus cystignathoides*): **Brownsville, Cameron Co.**
Cope (1878)	*Syrrhophus* (= *Eleutherodactylus*) *marnocki*: **near San Antonio, Bexar Co.**
Strecker (1910)	*Hyla versicolor chrysoscelis* (= *H. chryososcelis*): **Dallas, Dallas Co.**
Baird (1854)	*Helocaetes* (*Pseudacris*) *clarki*: **Galveston and Indianola**
Wright & Wright (1933)	*Pseudacris streckeri*: **unknown but likely Somerset, Bexar Co.**
Baird & Girard (1852)	*Rana* (= *Lithobates*) *areolata*: **Indianola, Calhoun Co.**
Baird (1859b)	*Rana* (= *Lithobates*) *berlandieri*: **Brownsville, Cameron Co.**
Strecker (1910)	*Scaphiopus hurteri*: **3.5 mi E Waco, McClellan Co.**
Cope (1863)	*Scaphiopus* (= *Spea*) *bombifrons*: **Llano Estacado**
Tortugas (6)	
Stejneger (1925)	*Graptemys versa*: **Austin, Travis Co.**
Baur (1893)	*Pseudemys texana*: **San Antonio, Bexar Co.**
Hartweg (1939)	*Pseudemys scripta gaigeae* (= *Trachemys gaigeae*): **Boquillas, Río Grande, Brewster Co.**
Agassiz (1857)	*Platythyra* (= *Kinosternon*) *flavescens*: **Río Blanco**
Agassiz (1857)	*Xerobates* (= *Gopherus*) *berlandieri*: **Lower Río Grande**
Agassiz (1857)	*Aspidonectes* (= *Apalone*) *emoryi*: **Brownsville, Cameron Co.**
Lagartijas (20)	
Baird (1859a [1858])	*Gerrhonotus infernalis*: **Devils River, Val Verde Co.**
Baird (1859a [1858])	*Crotaphytus reticulatus*: **Laredo and ringgold barracks, Starr Co.**
Stejneger (1893)	*Coleonyx brevis*: **Helotes, Bexar Co.**
Davis & Dixon (1858)	*Coleonyx reticulatus*: **Black Gap Wildlife Management Area, Brewster Co.**
Troschel (1852)	*Cophosaurus texanus*: **Guadalupe River at New Braunfels, Comal Co.**
Cope (1880)	*Holbrookia lacerata*: **Helotes, Bexar Co.**

Autor(es)	Nombre original: Localidad tipo
Baird & Girard (1852)	*Holbrookia propinqua*: **between San Antonio and Indianola**
Girard (1852)	*Phrynosoma modestum*: **between San Antonio and El Paso**
Stejneger (1916)	*Sceloporus disparilis* (= *S. grammicus disparilis*): **Lomita Ranch, 6 mi N. Hidalgo, Hidalgo Co.**
Wiegmann (1834)	*Sceloporus marmoratus* (= *variabilis*): **San Antono, Bexar Co.**
Stejneger (1904)	*Sceloporus merriami*: **East Painted Cave, near mouth Pecos River**
Smith (1934)	*Sceloporus olivaceus*: **near lower end of Arroyo Los Lomos, 3 mi SE Rio Grande City, Starr Co.**
Baird & Girard (1852)	*Sceloporus poinsettii*: **Río San Pedro of the Río Grande (= Devils River), Val Verde Co.**
Baird & Girard (1852)	*Urosaurus ornatus*: **Río San Pedro of the Río Grande (= Devils River), Val Verde Co.**
Baird & Girard (1852)	*Plestiodon obsoletum*: **Río San Pedro of the Río Grande (= Devils River), Val Verde Co.**
Bocourt (1879)	*Eumeces obtusirostris* (= *Plestiodon septentrionalis obtusirostris*): **unknown locality in Texas**
Cope (1880)	*Eumeces brevilineatus* (= *Plestiodon tetragrammus*): **Helotes Creek, Bexar Co.**
Baird & Girard (1852)	*Cnemidophorus* (=*Aspidoscelis*) *gularis*: **Indianola, Calhoun Co.**
McKinney et al. (1973)	*Cnemidophorus* (=*Aspidoscelis*) *laredoensis*: **Chacon Creek, at hwy 83, Laredo, Webb Co.**
Baird & Girard 1852	*Cnemidophorus* (=*Aspidoscelis*) *marmoratus*: **between San Antonio and El Paso**

Serpientes (24)

Kennicott *in* Baird (1859b)	*Arizona elegans*: **Lower Río Grande Valley**
Brown (1901b)	*Coluber* (=*Bogertophis*) *subocularis*: **Davis Mountains, Jeff Davis Co.**
Kennicott *in* Baird (1859b)	*Taeniophis* (= *Coniophanes*) *imperialis*: **Brownsville, Cameron Co.**
Smith (1941)	*Drymarchon erebennus* (= *D. melanurus erebennus*): **Eagle Pass, Maverick Co.**
Taylor (1941)	*Ficimia streckeri*: **3 mi E Río Grande City, Starr Co.**
Kennicott (1860)	*Heterodon kennerlyi*: **Lower Río Grande**
Baird & Girard (1853)	*Heterodon nasicus*: **Río Grande**
Stejneger (1893)	*Hypsiglena ochrorhyncha texana* (= *H. jani*): **between Laredo and Camargo**
Brown (1901a)	*Ophibolus alternus* (*Lampropeltis alterna*): **Davis Mountans, Jeff Davis Co.**
Kennicott *in* Baird (1859b)	*Dipsas* (= *Leptodeira*) *septentrionalis*: **Brownsville, Cameron Co.**
Baird & Girard (1853)	*Masticophis schotti*: **Eagle Pass, Maverick Co.**
Baird & Girard (1853)	*Masticophis ornatus* (= *M. taeniatus*): **between San Antonio and El Paso**
Baird & Girard (1853)	*Regina* (= *Nerodia*) *clarki*: **Indianola, Calhoun Co.**
Trapido (1941)	*Natrix* (= *Nerodia*) *harteri*: **Brazos River N of Palo Pinto, Palo Pinto Co.**
Yarrow (1880)	*Coluber* (= *Pantherophis*) *bairdi*: **near Ft. Davis, Jeff Davis Co.**
Baird & Girard (1852)	*Churchillia bellona* (= *Pituophis catenifer*): **Presidio del Norte, Texas**[1]
Baird & Girard (1853)	*Regina grahamii*: **Río Salado, 4 mi from San Antonio, Bexar Co.**
Baird & Girard (1853)	*Tantilla gracilis*: **Indianola, Calhoun Co.**
Kennicott (1860)	*Tantilla nigriceps*: **Indianola to Nueces**
Baird & Girard (1853)	*Elaps* (= *Micrurus*) *tener*: **Río San Pedro of the Río Grande (= Devils River), Val Verde Co.**
Baird & Girard (1853)	*Rena dulcis*: **between San Pedro and Commanche Springs**
Klauber (1939)	*Leptotyphlops humilis segeregus* (= *Rena segrega*): **Chalk Draw, Brewster Co.**
Baird & Girard (1853)	*Crotalus atrox*: **Indianola, Calhoun Co.**
Kennicott (1861)	*Caudisoma lepida* (= *Crotalus lepidus*): **Presidio del Norte and Eagle Pass, Maverick Co.**

[1] Aunque la localidad tipo reportada por Baird & Girard (1852) y Smith & Taylor (1950, 1966) es Presidio del Norte, Chihuahua, esto es un error; la localidad tipo correcta es Presidio del Norte, Texas.

mixtos. Esta provincia es continua con las planicies rojas del sur del Red River. La humedad es baja y hay una disminución de ésta de este a oeste, y la región frecuentemente es clasificada como semiárida. Las principales especies de pastos son pasto oso y pasto búfalo. Áreas de médanos de arena se caracterizan por la presencia de salvia de arena y encino resplandeciente. Hay planicies de mezquite con varias especies de árboles: mezquite, encino, olmo, almez y maple. Las principales especies de herpetofauna incluyen las tortugas *Terrapene ornata*, *Kinosternon flavescens*; las lagartijas *Crotaphytus collaris*, *Cophosaurus texanus*, *Holbrookia maculata*, *Phrynosoma cornutum*, *P. modestum*, *Sceloporus consobrinus*, *Plestiodon obsoletus*, *Aspidoscelis gularis* y *A. sexlineata*; 38 especies de serpientes, por ejemplo, *Arizona elegans*, *Lampropeltis getula*, *Masticophis flagellum*, *Nerodia erythrogaster*, *Pantherophis emoryi*, *Tantilla nigriceps*, *Thamnophis radix*, *Rena dissecta*, *Crotalus atrox* y *C. viridis*.

La única especie de salamandra es la ampliamente conocida *Ambystoma marvortium*. Anfibios representativos son *Anaxyrus cognatus*, *A. debilis*, *A. punctatus*, *A. speciosus*, *A. woodhousii*, *Acris crepitans*, *Lithobates blairi*, *Scaphiopus couchi*, *Spea bombifrons* y *S. multiplicata*.

Provincia Biótica de los Balcones

Esta provincial es única por su geología, con superficies de flujos de roca caliza del Cretáceo Temprano, salientes ígneas de granito, y sedimentos del Precámbrico dispersos a través de esta región. Por lo menos ocho ríos importantes corren a través de esta región. La orilla sur de esta provincia forma la frontera con el límite norte de la Provincia Tamaulipeca. Su frontera este es el límite de la Texana, la oeste es el límite de la Chihuahuense, y la norte es el límite de la Kansas. La precipitación en esta provincia disminuye de este a oeste. Su vegetación consiste grandemente de bosques de árboles cortos de encino, encino texano, mezquite y táscate, con un sotobosque de caqui texano, pasto oso, y pastos grama. Los bosques rivereños frecuentemente consisten de olmo, almez, roble y encinos grandes. A lo largo de la orilla sur de los bosques rivereños se observan cipreses (*Taxodium districhum*).

Salamandras características de esta región son todas las especies del género *Eurycea*, excepto por *E. quadridigitata*. Un *Plethodon*, la Salamandra Babosa de Occidente (*P. albagula*), también está presente. Hay aproximadamente 17 especies de anuros en esta provincia, pero únicamente 2, *Craugastor augusti* y *Eleutherodactylus marnocki*, parecen estar consistentemente asociados a esta provincia. Ambas especies habitan más ampliamente hacia el sur, principalmente en México. Especies de anuros representativas son *Anaxyrus debilis*, *A. punctatus*, *A. speciosus*, *A. woodhousii*, *Acris crepitans*, *Hyla chysoscelis*, *Pseudacris streckeri*, *Lithobates berlandieri*, *L. catebeianus* y *Scaphiopus couchi*.

Once especies de tortugas habitan en la Provincia Balconian. Sólo 3 acuáticas y 1 terrestre son comunes, éstas son: *Graptemys versa*, *Pseudemys texana*, *Terrapene ornata* y *Apalone spinifera*.

De las 16 lagartijas encontradas en esta región, la mayoría de ellas se distribuyen ampliamente hacia el oeste, pero algunas de ellas se filtran a las provincias Austroriparia y Tamaulipeca. Algunas especies características son: *Gerrhonotus liocephalus*, *Cophosaurus texanus* y *Sceloporus poinsettii*. Otras especies comúnmente observadas son *Crotaphytus collaris*, *Coleonyx brevis*, *Sceloporus consobrinus*, *S. olivaceus*, *Urosaurus ornatus*, *Plestiodon obsoletus*, *P. tetragrammus* y *Aspidoscelis gularis*.

Hay 43 especies de serpientes en esta provincia. Ninguna de ellas limitada a esta región, pero varias de ellas razonablemente comunes. Éstas son *Coluber constrictor*, *Diadophis punctatus*, *Gyalopion canum*, *Heterodon platirhinos*, *Lampropeltis getula*, *Masticophis flagellum*, *M. taeniatus*, *Nerodia erythrogaster*, *Pantherophis emoryi*, *P. obsoletus*, *Pituophis catenifer*, *Rhinochelius lecontei*, *Salvadora grahamiae*, *Sonora semiannulata*, *Tantilla nigriceps*, *Thamnophis cyrtopsis*, *T. proximus*, *Rena dulcis*, *Micrurus tener*, *Agkistrodon contortrix*, *Agkistrodon piscivorus*, *Crotalus atrox* y *C. molossus*.

Provincia Biótica Texana

Dice (1943) describió a esta provincia como el amplio ecotono entre los bosques del este y los pastizales del oeste. En Texas esta provincia está insertada entre las provincias Kansas, Balconian y Tamaulipeca del oeste, y la provincia Austroriparia del este. Ésta es una mezcla de bosques y praderas, con una abundante precipitación en el este y una casi deficiente cantidad de humedad en el oeste. Suelos arenosos albergan bosques de encino poste, encino blackjack y nogales americanos, mientras que suelos arcillosos albergan praderas de pastos altos. Este último actualmente es cultivado, mientras que las tierras forestales ahora albergan algunas plantaciones de pinos para cosecha y/o han sido taladas para pastura para ganado. Los principales ríos que irrigan esta provincia en el sur son los Ríos Brazos, Colorado, San Marcos y Guadalupe, mientras que los Ríos Red y Trinity irrigan la parte norte de esta provincia pasando antes por la Austroriparia.

La herpetofauna representa una mezcla de especies de anfibios y reptiles de este y oeste. Las especies de anfibios características del este son tres salamandras (*Ambystoma texana, A. tigrinum* y *Siren intermedia*), y ocho anuros (*Hyla cinerea, H. squirella, H. versicolor, Pseudacris crucifer, P. fouquettei, Gastrophryne carolinensis, Lithobates sphenocephala, Scaphiopus hurteri*). Especies de anfibios del oeste entrando a esta provincia son *Anaxyrus nebulifer, Pseudacris clarki, P. streckeri, Gastrophryne olivácea* y *Scaphiopus couchi*.

La fauna de reptiles también muestra una mezcla de especies de este y oeste, con especies de tortugas del este tales como *Chelydra serpentina, Graptemys pseudogeographica, Terrapene carolina, Kinosternon subrubrum* y *Sternotherus odoratus*; formas del oeste como *Pseudemys texana, Terrapene ornata* y *Apalone spinifera*. La única lagartija que parece ser característica de esta región es *Plestiodon septentrionalis*. Lagartijas del oeste que habitan en esta región son *Crotaphytus collaris, Phrynosoma cornutum, Sceloporus olivaceus* y *Plestiodon obsoletus*, mientras

que las especies del este son *Anolis carolinensis, Plestiodon fasciatus, P. laticeps* y *Scincella lateralis*.

De las 41 especies de serpientes que habitan en esta provincial, únicamente 14 son del este y las restantes 27 del oeste. Las únicas características a esta región pueden ser *Lampropeltis calligaster, Nerodia harteri, Regina grahamii, Tantilla gracilis, Tropidoclonion lineatum* y *Virginia valeriae*.

Provincia Biótica Austroriparia

De acuerdo con Blair, Dice (1943) configuró esta provincia desde la costa del Atlántico hasta el este de Texas. Blair (1950) señaló que la frontera oeste de esta provincia es aproximadamente una línea desde el oeste del condado Harris hacia el norte hasta el oeste del condado Red River. Posteriormente, Blair indicó que la frontera oeste es la orilla oeste del principal bosque de pino y madera dura, pero aquí no hay una separación fisiográfica que limite la orilla oeste de los bosques.

La vegetación de esta provincia está compuesta de las mismas especies de maderas duras y pinos que caracterizan la vegetación hacia el este hasta el Océano Atlántico. El pino de hojaslargas es el pino dominante en la parte de Texas de esta provincia. Las maderas duras de la parte sureste de este bosque (Big Thicket) en Texas están compuestas de liquidámbar, magnolia, encino blanco, encino de agua, tupelo, ciprés, mirto de cera, cornejo, haya y un sotobosque de palmeto y musgo español. El bosque del norte es menos pantanoso y contiene pino de incienso, pino amarillo, encino rojo, encino poste y encino blackjack. Hay aproximadamente 13 especies de salamandras, tales como *Amphiuma tridactylus, Necturus beyeri, Siren intermedia*, varias especies de tierras bajas de *Ambystoma* y *Notophthalmus viridescens*. Los anuros comunes a esta región consisten de *Anaxyrus fowleri, A nebulifer, Acris blanchardi, Hyla cinerea, H. squirella, H. versicolor, Pseudacris crucifer, P. foquettei, Gastrophryne carolinensis, Lithobates catesbeianus* y *L. clamitans*.

La fauna característica de tortugas consiste

de *Graptemys ouachitensis, Macrochelys temmin-ckii, Pseudemys concinna, Terrapene carolinensis, Kinosternon subrubrum* y *Sternotherus carinatus*. Lagartijas típicas son *Ophisaurus attenuatus, Anolis carolinensis, Plestiodon fasciatus, P. laticeps* y *Scincella lateralis*.

Serpientes características incluyen *Cemophora coccinea, Farancia abacura, Lampropeltis triangulum, Nerodia cyclopion, N. erythrogaster, N. fasciata, N. rhombifer, Opheodrys aestivus, Regina rigida, Micrurus tener, Agkistrodon contortrix, A. piscivorus, Crotalus horridus* y *Sistrurus miliarius*.

Provincia Biótica Tamaulipeca

Blair (1950) limitó esta provincia por su consistente vegetación, clima y precipitación, asi como por su característica fauna de vertebrados. Su frontera este comienza cerca de San Antonio y a lo largo de un tipo de vegetación bastante distintivo desde San Antonio extendiéndose hacia el sur-sureste hasta la costa del Golfo cerca de Rockport. La vegetación es predominantemente matorral espinoso, con su frontera norte en la zona de fallas de los Balcones desde Del Río hasta San Antonio, y su frontera sur cerca del Golfo de México. La vegetación se desarrolla principalmente sobre suelos de caliche y consiste de mezquites, varias especies de *Acacia* y *Mimosa*, cenizo, cepillo blanco, granjero, tasajillo, tuna, *Condalia* y *Castela*.

La zona de suelo arenoso desde cerca de Corpus Christi hasta Raymondsville alberga masas de encinos y pastos que no son considerados parte de la Provincia Tamaulipeca.

Una salamandra (*Notophthalmus meridionalis*) es considerada endémica a esta provincia. Posiblemente la *Siren* gigante también es endémica a esta provincia, y *Ambystoma marvortium* es común. La población de *Siren* gigante que habita el Valle del Río Bravo, en Brownsville, Texas, y Matamoros, Tamaulipas, fue tentativamente asignada a *Siren lacertina* por Flores-Villela y Brandon (1992), sin embargo, esta población en realidad parece representar una especie aún no descrita, por lo que en este momento es citada como *Siren* sp. indet. Algunos de los anuros más comunes son *Anaxyrus nebulifer, Rhinella marina, Eleutherodactylus cystignathoides, Smilisca baudinii, Leptodactylus fragilis, Hypopachus variolosus, Rhinophrynus dorsalis* y *Scaphiopus couchi*.

Algunos reptiles, especialmente tortugas y lagartijas son únicos a esta provincia. La Tortuga de Berlandier es endémica a esta provincial, y la Tortuga de Concha Blanda y la Tortuga de Oreja Amarilla son especies características. Lagartijas típicas son *Crotaphytus reticulatus, Coleonyx brevis, Holbrookia lacerata, H. propinqua, Phrynosoma cornutum, Sceloporus cyanogenys, S. grammicus, S. olivaceus, S. variabilis* y *Aspidoscelis laredoensis*. Siete de las 44 especies de serpientes que habitan en esta región, están más o menos limitadas a ella: *Coniophanes imperialis, Drymobius margaritiferus, Drymarchon melanurus, Ficimia streckeri, Leptodeira septentrionalis, Masticophis schotti* y *Tantilla atriceps*. Adicionalmente, 17 especies son típicas de esta provincia: *Arizona elegans, Coluber constrictor, Elaphe emoryi, Hypsiglena jani, Lampropeltis getula, L. triangulum, Masticophis flagellum, Pituophis catenifer, Rhinochelius lecontei, Salvadora grahamiae, Sonora semiannulata, Storeria dekayi, Thamnophis marcianus, T. proximus, Micrurus tener, Crotalus atrox* y *Sistrurus catenatus*.

Provincia Biótica Chihuahuense

Dice (1943) restringió esta provincial al Trans-Pecos de Texas, excluyendo las Montañas Guadalupe. Ésta alcanza su límite norte en el sur de Nuevo México. De acuerdo con Blair (1950), su límite este es el borde este de Toyah Basin, hacia el sur hasta el condado Crockett, donde sigue el Río Pecos hacia el sur hasta su unión con el Río Bravo. El límite oeste está en México. Esta provincia tiene más características fisiográficas que cualquier otra. Su parte noreste es aparentemente un bolsón antiguo, actualmente irrigado por el Río Pecos. Al sur y sureste del Toyah Basin, está la Meseta Stockton, la cual es la extensión del Trans-Pecos de rocas calcáreas del Cretáceo Temprano de la Meseta Edwards. Esta meseta comprende una distintiva característica fisiográfica en el sur del condado Pecos, este de Brewster, y la mayor parte de Terrell. De acuerdo con Blair, el antiguo cauce del Río Bravo entra al Trans-Pecos de Texas en el sur del condado de Terrell. El curso entero del Río Bravo en Texas, abajo del Cañón Boquillas, es un antiguo bolsón de lago (ver Sellars y Baker 1935). El pico más alto en la Provincia Chihuahuense es el Monte Livermore, 2563 m, de las Montañas Davis. El Pico Emory de las Montañas Chisos, y el Pico Chinati de las Montañas Chinati, exceden 2154 m.

El clima de la Provincia Chihuahuense está clasificado como árido, con un índice de deficiencia de humedad de −40 a −60. La vegetación es de plantas típicas y altamente diversificadas del Desierto Chihuahuense. Los pastos son búfalo, grama negra, tobosa y galleta. Los arbustos comunes son gobernadora, uña de gato, arbusto negro y varias especies de nopales. En algunos cañones someros huizache, mezquite, encinos, táscates y cedros predominan en el paisaje.

La única salamandra que comúnmente habita en aguajes someros de ranchos es *Ambystoma marvortium*, y los anuros característicos son *Anayxrus cognatus*, *A. debilis*, *A. punctatus*, *A. speciosus*, *Scaphiopus couchi*, *Spea multiplicata*, y en los cañones con arroyos se puede encontrar *Hyla arenicolor*.

La fauna de tortugas consiste de siete especies; dos son semi-terrestres: *Terrapene ornata* y *Kinosternon flavescens*. Los Ríos Pecos y Bravo albergan cuatro especies de tortugas acuáticas: *Pseudemys gorzugi*, *Trachemys gaigeae*, *T. scripta* (mascotas liberadas) y *Apalone spinifera*. Uno de los tributarios semi-permanentes del Río Bravo (Alamito Creek) hospeda la única población de *Kinosternon hirtipes* en esta provincia.

Aproximadamente 33 especies de lagartijas habitan en la porción texana de la Provincia Biótica Chihuahuense. De éstas, 2 *Coleonyx* son característicos a ella, así como 3 *Phrynosoma*, 4 *Sceloporus*, 2 *Plestiodon*, 9 *Aspidoscelis*, y *Crotaphytus collaris*, *Gambelia wislizenii*, *Cophosaurus texanus*, *Urosaurus ornatus* y *Uta stansburiana*.

Hay aproximadamente 42 especies de serpientes. Las especies venenosas son *Crotalus atrox*, *C. lepidus*, *C. molossus*, *C. scutulatus* y *C. viridis*. Otros géneros típicos son *Arizona*, *Coluber*, *Lampropeltis*, *Masticophis*, *Pantherophis*, *Pituophis*, *Rhinocheilus* y *Salvadora*. Especies raras incluyen *Tantilla cucullata*, *Trimorphodon vilkinsoni* y *Rena segrega*.

De las 219 especies nativas en Texas, 91 (41%) no se comparten con México. De estas especies no compartidas con México, 24 (86%) son salamandras, 18 (44%) son anuros, 15 (50%) son tortugas, 1 es un lagarto (100%), 9 (20%) son lagartijas, y 24 (32%) son serpientes.

El estatus de las 13 especies no compartidas de las salamandras de arroyo (*Eurycea*) distribuidas a lo largo de la pendiente de la orilla sur de la Meseta Edwards no está resuelto. Los genes de algunas especies muestran características intermedias, mientras que otros no. Algunas especies de taxa ciegos son muy distintivas, mientras que otros no. No espero ver una solución rápida a este problema. Salamandras que no son de arroyo del género *Eurycea* en Texas, *E. quadridigitata* se encuentran únicamente en la Provincia Biótica Austroriparia. Su distribución se extiende hacia el este hasta Florida y hacia arriba de la costa este hasta la frontera de Virginia. La Salamandra Babosa de Occidente (*Plethodon albagula*) es endémica a Texas, se le encuentra principalmente a lo largo de cañones y en fondos de arroyos cerca de la pendiente de la Meseta Edwards. La Salamandra de Conant (*Desmognathus conanti*) es conocida únicamente para algunas localidades en la Provincia Biótica Austroriparia en Texas, pero se extiende hacia el este hasta la Península de Florida. La mayoría del resto de las salamandras no compartidas con México (*Ambystoma maculatum*, *A. opacum*, *A. talpoidium*, *A. texanum*, *A. tigrinum*, *Amphiuma tridactylum*, *Necturus beyeri* y *Notophthalmus viridescens*) son salamandras del este o sureste de los Estados Unidos, y generalmente se limitan a la Provincia Austroriparia en Texas.

Las 4 salamandras compartidas con México son *Notophthalmus meridionalis*, *Ambystoma marvortium, Siren intermedia* y *Siren* sp. indet. El Tritón de Manchas Negras está compuesto de dos subespecies. La raza sureña pudo haber evolucionado como un relicto de un evento de dispersión del Pleistoceno, y es conocido únicamente para el norte de Veracruz y sur de Tamaulipas. La subespecie norteña habita desde aproximadamente el tercio sur de la costa de Texas hasta el límite sur de la Provincia Biótica Tamaulipeca en México, cerca del Trópico de Cáncer, al sur de Llera, Tamaulipas. La segunda especie, el Ajolote Tigre Rayado, puede ser un remanente de la alguna vez ampliamente distribuida Salamandra Tigre cuya distribución ha sido dividida en varios segmentos aislados, cada uno evolucionando en linajes independientes a través del tiempo. Su distribución incluye la mayor parte del noroeste de Texas, oeste de Nuevo México hacia el norte hasta la mayor parte de Nebraska, al sur a través de la mayor parte de Kansas y Oklahoma. Hacia el suroeste, la especie se extiende hasta los estados mexicanos de Coahuila y Chihuahua. La tercera especie compartida con México es *Siren intermedia*. Esta especie se distribuye principalmente en el sureste de Estados Unidos, se le encuentra en la Provincia Austroriparia y en áreas de la costa sur de la Provincia Biótica Texana y norte de la Tamaulipeca. De acuerdo con Petranka (1998), hay dos taxa en Texas, *S. i. nettingi* y *S. i. texana*. Hay una controversia considerable sobre el estatus de *S. i. texana*. La distribución de esta última se extiende ligeramente hasta México debajo de Brownsville, y hacia arriba sobre la costa de Texas hasta Kingsville. En la actualidad su genética es muy similar a *S. intermedia*, además es simpátrica con ella en varias localidades del sur de Texas, sugiriendo que ésta puede ser una especie. El estatus taxonómico de este taxón aún está por resolverse. En el caso particular de esta publicación y para ser consistente con el capítulo de Tamaulipas, la especie de *Siren* aún no descrita está considerada como la cuarta especie compartida con México con el nombre de *Siren* sp. indet.

Los anuros están más o menos igualmente divididos entre aquellos que no se comparten con México (44%) y los que se comparten con este país (56%). Uno de los no compartidos (*A. houstonensis*) es endémico a Texas y está en peligro de extinción. Ocho de las 17 especies restantes no compartidas, se encuentran en la Provincia Austroriparia. Éstas son Rana Cangrejo de Río (*Lithobates areolatus*), Rana de Invernadero (*Lithobates clamitans*), Rana de Ciénega (*Lithobates palustris*), Rana Leopardo del Sur de Estados Unidos (*Lithobates sphenocephalus*), Ranita de Oriente (*Gastrophryne carolinensis*), Sapo del Este de Texas (*Anaxyrus velatus*), Sapo Americano (*Anaxyrus americanus*) y Rana Pimienta (*Pseudacris crucifer*). Siete de las restantes 9 especies no compartidas se encuentran generalmente en la Austroriparia, pero se distribuyen más hacia el oeste hasta la Provincia Biótica Texana, tal como *Hyla chrysoscelis, Hyla cinerea, H. squirella, Hyla versicolor, Pseudacris fouquettei, P. streckeri* y *Scaphiopus hurteri. Lithobates grylio* está limitada en Texas a las marismas de la costa sureste, y *Lithobates blairi* se encuentra en el noroeste de Texas.

De las 23 especies nativas que se comparten con México, 8 se encuentran principalmente en la Provincia Tamaulipeca. Éstas son *Incilius nebulifer, Rhinella marina, Eleutherodactylus cystignathoides, Smilisca baudinii, Leptodactylus fragilis, Hypopachus variolosus, Lithobates berlandieri* y *Rhinophrynus dorsalis*.

Una de las restantes 15 especies es un generalista, pero también habita la Provincia Austroriparia, *Lithobates catesbeianus*. Ésta ha sido introducida en Arizona, Nuevo México, México y en el Desierto Chihuahuense de Texas, donde es considerada una especie invasiva. Dos especies se distribuyen principalmente en Norteamérica. *Anaxyrus woodhousii* es característico del este de Norteamérica, pero se extiende hasta las Grandes Planicies y en el Desierto Chihuahuense. *Anaxyrus cognatus*, que se distribuye principalmente en el centro de Norteamérica, pero también se encuentra en los pastizales del Desierto Chihuahuense a elevaciones más altas. Las restantes 12 especies tienen afinidades a los desiertos del suroeste de los Estados Unidos. Éstas son *Anaxyrus debilis, A. punctatus,*

A. speciosus, Craugastor augusti, Eleutherodactylus guttilatus, E. marnocki, Hyla arenicolor, Pseudacris clarki, Gastrophryne oivacea, Scaphiopus couchi, Spea bombifrons y *S. multiplicata*. Muchas de estas especies también habitan en las Provincias Bióticas de los Balcones, Chihuahuense y Tamaulipeca.

La única especie de Lagarto que habita en Texas es el Lagarto Americano (*Alligator mississippiensis*). Ésta habita principalmente en el sureste de los Estados Unidas, desde la costa de Carolina del Norte extendiéndose hacia el sur hasta la punta de Florida y hacia el oeste hasta la punta de Texas. Existen dientes de un Lagarto proveniente de Matamoros, Tamaulipas, México, en el Museo Nacional de los Estados Unidos, y un espécimen reportado para Brownville, Texas (USNM 3184) por Baird (1859b, 5). En la actualidad, los Lagartos son comunes en áreas del condado de Cameron, han sido observados en el Río Bravo cerca de Brownsville, y río arriba a lo largo de este río hasta a 85 km en línea recta al este de Puerto Isabel. No hay registros de especímenes para México, pero esta especie se encuentra en el Río Bravo, por lo que sólo es cuestión de tiempo para que alguno sea registrado en el lado del río que pertenece a México. Esta especie también ha sido observada en la boca del Cañón de Santa Elena, Big Bend National Park. Este último registro corresponde a un individuo liberado de una pequeña exhibición en Terlingua, Texas. Es interesante mencionar que *Crocodylus moreletii* habita en el Río San Fernando, Tamaulipas, a solo 150 m en línea recta al sur del Río Bravo.

Las 30 especies de tortugas compartidas/no-compartidas son difíciles de asignar a una provincia biótica o aún a una formación vegetal en Texas. Cinco especies son tortugas marinas, 21 son tortugas de agua dulce, 1 es de marismas, y 3 son terrestres.

Tres de las tortugas no compartidas con México son endémicas a Texas. Éstas son *Graptemys caglei, G. versa* y *Pseudemys texana*. La Tortuga de Cagle está limitada a la cuenta del Río Guadalupe, la Tortuga Mapa Texana a la cuenca del Río Colorado, y la Jicotea Texana a las cuencas de los Ríos Colorado y Brazos. Tres tortugas se encuentran únicamente en la cuenca del Río Mississippi, éstas son *Graptemys ouachitensis, G. pseudogeographica* y *Apalone mutica*. Cuatro se encuentran en la parte sureste de los Estados Unidos, *Macrochelys temminckii, Chrysemys dorsalis, Deirochelys reticularia* y *Kinosternon subrubrum*. *Chelydra serpentina, Pseudemys concinna* y *Sternotherus odoratus* se encuentran a través del este de los Estados Unidos, y por lo menos a través de la Provincia Biótica de los Balcones. *Sternotherus carinatus* se encuentra a través de la Provincia Austroriparia y en los estados de Louisiana, Arkansas y Oklahoma. *Malaclemys terrapin* habita en las marismas de Texas desde Corpus Christi hasta el Río Sabine.

Las 15 especies de tortugas compartidas con México incluyen a las 5 especies de tortugas marinas. También incluyen a 2 especies que se distribuyen a través de la mitad este de los Estados Unidos. Éstas son *Trachemys scripta* y *Apalone spinifera*. Dos de ellas están limitadas a la cuenca del Río Bravo (*Trachemys gaigeae* y *Pseudemys gorzugi*). Una especie, *Kinosternon hirtipes*, se limita a un tributario (Alameda Creek) del Río Bravo en el oeste de Texas. Otra especie, *Kinosternon flavescens*, se encuentra sobre la parte central y oeste de Texas. Una más, *Chrysemys picta*, se encuentra en los Ríos Grande y Pecos, y en el extremo noreste de Texas. Tres de estas tortugas compartidas son terrestres, éstas son: *Gopherus berlandieri*, limitada a la Provincia Tamaulipeca; *Terrapene ornata*, característica de las Grandes Planicies, pero distribuida también en el oeste de los Estados Unidos y norte de México; y *Terrapene carolina*, cuya distribución principal es el este de los Estados Unidos, pero también distribuida en el suroeste de Texas y en Tamaulipas, México.

De las 45 especies de lagartijas nativas a Texas, únicamente 9 no se comparten con México. Dos de estas 9 tienen una distribución muy pequeña en Texas y Nuevo México. La Lagartija de Arena (*Sceloporus arenicolus*) es conocida únicamente para las Dunas Mezcalero del sureste de Nuevo México y parte adyacente de Texas, y el Huico Gris (*Aspidoscelis dixoni*), que habita a lo largo de las terrazas del Río Bravo en los conda-dos de Presidio, Texas y Antelope Pass, suroeste de Nuevo México. La Lagartija de Arena es un especialista en el hábitat que ocupa, únicamente se le encuentra en o cerca de médanos activos donde el extremo de los médanos está un poco estabilizado por comunidades donde domina el encino de Havard. El Huico Gris es una espe-cie partenogenética resultado de la hibridación entre *A. marmorata* y *A. scalaris*. Tres scincidos, Lincer de Carbón (*Plestiodon anthracinus*), Lincer de Cinco Líneas (*Plestiodon fasciatus*) y Lincer de Cabeza-ancha (*Plestiodon laticeps*), se distri-buyen en la mitad este de los Estados Unidos, y en Texas generalmente están limitados a la Provincia Austroriparia. Otra especies, *Ophisau-rus attenuatus* (Lagartija de Cristal Delgada), se encuentra principalmente en praderas tanto cos-teras como de tierra adentro, a través de la mitad este de los Estados Unidos. Otro scincido está limitado al bioma de las Grandes Praderas, y en Texas está más o menos limitado a la Provincia Biótica Texana. La última lagartija, el Abaniquillo Verde (*Anolis carolinensis*), habita a través del su-reste de los Estados Unidos, en Texas ocupa las Provincias Austroriparia, de los Balcones, Texana y partes de la Chihuahuense y Tamaulipeca. Esta especie es comúnmente comercializada como mascota por lo que puede ser fácilmente trans-portada y probablemente se encuentre en México como especie introducida (escape de cautiverio).

De las 36 especies nativas que se comparten con México, más de la mitad de ellas (21) se encuentra en la Provincia Biótica Chihuahuense, éstas son: *Gerrhonotus infernalis, Crotaphytus collaris, Gambelia wislizenii, Coleonyx brevis, C. reticulatus, Cophosaurus texanus, Phrynosoma cornutum, P. hernandesi, P. modestum, Scelo-porus bimaculosus, S. consobrinus, S. cowlesi, S. merriami, S. poinsettii, Urosaurus ornatus, Uta stansburiana, Aspidoscelis exsanguis, A. inornata, A. marmorata, A. tesselata* y *A. uniparens.*

Siete de las especies compartidas se en-cuentran en la Provincia Biótica Tamaulipeca, *Crotaphytus reticulatus, Holbrookia propinqua, Sceloporus cyanogenys, S. grammicus, S. variabilis, Plestiodon tetragrammus* y *Aspidoscelis laredoen-*

sis. De las ocho restantes, *Holbrookia maculata,* en Texas, se distribuye en las Provincias Chi-huahuense y Kansas, y *Holbrookia lacerata* se encuentra sólo en las Provincias de los Balcones y Tamaulipeca. *Sceloporus olivaceus, Aspidoscelis gularis* y *A. sexlineatus,* son generalistas y en Texas se encuentran en todas las provincias bió-ticas excepto la Chihuahuense. *Scincella lateralis* habita en todas las provincias bióticas excepto en la Chihuahuense y Kansas. *Plestiodon obsoletus* sólo está ausente en las Provincias Austroriparia y Texana, y *Plestiodon multivirgatus* está ausente en la Provincia Biótica Kansas.

Hay 74 especies de serpientes nativas de Texas. De éstas, 50 se comparten con México y 24 no se comparten con este país. Las especies que no se comparten generalmente se encuen-tran en la parte este de los Estados Unidos y este de Texas, pero algunas de ellas pueden habitar en todas las provincias bióticas a excepción de la Chihuahuense y/o Tamaulipeca. Las espe-cies que más o menos se limitan a la Provincia Austroriparia son *Carphophis vermis, Farancia abacura, Nerodia cyclopion, N. fasciata, N. sipe-don, Pantherophis guttatus, Pituophis ruthveni, Regina rigida* y *Storeria occipitomaculata.* Las que habitan en todas las provincias bióticas excepto la Chihuahuense son *Heterodon platirhinos, Pantherophis obsoletus, Regina grahamii* y *Virginia striatula.* Dos especies son endémicas a Texas; estás son *Tantilla cucullata* y *Nerodia harteri.* La primera habita en la Provincia Chihuahuense, y *Nerodia harteri* está limitada al sistema de playas del Río Brazos y a la cuenca del Río Colorado. Tres especies, *Lampropeltis calligaster, Crotalus horridus* y *Sistrurus miliarius,* se encuentran en dos provincias, la Austroriparia y la Texana. Una especie, *Nerodia clarkii,* se encuentra únicamente a lo largo de la costa del Golfo de México, desde el Río Sabine hacia el sur hasta cerca de Corpus Christi. Una especie, *Cemophora coccinea,* con una dieta especializada en huevos de reptiles, tiene dos poblaciones en Texas. Una está limitada a la Provincia Austroriparia, la otra a la Tamaulipeca. *Thamnophis radix* se conoce únicamente para la Provincia Kansas, mientras que *Agkistrodon piscivorus* habita la mayor parte de Texas excepto por la Provincia Chihuahuense. Las dos últimas especies no compartidas con México, *Tropidoclonion lineatum* y *Virginia valeriae*, se distribuyen en varias localidades aisladas de partes de las Provincias Austroriparia y Texana.

De las 50 especies compartidas con México, 15 son generalistas en sus patrones de distribución y en general habitan sobre la mayor parte de Texas. Éstas son *Coluber constrictor, Diadophis punctatus, Lampropeltis getula, L. triangulum, Masticophis flagellum, Nerodia erythrogaster, N. rhombifer, Opheodrys aestivus, Pituophis catenifer, Storeria dekayi, Thamnophis proximus, T. sirtalis, Micrurus tener, Agkistrodon contortrix* y *Crotalus atrox*

Veintiuna de las especies compartidas habitan en la Provincia Biótica Chihuahuense, *Bogertophis subocularis, Gyalopion canum, Hypsiglena jani, Lampropeltis alterna, Masticophis taeniatus, Rhinochelius lecontei, Salvadora deserticola, S. grahamiae, Sonora semmiannulata, Tantilla atriceps, T. hobartsmithi, T. nigriceps, Thamnophis cyrtopsis, T. marcianus, Trimorphodon vilkinsoni, Rena dissecta, R. dulcis, R. segrega, Crotalus lepidus, C. molossus* y *C. scutulatus*. En Texas, algunas de estas serpientes también habitan en las partes adyacentes de otras provincias bióticas, tales como la de los Balcones y Tamaulipeca. Otras de las especies compartidas habitan únicamente en la Provincia Tamaulipeca, éstas son: *Coniophanes imperialis, Drymarchon melanurus, Drymobius margaritiferus, Ficimia streckeri, Leptodeira septentrionalis* y *Masticophis schotti*. Tres de las restantes cuatro especies compartidas pertenecen al Bioma de las Grandes Praderas. Una, *Crotalus viridis*, habita únicamente en el noroeste de Texas, en la Provincia Kansas. *Tantilla gracilis* habita en todas las provincias bióticas de Texas excepto en la Chihuahuense, y *Heterodon kennerlyi* sólo está ausente de la de los Balcones. La última serpiente compartida es *Sistrurus catenatus*. Ésta se distribuye en varias localidades aisladas de todas las provincias bióticas de Texas excepto por las Austroriparia y Balcones, y se considera rara en todas las otras provincias.

Composición por Unidades Políticas

Un análisis de las densidades de especies por condados (unidades políticas de tamaño desigual) revela más acerca de los herpetólogos residentes e inventarios intensivos de herpetofauna que acerca de la distribución de anfibios y reptiles en Texas. El mayor número de especies registradas se encuentra en condados como Bexar, Brazos, Dallas, Harris, Hays, McLennan, Tarrant y Travis. Cada uno de estos condados tiene una de las principales universidades estatales y uno o más herpetólogos residentes. Otros condados ricos en especies pueden ser el resultado de colectas incidentales de especies capturadas cuando se buscaba otra especie más deseable, o puede ser atribuido a sus diversidades de ambientes. Por ejemplo, el condado Bexar hospeda 97 especies de anfibios y reptiles, la mayor riqueza en el estado y 48% del total de la herpetofauna de Texas; el condado Bexar también es el más diverso en ambientes, y en él residen varios herpetólogos activos.

A todos los principales museos en los Estados Unidos y museos más pequeños en Texas se les solicitaron registros de sus anfibios y reptiles. Aproximadamente 90% respondieron con más de 1300 registros y 110,000 localidades para los condados texanos. Identificaciones cuestionables y registros distribucionales aislados fueron verificados siempre que fue posible, ya fuera visitando el museo o pidiendo prestamos de material para su examinación.

Los más de 13,000 registros para 254 condados de Texas revelan que 49 especies habitan en 60% de los condados de Texas, y 50 especies habitan sólo en 1.5% de los condados de este estado. La distribución de especies sugiere que 25% de los taxa son abundantes, 25% son raros, y 50% tienen una distribución promedio. Las salamandras comprenden 4.2% del total de los registros, anuros 22.9%, tortugas 11.6%, lagartijas 20.3%, y serpientes 41%. Aproximadamente 34% (4488) de los 13,284 registros han sido acumulados durante los pasados 47 años. No tengo el total de registros para los últimos 42 años, pero la mayoría se encuentra en *Herpetological Review* (ver Dixon 2000).

Problemas Taxonómicos

Sería negligente no discutir la cantidad excesiva de cambios taxonómicos que han ocurrido desde 1987. Las nuevas herramientas, tales como cariología, inmunología, electroforesis, ADN mitocondrial, y ADN nuclear cambian la perspectiva de la taxonomía tradicional. Estas nuevas herramientas, en conjunto con programas computacionales que pueden ver la "evidencia completa" en cuestión de segundos, crean una "falsa ilusión" para los taxónomos de la "vieja escuela." Biólogos trabajando con ADN pueden desarrollar miles de escenarios para evolucionistas, cladistas y fenetistas para ser analizados con cada segmento del gen. ¿Quién está en lo correcto? ¿Qué pasa con cualquier técnica mal aplicada? Para una discusión interesante de estos problemas se debe leer el artículo de James Lazell "Taxonomic Tyranny and the Exoteric" (Lazell 1992:14), o Pauly et al. (2009) "Taxonomic Freedom." Sus comentarios reflejan mis pensamientos perfectamente.

Problemas de Conservación

Por primera vez en la historia del Texas Parks and Wildlife Department (TPWD), los miembros de la comisión han tomado una posición para la regulación y control de la fauna no cinegética en el estado. En 1998, el Texas Parks and Wildlife Department emitió nuevas regulaciones en relación a la colecta, límites de posesión e intercambio de fauna no cinegética, la cual incluye anfibios y reptiles. Texas está muy atrasada en relación a otros estados en tomar acción para asegurar la protección de fauna no cinegética.

Un problema es la colecta comercial, la cual si se deja sin revisar, puede fácilmente acabar con poblaciones de tortugas, lagartijas y salamandras en ciertas áreas del estado. Debido al incremento en la demanda de ciertas especies de tortugas y otros reptiles y anfibios en Europa y Asia para mascotas y alimento, muchas de nuestras especies nativas han sufrido una marcada disminución en sus números.

Estoy razonablemente feliz de decir que el TPWD está haciendo un esfuerzo para contener este problema. Nosotros podemos ayudar a lograr esta meta. En 1983, el TPWD creó el Special Nongame and Endangered Species Conservation Fund. Este fondo puede ser utilizado para investigación sobre fauna no cinegética y especies en peligro de extinción, conservación, adquisición de hábitat y desarrollo, y divulgación sobre estas especies. El dinero del fondo se obtiene a través de donaciones privadas y venta de litografías, calcomanías y estampillas de fauna no cinegética. Para más información sobre los fondos o especies en peligro de extinción se puede contactar al TPWD.

Adicionalmente, los miembros de la comisión han ordenado al TPWD implementar la prohibición de la colecta de tortugas texanas en aguas públicas. El departamento permite que continúe la colecta comercial de Jicotea de Orejas Rojas, Tortuga Lagarto y dos especies de Tortuga de Concha Blanca en agua privada. Los miembros de la comisión también iniciaron un estudio de 5 años de las tortugas de Texas para determinar que tanto migran estas tortugas entre aguas públicas y privadas.

Otro problema es la carencia de consciencia pública para la protección del hábitat de ciertas especies. El Sapo de Houston es un excelente ejemplo. La explotación de su hábitat para campos de golf y desarrollos agrícolas lentamente ha socavado la habilidad de este sapo para sustentarse a sí mismo. Este problema se aplica a otras especies de Texas: Tritón de Manchas Negras, Ranita Chirriadora del Río Bravo, Tortuga de Cagle, Galápago de Berlandier, Reticulada de Collar, Camaleones y Palancacóatls.

Un incremento en el número de vehículos motorizados ha resultado en múltiples muertes de anfibios y reptiles de desplazamiento lento. Otro resultado del incremento del tráfico es la creación de más autopistas, siempre más amplias, todo a expensas del hábitat alguna vez utilizado por varias especies de vida silvestre. Lenta y predictivamente, hay menos y menos espacios verdes, menos animales, y menos que lamentan sus pérdidas. Esto también aplica a las tortugas marinas. Un incremento en la demanda de mariscos ha traído barcos pesqueros y redes adicionales que resultan en un incremento de tortugas marinas ahogadas. Filosóficamente, uno debe preguntarse ¿Dónde va a terminar todo esto? ¿Estamos haciendo muy poco demasiado tarde? ¿"La Primavera Silenciosa" de Rachel Carson (1962) está saliendo a la vista?

Diversidad de la Herpetofauna de los Estados de la Frontera México-Estados Unidos

GEOFFREY R. SMITH
Y JULIO A. LEMOS-ESPINAL

Los 3169 km de frontera que han separado a México de los Estados Unidos durante los últimos 169 años tuvieron su origen en la Guerra México—Estados Unidos, una guerra que inició sobre una disputa fronteriza. En 1846, México declaró que el Río Nueces era su frontera norte con Texas. Los Estados Unidos sostenían que la frontera era más al sur en el Río Bravo. Después de que las tropas mexicanas atacaron a las tropas estadounidenses que habían incursionado al sur del Río Nueces en mayo de 1846, los Estados Unidos le declararon la guerra a México. Después de un año y medio de combates, los Estados Unidos finalmente prevalecieron y demandaron una enorme franja del norte de México. En febrero 2, 1848 el Tratado de Guadalupe Hidalgo terminó con la guerra entre México y los Estados Unidos y creó una línea fronteriza separando a los dos países. El tratado obligó a México a renunciar a 26,418,000 km² de su frontera norte, más de la mitad de su territorio, a los Estados Unidos por 15 millones de dólares. Actualmente este territorio comprende los estados de California, Arizona, Nuevo México, y partes de Texas, Nevada, Colorado y Utah. La frontera internacional entre México y los Estados Unidos corre desde Imperial Beach, California y Tijuana, Baja California en el oeste hasta Matamoros, Tamaulipas y Brownswille, Texas en el este siguiendo la parte media del Río Bravo, "a lo largo

del canal más profundo" desde su desembocadura en el Golfo de México a una distancia de 2019 km hasta un punto ligeramente río arriba de El Paso y Ciudad Juárez. Posteriormente continúa hacia el oeste sobre tierra, en donde se encuentra señalada por monumentos tipo mojoneras, a través de una distancia de 858 km hasta el Río Colorado, en este tramo alcanza su mayor altitud en la intersección con la División Continental en la Sierra de San Luis de Chihuahua, México. La frontera continua siguiendo la parte media del Río Colorado hacia el norte por una distancia de 226 km, para posteriormente volver a dirigirse hacia el oeste por tierra siguiendo una línea marcada por monumentos tipo mojoneras a través de una distancia de 226 km en línea recta a lo largo de la división entre California y Baja California hasta llegar al Océano Pacífico (Hughes 2007; Levanetz 2008).

La región a lo largo de la frontera, está caracterizada por desiertos, cadenas montañosas escarpadas, abundante incidencia de rayos solares y dos grandes ríos, el Colorado y el Bravo, los cuales proporcionan aguas que dan vida a las tierras en gran parte áridas pero fértiles a lo largo de estos ríos en ambos países. Los estados norteamericanos a lo largo de la frontera, de oeste a este, son: California, Arizona, Nuevo México y Texas. Los estados mexicanos son: Baja California, Sonora, Chihuahua, Coahuila, Nuevo

León y Tamaulipas. En los Estados Unidos, Texas tiene la frontera más larga con México en comparación con los otros estados, y California tiene la más angosta. En México, Chihuahua tiene la frontera más larga con los Estados Unidos, y Nuevo León tiene la más angosta. Texas limita con cuatro estados mexicanos (Tamaulipas, Nuevo León, Coahuila y Chihuahua), el mayor número para cualquiera de los estados norteamericanos. Nuevo México y Arizona cada uno limita con dos estados mexicanos (Chihuahua y Sonora; Sonora y Baja California, respectivamente). California limita sólo con Baja California.

La superficie de los cuatro estados norteamericanos totaliza 1,726,665 km² (= 17.6 % del total de la superficie territorial de Estados Unidos); la superficie de los seis estados mexicanos totaliza 799,288 km² (= 40.5 % del total de la superficie territorial de México). Estos diez estados cubren una superficie total de 2,525,953 km², y quedan comprendidos entre 22°12′ y 42°00′ Latitud Norte, y entre 93°31′ y 124°26′ Longitud Oeste, a altitudes desde por debajo del nivel del mar (dolinas en Sonora) hasta 4421 m de altitud (Mount Whitney, California) (cuadro 13.1).

Este vasto territorio combinado con el amplio intervalo latitudinal, longitudinal y altitudinal hospeda una gran biodiversidad, donde los seis principales biomas terrestres de Norteamérica están representados (Tundra, Bosque de Coníferas, Pradera, Bosque Deciduo, Desierto y Bosque Tropical Lluvioso), junto con una increíble riqueza específica de vida silvestre. Anfibios y reptiles son dos de las clases de vertebrados mejor representadas en los estados de la frontera. La riqueza específica de herpetozoarios en los 10 estados de la frontera, incluyendo sus islas oceánicas, totaliza 648 especies: 186 anfibios (85 salamandras, 101 anuros); 462 reptiles (2 cocodrilos, 50 tortugas, 213 lagartijas, 1 anfisbénido y 196 serpientes). Siete de estas especies son marinas: 6 tortugas (*Caretta caretta, Chelonia mydas, Eretmochelys imbricata, Lepidochelys kempi, L. olivacea* y *Dermochelys coriacea*), y una serpiente (*Hydrophis platurus*). La riqueza específica de cada uno de los 10 estados de la frontera se muestra en los cuadros 13.2 y 13.3.

Doce de estas 648 especies fueron introducidas a la región desde África, Asia o Europa: *Xenopus laevis* (introducida a Baja California, California y Arizona); *Chamaeleo jacksonii* (California); *Cyrtopodion scabrum* (Texas); *Gehyra mutilata* (Baja California y California); *He-*

CUADRO 13.1 Comparación de la superficie territorial de los 10 estados de la frontera

	ST	%STo	LN max	LN min	LO max	LO min
California	423,970	16.8	42°	32°32′	−124°26′	−114°8′
Arizona	295,253	11.7	37°	31°20′	−114°49′	−109°2′
Nuevo México	315,194	12.5	37°	31°20′	−109°3′	−103°
Texas	692,248	27.4	36°30′	25°50′	−106°39′	−93°31′
Baja California	71,576	2.8	32°50′	27°42′	−117°7′	−112°45′
Sonora	185,430	7.3	32°3′	26°12′	−115°	−108°48′
Chihuahua	245,612	9.7	31°47′	25°38′	−109°7′	−103°18′
Coahuila	151,571	6.0	29°53′	24°32′	−103°58′	−99°51′
Nuevo León	64,924	2.6	27°49′	23°11′	−101°14′	−98°26′
Tamaulipas	80,175	3.2	27°40′	22°12′	−100°8′	−97°8′
TOTAL	2,525,953	100.0				

Note: Superficie Territorial (ST en km²); porcentaje de la Superficie Total (% STo) en relación al total de superficie de los 10 estados; Latitud Norte máxima y mínima (LN max y LN min, respectivamente); y Longitud Oeste máxima y mínima (LO max y LO min, respectivamente).

CUADRO 13.2 Riqueza específica de los 10 estados de la frontera

	S	E	I	CAU	ANU	CRO	TES	LAC	AMPH	OPH
California	194	45	21	46¹ (30)	31⁵ (3)	—	13⁵	54⁶ (10)	—	50⁴ (2)
Arizona	153	4	12	2¹	28³	—	10³	57⁴ (4)	—	56¹
Nuevo México	136	3	2	3 (2)	24¹	—	10	47¹ (1)	—	52
Texas	229	20	10	28 (14)	42¹ (1)	1	31¹ (3)	51⁶	—	76² (2)
Baja California	119	21	6	4	16³	—	7¹	51² (15)	1	40 (6)
Sonora	195	16	7	3	35²	—	16¹ (1)	69³ (13)	—	72¹ (2)
Chihuahua	174	3	2	4	34¹ (1)	—	13	51¹ (2)	—	72
Coahuila	133	7	1	4	20	—	11 (2)	49¹ (4)	—	49 (1)
Nuevo León	132	2	2	3	20	—	6	42¹ (1)	—	61 (1)¹
Tamaulipas	179	6	3	11 (2)	31	1	15	47² (2)	—	74¹ (2)

Nota: S = riqueza específica; E = endémicas; I = introducidas; CAU = caudata; ANU = anura; CRO = crocodilia; TES = testudines; LAC = lacertilia; AMPH = amphisbaenia; OPH = ophidia. Los superíndices indican el número de especies introducidas en un grupo dado. Los números entre paréntesis indican el número de especies endémicas en un grupo dado.

midactylus frenatus (Baja California, Sonora, Tamaulipas y Texas); *H. garnoti* (Texas); *H. turcicus* (a los 10 estados de la frontera); *Tarentola annularis* y *T. mauritanica* (California); *Podarcis sicula* (California); *Chalcides* sp. (Arizona); e *Indotyphlops braminus* (Sonora, Nuevo León, Tamaulipas, Texas, Arizona y California). Otras 3 especies fueron introducidas a la región desde el Caribe: *Eleutherodactylus coqui* (introducida a California); *Euhyas planirostris* (Texas); y *Anolis sagrei* (Texas).

Otras 19 especies que habitan en la región de los estados de la frontera o en el sur de México han sido introducidas a alguno de los estados de la frontera: *Ambystoma mavortium* (introducida a California); *Ensatina eschscholtzii* (Arizona); *Lithobates berlandieri* (Baja California, Sonora, Arizona y California); *L. catesbeianus* (Baja California, Sonora, Chihuahua, Nuevo México, Arizona y California); *L forreri* (Baja California); *L sphenocephalus* (California); *Chelydra serpentina* (Arizona y California); *Chrysemys picta* (California); *Graptemys pseudogeographica* (California); *Pseudemys nelsoni* (Texas); *Trachemys scripta* (Arizona y California); *Apalone spinifera* (Baja California, Sonora y Arizona); *Ctenosaura* sp. (Arizona); *C. pectinata* (Texas); *Sauromalus hispidus* (Sonora); *S. varius* (Baja California); *Aspidoscelis neomexicana* (Arizona); *Nerodia*

fasciata (California); *N. rhombifer* (California); y *N. sipedon* (California).

La riqueza de especies de la región, sin contar a las especies introducidas, totaliza 630 especies, de ellas, 233 (37%) se comparten entre los dos países; 193 (30.2%) habitan sólo al norte de la frontera en los Estados Unidos, incluyendo 72 especies endémicas a uno de los cuatro estados norteamericanos de la frontera; y 207 (32.8%) habitan sólo al sur de la frontera en México, incluyendo 55 especies endémicas a uno de los seis estados mexicanos de la frontera (cuadro 13.4).

Obtuvimos el listado de especies de anfibios y reptiles para los estados de la frontera México—Estados Unidos, utilizando cada uno de los listados de este libro (ver Anexo). Para analizar los patrones de especies compartidas entre los estados de la frontera, utilizamos el índice de análisis de agrupamiento jerárquico de Jaccard (ver Enderson et al. 2009). Estos análisis nos permitieron determinar cuáles estados muestran la mayor similitud. Realizamos estos análisis para varios grupos taxonómicos: todas las especies, anfibios, anuros, salamandras, reptiles, lagartijas, serpientes y tortugas. Adicionalmente repetimos estos análisis para grupos apropiados con una base de datos que excluyó especies marinas (seis tortugas y una serpiente) y una base de datos que excluyó tanto a especies marinas como

CUADRO 13.3 Riqueza específica por clase, orden y familia en los 10 estados de la frontera

	CA	AZ	NM	TX	BC	SO	CH	CO	NL	TA
CLASS AMPHIBIA										
Order Caudata	85	46	2	3	28	4	3	4	4	3	11
Ambystomatidae	11	4	1	1	6		2	3	1	1	
Amphiumidae	1				1						
Dicamptodontidae	2	2									
Plethodontidae	61	35	1	2	16	4	1	1	3	2	8
Proteidae	1				1						
Rhyacotritonidae	1	1									
Salamandridae	6	4			2						1
Sirenidae	2				2						2
Order Anura	101	31	28	24	42	16	35	34	20	20	31
Bufonidae	22	9	7	8	10	6	12	10	7	6	6
Craugastoridae	5		1	1	1		3	2	1	1	3
Eleutherodactylidae	9	1			4		1	2	3	3	5
Hylidae	25	4	6	5	10	2	6	5	4	2	6
Leiopelmatidae	1	1									
Leptodactylidae	2				1		1			1	2
Microhylidae	5		1	1	3		2	3	1	2	3
Pipidae	1	1	1			1					
Ranidae	24	12	8	6	8	5	8	9	2	1	2
Rhinophrynidae	1				1					1	1
Scaphiopodidae	6	3	4	3	4	2	2	3	2	3	3
CLASS REPTILIA										
Order Crocodilia	2				1						1
Crocodylidae	2				1						1
Order Testudines	50	13	10	10	31	7	16	13	11	6	15
Chelonidae	5	4			4	4	4				4
Chelydridae	2	1	1	1	2						
Dermochelyidae	1	1			1	1	1				1
Emydidae	23	4	3	5	16	1	4	4	5	2	4
Geoemydidae	1						1	1			
Kinosternidae	12	1	3	2	5		4	5	3	2	4
Testudinidae	4	1	2		1		1	2	2	1	1
Trionychidae	2	1	1	2	2	1	1	1	1	1	1
Order Squamata										
Suborder Lacertilia	213	54	57	47	51	51	69	51	49	42	47
Anguidae	16	3	1	1	2	3	1	4	3	3	5
Anniellidae	6	5				2					

	CA	AZ	NM	TX	BC	SO	CH	CO	NL	TA
Chamaeleonidae	1	1								
Corytophanidae	1									1
Crotaphytidae	12	5	4	2	3	5	4	2	4	2	2
Dactyloidae	4			2		1	1			1
Dibamidae	1									1
Eublepharidae	5	2	1	2	2	2	2	1	2	1	1
Gekkonidae	12	5	1	1	4	3	6	2	1	1	2
Helodermatidae	2	1	1	1			2	1			
Iguanidae	11	2	3		1	4	7	1			2
Lacertidae	1	1								
Phrynosomatidae	81	19	26	22	19	24	29	24	29	27	19
Scincidae	19	2	6	3	8	2	3	7	6	4	5
Teiidae	29	2	11	15	10	4	12	8	4	3	5
Xantusidae	11	6	3			2	2			1	2
Xenosauridae	1									1
Suborder Amphisbaenia	1					1				
Bipedidae	1					1				
Suborder Serpentes	196	50	56	52	76	40	72	72	49	61	74
Boidae	5	4	1			1	2	1			1
Colubridae	149	36	38	41	61	27	52	56	37	49	58
Elapidae	5	1	1	1	1	1	3	2	1	1	2
Leptotyphlopidae	6	1	2	2	3	1	1	3	3	2	3
Typhlopidae	1	1	1		1		1			1	1
Viperidae	30	7	13	8	10	10	13	10	8	8	9
Total	648	194	153	136	229	119	195	174	133	132	179

Note: CA = California; AR[4] = Arizona; NM = New Mexico; TX = Texas; BC = Baja California; SO = Sonora; CH = Chihuahua; CO = Coahuila; NL = Nuevo León; TA = Tamaulipas.

CUADRO 13.4 Número de especies nativas de los 10 estados de la frontera por grupo taxonómico

	Compartidas	No compartidas		Total
		US	México	
Salamandras	8	67 (46)	10 (2)	85
Anuros	38	35 (4)	25 (1)	98
Crocodrilos	—	1	1	2
Tortugas	20	16 (3)	13 (3)	49
Lagartijas	79	33 (15)	88 (37)	200
Amphisbaenidos	—	—	1	1
Serpientes	88	38 (4)	69 (12)	195
Total	**233**	**190**	**207**	**630**

Nota: Números entre paréntesis representan el número de especies endémicas a uno de los seis estados de México o cuatro de los Estados Unidos.

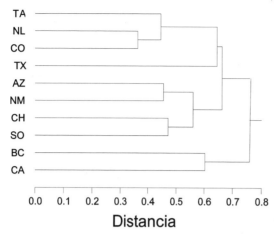

FIGURA 13.1 Dendograma generado para todas las especies de anfibios y reptiles de los estados de la frontera. Ver texto para explicación del método. Abreviaciones: AZ = Arizona, BC = Baja California, CA = California, CH = Chihuahua, CO = Coahuila, NL = Nuevo León, NM = Nuevo México, SO = Sonora, TA = Tamaulipas y TX = Texas.

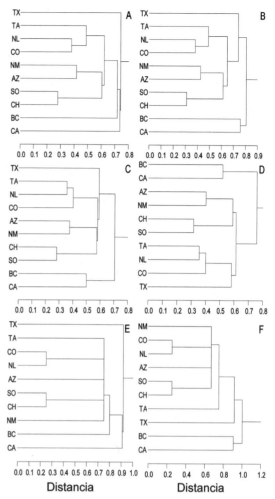

FIGURA 13.2 Dendogramas generados para (A) todos los anfibios, (B) todos los anfibios excluyendo a los introducidos, (C) anuros, (D) anuros excluyendo a los introducidos, (E) salamandras, y (F) salamandras excluyendo a las introducidas. Ver texto para explicación del método. Ver figura 13.1 para abreviaciones.

a especies introducidas. Además realizamos análisis similares sobre las provincias bióticas que se encuentran en cada uno de los diez estados (con base a Brown et al. 2007) para considerar que tan similar eran los estados en los hábitats o ambientes disponibles.

Para el análisis de todas las especies se obtuvieron varios grupos distintos de estados (figura 13.1). Texas se agrupa con Tamaulipas, Nuevo León y Coahuila. Arizona y Nuevo México se agrupan con Chihuahua y Sonora. Baja California y California se agrupan entre ellos. Al remover las especies marinas no se obtuvo ningún efecto cualitativo en el agrupamiento de estos estados, y tuvo muy poco efecto cuantitativo. Esto mismo se obtuvo al remover a las especies marinas y a las introducidas.

Para los anfibios el patrón de agrupamiento fue diferente al encontrado para todas las especies combinadas (figura 13.2A). La mayor diferencia fue en la ubicación de Texas, Baja California y California. Baja California se agrupó en un gran grupo formado por Tamaulipas, Nuevo León y Coahuila; Nuevo México y Arizona; y Sonora y Chihuahua. Posteriormente a este gran grupo se le une California y finalmente Texas.

Excluyendo a las especies introducidas (ninguno de los anfibios de la región es marino) resultó en grandes cambios en los resultados del análisis de agrupación. En este análisis Baja California y California formaron un par que se agrupo con un grupo formado por Tamaulipas, Nuevo León y Coahuila; Nuevo México y Arizona; y Sonora y Chihuahua con Texas.

El resultado del análisis de todas las especies de reptiles fue casi idéntico al análisis de todas las especies combinadas (figura 13.3A). Este patrón de similitud permanece igual cuando las

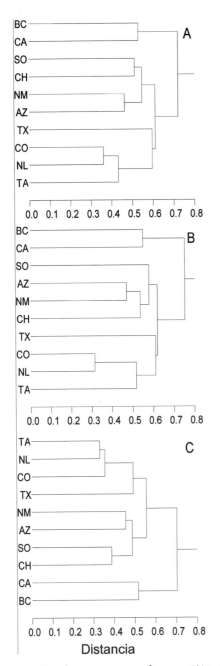

FIGURA 13.3 Dendogramas generados para (A) todos los reptiles, (B) lagartijas, y (C) serpientes. Ver texto para explicación del método. Ver figura 13.1 para abreviaciones.

especies marinas son excluidas del análisis, así como cuando tanto las especies marinas como las introducidas son excluidas del análisis.

Cuando se analiza la distribución de anuros, en el dendograma quedan colocados Arizona y Nuevo México como un par que se agrupa con el par de Chihuahua y Sonora (figura 13.2C). Posteriormente, estos dos grupos se unen a un grupo formado por Tamaulipas, Nuevo León y Coahuila. Texas se une a este grupo, y Baja California y California forman un par diferente del resto de los estados. Cuando se excluye a las especies introducidas, el único cambio es que Texas se agrupa más cercanamente con Tamaulipas, Nuevo León y Coahuila (figura 13.2D).

El análisis de salamandras produce un dendograma único (figura 13.2E). Coahuila y Nuevo Léon forman un par distintivo, así como Sonora y Chihuahua. Además, hay un gran grupo de estados igualmente similares que incluye a Tamaulipas, el par Coahuila y Nuevo León, Arizona, el par de Sonora y Chihuahua, y Nuevo México. Este gran grupo posteriormente se une con Baja California y después con California, para finalmente unirse con Texas. Varios cambios se producen cuando las salamandras introducidas son excluidas del análisis (figura 13.2F). En general hay una mayor resolución, con un grupo formado por Nuevo México, el par Coahuila y Nuevo León, Arizona, y el par de Sonora y Chihuahua formando un grupo con Tamaulipas. Texas se une a este grupo, y Baja California y California forman su propio par.

El análisis de las especies de lagartijas es generalmente similar al análisis de todas las especies de reptiles con excepción de Chihuahua que se agrupa más cercanamente con el par de Arizona y Nuevo México, y Sonora que se une al grupo de Chihuahua, Arizona y Nuevo México (figura 13.3B). Al excluir a las lagartijas introducidas no se cambia el resultado del análisis.

Las serpientes forman el mismo patrón de agrupamiento que el obtenido para todos los reptiles y para las lagartijas (figura 13.3C). Este patrón se mantiene para el análisis de serpientes excluyendo a las introducidas y a las marinas.

El análisis para las tortugas resulta en un grupo que incluye a Chihuahua, el par de Nuevo México y Arizona, el par de Baja California y California, y Sonora (figura 13.4A). Hay otro grupo de estados que incluye a Tamaulipas, Nuevo León y Coahuila que se une con el otro gran grupo de estados. Texas se mantiene sólo.

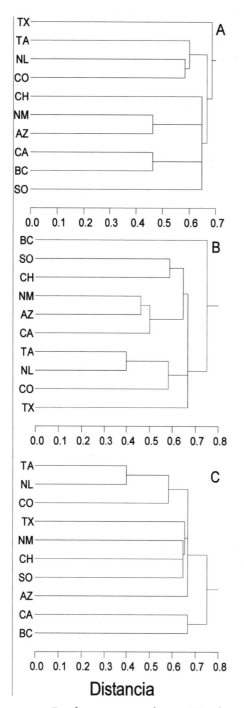

FIGURA 13.4 Dendograma generado para (A) todas las tortugas, (B) tortugas con las especies marinas excluidas, y (C) tortugas con las especies marinas e introducidas excluidas. Ver texto para explicación del método. Ver Figura 13.1 para abreviaciones.

Al excluir a las especies marinas se re-estructura el resultado del análisis (figura 13.4B). Hay un grupo de estados que incluye el grupo de California y el par de Arizona y Nuevo México con el par de Chihuahua y Sonora; otro grupo de estados incluye Tamaulipas, Nuevo León y Coahuila y Texas. Baja California ahora se mantiene fuera de estos grupos. Más cambios se producen cuando tanto las especies marinas como las introducidas son excluidas del análisis (figura 13.4C). Ahora, Baja California y California forman un par y se encuentran separados del resto de los estados. Los otros estados se agrupan en un subgrupo formado por Tamaulipas, Nuevo León y Coahuila; y otro subgrupo formado por Texas que se une con Nuevo México, Chihuahua y Sonora; posteriormente Arizona se agrupa con estos dos subgrupos.

El análisis de las provincias bióticas resulta en varios grupos distintivos (figura 13.5). Texas se agrupa con Nuevo México y Arizona, y Sonora y Chihuahua se agrupan entre sí. Estos dos grupos forman un grupo grande. Coahuila, Nuevo León y Tamaulipas forman otro grupo que se une con el grupo formado por Texas, Nuevo México, Arizona, Sonora y Chihuahua. Baja California y California forman un par separado del resto de los estados.

Nuestros resultados generaron varios patrones. Primero, es claro que, en general, ciertos conjuntos de estados tienden a ser más similares en composición de especies, que otros. En Particular, Arizona y Nuevo México se agrupan entre si frecuentemente, así como Sonora y Chihuahua. Este patrón de agrupamiento para estos estados también se encontró en el análisis realizado por Enderson et al. (2009) sobre los estados que rodean a Sonora. Esto era de esperarse ya que Arizona y Nuevo México comparten varias provincias bióticas en sus porciones norte, y Sonora y Chihuahua también comparten los hábitats de la Sierra Madre Occidental, incluyendo las tierras bajas de las Barrancas del Cobre. Estos dos pares de estados también tienden a agruparse entre sí, como era de esperarse, dado que comparten el Desierto Sonorense, el Desierto Chihuahuense, y los hábitats de las islas continentales. Tamaulipas, Coahuila y Nuevo León también se encuentran generalmente como un grupo de estados similares, y este grupo tiende a unirse con Texas en varios de los grupos taxonómicos que examinamos. De nuevo, este agrupamiento era de esperarse debido a su proximidad geográfica y a sus hábitats similares. Además, Tamaulipas, Coahuila y Nuevo León comparten los hábitats de la Sierra Madre Oriental, así como parte de la Planicie Costera que también es compartida con Texas. El último par que comúnmente se dio es entre Baja California y California. Dado el relativo aislamiento natural de Baja California y el hecho de que este estado comparte una frontera común así como muchos de sus hábitats con California, esto también era de esperarse. Una hipótesis de porqué se observan comúnmente estos patrones a través de los grupos taxonómicos es que ellos simplemente siguen las condiciones ambientales. Como ha sido enfatizado en muchos de los capítulos de este libro, las condiciones ambientales y hábitats encontrados en cada estado parecen estar condicionando fuertemente la distribución y presencia de las especies de anfibios y reptiles encontradas en estos estados. De hecho, muchos de los pares y grupos que observamos en nuestros análisis taxonómicos también aparecieron en el análisis de las similitudes entre provincias bióticas encontradas entre estos estados (ver figura 13.5). Por lo que, estados que comparten provincias bióticas tienden a compartir especies similares como se puede esperar. Por supuesto hay algunas excepciones; pero, en general, parece que la distribución de los hábitats a través de los estados de la frontera condiciona muchas de las similitudes en biodiversidad de estos estados.

Mientras que varios de los agrupamientos más similares se dan entre los estados de Estados Unidos (por ejemplo, Arizona y Nuevo México) o México (Sonora y Chihuahua; Tamaulipas, Coahuila y Nuevo León), hay uniones transfronterizas importantes en términos de diversidad compartida, como ha sido resaltado en cada uno de los capítulos al referirse a la herpetofauna estatal. El análisis de la biodiversidad transfronteriza compartida indica una fuerte necesidad de

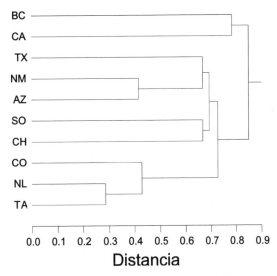

FIGURA 13.5 Dendograma para provincias bióticas (definidas y delimitadas por Brown et al. [2007]) presentes en los estados de la frontera. Ver texto para explicación del método. Ver figura 13.1 para abreviaciones.

hábitats, y por lo tanto de organismos, para ser conservados como parte de la solución transfronteriza, particularmente debido a que la rápida urbanización a lo largo de la frontera México—Estado Unidos puede impactar la cubierta terrestre y la hidrología (Biggs et al. 2010) y se espera que la región se vuelva más seca debido al cambio climático (Seager y Vecchi 2010).

Adicionalmente, esta biodiversidad transfronteriza compartida plantea preocupaciones sobre la construcción de una barrera en forma de muro diseñada para limitar el movimiento de la gente a través de la frontera. Esta barrera puede afectar a la herpetofauna si no es propiamente diseñada, y puede resultar en el aislamiento de poblaciones que cruzan la frontera. De hecho, Lasky et al. (2011) identificó 14 especies de anfibios y 21 de reptiles que están un riesgo potencial con la actual barrera pues representa un impedimento para su dispersión, sugiriendo que la expansión de esta barrera puede incrementar el número de especies en riesgo. Las consecuencias de la barrera fronteriza no son una preocupación exclusiva para la herpetofauna. Hay evidencias de que esta barrera puede afectar al jaguar (*Panthera onca*) a lo largo de la frontera

México—Estados Unidos (McCain y Childs 2008). Estudios en otras barreras fronterizas han demostrado que en algunas especies se dan efectos negativos en el flujo genético y acceso a hábitats adecuados a través de la frontera (ejem. Daleszczyk y Bunevich 2009; Kaczensky et al. 2011). Además, el extenso uso de luz a lo largo de la barrera fronteriza puede afectar la actividad de animales nocturnos (por ejemplo, Grigione y Mrykalo 2004), y debido a que muchas especies de anfibios y reptiles están activos en la noche, el uso de luz artificial en esta barrera puede tener consecuencias negativas en estas especies.

Segundo, para la mayoría de los grupos que examinamos la exclusión de especies introducidas tuvo relativamente poco impacto sobre el patrón existente de similitudes en biodiversidad de la herpetofauna. Para los reptiles excepto para las tortugas, la remoción de las especies introducidas no tuvo ningún efecto sobre el patrón de biodiversidad compartida que observamos. Hubo relativamente ligeros cambios en los patrones observados cuando excluimos a los anfibios introducidos del análisis de todos los anfibios, anuros y salamandras. En general la remoción de especies introducidas de anfibios cambio la ubicación de Texas; en los tres casos, la remoción de anfibios introducidos resultó en Texas moviéndose de afuera de un gran grupo a dentro de este grupo, y el patrón volviéndose más similar al patrón encontrado para el análisis de todas las especies combinadas (ver figura 13.1). La ausencia general de un mayor efecto de las especies introducidas en nuestro análisis puede ser debida al relativo bajo número de especies de reptiles y anfibios que han sido introducidas en México, especialmente comparado con la relativamente alta riqueza específica de herpetofauna nativa. Sin embargo, esto no significa que el continuar con la introducción de especies no nativas no vaya a afectar la diversidad y similitudes de diversidad en estos estados fronterizos (ver también Lavín-Murcio y Lazcano 2010). Por ejemplo, sólo se necesita una especie invasiva muy agresiva para reducir la diversidad nativa de un estado en particular (ejem. Rana Toro, Moyle 1973; Lawler et al. 1999; Casas-Andreu

et al. 2001; Luja y Rodríguez-Estrella 2010; ver también capítulos por Hollingsworth y Mahrdt y por Painter y Stuart en este libro). Además, aun sin los efectos negativos de una especie invasiva agresiva, el incremento de transporte antropogénico de especies alrededor del mundo puede resultar en homogenización de la diversidad, como ha sido mostrado para anfibios y reptiles (Smith 2006) y peces (Rahel 2000; Villeger et al. 2011) en otras partes del mundo.

En nuestro análisis, las tortugas como grupo se comportaron diferente. El análisis global de las tortugas generó un dendograma relativamente único a los otros análisis. Creemos que esto surgió de la inclusión de las tortugas marinas en este análisis en particular. Una vez que las tortugas marinas fueron removidas, el dendograma comenzó a ser más similar a los otros dendogramas de reptiles, aunque aún con algunas diferencias significativas, en particular la ubicación de California haciendo un par con el grupo formado por Arizona y Nuevo México, un patrón que no se presentó en ningún otro grupo taxonómico que analizamos. Igualmente, excluyendo a las especies de tortugas marinas y a las introducidas incrementó la similitud de los resultados a los encontrados en otros grupos taxonómicos. Sin embargo, la remoción de las especies introducidas resultó en el resurgimiento del par formado por Baja California y California. De acuerdo con esto, parece que de todos los grupos taxonómicos que examinamos, éste es el único en que las especies introducidas tienen un papel relativamente importante en los patrones de similitud en tortugas. Esto puede ser el resultado del pequeño número de tortugas en los estados de la frontera combinado con la amplia introducción que se ha dado de una o dos tortugas (ejem. *Apalone spinifera* y *Trachemys scripta*).

En conclusión, las similitudes de las herpetofaunas entre los estados de la frontera probablemente reflejan la distribución de los ambientes compartidos entre estados geográficamente próximos. Hay muchos enlaces herpetofaunísticos entre estados a través de la frontera México—Estados Unidos debido a estos ambientes compartidos; sin embargo, estos enlaces pueden estar amenazados por el muro fronterizo levantado por el gobierno de los Estados Unidos (Lasky et al. 2011) así como por otras amenazas en la región (una revisión de estas amenazas se presenta en Lavín-Murcio y Lazcano 2010). Se necesitan más estudios experimentales sobre los efectos de los cambios a lo largo de la frontera México—Estados Unidos sobre las poblaciones de anfibios y reptiles para entender plenamente las consecuencias y desarrollar actividades correctivas potenciales.

Appendix

Checklist of the Amphibians and Reptiles of the United States–Mexico Border States

Anexo

Listado de los anfibios y reptiles de los Estados de la Frontera Estados Unidos–México

Abbreviations / Abreviaciones: 1 = present; 0 = absent; **E** = state endemic; **M** = marine species; **IN** = Introduced; **BC** = Baja California; **SO** = Sonora;

CH = Chihuahua; **CO** = Coahuila; **NL** = Nuevo León; **TA** = Tamaulipas; **TX** = Texas; **NM** = New Mexico; **AZ** = Arizona; **CA** = California.

	SPANISH NAME	ENGLISH NAME	BC	SO	CH	CO	NL	TA	TX	NM	AZ	CA
CLASS AMPHIBIA												
Order Caudata												
Ambystomatidae												
Ambystoma californiense Gray, 1853	Salamandra Tigre de California	California Tiger Salamander	0	0	0	0	0	0	0	0	0	E
Ambystoma gracile (Baird, 1859)	Salamandra del Noroeste	Northwestern Salamander	0	0	0	0	0	0	0	0	0	1
Ambystoma macrodactylum Baird, 1850	Salamandra de Dedos-largos	Long-toed Salamander	0	0	0	0	0	0	0	0	0	1
Ambystoma maculatum (Shaw, 1802)	Ajolote Manchado	Spotted Salamander	0	0	0	0	0	0	1	0	0	0
Ambystoma mavortium Baird, 1850	Ajolote Tigre Rayado	Barred Tiger Salamander	0	1	1	1	1	0	1	1	1	IN
Ambystoma opacum (Gravenhorst, 1807)	Ajolote Marmoleado	Marbled Salamander	0	0	0	0	0	0	1	0	0	0
Ambystoma rosaceum Taylor, 1941	Ajolote Tarahumara	Tarahumara Salamander	0	1	1	0	0	0	0	0	0	0
Ambystoma silvense Webb, 2004	Salamandra del Bosque de Pino	Pine Woods Salamander	0	0	1	0	0	0	0	0	0	0
Ambystoma talpoideum (Holbrook, 1838)	Ajolote Topo	Mole Salamander	0	0	0	0	0	0	1	0	0	0
Ambystoma texanum (Matthes, 1855)	Ajolote Texano	Small-mouthed Salamander	0	0	0	0	0	0	1	0	0	0

	SPANISH NAME	ENGLISH NAME	BC	SO	CH	CO	NL	TA	TX	NM	AZ	CA
Ambystoma tigrinum (Green, 1825)	Ajolote Tigre del Este	Eastern Tiger Salamander	0	0	0	0	0	0	1	0	0	0

Amphiumidae

	SPANISH NAME	ENGLISH NAME	BC	SO	CH	CO	NL	TA	TX	NM	AZ	CA
Amphiuma tridactylum Cuvier, 1827	Amphiuma de Tres-dedos	Three-toed Amphiuma	0	0	0	0	0	0	1	0	0	0

Dicamptodontidae

	SPANISH NAME	ENGLISH NAME	BC	SO	CH	CO	NL	TA	TX	NM	AZ	CA
Dicamptodon ensatus (Eschscholtz, 1833)	Salamandra Gigante de California	California Giant Salamander	0	0	0	0	0	0	0	0	0	E
Dicamptodon tenebrosus (Baird & Girard, 1852)	Salamandra Gigante de la Costa	Coastal Giant Salamander	0	0	0	0	0	0	0	0	0	1

Plethodontidae

	SPANISH NAME	ENGLISH NAME	BC	SO	CH	CO	NL	TA	TX	NM	AZ	CA
Aneides ferreus Cope, 1869	Salamandra Nebulosa	Clouded Salamander	0	0	0	0	0	0	0	0	0	1
Aneides flavipunctatus (Strauch, 1870)	Salamandra Negra	Black Salamander	0	0	0	0	0	0	0	0	0	1
Aneides hardii (Taylor, 1941)	Salamandra de las Montañas Sacramento	Sacramento Mountains Salamander	0	0	0	0	0	0	0	E	0	0
Aneides iecanus (Cope, 1883)	Salamandra Negra del Monte Shasta	Shasta Black Salamander	0	0	0	0	0	0	0	0	0	E
✓ *Aneides lugubris* (Hallowell, 1849)	Salamandra Arborícola	Arboreal Salamander	1	0	0	0	0	0	0	0	0	1
Aneides niger Myers & Maslin, 1948	Salamandra Negra de Santa Cruz	Santa Cruz Black Salamander	0	0	0	0	0	0	0	0	0	E
Aneides vagrans Wake & Jackman, 1999	Salamandra Vagabunda	Wandering Salamander	0	0	0	0	0	0	0	0	0	E
Batrachoseps altasierrae Jockusch, Martínez-Solano, Hansen, & Wake 2012	Salamandra Delgada de las Montañas Greenhouse	Greenhouse Mountains Slender Salamander	0	0	0	0	0	0	0	0	0	E
Batrachoseps attenuatus (Eschscholtz, 1833)	Salamandra Delgada de California	California Slender Salamander	0	0	0	0	0	0	0	0	0	1
Batrachoseps bramei Jockusch, Martínez-Solano, Hansen, & Wake 2012	Salamandra Delgada de Fairview	Fairview Slender Salamander	0	0	0	0	0	0	0	0	0	E
Batrachoseps campi Marlow, Brode, & Wake, 1979	Salamandra de las Montañas Inyo	Inyo Mountains Salamander	0	0	0	0	0	0	0	0	0	E
Batrachoseps diabolicus Jockusch, Wake, & Yanev, 1998	Salamandra Delgada de Hell Hollow	Hell Hollow Slender Salamander	0	0	0	0	0	0	0	0	0	E
Batrachoseps gabrieli Wake, 1996	Salamandra Delgada de las Montañas San Gabriel	San Gabriel Mountains Slender Salamander	0	0	0	0	0	0	0	0	0	E

	SPANISH NAME	ENGLISH NAME	BC	SO	CH	CO	NL	TA	TX	NM	AZ	CA
Batrachoseps gavilanensis Jockusch, Yanev, & Wake, 2001	Salamandra Delgada de las Montañas Gavilán	Gabilan Mountains Slender Salamander	0	0	0	0	0	0	0	0	0	E
Batrachoseps gregarius Jockusch, Wake, & Yanev, 1998	Salamandra Delgada Gregaria	Gregarious Slender Salamander	0	0	0	0	0	0	0	0	0	E
Batrachoseps incognitus Jockusch, Yanev, & Wake, 2001	Salamandra Delgada de San Simeon	San Simeon Slender Salamander	0	0	0	0	0	0	0	0	0	E
Batrachoseps kawia Jockusch, Wake, & Yanev, 1998	Salamandra Delgada de Secoya	Sequoia Slender Salamander	0	0	0	0	0	0	0	0	0	E
Batrachoseps luciae Jockusch, Yanev, & Wake, 2001	Salamandra Delgada de las Montañas Santa Lucia	Santa Lucia Mountains Slender Salamander	0	0	0	0	0	0	0	0	0	E
✓ *Batrachoseps major* Camp, 1915	Salamandra Mayor	Garden Slender Salamander	1	0	0	0	0	0	0	0	0	1
Batrachoseps minor Jockusch, Yanev, & Wake, 2001	Salamandra Delgada Menor	Lesser Slender Salamander	0	0	0	0	0	0	0	0	0	E
Batrachoseps nigriventris Cope, 1869	Salamandra Delgada de Vientre-negro	Black-bellied Slender Salamander	0	0	0	0	0	0	0	0	0	E
Batrachoseps pacificus (Cope, 1865)	Salamandra Delgada de las Islas Canal	Channel Islands Slender Salamander	0	0	0	0	0	0	0	0	0	E
Batrachoseps regius Jockusch, Wake, & Yanev, 1998	Salamandra Delgada del Río Kings	Kings River Slender Salamander	0	0	0	0	0	0	0	0	0	E
Batrachoseps relictus Brame & Murray, 1968	Salamandra Delgada Relictual	Relictual Slender Salamander	0	0	0	0	0	0	0	0	0	E
Batrachoseps robustus Wake, Yanev, & Hansen, 2002	Salamandra de la Meseta Kern	Kern Plateau Salamander	0	0	0	0	0	0	0	0	0	E
Batrachoseps simatus Brame & Murray, 1968	Salamandra Delgada del Cañón Kern	Kern Canyon Slender Salamander	0	0	0	0	0	0	0	0	0	E
Batrachoseps stebbinsi Brame & Murray, 1968	Salamandra Delgada de Tehachapi	Tehachapi Slender Salamander	0	0	0	0	0	0	0	0	0	E
Bolitoglossa platydactyla (Gray, 1831)	Achoque de Tierra	Broad-footed Salamander	0	0	0	0	0	1	0	0	0	0
Chiropterotriton cracens Rabb, 1958	Salamandra Graciosa	Graceful Flat-footed Salamander	0	0	0	0	0	E	0	0	0	0
Chiropterotriton multidentatus (Taylor, 1938 [1939])	Salamandra Multidentada	Toothy Salamander	0	0	0	0	0	1	0	0	0	0
Chiropterotriton cf priscus Rabb, 1956	Salamandra Primitiva	Primeval Flat-footed Salamander	0	0	0	1	1	1	0	0	0	0
Desmognathus conanti Rossman, 1958	Salamandra de Conant	Spotted Dusky Salamander	0	0	0	0	0	0	1	0	0	0

	SPANISH NAME	ENGLISH NAME	BC	SO	CH	CO	NL	TA	TX	NM	AZ	CA
✓ *Ensatina eschscholtzii* Gray, 1850	Ensatina de Bahía Monterrey	Monterrey Ensatina	1	0	0	0	0	0	0	0	1	1
✓ *Ensatina klauberi* Dunn, 1929 [1928]	Ensatina Manchas-grandes	Large-blotched Ensatina	1	0	0	0	0	0	0	0	0	1
Eurycea chisholmensis Chippindale, Price, Wiens, & Hillis, 2000	Salamandra Salada	Salado Salamander	0	0	0	0	0	0	E	0	0	0
Eurycea latitans Smith & Potter, 1946	Salamandra Cascada de Cavernas	Cascade Caverns Salamander	0	0	0	0	0	0	E	0	0	0
Eurycea nana Bishop, 1941	Salamandra de San Marcos	San Marcos Salamander	0	0	0	0	0	0	E	0	0	0
Eurycea naufragia Chippindale, Price, Wiens, & Hillis, 2000	Salamandra de Georgetown	Georgetown Salamander	0	0	0	0	0	0	E	0	0	0
Eurycea neotenes Bishop & Wright, 1937	Salamandra de Texas	Texas Salamander	0	0	0	0	0	0	E	0	0	0
Eurycea pterophila Burger, Smith, & Potter, 1950	Salamandra de Helechos	Fern Bank Salamander	0	0	0	0	0	0	E	0	0	0
Eurycea quadridigitata (Holbrook, 1842)	Salamandra Enana	Dwarf Salamander	0	0	0	0	0	0	1	0	0	0
Eurycea rathbuni (Stejneger, 1896)	Salamandra Ciega Texana	Texas Blind Salamander	0	0	0	0	0	0	E	0	0	0
Eurycea robusta (Potter & Sweet, 1981)	Salamandra Ciega de Blanco	Blanco Blind Salamander	0	0	0	0	0	0	E	0	0	0
Eurycea sosorum Chippindale, Hillis, & Price, 1993	Salamandra de Arroyos de Barton	Barton Springs Salamander	0	0	0	0	0	0	E	0	0	0
Eurycea tonkawae Chippindale, Price, Wiens, & Hillis 2000	Salamandra de Tonkawe	Jollyville Plateau Salamander	0	0	0	0	0	0	E	0	0	0
Eurycea tridentifera Mitchell & Reddell, 1965	Salamandra Ciega de Comal	Comal Blind Salamander	0	0	0	0	0	0	E	0	0	0
Eurycea troglodytes Baker, 1957	Salamandra Agujero-morador	Valdina Farms Salamander	0	0	0	0	0	0	E	0	0	0
Eurycea waterlooensis Hillis, Chamberlain, Wilcox, & Chippindale, 2001	Salamandra Ciega de Austin	Austin Blind Salamander	0	0	0	0	0	0	E	0	0	0
Hydromantes brunus Gorman, 1954	Salamandra de Roca Caliza	Limestone Salamander	0	0	0	0	0	0	0	0	0	E
Hydromantes platycephalus (Camp, 1916)	Salamandra del Monte Lyell	Mount Lyell Salamander	0	0	0	0	0	0	0	0	0	E
Hydromantes shastae Gorman & Camp, 1953	Salamandra del Monte Shasta	Shasta Salamander	0	0	0	0	0	0	0	0	0	E
Plethodon albagula Grobman, 1944	Salamandra Babosa de Occidente	Western Slimy Salamander	0	0	0	0	0	0	E	0	0	0
Plethodon asupak Mead, Clayton, Nauman, Olson, & Pfrender, 2005	Salamandra de la Barra Scott	Scott Bar Salamander	0	0	0	0	0	0	0	0	0	E

	SPANISH NAME	ENGLISH NAME	BC	SO	CH	CO	NL	TA	TX	NM	AZ	CA
Plethodon dunni Bishop, 1934	Salamandra de Dunn	Dunn's Salamander	0	0	0	0	0	0	0	0	0	1
Plethodon elongatus Van Denburgh, 1916	Salamandra Del Norte	Del Norte Salamander	0	0	0	0	0	0	0	0	0	1
Plethodon neomexicanus Stebbins & Riemer, 1950	Salamandra de las Montañas Jemez	Jemez Mountains Salamander	0	0	0	0	0	0	0	E	0	0
Plethodon stormi Highton & Brame, 1965	Salamandra de las Montañas Siskiyou	Siskiyou Mountains Salamander	0	0	0	0	0	0	0	0	0	1
Pseudoeurycea bellii (Gray, 1850)	Ajolote de Tierra	Bell's Salamander	0	1	1	0	0	1	0	0	0	0
Pseudoeurycea cephalica (Cope, 1865)	Babosa	Chunky False Brook Salamander	0	0	0	0	0	1	0	0	0	0
Pseudoeurycea galeanae (Taylor, 1941)	Tlaconete Neoleonense	Galeana False Brook Salamander	0	0	0	1	1	1	0	0	0	0
Pseudoeurycea scandens Walker, 1955	Tlaconete Tamaulipeco	Tamaulipan False Brook Salamander	0	0	0	1	0	1	0	0	0	0

Proteidae

	SPANISH NAME	ENGLISH NAME	BC	SO	CH	CO	NL	TA	TX	NM	AZ	CA
Necturus beyeri Viosca, 1937	Salamandra de la Costa del Golfo	Gulf Coast Waterdog	0	0	0	0	0	0	1	0	0	0

Rhyacotritonidae

	SPANISH NAME	ENGLISH NAME	BC	SO	CH	CO	NL	TA	TX	NM	AZ	CA
Rhyacotriton variegatus Stebbins & Lowe, 1951	Salamandra del Torrente del Sur	Southern Torrent Salamander	0	0	0	0	0	0	0	0	0	1

Salamandridae

	SPANISH NAME	ENGLISH NAME	BC	SO	CH	CO	NL	TA	TX	NM	AZ	CA
✓ *Notophthalmus meridionalis* (Cope, 1880)	Tritón de Manchas Negras	Black-spotted Newt	0	0	0	0	0	1	1	0	0	0
Notophthalmus viridescens (Rafinesque, 1820)	Tritón de Oriental	Eastern Newt	0	0	0	0	0	0	1	0	0	0
Taricha granulosa (Skilton, 1849)	Tritón de Piel-rugosa	Rough-skinned Newt	0	0	0	0	0	0	0	0	0	1
Taricha rivularis (Twitty, 1935)	Tritón de Vientre-rojo	Red-bellied Newt	0	0	0	0	0	0	0	0	0	E
Taricha sierrae (Twitty, 1942)	Tritón de la Sierra	Sierra Newt	0	0	0	0	0	0	0	0	0	E
Taricha torosa (Rathke, 1833)	Tritón de California	California Newt	0	0	0	0	0	0	0	0	0	E

Sirenidae

	SPANISH NAME	ENGLISH NAME	BC	SO	CH	CO	NL	TA	TX	NM	AZ	CA
✓ *Siren intermedia* Barnes, 1826	Sirena Menor	Lesser Siren	0	0	0	0	0	1	1	0	0	0
✓ *Siren* sp. indet.			0	0	0	0	0	1	1	0	0	0

Order Anura

Bufonidae

	SPANISH NAME	ENGLISH NAME	BC	SO	CH	CO	NL	TA	TX	NM	AZ	CA
Anaxyrus americanus (Holbrook, 1836)	Sapo Americano	American Toad	0	0	0	0	0	0	1	0	0	0
✓ *Anaxyrus boreas* (Baird & Girard, 1852)	Sapo Occidental	Western Toad	1	0	0	0	0	0	0	1	0	1

	SPANISH NAME	ENGLISH NAME	BC	SO	CH	CO	NL	TA	TX	NM	AZ	CA
Anaxyrus californicus (Camp, 1915)	Sapo de Arroyo	Arroyo Toad	1	0	0	0	0	0	0	0	0	1
Anaxyrus canorus (Camp, 1916)	Sapo de Yosemite	Yosemite Toad	0	0	0	0	0	0	0	0	0	E
Anaxyrus cognatus (Say, *in* James, 1823)	Sapo de Espuelas	Great Plains Toad	1	1	1	1	1	1	1	1	1	1
Anaxyrus debilis (Girard, 1854)	Sapo Verde	Green Toad	0	1	1	1	1	1	1	1	1	0
Anaxyrus exsul (Myers, 1942)	Sapo Negro	Black Toad	0	0	0	0	0	0	0	0	0	E
Anaxyrus houstonensis (Sanders, 1953)	Sapo de Houston	Houston Toad	0	0	0	0	0	0	E	0	0	0
Anaxyrus kelloggi (Taylor, 1938 [1936])	Sapo Mexicano	Little Mexican Toad	0	1	0	0	0	0	0	0	0	0
Anaxyrus mexicanus (Brocchi, 1879)	Sapo Pie de Pala	Mexican Madre Toad	0	1	1	0	0	0	0	0	0	0
Anaxyrus microscaphus (Cope, 1867)	Sapo de Arizona	Arizona Toad	0	0	0	0	0	0	0	1	1	1
Anaxyrus punctatus (Baird & Girard, 1852)	Sapo de Puntos Rojos	Red-spotted Toad	1	1	1	1	1	1	1	1	1	1
Anaxyrus retiformis (Sanders & Smith, 1951)	Sapo Sonorense	Sonoran Green Toad	0	1	0	0	0	0	0	0	1	0
Anaxyrus speciosus (Girard, 1854)	Sapo Texano	Texas Toad	0	0	1	1	1	1	1	1	0	0
Anaxyrus velatus (Bragg & Sander, 1951)	Sapo del Este de Texas	East Texas Toad	0	0	0	0	0	0	1	0	0	0
Anaxyrus woodhousii (Girard, 1854)	Sapo de Woodhouse	Woodhouse's Toad	1	1	1	1	0	0	1	1	1	1
Incilius alvarius (Girard, 1859)	Sapo del Desierto de Sonora	Sonoran Desert Toad	1	1	1	0	0	0	0	1	1	1
Incilius marmoreus (Wiegmann, 1833)	Sapo Marmoleado	Marbled Toad	0	1	0	0	0	0	0	0	0	0
Incilius mazatlanensis (Taylor, 1940)	Sapo de Mazatlán	Sinaloa Toad	0	1	1	0	0	0	0	0	0	0
Incilius mccoyi Santos-Barrera & Flores-Villela, 2011	Sapo de McCoy	McCoy´s Toad	0	1	1	0	0	0	0	0	0	0
Incilius nebulifer (Girard, 1854)	Sapo Nebuloso	Gulf Coast Toad	0	0	0	1	1	1	1	0	0	0
Rhinella marina (Linnaeus, 1758)	Sapo Gigante	Cane Toad	0	1	1	1	1	1	1	0	0	0

Craugastoridae

	SPANISH NAME	ENGLISH NAME	BC	SO	CH	CO	NL	TA	TX	NM	AZ	CA
Craugastor augusti (Dugès, *in* Brocchi, 1879)	Sapo Ladrador	Barking Frog	0	1	1	1	1	1	1	1	1	0
Craugastor batrachylus (Taylor, 1940)	Rana Arborícola Tamaulipeca	Tamaulipan Arboreal Robber Frog	0	0	0	0	0	E	0	0	0	0
Craugastor decoratus (Taylor, 1942)	Rana Labradora	Adorned Robber Frog	0	0	0	0	0	1	0	0	0	0
Craugastor occidentalis (Taylor, 1941)	Rana Costeña	Taylor's Barking Frog	0	1	0	0	0	0	0	0	0	0

	SPANISH NAME	ENGLISH NAME	BC	SO	CH	CO	NL	TA	TX	NM	AZ	CA
Craugastor tarahumaraensis (Taylor, 1940)	Rana Ladradora Amarilla	Tarahumara Barking Frog	0	1	1	0	0	0	0	0	0	0
Eleutherodactylidae												
Eleutherodactylus coqui Thomas, 1966	Coquí	Coquí	0	0	0	0	0	0	0	0	0	IN
Eleutherodactylus cystignathoides (Cope, 1877)	Ranita Chirriadora del Río Bravo	Rio Grande Chirping Frog	0	0	0	0	1	1	1	0	0	0
Eleutherodactylus dennisi (Lynch, 1970)	Rana Chirradora	Dennis's Chirping Frog	0	0	0	0	0	1	0	0	0	0
Eleutherodactylus guttilatus (Cope, 1879)	Ranita Chirriadora de Manchas	Spotted Chirping Frog	0	0	0	1	1	1	1	0	0	0
Eleutherodactylus interorbitalis (Langebartel & Shannon, 1956)	Ranita de Lentes	Spectacled Chirping Frog	0	1	1	0	0	0	0	0	0	0
Eleutherodactylus longipes (Baird, in Emory, 1859)	Ranita Chirriadora de la Huasteca	Long-footed Chirping Frog	0	0	0	1	1	1	0	0	0	0
Eleutherodactylus marnockii (Cope, 1878)	Ranita Chirriadora de Escarpes	Cliff Chirping Frog	0	0	1	1	0	0	1	0	0	0
Eleutherodactylus verrucipes (Cope, 1885 [1884])	Ranita Orejona	Big-eared Chirping Frog	0	0	0	0	0	1	0	0	0	0
Euhyas planirostris (Cope, 1862)	Rana de Invernadero	Greenhouse Frog	0	0	0	0	0	0	IN	0	0	0
Hylidae												
Acris crepitans Baird, 1854	Rana Grillo Norteña	Northern Cricket Frog	0	0	0	1	0	0	0	1	1	0
Ecnomiohyla miotympanum (Cope, 1863)	Calate Arborícola	Small-eared Treefrog	0	0	0	1	1	1	0	0	0	0
Hyla arenicolor Cope, 1866	Ranita de las Rocas	Canyon Treefrog	0	1	1	1	0	0	1	1	1	0
Hyla chrysoscelis Cope, 1880	Ranita Gris de Cope	Cope's Gray Treefrog	0	0	0	0	0	0	1	0	0	0
Hyla cinerea (Schneider, 1799)	Ranita Verde	Green Treefrog	0	0	0	0	0	0	1	0	0	0
Hyla eximia Baird, 1854	Ranita de montaña	Mountain Treefrog	0	0	0	0	0	1	0	0	0	0
Hyla squirella Bosc, 1800	Ranita Ardilla	Squirrel Treefrog	0	0	0	0	0	0	1	0	0	0
Hyla versicolor Le Conte, 1825	Ranita Gris	Gray Treefrog	0	0	0	0	0	0	1	0	0	0
Hyla wrightorum Taylor, 1939	Ranita de Wright	Arizona Treefrog	0	1	1	0	0	0	0	1	1	0
Pachymedusa dacnicolor (Cope, 1864)	Ranita Verduzca	Mexican Leaf Frog	0	1	1	0	0	0	0	0	0	0
Pseudacris cadaverina (Cope, 1866)	Rana de Coro de California	California Treefrog	1	0	0	0	0	0	0	0	0	1
Pseudacris clarkii (Baird, 1854)	Rana de Coro Manchada	Spotted Chorus Frog	0	0	0	0	0	1	1	1	0	0
Pseudacris crucifer (Wied-Neuwied, 1838)	Rana Pimienta	Spring Peeper	0	0	0	0	0	0	1	0	0	0
Pseudacris fouquettei Lemmon, Lemmon, Collins, & Cannatella, 2008	Rana de Coro de Tierra Alta	Upland Chorus Frog	0	0	0	0	0	0	1	0	0	0

	SPANISH NAME	ENGLISH NAME	BC	SO	CH	CO	NL	TA	TX	NM	AZ	CA
✓ *Pseudacris hypochondriaca* (Hallowell, 1854)	Rana de Coro de Baja California	Baja California Treefrog	1	0	0	0	0	0	0	0	1	1
Pseudacris maculata (Agassiz, 1850)	Rana de Coro Boreal	Boreal Chorus Frog	0	0	0	0	0	0	0	1	0	0
Pseudacris regilla (Baird & Girard, 1852)	Rana de Coro del Pacífico Norte	Northern Pacific Treefrog	0	0	0	0	0	0	0	0	0	1
Pseudacris sierra (Jameson, Mackey, & Richmond, 1966)	Rana de Coro de la Sierra	Sierran Treefrog	0	0	0	0	0	0	0	0	0	1
Pseudacris streckeri Wright & Wright, 1933	Rana de Coro de Strecker	Strecker's Chorus Frog	0	0	0	0	0	0	1	0	0	0
Pseudacris triseriata (Wied-Neuwied, 1838)	Rana de Coro Occidental	Western Chorus Frog	0	0	0	0	0	0	0	0	1	0
Scinax staufferi (Cope, 1865)	Rana Arborícola Trompuda	Stauffer's Treefrog	0	0	0	0	0	1	0	0	0	0
✓ *Smilisca baudinii* (Duméril & Bibron, 1841)	Rana Arborícola Mexicana	Mexican Treefrog	0	1	1	1	1	1	1	0	0	0
✓ *Smilisca fodiens* (Boulenger, 1882)	Rana Chata	Lowland Burrowing Treefrog	0	1	0	0	0	0	0	0	1	0
Tlalocohyla smithi (Boulenger, 1902)	Ranita Enana Mexicana	Dwarf Mexican Treefrog	0	1	1	0	0	0	0	0	0	0
Trachycephalus typhonius (Linnaeus, 1758)	Rana Venulosa	Veined Treefrog	0	0	0	0	0	1	0	0	0	0

Leiopelmatidae

	SPANISH NAME	ENGLISH NAME	BC	SO	CH	CO	NL	TA	TX	NM	AZ	CA
Ascaphus truei Stejneger, 1899	Rana Coluda de la Costa	Coastal Tailed Frog	0	0	0	0	0	0	0	0	0	1

Leptodactylidae

	SPANISH NAME	ENGLISH NAME	BC	SO	CH	CO	NL	TA	TX	NM	AZ	CA
✓ *Leptodactylus fragilis* (Brocchi, 1877)	Ranita de Hojarasca	White-lipped Frog	0	0	0	0	1	1	1	0	0	0
Leptodactylus melanonotus (Hallowell, 1861)	Rana del Sabinal	Sabinal Frog	0	1	0	0	0	1	0	0	0	0

Microhylidae

	SPANISH NAME	ENGLISH NAME	BC	SO	CH	CO	NL	TA	TX	NM	AZ	CA
Gastrophryne carolinensis (Holbrook, 1836 [1835])	Ranita de Oriente	Eastern Narrow-mouthed Frog	0	0	0	0	0	0	1	0	0	0
Gastrophryne elegans (Boulenger, 1882)	Sapito Elegante	Elegant Narrow-mouthed Toad	0	0	0	0	0	1	0	0	0	0
✓ *Gastrophryne mazatlanensis* (Taylor, 1943)	Ranita Olivo de Mazatlán	Mazatlan Narrow-mouthed Toad	0	1	1	0	0	0	0	0	1	0
✓ *Gastrophryne olivacea* (Hallowell, 1857 [1856])	Ranita Olivo	Western Narrow-mouthed Toad	0	0	1	1	1	1	1	1	0	0
✓ *Hypopachus variolosus* (Cope, 1866)	Rana Manglera	Sheep Frog	0	1	1	0	1	1	1	0	0	0

Pipidae

	SPANISH NAME	ENGLISH NAME	BC	SO	CH	CO	NL	TA	TX	NM	AZ	CA
Xenopus laevis (Daudin, 1802)	Rana Africana con Garras	African Clawed Frog	IN	0	0	0	0	0	0	0	IN	IN

	SPANISH NAME	ENGLISH NAME	BC	SO	CH	CO	NL	TA	TX	NM	AZ	CA
Ranidae												
Lithobates areolatus (Baird & Girard, 1852)	Rana Cangrejo de Río	Crawfish Frog	0	0	0	0	0	0	1	0	0	0
Lithobates berlandieri (Baird, 1859)	Rana Leopardo de El Río Bravo	Rio Grande Leopard Frog	IN	IN	1	1	1	1	1	1	1	IN
Lithobates blairi (Mecham, Littlejohn, Oldham, Brown, & Brown, 1973)	Rana Leopardo de las Planicies	Plains Leopard Frog	0	0	0	0	0	0	1	1	1	0
Lithobates catesbeianus (Shaw, 1802)	Rana Toro	American Bullfrog	IN	IN	IN	1	0	1	1	IN	IN	IN
Lithobates chiricahuensis (Platz & Mecham, 1979)	Rana Leopardo Chiricahua	Chiricahua Leopard Frog	0	1	1	0	0	0	0	1	1	0
Lithobates clamitans (Latreille, 1801)	Rana Verde	Green Frog	0	0	0	0	0	0	1	0	0	0
Lithobates forreri (Boulenger, 1883)	Rana Leopardo de Forrer	Forrer's Leopard Frog	0	1	1	0	0	0	0	0	0	0
Lithobates grylio (Stejneger, 1901)	Rana Puerco	Pig Frog	0	0	0	0	0	0	1	0	0	0
Lithobates lemosespinali (Smith & Chiszar, 2003)	Rana Leopardo de Lemos-Espinal	Lemos-Espinal's Leopard Frog	0	0	E	0	0	0	0	0	0	0
Lithobates magnaocularis (Frost & Bagnara, 1976)	Rana Leopardo del Noroeste de México	Northwest Mexico Leopard Frog	0	1	1	0	0	0	0	0	0	0
Lithobates onca (Cope, 1875)	Rana Leopardo Relicto	Relict Leopard Frog	0	0	0	0	0	0	0	0	1	0
Lithobates palustris (Le Conte, 1825)	Rana de Ciénega	Pickerel Frog	0	0	0	0	0	0	1	0	0	0
Lithobates pipiens (Schreber, 1782)	Rana Leopardo Norteña	Northern Leopard Frog	0	0	0	0	0	0	0	1	1	1
Lithobates pustulosus (Boulenger, 1883)	Rana Rayas Blancas	White-striped Frog	0	1	1	0	0	0	0	0	0	0
Lithobates sphenocephalus (Cope, 1886)	Rana Leopardo del Sur de Estado Unidos	Southern Leopard Frog	0	0	0	0	0	0	1	0	0	IN
Lithobates tarahumarae (Boulenger, 1917)	Rana Tarahumara	Tarahumara Frog	0	1	1	0	0	0	0	0	1	0
Lithobates yavapaiensis (Platz & Frost, 1984)	Rana Leopardo de Yavapai	Lowland Leopard Frog	1	1	1	0	0	0	0	1	1	1
Rana aurora Baird & Girard, 1852	Rana Patas-rojas del Norte de Estados Unidos	Northern Red-legged Frog	0	0	0	0	0	0	0	0	0	1
Rana boylii Baird, 1854	Rana Patas-amarillas	Foothill Yellow-legged Frog	1	0	0	0	0	0	0	0	0	1
Rana cascadae Slater, 1939	Rana de las Montañas Cascada	Cascades Frog	0	0	0	0	0	0	0	0	0	1
Rana draytonii Baird & Girard, 1852	Rana Patas-rojas de California	California Red-legged Frog	1	0	0	0	0	0	0	0	0	1

	SPANISH NAME	ENGLISH NAME	BC	SO	CH	CO	NL	TA	TX	NM	AZ	CA
Rana muscosa Camp, 1917	Rana Patas-amarillas de las Montañas del Sur	Southern Mountain Yellow-legged Frog	0	0	0	0	0	0	0	0	0	E
Rana pretiosa Baird & Girard, 1853	Rana Manchada de Oregón	Oregon Spotted Frog	0	0	0	0	0	0	0	0	0	1
Rana sierrae Camp, 1917	Rana Patas-amarillas de Sierra Nevada	Sierra Nevada Yellow-legged Frog	0	0	0	0	0	0	0	0	0	1

Rhinophrynidae

✓ *Rhinophrynus dorsalis* Duméril & Bibron, 1841 | Sapo de Madriguera | Burrowing Toad | 0 | 0 | 0 | 0 | 1 | 1 | 1 | 0 | 0 | 0 |

✓ *Rhinophrynus dorsalis* Duméril & Bibron, 1841	Sapo de Madriguera	Burrowing Toad	0	0	0	0	1	1	1	0	0	0

Scaphiopodidae

	SPANISH NAME	ENGLISH NAME	BC	SO	CH	CO	NL	TA	TX	NM	AZ	CA
✓ *Scaphiopus couchi* Baird, 1854	Cavador	Couch's Spadefoot	1	1	1	1	1	1	1	1	1	1
Scaphiopus hurteri Strecker, 1910	Sapo de Espuelas de Hurter	Hurter's Spadefoot	0	0	0	0	0	0	1	0	0	0
✓ *Spea bombifrons* (Cope, 1863)	Sapo de Espuelas de los Llanos	Plains Spadefoot	0	0	1	0	1	1	1	1	1	0
✓ *Spea hammondii* (Baird, 1859)	Sapo de Espuelas Occidental	Western Spadefoot	1	0	0	0	0	0	0	0	0	1
Spea intermontana (Cope, 1883)	Sapo de Espuelas de la Gran Cuenca	Great Basin Spadefoot	0	0	0	0	0	0	0	0	1	1
✓ *Spea multiplicata* (Cope, 1863)	Sapo de Espuelas Mexicano	Mexican Spadefoot	0	1	1	1	1	1	1	1	1	0

CLASS REPTILIA

Order Crocodylia

Crocodylidae

	SPANISH NAME	ENGLISH NAME	BC	SO	CH	CO	NL	TA	TX	NM	AZ	CA
Alligator mississippiensis (Daudin, 1801)	Lagarto Americano	American Alligator	0	0	0	0	0	0	1	0	0	0
Crocodylus moreletii Duméril & Bibron, 1851	Cocodrilo de Pantano	Morelet's Crocodile	0	0	0	0	0	1	0	0	0	0

Order Testudines

Chelonidae

	SPANISH NAME	ENGLISH NAME	BC	SO	CH	CO	NL	TA	TX	NM	AZ	CA
Caretta caretta (Linnaeus, 1758)	Caguama	Loggerhead Sea Turtle	M	M	0	0	0	M	M	0	0	M
Chelonia mydas (Linnaeus, 1758)	Tortuga Blanca de Mar	Green Sea Turtle	M	M	0	0	0	M	M	0	0	M
Eretmochelys imbricata (Linnaeus, 1766)	Carey	Hawksbill Sea Turtle	M	M	0	0	0	M	M	0	0	M
Lepidochelys kempii (Garman, 1880)	Lora	Kemp's Ridley Sea Turtle	0	0	0	0	0	M	M	0	0	0
Lepidochelys olivacea (Eschscholtz, 1829)	Tortuga Golfina	Olive Ridley Sea Turtle	M	M	0	0	0	0	0	0	0	M

	SPANISH NAME	ENGLISH NAME	BC	SO	CH	CO	NL	TA	TX	NM	AZ	CA
Chelydridae												
Chelydra serpentina (Linnaeus, 1758)	Tortuga Lagarto	Snapping Turtle	0	0	0	0	0	0	1	1	IN	IN
Macrochelys temminckii (Troost *in* Harlan, 1835)	Tortuga Lagarto de Temminck	Alligator Snapping Turtle	0	0	0	0	0	0	1	0	0	0
Dermochelyidae												
Dermochelys coriacea (Vandelli, 1761)	Laúd	Leatherback Sea Turtle	M	M	0	0	0	M	M	0	0	M
Emydidae												
Actinemys marmorata (Baird & Girard, 1852)	Tortuga Occidental de Estanque	Western Pond Turtle	1	0	0	0	0	0	0	0	0	1
Chrysemys picta (Schneider, 1783)	Tortuga Pinta	Painted Turtle	0	0	1	0	0	0	1	1	1	IN
Chrysemys dorsalis Agassiz, 1857	Tortuga Pinta del Sur de Estados Unidos	Southern Painted Turtle	0	0	0	0	0	0	1	0	0	0
Deirochelys reticularia (Latreille, 1801)	Tortuga Pollo	Chicken Turtle	0	0	0	0	0	0	1	0	0	0
Graptemys caglei Haynes & McKown, 1974	Tortuga de Cagle	Cagle's Map Turtle	0	0	0	0	0	0	E	0	0	0
Graptemys ouachitensis Cagle, 1953	Tortuga de Ouachita	Ouachita Map Turtle	0	0	0	0	0	0	1	0	0	0
Graptemys pseudogeographica (Gray, 1831)	Falsa Tortuga Mapa	False Map Turtle	0	0	0	0	0	0	1	0	0	IN
Graptemys versa Stejneger, 1925	Tortuga Mapa Texana	Texas Map Turtle	0	0	0	0	0	0	E	0	0	0
Malaclemys terrapin (Schoepff, 1793)	Tortuga Espalda Diamante	Diamond-backed Terrapin	0	0	0	0	0	0	1	0	0	0
Pseudemys concinna (Le Conte, 1830)	Tortuga Oriental de Río	River Cooter	0	0	0	0	0	0	1	0	0	0
Pseudemys gorzugi Ward, 1984	Tortuga de Oreja Amarilla	Rio Grande Cooter	0	0	0	1	1	1	1	1	0	0
Pseudemys nelsoni Carr, 1938	Tortuga Vientre-rojo de Florida	Florida Red-bellied Turtle	0	0	0	0	0	0	IN	0	0	0
Pseudemys texana Bauer, 1893	Jicotea Texana	Texas Cooter	0	0	0	0	0	0	E	0	0	0
Terrapene carolina (Linnaeus, 1758)	Caja de Carolina	Eastern Box Turtle	0	0	0	0	0	1	1	0	0	0
Terrapene coahuila Schmidt & Owens, 1944	Caja de Cuatrociénegas	Coahuilan Box Turtle	0	0	0	E	0	0	0	0	0	0
Terrapene nelsoni Stejneger, 1925	Caja de Manchas	Spotted Box Turtle	0	1	1	0	0	0	0	0	0	0
Terrapene ornata (Agassiz, 1857)	Caja Ornamentada	Ornate Box Turtle	0	1	1	0	0	0	1	1	1	0
Trachemys gaigeae (Hartweg, 1939)	Jicotea de la Meseta Mexicana	Mexican Plateau Slider	0	0	1	1	0	0	1	1	0	0

	SPANISH NAME	ENGLISH NAME	BC	SO	CH	CO	NL	TA	TX	NM	AZ	CA
Trachemys nebulosa (Van Denburgh, 1895)	Jicotea del Vientre Negro	Black-bellied Slider	0	1	0	0	0	0	0	0	0	0
Trachemys scripta (Thunberg, *in* Schoepff, 1792)	Jicotea de Estanque	Pond Slider	0	0	0	1	1	1	1	1	IN	IN
Trachemys taylori (Legler, 1960)	Jicotea de Cuatrociénegas	Cuatrociénegas Slider	0	0	0	E	0	0	0	0	0	0
Trachemys venusta (Gray, 1855)	Jicotea de Agua	Mesoamerican Slider	0	0	0	0	0	1	0	0	0	0
Trachemys yaquia (Legler & Webb, 1970)	Jicotea del Yaqui	Yaqui Slider	0	E	0	0	0	0	0	0	0	0

Geoemydidae

	SPANISH NAME	ENGLISH NAME	BC	SO	CH	CO	NL	TA	TX	NM	AZ	CA
Rhinoclemmys pulcherrima (Gray, 1855)	Casco Rojo	Painted Wood Turtle	0	1	1	0	0	0	0	0	0	0

Kinosternidae

	SPANISH NAME	ENGLISH NAME	BC	SO	CH	CO	NL	TA	TX	NM	AZ	CA
Kinosternon alamosae Berry & Legler, 1980	Casquito de Álamos	Álamos Mud Turtle	0	1	0	0	0	0	0	0	0	0
Kinosternon arizonense Gilmore, 1922	Casquito de Arizona	Arizona Mud Turtle	0	1	0	0	0	0	0	0	1	0
Kinosternon durangoense Iverson, 1979	Casquito de Durango	Durango Mud Turtle	0	0	1	1	0	0	0	0	0	0
Kinosternon flavescens (Agassiz, 1857)	Casquito Amarillo	Yellow Mud Turtle	0	0	1	1	1	1	1	1	1	0
Kinosternon herrerai Stejneger, 1925	Casquito de Herrera	Herrera's Mud Turtle	0	0	0	0	0	1	0	0	0	0
Kinosternon hirtipes (Wagler, 1830)	Casquito de Pata Rugosa	Rough-footed Mud Turtle	0	0	1	1	0	0	1	0	0	0
Kinosternon integrum LeConte, 1854	Casquito de Fango Mexicana	Mexican Mud Turtle	0	1	1	0	1	1	0	0	0	0
Kinosternon scorpioides (Linnaeus, 1766)	Tortuga Escorpión	Scorpion Mud Turtle	0	0	0	0	0	1	0	0	0	0
Kinosternon sonoriense Le Conte, 1854	Casquito de Sonora	Sonora Mud Turtle	0	1	1	0	0	0	0	1	1	1
Kinsternon subrubrum (Lacépède, 1788)	Casquito Oriental	Eastern Mud Turtle	0	0	0	0	0	0	1	0	0	0
Sternotherus carinatus (Gray, 1855)	Tortuga Almizclera Espalda de Rasuradora	Razor-backed Musk Turtle	0	0	0	0	0	0	1	0	0	0
Sternotherus odoratus (Latreille, 1801)	Tortuga Almizclera	Eastern Musk Turtle	0	0	0	0	0	0	1	0	0	0

Family Testudinidae

	SPANISH NAME	ENGLISH NAME	BC	SO	CH	CO	NL	TA	TX	NM	AZ	CA
Gopherus agassizii (Cooper, 1861)	Tortuga de Desierto	Desert Tortoise	0	0	0	0	0	0	0	0	1	1
Gopherus berlandieri (Agassiz, 1857)	Tortuga de Berlandier	Texas Tortoise	0	0	0	1	1	1	1	0	0	0

	SPANISH NAME	ENGLISH NAME	BC	SO	CH	CO	NL	TA	TX	NM	AZ	CA
Gopherus flavomarginatus Legler, 1959	Tortuga del Bolsón de Mapimí	Bolson Tortoise	0	0	1	1	0	0	0	0	0	0
Gopherus morafkai Murphy, Berry, Edwards,Leviton, Lathrop, & Riedle, 2011	Tortuga del Desierto de Sonora	Sonoran Desert Tortoise	0	1	1	0	0	0	0	0	1	0

Trionychidae

	SPANISH NAME	ENGLISH NAME	BC	SO	CH	CO	NL	TA	TX	NM	AZ	CA
Apalone mutica (Le Sueur, 1827)	Tortuga de Concha Blanda Lisa	Smooth Softshell	0	0	0	0	0	0	1	1	0	0
Apalone spinifera (Le Sueur, 1827)	Tortuga de Concha Blanda	Spiny Softshell	IN	IN	1	1	1	1	1	1	IN	IN

Order Squamata

Suborder Lacertilia

Anguidae

	SPANISH NAME	ENGLISH NAME	BC	SO	CH	CO	NL	TA	TX	NM	AZ	CA
Abronia taeniata (Wiegmann, 1828)	Escorpión Arborícola de Bandas	Bromeliad Arboreal Alligator Lizard	0	0	0	0	0	1	0	0	0	0
Barisia ciliaris (Smith, 1942)	Escorpión de Montaña	Northern Alligator Lizard	0	0	1	1	1	1	0	0	0	0
Barisia levicollis Stejneger, 1890	Escorpión de Chihuahua	Chihuahuan Alligator Lizard	0	0	E	0	0	0	0	0	0	0
Elgaria cedrosensis (Fitch, 1934)	Lagartija Lagarto de Isla Cedros	Isla Cedros Alligator Lizard	E	0	0	0	0	0	0	0	0	0
Elgaria coerulea (Wiegmann, 1828)	Lagartija Lagarto del Norte de Estados Unidos	Northern Alligator Lizard	0	0	0	0	0	0	0	0	0	1
Elgaria kingii Gray, 1838	Lagartija Lagarto de Montaña	Madrean Alligator Lizard	0	1	1	0	0	0	0	1	1	0
Elgaria multicarinata (Blainville, 1835)	Lagartija Lagarto del Suroeste de Estados Unidos	Southern Alligator Lizard	1	0	0	0	0	0	0	0	0	1
Elgaria nana (Fitch, 1934)	Lagartija Lagarto de Los Coronados	Islas Los Coronados Alligator Lizard	E	0	0	0	0	0	0	0	0	0
Elgaria panamintina Stebbins, 1958	Lagartija Lagarto de la Sierra Panamint	Panamint Alligator Lizard	0	0	0	0	0	0	0	0	0	E
Gerrhonotus farri Bryson & Graham, 2010	Lagartija Escorpión de Farr	Farr's Alligator Lizard	0	0	0	0	0	E	0	0	0	0
Gerrhonotus infernalis Baird, 1859 (1858)	Cantil de Tierra	Texas Alligator Lizard	0	0	0	1	1	1	1	0	0	0
Gerrhonotus lugoi McCoy, 1970	Lagartija Escorpión de Lugo	Lugo's Alligator Lizard	0	0	0	E	0	0	0	0	0	0
Gerrhonotus parvus (Knight & Scudday, 1985)	Lagarto Pigmeo	Pygmy Alligator Lizard	0	0	0	0	E	0	0	0	0	0
Gerrhonotus taylori Tihen, 1954	Lagartija Escorpión de Taylor	Taylor's Alligator Lizard	0	0	1	0	0	0	0	0	0	0

	SPANISH NAME	ENGLISH NAME	BC	SO	CH	CO	NL	TA	TX	NM	AZ	CA
Ophisaurus attenuata Cope, 1880	Lagartija de Cristal Delegada	Slender Glass Lizard	0	0	0	0	0	0	1	0	0	0
Ophisaurus incomptus McConkey, 1955	Lagartija de Cristal de Cuello Plano	Plain-necked Glass Lizard	0	0	0	0	0	1	0	0	0	0

Anniellidae

	SPANISH NAME	ENGLISH NAME	BC	SO	CH	CO	NL	TA	TX	NM	AZ	CA
Anniella alexanderae Papenfuss & Parham, 2013	Lagartija Apoda de Alexander	Temblor Legless Lizard	0	0	0	0	0	0	0	0	0	E
Anniella campi Papenfuss and Parham, 2013	Lagartja Apoda de Camp	Southern Sierra Legless Lizard	0	0	0	0	0	0	0	0	0	E
Anniella geronimensis Shaw, 1940	Lagartija Sin Patas	Baja California Legless Lizard	E	0	0	0	0	0	0	0	0	0
Anniella grinnelli Papenfuss and Parham, 2013	Lagartija Apoda de Grinnell	Bakersfield Legless Lizard	0	0	0	0	0	0	0	0	0	E
Anniella pulchra Gray, 1852	Lagartija Apoda	California Legless Lizard	0	0	0	0	0	0	0	0	0	E
✓ *Anniella stebbinsi* Papenfuss & Parham, 2013	Lagartija Apoda del Sur de California	Southern California Legless Lizard	1	0	0	0	0	0	0	0	0	1

Chamaeleonidae

	SPANISH NAME	ENGLISH NAME	BC	SO	CH	CO	NL	TA	TX	NM	AZ	CA
Chamaeleo jacksonii (Boulenger, 1896)	Camaleón de Jackson	Jackson's Chameleon	0	0	0	0	0	0	0	0	0	IN

Corytophanidae

	SPANISH NAME	ENGLISH NAME	BC	SO	CH	CO	NL	TA	TX	NM	AZ	CA
Laemanctus serratus Cope, 1864	Lemancto Coronado	Serrated Casque-headed Lizard	0	0	0	0	0	1	0	0	0	0

Crotaphytidae

	SPANISH NAME	ENGLISH NAME	BC	SO	CH	CO	NL	TA	TX	NM	AZ	CA
Crotaphytus antiquus Axtell & Webb, 1995	Cachorón de Coahuila	Venerable Collared Lizard	0	0	0	E	0	0	0	0	0	0
Crotaphytus bicinctores Smith & Tanner, 1972	Cachorón de la Gran Cuenca	Great Basin Collared Lizard	0	0	0	0	0	0	0	0	1	1
✓ *Crotaphytus collaris* (Say, *in* James, 1823)	Cachorón de Collar	Eastern Collared Lizard	0	1	1	1	1	1	1	1	1	0
Crotaphytus dickersonae Schmidt, 1922	Cachorón Azul de Collar	Dickerson's Collared Lizard	0	E	0	0	0	0	0	0	0	0
Crotaphytus grismeri McGuire, 1994	Cachorón de Sierra de las Cucupas	Sierra de las Cucupas Collared Lizard	E	0	0	0	0	0	0	0	0	0
Crotaphytus insularis Van Denburgh & Slevin, 1921	Cachorón de Collar Negro	Isla Ángel de la Guarda Collared Lizard	E	0	0	0	0	0	0	0	0	0
✓ *Crotaphytus nebrius* Axtell & Montanucci, 1977	Cachorón de Sonora	Sonoran Collared Lizard	0	1	0	0	0	0	0	0	1	0
✓ *Crotaphytus reticulatus* Baird, 1859 (1858)	Reticulada de Collar	Reticulate Collared Lizard	0	0	0	1	1	1	1	0	0	0
✓ *Crotaphytus vestigium* Smith & Tanner, 1972	Cachorón de Baja California	Baja California Collared Lizard	1	0	0	0	0	0	0	0	0	1

	SPANISH NAME	ENGLISH NAME	BC	SO	CH	CO	NL	TA	TX	NM	AZ	CA
✓ *Gambelia copeii* (Yarrow, 1882)	Cachorón Leopardo de Baja California	Baja California Leopard Lizard	1	0	0	0	0	0	0	0	0	1
Gambelia sila Stejneger, 1890	Lagartija Leopardo de Nariz-chata	Blunt-nosed Leopard Lizard	0	0	0	0	0	0	0	0	0	E
✓ *Gambelia wislizenii* (Baird & Girard, 1852)	Lagartija Mata Caballo	Long-nosed Leopard Lizard	1	1	1	1	0	0	1	1	1	1

Dactyloidae

Anolis carolinensis (Voigt, 1832)	Abaniquillo Verde	Green Anole	0	0	0	0	0	0	1	0	0	0
Anolis nebulosus (Wiegmann, 1834)	Roño de Paño	Clouded Anole	0	1	1	0	0	0	0	0	0	0
Anolis sagrei Duméril & Bibron, 1837	Abaniquillo Costero Maya	Brown Anole	0	0	0	0	0	0	IN	0	0	0
Anolis sericeus Hallowell, 1856	Abaniquillo Punto Azul	Silky Anole	0	0	0	0	0	1	0	0	0	0

Dibamidae

Anelytropsis papillosus Cope, 1885	Lagartija Ciega Mexicana	Mexican Blind Lizard	0	0	0	0	0	1	0	0	0	0

Eublepharidae

✓ *Coleonyx brevis* Stejneger, 1893	Salamanquesa del Desierto	Texas Banded Gecko	0	0	1	1	1	1	1	1	0	0
Coleonyx fasciatus (Boulenger, 1885)	Geco de Bandas Negras	Black Banded Gecko	0	1	0	0	0	0	0	0	0	0
✓ *Coleonyx reticulatus* Davis & Dixon, 1958	Geco Reticulado	Reticulate Banded Gecko	0	0	0	1	0	0	1	0	0	0
✓ *Coleonyx switaki* (Murphy, 1974)	Geco Descalzo	Barefoot Banded Gecko	1	0	0	0	0	0	0	0	0	1
✓ *Coleonyx variegatus* (Baird, 1858)	Geco de Bandas Occidental	Western Banded Gecko	1	1	0	0	0	0	0	1	1	1

Gekkonidae

Cyrtopodion scabrum (Heyden, 1827)	Geco de Cola Rugosa	Rough-tailed Gecko	0	0	0	0	0	0	IN	0	0	0
Gehyra mutilata (Wiegmann, 1834)	Salamanquesa que Mutila	Stump-toed Gecko	0	0	0	0	0	0	0	0	0	IN
Hemidactylus frenatus Schlegel, in Dúmeril & Bribon, 1836	Besucona	Common House Gecko	0	IN	0	0	0	IN	IN	0	0	0
Hemidactylus garnoti Duméril & Bibron, 1836	Geco del Indo-Pacífico	Indo-Pacific House Gecko	0	0	0	0	0	0	IN	0	0	0
Hemidactylus turcicus (Linnaeus, 1758)	Geco del Mediterráneo	Mediterranean House Gecko	IN	IN	IN	IN	IN	IN	IN	IN	IN	IN
Phyllodactylus homolepidurus Smith, 1935	Salamanquesa de Sonora	Sonoran Leaf-toed Gecko	0	1	0	0	0	0	0	0	0	0
✓ *Phyllodactylus nocticolus* Dixon, 1964	Salamanquesa Insular	Peninsular Leaf-toed Gecko	1	1	0	0	0	0	0	0	0	1

	SPANISH NAME	ENGLISH NAME	BC	SO	CH	CO	NL	TA	TX	NM	AZ	CA
Phyllodactylus nolascoensis Dixon, 1964	Salamanquesa de San Pedro Nolasco	San Pedro Nolasco Gecko	0	E	0	0	0	0	0	0	0	0
Phyllodactylus partidus Dixon, 1966	Salamanquesa de Isla Partida Norte	Isla Partida Norte Leaf-toed Gecko	E	0	0	0	0	0	0	0	0	0
Phyllodactylus tuberculosus Wiegmann, 1835	Geco Panza Amarilla	Yellow-bellied Gecko	0	1	1	0	0	0	0	0	0	0
Tarentola annularis (Geoffroy de Saint-Hilaire, 1827)	Salamanquesa de Cruz	White-spotted Wall Gecko	0	0	0	0	0	0	0	0	0	IN
Tarentola mauritanica (Linnaeus, 1758)	Geco Moro	Moorish Gecko	0	0	0	0	0	0	0	0	0	IN

Helodermatidae

Heloderma exasperatum Bogert & Martín Del Campo, 1956	Escorpión del Río Fuerte	Rio Fuerte's Beaded Lizard	0	1	1	0	0	0	0	0	0	0
✓*Heloderma suspectum* Cope, 1869	Monstruo de Gila	Gila Monster	0	1	0	0	0	0	0	1	1	1

Iguanidae

Ctenosaura sp.	Garrobo	Spiny-tailed Iguana	0	0	0	0	0	0	0	0	IN	0
Ctenosaura acanthura (Shaw, 1802)	Garrobo de Noreste	Mexican Spiny-tailed Iguana	0	0	0	0	0	1	0	0	0	0
Ctenosaura conspicuosa Dickerson, 1919	Garrobo de Isla San Esteban	Isla San Esteban Spiny-tailed Iguana	0	E	0	0	0	0	0	0	0	0
Ctenosaura macrolopha Smith, 1972	Garrobo de Sonora	Sonora Spiny-tailed Iguana	0	1	1	0	0	0	0	0	0	0
Ctenosaura nolascensis Smith, 1972	Garrobo de Isla San Pedro Nolasco	Isla San Pedro Nolasco Spiny-tailed Iguana	0	E	0	0	0	0	0	0	0	0
Ctenosaura pectinata (Wiegmann, 1834)	Garrobo de Roca	Mexican Spiny-tailed Iguana	0	0	0	0	0	0	IN	0	0	0
✓*Dipsosaurus dorsalis* (Baird & Girard, 1852)	Cachorón Güero	Northern Desert Iguana	1	1	0	0	0	0	0	0	1	1
Iguana iguana (Linnaeus, 1758)	Iguana de Ribera	Common Green Iguana	0	0	0	0	0	1	0	0	0	0
✓*Sauromalus ater* Duméril, 1856	Cachorón de Roca	Common Chuckwalla	1	1	0	0	0	0	0	0	1	1
Sauromalus hispidus Stejneger, 1891	Iguana Espinosa de Pared	Spiny Chuckwalla	E	IN	0	0	0	0	0	0	0	0
Sauromalus varius Dickerson, 1919	Iguana de Pared de Piebald	Piebald Chuckwalla	IN	E	0	0	0	0	0	0	0	0

Lacertidae

Podarcis sicula (Rafinesque, 1810)	Lagartija Italiana de Pared	Italian Wall Lizard	0	0	0	0	0	0	0	0	0	IN

Phrynosomatidae

✓*Callisaurus draconoides* Blainville, 1835	Cachora Arenera	Zebra-tailed Lizard	1	1	0	0	0	0	0	1	1	1

	SPANISH NAME	ENGLISH NAME	BC	SO	CH	CO	NL	TA	TX	NM	AZ	CA
Callisaurus splendidus Dickerson, 1919	Cachora de Ángel de la Guarda	Ángel de la Guarda Zebra-tailed Lizard	E	0	0	0	0	0	0	0	0	0
Cophosaurus texanus Troschel, 1852	Lagartijón Sordo	Greater Earless Lizard	0	1	1	1	1	1	1	1	1	0
Holbrookia approximans Baird, 1859	Lagartija Sorda Manchada	Speckled Earless Lizard	0	1	1	1	1	0	0	0	0	0
Holbrookia elegans Bocourt, 1874	Lagartija Sorda Elegante	Elegant Earless Lizard	0	1	1	0	0	0	0	1	1	0
Holbrookia lacerata Cope, 1880	Lagartija Cola Punteada	Spot-tailed Earless Lizard	0	0	0	1	1	0	1	0	0	0
Holbrookia maculata Girard, 1851	Lagartija Sorda Pequeña	Common Lesser Earless Lizard	0	0	1	0	0	0	1	1	1	0
Holbrookia propinqua Baird & Girard, 1852	Lagartija Sorda Carinada	Keeled Earless Lizard	0	0	0	0	0	1	1	0	0	0
Petrosaurus mearnsi (Stejneger, 1894)	Lagarto de Roca Rayado	Banded Rock Lizard	1	0	0	0	0	0	0	0	0	1
Petrosaurus repens (Van Denburgh, 1895)	Lagarto de Roca de Hocico Corto	Central Baja California Banded Rock Lizard	1	0	0	0	0	0	0	0	0	0
Petrosaurus slevini (Van Denburgh, 1922)	Lagarto de Roca Bandeado	Slevin's Banded Rock Lizard	E	0	0	0	0	0	0	0	0	0
Phrynosoma blainvillii Gray, 1839	Camaleón de Blainville	Blainville's Horned Lizard	1	0	0	0	0	0	0	0	0	1
Phyrnosoma cerroense Stejneger, 1893	Camaleón de Isla Cedros	Cedros Island Horned Lizard	1	0	0	0	0	0	0	0	0	0
Phrynosoma cornutum (Harlan, 1825)	Camaleón Común	Texas Horned Lizard	0	1	1	1	1	1	1	1	1	0
Phrynosoma ditmarsi Stejneger, 1906	Camaleón de Roca	Rock Horned Lizard	0	E	0	0	0	0	0	0	0	0
Phrynosoma douglasii (Bell, 1828)	Camaleón Pigmeo de Cola Corta	Pygmy Short-horned Lizard	0	0	0	0	0	0	0	0	0	1
Phrynosoma goodei Stejneger, 1893	Camaleón de Sonora	Goode's Desert Horned Lizard	0	1	0	0	0	0	0	0	1	0
Phrynosoma hernandesi Girard, 1858	Camaleón Cuernitos de Hernández	Greater Short-horned Lizard	0	1	1	0	0	0	1	1	1	0
Phrynosoma mcallii (Hallowell, 1852)	Camaleón de Cola Plana	Flat-tailed Horned Lizard	1	1	0	0	0	0	0	0	1	1
Phrynosoma modestum Girard, 1852	Camaleón Modesto	Round-tailed Horned Lizard	0	1	1	1	1	1	1	1	1	0
Phrynosoma orbiculare (Linnaeus, 1758)	Camaleón de Montaña	Mountain Horned Lizard	0	1	1	1	1	1	0	0	0	0
Phrynosoma platyrhinos Girard, 1852	Camaleón del Desierto	Desert Horned Lizard	1	0	0	0	0	0	0	0	1	1
Phrynosoma solare Gray, 1845	Camaleón Real	Regal Horned Lizard	0	1	0	0	0	0	0	1	1	0
Sceloporus albiventris Smith, 1939	Bejore de Vientre Blanco	Western White-bellied Spiny Lizard	0	1	1	0	0	0	0	0	0	0

	SPANISH NAME	ENGLISH NAME	BC	SO	CH	CO	NL	TA	TX	NM	AZ	CA
Sceloporus arenicolus Degenhardt & Jones, 1972	Lagartija de Arena	Dunes Sagebrush Lizard	0	0	0	0	0	0	1	1	0	0
Sceloporus bimaculosus Phelan & Brattstrom, 1955	Cachora	Twin-spotted Spiny Lizard	0	0	1	1	0	0	1	1	1	0
Sceloporus cautus Smith, 1938	Espinosa Llanera	Shy Spiny Lizard	0	0	0	1	1	1	0	0	0	0
Sceloporus chaneyi Liner & Dixon, 1992	Espinosa de Chaney	Chaney's Bunchgrass Lizard	0	0	0	0	1	1	0	0	0	0
Sceloporus clarkii Baird & Girard, 1852	Bejori de Clark	Clark's Spiny Lizard	0	1	1	0	0	0	0	1	1	0
Sceloporus consobrinus Baird & Girard, 1853	Lagartija de las Cercas	Prairie Lizard	0	0	1	1	1	1	1	1	0	0
Sceloporus couchii Baird, 1859 (1858)	Espinosa de las Rocas	Couch's Spiny Lizard	0	0	0	1	1	0	0	0	0	0
Sceloporus cowlesi Lowe & Norris, 1956	Lagartija de Cerca del Noroeste	Southwestern Fence Lizard	0	1	1	0	0	0	1	1	1	0
Sceloporus cyanogenys Cope, 1885	Lagartija de Barba Azul	Bluechinned Rough-scaled Lizard	0	0	0	1	1	1	1	0	0	0
Sceloporus cyanostictus Axtell & Axtell, 1971	Lagarto con Espinas Azules	Blue Spiny Lizard	0	0	0	1	1	0	0	0	0	0
Sceloporus goldmani Smith, 1937	Lagartija de Pastizal de Goldman	Goldman's Bunchgrass Lizard	0	0	0	1	1	0	0	0	0	0
Sceloporus graciosus Baird & Girard, 1852	Lagartija Común de las Salvias	Common Sagebrush Lizard	0	0	0	0	0	0	0	1	1	1
Sceloporus grammicus Wiegmann, 1828	Lagartija de Árbol	Graphic Spiny Lizard	0	0	0	1	1	1	1	0	0	0
Sceloporus jarrovii Cope, *in* Yarrow, 1875	Lagartija de Yarrow	Yarrow´s Spiny Lizard	0	1	1	0	0	0	0	1	1	0
Sceloporus lemosespinali Lara-Góngora, 2004	Lagartija de Lemos-Espinal	Lemos-Espinal's Spiny Lizard	0	1	1	0	0	0	0	0	0	0
Sceloporus maculosus Smith, 1934	Lagartija Chata del Norte	Northern Snub-nosed Lizard	0	0	0	1	0	0	0	0	0	0
Sceloporus magister Hallowell, 1854	Lagartija del Desierto	Desert Spiny Lizard	1	1	0	0	0	0	0	1	1	1
Sceloporus merriami Stejneger, 1904	Lagartija de Cañón	Canyon Lizard	0	0	1	1	1	0	1	0	0	0
Sceloporus minor Cope, 1885	Lagartija Menor	Minor Lizard	0	0	0	1	1	1	0	0	0	0
Sceloporus nelsoni Cochran, 1923	Espinosa de Nelson	Nelson's Spiny Lizard	0	1	1	0	0	0	0	0	0	0
Sceloporus oberon Smith & Brown, 1941	Lagartija Menor Negra	Royal Lesser Minor Lizard	0	0	0	1	1	0	0	0	0	0
Sceloporus occidentalis Baird & Girard (1852)	Bejori de Cerca Occidental	Western Fence Lizard	1	0	0	0	0	0	0	0	0	1
Sceloporus olivaceus Smith, 1934	Espinosa de los Árboles	Texas Spiny Lizard	0	0	0	1	1	1	1	0	0	0
Sceloporus orcutti Stejneger, 1893	Espinosa del Granito	Granite Spiny Lizard	1	0	0	0	0	0	0	0	0	1

	SPANISH NAME	ENGLISH NAME	BC	SO	CH	CO	NL	TA	TX	NM	AZ	CA
Sceloporus ornatus Baird, 1859 (1858)	Espinosa Ornamentada	Ornate Spiny Lizard	0	0	0	1	1	0	0	0	0	0
Sceloporus parvus Smith, 1934	Lagartija de Panza Azul	Blue-bellied Lizard	0	0	0	1	1	1	0	0	0	0
✓ *Sceloporus poinsettii* Baird & Girard, 1852	Lagartija de las Grietas	Crevice Spiny Lizard	0	1	1	1	1	0	1	1	0	0
Sceloporus samcolemani Smith & Hall, 1974	Lagartija de Coleman	Coleman's Bunchgrass Lizard	0	0	0	1	1	0	0	0	0	0
Sceloporus scalaris Wiegmann, 1828	Lagartija Escamosa	Light-bellied Bunchgrass Lizard	0	0	0	0	0	1	0	0	0	0
Sceloporus serrifer Cope, 1866	Escamosa del Ocote	Rough-scaled Lizard	0	0	0	0	1	1	0	0	0	0
✓ *Sceloporus slevini* Smith, 1937	Espinosa de Pastizal de Slevin	Slevin's Bunchgrass Lizard	0	1	1	0	0	0	0	1	1	0
Sceloporus spinosus Weigmann, 1828	Chintete Espinoso	Eastern Spiny Lizard	0	0	0	1	1	1	0	0	0	0
Sceloporus torquatus Wiegmann, 1828	Espinosa de Collar	Torquate Lizard	0	0	0	0	1	1	0	0	0	0
Sceloporus tristichus Cope, *in* Yarrow, 1875	Lagartija de Cercas de la Meseta	Plateau Fence Lizard	0	0	0	0	0	0	0	1	1	0
Sceloporus uniformis Phelan & Brattstrom, 1955	Lagartija Espinosa de Espalda-amarilla	Yellow-backed Spiny Lizard	0	0	0	0	0	0	0	0	1	1
✓ *Sceloporus vandenburgianus* Cope, 1896	Septentrional de las Salvias	Southern Sagebrush Lizard	1	0	0	0	0	0	0	0	0	1
✓ *Sceloporus variabilis* Wiegmann, 1834	Panza Azul-Rosada	Rose-bellied Lizard	0	0	0	1	1	1	1	0	0	0
✓ *Sceloporus virgatus* Smith, 1938	Lagartija Rayada de la Meseta	Striped Plateau Lizard	0	1	1	0	0	0	0	1	1	0
Sceloporus zosteromus Cope, 1863	Bejori	Baja California Spiny Lizard	1	0	0	0	0	0	0	0	0	0
Uma exsul Schmidt & Bogert, 1947	Lagartija Arenera	Fringe-toed Sand Lizard	0	0	0	E	0	0	0	0	0	0
Uma inornata Cope, 1895	Arenera del Valle Coachella	Coachella Fringe-toed Lizard	0	0	0	0	0	0	0	0	0	E
✓ *Uma notata* Baird, 1858	Arenera del Desierto del Colorado	Colorado Desert Fringe-toed Lizard	1	0	0	0	0	0	0	0	0	1
Uma paraphygas Williams, Chrapliwy, & Smith, 1959	Arenera de Chihuahua	Chihuahua Fringe-toed Lizard	0	0	1	1	0	0	0	0	0	0
✓ *Uma rufopunctata* Cope, 1895	Arenera de Manchas Laterales	Yuman Desert Fringe-toed Lizard	0	1	0	0	0	0	0	0	1	0
Uma scoparia Cope, 1894	Arenera del Desierto de Mohave	Mohave Fringe-toed Lizard	0	0	0	0	0	0	0	0	1	1
Urosaurus bicarinatus (Duméril, 1856)	Roñito Arborícola	Tropical Tree Lizard	0	1	1	0	0	0	0	0	0	0
✓ *Urosaurus graciosus* Hallowell, 1854	Roñito de Matorral	Long-tailed Brush Lizard	1	1	0	0	0	0	0	0	1	1

	SPANISH NAME	ENGLISH NAME	BC	SO	CH	CO	NL	TA	TX	NM	AZ	CA
Urosaurus lahtelai Rau & Loomis, 1977	Roñito de Matorral Bajacaliforniano	Baja California Brush Lizard	E	0	0	0	0	0	0	0	0	0
✓ *Urosaurus nigricaudus* (Cope, 1864)	Roñito de Matorral Cola-negra	Black-tailed Brush Lizard	1	0	0	0	0	0	0	0	0	1
✓ *Urosaurus ornatus* (Baird & Girard, 1852)	Roñito Ornado	Ornate Tree Lizard	1	1	1	1	0	1	1	1	1	1
Uta antiqua Ballinger & Tinkle, 1968	Mancha-lateral de San Lorenzo	San Lorenzo Islands Side-blotched Lizard	E	0	0	0	0	0	0	0	0	0
Uta encantadae Grismer, 1994	Mancha-lateral Encantadas	Enchanted Side-blotched Lizard	E	0	0	0	0	0	0	0	0	0
Uta lowei Grismer, 1994	Mancha-lateral Muerta	Dead Side-blotched Lizard	E	0	0	0	0	0	0	0	0	0
Uta nolascensis Van Denburgh & Slevin, 1921	Mancha-lateral de San Pedro Nolasco	Isla San Pedro Nolasco Lizard	0	E	0	0	0	0	0	0	0	0
Uta palmeri Stejneger, 1890	Mancha-lateral de San Pedro Mártir	Isla San Pedro Mártir Side-blotched Lizard	0	E	0	0	0	0	0	0	0	0
✓ *Uta stansburiana* Baird & Girard, 1852	Mancha-lateral Común	Common Side-blotched Lizard	1	1	1	1	1	0	1	1	1	1
Uta tumidarostra Grismer, 1994	Mancha-lateral Narigona	Swollen-nosed Side-Blotched Lizard	E	0	0	0	0	0	0	0	0	0

Scincidae

	SPANISH NAME	ENGLISH NAME	BC	SO	CH	CO	NL	TA	TX	NM	AZ	CA
Chalcides sp.	Lincer Ocelado	Ocellated Skink	0	0	0	0	0	0	0	0	IN	0
Plestiodon anthracinus (Baird, 1850)	Lincer del Carbón	Coal Skink	0	0	0	0	0	0	1	0	0	0
Plestiodon bilineatus (Tanner, 1958)	Alicante de Dos Líneas	Two-lined Short-nosed Skink	0	0	1	0	0	0	0	0	0	0
✓ *Plestiodon callicephalus* (Bocourt, 1879)	Lincer de Barranco	Mountain Skink	0	1	1	0	0	0	0	1	1	0
Plestiodon dicei (Ruthven & Gaige, 1933)	Alicante de Cola Azul	Dice's Short-nosed Skink	0	0	0	1	1	1	0	0	0	0
Plestiodon fasciatus (Linnaeus, 1758)	Lincer de Cinco Líneas	Five-lined Skink	0	0	0	0	0	0	1	0	0	0
✓ *Plestiodon gilberti* (Van Denburgh, 1896)	Lincer de Gilbert	Gilbert's Skink	1	0	0	0	0	0	0	0	1	1
Plestiodon laticeps (Schneider, 1801)	Lincer de Cabeza-ancha	Broad-headed Skink	0	0	0	0	0	0	1	0	0	0
Plestiodon lynxe (Wiegmann, 1834)	Lincer de los Encinos	Oak Forest Skink	0	0	0	0	0	1	0	0	0	0
Plestiodon multilineatus (Tanner, 1957)	Eslaboncillo	Chihuahuan Skink	0	0	E	0	0	0	0	0	0	0
✓ *Plestiodon multivirgatus* Hallowell, 1856	Lincer Variable	Many-lined Skink	0	0	1	0	0	0	1	1	1	0
✓ *Plestiodon obsoletus* Baird & Girard, 1852	Lincer de Llanura	Great Plains Skink	0	1	1	1	1	1	1	1	1	0
Plestiodon parviauriculatus (Taylor, 1933)	Lincer Pigmeo Norteño	Northern Pygmy Skink	0	1	1	0	0	0	0	0	0	0

	SPANISH NAME	ENGLISH NAME	BC	SO	CH	CO	NL	TA	TX	NM	AZ	CA
Plestiodon septentrionalis Baird, 1859	Lincer Norteño	Prairie Skink	0	0	0	0	0	0	1	0	0	0
Plestiodon skiltonianus Baird & Girard, 1852	Lincer Occidental	Western Skink	1	0	0	0	0	0	0	0	1	1
Plestiodon tetragrammus Baird, 1859 (1858)	Lincer de Cuatro Líneas	Four-lined Skink	0	0	1	1	1	1	1	0	0	0
Scincella kikaapoa García-Vázquez, Canseco-Márquez, & Nieto Montes de Oca, 2010	Escíncela de Cuatrociénegas	Cuatrociénegas Little Skink	0	0	0	E	0	0	0	0	0	0
Scincella lateralis (Say, *in* James, 1823)	Escíncela de Tierra	Little Brown Skink	0	0	0	1	0	0	1	0	0	0
Scincella silvicola (Taylor, 1937)	Correlón	Taylor´s Ground Skink	0	0	0	1	1	1	0	0	0	0

Teiidae

	SPANISH NAME	ENGLISH NAME	BC	SO	CH	CO	NL	TA	TX	NM	AZ	CA
Ameiva undulata (Wiegmann, 1834)	Lagartija Metálica	Rainbow Ameiva	0	0	0	0	0	1	0	0	0	0
Aspidoscelis arizonae (Van Denburgh, 1896)	Huico de Arizona	Arizona Striped Whiptail	0	0	0	0	0	0	0	0	E	0
Aspidoscelis baccata (Van Denburgh & Slevin, 1921)	Huico de San Pedro Nolasco	San Pedro Nolasco Whiptail	0	E	0	0	0	0	0	0	0	0
Aspidoscelis burti (Taylor, 1938)	Huico Manchado del Cañón	Canyon Spotted Whiptail	0	1	0	0	0	0	0	0	0	0
Aspidosceles cana (Van Denburgh & Slevin 1921)	Huico de la Isla Salsipuedes	Isla Salsipuedes Whiptail	E	0	0	0	0	0	0	0	0	0
Aspidoscelis costata (Cope, 1878)	Huico Llanero	Western Mexico Whiptail	0	1	1	0	0	0	0	0	0	0
Aspidoscelis dixoni (Scudday, 1973)	Huico Gris	Gray-checkered Whiptail	0	0	0	0	0	0	1	1	0	0
Aspidoscelis estebanensis (Dickerson, 1919)	Huico de San Esteban	San Esteban Whiptail	0	E	0	0	0	0	0	0	0	0
Aspidoscelis exsanguis (Lowe, 1956)	Corredora de Chihuahua	Chihuahuan Spotted Whiptail	0	1	1	0	0	0	1	1	1	0
Aspidoscelis flagellicauda (Lowe & Wright, 1964)	Huico Manchado del Gila	Gila Spotted Whiptail	0	0	0	0	0	0	0	1	1	0
Aspidoscelis gularis (Baird & Girard, 1852)	Corredora Pinta Texana	Texas Spotted Whiptail	0	0	1	1	1	1	1	1	0	0
Aspidoscelis gypsi (Wright & Lowe, 1993)	Huico Blanco	Little White Whiptail	0	0	0	0	0	0	0	E	0	0
Aspidoscelis hyperythra (Cope, 1864)	Huico Garganta-naranja	Orange-throated Whiptail	1	0	0	0	0	0	0	0	0	1
Aspidoscelis inornata (Baird, 1859)	Huico Liso	Little Striped Whiptail	0	0	1	1	1	1	1	1	0	0
Aspidoscelis labialis (Stejneger, 1890)	Huico de Baja California	Baja California Whiptail	1	0	0	0	0	0	0	0	0	0

	SPANISH NAME	ENGLISH NAME	BC	SO	CH	CO	NL	TA	TX	NM	AZ	CA
√ *Aspidoscelis laredoensis* (McKinney, Kay, & Anderson, 1973)	Llanera de Laredo	Laredo Striped Whiptail	0	0	0	0	0	1	1	0	0	0
√ *Aspidoscelis marmorata* (Baird & Girard, 1852)	Huico Marmóreo	Western Marbled Whiptail	0	0	1	1	1	0	1	1	0	0
Aspidoscelis martyris (Stejneger, 1892)	Huico de San Pedro Mártir	San Pedro Mártir Whiptail	0	E	0	0	0	0	0	0	0	0
Aspidoscelis neomexicana (Lowe & Zweifel, 1952)	Huico de Nuevo México	New Mexico Whiptail	0	0	0	0	0	0	1	1	IN	0
Aspidoscelis opatae (Wright, 1967)	Huico Opata	Opata Whiptail	0	E	0	0	0	0	0	0	0	0
Aspidoscelis pai (Wright & Lowe, 1993)	Huico Rayado de Pai	Pai Striped Whiptail	0	0	0	0	0	0	0	0	E	0
√ *Aspidoscelis sexlineata* (Linnaeus, 1766)	Huico de Seis Líneas	Six-lined Racerunner	0	0	0	0	0	1	1	1	0	0
√ *Aspidoscelis sonorae* (Lowe & Wright, 1964)	Huico Manchado de Sonora	Sonoran Spotted Whiptail	0	1	1	0	0	0	0	1	1	0
√ *Aspidoscelis stictogramma* (Burger, 1950)	Huico Manchado Gigante	Giant Spotted Whiptail	0	1	0	0	0	0	0	1	1	0
√ *Aspidoscelis tesselata* (Say, *in* James, 1823)	Huico Teselado	Common Checkered Whiptail	0	0	1	1	0	0	1	1	0	0
√ *Aspidoscelis tigris* (Baird & Girard, 1852)	Huico Tigre	Tiger Whiptail	1	1	0	0	0	0	0	1	1	1
√ *Aspidoscelis uniparens* (Wright & Lowe, 1965)	Huico de la Pradera del Desierto	Desert Grassland Whiptail	0	1	1	0	0	0	1	1	1	0
√ *Aspidoscelis velox* (Springer, 1928)	Huico Rayado de la Meseta	Plateau Striped Whiptail	0	0	0	0	0	0	0	1	1	0
√ *Aspidoscelis xanthonota* (Duellman & Lowe, 1953)	Huico de Dorso-rojo	Red-backed Whiptail	0	1	0	0	0	0	0	0	1	0

Xantusidae

	SPANISH NAME	ENGLISH NAME	BC	SO	CH	CO	NL	TA	TX	NM	AZ	CA
Lepidophyma micropholis Walker, 1955	Lagartija Nocturna de Cueva	Cave Tropical Night Lizard	0	0	0	0	0	1	0	0	0	0
Lepidophyma sylvaticum Taylor, 1939	Lagartija Nocturna de Montaña	Madrean Tropical Night Lizard	0	0	0	0	1	1	0	0	0	0
Xantusia arizonae Klauber, 1931	Nocturna de Arizona	Arizona Night Lizard	0	0	0	0	0	0	0	0	E	0
Xantusia bezyi Papenfuss, Macey, & Schulte, 2001	Nocturna de Bezy	Bezy's Night Lizard	0	0	0	0	0	0	0	0	E	0
Xantusia gracilis Grismer & Galvan, 1986	Nocturna de Arenisca	Sandstone Night Lizard	0	0	0	0	0	0	0	0	0	E
√ *Xantusia henshawi* Stejneger, 1893	Nocturna de Granito	Granite Night Lizard	1	0	0	0	0	0	0	0	0	1
Xantusia jaycolei Bezy, Bezy, & Bolles, 2009	Lagartija Nocturna de Cole	Cole's Night Lizard	0	E	0	0	0	0	0	0	0	0
Xantusia riversiana Cope, 1883	Nocturna Insular	Island Night Lizard	0	0	0	0	0	0	0	0	0	E

	SPANISH NAME	ENGLISH NAME	BC	SO	CH	CO	NL	TA	TX	NM	AZ	CA
Xantusia sierrae Bezy, 1967	Nocturna de la Sierra	Sierra Night Lizard	0	0	0	0	0	0	0	0	0	E
Xantusia vigilis Baird, 1859	Nocturna del Desierto	Desert Night Lizard	0	1	0	0	0	0	0	0	1	1
Xantusia wigginsi Savage, 1952	Nocturna de Wiggins	Wiggins' Night Lizard	1	0	0	0	0	0	0	0	0	1

Xenosauridae

Xenosaurus platyceps King & Thompson, 1968	Xenosauro de Cabeza-plana	Flathead Knob-scaled Lizard	0	0	0	0	0	E	0	0	0	0

Order Squamata

Suborder Amphisbaenia

Bipedidae

Bipes biporus (Cope, 1894)	Dos Manos de Cinco Dedos	Five-toed Worm Lizard	1	0	0	0	0	0	0	0	0	0

Order Squamata

Suborder Serpentes

Boidae

Boa constrictor Linnaeus, 1758	Mazacoatl	Boa constrictor	0	1	1	0	0	1	0	0	0	0
Charina bottae (Blainville, 1835)	Boa de Hule del Norte	Northern Rubber Boa	0	0	0	0	0	0	0	0	0	1
Charina umbratica Klauber, 1943	Boa de Hule del Sur de California	Southern Rubber Boa	0	0	0	0	0	0	0	0	0	E
Lichanura orcutti Stejneger, 1889	Boa del Desierto	Desert Rosy Boa	0	0	0	0	0	0	0	0	0	1
Lichanura trivirgata Cope, 1861	Boa del Desierto Mexicana	Mexican Rosy Boa	1	1	0	0	0	0	0	0	1	1

Colubridae

Adelphicos newmanorum Taylor, 1950	Zacatera Roja	Newman's Earth Snake	0	0	0	0	1	1	0	0	0	0
Amastridium sapperi (Werner, 1903)	Zacatera Negra	Rusty-headed Snake	0	0	0	0	1	1	0	0	0	0
Arizona elegans Kennicott, 1859	Brillante Arenícola	Glossy Snake	1	1	1	1	1	1	1	1	1	1
Arizona pacata Klauber, 1946	Brillante Peninsular	Baja California Glossy Snake	1	0	0	0	0	0	0	0	0	0
Bogertophis rosaliae (Mocquard, 1899)	Ratonera de Baja California	Baja California Ratsnake	1	0	0	0	0	0	0	0	0	1
Bogertophis subocularis (Brown, 1901)	Ratonera de Trans-Pecos	Trans-Pecos Ratsnake	0	0	1	1	1	0	1	1	0	0
Carphophis vermis (Kennicott, 1859)	Culebra-lombriz de Occidente	Western Wormsnake	0	0	0	0	0	0	1	0	0	0
Cemophora coccinea (Blumenbach, 1788)	Serpiente Escarlata	Scarletsnake	0	0	0	0	0	0	1	0	0	0

	SPANISH NAME	ENGLISH NAME	BC	SO	CH	CO	NL	TA	TX	NM	AZ	CA
Chilomeniscus stramineus Cope, 1860	Arenera de Modelo Variable	Variable Sandsnake	1	1	0	0	0	0	0	0	1	0
Chionactis occipitalis (Hallowell, 1854)	Rostro de Pala Occidental	Western Shovel-nosed Snake	1	1	0	0	0	0	0	0	1	1
Chionactis palarostris (Klauber, 1937)	Rostro de Pala de Sonora	Sonoran Shovel-nosed Snake	0	1	0	0	0	0	0	0	1	0
Coluber constrictor Linnaeus, 1758	Corredora	North American Racer	0	0	0	1	1	1	1	1	1	1
Coniophanes fissidens (Günther 1858)	Panza Amarilla	Yellow-bellied Snake	0	0	0	0	0	1	0	0	0	0
Coniophanes imperialis (Baird & Girard, 1859)	Culebra de Raya Negra	Regal Black-striped Snake	0	0	0	0	0	1	1	0	0	0
Coniophanes piceivittis Cope, 1870 (1869)	Culebra Rayada	Cope's Black-striped Snake	0	0	0	0	0	1	0	0	0	0
Conopsis nasus Günther, 1858	Culebra de Nariz Grande	Large-nosed Earthsnake	0	0	1	0	0	0	0	0	0	0
Contia longicaudae Feldman & Hoyer, 2010	Culebra del Bosque de Cola-afilada	Forest Sharp-tailed Snake	0	0	0	0	0	0	0	0	0	1
Contia tenuis (Baird & Girard, 1852)	Culebra de Cola-afilada	Sharp-tailed Snake	0	0	0	0	0	0	0	0	0	1
Diadophis punctatus (Linnaeus, 1766)	Culebra de Collar	Ring-necked Snake	1	1	1	1	1	0	1	1	1	1
Drymarchon melanurus (Duméril, Bibron, & Duméril, 1854)	Palancacóatls	Central American Indigo Snake	0	1	1	1	1	1	1	0	0	0
Drymobius margaritiferus (Schlegel, 1837)	Petatillo	Speckled Racer	0	1	1	0	1	1	1	0	0	0
Farancia abacura (Holbrook, 1836)	Serpiente de Fango de Vientre-rojo	Red-bellied Mudsnake	0	0	0	0	0	0	1	0	0	0
Ficimia hardyi Mendoza-Quijano & Smith, 1993	Nariz de Gancho de Hardy	Hardy's Hook-nosed Snake	0	0	0	0	0	1	0	0	0	0
Ficimia olivacea Gray, 1849	Nariz de Gancho Huasteca	Brown Hook-nosed Snake	0	0	0	0	0	1	0	0	0	0
Ficimia streckeri Taylor, 1931	Naris de Gancho Tamaulipeca	Tamaulipan Hook-nosed Snake	0	0	0	0	1	1	1	0	0	0
Geophis dugesii Bocourt, 1883	Minadora de Dugès	Dugès' Earthsnake	0	1	1	0	0	0	0	0	0	0
Geophis latifrontalis Garman, 1884 (1883)	Minadora de San Luis Potosí	Potosí Earth Snake	0	0	0	0	0	1	0	0	0	0
Gyalopion canum Cope, 1861	Naricilla Chihuahuense	Chihuahuan Hook-nosed Snake	0	1	1	1	1	0	1	1	1	0
Gyalopion quadrangulare (Günther, 1893)	Naricilla del Desierto	Thornscrub Hook-nosed Snake	0	1	1	0	0	0	0	0	1	0
Heterodon kennerlyi Kennicott, 1860	Trompa de Cerdo Mexicana	Mexican Hog-nosed Snake	0	1	1	1	1	1	1	1	1	0
Heterodon nasicus Kennicott, 1860	Trompa de Cerdo Occidental	Western Hog-nosed Snake	0	0	0	0	0	0	1	1	0	0

	SPANISH NAME	ENGLISH NAME	BC	SO	CH	CO	NL	TA	TX	NM	AZ	CA
Heterodon platirhinos Latreille, 1801	Trompa de Cerdo Oriental	Eastern Hog-nosed Snake	0	0	0	0	0	0	1	1	0	0
Hypsiglena chlorophaea Cope, 1860	Nocturna Verde-oscuro	Desert Nightsnake	1	1	1	0	0	0	0	1	1	1
Hypsiglena jani (Dugès, 1865)	Nocturna de Chihuahua	Chihuahuan Nightsnake	0	0	1	1	1	1	1	1	1	0
Hypsiglena ochrorhyncha Cope, 1860	Nocturna Moteada	Coast Nightsnake	1	0	0	0	0	0	0	0	0	1
Hypsiglena slevini Tanner, 1943	Nocturna de Baja California	Slevin's Nightsnake	1	0	0	0	0	0	0	0	0	0
Imantodes cenchoa (Linnaeus, 1758)	Cordelilla Manchada	Blunthead Tree Snake	0	0	0	0	0	1	0	0	0	0
Imantodes gemmistratus (Cope, 1861)	Cordelilla Escamuda	Central American Tree Snake	0	1	1	0	0	0	0	0	0	0
Lampropeltis alterna (Brown, 1901)	Culebra Real Gris	Gray-banded Kingsnake	0	0	0	1	1	0	1	1	0	0
Lampropeltis calligaster (Harlan, 1827)	Culebra Real de Pradera	Prairie Kingsnake	0	0	0	0	0	0	1	0	0	0
Lampropeltis getula (Linnaeus, 1766)	Barila	Common Kingsnake	1	1	1	1	1	1	1	1	1	1
Lampropeltis herrerae Van Denburgh & Slevin, 1923	Culebra Real de Todos Santos	Islas Todos Santos Mountain Kingsnake	E	0	0	0	0	0	0	0	0	0
Lampropeltis knoblochi Taylor, 1940	Culebra Real de Chihuahua	Chihuahuan Mountain Kingsnake	0	1	1	0	0	0	0	1	1	0
Lampropeltis mexicana (Garman, 1884 [1883])	Culebra Real Roja	San Luis Potosí Kingsnake	0	0	0	1	1	1	0	0	0	0
Lampropeltis multifasciata (Bocourt, 1886)	Culebra Real de las Montañas Costeras	Coast Mountain Kingsnake	1	0	0	0	0	0	0	0	0	1
Lampropeltis pyromelana (Cope, 1867)	Culebra Real de Arizona	Arizona Mountain Kingsnake	0	0	0	0	0	0	0	1	1	0
Lampropeltis triangulum (Lacépède, 1788)	Falsa Coralillo	Milksnake	0	1	1	1	1	1	1	1	1	0
Lampropeltis zonata (Lockington, *in* Blainville, 1835)	Culebra Real de California	California Mountain Kingsnake	0	0	0	0	0	0	0	0	0	1
Leptodeira maculata (Hallowell, 1861 [1860])	Escombrera del Suroeste Mexicano	Southwestern Cat-eyed Snake	0	0	0	0	0	1	0	0	0	0
Leptodeira punctata (Peters, 1867 [1866])	Escombrera del Occidente	Western Cat-eyed Snake	0	1	0	0	0	0	0	0	0	0
Leptodeira septentrionalis (Kennicott, *in* Baird, 1859)	Escombrera Manchada	Northern Cat-eyed Snake	0	0	0	1	1	1	1	0	0	0
Leptodeira splendida Günther, 1895, *in* Salvini & Goldman, 1885–1902)	Escombrera Sapera	Splendid Cat-eyed Snake	0	1	1	0	0	0	0	0	0	0
Leptophis diplotropis (Günther, 1872)	Ratonera de la Costa del Pacífico	Pacific Coast Parrot Snake	0	1	1	0	0	0	0	0	0	0
Leptophis mexicanus Duméril, Bibron, & Duméril, 1854	Ranera Mexicana	Mexican Parrot Snake	0	0	0	0	1	1	0	0	0	0

	SPANISH NAME	ENGLISH NAME	BC	SO	CH	CO	NL	TA	TX	NM	AZ	CA
Masticophis bilineatus Jan, 1863	Látigo de Sonora	Sonoran Whipsnake	0	1	1	0	0	0	0	1	1	0
Masticophis flagellum Shaw, 1802	Chirrionera	Coachwhip	1	1	1	1	1	1	1	1	1	1
Masticophis fuliginosus Cope, 1895	Chirrionera de Baja California	Baja California Coachwhip	1	0	0	0	0	0	0	0	0	1
Masticophis lateralis Hallowell, 1853	Corredora Rayada	Striped Racer	1	0	0	0	0	0	0	0	0	1
Masticophis mentovarius (Duméril, Bibron, & Duméril, 1854)	Sabanera	Neotropical Whipsnake	0	1	1	0	0	1	0	0	0	0
Masticophis schotti Baird & Girard, 1853	Látigo de Schott	Schott's Whipsnake	0	0	0	1	1	1	1	0	0	0
Masticophis slevini Lowe & Norris, 1955	Látigo de San Esteban	Isla San Esteban Whipsnake	0	E	0	0	0	0	0	0	0	0
Masticophis taeniatus (Hallowell, 1852)	Látigo Rayada	Striped Whipsnake	0	0	1	1	1	0	1	1	1	1
Mastigodryas cliftoni (Hardy, 1964)	Lagartijera de Clifton	Clifton's Lizard Eater	0	1	1	0	0	0	0	0	0	0
Mastigodryas melanolomus (Cope 1868)	Lagartijera Común	Common Lizard Eater	0	0	0	0	0	1	0	0	0	0
Nerodia clarki (Baird & Girard, 1853)	Culebra de Agua de Clark	Saltmarsh Watersnake	0	0	0	0	0	0	1	0	0	0
Nerodia cycopion (Duméril, Bibron, & Duméril, 1854)	Culebra de Agua del Mississippi	Mississippi Green Watersnake	0	0	0	0	0	0	1	0	0	0
Nerodia erythrogaster (Forster, 1771)	Culebra de Agua de Vientre Plano	Plain-bellied Watersnake	0	0	1	1	1	1	1	1	0	0
Nerodia fasciata (Linnaeus, 1766)	Culebra de Agua del Sur de Estados Unidos	Southern Watersnake	0	0	0	0	0	0	1	0	0	IN
Nerodia harteri (Trapido, 1941)	Culebra de Agua de Harter	Harter's Watersnake	0	0	0	0	0	0	E	0	0	0
Nerodia rhombifer (Hallowell, 1852)	Culebra de Agua Diamantada	Diamond-backed Watersnake	0	0	0	1	1	1	1	0	0	IN
Nerodia sipedon (Linnaeus, 1758)	Culebra de Agua del Norte de Estados Unidos	Northern Watersnake	0	0	0	0	0	0	1	0	0	IN
Opheodrys aestivus (Linnaeus, 1766)	Estival Rugosa	Rough Greensnake	0	0	0	1	1	1	1	0	0	0
Opheodrys vernalis (Harlan, 1827)	Culebra Verde Lisa	Smooth Greensnake	0	0	1	0	0	0	1	1	0	0
Oxybelis aeneus (Wagler, 1824)	Bejuquilla Neotropical	Neotropical Vinesnake	0	1	1	0	1	1	0	0	1	0
Pantherophis bairdi (Yarrow, *in* Cope, 1880)	Ratonera de Bosque	Baird's Ratsnake	0	0	0	1	1	1	1	0	0	0
Pantherophis emoryi (Baird & Girard, 1853)	Ratonera de Emory	Great Plains Ratsnake	0	0	1	1	1	1	1	1	0	0

	SPANISH NAME	ENGLISH NAME	BC	SO	CH	CO	NL	TA	TX	NM	AZ	CA
Pantherophis guttatus (Linnaeus, 1766)	Ratonera Roja	Red Cornsnake	0	0	0	0	0	0	1	0	0	0
Pantherophis obsoletus (Say, 1823)	Ratonera Texana	Texas Ratsnake	0	0	0	0	0	0	1	0	0	0
Phyllorhynchus browni Stejneger, 1890	Culebra Ensillada	Saddled Leaf-nosed Snake	0	1	0	0	0	0	0	0	1	0
Phyllorhynchus decurtatus (Cope, 1868)	Culebra Nariz Moteada	Spotted Leaf-nosed Snake	1	1	0	0	0	0	0	0	1	1
Pituophis catenifer (Blainville, 1835)	Cincuate Casero	Gopher Snake	1	1	1	1	1	1	1	1	1	1
Pituophis deppei (Duméril, 1853)	Cincuate Mexicano	Mexican Bullsnake	0	1	1	1	1	1	0	0	0	0
Pituophis insulanus Klauber, 1946	Cincuate de Isla Cedros	Isla Cedros Gophersnake	E	0	0	0	0	0	0	0	0	0
Pituophis ruthveni Stull, 1929	Cincuate de Louisiana	Louisiana Pinesnake	0	0	0	0	0	0	1	0	0	0
Pituophis vertebralis (Blainville, 1835)	Cincuate de San Lucas	Baja California Gophersnake	1	0	0	0	0	0	0	0	0	0
Pliocercus bicolor Smith, 1941	Falsa Coral del Norte	Northern False Coral Snake	0	0	0	0	0	1	0	0	0	0
Pseudelaphe flavirufa (Cope, 1867 [1866])	Ratonera del Trópico	Tropical Ratsnake	0	0	0	0	0	1	0	0	0	0
Pseudoficimia frontalis (Cope, 1864)	Ilamacoa	False Ficimia	0	1	0	0	0	0	0	0	0	0
Regina grahami Baird & Girard, 1853	Regina de Graham	Graham's Crayfish Snake	0	0	0	0	0	0	1	0	0	0
Regina rigida (Say, 1825)	Regina Brillante	Glossy Crayfish Snake	0	0	0	0	0	0	1	0	0	0
Rhadinaea gaigeae Bailey, 1937	Hojarasquera de Gaige	Gaige's Pine Forest Snake	0	0	0	0	0	1	0	0	0	0
Rhadinaea hesperia Bailey, 1940	Culebra Rayada Occidental	Western Graceful Brown Snake	0	0	1	0	0	0	0	0	0	0
Rhadinaea laureata (Günther, 1868)	Hojarasquera Coronada	Crowned Graceful Brown Snake	0	0	1	0	0	0	0	0	0	0
Rhadinaea montana Smith, 1944	Hojarasquera de Nuevo León	Nuevo León Graceful Brown Snake	0	0	0	0	E	0	0	0	0	0
Rhinocheilus lecontei Baird & Girard, 1853	Culebra Nariz-larga	Long-nosed Snake	1	1	1	1	1	1	1	1	1	1
Salvadora bairdii Jan & Sordelli, 1860	Culebra Chata de Baird	Baird's Patch-nosed Snake	0	1	1	0	0	0	0	0	0	0
Salvadora deserticola Schmidt, 1940	Cabestrillo del Big-Bend	Big-Bend Patch-nosed Snake	0	1	1	0	0	0	1	1	1	0
Salvadora grahamiae Baird & Girard, 1853	Culebra Chata de Montaña	Eastern Patch-nosed Snake	0	1	1	1	1	1	1	1	1	0
Salvadora hexalepis (Cope, 1867)	Cabestrillo	Western Patch-nosed Snake	1	1	0	0	0	0	0	0	1	1

	SPANISH NAME	ENGLISH NAME	BC	SO	CH	CO	NL	TA	TX	NM	AZ	CA
Scaphiodontophis annulatus (Duméril, Bibron, & Duméril, 1854)	Culebra Añadida de Guatemala	Guatemala Neck-banded Snake	0	0	0	0	0	1	0	0	0	0
✓ *Senticolis triaspis* (Cope, 1866)	Culebra Ratonera Verde	Green Ratsnake	0	1	1	0	1	1	0	1	1	0
Sonora aemula (Cope, 1879)	Culebra de Tierra Cola Plana	Filed-tailed Groundsnake	0	1	1	0	0	0	0	0	0	0
✓ *Sonora semiannulata* Baird & Girard, 1853	Culebrilla de Tierra	Western Groundsnake	1	1	1	1	1	1	1	1	1	1
Spilotes pullatus (Linnaeus, 1758)	Voladora Mico	Tropical Tree Snake	0	0	0	0	0	1	0	0	0	0
✓ *Storeria dekayi* (Holbrook, 1836)	Culebra Parda	Dekay's Brownsnake	0	0	0	0	1	1	1	0	0	0
Storeria hidalgoensis Taylor, 1942	Culebra de Cuello Blanco	Mexican Yellow-bellied Brownsnake	0	0	0	1	1	1	0	0	0	0
Storeria occipitomaculata (Storer, 1839)	Serpiente de Vientre-Rojo	Red-bellied Snake	0	0	0	0	0	0	1	0	0	0
Storeria storerioides (Cope, 1865)	Culebra Parda Mexicana	Mexican Brownsnake	0	1	1	0	0	0	0	0	0	0
Sympholis lippiens Cope, 1862	Culebra Cola Corta Mexicana	Mexican Short-tailed Snake	0	1	1	0	0	0	0	0	0	0
✓ *Tantilla atriceps* (Günther, 1895, *in* Salvin & Godman, 1885–1902)	Culebrilla de Cabeza Negra	Mexican Black-headed Snake	0	0	0	1	1	1	1	0	0	0
Tantilla cucullata Minton, 1956	Culebra de Cabeza-negra de Trans-Pecos	Trans-Pecos Black-headed Snake	0	0	0	0	0	0	E	0	0	0
✓ *Tantilla gracilis* Baird & Girard, 1853	Culebra Cabeza-plana	Flat-headed Snake	0	0	0	1	0	0	1	0	0	0
✓ *Tantilla hobartsmithi* Taylor, 1936	Culebra Cabeza Negra del Norte	Smith's Black-headed Snake	0	1	1	1	1	0	1	1	1	1
✓ *Tantilla nigriceps* Kennicott, 1860	Culebra Cabeza Negra de los Llanos	Plains Black-headed Snake	0	0	1	1	1	1	1	1	1	0
✓ *Tantilla planiceps* (Blainville, 1835)	Culebra Cabeza Negra Occidental	Western Black-headed Snake	1	0	0	0	0	0	0	0	0	1
Tantilla rubra Cope, 1876 (1875)	Rojilla	Red Black-headed Snake	0	0	0	0	1	1	0	0	0	0
✓ *Tantilla wilcoxi* Stejneger, 1902	Centipedívora de Chihuahua	Chihuahuan Black-headed Snake	0	1	1	1	1	1	0	0	1	0
✓ *Tantilla yaquia* Smith, 1942	Culebra Cabeza Negra Yaqui	Yaqui Black-headed Snake	0	1	1	0	0	0	0	1	1	0
Thamnophis angustirostris (Kennicott, 1860)	Jarretera de Hocico-largo	Longnose Gartersnake	0	0	0	E	0	0	0	0	0	0
Thamnophis atratus Kennicott, 1860	Jarretera Acuática	Aquatic Gartersnake	0	0	0	0	0	0	0	0	0	1
Thamnophis couchii Kennicott, 1859	Jarretera de la Sierra	Sierra Gartersnake	0	0	0	0	0	0	0	0	0	1

	SPANISH NAME	ENGLISH NAME	BC	SO	CH	CO	NL	TA	TX	NM	AZ	CA
Thamnophis cyrtopsis (Kennicott, 1860)	Jarretera Cuello-negro	Black-necked Gartersnake	0	1	1	1	1	1	1	1	1	0
Thamnophis elegans (Baird & Girard, 1853)	Jarretera Terrestre Occidental	Terrestrial Gartersnake	1	0	1	0	0	0	0	1	1	1
Thamnophis eques (Reuss, 1834)	Jarretera Mexicana	Mexican Gartersnake	0	1	1	0	1	0	0	1	1	0
Thamnophis errans Smith, 1942	Jarretera Errante Mexicana	Mexican Wandering Gartersnake	0	0	1	0	0	0	0	0	0	0
Thamnophis exsul Rossman, 1969	Jarretera Mexicana Exiliada	Exile Mexican Gartersnake	0	0	0	1	1	1	0	0	0	0
Thamnophis gigas Fitch, 1940	Jarretera Gigante	Giant Gartersnake	0	0	0	0	0	0	0	0	0	E
Thamnophis hammondii (Kennicott, 1860)	Jarretera de Hammond	Two-striped Gartersnake	1	0	0	0	0	0	0	0	0	1
Thamnophis marcianus (Baird & Girard, 1853)	Sochuate	Checkered Gartersnake	1	1	1	1	1	1	1	1	1	1
Thamnophis melanogaster (Peters, 1864)	Jarretera Vientre-negro Mexicana	Mexican Black-bellied Gartersnake	0	1	1	0	0	0	0	0	0	0
Thamnophis mendax Walker, 1955	Jarretera Tamaulipeca de Montaña	Tamaulipan Montane Gartersnake	0	0	0	0	0	E	0	0	0	0
Thamnophis ordinoides (Baird & Girard, 1852)	Jarretera del Noroeste de Estados Unidos	Northwestern Gartersnake	0	0	0	0	0	0	0	0	0	1
Thamnophis proximus (Say, *in* James, 1823)	Jarretera Occidental	Western Ribbonsnake	0	0	0	1	1	1	1	1	0	0
Thamnophis pulchrilatus (Cope, 1885 [1884])	Jarretera Mexicana del Altiplano	Mexican Highland Gartersnake	0	0	0	0	1	1	0	0	0	0
Thamnophis radix (Baird & Girard, 1853)	Jarretera de las Planicies	Plains Gartersnake	0	0	0	0	0	0	1	1	0	0
Thamnophis rufipunctatus (Cope, 1875)	Jarretera Cabeza-angosta del Mogollon	Mogollon Narrow-headed Gartersnake	0	0	0	0	0	0	0	1	1	0
Thamnophis sirtalis (Linnaeus, 1758)	Jarretera Común	Common Gartersnake	0	0	1	0	0	0	1	1	0	1
Thamnophis unilabialis Tanner, 1985	Jarretera Cabeza-angosta de Chihuahua	Madrean Narrow-headed Gartersnake	0	1	1	0	0	0	0	0	0	0
Thamnophis validus (Kennicott, 1860)	Jarretera Mexicana del Pacífico	Mexican West Coast Gartersnake	0	1	1	0	0	0	0	0	0	0
Trimorphodon lambda Cope, 1886	Ilimacoa de Sonora	Sonoran Lyresnake	0	1	0	0	0	0	0	1	1	1
Trimorphodon lyrophanes (Cope, 1861)	Ilamacoa de California	California Lyresnake	1	0	0	0	0	0	0	0	0	1
Trimorphodon tau Cope, 1870	Falsa Nauyaca Mexicana	Mexican Lyresnake	0	1	1	0	1	1	0	0	0	0
Trimorphodon vilkinsonii Cope, 1886	Serpiente de Tetalura	Texas Lyresnake	0	0	1	0	0	0	1	1	0	0

	SPANISH NAME	ENGLISH NAME	BC	SO	CH	CO	NL	TA	TX	NM	AZ	CA
Tropidoclonion lineatum (Hallowell, 1856)	Serpiente Lineada	Lined Snake	0	0	0	0	0	0	1	1	0	0
Tropidodipsas fasciata Günther, 1858	Caracolera Anillada	Banded Snail Sucker	0	0	0	0	0	1	0	0	0	0
Tropidodipsas repleta Smith, Lemos-Espinal, Hartman, & Chiszar, 2005	Caracolera Repleta	Banded Blacksnake	0	1	1	0	0	0	0	0	0	0
Tropidodipsas sartorii Cope, 1863	Caracolera Terrestre	Terrestrial Snail Sucker	0	0	0	0	1	1	0	0	0	0
Virginia striatula (Linnaeus, 1766)	Serpiente Rugosa de Tierra	Rough Earthsnake	0	0	0	0	0	0	1	0	0	0
Virginia valeriae Baird & Girard, 1853	Serpiente Lisa de Tierra	Smooth Earthsnake	0	0	0	0	0	0	1	0	0	0

Elapidae

	SPANISH NAME	ENGLISH NAME	BC	SO	CH	CO	NL	TA	TX	NM	AZ	CA
Hydrophis platurus (Linnaeus, 1766)	Culebra del Mar	Yellow-bellied Seasnake	M	M	0	0	0	0	0	0	0	M
Micruroides euryxanthus (Kennicott, 1860)	Coralillo Occidental	Sonoran Coralsnake	0	1	1	0	0	0	0	1	1	0
Micrurus distans (Kennicott, 1860)	Coralillo Bandas Claras	West Mexican Coralsnake	0	1	1	0	0	0	0	0	0	0
Micrurus tamaulipensis Lavin-Murcio & Dixon, 2004	Coralillo de la Sierra de Tamaulipas	Sierra de Tamaulipas Coralsnake	0	0	0	0	0	E	0	0	0	0
Micrurus tener (Baird & Girard, 1953)	Coralillo Texano	Texas Coralsnake	0	0	0	1	1	1	1	0	0	0

Leptotyphlopidae

	SPANISH NAME	ENGLISH NAME	BC	SO	CH	CO	NL	TA	TX	NM	AZ	CA
Epictia goudotii (Dumeril & Bibron, 1844)	Culebra Lombriz	Black Threadsnake	0	0	0	0	0	1	0	0	0	0
Rena dissecta (Cope, 1896)	Culebrilla Ciega de Nuevo México	New Mexico Threadsnake	0	0	1	1	0	0	1	1	1	0
Rena dulcis (Baird & Girard, 1853)	Culebrilla Ciega Texana	Texas Threadsnake	0	0	0	1	1	1	1	0	0	0
Rena humilis Baird & Girard, 1853	Culebra Lombriz	Western Threadsnake	1	1	1	0	0	0	0	0	1	1
Rena myopica (Garman, 1884 [1883])	Culebrilla Ciega de Tampico	Tampico Threadsnake	0	0	0	0	1	1	0	0	0	0
Rena segrega (Klauber, 1939)	Culebra Lombriz de Trans-Pecos	Chihuahuan Threadsnake	0	0	1	1	0	0	1	1	0	0

Typhlopidae

	SPANISH NAME	ENGLISH NAME	BC	SO	CH	CO	NL	TA	TX	NM	AZ	CA
Indotyphlops braminus (Daudin, 1803)	Culebrilla Ciega	Brahminy Blindsnake	0	IN	0	0	IN	IN	IN	0	IN	IN

Viperidae

	SPANISH NAME	ENGLISH NAME	BC	SO	CH	CO	NL	TA	TX	NM	AZ	CA
Agkistrodon bilineatus (Günther, 1863)	Cantil	Cantil	0	1	1	0	0	0	0	0	0	0

	SPANISH NAME	ENGLISH NAME	BC	SO	CH	CO	NL	TA	TX	NM	AZ	CA
Agkistrodon contortrix (Linnaeus, 1766)	Mocasín	Copperhead	0	0	1	1	0	0	1	0	0	0
Agkistrodon piscivorus (Lacépède, 1789)	Boca de Algodón	Cottonmouth	0	0	0	0	0	0	1	0	0	0
Agkistrodon taylori Burger & Robertson, 1951	Metapil	Taylor's Cantil	0	0	0	0	1	1	0	0	0	0
Bothrops asper (Garman, 1884 (1883)	Cuatro Narices	Terciopelo	0	0	0	0	0	1	0	0	0	0
Crotalus angelensis Klauber, 1963	Cascabel de Isla Ángel de la Guarda	Isla Ángel de la Guarda Rattlesnake	E	0	0	0	0	0	0	0	0	0
Crotalus atrox Baird & Girard, 1853	Cascabel de Diamantes	Western Diamond-backed Rattlesnake	1	1	1	1	1	1	1	1	1	1
Crotalus basiliscus (Cope, 1864)	Saye	Mexican West Coast Rattlesnake	0	1	1	0	0	0	0	0	0	0
Crotalus caliginis Klauber, 1949	Cascabel de Isla Coronado	Islas Coronado Rattlesnake	E	0	0	0	0	0	0	0	0	0
Crotalus cerastes Hallowell, 1854	Víbora Cornuda	Sidewinder	1	1	0	0	0	0	0	0	1	1
Crotalus cerberus (Coues, 1875)	Cascabel Negra de Arizona	Arizona Black Rattlesnake	0	0	0	0	0	0	0	1	1	0
Crotalus enyo (Cope, 1861)	Cascabel de Baja California	Baja California Rattlesnake	1	0	0	0	0	0	0	0	0	0
Crotalus estebanensis (Klauber, 1949)	Cascabel de San Esteban	Isla San Esteban Black-tailed Rattlesnake	0	E	0	0	0	0	0	0	0	0
Crotalus horridus (Linnaeus, 1758)	Cascabel de Bosque	Timber Rattlesnake	0	0	0	0	0	0	1	0	0	0
Crotalus lepidus (Kennicott, 1861)	Cascabel Verde	Rock Rattlesnake	0	1	1	1	1	1	1	1	1	0
Crotalus lorenzoensis Radcliffe & Maslin, 1975	Cascabel de San Lorenzo	San Lorenzo Island Rattlesnake	E	0	0	0	0	0	0	0	0	0
Crotalus mitchellii (Cope, 1861)	Víbora Blanca	Speckled Rattlesnake	1	1	0	0	0	0	0	0	1	1
Crotalus molossus Baird & Girard, 1853	Cola Prieta	Black-tailed Rattlesnake	0	1	1	1	1	1	1	1	1	0
Crotalus muertensis Klauber, 1949	Cascabel de El Muerto	Isla El Muerto Rattlesnake	E	0	0	0	0	0	0	0	0	0
Crotalus oreganus Holbrook, 1840	Serpiente de Cascabel Occidental	Western Rattlesnake	1	0	0	0	0	0	0	0	1	1
Crotalus pricei Van Denburgh, 1895	Cascabel de Manchas-gemelas	Twin-spotted Rattlesnake	0	1	1	1	1	1	0	0	1	0
Crotalus ruber Cope, 1892	Cascabel Diamante Rojo	Red Diamond Rattlesnake	1	0	0	0	0	0	0	0	0	1
Crotalus scutulatus (Kennicott, 1861)	Chiauhcoatl	Mojave Rattlesnake	0	1	1	1	1	1	1	1	1	1
Crotalus stephensi Klauber, 1930	Cascabel de la Sierra Panamint	Panamint Rattlesnake	0	0	0	0	0	0	0	0	0	1

	SPANISH NAME	ENGLISH NAME	BC	SO	CH	CO	NL	TA	TX	NM	AZ	CA
Crotalus tigris Kennicott, *in* Baird, 1859	Cascabel Tigre	Tiger Rattlesnake	0	1	0	0	0	0	0	0	1	0
Crotalus totonacus Gloyd & Kauffeld, 1940	Tepocolcoatl	Totonacan Rattlesnake	0	0	0	0	1	1	0	0	0	0
Crotalus viridis (Rafinesque, 1818)	Serpiente de Cascabel de la Pradera	Prairie Rattlesnake	0	1	1	1	0	0	1	1	1	0
Crotalus willardi Meek, 1905	Cascabel de Nariz-afilada	Ridge-nosed Rattlesnake	0	1	1	0	0	0	0	1	1	0
Sistrurus catenatus (Rafinesque, 1818)	Cascabel de Massasagua	Massasagua Rattlesnake	0	0	0	1	1	1	1	1	1	0
Sistrurus miliarius (Linnaeus, 1766)	Cascabel Pigmea	Pygmy Rattlesnake	0	0	0	0	0	0	1	0	0	0
Total			119	195	174	133	132	179	229	136	153	194

References

Referencias Bibliográficas

1. Introduction/Introducción

Brennan, T. C., and A. T. Holycross. 2006. A field guide to amphibians and reptiles in Arizona. Phoenix: Arizona Game and Fish Department.

Degenhardt, W. G., C. W. Painter, and A. H. Price. 1996. Amphibians and reptiles of New Mexico. Albuquerque: University of New Mexico Press.

Dixon, J. R. 1987. Amphibians and reptiles of Texas, with keys, taxonomic synopses, bibliography, and distribution maps. College Station: Texas A&M University Press.

Grismer, L. L. 2002. Amphibians and reptiles of Baja California, including its Pacific Islands, and the islands in the Sea of Cortés. Berkeley: University of California Press.

Lemos-Espinal, J. A., and H. M. Smith. 2007a. Anfibios y reptiles del estado de Chihuahua, México/Amphibians and reptiles of the state of Chihuahua, Mexico. Mexico City: CONABIO.

———. 2007b. Anfibios y reptiles del estado de Coahuila, México/Amphibians and reptiles of the state of Coahuila, Mexico. Mexico City: CONABIO.

Stebbins, R. C., and S. M. McGinnis. 2012. Field guide to amphibians and reptiles of California. Berkeley: University of California Press.

2. Nomenclatural and Distributional Notes/Notas sobre nomenclatura y distribución

Adalsteinsson, S. A., W. R. Branch, S. Trape, L. J. Vitt, and S. B. Hedges. 2009. Molecular phylogeny, classification, and biogeography of snakes of the family Leptotyphlopidae (Reptilia, Squamata). Zootaxa 2244:1–50.

Baird, S. F., and C. Girard. 1853. Catalogue of North American reptiles in the Museum of the Smithsonian Institution. Part 1, Serpents. Washington, DC: Smithsonian Institution.

Barker, D. G. 1992. Variation, intraspecific relationships and biogeography of the Ridgenose Rattlesnake *Crotalus willardi*. In Biology of the pitvipers, ed. J. A. Campbell and E. D. Brodie Jr., 89–106. Tyler, TX: Selva.

Beltz, E. 2006. Scientific and common names of the reptiles and amphibians of North America—explained. http://ebeltz.net/herps/etymain.html.

Bocourt, M. F. 1881. Description d'un ophidian opotérodonte appartenant au genre *Catodon*. Bull. Soc. Philomathique Paris 4 (7): 81–82.

Burbrink, F. T., F. Fontanella, R. A. Pyron, T. J. Guther, and C. Jiménez. 2008. Phylogeography across a continent: The evolutionary and demographic history of the North American Racer (Serpentes: Colubridae: *Coluber constrictor*). Mol. Phylogenet. Evol. 47:272–288.

Burbrink, F. T., H. Yao, M. Ingrasci, R. W. Bryson Jr., T. J. Guiher, and S. Ruane. 2011. Speciation at the Mogollon Rim in the Arizona Mountain Kingsnake (*Lampropeltis pyromelana*). Mol. Phylogenet. Evol. 60:445–454.

Collins, J. T., and T. W. Taggart. 2008. A proposal to retain *Masticophis* as the generic name for the Coachwhip and Whipsnakes. J. Kans. Herpetol. 27:12.

———. 2009. Standard common and current

scientific names for North American amphibians, turtles, reptiles, and crocodilians. 6th ed. Lawrence, KS: Center for North American Herpetology.

Crother, B. I., ed. 2008. Scientific and standard English names of amphibians and reptiles of North America north of Mexico, with comments regarding confidence in our understanding. 6th ed. SSAR Herpetol. Circ. 37:1–84.

Daza, J. M., E. N. Smith, V. P. Pàez, and C. L. Parkinson. 2009. Complex evolution in the Neotropics: The origin and diversification of the widespread genus *Leptodeira* (Serpentes: Colubridae). Mol. Phylogenet. Evol. 53:653–667.

Devitt, T. J., T. J. LaDuc, and J. McGuire. 2008. The *Trimorphodon biscutatus* (Squamata: Colubridae) species complex revisited: A multivariate statistical analysis of geographic variation. Copeia 2008:370–387.

Dixon, J. R. 1964. The systematics and distribution of lizards of the genus *Phyllodactylus* in North and Central America. New Mex. State Univ. Res. Ctr. Sci. Bull. 64 (1):1–139.

———. 2000. Amphibians and reptiles of Texas, with keys, taxonomic synopses, bibliography, and distribution maps. 2nd ed. College Station: Texas A&M University Press.

Douglas M. E., M. R. Douglas, G. W. Schuett, D. D. Beck, and B. K. Sullivan. 2010. Conservation phylogenetics of helodermatid lizards using multiple molecular markers and a supertree approach. Mol. Phylogenet. Evol. 55:153–167.

Duellman, W. E. 1958. A monographic study of the colubrid snake genus *Leptodeira*. Bull. Amer. Mus. Nat. Hist. 114:1–152.

Farr, W. L., D. Lazcano, and P. A. Lavin-Murcio. 2013. New distributional records for amphibians and reptiles from the state of Tamaulipas, Mexico III. Herpetol. Rev. 44:631–645.

Feldman, C. R., and R. F. Hoyer. 2010. A new species of snake in the genus *Contia* (Squamata: Colubridae) from California and Oregon. Copeia 2010:254–267.

Feria-Ortiz, M., N. L. Manríquez-Morán, and A. Nieto-Montes De Oca. 2011. Species limits based on mtDNA and morphological data in the polytypic species *Plestiodon brevirostris* (Squamata: Scincidae). Herpetol. Monogr. 25:25–51.

Flores-Villela, O., and R. A. Brandon. 1992. *Siren lacertina* (Amphibia: Caudata) in northeastern Mexico and southern Texas. Ann. Carnegie Mus. 61:289–291.

García-Vázquez, Canseco-Márquez, and A. Nieto-Montes de Oca. 2010. A new species of *Scincella* (Squamata: Scincidae) from the Cuatro Ciénegas Basin, Coahuila, Mexico. Copeia 2010: 373–381.

Good, D. A. 1988. Phylogenetic relationships among gerrhonotine lizards: An analysis of external morphology. Univ. Calif. Publ. Zool. 121:1–139.

———. 1994. Species limits in the genus *Gerrhonotus* (Squamata: Anguidae). Herpetol. Monogr. 8:180–202.

Gray, J. E. 1845. Catalogue of the specimens of lizards in the collections of the British Museum. London: British Museum.

Grismer, L. L. 1994. The origin and evolution of the peninsular herpetofauna of Baja California, Mexico. Herpetol. Nat. Hist. 2:51–106.

———. 1999. An evolutionary classification of reptiles on islands in the Gulf of California. Herpetologica 55:446–469.

———. 2002. Amphibians and reptiles of Baja California including its Pacific Islands and the islands in the Sea of Cortés. Berkeley: University of California Press.

Grismer, L. L., and E. Mellink. 2005. Historical and ecological biogeography of the terrestrial herpetofauna of northern Baja California. In Biodiversity, ecosystems, and conservation in northern Mexico, ed. J. E. Cartron, G. Ceballos, and R. S. Felger, 167–178. New York: Oxford University Press.

Hansen, L., ed. 2008. Daniel Rolander's journal. Trans. J. Dobreff, C. Dahlman, D. Morgan, and J. Tipton. In The Linnaeus apostles—global science and adventure, vol. 3, book 3, Europe, North & South America, 1215–1576. London: IK Foundation. http://www.reptile-database.org.

Hedges, S. B., A. B. Marion, K. M. Lipp, J. Marin, and N. Vidal. 2014. A taxonomic framework for typhlopid snakes from the Caribbean and other regions (Reptilia, Squamata). Caribb. Herpetol. 49:1–61.

Jockusch, E. L., I. Martínez-Solano, R. W. Hansen, and D. B. Wake. 2012. Morphological and molecular diversification of slender salamanders (Caudata: Plethodontidae: *Batrachoseps*) in the southern Sierra Nevada of California with descriptions of two new species. Zootaxa 3190:1–30.

Keiser, E. D., Jr. 1982. *Oxybelis aeneus*. Cat. Am. Amphib. Rept. 305:1–4.

Klauber, L. M. 1940. The worm snakes of the genus *Leptotyphlops* in the United States and northern Mexico. Trans. San Diego Soc. Nat. Hist. 9:87–162.

———. 1946. The glossy snake, *Arizona*, with descriptions of new subspecies. Trans. San Diego Soc. Nat. Hist. 10:311–398.

Kluge, A. G. 1993. *Calabaria* and the phylogeny of erycine snakes. Zool. J. Linn. Soc. 107:293–351.

Köhler, G., and P. Heimes. 2002. Stachelleguane Lebensweise Pflege Zucht. Offenbach, Germany: Herpton.

LaDuc, T. J. 1996. A taxonomic revision of the *Adelphicos quadrivirgatum* species group (Serpentes: Colubridae). Master's thesis, University of Texas at El Paso.

Lavilla, E. O., J. A. Langone, J. M. Padial, and R. O. De Sá. 2010. The identity of the crackling, luminescent frog of Suriname (*Rana typhonia* Linnaeus, 1758) (Amphibia, Anura). Zootaxa 2671:17–30.

Lavín-Murcio, P. A., O. M. Hinojosa-Falcón, G. Herrera-Patiño, R. E. Núñez-Lara, and L. H. Vélez-Horta. 2005. Anfibios y reptiles de Tamaulipas: Un listado preliminar. In Biodiversidad tamaulipeca, ed. L. Barrientos, A. Correa, J. Horta, and J. Jiménez, 1:185–192. Tamaulipas: DGIT.

Lemos-Espinal, J. A., and H. M. Smith. 2007. Anfibios y reptiles del estado de Coahuila, México/ Amphibians and reptiles of the state of Coahuila, Mexico. Mexico City: CONABIO.

Lemos-Espinal, J. A., H. M. Smith, D. Chiszar, and G. Woolrich-Piña. 2004. Year 2003 snakes from Chihuahua and adjacent states of Mexico. Bull. Chic. Herpetol. Soc. 39:206–213.

Liner, E. A., and G. Casas-Andreu. 2008. Standard Spanish, English and scientific names of the amphibians and reptiles of Mexico. 2nd ed. SSAR Herpetol. Circ. 38:1–162.

Manier, M. K. 2004. Geographic variation in the Long-nosed Snake, *Rhinocheilus lecontei* (Colubridae): Beyond the subspecies debate. Biol. J. Linn. Soc. 83:65–85.

McCranie, J. R. 2011. The snakes of Honduras: Systematics, distribution and conservation. Salt Lake City, UT: SSAR.

Mendoza-Quijano, F., O. Flores-Villela, and J. W. Sites Jr. 1998. Genetic variation, species status, and phylogenetic relationships in Rose-bellied Lizards (*variabilis* group) of the genus *Sceloporus* (Squamata: Phrynosomatidae). Copeia 1998:354–366.

Mulcahy, D. G. 2007. Molecular systematics of Neotropical Cat-Eyed Snakes: A test of the monophyly of Leptodeirini (Colubridae: Dipsadinae) with implications for character evolution and biogeography. Biol. J. Linn. Soc. 92:483–500.

———. 2008. Phylogeography and species boundaries of the western North American Nightsnake (*Hypsiglena torquata*): Revisiting the subspecies concept. Mol. Phylogenet. Evol. 46:1095–1115.

Murphy, R. W., K. H. Berry, T. Edwards, A. E. Leviton, A. Lathrop, and J. D. Riedle. 2011. The dazed and confused identity of Agassiz's land tortoise, *Gopherus agassizii* (Testudines, Testudinidae) with the description of a new species, and its consequences for conservation. ZooKeys 113:39–71.

Pauly, G. B., D. M. Hillis, and D. C. Cannatella. 2009. Taxonomic freedom and the role of official lists of species names. Herpetologica 65:115–128.

Pyron, R. A., and F. T. Burbrink. 2009. Systematics of the Common Kingsnake (*Lampropeltis getula*; Serpentes: Colubridae) and the burden of heritage in taxonomy. Zootaxa 2241:22–32.

Reiserer, R. S., G. W. Schuett, and D. D. Beck. 2013. Taxonomic reassessment and conservation status of the beaded lizard, *Heloderma horridum* (Squamata: Helodermatidae). Amphib. Rept. Conserv. 7 (1): 74–96.

Rossman, D. A., N. B. Ford, and R. A. Seigel. 1996. The garter snakes: Evolution and ecology. Animal Natural History Series. Norman: University of Oklahoma Press.

Sanders, K. L., M. S. Y. Lee, Mumpuni, T. Bertozzi, and A. R. Rasmussen. 2013. Multilocus phylogeny and recent rapid radiation of the viviparous sea snakes (Elapidae: Hydrophiinae). Mol. Phylogenet. Evol. 66 (3): 575–591. doi:10.1016/j.ympev.2012.09.021.

Santos-Barrera, G., and O. Flores-Villela. 2011. A new species of toad of the genus *Incilius* from the Sierra Madre Occidental of Chihuahua, Mexico (Anura: Bufonidae). J. Herpetol. 45:211–215.

Savage, J. M., and B. I Crother. 1989. The status of *Pliocercus* and *Urotheca* (Serpentes: Colubridae), with a review of included species of coral snake mimics. Zool. J. Linn. Soc. 95:335–362.

Smith, H. M., and D. Chiszar. 2001a. *Pliocercus bicolor* Smith. Cat. Am. Amphib. Rept. 737:1–4.

————. 2001b. *Pliocercus* Smith. Cat. Am. Amphib. Rept. 735:1–9.

Streicher, J. W., C. L. Cox, J. A. Campbell, E. N. Smith, and R. O. de Sá. 2012. Rapid range expansion in the Great Plains Narrow-mouthed Toad (*Gastrophryne olivacea*) and a revised taxonomy for North American microhylids. Mol. Phylogenet. Evol. 64:645–653.

Taylor, E. H. 1950. Second contribution to the herpetology of San Luis Potosí. Univ. Kans. Sci. Bull. 33:441–457.

Tihen, J. A. 1954. Gerrhonotine lizards recently added to the American Museum collection, with further revision of the genus *Abronia*. Am. Mus. Novit. 1687:1–26.

Townsend, T. M., D. G. Mulcahy, B. P. Noonan, J. W. Sites Jr., C. A. Kuczynski, J. J. Wiens, and T. W. Reeder. 2011. Phylogeny of iguanian lizards inferred from 29 nuclear loci, and a comparison of concatenated and species-tree approaches for an ancient, rapid radiation. Mol. Phylogenet. Evol. 61:363–380.

Uetz, P. 2012. The reptile data base. Accessed November 17, 2014. http://reptile-database.org.

Utiger, U., B. Schätti, and N. Helfenberger. 2005. The oriental colubrine genus *Coelognathus* Fitzinger, 1843 and classification of Old and New World racers and rat snakes (Reptilia, Squamata, Colubridae, Colubrinae). Russ. J. Herpetol. 12 (1): 39–60.

Walker, J. M., and J. E. Cordes. 2011. Taxonomic implications of color pattern and meristic variation in *Aspidoscelis burti burti*, a Mexican whiptail lizard. Herpetol. Rev. 42:33–39.

Wilson, L. D., and J. D. Johnson. 2010. Distributional patterns of the herpetofauna of Mesoamerica, a biodiversity hotspot. In Conservation of Mesoamerican amphibians and reptiles, ed. L. D. Wilson, J. H. Townsend, and J. D. Johnson, 31–235. Eagle Mountain, UT: Eagle Mountain Publishing.

Wilson, L. D., V. Mata-Silva, and J. D. Johnson. 2013. A conservation reassessment of the reptiles of Mexico based on the EVS measure. Amphib. Rept. Conserv. 7:1–47.

Wood, D. A., R. N. Fisher, and T. R. Reeder. 2008. Novel patterns of historical isolation, dispersal, and secondary contact across Baja California in the Rosy Boa (*Lichanura trivirgata*). Mol. Phylogenet. Evol. 46:484–502.

Wood, D. A., A. G. Vandergast, J. A. Lemos-Espinal, R. N. Fisher, and A. T. Holycross. 2011.

Refugial isolation and divergence in the Narrow-headed Gartersnake species complex (*Thamnophis rufipunctatus*) as revealed by multilocus DNA sequence data. Mol. Ecol. 20:3856–3878.

3. Baja California

Adler, K. 1978. Herpetology in western North America during the nineteenth century. In Herpetological explorations of the Great American West, ed. K. Adler, 1:1–17. New York: Arno Press.

Aguirre-Léon, G., D. J. Morafka, and R. W Murphy. 1999. The peninsular archipelago of Baja California: A thousand kilometers of tree lizard genetics. Herpetologica 55:369–381.

Álvarez-Borrego, S. 2002. Physical oceanography. In A new island biogeography of the Sea of Cortés, ed. T. J. Case, M. L. Cody, and E. Ezcurra, 41–59. New York: Oxford University Press.

Arriaga, L., C. Aguilar, D. Espinosa-Organista, and R. Jiménez. 1997. Regionalización ecológica y biogeográfica de México. Mexico City: Taller de la Comisión Nacional para el Conocimiento y Uso de la Biodiversidad (CONABIO).

Ballinger, R. E., and D. W. Tinkle. 1968. A new species of *Uta* (Sauria: Iguanidae) from Salsipuedes Island, Gulf of California, Mexico. Occas. Pap. Mus. Zool. Univ. Mich. 656:1–6.

Barnard, F. L. 1970. Structural geology of the Sierra de Los Cucapas, northeastern Baja California, Mexico, and Imperial County, California. PhD diss., University of Colorado, Boulder.

Blainville, H. M. D. 1835. Description de quelques espèces de reptiles de la CalifFornie précédée de l'analyse d'un système général d'herpétologie et d'amphibiologie. Nouv. Ann. Mus. Hist. Natur. Paris 4:232–296.

Blair, C., F. R. Méndez de la Cruz, A. Ngo, J. Lindell, A. Lathrop, and R. W. Murphy. 2009. Molecular phylogenetics and taxonomy of leaf-toed geckos (Phyllodactylidae: *Phyllodactylus*) inhabiting the peninsula of Baja California. Zootaxa 2027:28–42.

Bostic, D. L. 1965. Home range of the teiid lizard *Cnemidophorus hyperythrus beldingi*. Southwest. Nat. 10:278–281.

————. 1966a. Food and feeding behavior of the teiid lizard *Cnemidophorus hyperythrus beldingi*. Herpetologica 22:12–31.

————. 1966b. Preliminary report on reproduction in the teiid lizard *Cnemidophorus hyperythrus beldingi*. Herpetologica 22:81–90.

———. 1966c. Thermoregulation and hibernation of the teiid lizard, *Cnemidophorus hyperythrus beldingi* (Sauria: Teiidae). Southwest. Nat. 11:275–289.

———. 1966d. Threat display in the lizards *Cnemidophorus hyperythrus* and *Cnemidophorus labialis*. Herpetologica 22:77–79.

———. 1968. Thermal relations, distribution, and habitat of *Cnemidophorus labialis* (Sauria: Teiidae). Trans. San Diego Soc. Nat. Hist. 15:21–30.

———. 1971. Herpetofauna of the Pacific coast of north central Baja California, Mexico, with a description of a new subspecies of *Phyllodactylus xanti*. Trans. San Diego Soc. Nat. Hist. 16:237–264.

———. 1975. A natural history guide to the Pacific coast of north central Baja California and adjacent islands. Vista, CA: Biological Education Expeditions.

Brattstrom, B. H. 1951. Two additions to the herpetofauna of Baja California, Mexico. Herpetologica 7:196.

Bury, R. B. 1983. Geographic distribution: *Anniella nigra argentea*. Herpetol. Rev. 14:83–84.

Caldwell, D. K. 1962. Sea turtles in Baja California waters (with special reference to those of the Gulf of California), and the description of a new subspecies of northeastern Pacific green turtle. Contr. Sci. Los Angeles Co. Mus. Nat. Hist. 61:1–31.

Campbell, J. A., and W. W. Lamar. 2004. The venomous reptiles of the Western Hemisphere. Ithaca, NY: Cornell University Press.

Carreño, A. L., and J. Helenes. 2002. Geology and ages of islands. In A new island biogeography of the Sea of Cortés, ed. T. J. Case, M. L. Cody, and E. Ezcurra, 14–40. New York: Oxford University Press.

Case, T. J. 1982. Ecology and evolution of the insular giant chuckwallas, *Sauromalus hispidus* and *S. varius*. In Iguanas of the world: Their behavior, ecology, and conservation, ed. G. M. Burghardt and A. S. Rand, 184–212. Park Ridge, NJ: Noyes Publications.

———. 1983. The reptiles: Ecology. In Island biogeography in the Sea of Cortés, ed. T. J. Case and M. L. Cody, 159–209. Berkeley: University of California Press.

Censky, E. J. 1986. *Sceloporus graciosus*. Cat. Am. Amphib. Rept. 386:1–4.

Clark, K. B., M. Dodero, A. Chavez, and J. Snapp-Cook. 2008. The threatened biological riches of Baja California's Colonet Mesa. Fremontia 36:3–10.

Clarkson, R. W., and J. C. de Vos Jr. 1986. The bullfrog, *Rana catesbeiana* Shaw, in the Lower Colorado River, Arizona–California. J. Herpetol. 20:42–49.

Clarkson, R. W., and J. C. Rorabaugh. 1989. Status of leopard frogs (*Rana pipiens* complex: Ranidae) in Arizona and southeastern California. Southwest. Nat. 34:531–538.

Cliff, F. S. 1954a. Snakes of the islands in the Gulf of California, Mexico. Trans. San Diego Soc. Nat. Hist. 12:67–98.

———. 1954b. Variation and evolution of the reptiles inhabiting the islands in the Gulf of California, Mexico. PhD diss., Stanford University.

Cliffton, K., D. O. Cornejo, and R. S. Felger. 1982. Sea turtles of the Pacific coast of Mexico. In Biology and conservation of sea turtles, ed. K. A. Bjorndal, 199–209. Washington, DC: Smithsonian Institution Press.

Cochran, D. M. 1961. Type specimens of reptiles and amphibians in the United States National Museum. Bull. U.S. Natl. Mus. 220:1–199.

Cody, M. L., R. Moran, J. Rebman, and H. Thompson. 2002. Plants. In A new island biogeography of the Sea of Cortés, ed. T. J. Case, M. L. Cody, and E. Ezcurra, 63–111. New York: Oxford University Press.

Cody, M. L., R. Moran, and H. Thompson. 1983. The plants. In Island biogeography of the Sea of Cortés, ed. T. J. Case and M. L. Cody, 49–97. Berkeley: University of California Press.

Cope, E. D. 1868. Sixth contribution to the herpetology of tropical America. Proc. Acad. Nat. Sci. Phila. 20:305–313.

Dall, W. H. 1909. Biographical memoir of William More Gabb 1839–1878. Natl. Acad. Sci. Biol. Mem. 6:347–361.

Delgadillo, J. 2004. El bosque de coníferas de la Sierra San Pedro Mártir, Baja California, México. Mexico City: Instituto Nacional de Ecología, SEMARNAT.

Devitt, T. J., S. E. Cameron Devitt, B. D. Hollingsworth, J. A. McGuire, and C. Moritz. 2013. Montane refugia predict population genetic structure in the Large-blotched Ensatina salamander. Mol. Ecol. 22 (6): 1–16.

Devitt, T. J., T. J. LaDuc, and J. A. McGuire. 2008. The *Trimorphodon biscutatus* (Squamata: Colubridae) species complex revisited: A multivariate statistical analysis of geographic variation. Copeia 2008:370–387.

Dickerson, M. C. 1919. Diagnoses of twenty-three new species and a new genus of lizards from Lower California. Bull. Am. Mus. Nat. Hist. 61:461–477.

Dixon, J. R. 1964. The systematics and distribution of lizards of the genus *Phyllodactylus* in North and Central America. New Mex. State Univ. Res. Ctr. Sci. Bull. 64 (1): 1–139.

———. 1966. Speciation and systematics of the gekkonid lizard genus *Phyllodactylus* of the islands in the Gulf of California. Proc. Calif. Acad. Sci., 4th ser., 33:415–452.

Fitch, H. S. 1934. New alligator lizards from the Pacific coast. Copeia 1934:6–7.

Funk, R. S. 1981. *Phrynosoma mcallii*. Cat. Am. Amphib. Rept. 281:1–2.

Gabb, W. M. 1882, Notes on the geology of Lower California. In Geological survey of California, Geology, vol. 2, The coast ranges, ed. J. D. Whitney, 137–148. Cambridge, MA: University Press, John Wilson and Son.

Grismer, L. L. 1982. A new population of slender salamander (*Batrachoseps*) from northern Baja California, Mexico. San Diego Herpetol. Soc. Newsltr. 4:3–4.

———. 1988. Geographic variation, taxonomy, and biogeography of the anguid genus *Elgaria* (Reptilia: Squamata) in Baja California, Mexico. Herpetologica 44:431–439.

———. 1989a. *Chionactis occipitalis*: Geographic distribution. Herpetol. Rev. 20:13.

———. 1989b. *Urosaurus g. graciosus*: Geographic distribution. Herpetol. Rev. 20:13.

———. 1990a. A new long-nosed snake (*Rhinocheilus lecontei*) from Isla Cerralvo, Baja California Sur, Mexico. Proc. San Diego Soc. Nat. Hist. 4:1–7.

———. 1990b. The relationships, taxonomy, and geographic variation of the *Masticophis lateralis* complex (Serpentes: Colubridae) of Baja California, Mexico. Herpetologica 46:66–77.

———. 1993. The insular herpetofauna of the Pacific coast of Baja California, Mexico. Herpetol. Nat. Hist. 1:1–10.

———. 1994a. Ecogeography of the peninsular herpetofauna of Baja California, Mexico, and its utility in historical biogeography. In Proceedings of the conference on herpetology of North American deserts, ed. J. W. Wright and P. Brown, 89–125. Southwest. Herpetol. Soc. Spec. Pub. Van Nuys, CA: Southern Herpetology Society.

———. 1994b. The origin and evolution of the peninsular herpetofauna of Baja California, Mexico. Herpetol. Nat. Hist. 2:51–106.

———. 1994c. Three new side-blotched lizards (genus *Uta*) from the Gulf of California. Herpetologica 50:451–474.

———. 1996. Geographic variation, taxonomy, and distribution of *Eumeces skiltonianus* and *E. lagunensis* (Squamata: Scincidae) in Baja California, Mexico. Amphibia-Reptilia 17:361–375.

———. 1999a. Checklist of amphibians and reptiles on islands in the Gulf of California, Mexico. Bull. So. Calif. Acad. Sci. 98 (2): 45–56.

———. 1999b. An evolutionary classification of reptiles on islands in the Gulf of California, Mexico. Herpetologica 55:446–469.

———. 1999c. Phylogeny, taxonomy, and biogeography of *Cnemidophorus hyperythrus* and C. *ceralbensis* (Squamata: Teiidae) in Baja California, Mexico. Herpetologica 55:28–42.

———. 2001. An evolutionary classification and checklist of amphibians and reptiles on the Pacific islands of Baja California, Mexico. Bull. So. Calif. Acad. Sci. 100:12–23.

———. 2002. Amphibians and reptiles of Baja California including its Pacific islands and the islands in the Sea of Cortés. Berkeley: University of California Press.

Grismer, L. L., K. R. Beaman, and H. E. Lawler. 1995. *Sauromalus hispidus*. Cat. Am. Amphib. Rept. 615:1–4.

Grismer, L. L., and B. D. Hollingsworth. 1996. *Cnemidophorus tigris* does not occur on Islas San Benito, Baja California, Mexico. Herpetol. Rev. 27:69–70.

———. 2001. A taxonomic review of the endemic alligator lizard *Elgaria paucicarinata* (Anguidae: Squamata) of Baja California, México, with a description of a new species. Herpetologica 57:488–496.

Grismer, L. L., and J. A. McGuire. 1993. The oases of central Baja California, México. Part I, A preliminary account of the relict mesophilic herpetofauna and the status of the oases. Bull. So. Calif. Acad. Sci. 92:2–24.

———. 1996. Taxonomy and biogeography of the *Sceloporus magister* complex (Squamata: Phrynosomatidae) in Baja California, Mexico. Herpetologica 52:416–427.

Grismer, L. L., J. A. McGuire, and B. D. Hollingsworth. 1994. A report on the herpetofauna of the Vizcaino Peninsula, Baja California, Mexico, with a discussion of its biogeographic and taxonomic implications. Bull. So. Calif. Acad. Sci. 93:45–80.

Grismer, L. L., and E. Mellink. 1994. The addition of *Sceloporus occidentalis* to the herpetofauna of Isla de Cedros, Baja California, Mexico, and its historical and taxonomic implications. J. Herpetol. 28:120–126.

———. 2005. Historical and ecological biogeography of the terrestrial herpetofauna of northern Baja California. In Biodiversity, ecosystems, and conservation in northern Mexico, ed. J. E. Cartron, G. Ceballos, and R. S. Felger, 167–178. New York: Oxford University Press.

Grismer, L. L., and J. R. Ottley. 1988. A preliminary analysis of geographic variation in *Coleonyx switaki* (Squamata: Eublepharidae), with a description of a new subspecies. Herpetologica 44:143–154.

Grismer, L. L., H. Wong, and P. Galina-Tessaro. 2002. Geographic variation and taxonomy of the sand snakes, *Chilomeniscus* (Squamata: Colubridae). Herpetologica 58:18–31.

Hafner, D. J., and B. R. Riddle. 1997. Biogeography of Baja California peninsular desert mammals. In Life among the muses: Papers in honor of James S. Findley, ed. T. L. Yates, W. L. Gannon, and D. E. Wilson, 39–65. Special Publication 3. Albuquerque: Museum of Southwestern Biology, University of New Mexico.

Hanna, G. D. 1922. The 1921 expedition of the California Academy of Sciences to the Gulf of California. Science 55:305–307.

Hastings, J. R., and R. R. Humphrey. 1969. Climatological data and statistics for Baja California. Univ. Ariz., Inst. Atmos. Phys., Tech. Rep. Meteorol. Climatol. Arid Regions 18:1–96.

Hastings, J. R., and R. M. Turner. 1965. Seasonal precipitation regimes in Baja California, México. Geogr. Ann., ser. A, 47:204–223.

Hazard, L. C., V. H. Shoemaker, and L. L. Grismer. 1998. Salt gland secretion by an intertidal lizard, *Uta tumidarostra*. Copeia 1998:231–234.

Heim, C. D., B. Alexander, R. W. Hansen, J. H. Valdez-Villavicencio, T. J. Devitt, B. D. Hollingsworth, J. A. Soto-Centeno, and C. R. Mahrdt. 2005. Geographic distribution: *Ensatina eschscholtzii klauberi*. Herpetol. Rev. 36:330–331.

Hollingsworth, B. D. 1998. The systematics of chuckwallas (*Sauromalus*) with a phylogenetic analysis of other iguanid lizards. Herpetol. Monogr. 11:38–191.

Hollingsworth, B. D., C. R. Mahrdt, L. L. Grismer, B. H. Banta, and C. K. Sylber. 1997. The occurrence of *Sauromalus varius* on a satellite islet of Isla Salsipuedes, Gulf of California, México. Herpetol. Rev. 28:26–28.

Humphrey, R. R. 1974. The boojum and its home. Tucson: University of Arizona Press.

Hunt, L. E. 1983. A nomenclatural rearrangement of the genus *Anniella* (Sauria: Anniellidae). Copeia 1983:79–89.

Klauber, L. M. 1924. *Tantilla eiseni* reported from Lower California. Copeia 131:62.

———. 1928. The *Trimorphodon* (Lyre Snake) of California, with notes on the species of the adjacent areas. Trans. San Diego Soc. Nat. Hist. 5:183–194.

———. 1931a. An addition to the fauna of Lower California. Copeia 1931:141.

———. 1931b. *Crotalus tigris* and *Crotalus enyo*, two little known rattlesnakes of the Southwest. Trans. San Diego Soc. Nat. Hist. 6:353–370.

———. 1931c. A new subspecies of the California Boa, with notes on the genus *Lichanura*. Trans. San Diego Soc. Nat. Hist. 6:305–318.

———. 1931d. Notes on the worm snakes of the Southwest, with descriptions of two new subspecies. Trans. San Diego Soc. Nat. Hist. 6:333–352.

———. 1931e. A statistical survey of the snakes of the southern border of California. Bull. Zool. Soc. San Diego 8:1–93.

———. 1932. The Flat-tailed Horned Toad in Lower California. Copeia 1932:100.

———. 1934. Annotated list of amphibians and reptiles of the southern border of California. Bull. Zool. Soc. San Diego 11:1–28.

———. 1935. *Phyllorhynchus*, the Leaf-nosed Snake. Bull. Zool. Soc. San Diego 12:1–31.

———. 1936. *Crotalus mitchellii*, the Speckled Rattlesnake. Trans. San Diego Soc. Nat. Hist. 8:149–184.

———. 1940a. The lyre snakes (genus *Trimorphodon*) of the United States. Trans. San Diego Soc. Nat. Hist. 9:163–194.

———. 1940b. Two new subspecies of *Phyllorhynchus*, the Leaf-nosed Snake, with notes on the genus. Trans. San Diego Soc. Nat. Hist. 9:195–214.

———. 1940c. The worm snakes of the genus

Leptotyphlops in the United States and northern Mexico. Trans. San Diego Soc. Nat. Hist. 9:87—162.

———. 1941. The Long-nosed Snake of the genus *Rhinocheilus*. Trans. San Diego Soc. Nat. Hist. 9:289–332.

———. 1943a. A desert subspecies of the snake *Tantilla eiseni*. Trans. San Diego Soc. Nat. Hist. 10:71–74.

———. 1943b. A new snake of the genus *Sonora* from Lower California, Mexico. Trans. San Diego Soc. Nat. Hist. 10:69–70.

———. 1944. The Sidewinder, *Crotalus cerastes*, with description of a new subspecies. Trans. San Diego Soc. Nat. Hist. 10:91–126.

———. 1945. The geckos of the genus *Coleonyx* with descriptions of new subspecies. Trans. San Diego Soc. Nat. Hist. 10:133–216.

———. 1946a. The Glossy Snake, *Arizona*, with descriptions of new subspecies. Trans. San Diego Soc. Nat. Hist. 10:311–398.

———. 1946b. The gopher snakes of Baja California with descriptions of new subspecies of *Pituophis catenifer*. Trans. San Diego Soc. Nat. Hist. 11:1–40.

———. 1949a. The relationships of *Crotalus ruber* and *Crotalus lucasensis*. Trans. San Diego Soc. Nat. Hist. 11:57–60.

———. 1949b. Some new and revived subspecies of rattlesnakes. Trans. San Diego Soc. Nat. Hist. 11:61–116.

———. 1951. The Shovel-nosed Snake, *Chionactis*, with descriptions of two new subspecies. Trans. San Diego Soc. Nat. Hist. 11:141–204.

———. 1956. Rattlesnakes: Their habits, life histories, and influence on mankind. Berkeley: University of California Press.

———. 1963. A new insular subspecies of the Speckled Rattlesnake. Trans. San Diego Soc. Nat. Hist. 13:73–80.

Lais, M. P. 1976. *Gerrhonotus cedrosensis*. Cat. Am. Amphib. Rept. 177:1.

Leaché, A. D., M. S. Koo, C. L. Spencer, T. J. Papenfuss, R. N. Fisher, and J. A. McGuire. 2009. Quantifying ecological, morphological, and genetic variation to delimit species in the coast horned lizard species complex (*Phrynosoma*). Proc. Natl. Acad. Sci. USA 106 (30): 12418–12423.

Leaché, A. D., and J. A. McGuire. 2006. Phylogenetic relationships of horned lizards (*Phrynosoma*) based on nuclear and mitochondrial data:

Evidence for a misleading mitochondrial gene tree. Mol. Phylogenet. Evol. 39:628–644.

Leaché, A. D., and D. G. Mulcahy. 2007. Phylogeny, divergence times and species limits of spiny lizards (*Sceloporus magister* species group) in western North American deserts and Baja California. Mol. Ecol. 16:5216–5233.

Leavitt, D. H., R. L. Bezy, K. A. Crandall, and J. W. Sites Jr. 2007. Multi-locus DNA sequence data reveal a history of deep cryptic vicariance and habitat-driven convergence in the Desert Night Lizard *Xantusia vigilis* species complex (Squamata: Xantusiidae). Mol. Ecol. 16:4455–4481.

Lindell, J. S., F. Méndez de la Cruz, and R. W. Murphy. 2005. Deep genealogical history without population differentiation: Discrepancy between mtDNA and allozyme divergence in the Zebra-tailed Lizard (*Callisaurus draconoides*). Mol. Phylogenet. Evol. 36:682–694.

———. 2008. Deep biogeographical history and cytonuclear discordance in the Black-tailed Brush Lizard (*Urosaurus nigricaudus*) of Baja California. Biol. J. Linn. Soc. 94:89–104.

Linsdale, J. M. 1932. Amphibians and reptiles from Lower California. Univ. Calif. Publ. Zool. 38:345–386c.

Linsdale, J. M., and J. L. Gressitt. 1937. Soft-shelled turtles in the Colorado River basin. Copeia 1937:222–225.

Lindsay, G. E. 1962. The Belvedere expedition to the Gulf of California. Trans. San Diego Soc. Nat. Hist. 13:1–44.

Lonsdale, P. 1989. Geology and tectonic history of the Gulf of California. In The eastern Pacific Ocean and Hawaii, ed. E. L. Winterer, D. M. Hussong, and R. W. Decker, 123–146. Boulder, CO: Geological Society of America.

Loomis, R. B. 1965. The Yellow-legged Frog, *Rana boylii*, from the Sierra San Pedro Martir, Baja California Norte, Mexico. Herpetologica 21:78–80.

Lovich, R. E., T. Akre, J. Blackburn, T. Robison, and C. Mahrdt. 2007. Geographic distribution: *Actinemys marmorata*. Herpetol. Rev. 38:216–217.

Lovich, R. E., L. L. Grismer, and G. Danemann. 2009. Conservation status of the herpetofauna of Baja California, México and associated islands in the Sea of Cortez and Pacific Ocean. Herpetol. Conserv. Biol. 4:358–378.

Lovich, R. E., and C. R. Mahrdt. 2008. Terrestrial herpetofauna of the Bahía de Los Angeles. In Bahía de Los Ángeles: Recursos naturales y comunidad: Línea base 2007, ed. G. Danemann and E. Ezcurra, 495–522. Ensenada, Baja California: D. R. Pronatura Noroeste, A.C.

Lovich, R. E., C. R. Mahrdt, and B. Downer. 2005. Geographic distribution: *Actinemys marmorata*. Herpetol. Rev. 36:200–201.

Lowe, C. H., and K. S. Norris. 1954. Analysis of the herpetofauna of Baja California, Mexico. Trans. San Diego Soc. Nat. Hist. 12:47–64.

Lynch, J. F., and D. B. Wake. 1974. *Aneides lugubris*. Cat. Am. Amphib. Rept. 159:1–2.

Mahrdt, C. R. 1975. The occurrence of *Ensatina eschscholtzii eschscholtzii* in Baja California, Mexico. J. Herpetol. 9:240–242.

Mahrdt, C. R., and R. E. Lovich. 2004. Geographic distribution: *Bufo californicus*. Herpetol. Rev. 35:280.

Mahrdt, C. R., R. E. Lovich, and S. J. Zimmitti. 2002. Natural history notes: *Bufo californicus*. Herpetol. Rev. 33:123–125.

Mahrdt, C. R., R. E. Lovich, S. J. Zimmitti, and G. D. Danemann. 2003. Geographic distribution: *Bufo californicus*. Herpetol. Rev. 34:256–257.

Mahrdt, C. R., R. H. McPeak, and L. L. Grismer. 1998. The discovery of *Ensatina eschscholtzii klauberi* (Plethodontidae) in the Sierra San Pedro Martir, Baja California, Mexico. Herpetol. Nat. Hist. 6:73–76.

Markham, C. G. 1972. Baja California's climate. Weatherwise 25:66–101.

Martínez-Isaac, R., and J. H. Valdez-Villavicencio. 2000. Geographic distribution: *Hemidactylus turcicus*. Herpetol. Rev. 31:254.

Martínez-Solano, I., A. Peralta-García, E. L. Jockusch, D. B. Wake, E. Vázquez-Domínguez, and G. Parra-Olea. 2012. Molecular systematics of *Batrachoseps* (Caudata, Plethodontidae) in southern California and Baja California: Mitochondrial-nuclear DNA disconcordance and the evolutionary history of *B. major*. Mol. Phylogenet. Evol. 63:131–149.

McGuire, J. A. 1994. A new species of collared lizard (Iguania: Crotaphytidae) from northeastern Baja California. Herpetologica 50:438–450.

———. 1996. Phylogenetic systematics of crotaphytid lizards (Reptilia: Iguania: Crotaphytidae). Bull. Carnegie Mus. Nat. Hist. 32:1–143.

McGuire, J. A., and L. L. Grismer. 1993. The taxonomy and biogeography of *Thamnophis hammondii* and *T. digueti* (Reptilia: Squamata: Colubridae) in Baja California, Mexico. Herpetologica 49:354–365.

Meigs, P. 1953. World distribution of arid and semi-arid homoclimates. In Reviews of research on arid zone hydrology, 202–210. Arid Zone Prog. 1. Paris: UNESCO.

———. 1966. Geography of coastal deserts. Arid Zone Res. 28:1–140.

Mellink, E. 1995. The potential effect of commercialization of reptiles from Mexico's Baja California peninsula and its associated islands. Herpetol. Nat. Hist. 3:95–99.

Mellink, E., and V. Ferreira-Bartrina. 2000. On the wildlife of wetlands of the Mexican portion of the Rio Colorado delta. Bull. So. Calif. Acad. Sci. 99:115–127.

Miller, R. R. 1946. The probable origin of the soft-shelled turtles in the Colorado River basin. Copeia 1946:46.

Mocquard, F. 1899. Contribution à la faune herpétologique de la Basse Californie. Nouv. Arch. Mus. Hist. Natur. Paris 4:297–343.

Morrone, J. J. 2005. Hacia una síntesis biogeográfica de México [Toward a synthesis of Mexican biogeography]. Rev. Mex. Biodivers. 76:207–252.

Mulcahy, D. G. 2008. Phylogeography and species boundaries of the western North American Night snake (*Hypsiglena torquata*): Revisiting the subspecies concept. Mol. Phylogenet. Evol. 46:1095–1115.

Murphy, R. W. 1974. A new genus and species of eublepharine gecko (Sauria: Gekkonidae) from Baja California, Mexico. Proc. Calif. Acad. Sci., 4th ser., 60:87–92.

———. 1975. Two new blind snakes (Serpentes: Leptotyphlopidae) from Baja California, Mexico, with a contribution to the biogeography of peninsular and insular herpetofauna. Proc. Calif. Acad. Sci., 4th ser., 40:93–107.

———. 1976. The evolution of a peninsular and insular herpetofauna: Drift based alternative hypothesis. Master's thesis, San Francisco State University.

———. 1983a. A distributional checklist of the amphibians and reptiles on the islands in the Sea of Cortés. Appendix 6.1 and 6.2. In Island biogeography in the Sea of Cortés, ed. T. J. Case and M. L. Cody, 429–437. Berkeley: University of California Press.

———. 1983b. Paleobiogeography and genetic differentiation of the Baja California herpetofauna. Occas. Pap. Calif. Acad. Sci. 137:1–48.

———. 1983c. The reptiles: Origin and evolution.

In Island biogeography in the Sea of Cortés, ed. T. J. Case and M. L. Cody, 130–158. Berkeley: University of California Press.

Murphy, R. W., and G. Aguirre-León. 2002a. A distributional checklist of the amphibians and reptiles on the islands in the Sea of Cortés. Appendix 8.2 and 8.4. In A new island biogeography of the Sea of Cortés, ed. T. J. Case, M. L. Cody, and E. Ezcurra, 580–599. New York: Oxford University Press.

———. 2002b. The non-avian reptiles: Origins and evolution. In A new island biogeography of the Sea of Cortés, ed. T. J. Case, M. L. Cody, and E. Ezcurra, 181–220. New York: Oxford University Press.

Murphy, R. W., J. Fu, A. Lathrop, J. V. Feltham, and V. Kovac. 2002. Phylogeny of the rattlesnakes (*Crotalus* and *Sistrurus*) inferred from sequences of five mitochondrial DNA genes. In Biology of the vipers, ed. G. W. Schuett, M. Hoggren, M. E. Douglas, and H. W. Greene, 71–92. Eagle Mountain, UT: Eagle Mountain Publishing.

Murphy, R. W., V. Kovac, O. Haddrath, G. S. Allen, A. Fishbein, and N. E. Mandrak. 1995. mtDNA gene sequence, allozyme, and morphological uniformity among red diamond rattlesnakes, *Crotalus ruber* and *Crotalus exsul*. Can. J. Zool. 73:270–281.

Murphy, R. W., and F. R. Méndez de la Cruz. 2010. The herpetofauna of Baja California and its associated islands: A conservation assessment and priorities. In Conservation of Mesoamerican amphibians and reptiles, ed. L. D. Wilson, J. H. Townsend, and J. D. Johnson, 238–273. Eagle Mountain, UT: Eagle Mountain Publishing.

Murphy, R. W., and J. R. Ottley. 1984. Distribution of amphibians and reptiles on islands in the Gulf of California. Ann. Carnegie Mus. 53:207–230.

Murphy, R. W., and H. M. Smith. 1985. Conservation of the name *Anniella pulchra* for the California legless lizard. Herpetol. Rev. 16:68.

———. 1991. *Anniella pulchra* Gray, 1852 (Reptilia, Squamata): Proposed designation of a neotype. Bull. Zool. Nomencl. 48:316–318.

Myers, E. A., J. A. Rodríguez-Robles, D. F. Denardo, R. E. Staub, A. Stropoli, S. Ruane, and F. T. Burbrink. 2013. Multilocus phylogeography assessment of the California Mountain Kingsnake (*Lampropeltis zonata*) suggests alternative patterns of diversification for the California Floristic Province. Mol. Ecol. 22 (21): 1–12.

Myers, N., R. A. Mittermeier, C. G. Mittermeier, G. A. B. da Fonseca, and J. Kent. 2001. Biodiversity hotspots for conservation priorities. Nature 403:853–858.

The Nature Conservancy. 2007. Conservation vision for Bahía San Quintín. The Nature Conservancy and Conservation Biology Institute. http://d2k78bk4kdhbpr.cloudfront.net/media /content/files/BSQreport0907small.pdf.

Nelson, E. W. 1922. Lower California and its natural resources. Mem. Natl. Acad. Sci. Wash. 16:1–194.

Oberbauer, T. A. 1992. Herpetology of Islas Coronados. In Natural history of the Coronado Islands, Baja California, Mexico, ed. L. Perry, 24–26. San Diego, CA: San Diego Association of Geologists.

Ottley, J. R. 1978. A new subspecies of the snake *Lichanura trivirgata* from Cedros Island, Mexico. Great Basin Nat. 38:411–416.

Ottley, J. R., and L. E. Hunt. 1981. Geographic distribution: *Crotalus viridis helleri*. Herpetol. Rev. 12:65.

Ottley, J. R., and E. E. Jacobsen. 1983. Pattern and coloration of juvenile *Elaphe rosaliae*, with notes on natural history. J. Herpetol. 17:189–190.

Ottley, J. R., R. W. Murphy, and G. V. Smith. 1980. The taxonomic status of the Rosy Boa *Lichanura roseofusca* (Serpentes: Boidae). Great Basin Nat. 40:59–62.

Papenfuss, T. J. 1982. The ecology and systematics of the amphisbaenian genus *Bipes*. Occas. Pap. Calif. Acad. Sci. 136:1–42.

Papenfuss, T. J., and J. F. Parham. 2013. Four new species of California legless lizards (*Anniella*). Breviora 536:1–17.

Pase, C. P. 1982. Sierran montane conifer forest. In Biotic communities of the American Southwest: United States and Mexico, ed. D. E. Brown, 49–51. Tucson: University of Arizona for the Boyce Thompson Southwestern Arboretum.

Peinado, M., F. Alcaraz, J. Delgadillo, and I. Aguado. 1994. Fitogeografía de la peninsula de Baja California, México. An. Jardín Botán. Madrid 51:255–277.

Peralta-García, A., and J. H. Valdez-Villavicencio. 2004. Geographic distribution: *Ensatina eschscholtzii eschscholtzii*. Herpetol. Rev. 5:279.

Pickwell, G. V., R. L. Bezy, and J. E. Fitch. 1983. Northern occurrences of the sea snake, *Pelamis platurus,* in the eastern Pacific, with a record of predation on the species. Calif. Fish Game 69:172–177.

Platz, J. E., R. W. Clarkson, J. C. Rorabaugh, and D. M. Hillis. 1990. *Rana berlandieri:* Recently introduced populations in Arizona and southeastern California. Copeia 1990:324–333.

Platz, J. E., and J. S. Frost. 1984. *Rana yavapaiensis,* a new species of leopard frog (*Rana pipiens* complex). Copeia 1984:940–948.

Pough, F. H. 1977. *Uma notata.* Cat. Am. Amphib. Rept. 197:1–2.

Radcliffe, C. W., and T. P. Maslin. 1975. A new subspecies of the Red Rattlesnake, *Crotalus ruber,* San Lorenzo Sur Island, Baja California Norte, Mexico. Copeia 1975:490–493.

Rau, C. S., and R. B. Loomis. 1977. A new species of *Urosaurus* (Reptilia, Lacertilia, Iguanidae) from Baja California, Mexico. J. Herpetol. 11:25–29.

Rebman, J. P., and N. C. Roberts. 2012. Baja California plant field guide. El Cajon, CA: Sunbelt Publications.

Roberts, N. 1984. Important riparian/wetland systems of peninsular Baja California: An overview. In California riparian systems: Ecology, conservation, and productive management, ed. R. E. Warner and K. M. Hendrix, 390–403. Berkeley: University of California Press.

———. 1987. Baja California plant field guide. La Jolla, CA: Natural History Publishing.

Rodríguez-Robles, J. A. 2000. Molecular systematics of New World gopher, bull, and pinesnakes (*Pituophis:* Colubridae), a transcontinental species complex. Mol. Phylogenet. Evol. 14:35–50.

Rodríguez-Robles, J. A., D. F. Denardo, and R. E. Staub. 1999. Phylogeography of the California Mountain Kingsnake, *Lampropeltis zonata* (Colubridae). Mol. Ecol. 8:1923–1934.

Rorabaugh, J. C., M. J. Sredl, V. Miera, and C. A. Drost. 2002. Continued invasion by an introduced frog (*Rana berlandieri*): Southwestern Arizona, southeastern California, and Río Colorado, México. Southwest. Nat. 47:12–20.

Ruíz-Campos, G., and J. H. Valdez-Villavicencio. 2012. Geographic distribution. *Xenopus laevis.* Herpetol. Rev. 43:99.

Samaniego-Herrera, A., A. Peralta-García, and A. Aguirre-Muñoz. 2007. Vertebrados de las islas Pacífico de Baja California: Guía de campo. Ensenada, Baja California: Grupo de Ecología y Conservación de Islas, A.C.

Sánchez-Pacheco, J. A., and E. Mellink. 2001. Geographic distribution: *Anniella geronimensis.* Herpetol. Rev. 32:192.

SARH (Secretaría de Agricultura y Recursos Hidráulicos). 1977. Inventario forestal del estado de Baja California. Resumen de la Publicación No. 3 (1968). Publicación No. 43. Mexico City: Dirección General de Inventarios Forestales, SARH.

Savage, J. M. 1952. Studies on the lizard family Xantusiidae. Vol. 1, The systematics status of the Baja California night lizards allied to *Xantusia vigilis,* with the description of a new subspecies. Am. Midl. Nat. 48:467–479.

———. 1960. Evolution of a peninsular herpetofauna. Syst. Zool. 9:184–212.

———. 1967. Evolution of the insular herpetofauna. In Proceedings of the symposium on the biology of the California Islands, ed. R. Philbrick, 219–227. Santa Barbara, CA: Santa Barbara Botanic Garden.

Schad, J. 1988. Parque Nacional San Pedro Martir (San Pedro Martir National Park) including Picacho del Diablo, topographic maps and visitor's guide to Baja's highest mountains: Featuring roads, trails, climbing routes, points of interest, natural features. San Diego, CA: Centra Publications.

Schmidt, K. P. 1922. The amphibians and reptiles of Lower California and the neighboring islands. Bull. Am. Mus. Nat. Hist. 46:607–707.

Shaw, C. E. 1940. A new species of legless lizard from San Geronimo Island, Lower California. Trans. San Diego Soc. Nat. Hist. 9:225–228.

———. 1945. The chuckwallas, genus *Sauromalus.* Trans. San Diego Soc. Nat. Hist. 10:296–306.

———. 1947. First records of the Red-Brown Loggerhead Turtle from the eastern Pacific. Herpetologica 4:55–56.

———. 1949. *Anniella geronimensis* on the Baja California peninsula. Herpetologica 5:27–28.

———. 1953. *Anniella pulchra* and *Anniella geronimensis,* sympatric species. Herpetologica 8:167–170.

Shreve, F. 1951. Vegetation and flora of the Sonoran Desert. Vol. 1, Vegetation. Carnegie Inst. Wash. Publ. 591:1–192.

Shreve, F., and I. R. Wiggins. 1964. Vegetation and flora of the Sonoran Desert. 2 vols. Stanford, CA: Stanford University Press.

Slevin, J. R. 1922. Expedition of the California Academy of Sciences to the Gulf of California in 1921. Proc. Calif. Acad. Sci. 7:55–72.

———. 1928. The amphibians of western North America: An account of the species known to inhabit California, Alaska, British Columbia, Washington, Oregon, Idaho, Utah, Nevada, Arizona, Sonora, and Lower California. Occas. Pap. Calif. Acad. Sci. 16:1–152.

Smith, H. M., and R. B. Smith. 1979. Synopsis of the herpetofauna of Mexico. Vol. 6, Guide to Mexican turtles, bibliographic addendum III. North Bennington, VT: John Johnson.

Smith, H. M., and W. W. Tanner. 1972. Two new subspecies of *Crotaphytus* (Sauria: Iguanidae). Great Basin Nat. 32:25–34.

Smith, H. M., and E. H. Taylor. 1950. Type localities of Mexican reptiles and amphibians. Univ. Kans. Sci. Bull. 33:313–380.

Soulé, M., and A. J. Sloan. 1966. Biogeography and distribution of the reptiles and amphibians on islands in the Gulf of California, México. Trans. San Diego Soc. Nat. Hist. 14:137–156.

Spencer, J. E., and P. A. Pearthree. 2001. Headward erosion versus closed-basin spillover as alternative causes of Neogene capture of the ancestral Colorado River by the Gulf of California. In The Colorado River: Origin and evolution: Grand Canyon, Arizona, ed. R. A. Young and E. E. Spamer, 215–219. Grand Canyon Association Monograph 12. Grand Canyon, AZ: Grand Canyon Association.

Stebbins, R. C. 1962. Amphibians of western North America. Berkeley: University of California Press.

Stejneger, L. 1890. Description of a new lizard from Lower California. Proc. U.S. Natl. Mus. 12:643–644.

———. 1891. Description of a new North American lizard of the genus *Sauromalus*. Proc. U.S. Natl. Mus. 14:409–411.

———. 1893. Annotated list of reptiles and batrachians collected by the Death Valley expedition in 1891, with descriptions of new species. North Am. Fauna 7:159–228.

Stock, J. M., and K. V. Hodges. 1989. Pre-Pliocene extension around the Gulf of California and the transfer of Baja California to the Pacific Plate. Tectonics 8:99–115.

Townsend, C. H. 1916. Voyage of the "Albatross" to the Gulf of California in 1911. Bull. Am. Mus. Nat. Hist. 35:399–707.

Turner, R. M., and D. E. Brown. 1982. Tropical-subtropical desertlands. In Biotic communities of the American Southwest: United States and Mexico, ed. D. E. Brown, 180–221. Tucson: University of Arizona for the Boyce Thompson Southwestern Arboretum.

Upton, D. E., and R. W Murphy. 1997. Phylogeny of the side-blotched lizards (Phrynosomatidae: *Uta*), based on mtDNA sequences: Support for a midpeninsular seaway in California. Mol. Phylogenet. Evol. 8:104–113.

Van Denburgh, J. 1895a. A review of the herpetology of Lower California. Part I—Reptiles. Proc. Calif. Acad. Sci., 2nd ser., 5:77–162.

———. 1895b. A review of the herpetology of Lower California. Part II—Batrachians. Proc. Calif. Acad. Sci., 2nd ser., 5:556–561.

———. 1896. Additional notes on the herpetology of Lower California. Proc. Calif. Acad. Sci., 2nd ser., 5:1004–1008.

———. 1905. The reptiles and amphibians of the islands of the Pacific coast of North America from the Farallons to Cape San Lucas and the Revilla Gigedos. Proc. Calif. Acad. Sci. 4:1–41.

———. 1922. The reptiles of western North America; an account of the species known to inhabit California and Oregon, Washington, Idaho, Utah, Nevada, Arizona, British Columbia, Sonora, and Lower California. Occas. Pap. Calif. Acad. Sci. 1–2:1–1028.

Van Denburgh, J., and J. R. Slevin. 1914. Reptiles and amphibians of islands of the west coast of North America. Proc. Calif. Acad. Sci. 4:129–152.

———. 1921a. A list of amphibians and reptiles of the peninsula of Lower California, with notes on the species in the California Academy. Proc. Calif. Acad. Sci. 11:27–72.

———. 1921b. Preliminary diagnoses of new species of reptiles from islands in the Gulf of California, Mexico. Proc. Calif. Acad. Sci. 11:95–98.

———. 1923. Preliminary diagnoses of four new snakes from Lower California, Mexico. Proc. Calif. Acad. Sci. 13:1–2.

Van Devender, T. 1990. Late Quaternary vegetation and climate of the Sonoran Desert, United States and Mexico. In Packrat middens: The last 40,000 years of biotic change, ed. J. L. Betancourt, T. R. Van Devender, and P. S. Martin, 134–165. Tucson: University of Arizona Press.

Velarde, E., B. D. Hollingsworth, and J. P. Rebman. 2008. Geographic distribution: *Sauromalus hispidus*. Herpetol. Rev. 39:368.

Vitt, L. J., and R. D. Ohmart. 1978. Herpetofauna of the lower Colorado River: Davis dam to the Mexican border. West. Found. Vertebr. Zool. 2:35–72.

Walker, J. M. 1981. A new subspecies of *Cnemidophorus tigris* from south Coronado Island, Mexico. J. Herpetol. 15:193–197.

Welsh, H. H., Jr. 1988. An ecogeographic analysis of the herpetofauna of the Sierra San Pedro Mártir region, Baja California: With a contribution to the biogeography of Baja California herpetofauna. Proc. Calif. Acad. Sci., 4th ser., 46:1–72.

Welsh, H. H., Jr., and R. B. Bury. 1984. Additions to the herpetofauna of the south Colorado Desert, Baja California, with comments on the relationships of *Lichanura trivirgata*. Herpetol. Rev. 15:53–56.

Wiggins, I. L. 1980. Flora of Baja California. Stanford, CA: Stanford University Press.

Winker, C. D., and S. M. Kidwell. 1986. Paleocurrent evidence for lateral displacement of the Pliocene Colorado River delta by the San Andreas fault system, southeastern California. Geology 14:788–791.

Zug, G. R., C. H. Ernst, and R. V. Wilson. 1998. *Lepidochelys olivacea*. Cat. Am. Amphib. Rept. 653:1–13.

Zweifel, R. G. 1952a. Notes on the lizards of the Coronados Islands, Baja California, Mexico. Herpetologica 8:9–11.

———. 1952b. Pattern variation and evolution of the Mountain Kingsnake, *Lampropeltis zonata*. Copeia 1952:152–168.

———. 1958. Results of the Puritan-American Museum of Natural History Expedition to western Mexico. 2, Notes on the reptiles and amphibians from the Pacific coastal islands of Baja California. Am. Mus. Novit. 1895:1–17.

———. 1975. *Lampropeltis zonata*. Cat. Am. Amphib. Rept. 174:1–4.

4. Sonora

Allen, J. A. 1893. List of mammals and birds collected in northeastern Sonora and northwestern Chihuahua, Mexico, on the Lumholtz Archaeological Expedition 1890–92. Bull. Am. Mus. Nat. Hist. 5:27–42.

Allen, M. J. 1933. Report on a collection of amphibians and reptiles from Sonora, Mexico, with the description of a new lizard. Occas. Pap. Mus. Zool. Univ. Mich. 259:1–15.

Avila, S., C. Robles Elias, and J. C. Rorabaugh. 2008. Geographic distribution: *Masticophis mentovarius*. Herpetol. Rev. 39:370.

Axtell, R. W., and R. R. Montanucci. 1977. *Crotaphytus collaris* from the eastern Sonoran Desert: Description of a previously unrecognized geographic race. Nat. Hist. Misc. Chic. Acad. Sci. 201:1–8.

Baird, S. F. 1859a (1858). Description of new genera and species of North American lizards in the Museum of the Smithsonian Institution. Proc. Acad. Nat. Sci. Phila. 10:253–256.

Baird, S. F. 1859b. Reptiles of the boundary, with notes by the naturalists of the survey. In United States and Mexican Boundary Survey, under the order of Lieut. Col. W. H. Emory, major First Cavalry, and United States commissioner. Vol. 2, pt. 2, U.S. 34th Cong., 1st Sess., Exec. Doc. 108, 1–35, plates 1–41. Washington, DC: Government Printing Office.

Baird, S. F., and C. Girard. 1853. Catalogue of North American reptiles in the Museum of the Smithsonian Institution. Part 1, Serpents. Washington, DC: Smithsonian Institution.

Ballinger, R. E., and D. W. Tinkle. 1972. Systematics and evolution of the genus *Uta* (Sauria: Iguanidae). Misc. Publ. Mus. Zool. Univ. Mich. 145:1–83.

Barbour, T. 1921. A new lizard from Guaymas, Mexico. Proc. New Engl. Zool. Club 7:79–80.

Berry, J. F., and J. M. Legler. 1980. A new turtle (genus *Kinosternon*) from Sonora, Mexico. Los Angeles Co. Mus. Contr. Sci. 325:1–12.

Bezy, R. L., K. B. Bezy, and K. Bolles. 2008. Two new species of night lizards (*Xantusia*) from Mexico. J. Herpetol. 42:680–688.

Bogert, C. M. 1943. A new box turtle from southeastern Sonora, Mexico. Am. Mus. Novit. 1226:1–7.

Bogert, C. M., and E. E. Dorson. 1942. A new lizard of the genus *Callisaurus* from Sonora. Copeia 1942 (3): 173–175.

Bogert, C. M., and J. A. Oliver. 1945. A preliminary analysis of the herpetofauna of Sonora. Bull. Am. Mus. Nat. Hist. 83:297–426.

Bonine, K. B., E. F. Enderson, and R. L. Bezy. 2006. Geographic distribution: *Imantodes gemmistratus*. Herpetol. Rev. 37:363.

Brattstrom, B. H. 1963. A preliminary review of the thermal requirements of amphibians. Ecology 44:238–255.

Brennan, T. C. 2008. Online field guide to the reptiles and amphibians of Arizona. http://www.reptilesofaz.org/.

Brennan, T. C., and A. T. Holycross. 2006. A field guide to amphibians and reptiles in Arizona. Phoenix: Arizona Game and Fish Department.

Brito-Castillo, L., M. A. Crimmins, and S. L. Swift. 2010. Clima. In Diversidad biologica de Sonora, ed. F. E. Molina Freaner and T. R. Van Devender, 73–96. Mexico City: Universidad Nacional Autónoma de México y CONABIO.

Brown, D. E., ed. 1982a. Biotic communities of the American Southwest: United States and Mexico. Desert Plants 4 (1–4): 1–342.

———. 1982b. Grasslands. Desert Plants 4 (1–4): 107–142.

———. 1982c. Sinaloan thornscrub. Desert Plants 4 (1–4): 101–107.

Brown, D. E., and C. H. Lowe. 1980. Biotic communities of the American Southwest. USDA Forest Service General Technical Report RM-78.

Burger, W. L., and M. M. Hensley. 1949. Notes on a collection of reptiles and amphibians from northwestern Sonora. Nat. His. Misc. 35:1–6.

Burquez, A., A. Martínez-Yrizar, R. S. Felger, and D. A. Yetman. 1999. Vegetation and habitat diversity at the southern edge of the Sonoran Desert. In Ecology of Sonoran Desert plants and plant communities, ed. R. H. Robichaux, 36–67. Tucson: University of Arizona Press.

Burquez, A., A. Martínez-Yrizar, and P. S. Martin. 1992. From the High Sierra Madre to the coast: Changes in vegetation along Highway 16, Maycoba-Hermosillo. In Geology and mineral resources of the northern Sierra Madre Occidental, Mexico, ed. K. F. Clark, J. Roldan-Quintana, and R. Schmidt, 239–252. El Paso, TX: El Paso Geological Society.

Burquez, A., M. E. Miller, and A. Martínez-Yrizar. 2002. Mexican grasslands, thornscrub, and transformation of the Sonoran Desert by invasive exotic buffelgrass (Pennisetum ciliare). In Invasive exotic species in the Sonoran region, ed. B. Tellman, 126–146. Tucson: University of Arizona Press and the Arizona-Sonora Desert Museum.

Bury, R. B., D. J. Germano, T. R. Van Devender, and B. E. Martin. 2002. The desert tortoise in Mexico: Distribution, ecology, and conservation. In The Sonoran Desert tortoise: Natural history, biology and conservation, ed. Thomas R. Van Devender, 86–108. Tucson: University of Arizona Press.

Carr, A. F. 1942. A new Pseudemys from Sonora, Mexico. Am. Mus. Novit. 1181:1–4.

Conant, R. 1968. Zoological exploration in Mexico: The route of Lieut. D. N. Cuch in 1853. Am. Mus. Novit. 2350:1–14.

Cope, E. D. 1861 (1860). Notes and descriptions of new and little-known species of American reptiles. Proc. Acad. Nat. Sci. Phila. 12:339–345.

———. 1867. On the reptilian and batrachia of the Sonoran Province of the Nearctic region. Proc. Acad. Nat. Sci. Phila. 18:300–314.

———. 1886 (1885). Thirteenth contribution to the herpetology of tropical America. Proc. Am. Philos. Soc. 23:271–287.

———. 1900 (1898). Crocodilians, lizards and snakes of North America. Ann. Rept. U.S. Nat. Mus.:151–1294.

Cowles, R. B., and C. M. Bogert. 1944. A preliminary study of the thermal requirements of desert reptiles. Bull. Am. Mus. Nat. Hist. 83:261–296.

Cragin, F. W. 1884. Notes on some southwestern reptiles in the cabinet of Washburn College. Bull. Washburn Lab. Nat. Hist. 1:6–8.

Davy, C. M., F. R. Méndez de la Cruz, A. Lathrop, and R. W. Murphy. 2011. Seri Indian traditional knowledge and molecular biology agree: No express train for island-hopping spiny-tailed iguanas in the Sea of Cortés. J. Biogeogr. 38:272–284.

Dibble, C. J., A. Boyd, J. A. Lemos-Espinal, G. R. Smith, and M. E. Ogle. 2007. Natural history notes: Callisaurus draconoides. Diet and clutch size. Herpetol. Rev. 38:75–76.

Dibble, C. J., G. R. Smith, and J. A. Lemos-Espinal. 2008. Diet and sexual dimorphism of the Desert Iguana Dipsosaurus dorsalis, from Sonora, México. West. N. Am. Nat. 68:521–523.

Dickerson, M. C. 1919. Diagnoses of twenty-three new species and a new genus of lizards from Lower California. Bull. Am. Mus. Nat. Hist. 41:461–477.

Dixon, J. R. 1964. Systematics and distribution of lizards of the genus Phyllodactylus in North and Central America. New Mex. State Univ. Res. Ctr. Sci. Bull. 64:1–139.

———. 1966. Speciation and systematics of the gekkonid lizard genus Phyllodactylus on the islands of the Gulf of California. Proc. Calif. Acad. Sci. 33:415–442.

Dixon, J. R., and R. H. Dean. 1986. Status of the southern populations of the night snake (Hypsiglena: Colubridae) exclusive of California and Baja California. Southwest. Nat. 31 (3): 307–318.

Enderson, E. F. 2010. The occurrence of Crotalus viridis in Sonora, Mexico. Sonoran Herpetol. 23:94–95.

Enderson, E. F., and R. L. Bezy. 2007a. Geographic distribution: *Geophis dugesii*. Herpetol. Rev. 38:103.

———. 2007b. Geographic distribution: *Lampropeltis triangulum sinaloae*. Herpetol. Rev. 38:487.

———. 2007c. Geographic distribution: *Leptodeira splendida ephippiata*. Herpetol. Rev. 38:220.

———. 2007d. Geographic distribution: *Pseudoficimia frontalis*. Herpetol. Rev. 38:105.

———. 2007e. Geographic distribution: *Syrrhophus interorbitalis*. Herpetol. Rev. 38:216.

Enderson, E. F., K. B. Bonine, and R. L. Bezy. 2006. Geographic distribution: *Gyalopion canum*. Herpetol. Rev. 37:362.

Enderson, E. F., S. F. Hale, and R. L. Bezy. 2007. Geographic distribution: *Kinosternon integrum*. Herpetol. Rev. 38:217.

Enderson, E. F., A. Quijada-Mascareñas, D. S. Turner, R. L. Bezy, and P. C. Rosen. 2010. Una sinopsis de la herpetofauna con comentarios sobre las prioridades en investigacion y conservacion. In The biological diversity of Sonora, Mexico, ed. F. E. Molina Freaner and T. R. Van Devender, 357–383. Mexico City: Universidad Nacional Autónoma de México.

Enderson, E. F., A. Quijada-Mascareñas, D. S. Turner, P. C. Rosen, and R. L. Bezy. 2009. The herpetofauna of Sonora, Mexico, with comparisons to adjoining states. Check List 5:632–672.

Enderson, E. F., T. R. Van Devender, and R. L. Bezy. 2014. Amphibians and reptiles of Yécora, Sonora and the Madrean Tropical Zone of the Sierra Madre Occidental in northwestern Mexico. Check List 10:913–926.

Felger, R. S., M. B. Johnson, and M. F. Wilson. 2001. The trees of Sonora, Mexico. New York: Oxford University Press.

Ferrusquia-Villafranca, I., and L. I. Gonzáles-Guzmán. 2005. Northern Mexico's landscape. Part II, The biotic setting across time. In Biodiversity, ecosystems, and conservation in Northern Mexico, ed. J. E. Cartron, G. Ceballos, and R. S. Felger, 39–51. New York: Oxford University Press.

Franklin, K., and F. Molina-Freaner. 2010. Consequences of buffelgrass pasture development for primary productivity, perennial plant richness, and vegetation structure in the drylands of Sonora, Mexico. Conserv. Biol. 24:1664–1673.

Girard, C. F. 1858. United States Exploring Expedition during the years 1838, 1839, 1840, 1841, 1842, under the command of Charles Wilkes, U.S.N. Vol. 20. Philadelphia: J. B. Lippincott.

Goldberg, S. R. 1997. Reproduction in the Western Coral Snake, *Micruroides euryxanthus* (Elapidae), from Arizona and Sonora, Mexico. Great Basin Nat. 57:363–365.

Goldman, E. A. 1951. Biological investigations in Mexico. Smithson. Misc. Coll. 115. Washington, DC: Smithsonian Institution.

González-Romero, A., and S. Álvarez-Cardenas. 1989. Herpetofauna de la región del Pinacate, Sonora, México: Un inventario. Southwest. Nat. 34:519–526.

Gould, S. J., and E. S. Vrba. 1982. Exaptation—a missing term in the science of form. Paleobiology 8:4–15.

Grismer, L. L. 2002. Amphibians and reptiles of Baja California including its Pacific islands and the islands in the Sea of Cortés. Berkeley: University of California Press.

Hardy, L. M., and R. W. McDiarmid. 1969. The amphibians and reptiles of Sinaloa, Mexico. Univ. Kans. Publ. Mus. Nat. Hist. 18:39–252.

Hartweg, N. 1938. *Kinosternon flavescens stejnegeri*, a new turtle from northern Mexico. Occas. Pap. Mus. Zool. Univ. Mich. 371:1–5.

Hastings, P. A., and L. T. Findley. 2007. Marine fishes of the Upper Gulf Biosphere Reserve, Northern Gulf of California. In Dry borders, great natural reserves of the Sonoran Desert, ed. R. S. Felger and B. Broyles, 364–382. Salt Lake City: University of Utah Press.

Heifetz, W. 1941. A review of the lizards of the genus *Uma*. Copeia 1941:99–111.

Heringhi, H. L. 1969. An ecological survey of the herpetofauna of Álamos, Sonora, Mexico. Master's thesis, Arizona State University, Tempe.

Howard, J. L. 1996. Paleocene to Holocene paleodeltas of ancestral Colorado River offset by San Andreas Fault system, southern California. Geology 24:783–786.

Humphrey, R. R. 1958. The desert grassland. Bot. Rev. 24:193–253.

Ingram, M. 2000. Desert storms. In A natural history of the Sonoran Desert, ed. S. J. Phillips and P. Wentworth Comus, 41–50. Tucson: Arizona–Sonora Desert Museum Press.

IPCC (Intergovernmental Panel on Climate Change). 2007. Climate change 2007: The physical science basis. Contribution of Working Group I to the Fourth Assessment Report of the Intergovernmental Panel on Climate Change, ed. S. Solomon, D. Qin, M. Manning, Z. Chen, M. Marquis, K. B. Avery, M. Tignor, and H. L. Miller. Cambridge: Cambridge University Press.

Iverson, J. B. 1981. Biosystematics of the *Kinoster-*

non hirtipes complex (Testudines: Kinosternidae). Tulane Stud. Zool. Bot. 23:1–74.

Jan, G. 1863. Elenco sistematico degli ofidi descriti e disegnati per l'iconografia generale. Milan: A. Lombardi.

Kellogg, R. 1932. Mexican tailless amphibians in the United States National Museum. Bull. U.S. Natl. Mus. 160:1–224.

Kennicott, R. 1860. Descriptions of new species of North American serpents in the Museum of the Smithsonian Institution, Washington. Proc. Acad. Nat. Sci. Phila. 12:328–338.

———. 1861. On three new forms of rattlesnakes. Proc. Acad. Nat. Sci. Phila. 13:206–208.

Klauber, L. M. 1937. A new snake of the genus *Sonora* from Mexico. Trans. San Diego Soc. Nat. Hist. 8:363–366.

———. 1945. The geckos of the genus *Coleonyx* with descriptions of new subspecies. Trans. San Diego Soc. Nat. Hist. 10 (11): 133–216.

———. 1949. Some new and revived subspecies of rattlesnakes. Trans. San Diego Society Nat. Hist. 11 (6): 61–116.

Lara-Góngora, G. 1986. New distributional records for some Mexican reptiles and amphibians. Bull. Md. Herpetol. Soc. 22:62–67.

———. 2004. A new species of *Sceloporus* (Reptilia: Sauria: Phrynosomatidae) of the *grammicus* complex from Chihuahua and Sonora, Mexico. Bull. Md. Herpetol. Soc. 40:1–41.

Larson, R. L. 1972. Bathymetry, magnetic anomalies, and plate tectonic history of the mouth of the Gulf of California. Bull. Geol. Soc. Am. 83:3345–3360.

Legler, J. M., and R. G. Webb. 1970. A new Slider Turtle (*Pseudemys scripta*) from Sonora, Mexico. Herpetologica 26:157–168.

Lemos-Espinal, J. A., D. Chiszar, H. M. Smith, and G. Woolrich-Piña. 2004. Selected records of 2003 lizards from Chihuahua and Sonora, México. Bull. Chic. Herpetol. Soc. 39:164–168.

Lemos-Espinal, J. A., and H. M. Smith. 2007. Anfibios y reptiles del estado de Chihuahua, México/Amphibians and reptiles of the state of Chihuahua, México. Mexico City: CONABIO.

———. 2009. Claves para los anfibios y reptiles de Sonora, Chihuahua y Coahuila, México/ Key to the amphibians and reptiles of Sonora, Chihuahua and Coahuila, México. Mexico City: CONABIO.

Lemos-Espinal, J. A., H. M. Smith, D. Chiszar, and G. Woolrich-Piña. 2004. Year 2003 snakes from Chihuahua and adjacent states of Mexico. Bull. Chic. Herpetol. Soc. 39:206–213.

Lemos-Espinal, J. A., H. M. Smith, J. R. Dixon, and A. Cruz. Forthcoming. Anfibios y reptiles de Sonora, Chihuahua y Coahuila, México/ Amphibians and reptiles of Sonora, Chihuahua and Coahuila, México. 2 vols. Mexico City: CONABIO.

Lemos-Espinal, J. A., H. M. Smith, D. Hartman, and D. Chiszar. 2004. Selected year 2003 amphibians and turtles from Chihuahua and Sonora, México. Bull. Chic. Herpetol. Soc. 39:107–109.

Liner, E. A., and G. Casas-Andreu. 2008. Standard Spanish, English and scientific names of the amphibians and reptiles of Mexico. 2nd ed. SSAR Herpetol. Circ. 38:1–162.

Lowe, C. H., and C. W. Howard. 1975. Viviparity and reproductive pattern in *Phrynosoma ditmarsi* in Sonora, Mexico. Southwest. Nat. 20:265–270.

Lowe, C. H., C. J. Jones, and J. W. Wright. 1968. A new plethodontid salamander from Sonora, Mexico. Contr. Sci. Los Angeles Co. Mus. Nat. Hist. 140:1–11.

Lowe, C. H., and K. S. Norris. 1955. Analysis of the herpetofauna of Baja California, Mexico. III, New and revived reptilian subspecies of Isla de San Esteban, Gulf of California, Sonora, Mexico, with notes on other satellite islands of Isla Tiburón. Herpetologica 11:89–96.

Lowe, C. H., M. D. Robinson, and V. D. Roth. 1971. A population of *Phrynosoma ditmarsi* from Sonora, Mexico. J. Ariz. Acad. Sci. 6:275–277.

Lowe, C. H., and W. H. Woodin. 1954. A new racer (genus *Masticophis*) from Arizona and Sonora, Mexico. Proc. Biol. Soc. Wash. 67:247–250.

Lumholtz, C. 1891. Report on explorations in northern Mexico. *J. Am. Geogr. Soc. N. Y.* 23:386–402.

Macedonia, J. M., A. K. Lappin, E. R. Loew, J. A. McGuire, P. S. Hamilton, M. Plasman, Y. Brant, J. A. Lemos-Espinal, and D. J. Kemp. 2009. Conspicuousness of Dickerson's Collared Lizard (*Crotaphytus dickersonae*) through the eyes of conspecifics and predators. Biol. J. Linn. Soc. 97:749–765.

Maldonado-Leal, B. G., P. L. Warren, T. R. Jones, V. L. Boyarksi, and J. C. Rorabaugh. 2009. Geographic distribution: *Hyla wrightorum*. Herpetol. Rev. 40:108.

Martin, P. S., D. Yetman, M. Fishbein, P. Jenkins, T. R. Van Devender, and R. K. Wilson. 1998.

Gentry's Río Mayo plants. Tucson: University of Arizona Press.

Martinez-Yrizar, A., R. S. Felger, and A. Burquez. 2010. Los ecosistemas terrestres: Un diverso capital natural. In Diversidad biológica de Sonora, ed. F. E. Molina Freaner and T. R. Van Devender, 129–156. Mexico City: Universidad Nacional Autónoma de México y CONABIO.

May, L. A. 1973. Resource reconnaissance of the Gran Desierto region, northwestern Sonora, Mexico. Master's thesis, University of Arizona, Tucson.

McKinney, C. O. 1969. Experimental hybridization between three populations of Uta stansburiana. Copeia 1969:289–292.

Merriam, R. H., and O. L. Bandy. 1965. Source of upper Cenozoic sediments in Colorado River delta region. J. Sediment. Petrol. 35:911–916.

Minckley, W. L., and P. C. Marsh. 2009. Inland fishes of the Greater Southwest: Chronicle of a vanishing biota. Tucson: University of Arizona Press.

Mulcahy, D. G. 2008. Phylogeography and species boundaries of the western North American Night Snake (Hypsiglena torquata): Revisiting the subspecies concept. Mol. Phylogenet. Evol. 46:1095–1115.

Mulcahy, D. G., J. E. Martinez-Gómez, G. Aguirre-Léon, J. A. Cervantes-Pasqualli, and G. R. Zug. 2014. Rediscovery of an endemic vertebrate from the remote Isla Revillagigedo in the eastern Pacific Ocean: The Clarion Nightsnake lost and found. PLOS ONE 9 (5): 1–5.

Murphy, R. W. 1976. The evolution of a peninsular and insular herpetofauna: Drift based alternative hypothesis. Master's thesis, San Francisco State University.

———. 1983a. A distributional checklist of the amphibians and reptiles on the islands in the Sea of Cortés. Appendix 6.1 and 6.2. In Island biogeography in the Sea of Cortés, ed. T. J. Case and M. L. Cody, 429–437. Berkeley: University of California Press.

———. 1983b. The reptiles: Origin and evolution. In Island biogeography in the Sea of Cortés, ed. T. J. Case and M. L. Cody, 130–158. Berkeley: University of California Press.

Murphy, R. W., and G. Aguirre-León. 2002a. A distributional checklist of the amphibians and reptiles on the islands in the Sea of Cortés. Appendix 8.2 and 8.4. In A new island biogeography of the Sea of Cortés, ed. T. J. Case, M. L.

Cody, and E. Ezcurra, 580–599. New York: Oxford University Press.

———. 2002b. The non-avian reptiles: Origins and evolution. In A new island biogeography of the Sea of Cortés, ed. T. J. Case, M. L. Cody, and E. Ezcurra, 181–220. New York: Oxford University Press.

Murphy, R. W., K. H. Berry, T. Edwards, A. E. Leviton, A. Lathrop, and J. D. Riedle. 2011. The dazed and confused identity of Agassiz's land tortoise, Gopherus agassizii (Testudines, Testudinidae) with the description of a new species, and its consequences for conservation. ZooKeys 113:39–71.

Nabhan, G. P. 2003. Singing the turtles to sea: The Comcáac (Seri) art and science of reptiles. Berkeley: University of California Press.

Nickerson, M. A., and H. L. Heringhi. 1966. Three noteworthy colubrids from southern Sonora, Mexico. Great Basin Nat. 26:136–140.

O'Brien, C., A. D. Flesch, E. Wallace, M. Bogan, S. E. Carrillo-Percástegui, S. Jacobs, and C. van Riper III. 2006. Biological inventory of the Río Aros, Sonora, México: A river unknown. Tucson: University of Arizona and Sonoran Desert Research Station.

Palacio-Baez, G., and E. Enderson. 2012. Geographic distribution: Incilius marmoreus (Marbled Toad). Herpetol. Rev. 43:613.Perrill, R. H. 1983. Geographic distribution: Phrynosoma ditmarsi. Herpetol. Rev. 14:123.

Peters, D. P. C., and K. M. Havstad. 2006. Cross-scale interactions, nonlinearities, and forecasting catastrophic events. J. Arid Environ. 65:196–206.

Pianka, E. R. 1986. Ecology and natural history of desert lizards. Princeton, NJ: Princeton University Press.

Quijada-Mascareñas, A., and E. F. Enderson. 2007. Geographic distribution: Ramphotyphlops braminus. Herpetol. Rev. 38:490.

Quijada-Mascareñas, A., E. F. Enderson, I. Parra-Salazar, and R. L. Bezy. 2007. Geographic distribution: Plestiodon obsoletus. Herpetol. Rev. 38:353.

Recchio, I. M., C. M. Rodriguez, and D. Lazcano. 2007. Geographic distribution: Geophis dugesii. Herpetol. Rev. 38:103–104.

Rorabaugh, J. C. 2008. An introduction to the herpetofauna of mainland Sonora, Mexico, with comments on conservation and management. J. Ariz.-Nev. Acad. Sci. 40:20–65.

————. 2010. Conservation of amphibians and reptiles in northwestern Sonora and southwestern Arizona. In Southwestern desert resources, ed. W. Halvorson, C. Schwalbe, and C. van Riper III, 181–204. Tucson: University of Arizona Press.

————. 2013. Geographic distribution: *Trimorphodon tau*. Herpetol. Rev. 44 (1): 111.

Rorabaugh, J. C., and E. F. Enderson. 2009. Lizards of Sonora, Mexico. In Lizards of the American Southwest, a photographic field guide, ed. L. L. C. Jones and R. E. Lovich, 508–514. Tucson, AZ: Rio Nuevo Publishers.

Rorabaugh, J. C., M. A. Gómez-Ramírez, C. E. Gutiérrez-González, J. E. Wallace, and T. R. Van Devender. 2011. Amphibians and reptiles of the Northern Jaguar Reserve and vicinity, Sonora, Mexico: A preliminary evaluation. Sonoran Herpetol. 24:123–131.

Rorabaugh, J. C., and A. King. 2013. Geographic distribution: *Apalone spinifera*. Herpetol. Rev. 44:104–105.

Rorabaugh, J. C., A. King, C. Gutiérrez-González, and M. Gómez-Ramírez. 2013. Geographic distribution: *Micrurus distans*. Herpetol. Rev. 44 (1): 110.

Rorabaugh, J. C., E. Soto Montoya, and M. M. Gómez-Sapiens. 2008. Geographic distribution: *Apalone spinifera*. Herpetol. Rev. 39:365.

Rorabaugh, J. C., and J. M. Servoss. 2006. Geographic distribution: *Rana berlandieri*. Herpetol. Rev. 37:102.

Rorabaugh, J. C., J. M. Servoss, V. L. Boyarski, E. Fernandez, D. Duncan, C. Robles Elías, and K. E. Bonine. 2013. A comparison of the herpetofaunas of Ranchos Los Fresnos and El Aribabi in northern Sonora, Mexico. In Proceedings: Merging science and management in a rapidly changing world: Biodiversity and management of the Madrean Archipelago III and 7th Conference on Research and Resource Management in the Southwestern Deserts, ed. G. J. Gottfried, P. F. Ffolliott, B. S. Gebow, L. G. Eskew, and L. C. Collins, 103–109. Fort Collins, CO: US Department of Agriculture, Forest Service, Rocky Mountain Research Station RMRS-P-67.

Rosen, P. C. 2007. Reptiles and amphibians of arid southwestern Arizona and northwestern Sonora. In Dry borders, great natural reserves of the Sonoran Desert, ed. R. S. Felger and B. Broyles, 310–337. Salt Lake City: University of Utah Press.

Rosen, P. C., and C. Melendez. 2010. Observations of the status of aquatic turtles and the occurrence of ranid frogs and other aquatic vertebrates in northwestern Mexico. In Southwestern desert resources, ed. W. Halvorson, C. Schwalbe, and C. van Riper III, 205–224. Tucson: University of Arizona Press.

Rosen, P. C., and A. Quijada-Mascareñas. 2009. Geographic distribution: *Aspidoscelis xanthonota*. Herpetol. Rev. 40:237.

Roth, V. D. 1971. Food habits of Ditmars' Horned Lizards with speculation on its type locality. J. Ariz. Acad. Sci. 6:278–281.

————. 1997. Ditmars' Horned Lizards (*Phrynosoma ditmarsi*) or the case of the lost lizard. Sonoran Herpetol. 10:2–6.

Russell, S. M., and G. Monson. 1998. The birds of Sonora. Tucson: University of Arizona Press.

Scarborough, R. 2000. The geologic origin of the Sonoran Desert. In A natural history of the Sonoran Desert, ed. S. J. Phillips and P. Wentworth Comus, 71–85. Tucson: Arizona-Sonora Desert Museum Press.

Schmidt, K. P. 1922. The amphibians and reptiles of Lower California and the neighboring islands. Bull. Am. Mus. Nat. Hist. 46:607–707.

Schwalbe, C. R., and C. H. Lowe. 2000. Amphibians and reptiles of the Sierra de Álamos. In The tropical deciduous forest of Álamos: Biodiversity of a threatened ecosystem in Mexico, ed. R. H. Robichaux and D. A. Yetman, 172–199. Tucson: University of Arizona Press.

Seminoff, J. A., and W. J. Nichols. 2007. Sea turtles of the Alto Golfo: A struggle for survival. In Dry borders, great natural reserves of the Sonoran Desert, ed. R. S. Felger and B. Broyles, 505–518. Salt Lake City: University of Utah Press.

Shannon, F. A. 1951. Notes on a herpetological collection from Oaxaca and other localities in Mexico. Proc. U.S. Nat. Mus. 101:465–484.

Sherbooke, W. C. 2003. Introduction to horned lizards of North America. Richmond: University of California Press.

Sherbrooke, W. C., B. E. Martin, and C. H. Lowe. 1998. Geographic distribution: *Phrynosoma ditmarsi*. Herpetol. Rev. 29:110–111.

Shreve, F. 1951. Vegetation of the Sonoran Desert. Publication 591. Washington, DC: Carnegie Institute of Washington.

Slevin, J. R. 1928. The amphibians of western North America. Occas. Pap. Calif. Acad. Sci. 16:1–152.

Smith, H. M. 1935a. Descriptions of new species

of lizards from Mexico of the genus *Uta*, with notes on other Mexican species. Univ. Kans. Sci. Bull. 22:157–183.

———. 1935b. Miscellaneous notes on Mexican lizards. Univ. Kans. Sci. Bull. 22:119–155.

———. 1935c. Notes on some Mexican lizards of the genus *Holbrookia*, with the descriptions of a new species. Univ. Kans. Sci. Bull. 22:185–201.

———. 1938. Remarks on the status of the subspecies of *Sceloporus undulatus*, with descriptions of new species and subspecies of the *undulatus* group. Occ. Pap. Mus. Zool. Univ. Mich. 387:1–17.

———. 1972. The Sonoran subspecies of the lizard *Ctenosaura hemilopha*. Great Basin Nat. 32:104–112.

Smith, H. M., and D. A. Langebartel. 1951. A new geographic race of Leaf-nosed Snake from Sonora, Mexico. Herpetologica 7:181–184.

Smith, H. M., J. A. Lemos-Espinal, and D. Chiszar. 2005. Amphibians and lizards from Sonora, Chihuahua and Coahuila. Bull. Chic. Herpetol. Soc. 40:45–51.

Smith, H. M., J. A. Lemos-Espinal, D. Hartman, and D. Chiszar. 2005. A new species of *Tropidodipsas* (Serpentes: Colubridae) from Sonora, Mexico. Bull. Md. Herpetol. Soc. 41:39–41.

Smith, H. M., J. A. Lemos-Espinal, and P. Heimes. 2005. 2005 Amphibians and reptiles from northwestern México. Bull. Chic. Herpetol. Soc. 40:206–212.

Smith, H. M., and W. W. Tanner. 1944. Description of a new snake from Mexico. Copeia 1944:131–136.

Smith, H. M., and E. H. Taylor. 1950. Type localities of Mexican reptiles and amphibians. Univ. Kans. Sci. Bull. 33:313–380.

Smith, P. W., and M. M. Hensley. 1958. Notes on a small collection of amphibians and reptiles from the vicinity of the Pinacate lava cap in northwestern Mexico. Trans. Kans. Acad. Sci. 61:64–76.

Stejneger, L. 1890. Part 5, Annotated list of reptiles and batrachians collected by Dr. C. Hart Merriam and Vernon Bailey on the San Francisco Mountain Plateau. In Results of a biological survey of the San Francisco Mountain region and desert of the Little Colorado, Arizona. North Am. Fauna 3:103–118.

———. 1893. Annotated list of the reptiles and batrachians collected by the Death Valley Expedition in 1891, with descriptions of new species. North Am. Fauna 7:159–228.

———. 1906. A new lizard of the genus *Phrynosoma* from Mexico. Proc. U.S. Natl. Mus. 29:575–567.

Stoleson, S. H., R. S. Felger, G. Ceballos, C. Raish, M. Wilson, and A. Búrquez. 2005. Recent history of natural resource use and population growth in northern Mexico. In Biodiversity, ecosystems, and conservation in northern Mexico, ed. J.-L. E. Cartron, G. Ceballos, and R. S. Felger, 52–86. New York: Oxford University Press.

Tanner, W. W. 1959. A new *Thamnophis* from western Chihuahua with notes on four other species. Herpetologica 15:165–172.

———. 1981. A new *Hypsiglena* from Tiburon Island, Sonora, Mexico. Great Basin Nat. 41 (1): 139–142.

Taylor, E. H. 1933. New species of skinks from Mexico. Proc. Biol. Soc. Wash. 46:175–182.

———. 1936a. Description of a new Sonora snake of the genus *Ficimia*, with notes on other Mexican species. Proc. Biol. Soc. Wash. 49:51–54.

———. 1936b. Notes on the herpetological fauna of the Mexican state of Sonora. Univ. Kans. Sci. Bull. 24:475–503.

———. 1937 (1936). Notes and comments on certain American and Mexican snakes of the genus *Tantilla*, with descriptions of new species. Trans. Kans. Acad. Sci. 39:335–348.

Townsend, C. H. 1916. Voyage of the "Albatross" to the Gulf of California in 1911. Bull. Am. Mus. Nat. Hist. 35:399–476.

Turner, R. M., and D. E. Brown. 1982. Sonoran desertscrub. In Biotic communities of the American Southwest—United States and Mexico, ed. D. E. Brown, 181–221. Desert Plants 4 (1–4): 1–342.

Van Denburgh, J. 1922. The reptiles of western North America. Occas. Pap. Calif. Acad. Sci. 10:1–1028.

Van Denburgh, J., and J. R. Slevin. 1921. Preliminary diagnoses of more new species from islands in the Gulf of California, Mexico. Proc. Calif. Acad. Sci., 4th ser., 11:395–398.

Van Devender, T. R., and E. F. Enderson. 2007. Geographic distribution: *Micrurus distans*. Herpetol. Rev. 38:488.

Van Devender, T. R., E. F. Enderson, D. S. Turner, R. A. Villa, S. F. Hale, G. M. Ferguson, and C. Hedgcock. 2013. Comparison of preliminary herpetofaunas of the Sierras la Madera (*Oposura*) and Bacadéhuachi with the mainland Sierra Madre Occidental in Sonora, Mexico. In Proceedings: Merging science and management

in a rapidly changing world: Biodiversity and
management of the Madrean Archipelago III
and 7th conference on research and resource
management in the southwestern deserts, ed.
G. J. Gottfried, P. F. Ffolliott, B. S. Gebow, L. G.
Eskew, and L. C. Collins, 110–118. RMRS-PJ-67.
Fort Collins, CO: US Department of Agriculture, Forest Service, Rocky Mountain Research
Station.

Van Devender, T. R., P. A. Holm, and C. H.
Lowe. 1989. Life history notes: *Pseudoeurycea
bellii sierraoccidentalis*. Habitat. Herpetol. Rev.
20:48–49.

Van Devender, T. R., C. H. Lowe, and P. A. Holm.
1989. Geographic distribution: *Pseudoeurycea
bellii sierraoccidentalis*. Herpetol. Rev. 20:48–49.

Van Devender, T. R., C. H. Lowe, and H. E. Lawler.
1994. Factors influencing the distribution of
the Neotropical Vine Snake *Oxybelis aeneus* in
Arizona and Sonora, Mexico. Herpetol. Nat.
Hist. 2:27–44.

Villa, R. A., P. T. Condon, T. A. Hare, S. Avilla-
Villegas, and D. G. Barker. 2007. Geographic
distribution: *Crotalus willardi willardi*. Herpetol.
Rev. 38:220.

Weiss, J. L., and J. T. Overpeck. 2005. Is the Sonoran Desert losing its cool? Glob. Change Biol.
11 (12): 2065–2077.

Wiewandt, T. A., C. H. Lowe, and M. W. Larson.
1972. Occurrence of *Hypopachus variolosus*
(Cope) in the short-tree forest of southern Sonora. Herpetologica 28:162–164.

Winter, K. E., G. R. Smith, J. A. Lemos-Espinal,
M. E. Ogle, and A. Boyde. 2007. Natural history
notes: *Leptodactylus melanonotus*. Diet. Herpetol.
Rev. 38:324.

Wood, D. A., A. G. Vandergast, J. A. Lemos-
Espinal, R. N. Fisher, and A. T. Holycross. 2011.
Refugial isolation and divergence in the Narrow-
headed Gartersnake species complex (*Tham-
nophis rufipunctatus*) as revealed by multilocus
DNA sequence data. Mol. Ecol. 20:3856–3878.

Zweifel, R. G. 1959. Variation in and distribution of lizards of western Mexico related to
Cnemidophorus sacki. Bull. Am. Mus. Nat. Hist.
117:61–116.

Zweifel, R. G., and K. S. Norris. 1955. Contribution to the herpetology of Sonora, Mexico:
Descriptions of new species of snakes (*Micruroides euryxanthus* and *Lampropeltis getulus*) and
miscellaneous collecting notes. Am. Midl. Nat.
54:230–249.

5. Chihuahua

Anderson, J. D. 1960. *Storeria storerioides* in western Mexico. Herpetologica 16:63–66.
———. 1961. The life history and systematics of
Ambystoma rosaceum. Copeia 1961:371–377.
———. 1962a. Egg laying and nesting in *Sceloporus scalaris slevini*. Herpetologica 18:162–164.
———. 1962b. *Eumeces brevilineatus* in Chihuahua and San Luis Potosí, Mexico. Herpetologica
18:56–57.
———. 1962c. A new subspecies of the Ridge-
nosed Rattlesnake, *Crotalus willardi*, from Chihuahua, Mexico. Copeia 1962:160–163.
———. 1972. Pattern polymorphism in the
Bunch-grass Lizard, *Sceloporus scalaris slevini*.
J. Herpetol. 6:80.

Baird, S. F. 1859. Reptiles of the boundary, with
notes by the naturalists of the survey. In United
States and Mexican Boundary Survey, under the
order of Lieut. Col. W. H. Emory, major First
Cavalry, and United States commissioner. Vol.
2, pt. 2, U.S. 34th Cong., 1st Sess., Exec. Doc.
108, 1–35, plates 1–41. Washington, DC: Government Printing Office.

Baird, S. F., and C. Girard. 1852. Reptiles. In
Stansbury's exploration and survey of the valley
of the Great Salt Lake of Utah, 1852, including
a reconnoissance of a new route through the
Rocky Mountains, 336–365. Philadelphia: Lippincott, Grambo.

Bogert, C. M., and J. A. Oliver. 1945. A preliminary
analysis of the herpetofauna of Sonora. Bull.
Am. Mus. Nat. Hist. 83:297–426.

Boulenger, G. A. 1917. Descriptions of new frogs of
the genus *Rana*. Ann. Mag. Nat. Hist., 8th ser.,
20:413–418.

Brennan, T. C., and A. T. Holycross. 2006. A field
guide to amphibians and reptiles in Arizona.
Phoenix: Arizona Game and Fish Department.

Chiszar, D., H. M. Smith, and J. A. Lemos-
Espinal. 1995. Two ethomorphological hypotheses regarding the arenicolous *Sceloporus undulatus speari*. J. Colo.-Wyo. Acad. Sci. 27:23.

Chrapliwy, P. S., K. Williams, and H. M. Smith.
1961. Noteworthy records of amphibians from
Mexico. Herpetologica 17:85–90.

Cole, C. J., and L. M. Hardy. 1981. Systematics of
North American colubrid snakes related to *Tan-
tilla planiceps* (Blainville). Bull. Am. Mus. Nat.
Hist. 171:199–284.

Conant, R., and J. T. Collins. 1998. A field guide

to reptiles and amphibians: Eastern and central North America. 3rd ed. exp. Boston: Houghton Mifflin.

Cope, E. D. 1879. Eleventh contribution to the herpetology of tropical America. Proc. Am. Philos. Soc. 18:261–277.

———. 1885. A contribution to the herpetology of Mexico. Proc. Am. Philos. Soc. 22:379–404.

———. 1886. Thirteenth contribution to the herpetology of tropical America. Proc. Am. Philos. Soc. 23:271–287.

———. 1887. Catalogue of batrachia and reptilia of Central America and Mexico. Bull. U.S. Natl. Mus. 32:1–98.

Degenhardt, W. G., C. W. Painter, and A. H. Price. 1996. Amphibians and reptiles of New Mexico. Albuquerque: University of New Mexico Press.

Dinerstein, E., D. Olson, J. Atchley, C. Loucks, S. Contreras-Balderas, R. Abell, E. Iñigo, E. Enkerlin, C. E. Williams, and G. Castilleja. 2000. Ecoregion-based conservation in the Chihuahuan Desert: A biological assessment and biodiversity vision. A collaborative effort by World Wildlife Fund (WWF), Comisión Nacional para el Conocimiento y Uso de la Biodiversidad (CONABIO), The Nature Conservancy, PRONATURA Noreste, e Instituto Tecnológico y de Estudios Superiores de Monterrey (ITESM). http://bva.colech.edu.mx/xmlui/bitstream /handle/1/1309/bio0164.pdf?sequence=1.

Dixon, J. R., R. K. Vaughan, and L. D. Wilson. 2000. The taxonomy of *Tantilla rubra* and allied taxa (Serpentes: Colubridae). Southwest. Nat. 45:141–153.

Domínguez, P., T. Álvarez, and P. Huerta. 1977. Colección de anfibios y reptiles del noroeste de Chihuahua México. Rev. Soc. Mex. Hist. Nat. 35:117–142.

Duellman, W. E. 2001. Hylid frogs of Middle America. Rev. and exp. ed. 2 vols. Ithaca, NY: Society for the Study of Amphibians and Reptiles.

Dugès, A. A. D. 1896. Reptiles y batracios de los Estados Unidos Mexicanos. La Naturaleza 2:479–485.

Frost, J. S., and J. T. Bagnara. 1976. A new species of leopard frog (*Rana pipiens* complex) from northwestern Mexico. Copeia 1976:332–338.

Girard, C. 1854. A list of North American bufonids, with diagnoses of new species. Proc. Acad. Nat. Sci. Phila. 7:86–88.

Goldman, E. A. 1951. Biological investigations in Mexico. Smithson. Misc. Coll. 115. Washington, DC: Smithsonian Institution.

Hardy, L. M. 1972. A systematic revision of the genus *Pseudoficimia* (Serpentes: Colubridae). J. Herpetol. 6:53–69.

Hardy, L. M., and R. W. McDiarmid. 1969. The amphibians and reptiles of Sinaloa, Mexico. Univ. Kans. Publ. Mus. Nat. Hist. 18:39–252.

INEGI (Instituto Nacional de Geografía, Estadística e Informatica—Dirección General de Geografía). 2004. Anuario estadístico: Chihuahua. Mexico City: INEGI.

Kellogg, R. 1932. Mexican tailless amphibians in the United States National Museum. Bull. U.S. Natl. Mus. 160:1–224.

Kennicott, R. 1860. Descriptions of North American serpents in the Museum of the Smithsonian Institution, Washington. Proc. Acad. Nat. Sci. Phila. 12:328–338.

Klauber, L. M. 1946. The Glossy Snake, *Arizona*, with descriptions of new subspecies. Trans. San Diego Soc. Nat. Hist. 10:163–194.

———. 1949. The subspecies of the Ring-nosed Rattlesnake, *Crotalus willardi*. Trans. San Diego Soc. Nat. Hist. 11:121–140.

Lara-Góngora, G. 2004. A new species of *Sceloporus* (Reptilia: Sauria: Phrynosomatidae) of the *grammicus* complex from Chihuahua and Sonora, Mexico. Bull. Md. Herpetol. Soc. 40:1–41.

Larson, E. T., F. van Breukelen, J. A. Lemos-Espinal, R. E. Ballinger, H. M. Smith, and D. Chiszar. 1998. Natural history notes: *Sceloporus bellii*. Pattern. Herpetol. Rev. 29:42–43.

Legler, J. M. 1959a. A new snake of the genus *Geophis* from Chihuahua, Mexico. Univ. Kans. Publ. Mus. Nat. Hist. 11:327–334.

———. 1959b. A new tortoise, genus *Gopherus*, from northcentral Mexico. Univ. Kans. Publ. Mus. Nat. Hist. 11:335–343.

Legler, J. M., and R. G. Webb. 1970. A new slider turtle (*Pseudemys scripta*) from Sonora, Mexico. Herpetologica 26:27–37.

Lemos-Espinal, J. A., D. L. Auth, D. Chiszar, and H. M. Smith. 2001a. Year 2000 amphibians taken in Chihuahua, Mexico. Bull. Md. Herpetol. Soc. 37:151–155.

———. 2001b. Year 2000 data on distribution and variation of some lizards of the *Sceloporus undulatus* complex in Chihuahua, Mexico. Bull. Chic. Herpetol. Soc. 37:29–31.

———. 2002a. Geographic distribution: *Am-*

bystoma tigrinum mavortium. Herpetol. Rev. 33:216–217.

———. 2002b. Geographic distribution: *Masticophis flagellum testaceus.* Herpetol. Rev. 33:69.

———. 2002c. Geographic distribution: *Rana forreri.* Herpetol. Rev. 33:63.

———. 2002d. Geographic distribution: *Thamnophis validus validus.* Herpetol. Rev. 33:325–326.

———. 2002e. Observations on the Chihuahua Fringe-toed Lizard, *Uma paraphygas.* Bull. Chic. Herpetol. Soc. 37:4–7.

———. 2002f. Year 2000 snakes from Chihuahua, Mexico. Bull. Chic. Herpetol. Soc. 37:51–55.

———. 2002g. Year 2001 snakes from Chihuahua, Mexico. Bull. Chic. Herpetol. Soc. 37:180–182.

Lemos-Espinal, J. A., D. Chiszar, C. Henke, and H. M. Smith. 1998. Natural history notes: *Phrynosoma cornutum.* Predation. Herpetol. Rev. 29:168.

Lemos-Espinal, J. A., D. Chiszar, M. J. Ingrasci, and H. M. Smith. 2004. Year 2002 turtles and snakes from Chihuahua, Mexico. Bull. Chic. Herpetol. Soc. 39:82–87.

Lemos-Espinal, J. A., D. Chiszar, and H. M. Smith. 1994a. The distribution of the Prairie Rattlesnake (*Crotalus v. viridis*) in Mexico. Bull. Md. Herpetol. Soc. 30:143–148.

———. 1994b. Results and their biological significance of a fall herpetological survey of the transmontane sand dunes of northern Chihuahua, Mexico. Bull. Md. Herpetol. Soc. 30:157–176.

———. 1997. Seasonal dorsal coloration variation in the lizard *Sceloporus undulatus speari* not confirmed. Bull. Chic. Herpetol. Soc. 32:173.

———. 2000. The lizard *Sceloporus merriami* [sic] in Chihuahua, Mexico. Bull. Md. Herpetol. Soc. 36:86–97.

———. 2001a. Distributional and variational data on some species of turtles and lizards from Chihuahua, Mexico. Bull. Chic. Herpetol. Soc. 36:201–208.

———. 2001b. The identity of *Sceloporus clarkii uriquensis* Tanner and Robison (Reptilia: Sauria). Bull. Md. Herpetol. Soc. 37:115–118.

———. 2001c. Natural history notes: *Sceloporus bellii.* Predation. Herpetol. Rev. 32:42–43.

———. 2002a. Geographic distribution: *Holbrookia elegans.* Herpetol. Rev. 33:225.

———. 2002b. The 2001 collection of *Sceloporus* (Reptilia: Sauria) from Chihuahua, Mexico. Bull. Chic. Herpetol. Soc. 37:163–167.

———. 2003a. Geographic distribution: *Rana magnaocularis.* Herpetol. Rev. 34:38.

———. 2003b. Knobloch's King Snake (*Lampropeltis pyromelana knoblochi*) of Mexico a species. Bull. Md. Herpetol. Soc. 39:53–58.

———. 2003c. Presence of the Río Fuerte Beaded Lizard (*Heloderma horridum exasperatum*) in western Chihuahua, Mexico. Bull. Md. Herpetol. Soc. 39:47–51.

———. 2004a. *Dryadophis cliftoni* (Serpentes: Colubridae) in Chihuahua, Mexico. Bull. Md. Herpetol. Soc. 40:77–80.

———. 2004b. Geographic distribution: *Crotalus basiliscus.* Herpetol. Rev. 35:83.

———. 2004c. Miscellaneous 2002 lizards from Chihuahua, Mexico. Bull. Chic. Herpetol. Soc. 39:1–7.

———. 2004d. Variation in *Procinura aemula,* the File-tailed Groundsnake of Mexico. Bull. Md. Herpetol. Soc. 40:61–69.

Lemos-Espinal, J. A., D. Chiszar, H. M. Smith, and C. Henke. 1999. The known distribution in 1998 of the members of the *undulatus* group of the lizard genus *Sceloporus* in Chihuahua, Mexico. Bull. Md. Herpetol. Soc. 35:152–163.

Lemos-Espinal, J. A., D. Chiszar, H. M. Smith, and G. Woolrich-Piña. 2004. Selected records of 2003 lizards from Chihuahua and Sonora, Mexico. Bull. Chic. Herpetol. Soc. 39:164–168.

Lemos-Espinal, J. A., P. Heimes, and H. M. Smith. 2007. Natural history notes: *Crotalus willardi amabilis.* Diet. Herpetol. Rev. 38:205.

Lemos-Espinal, J. A., G. R. Smith, and R. E. Ballinger. 2004. Diets of four species of horned lizards (genus *Phrynosoma*) from Mexico. Herpetol. Rev. 35:131–134.

Lemos-Espinal, J. A., G. R. Smith, R. E. Ballinger, and H. M. Smith. 2003. Ecology of *Sceloporus undulatus speari* (Sauria: Phrynosomatidae) from north-central Chihuahua, Mexico. J. Herpetol. 37:722–725.

Lemos-Espinal, J. A., G. R. Smith, H. M. Smith, and R. E. Ballinger. 2001. Diet of *Gambelia wislizenii* from Chihuahua, Mexico. Bull. Md. Herpetol. Soc. 36:115–118.

Lemos-Espinal, J. A., and H. M. Smith. 2007. Anfibios y reptiles del estado de Chihuahua, México/Amphibians and reptiles of the state of Chihuahua, México. Mexico City: CONABIO.

———. 2009. Claves para los anfibios y reptiles de Sonora, Chihuahua y Coahuila, México/ Key to the amphibians and reptiles of Sonora,

Chihuahua and Coahuila, México. Mexico City: CONABIO.

Lemos-Espinal, J. A., H. M. Smith, D. L. Auth, and D. Chiszar. 2001. The subspecies of *Sceloporus merriami* (Reptilia: Lacertilia) in Chihuahua and Durango, Mexico. Bull. Md. Herpetol. Soc. 37:123–129.

Lemos-Espinal, J. A., H. M. Smith, R. E. Ballinger, G. R. Smith, and D. Chiszar. 1997. A herpetological collection from northern Chihuahua, Mexico. Bull. Chic. Herpetol. Soc. 32:198–201.

———. 1998. A contribution to the superspecies concept of the lizard *Sceloporus undulatus belli*: A species. Southwest. Nat. 43:20–24.

Lemos-Espinal, J. A., H. M. Smith, and D. Chiszar. 2000a. Distributional records of anurans in Chihuahua, Mexico. Bull. Chic. Herpetol. Soc. 35:162–163.

———. 2000b. Geographic distribution: *Crotalus lepidus lepidus*. Herpetol. Rev. 31:113.

———. 2000c. New data on the geographic ranges of *Sceloporus belli* and *S. undulatus* in Chihuahua, Mexico (Reptilia: Sauria). Bull. Md. Herpetol. Soc. 36:133–138.

———. 2000d. New distributional and variational data on some species of lizards from Chihuahua, Mexico. Bull. Chic. Herpetol. Soc. 35:181–187.

———. 2000e. New distributional and variational data on some species of snakes from Chihuahua, Mexico. Bull. Chic. Herpetol. Soc. 35:19–24.

———. 2001a. Distributional and variational data on some species of turtles and lizards from Chihuahua, Mexico. Bull. Chic. Herpetol. Soc. 36:201–208.

———. 2001b. Geographic distribution: *Eleutherodactylus marnocki*. Herpetol. Rev. 32:270.

———. 2001c. Natural history notes: *Phrynosoma*. Protection. Herpetol. Rev. 32:41–42.

———. 2002a. Geographic distribution: *Boa constrictor imperator*. Herpetol. Rev. 32:277.

———. 2002b. Geographic distribution: *Hemidactylus turcicus turcicus*. Herpetol. Rev. 32:276.

———. 2002c. Geographic distribution: *Terrapene nelsoni klauberi*. Herpetol. Rev. 32:274.

———. 2002d. Miscellaneous 2001 lizards from Chihuahua, Mexico. Bull. Chic. Herpetol. Soc. 37:102–106.

———. 2003. 2001–2002 anurans, exclusive of *Rana*, from Durango and Chihuahua, Mexico. Bull. Md. Herpetol. Soc. 39:92–98.

———. 2004a. Introducción a los anfibios y reptiles del estado de Chihuahua, México/Introduction to the amphibians and reptiles of the state of Chihuahua, Mexico. Mexico City: CONABIO.

———. 2004b. A second record of *Pituophis deppei* (Deppe's Gopher Snake) in Chihuahua. Bull. Md. Herpetol. Soc. 40:81–83.

———. 2004c. Selected year 2003 amphibians and turtles from Chihuahua and Sonora, Mexico. Bull. Chic. Herpetol. Soc. 39:107–109.

———. 2004d. Year 2002 turtles and snakes from Chihuahua, Mexico. Bull. Chicago Herpetol. Soc. 39:82–87.

———. 2005. Apparent hybridization of *Bufo mazatlanensis* and *B. punctatus* (Anura: Bufonidae) in nature. Bull. Md. Herpetol. Soc. 41:42–44.

———. 2006a. *Agkistrodon b. bilineatus* in western Chihuahua, México. Bull. Md. Herpetol. Soc. 42:173–174.

———. 2006b. *Syrrhophus interorbitalis* (Amphibia: Anura) in Chihuahua, México. Bull. Md. Herpetol. Soc. 42:176—178.

Lemos-Espinal, J. A., H. M. Smith, D. Chiszar, and D. L. Auth. 2002. Geographic distribution: *Sceloporus merriami annulatus*. Herpetol. Rev. 32:276.

Lemos-Espinal, J. A., H. M. Smith, D. Chiszar, and G. Woolrich-Piña. 2004. Year 2003 Snakes from Chihuahua and adjacent state of Mexico. Bull. Chic. Herpetol. Soc. 39:206–213.

Lemos-Espinal, J. A., H. M. Smith, J. R. Dixon, and A. Cruz. Forthcoming. Anfibios y reptiles de Sonora, Chihuahua y Coahuila, México/Amphibians and reptiles of Sonora, Chihuahua and Coahuila, México. 2 vols. Mexico City: CONABIO.

Lemos-Espinal, J. A., H. M. Smith, D. Hartman, and D. Chiszar. 2004. Selected year 2003 amphibians and turtles from Chihuahua and Sonora, Mexico. Bull. Chic. Herpetol. Soc. 39:107–109.

Lemos-Espinal, J. A., J. M. Walker, and H. M. Smith. 2003. Natural history notes: *Cnemidophorus costatus barrancarum*. Habitat. Herpetol. Rev. 34:365–366.

Lemos-Espinal, J. A., R. G. Webb, D. Chiszar, and H. M. Smith. 2000. Geographic distribution: *Barisia imbricata ciliaris*. Herpetol. Rev. 31:112.

Liner, E. A., and G. Casas-Andreu. 2008. Standard Spanish, English and scientific names of the amphibians and reptiles of Mexico. 2nd ed. SSAR Herpetol. Circ. 38:1–162.

Mittleman, M. B. 1942. A summary of the iguanid genus *Urosaurus*. Bull. Mus. Comp. Zool. 91:103–181.

Owen, R. 1844. Characters of a new species of axolotl. Ann. Mag. Nat. Hist. 15 (88): 1–23.

Reynolds, L. P. 1982. Seasonal incidence of snakes in northeastern Chihuahua, Mexico. Southwest. Nat. 27:161–166.

Santos-Barrera G., and O. Flores-Villela. 2011. A new species of toad of the genus *Incilius* from the Sierra Madre Occidental of Chihuahua, Mexico (Anura: Bufonidae). J. Herpetol. 45:211–215.

Seidel, M. E. 2002. Taxonomic observations on extant species and subspecies of slider turtles, genus *Trachemys*. J. Herpetol. 36:285–292.

SEMARNAP (Secretaría del Medio Ambiente Recursos Naturales y Pesca). 1997. Programa de manejo del Área de Protección de Flora y Fauna Cañón de Santa Elena. Mexico City: SEMARNAP.

Shaffer, H. B., and M. L. McKnight. 1996. The polytypic species revisited: Genetic differentiation and molecular phylogenetics of the Tiger Salamander *Ambystoma tigrinum* (Amphibia: Caudata) complex. Evolution 50:417–433.

Smith, H. M. 1935a. Descriptions of new species of lizards from Mexico of the genus *Uta*, with notes on other Mexican species. Univ. Kans. Sci. Bull. 22:157–183.

———. 1935b. Notes on some Mexican lizards of the genus *Holbrookia*, with the descriptions of a new species. Univ. Kans. Sci. Bull. 22:185–201.

———. 1941. Notes on Mexican snakes of the genus *Masticophis*. J. Wash. Acad. Sci. 31:388–398.

———. 1942a. A resumé of Mexican snakes of the genus *Tantilla*. Zoologica 27:33–42.

———. 1942b. The synonymy of the garter snakes (*Thamnophis*), with notes on Mexican and Central American species. Zoologica 27:97–123.

Smith, H. M., and D. Chiszar. 2003. Distributional and variational data on the frogs of the genus *Rana* in Chihuahua, Mexico, including a new species. Bull. Md. Herpetol. Soc. 39:59–66.

Smith, H. M., D. Chiszar, and J. A. Lemos-Espinal. 1995. A new subspecies of the polytypic lizard species *Sceloporus undulatus* (Sauria: Iguanidae) from northern Mexico. Tex. J. Sci. 47:117–143.

Smith, H. M., J. A. Lemos-Espinal, and D. Chiszar. 2003. New subspecies of *Sceloporus merriami* (Reptilia: Lacertilia) and the derivation of its subspecies. Southwest. Nat. 48:700–705.

———. 2005. Amphibians and lizards from Sonora, Chihuahua and Coahuila. Bull. Chic. Herpetol. Soc. 40:45–51.

Smith, H. M., J. A. Lemos-Espinal, D. Chiszar, and M. J. Ingrasci. 2003. The Madrean Alligator Lizard of the Sierra del Nido, Chihuahua, Mexico (Reptilia: Sauria: Anguidae). Bull. Md. Herpetol. Soc. 39:99–102.

Smith, H. M., and E. H. Taylor. 1945. An annotated checklist and key to the snakes of Mexico. Bull. U.S. Natl. Mus. 187:1–239.

———. 1948. An annotated checklist and key to the amphibia of Mexico. Bull. U.S. Natl. Mus. 194:1–118.

———. 1950a. An annotated checklist and key to the reptiles of Mexico exclusive of snakes. Bull. U.S. Natl. Mus. 199:1–293.

———. 1950b. Type localities of Mexican reptiles and amphibians. Univ. Kans. Sci. Bull. 33: 313–380.

———. 1966. Herpetology of Mexico: Annotated checklists and keys to the amphibians and reptiles. A reprint of Bulletins 187, 194, and 199 of the U.S. National Museum with a list of subsequent taxonomic innovations. Ashton, MD: Eric Lundberg.

Smith, H. M., K. L. Williams, and E. O. Moll. 1963. Herpetological exploration of the Río Conchos, Chihuahua, Mexico. Herpetologica 19:205–215.

Stebbins, R. C. 2003. A field guide to western reptiles and amphibians. 3rd ed. Boston: Houghton Mifflin.

Stejneger, L. 1890. On the North American lizards of the genus *Barissia* of Gray. Proc. U.S. Natl. Mus. 13 (809): 183–185.

Tanner, W. W. 1957. A new skink of the *multivirgatus* group from Chihuahua. Great Basin Nat. 17:111–117.

———. 1961. A new subspecies of *Conopsis nasus* from Chihuahua, Mexico. Herpetologica 17:13–18.

———. 1985. Snakes of western Chihuahua. Great Basin Nat. 45:615–676.

———. 1987. Lizards and turtles of western Chihuahua. Great Basin Nat. 47:383–421.

———. 1988. *Eumeces multilineatus*. Cat. Am. Amphib. Rept. 446:1.

———. 1989. Amphibians of western Chihuahua. Great Basin Nat. 49:38–70.

———. 1990. *Thamnophis rufipunctatus*. Cat. Am. Amphib. Rept. 505:1–2.

Tanner, W. W., and W. G. Robison. 1959. A collection of herptiles from Urique, Chihuahua. Great Basin Nat. 19:75–82.

Taylor, E. H. 1940a. A new frog from the Tarahumara Mountains of Mexico. Copeia 1940:250–253.

———. 1940b. A new *Lampropeltis* from western Mexico. Copeia 1940:253–255.

———. 1941. A new ambystomid salamander from Chihuahua. Copeia 1941:143–146.

———. 1944. Two new species of crotalid snakes from Mexico. Univ. Kans. Sci. Bull. 30:47–56.

Taylor, E. H., and I. W. Knobloch. 1940. Report on an herpetological collection from the Sierra Madre Mountains of Chihuahua. Proc. Biol. Soc. Wash. 53:125–130.

Taylor, H. L., J. A. Lemos-Espinal, and H. M. Smith. 2003. Morphological characteristics of a newly discovered population of *Aspidoscelis tesselatus* (Squamata: Teiidae) from Chihuahua, Mexico, the identity of an associated hybrid, and a pattern of geographic variation. Southwest. Nat. 48:692–700.

Tihen, J. A. 1948. Two races of *Elgaria kingii* Gray. Trans. Kans. Acad. Sci. 51 (3): 299–301.

Trueb, L. 1969. *Pternohyla* Boulenger. Burrowing tree frogs. Cat. Am. Amphib. Rept. 77:1–4.

Uriarte-Garzón, P., and U. O. García-Vázquez. 2014. Primer registro de *Nerodia erythrogaster bogerti* (Conant, 1953) (Serpentes: Colubridae) para el estado de Chihuahua, México. Acta Zool. Mex., n.s., 30:221–225.

Van Devender, T. R. 1973a. Behavior and disruptive coloration in the New Mexico Gartersnake *Thamnophis sirtalis ornata*. Southwest. Nat. 18:247–248.

———. 1973b. Populations of *Ambystoma tigrinum* and *A. rosaceum* in Chihuahua, Mexico. J. Ariz. Acad. Sci. 8:84.

Van Devender, T. R., and C. H. Lowe Jr. 1977. Amphibians and reptiles of Yepómera, Chihuahua, Mexico. J. Herpetol. 11:41–50.

Van Devender, T. R., and W. Van Devender. 1985. Ecological notes on two Mexican skinks. Southwest. Nat. 20:279–282.

Villa, R. A., R. W. Bryson, and R. Ramírez-Chaparro. 2012. Geographic distribution: *Rhadinaea laureata*. Herpetol. Rev. 43:308–309.

Webb, R. G. 2001. Frogs of the *tarahumarae* group in western Mexico. In Mesoamerican herpetology: Systematics, zoogeography, and conservation, ed. J. D. Johnson, R. G. Webb, and O. A. Flores Villela, 20–43. El Paso: University of Texas at El Paso Special Publication.

Webb, R. G., and R. H. Baker. 1984. Terrestrial vertebrates of the Cerro Mohinora region, Chihuahua, Mexico. Southwest. Nat. 29:243–246.

Williams, K. L. 1994. *Lampropeltis triangulum*. Cat. Am. Amphib. Rept. 594:1–10.

Williams, K. L., P. S. Chrapliwy, and H. M. Smith. 1959. A new Fringe-footed Lizard (*Uma*) from Mexico. Trans. Kans. Acad. Sci. 62:166–172.

———. 1960. Snakes from northern Mexico. Chic. Acad. Sci. Nat. Hist. Misc. 177:1–8.

Williams, K. L., H. M. Smith, and P. S. Chrapliwy. 1960. Turtles and lizards from northern Mexico. Trans. Ill. Acad. Sci. 53:36–45.

Wilson, L. D. 1966. The range of the Rio Grande Racer in Mexico and the status of *Coluber oaxaca* (Jan). Herpetologica 22:42–47.

Wright, J. W. 1971. *Cnemidophorus neomexicanus*. Cat. Am. Amphib. Rept. 109:1–3.

Wright, J. W., and C. H. Lowe. 1993. Synopsis of the subspecies of the Little Striped Whiptail Lizard, *Cnemidophorus inornatus* Baird. Ariz.-Nev. Acad. Sci. 27:129–157.

6. Coahuila

Axtell, R. W. 1960. New subspecies of *Eumeces dicei* from the Sierra Madre of northeastern México. Copeia 1960:19–26.

Axtell, R. W., and C. A. Axtell. 1971. A new lizard (*Sceloporus jarrovii cyanostictus*) from the Sierra Madre of Coahuila, Mexico. Copeia 1971:89–98.

Axtell, R. W., and M. D. Sabath. 1963. *Crotalus pricei miquihuanus* from the Sierra Madre of Coahuila, Mexico. Copeia 1963:161–164.

Axtell, R. W., and R. G. Webb. 1995. Two new *Crotaphytus* from southern Coahuila and the adjacent states of central Mexico. Bull. Chic. Acad. Sci. 16:1–15.

Baird, S. F. 1859a (1858). Description of new genera and species of North American lizards in the Museum of the Smithsonian Institution. Proc. Acad. Nat. Sci. Phila.10:253–256.

———. 1859b. Reptiles of the boundary, with notes by the naturalists of the survey. In United States and Mexican Boundary Survey, under the order of Lieut. Col. W. H. Emory, major First Cavalry, and United States commissioner. Vol. 2, pt. 2, U.S. 34th Cong., 1st Sess., Exec. Doc. 108, 1–35, plates 1–41. Washington, DC: Government Printing Office.

Berlandier, J. L., and R. Chovell. 1850. Diario de viaje de la Comisión de Límites. México, Tip. de J. R. Navarro.

Castañeda-Gaytán, G., C. García-de la Peña, and D. Lazcano. 2004. Notes on herpetofauna 5: Herpetofauna of the sand dunes of Viesca, Coahuila, México: Preliminary list. Bull. Chic. Herpetol. Soc. 39:65–68.

Chrapliwy, P. S. 1956. Extensions of known range of certain amphibians and reptiles of Mexico. Herpetologica 12:121–124.

Chrapliwy, P. S., K. Williams, and H. M. Smith. 1961. Noteworthy records of amphibians from Mexico. Herpetologica 17:85–90.

Conant, R. 1975. A field guide to reptiles and amphibians: Eastern and central North America. 2nd ed. Boston: Houghton Mifflin.

Conant, R., and J. T. Collins. 1998. A field guide to reptiles and amphibians: Eastern and central North America. 3rd ed. exp. Boston: Houghton Mifflin.

Contreras-Arquieta, A. 1989. Variación morfológica y relaciones ecológicas y zoogeográficas de la herpetofauna del valle de Cuatro Ciénegas, Coahuila, México. Master's thesis, Universidad Autónoma de Nuevo León.

Cope, E. D. 1861. Descriptions of reptiles from tropical America and Asia. Proc. Acad. Nat. Sci. Phila. 1860:368–374.

———. 1892. A synopsis of the species of the teid genus *Cnemidophorus*. Trans. Amer. Philos. Soc. 17:27–52.

———. 1963. On *Trachycephalus*, *Scaphiopus* and other American batrachia. Proc. Acad. Nat. Sci. Phila. 15:43–54.

Dixon, J. R. 2000. Amphibians and reptiles of Texas, with keys, taxonomic synopses, bibliography, and distributional maps. 2nd ed. College Station: Texas A&M University Press.

Duellman, E. W., and R. G. Zweifel. 1962. A synopsis of the lizards of the sexlineatus group (genus *Cnemidophorus*). Bull. Am. Mus. Nat. Hist. 123:155–210.

Fugler, C. M., and R. G. Webb. 1956. Distributional notes on some reptiles and amphibians from southern and central Coahuila. Herpetologica 12:167–171.

Gadsden-Esparza, H., J. L. Estrada-Rodríguez, and S. V. Leyva-Pacheco. 2006. Checklist of amphibians and reptiles of the Comarca Lagunera in Durango-Coahuila, Mexico. Bull. Chic. Herpetol. Soc. 41:2–9.

Gadsden-Esparza, H., H. López-Corrujedo, J. L. Estrada-Rodríguez, and U. Romero-Méndez. 2001. Biología poblacional de la lagartija de arena *Uma exsul* (Sauria: Phrynosomatidae):

Implicaciones para su conservación. Bol. Soc. Herpetol. Méx. 9:51–66.

Gadsden-Esparza, H., F. R. Méndez de la Cruz, R. Gil-Martínez, and G. Casas-Andreu. 1993. Patrón reproductivo de una lagartija (*Uma paraphygas*) en peligro de extinción. Bol. Soc. Herpetol. Méx. 5:42–50.

Gadsden-Esparza, H., and L. E. Palacios-Orona. 1997. Seasonal dietary patterns of the Mexican Fringe-toed Lizard (*Uma paraphygas*). J. Herpetol. 31:1–9.

Gadsden-Esparza, H., L. E. Palacios-Orona, and G. A. Cruz-Soto. 2001. Diet of the Mexican Fringe-toed Lizard (*Uma exsul*). J. Herpetol. 35:493–496.

García-Vázquez, U. O., L. Canseco-Márquez, and A. Nieto-Montes de Oca. 2010. A new species of *Scincella* (Squamata: Scincidae) from the Cuatro Ciénegas Basin, Coahuila, Mexico. Copeia 2010:373–381.

Garman, S. 1883. The reptiles and batrachians of North America. Part 1, Ophidia. Mem. Mus. Comp. Zool. 8:i–xxxiv, 1–185.

———. 1887. Reptiles and batrachians from Texas and Mexico. Bull. Essex Inst. 19:119–138.

Garza-Tobón, D., and J. A. Lemos-Espinal. 2013a. Geographic distribution: *Ecnomiohyla miotympanum*. Herpetol. Rev. 44:103.

Garza-Tobón, D., and J. A. Lemos-Espinal. 2013b. Geographic distribution: *Lithobates catesbeianus*. Herpetol. Rev. 44:104.

Gloyd, H. K., and H. M. Smith. 1942. Amphibians and reptiles from the Carmen Mountains, Coahuila. Bull. Chic. Acad. Sci. 6:231–235.

Goldman, E. A. 1951. Biological investigations in Mexico. Smithson. Misc. Coll. 115. Washington, DC: Smithsonian Institution.

González-Alonso, H. A., F. Mendoza-Quijano, R. R. Montanucci, and E. A. Liner. 1988. Una colecta herpetológica en el norte de Coahuila. Escuela Nacional de Estudios Profesionales Iztacala, Coloquio de Investigaciones 8:26.

Iverson, J. B. 1981. Biosystematics of the *Kinosternon hirtipes* complex (Testudines: Kinosternidae). Tulane Stud. Zool. Bot. 23:1–74.

Kennicott, R. 1860. Descriptions of North American serpents in the Museum of the Smithsonian Institution, Washington. Proc. Acad. Nat. Sci. Phila. 12:328–338.

Legler, J. M. 1960. A new subspecies of slider turtle (*Pseudemys scripta*) from Coahuila, Mexico. Univ. Kans. Publ. Mus. Nat. Hist. 13:73–84.

———. 1990. The genus *Pseudemys* in Mesoamerica: Taxonomy, distribution and origins. In Life history and ecology of the slider turtle, ed. J. W. Gibbons, 82–105. Washington, DC: Smithsonian Institution Press.

Lemos-Espinal, J. A., and H. M. Smith. 2007. Anfibios y reptiles del estado de Coahuila, México/ Amphibians and reptiles of the state of Coahuila, México. Mexico City: CONABIO.

———. 2009. Claves para los anfibios y reptiles de Sonora, Chihuahua y Coahuila, México/ Key to the amphibians and reptiles of Sonora, Chihuahua and Coahuila, México. Mexico City: CONABIO.

Lemos-Espinal, J. A., G. R. Smith, and R. E. Ballinger. 2004. Diets of four species of horned lizards (genus *Phrynosoma*) from Mexico. Herpetol. Rev. 35:131–134.

Lemos-Espinal, J. A., H. M. Smith, D. Chiszar, and G. Woolrich-Piña. 2004. Year 2003 snakes from Chihuahua and adjacent state of Mexico. Bull. Chic. Herpetol. Soc. 39:206–213.

Lemos-Espinal, J. A., H. M. Smith, J. R. Dixon, and A. Cruz. Forthcoming. Anfibios y reptiles de Sonora, Chihuahua y Coahuila, México/ Amphibians and reptiles of Sonora, Chihuahua and Coahuila, México. Mexico City: CONABIO.

Liner, E. A., and G. Casas-Andreu. 2008. Standard Spanish, English and scientific names of the amphibians and reptiles of Mexico. 2nd ed. SSAR Herpetol. Circ. 38:1–162.

Liner, E. A., R. M. Johnson, and A. H. Chaney. 1977. A contribution to the herpetology of northern Coahuila, Mexico. Trans. Kans. Acad. Sci. 80:50–53.

Liner, E. A., R. R. Montanucci, A. González-Alonso, and F. Mendoza-Quijano. 1993. An additional contribution to the herpetology of northern Coahuila, Mexico. Bol. Soc. Herpetol. Méx. 5:9–11.

McCoy, C. J. 1970. A new alligator lizard (genus *Gerrhonotus*) from the Cuatro Ciénegas Basin, Coahuila, México. Southwest. Nat. 15:37–44.

———. 1984. Ecological and zoogeographic relationships of amphibians and reptiles of the Cuatro Ciénegas Basin. J. Ariz.-Nev. Acad. Sci. 19:49–59.

McCoy, C. J., and W. L. Minckley. 1969. *Sistrurus catenatus* (Reptilia: Crotalidae) from the Cuatro Ciénegas Basin, Coahuila, México. Herpetologica 25:152–153.

McGuire, J. A. 1996. Phylogenetic systematics of crotaphytid lizards (Reptilia: Iguanidae:

Crotaphytidae). Bull. Carnegie Mus. Nat. Hist. 32:1–143.

Mendoza-Quijano, F., E. A. Liner, R. R. Montanucci, and A. González-Alonso. 1993. Geographic distribution: *Sceloporus serrifer cyanogenys*. Herpetol. Rev. 24:155.

Montanucci, R. R. 1971. Ecological and distributional data on *Crotaphytus reticulatus* (Sauria, Iguanidae). Herpetologica 27:183–197.

Pinkava, D. J. 1979. Vegetation and flora of the Bolsón of Cuatro Ciénegas region, Coahuila, México: I. Bol. Soc. Bot. Méx. 38:35–73.

Powell, R., N. A. Laposha, D. D. Smith, and J. S. Parmerlee. 1984. New distributional records for some semiaquatic amphibians and reptiles from the Río Sabinas basin, Coahuila, Mexico. Herpetol. Rev. 15:78–79.

Rossman, D. A. 1963. The colubrid snake genus *Thamnophis*: A revision of the *sauritus* group. Bull. Fla. State Mus. Biol. Sci. 7:99–178.

———. 1969. A new natricine snake of the genus *Thamnophis* from northern Mexico. Occas. Pap. Mus. Zool. La. State Univ. 39:1–4.

Rossman, D. A., N. B. Ford, and R. A. Seigel. 1996. The gartersnakes: Evolution and ecology. Norman: University of Oklahoma Press.

Rzedowski, J. 1978. Vegetación de México. Mexico City: Editorial Limusa.

Savitzky, A. H., and J. T. Collins. 1971a. The ground snake *Sonora episcopa episcopa* in Coahuila, Mexico. J. Herpetol. 5:87–88.

———. 1971b. *Tantilla gracilis*, a snake new to the fauna of Mexico. J. Herpetol. 5:86–87.

Schmidt, K. P. 1921. New species of North American lizards of the genera *Holbrookia* and *Uta*. Am. Mus. Novit. 22:1–6.

Schmidt, K. P., and C. M. Bogert. 1947. A new Fringe-footed Sand Lizard from Coahuila, Mexico. Am. Mus. Novit. 1339:1–9.

Schmidt, K. P., and D. W. Owens. 1944. Amphibians and reptiles of northern Coahuila, Mexico. Publ. Field Mus. Nat. Hist. Zool. Ser. 29:97–115.

Smith, H. M. 1936. Two new subspecies of Mexican lizards of the genus *Sceloporus*. Copeia 1936:223–230.

———. 1938. Remarks on the status of the subspecies of *Sceloporus undulatus*, with descriptions of new species and subspecies of the *undulatus* group. Occas. Pap. Mus. Zool. Univ. Mich. 387:1–17.

———. 1942. Mexican herpetological miscellany. Proc. U.S. Natl. Mus. 92:349–395.

Smith, H. M., and B. C. Brown. 1941. New subspecies of *Sceloporus jarrovii* from Mexico. Publ. Field Mus. Nat. Hist. Zool. Ser. 24:253–257.

Smith, H. M., R. Conant, and D. Chiszar. 2003. Berlandier's herpetology of Tamaulipas, Mexico, 150 years ago. Newsl. Bull. Int. Soc. Hist. Bibliog. Herpetol. 4:19–30.

Smith, H. M., J. A. Lemos-Espinal, and D. Chiszar. 2003. New subspecies of *Sceloporus merriami* (Reptilia: Lacertilia) and the derivation of its subspecies. Southwest. Nat. 48:700–705.

———. 2005a. 2004 amphibians and lizards from Sonora, Chihuahua and Coahuila, Mexico. Bull. Chic. Herpetol. Soc. 40:45–51.

———. 2005b. 2004 snakes from Sonora, Chihuahua and Coahuila, Mexico. Bull. Chic. Herpetol. Soc. 40:66–70.

Villarreal, J. Á., and R. Valdés. 1992–1993. Vegetación de Coahuila, México. Rev. Man. Pastizales 6:9–18.

Ward, J. P. 1984. Relationships of the chrysemyd turtles of North America (Testudines: Emydidae). Spec. Publ. Mus. Tex. Tech. Univ. 21:1–50.

Webb, R. G., and J. M. Legler. 1960. A new softshell turtle (genus *Trionyx*) from Coahuila, Mexico. Univ. Kans. Sci. Bull. 40:21–30.

Williams, K. L., P. S. Chrapliwy, and H. M. Smith. 1959. A new Fringe-footed Lizard (*Uma*) from Mexico. Trans. Kans. Acad. Sci. 62:166–172.

———. 1961. Snakes from northern Mexico. Chic. Acad. Sci. Nat. Hist. Misc. 177:1–8.

Williams, K. L., H. M. Smith, and P. S. Chrapliwy. 1960. Turtles and lizards from northern Mexico. Trans. Ill. Acad. Sci. 53:36–45.

Wills, F. H. 1977. Distribution, geographic variation and natural history of *Sceloporus parvus* Smith (Sauria: Iguanidae). Master's thesis, Texas A&M University, College Station.

Wright, J. W., and C. H. Lowe. 1993. Synopsis of the subspecies of the Little Striped Whiptail Lizard, *Cnemidophorus inornatus* Baird. J. Ariz.-Nev. Acad. Sci. 27:129–157.

Zweifel, R. G. 1956. A survey of the frogs of the *augusti* group, genus *Eleutherodactylus*. Am. Mus. Novit. 1813:1–35.

———. 1958. The lizard *Eumeces tetragrammus* in Coahuila, Mexico. Herpetologica 14:175.

7. Nuevo León

Alanís-Flores G. J., G. Cano y Cano, and M. Rovalo-Merino. 1996. Vegetación y flora de Nuevo León, una guía botánico-ecológica. Monterrey: Impresora.

Baird, S. F. 1859 (1858). Description of new genera and species of North American lizards in the Museum of the Smithsonian Institution. Proc. Acad. Nat. Sci. Phila. 10:253–256.

Banda-Leal J., R. W. Bryson Jr., and D. Lazcano-Villarreal. 2002. New record of *Elgaria parva* (Lacertilia: Anguidae) from Nuevo León, Mexico. Southwest. Nat. 47:614–615.

———. 2005. Natural history notes: *Gerrhonotus parvus*. Maximum size. Herpetol. Rev. 36:449.

Beltz, E. 2006. Scientific and common names of the reptiles and amphibians of North America—explained. http://ebeltz.net/herps/etymain.html.

Bryson, R. W., Jr. 2005. Phylogenetic relationships of the *Lampropeltis mexicana* complex (Serpentes: Colubridae) as inferred from mitochondrial DNA sequences. Master's thesis, Sul Ross State University, Alpine, Texas.

Bryson, R. W., and D. Lazcano-Villareal. 2005. *Gerrhonotus parvus*. Reptilia (GB) 39:69–72.

Bryson, R. W., Jr., J. Pastorini, F. T. Burbrink, and M. R. J. Forstner. 2007. A phylogeny of the *Lampropeltis mexicana* complex (Serpentes: Colubridae) based on mitochondrial DNA sequences suggests evidence for species-level polyphyly within *Lampropeltis*. Mol. Phylogenet. Evol. 43:674–684.

Chaney, A. H., and E. A. Liner. 1986. Geographic distribution: *Rhadinaea montana*. Herpetol. Rev. 17:67.

———. 1990. Geographic distribution: *Rhadinaea montana* (Nuevo Leon Yellow-lipped Snake). Herpetol. Rev. 21: 23–24.

Chaney, A. H., E. A. Liner, and R. M. Johnson. 1982a. Geographic distribution: *Chiropterotriton prisca*. Herpetol. Rev. 13:51.

———. 1982b. Geographic distribution: *Sceloporus poinsettii poinsettii*. Herpetol. Rev. 13:52.

Cochran, D. M. 1961. Type specimens of reptiles and amphibians in the United States National Museum. Bull. U.S. Natl. Mus. 220:1–291.

Cole, C. J. 1965. The Regal Ringneck Snake (*Diadophis punctatus regalis* Baird and Girard) in Nuevo León, Mexico. Herpetologica 21:156.

Conant, R. 1968. Zoological exploration in Mexico—the route of Lieut. D. N. Couch in 1853. Am. Mus. Novit. 2350:1–14.

Conroy, C. J., R. W. Bryson Jr., D. Lazcano-Villarreal, and A. Knight. 2005. Phylogenetic placement of the Pygmy Alligator Lizard based on mitochondrial DNA. J. Herpetol. 39:142–147.

Contreras-Arquieta A., and D. Lazcano-Villarreal. 1995. Lista revisada de los reptiles del estado de Nuevo León, México. In Listado preliminar de la fauna silvestre del estado de Nuevo León, México, ed. S. Contreras-Balderas, F. González-Saldívar, D. Lazcano-Villarreal, and A. Contreras-Arquieta, 57–64. Mexico City: Consejo Consultivo para la Preservación y Fomento de la Flora y Fauna Silvestre de Nuevo León, Gobierno del Estado de Nuevo León.

Contreras-Lozano, J. A., D. Lazcano-Villarreal, A. J. Contreras-Balderas, and P. A. Lavín-Murcio. 2010. Notes on Mexican herpetofauna 14: An update to the herpetofauna of Cerro El Potosí, Galeana, Nuevo León, Mexico. Bull. Chic. Herpetol. Soc. 45:41–46.

Cope, E. 1885. A contribution of the herpetology of Mexico. Proc. Am. Philos. Soc. 22:379–404.

Duméril, A. M. C., G. Bibron, and A. H. A. Duméril. 1854. Erpétologie générale ou histoire naturelle complète des reptiles. Vol. 7, pt. 1. Paris: Roret.

Dunn, E. R. 1936. The amphibians and reptiles of the Mexican expedition of 1934. Proc. Acad. Nat. Sci. Phila. 88:474–475.

Gadow, H. 1906. A contribution to the study of evolution based upon the Mexican species of *Cnemidophorus*. Proc. Zool. Soc. Lond. 1906:277–375.

Girard, C. 1854. A list of the North American bufonids, with diagnoses of new species. Proc. Acad. Nat. Sci. Phila. 7:86–88.

Gloyd, H. K. 1940. The rattlesnakes, genera *Sistrurus* and *Crotalus*. Spec. Publ. Chic. Acad. Sci. 4: 104–118.

Günther, A. C. L. G. 1885–1902. Reptilia and Batrachia. In Biologia Centrali-Américana, ed. O. Salvin and F. D. Godman. London: R. H. Porter. Reprinted by SSAR in 1987.

Harris, H. S., and R. S. Simmons. 1978. A preliminary account of the rattlesnakes with the description of four new subspecies. Bull. Md. Herpetol. Soc. 14:105–211.

INEGI (Instituto Nacional de Geografía, Estadística e Informatica—Dirección General de Geografía). 2010. Anuario estadístico: Nuevo León. Mexico City: INEGI.

Johnson, R. M., E. A. Liner, and A. H. Chaney. 1982a. Geographic distribution: *Pseudoeurycea galeanae*. Herpetol. Rev. 13:51.

———. 1982b. Geographic distribution: *Sceloporus couchi*. Herpetol. Rev. 13:52.

Kellogg, R. 1932. Mexican tailless amphibians in the United States National Museum. Bull. U.S. Natl. Mus. 160:1–224.

Kennicott, R. 1860. Descriptions of new species of North American serpents in the Museum of the Smithsonian Institution, Washington. Proc. Acad. Nat. Sci. Phila. 12:328–338.

Knight, R. A., and J. F. Scudday. 1985. A new *Gerrhonotus* (Lacertilia: Anguidae) from the Sierra Madre Oriental, Nuevo León, Mexico. Southwest. Nat. 30 (1): 89–94.

Lazcano, D., A. Contreras-Balderas, J. I. González-Rojas, G. Castañeda, C. García de la Peña, and C. Solis-Rojas. 2004. Notes on Mexican herpetofauna 6: Herpetofauna of Sierra San Antonio Peña Nevada, Zaragoza, Nuevo León, Mexico: Preliminary list. Bull. Chic. Herpetol. Soc. 39:181–187.

Lazcano, D., A. J. A. Contreras-Lozano, J. Gallardo-Valdez, C. García de la Peña, and G. Castañeda. 2009. Notes on Mexican herpetofauna 11: Herpetological diversity in Sierra "Cerro de la Silla" (Saddleback Mountain), Nuevo León, Mexico: Preliminary list. Bull. Chic. Herpetol. Soc. 42:1–6.

Lazcano, D., A. Kardon, R. J. Muscher, and J. A. Contreras-Lozano. 2011. Notes on Mexican herpetofauna 16: Captive husbandry—propagation of the exile Mexican Garter Snake, *Thamnophis exsul* Rossman, 1969. Bull. Chic. Herpetol. Soc. 46:13–17.

Lazcano, D., M. A. Salinas-Camarena, and J. A. Contreras-Lozano. 2009. Notes on Mexican herpetofauna 12: Are roads in Nuevo León, Mexico, taking their toll on snake populations? Bull. Chic. Herpetol. Soc. 44:69–75.

Lazcano, D., A. Sánchez-Almazán, C. García de la Peña, G. Castañeda, and A. Contreras-Balderas. 2007. Notes on Mexican herpetofauna 9: Herpetofauna of a fragmented *Juniperus* forest in the State Natural Protected Area of San Juan y Puentes, Aramberri, Nuevo León, Mexico: Preliminary list. Bull. Chic. Herpetol. Soc. 42:21–27.

Liner, E. A. 1964. Notes on four small herpetological collections from Mexico. I, Introduction: Turtles and snakes. Southwest. Nat. 8:221–227.

———. 1966a. Notes on four small herpetological collections from Mexico. II, Amphibians. Southwest. Nat. 11:296–312.

———. 1966b. Notes on four small herpetological

collections from Mexico. III, Lizards. Southwest. Nat. 11:406–414.

———. 1991a. Mexico bound. Gulf Coast Herpetol. Soc. Newsl. (July): 12–15.

———. 1991b. Mexico bound II. Gulf Coast Herpetol. Soc. Newsl. 1:4–8.

———. 1992a. Mexico bound III. Gulf Coast Herpetol. Soc. Newsl. 2:12–21.

———. 1992b. Mexico bound IV. Gulf Coast Herpetol. Soc. Newsl. 3:5–7.

———. 1993. Mexico bound V. Gulf Coast Herpetol. Soc. Newsl. (July): 3–5.

———. 1994. Mexico bound VI. Part two. Gulf Coast Herpetol. Soc. Newsl. 5:2–3.

———. 1996a. Herpetological type material from Nuevo León, Mexico. Bull. Chic. Herpetol. Soc. 31 (9): 168–171.

———. 1996b. *Rhadinaea montana* Smithson. Cat. Am. Amphib. Rept. 640:1–2.

Liner, E. A., and A. H. Chaney. 1986. Natural history notes: *Crotalus lepidus lepidus*. Reproduction. Herpetol. Rev. 17:89.

———. 1987. Natural history notes: *Rhadinaea montana*. Habitat. Herpetol. Rev. 18:37.

———. 1990a. Geographic distribution: *Sceloporus torquatus mikeprestoni*. Herpetol. Rev. 21:22–23.

———. 1990b. Natural history notes: *Tantilla rubra rubra*. Arboreality. Herpetol. Rev. 21:20.

———. 1995a. Geographic distribution: *Cnemidophorus inornatus inornatus*. Herpetol. Rev. 26:154–155.

———. 1995b. Geographic distribution: *Cnemidophorus inornatus paulus*. Herpetol. Rev. 26:155.

Liner, E. A., A. H. Chaney, J. R. Dixon, and J. F. Scudday. 1990. Geographic distribution: *Thamnophis cyrtopsis pulchrilatus*. Herpetol. Rev. 21:42.

Liner, E. A., A. H. Chaney, and R. M. Johnson. 1982. Geographic distribution: *Elaphe subocularis*. Herpetol. Rev. 13:52–53.

Liner, E. A., and J. R. Dixon. 1992. A new species of the *Sceloporus scalaris* group from Cerro Peña Nevada, Nuevo León, México. Tex. J. Sci. 44:421–427.

———. 1994. *Sceloporus chaneyi*. Cat. Am. Amphib. Rept. 588:1.

Liner, E. A., and R. M. Johnson. 1973. Natural history notes: *Storeria occipitomaculata hidalgoensis*. Herpetol. Rev. 1:185.

Liner, E. A., R. M. Johnson, and A. H. Chaney. 1976. Amphibians and reptiles records and range extensions for Mexico. Herpetol. Rev. 7:177.

———. 1978. Geographic distribution: *Tantilla nigriceps fumiceps*. Herpetol. Rev. 9:22.

———. 1982. Geographic distribution: *Tantilla rubra rubra*. Herpetol. Rev. 13:53.

Liner, E. A., and R. E. Olson. 1973. Adults of the lizard *Sceloporus torquatus binocularis* Dunn. Herpetologica 29:53–55.

Liner, E. A., D. A. Rossman, and R. M. Johnson. 1973. *Gerrhonotus (Barisia) imbricatus ciliaris*. HISS News-J. 1:185.

Mattison, C. 1998. The Nuevo León Kingsnake. *Lampropeltis mexicana thayeri*. Reptilia Mag. 2:43–46.

Osborne S. T. 1983. Natural history notes: *Lampropeltis mexicana thayeri*. Coloration. Herpetol. Rev. 14:120.

Price, M., C. R. Harrison, and D. Lazcano-Villarreal. 2010. Geographic distribution: *Sceloporus cyanostictus*. Herpetol. Rev. 41:108.

Price, M., and D. Lazcano-Villarreal. 2010. Geographic distribution: *Sceloporus merriami australis*. Herpetol. Rev. 41:109.

Rabb, G. B. 1956. A new plethodontid salamander from Nuevo León, Mexico. Fieldiana Zool. 39:11–20.

Reese, R. W. 1971. Notes on a small herpetological collection from northeastern Mexico. J. Herpetol. 5:67–69.

Rossman, D. A., E. A. Liner, C. H. Treviño, and A. H. Chaney. 1989. Redescription of the garter snake *Thamnophis exsul* Rossman, 1969 (Serpentes: Colubridae). Proc. Biol. Soc. Wash. 102:507–514.

Salmon G. T., R. W. Bryson Jr., and D. Lazcano. 2001. Geographic distribution: *Lampropeltis mexicana*. Herpetol. Rev. 35:292.

Salmon, G. T., E. A. Liner, J. E. Forks, and D. Lazcano. 2004. Geographic distribution: *Lampropeltis alterna*. Herpetol. Rev. 35:292.

Smith, H. M. 1934. Descriptions of new lizards of the genus *Sceloporus* from Mexico and southern United States. Trans. San Diego Soc. Natl. Hist. 37:263–285.

———. 1942. Summary of the collections of snakes and crocodilians made in Mexico under the Walter Rathbone Bacon Traveling Scholarship. Proc. U.S. Natl. Mus. 93 (3169): 393–504.

———. 1944. Snakes of the Hoogstraal Expeditions to northern Mexico. Zool. Ser. Field Mus. Nat. Hist. 29 (8): 135–152.

———. 1951. A new species of *Leiolopisma* (Reptilia: Sauria) from Mexico. Univ. Kans. Sci. Bull. 34 (3): 195–200.

Smith, H. M., and W. P. Hall. 1974. Contributions to the concepts of reproductive cycles and the systematics of the *scalaris* group of the lizard genus *Sceloporus*. Great Basin Nat. 34 (2): 97–104.

Smith, H. M., and E. H. Taylor. 1945. An annotated checklist and key to the snakes of Mexico. U.S. Natl. Mus. Bull. 187:i–iv, 1–239.

———. 1948. An annotated checklist and key to the amphibia of Mexico. U.S. Natl. Mus. Bull. 194:i–iv, 1–118.

———. 1950a. An annotated checklist and key to the reptiles of Mexico exclusive of the snakes. U.S. Natl. Mus. Bull. 199:1–253.

———. 1950b. Type localities of Mexican reptiles and amphibians. Univ. Kans. Sci. Bull. 33:313–380.

Taylor, E. H. 1940a. New species of Mexican anuran. Univ. Kans. Sci. Bull. 26:385–405.

———. 1940b. Two new anuran amphibians from Mexico. Proc. U.S. Natl. Mus. 89:43–47.

———. 1941. Two new species of Mexican plethodontid salamanders. Proc. Biol. Soc. Wash. 54:81–85.

Taylor, E. H., and H. M. Smith. 1945. Summary of the collections of amphibians made in Mexico under the Walter Rathbone Bacon Traveling Scholarship. Proc. U.S. Natl. Mus. 95:521–613.

Treviño-Saldaña, C. H. 1988. A new montane lizard (*Sceloporus jarrovi cyaneus*) from Nuevo León, Mexico. Rev. Biol. Trop. 36:407–411.

Velazco-Macías, C. G. 2009. Flora del estado de Nuevo León, México: Diversidad y análisis espacio-temporal. PhD diss., Universidad Autónoma de Nuevo León.

Velazco-Macías, C. G., R. Foroughbakhch, M. A. Álvarado-Vázquez, and G. J. Alanís-Flores. 2008. La familia Nymphaeaceae en el estado de Nuevo León. J. Bot. Res. Inst. Tex. 2:593–603.

Zertuche, J. J., and C. H. Treviño. 1978. Una nueva subespecie de *Crotalus lepidus* encontrada en Nuevo León. Monterrey, Nuevo León, Mexico: Resúmenes del Segundo Congreso Nacional de Zoología.

8. Tamaulipas

Adams, R. P. 2011. Junipers of the world: The genus *Juniperus*. 3rd ed. Bloomington, IN: Trafford Publishing.

Agassiz, J. L. R. 1857. Contributions to the natural history of the United States of America. Vol. 1. Boston: Little, Brown.

Alcántara, O., I. Luna, and A. Velázquez. 2002. Altitudinal distribution of Mexican cloud forest based upon preferential characteristic genera. Plant Ecol. 161:167–174.

Álvarez, T. 1963. The recent mammals of Tamaulipas, Mexico. Univ. Kans. Publ. Mus. Nat. Hist. 14:365–473.

Auth, D. L., H. M. Smith, B. C. Brown, and D. Lintz. 2000. A description of the Mexican amphibian and reptile collection of the Strecker Museum. Bull. Chic. Herpetol. Soc. 35:65–85.

Auth, D. L., H. M. Smith, B. C. Brown, D. Lintz, and D. Chiszar. 2000. Further observations on Iverson's Blind Snake in Tamaulipas, Mexico. Bull. Md. Herpetol. Soc. 36:1–4.

Axtell, R. W. 1958a. A monographic revision of the iguanid genus *Holbrookia*. PhD diss., University of Texas at Austin.

———. 1958b. A northward range extension for the lizard *Anelytropsis papillosus*, with notes on the distribution and habits of several other Mexican lizards. Herpetologica 14:189–191.

———. 1960. A new subspecies of *Eumeces dicei* from the Sierra Madre of northeastern Mexico. Copeia 1960:19–26.

Axtell, R. W., and A. O. Wasserman. 1953. Interesting herpetological records from southern Texas and northern Mexico. Herpetologica 9:1–6.

Axtell, R. W., and R. G. Webb. 1995. Two new *Crotaphytus* from southern Coahuila and the adjacent states of east-central Mexico. Bull. Chic. Acad. Sci. 16:1–15.

Bailey, J. R. 1937. New forms of *Coniophanes* Hallowell, and the status of *Dromicus clavatus* Peters. Occas. Pap. Mus. Zool. Univ. Mich. 362:1–6.

———. 1939. A systematic revision of the snakes of the genus *Coniophanes*. Pap. Mich. Acad. Sci. Arts Ltr. 24:1–48.

Bailey, J. W. 1928. A revision of the lizards of the genus *Ctenosaura*. Proc. U.S. Natl. Mus. 73:1–55.

Bailey, J. W., J. R. Dixon, R. Hudson, and M. R. J. Forstner. 2008. Minimal genetic structure in the Rio Grande Cooter (*Pseudemys gorzugi*). Southwest. Nat. 53:406–411.

Baird, S. F. 1854. Descriptions of new genera and species of North American frogs. Proc. Acad. Nat. Sci. Phila. 7:59–62.

———. 1859a (1858). Description of new genera and species of North American lizards in the

Museum of the Smithsonian Institution. Proc. Acad. Nat. Sci. Phila. 10:253–256.

———. 1859b. Reptiles of the boundary, with notes by the naturalists of the survey. In United States and Mexican Boundary Survey, under the order of Lieut. Col. W. H. Emory, major First Cavalry, and United States commissioner. Vol. 2, pt. 2, U.S. 34th Cong., 1st Sess., Exec. Doc. 108, 1–35, plates 1–41. Washington, DC: Government Printing Office.

Baird, S. F., and C. Girard. 1853. Catalogue of North American reptiles in the Museum of the Smithsonian Institution. Part 1, Serpents. Washington, DC: Smithsonian Institution.

Baker, R. H., and R. G. Webb. 1966. Notas acerca de los anfibios, reptiles y mamiferos de La Pesca, Tamaulipas. Rev. Soc. Mex. Hist. Nat. 27:179–190.

Ballinger, R. E., G. R. Smith, and J. A. Lemos-Espinal. 2000. Xenosaurus platyceps. Cat. Am. Amphib. Rept. 715:1–2.

Berlandier, J. L. 1980. Journey to Mexico during the years 1826 to 1834. 2 vols. Austin: Texas State Historical Association in cooperation with the Center for Studies in Texas History, University of Texas at Austin.

Bezy, R. L. 1984. Systematics of xantusiid lizards of the genus Lepidophyma in northeast Mexico. Contr. Sci. 349:1–16.

Blair, W. F. 1950. The biotic provinces of Texas. Tex. J. Sci. 2:93–117.

Bocourt, M. F. 1869. Descriptions de quelques reptiles et poissons nouveaux appurtenant à la faune tropicale de l'Amérique. Nouv. Arch. Mus. Hist. Natur. Paris 5:19–24.

Brown, B. C., and L. M. Brown. 1967. Notable records of Tamaulipan snakes. Tex. J. Sci. 19:323–326.

Brown, B. C., and H. M. Smith. 1942. A new subspecies of Mexican coral snake. Proc. Biol. Soc. Wash. 55:63–66.

Bryson, R. W., Jr., and M. R. Graham. 2010. A new alligator lizard from northeastern Mexico. Herpetologica 66:92–98.

Burchfield, P. M. 1982. Additions to the natural history of the crotaline snake Agkistrodon bilineatus taylori. J. Herpetol. 16:376–382.

———. 2004. Report on the Mexico/United States of America population restoration project for the Kemp's Ridley Sea Turtle, on the coast of Tamaulipas and Veracruz, Mexico 2004. Unpublished report, Albuquerque, NM, U.S. Fish and Wildlife Service.

Burger, W. L., and W. B. Robertson. 1951. A new subspecies of the Mexican Moccasin Agkistrodon bilineatus. Kans. Univ. Sci. Bull. 24:213–218.

Burt, C. E. 1932. Some Mexican herpetological records. Herpetologica 1932:158.

Casas Andreu, G. 1974. Los habitos de anidacion de Lepidochelys olivacea y Lepidochelys kempi (Reptilia: Testudines) en las castas de Jalisco, Oaxaca y Tamaulipas, Mexico. Mexico City: VI Congreso Latinoamerica de Zoologia.

Chávez, H., M. Contreras, and E. T. P. Hernández. 1967. Aspectos biologicos y proteccion de la Tortuga lora, Lepidochelys kempi (German), en las coasta de Tamaulipas, Mexico. Inst. Nac. Invest. Biol. Pesq. 17:1–40.

———. 1968. On the coast of Tamaulipas. Int. Turtle Tortoise Soc. J. 2:20–29.

Chiszar, D., R. Conant, and H. M. Smith. 2003. Observations of the rattlesnake Crotalus atrox by Berlandier. Bull. Chic. Herpetol. Soc. 38:138–142.

Clay, W. M. 1938. A new water snake of the genus Natrix from Mexico. Ann. Carnegie Mus. 27:251–253.

Cochran, D. M. 1961. Type specimens of reptiles and amphibians in the U.S. National Museum. Bull. U.S. Natl. Mus. 220:1–216.

CONABIO (Comisión Nacional para el Conocimiento y Uso de la Biodiversidad). 2008. Manglares de México. Mexico City: CONABIO.

Conant, R. 1965. Miscellaneous notes and comments on toads, lizards, and snakes from Mexico. Am. Mus. Novit. 2205:1–38.

———. 1968. Zoological exploration in Mexico—the route of Lieut. D. N. Couch in 1853. Am. Mus. Novit. 2350:1–14.

———. 1969. A review of the water snakes of the genus Natrix in Mexico. Bull. Am. Mus. Nat. Hist. 142:1–140.

Cope, E. D. 1863. On Trachycephalus, Scaphiopus and other American batrachia. Proc. Acad. Nat. Sci. Phila. 15:43–54.

———. 1880. On the zoological position on Texas. Bull. U.S. Natl. Mus. 17:1–51.

———. 1887. Catalogue of batrachians and reptiles of Central America and Mexico. Bull. U.S. Natl. Mus. 32:1–98.

———. 1895. On some new North American snakes. Bull Am. Natl. Mus. 29:676–680.

———. 1900. The crocodilians, lizards, and snakes of North America. Ann. Rept. U.S. Natl. Mus. 1898:153–1270.

Cram, S., I. Sommer, L. M. Morales, O. Oropeza,

E. Carmona, and F. González-Medrano. 2006. Suitability of the vegetation types in Mexico's Tamaulipas state for the siting of hazardous waste treatment plants. J. Environ. Manage. 80:13–24.

Davis, W. B. 1953a. Northernmost record of the frog *Rhinophrynus dorsalis* in Mexico. Copeia 1953:65.

———. 1953b. The turtle *Kinosternon cruentatum cruentatum* in Tamaulipas. Copeia 1953:65.

Degenhardt, W. G., C. W. Painter, and A. H. Price. 1996. Amphibians and reptiles of New Mexico. Albuquerque: University of New Mexico Press.

De Queiroz, K. 1995. Checklist and key to the extant species of Mexican iguanas (Reptilia: Iguaninae). Publ. Espec. Mus. Zool. 9:1–48.

Dice, L. R. 1937. Mammals of the San Carlos Mountains and vicinity. In The geology and biology of the San Carlos Mountains, Tamaulipas, Mexico, ed. L. B. Kellum, 243–268. Ann Arbor: University of Michigan Press.

Dixon, J. R. 1959. Geographic variation and distribution of the long-tailed group of the Glossy Snake, *Arizona elegans* Kennicott. Southwest. Nat. 4:20–29.

———. 1962. Three additional specimens of the night snake *Hypsiglena dunklei*. Herpetologica 18:134–135.

———. 1969. Taxonomic review of the Mexican skinks of the *Eumeces brevirostris* group. Contr. Sci. 168:1–30.

Dixon, J. R., and R. H. Dean. 1986. Status of the southern populations of the night snake (*Hypsiglena*: Colubridae) exclusive of California and Baja California. Southwest. Nat. 31:307–318.

Dixon, J. R., and R. A. Thomas. 1974. A dichromatic population of the snake, *Geophis latifrontalis*, with comments on the status of *Geophis semiannulatus*. J. Herpetol. 8:271–273.

Dixon, J. R., and R. K. Vaughan. 2003. The status of Mexican and southwestern United States blind snakes allied with *Leptotyphlops dulcis* (Serpentes: Leptotyphlopidae). Tex. J. Sci. 55:3–24.

Dixon, J. R., R. K. Vaughan, and L. D. Wilson. 2000. The taxonomy of *Tantilla rubra* and allied taxa (Serpentes: Colubridae). Southwest. Nat. 45:141–153.

Duellman, W. E. 1958. A monographic study of the colubrid snake genus *Leptodeira*. Bull. Am. Mus. Nat. Hist. 114:1–152.

Dugès, A. A. D. 1896. Reptiles y batraciios de los Estados Unidos Mexicanos. Naturaleza 2: 479–485.

Duméril, A. H. A., P. Brocchi, and M. F. Mocquard. 1870–1909. Études sur les reptiles. In Mission scientifique au Mexique et dans l'Amérique Centrale, recherches zoolgiques. Paris: Imprimerie Impériale. Repr., New York: Arno Press, 1978.

Dundee, H. A., and E. A. Liner. 1997. *Trimorphodon tau tau*. Mexico, Tamaulipas. Herpetol. Rev. 28:211.

Dunkle, D. H. 1935. Note on *Crocodylus moreletti* A. Duméril from Mexico. Copeia 1935:182.

Dunkle, D. H., and H. M. Smith. 1937. Notes on some Mexican ophidians. Occas. Pap. Mus. Zool. Univ. Mich. 363:1–15.

Farr, W. L. 2011. The distribution of *Hemidactylus frenatus* in México. Southwest. Nat. 56:267–274.

Farr, W. L., T. Burkhardt, and D. Lazcano. 2011. *Tantilla rubra* (Red Blackhead Snake): Maximum size. Herpetol. Rev. 42:445.

Farr, W. L., A. Godambe, and D. Lazcano. 2010. *Spea multiplicata* (Mexican Spadefoot): Predation. Herpetol. Rev. 41:209.

Farr, W. L., P. A. Lavín Murcio, and D. Lazcano. 2007. New distributional records for amphibians and reptiles from the state of Tamaulipas, Mexico. Herpetol. Rev. 38:226–233.

Farr, W. L., and D. Lazcano. 2011. Natural history notes: *Drymobius margaritiferus*. Defensive behavior-thanatosis. Herpetol. Rev. 42:613.

Farr, W. L., D. Lazcano, and P. A. Lavín Murcio. 2009. New distributional records for amphibians and reptiles from the state of Tamaulipas, Mexico II. Herpetol. Rev. 40:459–467.

———. 2013. New distributional records for amphibians and reptiles from the state of Tamaulipas, Mexico III. Herpetol. Rev. 44:631–645.

Ferrusquia-Villafranca, I. 1993. Geology of Mexico: A synopsis. In Biological diversity of Mexico: Origins and distribution, ed. T. P. Ramamoorthy, R. Bye, A. Lot, and J. Fa, 3–107. Oxford: Oxford University Press.

Flores-Benabib, J., and O. Flores-Villela. 2008. Nuevo registro de *Leptotyphlops goudotii* en Tamaulipas, Mexico. Bol. Soc. Herpetol. Méx. 16:13–14.

Flores-Villela, O. 1993. Herpetofauna mexicana: Lista anotada de las especies de los anfibios y reptiles de México, cambios taxonómicos recientes y nuevas especies. Carnegie Mus. Nat. Hist. Spec. Publ. 17:1–73.

Flores-Villela, O., and R. A. Brandon. 1992. *Siren lacertina* (Amphibia: Caudata) in northeastern

Mexico and southern Texas. Ann. Carnegie Mus. 61:289–291.

Flores-Villela, O., and H. A. Pérez-Mendoza. 2006. Herpetofaunas estatales de México. In Inventarios herpetofaunísticos de México: Avances en el conocimiento de su biodiversidad, ed. A. Ramírez-Bautista, L. Canseco-Márquez, and F. Mendoza-Quijano, 327–346. Publication no. 3. Mexico City: Sociedad Herpetológica Mexicana.

Gaige, H. T. 1937. Some amphibians and reptiles from Tamaulipas. In The geology and biology of the San Carlos Mountains, Tamaulipas, Mexico, ed. L. B. Kellum, 301–304. Ann Arbor: University of Michigan Press.

García-Padilla, E., J. I. Cumpián-Medellín, N. F. Vargas Gonzàlez, and A. Hernàndez Maldonado. 2011. Natural history notes: *Pantherophis bairdi*. Diet and foraging behavior. Herpetol. Rev. 42:300.

García-Padilla, E., and W. L. Farr. 2010. *Anelytropsis papillosus* (Mexican Blind Lizard) Mexico, Tamaulipas. Herpetol. Rev. 41:511.

Garman, S. 1883. The reptiles and batrachians of North America. Part I, Ophidia. Mem. Mus. Comp. Zool. 8:i–xxxiv, 1–185.

Geiser, S. W. 1948. Naturalist of the frontier. Dallas, TX: Southern Methodist University Press.

Girard, C. F. 1854. A list of North American bufonids with diagnoses of new species. Proc. Acad. Nat. Sci. Phila. 7:86–88.

Gloyd, H. K. 1940. The rattlesnakes, genera *Sistrurus* and *Crotalus*. A study in zoogeography and evolution. Spec. Publ. Chic. Acad. Sci. 4:1–270.

Goin, C. J. 1957. Description of a new salamander of the genus *Siren* from the Rio Grande. Herpetologica 13:37–42.

Goldman, E. A. 1951. Biological investigations in Mexico. Smithson. Misc. Coll. 115. Washington, DC: Smithsonian Institution.

Goldman, E. A., and R. T. Moore. 1945. The biotic provinces of Mexico. J. Mammal. 26:347–360.

Gómez-Hinostrosa, R., and H. M. Hernández. 2000. Diversity, geographical distribution, and conservation of Cactaceae in the Mier y Noriega region, Mexico. Biodivers. Conserv. 9:403–418.

Gray, J. E. 1849. Description of a new species of box turtle from Mexico. Proc. Zool. Soc. Lond. 16:16–17.

Greene, H. W. 1970. Mode of reproduction in lizards and snakes of the Gómez Farías region, Tamaulipas, Mexico. Copeia 1970:565–568.

Günther, A. C. L. G. 1885–1902. Reptilia and batrachia. In Biologia Centrali-Americana, ed. O. Salvin and F. D. Godman. London: R. H. Porter. Reprinted by SSAR in 1987.

Harlan, R. 1825. Description of two species of Linnean Lacerta, not before described, and construction of the new genus *Cyclura*. J. Acad. Nat. Sci. Phila. 4:242–251.

Heim, A. 1940. The front ranges of the Sierra Madre Oriental, Mexico, from Ciudad Victoria to Tamazunchale. Eclogae Geol. Helv. 33: 313–352.

Hendricks, F. S., and A. M. Landry. 1976. Vertebrate fauna of the San Carlos Mountains, Tamaulipas, Mexico: Identification keys, annotated checklist of the species, and field study techniques. Unpublished Wildlife and Fisheries Sciences Report (Class 300), Texas A&M University.

Henrickson, J., and R. M. Straw. 1976. A gazetteer of the Chihuahuan Desert region. A supplement to the Chihuahuan Desert flora. Los Angeles: California State University.

Horowitz, S. B. 1955. An arrangement of the subspecies of the horned toad, *Phrynosoma orbiculare* (Iguanidae). Am. Midl. Nat. 54:204–218.

Iverson, J. B., and J. F. Berry. 1979. The mud turtle genus *Kinosternon* in northeastern Mexico. Herpetologica 35:318–324.

Jackson, M. K. 1971. Another *Tropidodipsas fasciata fasciata* (Colubridae) from Tamaulipas, Mexico. Southwest. Nat. 16:124.

Jan, G. 1863. Enumerazione sistematico degli ofidi appartenenti al gruppo Coronellidae. Arch. Zool. Anat. Fish. 2:211–330.

Jiménez-Ramos, D., E. Pérez-Ramos, and J. A. Vargas-Contreras. 1999. Geographic distribution: *Cnemidophorus sexlineatus*. Herpetol. Rev. 30:109–110.

Johnston, M. C. 1963. Past and present grasslands of southern Texas and northeastern Mexico. Ecology 44:456–466.

Jones, F. A., and D. L. Gorchov. 2000. Patterns of abundance and human use of the vulnerable understory palm, *Chamaedorea radicalis* (Arecaceae), in a montane cloud forest, Tamaulipas, Mexico. Southwest. Nat. 45:421–430.

Kellogg, R. 1932. Mexican tailless amphibians in the United States National Museum. U.S. Natl. Mus. Bull. 160:1–224.

Kellum, L. B. 1937. The geology and biology of the San Carlos Mountains, Tamaulipas, Mexico. Ann Arbor: University of Michigan Press.

Kennicott, R. 1859. *Dipsas septentrionalis.* In United States and Mexican Boundary Survey, under the order of Lieut. Col. W. H. Emory, major First Cavalry, and United States commissioner. Vol. 2, Part 2, U.S. 34th Cong., 1st Sess., Exec. Doc. 108, 16. Washington, DC: Government Printing Office.

———. 1860. Descriptions of new species of North American serpents in the Museum of the Smithsonian Institution, Washington. Proc. Acad. Nat. Sci. Phila. 12:328–338.

King, F. W., and F. G. Thompson. 1968. A review of the American lizards of the genus *Xenosaurus* Peters. Bull. Fla. State Mus. 12:93–123.

Lavin-Murcio, P. A. 1998. An ecological analysis of the herpetofauna of a cloud forest community in the "El Cielo" Biosphere Reserve, Tamaulipas, Mexico. PhD diss., Texas A&M University, College Station.

Lavin-Murcio, P. A., and J. R. Dixon. 2004. A new species of coral snake (Serpentes, Elapidae) from the Sierra de Tamaulipas, Mexico. Phyllomedusa 3:3–7.

Lavin-Murcio, P. A., O. M. Hinojosa-Falcón, G. Herrera-Patiño, R. E. Nuñez-Lara, and L. H. Vélez-Horta. 2005. Anfibios y reptiles de Tamaulipas: Un listado preliminar. In Biodiversidad tamaulipeca, ed. L. Barrientos, A. Correa, J. Horta, and J. Jiménez, 1:185–192. Tamaulipas: DGIT.

Lavin-Murcio, P. A., X. M. Sampablo Angel, O. M. Hinojosa Falcon, J. R. Dixon, and D. Lazcano. 2005. La herpetofauna. In Historia natural de la Reserva de la Biosfera "El Cielo" Tamaulipas, Mexico, ed. G. Sánchez-Ramos, P. Reyes-Castillo, and R. Dirzo, 489–509. Hong Kong: Universidad Autónoma de Tamaulipas.

Lazcano, D., W. L. Farr, P. A. Lavín-Murcio, J. A. Contreras-Lozano, A. Kardon, S. Narvaez-Torres, and J. A. Chávez-Cisneros. 2009. Notes on Mexican herpetofauna 13: DORs in the municipality of Aldama Tamaulipas, Mexico. Bull. Chic. Herpetol. Soc. 44:181–195.

Lemos-Espinal, J. A., and I. Rojas-González. 2000. Observations on neonate size and sex ratio of the crevice-dwelling lizard *Xenosaurus platyceps.* Herpetol. Rev. 31:153.

Lemos-Espinal, J. A., G. R. Smith, and R. E. Ballinger. 1997. Natural history of *Xenosaurus platyceps*, a crevice-dwelling lizard from Tamaulipas, Mexico. Herpetol. Nat. Hist. 5:181–186.

———. 2003. Diets of three species of knob-scaled lizards (Genus *Xenosaurus*) from Mexico. Southwest. Nat. 48:119–122.

Liner, E. A. 1960. A new subspecies of false coral snake (*Pliocercus elapoides*) from San Luis Potosi, Mexico. Southwest. Nat. 5:217–220.

———. 1964. Notes on four small herpetological collections from Mexico. I. Introduction, turtles and snakes. Southwest. Nat. 8:221–227.

———. 1966a. Notes on four small herpetological collections from Mexico. II. Amphibians. Southwest. Nat. 11:296–298.

———. 1966b. Notes on four small herpetological collections from Mexico. III. Lizards. Southwest. Nat. 11:406–408.

Liner, E. A., and J. R. Dixon. 1992. A new species of the *Sceloporus scalaris* group from Cerro Pena Nevada, Nuevo León, Mexico (Sauria: Iguanidae). Tex. J. Sci. 44:421–427.

Linnaeus, C. 1758. Systema naturae per regna tria naturae, secundum classes, ordines, genera, species cum characteribus, differentiis, synonymis, locis. 10th ed. Vol. 1. Stockholm: L. Salvius.

———. 1766. Systema naturae per regna tria naturae, secundum classes, ordines, genera, species cum characteribus, differentiis, synonymis, locis. 12th ed. Vol. 1. Stockholm: L. Salvius.

Little, E. L., Jr. 1976. Atlas of United States trees. Vol. 3, Minor western hardwoods. USDA Misc. Publ. 1314:1–13.

———. 1977. Atlas of United States trees. Vol. 4, Minor eastern hardwoods. USDA Misc. Publ. 1342:1–17.

Loveridge, A. 1924. A new snake of the genus *Lampropeltis.* Occas. Pap. Bost. Soc. Nat. Hist. 5:137–139.

Lynch, J. D. 1963. The status of *Eleutherodactylus longipes* (Baird) of Mexico (Amphibia: Leptodactylidae). J. Ohio Herpetol. Soc. 3:580–581.

———. 1965. A record of the hylid genus *Pseudacris* in Mexico. J. Ohio Herpetol. Soc. 5:31.

———. 1967. Synonymy, distribution and variation in *Eleutherodactylus decoratus* of Mexico (Amphibia: Leptodactylidae). Trans. Ill. State Acad. Sci. 60:299–304.

———. 1970. A taxonomic revision of the leptodactylid frog genus *Syrrhophus* Cope. Univ. Kans. Pub. Mus. Nat. Hist. 20:1–45.

Martin, P. S. 1952. A new subspecies of the iguanid lizard *Sceloporus serrifer* from Tamaulipas, Mexico. Occas. Pap. Mus. Zool. Univ. Mich. 543:1–7.

———. 1955a. Herpetological records from the Gómez Farías region of southwest Tamaulipas, Mexico. Copeia 1955:173–180.

———. 1955b. Zonal distribution of vertebrates in a Mexican cloud forest. Am. Nat. 89:347–361.

———. 1958. A biogeography of reptiles and amphibians in the Gómez Farías region, Tamaulipas, Mexico. Misc. Publ. Mus. Zool. Univ. Mich. 101:1–102.

———. 2005. El Cielo in 1948: Discovery of a tropical cloud forest. In Historia natural de la Reserva de la Biosfera "El Cielo" Tamaulipas, Mexico, ed. G. Sánchez-Ramos, P. Reyes-Castillo, and R. Dirzo, 24–37. Hong Kong: Universidad Autónoma de Tamaulipas.

Martin, P. S., and B. E. Harrell. 1957. The Pleistocene history of temperate biotas in Mexico and eastern United States. Ecology 38:468–480.

Martin, P. S., C. Robins, and W. Heed. 1954. Birds and biogeography of the Sierra de Tamaulipas, an isolated pine-oak habitat. Wilson Bull. 66:38–57.

Mecham, J. S. 1968a. *Notophthalmus meridionalis*. Cat. Am. Amphib. Rept. 74:1–2.

———. 1968b. On the relationships between *Notophthalmus meridionalis* and *Notophthalmus kallerti*. J. Herpetol. 2:121–127.

Mecham, J. S., and R. W. Mitchell. 1983. *Siren intermedia* in Mexico. Herpetol. Rev. 14:55.

Morafka, D. J. 1977. A biogeographical analysis of the Chihuahuan Desert through its herpetofauna. The Hague, Netherlands: Dr. W. Junk Publishers.

Morrone, J. J., D. Espinosa-Organista, and J. Llorente-Bousquets. 2002. Mexican biogeographic provinces: Preliminary scheme, general characterizations, and synonymies. Acta Zool. Mex., n.s., 85:83–108.

Müller, L. 1936. Beiträge zur kenntnis der schildkrötenfauna von Mexico. Aool. Anz. 113:97–114.

Nelson, Craig E., and M. A. Nickerson. 1966. Notes on some Mexican and Central American amphibians and reptiles. Southwest. Nat. 11:128–131.

Olson, R. E. 1986. A new subspecies of *Sceloporus torquatus* from the Sierra Madre Oriental, Mexico. Bull. Md. Herpetol. Soc. 22:167–170.

———. 1987. Taxonomic revision of the lizards *Sceloporus serrifer* and *cyanogenys* of the Gulf Coastal Plain. Bull. Md. Herpetol. Soc. 23:158–167.

———. 1990. *Sceloporus torquatus:* Its variation and zoogeography. Bull. Chic. Herpetol. Soc. 25:117–127.

Ortenburger, A. I. 1923. A note on the genera *Coluber* and *Masticophis* and a description of a new species of *Masticophis*. Occas. Pap. Mus. Zool. Univ. Mich. 139:1–14.

Paulissen, M. A., and J. M. Walker. 1998. *Cnemidophorus laredoensis*. Cat. Am. Amphib. Rept. 673:1–5.

Pérez-Ramos, E., A. Nieto-Montes de Oca, J. A. Vargas-Contreras, J. E. Cordes, M. A. Paulissen, and J. M. Walker. 2010. *Aspidoscelis sexlineata* (Sauria: Teiidae) in Mexico: Distribution, habitat, morphology, and taxonomy. Southwest. Nat. 55:419–425.

Perry, J. P. 1991. The pines of Mexico and Central America. Portland, OR: Timber Press.

Peters, J. A. 1948. The northern limit of the range of *Laemanctus serratus*. Nat. Hist. Misc. Chic. Acad. Sci. 27:1–3.

Plotkin, P. T. 2007. Biology and conservation of ridley sea turtles. Baltimore: John Hopkins University Press.

Rabb, G. B. 1958. On certain Mexican salamanders of the plethodontid genus *Chiropterotriton*. Occas. Pap. Mus. Zool. Univ. Mich. 587:1–37.

Reese, R., and I. L. Firschein. 1950. Herpetological results of the University of Illinois field expedition, spring 1949. II, Amphibia. Trans. Kans. Acad. Sci. 53:44–55.

Reilly, S. M. 1990. Biochemical systematics and evolution of the eastern North American newts, genus *Notophthalmus* (Caudata: Salamandridae). Herpetologica 46:51–59.

Rojas-González, R. I., J. J. Zuñiga-Vega, and J. A. Lemos-Espinal. 2008. Reproductive variation of the lizard *Xenosaurus platyceps:* Comparing two populations of contrasting environments. J. Herpetol. 42:332–336.

Rose, J. A. 1967. A check list of the New World venomous coral snakes (Elapidae), with descriptions of new forms. Am. Mus. Novit. 2287:1–60.

Rossman, D. A. 1969. A new natricine snake of the genus *Thamnophis* from northern Mexico. Occas. Pap. Mus. Zool. La. State Univ. 39:1–4.

———. 1992a. The black-necked garter snake (*Thamnophis cyrtopsis*): Polytypic species or cryptic species complex? Abstracts, annual meeting of the American Society of Ichthyologists and Herpetologists, Urbana, Illinois.

———. 1992b. Taxonomic status and relationships of the Tamaulipan montane garter snake, *Thamnophis mendax* Walker, 1955. Proc. La. Acad. Sci. 55:1–14.

Rossman, D. A., and M. J. Fouquette Jr. 1963. Noteworthy records of Mexican amphibians and reptiles in the Florida State Museum and the Texas Natural History Collection. Herpetologica 19:185–201.

Ruthven, A. G., and H. T. Gaige. 1933. A new skink from Mexico. Occas. Pap. Mus. Zool. Univ. Mich. 260:1–3.

Sampablo-Brito, X., and J. R. Dixon. 1998. Geographic distribution: *Gastrophryne elegans*. Herpetol. Rev. 29:48.

Seidel, M. E. 1976. Geographic distribution: *Kinosternon flavescens flavescens*. Herpetol. Rev. 7:122.

Seidel, M. E., J. N. Stuart, and W. G. Degenhardt. 1999. Variation and species status of slider turtles (Emydidae: *Trachemys*) in the southwestern United States and adjacent Mexico. Herpetologica 55:470–487.

Selander, R. K., R. F. Johnston, B. J. Wilks, and G. G. Raun. 1962. Vertebrates from the barrier islands of Tamaulipas, Mexico. Univ. Kans. Publ. Mus. Nat. Hist. 12:309–345.

Shannon, F. A., and H. M. Smith. 1950 (1949). Herpetological results of the University of Illinois field expedition, spring 1949. I, Introduction, testudines, serpentes. Trans. Kans. Acad. Sci. 52:499–514.

Shaw, G. 1802. General zoology of systematic natural history. Vol. 3. London: Thomas Davidson.

Sites, J. W., Jr., and J. R. Dixon. 1981. A new subspecies of the iguanid lizard, *Sceloporus grammicus* from northeastern Mexico, with comments on its evolutionary implications and the status of *S. g. disparilis*. J. Herpetol. 15:59–69.

———. 1982. Geographic variation in *Sceloporus variabilis*, and its relationship to *S. teapensis* (Sauria: Iguanidae). Copeia 1982:14–27.

Smith, H. M. 1943. Another Mexican snake of the genus *Pliocercus*. J. Wash. Acad. Sci. 33:344–345.

Smith, H. M., and T. Álvarez. 1976. Possible intraspecific sympatry in the lizard species *Sceloporus torquatus*, and its relationship with *S. cyanogenys*. Trans. Kans. Acad. Sci. 77:219–224.

Smith, H. M., and R. A. Brandon. 1968. Data nova herpetologica Mexicana. Trans. Kans. Acad. Sci. 71:49–61.

Smith, H. M., and D. Chiszar. 2003. Observations by Berlandier 1827–1834 on the crocodilians of Texas and northeastern Mexico. Bull. Chic. Herpetol. Soc. 38:155–157.

Smith, H. M., R. Conant, and D. Chiszar. 2003. Berlandier's herpetology of Tamaulipas, Mexico, 150 years ago. Newsl. Bull. Int. Soc. Hist. Bibliog. Herpetol. 4:19–30.

Smith, H. M., and J. R. Dixon. 1987. The amphibians and reptiles of Texas: A guide to records needed for Mexico. Bull. Md. Herpetol. Soc. 23:154–157.

Smith, H. M., R. L. Holland, and R. Spiering. 1976. Observations on a species of salamander (*Pseudoeurycea*) from Tamaulipas, Mexico. Bull. Md. Herpetol. Soc. 12:33–36.

Smith, H. M., and L. E. Laufe. 1946. A summary of Mexican lizards of the genus *Ameiva*. Univ. Kans. Sci. Bull. 31:7–73.

Smith, H. M., and R. B. Smith. 1979. Synopsis of the herpetofauna of Mexico. Vol. 6, Guide to Mexican turtles, bibliographic addendum III. North Bennington, VT: John Johnson.

Smith, H. M., and E. H. Taylor. 1945. An annotated checklist and key to the snakes of Mexico. U.S. Natl. Mus. Bull. 187:i–iv, 1–239.

———. 1948. An annotated checklist and key to the amphibia of Mexico. U.S. Natl. Mus. Bull. 194:i–iv, 1–118.

———. 1950a. An annotated checklist and key to the reptiles of Mexico exclusive of the snakes. U.S. Natl. Mus. Bull. 199: i–iv, 1–253.

———. 1950b. Type localities of Mexican reptiles and amphibians. Univ. Kans. Sci. Bull. 33: 313–380.

———. 1966. Herpetology of Mexico: Annotated checklist and keys to the amphibians and reptiles. A reprint of Bulletins 187, 194 and 199 of the U.S. National Museum with a list of subsequent taxonomic innovations. Ashton, MD: Eric Lundberg.

Smith, H. M., F. van Breukelen, D. L. Auth, and D. Chiszar. 1998. A subspecies of the Texas blind snake (*Leptotyphlops dulcis*) without supraoculars. Southwest. Nat. 43:437–440.

Smith, P. W., and W. L. Burger. 1950. Herpetological results of the University of Illinois field expedition, spring 1949. III, Sauria. Trans. Kans. Acad. Sci. 53:165–175.

Smith, P. W., H. M. Smith, and J. E. Werler. 1952. Notes on a collection of amphibians and reptiles from eastern Mexico. Tex. J. Sci. 4: 251–260.

Stejneger, L. H. 1893. Annotated list of the reptiles and batrachians collected by the Death Valley Expedition in 1891, with descriptions of new species. North Am. Fauna 7:159–228.

———. 1925. New species and subspecies of American turtles. J. Wash. Acad. Sci. 15: 462–463.

Stejneger, L. H., and T. Barbour. 1933. A checklist of North American amphibians and reptiles. 3rd ed. Cambridge, MA: Harvard University Press.

Taylor, E. H. 1935. A taxonomic study of the cosmopolitan scincoid lizards of the genus *Eumeces*. Univ. Kans. Sci. Bull. 23:1–643.

————. 1938. On the Mexican snakes of the genera *Trimorphodon* and *Hypsiglena*. Univ. Kans. Sci. Bull. 25:357–383.

————. 1940. A new eleutherodactylid frog from Mexico. Proc. New Engl. Zool. Club 18:13–16.

Terán-Juárez, S. A. 2008. *Anguis incomptus* (Sauria: Anguidae), una adicion a la herpetofauna de Tamaulipas, Mexico. Acta Zool. Mex. 24:235–238.

Tunnell, J. W., Jr. 2002. Conservation issues and recommendations. In The Laguna Madre of Texas and Tamaulipas, ed. J. W. Tunnell Jr. and F. W. Judd, 275–288. College Station: Texas A&M University Press.

Tunnell, J. W., Jr., and F. W. Judd. 2002. The Laguna Madre of Texas and Tamaulipas. College Station: Texas A&M University Press.

Velasco, A. L. 1892. Geografía y estadistica del estado de Tamaulipas. In Geografía y estadística de la República Méxicana, vol. 12, 1–204. Mexico City: Oficina Tip, De la Secretaria de Fomento.

Walker, C. F. 1955a. A new gartersnake (*Thamnophis*) from Tamaulipas. Copeia 1955:110–113.

————. 1955b. A new salamander of the genus *Pseudoeurycea* from Tamaulipas. Occas. Pap. Mus. Zool. Univ. Mich. 567:1–8.

————. 1955c. Two new lizards of the genus *Lepidophyma* from Tamaulipas. Occas. Pap. Mus. Zool. Univ. Mich. 564:1–10.

Walker, J. M. 1986. The taxonomy of parthenogenetic species of hybrid origin: Cloned hybrid populations of *Cnemidophorus* (Sauria: Teiidae). Syst. Zool. 35:427–440.

————. 1987a. Distribution and habitat of a new major clone of a parthenogenetic whiptail lizard (genus *Cnemidophorus*) in Texas and Mexico. Tex. J. Sci. 39:313–334.

————. 1987b. Distribution and habitat of the parthenogenetic whiptail lizard, *Cnemidophorus laredoensis* (Sauria: Teiidae). Am. Midl. Nat. 117:319–332.

Walker, J. M., R. M. Abuhteba, and M. A. Paulissen. 1990. Additions to the distributional ecology of two parthenogenetic clonal complexes in the subgroup *Cnemidophorus laredoensis* (Sauria: Teiidae) in Texas and Mexico. Tex. J. Sci. 42:129–135.

Walker, J. M., S. E. Trauth, J. E. Cordes, and J. M. Britton. 1986. Geographic distribution: *Cnemidophorus laredoensis*. Herpetol. Rev. 17:27–28.

Watkins-Colwell, G. J., H. M. Smith, and D. Chiszar. 1996. Geographic distribution: *Sceloporus chaneyi*. Herpetol. Rev. 27:153.

Webb, R. G. 1960. Notes on some amphibians and reptiles from northern Mexico. Trans. Kans. Acad. Sci. 63:289–298.

Werler, J. E. 1951. Miscellaneous notes on the eggs and young of Texan and Mexican reptiles. Zoologica 56:37–48.

Werler, J. E., and D. M. Darling. 1950. *Rhinocheilus lecontei tessellatus* Garman in Tamaulipas, Mexico. Herpetologica 6:112.

Werler, J. E., and H. M. Smith. 1952. Notes on a collection of reptiles and amphibians from Mexico, 1951–1952. Tex. J. Sci. 4:551–573.

Wiens, J. J., and T. A. Penkrot. 2002. Delimiting species using DNA and morphological variation and discordant species limits in spiny lizards (*Sceloporus*). Syst. Biol. 51:69–91.

Wiens, J. J., and T. W. Reeder. 1997. Phylogeny of the spiny lizards (*Sceloporus*) based on molecular and morphological evidence. Herpetol. Monogr. 11:1–101.

Wilson, L. D., and G. R. Zug. 1991. *Lepidochelys kempii*. Cat. Am. Amphib. Rept. 509:1–8.

Wolterstorff, W. 1930. Beiträge zur herpetologie Mexikos. II, Zur systematik und biologie der urodelen Mexikos. Abh. Ber. Mus. Natur-u. Heimatkunde Magdeburg 6:129–149.

9. California

Adler, K. 1978. Herpetology in western North America during the nineteenth century. In Herpetological explorations of the Great American West, ed. K. Adler, 1:1–17. New York: Arno Press.

————. 1989. Contributions to the history of herpetology. Vol. 1. Salt Lake City, UT: Society for the Study of Amphibians and Reptiles.

————. 2007. Contributions to the history of herpetology. Vol. 2. Salt Lake City, UT: Society for the Study of Amphibians and Reptiles.

Ashton, K. G., and A. de Queiroz. 2001. Molecular systematic of the Western Rattlesnake, *Crotalus viridis* (Viperidae), with comments on the utility of the d-loop in phylogenetic studies of snakes. Mol. Phylogenet. Evol. 21:176–189.

Baird, S. F. 1854. Descriptions of new genera and species of North American frogs. Proc. Acad. Nat. Sci. Phila. 7:59–62.

————. 1859a (1858). Descriptions of new genera and species of North American lizards in the Museum of the Smithsonian Institution. Proc. Acad. Nat. Sci. Phila. 10:253–256.

————. 1859b. Report upon reptiles collected on the survey. In Reports of explorations and surveys, to ascertain the most practicable and economical route for a railroad from the Mississippi River to the Pacific Ocean, ed. United States War Department, vol. 10, pt. 4, no. 4, 9–13. Washington, DC: Department of the Interior.

Baird, S. F., and C. Girard. 1852a. Characteristics of some new reptiles in the Museum of the Smithsonian Institution. Part 2. Proc. Acad. Nat. Sci. Phila. 6:125–129.

———. 1852b. Descriptions of new species of reptiles, collected by the U.S. exploring expedition under the command of Capt. Charles Wilkes, U.S.N. First part.—including the species from the western coast of America. Proc. Acad. Nat. Sci. Phila. 6:174–177.

———. 1853. Catalogue of North American reptiles in the Museum of the Smithsonian Institution. Part 1, Serpents. Washington, DC: Smithsonian Institution.

Barbour, M. G., T. Keeler-Wolf, and A. A. Schoen-herr. 2007. Terrestrial vegetation of California. Berkley: University of California Press.

Barrows, C. W., and M. Fisher. 2009. Coachella Fringe-toed Lizard: *Uma scoparia*, Cope 1894. In Lizards of the American Southwest: A photographic field guide, ed. L. L. C. Jones and R. E. Lovich, 266–269. Tucson, AZ: Rio Nuevo Publishers.

Benson, L. 1957. Plant classification. Boston: D. C. Heath.

Bezy, R. L. 1967. A new night lizard (*Xantusia vigilis sierrae*) from the southern Sierra Nevada in California. J. Ariz. Acad. Sci. 4 (3): 163–167.

———. 2009. Sierra Night Lizard: *Xantusia sierrae* Bezy, 1967. In Lizards of the American Southwest: A photographic field guide, ed. L. L. C. Jones and R. E. Lovich, 432–435. Tucson, AZ: Rio Nuevo Publishers.

Blainville, H. M. D. 1835. Description de quelques espèces de reptiles de la Californie précédée de l'analyse d'un système général d'herpétologie et d'amphibiologie. Nouv. Ann. Mus. Hist. Natur. Paris 4:232–296.

Bocourt, M. F. 1879. Études sur les reptiles. In Recherches zoologiques pour server à l'histoire et la faune de l'Amérique centrale et du Mexique. Mission scientifique au Mexique et dans l'Amérique centrale, recherches zoologiques, ed. A. H. A. Duméril, M. F. Bocourt, and M. F. Mocquard, pt. 3, sec. 1, 1–1012. Paris: Imprimerie Nationale.

Brame, A. H., Jr. and K. F. Murray. 1968. Three new slender salamanders (*Batrachoseps*) with a discussion of relationships and speciation within the genus. Bull. Los Angeles Co. Mus. Nat. Hist. 4:1–35.

Brattstrom, B. H. 1953. An ecological restriction of the type locality of the Western Worm Snake, *Leptotyphlops h. humilis*. Herpetologica 8:180–181.

Bury, R. B. 2005. *Dicamptodon ensatus* Eschscholtz, 1833: California Giant Salamander. In Amphibian declines: The conservation status of United States species, ed. M. Lannoo, 653–654. Berkeley: University of California Press.

Camp, C. L. 1915. *Batrachoseps major* and *Bufo cognatus californicus*, new amphibian from southern California. Univ. Calif. Publ. Zool. 12:327–334.

———. 1916a. Description of *Bufo canorus*: A new toad from the Yosemite National Park. Univ. Calif. Publ. Zool. 17:59–62.

———. 1916b. *Spelerpes platycephalus*, a new alpine salamander from the Yosemite National Park. Univ. Calif. Publ. Zool. 17:11–14.

———. 1917. Notes of the systematic status of the toads and frogs of California. Univ. Calif. Publ. Zool. 17 (9): 115–125.

Campbell, J. A., and W. W. Lamar. 2004. The venomous reptiles of the Western Hemisphere. Ithaca, NY: Cornell University Press.

Cochran, D. M. 1961. Type specimens of reptiles and amphibians in the U.S. National Museum. Bull. U.S. Natl. Mus. 220:1–199.

Cooper, J. G. 1861. New Californian animals. Proc. Calif. Acad. Sci. 2:120.

Cope, E. D. 1865. Third contribution to the herpetology of Tropical America. Proc. Acad. Nat. Sci. Phila. 17:185–198.

———. 1866. On the structures and distribution of the genera of the arciferous *Anura*. Proc. Acad. Nat. Sci. Phila., 2nd ser., 6:67–112.

———. 1869. A review of the species of the Plethodontidae and Desmognathidae. Proc. Acad. Nat. Sci. Phila. 21:93–118.

———. 1883 (1884). Notes on the geographical distribution of batrachia and reptilia in western North America. Proc. Acad. Nat. Sci. Phila. 35:10–35.

———. 1889. The batrachia of North America. Bull. U.S. Natl. Mus. 34:1–525.

———. 1892. A critical review of the characters and variation of the snakes of North America. Proc. U.S. Natl. Mus. 14:589–694.

———. 1894. On the iguanian genus *Uma* Baird. Am. Nat. 28:434–435.

———. 1895. On the species *Uma* and *Xantusia*. Am. Nat. 29:938–939.

———. 1896. On two new species of lizards from southern California. Am. Nat. 30:883–836.

Davidson, C., and G. M. Fellers. 2005. *Bufo canorus* Camp, 1916: Yosemite Toad. In Amphibian declines: The conservation status of United States species, ed. M. Lannoo, 400–401. Berkeley: University of California Press.

Devitt, T. J., S. E. Cameron Devitt, B. D. Hollingsworth, J. A. McGuire, and C. Moritz. 2013. Montane refugia predict population genetic structure in the Large-blotched Ensatina salamander. Mol. Ecol. 22 (6): 1–16.

Dixon, J. R. 1964. The systematics and distribution of lizards of the genus *Phyllodactylus* in North and Central America. New Mex. State Univ. Res. Ctr. Sci. Bull. 64:1–139.

Douglas, M. E., M. R. Douglas, G. W. Schuett, L. W. Porras, and A. T. Holycross. 2002. Phylogeography of the Western Rattlesnake (*Crotalus viridis*) complex, with emphasis on the Colorado Plateau. In The biology of the vipers, ed. G. W. Schuett, M. Höggren, M. E. Douglas, and H. W. Green, 11–50. Eagle Mountain, UT: Eagle Mountain Publishing.

Dunn, E. R. 1926. The salamanders of the family Plethodontidae. Northampton, MA: Smith College.

———. 1929. A new salamander from southern California. Proc. U.S. Natl. Mus. 74:1–3.

Eschscholtz, J. E. 1833. Zoologischer Atlas, enthaltend Abbildungen und Beschreibungen neuer Thierarten, während des Flottcapitains von Kotzebue zweiter Reise um die Welt, auf der Russisch-Kaiserlichen Kreigsschlupp Predpriaetië in den Jahren 1823–1826. Berlin: G. Reimer.

Espinoza, R. E. 2009. Mohave Fringe-toed Lizard: *Uma inornata*, Cope 1895. In Lizards of the American Southwest: A photographic field guide, ed. L. L. C. Jones and R. E. Lovich, 278–281. Tucson, AZ: Rio Nuevo Publishers.

Feldman, C. R., and R. F. Hoyer. 2010. A new species of snake in the genus *Contia* (Squamata: Colubridae) from California and Oregon. Copeia 2010:254–267.

Feldman, C. R., and G. S. Spicer. 2006. Comparative phylogeography of woodland reptiles in California: Repeated patterns of cladogenesis and population expansion. Mol. Ecol. 15:2201–2222.

Fellers, G. M. 2005. *Bufo exsul* Myers, 1942: Black Toad. In Amphibian declines: The conservation status of United States species, ed. M. Lannoo, 406–408. Berkeley: University of California Press.

Fellers, G. M., and C. A. Drost. 1993. Disappearance of the Cascades Frog, *Rana cascadae*, at the southern end of its range, California, USA. Biol. Conserv. 65:177–181.

———. 2009. Island Night Lizard: *Xantusia riversiana* Cope, 1883. In Lizards of the American Southwest: A photographic field guide, ed. L. L. C. Jones and R. E. Lovich, 428–431. Tucson, AZ: Rio Nuevo Publishers.

Fitch, H. S. 1934. A shift of specific names in the genus *Gerrhonotus*. Copeia 1934 (4): 172–173.

———. 1940. A biogeographical study of the *ordinoides* artenkreis of garter snakes (genus *Thamnophis*). Univ. Calif. Publ. Zool. 44:1–150.

Fontanella, F. M., C. R. Feldman, M. E. Siddall, and F. T. Burbrink. 2008. Phylogeography of *Diadophis punctatus*: Extensive lineage diversity and repeated patterns of historical demography in a trans-continental snake. Mol. Phylogenet. Evol. 46:1049–1070.

Fouquette, M. J. 1968. Remarks on the type specimen of *Bufo alvarius* Girard. Great Basin Nat. 28 (2): 70–72.

Frost, D. R. 2011. Amphibian species of the world: An online reference. Version 5.5, January 31. New York: American Museum of Natural History. http://research.amnh.org/vz/herpetology/amphibia/.

Germano, D. J. 2009. Blunt-nosed Leopard Lizard: *Gambelia sila* (Stejneger). In Lizards of the American Southwest: A photographic field guide, ed. L. L. C. Jones and R. E. Lovich, 120–123. Tucson, AZ: Rio Nuevo Publishers.

Girard, C. 1858. United States Exploring Expedition during the years 1838, 1839, 1840, 1841, 1842, under the command of Charles Wilkes, U.S.N. Vol. 20. Philadelphia: J. B. Lippincott.

———. 1859. Reptiles of the boundary. In Report on the United States and Mexican Boundary Survey, by S. F. Baird. Vol. 2. Washington, DC: Department of the Interior.

Gorman, J. 1954. A new species of salamander from central California. Herpetologica 10: 153–158.

Gorman, J., and C. L. Camp. 1953. A new cave species of salamander of the genus *Hydromantes* from California, with notes on habits and habitat. Copeia 1953:39–43.

Goudey, C. B., and D. W. Smith. 1994. Ecological units of California. San Francisco: USDA Forest Service and Natural Resource Conservation Service, Pacific Southwest Region

Gray, J. E. 1839. Reptiles. In The zoology of Captain Beechey's voyage to the Pacific and Behring's Straits performed in his majesty's ship *Blossom*, by J. Richardson et al., 93–97. London: Henry G. Bohn.

———. 1850. Catalogue of the species of amphibians in the collection of the British Museum. Part 2, Batrachia Gradientia, etc. London: Taylor and Francis.

———. 1852. Description of several new genera of reptiles, principally from the collection of the H.M.S. *Herald*. Ann. Mag. Nat. Hist. 10:437–440.

———. 1853. On a new species of salamander from California. Proc. Zool. Soc. London 21:11.

Grinnell, J., and C. L. Camp. 1917. A distributional list of the amphibians and reptiles of California. Univ. Calif. Publ. Zool. 17 (10): 127–208.

Grismer, L. L. 2002. Amphibians and reptiles of Baja California, including its Pacific islands, and the islands in the Sea of Cortés. Berkeley: University of California Press.

Grismer, L. L., and M. A. Galvan. 1986. A new night lizard (*Xantusia henshawi*) from a sandstone habitat in San Diego County, California. Trans. San Diego Soc. Nat. Hist. 21:155–165.

Hall, W. P., and H. M. Smith. 1979. Lizards of the *Sceloporus orcutti* complex of the Cape Region of Baja California. Breviora 452:1–26.

Hallowell, E. 1849. Description of a new species of salamander from Upper California. Proc. Acad. Nat. Sci. Phila. 4:126.

———. 1852. Descriptions of new species of reptiles inhabiting North America. Proc. Acad. Nat. Sci. Phila. 6:177–184.

———. 1853. On some new reptiles from California. Proc. Acad. Nat. Sci. Phila. 6:236–238.

———. 1854. Descriptions of new reptiles from California. Proc. Acad. Nat. Sci. Phila. 7:91–97.

Hanes, T. L. 1971. Succession after fire in the chaparral of southern California. Ecol. Monogr. 41:27–52.

Hansen, R. W., R. H. Goodman, and D. B. Wake. 2005. *Batrachoseps gabriela* Wake, 1996: San Gabriel Mountains Salamander. In Amphibian declines: The conservation status of United States species, ed. M. Lannoo, 672–673. Berkeley: University of California Press.

Hansen, R. W., and D. B. Wake. 2005. *Batrachoseps campi* Marlow, Brode, and Wake, 1979: Inyo Mountains Salamander. In Amphibian declines: The conservation status of United States species, ed. M. Lannoo, 669–671. Berkeley: University of California Press.

Hansen, R. W., D. B. Wake, and G.M. Fellers. 2005. *Batrachoseps pacificus* (Cope, 1865): Channel Islands Slender Salamander. In Amphibian declines: The conservation status of United States species, ed. M. Lannoo, 685–686. Berkeley: University of California Press.

Hickman, J. C., ed. 1993. The Jepson manual. Berkeley: University of California Press.

Holbrook, J. E. 1842. North American herpetology; or, A description of the reptiles inhabiting the United States. Philadelphia: J. Dobson.

Jackman, T. R. 1999. Molecular and historical evidence for the introduction of clouded salamanders (genus *Aneides*) to Vancouver Island, British Columbia, Canada, from California. Can. J. Zool. 76:1570–1579.

Jameson, D. L., J. P. Mackey, and R. C. Richmond. 1966. The systematics of the Pacific Tree Frog, *Hyla regilla*. Proc. Calif. Acad. Sci., 4th ser., 33 (19): 551–620.

Jennings, M. R. 1988 (1987). A biography of Dr. Charles Elisha Boyle, with notes on his 19th century natural history collection from California. Wasmann J. Biol., San Francisco 45: 58–68.

Jensen, H. A. 1947. A system for classifying vegetation in California. Calif. Fish Game 33:199–266.

Jepson, W. L. 1925. A manual of the flowering plants of California. Berkeley: University of California Press.

Jockusch, E. L., I. Martínez-Solano, R. W. Hansen, and D. B. Wake. 2012. Morphological and molecular diversification of slender salamanders (Caudata: Plethodontidae: *Batrachoseps*) in the southern Sierra Nevada of California with descriptions of two new species. Zootaxa 3190:1–30.

Jockusch, E. L., D. B. Wake, and K. P. Yanev. 1998. New species of slender salamanders, *Batrachoseps* (Amphibia: Plethodontidae), from the Sierra Nevada of California. Contr. Sci. Los Angeles Co. Mus. Nat. Hist. 472:1–17.

Jockusch, E. L., K. P. Yanev, and D. B. Wake. 2001. Molecular phylogenetic analysis of slender salamanders, genus *Batrachoseps* (Amphibia: Plethodontidae), from central coastal California with descriptions of four new species. Herpetol. Monogr. 15:54–99.

Kennicott, R. 1859. In Report upon reptiles collected on the survey, by S. F. Baird. In Reports of explorations and surveys, to ascertain the most practicable and economical route for a railroad from the Mississippi River to the Pacific Ocean, ed. United States War Department, vol. 10, pt. 4, no. 4, 9–13. Washington, DC: Department of the Interior.

———. 1860a. Descriptions of new species of North American serpents in the Museum of the Smithsonian Institution, Washington. Proc. Acad. Nat. Sci. Phila. 12:328–338.

———. 1860b. In Report upon reptiles collected on the survey, by J. G. Cooper, 292–396. In Reports of explorations and surveys, to ascertain the most practicable and economical route for a railroad from the Mississippi River to the Pacific Ocean, vol. 12, bk. 2, pt. 3, no. 4. Washington, DC: Department of the Interior.

———. 1861. On three new forms of rattlesnakes. Proc. Acad. Nat. Sci. Phila. 13:206–208.

Klauber, L. M. 1930. New and renamed subspecies of Crotalus confluentus Say, with remarks on related species. Trans. San Diego Soc. Nat. Hist. 6:95–144.

———. 1932. The Flat-tailed Horned Toad in Lower California. Copeia 1932 (2): 100.

———. 1943. The subspecies of the rubber snake, Charina. Trans. San Diego Soc. Nat. Hist. 10 (7): 83–90.

———. 1956. Rattlesnakes: Their habits, life histories, and influence on mankind. Berkeley: University of California Press.

———. 1972. Rattlesnakes: Their habits, life histories, and influence on mankind. 2nd ed. Berkeley: University of California Press.

Kruckeberg, A. R. 2006. Introduction to California soils and plants: Serpentine, vernal pools, and other geological wonders. Berkeley: University of California Press.

Kuchta, S. R. 2007. Contact zones and species limits: Hybridization between the lineages of the California Newt, Taricha torosa, in the southern Sierra Nevada. Herpetologica 63:332–350.

Kuchta, S. R., D. S. Parks, R. L. Mueller, and D. B. Wake. 2009. Closing the ring: Historical biogeography of the salamander ring species Ensatina eschscholtzii. J. Biogeogr. 36:82–995.

Kuchta, S. R., D. S. Parks, and D. B. Wake. 2009. Pronounced phylogeographic structure on a small spatial scale: Geomorphological evolution and lineage history in the salamander ring species Ensatina eschscholtzii in central coastal California. Mol. Phylogenet. Evol. 50:240–255.

Lockington, W. N. 1876. Description of a new genus and species of colubrine snake. Proc. Calif. Acad. Sci. 7:52–53.

Lovich, R. E. 2001. Phylogeography of the night lizard, Xantusia henshawi, in southern California: Evolution across fault zones. Herpetologica 57:470–487.

Mahoney, M. J., D. S. M. Parks, and G. M. Fellers. 2003. Uta stansburiana and Elgaria multicarinata on the California Channel Islands: Natural dispersal or artificial introduction? J. Herpetol. 37:586–591.

Mahrdt, C. R., and K. R. Beaman. 2009. Panamint Alligator Lizard: Elgaria panamintina (Stebbins, 1958). In Lizards of the American Southwest: A photographic field guide, ed. L. L. C. Jones and R. E. Lovich, 488–491. Tucson, AZ: Rio Nuevo Publishers.

Marlow, R. W., J. M. Brode, and D. B. Wake. 1979. A new salamander, genus Batrachoseps, from the Inyo Mountains of California, with a discussion of relationships in the genus. Contr. Sci. Los Angeles Co. Nat. Hist. Mus. 308:1–17.

Mayer, K. E., and W. F. Laudenslayer Jr. 1988. A guide to the wildlife habitats of California. Sacramento: California Department of Forestry and Fire Protection.

McGuire, J. A. 1996. Phylogenetic systematics of crotaphytid lizards (Reptilia: Iguania: Crotaphytidae). Bull. Carnegie Mus. Nat. Hist. 32:1–143.

Mead, L. S., D. R. Clayton, R. S. Nauman, D. H. Olson, and M. E. Pfrender. 2005. Newly discovered populations of salamanders from Siskiyou County California represent a species distinct from Plethodon stormi. Herpetologica 61:158–177.

Meik, J. M. 2008. Morphological analysis of the contact zone between the rattlesnakes Crotalus mitchellii stephensi and Crotalus m. pyrrhus. In The biology of the rattlesnakes, ed. W. K. Hayes, K. R. Beaman, M. D. Cardwell, and S. P. Bush, 39–46. Loma Linda, CA: Loma Linda University Press.

Moritz, C., C. J. Schneider, and D. B. Wake. 1992. Evolutionary relationships within the Ensatina eschscholtzii complex confirm the ring species interpretation. Syst. Biol. 41:273–291.

Munz, P. A., and D. D. Keck. 1959. A California flora. Berkeley: University of California Press.

Murphy, R. W., K. H. Berry, T. Edwards, A. E. Leviton, A. Lathrop, and J. D. Riedle. 2011. The dazed and confused identity of Agassiz's land tortoise, *Gopherus agassizii* (Testudines, Testudinidae) with the description of a new species and its consequences for conservation. ZooKeys 113:39–71.

Murphy, R. W., and H. M. Smith. 1991. *Anniella pulchra* Gray, 1852 (Reptilia, Squamata): Proposed designation of a neotype. Bull. Zool. Nomencl. 48:316–318.

Myers, E. A., J. A. Rodríguez-Robles, D. F. Denardo, R. E. Staub, A. Stropoli, S. Ruane, and F. T. Burbrink. 2013. Multilocus phylogeography assessment of the California mountain kingsnake (*Lampropeltis zonata*) suggests alternative patterns of diversification for the California Floristic Province. Mol. Ecol. 22:1–12.

Myers, G. S. 1942. The black toad of Seep Springs Valley, Inyo County, California. Occ. Pap. Mus. Zool. Univ. Mich. 460:1–19.

Myers, G. S., and T. P. Maslin Jr. 1948. The California plethodontid salamander, *Aneides flavipunctatus* (Strauch), with a description of a new subspecies and notes on other western *Aneides*. Proc. Biol. Soc. Wash. 61:127–138.

Nassbaum, R. A. 1976. Geographic variation and systematic of salamanders of the genus *Dicamptodon* Strauch (Ambystomatidae). Misc. Publ. Mus. Zool. Univ. Mich. 149:1–94.

Papenfuss, T. J., and J. F. Parham. 2013. Four new species of California legless lizards (*Anniella*). Breviora 536:1–17.

Parham, J. F., and T. J. Papenfuss. 2009. High genetic diversity among fossorial lizard populations (*Anniella pulchra*) in rapidly developing landscape (Central California). Conserv. Genet. 10:169–176.

Pauly, G. B., S. R. Ron, and L. Lerum. 2008. Molecular and ecological characterization of extralimital populations of Red-legged Frogs from western North America. J. Herpetol. 42:668–679.

Pereira, R. J., and D. B. Wake. 2009. Genetic leakage after adaptive and nonadaptive divergence in *Ensatina eschscholtzii* ring species. Evolution 63:2288–2301.

Phelan, R. L., and B. H. Brattstrom. 1955. Geographic variation in *Sceloporus magister*. Herpetologica 11 (1): 1–14.

Pook, C. E., W. Wüster, and R. S. Thorpe. 2000. Historical biogeography of the Western Rattle-

snake (Serpentes: Viperidae: *Crotalus viridis*), inferred from mitochondrial DNA sequence information. Mol. Phylogenet. Evol. 15:269–282.

Quinn, R. D., and S. C. Keeley. 2006. Introduction to California chaparral. Berkeley: University of California Press.

Rathke, M. H. 1833. *Triton torosus*. Eschscholtz, J. F. v. ed., Zoologischer Atlas, enthaltend Abbildungen und Beschreibungen neuer Thierarten, während des Flottcapitains von Kotzebue zweiter Reise um die Welt, auf der Russisch-Kaiserlichen Kreigsschlupp Predpriaetië in den Jahren 1823–1826. Fünftes Heft:12–14. Berlin: G. Reimer.

Recuero, E., Í. Martínez-Solano, G. Parra-Olea, and M. García-París. 2006a. Corrigendum to "Phylogeography of *Pseudacris regilla* (Anura: Hylidae) in western North America, with a proposal for a new taxonomic rearrangement" [Mol. Phylogenet. Evol. 39 (2006) 293–301]. Mol. Phylogenet. Evol. 41:511.

———. 2006b. Phylogeography of *Pseudacris regilla* (Anura: Hylidae) in western North America, with a proposal for a new taxonomic rearrangement. Mol. Phylogenet. Evol. 39: 293–301.

Richmond, J. Q., and E. L. Jockusch. 2007. Body size evolution simultaneously creates and collapses species boundaries in a clade of scincid lizards. Proc. Roy. Soc. 274:1701–1708.

Richmond, J. Q., and T. W. Reeder. 2002. Evidence for parallel ecological speciation in scincid lizards of the *Eumeces skiltonianus* species group (Squamata: Scincidae). Evolution 56:1498–1513.

Rissler, L. J., and J. J. Apodaca. 2007. Adding more ecology into species delimitation: Ecological niche models and phylogeography help define cryptic species in the Black Salamander (*Aneides flavipunctatus*). Syst. Biol. 56:924–942.

Rissler, L. J., R. J. Hijmans, C. H. Graham, C. Moritz, and D. B. Wake. 2006. Phylogeographic lineages and species comparisons in conservation analyses: A case study of California herpetofauna. Am. Nat. 167:655–666.

Rodríguez-Robles, J. A., and J. M. De Jesús-Escobar. 2000. Molecular systematics of New World gopher, bull, and pinesnakes (*Pituophis*; Colubridae), a transcontinental species complex. Mol. Phylogenet. Evol. 14:35–50.

Rodríguez-Robles, J. A., D. F. Denardo, and R. E. Staub. 1999. Phylogeography of the California

Mountain Kingsnake, *Lampropeltis zonata* (Colubridae). Mol. Ecol. 8:1923–1934.

Rodríguez-Robles, J. A., D. A. Good, and D. B. Wake. 2003. Brief history of herpetology in the Museum of Vertebrate Zoology, University of California, Berkeley, with a list of type specimens of recent amphibians and reptiles. Univ. Calif. Publ. Zool. 131:1–119.

Rodríguez-Robles, J. A., G. R. Stewart, and T. J. Papenfuss. 2001. Mitochondrial DNA-based phylogeography of North American Rubber Boas, *Charina bottae* (Serpentes: Boidae). Mol. Phylogenet. Evol. 18:227–237.

Rossman, D. A., N. B. Ford, and R. A. Seigel. 1996. The garter snakes: Evolution and ecology. Norman: University of Oklahoma Press.

Schmidt, K. P. 1953. A check list of North American amphibians and reptiles. 6th ed. Chicago: American Society of Ichthyologists and Herpetologists.

Schneider, G. E. 1986. Geographic variation in the contact zone of two subspecies of the Pacific Rattlesnake, *Crotalus viridis oreganus* and *Crotalus viridis helleri*. Master's thesis, University of California, Santa Barbara.

Schoenherr, A. A. 1992. A natural history of California. Berkeley: University of California Press.

Schoenherr, A. A., C. R. Feldmeth, and M. J. Emerson. 1999. Natural history of the islands of California. Berkeley: University of California Press.

Shaffer, H. B., G. M. Fellers, S. R. Voss, J. C. Olive, and G. B. Pauly. 2004. Species boundaries, phylogeography and conservation genetics of the Red-legged Frog (*Rana aurora/draytonii*) complex. Mol. Ecol. 13:2667–2677.

Shaffer, H. B., and A. E. Leviton. 1956. Holotype specimens of reptiles and amphibians in the collection of the California Academy of Sciences. Proc. Calif. Acad. Sci. 28:529–560.

Shaffer, H. B., G. B. Pauly, J. C. Oliver, and P. C. Trenham. 2004. The molecular phylogenetics of endangerment cryptic variation and historical phylogeography of the California Tiger Salamander, *Ambystoma californiense*. Mol. Ecol. 13:3033–3049.

Shaffer, H. B., and P. C. Trenham. 2005. *Ambystoma californiense* Gray, 1853: California Tiger Salamander. In Amphibian declines: The conservation status of United States species, ed. M. Lannoo, 605–608. Berkeley: University of California Press.

Slevin, J. R. 1928. The amphibians of western North America: An account of the species known to inhabit California, Alaska, British Columbia, Washington, Oregon, Idaho, Utah, Nevada, Arizona, Sonora, and Lower California. Occas. Pap. Calif. Acad. Sci. 16:1–152.

Smith, H. M., and E. H. Taylor. 1950. Type localities of Mexican reptiles and amphibians. Univ. Kans. Sci. Bull. 33:313–380.

Spickler, J. C., S. C. Sillett, S. B. Marks, and H. H. Welsh. 2006. Evidence of a new niche for the North American salamander: *Aneides vagrans* residing in the canopy of old-growth redwood forest. Herpetol. Conserv. Biol. 1:16–26.

Spinks, P. Q., and H. B. Shaffer. 2005. Range-wide molecular analysis of the Western Pond Turtle (*Emys marmorata*): Cryptic variation, isolation by distance, and their conservation implications. Mol. Ecol. 14:2047–2064.

Stebbins, G. L., and J. Major. 1965. Endemism and speciation in the California flora. Ecol. Monogr. 35:1–35.

Stebbins, R. C. 1958. A new alligator lizard from the Panamint Mountains, Inyo County, California. Am. Mus. Novit. 1883:1–27.

———. 2003. A field guide to western reptiles and amphibians. 3rd ed. Boston: Houghton Mifflin.

Stebbins, R. C., and C. H. Lowe Jr. 1951. Subspecific differentiation in the Olympic Salamander, *Rhyacotriton olympicus*. Univ. Calif. Publ. Zool. 50:465–484.

Stebbins, R. C., and S. M. McGinnis. 2012. Field guide to amphibians and reptiles of California. Berkeley: University of California Press.

Stejneger, L. 1889. Diagnosis of a new species of snake (*Lichanura orcutti*), from San Diego County, California. West Am. Sci. 6:83.

———. 1890. Annotated list of reptiles and batrachians collected by Dr. C. Hart Merriam and Vernon Bailey on the San Francisco Mountain plateau and desert of the Little Colorado, Arizona, with descriptions of new species. North Am. Fauna 3:103–118.

———. 1893a. Annotated list of the reptiles and batrachians collected by the Death Valley Expedition in 1891, with descriptions of new species. North Am. Fauna 7:159–228.

———. 1893b. Identification of a new California lizard. Proc. U.S. Natl. Mus. 16:467.

———. 1894. Description of *Uta mearnsi*, a new lizard from California. Proc. U.S. Natl. Mus. 17:589–591.

———. 1902. *Gerrhonotus caeruleus* versus *Gerrhonotus burnettii*. Proc. Biol. Soc. Wash. 15:37.

Stephens, F. 1905. Life areas of California. Trans. San Diego Soc. Nat. Hist. 1:1–8.

Stinson, M. L. 1984. Biology of sea turtles in San Diego Bay, California, and in the northeastern Pacific Ocean. Master's thesis, San Diego State University.

Storer, T. I. 1925. A synopsis of the amphibia of California. Univ. Calif. Publ. Zool. 27:1–342.

Strauch, A. 1870. Revision der salamandriden Gattungen. Mem. Acad. Sci., 7th ser., 16 (4):1–110.

Twitty, V. C. 1935. Two new species of *Triturus* from California. Copeia 1935:73–80.

———. 1942. The species of California *Triturus*. Copeia 1942:65–76.

Van Denburgh, J. 1896. Description of a new lizard (*Eumeces gilberti*) from the Sierra Nevada of California. Proc. Calif. Acad. Sci. 6:350–352.

———. 1897. The reptiles of the Pacific Coast and Great Basin: An account of the species known to inhabit California, and Oregon, Washington, Idaho, and Nevada. Occas. Pap. Calif. Acad. Sci. 5:1–236.

———. 1905. The reptiles and amphibians of the islands of the Pacific coast of North America from the Farallons to Cape San Lucas and the Revilla Gigedos. Proc. Calif. Acad. Sci., 3rd ser., 4 (1):1–41.

———. 1916. Four species of salamanders new to the state of California, with a description of *Plethodon elongates*, a new species, and notes on other salamanders. Proc. Calif. Acad. Sci., 4th ser., 6:215–221.

———. 1922. The reptiles of western North America: An account of the species known to inhabit California, and Oregon, Washington, Idaho, Utah, Nevada, Arizona, British Columbia, Sonora, and Lower California. Volume 1, Lizards. Volume 2, Snakes and turtles. San Francisco: California Academy of Sciences.

Van Dyke, E. C. 1919. The distribution of insects in western North America. Ann. Entomol. Soc. Am. 12:1–12.

———. 1929. The influence which geographical distribution has had on the production of the insect fauna of North America. 4th Int. Cong. Entomol. Proc. 2:556–566.

Vredenburg, V. T., R. Bingham, R. Knapp, J. A. T. Morgan, C. Moritz, and D. B. Wake. 2007. Concordant molecular and phenotypic data delineate new taxonomy and conservation priorities for the endangered Mountain Yellow-legged Frog. J. Zool. Lond. 271:361–374.

Wake, D. B. 1996. A new species of *Batrachoseps* (Amphibia: Plethodontidae) from the San Gabriel Mountains, southern California. Contr. Sci. Los Angeles Co. Nat. Hist. Mus. 463:1–12.

Wake, D. B., and T. R. Jackman. 1999. Description of a new species of plethodontid salamander from California. Can. J. Zool. 76:1579–1580.

Wake, D. B., and E. L. Jockusch. 2000. Detecting species borders using diverse data sets: Examples from plethodontid salamanders in California. In The biology of plethodontid salamanders, ed. R. C. Bruce, R. G. Jaeger, and L. D. Houck, 95–119. New York: Kluwer Academic/Plenum Publishers.

Wake, D. B., and T. J. Papenfuss. 2005a. *Hydromantes brunus* Gorman, 1954: Limestone Salamander. In Amphibian declines: The conservation status of United States species, ed. M. Lannoo, 781–782. Berkeley: University of California Press.

———. 2005b. *Hydromantes platycephalus* (Camp, 1916): Mt. Lyell Salamander. In Amphibian declines: The conservation status of United States species, ed. M. Lannoo, 783–784. Berkeley: University of California Press.

Wake, D. B., K. P. Yanev, and R. W. Hansen. 2002. New species of slender salamander, genus *Batrachoseps*, from the southern Sierra Nevada of California. Copeia 2002:1016–1028.

Wiegmann, A. F. A. 1828. Beyträge zur Amphibienkunde. Isis von Oken 21:364–383.

Wieslander, A. E. 1935. A vegetation type map of California. Madrono 19:161–163.

10. Arizona

Arizona atlas and gazetteer. 2008. 7th ed. Yarmouth, ME: DeLorme Publishing.

Baird, S. F. 1859. Reptiles of the boundary, with notes by the naturalists of the survey. In United States and Mexican Boundary Survey, under the order of Lieut. Col. W. H. Emory, major First Cavalry, and United States commissioner. Vol. 2, pt. 2, U.S. 34th Cong., 1st Sess., Exec. Doc. 108, 1–35, plates 1–41. Washington, DC: Government Printing Office.

Baird, S. F., and C. Girard. 1852. Characteristics of some new reptiles in the Museum of the Smithsonian Institution. Part 2. Proc. Acad. Nat. Sci. Phila. 6:125–129.

————. 1853. Catalogue of North American reptiles in the Museum of the Smithsonian Institution. Part 1, Serpents. Washington, DC: Smithsonian Institution.

Bogert, C. M., and R. Martín Del Campo. 1956. The Gila Monster and its allies. The relationships, habits, and behavior of the lizards of the family Helodermatidae. Bull. Am. Mus. Nat. Hist. 109:1–238.

Brown, D. E. 1994a. Arctic-boreal grasslands. In Biotic communities: Southwestern United States and northwestern Mexico, ed. D. E. Brown, 108–111. Salt Lake City: University of Utah Press.

————. 1994b. Chihuahuan desertscrub. In Biotic communities: Southwestern United States and northwestern Mexico, ed. D. E. Brown, 169–179. Salt Lake City: University of Utah Press.

————. 1994c. Great Basin conifer woodland. In Biotic communities: Southwestern United States and northwestern Mexico, ed. D. E. Brown, 52–57. Salt Lake City: University of Utah Press.

————. 1994d. Madrean evergreen woodland. In Biotic communities: Southwestern United States and northwestern Mexico, ed. D. E. Brown, 59–65. Salt Lake City: University of Utah Press.

————. 1994e. Plains and Great Basin grasslands. In Biotic communities: Southwestern United States and northwestern Mexico, ed. D. E. Brown, 115–121. Salt Lake City: University of Utah Press.

————. 1994f. Semidesert grassland. In Biotic communities: Southwestern United States and northwestern Mexico, ed. D. E. Brown, 123–131. Salt Lake City: University of Utah Press.

Brown, D. E., N. Carmony, H. Shaw, and W. L. Minckley. 2009. Arizona wildlife: The territorial years 1863–1912. Phoenix: Arizona Game and Fish Department.

Brown, D. E., N. B. Carmony, and R. M. Turner. 1981. Drainage map of Arizona showing perennial streams and some important wetlands. Phoenix: Arizona Game and Fish Department.

Brown, D. E., and C. H. Lowe. 1980. Biotic communities of the Southwest. Color map. USDA Forest Service, General Technical Report RM-78.

Chronic, H. 1983. Roadside geology of Arizona. Missoula, MT: Mountain Press Publishing.

Cochran, D. M. 1961. Type specimens of reptiles and amphibians in the United States National Museum. Bull. U.S. Natl. Mus. Bull. 220:1–291.

Cope, E. D. 1860. Catalogue of the Colubridae in the Museum of the Academy of Natural Sciences of Philadelphia, with notes and descriptions of new species. Part II. Proc. Acad. Nat. Sci. Phila. 12:241–266.

————. 1866. On the structure and distribution of the genera of the arciferous Anura. J. Acad. Nat. Sci. Phila., 2nd ser., 6:67–112.

————. 1867 (1866). On the reptilia and batrachia of the Sonoran Province of the Nearctic region. Proc. Acad. Nat. Sci. Phila. 18:300–314.

————. 1869. [Protocol of the March 9, 1869 meeting]. Proc. Acad. Nat. Sci. Phila. 21:5.

————. 1892. A critical review of the characters and variations of the snakes of North America. Proc. U.S. Natl. Mus. 14:589–694.

————. 1895. On the species of *Uma* and *Xantusia*. Am. Nat. 29:938–939.

Coues, E. 1871. The Yellow-headed Black Bird. Am. Nat. 5:200.

————. 1875. Synopsis of the reptiles and batrachians of Arizona: With critical and field notes, and an extensive synonymy. In Report upon geographical and geological explorations and surveys west of the one hundredth meridian, ed. G. M. Wheeler, 5:585–633. Washington, DC: US Government Printing Office.

Curtis, C. A. 1902. Coues at his first army post. Bird Lore 4:5–9.

Davis, G. P. 1982. Man and wildlife in Arizona: The American exploration period 1824–1865. Phoenix: Arizona Game and Fish Department.

Degenhardt, W. G., C. W. Painter, and A. H. Price. 1996. Amphibians and reptiles of New Mexico. Albuquerque: University of New Mexico Press.

Duellman, W. E., and C. H. Lowe. 1953. A new lizard of the genus *Cnemidophorus* from Arizona. Nat. Hist. Misc. Chic. Acad. Sci. 120:1–8.

Edwards, T., K. E. Bonine, C. Ivanyi, and R. Prescott. 2005. The molecular origins of Spiny-tailed Iguanas (*Ctenosaura*) on the grounds of the Arizona-Sonora Desert Museum. Sonoran Herpetol. 18:122–125.

Ewan, J. 1981. Biographical dictionary of Rocky Mountain naturalists: A guide to the writings and collections of botanists, zoologists, geologists, artists and photographers 1682–1932. Utrecht: Bohn, Scheltema and Holkema.

Fowlie, J .A. 1965. The snakes of Arizona. Fallbrook, CA: Azul Quinta Press.

Garfin, G., and R. Emanuel. 2006. Arizona watershed stewardship guide: Arizona weather and climate. Tucson: University of Arizona Cooperative Extension.

Girard, C. 1854. A list of the North American bufonids, with diagnoses of new species. Proc. Acad. Nat. Sci. Phila. 7:86–88.

Gloyd, H. K. 1936. The subspecies of *Crotalus lepidus*. Occas. Pap. Mus. Zool. Univ. Mich. 337:1–5.

Gorman, J. 1960. *Treetoad studies, 1. Hyla californiae, new species.* Herpetologica 16:214–222.

Gray, J. E. 1838. Catalogue of the slender-tongued saurians, with descriptions of many new genera and species. Part 3. Ann. Mag. Nat. Hist. 1:390.

———. 1845. Catalogue of the specimens of lizards in the collection of the British Museum. London: Taylor and Francis.

Hallowell, E. 1852. Descriptions of new species of reptiles inhabiting North America. Proc. Acad. Nat. Sci. Phila. 6:177–182.

———. 1854. Descriptions of new reptiles from California. Proc. Acad. Nat. Sci. Phila. 7:91–97.

Holycross, A. T. 2003. Desert Massasauga (*Sistrurus catenatus edwardsii*). Sonoran Herpetol. 16:30–32.

Kauffeld, C. 1957. Snakes and snake hunting. Garden City, NY: Hanover House.

Kearney, T. H., and R. H. Peebles. 1951. Arizona flora. Berkeley: University of California Press.

Kennicott, R. 1860. Descriptions of new species of North American serpents in the Museum of the Smithsonian Institution, Washington. Proc. Acad. Nat. Sci. Phila. 12:328–338.

———. 1861. On three new forms of rattlesnakes. Proc. Acad. Nat. Sci. Phila. 13:206–207.

Klauber, L. M. 1931. A new species of *Xantusia* from Arizona, with a synopsis of the genus. Trans. San Diego Soc. Nat. Hist. 7:1–16.

———. 1945. The geckos of the genus *Coleonyx* with description of new subspecies. Trans. San Diego Soc. Nat. Hist. 10:133–216.

———. 1946. The Glossy Snake, *Arizona*, with descriptions of new subspecies. Trans. San Diego Soc. Nat. Hist. 10:311–398.

Le Conte, J. 1854. Description of four new species of *Kinosternum*. Proc. Acad. Nat. Sci. Phila. 7:180–190.

Lowe, C. H. 1955. A new species of whiptail lizard (genus *Cnemidophorus*) from the Colorado Plateau of Arizona, New Mexico, Colorado, and Utah. Breviora 47:1–7.

Lowe, C. H., and D. E. Brown. 1994. Introduction. In Biotic communities: Southwestern United States and northwestern Mexico, ed. D. E. Brown, 8–16. Salt Lake City: University of Utah Press.

Lowe, C. H., C. R. Schwalbe, and T. B. Johnson. 1986. Venomous reptiles of Arizona. Phoenix: Arizona Game and Fish Department.

Lowe, C. H., and J. W. Wright. 1964. Species of the *Cnemidophorus exsanguis* group of whiptail lizards. J. Ariz. Acad. Sci. 3:78–80.

Mearns, E. A. 1907. Mammals of the Mexican boundary of the United States. Bull. U.S. Natl. Mus. 56:1–530.

Meek, S. E. 1905. An annotated list of a collection of reptiles from southern California and northern lower California. Field Columbian Mus. Zool. Ser. 7:1–19.

Minckley, W. L., and D. E. Brown. 1994. Warm temperate wetlands. In Biotic communities: Southwestern United States and northwestern Mexico, ed. D. E. Brown, 249–267. Salt Lake City: University of Utah Press.

Mulcahy, D. G. 2006. Historical biogeography of North American Nightsnakes and their relationships among the dipsadines: Evidence for vicariance associated with Miocene formations of northwestern Mexico. PhD diss., Utah State University, Logan.

———. 2008. Phylogeography and species boundaries of the western North American Nightsnake (*Hypsiglena torquata*). Mol. Phylogenet. Evol. 46:1095–1115.

Murphy, R. W., K. H. Berry, T. Edwards, A. E. Leviton, A. Lathrop, and J. D. Riedle. 2011. The dazed and confused identity of Agassiz's land tortoise, *Gopherus agassizii* (Testudines, Testudinidae) with the description of a new species, and its consequences for conservation. ZooKeys 113:39–71.

Nations, D., and E. Stump. 1981. Geology of Arizona. Dubuque, IA: Kendall Hunt Publishing.

Papenfuss, T. J., J. R. Macey, and J. A. Schulte II. 2001. A new lizard in the genus *Xantusia* from Arizona. Sci. Pap. Nat. Hist. Mus. Univ. Kans. 23:1–9.

Pase, C. P., and D. E. Brown. 1994a. Alpine tundra. In Biotic communities: Southwestern United States and northwestern Mexico, ed. D. E. Brown, 27–33. Salt Lake City: University of Utah Press.

———. 1994b. Interior chaparral. In Biotic communities: Southwestern United States and northwestern Mexico, ed. D. E. Brown, 95–99. Salt Lake City: University of Utah Press.

———. 1994c. Rocky Mountain (Petran) and Madrean montane conifer forest. In Biotic communities: Southwestern United States and northwestern Mexico, ed. D. E. Brown, 43–48. Salt Lake City: University of Utah Press.

———. 1994d. Rocky Mountain (Petran) sub-alpine conifer forest. In Biotic communities: Southwestern United States and northwestern Mexico, ed. D. E. Brown, 37–39. Salt Lake City: University of Utah Press.

Platz, J. E., and J. S. Frost. 1984. *Rana yavapaiensis:* A new species of leopard frog (*Rana pipiens* complex) from Arizona. Copeia 1984:940–948.

Platz, J. E., and J. S. Mecham. 1979. *Rana chiricahuensis:* A new species of leopard frog (*Rana pipiens* complex) from Arizona. Copeia 1979:383–390.

Richmond, C. W. 1918. In memoriam: Edgar Alexander Mearns. Auk 35:1–18.

Ruthven, A. G. 1907. A collection of reptiles and amphibians from southern New Mexico and Arizona. Am. Mus. Nat. Hist. 23:438–604.

Sanders, O., and H. M. Smith. 1951. Geographic variation in toads of the debilis group of *Bufo*. *Field and Lab. So. Meth. Univ. 19:153*.

Schmidt, K. P. 1921. New species of North American lizards of the genera *Holbrookia* and *Uta*. Am. Mus. Novit. 22:1.

———. 1953. A checklist of North American amphibians and reptiles. 6th ed. Chicago: University of Chicago Press.

Sellers, W. D. 1964. The climate of Arizona. In Arizona climate, ed. C. R. Green and W. D. Sellers, 5–51. Tucson: University of Arizona Press.

Sellers, W. D., and R. H. Hill, eds. 1974. Arizona climate 1931–1972. 2nd ed. rev. Tucson: University of Arizona Press.

Smith, H. M. 1937. A synopsis of the *scalaris* group of the lizard genus *Sceloporus*. Occas. Pap. Mus. Zool. Univ. Mich. 361:1–8.

Smith, H. M., and E. H. Taylor. 1948. An annotated checklist and key to the amphibia of Mexico. Bull. U.S. Natl. Mus. 194:1–118.

———. 1950. Type localities of Mexican reptiles and amphibians. Univ. Kans. Sci. Bull. 33: 313–380.

Springer, S. 1928. An annotated list of the lizards of Lee's Ferry, Arizona. Copeia 1928:100–104.

Stebbins, R. C. 1954. *Reptiles and amphibians of western North America*. New York: McGraw-Hill.

———. 2003. *A field guide to western reptiles and amphibians. 3rd ed.* Boston: Houghton Mifflin.

Stejneger, L. 1890. On a new genus and species of colubrine snakes from North America. Proc. U.S. Natl. Mus. 13:151–155.

———. 1902 (1903). The reptiles of the Huachuca Mountains, Arizona. Proc. U.S. Natl. Mus. 25:149–158.

Stickel, W. H. 1943. The Mexican snakes of the genera *Sonora* and *Chionactis* with notes on the status of other colubrid genera. Proc. Biol. Soc. Wash. 56:109–128.

Swarth, H. S. 1921. The type locality of *Crotalus willardi* (Meek). Copeia 1921:83.

Taylor, E. H. 1939 (1938). Frogs of the *Hyla eximia* group in Mexico with descriptions of two new species. Univ. Kans. Sci. Bull. 25:436.

Turner, R. M. 1994a. Great Basin desertscrub. In Biotic communities: Southwestern United States and northwestern Mexico, ed. D. E. Brown, 145–155. Salt Lake City: University of Utah Press.

———. 1994b. Mohave desertscrub. In Biotic communities: Southwestern United States and northwestern Mexico, ed. D. E. Brown, 157–167. Salt Lake City: University of Utah Press.

Turner, R. M., and D. E. Brown. 1994. Sonoran desertscrub. In Biotic communities: Southwestern United States and northwestern Mexico, ed. D. E. Brown, 181–221. Salt Lake City: University of Utah Press.

Van Denburgh, J. 1895. Description of a new rattlesnake (*Crotalus pricei*) from Arizona. Proc. Calif. Acad. Sci. 2:856–857.

———. 1896. A list of some reptiles from southeastern Arizona, with a description of a new species of *Cnemidophorus*. Proc. Calif. Acad. Sci., 2nd ser., 6:344.

Van Denburgh, J., and J. R. Slevin. 1913. A list of reptiles and amphibians of Arizona, with notes on species in the collection of the Academy. Calif. Acad. Sci. 3:391–454.

Webb, R. and R. Axtell. 1986. Type and type-locality of *Sceloporus jarrovii* Cope, with travel-routes of Henry W. Henshaw in Arizona in 1873 and 1874. J. Herpetol. 20:32–41.

Western Regional Climate Center. 2001. Climate of Arizona. Accessed December 11, 2014. http://www.wrcc.dri.edu/narratives/ARIZONA.htm.

Wheeler, G. M. 1875. Report upon geographical explorations and surveys west of the one hundredth meridian. Vol. 5. Washington, DC: US Government Printing Office.

Woodin, W. H. 1953. Notes on some reptiles from the Huachuca area of southern Arizona. Bull. Chic. Acad. Sci. 9:285–296.

———. 1962. *Ficimia quadrangularis*, a snake new to the fauna of the United States. Herpetologica 18:152–153.

Wright, J. W., and C. H. Lowe. 1965. The rediscovery of *Cnemidophorus arizonae* Van Denburgh. J. Ariz. Acad. Sci. 3:164–168.

———. 1993. Synopsis of the subspecies of the Little Striped Whiptail lizard, *Cnemidophorus inornatus* Baird. J. Ariz.-Nev. Acad. Sci. 27:129–157.

Yarrow, H. C. 1875. Report upon the collections of batrachians and reptiles made in portions of Nevada, Utah, California, Colorado, New Mexico, and Arizona during the years of 1871, 1872, 1873, and 1874. In Report upon geographical and geological explorations and surveys west of the one hundredth meridian, by G. M. Wheeler, 5:509–584. Washington, DC: US Government Printing Office.

11. New Mexico/ Nuevo México

Abert, J. W. 1846. Report of the expedition led by Lieutenant Abert on the upper Arkansas and through the country of the Comanche Indians, in the fall of the year 1845. 29th Cong., 1st Sess., Sen. Exec. Doc. VIII (438), 1–75.

———. 1848. Report of Lieut. J. W. Abert of his examination of New Mexico in the years 1846–1847. In Notes of a military reconnaissance, from Fort Leavenworth, in Missouri, to San Diego, in California, including parts of the Arkansas, Del Norte, and Gila Rivers, by W. H. Emory, 417–546. 30th Cong., 1st Sess., House of Rep. Exec. Doc. (41), Washington, DC.

Adler, K. 1989. Herpetologists of the past. In Contributions to the history of herpetology, ed. K. Adler, 5–141. Oxford, OH: Society for the Study of Amphibians and Reptiles.

Altenbach, M. J., and C. W. Painter. 1998. A bibliography and review of the Jemez Mountains Salamander, *Plethodon neomexicanus*, 1913–1998. New Mex. Nat. Notes 1:46–82.

Axtell, R. W. 1988. *Phrynosoma modestum*. Interpretive atlas of Texas lizards, 6:1–18. Privately printed.

Bailey, V. 1905. Biological survey of Texas. North Am. Fauna 25:1–222.

———. 1913. Life zones and crop zones of New Mexico. North Am. Fauna 35:1–100.

Baird, S. F. 1850. Revision of North American tailed-batrachia, with descriptions of new genera and species. J. Acad. Nat. Sci. Phila., 2nd ser., 1:281–294.

Baird, S. F., and C. Girard. 1852. Characteristics of some new reptiles in the Museum of the Smithsonian Institution. Parts 1–3. Proc. Acad. Nat. Sci. Phila. 6:68–70, 125–129, 173.

———. 1853. Catalogue of North American reptiles in the Museum of the Smithsonian Institution. Part 1, Serpents. Washington, DC: Smithsonian Institution.

Baltosser, W. H., and T. L. Best. 1990. Seasonal occurrence and habitat utilization by lizards in southwestern New Mexico. Southwest. Nat. 35:377–384.

Beck, D. D. 2005. Biology of Gila Monsters and Beaded Lizards. Berkeley: University of California Press.

Bettison, C. A., R. Shock, R. D. Jennings, and D. Miller. 1999. New identifications of naturalistic motifs on Mimbres pottery. In Sixty years of Mogollon archeology: Papers from the Ninth Mogollon Conference, Silver City, New Mexico. Tucson, AZ: SRI Press.

Best, T. L., H. C. James, and F. H. Best. 1983. Herpetofauna of the Pedro Armendariz lava field, New Mexico. Tex. J. Sci. 35 (3): 245–255.

Brown, D. E., ed. 1994a. Alpine and subalpine grasslands. In Biotic communities: Southwestern United States and northwestern Mexico, ed. D. E. Brown, 109–111. Salt Lake City: University of Utah Press.

———. 1994b. Biotic communities: Southwestern United States and northwestern Mexico. Salt Lake City: University of Utah Press.

———. 1994c. Chihuahuan desertscrub. In Biotic communities: Southwestern United States and northwestern Mexico, ed. D. E. Brown, 169–179. Salt Lake City: University of Utah Press.

———. 1994d. Great Basin conifer woodland. In Biotic communities: Southwestern United States and northwestern Mexico, ed. D. E. Brown, 52–57. Salt Lake City: University of Utah Press.

———. 1994e. Great Basin montane scrubland. In Biotic communities: Southwestern United States and northwestern Mexico, ed. D. E. Brown, 83–84. Salt Lake City: University of Utah Press.

———. 1994f. Madrean evergreen woodland. In Biotic communities: Southwestern United

States and northwestern Mexico, ed. D. E. Brown, 59–65. Salt Lake City: University of Utah Press.

———. 1994g. Montane meadow grassland. In Biotic communities: Southwestern United States and northwestern Mexico, ed. D. E. Brown, 113–114. Salt Lake City: University of Utah Press.

———. 1994h. Plains and Great Basin grasslands. In Biotic communities: Southwestern United States and northwestern Mexico, ed. D. E. Brown, 115–121. Salt Lake City: University of Utah Press.

———. 1994i. Semidesert grassland. In Biotic communities: Southwestern United States and northwestern Mexico, ed. D. E. Brown, 123–131. Salt Lake City: University of Utah Press.

———. 1994j. Subalpine scrub. In Biotic communities: Southwestern United States and northwestern Mexico, ed. D. E. Brown, 81. Salt Lake City: University of Utah Press.

Brown, D. E., and C. H. Lowe. 1980. Biotic communities of the Southwest (map). USDA Forest Service, Rocky Mountain Forest and Range Experiment Station, Gen. Tech. Rep. RM-78.

Brown, T. L. 2002. A herpetological survey conducted in Mills Canyon, Harding and Mora Counties, New Mexico. New Mex. Herpetol. Soc. Newsl. 39:2–4.

———. 2003. A herpetological survey of Chloride Canyon in the Black Range, Sierra County, New Mexico. Unpublished report, New Mexico Herpetological Society.

Brown, T. L., and J. N. Stuart. 1990. Canadian River Canyon, Harding, Mora, and San Miguel Counties. Special Publication No. 3, New Mexico Herpetological Society.

Burkett, D. 2000. Amphibians and reptiles of White Sands Missile Range. Unpublished report, US Dept. of the Army, CSTE-DTC-WE-ES-ES.

Burkett, D., and D. Black. 1999. Baseline herpetofauna survey report: San Andres National Wildlife Refuge. Unpublished report, San Andres NWR, US Fish and Wildlife Service.

Byers, M., D. S. Sias, and J. N. Stuart. 2007. The introduced Mediterranean gecko (*Hemidactylus turcicus*) in north-central New Mexico. Bull. Chic. Herpetol. Soc. 42:18–19.

Campbell, H. 1953. Observations of snakes DOR in New Mexico. Herpetologica 9:157–160.

———. 1956. Snakes found dead on the roads of New Mexico. Copeia 1956:124–125.

Carpenter, C. C., and J. J. Krupa. 1989. Oklahoma herpetology: An annotated bibliography. Norman: University of Oklahoma Press.

Cockerel, T. D. A. 1896. Reptiles and batrachians of Mesilla Valley, New Mexico. Am. Nat. 30:325–327.

Collins J. T. 2000. A survey of the amphibians, turtles, and reptiles of the eastern portion of the Kiowa National Grassland of New Mexico (Union County) and the Rita Blanca National Grassland of adjacent Oklahoma and Texas. Report to USDA, Forest Service. Lawrence, KS: JTC Enterprises.

Cope, E. D. 1883. Notes on the geographical distribution of batrachia and reptilia in western North America. Proc. Acad. Sci. Phila. 35: 10–35.

———. 1896. The geographical distribution of batrachia and reptilia in western North America. Am. Nat. 30:886–902, 1003–1026.

———. 1875. In Report upon the collection of batrachians and reptiles made in portions of Nevada, Utah, California, Colorado, New Mexico and Arizona, during the years 1871, 1872, 1873, and 1874, by H. C. Yarrow. In Report upon geographical and geological explorations and surveys west of the one hundredth meridian, by G. M. Wheeler, 5:509–584, plates 16–25. Washington, DC: US Government Printing Office.

Coues, E. 1875. Synopsis of the reptiles and batrachians of Arizona, with critical and field notes, and an extensive synonymy. In Report upon geographical and geological explorations and surveys west of the one hundredth meridian, by G. M. Wheeler, 5:585–633. Washington, DC: US Government Printing Office.

Creusere, F. M., and W. G. Whitford. 1976. Ecological relationships in a desert anuran community. Herpetologica 32:7–18.

Crother, B. I., ed. 2008. Scientific and standard English names of amphibians and reptiles of North America north of Mexico, with comments regarding confidence in our understanding. 6th ed. SSAR Herpetol. Circ. 37:1–84.

Cummer, M. R., B. L. Christman, and M. A. Wright. 2003. Investigations of the status and distribution of amphibians and reptiles on the Valles Caldera National Preserve, Sandoval County, New Mexico. Unpublished report to Valles Caldera Trust, Los Alamos, NM.

Cushing, F. H. 1883. My adventures in Zuñi, III. Century Mag. 26:28–47.

Dall, W. H. 1914. Spencer Fullerton Baird, a biography. Philadelphia: J. B. Lippincott.

Davis, C. O. 1995. Treasured earth: Hattie Cosgrove's Mimbres archeology in the American Southwest. Tucson, AZ: Sanpete Publications.

Degenhardt, W. G., and J. L. Christiansen. 1974. Distribution and habitats of turtles in New Mexico. Southwest. Nat. 19:21–46.

Degenhardt, W. G., and K. L. Jones. 1972. A new sagebrush lizard, *Sceloporus graciosus*, from New Mexico and Texas. Herpetologica 28:212–217.

Degenhardt, W. G., C. W. Painter, and A. H. Price. 1996. Amphibians and reptiles of New Mexico. Albuquerque: University of New Mexico Press.

Dixon, J. R. 1987. Amphibians and reptiles of Texas, with keys, taxonomic synopses, bibliography, and distribution maps. College Station: Texas A&M University Press.

Ellis, M. M. 1917. Amphibians and reptiles from the Pecos Valley. Copeia 1917:39–40.

Emory, W. H. 1848. Notes of a military reconnaissance, from Fort Leavenworth, in Missouri, to San Diego, in California, including parts of the Arkansas, Del Norte, and Gila Rivers. 30th Cong., 1st Sess., House of Rep. Exec. Doc. (41), Washington, DC.

———. 1857. Report of the United States and Mexican Boundary Survey, made under the direction of the secretary of the Interior. Vol. 1, 34th Cong., 1st Sess., House of Rep. Exec. Doc. (135), Washington, DC.

———. 1859. Report of the United States and Mexican Boundary Survey, made under the direction of the secretary of the Interior. Vol. 2, 34th Cong., 1st Sess., Senate Exec. Doc. (108), Washington, DC.

Fewkes, J. W. 1986. Hopi Snake ceremonies: An eyewitness account by Jesse Walter Fewkes: Selections from Bureau of American Ethnology annual reports nos. 16 and 19 for the years 1894–95 and 1897–98. Repr., Albuquerque, NM: Avanyu Publishing.

Fitzgerald, L. A., and C. W. Painter. 2000. Rattlesnake commercialization: Long-term trends, issues, and implications for conservation. Wildlife Soc. Bull. 28:235–253.

———. 2009. Dunes Sagebrush Lizard *Sceloporus arenicolus* Degenhardt and Jones, 1972. In Lizards of the American Southwest, ed. L. L. C. Jones and R. E. Lovich, 198–201. Tucson, AZ: Rio Nuevo Publishers.

Fitzgerald, L. A., C. W. Painter, A. Reuter, and C. Hoover. 2004. Collection, trade, and regulation of reptiles and amphibians of the Chihuahuan Desert ecoregion. TRAFFIC North America. Washington, DC: World Wildlife Fund.

Fox, T. S., T. K. Haarmann, and D. C. Keller. 1999. Amphibians and reptiles of Los Alamos County, New Mexico. LA-13626-MS. Los Alamos, NM: Los Alamos National Laboratory.

Fritts, T. H., R. D. Jennings, and N. J. Scott Jr. 1984. A review of the leopard frogs of New Mexico. Unpublished report, Santa Fe: New Mexico Dept. Game and Fish.

Garman, S. 1887. Reptiles and batrachians from New Mexico and Texas. Bull. Essex Inst. 19: 1–20.

Gehlbach, F. R. 1965. Herpetology of the Zuni Mountain region, northwestern New Mexico. Bull. U.S. Natl. Mus. 116:243–332.

Hallowell, E. 1852a. Descriptions of new species of reptiles inhabiting North America. Proc. Acad. Nat Sci. Philadelphia 6:177–182.

———. 1852b (1853). On a new genus and three new species of reptiles inhabiting North America. Proc. Acad. Sci. Phila. 6:206–209.

Harris, A. H., ed. 1963. Ecological distribution of some vertebrates in the San Juan Basin, New Mexico. Mus. New Mex. Pap. Anthropol. 8:1–64.

Harris, H. S., Jr., and R. S. Simmons. 1976. The paleogeography and evolution of *Crotalus willardi*, with a formal description of a new subspecies from New Mexico, United States. Bull. Md. Herpetol. Soc. 12:1–22.

Henderson, J., and J. P. Harrington. 1914. Ethnozoology of the Tewa Indians. Bull. Bur. Am. Ethnol. Smithson. Inst. 56:1–76.

Herrick, C. L., J. Terry, and H. N. Herrick. 1899. Notes on a collection of lizards from New Mexico. Bull. Sci. Lab. Denison Univ. 11:117–148.

Hough, W. 2010. The Moki Snake dance: A popular account of that unparalleled dramatic pagan ceremony of the Pueblo Indians of Tusavan, Arizona, with incidental mention of their life and customs. Repr., Charleston, SC: BiblioBazaar.

James, E. 1823. An account of an expeditions from Pittsburgh to the Rocky Mountains, performed in the years 1819 and '20 by the order of the Hon. J. C. Calhoun, sec'y of war: Under the command of Major Stephen H. Long. 2 vols. Philadelphia: H. C. Carey and I. Lea.

Jones, K. L. 1970. An ecological survey of the reptiles and amphibians of Chaco Canyon National Monument, San Juan County, New Mexico.

Master's thesis, University of New Mexico, Albuquerque.

Kennerly, C. B. R. 1856. Report on the zoology of the expedition. No. 1. Field notes and explanations. Pacific R.R. Reports, 35th Parallel, vol. 4, pt. 6, 1–17.

Klauber, L. M. 1939. Studies of reptile life in the arid southwest. Part 1, Night collecting on the desert with ecological statistics. Bull. Zool. Soc. San Diego 14:6–64.

———. 1972. Rattlesnakes: Their habits, life histories, and influence on mankind. 2nd ed. 2 vols. Berkeley: University of California Press.

Krupa, J. J. 1998. Amphibians and reptiles of Rattlesnake Springs, Carlsbad Caverns National Park and the impact of the non-native bullfrog on the herpetofauna. Unpublished report to Carlsbad Caverns National Park.

Lannoo, M., ed. 2005. Amphibian declines: The conservation status of United States species. Berkeley: University of California Press.

Lewis, T. H. 1950. The herpetofauna of the Tularosa Basin and Organ Mountains of New Mexico with notes on some ecological features of the Chihuahuan Desert. Herpetologica 6:1–10.

Little, E. L., Jr., and J. G. Keller. 1937. Amphibians and reptiles of the Jornada Experimental Range, New Mexico. Copeia 1937:216–222.

Lowe, C. H., Jr. 1956. A new species and subspecies of whiptailed lizards (genus *Cnemidophorus*) of the inland Southwest. Bull. Chic. Acad. Sci. 10:137–150.

Lowe, C. H., Jr., and K. S. Norris. 1956. A subspecies of the lizard *Sceloporus undulatus* from the White Sands of New Mexico. Herpetologica 12:125–127.

Lowe, C. H., Jr., and J. W. Wright. 1964. Species of the *Cnemidophorus exsanguis* subgroup of whiptail lizards. Ariz. Acad. Sci. 3:78–80.

Lowe, C. H., Jr., and R. G. Zweifel. 1952. A new species of whiptailed lizard (genus *Cnemidophorus*) from New Mexico. Bull Chic. Acad. Sci. 9:229–247.

MacBride, T. H. 1905. The Alamogordo Desert. Science, n.s., 21:90–97.

Martin-Bashore, T. E. 1997. Study of species composition, diversity, and relative abundance of reptiles and amphibians from six vegetative community associations on Otero Mesa, McGregor Range, Fort Bliss. Unpublished report to US Army Air Defense Artillery Center and Ft. Bliss, Texas. COMPA Industries, Albuquerque, NM.

Mecham, J. S. 1979. The biogeographical relationships of the amphibians and reptiles of the Guadalupe Mountains. In Biological investigations in the Guadalupe Mountains National Park, Texas, ed. H. H. Genoways and R. J. Baker, 169–179. USDI, Natl. Park Serv., Proc. Trans. Ser. 4:1–442.

Miller, R. R. 1946. The probable origin of the soft-shelled turtle in the Colorado River basin. Copeia 1946:46.

Moir, W. H., and H. M. Smith. 1970. Occurrence of an American salamander, *Aneides hardyi* (Taylor), in tundra habitat. Arctic Alpine Res. 2:155–156.

Mosauer, W. 1932. The amphibians and reptiles of the Guadalupe Mountains of New Mexico and Texas. Occas. Pap. Mus. Zool. Univ. Mich. 246:1–18.

Mulcahy, D. G. 2006. Historical biogeography of North American Nightsnakes and their relationships among the dipsadines: Evidence for vicariance associated with Miocene formations of northwestern Mexico. PhD diss., Utah State University, Logan.

———. 2008. Phylogeography and species boundaries of the western North American Nightsnake (*Hypsiglena torquata*). Mol. Phylogenet. Evol. 46:1095–1115.

Murphy, R. W., K. H. Berry, T. Edwards, A. E. Leviton, A. Lathrop, and J. D. Riedle. 2011. The dazed and confused identity of Agassiz's Land Tortoise, *Gopherus agassizi* (Testudines, Testudinidae) with the description of a new species, and its consequences for conservation. ZooKeys 113:39–71.

New Mexico Department of Game and Fish. 2006. Comprehensive wildlife conservation strategy for New Mexico. Santa Fe: New Mexico Department of Game and Fish.

Painter, C. W. 1985. Herpetology of the Gila and San Francisco River drainages of southwestern New Mexico. Unpublished report, New Mexico Department of Game and Fish, Santa Fe.

———. 2009. Gray Checkered Whiptail (unisexual) *Aspidoscelis dixoni* (Scudday, 1973). In Lizards of the American Southwest, ed. L. L. C. Jones and R. E. Lovich, 334–337. Tucson, AZ: Rio Nuevo Publishers.

Painter, C. W., P. W. Hyder, and G. Swinford. 1992. Three species new to the herpetofauna of New Mexico. Herpetol. Rev. 23:64.

Painter, C. W., and C. M. Milensky. 1993. Geographic distribution: *Crotalus tigris*. Herpetol. Rev. 24:155–156.

Parke, J. G. 1855. Report of exploration for that portion of a railroad route, near the thirty-second parallel of north latitude, lying between Dona Ana, on the Rio Grande, and Pima Villages, on the Gila. In Reports of exploration and surveys, to ascertain the most practicable and economical route for a railroad from the Mississippi River to the Pacific Ocean. Vol. 2 (6), 33rd Cong., 2nd Sess., House of Rep. Exec. Doc (91), 1–28. Washington, DC.

Pase, C. P. 1994. Alpine tundra. In Biotic communities: Southwestern United States and northwestern Mexico, ed. D. E. Brown, 27–33. Salt Lake City: University of Utah Press.

Pase, C. P., and D. E. Brown. 1994a. Interior chaparral. In Biotic communities: Southwestern United States and northwestern Mexico, ed. D. E. Brown, 95–99. Salt Lake City: University of Utah Press.

———. 1994b. Rocky Mountain (Petran) and Madrean montane conifer forest. In Biotic communities: Southwestern United States and northwestern Mexico, ed. D. E. Brown, 43–48. Salt Lake City: University of Utah Press.

———. 1994c. Rocky Mountain (Petran) subalpine conifer forest. In Biotic communities: Southwestern United States and northwestern Mexico, ed. D. E. Brown, 37–39. Salt Lake City: University of Utah Press.

Phelan, R. L., and B. H. Brattstrom. 1955. Geographic variation in *Sceloporus magister*. Herpetologica 11 (1): 1–14.

Pope, J. 1854 (1855). Report of exploration of a route for the Pacific railroad, near the 32nd parallel of north latitude, from the Red River to the Rio Grande. In Reports of exploration and surveys, to ascertain the most practicable and economical route of a railroad from the Mississippi River to the Pacific Ocean. Vol. 2 (6), 33rd Cong., 2nd Sess., House of Rep. Exec. Doc. (91), 1–185. Washington, DC.

Price, A. H., and J. L. LaPointe. 1990. Activity patterns of a Chihuahuan Desert snake community. Ann. Carnegie Mus. 59:15–23.

Ruthven, A. G. 1907. A collection of reptiles and amphibians from southern New Mexico and Arizona. Bull. Am. Mus. Nat. Hist. 23:483–603.

Schmidt, K. P. 1921. New species of North American lizards of the genera *Holbrookia* and *Uta*. Am. Mus. Novit. 22:1–6.

Sitgreaves, L. 1853. Report of an expedition down the Zuñi and Colorado Rivers, accompanied by maps, sketches, views, and illustrations. U.S. 32nd Cong., 2nd Sess., Senate Exec. Doc. 59, 1–198.

Smith, H. M., E. L. Bell, J. S. Applegarth, and D. Chiszar. 1992. Adaptive convergence in the lizard superspecies *Sceloporus undulatus*. Bull. Md. Herpetol. Soc. 28:123–149.

Smith, H. M., and R. B. Smith. 1973. Synopsis of the herpetofauna of Mexico. Vol. 2, Analysis of the literature exclusive of the Mexican axolotol. North Bennington, VT: John Johnson.

———. 1976. Synopsis of the herpetofauna of Mexico. Vol. 3, Source analysis and index for Mexican reptiles. North Bennington, VT: John Johnson.

———. 1979 (1980). Synopsis of the herpetofauna of Mexico. Vol. 4, Guide to Mexican turtles, bibliographic addendum III. North Bennington, VT: John Johnson.

———. 1993. Synopsis of the herpetofauna of Mexico. Vol. 7, Bibliographic addendum IV and index, bibliographic addenda II–IV 1979–1991. Niwot: University Press of Colorado.

Smith, H. M., and E.D. Taylor. 1950. Type localities of Mexican reptiles and amphibians. Univ. Kans. Sci. Bull. 33:313–380.

Stebbins, R. C., and W. J. Riemer. 1950. A new species of plethodontid salamander from the Jemez Mountains of New Mexico. Copeia 1950:73–80.

Stejneger, L. 1890. Part 5, Annotated list of reptiles and batrachians collected by Dr. C. Hart Merriam and Vernon Bailey on the San Francisco Mountain Plateau and Desert of the Little Colorado, Arizona, with descriptions of new species. In Results of a biological survey of the San Francisco Mountain region and Desert of the Little Colorado, Arizona. North Am. Fauna 3:103–118.

Stone, W., and J. A. G. Rehn. 1903. On the terrestrial vertebrates of portions of southern New Mexico and western Texas. Proc. Acad. Sci. Nat. Sci. Phila. 55:16–34.

Stuart, J. N. 2000. Additional notes on native and non-native turtles of the Rio Grande drainage basin, New Mexico. Bull. Chic. Herpetol. Soc. 35:229–235.

———. 2005. A supplemental bibliography of herpetology in New Mexico. Museum of Southwestern Biology, University of New Mexico.

Accessed November 21, 2014. http://inram.msb
.unm.edu/herpetology/publications/stuart
_supl_biblio.pdf.

Stuart, J. N., and T. L. Brown. 1988. Notes on amphibians and reptiles of the Dry Cimarron River
Valley and vicinity, Union County, New Mexico
and Cimarron County, Oklahoma. New Mex.
Herpetol. Soc. Spec. Publ. 2:1–11.

Stuart, J. N., and C. W. Painter. 1988. Geographic
distribution: *Chelydra serpentina serpentina*.
Herpetol. Rev. 19:21.

Stuart, J. N., and J. A. White. 1987. Notes on amphibians and reptiles from spring run areas of
southern Eddy County, New Mexico. New Mex.
Herpetol. Soc. Spec. Publ. 1:1–24.

Tanner, D. L. 1975. Lizards of the New Mexican
Llano Estacado and its adjacent river valleys.
Stud. Nat. Sci. East. New Mex. Univ. 2:1–39.

Taylor, E. H. 1941. A new plethodontid salamander from New Mexico. Proc. Biol. Soc. Wash.
54:77–79.

Townsend, C. H. T. 1893. On the life zones of the
Organ Mountains and adjacent region in southern New Mexico, with notes on the fauna of the
range. Science 22:313–315.

Turner, R. M. 1994. Great Basin desertscrub. In
Biotic communities: Southwestern United
States and northwestern Mexico, ed. D. E.
Brown, 145–155. Salt Lake City: University of
Utah Press.

Van Denburgh, J. 1924. Notes on the herpetology
of New Mexico, with a list of species known
from that state. Proc. Calif. Acad. Sci. 13:
189–230.

Webb, R. G. 1970. Reptiles of Oklahoma. Norman:
University of Oklahoma Press.

———. 1988. Type and type locality of *Sceloporus
poinsetti* Baird and Girard (Sauria: Iguanidae).
Tex. J. Sci. 40:407–415.

Western Regional Climate Center. 2011. Climate
of New Mexico. Accessed December 11, 2014.
http://www.wrcc.dri.edu/narratives/NEW
MEXICO.htm.

Whipple, A. W. 1854 (1856). Itinerary. In Reports
of exploration and surveys, to ascertain the most
practicable and economical route of a railroad
from the Mississippi River to the Pacific Ocean.
Vol. 2 (6), 33rd Cong., 2nd Sess., House of Rep.
Exec. Doc. (91), 1–136. Washington, DC.

Whitford, W. G., and F. M. Creusere. 1977. Seasonal and yearly fluctuations in Chihuahuan
Desert lizard communities. Herpetologica
33:54–65.

Wislizenus, A. M. D. 1848. Memoir of a tour to
northern Mexico, connected with Col. Doniph-
an's expedition, in 1846 and 1847. 30th Cong.,
1st Sess., Sen. Misc. Doc. (26). Washington, DC.

Woodward, B. D. 1982. Tadpole competition in a
desert anuran community. Oecologica 54:
96–100.

———. 1983. Predator-prey interactions and
breeding pond use by temporary-pond species in a desert anuran community. Ecology
64:1549–1555.

———. 1987. Clutch parameters and pond use in
some Chihuahuan Desert anurans. Southwest.
Nat. 32:13–19.

Wright, J. W., and C. H. Lowe Jr. 1993. Synopsis
of the subspecies of the Little Striped Whiptail,
Cnemidophorus inornatus Baird. J. Ariz.-Nev.
Acad. Sci. 27:129–157.

Yarrow, H. C. 1875. Report upon the collections of
batrachians and reptiles made in portions of Nevada, Utah, California, Colorado, New Mexico,
and Arizona, during the years 1871, 1872, 1873,
and 1874. In Report upon geographical and
geological explorations and surveys west of the
one hundredth meridian, ed. G. M. Wheeler,
5:509–584, plates 16–25. Washington, DC: US
Government Printing Office.

Zweifel, R. G. 1968. Reproductive biology of
anurans of the arid Southwest, with emphasis
on adaptation of embryos to temperature. Bull.
Am. Mus. Nat. Hist. 140:1–64.

12. Texas

Agassiz, L. 1857. Contributions to the natural
history of the United States of America. Vol. 1.
Boston: Little, Brown.

Baird, S. F. 1854. Descriptions of new genera and
species of North American frogs. Proc. Acad.
Nat. Sci. Phila. 7:59–62.

———. 1859a (1858). Description of new genera
and species of North American lizards in the
Museum of the Smithsonian Institution. Proc.
Acad. Nat. Sci. Phila. 10:253–256.

———. 1859b. Reptiles of the boundary, with
notes by the naturalists of the survey. In Report
of the United States and Mexican Boundary Survey, under the order of Lieut. Col. W. H. Emory,
major First Cavalry, and United States commissioner. Vol. 2, pt. 2, U.S. 34th Cong., 1st Sess.,
Exec. Doc. 108, 1–35, plates 1–41. Washington,
DC: Government Printing Office.

Baird, S. F., and C. Girard. 1852a. Characteristics

of some new reptiles in the Museum of the Smithsonian Institution. Proc. Acad. Nat. Sci. Phila. 6:68–70.

———. 1852b. Characteristics of some new reptiles in the Museum of the Smithsonian Institution. Second part. Proc. Acad. Nat. Sci. Phila. 6:125–129.

———. 1852c. Characteristics of some new reptiles in the Museum of the Smithsonian Institution. Third part. Proc. Acad. Nat. Sci. Phila. 6:173.

———. 1852d. Reptiles. In Stansbury's exploration and survey of the valley of the Great Salt Lake of Utah, 1852, including a reconnoissance of a new route through the Rocky Mountains, 336–365. Philadelphia: Lippincott, Grambo.

———. 1853. Catalogue of North American reptiles in the Museum of the Smithsonian Institution. Part 1, Serpents. Washington, DC: Smithsonian Institution.

———. 1854. Reptiles. In Exploration of the Red River of Louisiana in the year 1852, ed. R. B. Marcy and G. B. McClellan, 188–215. Washington, DC: United States War Department.

Baker, J. K. 1957. *Eurycea troglodytes*: A new blind cave salamander from Texas. Tex. J. Sci. 9:328.

Baur, G. 1893. Notes on the classification and taxonomy of the Testudinata. Proc. Am. Philos. Soc. 31:210–225.

Berlandier, J. L. 1980. Journey to Mexico during the years 1826–1834. Trans. S. M. Ohlendorf, J. M. Bigelow, and M. M. Standifer. 2 vols. Austin: Texas State Historical Association.

Bishop, S. C. 1941. Notes on salamanders with descriptions of several new forms. Occas. Pap. Mus. Zool. Univ. Mich. 451:1–21.

Bishop, S. C., and M. R. Wright. 1937. A new neotenic salamander from Texas. Proc. Biol. Soc. Wash. 50:141–144.

Blair, W. F. 1950. The biotic provinces of Texas. Tex. J. Sci. 2:93–117.

Bocourt, M. F. 1879. Études sur les reptiles. In Recherches zoologiques pour servir à l'histoire de la faune de l'Amérique centrale et du Mexique. Mission scientifique au Mexique et dans l'Amérique centrale, recherches zoologiques, ed. A. H. A. Duméril, M. F. Bocourt, and M. F. Mocquard, pt. 3, sec. 1, 1–1012. Paris: Imprimerie Nationale.

Bragg, A. N., and O. Sanders. 1951. A new subspecies of the *Bufo woodhousii* group of toads (Salientia: Bufonidae). Wasmann J. Biol. 9:363–378.

Brown, A. E. 1901a. A new species of *Ophiobolus*
from western Texas. Proc. Acad. Nat. Sci. Phila. 53 (3): 612–613.

———. 1901b. A review of the genera and species of American snakes, north of Mexico. Proc. Acad. Nat. Sci. Phila. 53:492.

Brown, B. C. 1950. An annotated check list of the reptiles and amphibians of Texas. Waco, TX: Baylor University Studies.

Carson, R. 1962. Silent spring. New York: Houghton Mifflin.

Chippindale, P. T., A. H. Price, and D. M. Hillis. 1993. A new species of perennibranchiate salamander (*Eurycea*: Plethodontidae) from Austin, Texas. Herpetologica 49:248–259.

Chippindale, P. T., A. H. Price, J. J. Wiens, and D. M. Hillis. 2000. Phylogenetic relationships and systematic revision of Central Texas hemidactyline plethodontid salamanders. Herpetol. Monogr. 14:1–81.

Cope, E. D. 1863. On *Trachycephalus*, *Scaphiopus* and other Batrachia. Proc. Acad. Nat. Sci. Phila. 15:43–54.

———. 1878. A new genus of Cystignathidae from Texas. Am. Nat. 12:252–253.

———. 1880. On the zoological position of Texas. Bull. U.S. Natl. Mus. 17:1–51.

Davis, W. B., and J. R. Dixon. 1958. A new *Coleonyx* from Texas. Proc. Biol. Soc. Wash. 71:149–152.

Dice, L. R. 1943. The biotic provinces of North America. Ann Arbor: University of Michigan Press.

Dixon, J. R. 1987. Amphibians and reptiles of Texas, with keys, taxonomic synopses, bibliography, and distribution maps. College Station: Texas A&M University Press.

———. 2000. Amphibians and reptiles of Texas, with keys, taxonomic synopses, bibliography, and distribution maps. 2nd ed. College Station: Texas A&M University Press.

Flores-Villela, O., and R. A. Brandon. 1992. *Siren lacertina* (Amphibia: Caudata) in northeastern Mexico and southern Texas. Ann. Carnegie Mus. 61:289–291.

Girard, C. 1852. A monographic essay on the genus *Phrynosoma*. In Stansbury's exploration and survey of the valley of the Great Salt Lake of Utah, 1852, including a reconnoissance of a new route through the Rocky Mountains, 354–365. Philadelphia: Lippincott, Grambo.

———. 1854. A list of North American bufonids, with diagnoses of new species. Proc. Acad. Nat. Sci. Phila. 7:86–88.

———. 1858. Herpetology in United States Exploring Expedition during the years 1838, 1840, 1841,
1842, under the command of Charles Wilkes,
U.S.N. Vol. 20. Philadelphia: J. B. Lippincott.

———. 1859. Herpetological notices. Proc. Acad.
Nat. Sci. Phila. 11:169–170.

Grobman, A. B. 1944. The distribution of the sal-
amanders of the genus *Plethodon* in the eastern
United States and Canada. Ann. N. Y. Acad.
Sci. 45:261–316.

Hartweg, N. 1939. A new American *Pseudemys*.
Occas. Pap. Mus. Zool. Univ. Mich. 397:1–4.

Hillis, D. M., D. A. Chamberlain, T. P. Wilcox, and
P. T. Chippindale. 2001. A new species of subterranean blind salamander (Plethodontidae:
Hemidactyliini: *Eurycea*: *Typhlomolge*). Herpetologica 57:266–280.

Kennicott, R. 1859. In Reptiles of the boundary,
with notes by the naturalists of the survey,
by S. F. Baird. In United States and Mexican
Boundary Survey, under the order of Lieut. Col.
W. H. Emory, major First Cavalry, and United
States commissioner. Vol. 2, pt. 2, U.S. 34th
Cong., 1st Sess., Exec. Doc. 108, 1–35, plates 1–41
(*Arizona elegans*, 18–19, plate 13). Washington,
DC: Government Printing Office.

———. 1860. Descriptions of new species of
North American serpents in the Museum of
the Smithsonian Institution, Washington. Proc.
Acad. Nat. Sci. Phila. 12:328–338.

———. 1861. On three new forms of rattlesnakes.
Proc. Acad. Nat. Sci. Phila. 13:206–207.

Klauber, L. M. 1939. A new subspecies of the
Western Worm Snake. Trans. San Diego Soc.
Nat. Hist. 9 (14): 67–68.

Lazell, J. 1992. Taxonomic tyranny and the exoteric. Herpetol. Rev. 23:14.

Matthes, B. 1885. Die Hemibatrachier im All-
gemeinen und die Hemibatrachier von Nor-
damerika im Speciellem. Allegemeine deutsche
naturhistorische, n.s., 1:249–280.

McKinney, C. O., F. R. Kay, and R. A. Anderson.
1973. A new all-female species of the genus *Cnemidophorus*. Herpetologica 29:361–366.

Mitchell, R. W., and J. R. Reddell. 1965. *Eurycea
tridentifera*, a new species of Troglobitic Salamander from Texas and a reclassification of
Typhlomolge rathbuni. Tex. J. Sci. 17:12–27.

Pauly, G. B., D. M. Hillis, and D. C. Cannatella.
2009. Taxonomic freedom and the role of
official lists of species names. Herpetologica
65:115–128.

Petranka, J. W. 1998. Salamanders of the United
States and Canada. Washington, DC: Smithsonian Institution Press.

Raun, G. G., and F. R. Gehlbach. 1972. Amphibians and reptiles in Texas. Dallas Mus. Nat. Hist.
Bull. 2:1–61.

Sanders, O. 1953. A new species of toad, with a
discussion of morphology of the bufonid skull.
Herpetologica 9:25–47.

Sellards, E. H., and C. L. Baker. 1934. The geology
of Texas. Vol. 2. Structural and economic geology. University of Texas Bulletin 3401.

Smith, H. M. 1934. Descriptions of new lizards of
the genus *Sceloporus* from Mexico and southern
United States. Trans. Kans. Acad. Sci. 37:263–285.

———. 1941. A review of the subspecies of the Indigo Snake (*Drymarchon corais*). J. Wash. Acad.
Sci. 31:466–481.

Smith, H. M., and F. E. Potter Jr. 1946. A third
neotenic salamander of the genus *Eurycea* from
Texas. Herpetologica 3:105–109.

Smith, H. M., and E. H. Taylor. 1950. Type localities of Mexican reptiles and amphibians. Univ.
Kans. Sci. Bull. 33:313–380.

———. 1966. Herpetology of Mexico: Annotated checklist and keys to the amphibians and
reptiles. A reprint of Bulletins 187, 194 and 199
of the U.S. National Museum with a list of subsequent taxonomic innovations. Ashton, MD:
Eric Lundberg.

Stejneger, L. H. 1893. Annotated list of the reptiles
and batrachians collected by the Death Valley
Expedition in 1891, with descriptions of new
species. North Am. Fauna 7:159–228.

———. 1896. Description of a new genus and
species of blind tailed batrachians from the
subterranean waters of Texas. Proc. U.S. Natl.
Mus. 18:619–621.

———. 1904. A new lizard from the Rio Grande
Valley, Texas. Proc. Biol. Soc. Wash. 17:17–20.

———. 1915. A new species of tailless batra-
chian from North America. Proc. Biol. Soc.
Wash. 28:131–132.

———. 1916. A new lizard of the genus *Sceloporus*
from Texas. Proc. Biol. Soc. Wash. 29:227–230.

———. 1925. New species and subspecies of
American turtles. J. Wash. Acad. Sci. 15 (20):
462–463.

Strecker, J. K. 1910. Description of a new Solitary
Spadefoot (*Scaphiopus hurterii*) from Texas, with
other herpetological novelties. Bull. U.S. Natl.
Mus. 23:115–122.

————. 1915. Reptiles and amphibians of Texas. Baylor Univ. Bull. 18:1–82.

Taylor, E. H. 1931. Notes on two new specimens of the rare snake *Ficimia cana*, and the description of a new species of *Ficimia* from Texas. Copeia 1931 (4): 4–7.

Thomas, R. A. 1974. A checklist of Texas amphibians and reptiles. Tech. Ser. Tex. Parks Wildlife Dept. 17:1–15.

————. 1976. A checklist of Texas amphibians and reptiles. Tech. Ser. Tex. Parks Wildlife Dept. 17:1–16.

Trapido, H. 1941. A new species of *Natrix* from Texas. Am. Midl. Nat. 25:673–680.

Troschel, F. H. 1852. *Cophosaurus texanus*, neue Eidechsen-gatlung aus Texas. Arch. Naturg. 1 (1850): 388–394.

Wiegmann, A. F. A. 1834. Herpetologia mexicana, seu descriptio amphibiorum novae hispaniae, quae itineribus comitis de Sack, Ferdinandi Deppe et Chr. Guil. Schiede im Museum Zoologicum Berolinense Pervenerunt. Pars prima, saurorum species. Berlin: Lüderitz.

Wright, A. A., and A. H. Wright. 1933. Handbook of frogs and toads of the United States and Canada. Ithaca, NY: Comstock Publishing Associates.

Yarrow, H. C. 1880. *Coluber bairdi* sp. nov. In On the zoological position of Texas, by E. D. Cope. Bull. U.S. Natl. Mus. 17:1–51.

13. Herpetofaunal Diversity of the United States–Mexico Border States/ Diversidad de la herpetofauna de los estados de la frontera México–Estados Unidos

Biggs, T. W., E. Atkinson, R. Powell, and L. Ojeda-Revah. 2010. Land cover following rapid urbanization on the US-Mexico border: Implications for conceptual models of urban watershed processes. Landscape Urban Plan. 96:78–87.

Brown, D. E., P. J. Unmack, and T. C. Brennan. 2007. Digitized map of biotic communities for plotting and comparing distributions of North American animals. Southwest. Nat. 52:610–616.

Casas Andreu, G., X. Aguilar Miguel, and R. Cruz Aviva. 2001. La introducción y el cultivo de la Rana Toro (*Rana catesbeiana*). Un atentado a la biodiversidad de México? Cienc. Ergo Sum 8:62–67.

Daleszczyk, K., and A. N. Bunevich. 2009. Population viability analysis of European bison populations in Polish and Belarusian parts of Bialowieza Forest with and without gene exchange. Biol. Conserv. 142:3068–3075.

Enderson, E. F., A. Quijada-Mascareñas, D. S. Turner, P. C. Rosen, and R. L. Bezy. 2009. The herpetofauna of Sonora, Mexico, with comparisons to adjoining states. Check List 5:632–672.

Grigione, M. M., and R. Mrykalo. 2004. Effects of artificial night lighting on endangered ocelots (*Leopardus paradalis*) and nocturnal prey along the United States–Mexico border: A literature review and hypotheses of potential impacts. Urban Ecosyst. 7:65–77.

Hughes, C. W. 2007. "La Mojonera" and the marking of California's U.S.-Mexico boundary line, 1849–1851. J. San Diego Hist. 53:126–147.

Kaczensky, P., R. Kuehn, B. Lhagvasuren, S. Pietsch, W. Yang, and C. Walzer. 2011. Connectivity of the Asiatic wild ass population in the Mongolian Gobi. Biol. Conserv. 144:920–929.

Lasky, R. J., W. Jetz, and T. H. Keitt. 2011. Conservation biogeography of the US-Mexico border: A transcontinental risk assessment of barriers to animal dispersal. Divers. Distrib. 17:673–687.

Lavín-Murcio, P. A., and D. Lazcano. 2010. Geographic distribution and conservation of the herpetofauna of northern Mexico. In Conservation of Mesoamerican amphibians and reptiles, ed. L. D. Wilson, J. H. Townsend, and J. D. Johnson, 275–301. Eagle Mountain, UT: Eagle Mountain Publishing.

Lawler, S. P., D. Dritz, T. Strange, and M. Holyoak. 1999. Effects of introduced mosquitofish and bullfrogs on the threatened California Red-legged Frog. Conserv. Biol. 13:613–620.

Levanetz, J. 2008. A compromised country: Redefining the U.S.-Mexico border. J. San Diego Hist. 54:39–42.

Luja, V. H., and R. Rodríguez-Estrella. 2010. The invasive bullfrog, *Lithobates catesbeianus* in oases of Baja California Sur, Mexico: Potential effects in a fragile ecosystem. Biol. Invasions 12:2979–2983.

McCain, E. B., and J. L. Childs. 2008. Evidence of resident jaguars (*Panthera onca*) in the southwestern United States and the implications for conservation. J. Mammal. 89:1–10.

Moyle, P. B. 1973. Effects of introduced bullfrogs, *Rana catesbeiana*, on the native frogs of the San Joaquin Valley, California. Copeia 1973:18–22.

Rachel, F. J. 2000. Homogenization of fish faunas across the United States. Science 288:854–856.

Seager, R., and G. A. Vecchi. 2010. Greenhouse warming and the 21st century hydroclimate of southwestern North America. Proc. Nat. Acad. Sci. 107:21, 277–21, 282.

Smith, K. G. 2006. Patterns of nonindigenous herpetofaunal richness and biotic homogeni-zation among Florida counties. Biol. Conserv. 127:327–335.

Villeger, S., S. Blanchet, O. Beauchard, T. Ober-dorft, and S. Brosse. 2011. Homogenization patterns of the world's freshwater fish faunas. Proc. Nat. Acad. Sci. 108:18003–18008.

Index

Note: Page numbers in *italics* refer to tables and graphs.

Owens, D. W., 66, 67, 70
Oxybelis aeneus (Bejuquilla Neotropical; Neotropical
Vinesnake), *plates 466–467*; of Arizona, 157, 159,
162; Brown's collection of, 148; of Chihuahua, 58,
59, 60, 65; distribution of, 452; nomenclature
notes for, 13; of Nuevo León, 95, 97, 100; of Sonora,
38, 47, 50; of Tamaulipas, 113, 114, 115

Pachymedusa, 62
Pachymedusa dacnicolor (Ranita Verduzca; Mexican
Leaf Frog), *plate 95*, 46, 51, 54, 58, 59, 433
Pacific Coast Parrot Snake. See *Leptophis diplotropis*
Pai Striped Whiptail. See *Aspidoscelis pai*
Painted Turtle. See *Chrysemys picta*
Painted Wood Turtle. See *Rhinoclemmys pulcherrima*
Painter, Charles W., 167–68, 178, 179
Palacio-Baez, G., 38
Palacios-Orona, L. E., 69
Palmer, Edward, 103
Panamint Alligator Lizard. See *Elgaria panamintina*
Panamint Rattlesnake. See *Crotalus stephensi*
Panthera onca (jaguars), 60, 204
Pantherophis bairdi (Ratonera de Bosque; Baird's
Ratsnake), *plate 468*; of Chihuahua, 64; of Coa-
huila, 74, 75; distribution of, 452; of Nuevo León,
93, 94, 99; of Tamaulipas, 106, 111, 116, 117, 118; of
Texas, 186, 190
Pantherophis emoryi (Ratonera de Emory; Great Plains
Ratsnake), *plate 469*; of Chihuahua, 61; of Coa-
huila, 74, 75; distribution of, 452; of New Mexico,
173, 175, 176; of Nuevo León, 93, 94, 99; of Tamau-
lipas, 111, 112, 113, 115, 117, 119, 120; of Texas, 187
Pantherophis guttatus (Ratonera Roja; Red Corn-
snake), *plate 470*, 193, 453
Pantherophis obsoletus (Ratonera Texana; Texas
Ratsnake), *plate 472*, 187, 193, 453
Papenfuss, T. J., 127, 136, 137, 152
Parham, J. F., 127, 136, 137
Parke, John G., 166
Parmerlee, John S., 68
Pauly, G. B., 194
Pelamis platura, 10
Peninsular Leaf-toes Gecko. See *Phyllodactylus nocti-
colus*
Pennell, Francis W., 85
Pérez-Mendoza, H. A., 105
Pérez-Ramos, E., 105
Perrill, R. H., 38
Peters, James A., 104
Petranka, J. W., 190
petroglyphs, 52–53
Petrosaurus mearnsi (Lagarto de Roca Rayado; Banded

Rock Lizard), *plate 265*; of Baja California, 29, 30,
33; of California, 124, 127, 141, 142; distribution of,
443
Petrosaurus repens (Lagarto de Roca de Hocico Corto;
Central Baja California Banded Rock Lizard), *plate
266*, 16, 30, 32, 33, 443
Petrosaurus slevini (Lagarto de Roca Bandeado;
Slevin's Banded Rock Lizard), *plate 267*, 16, 31, 443
Phelan, R. L., 127, 169
Phenacosaurus, 7
Phimothyra decurtata, 20
Phrynosoma, 19, 42, 190
Phrynosoma blainvillii (Camaleón de Blainville;
Blainville's Horned Lizard): of Baja California, 24,
26, 33; of California, 127, 142; distribution of, 443;
of Texas, 195
Phrynosoma cerroense, 16, 20, 26, 30, 32, 33
Phrynosoma cornutum (Camaleón Común; Texas
Horned Lizard), *plate 269*; of Arizona, 157, 162; of
Chihuahua, 61; of Coahuila, 67, 73, 78; distribu-
tion of, 443; of New Mexico, 165, 173, 174; of Nuevo
León, 93, 94, 97, 99; of Sonora, 47; of Tamaulipas,
111, 113, 115, 117; of Texas, 184, 188, 189, 192
Phrynosoma ditmarsi (Camaleón de Roca; Rock
Horned Lizard), *plate 270*; distribution of, 443; of
Sonora, 36, 38, 40, 42, 46, 50
Phrynosoma douglasii (Camaleón Pigmeo de Cola
Corta; Pygmy Short-horned Lizard), 133, 443
Phrynosoma goodei (Camaleón de Sonora; Goode's
Desert Horned Lizard), *plate 271*, 40, 45, 155, 156,
162, 443
Phrynosoma hernandesi (Camaleón Cuernitos de
Hernández; Greater Short-horned Lizard), *plate
272*; of Arizona, 149, 157, 158, 159–60, 162; of Chihuahua, 60; of Coahuila, 80; distribution of, 443;
of New Mexico, 165, 173, 174, 175, 176; of Sonora,
42, 46; of Texas, 192
Phrynosoma mcallii (Camaleón de Cola Plana; Flat-
tailed Horned Lizard), *plate 273*; of Arizona, 150,
155, 161, 162; of Baja California, 27, 28; of California, 123, 127, 141, 142; distribution of, 443; Rorabaugh's study of, 150
Phrynosoma modestum (Camaleón Modesto; Round-
tailed Horned Lizard), 12, *plate 274*; of Arizona,
157, 162; of Chihuahua, 61, 65; of Coahuila, 73, 78;
distribution of, 443; of New Mexico, 169, 173, 174,
175; nomenclature notes for, 12; of Nuevo León, 93,
94, 99; of Sonora, 47; of Tamaulipas, 111, 112, 119,
120; of Texas, 184, 186, 192
Phrynosoma orbiculare (Camaleón de Montaña; Mountain Horned Lizard), *plate 275*; of Chihuahua, 59,
60; of Coahuila, 77, 81; distribution of, 443; nomen-

Snapping Turtle. See *Chelydra serpentina*

Snyder, J. O., 103

Sonora, 34–51; amphibians of, 34–35, 36–39, 40, 42, 45, 48, *201*; anurans of, 35, 37, 39, 48, 49, *201*; biotic provinces of, *203*; Chihuahuan Desert assemblage, 47; grassland assemblage, 47; marine assemblage, 47–48; Montane/foothills assemblage, 46–47; Sonoran Desert assemblage, 45–46; subtropical assemblage, 46; as border state, 196; clustering present in, 198, 201, *201*, 202, 202, *203*, 203, 204; crocodilians of, 39, 48–49; distribution of, 454; endemic species of, 45, 50–51, *198*; frogs of, 47, 49; introduced species of, 35, 47, 48, 49, 50, 198, 198, *198*; lizards of, 35, 36, 37, 39, 40–41, 42, 45, 48, 49, *202*; physiographic characteristics of, 39, 42–45; previous studies of, 35–39; richness of species in, 197–98, *198*, *199–200*; salamanders of, 35, 39, 40, 48, 201, 201, 202; sea turtles of, 38, 39, 48, 49; shared/not shared species of, 50; snakes of, 35, 36, 37, 39, 41, 42, 47, 48, 49–50, *202*; summary of herpetofauna, 35, 48; surface area in, *197*; terrain and climate of, 34; turtles of, 35, 36, 37, 38, 39, 40, 47, 48, 49, 202, 203

"Sonora, Mexico" localities, 150

Sonora aemula (Culebra de Tierra Cola Plana; Filed-tailed Groundsnake), *plates 500–501*, 37, 46, 59, 60, 454

Sonora Mud Turtle. See *Kinosternon sonoriense*

Sonora palarostris, 41

Sonora semiannulata (Culebrilla de Tierra; Western Groundsnake), *plates 502–512*; of Arizona, 145, 152, 155, 156, 157, 158, 159, 162; of Baja California, 24, 29, 30, 32, 33; of California, 133, 143; of Chihuahua, 61; of Coahuila, 68, 73, 78; of New Mexico, 173, 174; of Nuevo León, 93, 94, 97, 99; of Sonora, 35, 48; of Tamaulipas, 111, 112; of Texas, 187, 189, 193

Sonora Spiny-tailed Iguana. See *Ctenosaura macrolopha*

Sonoran Collared Lizard. See *Crotaphytus nebrius*

Sonoran Coralsnake. See *Micruroides euryxanthus*

Sonoran Desert Toad. See *Incilius alvarius*

Sonoran Desert Tortoise. See *Gopherus morafkai*

Sonoran Green Toad. See *Anaxyrus retiformis*

Sonoran Leaf-toes Gecko. See *Phyllodactylus homolepidurus*

Sonoran Lyresnake. See *Trimorphodon lambda*

Sonoran Mountain Kingsnake, 13

Sonoran Shovelnosed Snake. See *Chionactis palarostris*

Sonoran Spotted Whiptail. See *Aspidoscelis sonorae*

Sonoran Whipsnake. See *Masticophis bilineatus*

Soulé, Michael, 18

Southern Alligator Lizard. See *Elgaria multicarinata*

Southern California Legless Lizard. See *Anniella stebbinsi*

Southern Leopard Frog. See *Lithobates sphenocephalus*

Southern Mountain Yellow-legged Frog. See *Rana muscosa*

Southern Painted Turtle. See *Chrysemys dorsalis*

Southern Rubber Boa. See *Charina umbratica*

Southern Sagebrush Lizard. See *Sceloporus vandenburgianus*

Southern Sierra Legless Lizard. See *Anniella campi*

Southern Torrent Salamander. See *Rhyacotriton variegatus*

Southern Watersnake. See *Nerodia fasciata*

Southwestern Cat-eyed Snake. See *Leptodeira maculata*

Southwestern Fence Lizard. See *Sceloporus cowlesi*

Southwestern Naturalist, 183

Spea, 62

Spea bombifrons (Sapo de Espuelas de los Llanos; Plains Spadefoot), *plate 134*; of Arizona, 156, 157, 158; of Chihuahua, 61; of Coahuila, 80; distribution of, 436; of New Mexico, 173, 174, 177; of Nuevo León, 97, 98, 99; of Sonora, 49; of Tamaulipas, 110, 112; of Texas, 185, 187, 191

Spea hammondii (Sapo de Espuelas Occidental; Western Spadefoot), *plate 135*, 24, 26, 33, 127, 142, 436

Spea intermontana (Sapo de Espuelas de la Gran Cuenca; Great Basin Spadefoot): of Arizona, 157, 158, 159, 161, 163; of California, 133; distribution of, 436

Spea multiplicata (Sapo de Espuelas Mexicano; Mexican Spadefoot), *plate 136*; of Arizona, 156–57, 158, 159, 161, 162; of Chihuahua, 62; of Coahuila, 67, 73, 75, 76, 78, 79; distribution of, 436; of New Mexico, 173, 174, 175, 177; of Nuevo León, 87, 93, 94, 95, 97, 98, 99; of Sonora, 47, 48; of Tamaulipas, 116, 117, 119; of Texas, 187, 189, 191

Special Nongame and Endangered Species Conservation Fund, 194

Speckled Earless Lizard. See *Holbrookia approximans*

Speckled Racer. See *Drymobius margaritiferus*

Speckled Rattlesnake. See *Crotalus mitchellii*

Spectacled Chirping Frog. See *Eleutherodactylus interorbitalis*

Spelerpes platycephalus, 126

Sphargis coriacea schlegelii, 40

Spilotes pullatus (Voladora Mico; Tropical Tree Snake), 113, 114, 115, 454

Spiny Chuckwalla. See *Sauromalus hispidus*

Spiny Softshell. See *Apalone spinifera*

Spiny-tailed Iguana. See *Ctenosaura* sp.

Trimorphodon, 19

Trimorphodon biscutatus, 10

Trimorphodon lambda (Ilimacoa de Sonora; Sonoran Lyresnake), plate 540; of Arizona, 156, 157, 158, 159, 162; of California, 141, 142; distribution of, 455; of New Mexico, 174, 175, 176; and Pliocercus bicolor, 10; of Sonora, 41, 48

Trimorphodon lyrophanes (Ilamacoa de California; California Lyresnake), plate 541; of Baja California, 24, 26, 29, 30, 32, 33; of California, 140, 143; distribution of, 455; and Pliocercus bicolor, 10

Trimorphodon paucimaculatus, 10

Trimorphodon quadruplex, 10

Trimorphodon tau (Falsa Nauyaca Mexicana; Mexican Lyresnake), plate 542; of Chihuahua, 60; distribution of, 455; of Nuevo León, 93, 95, 97, 100; of Sonora, 38, 46; of Tamaulipas, 105, 111, 112, 115, 117, 119, 120

Trimorphodon vilkinsonii (Serpiente de Tetalura; Texas Lyresnake): of Chihuahua, 57, 62; of Coahuila, 81; distribution of, 455; of New Mexico, 173, 174, 175; and Pliocercus bicolor, 10; of Texas, 190, 193

Trionychidae, 199, 439

Trionyx, 183

Trionyx ater, 70

Triton ensatus, 126

Triton torosa, 126

Triturus rivularis, 126

Triturus sierrae, 126

Tropical Ratsnake. See Pseudelaphe flavirufa

Tropical Tree Lizard. See Urosaurus bicarinatus

Tropical Tree Snake. See Spilotes pullatus

Tropidoclonion lineatum (Serpiente Lineada; Lined Snake), plate 543, 173, 175, 188, 193, 456

Tropidodipsas fasciata (Caracolera Anillada; Banded Snail Sucker), plate 544; distribution of, 456; nomenclature notes for, 13; of Tamaulipas, 104, 111, 112, 113, 115

Tropidodipsas repleta (Caracolera Repleta; Banded Blacksnake), plate 545; of Chihuahua, 55, 60; distribution of, 456; nomenclature notes for, 13; of Sonora, 38, 41, 47

Tropidodipsas sartorii (Caracolera Terrestre; Terrestrial Snail Sucker), plate 546; distribution of, 456; of Nuevo León, 97, 100; protected status of, 121; of Tamaulipas, 111, 112, 113, 115, 117, 119, 120, 121

Troschel, Franz H., 181, 186

Trueb, L., 63

Tunnell, J. W., Jr., 110

Turner, R. M., 22

turtles: of Arizona, 151, 161, 202, 203, 205; of Baja California, 16, 19, 20, 32, 202, 203; of California, 122, 125, 127, 129, 142, 202, 203, 205; of Chihuahua, 53, 54, 62, 63, 202, 203; clustering patterns of, 198, 202, 202, 203, 204, 205; of Coahuila, 66, 67, 68, 69, 70, 79, 80, 202, 203; legal protection of, under Mexican law, 48; of New Mexico, 164, 165, 166, 167, 177, 178, 202, 203, 205; of Nuevo León, 89, 98, 99, 202, 203; shared/not shared species of, 200; of Sonora, v, 35, 36, 37, 38, 39, 40, 47, 48, 49, 202, 203; species richness of, 197; of Tamaulipas, 105, 106, 107, 111, 112, 115, 117, 119, 120, 202, 203; of Texas, 182, 183, 184, 185, 187, 188, 189–90, 191–92, 194, 195, 203. See also specific species

Twin-spotted Rattlesnake. See Crotalus pricei

Twin-spotted Spiny Lizard. See Sceloporus bimaculosus

Twitty, V. C., 126

Two-lined Shortnosed Skink. See Plestiodon bilineatus

Two-striped Gartersnake. See Thamnophis hammondii

Typhlomolge, 185

Typhlomolge rathbuni, 185

Typhlopidae, 200, 456

Uetz, Peter, 9

Uma exsul (Lagartija Arenera; Fringe-toed Sand Lizard), plate 331, 67, 70, 73, 74, 81, 445

Uma inornata (Arenera del Valle Coachella; Coachella Fringe-toed Lizard), plate 332, 128, 141, 445

Uma notata (Arenera del Desierto del Colorado; Colorado Desert Fringe-toed Lizard), plate 333, 27, 28, 128, 141, 445

U. n. cowlesi, 40

Uma paraphygas (Arenera de Chihuahua; Chihuahua Fringetoed Lizard), plate 334; of Chihuahua, 54, 56, 61, 65; of Coahuila, 73, 74, 82; distribution of, 445

Uma rufopunctata (Arenera de Manchas Laterales; Yuman Desert Fringe-toed Lizard), plate 335; of Arizona, 151, 155, 161, 162; distribution of, 445; of Sonora, 40, 45

Uma scoparia (Arenera del Desierto de Mohave; Mohave Fringe-toed Lizard), plate 336, 128, 141, 155, 161, 162, 445

United States: border with Mexico, 17, 35; shared/not shared species of, 200. See also Arizona; California; New Mexico; Texas

United States Biological Survey, 166

United States Bureau of Fisheries, 36

United States Exploring Expedition, 123

United States Fish and Wildlife Service, 34

United States National Museum (USNM), 34, 67, 84, 146

Universidad Autónoma de Nuevo León (UANL), 87

Universidad Nacional Autónoma de México, 34

University of Arizona, 34, 149
University of Illinois Mexican Expedition, 104
University of Illinois Museum of Natural History
    (UIMNH), 54, 68, 85
University of Michigan Museum of Zoology
    (UMMZ), 85, 103
University of New Mexico (UNM), 167
Upland Chorus Frog. See *Pseudacris fouquettei*
urbanization, 204
Urosaurus bicarinatus (Roñito Arborícola; Tropical
    Tree Lizard), *plate 337*, 46, 59, 445
U. b. tuberculatus, 56
Urosaurus graciosus (Roñito de Matorral; Long-tailed
    Brush Lizard), *plates 338–339*; of Arizona, 155, 156,
    162; of Baja California, 27, 28; of California, *128*,
    141, 142; distribution of, 445; of Sonora, 45
Urosaurus lahtelai (Roñito de Matorral Bajacalifor-
    niano; Baja California Brush Lizard), *plate 340*, 16,
    20, 30, 446
Urosaurus nigricaudus (Roñito de Matorral Cola-
    negra; Black-tailed Brush Lizard), *plate 341*; of Baja
    California, 24, 29, 30, 32, 33; of California, 141, 142;
    distribution of, 446
Urosaurus ornatus (Roñito Ornado; Ornate Tree
    Lizard), *plate 342*; of Arizona, 155, 156, 157, 158, 159,
    160, 161, 162; of Baja California, 27, 28; of Cali-
    fornia, 141, 142; of Chihuahua, 56, 59, 62, 65; of
    Coahuila, 74; distribution of, 446; of New Mexico,
    169, 173, 174, 175; of Sonora, 35, 40, 48; of Tamau-
    lipas, 111, 112; of Texas, 181, *186*, 187, 190, 192; *U. o.
    caeruleus*, 56; *U. o. levis*, 169; *U. o. linearis*, 40; *U. o.
    schotti*, 40; *Urosaurus unicus*, 56; *Ursus horribilis*
    (grizzly bears), 61
Uta, 28
Uta antiqua (Mancha-lateral de San Lorenzo; San Lo-
    renzo Islands Side-blotched Lizard), 16, 20, 31, 446
Uta caerulea, 56
Uta encantadae (Mancha-lateral Encantadas; En-
    chanted Sideblotched Lizard), 16, 20, 27, 28, 446
Uta lowei (Mancha-lateral Muerta; Dead Side-
    blotched Lizard), 16, 20, 27, 28, 446
Uta mearnsi, 127
Uta nolascensis (Mancha-lateral de San Pedro No-
    lasco; Isla San Pedro Nolasco Lizard), *plate 343*, 35,
    40, 45, 50, 446
Uta ornata var. *linearis*, 40
Uta palmeri (Mancha-lateral de San Pedro Mártir; Isla
    San Pedro Mártir Side-blotched Lizard), 35, 40, 45,
    50, *plate 344*, 446
Uta schottii, 40
Uta slevini, 20
Uta stansburiana (Mancha-lateral Común; Common

Sideblotched Lizard), *plates 345–46*; of Arizona, 155,
    156, 157, 158, 161, 162; of Baja California, 24, 26, 29,
    30, 32, 33; of California, 142, 143; of Chihuahua,
    59, 61; of Coahuila, 67, 73; distribution of, 446;
    of New Mexico, 169, 173, 174, 175; of Nuevo León,
    93, 94, 97, 99; of Sonora, 36, 41, 45; of Texas, 190,
    192; *U. s. stejnegeri*, 169; *U. s. taylori*, 36, 41
Uta taylori, 36, 41
Uta tumidarostra (Mancha-lateral Narigona; Swollen-
    nosed Side-Blotched Lizard), *plate 347*, 16, 20, 27,
    28, 446
Utinger, Urs, 9

Valdina Farms Salamander. See *Eurycea troglodytes*
Van Denburgh, John: on Arizonan species, 147,
    151, *152*; on Baja Californian species, 17, 20, 24;
    on Californian species, 124, *126*, *128*; "A List of
    Reptiles and Amphibians of Arizona, with Notes
    on Species in the Collection of the Academy", 147;
    on New Mexican species, 166; on Sonoran species,
    34, 36, 38, 40
Van Devender, Thomas R., 54
Van Devender, Wayne, 54
Vance, Thomas, 183
Variable Sandsnake. See *Chilomeniscus stramineus*
Veined Treefrog. See *Trachycephalus typhonius*
Velasco, A. L., 101
Venerable Collared Lizard. See *Crotaphytus antiquus*
Venomous Reptiles of Arizona (Lowe, Schwalbe, and
    Johnson), 149
Villa, Robert A., 38
Viperidae, 200, 456–58
Virginia striatula (Serpiente Rugosa de Tierra; Rough
    Earthsnake), *plate 547*, 193, 456
Virginia valeriae (Serpiente Lisa de Tierra; Smooth
    Earthsnake), 188, 193, 456
Vitt, Laurie J., 149
viviparity, 42
Vorhies, Charles T., 148
Vuilleumier, Francois, 54

Wagner, Hellmuth, 104
Wake, David B., 125, *126*, 130
Walker, Charles F., 103, 104, *107*, *108*
Walker, J., 7, 24, 183
Wandering Gartersnake. See *Thamnophis elegans*
Wandering Salamander. See *Aneides vagrans*
Ward, J. P., 70
Warren, P., 178
Webb, R., 40, 54, 63, 68, 70, 101
Webb, Thomas H., 53, 67–68, 84
Welsh, Hartwell H., 18, 26

Índice

Nota: Los números de páginas en *cursivas* se refieren a cuadros y gráficas.

California Red-legged Frog), *Imagen 129*; de Baja California, 230, 233, 240; de California, *341, 345, 358*; distribución de, 435

Rana Grillo Norteña. Véase *Acris crepitans*

Rana Labradora. Véase *Craugastor decoratus*

Rana Ladradora Amarilla. Véase *Craugastor tarahumaraensis*

Rana lemosespinali, 264

Rana Leopardo Chiricahua. Véase *Lithobates chiricahuensis*

Rana Leopardo de El Río Bravo. Véase *Lithobates berlandieri*

Rana Leopardo de Forrer. Véase *Lithobates forreri*

Rana Leopardo de las Planicies. Véase *Lithobates blairi*

Rana Leopardo de Lemos-Espinal. Véase *Lithobates lemosespinali*

Rana Leopardo de Yavapai. Véase *Lithobates yavapaiensis*

Rana Leopardo del Noroeste de México. Véase *Lithobates magnaocularis*

Rana Leopardo del Sur de Estado Unidos. Véase *Lithobates sphenocephalus*

Rana Leopardo Norteña. Véase *Lithobates pipiens*

Rana Leopardo Relicto. Véase *Lithobates onca*

Rana luteiventris, 348, 349

Rana Manchada de Oregón. Véase *Rana pretiosa*

Rana Manglera. Véase *Hypopachus variolosus*

Rana muscosa (Rana Patas-amarillas de las Montañas del Sur; Southern Mountain Yellow-legged Frog), *Imagen 130, 339, 341, 355, 358*, 436

Rana Patas-amarillas. Véase *Rana boylii*

Rana Patas-amarillas de las Montañas del Sur. Véase *Rana muscosa*

Rana Patas-amarillas de Sierra Nevada. Véase *Rana sierrae*

Rana Patas-rojas de California. Véase *Rana draytonii*

Rana Patas-rojas del Norte de Estados Unidos. Véase *Rana aurora*

Rana Pimienta. Véase *Pseudacris crucifer*

Rana pretiosa (Rana Manchada de Oregón; Oregon Spotted Frog), 348, 349, 436

Rana Puerco. Véase *Lithobates grylio*

Rana Rayas Blancas. Véase *Lithobates pustulosus*

Rana sierrae (Rana Patas-amarillas de Sierra Nevada; Sierra Nevada Yellow-legged Frog), *339, 341*, 352, 436

Rana Tarahumara. Véase *Lithobates tarahumarae*

Rana Toro. Véase *Lithobates catesbeianus*

Rana typhonia, 212

Rana Venulosa. Véase *Trachycephalus typhonius*

Rana Verde. Véase *Lithobates clamitans*

Rana yavapaiensis, 367

ranas: de Arizona, 366; de Baja California, 222, 232; de California, 336, 339, *341, 345, 347, 348*, 352; de Chihuahua, 270, 272; de Nuevo León, 308, 309, 310; de Nuevo México, 384, 395–96, 397; de Sonora, 255, 257; de Tamaulipas, 334; de Texas, 399, 400, 402, 405, 406, 407, 408, 409, 412, 413. *Véase también especies específicas*

Rancho El Aribabi, 255

Rancho Los Fresnos, 255

Ranera Mexicana. Véase *Leptophis mexicanus*

Ranidae, 257, 272, *418*, 435–36

Ranita Ardilla. Véase *Hyla squirella*

Ranita Chirriadora de Escarpes. Véase *Eleutherodactylus marnockii*

Ranita Chirriadora de la Huasteca. Véase *Eleutherodactylus longipes*

Ranita Chirriadora de Manchas. Véase *Eleutherodactylus guttilatus*

Ranita Chirriadora del Río Bravo. Véase *Eleutherodactylus cystignathoides*

Ranita de Hojarasca. Véase *Leptodactylus fragilis*

Ranita de las Rocas. Véase *Hyla arenicolor*

Ranita de Lentes. Véase *Eleutherodactylus interorbitalis*

Ranita de Montaña. Véase *Hyla eximia*

Ranita de Oriente. Véase *Gastrophryne carolinensis*

Ranita de Wright. Véase *Hyla wrightorum*

Ranita Enana Mexicana. Véase *Tlalocohyla smithi*

Ranita Gris. Véase *Hyla versicolor*

Ranita Gris de Cope. Véase *Hyla chrysoscelis*

Ranita Olivo. Véase *Gastrophryne olivacea*

Ranita Olivo de Mazatlán. Véase *Gastrophryne mazatlanensis*

Ranita Orejona. Véase *Eleutherodactylus verrucipes*

Ranita Verde. Véase *Hyla cinerea*

Ranita Verduzca. Véase *Pachymedusa dacnicolor*

Rathke, M. H., *341*

Ratonera de Baja California. Véase *Bogertophis rosaliae*

Ratonera de Bosque. Véase *Pantherophis bairdi*

Ratonera de Emory. Véase *Pantherophis emoryi*

Ratonera de la Costa del Pacífico. Véase *Leptophis diplotropis*

Ratonera de Trans-Pecos. Véase *Bogertophis subocularis*

Ratonera del Trópico. Véase *Pseudelaphe flavirufa*

Ratonera Roja. Véase *Pantherophis guttatus*

Ratonera Texana. Véase *Pantherophis obsoletus*

Rau, C. S., *226, 237*

Raun, Gerald, 400, 401, 402

Xantusia jaycolei (Lagartija Nocturna de Cole; Cole's Night Lizard), 246, *248*, 253, 259, 448
Xantusia riversiana (Nocturna Insular; Island Night Lizard), *Imagen 387, 342, 354, 355, 358*, 448; *X. r. reticulata*, 355; *X. r. riversiana*, 355
Xantusia sierrae (Nocturna de la Sierra; Sierra Night Lizard), *342, 352, 353*, 449
Xantusia vigilis (Nocturna del Desierto; Desert Night Lizard), *Imagen 388*; de Arizona, 373, 374, 376; de Baja California, *226*; de California, *342, 358*; de Sonora, 253; distribución de, 449
Xantusia wigginsi (Nocturna de Wiggins; Wiggins' Night Lizard), *Imagen 389*; de Baja California, 225, *226, 230, 233, 235, 237, 239, 240*; de California, *356, 357*; distribución de, 449
Xantusidae, *419*, 448–49
Xenopus laevis (Rana Africana con Garras; African Clawed Frog): de Arizona, 377, 379; de Baja California, 222, 240; de California, 359; de Texas, 400; distribución de, 434; en todos los estados de la frontera, 416
Xenosauridae, *419*, 449

Xenosauro de Cabeza-plana. Véase *Xenosaurus platyceps*
Xenosaurus platyceps (Xenosauro de Cabeza-plana; Flathead Knob-scaled Lizard), *Imagen 390*; de Tamaulipas, 318, 319, *321*, 331; distribución de, 449
Xerobates agassizii, 341
Xerobates berlandieri, 403

Yarrow, Henry, 362, 367, 404

Zacatera Negra. Véase *Amastridium sapperi*
Zacatera Roja. Véase *Adelphicos newmanorum*
Zacholus zonatus, *342*
Zamenis conirostris, *321*
Zertuche, J. J., *299*
Zoological Museum of Berlin, 337
Zoological Society of San Diego, 224
Zug, G. R., 318
Zweifel, Richard G.: sobre especies de Baja California, 224, 233; sobre especies de Coahuila, 278, *281*; sobre especies de Nuevo México, 384, *387*; sobre especies de Sonora, 244